ANNUAL REVIEW OF BIOCHEMISTRY

EDITORIAL COMMITTEE (1989)

ANNUAL REVIEW OF BIOCHEMISTRY

VOLUME 58, 1989

CHARLES C. RICHARDSON, *Editor*
Harvard Medical School

JOHN N. ABELSON, *Associate Editor*
California Institute of Technology

PAUL D. BOYER, *Associate Editor*
University of California, Los Angeles

ALTON MEISTER, *Associate Editor*
Cornell University Medical College

ANNUAL REVIEWS INC.   4139 EL CAMINO WAY   P.O. BOX 10139   PALO ALTO, CALIFORNIA 94303-0897

Ⓡ ANNUAL REVIEWS INC.
Palo Alto, California, USA

International Standard Serial Number: 0066–4154
International Standard Book Number: 0–8243–0858-1
Library of Congress Catalog Card Number: 32–25093

∞ The paper used in this publication meets the minimum requirements of American National Standard for Information Sciences—Permanence of Paper for Printed
Library Materials, ANSI Z39.48-1984.

Annual Reviews Inc. and the Editors of its publications assume no responsibility
for the statements expressed by the contributors to this *Review*.

Typesetting by Kachina Typesetting Inc., Tempe, Arizona; John Olson, President
Typesetting coordinator, Typesetting by Kachina Typesetting Inc., Tempe, Arizona;
John Olson, President Typesetting coordinator, Janis Hoffman

PRINTED AND BOUND IN THE UNITED STATES OF AMERICA

Annual Review of Biochemistry
Volume 58, 1989

CONTENTS

SOME RELATED ARTICLES IN OTHER *ANNUAL REVIEWS*

From the *Annual Review of Biophysics and Biophysical Chemistry*, Volume 18 (1989)

Protein-Mediated Membrane Fusion, T. Stegmann, R. W. Doms, and A. Helenius
Physical Studies of Protein-DNA Complexes by Footprinting, T. D. Tullius
Thermodynamic Problems of Protein Structure, P. L. Privalov
Escherichia coli *Aspartate Transcarbamoylase*, N. M. Allewell

From the *Annual Review of Cell Biology*, Volume 5 (1989)

The Chloroplast Chromosomes in Land Plants, M. Sugiura
Simple and Complex Cell Cycles, F. Cross, H. Weintraub, J. Roberts
Initiation of Eukaryotic DNA Replication in Vitro, B. Stillman
The Interleukin 2 Receptor, K. A. Smith
Lysosomal Targeting in Animal Eukaryotes, S. Kornfeld

From the *Annual Review of Genetics*, Volume 22 (1988)

Biological Regulation by Antisense RNA in Prokaryotes, R. W. Simons and N. Kleckner
Spliceosomal snRNAs, C. Guthrie and B. Patterson
DNA Double-Chain Breaks in Recombination of Phage λ and of Yeast, D. S. Thaler and F. W. Stahl
The Mechanism of Conservative Site-Specific Recombination, N. L. Craig
The Heat-Shock Proteins, S. Lindquist and E. A. Craig

From the *Annual Review of Immunology*, Volume 7 (1989)

The Structure, Function, and Molecular Genetics of the Delta/Gamma T Cell Receptor, D. H. Raulet

From the *Annual Review of Medicine*, Volume 40 (1989)

Thrombolytic Therapy in Acute Myocardial Infarction, M. L. Simoons
Phosphoinositide Metabolism and Hormone Action, K. J. Catt and T. Balla

(*continued*)   vii

From the *Annual Review of Microbiology,* Volume 43 (1989)

Scrapie Prions, S. B. Prusiner
Structural Domains and Organizational Conformation Involved in the Sorting and Transport of Influenza Virus Transmembrane Proteins, D. P. Nayak and M. A. Jabbar
Role of the DNA/Membrane Complex in Prokaryotic DNA Replication, W. Firshein
Biochemistry of Acetate Catabolism in Anaerobic Chemotrophic Bacteria, R. K. Thauer, D. Möller-Zinkhan, and A. M. Spormann
DNA Packaging in dsDNA Bacteriophages, L. W. Black
Transcription and Reverse Transcription of Retrotransposons, J. D. Boeke and V. G. Corces
Genetics and Regulation of Bacterial Lipid Metabolism, T. Vanden Boom and J. E. Cronan, Jr.

From the *Annual Review of Neuroscience,* Volume 12 (1989)

Structure of the Adrenergic and Related Receptors, B. F. O'Dowd, R. J. Lefkowitz, and M. G. Caron
Biochemistry of Altered Brain Proteins in Alzheimer's Disease, D. J. Selkoe
Spider Toxins, H. Jackson and T. N. Parks
Acute Regulation of Tyrosine Hydroxylase by Nerve Activity and by Neurotransmitters via Phosphorylation, R. E. Zigmond, M. A. Schwarzschild, and A. R. Rittenhouse
Axonal Growth-Associated Proteins, J. H. P. Skene

From the *Annual Review of Physical Chemistry,* Volume 40 (1989)

Protein Translocation in Proteins, R. A. Copeland and S. I. Chan

From the *Annual Review of Physiology,* Volume 51 (1989)

Homologies Between Sugar Transporters from Eukaryotes and Prokaryotes, S. A. Baldwin and P. J. F. Henderson

From the *Annual Review of Plant Physiology and Plant Molecular Biology,* Volume 40 (1989)

Physiology and Molecular Biology of Phenylpropenoid Metabolism, K. Hahlbrock and D. Scheel
Biosynthesis and Degradation of Starch in Higher Plants, E. Beck and P. Ziegler

ANNUAL REVIEWS INC. is a nonprofit scientific publisher established to promote the advancement of the sciences. Beginning in 1932 with the *Annual Review of Biochemistry,* the Company has pursued as its principal function the publication of high quality, reasonably priced *Annual Review* volumes. The volumes are organized by Editors and Editorial Committees who invite qualified authors to contribute critical articles reviewing significant developments within each major discipline. The Editor-in-Chief invites those interested in serving as future Editorial Committee members to communicate directly with him. Annual Reviews Inc. is administered by a Board of Directors, whose members serve without compensation.

For the convenience of readers, a detachable order form/envelope is bound into the back of this volume.

Arthur Kornberg

Annu. Rev. Biochem. 1989. 58:1–30

NEVER A DULL ENZYME

Arthur Kornberg

Department of Biochemistry, Stanford University School of Medicine, Stanford, California 94305

CONTENTS

GROWING UP WITHOUT SCIENCE (1918–1942)

The current generation of scientists may be suprised to know that I had no formal research training. I was well started in a career of clinical medicine until World War II placed me in the National Institutes of Health (NIH) where I soon became an eager investigator of rat nutrition. Three years later, in 1945, I responded to the lure of enzymes and have remained faithful to them ever since.

Science was unknown in my family and circle of friends. Once, in 1947,

0066-4154/89/0701-0001$02.00

when I was in the biochemistry department of Washington University in St. Louis, working under the guidance of Carl and Gerty Cori, Gerty told me that Carl had collected beetles and butterflies in his youth, and then asked: "Arthur, what did you collect?" "Matchbook covers," was my sheepish response. What else? They were the dominant flora in the Brooklyn streets where I played and in the subways where my father often risked being trampled when he stooped to add one more to my collection.

My early education in grade school and Abraham Lincoln High School in Brooklyn was distinguished only by "skipping" a few grades and finishing three years ahead of schedule. I recall nothing inspirational from teachers or courses except encouragement to get good marks. I remember the glow of my chemistry teacher when I received a grade of 100 in New York State Regents examination. It was the first time, in more than twenty years of teaching, a student of his had gotten a perfect grade. Once when I boasted about this to my wife, Sylvy, she remarked that she too had gotten 100, not only in chemistry, but also in algebra and geometry.

I chose the cachet of City College in uptown Manhattan over nearby Brooklyn College, even though commuting from Bath Beach (near Coney Island) meant three hours a day in crowded subways. Competition among a large body of bright and highly motivated students was fierce in all subjects. I carried over my high school interest in chemistry, but the prospects for employment in college teaching or industry were dismal. For lack of graduate studies or research laboratories at City College then, these possibilities barely existed. At age 19 in 1937, with a Bachelor of Science degree, and no jobs to be had in the depths of the Great Depression, I welcomed the haven that medical school would provide for four more years.

Throughout college I worked evenings, weekends, and school holidays as a salesman in men's furnishings stores. This left little time for study or sleep and none for leisure. With these earnings, a New York State Regents Scholarship of $100 a year, no college tuition, and frugal living, I saved enough to see myself through the first half of medical school at the University of Rochester.

I enjoyed medical school and the training to become a doctor. Among my courses, biochemistry seemed rather dull. The descriptive emphasis on the constituents of tissues, blood, and urine reflected biochemistry in the United States in the 1930s. The dynamism of cellular energy exchanges and macromolecules was still unknown, and the importance of enzymes had not penetrated my course or textbook. By contrast, anatomy and physiology presented integrated and awesome structures and functions. The aberrations presented in pathology and bacteriology were absorbing, as were the responsibilities to diagnose and treat patients during the clinical years.

Did I as a medical student consider a career in research? Not really. I expected to practice internal medicine, preferably in an academic setting; the idea of spending a significant fraction of my future days in the laboratory had no appeal. The medical school of the University of Rochester granted some students fellowships to take a year out for research. I had hoped but failed to get such an award from any of the departments. In those years, ethnic and religious barriers were formidable, even within the enlightened circle of academic science.

I did some research on my own, which grew out of curiosity about jaundice. I had noticed a slightly yellow discoloration of the whites of my eyes, and found that my blood bilirubin level was elevated and my tolerance to injected bilirubin reduced. I made similar measurements on as many medical students and patients as I could. I collected samples at odd moments and did the analyses on a borrowed bench, late at night and on weekends. The report I published (1) called attention to the frequent occurrence of high bilirubin levels and reduced capacity to eliminate bilirubin, now recognized as signs of the benign familial trait called Gilbert's Disease.

Looking back, I realized that I enjoyed collecting data. I kept on collecting bilirubin measurements during my internship year and started setting up to do more analyses in the small sickbay of a Navy ship soon after I joined it. A lucky consequence was that the publication of my student work on jaundice attracted attention and led to my transfer from sea duty to do research at the NIH, a rare assignment at that time.

JOINING THE VITAMIN HUNTERS (1942–1945)

The Nutrition Laboratory at NIH to which I was assigned in the fall of 1942 as a commissioned officer in the U. S. Public Health Service had been started by Joseph Goldberger (1874–1929). He was among the first to recognize that a vitamin deficiency can cause an epidemic disease, and in tracking the missing vitamin in the diets of pellagra patients, he emerged as one of the greatest of the vitamin hunters. W. H. (Henry) Sebrell, whom he had trained, was now chief of the laboratory and my senior boss. The laboratory had moved in 1938 from downtown Washington to suburban Bethesda, Maryland, but some of Goldberger's animal caretakers, kitchen staff, and diet notebooks, as well as his aura were still around.

My initial project as a nutritionist was to find out why rats fed a purified ("synthetic") diet containing a sulfa drug developed a severe blood disorder in a few weeks and died. A stock animal ration or inclusion of a yeast or liver supplement in the purified diet was effective in preventing and curing the disease (2). After other vitamin hunters (3) with the use of a microbial assay had succeeded in isolating folic acid and made it available to us, we could show that an induced deficiency of this vitamin was responsible for the sulfa drug effect.

It seemed clear that sulfa drugs, as analogues of para-aminobenzoic acid (PABA), a component of folic acid, were preventing bacteria from synthesizing this essential constituent and thus preventing their growth. We also knew that many animals rely on their intestinal bacteria for an adequate supply of folic acid and other vitamins, including vitamin K. I was therefore puzzled by a report (4) that PABA prevented the sulfa drug from producing a vitamin K deficiency even when given by injection. This result was taken to mean that the sulfa drug was not exerting its toxic effect on the intestinal bacteria, but somewhere else in the body.

I repeated the experiments with PABA injections and sulfa drugs, and developed a method to measure the amounts of vitamin K and PABA in the intestinal contents and feces. Ample quantities of vitamin K were produced by the intestinal bacteria of rats on a purified diet and this production was eliminated by sulfa drugs. As for rats injected with PABA, high levels of this substance accumulated in the intestinal contents, amounts sufficient to offset the action of the sulfa drug taken in the diet. These findings were reported in my first contribution to the *Journal of Biological Chemistry* (5), one of a very few biochemical papers from the NIH.

With the isolation of folic acid, it was apparent that virtually all the vitamins had been discovered. But we did not understand what most of the vitamins did in the body. How was folic acid serving in the growth of blood cells? What clues did the structure of folic acid offer to understanding its precise metabolic function? Could this understanding explain why sulfa drugs kill bacteria but not animal cells?

The answers to these questions, as well as to similar questions about the functions of the other vitamins, would be answered in the next two decades by enzymology. Just as the microbe hunters, who led the way in the first two decades of this century, were succeeded in the 1920s and 1930s by the vitamin hunters, so the latter would be overrun in the next two decades by the enzyme hunters.

I had come to nutrition in its twilight, decades late for the excitement and adventures of the early vitamin hunters who had solved the riddles of diseases that had plagued the world for centuries. My envy of their exploits impelled me to search for a new frontier. The discoveries of each of the vitamins— nicotinic acid, riboflavin, and thiamine in intermediary metabolism, and folic acid in nucleotide biosynthesis—became part of my heritage as I went on to learn about their biochemical functions. The rush to biochemistry depopulated the ranks of nutrition. How tragic that diet remains to this day as controversial as politics and the science of nutrition is in disarray.

FROM RATS TO ENZYMES (1945–1947)

By 1945, with the war over, I had become bored with feeding rats variations of purified diets. I was excited reading for the first time about enzymes, coenzymes, and ATP, in papers by Otto Warburg, Otto Meyerhof, Carl Cori, Herman Kalckar, and Fritz Lipmann. I had learned nothing about these things or people in medical school. While at NIH, I was startled and fascinated by a seminar in which Edward Tatum described his and George Beadle's work with *Neurospora* mutants and their one gene-one enzyme hypothesis. I knew even less about genetics than about biochemistry.

Fortunately, I was able to persuade Dr. Sebrell to let me quit my nutritional work and go to a laboratory where I could learn about ATP and enzymes. Immediately, I apprenticed myself to Bernard Horecker, a friend at NIH, who had been studying effects of DDT on cockroaches and was returning to the subject of his doctoral dissertation, the cytochromes of cellular respiration. Bernie introduced me to succinoxidase, cytochrome c, and the Beckman Model DU spectrometer. The unsolved problem of oxidative phosphorylation seemed to me to be the most important thing to do in biochemistry.

While still in uniform, I spent the year 1946 with Severo Ochoa at New York University Medical School; it was one of the happiest and most exhilarating in my life. Never had my learning curve been so sharply exponential and sustained. And in the few waking hours outside the laboratory, Sylvy and I discovered the theater, music, and museums that are the heartthrob of New York. Despite my being a native of Brooklyn and having attended City College in uptown Manhattan, and despite Sylvy's many visits from Rochester where she grew up and studied biochemistry, we were strangers to the city.

My mission from Ochoa was to purify heart muscle aconitase. This was my first solo stab at enzyme purification. We expected to resolve the activity into two enzymes to account for the successive subtraction and readdition of a water molecule that converts citric to isocitric acid. Despite repeated failure (aconitase proved to be one enzyme), this immersion in enzymology was intoxicating. Aside from the fascination of seeing an enzyme in action, the pace of the experimental work was breathtaking. By coupling aconitase action to isocitrate dehydrogenase, spectrophotometric assays could be performed in a few minutes, and many ideas could be tested and discarded in the course of a day. Late evenings were occupied preparing a series of protocols for the following day. What a contrast with the tedious pace of nutritional experiments on rats.

In my work on aconitase, I learned the philosophy and practice of enzyme purification. To attain the goal of a pure protein, the notebook record of an enzyme purification should withstand the scrutiny of an auditor or bank examiner. Not that I ever regarded the enterprise as a business or banking operation. Rather, it often seemed like the ascent of an uncharted mountain: the logistics resembled supplying successively higher base camps; protein fatalities and confusing contaminants resembled the adventure of unexpected storms and hardships. Gratifying views along the way fed the anticipation of what would be seen from the top. The ultimate reward of a pure enzyme was tantamount to the unobstructed and commanding view from the summit. Beyond the grand vista and thrill of being there first, there was no need for descent, but rather the prospect of ascending even more inviting mountains, each with the promise of even grander views.

I was luckier in my second attempt at enzyme purification when I joined Ochoa and Alan Mehler, his first graduate student, in purifying the liver malic enzyme, the enzyme that converts malic to lactic acid (6). Mehler was already on the scene when I arrived in Ochoa's lab and became my indefatigable and devoted tutor. Having always been the youngest in my class, it was a shock to find that I was so far behind someone four years my junior.

To let me pursue my training and the problem of aerobic phosphorylation, the NIH extended my stay with Ochoa to a full year, and allowed me another six months in the laboratory of Carl and Gerty Cori at the Washington University Medical School in St. Louis. Right after the war, the Cori laboratory was the mecca of enzymology. There I joined a young Swedish visitor, Olov Lindberg, who was investigating a striking observation made six years earlier by Ochoa when he worked in the Cori laboratory. Liver particles metabolizing pyruvic and related acids produced inorganic pyrophosphate (PP), a compound previously unknown as a cellular constituent. We began by ruling out the possibility that PP was released from an unstable form of ATP, but then found little else to guide us.

Later, while trying to enhance the levels of respiration and coupled aerobic phosphorylation by kidney particles, we observed a strong stimulation by NAD, and discovered that the effect could be traced to AMP generated by its hydrolysis.

$$\text{Nicotinamide-ribose-P-P-ribose-adenine} + H_2O \rightarrow \text{Nicotinamide-ribose-P} + \text{AMP} \qquad 1.$$
$$\text{(NAD)} \qquad\qquad\qquad\qquad \text{(NicRP)}$$

The AMP produced by NAD cleavage stimulated the reaction because it served as an acceptor of inorganic phosphate to form ATP. This mundane result marked the end of my search for the source of ATP in aerobic phosphorylation. The search had been doomed from the start because I was committed to finding discrete soluble enzymes that linked the synthesis of

ATP to respiration. As the late Albert Lehninger recognized a few years later, these enzymes are firmly embedded in mitochondria.

MY ROOKIE YEAR (1948)

Nineteen forty-eight, the year I set up my own biochemistry lab, was a great year for me. When I returned to the NIH, my former laboratory space in the Nutrition Division (in Building 4) was occupied. Just about then, one of the frequent organizational convulsions in the Industrial Hygiene Division (in Building 2) threatened Bernie Horecker and Leon Heppel, a close friend and medical school classmate, with a transfer to Cincinnati. Fortunately, Henry Sebrell agreed to let me start an Enzyme Section that would include the three of us in a few laboratory rooms in Building 3. (Considering the present mammoth size of the NIH, covering 300 acres and employing 13,000 people, it is hard to believe that in 1947 there were only six small buildings and that the research emphasis was still on infectious disease, dominated by a small corps of commissioned medical officers.)

I continued the work on the rabbit kidney enzyme that Lindberg and I had discovered in St. Louis and established that it cleaves NAD at the pyrophosphate linkage (7). However, the enzyme was firmly attached to tissue particles and there was little hope of obtaining it in pure form. At the suggestion of the late Sidney Colowick and Oliver Lowry, I looked for and found a similar enzyme activity in potatoes, from which it could readily be extracted in a free, soluble form (8).

The purified enzyme cleaved not only NAD, but all nucleotides with a pyrophosphate bond. I called the enzyme *nucleotide pyrophosphatase*. By using the enzyme to cleave NADP, I could show that the position of the extra phosphate, then unknown, was part of the AMP moiety on carbon 2 of the ribose. Best of all, having isolated NicRP from NAD cleavage, I wondered whether it might serve in the synthesis of NAD. It did! Enzymes purified from yeast and liver condensed NicRP and ATP to produce not only NAD, but PP as well, the first clue to the origin of PP after years of speculation. The reaction was readily reversible and could support a vigorous exchange of PP with ATP (9).

$$NicRP + PPPRA \rightleftarrows NicRPPRA + PP \qquad\qquad 2.$$
$$(ATP)(NAD)$$

This mechanism immediately led us to the discovery of the enzyme that synthesizes flavin adenine dinucleotide (FAD) from riboflavin phosphate and ATP (10). In the ensuing years, the mechanism of nucleotidyl transfer from a nucleoside triphosphate for the biosynthesis of coenzymes was discovered again and again in the biosynthesis of proteins, lipids, carbohydrates, and nucleic acids. A variety of phosphoric, carboxylic, and sulfuric acids (XO^-) accept a nucleotidyl group from a nucleoside triphosphate (PPPRN) to generate an activated form of XO^- with the release of PP.

$$XO^- + PPPRN \rightleftarrows XO-PRN + PP \qquad\qquad 3.$$
$$\downarrow$$
$$2\ Pi$$

Hydrolysis of PP by a strong and ubiquitous inorganic pyrophosphatase drives these reversible condensations toward biosynthesis (11).

What a wondrous enzyme, the humble potato pyrophosphatase! It helped solve an aspect of NADP structure, set up the discovery of coenzyme biosynthesis, and with it a major theme in biochemistry, and then led me on to the enzymes that assemble DNA, genes, and chromosomes.

OROTIC ACID IS ON THE MAIN TRACK (1953–1955)

In 1955, two years after the historic Watson and Crick reports (12) of the double helix and its implications for replication, I found an enzyme that synthesizes DNA chains from simple building blocks. Based on this chronology, it is commonly assumed that the Watson-Crick discovery spurred me to search for the enzymes of replication. But that is not the way it happened. In 1953, DNA was far from the center of my interests. The significance of the double helix did not intrude on my work until 1956, when the enzyme that assembles the nucleotide building blocks into a DNA chain was already in hand.

My interest in the replication of DNA, the focus of my research for the past 33 years, developed primarily from a fascination with enzymes. Having found an enzyme that incorporates a nucleotide into a coenzyme, I began, around 1950, to wonder about enzymes that might assemble the many nucleotides that make up the chains of nucleic acids, particularly RNA. But first we had to know the building blocks of the nucleic acids. It was not at all obvious in 1950 what they might be. Was the backbone assembled first and were the bases attached later? Was each link added to the chain as a single nucleotide? If so, was the phosphate in each component nucleotide initially attached to carbon number three or five, or to either one randomly or in a cyclic form to both?

In anticipating what the building block might be, I was influenced by what I had learned from the biosynthesis of coenzymes. I also felt that in searching for the form of the nucleotide that might serve as a building block for RNA and DNA, it would help to know how a nucleotide itself is built from simpler molecules, and thus what its nascent form might be. Inasmuch as Jack Buchanan and Bob Greenberg were already pursuing purine biosynthesis, I decided to go after the pyrimidines.

During a brief interlude, I acted on the hunch that biosynthesis of the phosphodiester bond, accessible in phospholipids, might offer a model for building the backbone of nucleic acids. In exploratory experiments with [^{32}P]-α-glycerophosphate and [^{14}C]-phosphoryl choline, I could find no evidence for their condensation to form the diester (glycerophosphoryl choline), but I did stumble on the formation of phosphatidic acid and phosphatidyl choline in the cell-free extract (13). I worked out the enzymatic synthesis of phosphatidic acid (14), the key precursor of phospholipids, but still was eager to get away from greasy molecules and return to pyrimidines and the aqueous phase. In the future, I would not rely on intuition about model systems, but would head toward an objective directly.

Osamu Hayaishi came as a postdoctoral fellow in 1950 experienced in the use of soil bacterial enrichment cultures. Among the huge variety of species in soil, at least one can be found that will respond to virtually every natural organic compound and use it as a source of carbon and energy. Believing too that reversibility of metabolic pathways might provide clues to biosynthesis, we examined the breakdown of uracil and thymine in extracts of bacteria isolated from soil by aerobic enrichment on these pyrimidines. Uracil and thymine were converted to the corresponding barbiturates, not at all promising as biosynthetic precursors (15). But the next year, during my first visit to California, H. A. Barker helped me find an anaerobe in San Francisco Bay mud that consumed orotic acid. Back at NIH, with the participation of Irving Lieberman, who had been a student of Barker, studies of this organism identified as metabolic products dihydroorotic acid and carbamyl aspartate (16), which later proved to be intermediates in the biosynthesis of orotic acid.

Orotic acid was known from intact cell studies to be a precursor of nucleic acid pyrimidines, but it was uncertain whether it was on the main track or connected to it by a spur. With orotic acid tagged in its carboxyl group, the release of CO_2 to form uracil might lead us to the enzyme that took orotic "up" to nucleic acid. CO_2 release by extracts from yeast or liver was terribly feeble, yet showed a tantalizing requirement for ATP and ribose 5P. One happy day, instead of using extracts of either yeast or liver, I combined them. The reaction was explosive, hundreds of times greater than before, one of those rare moments in a scientific lifetime.

The enzyme abundant in liver extracts transferred a PP group from ATP to carbon 1 of ribose 5P to produce the novel phosphoribosyl pyrophosphate (PRPP) (17), later recognized as the key precursor of purine nucleotides, histidine, tryptophan, and NAD. The enzyme in yeast extracts (actually two enzymes) formed orotidine 5P, which then was decarboxylated to UMP (18), the direct precursor of all the nucleic acid pyrimidines (Figure 1).

The transfer of pyrophosphate to ribose 5P entails an attack on the middle phosphate of ATP, as Gobind Khorana showed during one of his whirlwind and productive visits to my lab (19). (Other examples of this unusual reaction are the synthesis of thiamine PP, and guanosine tetraphosphate.) PRPP synthetase remains one of my favorite enzymes. As I wrote in the 1975 Festschrift for Ochoa (*Reflections on Biochemistry,* Pergamon Press): "Most of us anticipated that ribosyl activation for nucleotide biosynthesis would use the same device of phosphorylation, so well known for glucose. But the novelty of pyrophosphorylation used by this enzyme (coupled with elimination of inorganic pyrophosphate upon subsequent condensations) established my unalloyed awe for the ingenuity and fitness of an enzyme."

Knowing that PRPP enables a free pyrimidine (orotic acid) to be converted directly to a nucleotide, we sought and found enzymes that used PRPP to convert free purines (adenine, hypoxanthine, guanine) directly to nucleotides (20). Yet, I also knew from Buchanan's and Greenberg's studies (21, 22) that a purine ring is assembled from the very outset attached to ribose phosphate (later shown to be derived from PRPP). These facts, coupled with the knowledge that nucleotides can be formed from nucleosides by kinases, made it clear to me that cells have alternate pathways to the biosynthesis of nucleotides: salvage of preformed bases and nucleosides, and de novo routes from smaller molecules (e.g. sugar phosphates, amino acids, ammonia, one-carbon units). We have since realized that the role of salvage pathways can be as vital as the de novo pathways even under normal conditions when the de novo routes are not blocked by mutation, drugs, disease, or excessive traffic (23).

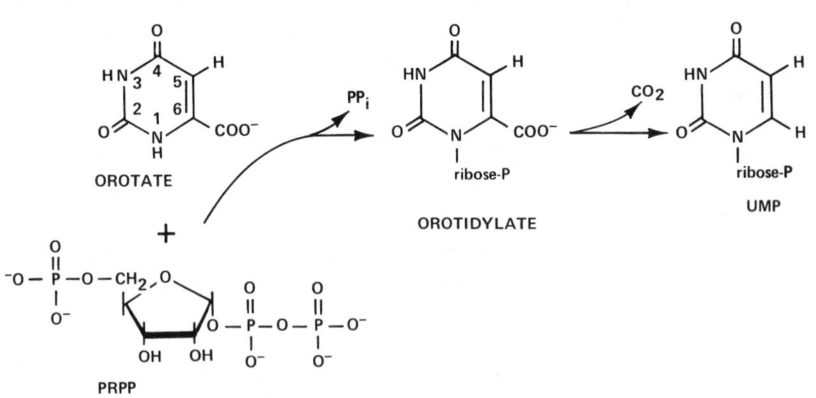

Figure 1   Condensation of orotic acid with PRPP produces the nucelotide, orotidylate (orotidine 5P), which upon decarboxylation generates uridine 5P (UMP).

DISCOVERY OF DNA POLYMERASE (1955–1959)

Having learned how the likely nucleotide building blocks of nucleic acids are synthesized and activated in cells, it seemed natural that in 1954 I would look for the enzymes that assemble them into RNA and DNA. Such an attempt might have been considered by some as audacious. Synthesis of starch and fat, once regarded as impossible outside the living cell, had been achieved with enzymes in the test tube. But, the monotonous array of sugar units in starch or the acetic acid units in fat was a far cry from the assembly of DNA, thousands of times larger and genetically precise.

Yet, I was only following the classical biochemical traditions practiced by my teachers. It always seemed to me that a biochemist devoted to enzymes could, if persistent, reconstitute any metabolic event in the test tube as well as the cell does it. In fact better! Without the constraints under which an intact cell must operate, the biochemist can manipulate the concentrations of substrates and enzymes and arrange the medium around them to favor the reaction of his choice.

I have adhered to the rule that all chemical reactions in the cell proceed through the catalysis and control of enzymes. Once, in a seminar on the enzymes that degrade orotic acid (16), I realized that my audience in the Washington University chemistry department was drifting away. In a last-ditch attempt to gain their attention, I pronounced loudly that every chemical event in the cell depends on the action of an enzyme. At that point, Joseph Kennedy, the brilliant young chairman, awoke: "Do you mean to tell us that something as simple as the hydration of carbon dioxide (to form bicarbonate) needs an enzyme?" The Lord had delivered him into my hands. "Yes, Joe, cells have an enzyme, called carbonic anhydrase. It enhances the rate of that reaction more than a million-fold."

By 1954, the rapidly growing *Escherichia coli* cell had become a favored object of biochemical and genetic studies, and for me had replaced yeast and animal tissues as the preferred source of enzymes. To explore the synthesis of RNA, Uri Littauer, a postdoctoral fellow, and I prepared [^{14}C-adenine]-ATP and maintained it as ATP with a regenerating system. Upon incubation with an *E. coli* extract, a small but significant amount of the radioactivity was incorporated into an acid-insoluble form, presumably RNA, and we proceeded eagerly to purify the activity responsible.

I also pursued the synthesis of DNA. Here, I had the invaluable help of Morris Friedkin, who had synthesized ^{14}C-thymidine and was studying its uptake into the DNA of rabbit bone marrow or onion root tip cells. Disinclined to work with cell-free extracts, he generously saved the spent reaction fluid from which I recovered radioactive thymidine to use in trials with extracts of *E. coli*.

The results were mixed. Very little thymidine was incorporated into the acid-insoluble form indicative of DNA, only about 50 cpm out of the million with which we started. On the other hand, 5–10% of the thymidine was converted to novel soluble forms that resembled the phosphorylated states of the nucleotide building blocks, possibly better precursors than thymidine for DNA synthesis.

At this juncture, Herman Kalckar on a visit to St. Louis brought us the startling and unsettling news that Ochoa and Marianne Grunberg-Manago, a postdoctoral fellow, had just discovered the enzymatic synthesis of RNA. It was for them a totally unexpected finding made while exploring aerobic phosphorylation in extracts of *Azotobacter vinelandii*. They observed an exchange of phosphate into ADP and the reversible conversion of ADP (or other nucleoside diphosphates) into RNA-like chains (24) and they named the enzyme polynucleotide phosphorylase.

On the strength of this new information, we shifted to using ADP rather than ATP in our studies with *E. coli*. The rate and extent of reaction were far greater and we readily purified the enzyme involved (25). We had made a classic blunder. Accounting for a phenomenon does not insure that it is the only or the best explanation of it. In this instance, we were diverted from the discovery of RNA polymerase, which depends on ATP. By switching to ADP, we tracked the synthetic activity of polynucleotide phosphorylase and missed the key enzyme for gene transcription.

Ten months passed before I repeated the experiment of converting radioactive thymidine to an acid-insoluble form. Once again, only a tiny amount of this presumed precursor was converted. But several things were different. For one, the radioactivity of the thymidine happened to be three times as great and so the results seemed more impressive. For another, believing I had lost out on the synthesis of RNA, the synthesis of DNA became a more precious goal. Finally, I exposed the product this time to pancreatic DNase and found that it became acid-soluble, a strong indication that it was DNA.

Even before I calculated the DNase results, I stopped to tell Bob Lehman about them. Although his postdoctoral problem was well started, he was eager to switch to DNA synthesis. Progress was rapid. Bob soon found that thymidine phosphate was a far better precursor than thymidine and later showed that thymidine triphosphate was much better still. With improvements in the assay of DNA synthesis by these crude extracts, our goal was to purify the enzyme that assembled nucleotides into a DNA chain, the enzyme we would name DNA polymerase (26, 27).

The most complex and revealing insights into the reaction would come from exploring the function of the DNA that I had included in the reaction mixture in my earliest attempt to incorporate thymidine into DNA. Some assume that DNA was included to serve as a template and that its primer role emerged many years later. Not so. I added DNA expecting that it would serve as a primer for growth of a DNA chain, because I was influenced by the Cori work on the growth of carbohydrate chains by glycogen phosphorylase. I never thought that I would discover a phenomenon utterly unprecedented in biochemistry: an absolute dependence of an enzyme for instruction by its substrate serving as a template.

I had added DNA for another reason. Nuclease action in the extracts was rampant, and I wanted a pool of DNA to surround the newly incorporated thymidine and protect at least some of it. Only later did Lehman and I learn with elation that the added DNA fulfilled two other essential roles. It indeed served as a template and also as a source of the missing nucleotides. The DNA was cleaved by DNases in the extract to nucleotides. These were converted by ATP and five kinases in the extract to the di- and triphosphates of the A, G, C, and T deoxyribonucleotides, which were then still unknown.

Maurice Bessman, Steve Zimmerman, and Julius Adler joined Bob Lehman, Sylvy, Ernie Simms (my research assistant), and me and occasionally one or two others, all in a small laboratory, only about 20 by 20 feet. Crowded and excited, we shared ideas, reagents, and data. The sum of our efforts was far greater than if we had been diluted into a larger room or separated by walls.

CREATION OF LIFE IN THE TEST TUBE (1960–1967)

With purified DNA polymerase, we could show that the DNA product reflected the base composition of the template and the frequencies of the 16 possible dinucleotides. The "nearest-neighbor" sequence method, which we devised to determine the dinucleotide frequencies, also revealed that the two strands of the double helix have opposite polarities, a structural feature that had not been experimentally demonstrated up to that time (28).

We also made the unexpected discovery that the enzyme, in the apparent absence of any template would, after a considerable delay, make DNA-like polymers of simple composition (29, 30): the alternating copolymers polydA·dT and polydG·dC and the homopolymer pairs of polydA with polydT and of polydG with polydC. These polymers, once made, proved to be superior templates and have been widely used in DNA chemistry and biology. Generation of the polymers de novo could be ascribed to the reiterative replication of short sequences in the immeasurably small amounts of DNA that contaminate a polymerase preparation (31, 32).

For more than 10 years, I had to find excuses at the end of every seminar to explain why the DNA product had no biologic activity. If the template had been copied accurately, why were we unsuccessful in all our attempts to multiply the transforming factor activity of DNA from *Pneumococcus, Hemophilus,* and *Bacillus* species? Finally, with the arrival of ligase in 1967, a crucial test could be made. [The enzyme had been discovered that year in five laboratories: those of Martin Gellert, Charles Richardson, and Jerard Hurwitz, in Lehman's next door, and in mine by Nicholas Cozzarelli.] Mehran Goulian and I could replicate the single-stranded circle of phage ϕX174 with DNA polymerase and then seal the complementary product with ligase. The circular product was isolated and then replicated to produce a circular copy of the original viral strand, which could be assayed for infectivity in *E. coli* (33). We found the completely synthetic viral strand to be as infectious as that of the phage DNA with which we started!

After so many years of trying, we had finally done it. We had gotten DNA polymerase to asemble a 5000-nucleotide DNA chain with the identical form, composition, and genetic activity of DNA from a natural virus. All the enzyme needed was the four common building blocks: A, G, T, and C. At that moment, it seemed there were no major impediments to the synthesis of DNA, genes, and chromosomes. The way was open to create novel DNA and genes by manipulating the building blocks and their templates.

In a very small way, we were observers of something akin to what those at Alamogordo on a July day in 1945 witnessed in the explosive force of the atomic nucleus. Harnessing the enzymic powers of the cellular nucleus had neither the dramatic staging of light and sound nor the stunningly apparent global consequences. Yet, this demonstration of our power with enzymes that build and link DNA chains would soon help others forge a different revolution, the engineering of genes and modification of species.

A hundred newspaper and television reporters and photographers came to a press conference called by the Stanford News Bureau on December 14, 1967, because of many inquiries about the paper we had just published in that month's issue of the *Proceedings of the National Academy of Sciences* (34). The title was: "Enzymatic Synthesis of DNA, XXIV. Synthesis of Infectious Phage ϕX174 DNA." To the editors who sent the newsmen, it seemed that a virus had been synthesized and life created in the test tube.

At the news conference, I tried to explain why the definition of life and a living molecule is so elusive. Afterwards, I overheard a reporter on the telephone to his office: "It's not what we expected. They haven't made a virus. It's only a molecule, a short chain of DNA. They've been making DNA in the test tube for 12 years." Hairy little monsters had not been created in the test tube. Yet, the story rated banner headlines worldwide and a newspaper article on January 4, 1968, was titled: "Creation of Life Rates Best of Science Stories in 1967." In smaller type: "Human Heart Transplant Second."

PROOFREADING AND EDITING BY A REPLICATING ENZYME (1967–1971)

Knowing that DNA polymerase synthesized a chain in the 5' to 3' direction, it made no sense to me then that the enzyme degraded the very 3'-end of the chain it would normally be extending. In the absence of the nucleotide building blocks needed for synthesis, nucleotide units were cleaved slowly and serially from the 3' end of a DNA chain. Then a simple fact about the nuclease gave us our best clue. Douglas Brutlag observed that the degrading activity was far more potent on a single strand of DNA than on the usual double-stranded form. This preference became extreme when the temperature of the reaction was lowered, presumably because the ends of duplex DNA are less frayed at lower temperatures.

Why should a loose primer end be a substrate for degradation by a synthesizing enzyme? We prepared a variety of duplex DNAs in which a few residues at the primer end of a chain were not matched to the other strand. The mismatched residues were removed immediately, after which the others were removed far more slowly. When deoxynucleoside triphosphates were supplied to permit chain extension, the mismatched residues were still removed quickly, but now the subjacent nucleotides remained intact and were extended by synthesis (35).

Thus, the enzyme removes all mismatched units, permitting fresh units to be added to the growing chain end only when it is correctly matched to the template chain. We could infer that if the synthesizing enzyme were to make a rare mistake during elongation of a chain, such as inserting a C opposite an A (estimated to happen once in 10,000 times), it would remove the mismatched C before proceeding with extension of the chain. This astonishing proofreading ability of the enzyme, coupled with its fine discrimination in the initial choice of correct building blocks during synthesis, reduces errors in the overall process of replication to one in 10 million.

Having finally made sense of why an activity that degrades DNA is part of the very enzyme that makes it, we were unprepared for the paradoxical observation by Lehman (36) that the nuclease action of DNA polymerase on double-stranded DNA was enhanced tenfold when all four building blocks required for synthesis (i.e. A, T, G, and C) were present. How could synthesis be enhancing degradation? After all, we had observed earlier that synthesis extends the primer end of a chain and thereby protects it from nuclease action.

The solution came from Edward Reich and his colleagues (37) and from Murray Deutscher in my laboratory (38), showing that nuclease activity in DNA polymerase persists even with the 3'-end blocked by an analogue or phosphate. A separate domain in the enzyme removes nucleotides from the 5'-end of a chain.

Now we could explain how the four building blocks, and the synthesis they make possible, enhance nuclease action by DNA polymerase (39). By removal of DNA from the 5'-end at a nick, a stretch of template becomes exposed for pairing with a substrate nucleotide and further synthesis. In its synthetic progress along the template, the polymerase is brought up to a 5'-end of the chain, which it then degrades. It was immediately obvious how this "nick translation" by polymerase could be useful in the repair of lesions in DNA (40), and as we recognized some years later, could perform an essential step in replication by removing the RNA that initiates the start of a DNA chain.

DNA POLYMERASE UNDER INDICTMENT (1970–1972)

DNA polymerase was called a "red herring" and charged by *Nature New Biology* in a series of editorials with masquerading as a replication enzyme (41). The replicative role of DNA polymerase was questioned because of the Cairns mutant of *E. coli* (42), which appeared to lack the enzyme and yet grew and multiplied at a normal rate. In addition to the apparent dispensability of DNA polymerase for cell multiplication and its more estimable qualifications as a repair enzyme, genes were being discovered (designated *dnaA, dnaB, dnaC,* etc) that strongly implicated many other proteins as essential for a replication process far more complex than had been imagined.

The rising skepticism about the importance of DNA polymerase was fanned by the *Nature New Biology* vendetta. Not only was the enzyme attacked, but the basic mechanism, the building blocks, and the assays of DNA synthesis were judged to have prevented the discovery of the true DNA-replicating enzymes. At this juncture, my middle son, Tom, entered the fray. (I was at the Molecular Biology Laboratory in Cambridge, England, in the second half of a sabbatical year devoted to membranes.) An injury to his hand prevented him from continuing his career as a cellist at the Juilliard School, and he was disturbed by disparaging comments about DNA polymerase in his biology course at Columbia College where he was also a full-time student. Despite lack of laboratory experience, he found, within a few weeks, a DNA polymerase in *E. coli* cells, distinct from the one I had discovered. The new activity, named DNA polymerase II (pol II) (43), was clearly different from the "classic" DNA polymerase (pol I) and from still another, DNA polymerase III (pol III) (44), which he discovered in the course of purifying pol II. Subsequently, he and Malcolm Gefter located the gene for pol III and showed that conditionally lethal mutations in this gene blocked DNA replication (45). Pol III, in a far more elaborate form, was to gain recognition as the central enzyme of DNA replication in *E. coli.*

All three polymerases, although differing significantly in structure, proved to be virtually identical in their mechanisms of DNA synthesis, proofreading, and use of the same building blocks. The maligned polymerase (pol I) became the prototype for all DNA polymerases in plants, animals, and viruses, as well as in *E. coli.* The gloomy prophecies of *Nature New Biology* soon disappeared, as did the magazine itself.

HOW DNA CHAINS ARE STARTED (1971–1975)

Despite the excitement over the synthesis of a chain of infectious viral DNA, I had felt a certain uneasiness. One of the inferences drawn from the replication of a single-stranded, circular template was that DNA polymerase I could start a new chain. Yet we were never able to find direct proof of this. Moreover, we had observed that replication of the circular template was far more efficient if a small amount of boiled *E. coli* extract was present. Although it seemed unlikely that a random fragment of DNA in the extract would match the viral DNA template accurately enough to serve as a primer, this possibility became a reality. DNA polymerase removed the unmatched regions of the fragment by proofreading at the 3' end; with generous editing at the 5' end, no trace of the fragment remained in the synthetic product.

We were left with the question of how DNA chains are started, how a single-stranded, circular viral DNA is converted to the duplex form upon entering the cell, how nascent chains are initiated in the replication of virtually all chromosomes. Indeed, Reiji Okazaki had shown earlier (46) that chains are started not just once, at the beginning of the chromosome, but repeatedly in staccato fashion during the progress of replication.

After several years of unproductive attempts, I recognized a basic flaw in our work, how hopeless it was to answer the question about chain starts with the DNA we were using as template and primer. A tenet I was taught and to which I had faithfully adhered is that one must purify an enzyme to understand what it does. An aphorism attributed to me, but actually due to Efraim Racker, is: "Don't waste clean thinking on dirty enzymes." Another basic tenet of enzymology, too obvious, it would seem, to mention, is that one must provide the enzyme, clean or not, with a pure substrate.

The blunder we had made for too many years was accepting the DNA extracted from bacterial and animal cells as an adequate substrate for the enzymes of replication. The huge chromosomes are very fragile and the mechanical forces of flow and mixing used during isolation are violent enough to reduce the DNA to a heterogeneous collection of damaged fragments. In short, we were giving our relatively clean enzyme a very "dirty" substrate.

When I finally recognized the futility of searching for the replication enzymes with bacterial DNA, let alone animal DNA, I also realized that we had been ignoring a proper DNA substrate, the chromosome of a tiny bacteriophage. I recalled belatedly the virtues of the intact, clean, phage chromosome that four years earlier had served us in demonstrating the synthesis of infectious DNA by DNA polymerase I. As small, single-stranded circles, we could actually view them with an electron microscope and verify that in a purified sample, they were intact, homogeneous, and uncontaminated by fragments of the bacterial host DNA. We also knew that immediately upon entering the cell, the phage DNA is converted by bacterial enzymes to a double-stranded circle, an event we could easily assess. Probing how a new phage circle is started and completed might illuminate the intricate enzymatic machinery the cell uses to replicate its own chromosome.

We could use the chromosome of either of the two classes of small phages, the icosahedral ϕX174 or the filamentous M13. Luckily, as events later proved, I chose to work on M13 and within a week or two switched the efforts of my entire group to the various stages of its life cycle. I have sometimes felt wistful reflecting on the boldness of that move and the exciting events in the weeks that followed.

Being preoccupied with the initial event in M13 replication enabled me to connect three otherwise unrelated facts and arrive at an idea as to how a DNA chain might get started. For one, RNA polymerase, unlike DNA polymerase, can start chains. Furthermore, DNA polymerase, while routinely excluding ribonucleotides in assembling a DNA chain, does accept an RNA chain end matched to a DNA template as a primer for extension in DNA synthesis. Finally, DNA polymerase I has an editing function, which can remove something foreign, like thymine dimers and RNA from the start of a DNA chain, and replace it with proper DNA.

Might RNA polymerase make a short piece of RNA on single-stranded M13, which DNA polymerase could use to start a DNA chain? Then, when the enzyme had come full circle in copying the available template, its editing system would erase the RNA and synthesize DNA in its place. We could test this hypothesis by using rifampicin to inhibit RNA polymerase in vivo (47). When Doug Brutlag did so, the M13 circle was not replicated! We went on to show with cell extracts and partially purified enzymes that RNA polymerase initiates DNA replication by forming a primer RNA for covalent attachment of the deoxyribonucleotide that starts the new DNA chain (48).

There was one discordant note. Rifampicin did not prevent the conversion of the phage ϕX174 single-stranded circle to its duplex form, nor did it interrupt the ongoing replication of the E. coli chromosome, despite the repeated initiations of DNA strands presumed to be occurring at the growing fork. Either RNA priming was of limited significance, or another mode of RNA synthesis, independent of RNA polymerase, existed. As we probed the initiation on ϕX174 circles, it became clear that this virus, instead of relying on RNA polymerase, exploits the extraordinarily complex assembly of initiation proteins that the cell uses for replicating its own chromosome.

PRIMOSOMES, HELICASES, AND REPLISOMES (1972–)

After a 10-year drought of discoveries of new enzymes, they now came in a torrent, and in 1972 a bright and boistrous group, led by Randy Schekman and

Bill Wickner, was there to collect and sort them. Upon fractionating the components responsible for conversion of the single-stranded circle of ϕX174 to the duplex form, we could separate them into two groups, one that primed the start of a chain and the other that extended it. Then, as we tried to purify each of these fractions, they splintered into many separate components. The joy of uncovering the trails to so many novel proteins soon gave way to the discouragement of being unable to track down any one of them. We judged there might be as many as eight different proteins, all scarce, that were needed to make the tiny bit of RNA that primed the synthesis of a DNA chain; the DNA polymerase seemed just as complex. What were all these proteins and what was each of them doing?

A major assist came from New Haven sewage via Nigel Godson, who discovered in it a new phage (G4) that resembled ϕX174 (49). Replication of G4 DNA required only only three of the eight fractions needed by ϕX174 (50). One of the fractions was purified easily because it withstood heating. It was the single-strand binding protein (SSB) (51), which coats the DNA, except at a region that forms a duplex, hairpinlike structure and is used by the second fraction as a template to synthesize a stretch of primer RNA. This protein, known to complement the deficiency of cell extracts of dnaG mutants, was named primase (52). The third fraction was DNA polymerase III in a complex, unstable form (53) with many auxiliary units that clamp the polymerase to the template and enable it to replicate great lengths of DNA with astonishing speed and accuracy. Replication of G4 DNA thus provided us with our best assays for purifying each of these components: SSB, primase, and the super polymerase we now called DNA polymerase III holoenzyme.

We could now return to ϕX174 and begin to explain the molecular operations of the multiprotein assembly (called a primosome) (54) that starts a DNA chain. The image of a locomotive seemed helpful for a time in accounting for primosome actions. The engine, protein n' (55) (its gene unknown to this day), powered by ATP energy, is a helicase (56, 57); it unzippers the DNA duplex and is equipped with a cowcatcher to remove SSB in its path. Another protein, dnaB, is both helicase (58, 59) and engineer (60), using ATP to locate or shape a section of DNA track upon which primase will find it possible to lay down a short stretch of RNA, which will then attract DNA polymerase to start a DNA chain. The primosome is translocated on DNA in only one direction, the one that keeps it at the advancing fork of a replicating chromosome (61).

Another awesome property of the primosome is the persistence of its attachment to the ϕX174 DNA circle even after the duplex circle has been completed (62). The attached primosome directs the next replication event and is used over and over again as part of a stamping machine to generate the many duplex forms needed for transcription into messages for the 10 proteins encoded by the phage DNA.

To multiply duplex circles (Figure 2), the phage employs a single gene (gene A) of its own and relies on host proteins to do the rest. The isolated gene A protein breaks the viral strand backbone at a particular diester bond and becomes covalently attached to the 5'-end of the break (63). In so doing, it makes the 3'-end available as a primer for DNA polymerase III holoenzyme replication. For lack of exposed template at this nick, the phage exploits the host rep protein, a helicase, to unzipper the duplex (64, 65). This was the first occasion in which a helicase was seen in its role of opening of a duplex for an advancing replication fork. Coordinate helicase and polymerase actions generate single-stranded viral circles repeatedly by a mechanism called rolling-circle replication (66). Each completed circle is then replicated to form a duplex in the same way as the one originally injected by the phage.

Based on these insights gained from the proteins involved in the replication of the small phages, we now propose that progress of replication of the host chromosome (Figure 3) (67) depends on helicases, SSB, and topoisomerases to prepare the templates for the continuous synthesis of a leading strand and the discontinuous synthesis of the other. We regard the polymerase III holoenzyme as a rather loose assembly of many auxiliary units attached

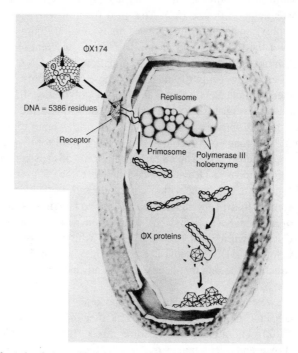

Figure 2   Life cycle of phage ϕX174 in *E. coli*. Conversion of the viral circle to a duplex is followed by multiplication of the duplexes by a rolling-circle mechanism.

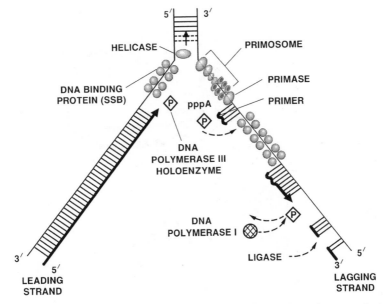

Figure 3   Scheme for enzymes operating at a replication fork in *E. coli:* continuous synthesis of a leading strand and RNA-primed discontinuous synthesis of the other strand.

asymmetrically to a pair of polymerase cores. One arm appears suited to the highly processive synthesis of the leading strand and the other arm to the discontinuous synthesis of the other strand (68, 69).

We now wonder whether essentially concurrent replication of both strands might be achieved were priming of nascent fragments of discontinuous synthesis integrated with continuous strand synthesis. A replisome comprising the holoenzyme, primosome, and helicases might periodically generate a loop in the lagging strand template to place it in the same orientation as the leading strand at the fork (Figure 4) (67).

We had been aware that the primosome and the primase in it moved in opposite directions on a DNA chain. With additional knowledge, the difficulty grew. The isolated dnaB component was found to move in the direction of the primosome (58), whereas the n' component by itself moved in the opposite (chain elongation) direction (57). A smoldering annoyance had become a serious paradox. How could integral parts of the primosome locomotive move in opposite directions on a fixed track of DNA? The analogy would have to be scrapped. Instead, it now makes more sense to invert the relative movements of the primosome and DNA and regard the primosome machine as fixed in place with the DNA chains being drawn through it (Figure 4) (57). This view also offers an attractive mechanism for coordinating the

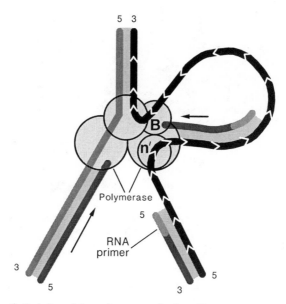

Figure 4   Hypothetical view of the replisome as a fixed machine through which a loop of DNA is pulled to provide for essentially concurrent replication of both strands. As the DNA is pulled through the primosome, the n' protein is translocated on the DNA in the direction of chain elongation and the B protein in the opposite direction.

helicase, priming, and replication actions at the advancing fork of a chromosome.

INITIATION OF CHROMOSOMES (1979–)

An aspect of replication that has long intrigued me is how an increase in *E. coli* cell mass triggers the initiation of replication that commits the cell to start a new cycle. What is the biochemistry of the replication switch, which in *E. coli* regulates the cell cycle and in eukaryotes responds to signals that turn the embryonic cell to a quiescent adult, or the quiescent cell to proliferation? With the cloning of the highly conserved, unique, 245–base pair chromosomal origin (*oriC*) in plasmids, we were afforded a substrate with which to seek the comparably complex, conserved multiprotein system that uses *oriC* to initiate a cycle of replication. After 10 man-years of fruitless effort to obtain a cell-free initiation system, Bob Fuller and Jon Kaguni were the ones who finally succeeded (70). Two apparently illogical maneuvers were essential. One was the inclusion of a hydrophilic polymer (e.g. polyethylene glycol) at high levels which, as we later realized, acted by a "macromolecular crowd-ing" effect, which concentrates the numerous proteins and DNA into a small volume. The other was subjecting an inert lysate to a refined ammonium sulfate fractionation, a trick that had worked for me 30 years earlier in the discovery of the yeast enzyme that converts NAD to NADP (71). In the active fraction the numerous required proteins are concentrated and potent inhibitors are excluded.

The proteins that we found responsible for initiating replication at *oriC* include *initiation proteins* (particularly dnaA protein) that recognize super-coiled *oriC*, alter its structural conformation, and lead to its further opening by dnaB helicase action, *specificity proteins* that suppress potential origins elsewhere on the chromosome, and the *replication proteins* that prime and elongate chains on the opened plasmid and propel two forks in opposite directions (72, 73).

The motif we are finding in the mechanism for initiation of the *E. coli* chromosome (Figure 5) seems to apply to a wide variety of bacteria and their plasmids and phages (74). Control of dnaA protein by ATP-ADP cycling (75) and by binding to the acidic headgroups of a fluid membrane (76), and activation of an inert origin in an overly relaxed supercoil by nearby transcrip-tion (77) are among the factors already discovered that influence ini-tiation, and likely more will be found as we explore this crucial event in the cell cycle.

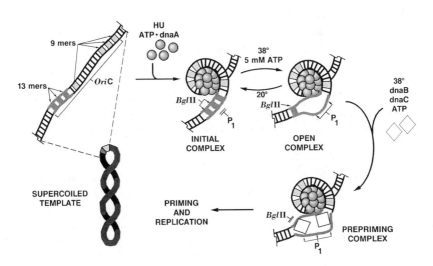

Figure 5 Scheme for early events in initiation of replication at the origin of the *E. coli* chromosome operating in a plasmid.

SPORULATION AND GERMINATION (1962–1970)

Those familiar with my research are aware of my intense concentration on a single subject—the enzymatic synthesis of DNA—and the blinders I have worn to maintain this focus. Nearly forgotten now by us all are the eight years, in the midst of the DNA work, when half of my research effort was devoted to an arcane subject, the development of spores (78).

During my tenure as chairman of a department of microbiology (1953–1959), I had become interested again in spores as agents of disease: anthrax, tetanus, botulism. I constantly remembered with deep anguish the ghastly *Clostridium perfringens* (gas gangrene) spore that killed my mother within a day of a "routine" gall bladder operation in 1939. Now I could look beyond the "bad" spores to the vast array of innocent species whose mysterious biology and biochemistry fascinated me.

How is a spore made and how does its chemical organization endow it with astonishing abilities: dormancy and resistance to extremes of heat, desiccation, disinfectants, and ultraviolet rays lethal to the cell? How does a spore, after years of hibernation, respond instantly to a substance that signals conditions are right for growth into a new cell? I believed then, and still do, that this knowledge would contribute in a major way toward understanding the embryonic development of animals and their response to environmental stresses.

During the eight-year period in which a succession of students and postdoctoral fellows (including Pieter Bonsen, Pierre Chambon, Murray Deutscher, Arturo Falaschi, David Nelson, Tuneko Okazaki, Peter Setlow, Jim Spudich, Henrique Tono, and Jim Vary) worked on spores, we published 26 papers and still were making little progress toward answering the global questions that had attracted me.

I abandoned the spore work when I came to realize how much more complex the problems were than I had imagined and that hardly anyone else in the world seemed to care. The little research on spores, then and even now, was largely of a practical nature: how to destroy spores in food canning or how to use them as pesticides on crops.

Beyond the discouragement and loneliness of working on a tough problem in an unfashionable area was the distraction of having the other half of my research group of eight or so engaged in the more glamorous and productive work on DNA replication. Because of my own ambivalence, I offered no resistance on occasions when a member of the sporulation group defected to the replication team. Finally, after this eight-year siege, I too gave up. Eventually only Peter Setlow has continued a biochemical interest in spores.

Progress in science depends on how vigorously the field is cultivated. In contrast with sporulation, interest in cancer is enormous. Hundreds of laboratories worldwide attract many thousands of scientists, including the brightest, to unravel the processes responsible for malignant growth. Yet studies of sporulation are also deserving of resources and talents. Sporulation, dormancy, and germination are fundamental processes in nature, more accessible to incisive examination, and, if better understood, might yield as much information relevant to the cancer process as some of the massive programs on tumor-bearing animals.

SCIENTIST, TEACHER, AUTHOR, CHAIRMAN: IN WHAT ORDER?

In May 1988, when my former and present students and colleagues gathered in San Francisco for a gala 70th birthday party, I thought it would be fun to select from the 30 or more enzymes I had worked with, the 10 that I favored most. I was surprised to find that 6 of the 10 were discovered in the brief period from 1948 to 1955: nucleotide pyrophosphatase (8), NAD synthetase (9), phosphatidic acid synthetase (14), PRPP synthetase (17), polyphosphate synthetase (79), and DNA polymerase (26). That left only 4 to be selected from more than 20 enzymes that appeared in the next 30+ years. Inasmuch as some of the enzymes omitted from the top 10 are far more deserving of selection than some of the chosen ones, it is clear that the basis for the choice of the first 6 was largely sentimental. I was most attached to those enzymes that came during the time of my life when I collected the data myself, from conception to delivery.

In my marriage to enzymes, I have found a level of complexity that suits me. I feel ill at ease grappling with the operations of a cell, let alone those of a multicellular creature. I also feel inadequate in probing the fine chemistry of small molecules. Becoming familiar with the personality of an enzyme performing in a major synthetic pathway is just right. To gain this intimacy, the enzyme must first be purified and I have never felt unrewarded for any effort expended this way.

I once shocked the Dean of the Washington University School of Medicine by telling him that my prime interest as Chairman of the Department of Microbiology was to do and foster research rather than teach. It has never been otherwise. Experiments are far more consuming and fulfilling for me than any form of teaching. Still, I have enjoyed a rather modest amount of formal lecture and laboratory instruction and have done it conscientiously. For the student, didactic teaching fails without the infusion of scientific skepticism and a fervor for new knowledge, and these things are naturally conveyed by someone dedicated to research. For me, some 10 lectures a year freshen my awareness of basic subjects, and on one occasion the preparation of a laboratory exercise on DNA opened a major avenue for my experimental work.

The most rewarding teaching for me has been in the intimate, daily contact with graduate and postdoctoral students. Well over a hundred of them spent from two to five years in my laboratory and were exposed to my tastes and my obsession with the use of time. I felt closest to those who shared my devotion to enzymes and my concern with the productive use of our most precious resource: each of the hours and days that so quickly stretch into the few years of a creative life. I recall in 1948 relating to Sidney Colowick and Ollie Lowry (both senior to me in age and experience) my failure in purifying an enzyme by a certain procedure. "I wasted a whole afternoon trying that," I said. Colowick turned to Lowry and said with mock gravity: "Imagine, Ollie, he wasted a *whole* afternoon."

Imagination or hard work? At either extreme—speculating about complex phenomena or doggedly collecting data—success may come on occasion and draw acclaim. But the most consistent approach for acquiring a biochemical understanding of nature lies in between. The novel is yet to be written that captures the creative and artistic essence of scientific discoveries and dispels images of the scientist as dreamer, walking in the woods awaiting a flash of insight or of the scientist as engineeer, at an instrument panel executing a precisely planned experiment. Some intermediate ground, hard work with a touch of fantasy, is what I have sought for myself and my students.

If asked to name varieties of mental torture, most scientists would place writing near the top of the list. As a result, scientific papers are usually put off or dashed off and demean the quality and value of the work they describe. Writing a paper is an integral part of the research and surely deserves the small fraction, say five percent, of the time spent finding the thing worth reporting. Yet, I feel uneasy seeing students and colleagues writing at their desks during "working hours" rather than busy at the laboratory bench. Whereas taking time to prepare a scientific report is unavoidable, writing a book always seemed an unconscionable abdication from research until I wrote one.

Writing *DNA Synthesis,* a 400-page book (80), was a surprise in many ways. First, the effort was far greater than I imagined. Very little from lecture notes and reprints could be lifted and placed in the right context and still remain readable. I was also surprised by the pleasure I found in reworking and polishing sentences and paragraphs for brevity and clarity, a satisfaction I had never found in crossword puzzles or other word games. Best of all, I could present my work, views, and excitement about the enzymology of DNA replication to an unexpectedly wide audience. The book—adopted as a text for some courses—became the reference source for writers of reviews of DNA replication and for authors of textbooks of biology and biochemistry.

The sequel to *DNA Synthesis,* entitled *DNA Replication* (23), came out in 1980. Twice the size of its predecessor, it was really a new book in scope and organization as well as in expanded contents. Despite its inflated size, it was a better book and found a wider readership. However, progress in this field is so rapid that revisions are needed annually. As an experiment in publishing, I assembled a 273-page "1982 Supplement to DNA Replication" (67), which extended the life of its parent. The publishers objected to "1982" in the title and correctly saw it as an advertisement of obsolescence. There have been no further supplements.

Having regarded teaching and book writing as deviant activities for a dedicated scientist, then surely the administrative work of a departmental chairman should be beyond the pale. Yet, I served as chairman for more than 20 years and never found it a serious intrusion on my time or attention. On the contrary, the benefits of creating and maintaining a collegial and stimulating scientific circle were well worth the investment I made. With excellent administrative assistance and the eager participation of my faculty colleagues, direction of departmental activities took no more time than being a conscientious member of the department.

Involvement in medical school and university affairs is a far different matter. I never found the skills and patience to function at these levels. For me, the most burdensome feature of being a departmental chairman was the obligated service on the Executive Committee of the Medical School, preoccupied with budgets, promotions, interdepartmental feuds, and salaries. In 6 years at Washington University and 10 at Stanford, I cannot recall a deliberate discussion of science or educational policy. No wonder I had no interest in being the dean of a medical or graduate school on occasions when this possibility was raised.

Increasingly conspicuous in current scientific life are the extramural administrative and educational activities, which, with the attendant travel, may consume half the time of prominent members of a science faculty. Lectures and visiting professorships, scientific meetings and society councils, government panels and advisory boards, consultantships in industry—all are prestigious, diverting, less demanding than research, and terribly tempting. I have done less than most, but have been unable to resist participating, particularly in writing essays (81–83), testifying for federal support of research and training, and most recently in the founding and development of a biotechnology enterprise (the DNAX Research Institute, Inc., later acquired by the Schering-Plough Corp.) with the mission of applying the techniques of molecular and cellular biology to the therapy of diseases of the immune system.

All these nonresearch activities, in and out of the university, fail to give me a deep sense of personal achievement. In research, it is up to me to select a corner of the giant jigsaw puzzle of nature and then find and fit a missing piece. When after false starts and fumbling, a piece falls into place and provides clues for more, I take pleasure in having done something creative. By contrast, in my other activities, which are just as personal, all I do, it seems, is try to behave in a commonsensical, fair, and responsible way, as anyone else would. With research so dominant over my teaching, writing, and administrative activities, in sharply descending order of importance to me, I sometimes wonder whether, valued for their contributions to science, this order might be inverted.

Beginning with administration, consider the creation and management of the Stanford Biochemistry Department. It was started in St. Louis as the Microbiology Department. From there, Paul Berg, Bob Lehman, Dave Hogness, Dale Kaiser, and Mel Cohn moved with me to Stanford in 1959 to be joined by Buzz Baldwin and a few years later by George Stark and Lubert Stryer. In polls of peers, the Department has been accorded a top rating for many years, and is regarded as a major source of discoveries basic to recombinant DNA and the genetic engineering revolution. More than 500 people trained in the department now staff and direct departments of biochemistry and molecular biology all over the world. The organization and development of this notable faculty and its preservation, largely intact, against strong and attractive centrifugal forces, I would have to admit is a unique achievement.

As for writing, the monographs on DNA replication, with more than 40,000 copies sold, have made it easier for others to enter and work in this field. More than offering a readable account of a forbiddingly specialized area of biochemistry, these books have helped revive an appreciation that enzymology provides a direct route toward solving biologic problems and creates reagents for the analysis and synthesis of a great variety of compounds for all branches of biologic science.

With regard to teaching, assumption of credit for the success of a student has always puzzled me. There simply are no controls in these experiments. How do I know, given a motivated, gifted student, whether I have been a help or hindrance? Nevertheless, having involved myself in the daily scientific lives of my students, I may have guided some of them in directions that attract me and thereby diverted them from a career in biology or chemistry to the love and pursuit of biochemistry and enzymes. These progeny now include illustrious figures in science who have spread this gospel to a widening circle of "grandstudents" and "great-grandstudents."

Finally, even were I forced to agree that my activities in administration, writing, and teaching had a singular quality, I would have to concede further that my discoveries in science did not. Very likely, they would have been made by others soon after. Yet in the last analysis, I will argue that for me it was the research that mattered most, because all my attitudes and activities were shaped by it.

ACKNOWLEDGMENTS

To the gifted students, postdoctoral fellows, and research associates (notably LeRoy Bertsch), including many whose work I have not cited here, I am deeply grateful. I also thank the NIH and NSF for 35 years of generous and uninterrupted research support. In preparing this memoir, I have borrowed extensively from *For the Love of Enzymes: The Odyssey of a Biochemist* (Harvard University Press, 1989) and I am indebted to the publishers for their permission to do so. Finally, I dedicate this autobiographical account to the memory of Sylvy for 43 years of love and devotion, for unwavering support for me to start and sustain a career in science, and for a happy family life with our three sons: Roger, Tom, and Ken.

Literature Cited

1. Kornberg, A. 1942. *J. Clin. Invest.* 21:299–308
2. Kornberg, A., Daft, F. S., Sebrell, W. H. 1943. *Science* 98:20–22
3. Hutchings, B. L., Stokstad, E. L. R., Bohonos, N., Slobodkin, N. H. 1944. *Science* 99:371
4. Black, S., Overman, R. S., Elvehjem, C. A., Link, K. P. 1942. *J. Biol. Chem.* 145:137–43
5. Kornberg, A., Daft, F. S., Sebrell, W. H. 1944. *J. Biol. Chem.* 155:193–200
6. Ochoa, S., Mehler, A. H., Kornberg, A. 1947. *J. Biol. Chem.* 167:871–72
7. Kornberg, A., Lindberg, O. 1948. *J. Biol. Chem.* 176:665–77
8. Kornberg, A., Pricer, W. E. 1950. *J. Biol. Chem.* 182:763–77
9. Kornberg, A. 1950. *J. Biol. Chem.* 182:779–93
10. Schrecker, A. W., Kornberg, A. 1950. *J. Biol. Chem.* 182:795–803
11. Kornberg, A. 1957. *Adv. Enzymol.* 18:191–240
12. Watson, J. D., Crick, F. H. C. 1953. *Nature* 171:737–38, 964–67
13. Kornberg, A., Pricer, W. E. 1952. *J. Am. Chem. Soc.* 74:1617
14. Kornberg, A., Pricer, W. E. 1953. *J. Biol. Chem.* 204:345–57
15. Hayaishi, O., Kornberg, A. 1952. *J. Biol. Chem.* 197:717–32
16. Lieberman, I., Kornberg, A. 1954. *J. Biol. Chem.* 207:911–24
17. Kornberg, A., Lieberman, I., Simms, E. 1955. *J. Biol. Chem.* 215:389–402
18. Lieberman, I., Kornberg, A., Simms, E. 1955. *J. Biol. Chem.* 215:403–15
19. Khorana, H. G., Fernandes, J. F., Kornberg, A. 1958. *J. Biol. Chem.* 230:941–48
20. Kornberg, A., Lieberman, I., Simms, E. 1955. *J. Biol. Chem.* 215:417–27
21. Buchanan, J. M., Wilson, D. W. 1953. *Fed. Proc.* 12:646
22. Greenberg, G. R. 1953. *Fed. Proc.* 12:651
23. Kornberg, A. 1980. *DNA Replication.* San Francisco: Freeman. 724 pp.
24. Grunberg-Manago, M., Ochoa, S. 1955. *J. Am. Chem. Soc.* 77:3165–66
25. Littauer, U. Z., Kornberg, A. 1957. *J. Biol. Chem.* 226:1077–92
26. Kornberg, A., Lehman, I. R., Bessman, M., Simms, E. 1956. *Biochem. Biophys. Acta* 21:197–98
27. Kornberg, A. 1957. *The Chemical Basis of Heredity.* ed. W. D. McElroy, B. Glass, pp. 579–608. Baltimore: Johns Hopkins
28. Josse, J., Kaiser, A. D., Kornberg, A. 1961. *J. Biol. Chem.* 236:864–75
29. Schachman, H. K., Adler, J., Radding, C. M., Lehman, I. R., Kornberg, A. 1960. *J. Biol. Chem.* 235:3242–49
30. Radding, C. M., Josse, J., Kornberg, A. 1962. *J. Biol. Chem.* 237:2869–76
31. Radding, C. M., Kornberg, A. 1962. *J. Biol. Chem.* 237:2877–82
32. Kornberg, A., Bertsch, L., Jackson, J. F., Khorana, H. G. 1964. *Proc. Natl. Acad. Sci. USA* 51:315–23
33. Goulian, M., Kornberg, A. 1967. *Proc. Natl. Acad. Sci. USA* 58:1723–30
34. Goulian, M., Kornberg, A., Sinsheim-

30 KORNBERG

er, R. L. 1967. *Proc. Natl. Acad. Sci. USA* 58:2321–28
35. Brutlag, D., Kornberg, A. 1972. *J. Biol. Chem.* 247:241–48
36. Lehman, I. R. 1967. *Annu. Rev. Biochem.* 36:645–68
37. Klett, R. P., Cerami, A., Reich, E. 1968. *Proc. Natl. Acad. Sci. USA* 60:943–50
38. Deutscher, M. P., Kornberg, A. 1969. *J. Biol. Chem.* 244:3029–37
39. Kornberg, A. 1969. *Science* 163:1410–18
40. Kelly, R. B., Atkinson, M. R., Huberman, J. A., Kornberg, A. 1969. *Nature* 224:495–501
41. Editorials. 1971. *Nature New Biol.* 229:65–66; 230:258; 233:97–98
42. De Lucia, P., Cairns, J. 1969. *Nature* 224:1164–66
43. Kornberg, T., Gefter, M. L. 1970. *Biochem. Biophys. Res. Commun.* 40:1348–55
44. Kornberg, T., Gefter, M. L. 1971. *Proc. Natl. Acad. Sci. USA* 68:761–64
45. Gefter, M. L., Hirota, Y., Kornberg, T., Wechsler, J. A., Barnoux, C. 1971. *Proc. Natl. Acad. Sci. USA* 68:3150–53
46. Sugimoto, K., Okazaki, T., Okazaki, R. 1968. *Proc. Natl. Acad. Sci. USA* 60:1356–62
47. Brutlag, D., Schekman, R., Kornberg, A. 1971. *Proc. Natl. Acad. Sci. USA* 68:2826–29
48. Wickner, W., Brutlag, D., Schekman, R., Kornberg, A. 1972. *Proc. Natl. Acad. Sci. USA* 69:965–69
49. Godson, G. M. 1974. *Virology* 58:272–89
50. Zechel, K., Bouché, J.-P., Kornberg, A. 1975. *J. Biol. Chem.* 250:4684–89
51. Weiner, J. H., Bertsch, L. L., Kornberg, A. 1975. *J. Biol. Chem.* 250:1972–80
52. Bouché, J.-P., Rowen, L., Kornberg, A. 1978. *J. Biol. Chem.* 253:765–69
53. Wickner, W., Kornberg, A. 1973. *Proc. Natl. Acad. Sci. USA* 70:3679–83
54. Arai, K., Low, R., Kobori, J., Shlomai, J., Kornberg, A. 1981. *J. Biol. Chem.* 256:5273–80
55. Shlomai, J., Kornberg, A. 1980. *Proc. Natl. Acad. Sci. USA* 77:799–803
56. Lee, M. S., Marians, K. J. 1987. *Proc. Natl. Acad. Sci. USA* 84:8345–49
57. Lasken, R. S., Kornberg, A. 1988. *J. Biol. Chem.* 263:5512–18

58. LeBowitz, J. H., McMacken, R. 1986. *J. Biol. Chem.* 261:4738–78
59. Baker, T. A., Funnell, B. E., Kornberg, A. 1987. *J. Biol. Chem.* 262:6877–85
60. Arai, K., Kornberg, A. 1981. *J. Biol. Chem.* 256:5267–72
61. Arai, K., Kornberg, A. 1981. *Proc. Natl. Acad. Sci. USA* 78:69–73
62. Low, R. L., Arai, K., Kornberg, A. 1981. *Proc. Natl. Acad, Sci. USA* 78:1436–40
63. Eisenberg, S., Kornberg, A. 1979. *J. Biol. Chem.* 254:5328–32
64. Scott, J. F., Kornberg, A. 1978. *J. Biol. Chem.* 253:3292–97
65. Kornberg, A., Scott, J. F., Bertsch, L. L. 1978. *J. Biol. Chem.* 253:3298–304
66. Gilbert, W., Dressler, D. 1968. *Cold Spring Harbor Symp. Quant. Biol.* 33:473–84
67. Kornberg, A. 1982. *1982 Supplement to DNA Replication.* San Francisco: Freeman. 273 pp
68. Maki, H., Maki, S., Kornberg, A. 1988. *J. Biol. Chem.* 263:6570–78
69. McHenry, C. S. 1988. *Annu. Rev. Biochem.* 57:519–50
70. Fuller, R. S., Kaguni, J. M., Kornberg, A. 1981. *Proc. Natl. Acad. Sci. USA* 78:7370–74
71. Kornberg, A. 1950. *J. Biol. Chem.* 182:805–13
72. Funnell, B. E., Baker, T. A., Kornberg, A. 1987. *J. Biol. Chem.* 262:10327–34
73. Bramhill, D., Kornberg, A. 1988. *Cell* 52:743–55
74. Bramhill, D., Kornberg, A. 1988. *Cell.* In press
75. Sekimizu, K., Bramhill, D., Kornberg, A. 1988. *J. Biol. Chem.* 263:7124–30
76. Yung, B. Y., Kornberg, A. 1988. *Proc. Natl. Acad. Sci. USA* 54:915–18
77. Baker, T. A., Kornberg, A. 1988. *Cell* 55:113–23
78. Kornberg, A., Spudich, J. A., Nelson, D. L., Deutscher, M. P. 1968. *Annu. Rev. Biochem.* 37:51–78
79. Kornberg, A., Kornberg, S. R., Simms, E. 1956. *Biochem. Biophys. Acta* 20:215–27
80. Kornberg, A. 1974. *DNA Synthesis.* San Francisco: Freeman. 399 pp
81. Kornberg, A. 1976. *New Engl. J. Med.* 294:1212–16
82. Kornberg, A. 1984. In *Medicine, Science and Society,* ed. K. J. Isselbacher, pp. 6–17. New York: Wiley
83. Kornberg, A. 1987. *Biochemistry* 26:6888–91

Annu. Rev. Biochem. 1989. 58:31–44

THE PROTEIN KINASE C FAMILY: HETEROGENEITY AND ITS IMPLICATIONS

Ushio Kikkawa, Akira Kishimoto, and Yasutomi Nishizuka

Department of Biochemistry, Kobe University School of Medicine, Kobe 650, Japan

CONTENTS

PERSPECTIVES AND SUMMARY

The physiological importance of protein kinase C (PKC) activation is widely appreciated and well documented. It is now clear that there is more than one species of PKC molecule, and several discrete subspecies have been defined. These proteins are derived both from multiple genes and from alternative splicing of a single RNA transcript, yet possess a primary structure containing conserved structural motifs with a high degree of sequence homology. The enzyme subspecies purified from tissues show subtle differences in their mode of activation, sensitivity of Ca^{2+}, and catalytic activity toward endogenous substrates. In the brain tissues, for example, at least seven subspecies can be distinguished, one of which is expressed only in the central nervous tissue. Biochemical and immunocytochemical studies with subspecies-specific antibodies suggest that the PKC subspecies may be differently located in particu-

0066-4154/89/0701-0031$02.00

lar cell types and at limited intracellular locations. Many cell types so far examined express more than one subspecies in variable ratios, and their intracellular distribution may depend on the state of activation of the cells. Interestingly, in response to extracellular signals, these subspecies are frequently down-regulated at different rates. Although, at present, there is little evidence to discuss specific functions of each PKC subspecies, the members of the enzyme family may have distinct roles in the processing and modulation of a variety of physiological and pathological cellular responses.

MOLECULAR HETEROGENEITY

Initially, four cDNA clones that encode α- βI-, βII, and γ-subspecies were found in the bovine (1, 2), rat (3–10), rabbit (11, 12), and human brain (2, 13), and later in human spleen (14) cDNA libraries. The integrated nomenclature used herein for the PKC subspecies known at present is as described elsewhere (5, 15). The genes for the α-, β- (βI plus βII), and γ-subspecies are located on different chromosomes (2). Partial genomic analysis has indicated that βI- and βII-subspecies are derived from a single RNA transcript by alternative splicing (13, 15, 16). Recently, another group of cDNA clones, encoding at least three further subspecies designated δ-, ϵ-, and ζ-PKC, have been isolated from a rat brain library by using a mixture of α-, βII, and γ-cDNAs as probes under low stringency conditions (17, 18). A cDNA clone encoding ϵ-PKC was also found more recently in the rabbit brain library (19). Another cDNA clone, isolated previously from a rat cDNA library (10), may encode a part of ϵ-PKC. These subspecies have a common structure closely related to, but clearly distinct from, the four subspecies initially described.

Figure 1 shows schematically the structures of PKC subspecies cloned to date. They are all composed of a single polypeptide chain, with each in the group of α-, βI-, βII- and γ-subspecies having four conserved (C_1–C_4) and five variable (V_1–V_5) regions (2, 6). The βI- and βII-subspecies differ from each other only in a short range of about 50 amino acid residues at their carboxyl-terminal end regions V_5, and even in this area possess a high degree of sequence homology (4, 6, 8, 11, 13–16). Although the second group of δ-, ϵ-, and ζ-subspecies lack the region C_2, the molecular mass of the enzyme molecules of δ- and ϵ-subspecies is, nevertheless, similar. Table 1 summarizes these members of the PKC family.

The conserved region C_1 of the members of the family so far identified, except for ζ-PKC, contains a tandem repeat of a cysteine-rich sequence, Cys-X_2-Cys-$X_{13(14)}$-Cys-X_2-Cys-X_7-Cys-X_7-Cys, where X represents any amino acid. ζ-PKC contains only one set of the cysteine-rich sequence, and has a relatively small molecular mass (Y. Ono et al, in preparation). The cysteine-rich sequence resembles the consensus sequence of a "cysteine-zinc

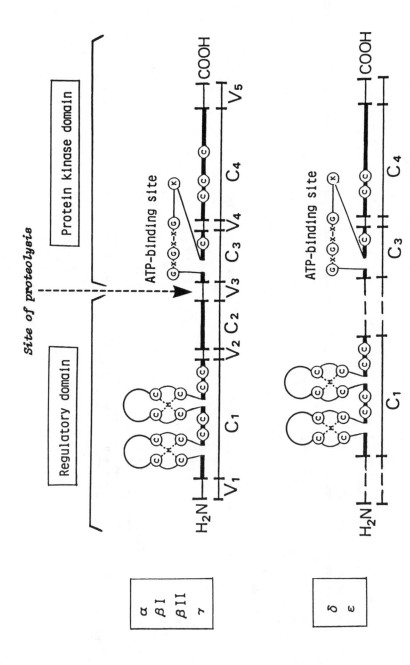

Figure 1 Common structure of protein kinase C subspecies. C, G, K, X, and M represent cysteine, glycine, lysine, any amino acid, and metal, respectively. Other details are described in the text. Adapted from Ref. 100.

Table 1 Subspecies of protein kinase C from mammalian tissues

Subspecies	α	βI	βII	γ	δ	ε	ζ[a]
Amino acid residues	672	671	673	697	673	737	592
Calculated molecular weight	76,799	76,790	76,933	78,366	77,517	83,474	67,740
Chromatographic subfraction	type III	—type II—	—type II—	type I	—	—	—
Activators[b]	PS+DG+Ca^{2+} AA+Ca^{2+}	—PS+DG+Ca^{2+}—	—PS+DG+Ca^{2+}—	PS+DG+Ca^{2+} AA	PS+DG+(Ca^{2+})	PS+DG+(Ca^{2+})	—
Tissue expression	Universal	Some tissues & cells	Many tissues & cells	Brain & spinal cord only	Many tissues[c]	Brain only[c]	Some tissues
Chromosome location (human)[d]	17	—16—	—16—	19	—	—	—
References (molecular cloning)	1, 2, 6, 7,11	4, 8, 11, 13–16	2, 4, 8, 11	2, 6, 7, 8, 12	17, 18	10, 17–19	17, 18

[a] The complete sequence will be described elsewhere (Y. Ono et al., in preparation).
[b] The activators for each subspecies are determined with calf thymus H1 histone as phosphate acceptor. PS, phosphatidylserine; DG, diacylglycerol; AA, arachidonic acid.
[c] The tissue expression is estimated by Northern blot analysis (Refs. 10, 18, 19).
[d] From Ref. 2.

DNA-binding finger" that is found in many metallo-proteins and DNA-binding proteins that are related to transcriptional regulation (20, 21). There is no evidence presently, however, indicating that PKC will bind to DNA. The conserved regions C_1 and C_2 apparently include the regulatory domain, but as yet the sites involved in Ca^{2+}-, diacylglycerol-, and phospholipid-binding have not been identified with any certainty. Rather, the mechanism of PKC activation appears to be a complex interaction between these components.

The carboxyl-terminal half containing the regions C_3 and C_4 appears to be the protein kinase domain, as it shows large clusters of sequence homology with many other protein kinases (1–4, 6–8, 10–12, 18, 19). The conserved region C_3 has an ATP-binding sequence, Gly-X-Gly-X-X-Gly——Lys (22). While the conserved region C_4 contains an additional sequence similar to the ATP-binding site, Gly-X-Gly-X-X-Gly——(X), the significance of this repeat remains unknown (6). The regulatory and protein kinase domains are cleaved by limited proteolysis catalyzed by a Ca^{2+}-dependent neutral protease, calpain (23, 24). The physiological significance of this proteolysis will be discussed below.

While the genomic structure of mammalian PKC has not yet been determined, a *Drosophila* genomic clone has been sequenced that shows about 60% homology to the mammalian PKC (25). This insect gene spans about 20 kilobases of chromosomal DNA with at least 14 exons.

DIFFERENTIAL REGIONAL EXPRESSION

Early studies using autoradiographic analysis after tritiated phorbol-12,13-dibutyrate-binding have shown an uneven distribution of PKC in the brain (26–28). Northern blot analysis with specific oligonucleotide probes has also suggested a tissue-specific expression of some PKC subspecies (2, 8, 11). In situ hybridization analysis has shown a distinct distribution of the transcripts of PKC subspecies in the rat brain and spleen (29). The immunocytochemical approach to identifying PKC also suggests a different regional distribution of the enzyme (30, 31). Using a combination of biochemical, immunological, and cytochemical procedures with subspecies-specific antibodies, the relative activity and individual pattern of expression of multiple PKC subspecies in several cells and tissues have recently been examined extensively and clarified in more detail (32–50).

PKC with γ-sequence is apparently expressed solely in the brain and spinal cord, and is not found in other tissues and cell types (32, 36–40). Immunocytochemical studies with γ-subspecies–specific antibodies indicate that the highest enzyme activity is associated with the hippocampus, cerebral cortex, amygdaloid complex, cerebellar cortex, and spinal cord (34, 39). This subspecies appears to be localized in various neuronal cells, particularly in the pyramidal cells of hippocampus, as well as in the cell bodies, dendrites, and axons of Purkinje cells (34, 39). Immunoelectron microscopic analysis reveals that the γ-subspecies is associated with most membranous structures present throughout the cell, except for the nucleus (44). Mitochondria generally lack or poorly express PKC. The γ-subspecies develops postnatally and reaches maximum activity, in the rat, around three weeks after birth (40, 46). Although it seems premature to discuss the relationship of this PKC subspecies with any specific functions, it may play a role in specialized neuronal processes, such as long-term potentiation in the hippocampus.

PKCs with βI- and βII-sequence also display differential expression in the brain and many other tissues, including endocrine tissues such as the pituitary gland (50) and pancreatic islets (A. Ito et al, in preparation). Normally, the activity of βII-subspecies far exceeds that of the βI-subspecies. These two subspecies are very similar, differing from each other only in the region V_5 as noted above. However, cytochemical analysis with polyclonal antibodies raised against synthetic peptides specific to each of the PKC subspecies indicates that, in certain tissues, a clear, distinct cellular expression is apparent (43). In the rat cerebellar cortex, for example, PKC with βI-sequence is localized mainly in the granular layer, whereas βII-subspecies is found primarily in the molecular layer, apparently in the presynaptic nerve endings that terminate on the dendrites and cell body of the Purkinje cells. Immunoelectron microscopic analysis shows that, in some cells, βII-subspecies is associated primarily with the Golgi apparatus membrane, inferring a possible involvement in protein processing or transport (C. Tanaka et al, unpublished).

On the other hand, PKC with α-sequence is widely distributed in many tissues and cell types (32, 33, 37, 38, 41, 42). Most tissues, including liver, kidney, spleen, and testis, additionally contain β-subspecies in variable ratios (38). The sciatic nerve has predominantly α-subspecies, and a very small amount of β-subspecies (41). Some tissues such as heart, lung, and platelets appear to possess several undefined subspecies (38). In general, one cell type contains more than one subspecies of PKC. For instance, recent observation in this laboratory has shown that T lymphocytes, and various cloned cell lines express α-, βI, and βII-subspecies in different ratios (42). These subspecies, by light microscopy resolution, apparently show a distinct intracellular location, depending upon the state of differentiation or proliferation. Cultured fibroblasts express the α- but not β-subspecies (33).

At present, the distribution and biochemical properties of the enzymes encoded by δ-, ε-, and ζ-subspecies has not been determined, but Northern blot analysis suggests that these members of the family are expressed in several tissues and organs in different ratios (10, 18, 19).

INDIVIDUAL ENZYMOLOGICAL PROPERTIES

Direct enzymatic analysis of each PKC subspecies is a prerequisite for determining the potential roles of this enzyme family in signal transduction. The variable regions seen in each subspecies may play key roles in governing the individual enzymological characteristics and possibly localization and function. The diversity of the sequence in these regions allows separation of this enzyme into several subfractions upon chromatography on a hydroxyapatite column (5, 15, 32, 36–38, 47, 51, 52). The structure and genetic identity of some of these subfractions have been determined by comparison with the enzymes expressed in COS 7 cells after transfection with individual PKC cDNA-containing plasmids (5, 15). To date, three subfractions, types I, II, and III (51), have been shown to correspond to γ-, β- (βI and βII), and α-subspecies, respectively. PKCs with βI- and βII-sequence show nearly identical kinetic and catalytic properties, and can be distinguished from each other only by immunochemical procedures (15, 32, 43). The subspecies having δ-, ϵ-, and ζ-sequence expressed in COS 7 cells can be partially separated upon chromatography (18), but their correspondence to the subfractions chromatographically obtained from tissues remains to be clarified.

The PKC subfractions obtained thus far from various tissues exhibit subtle differences in enzymatic properties (38, 50–55). PKC with γ-sequence (type I) shows less activation by diacylglycerol, but is significantly activated by the micromolar range of free arachidonic acid (50). Activation by arachidonic acid does not require Ca^{2+}, nor does it depend on phospholipid and diacylglycerol. PKCs with βI- and βII-sequence (type II) show substantial activity without added Ca^{2+} in the presence of diacylglycerol and phospholipid, but respond much less well to arachidonic acid. PKC with α-subspecies (type III) shows properties apparently similar to γ-subspecies, and responds to high concentrations of free arachidonic acid only when Ca^{2+} is increased. It is possible that some PKC subspecies may be activated at different phases of cellular responses by a series of phospholipid metabolites, such as diacylglycerol, arachidonic acid (53, 54, 56–61), and lipoxin A (61), that successively appear subsequent to stimulation of the receptor.

In addition to these well-defined PKC subspecies, structurally undefined enzymes obtained from some tissues such as heart, lung, and platelets respond to phospholipid, diacylglycerol, and Ca^{2+} in different ways (38). For instance, one enzyme obtained from human platelets is not sensitive to Ca^{2+} (61a). This enzyme activity is reminiscent of the previously described rabbit reticulocyte protease-activated kinase II, which is capable of phosphorylating ribosomal S6 protein (62). The exact relation of this enzyme to the members of the PKC family deduced from the cDNA analysis remains unclear.

It is worth noting, however, that the dependency of PKC on Ca^{2+}, diacylglycerol, and phospholipid, when assayed in in vitro systems, varies markedly with the phosphate acceptor protein employed (63, 64), and, in the most extreme case, neither Ca^{2+} nor phospholipid is needed for the enzymatic activity with protamine as phosphate acceptor (23, 24). Thus, comparison of the members of the family is possible only under limited, defined conditions. In general, there is a problem in protein phosphorylation research that, following disruption of the cell, most protein kinases show an activity to phosphorylate many proteins, which may or may not represent physiological substrates (65). Therefore, the facts, that PKC subspecies may display different responsiveness to activators in vivo, and that PKC subspecies are differentially distributed both in various cell types and within the same cell, greatly complicate the search for physiological substrates for each member of the PKC family. Presumably, the PKC subspecies show different preferences for substrate proteins that are located in specific intracellular compartments. For instance, the EGF (epidermal growth factor) receptor of the A431 epidermoid carcinoma cells appears to be phosphorylated most rapidly by the ubiquitous α-subspecies of PKC, whereas it is more slowly phosphorylated by the brain-specific γ-subspecies (66). Thus, important questions remain to be answered concerning both the in vivo mechanism of activation of PKC, and the true physiological target proteins for each subspecies.

LIMITED PROTEOLYSIS BY CALPAIN

PKC was originally detected as an undefined protein kinase, which was present in many tissues and could be activated by limited proteolysis with calpain (23, 24). This hydrolytic cleavage occurs at one or two specific sites in the region V_3, resulting in the release of a catalytically fully active fragment, which subsequently is rapidly removed from the cell (A. Kishimoto et al, submitted for publication). Note that calpain I, which is active at the micromolar range of Ca^{2+} (67–69), cleaves PKC in the presence of phosphatidylserine plus diacylglycerol or TPA (12-O-tetradecanoylphorbol-13-acetate), implying that the activated form of PKC is a target of the calpain action (70). Quantitative analysis shows that γ-subspecies (type I) is very susceptible, whereas α-subspecies (type III) is relatively resistant to calpain I, and there is no obvious common sequence around the sites of the cleavage (A. Kishimoto et al, submitted for publication).

Although the limited proteolysis of PKC may generate a catalytically fully active fragment, previously called protein kinase M, and regulatory fragment (23, 24, 70–73), the physiological significance of this proteolysis has not yet been unequivocally established. Two alternative possibilities may be considered. Firstly, this proteolysis is a process to activate the PKC molecule, and the resulting protein kinase fragment may play some roles in the control of cellular function (74–82). It is also possible that the regulatory fragment has some roles, since it contains a DNA-binding motif as described above. However, no obvious evidence has been available thus far suggesting this first possibility. Secondly, in contract, the limited proteolysis is a process to initiate the degradation of PKC, eventually depleting the enzyme from the cell. Recent reports from several laboratories (83–96) have shown that, in a variety of tissues and cell types, TPA, which induces persistent activation of PKC due to its stable properties, elicits the translocation of the enzyme from the soluble fraction to the membrane, and subsequent depletion, termed down-regulation of PKC. In fact, the catalytically active fragment, protein kinase M, is not always recovered from the cell, and is probably degraded further by the action of other proteases. A recent analysis has clearly shown that, upon treatment with TPA, some PKC subspecies coexpressed in a single cell type, such as KM3 and HL-60 cell lines, disappear quickly at different rates (97), which may reflect the specificity of the calpain action described above. It is more likely, then, that the limited proteolysis of the PKC molecules by calpain, particularly by calpain I, is directly related to the initiation of degradation of the enzyme, and that various subspecies of PKC are depleted from the cell at different rates due to the proteolysis at their variable region V_3.

DUAL ACTION AND PROSPECTIVES

Several physiological functions have been assigned to PKC, including involvement in secretion and exocytosis, modulation of ion conductance, interaction and down-regulation of receptors, smooth muscle contraction, gene expression, and cell proliferation (98–100). One can immediately anticipate, therefore, that the members of the family may each have different roles in the control of cellular functions. However, little is known at present to allow a discussion of the possible biological roles of individual PKC subspecies in biochemical terms.

It is now well recognized that synergistic interaction between PKC and Ca^{2+} pathways underlies a variety of cellular responses to external stimuli (98–102). In the Ca^{2+} pathway, inositol-1,4,5-trisphosphate is shown to mobilize Ca^{2+} from its internal store. A large body of evidence has accumulated to indicate that PKC has a dual action, providing both positive forward control, as well as negative feedback control, over various steps of cell signaling processes (98–100). In short-term responses, for instance, a major role of PKC appears to lie in decreasing Ca^{2+} concentrations in a manner given schematically in Figure 2. Persistent and high concentrations of Ca^{2+} have deleterious effects, and frequently lead ultimately to the death of the cell. The appearance of second messengers, such as diacylglycerol, is normally very transient, since a positive signal is usually followed by an immediate negative feedback control. A number of reports have suggested that in various cell types, PKC has a function to activate the Ca^{2+}-transport ATPase and the Na^+/Ca^{2+} exchange protein, both of which remove Ca^{2+} from the cytosol (98, 99). PKC often inhibits the receptor-mediated hydrolysis of inositol phospholipids, thereby blocking the activation of the Ca^{2+}-signaling pathway. The recent observations that TPA inhibits the frequency of repetitive Ca^{2+} transients recorded from single cells (103) could be explained by such an inhibitory action of PKC.

Although it sounds paradoxical, such a negative feedback role of PKC is not confined to short-term responses, but may be extended to long-term responses such as cell proliferation. The receptor for EGF has repeatedly been shown to be phosphorylated by PKC, resulting in a rapid decrease in high-affinity binding of EGF as well as inhibition of the ligand-induced tyrosine phosphorylation (104). It is plausible that the treatment of cells with TPA frequently causes depletion of the PKC molecule itself as discussed in the previous section, and thereby relieves the cell from down-regulation of the growth factor receptor, so that uncontrolled cell proliferation might occur in the presence of a mitogenic stimulus (Figure 2). Thus, the tumor-promoter, TPA, again provides a dual effect, furnishing a positive short-term activation of PKC, and then a negative action to initiate the degradation of the enzyme over a long time course.

Such a dual effect of PKC is also seen in T lymphocyte proliferation, where the synergistic action of TPA and Ca^{2+} ionophore has been demonstrated (105). Here, the activated PKC also has a negative feedback action in these cells, as it phosphorylates and thereby induces down-regulation of the T cell receptor, preventing any further stimulation by antigen to proliferation (106). At present, such a sequential intracellular event has not been fully substantiated on a firm biochemical basis, but it will be attractive to investigate the possible dual effect of PKC and of TPA for the action of many growth factors.

Obviously, the negative feedback role of PKC emphasized above does not exclude the existence of a positive forward action of the enzyme. Plausible evidence seems to indicate a possible involvement of PKC in gene expression, such as induction of the interleukin 2 receptor and some protooncogene activation (98, 99). Recent studies with fibroblasts overproducing a PKC subspecies suggest such a positive forward action of this enzyme in growth regulation (107, 108). Several lines of evidence also suggest that PKC may have a crucial role in modulating many membrane functions, including ion conductance and cross-talks of various receptors (98–100). These aspects are beyond the scope of the present review, but it may be pointed out that, based on the mode of activation of PKC as well as on its immunocytochemical

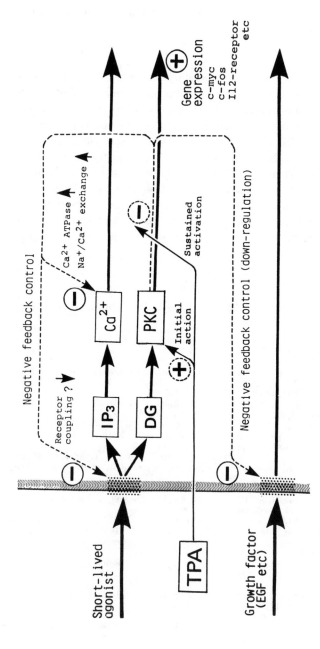

Figure 1   Common structure of protein kinase C subspecies. C, G, K, X, and M represent cysteine, glycine, lysine, any amino acid, and metal, respectively. Other details are described in the text. Adapted from Ref. 100.

Figure 2.   Dual action of protein kinase C and TPA. This schema is probably oversimplified, but emphasizes a negative feedback action of PKC on membrane functions. Unlike transiently produced diacylglycerol, TPA appears to show a dual effect with a positive action to immediately activate PKC, and with a negative action to initiate the degradation of the enzyme during sustained activation. The detailed explanation is as given in the text. DG represents diacylglycerol, and IP_3 indicates inositol-1,4,5-trisphosphate. Adapted from Ref. 100.

distribution, the principal targets of the enzyme family, if not all, must be located in the membrane.

ACKNOWLEDGMENTS

Some of the work reported from this laboratory was carried out in collaboration with the Biotechnology Laboratories of Takeda Chemical Industries and the Department of Pharmacology, Kobe University School of Medicine. The great efforts of all collaborators cited in each article are cordially acknowledged. The investigations carried out in this laboratory were supported in part by research grants from the Research Fund of the Ministry of Education, Science and Culture, Japan; Muscle Dystrophy Association, United States; Juvenile Diabetes Foundation International, United States; Yamanouchi Foundation for Research on Metabolic Disorders; Merck Sharp & Dohme Research Laboratories; Ajinomoto Central Research Laboratories; Meiji Institute Health Sciences; and New Lead Research Laboratories of Sankyo Company.

Literature Cited

1. Parker, P. J., Coussens, L., Totty, N., Rhee, L., Young, S., et al. 1986. *Science* 233:853–59
2. Coussens, L., Parker, P. J., Rhee, L., Yang-Feng, T. L., Chen, E., et al. 1986. *Science* 233:859–66
3. Ono, Y., Kurokawa, T., Kawahara, K., Nishimura, O., Marumoto, R., et al. 1986. *FEBS Lett.* 203:111–15
4. Ono, Y., Kurokawa, T., Fujii, T., Kawahara, K., Igarashi, K., et al. 1986. *FEBS Lett.* 206:347–52
5. Kikkawa, U., Ono, Y., Ogita, K., Fujii, T., Asaoka, Y., et al. 1987. *FEBS Lett.* 217:227–31
6. Kikkawa, U., Ogita, K., Asaoka, Y., Shearman, M. S., Fujii, T., et al. 1987. *FEBS Lett.* 223:212–16
7. Ono, Y., Fujii, T., Igarashi, K., Kikkawa, U., Ogita, K., et al. 1988. *Nucleic Acids Res.* 16:5199–200
8. Knopf, J. L., Lee, M.-H., Sultzman, L. A., Kriz, R. W., Loomis, C. R., et al. 1986. *Cell* 46:491–502
9. Makowske, M., Birnbaum, M. J., Ballester, R., Rosen, O. M. 1986. *J. Biol. Chem.* 261:13389–92
10. Housey, G. M., O'Brian, C. A., Johnson, M. D., Kirschmeier, P., Weinstein, I. B. 1987. *Proc. Natl. Acad. Sci. USA* 84:1065–69
11. Ohno, S., Kawasaki, H., Imajoh, S., Suzuki, K., Inagaki, M., et al. 1987. *Nature* 325:161–66
12. Ohno, S., Kawasaki, H., Konno, Y., Inagaki, M., Hidaka, H., et al. 1988. *Biochemistry* 27:2083–87
13. Coussens, L., Rhee, L., Parker, P. J., Ullrich, A. 1987. *DNA* 6:389–94
14. Kubo, K., Ohno, S., Suzuki, K. 1987. *FEBS Lett.* 223:138–42
15. Ono, Y., Kikkawa, U., Ogita, K., Fujii, T., Kurokawa, T., et al. 1987. *Science* 236:1116–20
16. Kubo, K., Ohno, S., Suzuki, K. 1987. *Nucleic Acids Res.* 15:7179–80
17. Ono, Y., Fujii, T., Ogita, K., Kikkawa, U., Igarashi, K., et al. 1987. *FEBS Lett.* 226:125–28
18. Ono, Y., Fujii, T., Ogita, K., Kikkawa, U., Igarashi, K., et al. 1988. *J. Biol. Chem.* 263:6927–32
19. Ohno, S., Akita, Y., Konno, Y., Imajoh, S., Suzuki, K. 1988. *Cell* 53:731–41
20. Berg, M. J. 1986. *Science* 232:485–87
21. Johnston, M. 1987. *Nature* 328:353–55
22. Hunter, T., Cooper, J. A. 1985. *Annu. Rev. Biochem.* 54:897–930
23. Takai, Y., Kishimoto, A., Inoue, M., Nishizuka, Y. 1977. *J. Biol. Chem.* 252:7603–9
24. Inoue, M., Kishimoto, A., Takai, Y., Nishizuka, Y. 1977. *J. Biol. Chem.* 252:7610–16
25. Rosenthal, A., Rhee, L., Yadegari, R., Paro, R., Ullrich, A., et al. 1987. *EMBO J.* 6:433–41
26. Nagle, D. S., Blumberg, P. M. 1983. *Cancer Lett.* 18:35–40

27. Worley, P. F., Baraban, J. M., De Souza, E. B., Snyder, S. H. 1986. *Proc. Natl. Acad. Sci. USA* 83:4053–57
28. Worley, P. F., Baraban, J. M., Snyder, S. H. 1986. *J. Neurosci.* 6:199–207
29. Brandt, S. J., Niedel, J. E., Bell, R. M., Young, W. S. III. 1987. *Cell* 49:67–63
30. Girard, P. R., Mazzei, G. J., Wood, J. G., Kuo, J. F. 1985. *Proc. Natl. Acad. Sci. USA* 82:3030–34
31. Wood, J. G., Girard, P. R., Mazzei, G. J., Kuo, J. F. 1986. *J. Neurosci.* 6:2571–77
32. Shearman, M. S., Naor, Z., Kikkawa, U., Nishizuka, Y. 1987. *Biochem. Biophys. Res. Commun.* 147:911–19
33. McCaffrey, P. G., Rosner, M. R., Kikkawa, U., Sekiguchi, K., Ogita, K., et al. 1987. *Biochem. Biophys. Res. Commun.* 146:140–46
34. Kitano, T., Hashimoto, T., Kikkawa, U., Ase, K., Saito, N., et al. 1987. *J. Neurosci.* 7:1520–25
35. Mochly-Rosen, D., Basbaum, A. I., Koshland, D. E. Jr. 1987. *Proc. Natl. Acad. Sci. USA* 84:4660–64
36. Huang, F. L., Yoshida, Y., Nakabayashi, H., Huang, K.-P. 1987. *J. Biol. Chem.* 262:15714–20
37. Huang, F. L., Yoshida, Y., Nakabayashi, H., Knopf, J. L., Young, W. S. III, et al. 1987. *Biochem. Biophys. Res. Commun.* 149:946–52
38. Kosaka, Y., Ogita, K., Ase, K., Nomura, H., Kikkawa, U., et al. 1988. *Biochem. Biophys. Res. Commun.* 151:973–81
39. Saito, N., Kikkawa, U., Nishizuka, Y., Tanaka, C. 1988. *J. Neurosci.* 8:369–83
40. Hashimoto, T., Ase, K., Sawamura, S., Kikkawa, U., Saito, N., et al. 1988. *J. Neurosci.* 8:1678–83
41. Shearman, M. S., Kosaka, Y., Ase, K., Kikk.wa, U., Nishizuka, Y. 1988. *Biochem. Soc. Trans.* 16:307–8
42. Shearman, M. S., Berry, N. S., Oda, T., Ase, K., Kikkawa, U., et al. 1988. *FEBS Lett.* 234:387–91
43. Ase, K., Saito, N., Shearman, M. S., Kikkawa, U., Ono, Y., et al. 1988. *J. Neurosci.* 8:3850–56
44. Kose, A., Saito, N., Ito, H., Kikkawa, U., Nishizuka, Y., et al. 1988. *J. Neurosci.* 8:4262–68
45. Hidaka, H., Tanaka, T., Onda, K., Higashihara, M., Watanabe, M., et al. 1988. *J. Biol. Chem.* 263:4523–26
46. Yoshida, Y., Huang, F. L., Nakabayashi, H., Huang, K.-P. 1988. *J. Biol. Chem.* 263:9068–73
47. Pelosin, J.-M., Vilgrain, I., Chambaz, E. M. 1987. *Biochem. Biophys. Res. Commun.* 147:382–91
48. Makowske, M., Ballester, R., Cayre, Y., Rosen, O. M. 1988. *J. Biol. Chem.* 263:3402–10
49. Kraft, A. S., Reeves, J. A., Ashendel, C. L. 1988. *J. Biol. Chem.* 263:8437–42
50. Naor, Z., Shearman, M. S., Kishimoto, A., Nishizuka, Y. 1988. *Mol. Endocrinol.* 2:1043–48
51. Huang, K.-P., Nakabayashi, H., Huang, F. L. 1986. *Proc. Natl. Acad. Sci. USA* 83:8535–39
52. Jaken, S., Kiley, S. C. 1987. *Proc. Natl. Acad. Sci. USA* 84:4418–22
53. Sekiguchi, K., Tsukuda, M., Ogita, K., Kikkawa, U., Nishizuka, Y. 1987. *Biochem. Biophys. Res. Commun.* 145:797–802
54. Sekiguchi, K., Tsukuda, M., Ase, K., Kikkawa, U., Nishizuka, Y. 1988. *J. Biochem.* 103:759–65
55. Asaoka, Y., Kikkawa, U., Sekiguchi, K., Shearman, M. S., Kosaka, Y., et al. 1988. *FEBS Lett.* 231:221–24
56. McPhail, L. C., Clayton, C. C., Snyderman, R. 1984. *Science* 224:622–25
57. Murakami, K., Routtenberg, A. 1985. *FEBS Lett.* 192:189–93
58. Murakami, K., Chan, S. Y., Routtenberg, A. 1986. *J. Biol. Chem.* 261:15424–29
59. Nishikawa, M., Hidaka, H., Shirakawa, S. 1988. *Biochem. Pharmacol.* 37:3079–89
60. Wooten, M. W., Wrenn, R. W. 1988. *Biochem. Biophys. Res. Commun.* 153:67–73
61. Hansson, A., Serhan, C. N., Haeggström, J., Ingelman-Sundberg, M., Samuelsson, B. 1986. *Biochem. Biophys. Res. Commun.* 134:1215–22
61a. Tsukuda, M., Asaoka, Y., Sekiguchi, K., Kikkawa, U., Nishizuka, Y. 1988. *Biochem. Biophys. Res. Commun.* 155:1387–95
62. Gonzatti, M. I., Traugh, J. A. 1986. *J. Biol. Chem.* 261:15266–72
63. Bazzi, M. D., Nelsestuen, G. L. 1987. *Biochemistry* 26:1974–82
64. Dell, K. R., Walsh, M. P., Severson, D. L. 1988. *Biochem. J.* 254:455–62
65. Krebs, E. G., Beavo, J. A. 1979. *Annu. Rev. Biochem.* 48:923–59
66. Ido, M., Sekiguchi, K., Kikkawa, U., Nishizuka, Y. 1987. *FEBS Lett.* 219:215–18
67. Mellgren, R. L. 1980. *FEBS Lett.* 109:129–33
68. Murachi, T., Tanaka, K., Hatanaka, M., Murakami, T. 1981. *Adv. Enzyme Regul.* 19:407–24
69. Kishimoto, A., Kajikawa, N., Tabuchi, H., Shiota, M., Nishizuka, Y. 1981. *J. Biochem.* 90:1787–93

70. Kishimoto, A., Kajikawa, N., Shiota, M., Nishizuka, Y. 1983. *J. Biol. Chem.* 258:1156–64
71. Hoshijima, M., Kikuchi, A., Tanimoto, T., Kaibuchi, K., Takai, Y. 1986. *Cancer Res.* 46:3000–4
72. Lee, M.-H., Bell, R. M. 1986. *J. Biol. Chem.* 261:14867–70
73. Huang, K.-P., Huang, F. L. 1986. *Biochem. Biophys. Res. Commun.* 139: 320–26
74. Melloni, E., Pontremoli, S., Michetti, M., Sacco, O., Sparatore, B., et al. 1985. *Proc. Natl. Acad. Sci. USA* 82: 6435–39
75. Tapley, P. M., Murray, A. W. 1984. *Biochem. Biophys. Res. Commun.* 122: 158–64
76. Tapley, P. M., Murray, A. W. 1985. *Eur. J. Biochem.* 151:419–23
77. Melloni, E., Pontremoli, S., Michetti, M., Sacco, O., Sparatore, B., et al. 1986. *J. Biol. Chem.* 261:4101–5
78. Chida, K., Kato, N., Kuroki, T. 1986. *J. Biol. Chem.* 261:13013–18
79. Fabbro, D., Regazzi, R., Costa, S. D., Borner, C., Eppenberger, U. 1986. *Biochem. Biophys. Res. Commun.* 135: 65–73
80. Guy, G. R., Gordon, J., Walker, L., Michell, R. H., Brown, G. 1986. *Biochem. Biophys. Res. Commun.* 135: 146–53
81. Fournier, A., Murray, A. W. 1987. *Nature* 330:767–69
82. Pontremoli, S., Melloni, E., Damiani, G., Salamino, F., Sparatore, B., et al. 1988. *J. Biol. Chem.* 263:1915–19
83. Rodriguez-Pena, A., Rozengurt, E. 1984. *Biochem. Biophys. Res. Commun.* 120:1053–59
84. Blackshear, P. J., Witters, L. A., Girard, P. R., Kuo, J. F., Quamo, S. N. 1985. *J. Biol. Chem.* 260:13304–15
85. Ballester, R., Rosen, O. M. 1985. *J. Biol. Chem.* 260:15194–99
86. Gainer, H. St. C., Murray, A. W. 1985. *Biochem. Biophys. Res. Commun.* 126: 1109–13
87. Katakami, Y., Nakao, Y., Matsui, T., Koizumi, T., Kaibuchi, K., et al. 1986. *Biochem. Biophys. Res. Commun.* 135: 355–62
88. Stabel, S., Rodriguez-Pena, A., Young, S., Rozengurt, E., Parker, P. J. 1987. *J. Cell. Physiol.* 130:111–17
89. Young, S., Parker, P. J., Ullrich, A., Stabel, S. 1987. *Biochem. J.* 244:775–79
90. Young, S., Rothbard, J., Parker, P. J. 1988. *Eur. J. Biochem.* 173:247–52
91. vonRuecker, A. A., Rao, G., Bidlingmaier, F. 1988. *Biochem. Biophys. Res. Commun.* 151:997–1003
92. Borner, C., Eppenberger, U., Wyss, R., Fabbro, D. 1988. *Proc. Natl. Acad. Sci. USA* 85:2110–14
93. Helper, J. R., Earp, H. S., Harden, T. K. 1988. *J. Biol. Chem.* 263:7610–19
94. Girard, P. R., Stevens, V. L., Blackshear, P. J., Merrill, A. H. Jr., Wood, J. G., et al. 1987. *Cancer Res.* 47:2892–98
95. Shoji, M., Girard, P. R., Charp, P. A., Koeffler, H. P., Vogler, W. R., et al. 1987. *Cancer Res.* 47:6363–70
96. Ewald, D. A., Matthies, H. J. G., Perney, T. M., Walker, M. W., Miller, R. J. 1988. *J. Neurosci.* 8:2447–51
97. Ase, K., Berry, N., Kikkawa, U., Kishimoto, A., Nishizuka, Y. 1988. *FEBS Lett.* 236:396–400
98. Nishizuka, Y. 1986. *Science* 233:305–12
99. Kikkawa, U., Nishizuka, Y. 1986. *Annu. Rev. Cell Biol.* 2:149–78
100. Nishizuka, Y. 1988. *Nature.* 334:661–65
101. Nishizuka, Y. 1984. *Nature.* 308:693–98
102. Berridge, M. J. 1987. *Annu. Rev. Biochem.* 56:159–93
103. Woods, N. M., Cuthbertson, K. S. R., Cobbold, R. H. 1987. *Biochem. J.* 246:619–23
104. Schlessinger, J. 1986. *J. Cell Biol.* 103:2067–72
105. Kaibuchi, K., Takai, Y., Nishizuka, Y. 1983. *J. Biol. Chem.* 260:1366–69
106. Cantrell, D. A., Davies, A. A., Crumpton, M. J. 1985. *Proc. Natl. Acad. Sci. USA* 82:8158–62
107. Housey, G. M., Johnson, M. D., Hsiao, W. L. W., O'Brian, C. A., Murphy, J. P., et al. 1988. *Cell* 52:343–54
108. Persons, D. A., Wilkinson, W. Q., Bell, R. M., Finn, O. J. 1988. *Cell* 52:447–58

Annu. Rev. Biochem. 1989. 58:45–77

HEMOPOIETIC CELL GROWTH FACTORS AND THEIR RECEPTORS

Nicos A. Nicola

The Walter and Eliza Hall Institute of Medical Research, P.O. Royal Melbourne Hospital, 3050, Victoria, Australia

CONTENTS

PERSPECTIVES AND SUMMARY

Hemopoietic growth factors form a family of glycosylated extracellular proteins that regulate the production and functional activity of hemopoietic cells. Most of these factors were initially described as activities produced by feeder cells or present in crude cell-conditioned media that stimulated the proliferation of white blood cell precursor cells. Over the last decade many of these growth factors have been purified to homogeneity and their genes cloned, so that it has now become possible to test their clinical efficacy.

45

Unlike classical hormones, hemopoietic growth factors are generally produced at multiple sites in the body by a few different cell types and act on specific target cells. Their production is elevated by immunological reactions and products of infectious agents that signal a requirement for the recruitment of white blood cells involved in host defense.

The number of hemopoietic growth factors that are well defined continues to expand at a rapid rate and includes the interleukins 1, 2, 4, 5, 6, and 7 with primary activities on lymphoid cells, the granulocyte-macrophage colony-stimulating factors (G-CSF, M-CSF, GM-CSF, and Multi-CSF or interleukin-3) active on neutrophils, macrophages, eosinophils, megakaryocytes, and mast cells, and erythropoietin active in the production of erythroid cells. Since all of these growth factors cannot be reviewed adequately here, a major emphasis will be placed on those hemopoietic growth factors regulating the production of granulocytes and macrophages (the CSFs, IL-5, and IL-6).

The cellular specificity and different levels of action of the hemopoietic growth factors will be discussed based on studies performed in vitro, in vivo in experimental animals, and in current clinical trials. The molecular properties of the growth factors, their receptors, and the growth factor genes will also be described as well as the regulation of the expression of the genes and gene products. Finally, the possible relationships of growth factor or receptor expression to the development of leukemias and other cancers will be discussed.

BIOLOGICAL ACTIONS OF THE HEMOPOIETIC GROWTH FACTORS IN VITRO

Enhancement of Cell Survival

When bone marrow cells or purified hemopoietic progenitor cells are cultured in vitro under conditions that exclude the endogenous production of growth factors by accessory cells, progenitor cells committed to different cell lineages die with a half-life varying from 9 to 24 hrs (1, 2). In contrast, in the presence of CSFs, such progenitor cells survive and proliferate. The life spans of blood neutrophils, eosinophils, and monocytes in vitro are also extended by the appropriate growth factors (GM-CSF, G-CSF; Multi-CSF, GM-CSF, IL-5; GM-CSF, M-CSF, respectively), although such cells eventually die even in the presence of the appropriate factor (3, 4). The concentrations of CSFs required for maintaining cell viability are generally considerably lower than those required for proliferation or functional activation (3), and there is some indication from factor-dependent cell lines that the mechanisms maintaining cell viability [maintenance of glucose transport and ATP levels, (5, 6)] may differ from those stimulating cell proliferation.

Stimulation of Cell Proliferation

The continuous presence of CSFs in the concentration range of 1–100 pM is absolutely required for the proliferation of hemopoietic progenitor cells, some leukemic cells, and cell lines in vitro. If the progenitor cells are not in cell cycle, the appropriate CSF will induce entry into the S-phase of the cell cycle within 3–12 hrs (7, 8). Nevertheless, the onset of cell proliferation is highly asynchronous, with lag times of several days for some types of cells. Increasing concentrations of a given CSF or combinations of different CSFs can reduce this lag time and make the cultures more synchronous (9, 10) but, in general, most cultures still retain a significant degree of asynchrony.

The dose-response curves for clonal cell proliferation with CSF are sigmoid when the CSF dose is plotted on a log scale, as is expected for a process mediated by saturable binding to cellular receptors. However, the relationship between CSF dose and response is not a simple one, because there is evidence of clonal heterogeneity in the responsiveness of different types of progenitor cells to CSFs (11). Moreover, as the concentration of CSF is increased, there is a direct effect on shortening the cell cycle time of progenitor cells to a minimum of 8–12 hrs (12) and in increasing the number of progeny cells generated (12), presumably by delaying the onset of terminal differentiation of colony cells to postmitotic granulocytes and macrophages.

The variability of the onset of proliferation, the doubling time, the final colony size, and the CSF-responsiveness of different progenitor cells are at least partly due to heterogeneity in the state of differentiation and the type of differentiation commitment displayed by the progenitor cell. Progenitor cells that are hyper-responsive to GM-CSF and form macrophage colonies can be separated, based on their smaller size, from less responsive progenitor cells that form larger granulocyte-macrophage colonies (13). Similarly, cluster-forming cells (with limited proliferative potential) can be separated from colony-forming cells which can, in turn, be separated from the more primitive stem cells that generate colony-forming cells (14). The result of this is that there is a distinct concentration dependence in the types of committed progenitor cells whose proliferation is stimulated by each CSF (11).

At low concentrations, G-CSF and M-CSF are relatively selective proliferative stimuli for neutrophil or macrophage precursor cells, respectively. However, at higher concentrations G-CSF also stimulates macrophage and granulocyte-macrophage progenitor cells and M-CSF stimulates granulocyte-macrophage and granulocyte progenitor cells. For GM-CSF, the order of decreasing responsiveness among progenitor cells is macrophage > neutrophil > neutrophil-macrophage > eosinophil > megakaryocyte, and for Multi-CSF the order is macrophage > neutrophil-macrophage = neutrophil = eosinophil = megakaryocyte = mast cell > erythroid > multipotential (11).

Amongst nonlymphoid cells, IL-5 and IL-6 are relatively specific stimuli for progenitors of eosinophils or neutrophils and macrophages, respectively (11, 15, 10) (Figure 1).

The proliferative actions of the CSFs depend on the stage of differentiation or maturity of the responding cells. In the simplest case hemopoietic progenitor cells can differentiate to mature cells, which are incapable of further proliferation (for example neutrophils, eosinophils, megakaryocytes, or erythroid cells), although some mature cells such as macrophages retain the capacity to proliferate in response to M-CSF or high concentrations of GM-CSF or Multi-CSF (16, 17). In other cases, progenitor cells still capable of extensive proliferation gain or lose the capacity to respond to a particular type of CSF, so that optimum expansion of some cell lineages requires synergistic or additive interactions of CSFs and other factors.

This type of synergistic interaction was first noted in the erythroid cell lineage where it was documented that the proliferation of early erythroid precursor cells (BFU-E) is stimulated by Multi-CSF and GM-CSF (and to a lesser extent G-CSF), but results in the production of late erythroid precursor cells (CFU-E), which lose responsiveness to GM-CSF, G-CSF, and possibly Multi-CSF, but gain responsiveness to erythropoietin (18–20). In this cell system erythropoietin provides an acute expansion of erythroid cells from CFU-E, but GM-CSF and Multi-CSF are also required for a larger or sustained production of erythroid cells from more immature erythroid precursor cells (BFU-E). Similarly, maximal production of eosinophils in vitro requires the combined actions of initially Multi-CSF and GM-CSF and subsequently IL-5 (21), while megakaryocyte production is enhanced by the combined actions of initially Multi-CSF and GM-CSF, and possibly thrombopoietin (22). It is possible that similar situations also exist in the production of macrophages (23) and neutrophils (24).

Synergistic interactions of CSFs and other factors on hemopoietic stem cells are even more complex because of the heterogeneity of such cells. The most primitive stem cells detected in vitro are probably those derived from the bone marrow of mice treated with 5-fluorouracil, which generate macroscopic colonies in vitro (25). For maximal expression of the proliferative capacity of these cells, several factors are required to be present simultaneously, including hemopoietin-1 (probably identical to interleukin-1), Multi-CSF, and M-CSF (26, 27). However, other factors active in such assays (28) and various CSF combinations (24) have also been described to produce macroscopic colonies. Multipotential stem cells detected by the in vivo spleen colony–forming assay proliferate in vitro in response to Multi-CSF but not other CSFs, although this proliferative response declines with progressively more ancestral stem cells (29, 30). Multipotential cells giving rise to colonies of blast cells in vitro that have a high content of committed progenitor cells are

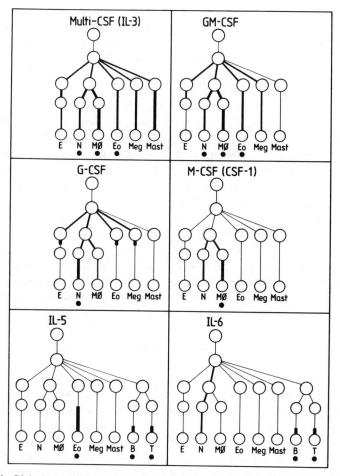

Figure 1 Biological specificities of the hemopoietic growth factors. The hemopoietic tree in each panel represents the self-renewing multipotential stem cell at the top with successive restriction to individual cell lineages going down the diagram. Heavy lines indicate a proliferative effect of the factor even at low concentrations, medium lines a proliferative effect at high concentrations only, and light lines indicate no effect. Abbreviations are E, erythroid, N, neutrophil, MØ, monocyte/macrophage, Eo, eosinophil, Meg, megakaryocyte, and Mast, mast cells. Dots under the abbreviation indicate the ability of the factor to stimulate functional activities of the mature cells.

stimulated by Multi-CSF and to a lesser extent by GM-CSF (31). Initiation of proliferation of such cells is enhanced by IL-6 (10). Finally, multipotential cells giving rise to colonies containing multiple lineages of differentiated hemopoietic cells are stimulated by Multi-CSF, and their initial proliferation can be stimulated more weakly by GM-CSF and G-CSF (18–20). These data suggest a stem cell hierarchy with the most ancestral to most mature stem cells being stimulated by factors in the order: unknown factors > interleukin-1 > Multi-CSF > IL-6, GM-CSF > G-CSF, with combinations of factors often required for maximal proliferative expansion.

Induction of Cellular Differentiation

It is a characteristic of the action of all the CSFs on hemopoietic progenitor cells that the stimulation of cellular proliferation is coupled to the induction of terminal differentiation. It is of course possible that this coupling is genetically determined in the responding cells as a consequence of cell division and that CSFs do not directly influence differentiative events. However, there are some lines of evidence that suggest the proliferative actions of CSFs can be discriminated from effects on cell differentiation.

Different cell lines have been described that are absolutely dependent on CSFs for their continuous growth in vitro and that show no signs of terminal differentiation or clonal extinction (32–34). The majority of these are dependent for their growth on Multi-CSF, a few on Multi-CSF or GM-CSF, and even fewer also respond to G-CSF, IL-4, erythropoietin, or M-CSF (32–34). In some of these cell lines, transfer of the cells from Multi-CSF to a different CSF (G-CSF or erythropoietin) results in the induction of terminal differentiation (33, 34). This would seem to imply that the proliferative effects of a CSF need not always be coupled to a predetermined capacity of the cell to differentiate.

In studies where the daughter cells of a single progenitor cell were separated and exposed to different concentrations of GM-CSF or GM-CSF versus M-CSF, Metcalf (12, 35) found that the fate of the daughter cells was influenced by the type or amount of CSF that they were exposed to. In these experiments evidence was found both for precommitment in some of the daughter cells and for a direct influence of the CSF on the differentiation outcome of other daughter cells. It is difficult in these experiments to rule out selective expansion of precommitted cells by the different CSFs, but these results may again suggest that individual progenitor cells do not differentiate only as a consequence of cell proliferation.

Perhaps the most convincing demonstrations of the differentiating-inducing actions of the CSFs come from studies on cell lines that are independent of CSF for proliferation.

Amongst the CSFs, G-CSF has the most potent capacity to induce terminal differentiation and clonal suppression in myeloid leukemic and nonleukemic cell lines. G-CSF induces differentiation in the myelomonocytic leukemic cell line WEHI-3B D$^+$ with the same dose-response relationship as on normal progenitor cells (36), and this eventually results in a profound suppression of self-renewing cell divisions (37). The effects of G-CSF are, however, slow so that initially there is little effect on cell proliferation. Subsequently, individual daughter cells become irreversibly committed in an asymmetric process to terminal differentiation (38). Similar effects of G-CSF have been observed on the monocytic leukemic cell line, M1, although much higher concentrations of G-CSF are required (39), and the suppression of proliferation is not as profound as seen with other factors (40). G-CSF also induces differentiation and complete clonal suppression of a Multi-CSF-dependent cell line, 32D cl.3, an action that is competitive with the proliferative stimulus exerted by Multi-CSF (34). In some of these experiments evidence has been obtained that the cells make the decision to leave the self-renewing proliferative pathway and enter the differentiation commitment pathway during the G1 phase of the cell cycle, and it is during this phase that G-CSF exerts its differentiation-inducing actions (39). To some extent, the pathway of differentiation may be genetically determined by the cell, since in the cases cited above G-CSF leads to neutrophilic differentiation in 32D cl.3 cells, to monocytic differentiation in M1 cells, and monocytic/neutrophil differentiation in WEHI-3B D$^+$ cells.

GM-CSF has a weaker differentiation-inducing action than G-CSF on WEHI-3B D$^+$ cells (41), but GM-CSF, Multi-CSF, and M-CSF have no differentiating actions on the other cell lines. Both GM-CSF and G-CSF have weak differentiation-inducing activities on the human promyeleocytic leukemic cell line HL-60 showing, on prolonged exposure, a suppression of clonal proliferation (42, 43). However, there is little evidence of CSF-induced differentiation of primary human myeloid leukemic cells (42, 43), except possibly with M-CSF acting on some subclasses of myeloid leukemia (44).

The variable responses seen with different types of cell lines to the differentiation-inducing actions of CSFs may result from a number of different lesions in these cells. Some are known not to display one or another type of CSF receptor (45), while others do display receptors but the cells may not be able to respond to the receptor-generated signals. Yet other cell lines have been described that produce differentiation inhibitors that negate the actions of differentiation inducers (46).

Activation of Functional Activities of Mature Hemopoietic Cells

With the exception of M-CSF, CSFs do not generally stimulate proliferation of mature cells (those in the blood and tissues). However, they probably prolong the life span of the mature cells, and either directly enhance functional activities associated with host defense or prime the cells so that they are more readily activated by products derived from infectious agents or factors generated in immunological reactions.

GM-CSF is directly chemotactic for neutrophils (47), and high concentrations of GM-CSF lead to their immobilization (migration inhibition) (48). It also increases the capacity of neutrophils to phagocytose and kill intracellularly yeast, parasites, and bacteria (49, 50), to participate in antibody-dependent cell killing (51), presumably through binding the Fc portion of both IgG and IgA (52), and to adhere to cell surfaces by inducing the expression of cell surface adhesion molecules (53). GM-CSF also primes neutrophils so that they respond more readily to bacterial products (formyl peptides) and immunological products (complement and leukotrienes) by releasing reactive oxygen metabolites (superoxide) that have a bacteriocidal activity (54). Some of these activities on neutrophils are shared by G-CSF, including the stimulation of antibody-dependent cell killing (51), IgA-mediated phagocytosis (52), and the priming of neutrophils to respond to bacterial chemotactic peptides by generating superoxide anion (55). In some cases the effects of G-CSF and GM-CSF on neutrophils have been shown to be additive (51), although it is not clear whether this is mediated at the single cell level or by recruiting additional populations of neutrophils. G-CSF and GM-CSF may also enhance the general immunological response by stimulating the release from neutrophils of other mediators of inflammatory responses: arachidonic acid by G-CSF in the presence of chemotactic peptides (56) and interleukin-1 by GM-CSF (57). The priming actions of GM-CSF and G-CSF on neutrophils for their response to chemotactic peptides like N-formyl-methionine-leucine-phenylalanine (FMLP) may be mediated by their up-regulation of FMLP receptors (58).

GM-CSF, but not G-CSF, has similar stimulating activities on eosinophils. GM-CSF enhances the capacity of eosinophils to adhere to antibody-coated parasites and to kill them (59, 60), to produce superoxide anion (59) and release leukotriene C4 in response to calcium-mobilizing agents (60), and to kill antibody-coated tumor cells (61). IL-5 shares with GM-CSF the ability to stimulate eosinophil-mediated killing of antibody-coated tumor cells and phagocytosis of serum-opsonized yeast (15).

Macrophages can be stimulated to proliferate by M-CSF and to a lesser extent by GM-CSF and Multi-CSF (17), but both GM-CSF and Multi-CSF synergize with M-CSF to increase the proliferative response of macrophages. M-CSF and GM-CSF stimulate the release of plasminogen activator, prostaglandin E, interferons, interleukin-1, and tumor necrosis factor by macrophages (62), increase the resistance of macrophages to viral infections (63), and increase the capacity of macrophages to kill intracellular parasites and yeast (64).

These actions of the growth factors on mature circulating and tissue cells would be expected to enhance their ability to resolve infections at local sites and repair tissues. The release of CSFs at local sites of infection (see later) would attract, either directly or indirectly, neutrophils, eosinophils, or macrophages, immobilize them, and either directly activate these cells for enhanced adherence to microorganisms as well as ingestion and killing or prime them so that they more readily respond to bacterial products to produce toxic metabolites. The induced release of other cytokines would recruit further cells and enhance the functional capacity of these cells at the site of infection.

ACTIONS OF HEMOPOIETIC GROWTH FACTORS IN VIVO

With the molecular cloning of the colony-stimulating factors and their high-level expression in mammalian, yeast, or bacterial cells, it has recently become possible to determine the efficacy of CSFs injected in vivo in a number of animal models and in phase I/II clinical trials of hemopoietically comprised patients.

The activity of CSFs in vivo is in part dependent on the dose and route of administration because of relatively rapid pharmacokinetics usually involving clearance through the liver and degradation in the kidneys (65). These clearance mechanisms do not seem to involve recognition of the carbohydrate moieties of CSFs, since nonglycosylated CSFs have similar clearance kinetics and similar activity in vivo (65, 66), and both types of CSF have access to cells in hemopoietic organs such as the bone marrow and spleen.

In general the actions of the CSFs in vivo are predictable from their activities in vitro, although their quantitative effects are less so. G-CSF caused a dramatic, selective, and reversible rise in circulating neutrophil levels in mice, hamsters, monkeys, and man (67–71), and a rise in spleen cellularity and content of early progenitor cells in mice. GM-CSF caused a more modest increase in blood neurophils, monocytes, and eosinophils of mice, monkeys, and man (72–78), but resulted in a marked elevation of 'activated' macrophages, neutrophils, and eosinophils in the peritoneal cavity and an infiltration of these cells into several other organs in mice (72). The spleen weight and content of progenitor cells were also increased in GM-CSF-injected mice. Multi-CSF and M-CSF had the least dramatic effects on circulating blood cell levels, but Multi-CSF caused a redistribution of hemopoiesis from the marrow to the spleen, and resulted in increased stem cell, progenitor cell, megakaryocyte, and mast cell levels in the spleen and increased macrophage, eosinophil, and neutrophil levels in the peritoneal cavity of mice (66, 79, 80). M-CSF administration to mice resulted in increased cell cycling rates of early progenitor cells in the bone marrow of mice, and its action was synergistic with Multi-CSF and GM-CSF (81).

The actions of CSFs in normal animals can thus be characterized as having modest effects in the bone marrow, relocating the site of hemopoiesis to the spleen, causing mobilization of mature cells into the blood or into peripheral organs, causing activation of mature cells, and being reversible after CSF withdrawal.

The ability of CSFs to assist in hemopoietic recovery or augment hemopoietic responses in compromised animals has also been assessed in vivo. G-CSF administered to cyclophosphamide-treated mice prevented the depression of blood neutrophil levels and enhanced the rate of hemopoietic recovery in the spleen resulting in a profound protection of these mice from lethal doses of a variety of bacteria (82). This advanced hemopoietic recovery after bone marrow ablation has also been seen in monkeys (68) and man (69–71). GM-CSF was shown to synergize with interleukin-1 in exerting a radioprotective effect on lethally irradiated mice (83), and to significantly shorten the period of neutropenia in a monkey model of autologous bone marrow transplantation following irradiation (84). Multi-CSF increased the recovery of multipotential hemopoietic stem cells in the spleens of irradiated mice and increased the cell cycling status of progenitor cells in the bone marrow and spleen of mice pretreated with cyclophosphamide (85, 86). M-CSF also increased the cycling status of hemopoietic progenitor cells in the bone marrow of mice pretreated with the myelosuppressive agent, lactoferrin (81).

These actions of the CSFs in augmenting hemopoietic responses or hemopoietic recovery in models of myelosuppression have indicated that they should find clinical use in immunocompromised or hematologically impaired patients. Indeed the first phase I/II clinical trials with G-CSF and GM-CSF have now been reported. GM-CSF and G-CSF have been used in patients with the acquired immunodeficiency syndrome (AIDS) (74), in patients receiving chemotherapy for cancer (69–71, 75, 76), in patients receiving bone marrow transplants after bone marrow ablation (76), and in patients with anemias associated with myelodysplastic syndromes (75, 78). The results in these patients have been very encouraging, with the animal models serving as a good guide to the effectiveness of the CSFs in patients: increased circulating blood cell levels, increased hemopoietic recovery rates, and reduced numbers of days of neutropenia.

Compared to other cytokines, the side effects of CSF treatment have been less severe. Nevertheless, some caution needs to be exercised in the use of the CSFs in clinical medicine, since the side effects include bone pain, phlebitis, edema, rashes, fever, and chills ranging from mild to severe at high doses. Some of these side effects may be minimized by careful selection of dose, site, mode of delivery, and duration of administration of the CSFs. Further caution is suggested by two murine models of chronic GM-CSF administration: transgenic mice bearing a constitutively active GM-CSF gene (87) and mice reconstituted with bone marrow cells harboring a retrovirally activated GM-CSF gene (88). In these mice, chronically activated macrophages, neutrophils, and eosinophils invade various organs resulting in a lethal syndrome characterized by extensive tissue damage in the eyes, muscles, liver, and lungs. Finally, the CSFs are able to stimulate the proliferation of some types of myeloid leukemic cells (89–92), so that there should be careful consideration before using them in patients with myelodysplastic syndromes or myeloid leukemias.

CONTROL OF THE PRODUCTION OF HEMOPOIETIC GROWTH FACTORS

Little is known about the steady-state production of hemopoietic growth factors that maintain the massive daily production of white blood cells from the bone marrow. Normal animals usually have very low or undetectable levels of the CSFs and interleukins in serum. Similarly, it has been difficult to detect these factors in long-term bone marrow culture supernatants or even show an effect of exogenously added CSFs in such cultures. In part, these difficulties might arise from the steady-state utilization rates of CSFs being equal to their rate of production in the bone marrow environment (93, 94) or to the sequestration of CSFs in the extracellular matrix, in particular by binding to heparin sulfate and other glycosaminoglycans (95, 96). However, cloned bone marrow stromal cell lines have been shown to produce only M-CSF and IL-6 constitutively and to require induction by IL-1 to produce GM-CSF and G-CSF (97, 98). The role of CSFs and interleukins in maintaining steady-state hemopoiesis thus remains unresolved.

When mice are infected with bacteria or parasites or injected with bacterial endotoxin, there is a rapid (3–6 hrs) and dramatic rise in the serum levels of M-CSF, G-CSF, IL-5, and IL-6, and a small rise in the serum level of GM-CSF (62, 99, 100). However, there have been virtually no situations in mice that have resulted in detectable circulating levels of Multi-CSF even under those conditions that result in Multi-CSF production in vitro.

Studies in vitro have suggested that fibroblasts (101), activated macrophages (102), bone marrow stromal cells (98, 103), and the epithelial cells of the endometrium of pregnant uterus are the main cellular sources of M-CSF (104); that T-lymphocytes (105), endothelial cells (106), fibroblasts (107), and activated macrophages (108) are the main cellular sources of GM-CSF; macrophages (109), endothelial cells (106), and fibroblasts (110) are the main cellular sources of G-CSF; macrophages and T-lymphocytes are the main cellular sources of IL-6 (10, 111); and T-lymphocytes are the main cellular sources of IL-5 and Multi-CSF (100, 105) (Figure 2). The complex interrelationships of hemopoietic growth factor-inducing signals are shown in Figure 2. This figure demonstrates the central role of T-lymphocytes and macrophages in responding to foreign antigens or bacterial products, respectively, by directly secreting growth factors or generating signals (interferon-γ, interleukin-1, tumor necrosis factor, GM-CSF, Multi-CSF) that result in growth factor secretion by other cells. Since the cells shown in Figure 2 are widely distributed throughout the tissues, this network system seems to be designed to respond rapidly to local infections by secretion of hemopoietic growth factors, which then serve to attract, immobilize, and activate hemopoietic cells inolved in host defense at the local site of infection. These growth factors are short-lived so that, with resolution of the infection, the signals for their production are eliminated and eventually the growth factors themselves are cleared from the site.

At the molecular level, the regulation of hemopoietic growth factor production in T-lymphocytes is probably the best understood. Activation of the T-cell antigen receptor (either directly by antigen or by antibodies against the receptor or by lectins) results in the intracellular production of inositol trisphosphate and the release of intracellular calcium, ultimately leading to activation of protein kinase C (112). By a mechanism that is not understood,

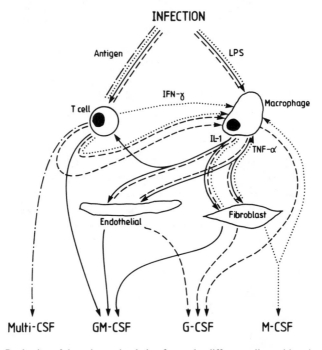

Figure 2   Production of the colony-stimulating factors by different cells, and its stimulation by intercellular mediator molecules. Infection results in the release of antigens and lipopolysaccharides that act on the central control cells, T-lymphocytes, and macrophages, respectively. These cells produce CSFs, but also release mediators (interferon-γ, interleukin-1, and tumor necrosis factor) that stimulate CSF production from fibroblasts and endothelial cells that are widely spread in the tissues.

activation of the T-cell receptor results in the accumulation of mRNA for a number of different growth factors, including GM-CSF, Multi-CSF, IL-2, IL-4, IL-5, IL-6, and interferon-γ, although there is a clonal variation in the ability of different T-cell clones to produce these factors (105). The accumulation of these mRNAs is independent of protein systhesis, since it is not inhibited by cycloheximide (105), but there is evidence that specific nuclear DNA-binding proteins recognize certain elements in the 5' region of a number of inducible T-cell genes and increase their rate of transcription (113, 114). In addition several inducible T-cell genes code for an A/U-rich sequence in the 3' untranslated region of the mRNAs that is ribonuclease sensitive and confers instability on the mRNA (115). Accumulation of such mRNAs may thus also involve proteins that stabilize this region of the mRNA (115, 116). One possibility is that protein kinase C or other kinases phosphorylate pre-existing nuclear proteins and activate their ability to bind to DNA promoter regions or to their mRNAs.

In addition to antigen, interleukin-1 and interleukin-2 have also been shown to induce the production of growth factors in T-cells. Interleukin-2 appears to do this by a different mechanism from that of antigen, since only low levels of GM-CSF and interferon-γ are produced, with no production of Multi-CSF, and this production is not inhibited by cyclosporin (117).

It is not clear if the basic mechanisms of growth factor induction are similar in different cells. GM-CSF production by endotoxin or otherwise-activated macrophages has been reported to involve mainly increased mRNA stability by a cycloheximide-sensitive process (108), in contrast to the situation observed in activated T-cells. On the other hand, production of GM-CSF, G-CSF, and M-CSF by endotoxin-treated endothelial cells is mediated, at least in part, by an increased rate of transcription (106). The conservation of certain small DNA sequences at the 5' ends of a number of different inducible hemopoietic growth factor genes (118) suggests that their transcription may be coordinately regulated by a set of nuclear DNA-binding factors. While some of these factors are now being identified (113, 114), further experiments are needed to determine their range of actions in different cells.

All of the hemopoietic growth factors are synthesized as precursor molecules with a hydrophobic leader sequence of 25–32 amino acids adjacent to a signal cleavage site (Table 1). For most of these factors there is little evidence for control of secretion at the level of glycosylation, transport, or signal peptide cleavage. However, there is clear evidence for M-CSF of alternative mRNA splicing that is differently regulated in different cell types and results in the production of at least two distinctly different molecules (119, 120). Both of these molecules contain, in additon to the N-terminal hydrophobic leader sequence, a C-terminal sequence with a typical transmembrane insertion and stop sequence. At least one of these forms of M-CSF is expressed at

Table 1 Molecular properties of hemopoietic growth factors

Growth factor	Other names	Species	Sub-units	Leader[a]	Mature protein[b] M_r (x 10^{-3}) (N)
GM-CSF	MGI-1GM	murine	1	25	14.4 (124)
	CSF-α, pluripoietinα	human	1	25	14.7 (127)
G-CSF	MGI-1G	murine	1	30	19.1 (178)
	CSF-β, pluripoietin	human	1	30	18.6 (174)
M-CSF	MGI-1M, CSF-1	murine	2	31	21(189), 18(158)[e]
	CSF-1	human	2	32	21(189), 18(158)
Multi-CSF	IL-3	murine	1	26	16.2 (140)
	IL-3	human	1	19	15.4 (133)
IL-5	EDF, BCGF II, TRF	murine	2	21	13.3 (112)
		human	2	22	13.2 (112)
IL-6	BSF-2, IFN-β_2, HP1	murine	1	24	21.7 (187)
	HSF, 26KD	human	1	29	20.8 (184)

[a] Number of amino acids in putative leader sequence
[b] molecular weight and number of amino acids (N) in the protein core of the mature protein
[c] ½ cystine amino acid positions are linked with a dash where known; [i] intramolecular; [o] intermolecular
[d] positions of potential N-linked glycosylation sites (N) or known O-glycosylation sites (O)
[e] at least two different forms of M-CSF are produced from a single gene, both involving N-terminal and separate C-terminal processing as well as an additional sequence between positions 149 and 150.

the cell surface of the synthesizing cell but can be released into the medium by proteolysis (121). Control of secretion of M-CSF could thus be mediated by specific protease levels, perhaps providing a mechanism for rapid release of M-CSF in certain situations.

MOLECULAR STRUCTURES OF THE HEMOPOIETIC GROWTH FACTORS

All of the hemopoietic growth factors discussed here are relatively small glycoproteins containing intra- or intermolecular disulfide bonds. Although these factors are N or O glycosylated (Table 1), the carbohydrate moieties in no case are necessary for their full range of biological activities. Rather, studies with nonglycosylated or alternatively glycosylated recombinant hemopoietic growth factors have suggested that the main role of the carbohydrate is to enhance the solubility, stability, and resistance to proteolysis of the factors. Indeed these glycosylated and disulfide-linked structures are often remarkably stable to denaturants and proteases, a property that would be expected to be of value in the extracellular environment of an inflammatory reaction.

With the exception of M-CSF and IL-5, which are composed of two identical, disulfide-linked subunits, all the other factors (GM-CSF, G-CSF, Multi-CSF, and IL-6) are single subunit proteins. Despite their similar size

Disulfides[c]	Glycosylation[d] sites		Glycosylated M_r ($\times$ 10^{-3})	Refs.
51–93, 85–118 i	N	58, 67+0	18–25	122–124
54–96, 88–121 i	N	19, 29+0	18–30	123, 125, 126
42–48, 70–80 i	O	139	25	127–129
36–42, 64–74 i	O	133	20	130–132
7, 31, 48, 90, 102, 139, 146 o	N	122, 140+0	70–90, 45–50	101, 133, 134
7, 31, 48, 90, 102, 139, 146 o	N	122, 140+0	70–90, 45–50	119–121, 134, 135
17–79, 80–140 i	N	16, 51, 86+0	18–30	136–139
16–84 i	N	15, 70+0	15–30	140, 141
41, 82 o	N	25, 54, 68	32–62	100, 118, 142, 143
41, 83 o	N	25, 68	46	100, 118, 144
43, 49, 72, 82 i	O	143	22–29	145, 146
44, 50, 73, 83 i	N	45, 143+0	19–21	147, 148

and general molecular nature as well as their overlapping biological specificities, each of the factors is encoded by a single unique gene, binds to a unique cellular receptor exhibiting no cross-reactivity, and shows no extended sequence homology with any of the other factors. Relatively weak areas of homology have been noted between some of the factors (G-CSF and IL-6; GM-CSF and IL-4), but these must reflect a very distant evolutionary relationship (100, 118, 148). Amino acid sequences for all the growth factors discussed here have been determined for both human and murine species. Some of the factor sequences have been remarkably well conserved (M-CSF, G-CSF, IL-5, ~75%) and retain some level of cross-species biological reactivity (149), while others show limited conservation (GM-CSF, IL-6 ~50%) or very little conservation (Multi-CSF ~28%) and no or very little cross-species reactivity at the biological level.

Amongst the hemopoietic growth factors, M-CSF is unique in that two distinct molecular forms of the factor can be produced by alternative mRNA splicing and have been shown to occur naturally. M-CSF produced by the smaller (1.6 kb) mRNA is synthesized as a 250-amino-acid subunit precursor containing a 32-amino-acid leader sequence, which is cleaved by a signal peptidase, and a second hydrophobic transmembrane-spanning sequence of 23 amino acids beginning at residue 166 and followed by three positive residues, which cause insertion of the polypeptide into membranes during its synthesis. The glycosylated disulfide-linked homodimer (68 kd) is presented at the cell surface as an integral membrane protein and may subsequently be proteolytically cleaved on the extracellular side near residue 158 to produce a soluble factor of 44 kd. On the other hand, the longer (4 kb) mRNA produces a 554-amino-acid subunit precursor, which contains an additional 298 amino acids interposed between amino acids 149 and 150 of the smaller protein. This additional sequence in the larger form of M-CSF appears to provide an alternate proteolytic processing site between residues 191 and 218 since this precursor protein is cleaved more efficiently than the short precursor, and processing may occur in the secretory compartments rather than at the cell surface. The longer form of M-CSF is secreted as a soluble factor of 70 kd (119–121, 133–135).

The existence of membrane-bound forms of M-CSF as well as at least two different forms of secreted M-CSF raises interesting questions about differental functions of M-CSF. It is known that the two types of mRNA are expressed differentially in different cell types, and that during pregnancy the longer mRNA is preferentially expressed in uterine cells (104). However, no evidence has been presented that the different forms of M-CSF differ in their spectrum of biological activities. In fact, both forms of secreted CSF are fully active in stimulating macrophage proliferation, suggesting that the information for this function resides between residues 1 and 158. This also suggests that the two additional potential N-glycosylation sites and additional cysteine residues in the inserted sequence in the longer form of M-CSF are not required for the correct folding and dimerization of the active form of M-CSF.

C-DNA cloning of G-CSF, GM-CSF, and Multi-CSF sequences has also suggested the possibility of alternate splicing or initiation producing different molecular forms of these factors. For human G-CSF the two forms differ by a tripeptide insertion between positions 35 and 36 with the longer form of G-CSF having a somewhat lower specific biological activity (150). For murine GM-CSF and Multi-CSF, alternative sites of transcriptional initiation have been suggested to provide the possibility of integral membrane forms of these factors (151). However, in all of these cases no direct evidence has been found for the existence of these alternate forms of the factors or for the predicted mRNAs in any cell type. The possibility must be considered that these alternate cDNA clones might have arisen as cloning artifacts.

Little is known about the detailed conformational structures of any of the growth factors discussed here or of the structural elements involved in defining the biologically active sites. Deletion and mutagenesis analysis of murine GM-CSF suggested that mutant molecules predicted to truncate two α-helical regions in the native structure near the N-terminal resulted in loss of biological activity (152). However, proteolysed murine GM-CSF molecules or synthetic human GM-CSF molecules with deletions of up to 14 N-terminal amino acids retained full biological activity (153). Similarly, deletion of the C-terminal six amino acids had little effect on biological activity, and it was suggested that the active site is between residues 14 and 96 (153). A similar synthetic approach to defining the active site of Multi-CSF has suggested that the N-terminal 16 amino acids are not required for activity, and that the active site is between residues 17 and 19, which form the first disulfide loop (154). It is clearly of some importance to determine the detailed conformations of these molecules by X-ray crystallography before the active sites can be mapped in any detail.

THE GENES FOR HEMOPOIETIC GROWTH FACTORS

The hemopoietic growth factors, because of their extremely high specific biological activities (10^8–10^9 Units/mg), are present in most natural sources in vanishingly small mass amounts (1–10 ng/ml in conditioned media from the best cellular sources). This has meant that natural sources of these factors are not practical avenues for the mass production of growth factors for animal or clinical trials, and that the biochemical purification of such factors was in each case a major undertaking. These considerations have resulted in a major effort by several groups over the last four years that has seen the C-DNA and genomic cloning of murine and human forms of all the growth factors discussed here. In some cases (murine GM-CSF, human G-CSF, IL-6), the C-DNAs were identified by probing libraries with oligonucleotides based on partial amino acid sequences of the purified natural proteins, while in other cases (murine and human Multi-CSF, human GM-CSF, murine IL-5), the C-DNAs were identified by direct expression of the biological activity from full-length C-DNAs in mammalian expression vectors. These strategies allowed the first identification of a full coding sequence for these growth factors and in some cases allowed isolation of the corresponding growth factor C-DNA from other species by cross-hybridization (Table 2).

The hemopoietic growth factors are encoded by single genes in both mouse and man. In the mouse, GM-CSF and Multi-CSF have been mapped close together on chromosome 11, which also contains the G-CSF and IL-5 genes as well as the gene for the M-CSF receptor/c-fms (157). However, none of these genes seems to be associated with the Evi-2 locus, a viral integration site on chromosome 11 involved in murine myeloid leukemias. Part of murine chromosome 11 is homologous to the long arm of human chromosome 5, and the human GM-CSF, Multi-CSF, M-CSF, IL-5 genes, as well as the M-CSF receptor gene have all been mapped to this region of chromosome 5. Interestingly, deletion of the long arm of chromosome 5 has been observed in a group of patients suffering from the 5q- syndrome. These patients often have myelodysplastic syndromes characterized by a refractory anemia, and may develop myeloid leukemias after cytotoxic therapy (158), but it is difficult at present to correlate this syndrome with the loss of one or more of the growth factor or receptor genes. The tight linkage of the GM-CSF and Multi-CSF genes in man (about 10 kb apart) is also of interest, and may reflect their common expression in T-cell lines. Another part of chromosome 11 is

Table 2 Genes for hemopoietic growth factors

Growth factor	Species	Chromosomal location	Number of exons	Gene[a] size (kb)	mRNA[b] sizes (kb)	Alternate[c] splicing described?	Protein[d] homology	Refs.
GM-CSF	murine	11	4	2.5	0.78	yes		124, 151, 155–157
	human	5q21–q31	4	2.5	0.78	no	52%	126, 156, 158
G-CSF	murine	11	5	2.2	1.6	no		131, 132, 157–159
	human	17q21–q22	5	2.2	1.7	yes	78%	150, 160
M-CSF	murine	—	—	—	4.5, 3.8, 2.3, 1.4	yes		133
	human	5q33.1	10	21	4.5, 3.3, 2.6, 1.6	yes	82%	119, 120, 135, 158, 161
Multi-CSF	murine	11	5	2.2	1.0	yes		138, 139, 157, 162
	human	5q21–q31	5	2.5	1.0	no	28%	140, 141, 163
IL-5	murine	11	4	3.8	1.7	no		100, 118, 143, 164
	human	5q31	4	1.9	—	no	67%	100, 118, 144, 165, 166
IL-6	murine	—	—	—	1.3	no		146
	human	7p15	5	4.7	1.3	yes	42%	147, 148, 167, 168

[a] From the beginning of exon 1 to the end of the final exon.
[b] Sizes of the mature cytoplasmic messages.
[c] Alternative splicing or alternative initiation sites for transcription.
[d] Between species.

homologous to human chromosome 17, and it is on this chromosome that the human G-CSF gene is found at q21–q22. Human neutrophilic promyelocytic leukemias are commonly associated with a translocation between chromosomes 15 and 17, but the G-CSF gene does not appear to be altered by this translocation (160). Human IL-6 is not linked to the other hemopoietic growth factors and is on chromosome 7 (Table 2).

The structures of the growth factor genes have shown some common structural elements at their 5' ends. At a position 73–183 nucleotides upstream of the TATA box is a common decanucleotide

GA A   C C
AGGGTTCTAT   (CK1)

in the GM-CSF, Multi-CSF, G-CSF, IL-5, and IL-6 genes of mouse and man. A second 5' conserved sequence of

A
TCAGGTA   (CK2)

is found in the GM-CSF and Multi-CSF genes just 3' to the decanucleotide sequence (100). Nuclear-binding proteins have been identified by gel retardation experiments for both these sites of the human GM-CSF gene. The first binds to the CK1 region and is distributed in many cells, including those that do not express GM-CSF, while the second binds to the CK2 region, is more restricted in its expression, and is inducible by phorbol esters in a bladder carcinoma cell line which subsequently expresses GM-CSF mRNA and protein (113). On the other hand, DNase I footprinting studies of the human GM-CSF gene have identified a different nuclear protein–binding region just upstream of the TATA box, and a 90-bp region of the gene including this site, but not the CK1 or CK2 sites, is induced to stimulate transcription of a reporter CAT gene in stimulated T-cells (114).

Alternate sites of transcriptional initiation have been described for a number of the growth factor genes, including GM-CSF, Multi-CSF, and IL-6. In these cases, the less common initiation site, which may be up to 10 kb 5' to the more common site, would result in the production of an altered protein with a hydrophilic domain N-terminal to the hydrophobic leader sequence, but further study will be required to determine if tissue-specific transcriptional initiation can be used to produce functionally different molecules.

Alternative splicing of the primary mRNA transcripts has also been shown to occur for some of the hemopoietic growth factors. The most dramatic of these occurs for M-CSF where use of alternative splice sites results in either a 4 kb or 1.6 kb mRNA. These mRNAs have common initiation and termination codons but use alternate noncoding exons at their 3' ends and, importantly, the longer mRNA contains an additional in frame insertion of a coding region for 298 amino acids at amino acid position 149. These alternately spliced mRNAs are differently expressed in different cells (161), but the functional importance of the altered proteins is not known.

A common feature of several growth facteor mRNAs is an A/U-rich sequence in the 3'-untranslated region (115, 116). This is a conserved region in the mRNAs for GM-CSF, Multi-CSF, G-CSF, and several other transiently expressed lymphokines (IL-1, IL-2, TNF, IFN) and oncogenes (cfos, sis, myc, myb). When this region of GM-CSF mRNA was inserted into the 3' region of normally long-lived β-globin mRNA and inserted into fibroblasts the globin RNA was unstable, suggesting that the A/U region conferred nuclease sensitivity to the mRNA (169).

RECEPTORS FOR HEMOPOIETIC GROWTH FACTORS

Specific, high-affinity receptors unique for each type of hemopoietic growth factor have been identified on several cell lines and on normal hemopoietic tissues as well as primary myeloid leukemic cells. The cellular distribution of the receptors matches the known biological specificities of each growth factor, and these receptors are generally low in number. M-CSF receptors are essentially restricted to cells of the monocyte-macrophage lineage, possibly with small numbers of receptors on some neutrophils, and receptor numbers increase from a few hundred on bone marrow monocytes and their precursors to several thousand on mature, 'activated' peritoneal exudate macrophages (170). Similarly, G-CSF receptors are essentially restricted to cells of the neutrophilic granulocyte cell lineage, with small numbers of receptors on monocyte/macrophages and receptor numbers increasing with cellular differentiation to a few hundred per cell (171, 149). On the other hand, GM-CSF and Multi-CSF receptors are more uniformly distributed on neutrophilic and eosinophilic granulocytes and monocyte/macrophages and generally decrease to a few hundred or less per cell as the cells become more mature (172, 173). The distributions of IL-5 and IL-6 receptors have not yet been determined. Quantitative autoradiographic analyses have suggested that at least some cells must simultaneously display at least four CSF receptors on their cell surface (173). Analysis of a variety of primary human acute myeloid and chronic myeloid leukemic cells indicated that most often these cells retain functional receptors for G-CSF, GM-CSF, and probably Multi-CSF and M-CSF (149, 174).

Some molecular and binding characteristics of the CSF receptors are shown in Table 3. The M-CSF receptor has been identified by chemical cross-linking and affinity purification as a single-chain glycoprotein tyrosine kinase (175,

176). M-CSF binding to its receptor stimulates its tyrosine kinase activity and results in autophosphorylation of tyrosine 969 on the intracellular side. The M-CSF receptor is almost certainly the cellular fms oncogene, since both proteins are tyrosine kinases, have the same size, and have a similar cellular distribution. M-CSF binds to the chicken viral fms oncogene, and some antibodies to this viral oncogene precipitate M-CSF/receptor complexes with associated tyrosine kinase activity (177). Finally, 3T3 fibroblasts transfected with an expressable c-fms construct become responsive to M-CSF for enhanced proliferation (178). c-fms or the M-CSF receptor has been molecularly cloned and is clearly homologous to the receptor for platelet-derived growth factor (PDGF) (179, 180). It contains a 490-amino-acid extracellular domain at the N-terminal end with limited homology to the immunoglobulin supergene family and a 536-amino-acid intracellular domain containing a typical but interrupted tyrosine kinase domain and a tyrosine phosphorylation site. Two point mutations in the extracellular domain are likely to be the main causes of the transforming potential of v-fms. However, the deletion of 40 amino acids at the C-terminal end of v-fms relative to c-fms results in the deletion of a tyrosine autophosphorylation site, and this deletion also enhances the transforming potential of v-fms (178).

On intact cells or cell membranes, M-CSF receptors display a single class of binding affinity, which at 2°C is characterized by an almost unmeasurably slow kinetic dissociation rate (181). At 37°C, dissociation of M-CSF is measurable, and kinetic measurements suggest that the true equilibrium dissociation constant would be about 400 pM. However, at this temperature, binding of M-CSF is rapidly followed by internalization and degradation of M-CSF, the degradation rate being rapid in peritoneal exudate macrophages but considerably slower in bone marrow–derived monocytes (182, 183). Binding of M-CSF increases the rate of internalization of receptors and causes down-regulation of surface M-CSF receptors. Besides the cognate ligand, a number of other agents unable to bind to the M-CSF binding site are able to indirectly 'down-regulate' the M-CSF receptor on macrophages or monocytes at 37°C. These include bacterial lipopolysaccharides, phorbol esters, interferon, GM-CSF, and Multi-CSF (173, 184, 185).

The G-CSF receptor has been identified by chemical cross-linking as a single-chain glycoprotein of about 150,000 in molecular weight (186). Binding to a variety of murine and human cell types is associated with a single class of affinity site with apparent K_D of about 100 pM, but again binding at 0–4°C is nearly irreversible (45, 149, 171). Kinetic results at 37°C indicate a true equilibrium dissociation constant of about 300 pM, but kinetic interactions involved with internalization and degradation of ligand mean that the steady-state apparent dissociation constant is considerably lower than this (20–100 pM) (94) and closer to the concentrations that are needed to exert

Table 3 Properties of hemopoietic growth factor receptors

Receptor[a] M_r (× 10^{-3})	K_D (pM)[b] Eq.	SS	Ave. receptor no[c] per + ve BM cell	Down-regulated by[d]	Refs.
m M-CSF	400	—	4600	Multi, GM, TPA, LPS, IFN	172, 179–185
h M-CSF	1000	—	—	—	179, 197, 198
m G-CSF	518	25	185	Multi, GM, TPA	186, 45, 149, 171, 94, 184
h G-CSF	500	—	50–300	TPA, FMLP, GM, LPS	187, 197
m Multi-CSF	244	34	97 (486)	Multi	94, 173, 184, 186–189
h Multi-CSF	100	—	100	—	197
m GM-CSF	189	31	176	Multi, TPA	94, 184, 190, 191
h GM-CSF	250	—	200–500	TPA, FMLP	192–194, 197
m IL-6	10, 1000	—	—	—	195, 196
h IL-6	—	—	—	—	—

Receptor M_r (× 10^{-3}) values in first column:
m M-CSF 165, h M-CSF 165, m G-CSF 150, h G-CSF —, m Multi-CSF 75, 55, 130, h Multi-CSF —, m GM-CSF 55, 115, h GM-CSF 85, m IL-6 60, h IL-6 —

[a] Determined by chemical cross-linking except for the M-CSF and IL-6 receptors, which have been cloned.
[b] Determined by the best kinetic estimates where available (Eq.) or the steady-state constant at 37°C (SS).
[c] Average receptor number per positive bone marrow cell.
[d] Abbreviations used: Multi, Multi-CSF; GM, GM-CSF; TPA, tetradecanoyl phorbol acetate; LPS, lipopolysaccharide; IFN, interferon; FMLP, F-Met-Leu-Phe.

maximal biological effects (10–50 pM). On murine bone marrow cells at 37°C, G-CSF, GM-CSF, and Multi-CSF cause 'down-regulation' of surface G-CSF receptors (184). On human neutrophils, G-CSF, GM-CSF, bacterial lipopolysaccharides, and chemotactic peptides all cause G-CSF receptor down-regulation (187).

Multi-CSF receptors have been identified by chemical cross-linking either as a single-chain protein of M_r 75,000, two noncovalently linked chains of M_r 75,000 and 55,000, or a third chain of M_r 113,000 (173, 186, 188). Despite this complex situation, only a single class of binding site has been described with apparent K_Ds varying from 100 to 1000 pM (173, 188, 189). Kinetic measurements at 37°C, as for the other CSFs, indicate a true equilibrium dissociation constant of about 300 pM but a lower apparent dissociation constant under steady-state conditions associated with internalization and degradation (94). Unlike the other CSFs, however, only Multi-CSF itself has been shown to down-regulate its receptor (184).

Chemical cross-linking of GM-CSF receptors on murine cells identified a receptor of M_r 51,000 (190) or 130,000 (191). Some studies have demonstrated two affinity classes of binding sites on murine (190) or human cells (N. A. Nicola, unpublished) with apparent K_Ds of 30 pM and 1 nM or a single class of apparent K_D varying from 30 pM to 1 nM (192–194). The high-affinity class mediates GM-CSF internalization and degradation at 37°C as for the other CSFs, and this results in a reduction of the equilibrium dissociation constant from about 200 pM to an apparent dissociation constant in the steady-state of about 30 pM (94; N. A. Nicola, unpublished). On murine bone marrow cells the GM-CSF receptor is 'down-regulated' by GM-CSF, Multi-CSF, and phorbol esters, while on human neutrophils the GM-CSF receptor is down-regulated by GM-CSF and chemotactic peptides (184, 194).

A similar situation exists for the IL-6 receptor with two affinity classes of apparent K_D 17 pM and 1 nM on U266 cells (195, 196). The IL-6 receptor has been cloned by direct expression and selection using biotinylated IL-6 and fluorescent avidin. Transfection of the cloned cDNA into a human T-cell line has demonstrated that the two affinity classes arise from a single gene product. The cloned receptor contains an 82-amino-acid intracellular domain with no homology to tyrosine kinase domains, but the 340-amino-acid ligand-binding extracellular domain contains regions with some homology to the immunoglobulin supergene family and M-CSF and PDGF receptors (196).

The common features of CSF-receptor interactions can be summarized as follows. In most cases the binding of CSF to its receptor at low temperatures is characterized by a tight interaction that seriously retards dissociation. This may reflect a conformational change in the receptor induced by ligand binding. Nevertheless, at 37°C this does not result in the same type of irreversible binding, but rather results in events that, among other things, lead to internalization of the CSF-receptor complex and ultimately to CSF degradation. The effects of these events are to result in down-regulation of cell-surface receptors and, under steady-state conditions, to increase the level of receptor occupancy at a given CSF concentration as well as to result in a rate of CSF utilization. Each CSF-receptor is also characterized by a unique pattern of noncognate ligands, including other CSFs that can cause trans-down-regulation. Elsewhere we have argued that this pattern may be consistent with trans-activation of CSF receptors (173, 190), but neither the mechanism of these trans effects nor their biological consequences are known. As other CSF receptors are cloned, it may be possible to dissect these mechanisms and their consequences in more detail.

SIGNAL TRANSDUCTION BY HEMOPOIETIC GROWTH FACTORS

As for most other growth factors, little is known about the detailed signal transduction pathways mediating the biological effects of hemopoietic growth factors. The demonstrated M-CSF-dependent tyrosine kinase activity of the M-CSF receptor (177–180), as well as its homology to the PDGF receptor, suggests that some pathways of cellular activation may be common to several types of ligand-activated tyrosine kinases. Indeed, both receptors result in the increased expression of some common genes in their respective cell types, including transient expression of c-fos, sustained expression of c-myc, and early elevation of JE and KC mRNAs (199). Moreover, the M-CSF receptor can transmit proliferative signals in fibroblasts either as the constitutively active v-fms or the normal c-fms molecule stimulated by M-CSF (178). Conversely, the epidermal growth factor (EGF) receptor can transmit proliferative signals in CSF-dependent cell lines either constitutively as the v-erb B product or as the normal c-erb B product stimulated by EGF (200, 201).

Despite the fact that the Multi-CSF receptor may be too small to encode a protein tyrosine kinase (though this has not yet been proved), there is evidence that Multi-CSF also stimulates tyrosine phosphorylation on cellular substrates in factor-dependent cell lines (202, 203). Two of these phosphorylation substrates (a membrane-bound 150-kd glycoprotein and a 33-kd protein) are also constitutively phosphorylated in cell line variants, which become independent of Multi-CSF for their growth (202, 203). Consistent with this, it has been shown that transfection of factor-dependent cell lines growing in the presence of Multi-CSF with retroviruses containing the constitutively active tyrosine kinase v-src results in the cells becoming independent of Multi-CSF for growth. Somewhat surprisingly, however, addition of Multi-CSF or GM-CSF to these cell lines results in a reduction of tyrosine kinase activity (204).

For some growth factor/receptor systems it has been suggested that the receptors are coupled to a GTP-binding protein (G-protein) that activates a phosphatidyl inositol–specific phospholipase C leading to rises in intracellular calcium and activation of protein kinase C (205). Some authors have found that Multi-CSF activates protein kinase C in factor-dependent cell lines and macrophages (206, 207), and that phorbol esters, which activate protein kinase C independently of ligand, can partially replace the requirement for CSFs for proliferation (208). In addition, v-fms–transformed cells or hemopoietic cells transformed with the EGF receptor and stimulated with EGF show evidence for enhanced activity of a phosphatidyl inositol–specific phospholipase C (200, 209). However, most studies have failed to detect enhanced activity of this enzyme induced directly by Multi-CSF or M-CSF in responsive cells (207, 210). Nevertheless, the use of several different inhibitors or activators of G-proteins has suggested that these proteins may be involved in the differential regulation of Multi-CSF and M-CSF proliferative responses (211, 212).

CSFs have been shown to regulate many other enzyme functions in responsive cells, including ATP levels, glucose transport, and the Na^+/K^+ ATPase system (5, 6, 213, 214). However, the relationships of these functions to cellular proliferation are unknown. Indeed the wide variety of cellular responses induced by the CSFs makes it difficult to correlate any of the observed signal transduction pathways to one or the other specific response, including those of proliferation and differentiation.

ONCOGENIC POTENTIAL OF HEMOPOIETIC GROWTH FACTORS

The involvement of the hemopoietic growth factors and their receptors in regulating blood cell proliferation raises the possibility that dysregulation of these molecules may be involved in the generation of leukemias or other cancers. The close relationship of platelet-derived growth factor to the v-sis oncogene (215), as well as of the epidermal growth factor receptor to the v-erb B oncogene (216), serves to reinforce this argument.

A clear parallel exists in the hemopoietic system with the sequence relationship between the M-CSF receptor and the v-fms oncogene. Unlike the v-erb B oncogene, v-fms retains the M-CSF–binding domain (although its binding characteristics may be altered), but like v-erb B, v-fms also shows a C-terminal deletion that removes a tyrosine autophosphorylation site (179). Despite its retention of the ligand-binding domain, v-fms, like v-erb B, acts as a constitutively activated tyrosine kinase and stimulates its autophosphorylation on tyrosine. The ability of v-fms to cause fibrosarcomas suggests that fibroblasts can respond to proliferative signals generated by an activated M-CSF receptor. This has been confirmed by the ability of cotransfection of the M-CSF receptor gene and the M-CSF gene to cause fibroblast transformation, while transfection of the M-CSF receptor alone was ineffective (121). Evidence that c-fms may also be involved in leukemic development comes from the observation that a common viral integration site (fim-2) involved in Friend leukemia virus-induced myeloid leukemias has been mapped to the 5' end of the c-fms/M-CSF receptor gene (217). Moreover, v-fms–transfected bone marrow cells injected into irradiated mice caused a myeloproliferative disease as well as erythroid leukemias and B-cell lymphomas (218). Similarly, v-fms transfection into a M-CSF-dependent cell line, BAC 1.2 F5, or a Multi-CSF-dependent cell line (FDCP-1) in both cases resulted in factor-independent growth and tumorigenicity (219, 220). Nevertheless, stimulation of normal CSF-dependent pathways is usually not sufficient to induce tumorigenicity. Transfection of BAC 1.2 F5 cells with an active M-CSF gene results in factor-independence but not tumorigenicity, indicating that altered regulation of M-CSF receptor activities is required to induce tumorigenicity (221).

Simple uncontrolled expression of a hemopoietic growth factor gene by a responsive cell would not be expected to result in leukemic transformation, since normal cells respond to these factors by coupling the proliferative response to a self-limiting terminal differentiation response. In fact, some of the CSFs have an anti-leukemic action in enforcing terminal differentiation of leukemic cells (222). As a result of these actions retroviral infection of primary hemopoietic cells with expressable CSF genes does not result in the development of tumorigenic cells either in vitro or in vivo (223, 88). Nor do transgenic mice bearing a constitutively activated GM-CSF gene develop myeloid leukemias (87).

However, there is ample evidence that constitutive activation of CSF genes may be involved as a secondary event in leukemia development. CSF-dependent immortal cell lines appear to have a primary defect in their capacity to differentiate but are rarely leukemogenic when transplanted in vivo. If these cell lines are transfected with retrovirally activated GM-CSF or Multi-CSF genes they become factor-independent for growth and leukemogenic when injected in vivo (223–225). Similarly, the development of some murine and human leukemias in vivo is associated with the constitutive activation by genetic rearrangement of Multi-CSF, GM-CSF, or M-CSF genes, presumably as a secondary event that confers a growth advantage to the affected cell clones (226–230). Nevertheless, the majority of human myeloid leukemias retain an absolute dependence on CSFs for their growth at least in vitro (62), suggesting that CSF levels are normally not limiting for leukemia development. A possible exception to this may be seen in the case of human myelomas where several cases were shown to be producing IL-6 constitutively and were inhibited in their proliferation in vitro by antibodies to IL-6 (231). In this case IL-6 appears to be required to interact with IL-6 receptors at the cell surface to stimulate cell proliferation. The insensitivity of other putative autocrine growth factor–dependent leukemias to antibody against the growth factor has led to the suggestion that intracellular activation might occur through the formation of growth factor/receptor complexes immediately after intracellular synthesis (224).

These results strongly suggest that at least two, and probably more, events are involved in the generation of myeloid leukemias. The first is an immortalization event (in the murine system sometimes associated with translocation at the c-myb locus) that blocks terminal differentiation and promotes self-renewing proliferation. The second is some event that results in autonomy from environmental stimulators of cell proliferation. This can occur at several levels involving constitutive activation of CSF receptors, CSF genes, or intracellular signaling pathways that are mimicked by the v-abl oncogene, cytoplasmic oncogenes such as v-ras and src, or nuclear oncogenes such as myb, myc, fos, and others (232–236).

FUTURE PROSPECTS

The molecular cloning and expression of the hemopoietic growth factors has made these factors accessible to a wide range of investigators and to the clinic. The factors are proving their usefulness in a variety of conditions associated with an impaired immune or hemopoietic system, including infections, AIDS, anemias, cancer chemotherapy, and bone marrow transplantation. They may also have a role in the treatment of leukemias and myelodysplastic syndromes, although the proliferative effects of the factors need to be carefully considered. It is to be hoped that the rapidly expanding number of clinical trials will be carefully monitored in order to improve the delivery of the factors and to minimize their side effects. It is also to be hoped that basic laboratory studies will soon provide more detail on the structures of the growth factors and their receptors, their mechanism of action, and their relationship to the development of hemopoietic disorders.

ACKNOWLEDGMENTS

The authors's work was supported by the National Health and Medical Research Council, Canberra, Australia; the Anti-Cancer Council of Victoria, Australia; the J. D. and L. Harris Trust Fund, and the National Institutes of Health, Bethesda, Md., Grant CA-22556.

Literature Cited

1. Metcalf, D., Merchav, S. 1982. *J. Cell. Physiol.* 112:411–18
2. Nicola, N. A., Metcalf, D. 1982. *J. Cell. Physiol.* 112:257–64
3. Begley, C. G., Lopez, A. F., Nicola, N. A., Warren, D. J., Vadas, M. A., et al. 1986. *Blood* 68:162–66
4. Rothenberg, M. E., Owen, W. F., Silberstein, D. S., Woods, J., Soberman, R. J., et al. 1988. *J. Clin. Invest.* 81:1986–92
5. Whetton, A. D., Dexter, T. M. 1983. *Nature* 303:629–31
6. Whetton, A. D., Bazill, G. W., Dexter, T. M. 1984. *EMBO J.* 3:409–13
7. Moore, M. A. S., Williams, N. 1973. In *Hemopoiesis in Culture*, ed. W. A. Robinson, pp. 17–27. Washington: DHEW Publ. No. 74–205
8. Tushinski, R. J., Stanley, E. R. 1983. *J. Cell. Physiol.* 116:67–75
9. Metcalf, D. 1970. *J. Cell. Physiol.* 76:89–100
10. Ikebuchi, K., Wong, G. G., Clark, S. C., Ihle, J. N., Hirai, Y., Ogawa, M. 1987. *Proc. Natl. Acad. Sci. USA* 84:9035–39
11. Metcalf, D. 1987. *Proc. R. Soc. London, Ser. B* 230:389–423
12. Metcalf, D. 1980. *Proc. Natl. Acad. Sci. USA* 77:5327–30
13. Metcalf, D., MacDonald, H. R. 1975. *J. Cell. Physiol.* 85:643–54
14. Nicola, N. A., Johnson, G. R. 1982. *Blood* 60:1019–29
15. Lopez, A. F., Begley, C. G., Williamson, D. J., Warren, D. J., Vadas, M. A., Sanderson, C. J. 1986. *J. Exp. Med.* 163:1085–99
16. Lin, H-S., Stewart, C. L. 1974. *J. Cell. Physiol.* 83:369–78
17. Hamilton, J. A., Vairo, G., Nicola, N. A., Burgess, A. W., Metcalf, D., Lingelbach, S. R. 1988. *Blood* 71:1574–80
18. Metcalf, D., Johnson, G. R., Burgess, A. W. 1980. *Blood* 55:138–47
19. Ihle, J. N., Keller, J., Oroszlan, S., Henderson, L. E., Copeland, T. P., et al. 1983. *J. Immunol.* 131:282–87
20. Metcalf, D., Nicola, N. A. 1983. *J. Cell. Physiol.* 116:198–206
21. Warren, D. J., Moore, M. A. 1988. *J. Immunol.* 140:94–99
22. Robinson, B. E., McGrath, H. E., Quesenberry, P. J. 1987. *J. Clin. Invest.* 79:1648–52
23. Chen, B. D., Clark, C. R., Chou, T. 1988. *Blood* 71:997–1002
24. McNiece, I. K., Stewart, F. M., Deacon, D. M., Quesenberry, P. J. 1988. *Exp. Hematol.* 16:383–88
25. Bradley, T. R., Hodgson, G. S. 1979. *Blood* 54:1446–50
26. Bartelmez, S., Stanley, E. R. 1985. *J. Cell. Physiol.* 122:370–78
27. McNiece, I. K., Kriegler, A. B., Hapel, A. J., Fung, M. C., Young, I. G., et al. 1984. *Cell. Biol. Int. Rep.* 8:812
28. Quesenberry, P., Song, Z. X., McGrath, E., McNiece, I., Shadduck, R., et al. 1987. *Blood* 69:827–35
29. Migliaccio, A. R., Visser, J. W. M. 1986. *Exp. Hematol.* 14:1043–48
30. Muller-Sieburg, C. E., Townsend, K., Weissman, I. L., Rennick, D. 1988. *J. Exp. Med.* 167:1825–40
31. Koike, K., Ogawa, M., Ihle, J. N., Miyake, T., Shimizu, T., et al. 1987. *J. Cell. Physiol.* 131:458–64
32. Dexter, T. M., Garland, J., Scott, D., Scolnick, E., Metcalf, D. 1980. *J. Exp. Med.* 152:1036–47
33. Hara, K., Suda, T., Suda, J., Eguchi, M., Ihle, J. N., et al. 1988. *Exp. Hematol.* 16:256–61
34. Valtieri, M., Tweardy, D. J., Caracciola, D., Johnson, K., Mavilio, F., et al. 1987. *J. Immunol.* 138:3829–35
35. Metcalf, D., Burgess, A. W. 1982. *J. Cell. Physiol.* 111:275–83
36. Nicola, N. A., Metcalf, D., Matsumoto, M., Johnson, G. R. 1983. *J. Biol. Chem.* 258:9017–23
37. Metcalf, D., Nicola, N. A. 1982. *Int. J. Cancer* 30:773–80
38. Metcalf, D. 1982. *Int. J. Cancer* 30:203–10
39. Tsuda, H., Neckers, L. M., Pluznik, D. M. 1986. *Proc. Natl. Acad. Sci. USA* 83:4317–21
40. Metcalf, D., Hilton, D. J., Nicola, N. A. 1988. *Leukemia* 2:216–21
41. Metcalf, D. 1982. *Int. J. Cancer* 24:616–23
42. Begley, C. G., Metcalf, D., Nicola, N. A. 1987. *Int. J. Cancer* 39:99–105
43. Hara, T., Umeda, T., Niijima, T., Okabe, T. 1985. *J. Cancer Res. Clin. Oncol.* 109:103–6
44. Miyauchi, J., Wang, C., Kelleher, C. A., Wong, G. G., Clark, S. C., et al. 1988. *J. Cell. Physiol.* 135:55–62
45. Nicola, N. A., Metcalf, D. 1984. *Proc. Natl. Acad. Sci. USA* 81:3765–69
46. Okabe-Kado, J., Kasukabe, T., Honma, Y., Hayashi, M., Hozumi, M. 1988. *J. Biol. Chem.* 263:10994–99
47. Wang, J. M., Colella, S., Allavena, P., Mantovani, A. 1987. *Immunology* 60:439–44
48. Weisbart, R. H., Golde, D. W., Clark,

S. C., Wong, G. G., Gasson, J. C. 1985. *Nature* 314:361–63
49. Fleischmann, J., Golde, D. W., Weisbart, R. H., Gasson, J. C. 1986. *Blood* 68:708–11
50. Villalta, F., Kierszenbaum, F. 1986. *J. Immunol.* 137:1703–7
51. Lopez, A. F., Nicola, N. A., Burgess, A. W., Metcalf, D., Battye, F. L., et al. 1983. *J. Immunol.* 131:2983–88
52. Weisbart, R. H., Kacena, A., Schuh, A., Golde, D. W. 1988. *Nature* 332:647–48
53. Arnaout, M. A., Wang, E. A., Clark, S. C., Sieff, C. A. 1986. *J. Clin. Invest.* 78:597–601
54. Weisbart, R. H., Kwan, L., Golde, D. W., Gasson, J. C. 1987. *Blood* 69:18–21
55. Yuo, A., Kitagawa, S., Okabe, T., Urabe, A., Komatsu, Y., et al. 1987. *Blood* 70:404–11
56. Avalos, B. R., Hedzat, C., Baldwin, G. C., Golde, D. W., Gasson, J. C., DiPersio, J. F. 1988. *Blood Suppl.* Abstr. Am. Soc. Hematol.
57. Lindemann, A., Riedel, D., Oster, W., Meuer, S. C., Blohm, D., et al. 1988. *J. Immunol.* 140:837–39
58. Weisbart, R. H., Golde, D. W., Gasson, J. C. 1986. *J. Immunol.* 137:3584–87
59. Vadas, M. A., Varigos, G., Nicola, N., Pincus, S., Dessein, A., et al. 1983. *Blood* 61:1232–41
60. Silberstein, D. S., Owen, W. F., Gasson, J. C., DiPersio, J. F., Golde, D. W., et al. 1986. *J. Immunol.* 137:3290–94
61. Vadas, M. A., Nicola, N. A., Metcalf, D. 1983. *J. Immunol.* 130:795–99
62. Metcalf, D. 1984. *The Hemopoietic Colony Stimulating Factors.* Amsterdam: Elsevier
63. Hammer, S. M., Gillis, J. M., Groopman, J. E., Rose, R. M. 1986. *Proc. Natl. Acad. Sci. USA* 83:8734–38
64. Weiser, W. Y., Van Niel, A., Clark, S. C., Dand, J. R., Remold, H. G. 1987. *J. Exp. Med.* 166:1436–46
65. Metcalf, D., Nicola, N. A. 1988. *Proc. Natl. Acad. Sci. USA* 85:3160–64
66. Metcalf, D., Begley, C. G., Nicola, N. A., Johnson, G. R. 1987. *Exp. Hematol.* 15:288–95
67. Cohen, A. M., Szebo, K. M., Inoue, H., Mines, D., Boone, T. C., et al. 1987. *Proc. Natl. Acad. Sci. USA* 84:2484–88
68. Welte, K., Bonilla, M. A., Gillio, A. P., Boone, T. C., Potter, G. K., et al. 1987. *J. Exp. Med.* 165:941–48
69. Bronchud, M. H., Scarffe, J. H.,

Thatcher, N., Crowther, D., Souza, L. M., et al. 1987. *Br. J. Cancer* 56:809–13
70. Morstyn, G., Souza, L. M., Keech, J., Sheridan, W., Campbell, L., et al. 1988. *Lancet* 1:667–72
71. Gabrilove, J. L., Jakubowski, A., Scher, H., Sternberg, C., Wong, G., et al. 1988. *New Engl. J. Med.* 318:1414–22
72. Metcalf, D., Begley, C. G., Williamson, D. J., Nice, E. C., DeLamarter, J., et al. 1987. *Exp. Hematol.* 15:1–9
73. Donahue, R. E., Wang, E. A., Stone, D. K., Kamen, R., Wong, G. G., et al. 1986. *Nature* 321:827–75
74. Groopman, J. E., Mitsuyasu, R. T., DeLeo, M. J., Oette, D. H., Golde, D. W. 1987. *New Engl. J. Med.* 317:593–98
75. Vadhan-Raj, S., Buescher, S., LeMaistre, A., Keating, M., Walters, R., et al. 1988. *Blood* 72:134–41
76. Brandt, S. J., Peters, W. P., Atwater, S. K., Kurtzberg, J., Borowitz, M. J., et al. 1988. *New Engl. J. Med.* 318:869–76
77. Lieschke, G. J., Maher, D., Cebon, J., O'Connor, M., Green, M., et al. 1988. *Ann. Intern. Med.* In press
78. Antin, J. H., Smith, B. R., Holmes, W., Rosenthal, D. S. 1988. *Blood* 72:705–13
79. Kindler, V., Thorens, B., DeKossodo S., Allet, B., Eliason, J. F., et al. 1986. *Proc. Natl. Acad. Sci. USA* 83:1001–5
80. Metcalf, D., Begley, C. G., Johnson, G. R., Nicola, N. A., Lopez, A. F., Williamson, D. J. 1986. *Blood* 68:46–57
81. Broxmeyer, H. E., Williams, D. E., Hangoc, G., Cooper, S., Gillis, S., et al. 1987. *Proc. Natl. Acad. Sci. USA* 84:3871–75
82. Matsumoto, M., Matsubara, S., Matsuno, T., Tamura, M., Hattori, K., et al. 1987. *Infect. Immun.* 55:2715–20
83. Neta, R., Oppenheim, J. J., Douches, S. D. 1988. *J. Immunol.* 140:108–11
84. Monroy, R. L., Skelly, R. R., MacVittie, T. J., Davis, T. A., Sauber, J. I., et al. 1987. *Blood* 70:1696–99
85. Kindler, V., Thorens, B., Vassalli, P. 1987. *Eur. J. Immunol.* 17:1511–14
86. Broxmeyer, H. E., Williams, D. E., Cooper, S. 1987. *Leuk. Res.* 11:201–5
87. Lang, R. A., Metcalf, D., Cuthbertson, R. A., Lyons, I., Stanley, E., et al. 1987. *Cell* 51:675–86
88. Johnson, G. R., Gonda, T. J., Metcalf, D., Hariharan, I. K., Cory, S. 1988. *EMBO J.* In press
89. Begley, C. G., Metcalf, D., Nicola, N. A. 1987. *Leukemia* 1:1–8

90. Kelleher, C., Miyauchi, J., Wong, G., Clark, S., Minden, M. D., McCulloch, E. A. 1987. *Blood* 69:1498–503

91. Vellenga, E., Ostapovicz, D., O'Rourke, B., Griffin, J. D. 1987. *Leukemia* 1:584–89

92. Miyauchi, J., Wang, C., Kelleher, C. A., Wong, G. G., Clark, S. C., et al. 1988. *J. Cell. Physiol.* 135:55–62

93. Tushinski, R. J., Oliver, I. T., Guilbert, L. J., Tynan, P. W., Warner, J. R., Stanley, E. R. 1982. *Cell* 28:71–81

94. Nicola, N. A., Peterson, L., Hilton, D. J., Metcalf, D. 1988. *Growth Factors.* In press

95. Gordon, M. Y., Riley, G. P., Watt, S. M., Greaves, M. F. 1987. *Nature* 326:403–5

96. Roberts, R., Gallagher, J., Spooncer, E., Allen, T. D., Bloomfield, F., Dexter, T. M. 1988. *Nature* 332:376–78

97. Rennick, D., Yang, G., Gemmell, L., Lee, F. 1987. *Blood* 69:682–91

98. Fibbe, W. E., Van Damme, J., Billiau, A., Goselink, H. M., Voogt, P. J., et al. 1988. *Blood* 71:430–35

99. Cheers, C., Haigh, A. M., Kelso, A., Metcalf, D., Stanley, E. R., Young, A. M. 1988. *Infect. Immun.* 56:247–51

100. Sanderson, C. J., Campbell, H. D., Young, I. G. 1988. *Immunol. Rev.* 102:29–50

101. Stanley, E. R., Heard, P. M. 1977. *J. Biol. Chem.* 252:4305–12

102. Rambaldi, A., Young, D. C., Griffin, J. D. 1987. *Blood* 69:1409–13

103. Gualtieri, R. J., Liang, C. M., Shadduck, R. K., Waheed, A., Bank, J. 1987. *Exp. Hematol.* 15:883–89

104. Pollard, J. W., Bartocci, A., Arceci, R., Orlofsky, A., Ladner, M. B., Stanley, E. R. 1987. *Nature* 330:484–86

105. Kelso, A., Gough, N. M. 1987. In *The Lymphokines*, ed. D. R. Webb, D. V. Goeddel, 3:209–38. New York: Academic

106. Seelentag, W. K., Mermod, J-J., Montesano, R., Vassalli, P. 1987. *EMBO J.* 6:2261–65

107. Koury, M. J., Pragnell, I. B. 1987. *Nature* 299:638–40

108. Thorens, B., Mermod, J-J., Vassalli, P. 1987. *Cell* 48:671–79

109. Metcalf, D., Nicola, N. A. 1985. *Leuk. Res.* 9:35–50

110. Koeffler, H. P., Gasson, J., Ranyard, J., Souza, L., Shepherd, M., Munker, R. 1987. *Blood* 70:55–59

111. Gauldie, J., Richards, C., Harnish, D., Lansdorp, P., Baumann, H. 1987. *Proc. Natl. Acad. Sci. USA* 84:7251–55

112. Hadden, J. W. 1988. *Immunol. Today* 9:235–39

113. Shannon, M. F., Gamble, J. R., Vadas, M. A. 1988. *Proc. Natl. Acad. Sci. USA* 85:674–78

114. Nimer, S. D., Marita, E. A., Martis, M. J., Wachsman, W., Gasson, J. C. 1988. *Mol. Cell. Biol.* 8:1979–84

115. Shaw, G., Kamen, R. 1986. *Cell* 46:659–67

116. Caput, D., Beutler, B., Hartog, K., Thayer, R., Brown-Shimer, S., Cerami, A. 1986. *Proc. Natl. Acad. Sci. USA* 83:1670–74

117. Kelso, A., Metcalf, D., Gough, N. M. 1986. *J. Immunol.* 136:1718–25

118. Yokota, T., Arai, N., De Vries, J., Spits, H., Banchereau, J., et al. 1988. *Immunol. Rev.* 102:137–87

119. Ladner, M. B., Martin, G. A., Noble, J. A., Nikoloff, D. M., Tal, R., et al. 1987. *EMBO J.* 6:2693–98

120. Wong, G. G., Temple, P. A., Leary, A. C., Witek-Giannotti, J. S., Yang, Y-C., et al. 1987. *Science* 235:1504–8

121. Rettenmier, C. W., Roussel, M. F., Ashmun, R. A., Ralph, P., Price, K., Sherr, C. J. 1987. *Mol. Cell. Biol.* 7:2378–87

122. Burgess, A. W., Camakaris, J., Metcalf, D. 1977. *J. Biol. Chem.* 252:1998–2003

123. Shrimser, J. L., Rose, K., Simona, M. G., Wingfield, P. 1987. *Biochem. J.* 247:195–99

124. Gough, N. M., Gough, J., Metcalf, D., Kelso, A., Grail, D., et al. 1984. *Nature* 309:763–67

125. Gasson, J. C., Weisbart, R. H., Kaufman, S. E., Clark, S. C., Hewick, R. M., et al. 1984. *Science* 226:1339–42

126. Wong, G. G., Witek, J. S., Temple, P. A., Wilkens, K. M., Leary, A. C., et al. 1985. *Science* 228:810–15

127. Nicola, N. A., Metcalf, D., Matsumoto, M., Johnson, G. R. 1983. *J. Biol. Chem.* 258:9017–23

128. Tsuchiya, M., Asano, S., Kaziro, Y., Nagata, S. 1986. *Proc. Natl. Acad. Sci. USA* 83:7633–37

129. Oheda, M., Hase, S., Ono, M., Ikenaka, T. 1988. *J. Biochem.* 103:544–46

130. Nomura, H., Imazeki, I., Oheda, M., Kubota, N., Tamura, M., et al. 1986. *EMBO J.* 5:871–76

131. Nagata, S., Tsuchiya, M., Asano, S., Kaziro, Y., Yamazaki, T., et al. 1986. *Nature* 319:415–18

132. Souza, L. M., Boone, T. C., Gabrilove, J., Lai, P. H., Zsebo, K. M., et al. 1986. *Science* 232:61–65

133. De Lamarter, J. F., Hession, C., Semon, D., Gough, N. M., Rothenbuhler, R., Mermod, J-J. 1987. *Nucleic Acids Res.* 15:2389–90

134. Das, S. K., Stanley, E. R. 1982. *J. Biol. Chem.* 257:13679–84
135. Kawasaki, E. S., Ladner, M. B., Wang, A. M., Arsdell, J. V., Warren, M. K., et al. 1985. *Science* 230:291–96
136. Ihle, J. N., Keller, J., Henderson, L., Klein, F., Palaszynski, E. 1982. *J. Immunol.* 129:2431–36
137. Cutler, R. L., Metcalf, D., Nicola, N. A., Johnson, G. R. 1983. *J. Biol. Chem.* 260:6579–84
138. Fung, M. C., Hapel, A. J., Ymer, S., Cohen, D. R., Johnson, R. M., et al. 1984. *Nature* 307:233–37
139. Yokota, T., Lee, F., Rennick, D., Hall, C., Arai, N., et al. 1984. *Proc. Natl. Acad. Sci. USA* 81:1070–74
140. Yang, Y. C., Ciarletta, A. B., Temple, P. A., Chung, M. P., Kovacic, S., et al. 1986. *Cell* 47:3–10
141. Dorssers, L., Burger, H., Bot, F. J., Delwel, R., Geurts Van Kessel, A.H.M., et al. 1987. *Gene* 55:115–24
142. McKenzie, D. T., Filutowicz, H. I., Swain, S. L., Dutton, R. W. 1987. *J. Immunol.* 139:2661–68
143. Kinashi, T., Harada, N., Severinson, E., Tanobe, T., Sideras, P., et al. 1986. *Nature* 324:70–73
144. Azuma, C., Tanabe, T., Konishi, M., Kinashi, T., Noma, T., et al. 1986. *Nucleic Acids Res.* 14:9149–58
145. Van Snick, J., Cayphas, S., Vink, A., Uyttenhove, C., Coulie, P. G., et al. 1986. *Proc. Natl. Acad. Sci. USA* 83:9679–83
146. Van Snick, J., Cayphas, S., Szikora, J-P., Renauld, J-C., Van Roost, E., et al. 1988. *Eur. J. Immunol.* 18:193–97
147. Hirano, T., Taga, T., Yasukawa, K., Nakajima, K., Nakano, N., et al. 1987. *Proc. Natl. Acad. Sci. USA* 84:228–31
148. Hirano, T., Yasukawa, K., Harada, H., Taga, T., Watanabe, Y., et al. 1986. *Nature* 324:73–76
149. Nicola, N. A., Begley, C. G., Metcalf, D. 1985. *Nature* 314:625–28
150. Nagata, S., Tsuchiya, M., Asano, S., Yamamoto, O., Hirata, Y., et al. 1986. *EMBO J.* 5:575–81
151. Gough, N. M., Metcalf, D., Gough, J., Grail, D., Dunn, A. R. 1985. *EMBO J.* 4:645–53
152. Gough, N. M., Grail, D., Gearing, D. P., Metcalf, D. 1987. *Eur. J. Biochem.* 169:353–58
153. Clark-Lewis, I., Lopez, A. F., To, L. B., Vadas, M. A., Schrader, J. W., et al. 1988. *J. Immunol.* 141:881–89
154. Clark-Lewis, I., Rebersold, R., Ziltener, H. J., Schrader, J. W., Hood, L. E., Kent, S.B.H. 1986. *Science* 231:134–39
155. Stanley, E., Metcalf, D., Sobieszcuk, P., Gough, N. M., Dunn, A. R. 1985. *EMBO J.* 4:2569–73
156. Miyatake, S., Otsuka, T., Yokota, T., Lee, F., Arai, K. 1985. *EMBO J.* 4:2561–68
157. Buchberg, A. M., Bedigian, H. G., Taylor, B. A., Brownell, E., Ihle, J. N., et al. 1988. *Oncogene Res.* 2:149–65
158. Le Beau, M. M., Pettenati, M. J., Lemons, R. S., Diaz, M. O., Westbrook, C. A., et al. 1986. *Cold Spring Harbor Symp. Quant. Biol.* 51 (Pt. 2):899–909
159. Tsuchiya, M., Kaziro, Y., Nagata, S. 1987. *Eur. J. Biochem.* 165:7–12
160. Simmers, R-N., Webber, L. N., Shannon, M. F., Garson, O. M., Wong, G., et al. 1987. *Blood* 70:330–32
161. Rajavashisth, T. B., Eng, R., Shadduck, R. K., Waheed, A., Ben-Avram, C. M., et al. 1987. *Proc. Natl. Acad. Sci. USA* 84:1157–61
162. Miyatake, S., Yokota, T., Lee, F., Arai, K. 1985. *Proc. Natl. Acad. Sci. USA* 82:316–20
163. Le Beau, M. M., Epstein, N. D., O'Brien, S. J., Nienhuis, A. W., Yang, Y. C., et al. 1987. *Proc. Natl. Acad. Sci. USA* 84:5913–17
164. Campbell, H. D., Sanderson, C. J., Wang, Y., Hort, Y., Martinson, M. E., et al. 1986. *Eur. J. Biochem.* 174:345–52
165. Campbell, H. D., Tucker, W.Q.J., Hort, Y., Martinson, M. E., Mayo, G., et al. 1987. *Proc. Natl. Acad. Sci. USA* 84:6629–33
166. Sutherland, G. R., Baker, E., Callen, D. F., Campbell, H. D., Young, I. G., et al. 1988. *Blood* 71:1050–52
167. Sutherland, G. R., Baker, E., Callen, D. F., Hyland, V. J., Wong, G., et al. 1988. *Hum. Genet.* 79:335–37
168. Yasukawa, K., Hirano, T., Watanabe, Y., Muratani, K., Matsuda, T., et al. 1987. *EMBO J.* 6:2939–45
169. Wreschner, D. H., Rechavi, G. 1988. *Eur. J. Biochem.* 172:333–40
170. Byrne, P. V., Guilbert, L. J., Stanley, E. R. 1981. *J. Cell. Biol.* 91:848–53
171. Nicola, N. A., Metcalf, D. 1985. *J. Cell. Physiol.* 124:313–21
172. Nicola, N. A., Metcalf, D. 1986. *J. Cell. Physiol.* 128:180–88
173. Nicola, N. A. 1987. *Immunol. Today* 8:134–39
174. Kelleher, C. A., Wong, G. G., Clark, S. C., Schendel, P. F., Minden, M. D., McCulloch, E. A. 1988. *Leukemia* 2:211–15
175. Morgan, C. J., Stanley, E. R. 1984. *Biochem. Biophys. Res. Commun.* 119:35–41

176. Yeung, Y. G., Jubinsky, P. T., Sengupta, A., Yeung, D.C.Y., Stanley, E. R. 1987. *Proc. Natl. Acad. Sci. USA* 84:1268–71

177. Sherr, C. J., Rettenmier, C. W., Sacca, R., Roussel, M. F., Look, A. T., Stanley, E. R. 1985. *Cell* 41:665–67

178. Roussel, M. F., Dull, T. J., Rettenmier, C. W., Ralph, P., Ullrich, A., Sherr, C. J. 1987. *Nature* 325:549–51

179. Coussens, L., Van Beveren, C., Smith, D., Chen, E., Mitchell, R. L., et al. 1986. *Nature* 320:277–80

180. Rothwell, V. M., Rohrschneider, L. R. 1987. *Oncogene Res.* 1:311–24

181. Stanley, E. R., Guilbert, L. J. 1981. *J. Immunol. Methods* 42:253–84

182. Guilbert, L. J., Stanley, E. R. 1986. *J. Biol. Chem.* 261:4024–32

183. Chen, B. D., Chou, T. H., Clark, C. R. 1987. *Br. J. Hematol.* 67:381–86

184. Walker, F., Nicola, N. A., Metcalf, D., Burgess, A. W. 1985. *Cell* 43:269–76

185. Chen, B. D. 1986. *J. Immunol.* 136:174–80

186. Nicola, N. A., Peterson, L. 1986. *J. Biol. Chem.* 261:12384–89

187. Nicola, N. A., Vadas, M. A., Lopez, A. F. 1986. *J. Cell. Physiol.* 128:501–9

188. Park, L. S., Friend, D., Gillis, S., Urdal, D. L. 1986. *J. Biol. Chem.* 261:4177–83

189. Palaszynski, E. W., Ihle, J. N. 1984. *J. Immunol.* 132:1872–78

190. Walker, F., Burgess, A. W. 1985. *EMBO J.* 4:933–39

191. Park, L. S., Friend, D., Gillis, S., Urdal, D. L. 1986. *J. Biol. Chem.* 261:205–10

192. Gasson, J. C., Kaufman, S. E., Weisbart, R. H., Tomonaga, M., Golde, D. W. 1986. *Proc. Natl. Acad. Sci. USA* 83:669–73

193. Park, L. S., Friend, D., Gillis, S., Urdal, D. L. 1986. *J. Exp. Med.* 164:251–62

194. DiPersio, J., Billing, P., Kaufman, S., Eghtedsay, P., Williams, R. E., Gasson, J. C. 1988. *J. Biol. Chem.* 263:1834–41

195. Coulie, P. G., Vanhecke, A., Van Damme, J., Cayphas, S., Poupart, P., et al. 1987. *Eur. J. Immunol.* 17:1435–40

196. Yamasaki, K., Taga, T., Hirata, Y., Yawata, H., Kawanishi, Y., et al. 1988. *Science* 241:825–28

197. Urdal, D. L., Park, L. S. 1988. *Behring Inst. Mitt.* 83:27–39

198. Motoyoshi, K., Suda, T., Kusomoto, K., Takaku, F., Miura, Y. 1982. *Blood* 60:1378–91

199. Orlofsky, A., Stanley, E. R. 1987. *EMBO J.* 6:2947–52

200. Pierce, J. H., Ruggiero, M., Fleming, T. P., DiFiore, P. P., Greenberger, J. S., et al. 1988. *Science* 239:628–31

201. von Ruden, T., Wagner, E. F. 1988. *EMBO J.* 7:2749–56

202. Koyasu, S., Tojo, A., Miyajima, A., Akiyama, T., Kasuga, M., et al. 1987. *EMBO J.* 6:3979–84

203. Garland, J. M. 1988. *Leukemia* 2:94–102

204. Watson, J. D., Jenkins, D. R., Eszes, M., Leung, E. 1988. *J. Immunol.* 140:501–7

205. Berridge, M. J., Irvine, R. F. 1984. *Nature* 312:315–21

206. Farrar, W. L., Thomas, T. P., Anderson, W. B. 1985. *Nature* 315:235–37

207. Whetton, A. D., Monk, P. N., Consalvey, S. D., Huang, S. J., Dexter, T. M., Downes, C. P. 1988. *Proc. Natl. Acad. Sci. USA* 85:3284–88

208. Whetton, A. D., Heyworth, C. M., Dexter, T. M. 1986. *J. Cell. Sci.* 84:93–104

209. Jackowski, S., Rettenmier, C. W., Sherr, C. J., Rock, C. D. 1986. *J. Biol. Chem.* 261:4978–85

210. Whetton, A. D., Monk, P. N., Consalvey, S. D., Downes, C. P. 1986. *EMBO J.* 5:3281–86

211. Dexter, T. M., Whetton, A. D., Heyworth, C. M. 1985. *Blood* 65:1544–48

212. He, Y. X., Hewlett, E., Temeles, D., Quesenberry, P. 1988. *Blood* 71:1187–95

213. Hamilton, J. A., Vairo, G., Lingelbach, S. R. 1988. *J. Cell. Physiol.* 134:405–12

214. Vairo, G., Hamilton, J. A. 1988. *J. Cell. Physiol.* 134:13–24

215. Doolittle, R. F., Hunkapiller, M. W., Hood, L. E., Devare, S. G., Robbins, K. C., et al. 1983. *Science* 221:275–77

216. Downward, J., Yarden, Y., Mayes, E., Scrace, G., Totty, N., et al. 1984. *Nature* 307:521–27

217. Gisselbrecht, S., Fichelson, S., Sola, B., Borderaux, D., Hampe, A., et al. 1987. *Nature* 329:259–61

218. Heard, J. M., Roussel, M. F., Rettenmier, C. W., Sherr, C. J. 1987. *Cell* 51:663–73

219. Wheeler, E. F., Askew, D., May, S., Ihle, J. N., Sherr, C. J. 1987. *Mol. Cell. Biol.* 7:1673–80

220. Wheeler, E. F., Rettenmier, C. W., Look, A. T., Sherr, C. J. 1986. *Nature* 324:377–80

221. Roussel, M. F., Rettenmier, C. W., Sherr, C. J. 1988. *Blood* 71:1218–25

222. Nicola, N. A., Metcalf, D. 1986. *Cancer Surveys* 4:789–815

223. Wong, P. M. C., Chung, S., Nienhuis, A. W. 1987. *Genes & Dev.* 1:358–65

224. Lang, R. A., Metcalf, D., Gough, N. M., Dunn, A. R., Gonda, T. J. 1985. *Cell* 43:531–42

225. Laker, C., Stocking, C., Bergholz, U., Hess, N., DeLamarter, J. F., Ostertag, W. 1987. *Proc. Natl. Acad. Sci. USA* 84:8458–62

226. Hapel, A. J., Young, I. G. 1987. *Anticancer Res.* 7:661–67

227. Young, D. C., Griffin, J. D. 1986. *Blood* 68:1178–81

228. Cheng, G. Y. M., Kelleher, C. A., Miyauchi, J., Wang, C., Wong, G., et al. 1988. *Blood* 71:204–8

229. Schrader, J. W. 1986. *Annu. Rev. Immunol.* 4:205–30

230. Baumbach, W. R., Stanley, E. R., Cole, M. D. 1987. *Mol. Cell. Biol.* 7:664–71

231. Kawano, M., Hirano, T., Matsuda, T., Taga, T., Horii, Y., et al. 1988. *Nature* 332:83–85

232. Ihle, J. N. 1988. *Lymphokines* 15:127–61

233. Klingler, K., Johnson, G. R., Walker, F., Nicola, N. A., Decker, T., Ostertag, W. 1987. *J. Cell. Physiol.* 132:22–32

234. Overell, R. W., Watson, J. D., Gallis, B., Weiser, K. E., Cosman, D., Widmer, M. B. 1987. *Mol. Cell. Biol.* 7:3394–401

235. Cook, W. D., Metcalf, D., Nicola, N. A., Burgess, A. W., Walker, F. 1985. *Cell* 41:677–83

236. Metcalf, D., Roberts, T. M., Cherington, V., Dunn, A. R. 1987. *EMBO J.* 6:3703–9

Annu. Rev. Biochem. 1989. 58:79–110

BIOCHEMISTRY OF OXYGEN TOXICITY[1]

Enrique Cadenas

Department of Pathology II, University of Linköping, S-581 85 Linköping, Sweden

CONTENTS

PERSPECTIVES AND SUMMARY

The biochemistry of reactive O_2 intermediates is an important field with practical implications, because while O_2 is an essential component for living organisms, the formation of reactive O_2 intermediates seems to be com-

---

[1]Abbreviations used: EPR, electron spin resonance; DMPO, 5,5'-dimethyl-1-pyrroline-*N*-oxide; PBN, phenyl-*t*-butyl nitrone; DBAS, dibromoanthracene-sulfonate; OXANO, 2-ethyl-2,5,5-trimethyl-3-oxazolidinoxyl; MNP,2-methyl-2-nitrosopropane; Mb^{III}, metmyoglobin; Mb^{IV}, ferrylmyoglobin; ·Mb^{IV}, radical form of ferrylmyoglobin

0066-4154/89/0701-0079$02.00

monplace in aerobically metabolizing cells. The stability of a variety of reactive O_2 intermediates, whether of free radical character or not, differs substantially, but in some instances their interaction with the intracellular environment sets the basis for the pathophysiology of several disease states.

Reactive O_2 intermediates in biochemistry is a very wide subject impossible to cover in a single review. This work surveys only some aspects of the biochemistry of reactive O_2 species and some defense mechanisms, which are subjected to physiological control. Within a limited space, emphasis is given to the biological generation of $^1O_2$, because there has been a recent surge of interest in this area. Certain particular features of the biochemistry of O_2 toxicity—relevant to some pathological situations—are listed here, including the metabolism of hydroperoxides, the redox and addition chemistry of quinoid compounds, and the redox transitions of myoglobin, which might be of future significance in cardiac disorders that might be related to ischemia/reperfusion-mediated tissue injury.

CHEMISTRY OF MOLECULAR OXYGEN

The chemistry of molecular O_2 has been discussed several times, and some excellent review articles are available (1–4). However, a short overview stressing some points seems essential. The kinetic inertness of dioxygen in solution can be explained by its electronic structure. Although molecular O_2 contains an even number of electrons, it has two unpaired electrons in its molecular orbitals, and is said to be in a triplet ground state. These two electrons have the same spin quantum number, and if O_2 attempts to oxidize another atom or molecule by accepting a pair of electrons from it, both new electrons must be of parallel spin so as to fit into the vacant spaces of the orbitals. Usually, a pair of electrons in an atomic or molecular orbital would have antiparallel spins. This imposes an important restriction on oxidation by O_2. Molecular O_2 is usually constrained to one electron transfer reaction at a time. Below are described three types of reactions. These lead to the formation of superoxide radical and related species, peroxyl radicals, and singlet molecular oxygen.

One-Electron Transfer to O_2: Formation of $O_2^-\cdot$ and Related Species

The one-electron reduction of molecular O_2 to superoxide radical, $O_2^-\cdot$, a species that has received considerable attention among the oxy-radicals generated in biological systems, is thermodynamically unfavorable $[E_0(O_2/O_2^-\cdot) = -0.33$ V] (5). A way to overcome this spin restriction lies in the interaction of the O_2 molecule with another paramagnetic center to participate in exchange coupling. Transition metals, such as Fe or Cu, frequently have unpaired electrons and are excellent catalysts of O_2 reduction.

$$\text{Fe}^{II} + O_2 \rightarrow \text{Fe}^{III} + O_2^- \cdot \qquad\qquad 1.$$

In addition to reaction 1, $O_2^- \cdot$ is produced when electrophiles, e.g. quinoid compounds, are reduced to semiquinones by cellular electron-transfer flavoproteins, which subsequently reduce O_2 via one electron, within a process termed *redox cycling*.

In aqueous, neutral pH solutions $O_2^- \cdot$ decays to H_2O_2 through disproportionation (reaction 2; $k_2 = 4.5 \times 10^5$ $M^{-1}s^{-1}$). The protonation of $O_2^- \cdot$ produces the hydroperoxyl radical $[O_2^- \cdot + H^+ \rightarrow \text{HOO}\cdot]$, a far more reactive species. The chemistry of $O_2^- \cdot$ in aprotic media serves as a model for the chemical environment in biological membranes (1, 3), and it encompasses four basic modes of action of $O_2^- \cdot$: electron transfer (the most common mode in biosystems), nucleophilic substitution, deprotonation, and H-atom abstraction.

$$O_2^- \cdot + O_2^- \cdot + 2H^+ \rightarrow H_2O_2 + O_2 \qquad\qquad 2.$$

Both types of chemistry of $O_2^- \cdot$ are important, because the question still remains whether the internal reactions in biological systems take place in an essentially aqueous or nonaqueous environment. This different chemistry would contemplate alternation in electron-donating and accepting behavior on proton gain or loss, and will, of course, be of significance in the chain of events connecting O_2 and H_2O.

The decomposition of H_2O_2 can be a source of hydroxyl radical, $\text{HO}\cdot$. This species can only diffuse 5–10 molecular diameters from its site of formation before it reacts (6), and it is responsible for most biological damage. The accepted concept of an $O_2^- \cdot$-driven Fenton-type reaction (7, 8) requires both $O_2^- \cdot$ and H_2O_2 as precursors for $\text{HO}\cdot$; the reaction necessitates an intermediate catalyst, such as a transition metal chelate, which is reduced by $O_2^- \cdot$ and reacts with H_2O_2 in a "Fenton-like" reaction to produce $\text{HO}\cdot$; thus, the generation of $\text{HO}\cdot$ upon reaction of Fe^{II} or Cu^I with H_2O_2,

$$\text{Fe}^{III} + O_2^- \cdot \rightarrow \text{Fe}^{II} + O_2 \qquad\qquad 3.$$

$$\text{FE}^{II} + H_2O_2 \rightarrow \text{Fe}^{III} + \text{HO}^- + \text{HO}\cdot \qquad\qquad 4.$$

might be consistent with the concept of site-specific damage. These observations are rationalized in terms of the well-known reaction sequences described above and give as net balance (Haber-Weiss reaction):

$$O_2^- \cdot + H_2O_2 \rightarrow O_2 + \text{HO}^- + \text{HO}\cdot \qquad\qquad 5.$$

The likelihood of reactions 3–4 will depend, of course, on the availability of transition metals in their complexed forms in the cell to catalyze the decomposition of H_2O_2 (4). It is likely that iron released from hemoglobin by peroxides is the true generator of HO· in the Fenton reaction (9). Where metal catalysts are involved in the generation of HO·, other alternative oxidizing species should be considered, such as the ferryl radical ($[Fe-OH]^{3+}$ or $[Fe=O]^{2+}$) and the perferryl radical ($[Fe=O]^{3+}$ or $[Fe^{3+}-O_2^-\cdot]$).

Transition metal complexes may play two roles in biosystems (10): on the one hand, as protectors against $O_2^-\cdot$ damage and, on the other hand, as sensitizers of toxic effects of $O_2^-\cdot$. The expression of one or another role could be accounted for by the kinetic properties of the metal ligands as well as the steady-state concentrations of the required O_2 species. The former role is supported by the disproportionation of $O_2^-\cdot$ catalyzed by several copper ligands (L-Cu) through a ping-pong mechanism and analogous to a superoxide dismutase reaction (reactions 6 and 7).

$$L\text{-}Cu^{II} + O_2^-\cdot \rightarrow L\text{-}Cu^{I} + O_2 \qquad\qquad 6.$$

$$L\text{-}Cu^{I} + O_2^-\cdot + 2\,H^+ \rightarrow L\text{-}Cu^{II} + H_2O_2 \qquad\qquad 7.$$

The latter role is accounted for by the production of HO· within an $O_2^-\cdot$-driven Fenton cycle as illustrated in reactions 3–4 above, where HO· is generated either in the bulk solution (non site-specific mechanism of damage) or close to the biological target when the metal ion or its complex is bound to the target, which might serve as an effective ligand for, e.g. Cu (site-specific mechanism of damage) (10). In the latter instances, the stereo- and regioselectivity of production of HO· in the vicinity of DNA may control the type of chemistry observed. The HO· is unique in being the only oxy-radical that, due to a combination of high electrophilicity, high thermochemical reactivity, and a mode of production that can occur near DNA, can both add to DNA bases and abstract H-atoms from the DNA helix (11).

A semiquinone/quinone couple with a reduction potential between −0.33 volts and 0.46 volts can theoretically bring about a metal-independent decomposition of H_2O_2 with formation of HO·. A discussion on the kinetic restraints as well as the likelihood of this reaction in biological systems has been offered (12). Furthermore, it has been shown that reaction 8 actually takes place, whereby the semiquinone forms of menadione, ubiquinone$_0$, ubiquinone$_1$, and ubiquinone$_4$ were able to reduce H_2O_2 to HO· within a reaction in which the participation of contaminating transition metals was excluded (13).

$$8.$$

In summary, three important species in the oxidation cycle concerning peroxides are free radicals: $O_2^- \cdot$, $HO_2 \cdot$, and $HO \cdot$, and there is, of course, a growing awareness of the possibility of interconversion of some of these species into each other (12). Only a few applications of EPR measurements seem to have been made in the studies presented by biochemists and physiologists. However, this has not prevented the postulation of several adverse reactions in in vivo systems. The ability of scavengers to distinguish $HO \cdot$ originating from the Fe-catalyzed Haber-Weiss reaction has been recently discussed in terms of different assays for $HO \cdot$ (14). In addition to EPR with spin trapping techniques (15), HPLC with electrochemical detection of hydroxylation products of aromatic compounds has been shown as a sensitive method for $HO \cdot$ detection (16, 17). Most of the biological research has been carried out in aqueous media in which it is likely that not only $O_2^- \cdot$ but also its conjugate acid, $HO_2 \cdot$, and $HO \cdot$, are the reactive species (18). Yet a biological system cannot be entirely defined on the basis of free radical reactions in water. Certain considerations need be taken into account such as the lack of homogeneous distribution of free radicals (probably determined by hydrophobic areas), site of formation of free radicals, as well as the mobility of target molecules, including those either "freely" mobile or rigidly held in a matrix (19). Therefore, in experimental models containing isolated cells, a correlation between rate constants and concentrations of injected scavenger cannot be expected, because of the differential concentrations that may occur within the cell.

Quenching of Carbon-Centered Radicals by O_2: Formation of Peroxyl Radicals

The reactions of molecular O_2 are constrained to one-electron transfers. A type of one-electron transfer reaction to O_2 to yield $O_2^- \cdot$ [involving transition metals (reaction 1) or a redox cycling process] as primary species has been described above. A second—no less important—type of one-electron transfer that contributes to the formation of oxy-radicals involves the quenching of carbon-centered radicals by molecular O_2 leading to the formation of peroxyl radicals (reaction 9; $k \approx 10^9$ $M^{-1}s^{-1}$). Hydroperoxides are subsequently formed within a chain reaction in which the chain carriers are the carbon radical and the peroxyl radical (20).

$$-\overset{|}{\underset{|}{C}} \cdot \longrightarrow -\overset{|}{\underset{|}{C}}OO \cdot \qquad\qquad 9.$$

The biological generation of secondary peroxyl radicals and the recombination reactions in which they can participate, yielding electronically excited states, are described below.

Formation of Singlet Molecular Oxygen

An alternative way of increasing the reactivity of O_2 is to "move" one of the unpaired electrons in a way that alleviates the spin restriction. Singlet molecular O_2 ($^1\Delta g$; $^1O_2$) has no unpaired electrons [the other singlet O_2 (Σg^+) usually decays to the ($^1\Delta g$) state before it has time to react with anything]. The description $(\pi^*_x)^2,(\pi^*_y)^0$, though not strictly correct for the gas-phase molecule, was suggested as suitable for $^1\Delta O_2$ in aqueous solution (21). At variance with the triplet ground state of molecular O_2 ($^3\Sigma$),$-^1\Delta O_2$ can react both as a nucleophile and as an electrophile (21). The thermodynamic considerations and mechanism of reactions involving $^1O_2$ and $O_2^- \cdot$ have been discussed (22).

Although $^1O_2$ generation in chemical systems is well defined [e.g. see (23)], only recently was it detected in biosystems, where its formation has been attributed to phagocytosis, photosensitized reactions, and recombination of peroxyl radicals (e.g. arising from lipid peroxidation), in peroxidase-catalyzed reactions, and in various medical conditions. The detection of $^1O_2$ in biological systems (24) has been aided by the use of "specific" $^1O_2$ scavengers and quenchers, solvent (D_2O) effects, the detection of specific products of cholesterol oxidation (5-α-hydroperoxide adduct, distinct from those adducts formed upon reaction with HO·) or chemical trapping by the water-soluble 9,10-bis(ethylene) anthracene disulfate (which yields specifically an endoperoxide as oxidation product) (25), and the observation of low-level chemiluminescence. The latter approach follows the decay of $^1O_2$ via its luminescence, offering a direct way to obtain rate constants for quenching $^1O_2$ and for determining yields of $^1O_2$, and it is expressed as $^1O_2$ monomol emission (due to $^1\Delta O_2$ itself; 1268 nm) and dimol emission (resulting from the dimer $^1\Delta O_2 \cdots ^1\Delta O_2$; 634 nm) (reactions 10

$$^1O_2 \rightarrow {}^3O_2 + h\nu_{1268 \text{ nm}} \qquad\qquad 10.$$

$$2\ {}^1O_2 \rightarrow 2\ {}^3O_2 + h\nu_{634 \text{ nm}} \qquad\qquad 11.$$

and 11). Although the intensity in the visible region is higher than that in the infrared region, dimol emission would require collision involving orbital overlap, a process restricted by the short lifetime of $^1O_2$. Moreover, observation of $^1O_2$ monomol emission in biosystems is possible because of the recent development of sensitive infrared spectroscopic techniques to detect $^1O_2$ in solution (25–28). The identification of a 1268-nm emission band, along with isotope effects, lends strong support to the generation of $^1O_2$ during enzymic reactions because, apart from $^1O_2$, there is no other molecular electronic state capable of emitting in the 1268-nm region. Chemiluminescence, however, implies solely the relaxation of an electronically excited state (P*) to the ground state with emission of a photon [A + B → P* → P + hv] (a process often observed in connection with certain oxidative experimental models mimicking biological reactions) (29–31), and if aided by spectral analysis, it can point to the involvement of a particular excited species in a system.

BIOCHEMISTRY OF OXYGEN INTERMEDIATES

O_2^- · seems to be a commonplace product of O_2 reduction and, in many cases, the precursor for H_2O_2 (18). O_2^- · is not, though, a major product of O_2 reduction, as judged by the tetravalent reduction of O_2 to H_2O by cytochrome oxidase and other water-producing oxidases (32). The production of O_2^- · in biosystems does constitute a minor pathway, and an upper limit of about 5% could be set for the fraction of total O_2 reduced to O_2^- · (33). The question as to whether the cytochrome oxidase-O_2 reaction produces O_2^- · is still unresolved: superoxide dismutase does not inhibit the oxidation of cytochrome c by cytochrome oxidase, indicating that either O_2^- · production is bypassed by a two-electron transfer, or if produced, the O_2^- · is not released from the active site of the enzyme (32). Moreover, it has been recently suggested that cytochrome oxidase is a O_2^- · scavenger by itself and has a built-in defense mechanism against the potentially hazardous O_2^- ·, whether it is produced by accident or as an intermediate (32).

The electron-transfer chain of mitochondria is a well-documented source of H_2O_2 from disproportionating O_2^- ·, and several components of complex I, II, and III exhibit thermodynamic properties appropriate for the reduction of O_2 to O_2^- · (33–35). A novel and more active O_2^- · generator in heart mitochondria has been recentlay described (36), which (a) exhibits high affinity for O_2, (b) is not involved in energy-linked respiration, (c) requires NADH to initiate its generation, and (d) releases the formed O_2^- · entirely into the extramitochondrial space. This mitochondrial generator of O_2^- · and the exogenous NADH oxidoreductase of heart mitochondria (localized at the cytosolic face of the inner membrane) are identical (37). Liver mitochondria exhibits no oxidation of external NADH, thus pointing to the organ-specific character of the heart mitochondrial activity (37). The latter may be of relevance in the one-electron activation of electrophilic foreign substances, such as adriamycin, thus explaining the selective cardiotoxicity of this antitumor drug (38).

The H_2O_2 concentration in liver, originating from various sources, was estimated to be about $10^{-7}–10^{-9}$ M and that of O_2^- · about 10^{-11} M (39). This figure for H_2O_2 steady-state concentration is expected to be higher in organs with less effective mechanisms than liver for disposal of H_2O_2. Thus, in most normal cells the ratio of the steady-state concentrations of H_2O_2 and $O_2^- \cdot$ is represented by $[H_2O_2]/[O_2^- \cdot] \approx 10^3$ (33).

Antioxidant Defenses

A modified terminology has been proposed, which introduces a differentiation between primary and secondary antioxidant defenses (40). The former include the activities of catalase, superoxide dismutase, glutathione peroxidase, and DT-diaphorase, as well as small molecules such as ascorbic acid, α-tocopherol, GSH, β-carotene, and uric acid. The latter include proteolytic-(40) and lipolytic (41) enzymes, as well as the DNA-repair systems (42).

1. PRIMARY ANTIOXIDANT DEFENSES   The enzymic antioxidant defenses have been extensively reviewed. The interaction of free radicals and electronically excited states with small molecules, such as β-carotene, uric acid, GSH, ascorbic acid, and α-tocopherol, mostly considered as chain-breaking antioxidants, reveals several new features that may be relevant in biosystems. Furthermore, some of the anticarcinogenic properties of antioxidants can be attributed, at least partly, to their free radical–scavenging abilities. Antioxidants located in cytosolic fraction or extracellular fluids, as well as those located in membrane lipids, act as potential anticarcinogens by interrupting the free radical process responsible for carcinogenesis (43). The continuously growing information on free radical reaction rate constants (see 44) obtained by stationary- or pulse-radiolysis studies, as well as the steady accumulation of reduction potentials of many free-radical/molecule couples, is providing a useful framework for the design of in vitro models with potential relevance or applications in vivo (44).

Vitamin E   Vitamin E or α-tocopherol is known to react with a variety of free radicals at fairly high rates, yielding the corresponding chromanoxyl radical (Vit-E—O·). The reaction is formally regarded as an H-atom transfer (reaction 12); however, another possibility regards electron transfer (reaction 13) ($k_{13} = 5.7 \times 10^6\ M^{-1}s^{-1}$) (45, 46), particularly in nonpolar media such as lipids and membranes, followed by deprotonation of the antioxidant radical cation in a reaction with H_2O immersed in the lipid phase (reaction 14).

$$\text{Vit-E-OH} + \text{ROO·} \rightarrow \text{Vit-E-O·} + \text{ROOH} \qquad 12.$$

$$\text{Vit-E-OH} + \text{ROO·} \rightarrow \text{Vit-E-OH·}^+ + \text{ROO}^- \qquad 13.$$

$$\text{Vit-E-OH·}^+ + H_2O \rightarrow \text{Vit-E-O·} + H_3O^+ \qquad 14.$$

A general mechanism as that illustrated in reactions 12–14 does not explain satisfactorily, though, why the inhibition of lipid peroxidation of membranes by α-tocopherol is not accompanied by the expected accumulation of lipid hydroperoxides. It has been proposed (47) that free radical scavengers as α-tocopherol can function as a molecular "channel" via which the free radicals can leave the hydrocarbon zone of the membrane as within the following three-step mechanism: (a) the inhibitor interrupts the propagation chain in the membrane by scavenging carbon-centered or unoxygenated free radicals (reaction 15); (b) the unoxygenated radicals reach the side hydrocarbon chains of the antioxidant by intermolecular H abstraction, and (c) the aroxyl radicals are formed by the intramolecular rearrangement of the radical in the antioxidant molecule (reaction 16) (47).

$$R\cdot \ + \ \text{Vit-E-OH} \rightarrow RH \ + \ \cdot\text{Vit-E-OH} \hspace{3cm} 15.$$

$$\cdot\text{Vit-E-OH} \rightarrow \text{Vit-E-O}\cdot \hspace{5cm} 16.$$

The vitamin E chromanoxy radical, formed by either mechanism outlined above, is known to be long-lived and to react (a) with itself (reaction 17) to form a set of molecular products ($2k = 3.5 \times 10^2 \ M^{-1}s^{-1}$) (48) as well as (b) with physiological antioxidants (e.g. ascorbate, AH^-), which contribute to its recovery (reaction 18).

$$2 \ \text{Vit-E-O}\cdot \rightarrow \text{Products} \hspace{5cm} 17.$$

$$\text{Vit-E-O}\cdot \ + \ AH^- \rightarrow \text{Vit-E-OH} \ + \ A\cdot^- \hspace{3cm} 18.$$

Examples of these antioxidants are ascorbic acid (45), uric acid (49), and serotonine as well as other hydroxyindole derivatives present in the nerve cells in high concentrations (M. G. Simic et al, unpublished). GSH cannot recover vitamin E-chromanoxyl radical, and the glutathionyl radical (GS·) does not react with the α-tocopherol analogue, Trolox C. GSH, however, has been proposed to establish a link between H-atom transfer and electron-transfer reactions in a synergistic fashion (44). Recovery of Trolox radical (generated by pulse radiolysis) can be accomplished by $O_2^-\cdot$ (reaction 19), with a bimolecular rate constant of $k_{19} = 1.2 \times 10^8 \ M^{-1}s^{-1}$ (E. Cadenas, unpublished results).

$$19.$$

α-Tocopherol is also known to react with electronically excited states such as $^1O_2$ and triplet carbonyl compounds ($^3[>C=O]*$). The interaction with the former proceeds at a rate constant of 6.7×10^8 $M^{-1}s^{-1}$ (50) within a reaction that involves both oxidation of the phenol and quenching of $^1O_2$ by the phenol without reaction. This reactivity of α-tocopherol toward $^1O_2$ is about 50-fold lower than that of the diffusion-controlled quencher, β-carotene. The interaction of α-tocopherol with triplet carbonyl compounds proceeds efficiently ($k = 6.7 \times 10^9$ $M^{-1}s^{-1}$) and, apparently, could be explained in terms of H-atom abstraction from the $-HO$ group in the α-tocopherol molecule (reaction 20) (51).

$$\text{Vit-E-OH} + {}^3[>C=O]* \rightarrow \text{Vit-E-O} \cdot + >C \cdot\text{-OH} \qquad 20.$$

Ascorbic acid Ascorbic acid seems to be involved in the H transfer-mediated recovery of vitamin E radical (reaction 18) as well as its direct reaction with peroxyl radicals (reaction 21) and $^1O_2$ (52). Ascorbic acid quenching of $^1O_2$ [$k = 1.6 \times 10^8$ $M^{-1}s^{-1}$ (53)] is predominantly chemical in nature (54). Triplet carbonyls, e.g. triplet benzophenone, promote H abstraction from the enediol groups of ascorbate ($k_R = 1.2 \times 10^9$ $M^{-1}s^{-1}$) (reaction 22) (51).

$$\text{ROO} \cdot + \text{AH}^- \rightarrow \text{ROOH} + \text{AH} \cdot^- \qquad 21.$$

$$^3[>C=O]* + \text{AH}^- \rightarrow >C \cdot\text{-OH} + \text{A} \cdot^- \qquad 22.$$

Glutathione GSH undergoes rather readily interactions with free radicals, as those produced during H_2O radiolysis; thus, a protective role of GSH might imply a repair mechanism of critical biological molecules damaged during irradiation (reaction 23) (55).

$$\text{GS}^- + \text{ROO} \cdot \rightarrow \text{GS} \cdot^- + \text{ROO}^- \qquad 23.$$

The GSH/free radical interactions have been well characterized and, as a rule, the thiyl radical (GS$\cdot$) is the main product, though not the unique consequence of these interactions. GS$\cdot$, unless stabilized, undergoes several reactions, which, in addition to its quenching by O_2 (reaction 24), include diradical annihilation (reaction 25), oxidation of GSH (reaction 26), and reaction with CSSG$\cdot^-$ (reaction 27).

$$\text{GS} \cdot + O_2 \rightarrow \text{GSOO} \cdot \qquad 24.$$

$$\text{GS} \cdot + \text{GS} \cdot \rightarrow \text{GSSG} \qquad 25.$$

$$GS \cdot + GS^- \rightarrow [GSSG \cdot]^- \qquad \qquad 26.$$

$$GS \cdot + GSSG \cdot^- \rightarrow GS^- + GSSG \qquad \qquad 27.$$

A new competitive method has been developed based on the oxidation properties of thiyl radicals (56), which allows the calculation of the bimolecular rate constants for the addition of O_2 to GS· (reaction 24). The values obtained were lower than previously reported, and they could be enhanced by considering the reversibility of reaction 24 (56, 57). The sulphenyl-peroxo radicals (GSOO·) were reported to be responsible for lysozyme and trypsin inactivation (58), thus indicating that GSOO· can be harmful to biomolecules; in addition, their recombination reaction yields electronically excited states, possibly $^1O_2$ (59).

The thiol also reacts with electronically excited states, such as $^1O_2$ and triplet carbonyl compounds. GSH, in its anionic form (GS$^-$) and at pH 7, inactivates $^1O_2$ with a rate constant of $2.9 \times 10^6 \ M^{-1}s^{-1}$ as calculated from time-resolved measurements of the decay of $^1O_2$ phosphorescence at 1268 nm in D_2O (60). Other thiols, such as cysteine and N-acetyl-cysteine, also undergo this reaction with similar rate constants. Methionine reacts with $^1O_2$ at higher rates ($k = 1.4 \times 10^7 \ M^{-1}s^{-1}$) in a pD-independent manner (60), and it decreases the $^1O_2$-induced damage in plasmid DNA pBR322 (61). Thus, sulfur compounds might be considered as possible candidates for protection against the side effects associated with skin photodynamic therapy. Although $^1O_2$ is known to react via electron transfer with certain electron donors with low oxidation potentials, no evidence was found for the formation of GS· upon reaction of $^1O_2$ with GSH. Addition of $^1O_2$ to GS$^-$ will likely involve a process that might eventually lead to disulfide formation via reactions 28–30 (60).

$$GS^- + {}^1O_2 \rightarrow GS\text{-}SO_2^- \qquad \qquad 28.$$

$$GS\text{-}SO_2^- + H_2O \rightarrow [GS\text{-}SO^+] + 2 \ HO^- \qquad \qquad 29.$$

$$[GS\text{-}SO^+] + GS^- \rightarrow GSSG + 1/2 \ O_2 \qquad \qquad 30.$$

The quenching of triplet carbonyls by GSH ($k_Q = 6.7 \times 10^8 M^{-1}s^{-1}$) requires its $-SH$ group and proceeds by a charge-transfer complex followed by H-atom abstraction from the S-H bond and/or from the α-carbon atom. The high k_Q values of GSH toward triplet carbonyls, along with the large concentration of the former in the cellular milieu, makes GSH an efficient defense against possible damage by these species (51).

Uric acid   Uric acid has been defined as one of the most important anti-oxidants (62), and it might trap peroxyl radicals in an aqueous phase more effectively than ascorbic acid (43). However, it has been observed that urate increases the inactivation of the enzyme alchohol dehydrogenase (63), probably due to a radiation-induced formation of a short-lived toxin when $O_2^- \cdot$ is generated in the presence of the protein and urate (44). This argument cannot distinguish whether enzyme inactivation is due to the formation of $HO_2\cdot$ or urate radical or a more stable product requiring urate at some stage.

2. SECONDARY ANTIOXIDANT DEFENSES   The suggested differentiation between primary and secondary antioxidant defenses is mainly based on the observation that in spite of the impressive array of primary antioxidant defenses, cellular proteins and enzymes are still susceptible to oxidative damage and inactivation in vivo (40). The oxidatively damaged proteins seem to be removed by novel, ATP and Ca^{2+}-independent proteases, which might be considered as secondary antioxidant defenses (40). Various oxy-radicals, oxidizing species of non–free radical character, and mixed-function oxidation systems cause oxidative damage of proteins, reflected as protein modification, loss of catalytic activity and protein fragmentation, and increased proteolytic susceptibility. The latter effect appears to be supported by denaturation and increased hydrophobicity of the oxidatively damaged protein. The formation of protein-peroxyl radicals seems to be involved in the sequence of steps including main-chain and side-chain reactions and leading to protein fragmentation products with new carbonyl moieties and amino moieties (64–69). The higher content of carbonyl derivatives originating from protein damage indicate that the latter process is involved in aging (70), O_2 toxicity, killing of bacteria by neutrophiles (71), and in the marking of enzymes for proteolytic degradation.

Oxidative damage of human low-density lipoprotein, however, seems to proceed by the sequence lipid peroxidation $\rightarrow$ protein peroxidation, with intermediate formation of lipophilic aldehydes—carbonyls—of medium-chain length. The latter, which seem to be required for lipoprotein damage, diffuse within the lipid core to the apoprotein B, further reacting with amino acid residues (72).

METABOLISM OF HYDROPEROXIDES

Hydroperoxide metabolism in mammalian cells has been reviewed in detail several years ago (39), mainly in terms of enzymes utilizing free and bound hydroperoxides, e.g. catalase, glutathione peroxidase, heme peroxidases, cytochrome P_{450}, and cytochrome oxidase, but also providing a complete overview of the cellular regulation of these enzyme activities.

In the section on CHEMISTRY OF MOLECULAR OXYGEN it has been pointed out that reactions of molecular O_2 are constrained to one-electron transfers and that quenching of carbon-centered radicals by O_2 to yield peroxyl radicals was one way in which an antibonding orbital of the diatomic O_2 participates in chemical bonding with another radical. The formation of peroxyl radicals (reaction 9) leads to the production of organic hydroperoxides. One aspect related to the metabolism of hydroperoxides could be formally referred to as the enzyme-catalyzed disproportionation of hydroperoxides involving certain peroxidase activities, in which $^1O_2$ seems to be one of the products. Another aspect covers mainly the disproportionation of peroxyl radicals, in itself a source of electronically excited states, a process that might be of relevance in vivo.

Disproportionation of Hydroperoxides

The metabolism of hydroperoxides by several hemoproteins or halide-dependent enzymic reactions can proceed, at least formally, as a hydroperoxide disproportionation with formation of dioxygen in an excited state, $^1O_2$. The generation of $^1O_2$ as a consequence of an enzymic reaction could represent a key step in the metabolism of hydroperoxides. Examples are given by the halide-mediated H_2O_2 decomposition by lacto-, myelo-, and chloroperoxidase (26, 27), the prostaglandin-endoperoxide synthase reaction (73), and the metabolism of hydroperoxides by cytochrome P_{450} (in the absence of electron donors and hydroxylatable substrates) (74). Also, the H_2O_2-mediated oxidation of metmyoglobin (Mb^{III}) to ferrylmyoglobin (Mb^{IV}) or its transient free radical form ($\cdot Mb^{IV}$) proceeds with formation of electronically excited states and promotion of cholesterol oxidation to the 5-α-hydroperoxide adduct, the latter product being a specific probe for $^1O_2$ (75). A short survey on the molecular mechanisms and the physiological role for $^1O_2$ formation is given below; more detailed discussions are found in the respective references.

MYELO-, CHLORO-, AND LACTOPEROXIDASES $^1O_2$ monomol emission (1268 nm) was observed during the lacto-, myelo-, and chloroperoxidase-H_2O_2-halide systems (26, 27), and hypothetically, during the decomposition of H_2O_2 by catalase (1642 nm) (76), though in the latter case emission has been also attributed to thermal radiation (77). The mechanism for the generation of $^1O_2$ during these peroxidase reactions proceeds in a two-step reaction as a disproportionation of H_2O_2 (reactions 31a,b) in a fashion suggestive of the classic inorganic H_2O_2/OCl^- reaction leading to the generation of $^1O_2$. The requirement of halide serves as a trigger of H_2O_2 decomposition with formation of a hypohalous acid; a second molecule of H_2O_2 is decomposed by the latter with generation of $^1O_2$.

$$X^- + H^+ + H_2O_2 \rightarrow HO-X + H_2O \qquad\qquad 31a.$$

$$HO-X + H_2O_2 \rightarrow H^+ + X^- + H_2O + {}^1O_2 \qquad\qquad 31b.$$

PROSTAGLANDIN-ENDOPEROXIDE SYNTHASE   The prostaglandin-endoperoxide synthase-catalyzed conversion of PGG_2 to PGH_2 involves the generation of a potent oxidant (78), which may account for both the rapid loss of activity of the enzymes involved in the initial steps of the prostaglandin biosynthesis and the cooxidation of chemicals. The enzymic conversion of PGG_2 to PGH_2 is associated with photoemission in the visible region with a spectral distribution resembling the pattern of ${}^1O_2$ dimol emission (its formation rationalized as in reactions 32a,b) (73), but not in the infrared region, that is, at 1268 nm, ascribed to ${}^1O_2$ monomol emission (79). This discrepancy could be accounted for in terms of the formation of excited species—still to be identified—other than ${}^1O_2$ and emitting in the region of the ${}^1O_2$ dimol emission bands (reaction 11) (79). However, depending on the resolution of the spectral analysis of visible emission (73, 79), such an explanation may carry in itself a trifling connotation. The visible photoemission cannot be ascribed uniquely to ${}^1O_2$, for a large fraction of it occurs below 600 nm with a spectral pattern difficult to attribute to any particular excited state. Mechanistic considerations do not apply in this context—although they are of relevance in terms of quantum yield—for a mechanism as that encompassed in reactions 32a,b or the recombination of peroxyl radicals (see below) can lead to the formation of the same excited species.

$$Fe^{3+} + PGG_2 \rightarrow [Fe=O]^{3+} + PGH_2 \qquad\qquad 32a.$$

$$[Fe=O]^{3+} + PGG_2 \rightarrow Fe^{3+} + PGH_2 + O_2 \qquad\qquad 32b.$$

The function of ${}^1O_2$ in cooxygenation systems, however, remains uncertain, and it could only account for a fraction of the cooxidized molecular products formed. The pathological aspects of the generation of such a powerful oxidant during the prostaglandin hydroperoxidase reaction has been related to inflammation (80), because of the effect of certain antiinflammatory agents, which act as free radical scavengers.

CYTOCHROME P_{450} The heterolytic cleavage of the O-O bond of the hydroperoxides by the oxene transferase activity of cytochrome P_{450} proceeds under formation of a transient $[Fe=O]^{3+}$ species (81), which upon reaction with a second hydroperoxide molecule yields, in part, ${}^1O_2$. An analogous reaction occurs with chloroperoxidase and H_2O_2 through a scrambling mechanism: the evolving dioxygen is derived from two different H_2O_2 molecules.

Chloroperoxidase closely resembles cytochrome P_{450} and probably also contains an axial thiolate ligand as postulated for cytochrome P_{450}. The metabolism of different oxene donors by cytochrome P_{450} proceeds with photoemission, which resembles $^1O_2$ dimol emission (reactions 33a,b) (74).

$$P_{450}\text{-}Fe^{3+} + ROOH \rightarrow P_{450}\text{-}[Fe{=}O]^{3+} + ROH \qquad\qquad 33a.$$

$$P_{450}\text{-}[Fe{=}O]^{3+} + ROOH \rightarrow P_{450}\text{-}Fe^{3+} + ROH + O_2 \qquad\qquad 33b.$$

The chemiluminescence arising from hydroperoxide breakdown catalyzed by cytochrome P_{450} can involve either a heterolytic or homolytic scission of the hydroperoxide. In the former case, the generation of excited states occurs in a *primary fashion* within a reaction that proceeds under anaerobic conditions (74). In the latter mechanism, the generation of excited states occurs in a *secondary fashion* as a consequence of free radical recombination reactions during lipid peroxidation (82) (see below).

MYOGLOBIN   The redox transitions of oxymyoglobin ($Mb^{II}O_2$) involve either (*a*) its one-electron oxidation expressed as electron transfer to O_2 with concomitant formation of products of dioxygen reduction (83) or (*b*) its oxidation—as well as that of metmyoglobin, Mb^{III}—by H_2O_2 or lipid hydroperoxides to one-oxidation equivalent over Fe^{III}, that is, to an intermediate hemoprotein similar to compound II of peroxidases and termed ferrylmyoglobin (Mb^{IV}) (75, 84–86). The exact molecular mechanism of H_2O_2-mediated oxidation of Mb^{III} to Mb^{IV} or its transient free radical form, $\cdot Mb^{IV}$, is ambiguous, and a *homolytic scission* (reaction 34a,b; HX- denoting an amino acid residue of the apoprotein)

$$Mb^{III}(HX\text{-}Fe^{III}) + H_2O_2 \rightarrow Mb^{IV}(HX\text{-}Fe^{IV}\text{-}OH) + HO\cdot \qquad 34a.$$

$$Mb^{IV}(HX\text{-}Fe^{IV}\text{-}OH) + HO\cdot \rightarrow \cdot Mb^{IV}(\cdot X\text{-}Fe^{IV}\text{-}OH) + H_2O \qquad 34b.$$

or a *heterolytic scission* (reaction 35) of the peroxide,

$$Mb^{III}(HX\text{-}Fe^{III}) + H_2O_2 \rightarrow \cdot Mb^{IV}(\cdot X\text{-}Fe^{IV}\text{-}OH) + H_2O \qquad 35.$$

as well as a *concerted mechanism* (reaction 36a), have been proposed.

$$Mb^{III}(HX\text{-}Fe^{III}) + H_2O_2 \rightarrow Mb^{V}(HX\text{-}Fe^{V}{=}O) + H_2O \qquad 36a.$$

Analysis of cholesterol oxidation products during the oxidation of Mb^{III} by H_2O_2 resulted in the formation of the 5-α-hydroperoxide adduct, indicative of

$^1O_2$ formation (75). The formation of $^1O_2$ can be rationalized in connection with a concerted mechanism (reaction 36a followed by reaction 36b), in a manner as similarly proposed for the reaction of different oxene donors with cytochrome P_{450} (reactions 33a,b) (74).

$$Mb^V(HX\text{-}Fe^V\text{=}O) + H_2O_2 \rightarrow Mb^{III}(HX\text{-}Fe^{III}) + H_2O + {}^1O_2 \qquad 36b.$$

The formation of HO· during the oxidation of Mb^{III} by H_2O_2 (reaction 34a) seems unlikely, for HO· could be detected neither by spin trapping methods with DMPO or DMPO-ethanol (87) nor by the isolation of the 7-α-hydroperoxide adduct of cholesterol (indicative of HO·) (75). The formation of hydroxylation products of salicylate—identified by HPLC with electrochemical detection and otherwise indicative of HO· occurrence (16, 17)—is ambiguous in systems containing Mb^{IV} or ·Mb^{IV}, for the ferryl heme can hydroxylate salicylate to a similar pattern of dihydroxy benzoic acid derivatives (16).

In spite of the ambiguity regarding the molecular mechanism for the formation of Mb^{IV} (reactions 34–36), its formulation as $[Fe^{IV}\text{-}OH]$ or $[Fe^{IV}\text{=}O]$ (88–90), as well as that of the EPR-detectable transient free radical form of the protein (formally represented as ·$X\text{-}Fe^{IV}\text{-}OH$) (85, 91, 92), has been substantiated by a variety of studies. Moreover, the formation of the transient free radical form of Mb^{IV} bears the important implication that, on the one hand, it bridges the discrepancy between the retention of one-oxidizing equivalent in the ferryl heme complex and the observed catalysis of two-electron oxidations and, on the other hand, sets an example of a reversible univalent oxidation state in a biological system, not involving valency changes in the transition metal (91).

Although the cytotoxic activity of H_2O_2 in biological systems is believed to be due to its capacity to be reduced to HO·, the reaction of H_2O_2—or lipid hydroperoxides (75)—with Mb^{III} leads to the powerful oxidant Mb^{IV} or ·Mb^{IV}, which is more selective and perhaps more relevant than HO· in vivo. The ferryl complex in Mb^{IV} or its transient free radical form is regarded as an oxidizing agent capable of promoting peroxidation of fatty acids, oxidation of β-carotene, ascorbic acid, methional, uric acid, and phenols, as well as epoxidation of styrene. Mb^{IV} and ·Mb^{IV} can intiate membrane lipid peroxidation in vitro, though its participation in some pathological situations in vivo remains to be established. A comprehensive sequence of reactions to account for lipid peroxidation initiated by ·Mb^{IV} includes (93, 94): initial H abstraction from the unsaturated fatty acid by ·Mb^{IV} (reaction 37), followed by quenching of the alkyl fatty acid radical by O_2 (reaction 38), and subsequent propagation reactions (reaction 39). The formed hydroperoxide (ROOH) can further react with Mb^{IV} to yield the corresponding peroxyl radical (ROO·) and Mb^{III} (reaction 40).

$$\cdot Mb^{IV}(\cdot X\text{-}Fe^{IV}{=}O) + RH \rightarrow Mb^{IV}(HX\text{-}Fe^{IV}{=}O) + R\cdot \qquad 37.$$

$$R\cdot + O_2 \rightarrow ROO\cdot \qquad 38.$$

$$ROO\cdot + RH \rightarrow ROOH + R\cdot \qquad 39.$$

$$Mb^{IV}(HX\text{-}Fe^{IV}{=}O) + ROOH \rightarrow Mb^{III}(HX\text{-}Fe^{III}) + ROO\cdot + HO^- \quad 40.$$

The reduction of Mb^{IV} or its transient free radical form is a complex process, involving steps, not necessarily irreversible, with both multiple components at different stages of oxidation and reactions, and it is expectedly dependent on the reduction potential of the couples involved. Although two-electron reduced products are isolated, these reactions proceed in one-electron transfer steps. Several conventional electron donors have been described to promote the conversion $Mb^{IV} \rightarrow Mb^{III}$. Recent investigations have stressed the role of GSH (95), quinone-thioether derivatives (96), and ascorbate (97).

1. Glutathione reduces Mb^{IV} to Mb^{III} with concomitant oxidation to its disulfide. However, GSH cannot reduce Mb^{III} to Mb^{II} or $Mb^{II}O_2$, a process probably restricted by an unfavorable reduction potential. The GSH-mediated reduction of Mb^{IV} is a one-electron transfer process (reaction 41), and GS· is an expected intermediate in the formation of GSSG (95). GS·, unless stabilized, decays to a variety of products, such as those partly summarized in reactions 24–27 (55).

$$Mb^{IV}(HX\text{-}Fe^{IV}\text{-}OH) + GS^- + H^+ \rightarrow Mb^{III}(HX\text{-}Fe^{III}) + H_2O + GS^-\cdot \quad 41.$$

The reduction of Mb^{IV} or $\cdot Mb^{IV}$ to Mb^{III} by GSH facilitates the further reduction of H_2O_2 to H_2O, within a process of cycling characteristics (reaction 35 $\Leftrightarrow$ reaction 41) and in a fashion similar to a glutathione peroxidase-like activity (95).

2. Quinone-thioether derivatives, e.g. glutathionyl-menadiol, at variance with GSH, facilitate the two-electron reduction of Mb^{IV} (or its free radical form, $\cdot Mb^{IV}$) to Mb^{II} or $Mb^{II}O_2$, depending on whether the assay is carried out in the presence or absence of O_2 (96). Although the overall reaction can be considered a two-electron transfer process, it proceeds likely in two one-electron transfer steps: $\cdot Mb^{IV} \rightarrow Mb^{IV} \rightarrow Mb^{III}$ (reactions 42a,b)

$$\text{(naphthalene-OH,CH}_3\text{,SG,OH)} + \cdot Mb^{IV}(\text{X–Fe}^{IV}\text{–OH}) \longrightarrow \text{(naphthalene-O}^\cdot\text{,CH}_3\text{,SG,O}^-) + Mb^{IV}(\text{HX–Fe}^{IV}\text{–OH}) + H^+ \qquad 42a.$$

$$\text{(naphthalene-O}^\cdot\text{,CH}_3\text{,SG,O}^-) + Mb^{IV}(\text{HX–Fe}^{IV}\text{–OH}) + H^+ \longrightarrow \text{(naphthoquinone-O,CH}_3\text{,SG,O)} + Mb^{III}(\text{HX–Fe}^{III}) + H_2O \qquad 42b.$$

or $Mb^{IV} \rightarrow Mb^{III} \rightarrow Mb^{II}$ (reaction 43a,b).

$$\text{(naphthalene-OH,CH}_3\text{,SG,OH)} + Mb^{IV}(\text{HX–Fe}^{IV}\text{–OH}) \longrightarrow \text{(naphthalene-O}^\cdot\text{,CH}_3\text{,SG,O}^-) + Mb^{III}(\text{HX–Fe}^{III}) + H_2O + H^+ \qquad 43a.$$

$$\text{(naphthalene-O}^\cdot\text{,CH}_3\text{,SG,O}^-) + Mb^{III}(\text{HX–Fe}^{III}) \longrightarrow \text{(naphthoquinone-O,CH}_3\text{,SG,O)} + Mb^{II}(\text{HX–Fe}^{II}) \qquad 43b$$

Mb^{III} is also reduced to Mb^{II} or $Mb^{II}O_2$ by quinone-thioether derivatives (reaction 44); in this case a glutathione-semiquinone intermediate product is expected, as demonstrated by EPR with the spin trapping technique (98).

$$\text{(naphthalene-OH,CH}_3\text{,SG,OH)} + Mb^{III}(\text{HX–Fe}^{III}) \longrightarrow \text{(naphthalene-O}^\cdot\text{,CH}_3\text{,SG,O}^-) + Mb^{II}(\text{HX–Fe}^{II}) + 2H^+ \qquad 44.$$

3. Ascorbate, like quinone-thioether derivatives, reduces Mb^{IV} and $\cdot Mb^{IV}$ to Mb^{III} (reactions 45a,b)

$$\text{(ascorbate: O=,O,CH–CH}_2\text{OH,OH,HO,O}^-) + \cdot Mb^{IV}(\text{X–Fe}^{IV}\text{–OH}) \longrightarrow \text{(dehydroascorbate: O=,O,CH–CH}_2\text{OH,OH,O,O)} + Mb^{IV}(\text{HX–Fe}^{IV}\text{–OH}) \qquad 45a.$$

$$\text{(ascorbate: O=,O,CH–CH}_2\text{OH,OH,O,O)} + Mb^{IV}(\text{HX–Fe}^{IV}\text{–OH}) \longrightarrow \text{(dehydroascorbate: O=,O,CH–CH}_2\text{OH,OH,O,O)} + Mb^{III}(\text{HX–Fe}^{III}) + H_2O + 2H^+ \qquad 45b.$$

and, albeit at slower rates, to Mb^{II} or $Mb^{II}O_2$ (reaction 46) (97).

$$\text{(ascorbate: O=,O,CH–CH}_2\text{OH,OH,HO,O}^-) + Mb^{III}(\text{HX–Fe}^{III}) \longrightarrow \text{(dehydroascorbate: O=,O,CH–CH}_2\text{OH,OH,O,O)} + Mb^{II}(\text{HX–Fe}^{II}) + H^+ \qquad 46.$$

The effect of ascorbate on the $Mb^{III} \Leftrightarrow Mb^{IV}$ conversions is indicative of the balance between (a) oxidation of Mb^{III} by H_2O_2 (reaction 35) and (b) re-reduction of Mb^{IV} by ascorbate (reactions 45–46). The apparent rate of re-reduction is reflecting the decrease in H_2O_2 in the reaction mixture. Only upon total oxidation of ascorbate, the remaining H_2O_2 is capable of oxidizing Mb^{III} (reaction 35). The consequence of the above reactions is a continuous cycling $Mb^{III} \Leftrightarrow Mb^{IV}$, during which Mb^{III} is oxidized by H_2O_2 and the formed Mb^{IV} is reduced by ascorbate, the overall effect similar to that of an ascorbic acid oxidase (97).

Disproportionation of Peroxyl Radicals: a Possible Source of Electronically Excited States in Biological Systems

The generation of electrophilic peroxyl radicals from the corresponding hydroperoxides occurs in a variety of biochemical models, such as lipid peroxidation, the lipoxygenase reaction, the conversion of linoleate hydroperoxide to different products catalyzed by hematin, and organic hydroperoxide/hematin mixtures. Reactions 47–51 list the possible sources of peroxyl radicals; however, no evidence was found for the direct production of peroxyl

$$ROOH + Fe^{3+} \rightarrow ROO\cdot + Fe^{2+} + H^+ \qquad\qquad 47.$$

$$ROOH + Fe^{2+} \rightarrow RO\cdot + Fe^{3+} + HO^- \qquad\qquad 48.$$

$$ROOH + RO\cdot \rightarrow + ROO\cdot + ROH \qquad\qquad 49.$$

$$ROOH + ROO\cdot \rightarrow ROO\cdot + ROOH \qquad\qquad 50.$$

$$ROOH + Fe^{4+}\text{-}OH \rightarrow ROO\cdot + Fe^{3+} + H_2O \qquad\qquad 51.$$

radicals originating from a mechanism as that shown in reaction 47, assuming that the reaction either does not occur or occurs at rates far slower than that involving the reductive formation of alkoxyl radicals (reaction 48). The metal-ion and lipoxygenase-catalyzed breakdown of peroxidized fatty acids produces initially alkoxyl radicals ($RO\cdot$) by one-electron reductive cleavage of the hydroperoxide (reaction 48) (99). Subsequent reactions of the alkoxyl radical yield carbon-centered radicals and peroxyl radicals (reaction 49) as well as a further species tentatively identified as an acyl-[RC(O)$\cdot$]-radical adduct (99). Peroxyl radicals can also be formed upon H abstraction of a hydroperoxide by a ferryl-hydroxo complex (Fe^{4+}-OH) (reaction 51). The ubiquitous nature of the latter heme complex, even when at relatively low levels, implies that it could be a quantitatively important catalyst of hydroperoxide metabolism in mammalian tissues (100).

Independent of the molecular mechanism supporting the formation of peroxyl radicals, their participation in recombination reactions is known to proceed via a tetroxide intermediate with elimination of a ketone, an alcohol, and O_2 (reaction 52).

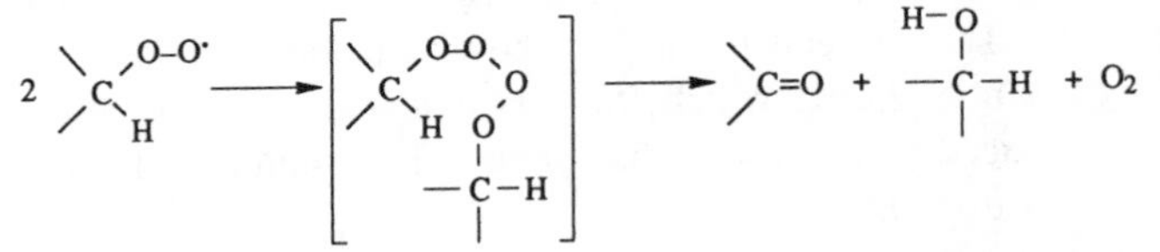

$$2 \quad \underset{/}{\overset{\backslash}{C}}\underset{H}{\overset{O-O\cdot}{}} \longrightarrow \left[\underset{/}{\overset{\backslash}{C}}\underset{H}{\overset{O-O}{}}\underset{O}{\overset{}{}}\underset{-C-H}{\overset{}{}} \right] \longrightarrow \underset{/}{\overset{\backslash}{C}}{=}O \; + \; \overset{H-O}{\underset{-C-H}{\overset{|}{}}} \; + \; O_2 \qquad 52.$$

The disproportionation of peroxyl radicals (ROO·) is concerted (101): (a) since the reaction exothermicity is ~115–150 kcal and the luminescence state of the carbonyl is commonly 75–80 kcal, a concerted reaction is virtually required to concentrate the energy in the ketone fragment, at least for those events leading to chemiluminescence; (b) the termination rate constant for secondary peroxyl radicals is 10^3-fold faster than for tertiary peroxyl radicals, in which there is no α hydrogen; (c) the termination rate constant decreases by 1.9 with a deuteration. A rationalization of this sequence of events can be explained using the Russell diagram (Figure 1) (101, 102).

To obtain an excited state, the reaction exothermicity must be concentrated in the carbonyl fragment. The concerted fragmentation then gives singlet ground state alcohol and triplet ground state O_2. To satisfy selection rules, the excited carbonyl must be of triplet multiplicity (101). This product distribution resulting from secondary ROO· recombination reactions is conventionally written as in reaction 53.

$$\text{ROO·} + \text{ROO·} \rightarrow \text{ROH} + {}^3\text{RO*} + {}^3O_2 \qquad 53.$$

The formation of ${}^1O_2$ during recombination of ROO· can be rather explained by a more efficient quenching of the triplet carbonyl by the O_2 eliminated in the reaction and retained in the solvent cage. The rate of triplet-triplet transfer quenching in the solvent cage has been estimated to be

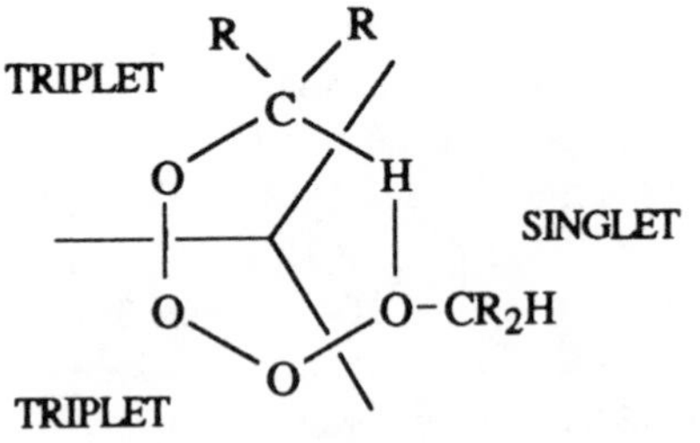

Figure 1 Russell diagram

$\sim 10^{11} s^{-1}$, which could lead to the formulation of reaction 54. However, the formation of $^1O_2$ during disproportionation of secondary ROO· is usually written as in equation 55 (102, 103).

$$RO^* + O_2 \text{ [cage]} \rightarrow RO + {}^1O_2 \ (k_Q = 10^{11}) \qquad 54.$$

$$ROO· + ROO· \rightarrow ROH + RO + {}^1O_2 \qquad 55.$$

Soybean lipoxygenase-3 degrades a mixture of 13-hydroperoxylinoleic acid and linoleic acid to produce modest amounts of $^1O_2$ (104), accounted for in terms of the bimolecular reaction of peroxyl radicals as first proposed by Russell (102). A similar mechanism seems to explain the formation of $^1O_2$ during the Fe^{III}-bleomycin complex-catalyzed degradation of hydroperoxides of long-chain unsaturated fatty acids (13-hydroperoxylinoleate and 15-hydroperoxyarachidonate) (105). The hydroperoxide structural requirements for $^1O_2$ production by Fe^{III}-bleomycin complex are identical to those heme-containing catalysts (100).

The degradation of linoleate, linolenate, and arachidonate by soybean lipoxygenase also produces photoemission in the visible region of the spectrum, and both the intensity of emission and amount of O_2 consumed are linearly related to the number of double bonds present in the fatty acid substrate (106). Photoemission is due to formation of triplet carbonyl compounds (107) as attested by the excitation of different acceptors resulting in sensitized emission. DBAS- and chlorophyll-a-sensitized emission from the arachidonate/lipoxygenase reaction is about 10^3-fold higher than in the absence of sensitizers. Since both acceptors have different energy requirements to populate their fluorescence state, two possible mechanisms for the generation of triplet carbonyls can be envisaged: (a) free radical interactions yielding conjugated carbonyl compounds (reaction 53) having sufficient energy to promote chlorophyll-a-sensitized emission [$\Delta E(S_1)_{\text{chlorophyll-}a} = 41.5$ kcal $\times$ mol^{-1}] and (b) breakdown of a possible dioxetane intermediate from lipid hydroperoxides (reaction 56) yielding unconjugated carbonyls, having enough energy to promote both DBAS- and

$$ 56. $$

chlorophyll-a-sensitized emission [$\Delta E(S_1)_{\text{DBAS}} = 70.2$ kcal $\times$ mol^{-1}]. Therefore, it can be expected that a variety of excited species of different multiplicities as well as distinct mechanisms are involved in the lipoxygenase reaction (107).

ELECTRON-TRANSFER AND NUCLEOPHILIC ADDITION REACTIONS OF QUINOID COMPOUNDS

The overall expression of toxicity of several xenobiotics can be considered as a function of their electrophilic character and/or their activation by electron transfer. The electrophilic character of several xenobiotics, e.g. quinones, allows their reaction with thiols or other cellular nucleophiles, whereas the one- or two-electron reduction of xenobiotics by flavo-enzymes is a source of oxy-radicals. Both mechanisms can be regarded in chemical terms as a single critical cellular bimolecular reaction, thereafter propagating its effects to other biochemical pathways and leading to the expression of toxicity. These reactions are generally proposed in the oxidative metabolism of quinoid compounds that are frequently used as chemical models of antitumor drugs possessing quinoid structure.

Electron Transfer to Quinoid Compounds

Many quinones possess the ability to redox cycle with microsomes or mitochondria or in whole cell preparations. The basis of redox cycling requires the initial reduction of quinones via a one- or two-electron transfer mechanism, which can be achieved by a variety of enzymic systems. By relating the rate of free radical formation over the rate of enzyme reaction, the enzymic reduction of quinones can be classified as a one-electron transfer mechanism (reaction 57), a two-electron transfer mechanism (reaction 58), and a mixed-type mechanism (108). The formation of semiquinones as in reaction 57 can be accomplished by among others, NADH-cytochrome b_5 reductase, NADPH-cytochrome P_{450} reductase, and chloroplast ferredoxin-$NADP^+$ reductase. The formation of hydroquinones as in reaction 58 can be catalyzed by DT-diaphorase, an enzyme considered as a cellular protective device against semiquinone and $O_2^- \cdot$ formation (109). A carbonyl transferase has been also shown as a two-electron transfer flavoprotein in human liver (110). The products of reactions 57 and 58

$$57.$$

$$58.$$

are prone to undergo one-electron transfer reactions to O_2, with formation of $O_2^- \cdot$ and quinone and semiquinone products, respectively.

In addition to the one-electron enzymic reduction of quinones (reaction 57), semiquinone intermediates can be generated nonenzymically by (a) comproportionation reactions, (b) oxidation of hydroquinones by either O_2, $O_2^- \cdot$, or excited molecules, and (c) reduction of quinones by $O_2^- \cdot$, radiolysis, and reaction of triplet quinone with suitable electron donors (111–114).

Nucleophilic Addition Reactions of Quinoid Compounds

Quinoid compounds are electrophiles by nature and capable of reacting with cellular nucleophiles, among them, thiol-containing compounds. The reaction between quinoid compounds and sulfur nucleophiles is a 1,4-reductive addition of the Michael type (115). The electrophilic center is represented by the activated or polarized double bond contained in the quinoid compound, whereas the polarizability of the compound that furnishes the nucleophilic sulfur atom is determined by both the electronegativity and the size of the atom bearing the nonbonding electrons. A model nucleophilic addition reaction is shown, for the case of 2-methyl-1,4-naphthoquinone with GSH, in reaction 59.

The possible reactions to which glutathionyl-hydroquinones are subjected include: cross-oxidation, autoxidation, disproportionation, free radical interactions, oxidative elimination, and enzymic reduction (116). The occurrence of glutathionyl-semiquinone intermediates (98) indicates that quinone-thioether derivatives can participate in one-electron transfer processes, such as cross-oxidation, autoxidation, or disproportionation reactions. The quinone-thioether derivatives can also redox cycle upon their effective two-electron reduction by DT-diaphorase (117) (reaction 60). Although autoxidation of the quinol-thioether in reaction 60 is written as a two-electron transfer reaction, it actually takes place as a one-electron transfer step with formation of a glutathionyl-semiquinone intermediate.

Thus, reaction 59 not only bears the implication of "depletion of cellular thiols," but it yields a compound that can be enzymically reduced at rates comparable to those of the unconjugated quinoids. The effect of the glutathionyl-substituent on the reduction potential of the quinone is ambivalent, and it is an expression of the overall effect exerted by the substituents, including those involved in the thioether linkage. Although the glutathionyl substituent exerts negligible (118) or small (117) changes in the reduction potential of the quinone, it causes significant alterations in the overall electron density of the compound, for it autoxidizes at far higher rates than the unconjugated parent quinone (117, 119). This observation, along with the described reduction of quinone-thioether derivatives by DT-diaphorase, brings forward an enzymic reduction ⇔ autoxidation cycle (reaction 60), which could be more harmful to the cell than those cycles involving quinones lacking a glutathionyl substituent.

The Relative Contribution of Redox Cycling and Sulfhydryl Arylation to Quinone Cytotoxicity

The relative significance of both mechanisms for quinone cytotoxicity, that is, redox cycling and sulfhydryl arylation reactions, has been amply discussed (120–122). Although the metabolism of quinones by isolated hepatocytes reveals that the toxicological properties of these compounds are mainly supported by the redox and addition chemistry of quinones (120), their connection to other—no less important—cytotoxic effects, such as inhibition of mitochondrial functions and critical enzymic activities, and alterations of intracellular Ca^{2+} balance (121), remains to be assessed.

Cytotoxicity depends primarily on the chemistry of the particular quinone under study, and a general scheme applying to quinone cellular metabolism cannot be drawn. The inborn physicochemical properties of the quinone are a major determinant in the expression of toxicity as well as in the type of bimolecular reaction that will trigger the alteration of biochemical pathways and propagation of unbalanced events leading to cell impairment and death. Thus, on the one hand, less-substituted p-benzoquinone derivatives, such as 2-methyl-, 2-Br-, 2,6-dimethyl-, 2,5-dimethyl, and 2,3,5-trimethyl-p-benzoquinone, depleted hepatocyte GSH as a result of arylation and—without apparently producing oxidative stress—were highly toxic to the cells. On the other hand, the tetramethyl-substituted p-benzoquinone (duroquinone) promoted GSH oxidation without arylation and was only toxic to the cells when they were compromised by inhibiting their glutathione reductase and/or catalase activities (123). Similar observations were described for the effect of several naphthoquinone derivatives with different substituents. 1,4-Naphthoquinone, 2-methyl-1,4-naphthoquinone, and 5-hydroxy-1,4-naphthoquinone seemed to exert their toxicity by means of both redox cycling and sulfhydryl arylation, whereas 2-hydroxy- and 2,3-dimethoxy-1,4-naphtho-quinone, which cannot react with sulfur nucleophiles due to the electron-donating properties of their substituents, seemed to exert their toxicity mainly by enzymic reduction followed by the formation of oxy-radicals (121, 122, 124). GSH depletion and GSH oxidation are commonly observed within the former group of naphthoquinone derivatives, whereas only GSSG formation, probably due to the oxidation of GSH by oxy-radicals, is found with the latter group of naphthoquinone derivatives. Oxy-radicals generated during redox cycling of naphthoquinones also stimulate a tyrosine-specific protein phosphorylation in the rat liver plasma membrane (125). It can be concluded that while oxidative processes (redox cycling $\rightarrow$ GSH oxidation) may be cause for quinone toxicity, the addition to nucleophilic intracellular thiols also contributes significantly to the cytotoxic effect of certain quinones.

The glutathionyl-hydroquinone conjugate is of a rather hydrophilic char-acter and, thus, it will display its chemical reactivity (summarized above) in the aqueous environment represented by the cytosol. However, given the cellular distribution of GSH, the reductive addition reaction (reaction 59) should take place in the hydrophilic milieu, whereas the electrophilic counter-part—the quinoid compound—depending on its partition coefficient, tends to be largely distributed in the hydrophobic milieu. The partition coefficient of quinoids has been shown to correlate well with their cytotoxicity (126). The expression of the chemical reactivity of quinone-thioether derivatives is assumed to be influenced by the balance in the biological milieu between potential activating and detoxifying activities, including (a) cellular reduc-tases, such as DT-diaphorase, (b) conjugating enzymes, such as UDP-glucuronosyl-transferase, and (c) the intracellular O_2 concentration. Neither of the latter are necessarily evenly distributed in the heterogeneous environ-ment of the cell, and this may give rise to local alterations of this balance, which, eventually, may be expressed as an impairment of vital cell functions.

Another aspect of quinone-mediated cytotoxicity is the formation of elec-tronically excited states accompanying the oxidative stress induced by their infusion into perfused liver or metabolism by isolated hepatocytes or micro-somal fractions (127, 128). The intensity of photoemission elicited during 2-methyl-1,4-naphthoquinone metabolism—which might be considered as indicative of cellular oxidative stress—is inversely related to the operativity of the two-electron reductive pathway of 2-methyl-1,4-naphthoquinone medi-ated by DT-diaphorase (128, 129). This indicates the protective role of the latter enzyme against oxidative stress, as previously suggested (109). A link between the production of $O_2^-\cdot$ by redox cycling of menadione and the formation of excited species is still missing. The possibility also remains to be established on the contribution to photoemission of an excited state of the quinone, menadione, which is a well-known electron affinity radiosensitizer of hypoxic cells (130) and sensitizes the photooxidation of nucleic acid and protein constituents (131).

Photoemission during hydro- or semiquinone oxidation (132) seems to be contributed in some cases by the triplet excited state of the quinone (133–136), and sets an example of formation of electronically excited states by electron transfer. The formation of triplet quinones might bear a cytotoxic aspect on its own, for triplet carbonyl compounds are known to possess a similar reactivity to that of alkoxyl radicals, at least in terms of H-atom abstraction. In some instances, the formation of an excited quinone compound of triplet multiplicity seems to require the oxidation of the semiquinone by $HO\cdot$ (133, 134). This mechanism would be preceded by the reduction of H_2O_2 to $HO\cdot$ by semiquinone intermediates (reaction 8) (12, 13). The semiquinone-mediated reduction of H_2O_2 explains the mitochondrial formation of $HO\cdot$, which is not sensitive to a variety of effective iron chelators, but appears to depend on the steady-state concentrations of ubisemiquinone radicals. Since H_2O_2 may be regarded as a quasi-physiological by-product of mitochondrial respiration (33), its relatively low permeability constant (35) would favor an efficient bimolecular collision with ubisemiquinone radicals, which have been documented for mitochondrial ubiquinone as an essential intermediate step of electron transfer in the respiratory chain. It was suggested that such an event, yielding steady-state concentrations of $HO\cdot$, has to be considered a realistic process of biological significance (13).

The Role of Superoxide Dismutase in Quinone Metabolism

At a cellular level, there seems to be a relationship between $O_2^- \cdot$-generating hydroquinones and superoxide dismutase activity (137). The observed inhibition of superoxide dismutase by quinones (138), along with the superoxide dismutase-mediated decrease in O_2 uptake and H_2O_2 production by hepatocytes supplemented with menadione (139), cannot be entirely explained in terms of the classical disproportionation of $O_2^- \cdot$ (18) or semiquinones (140) catalyzed by the enzyme, the latter reaction proceeding with a rate constant 3–4 orders of magnitude lower than that for the disproportionation of $O_2^- \cdot$.

The sequential activities DT-diaphorase and superoxide dismutase may serve as a main detoxication pathway for electrophilic quinones. These complimentary activities require, on the one hand, the occurrence of organic compounds that break down with formation of semiquinones and $O_2^- \cdot$, a process brought about by DT-diaphorase and, on the other hand, the reduction of semiquinones by $O_2^- \cdot$, a process mediated by superoxide dismutase. The role of superoxide dismutase in connection with DT-diaphorase consists, therefore, in inhibition of semiquinone autoxidation by facilitating its reduction at the expense of $O_2^- \cdot$. This activity has been termed superoxide:semiquinone oxidoreductase (141), its overall mechanism formulated as $Q^{-} + O_2^- \cdot + 2H^+ \Leftrightarrow QH_2 + O_2$. The autoxidation of several glutathionyl-substituted hydroquinones is likewise prevented by superoxide dismutase. This means that the sequences DT-diaphorase $\rightarrow$ superoxide dismutase or sulfur-reductive addition $\rightarrow$ superoxide dismutase provide an efficient means for protection against quinone toxicity. Not all quinones can be reduced at the expense of O_2^- · within a reaction involving superoxide dismutase: the autoxidation of un- and methyl-substituted p-benzoquinones is enhanced by superoxide dismutase, because the enzyme displaces the equilibrium of the autoxidation reaction $[Q \cdot ^- + O_2 \Leftrightarrow Q + O_2^- \cdot]$ towards the right.

OXIDATIVE PROCESSES IN TISSUE INJURY

Many in vitro experimental models have implicated oxy-radicals and other O_2-derived species, together with the chain reactions of lipid peroxidation, as important causative agents of aging and of various diseases (4). Accumulating evidence suggests the participation of oxy-radicals, particularly O_2^- ·, as a likely link in the pathophysiology of two disease states: ischemia/reperfusion injury and inflammation. While the mechanisms of O_2^- · formation in these two pathological situations are biochemically distinct, it has been recently discussed that logical links would be apparent, whereby ischemia/reperfusion-induced injury would lead to inflammation and, conversely, whereby an inflamed tissue would become vulnerable to ischemic injury (142).

The apparent occurrence of oxy-radicals in several pathological disorders led rapidly to the development of: (a) a therapeutic approach, implying the medical benefits of oxy-radical scavengers in a multitude of clinical situations; however, discrepancies found in the analysis of the therapeutic literature make correctly designed clinical trails an urgent need (143). (b) In vivo (or closely-related) experimental models that would allow the rapid and objective assessment of oxy-radical occurrence during, for example, ischemia/reperfusion cellular injury, correlating with the application of diverse techniques to monitor in vivo cellular oxidative changes and to test different drugs.

The exposure of tissues to a temporary restriction in O_2 supply (e.g. through an infarction or surgical intervention), followed by a restitution of normal blood flow, may result in oxidative stress due to the formation of oxy-radicals, although the pathogenic events that give rise to tissue damage during reoxygenation are not yet clearly defined. The development of clinical models for the evaluation of oxy-radicals has led to the application of EPR with spin trapping techniques as well as low-level chemiluminescence mainly on ischemia/reperfusion-induced injury. EPR with spin trapping techniques were applied to the intestinal and cardiac ischemia/reperfusion model in vivo using OXANO (2-ethyl-2,5,5-trimethyl-3-oxazolidinoxyl) (which seems rather specific for O_2^- ·) (144), PBN (phenyl-t-butyl nitrone), MNP (2-methyl-2-nitrosopropane), or DMPO (5,5'-dimethyl-1-pyrroline-N-oxide) (145–147) as spin traps, respectively. The use of spin traps indicated that oxy-radicals are important by-products of abnormal oxidative metabolism during regional ischemic heart followed by reperfusion: a carbon-centered adduct and an oxy-radical adduct have been observed in ischemia and reperfusion, respectively (145, 146). The formation of the oxy-radical adduct is observed in the early moments of reperfusion, is sensitive to superoxide dismutase, and is related to the severity of ischemia (145, 146).

In the absence of spin traps, EPR analysis of the ischemia/reperfusion model in the isolated rabbit heart has provided an almost-quantitative assessment of free radical production in this model (148, 149). Interestingly, some of the EPR signals are similar to that observed upon supplementation of myoglobin with H_2O_2 (150), which leads to the formation of ferrylmyoglobin (reactions 34b, 35). The oxidizing character of ferrylmyoglobin may play a role in tissue damage accompanying reperfusion of ischemic muscle tissue. Residual or increased steady-state concentrations of H_2O_2 during ischemia could favor the oxidation of met- to ferryl-myoglobin. Thus, myoglobin may play an important catalytic role as an electron sink in protecting, for example, unsaturated fatty acids in membranes from oxidative attack, when functioning within a catalytic cycle supported by, e.g. ascorbic acid (97). Such an action might be a determinant of tissue damage associated with the intra- or extracellular formation of oxidants during reperfusion of muscle following episodes of hypoxia or ischemia. In fact, ascorbate prevents damage to rat hearts perfused with O_2 after a brief period of ischemia (A. Arduini, D. Galaris, P. Hochstein, unpublished). Although ascorbate may also react with peroxyl and alkoxyl radicals, its protective action in the heart is consistent with a role in the redox cycling of myoglobin (see above). This concept provides a new rationale for the data that sufficient vitamin C status avoids an increased risk in human ischemic heart disease (151). An EPR signal corresponding to coenzyme Q free radical in myocardium was lowered during ischemia and turned to control levels upon reperfusion (152). Furthermore, the autoxidation of catecholamines has been proposed as a source of oxy-radicals in heart ischemia-reperfusion injury. However, this autoxidation per se is extremely slow at physiological pH and, therefore, unlikely to contribute as a primary source of tissue injury; these oxy-radicals are likely to arise from catalyzed oxidations through enzymic systems and/or metals ions (153).

The observation of an allopurinol- and superoxide dismutase-sensitive spontaneous chemiluminescence (154) or sensitized chemiluminescence (155) signal from intestine following reperfusion was observed simultaneously by two groups. The observation of chemiluminescence indicates only that an electronically excited state is being produced during reperfusion, thereby pointing to the requirement of O_2 for its production. Further considerations should be taken into account and remain to be established: (*a*) the nature of the excited species contributing to photoemission. This is difficult to solve by spectral analysis due to the low intensity of the signal, and it could not be expected that a single species would be responsible for the emission in this complex experimental model. (*b*) The metabolic pathway(s) supporting the generation of excited states and a possible link to free radical intermediates. The sensitivity of chemiluminescence to allopurinol (154) and superoxide dismutase (155) might indicate the participation of $O_2^- \cdot$ in some step leading to the formation of an excited product, its relaxation to the ground state being detected as photoemission. Yet it is known that several molecular mechanisms, involving at any early stage $O_2^- \cdot$, can support excited state(s) formation (29). (*c*) The formation of electronically excited species can bear a relevance of its own: the reactivity of both triplet carbonyls and $^1O_2$ has been thoroughly described and, in the former case, it could mimic that of alkoxyl radicals as they mediate H-atom abstraction.

The attempts to use both EPR and photon-counting techniques indicate a step forward in the direct evaluation of free radicals and excited species in the above-described disease states, rather than the indirect approach based on the prevention of damage (generally assessed by histological techniques) by free radical scavengers. Monitoring luminescence from in vivo models has shown that this technique is useful for the noninvasive evaluation of cellular oxidative stress, as indicated by the increased emission in tumor-bearing mice (156), from α-tocopherol- and selenium-deficient rat liver (157), and from in situ liver upon chloroform anesthesia (158) and by the decreased emission of in situ liver by certain flavonoid antioxidants (159). The field of organ conservation, interesting for organ transplantation surgery, might be a future area where the chemiluminescence technique could be used as a noninvasive assay to monitor cellular oxidative conditions.

ACKNOWLEDGMENTS

Supported by grant 2703-B89-01XA from the Swedish Cancer Foundation and grants 7697 and 4481 from the Swedish Medical Research Council.

Literature Cited

1. Frimer, A. A. 1983. In *The Chemistry of Functional Groups, Peroxides,* ed. S. Patai, pp. 429–61. New York: Wiley
2. Green, M. J., Hill, H. A. O. 1984. *Methods Enzymol.* 105:3–22
3. Sawyer, D. T., Roberts, J. L. Jr., Calderwood, T. S., Sugimoto, H., McDowell, M. S. 1985. *Philos. Trans. R. Soc. London, Ser. B.* 311:483–503
4. Halliwell, B., Gutteridge, J. M. C. 1985. *Mol. Asp. Med.* 8:89–193
5. Ilan, Y. A., Czapski, G., Meisel, D. 1976. *Biochim. Biophys. Acta* 430:209–24

6. Pryor, W. A. 1986. *Annu. Rev. Physiol.* 48:657–63
7. Goldstein, S., Czapski, G. 1983. *J. Am. Chem. Soc.* 105:7276–80
8. Rush, J. M., Bielski, B. H. J. 1985. *J. Phys. Chem.* 89:5062–66
9. Gutteridge, J. M. C. 1986. *FEBS Lett.* 201:291–95
10. Goldstein, S., Czapski, G. 1986. *J. Free Radic. Biol. Med.* 2:3–11
11. Pryor, W. A. 1988. *Free Radic. Biol. Med.* 4:219–23
12. Koppenol, W. H., Butler, J. 1985. *Adv. Free Radic. Biol. Med.* 1:91–131

13. Nohl, H., Jordan, W. 1987. *Bioorg. Chem.* 15:374–82
14. Winterbourn, C. C. 1987. *Free Radic. Biol. Med.* 3:33–40
15. Mottley, C., Connor, H. D., Mason, R. P. 1986. *Biochem. Biophys. Res. Commun.* 141:622–28
16. Floyd, R. A., Henderson, R., Watson, J. J., Wong, P. K. 1986. *J. Free Radic. Biol. Med.* 2:13–18
17. Radzik, D. M., Roston, D. A., Kissinger, P. T. 1983. *Anal. Biochem.* 131:458–64
18. Fridovich, I. 1976. In *Free Radicals in Biology,* ed. W. A. Pryor, I:239–75. New York: Academic
19. Williams, R. J. P. 1985. *Philos. Trans. R. Soc. London, Ser. B.* 311:593–603
20. Dunford, H. B. 1987. *Free Radic. Biol. Med.* 3:405–21
21. Symons, M. C. R. 1985. *Philos. Trans. R. Soc. London, Ser. B.* 311:451–72
22. Koppenol, W. H. 1976. *Nature* 262:420–21
23. Khan, A. U. 1976. *J. Phys. Chem.* 80:2219–28
24. Foote, C. S., Shook, F. C., Akaberli, R. B. 1984. *Methods Enzymol.* 105:36–47
25. Di Mascio, P., Sies, H. 1988. *J. Am. Chem. Soc.* 111:In press
26. Khan, A. U. 1984. *J. Photochem.* 25:327–34
27. Kanofsky, J. R. 1984. *J. Photochem.* 25:105–13
28. Khan, A. U. 1981. *J. Am. Chem. Soc.* 103:6516–17
29. Cadenas, E. 1984. *Photochem. Photobiol.* 40:823–30
30. Cilento, G. 1984. *Pure & Appl. Chem.* 56:1179–90.
31. Duran, N., Cadenas, E. 1987. *Rev. Chem. Intermed.* 8:147–87
32. Naqui, A., Chance, B., Cadenas, E. 1986. *Annu. Rev. Biochem.* 55:137–66
33. Boveris, A., Cadenas, E. 1982. In *Superoxide Dismutase,* ed. L. W. Oberley, II:15–30. Boca Raton: CRC
34. Cadenas, E., Boveris, A., Ragan, C. I., Stoppani, A. O. M. 1977. *Arch. Biochem. Biophys.* 180:248–57
35. Nohl, H., Jordan, W., Youngman, R. J. 1986. *Adv. Free Radic. Biol. Med.* 2:211–79
36. Nohl, H. 1987. *FEBS Lett.* 214:269–73
37. Nohl, H. 1987. *Eur. J. Biochem.* 169:585–91
38. Nohl, H. 1988. *Biochem. Pharmacol.* 37:2633–38
39. Chance, B., Sies, H., Boveris, A. 1979. *Physiol. Rev.* 59:527–605
40. Davies, K. J. A. 1986. *J. Free Radic. Biol. Med.* 2:155–73

41. Sevanian, A., Kim, E. 1985. *J. Free Radicals Biol. Med.* 1:263–71
42. Demple, B., Halbrook, J. 1983. *Nature* 304:466–68
43. Terao, J., Matsushita, S. 1988. *Free Radic. Biol. Med.* 4:205–8
44. Willson, R. L., Dunster, C. A., Forni, L. G., Gee, C. A., Kittridge, K. J. 1985. *Philos. Trans. R. Soc. London, Ser. B.* 311:545–63
45. Willson, R. L. 1985. In *Oxidative Stress,* ed. H. Sies, pp. 41–72. London: Academic
46. Simic, M. G., Hunter, E. P. L. 1985. In *Chemical Changes in Food During Processing,* pp. 107–19. Westport, CT: AVI
47. Ivanov, I. I. 1985. *J. Free Radic. Biol. Med.* 1:247–53
48. Simic, M. G. 1980. In *Autoxidation in Food and Biological Systems,* ed. M. G. Simic, M. Carel, pp. 17–26. New York: Plenum
49. Simic, M. G., Jovanovic, S. V. 1987. Submitted.
50. Foote, C. S., Ching, T.-Y., Geller, G. G. 1974. *Photochem. Photobiol.* 20:511–13
51. Encinas, M. V., Lissi, E. A., Olea, A. F. 1985. *Photochem. Photobiol.* 42:347–52
52. Bendich, A., Burton, G. W., Machlin, L. J., Scandurra, O., Wayner, D. D. M. 1986. *Adv. Free Radic. Biol. Med.* 2:419–44
53. Rougée, M., Bensasson, R. V. 1986. *C. R. Acad. Sci. Paris* 20:1223–26
54. Chou, P. T., Khan, A. U. 1983. *Biochem. Biophys. Res. Commun.* 115:932–37
55. Quintiliani, M., Badielo, R., Tamba, M., Esfandi, A., Gorin, G. 1977. *Int. J. Radiat. Biol.* 32:195–202
56. Mönig, J., Asmus, K.-D., Forni, L. G., Willson, R. L. 1987. *Int. J. Radiat. Biol.* 52:589–692
57. Tamba, M., Simone, G., Quintiliani, M. 1986. *Int. J. Radiat. Biol.* 50:595–600
58. Simone, G., Bremer, J. C. M., Tamba, M. 1987. *Int. Congr. Oxyg. Radic.* 4th, La Jolla, pp. 28–31 Abstr.
59. Wefers, H., Sies, H. 1983. *Eur. J. Biochem.* 137:29–36
60. Rougée, M., Bensasson, R. V., Land, E. J., Pariente, R. 1988. *Photochem. Photobiol.* 47:485–89
61. Wefers, H., Schulte-Frohlinde, D., Sies, H. 1987. *FEBS Lett.* 211:49–52
62. Ames, B. N., Cathcart, R., Schwiers, E., Hochstein, P. 1981. *Proc. Natl. Acad. Sci. USA* 78:6858–62
63. Kittridge, K., Willson, R. L. 1984. *FEBS Lett.* 170:162–64

64. Davies, K. J. A. 1987. *J. Biol. Chem.* 262:9895–901
65. Davies, K. J. A., Delsignore, M. E., Lin, S. W. 1987. *J. Biol. Chem.* 262:9902–7
66. Davies, K. J. A., Delsignore, M. E. 1987. *J. Biol. Chem.* 262:9908–13
67. Davies, K. J. A., Lin, S. W., Pacifici, R. E. 1987. *J. Biol. Chem.* 262:9914–20
68. Davies, K. J. A., Goldberg, A. L. 1987. *J. Biol. Chem.* 262:8220–26
69. Davies, K. J. A., Goldberg, A. L. 1987. *J. Biol. Chem.* 262:8227–34
70. Oliver, C. N., Ahn, B.-W., Moerman, E. J., Goldstein, S., Stadtman, E. R. 1987. *J. Biol. Chem.* 262:5488–91
71. Oliver, C. N. 1987. *Arch. Biochem. Biophys.* 253:62–72
72. Esterbauer, H. 1987. *J. Lipid Res.* 28: 495–509
73. Cadenas, E., Sies, H., Nastainczyk, W., Ullrich, V. 1983. *Hoppe-Seyler's Z. Physiol. Chem.* 364:519–28
74. Cadenas, E., Sies, H., Graf, H., Ullrich, V. 1983. *Eur. J. Biochem.* 130:117–21
75. Galaris, D., Mira, D., Sevanian, A., Cadenas, E., Hochstein, P. 1988. *Arch. Biochem. Biophys.* 262:221–31
76. Khan, A. U. 1983. *J. Am. Chem. Soc.* 105:7195–97
77. Kanofsky, J. R. 1984. *J. Am. Chem. Soc.* 106:4277–78
78. Marnett, L. J., Reed, G. A. 1979. *Biochemistry* 18:2923–29
79. Kanofsky, J. R. 1988. *Photochem. Photobiol.* 47:605–9
80. Kuehl, F. A., Humes, J. L., Ham, E. A., Egan, R. W., Dougherty, H. W. 1980. In *Advances in Prostaglandin and Thromboxane Research,* ed. B. Samuelsson, P. V. Ramwell, R. Paoletti, pp. 77–86. New York: Raven
81. Ullrich, V. 1984. In *Oxygen Radicals in Chemistry and Biology,* ed. W. Bors, M. Saran, D. Tait, pp. 391–404. Berlin: de Gruyter
82. Cadenas, E., Sies, H. 1982. *Eur. J. Biochem.* 124:349–56
83. Caughey, W. S., Watkins, J. A. 1985. In *Handbook of Methods for Oxygen Radical Research,* ed. R. A. Greenwald, pp. 95–104. Boca Raton: CRC
84. George, P., Irvine, D. H. 1952. *Biochem. J.* 52:511–17
85. King, K. N., Winfield, M. E. 1963. *J. Biol. Chem.* 238:1520–28
86. Yonetani, T., Schleyer, H. 1967. *J. Biol. Chem.* 242:1974–79
87. Harada, K., Tamura, M., Yamazaki, I. 1986. *J. Biochem.* 100:499–504
88. La Mar, G. N., de Ropp, J. S., Latos-Grazinsky, L., Balch, A. L., Johnson,

R. B., et al. 1983. *J. Am. Chem. Soc.* 105:782–87
89. Schonbaum, G. R., Lo, S. 1972. *J. Biol. Chem.* 247:3353–60
90. Roberts, J. E., Hoffman, B. M., Rutter, R., Hager, L. P. 1981. *J. Am. Chem. Soc.* 103:7654–56
91. Gibson, J. F., Ingram, D. J. E., Nichols, P. 1958. *Nature* 181:1398–99
92. Wittemberg, J. B. 1978. *J. Biol. Chem.* 253:5694–95
93. Kanner, J., Harel, S. 1985. *Arch. Biochem. Biophys.* 237:314–21
94. Kanner, J., Harel, S. 1985. *Lipids* 20:625–28
95. Galaris, D., Cadenas, E., Hochstein, P. 1989. *Free Radic. Biol. Med.* In press
96. Buffinton, G., Mira, D., Galaris, D., Hochstein, P., Cadenas, E. 1988. *Chem.-Biol. Interact.* 66:205–22
97. Galaris, D., Cadenas, E., Hochstein, P. 1989. Submitted
98. Gant, T. W., d'Arcy-Doherty, M., Odowole, D., Sales, K. D., Cohen, G. M. 1986. *FEBS Lett.* 201:296–300
99. Davies, M. J., Slater, T. F. 1987. *Biochem. J.* 245:167–73
100. Dix, T. A., Fontana, R., Panthani, A., Marnett, L. J. 1985. *J. Biol. Chem.* 260:5358–65
101. Kellogg, R. E. 1969. *J. Am. Chem. Soc.* 91:5433–36
102. Russell, G. A. 1957. *J. Am. Chem. Soc.* 79:3871–77
103. Howard, J. A., Ingold, K. U. 1968. *J. Am. Chem. Soc.* 90:1056–58
104. Kanofsky, J. R., Axelrod, B. 1986. *J. Biol. Chem.* 261:1099–104
105. Kanofsky, J. R. 1987. See Ref. 58, pp. 68–69
106. Schulte-Herbrüggen, T., Cadenas, E. 1985. In *Free Radicals in Liver Injury,* ed. G. Poli, K. H. Cheeseman, M. U. Dianzani, T. F. Slater, pp. 91–98. Oxford: IRL
107. Schulte-Herbrüggen, T., Cadenas, E. 1985. *Photobiochem. Photobiophys.* 10:35–51
108. Yamazaki, I. 1977. In *Free Radicals in Biology,* ed. W. A. Pryor, 3:183–218. New York: Academic
109. Lind, C., Hochstein, P., Ernster, L. 1982. *Arch. Biochem. Biophys.* 216: 178–85
110. Wermuth, B., Platts, K. L., Seidel, A., Oesch, F. 1986. *Biochem. Pharmacol.* 35:1277–82
111. Cadenas, E. 1987. *Chem. Scr.* 27A: 113–15
112. Amouyal, E., Bensasson, R. 1977. *J. Chem. Soc., Faraday Trans. 1.* 73:1561–68

113. Amoyal, E., Bensasson, R. 1976. *J. Chem. Soc., Faraday Trans. 1.* 72: 1274–87
114. Swallow, A. J. 1982. In *Function of Quinones in Energy Conserving Systems,* ed. B. L. Trumpower, pp. 59–72. London: Academic
115. Finley, K. T. 1974. In *The Chemistry of Quinonoid Compounds,* ed. S. Patai, pp. 877–1144. London: Wiley
116. Brunmark, A., Cadenas, E. 1989. In *Handbook of Glutathione: Metabolism and Physiological Functions,* ed. J. Viña. Boca Raton: CRC. In press
117. Buffinton, G., Öllinger, K., Brunmark, A., Cadenas, E. 1989. *Biochem. J.* 257:561–71
118. Wilson, I., Wardman, P., Tai-Shun, L., Sartorelli, A. C. 1986. *J. Med. Chem.* 29:1381–84
119. Brunmark, A., Cadenas, E. 1989. *Chem.-Biol. Interact.* In press
120. Orrenius, S. 1985. *Philos. Trans. R. Soc. London, Ser. B.* 311:673–77
121. Bellomo, G., Mirabelli, F., DiMonte, D., Richelmi, P., Thor, H., et al. 1987. *Biochem. Pharmacol.* 36:1313–20
122. Gant, T. W., Ramakrishna Rao, D. N., Mason, R. P., Cohen, G. M. 1988. *Chem.-Biol. Interact.* 65:157–73
123. Rossi, L., Moore, G. A., Orrenius, S., O'Brien, P. J. 1986. *Arch. Biochem. Biophys.* 251:25–35
124. DiMonte, D., Bellomo, G., Thor, H., Nicoreta, P., Orrenius, S. 1984. *Arch. Biochem. Biophys.* 235:334–42
125. Chan, T. M., Chen, E., Tatoyan, A., Shargill, N. S., Pleta, M., Hochstein, P. 1986. *Biochem. Biophys. Res. Commun.* 139:439–45
126. Prough, R. A., Gettings, S. D., Lubet, R. A., Nims, R. I., Santore, K. S., Powis, G. 1987. *Chem. Scr.* 27A:99–104
127. Wefers, H., Sies, H. 1983. *Arch. Biochem. Biophys.* 224:568–78
128. Wefers, H., Komai, T., Talalay, P., Sies, H. 1984. *FEBS Lett.* 169:63–66
129. Prochaska, H. J., Talalay, P., Sies, H. 1987. *J. Biol. Chem.* 262:1931–34
130. Adams, G. E. 1977. In *Cancer. A Comprehensive Treatise,* ed. F. F. Becker, 6:183–333. New York: Plenum
131. Murali Krishna, C., Decarroz, C., Wagner, J. R., Cadet, J., Riesz, P. 1987. *Photochem. Photobiol.* 46:175–82
132. Stauf, J., Bertolmes, P. 1970. *Angew. Chem.* 9:307–8
133. Brunmark, A., Cadenas, E. 1987. *Free Radic. Biol. Med.* 3:169–82
134. Ginsberg, M., Cadenas, E. 1985. *Photobiochem. Photobiophys.* 9:223–32
135. Hoffman, M. E., Ciampi, D. B., Duran, N. 1987. *Experientia* 43:217–20

136. Villablanca, M., Cilento, G. 1985. *Photochem. Photobiol.* 42:591–97
137. Hassan, H. M., Fridovich, I. 1979. *Arch. Biochem. Biophys.* 196:385–95
138. Smith, M. T., Evans, C. G. 1985. *Biochem. Pharmacol.* 33:3109–10
139. Ross, D., Thor, H., Orrenius, S., Moldéus, P. 1985. *Chem.-Biol. Interact.* 55:177–84
140. Butler, J., Hoey, B. M. 1986. *J. Free Radic. Biol. Med.* 2:77–81
141. Cadenas, E., Mira, D., Brunmark, A., Lind, C., Segura-Aguilar, J., Ernster, L. 1988. *Free Radic. Biol. Med.* 5:71–79
142. McCord, J. M. 1986. *Adv. Free Radic. Biol. Med.* 2:235–46
143. Greenwald, R. A. 1985. *J. Free Radic. Biol. Med.* 1:173–77
144. Nilsson, U. A., Lundgren, O., Haglind, E., Byluna-Fellenius, A.-C. 1987. See Ref. 58, pp. 150–53
145. Arroyo, C. M., Kramer, J. H., Dickens, B. F., Weglicki, W. B. 1987. *FEBS Lett.* 221:101–4
146. Kramer, J. H., Arroyo, C. M., Dickens, B. F., Weglicki, W. B. 1987. *Free Radic. Biol. Med.* 3:153–59
147. Baker, J. E., Felix, C. C., Olinger, G. N., Kalyanaraman, B. 1988. *Proc. Natl. Acad. Sci. USA* 85:2786–89
148. Zweir, J. L., Flaherty, J. T., Weisfeld, M. L. 1987. *Proc. Natl. Acad. Sci. USA* 84:1404–7
149. Limm, W., Mugiishi, M., Piette, L. H., McNamara, J. J. 1987. See Ref. 58, pp. 123–26
150. Harada, K., Yamazaki, I. 1987. *J. Biochem.* 101:283–86
151. Gey, K. F., Stahelin, H. B., Puska, P., Evans, A. 1987. *Ann. NY Acad. Sci.* 498:110–20
152. Nakazawa, H. K., Ban, K., Ichinori, K., Okino, H., Masuda, T., Aoki, N. 1987. See Ref. 58, pp. 127–29
153. Jewett, S. L., Eddy, L. J., Hochstein, P. 1989. *Free Radic. Biol. Med.* In press
154. Turrens, J. F., Giulivi, C., Pinus, C., Roldan, E., Boveris, A. 1987. See Ref. 58, pp. 64–65
155. Morris, J. B., Bulkley, G. B., Haglund, U., Cadenas, E., Sies, H. 1987. See Ref. 58, p. 67
156. Boveris, A., Llesuy, S. F., Fraga, C. G. 1985. *J. Free Radic. Biol. Med.* 1:131–38
157. Fraga, C. G., Martino, V. S., Ferraro, G. E., Coussio, J. D., Boveris, A. 1987. *Biochem. Pharmacol.* 36:717–20
158. Cohen, P. J., Chance, B. 1986. *Biochim. Biophys. Acta* 884:517–19
159. Fraga, C. G., Arias, R. F., Llesuy, S. F., Koch, O. R., Boveris, A. 1987. *Biochem. J.* 242:383–86

Annu. Rev. Biochem. 1989. 58:111–36

ATP SYNTHASE (H$^+$-ATPase): Results by Combined Biochemical and Molecular Biological Approaches[1]

Masamitsu Futai, Takato Noumi, and Masatomo Maeda

Department of Organic Chemistry and Biochemistry, The Institute of Scientific and Industrial Research, Osaka University, Ibaraki, Osaka 567, Japan

CONTENTS

[1]Abbreviations used: DCCD, dicyclohexylcarbodiimide; Pi, inorganic phosphate; PS3, the thermophilic bacterium PS3; Nbf-Cl, 7-chloro-4-nitrobenzofurazan; NBD-Cl, 7-chloro-4-nitrobenzoxadiazole; OSCP, oligomycin-sensitivity conferring protein; EF$_1$, *Escherichia coli* F$_1$; CF$_1$, chloroplast F$_1$; EF$_0$, *Escherichia coli* F$_0$; ^{125}I-TID, 3-(trifluoro-methyl)-3-(m ^{125}I iodophenyl)diazirine; AP$_3$-PL, adenosine triphosphopyridoxal; AP$_2$-PL, adenosine diphosphopyridoxal; FSBI, 5'-*p*-sulfonyl benzoylinosine; FSBA, 5'-*p*-sulfonyl benzoyladenosine; BzATP, 3'-O-(4-benzoyl)-benzoyl ATP.

    Coverage of literature is mainly until the spring of 1988.

0066-4154/89/0701-0111$02.00

INTRODUCTION

ATP synthase, or H^+-ATPase (F_0F_1), in membranes of mitochondria, chloroplasts, and bacteria, synthesizes ATP coupled with an electrochemical gradient of protons generated by the electron transfer chain. The enzyme can also hydrolyze ATP and form an electrochemical gradient of protons. The F_1 ATPase component was identified many years ago as a coupling factor (1), and the synthase as complex V (2) of oxidative phosphorylation. The synthase has now been well defined both genetically and biochemically. The enzymes from different sources have essentially the same structure, although eukaryotic enzymes have more complicated structures than those of bacteria (Table 1). The catalytic portion of the enzyme, F_1, consists of five subunits, α, β, γ, δ, and ϵ, and can be detached from membranes as a soluble ATPase. The integral membrane sector, F_0, functions as a proton pathway in ATP synthesis and hydrolysis. F_0 becomes a passive proton pathway when F_1 is removed, and *Escherichia coli* F_0 has three subunits, *a*, *b*, and *c*. F_0F_1 and its individual subunits have been purified, and the F_0 and F_1 have been reconstituted. First the *unc* operon of *E. coli* (3, 4) coding for the entire F_0F_1 complex, and later gene clusters for F_0F_1s of various organisms, have been cloned and sequenced, and the primary structures of the subunits of F_0F_1 have been determined (4, 5).

Studies of this enzyme are now at the molecular biological level, and are focused on details of the mechanism of catalysis coupled with proton translocation and the complicated subunit structure. Functional domains or amino acid residues have been studied by analyzing altered kinetics, assem-

Table 1 Subunit structures of H^+–ATPase from *E. coli*, chloroplasts, and mitochondria[a]

E. coli	Chloroplasts	Mitochondria
α	α	α
β	β	β
γ	γ	γ
δ	δ	OSCP
ϵ	ϵ	δ
		ϵ
	subunit II	
a	subunit IV	subunit 6
b	subunit I	subunit 4 (subunit *b*)
c	subunit III	subunit 9
		subunit 8 (A6L, *aap1*)
		ATPase inhibitor protein
		9-kd protein, 15-kd protein
		Factor B, F6, subunit *d*

[a] Equivalent subunits are shown. See text and references for details.

bly, or other properties of the enzyme after its chemical or genetic modification. A combined approach seems best at present, since the crystal structure of F_0F_1 is unknown. Homologous regions in the primary structures of individual subunits from different sources or in other ion-motive ATPases and enzymes having nucleotide-binding sites have also been important in assessing functional domains and amino acid residues. Results on the F_0 sector are of interest for its comparison with other proteins having ion pathways. Furthermore, from information on the similarities of F_0F_1 with other ATPases, it is now possible to discuss the evolution of ATPases.

F_0F_1s from various sources have been studied extensively and results have been reviewed many times. Thus here we review only results on the mechanism and assembly of F_0F_1, with special reference to recent molecular biological studies. For other aspects of the enzyme, the reader should refer to recent excellent review articles (6–18).

FROM GENE CLUSTERS FOR F_0F_1 TO PRIMARY STRUCTURES OF SUBUNITS

Gene Clusters for F_0F_1 and Their Diversity

BACTERIAL GENOMES   *E. coli* mutants defective in F_0F_1 were isolated and mapped in the *unc* operon located at approximately 83 min on the genetic map (3, 4). The *unc* operon was cloned and its entire DNA sequence was determined (4, 5). The order of the structural genes coding for the subunits of F_0F_1 is *uncB (a)*, *uncE (c)*, *uncF (b)*, *uncH (δ)*, *uncA (α)*, *uncG (γ)*, *uncD (β)*, and *uncC (ε)*. Upstream of the *uncB* gene, there is an open reading frame (*uncI*) whose role is still unknown. The functional promoter is located upstream of *uncI*, two weak promoters are present in the *uncI* reading frame, and a terminator is located downstream of *uncC*. Translational regulations seem to determine the amounts of synthesis of subunits giving the stoichiometry of F_0F_1 (4, 19). The gene cluster of thermophilic bacterium PS3, which has exactly the same cistron organization as that of the *unc* operon, was recently sequenced (20).

In *Rhodopseudomonas blastica* (21) and *Rhodospirillum rubrum* (22), genes for subunits of F_1 form an operon with the same gene order as that of the *unc* operon, except that in *R. blastica*, there is an unknown reading frame between the genes for the γ and β subunits. In cyanobacteria *Synechococcus* 6301 (23) and *Anabaena* sp. (24, 25), F_0F_1 is encoded by two gene clusters: *atp1* consists of gene I (related to *E. coli uncI*), *a*, *c*, *b*, and *b'* (a diverged form of *b*), and three F_1 genes (α, γ, and δ) in the same order as in the *unc* operon; and *atp2*, which is at least 15 kb from *atp1*, consists of genes for the β and ε subunits in this order. It is very interesting that F_0F_1 is not coded by a single operon in these organisms.

ORGANELLAR GENOMES   Organellar genomes code for some of the subunits of F_0F_1 in eukaryotes. Analogies of organellar and *E. coli* subunits were found by comparison of their primary structures, hydropathy plots, and predicted structures (Table 1). As in cyanobacteria, in chloroplasts the genes for the β and ϵ subunits form a single transcriptional unit. In the chloroplast genome, the stop codon of the β subunit gene often overlaps the initiation codon of the ϵ subunit gene (26–30). In chloroplast genes the IV (*a*), III (*c*), I (*b*), and α subunits form a single gene cluster in the same order as in *E. coli* or cyanobacteria, although the gene for subunit I has an intron (31–36). The gene (*rps2*) for a protein analogous to the S2 protein of the *E. coli* 30 S ribosome was found upstream of the subunit IV gene.

Yeast mitochondrial DNA has genes for subunits 9 (*oli1* gene) (*c*), 8 (*aap1*), and 6 (*oli2*) (*a*) (Table 1) (37), while mammalian mitochondrial DNA contains genes for only subunits 6 and A6L (subunit 8) (38–40), and *Neurospora crassa* mitochondrial DNA has a pseudogene for the subunit 9 (41). Four DNA segments carrying the α subunit coding sequences were located at homologous recombination sites of the plant mitochondrial genome: *Oenothera* has one functional gene (42), whereas the pea (43) and maize (44) have two functional genes. In pea mitochondria, the gene for subunit 9 is located about 2 kb upstream of the α subunit gene and is transcribed in the opposite direction to the latter (45). Interestingly, mitochondria of *Petunia hybrida* have two different transcriptionally active genes for subunit 9 (46, 47). The diverse gene organizations are of interest with regard to the evolution of mitochondria and chloroplasts.

NUCLEAR GENOMES   Complementary DNAs of nuclear genes have been cloned and sequenced for mitochondrial β subunits of human (48), rat (49), bovine (50), and yeast (51), the α subunit of yeast (52), the γ subunit (53) of chloroplasts, and the *c* subunits of *N. crassa* (54), bovine (55), and human (56), the bovine subunits *b* and *d* (57), and the yeast subunit 4 (58). Nuclear genomic structures were determined for the β subunits of *Nicotiana plumbaginifolia* (59) and human (60); the lengths and numbers of introns and exons in the two genes were found to be different. The bovine and human genomes have two sets of genes for the *c* subunit, which are expressed tissue specifically (55, 56). Interestingly, the fusion protein constructed from the targeting sequence of the subunit 9 of *N. crassa* and a reading frame for yeast subunit 8 (coded by the mitochondrial genome) could be transported into yeast mitochondria, and subunit 8 cleaved in the matrix could form active F_0 (61). This finding provided a way to introduce directed-mutagenesis in subunit 8.

The regulation of the coordinated expressions of nuclear and organellar genomes for synthesis of the F_0F_1 complex is interesting. Molecular biological approaches can now be applied to different F_0F_1 ATPases, since cloned DNA segments are available.

Primary Structures of Subunits of F_0F_1

The primary structures of subunits of F_0F_1 from various sources were determined by DNA and protein sequencing. The subunits of F_1 have similar polarities to those of soluble proteins. Of the subunits, β subunits have the most conserved primary structures, and 42.0% of the amino acid residues are identical in the 20 β subunits sequenced (20–23, 25–33, 48–51, 59, 62–65). The β subunit has limited homology with the subunits of ATPases from *Sulfolobus acidocaldarius* (66) and carrot vacuoles (67). The *S. acidocaldarius* ATPase seems to be similar to the F_0F_1 type (68), although *Halobacterium halobium,* another archaebacterium, may have a different type of ATP synthase (69). The amino acid sequences of the 16 α subunits so far obtained (20–24, 30, 32, 33, 43, 52, 70–74) show 30.0% identity when aligned to obtain maximal homology. Three regions (residues 148–186, 257–303, and 322–378 of the EF$_1$ α subunit) are conserved to an even higher degree, suggesting that they are important for catalysis or subunit-subunit interaction(s) (75). The γ subunits (EF$_1$ γ, 286 residues) so far sequenced (20–24, 53, 63, 71, 76) have only 33 identical residues. The carboxyl-terminal regions of the γ subunits are highly conserved in eight species and seem to be essential for catalysis and assembly. The γ subunit of chloroplast (53) has the highest homology (55%) with that of cyanobacteria (23, 24) and lower homology with that of *E. coli* (33%) (70, 76), supporting the idea of a close evolutionary relationship between chloroplasts and cyanobacteria. Close similarity of the F_0F_1s of chloroplasts and cyanobacteria is also found in their gene organization and in the homologies of other subunits. The γ subunits of chloroplasts and cyanobacteria have an extra-domain (about 40 amino acid residues), which is not present in the γ subunits from other species. The δ subunits show limited homologies, mainly in their carboxyl-terminal regions, and the ϵ subunits show homologies in their amino-terminal regions (20–23, 25–28, 31, 32, 63, 65, 71). The entire sequences of the δ subunit of bacteria and bovine OSCP show considerable homology, suggesting that these two proteins have the same function (77, 78). OSCP is also homologous to the *b* subunit of *E. coli* (78). The bovine δ subunit corresponds to the bacterial ϵ subunit (77).

ATPase inhibitor proteins have been reviewed (79). The primary structures of the yeast (80, 81) and bovine (82) ATPase inhibitor proteins have been determined. This protein binds directly to F_1 as well as to F_0F_1, and this interaction is stabilized by two proteins (9 kd and 15 kd) found in yeast mitochondria (83, 84). Interestingly, these two proteins have significant homologies with the ATPase inhibitor (85). The ϵ subunit of EF$_1$ also has limited homology with the inhibitor protein (85). Consistent with this homology, the ϵ subunit has intrinsic inhibitory activity (86) for F_1 ATPase and alters the conformation of the β subunit (87).

The homologies of the *a* subunits (subunit 6) of different species are low, although their hydropathy patterns are similar (4, 5). Subunit a has conserved polar residues in the carboxyl-terminal region. The *E. coli* subunit b is a hydrophilic protein with a polar residue content of 48%, but it has a hydrophobic stretch of 23 residues near its amino-terminus. This subunit has homology with subunit I of spinach or pea chloroplasts (19% identity) (33) and *Synechococcus b* (27% identity) (23). The bovine subunit b (57) and yeast subunit 4 (58) correspond to the *E. coli* subunit b. Subunit c (subunit 9 or III) is also called proteolipid or DCCD-binding protein, as it is extremely hydrophobic. DCCD binds to a Glu or Asp residue located in the center of the second hydrophobic stretch from the amino-terminus (4, 88). Two hydrophobic stretches of the subunit, possibly α helices, traverse the membrane in an antiparallel manner like a hairpin, and a central polar region is exposed to the mitochondrial matrix or bacterial cytoplasm. Recently, the proteolipid subunit of chromaffin granule H^+-ATPase was sequenced (89). Its molecular weight is twice that of the c subunit. Of the two halves, the amino-terminal half of the protein matches the c subunits of *E. coli*, chloroplasts, and yeast better, whereas the carboxyl half matches bovine and *N. crassa* subunits better. These findings suggest that the proteolipid and the c subunit evolved from the same ancestral gene.

Mitochondrial H^+-ATPase has other subunits such as yeast subunit 8, which is the same as mammalian A6L (40, 90), subunit d (57), Factor B (91), and Factor 6 (92) (Table 1). Subunit 8 is a highly hydrophobic F_0 subunit spanning the membrane only once with the carboxyl-terminus exposed to the matrix (93). Subunit II of chloroplast F_0 is coded by the nuclear genome and has no known counterpart in *E. coli* (34).

SUBUNIT-SUBUNIT INTERACTIONS AND ASSEMBLY OF F_0F_1

Higher-Ordered Structure of F_1

F_0F_1 can be solubilized and purified in the presence of detergents, while F_1 can be solubilized and purified without detergent (4). The subunit stoichiometry of *E. coli* F_0F_1 has been established as $\alpha_3\beta_3\gamma\delta\epsilon ab_2c_{6-10}$ (4). Crystals of F_1 have been obtained, and show a pseudohexagonal structure (4, 5, 94), but no information is available on the positions of amino acid residues.

Models of F_1 attached to F_0 through the δ and ϵ subunits have been proposed based on the interactions of the subunits and the reconstitution experiments discussed below (4, 5). Capaldi and coworkers recently proposed a similar model from analysis of negatively stained F_0F_1 from *E. coli* (95): a sphere of about 9 nm attached to the bilayer through a stalk (4.5 nm). Consistent with this model, a photoaffinity phospholipid analogue labeled the α and β subunits of *E. coli* membranes poorly, and the a and b subunits strongly (96). The close interaction of subunits α and δ was suggested by disulfide cross-linkage between these subunits (97). The δ subunit in F_1 was heavily labeled by hydrophobic and water-soluble aldehydes, but the same subunit in the F_0F_1 preparation was not labeled, possibly because the reactive residue(s) in the δ subunit was shielded by F_0 subunits (96). However, the δ subunit may not actually form a stalk, because in the absence of OSCP (δ), mitochondrial F_1 shows specific and saturable binding to submitochondrial particles depleted of F_1 (98), and OSCP increases the binding affinity of F_1 to F_0 (99). OSCP binds to F_1 in a molar ratio of 1 : 1 and cross-links to the α and β subunits with zero-length cross-linkers, suggesting that it binds to the interfaces of the α and β subunits (100). Similarly, CF_1 without the δ subunit could bind to chloroplasts depleted of CF_1, but the photophosphorylation activity of the reconstituted membranes was lower than that of those reconstituted with complete CF_1 (101). These results suggest that δ or OSCP stabilizes the linkage between F_1 and F_0.

A model stressing the interaction of the β subunit with the F_0 subunit has also been presented based on the results of cross-linking experiments (102). The α subunit of PS3 F_0F_1 in liposomes was labeled with a radioactive photoaffinity phospholipid analogue (103), indicating that part of the α subunit is close to the phospholipid bilayer. This finding differs from that with *E. coli* membranes (96), possibly due to a difference in reactivity of amino acid residues in the two bacteria. The actual higher-ordered structure of F_0F_1 may have the features of both the models discussed above. In most models of F_1, three α and three β subunits are arranged alternately (4, 5). This arrangement is supported by α-β cross-linking (104) and electron microscopic study of F_1 bound to monoclonal antibody specific for the α subunit (105). However, β-β (106) or α-α (104) cross-linkings were also observed, supporting a model of F_1 in which a triangular set of three α subunits is located above a triangular set of three β subunits in a staggered conformation (107).

Subunit-Subunit Interactions in F_1

Complexes of α, β, and γ subunits reconstituted from isolated individual subunits of *E. coli* (108, 109), PS3 (110), and *Salmonella typhimurium* (111) showed ATPase activity. It is noteworthy that the γ subunit is necessary for reconstitution of a complex with ATPase activity, although the catalytic site is on the β subunit or at the interface of the α and β subunits (4). Mixtures of the α and β subunits of *E. coli* and the γ subunit of PS3 or of the α and β subunits of PS3 and *E. coli* γ also formed complexes with ATPase activity (112), suggesting that homologous regions of the γ subunits of these two bacteria (38% identical residues) are sufficient for forming the $\alpha\beta\gamma$ complex. Similarly, the β subunit of *E. coli* (113) or chloroplasts (114) could replace the same subunit of *R. rubrum* chromatophores for ATP hydrolysis and synthesis, and the genes for the α and β subunits of *Bacillus megaterium* complemented *E. coli* mutants (115). Consistent with the results of reconstitution experiments, genetic loss of the EF_1 γ subunit resulted in loss of assembly of F_1 on membranes or the cytoplasm (116–118). Studies on the properties of membranes of mutants defective in the γ subunit suggested the importance of the carboxyl-terminal (Gln-269—carboxyl-terminus) and amino-terminal (Lys-21—Ala-27) regions of the subunit for formation of a stable F_1 complex with ATPase activity (116, 117). However, the γ subunit lacking four residues (Ala-Ala-Ala-Val) from the carboxyl-terminus could form active F_0F_1, suggesting that these residues are dispensable (119), although two of these Ala residues (underlined above) are conserved in different species.

Single amino acid substitutions in the α or β subunit of *E. coli* often resulted in loss of the normal F_0F_1 complex in membranes. The *uncD* alleles of such mutants were cloned and their DNA sequences were determined (120): βGlu-41$\rightarrow$ Lys, βGlu-185 $\rightarrow$ Lys, βGly-223 $\rightarrow$ Asp, βSer-292 $\rightarrow$ Phe, βGln-361 $\rightarrow$ end, and βGln-397 $\rightarrow$ end mutations were identified (Table 2). In the above missense mutants, the α and β subunits were partially assembled and/or no minor subunits were found in the membranes, suggesting that altered residues or the regions in their vicinities are essential for subunit-subunit interaction(s). The importance of βGlu-185 was further supported by studies on isolated β subunits with βGlu-185 $\rightarrow$ Gln or Lys mutations (121): only trace amounts of $\alpha\beta\gamma$ complexes could be reconstituted from either of these two mutant β subunits and wild-type α and γ subunits, although the wild-type and mutant subunits had similar K_d values for ATP and aurovertin. Three mutations in the α subunits, αGly-29 $\rightarrow$ Asp (122), αGlu-299 $\rightarrow$ Lys (123), and αAla-285 $\rightarrow$ Val (75), decreased the amount of membrane-bound F_1 and made the membranes permeable to protons, suggesting that these residues may be in the region(s) of the subunit that is important for the interaction between F_1 and F_0.

The $\alpha\beta\gamma$ complex of *E. coli* cannot bind to F_0 unless the δ and ϵ subunits are added (109). Supporting this finding, mutants of the Shine-Dalgarno sequence (SD^-) of the *uncC* gene for ϵ and a nonsense mutant (ϵGln-72 $\rightarrow$ end) (124) had F_1-ATPase activity only in the cytoplasmic fraction, and naturally could not grow by oxidative phosphorylation. Analyses of the SD^- mutant with different recombinant plasmids carrying truncated *uncC* genes indicated that the amino-terminal half (78–80 residues) of the ϵ subunit was sufficient for forming active F_1 capable of binding to F_0. Consistent with the conclusion that the amino-terminal region of the ϵ subunit is important, membranes of a mutant (ϵGly-48 $\rightarrow$ Asp) showed no activity for ATP-dependent H^+ translocation (125). Interestingly, pseudorevertants of the mutant had an ϵPro-47 $\rightarrow$ Ser or $\rightarrow$ Thr replacement, suggesting that the peptide oxygen between ϵPro-47 and ϵGly-48 is structurally involved in

Table 2 Important amino acid residues and regions in F$_1$-ATPase[a]

Subunit	Amino acid residues and regions	Comment	Refs.
α	Gly-29, Ala-285, Glu-299, Ser-347, Gly-502,	assembly mutation	75, 122, 123
	Lys-175, Asp-261, Pro-281, Gly-351, Ser-373, Ser-375, Arg-376	catalysis mutation	75, 123, 226, 242
	Lys-201	AP$_3$PL, AP$_2$PL	236, 244
β	Glu-41, Glu-185, Gly-214, Gly-223, Ser-292	assembly mutation	120, 121, 229
	Gly-142, Gly-149, Ala-151, Gly-154, Lys-155, Ser-174, Met-209, Asp-242, Arg-246, Thr-285	catalysis mutation	195, 208, 229, 230, 238, 240, 241, 242,
	Lys-155	AP$_3$PL	236
	Lys-155, Glu-192	cross-linking	106
	Lys-155, Tyr-297	Nbf-Cl	235, 252
	Glu-181, Glu-192	DCCD	246–248, 250
	Arg-281	phenylglyoxal	265, 266
	Thr-287(Lys), Ile-290, Tyr-297	8-azido-ATP	245
	Tyr-331, Tyr-354	2-azido-ATP	187, 190, 191
	Leu-328, Ile-330, Tyr-331, Pro-332	2-azido-ADP	267
	Tyr-331, Asp-338	BzATP	189
	Tyr-331	FSBI	188
	Tyr-354, Ser-413(His)	FSBA	251, 268
γ	Lys-21 $\sim$ Ala-27	assembly mutation	116
	Gln-269 $\sim$ carboxyl-terminus	catalysis mutation	117
ϵ	Gly-48	H$^+$-translocation	125
	amino-terminus $\sim$ Val-78	functional region	124

[a] Available information from *E. coli* mutant studies and chemical modifications is summarized. Results for chloroplasts and mitochondria are included using the *E. coli* residue numbering system. Lys residue of mitochondria modified by 8-azido-ATP is not conserved in *E. coli* and corresponds to Thr-287. Similarly His residue of mitochondria modified by FSBA corresponds to Ser-413 of *E. coli*. See text and references for details of the roles of amino acid residues.

intersubunit hydrogen bonding or more directly in H$^+$ movement through F$_1$. It is notable in this regard that CF$_1$ lacking ϵ binds to F$_0$ but does not block passive H$^+$ conduction (126). The EF$_1$ ϵ subunit binds tightly to the γ subunit (127), and is also cross-linked to the carboxyl-terminal region of the β subunit (128).

Topology of F$_0$ Subunits

SUBUNIT a (IV or 6) Genetic studies (129) and reconstitution experiments (130) indicated that three subunits (a, b, and c) are required for a functional EF$_0$. The F$_0$ sector was reviewed recently (131). Five to seven hydrophobic segments, possibly α helixes, which span the membrane, were predicted for the subunit a. The portions connecting these helixes may also be embedded in the membrane to some extent, because they are not accessible to proteolytic enzymes (132). Studies on mutants suggested that at least some of the F_1-binding sites could be formed without the region between residue 111 and the carboxyl-terminus of the a subunit (133).

SUBUNIT b (I or 4)   The hydrophobic amino-terminal region of the *E. coli b* subunit may be embedded in the membrane as a helical domain, and the two hydrophilic helical domains may be located in the cytoplasm (4, 131). Consistent with this model, treatment with proteases had no effect on the H^+ permeability of the F_0 portion, but abolished its functional binding of F_1 (132). The hydrophobic photoreactive reagent ^{125}I-TID labeled amino-terminal residues (from Leu-3 to Trp-26) (134). A photoreactive phospholipid analogue, 1-palmitoyl-2-(2-azido-4-nitro)benzoyl-*sn*-glycerol-3-(^3H)phos-phocholine, labeled bAsn-2, bCys-21, and bTrp-26 of the b subunit almost exclusively. These results strongly suggest that the amino-terminal region traverses the lipid bilayer and that bCys-21 may be close to the surface of the bilayer. Studies on pseudorevertants of a nonsense mutant (bTrp-26 → end) showed that position 26 could be replaced by acidic or basic amino acid residues (135). Membranes of the revertants showed impaired H^+ permeability on removal of F_1, suggesting that bTrp-26, located near the surface of the lipid bilayer, stabilizes the H^+ pathway.

Consistent with the intramembrane location of the amino-terminal region of

Table 3   Mutations in *E. coli* F_0^a

Subunit	Number of residues	Mutation (missense or nonsense)	Phenotype
a	271	Pro-143, Ser-206, Arg-210, Glu-219, His-245, Gln-252 → end	Defective H^+-translocation
		Pro-240	Suppression of bGly-9 mutation
b	156	Gly-131, Glu-155 → end	Defective H^+ translocation No F_1 binding
		Gly-9	Defective H^+ translocation
c	79	Ala-21, Ala-25, Asp-61, Pro-64	Defective H^+ translocation
		Gly-23, Leu-31, Gly-58	Loss of c subunit (Defective H^+ translocation, No F_1 binding)
		Gln-42	Impaired catalysis
		Ala-62	Suppression of bGly-9 mutation

[a] Important mutations are summarized. Amino acid residues shown are those replaced in mutations. Although many mutations show defective H^+ translocation, only a few residues (aSer-206, aArg-210, aGlu-219, aHis-245, cAsp-61) may actually participate in the H^+ pathway. See text and references for detailed discussion.

the b subunit, the mutation "bGly-9 → Asp" blocked H$^+$ permeability without causing significant loss of F$_1$ binding to F$_0$ (136). This mutation is suggested to affect the interaction with other F$_0$ subunits (136, 137). The mutation bGly-131 → Asp impaired H$^+$ permeability and F$_1$-binding activity (136, 138). The bGly-9 → Asp mutant was complemented with a plasmid carrying the mutant gene with bGly-131 → Asp substitution (138), confirming the requirement of two copies of the b subunit in F$_0$F$_1$ (4, 131). Recent analyses of strains carrying b subunits of various lengths suggested that the entire carboxyl-terminal region is necessary for formation of a functional F$_0$ (139): loss of a carboxyl-terminal residue resulted in significant reduction of F$_1$-binding and H$^+$-translocation, and loss of two residues abolished both activities almost completely. Consistent with these findings, the carboxyl-terminal region of the subunit b was shown to be necessary for proper reconstitution of F$_0$ in liposomes (140).

SUBUNIT c (9 or III)   As a membrane protein, the conformation of the c subunit is significantly influenced by phospholipids, and antibodies specific for the c subunit only embedded in liposomes have been identified (141). Consistent with its hydrophobicity and high copy number in F$_0$F$_1$, the subunit 9 or III, could form a supramolecular structure of about 12 copies even in the presence of a detergent (142). The oligomeric structure of the subunit appeared to be rigid, because little change in the labeling pattern of the subunit with ^{125}I-TID was observed upon energization or addition of inhibitors (143). Combined genetic and chemical modification studies support a hairpin model for the c subunit: two helixes (residues 15–40 and 51–76 of the EF$_0$ subunit) traverse the membrane in an antiparallel fashion, and a central hydrophilic sequence (residues 41–50) is exposed and possibly forms the F$_1$-binding region (4, 131). Mutations of cIle-28 → Val or Thr rendered cAsp-61 inaccessible to DCCD (144), and substitution of Glu-42 for Gln impaired coupling between F$_0$ and F$_1$ (145). The substitutions cAsp-61 → Gly or Asn (134, 146), cAla-21 → Val (134), cAla-25 → Thr (147), and cPro-64 → Leu (148) blocked H$^+$ permeability. These latter three residues may be important for maintaining the structure of the c subunit or for its interaction with the b and a subunits. The mutation cPro-64 → Leu was suppressed by a second mutation cAla-20 → Pro (148), suggesting that a Pro residue may be required in either of the two helixes embedded in the membrane. The three mutations cGly-23 → Asp (149), cGly-58 → Asp (150), and cLeu-31 → Phe (149) resulted in loss of the c subunit from membranes. Subunit 9 has modulatory effects on the assembly of respiratory complexes in yeast mitochondria, as mutants that failed to integrate subunit 9 into F$_0$F$_1$ showed a pleiotropic effect with marked reduction in the syntheses of respiratory complexes, notably cytochrome oxidase (151). In this regard it is noteworthy that a supramolecular complex structure formed from F_0F_1 and respiratory components has been proposed for bovine mitochondria (152).

F_0 SUBUNIT INTERACTIONS   The close interactions of the three EF_0 subunits were shown by cross-linking and genetic experiments. Hydrophobic cross-linking reagents have been shown to associate with the a and b subunits (102, 153). The a subunit also appears to associate strongly with the c subunit, since an a-c complex has been isolated (154). Furthermore, labeling of liposomes in which various F_0 subunits were embedded with a photoactivatable carbene-generating reagent has suggested that the conformation of the c subunit in F_0 is mainly determined by its interactions with the a subunit (155).

Intramembranous interactions between b and a or c subunits were shown by suppression of the mutation in the b subunit (bGly-9 → Asp) with that of the a (aPro-240 → Ala or Leu) (156) or c (cAla-62 → Ser) subunits (157). Close interaction(s) between F_0 subunits was also suggested by analyses of oligomycin-resistant mutants of yeast and $N.$ $crassa$ (88, 134). Radiochemical labeling studies have suggested that the primary binding site for oligomycin lies within subunit 9 (158, 159). However, oligomycin-resistant mutations were mapped at amino acid residues of both subunit 9 (88, 160) and 6 (88, 161), suggesting that these two subunits interact directly and form an oligomycin-binding site.

Assembly of F_0F_1

As described above, EF_1 (109) and EF_0 (130) could be reconstituted from their isolated subunits. For study of the assembly of F_0F_1 in vivo, plasmids carrying parts of the unc operon were introduced into a strain in which the unc operon was deleted. Consistent with the results of reconstitution experiments, subunits of F_1 could be synthesized and assembled into a complex in the absence of F_0 subunits (162), and similarly active F_0 could be formed from the a, b, and c subunits without F_1 subunits (163, 164). Thus the F_1 and F_0 sectors can be assembled independently, at least when the corresponding subunits are coded by recombinant plasmids. Recently, we obtained a mutant showing temperature-sensitive initiation of translation of the $uncG$ gene for the γ subunit (118). When this strain was grown at 42°C, its membranes contained about 2% of the amount of F_1-ATPase, and about 40% of the amount of F_0 of wild-type membranes. Thus F_0 could be assembled in membranes without F_1 subunits at 42°C, supporting results with the plasmid-coded system.

On the other hand, membranes of the polarity mutant of the $uncD$ gene for the β subunit did not have a functional F_0 or the α, γ, and δ subunits of F_1 (165), although the mutation affected only the two most distal genes ($uncD$ and $uncC$) to the promoter. From analyses of similar mutants, Cox et al (165) suggested that the α and β subunits were essential for insertion of the b

subunits into membranes and thus for assembly of F$_0$. Their model may be applicable to the *unc* operon coded by the *E. coli* chromosome, but appropriate explanations must be made for assembly mutants that cannot be included in their model.

CATALYTIC MECHANISM AND AMINO ACID RESIDUES

Nucleotide-Binding Sites

The catalytic mechanism can be studied using purified F$_1$-ATPase. Six nucleotide-binding sites were found in bovine F$_1$ (166, 167), and classified as three catalytic (exchangeable) and three noncatalytic (nonexchangeable) sites according to their abilities to exchange bound nucleotide rapidly during hydrolysis of Mg-ATP. Once filled, the three noncatalytic sites do not exchange nucleotide with that in the medium, at least during several hundred catalytic turnovers (168). Based on similar criteria, EF$_1$ also has three noncatalytic and three catalytic nucleotide-binding sites (169, 170). When bovine F$_1$ that had been stored in buffer containing ammonium sulfate was precipitated, rinsed, and desalted, the enzyme retained three bound adenine nucleotides, two at noncatalytic sites and one at a catalytic site (168). The vacant noncatalytic site is highly specific for adenine nucleotide (171). Only this noncatalytic site bound and released the ligand when F$_1$ was incubated in the presence of EDTA at pH 8, suggesting that the three noncatalytic sites have asymmetric properties. During the half-time for dissociation of the nucleotide from the noncatalytic site, the catalytic site turned over approximately 10^5 times. The rat liver enzyme has been reported to have four nucleotide-binding sites, one of which is specific for ADP (172). Possibly two additional sites were not detected. The roles of the noncatalytic sites are currently unknown. These sites may play structural roles and may be essential for proper assembly of subunits, as previously suggested (109). CF$_1$ also has noncatalytic and catalytic sites (173, 174), and analysis with 2-azido ATP suggested the presence of more than three, possibly six, nucleotide-binding sites (175), although a widely held view is that there are only three binding sites (173, 174).

The locations of the nucleotide-binding sites have been studied with isolated α and β subunits and photoreactive nucleotide analogues. Purified α (109) and β (176–178) subunits of *E. coli* or PS3 (179) have nucleotide-binding sites: the α and β subunits have higher-affinity and lower-affinity sites, respectively. Analysis of the binding of trinitrophenyl ATP suggested that the site in the β subunit has to be filled for assembly of this subunit with other subunits (178). The isolated α subunit shows a large conformational change upon binding of nucleotide, detected by changes of physical parameters (180) and sensitivity to trypsin (181). The dissociation rate constant of the α-ATP complex is about 10^6-fold less than that of most protein-ligand complexes (180). These results suggest that the properties of the binding site in the α subunit may change upon assembly, if the site forms at least part of the catalytic site. The presence of a nucleotide-binding site at the interface between the α and β subunits has been suggested (182–184). The presence of one (185) or two (186) nucleotide-binding sites and ATPase activity (185) in the isolated β subunit of *R. rubrum* have also been reported.

Using the same procedure to load the catalytic and noncatalytic sites with ADP or ATP, Cross et al found that βTyr-368 (noncatalytic) and βTyr-345 (catalytic) of the bovine β subunit were labeled with 2-azido-ATP (187). βTyr-345 was also labeled with FSBI (188) and BzATP (189). The 2-azido-ATP–binding catalytic and noncatalytic sites of other F_1 were determined: namely, βTyr-331 and βTyr-354, respectively, in *E. coli* (190), and βTyr 362 and βTyr-385, respectively, in chloroplast (191). Since the two sites can be occupied simultaneously, it is unlikely that the two residues were in a single site and that the different labelings reflected two conformational states of the β subunit. Another group of workers also detected two 2-azido-ADP–binding sites in the mitochondrial β subunit (192), but they found that two peptides, Pro-320—Met-358 and Gly-72—Arg-83, were labeled. These results suggest that at least parts of both nucleotide-binding sites are located in the β subunit, and that one type of site may be shared between the α and β subunits. The presence of a Pi-binding site in the CF_1 β subunit was suggested from studies with 4-azido-2-nitrophenyl phosphate (193).

Mechanism of ATP Hydrolysis by F_1-ATPase

No covalent phospho-enzyme intermediate is involved in hydrolysis of ATP by F_1-ATPase, as a direct in-line displacement reaction occurred yielding ADP and Pi from ATP and H_2O (194). In this regard, F_1-ATPase is different from other ion-motive ATPases, such as Na^+/K^+-ATPase and Ca^{2+}-ATPase, which have aspartyl phosphate intermediates. Consistent with these findings, attempts to detect such an intermediate with a mutant (β-Thr-285 → Asp to give a sequence like Na^+/K^+-ATPase) F_1-ATPase under similar conditions to those described for ion-motive ATPase were unsuccessful (195). Boyer and colleagues showed that F_1 catalyzes exchange between oxygen of water and oxygen of phosphate during steady-state hydrolysis of ATP at low concentration (196). However, this exchange was not observed with a high concentration of ATP. These results suggest that ATP hydrolysis and synthesis at the catalytic site are reversible at low ATP concentration, but that at high ATP concentration the reaction proceeds only in the direction of ATP hydrolysis, possibly due to the positive catalytic cooperatively. Consistent with this idea, synthesis of the enzyme-bound ATP was shown with purified F$_1$ (197–199). An oxygen exchange experiment suggested that ATP binding at one catalytic site increases the rate of release of Pi and ADP from another catalytic site at least 2×10^4-fold (200).

Penefsky and coworkers analyzed the hydrolysis of ATP during single site (uni-site) and steady-state (multi-site) catalyses, and suggested that the equilibrium between ATP and ADP + Pi at the high-affinity catalytic site during uni-site hydrolysis is close to unity, and can occur without energy input (201). Upon addition of excess ATP, the rate of release of Pi from the uni-site of bovine F$_1$ is accelerated and the rate of release of Pi in multi-site hydrolysis is about 10^6-fold that in uni-site hydrolysis (202). Kinetic evidence indicated that the dissociation of Pi from labeled ATP at the catalytic site is actually accelerated by the binding of a single ATP to the other catalytic site(s). (203). From kinetic analysis of normal and chemically modified F$_1$, the high-affinity site was proposed not to be a normal catalytic site (204, 205). However, Penefsky recently showed that hydrolysis of ATP bound to the high-afinity site during uni-site hydrolysis preceded the hydrolysis of ATP bound to the additional catalytic sites, supporting the idea of alternation between the catalytic sites (206). A similar mechanism was demonstrated for EF$_1$ (207, 208). Positive cooperativity of ATP hydrolysis has also been shown using fluorescent nucleotide analogues (209, 210). Sodium azide inhibited multi-site hydrolysis of EF$_1$ more than 90%, but affected the rate of uni-site hydrolysis only slightly (211), suggesting that this inhibitor blocks catalytic cooperativity. Similarly, DCCD causes more inhibition of multi-site hydrolysis of ATP by F$_1$ (212), and inhibits ATP hydrolysis more than ATP synthesis in submitochondrial particles (213).

Most of these results support the binding change mechanism (196) for ATP synthesis. According to this mechanism, tight-binding of ATP occurs at the catalytic site, and the electrochemical gradient of protons decreases the affinity for ATP and increases the affinities for ADP and Pi. Cooperativity between the three (or two) catalytic sites is essential for the mechanism. Experiments with submitochondrial particles support this conclusion: an electrochemical gradient of protons can release tightly bound ATP from the catalytic site at a similar rate to ATP synthesis during oxidative phosphorylation (214), and positive cooperativity of ATP synthesis and negative cooperativity of ADP binding have been shown (215). Positive cooperativity of photophosphorylation has also been shown, with the results explained by the sequential participations of at least two catalytic sites (216).

Considerations of the cooperative catalysis of F$_1$ carried by the β subunits or interfaces of the α and β subunits (possibly three copies) led to the suggestion of a catalytic mechanism involving rotation of a core of minor subunits (γ, δ, ϵ) and change in the position of this core relative to the different β subunits or the interfaces of the α and β subunits (217–219). Attempts have been made to test the rotation model by cross-linking the γ subunit, or another minor subunit with the α and/or β subunit (220, 221), but two opposing conclusions have been obtained. We think that it is difficult to draw conclusions unless rotation is clearly defined by some physical means. The requirement of three normal β subunits for catalysis was suggested by analyzing hybrid $\alpha\beta\gamma$ complexes reconstituted from mixtures of the mutant and wild-type β subunits (208, 222). We have discussed a model of three alternating β subunits during steady-state catalysis, as it seems to be more convincing than other models. However, other models for catalysis could not be completely excluded. Wang and coworkers suggested that three β subunits are permanently nonequivalent and proposed a model with one β subunit for steady-state catalysis and two for regulatory roles (223, 224). They also proposed the occasional switching of the roles of individual β subunits. Gautheron and coworkers suggested a model with two catalytic sites and a single regulatory site (225).

Essential Amino Acid Residues for Catalysis by the F_1 Sector

RESIDUES FOUND BY RANDOM MUTAGENESIS   The roles of the five subunits and their amino acid residues in catalysis have been studied by using F_0F_1 with mutationally (random and directed mutagenesis) altered subunits or wild-type F_0F_1 that is chemically modified or labeled with reactive nucleotide analogues (Table 2). The F_1s from αSer-373 $\rightarrow$ Phe (226) and αArg-376 $\rightarrow$ Cys (75) mutants of the α subunit showed 10^{-3}-fold and 10^{-4}-fold lower rates, respectively, of multi-site hydrolysis of ATP than wild-type F_1, but similar rates of uni-site hydrolysis to wild-type F_1. Similar results were obtained with αGly-351 $\rightarrow$ Asp and αSer-375 $\rightarrow$ Phe mutants (123). These results suggest that the region around residues 350–380, which is highly conserved, is essential for the conformational transmission leading to positive catalytic cooperativity. Altered conformational interactions between the α and β subunits of the αPhe-373 mutant were suggested (16, 227, 228).

The F_1s from βArg-246 $\rightarrow$ His (208) or Cys (207, 229) and βMet-209 $\rightarrow$ Ile (207, 229) mutants showed less than 1% of the rate of Pi release in the multi-site ATPase activity of the wild-type enzyme, while their rates of Pi release in uni-site catalysis were more than 10 times that of wild-type F_1. These results suggest that βArg-246 and βMet-209 or the regions in their vicinities may be important in catalytic cooperativity and also in Pi binding. The F_1 from the mutant βSer-174 $\rightarrow$ Phe had about 10% of the multi-site Mg^{2+}-dependent ATPase activity of the wild-type and showed an abnormal ratio of Ca^{2+}- to Mg^{2+}-dependent ATPase activity: this ratio was about 0.8 for the wild-type F_1, whereas it was more than 3 for the mutants (230). Thus βSer-174 or its vicinity may also be closely related to catalysis.

CONSERVED SEQUENCES   The sequence G-X-X-X-X-G-K-T/S is found in the β subunits, adenylate kinase, p21 *ras* protein, and other nucleotide-binding proteins (231). In crystalline adenylate kinase, residues 15–23 (G-G-P-G-S-G-K-G-T) form a flexible loop structure between an α-helix and β-sheet, and Lys-21 (K) in the loop may be close to the α-phosphate (232) or γ-phosphate (233) of ATP. The conserved sequence corresponds to residues 149–156 (G-G-A-G-V-G-K-T) of the EF$_1$ β subunit. Models of the catalytic site in the β subunit have been proposed based on the crystalline structure of adenylate kinase (49, 234). Corresponding Lys residues of bovine and *E. coli* F$_1$ were modified with Nbf-Cl (235) and AP$_3$-PL (236), respectively, suggesting that the Lys is located near the γ phosphoryl group of ATP. A similar conclusion was reached on the basis of energy transfer experiments between a fluorescent ATP analogue and NBD attached to the corresponding Lys of bovine F$_1$ (237). The EF$_1$ of the mutant βAla-151 $\rightarrow$ Val isolated by random mutagenesis showed much faster uni-site hydrolysis of ATP than the wild-type and had about 6% of the wild-type multi-site activity (238). It is interesting that normal and oncogenic p21 *ras* proteins have Gly and Val residues, respectively, at the position corresponding to Ala-151 of the β subunit (239). Like the mutant F$_1$, the oncogenic p21 *ras* protein has lower GTPase activity (about 10%) than normal.

This region has been studied by site-directed mutagenesis: the rates of multi-site ATP hydrolysis by EF$_1$ with βGln-155 and βGlu-155 were 20 and 34%, respectively, of that of the wild-type (βLys-155) (240, 241). However, similar rates of release of Pi in uni-site catalysis were observed with EF$_1$ with βLys-155, βGln-155, and βGlu-155, although the binding affinities for ATP of the mutant F$_1$s were considerably lower than that of the wild-type. These results suggest that βLys-155 is not directly involved in the cleavage or formation of a phospho-diester bond (241), but is located close to the catalytic site. PS3 F$_1$ lost multi-site ATPase activity completely on replacement of the corresponding Lys residue by Ile, although the mutant subunit bound ATP (242). Garboczi et al recently synthesized a 50-amino-acid peptide (corresponding to residues Asp-141—Thr-190 of the rat liver β subunit) including the conserved sequence, and found that it was precipitated with ATP and enhanced the fluorescence of 2'(3')-O-(2,4-6-trinitrophenyl)-adenosine 5'-triphosphate (243).

These results further support the idea that the conserved sequence (G-X-X-X-X-G-K-T) is located close to the catalytic site and undergoes conformational movement essential for the catalysis step. In addition, AP$_3$-PL bound to βLys-155 and αLys-201 of EF$_1$ (236), and AP$_2$-PL bound to αLys-201 of the isolated α subunit (244). Thus βLys-155 residues may be close to αLys-201 or in a region in its vicinity. The α subunit has a similar conserved sequence, and replacement of the corresponding lysine residue of

PS3 F_1 by Ile gave an α subunit that reassembled only weakly with other subunits (242).

A sequence of 10 amino acids (I-C-S-<u>D</u>-K-T-G-T-L-T) of ion-motive ATPases such as Na^+/K^+-ATPase is similar to the sequence of β subunits such as that of *E. coli* (I-T-S-<u>T</u>-K-T-G-S-I-T) (residues 282–291), as pointed out earlier (7). The Asp (<u>D</u>) residue phosphorylated in ion-motive ATPases corresponds to βThr-285 of *E. coli* (<u>T</u>). When a βThr-285 → Asp mutation was introduced into the *E. coli* β subunit (195), the uni- and multi-site ATPase activities of the $\alpha\beta\gamma$ complex with mutant β subunits were about 20 and 30% of those with the wild-type subunit. The rate of ATP binding of the mutant complex under uni-site conditions was about one-tenth of that of the wild-type complex. These results suggest that βThr-285, or the region in its vicinity, is essential for normal catalysis of the H^+-ATPase. Consistent with this notion, 8-azido-ATP bound to residues of bovine F_1 corresponding to βThr-287, βIle-290, and βTyr-297, respectively, of the *E. coli* β subunit (245). The mutant (βAsp-285) complex could not form a phosphoenzyme under conditions in which ion-motive ATPase is phosphorylated, suggesting that the mechanism of F_1-ATPase cannot be changed simply by replacing one amino acid residue (195).

CHEMICAL MODIFICATIONS AND MUTAGENESIS   DCCD reacted with homologous residues of the bovine (246) and *E. coli* (βGlu-192) (247) β subunit, but with a different Glu residue of PS3 F_1 (corresponding to EF_1 β Glu-181) (248). Oligonucleotide-directed mutagenesis was applied to these residues of EF_1 (249). The βGln-181 mutation resulted in strong impairment of ATPase and ATP-driven H^+-pumping activity, although low levels of both activities were detectable. The βGln-192 mutation resulted in partial inhibition of oxidative phosphorylation without affecting H^+ pumping activity, although the membrane ATPase activity was reduced by 78%. Thus the carboxyl side chain at residue βGlu-181 is important, although not absolutely necessary, and βGlu-192 may not be involved in catalysis. On the other hand, replacements of Glu residues in the PS3 β subunit corresponding to βGlu-181 and βGlu-192 of *E. coli* by Gln caused complete loss of ATPase activity (250). The reason for this marked difference between results is unknown. FSBA bound to a Tyr residue of bovine F_1 corresponding to βTyr-354 of EF_1 (251). However, genetic replacement of βTyr-354 by Phe did not affect ATPase activity significantly (240). Similarly, the mutation Tyr-297 → Phe, which corresponds to the Nbf-Cl–binding site of bovine F_1 (Tyr-311) (252), had only a slight effect on ATPase activity (240). Thus these two Tyr residues may not have direct roles in catalysis, although they may be close to the catalytic residue(s): the NBD moiety bound to Tyr-311 of bovine F_1 could be transferred to P^1-(5'-adenosyl)-P^2-N-(2-mercaptoethyl)diphosphoramidate at the catalytic site, indicating that the Tyr-311 residue is close to the γ-phosphate of ATP (253). Tyr-362 and Asp-369 of CF$_1$ may also be near the catalytic site, because they were labeled with BzATP (189). It will be of interest to apply directed mutagenesis to these residues and also to those modified by other reagents (Table 2).

Possible Amino Acid Residues in the H$^+$ Pathway

A passive proton pathway of PS3 F$_0$ incorporated into liposomes was sealed when the δ, ϵ, and γ subunits were added, but not when only the δ and ϵ subunits were added (110). The role of chloroplast δ as a gate to the F$_0$ proton pathway has been proposed (101, 254). Participation of the γ subunit in the gating of protons was shown in CF$_1$ (255). It is noteworthy that CF$_1$ lacking ϵ binds to F$_0$, but does not block passive proton conduction (126). These results suggest that the proton gate may be formed by the close interaction of these three subunits.

Mutation in the a, b, or c subunit often resulted in F$_0$ defective in H$^+$ translocation (Table 3). However, only a few of the amino acids involved in these mutations may actually participate in the H$^+$ pathway, as discussed below. A carboxyl residue (cAsp-61 in E. $coli$ and Glu in other organisms) in the middle of a putative membrane-spanning α helix of the c subunit seems to be essential for the H$^+$ pathway (131, 134), as its mutational replacement by Gly or Asn, or its binding to DCCD, rendered F$_0$ impermeable to H$^+$ (134, 146). However, the genetic replacement resulted in altered assembly of F$_o$ and loss of H$^+$ permeability: the b subunit of the mutant membranes was more readily solubilized by detergents (256) and binding of the F$_1$-ATPase to the mutant membranes was inhibited 50% (146). Therefore, cAsp-61 itself or a neighboring residue(s) may also be related to the interaction of F$_0$ and F$_1$ or the conformation of the entire F$_0$ complex. It may not be easy to construct an H$^+$ pathway from c subunits only, because there are not enough charged residues to form a transmembrane H$^+$ pathway. The high copy number of the c subunit in F$_0$ may also result in difficulty in constructing the actual H$^+$ pathway.

The two putative α helices of the c subunit are connected by a hydrophilic loop domain. Mutations in this domain in E. $coli$ (cGln-42 $\rightarrow$ Glu) (145) or yeast (Arg-39 $\rightarrow$ Met) (257) rendered the ATPase complex nonfunctional. Analyses of both intragenic and extragenic revertants of yeast indicated that there is a strict requirement for a positively charged residue in this loop region, possibly for interaction with other subunits.

Initially the E. $coli$ a subunit was suggested to play only a structural role(s) in formation of F$_0$ and not to have any direct part in the function of F$_0$F$_1$ (218). However, membranes of two missense mutants (aSer-206 $\rightarrow$ Leu and aHis-245 $\rightarrow$ Tyr) (258) showed impaired H$^+$ conductivity, whereas they had normal F_1-binding activity. Thus these residues may participate directly in H^+ conduction. Furthermore, membranes of the aArg-210 → Gln and aHis-245 → Leu mutants were also impermeable to H^+ (259) and had about 50% of the ATPase activity of the wild-type. aHis-245 occupies the position of a Glu residue conserved in the mitochondrial subunit 6, and subunit 6 has histidine at a position corresponding to aGlu-219 of *E. coli* (258). Interestingly, aGlu-219 mutations (Glu-219 → Asp, His, Gln, or Leu) affected the ability of the enzyme to translocate protons, but not F_0F_1 assembly (260). Furthermore, complementary interaction between aGlu-219 → His and aHis-245 → Glu mutations was observed, suggesting that these two sites may participate in H^+ translocation. Mutations were introduced into the two highly conserved residues aGlu-196 (261, 262) and aPro-190 (262), and findings indicated that these two residues are not obligatory components of the proton channel, but that they affect the H^+ pathway indirectly (262). Similarly an aLys-203 → Ile mutation had no effect on H^+ translocation (260). Membranes of an aGln-252 → end mutant showed no H^+ permeability, suggesting that the carboxyl-terminal region may have pertinent residues, possibly the conserved aTyr-263 and aGln-252 residues, for H^+ translocation (133). These mutant studies also indicated the locations of conservd polar residues in one side of the α helix embedded in the membranes. A missense mutant aPro-143 → Ser had an impaired H^+ pathway, possibly due to altered F_0F_1 interaction (133).

The actual structure of the H^+ pathway in F_0 is still unknown. This H^+ pathway may be formed from residues of different subunits, a possibility supported by the close interactions between subunits. In this regard, Cox et al proposed a model (263) in which the conserved amino acid residues between aGlu-196 and aGly-218 of the a subunit form a trans-membrane H^+ pathway with cAsp-61 of the c subunit. However, the similar sequence in bacteriorhodopsin is not related to the H^+ pathway (264). Thus further studies are required to determine the H^+ pathway in F_0 and the mechanism of coupling of its H^+ translocation with ATP hydrolysis or synthesis.

CONCLUSION

This article reviews recent results on H^+-ATPase (F_0F_1) from mitochondria, chloroplasts, and bacteria. The mechanism of this enzyme, the reversibility at the catalytic site, and the positive catalytic cooperativity have been firmly established. The subunit structure and the amino acid sequences of the subunits have also been established, although further studies may be required on mitochondrial subunits. In this article important (or essential) amino acid residues for catalysis, H^+ translocation, and subunit-subunit interaction(s) have been discussed. Once the crystal structure has been determined, determinations of the locations of these residues in the higher-ordered structure of F_0F_1 will be of great importance. However, we are still far from understanding this enzyme. We cannot answer exactly the naive question of why this enzyme is so complicated. Combined studies on chemical modification, affinity labeling, and mutagenesis (random or directed) will deepen our understanding of this enzyme, and especially of the role of protons at the molecular level during catalysis. Molecular biological studies on the enzyme have so far mainly been limited to the enzyme of *E. coli,* but extensions of studies to other organisms have started, as evidenced by the wealth of genetic studies.

ACKNOWLEDGMENTS

We thank our colleagues for sending us preprints and reprints, and apologize for not citing all of these, because of limitation of space and the scope of this article. The work carried out in our laboratory was supported in part by grants from the Ministry of Education, Science, and Culture of Japan, the Science and Technology Agency of the Japanese Government, and Mitsubishi Foundation.

Literature Cited

1. Racker, E. 1976. *A New Look at Mechanism in Bioenergetics.* New York: Academic
2. Hatefi, Y. 1985. *Annu. Rev. Biochem.* 54:1015–69
3. Downie, J. A., Gibson, F., Cox, G. B. 1979. *Annu. Rev. Biochem.* 48:103–31
4. Futai, M., Kanazawa, H. 1983. *Microbiol. Rev.* 47:285–312
5. Walker, J. E., Saraste, M., Gay, N. J. 1984. *Biochim. Biophys. Acta* 768:164–200
6. Fillingame, R. H. 1980. *Annu. Rev. Biochem.* 49:1079–113
7. Pedersen, P. L., Carafoli, E. 1987. *Trends Biochem. Sci.* 12:146–50, 186–89
8. Strotmann, H., Bickel-Sandkötter, S. 1984. *Annu. Rev. Plant Physiol.* 35:97–120
9. Cross, R. L. 1981. *Annu. Rev. Biochem.* 50:681–714
10. Vignais, P. V., Lunardi, J. 1985. *Annu. Rev. Biochem.* 54:977–1014
11. Tsong, T. Y., Astumian, R. D. 1988. *Annu. Rev. Physiol.* 50:273–90
12. Avron, M. 1987. In *Photosynthesis,* ed. J. Amesz, pp. 159–73. Holland: Elsevier
13. Nalin, C. M., Nelson, N. 1987. *Curr. Top. Bioenerg.* 15:273–94
14. Pfanner, N., Neupert, W. 1987. *Curr. Top. Bioenerg.* 15:177–219
15. Kagawa, Y. 1984. *New Comp. Biochem.* 9:149–86
16. Senior, A. E. 1988. *Physiol. Rev.* 68:177–231
17. Amzel, L. M., Pedersen, P. L. 1983. *Annu. Rev. Biochem.* 52:801–24
18. Wang, J. H. 1983. *Annu. Rev. Biophys. Bioeng.* 12:21–34
19. McCarthy, J. E. G. 1988. *J. Bioenerg. Biomembr.* 20:19–39
20. Ohta, S., Yohda, M., Ishizuka, M., Hirata, H., Hamamoto, T., Kagawa, Y., et al. 1988. *Biochim. Biophys. Acta* 933:141–55
21. Tybulewicz, V. L. J., Falk, G., Walker, J. E. 1984. *J. Mol. Biol.* 179:185–214
22. Falk, G., Hampe, A., Walker, J. E. 1985. *Biochem. J.* 228:391–407
23. Cozens, A. L., Walker, J. E. 1987. *J. Mol. Biol.* 194:359–83
24. McCarn, D. F., Whitaker, R. A., Alam, J., Vrba, J. M., Curtis, S. E. 1988. *J. Bacteriol.* 170:3448–58
25. Curtis, S. E. 1987. *J. Bacteriol.* 169:80–86
26. Zurawski, G., Bottomley, W., Whitfeld, P. R. 1982. *Proc. Natl. Acad. Sci. USA* 79:6260–64
27. Krebbers, E. T., Larrinua, I. M., McIntosh, L., Bogorad, L. 1982. *Nucleic Acids Res.* 10:4985–5002
28. Shinozaki, K., Deno, H., Kato, A.,

Sugiura, M. 1983. *Gene* (Amsterdam) 24:147–55
29. Zurawski, G., Clegg, M. T. 1984. *Nucleic Acids Res.* 12:2549–59
30. Howe, C. J., Fearnley, I. M., Walker, J. E., Dyer, T. A., Gray, J. C. 1985. *Plant Mol. Biol.* 4:333–46
31. Cozens, A. L., Walker, J. E., Phillips, A. L., Huttly, A. K., Gray, J. C. 1986. *EMBO J.* 5:217–22
32. Ohyama, K., Fukuzawa, H., Kohchi, T., Shirai, H., Sano, T., et al. 1986. *Nature* 322:572–74
33. Hudson, G. S., Mason, J. G., Holton, T. A., Koller, B., Cox, G. B., et al. 1987. *J. Mol. Biol.* 196:283–98
34. Hennig, J., Herrmann, R. G. 1986. *Mol. Gen. Genet.* 203:117–28
35. Bird, C. R., Koller, B., Auffret, A. D., Huttly, A. K., Howe, C. J., et al. 1985. *EMBO J.* 4:1381–88
36. Deno, H., Shinozaki, K., Sugiura, M. 1984. *Gene* 32:195–201
37. Ooi, B. G., Lukins, H. B., Linnane, A. W., Nagley, P. 1987. *Nucleic Acids Res.* 15:1965–77
38. Anderson, S., Bankier, A. T., Barrell, B. G., de Bruijn, M. H. L., Coulson, A. R., et al. 1981. *Nature* 290:457–65
39. Bibb, M. J., van Etten, R. A., Wright, C. T., Walberg, M. W., Clayton, D. A. 1981. *Cell* 26:167–80
40. Fearnley, I. M., Walker, J. E. 1986. *EMBO J.* 5:2003–8
41. van der Boogaart, P., Samallo, J., Agsteribbe, E. 1982. *Nature* 298:187–89
42. Schuster, W., Brennicke, A. 1986. *Mol. Gen. Genet.* 204:29–35
43. Morikami, A., Nakamura, K. 1987. *J. Biochem.* 101:967–76
44. Isaac, P. G., Brennicke, A., Dunbar, S. M., Leaver, C. J. 1985. *Curr. Genet.* 10:321–28
45. Morikami, A., Nakamura, K. 1987. *Nucleic Acids Res.* 15:4692
46. Rothenberg, M., Hanson, M. R. 1987. *Mol. Gen. Genet.* 209:21–27
47. Young, E. G., Hanson, M. R., Dierks, P. M. 1986. *Nucleic Acids Res.* 14:7995–8006
48. Ohta, S., Kagawa, Y. 1986. *J. Biochem.* 99:135–41
49. Garboczi, D. N., Fox, A. H., Gerring, S. L., Pedersen, P. L. 1988. *Biochemistry* 27:553–60
50. Breen, G. A. M., Holmans, P. L., Garnett, K. E. 1988. *Biochemistry* 27:3955–61
51. Takeda, M., Vassarotti, A., Douglas, M. G. 1985. *J. Biol. Chem.* 260:15458–65
52. Takeda, M., Chen, W.-J., Saltzgaber,

J., Douglas, M. G. 1986. *J. Biol. Chem.* 261:15126–33
53. Miki, J., Maeda, M., Mukohata, Y., Futai, M. 1988. *FEBS Lett.* 232:221–26
54. Viebrock, A., Perz, A., Sebald, W. 1982. *EMBO J.* 1:565–71
55. Gay, N. J., Walker, J. E. 1985. *EMBO J.* 4:3519–24
56. Farrell, L. B., Nagley, P. 1987. *Biochem. Biophys. Res. Commun.* 144:1257–64
57. Walker, J. E., Runswick, M. J. 1987. *J. Mol. Biol.* 197:89–100
58. Velours, J., Durrens, P., Aigle, M., Guerin, B. 1988. *Eur. J. Biochem.* 170:637–42
59. Boutry, M., Chua, N.-H. 1985. *EMBO J.* 4:2159–65
60. Ohta, S., Tomura, H., Matsuda, K., Kagawa, Y. 1988. *J. Biol. Chem.* 263:11257–62
61. Gearing, D. P., Nagley, P. 1986. *EMBO J.* 5:3651–55
62. Runswick, M. J., Walker, J. E. 1983. *J. Biol. Chem.* 258:3081–89
63. Kanazawa, H., Kayano, T., Kiyasu, T., Futai, M. 1982. *Biochem. Biophys. Res. Commun.* 105:1257–64
64. Kobayashi, K., Nakamura, K., Asahi, T. 1987. *Nucleic Acids Res.* 15:7177
65. Hawthorne, C. A., Brusilow, W. S. A. 1988. *Biochem. Biophys. Res. Commun.* 151:926–31
66. Denda, K., Konishi, J., Oshima, T., Date, T., Yoshida, M. 1988. *J. Biol. Chem.* 263:6012–15
67. Zimniak, L., Dittrich, P., Gogarten, J. P., Kibak, H., Taiz, L. 1988. *J. Biol. Chem.* 263:9102–12
68. Lübben, M., Schäfer, G. 1987. *Eur. J. Biochem.* 164:533–40
69. Mukohata, Y., Yoshida, M. 1987. *J. Biochem.* 102:797–802
70. Kanazawa, H., Kayano, T., Mabuchi, K., Futai, M. 1981. *Biochem. Biophys. Res. Commun.* 103:604–12
71. Walker, J. E., Fearnley, I. M., Gay, N. J., Gibson, B. W., Northrop, F. D., et al. 1985. *J. Mol. Biol.* 184:677–701
72. Deno, H., Shinozaki, K., Sugiura, M. 1983. *Nucleic Acids Res.* 11:2185–91
73. Braun, C. J., Levings, C. S. III. 1985. *Plant Physiol.* 79:571–77
74. Weeks, D. L., Melton, D. A. 1987. *Proc. Natl. Acad. Sci. USA* 84:2798–802
75. Soga, S., Noumi, T., Takeyama, M., Maeda, M., Futai, M. 1989. *Arch Biochem. Biophys.* In press
76. Saraste, M., Gay, N. J., Eberle, A., Runswick, M. J., Walker, J. E. 1981. *Nucleic Acids Res.* 9:5287–96

77. Walker, J. E., Runswick, M. J., Saraste, M. 1982. *FEBS Lett.* 146:393–96
78. Ovchinnikov, Y. A., Modyanov, N. N., Grinkevich, V. A., Aldanova, N. A., Trubetskaya, O. E., et al. 1984. *FEBS Lett.* 166:19–22
79. Schwerzmann, K., Pedersen, P. L. 1986. *Arch. Biochem. Biophys.* 250:1–18
80. Matsubara, H., Hase, T., Hashimoto, T., Tagawa, K. 1981. *J. Biochem.* 90:1159–65
81. Dianoux, A. C., Hoppe, J. 1987. *Eur. J. Biochem.* 163:155–60
82. Frangione, B., Rosenwasser, E., Penefsky, H. S., Pullman, M. E. 1981. *Proc. Natl. Acad. Sci. USA* 78:7403–7
83. Hashimoto, T., Yoshida, Y., Tagawa, K. 1983. *J. Biochem.* 94:715–20
84. Hashimoto, T., Yoshida, Y., Tagawa, K. 1984. *J. Biochem.* 95:131–36
85. Yoshida, Y., Wakabayashi, S., Matsubara, H., Hashimoto, T., Tagawa, K. 1984. *FEBS Lett.* 170:135–38
86. Klionsky, D. J., Brusilow, W. S. A., Simoni, R. D. 1984. *J. Bacteriol.* 160:1055–60
87. Dunn, S. D., Zadorozny, V. D., Tozer, R. G., Orr, L. E. 1987. *Biochemistry* 26:4488–93
88. Sebald, W., Hoppe, J. 1981. *Curr. Top. Bioenerg.* 12:1–64
89. Mandel, M., Moriyama, Y., Hulmes, J. D., Pan, Y.-C. E., Nelson, H., et al. 1988. *Proc. Natl. Acad. Sci. USA* 85:5521–24
90. Higuti, T., Negama, T., Takigawa, M., Uchida, J., Yamane, T., et al. 1988. *J. Biol. Chem.* 263:6772–76
91. Huang, Y., Kantham, L., Sanadi, D. R. 1987. *J. Biol. Chem.* 262:3007–10
92. Walker, J. E., Gay, N. J., Powell, S. J., Kostina, M., Dyer, M. R. 1987. *Biochemistry* 26:8613–19
93. Farrell, L. B., Nero, D., Meltzer, S., Braidotti, G., Devenish, R. J., Nagley, P. 1988. In *The Molecular Structure, Function and Assembly of ATP Synthase*, ed. S. Maruzuki. In press
94. Amzel, L. M., McKinney, M., Narayanan, P., Pedersen, P. L. 1982. *Proc. Natl. Acad. Sci. USA* 79:5852–56
95. Gogol, E. P., Lücken, U., Capaldi, R. A. 1987. *FEBS Lett.* 219:274–78
96. Aggeler, R., Zhang, Y.-Z., Capaldi, R. A. 1987. *Biochemistry* 26:7107–13
97. Bragg, P. D., Hou, C. 1986. *Biochim. Biophys. Acta* 851:385–94
98. Dupuis, A., Vignais, P. V. 1987. *Biochemistry* 26:410–18
99. Dupuis, A., Issartel, J.-P., Lunardi, J.,

Satre, M., Vignais, P. V. 1985. *Biochemistry* 24:728–33
100. Dupuis, A., Lunardi, J., Issartel, J.-P., Vignais, P. V. 1985. *Biochemistry* 24:734–39
101. Engelbrecht, S., Junge, W. 1987. *FEBS Lett.* 219:321–25
102. Aris, J. P., Simoni, R. D. 1983. *J. Biol. Chem.* 258:14599–609
103. Gao, Z., Bäuerlein, E. 1987. *FEBS Lett.* 223:366–70
104. Bragg, P. D., Hou, C. 1986. *Arch. Biochem. Biophys.* 244:361–72
105. Lünsdorf, H., Ehrig, K., Friedl, P., Schairer, H. U. 1984. *J. Mol. Biol.* 173:131–36
106. Joshi, V. K., Wang, J. H. 1987. *J. Biol. Chem.* 262:15721–25
107. Tiedge, H., Schäfer, G., Mayer, F. 1983. *Eur. J. Biochem.* 132:37–45
108. Futai, M. 1977. *Biochem. Biophys. Res. Commun.* 79:1231–37
109. Dunn, S. D., Futai, M. 1980. *J. Biol. Chem.* 255:113–18
110. Yoshida, M., Okamoto, H., Sone, N., Hirata, H., Kagawa, Y. 1977. *Proc. Natl. Acad. Sci. USA* 74:936–40
111. Hsu, S.-Y., Senda, M., Kanazawa, H., Tsuchiya, T., Futai, M. 1984. *Biochemistry* 23:988–93
112. Futai, M., Kanazawa, H., Takeda, K., Kagawa, Y. 1980. *Biochem. Biophys. Res. Commun.* 96:227–34
113. Gromet-Elhanan, Z., Khananshvili, D., Weiss, S., Kanazawa, H., Futai, M. 1985. *J. Biol. Chem.* 260:12635–40
114. Richter, M. L., Gromet-Elhanan, Z., McCarty, R. E. 1986. *J. Biol. Chem.* 261:12109–13
115. Hawthorne, C. A., Brusilow, W. S. A. 1986. *J. Biol. Chem.* 261:5245–48
116. Kanazawa, H., Hama, H., Rosen, B. P., Futai, M. 1985. *Arch. Biochem. Biophys.* 241:364–70
117. Miki, J., Takeyama, M., Noumi, T., Kanazawa, H., Maeda, M., Futai, M. 1986. *Arch. Biochem. Biophys.* 251:458–64
118. Miki, J., Maeda, M., Futai, M. 1988. *J. Bacteriol.* 170:179–83
119. Miki, J., Iwamoto, A., Maeda, M., Futai, M. 1989. Submitted
120. Noumi, T., Oka, N., Kanazawa, H., Futai, M. 1986. *J. Biol. Chem.* 261:7070–75
121. Noumi, T., Azuma, M., Shimomura, S., Maeda, M., Futai, M. 1987. *J. Biol. Chem.* 262:14978–82
122. Maggio, M. B., Parsonage, D., Senior, A. E. 1988. *J. Biol. Chem.* 263:4619–23
123. Maggio, M. B., Pagan, J., Parsonage,

D., Hatch, L., Senior, A. E. 1987. *J. Biol. Chem.* 262:8981–84
124. Kuki, M., Noumi, T., Maeda, M., Amemura, A., Futai, M. 1988. *J. Biol. Chem.* 263: In press
125. Cox, G. B., Hatch, L., Webb, D., Fimmel, A. L., Lin, Z.-H., et al. 1987. *Biochim. Biophys. Acta* 890:195–204
126. Richter, M. L., Patrie, W. J., McCarty, R. E. 1984. *J. Biol. Chem.* 259:7371–73
127. Dunn, S. D. 1982. *J. Biol. Chem.* 257:7354–59
128. Tozer, R. G., Dunn, S. D. 1987. *J. Biol. Chem.* 262:10706–11
129. Friedl, P., Hoppe, J., Gunsalus, R. P., Michelsen, O., von Meyenburg, K., Schairer, H. U. 1983. *EMBO J.* 2:99–103
130. Schneider, E., Altendorf, K. 1985. *EMBO J.* 4:515–18
131. Schneider, E., Altendorf, K. 1987. *Microbiol. Rev.* 51:477–97
132. Hoppe, J., Friedl, P., Schairer, H. U., Sebald, W., von Meyenburg, K., et al. 1983. *EMBO J.* 2:105–10
133. Eya, S., Noumi, T., Maeda, M., Futai, M. 1988. *J. Biol. Chem.* 263:10056–62
134. Hoppe, J., Sebald, W. 1984. *Biochim. Biophys. Acta.* 768:1–27
135. Jans, D. A., Hatch, L., Fimmel, A. L., Gibson, F., Cox, G. B. 1984. *J. Bacteriol.* 160:764–70
136. Porter, A. C. G., Kumamoto, C., Aldape, K., Simoni, R. D. 1985. *J. Biol. Chem.* 260:8182–87
137. Jans, D. A., Fimmel, A. L., Hatch, L., Gibson, F., Cox, G. B. 1984. *Biochem. J.* 221:43–51
138. Jans, D. A., Hatch, L., Fimmel, A. L., Gibson, F., Cox, G. B. 1985. *J. Bacteriol.* 162:420–26
139. Takeyama, M., Noumi, T., Maeda, M., Futai, M. 1988. *J. Biol. Chem.* 263: In press
140. Steffens, K., Schneider, E., Deckers-Hebestreit, G., Altendorf, K. 1987. *J. Biol. Chem.* 262:5866–69
141. Deckers-Hebestreit, G., Steffens, K., Altendorf, K. 1986. *J. Biol. Chem.* 261: 14878–81
142. Fromme, P., Boekema, E. J., Gräber, P. 1987. *Z. Naturforsch. Teil C* 42: 1239–45
143. Weber, H., Junge, W., Hoppe, J., Sebald, W. 1986. *FEBS Lett.* 202:23–26
144. Hoppe, J., Schairer, H. U., Sebald, W. 1980. *Eur. J. Biochem.* 112:17–24
145. Mosher, M. E., White, L. K., Hermolin, J., Fillingame, R. H. 1984. *J. Biol. Chem.* 260:4807–14
146. Fillingame, R. H., Peters, L. K., White, L. K., Mosher, M. E., Paule, C. R. 1984. *J. Bacteriol.* 158:1078–83

147. Fimmel, A. L., Jans, D. A., Hatch, L., James, L. B., Gibson, F., Cox, G. B. 1985. *Biochim. Biophys. Acta* 808:252–58
148. Fimmel, A. L., Jans, D. A., Langman, L., James, L. B., Ash, G. R., et al. 1983. *Biochem. J.* 213:451–58
149. Jans, D. A., Fimmel, A. L., Langman, L., James, L. B., Downie, J. A., et al. 1983. *Biochem. J.* 211:717–26
150. Mosher, M. E., Peters, L. K., Fillingame, R. H. 1983. *J. Bacteriol.* 156: 1078–92
151. Jean-Francois, M. J. B., Hadikusumo, R. G., Watkins, L. C., Lukins, H. B., Linnane, A. W., Marzuki, S. 1986. *Biochim. Biophys. Acta* 852:133–43
152. Ozawa, T., Nishikimi, M., Suzuki, H., Tanaka, M., Shimomura, Y. 1987. *Bioenergetics: Structure and Function of Energy Transducing Sytems*, ed. T. Ozawa, S. Papa, pp. 101–19. Tokyo/Berlin: Japan Sci. Soc. Press/Springer-Verlag
153. Hermolin, J., Gallant, J., Fillingame, R. H. 1983. *J. Biol. Chem.* 258:14550–55
154. Schneider, E., Altendorf, K. 1984. *Proc. Natl. Acad. Sci. USA* 81:7279–83
155. Steffens, K., Hoppe, J., Altendorf, K. 1988. *Eur. J. Biochem.* 170:627–30
156. Kumamoto, C. A., Simoni, R. D. 1986. *J. Biol. Chem.* 261:10037–42
157. Kumamoto, C. A., Simoni, R. D. 1987. *J. Biol. Chem.* 262:3060–64
158. Enns, R. K., Criddle, R. S. 1977. *Arch. Biochem. Biophys.* 182:587–600
159. Sebald, W., Gatti, D., Weber, H., Hoppe, J. 1985. In *Bioenergetics*, ed. E. Quagliariello, et al, 1:27–34. Amsterdam/New York: Elsevier
160. Nagley, P., Hall, R. M., Ooi, B. G. 1986. *FEBS Lett.* 195:159–63
161. John, U. P., Nagley, P. 1986. *FEBS Lett.* 207:79–83
162. Klionsky, D. J., Simoni, R. D. 1985. *J. Biol. Chem.* 260:11200–6
163. Aris, J. P., Klionsky, D. J., Simoni, R. D. 1985. *J. Biol. Chem.* 260:11207–15
164. Fillingame, R. H., Porter, B., Hermolin, J., White, L. K. 1986. *J. Bacteriol.* 165:244–51
165. Cox, G. B., Gibson, F. 1987. *Curr. Top. Bioenerg.* 15:163–75
166. Cross, R. L., Nalin, C. M. 1982. *J. Biol. Chem.* 257:2874–81
167. Weber, J., Lücken, U., Schäfer, G. 1985. *Eur. J. Biochem.* 148:41–47
168. Kironde, F. A. S., Cross, R. L. 1986. *J. Biol. Chem.* 261:12544–49
169. Wise, J. G., Duncan, T. M., Latchney, L. R., Cox, D. N., Senior, A. E. 1983. *Biochem. J.* 215:343–50
170. Issartel, J.-P., Lunardi, J., Vignais, P. V. 1986. *J. Biol. Chem.* 261:895–901

171. Kironde, F. A. S., Cross, R. L. 1987. *J. Biol. Chem.* 262:3488–95
172. Williams, N., Hullihen, J., Pedersen, P. L. 1987. *Biochemistry* 26:162–69
173. McCarty, R. E., Hammes, G. G. 1987. *Trends Biochem. Sci.* 12:234–37
174. Leckband, D., Hammes, G. G. 1987. *Biochemistry* 26:2306–12
175. Xue, Z., Zhou, J.-M., Melese, T., Cross, R. L., Boyer, P. D. 1987. *Biochemistry* 26:3749–53
176. Hirano, M., Takeda, K., Kanazawa, H., Futai, M. 1984. *Biochemistry* 23:1652–56
177. Issartel, J.-P., Vignais, P. V. 1984. *Biochemistry* 23:6591–95
178. Rao, R., Al-Shawi, M. K., Senior, A. E. 1988. *J. Biol. Chem.* 263:5569–73
179. Ohta, S., Tsuboi, M., Oshima, T., Yoshida, M., Kagawa, Y. 1980. *J. Biochem.* 87:1609–17
180. Dunn, S. D. 1980. *J. Biol. Chem.* 255:11857–60
181. Senda, M., Kanazawa, H., Tsuchiya, T., Futai, M. 1983. *Arch. Biochem. Biophys.* 220:398–404
182. Williams, N., Coleman, P. S. 1982. *J. Biol. Chem.* 257:2834–41
183. Lübben, M., Lücken, U., Weber, J., Schäfer, G. 1984. *Eur. J. Biochem.* 143:483–90
184. Noumi, T., Tagaya, M., Miki-Takeda, K., Maeda, M., Fukui, T., Futai, M. 1987. *J. Biol. Chem.* 262:7686–92
185. Harris, D. A., Boork, J., Baltscheffsky, M. 1985. *Biochemistry* 24:3876–83
186. Khananshvili, D., Gromet-Elhanan, Z. 1985. *Proc. Natl. Acad. Sci. USA* 82:1886–90
187. Cross, R. L., Cunningham, D., Miller, C. G., Xue, Z., Zhou, J.-M., Boyer, P. D. 1987. *Proc. Natl. Acad. Sci. USA* 84:5715–19
188. Bullough, D. A., Allison, W. S. 1986. *J. Biol. Chem.* 261:14171–77
189. Admon, A., Hammes, G. G. 1987. *Biochemistry* 26:3193–97
190. Wise, J. G., Hicke, B. J., Boyer, P. D. 1987. *FEBS Lett.* 223:395–401
191. Xue, Z., Miller, C. G., Zhou, J.-M., Boyer, P. D. 1987. *FEBS Lett.* 223:391–94
192. Lunardi, J., Garin, J., Issartel, J.-P., Vignais, P. V. 1987. *J. Biol. Chem.* 262:15172–81
193. Pougeois, R., Lauquin, G. J.-M., Vignais, P. V. 1983. *Biochemistry* 22:1241–45
194. Webb, M. R., Grubmeyer, C., Penefsky, H. S., Trentham, D. R. 1980. *J. Biol. Chem.* 255:11637–39
195. Noumi, T., Maeda, M., Futai, M. 1988. *J. Biol. Chem.* 263:8765–70
196. Boyer, P. D. 1979. In *Membrane Bioenergetics*, ed. C.-P. Lee, G. Shatz, L. Ernster, pp. 461–79. Reading, Mass: Addison-Wesley
197. Feldman, R. I., Sigman, D. S. 1982. *J. Biol. Chem.* 257:1676–83
198. Sakamoto, J., Tonomura, Y. 1983. *J. Biochem.* 93:1601–14
199. Kandpal, R. P., Stempel, K. E., Boyer, P. D. 1987. *Biochemistry* 26:1512–17
200. O'Neal, C. C., Boyer, P. D. 1984. *J. Biol. Chem.* 259:5761–67
201. Grubmeyer, C., Cross, R. L., Penefsky, H. S. 1982. *J. Biol Chem.* 257:12092–100
202. Cross, R. L., Grubmeyer, C., Penefsky, H. S. 1982. *J. Biol. Chem.* 257:12101–5
203. Beharry, S., Gresser, M. J. 1987. *J. Biol. Chem.* 262:10630–37
204. Bullough, D. A., Verburg, J. G., Yoshida, M., Allison, W. S. 1987. *J. Biol. Chem.* 262:11675–83
205. Milgrom, Ya. M., Murataliev, M. B. 1987. *FEBS Lett.* 222:32–36
206. Penefsky, H. S. 1988. *J. Biol. Chem.* 263:6020–22
207. Duncan, T. M., Senior, A. E. 1985. *J. Biol. Chem.* 260:4901–7
208. Noumi, T., Taniai, M., Kanazawa, H., Futai, M. 1986. *J. Biol. Chem.* 261:9196–201
209. Grubmeyer, C., Penefsky, H. S. 1981. *J. Biol. Chem.* 256:3728–34
210. Matsuoka, I., Watanabe, T., Tonomura, Y. 1981. *J. Biochem.* 90:967–89
211. Noumi, T., Maeda, M., Futai, M. 1987. *FEBS Lett.* 213:381–84
212. Tommasino, M., Capaldi, R. A. 1985. *Biochemistry* 24:3972–76
213. Matsuno-Yagi, A., Hatefi, Y. 1984. *Biochemistry* 23:3508–14
214. Penefsky, H. S. 1985. *J. Biol. Chem.* 260:13735–41
215. Matsuno-Yagi, A., Hatefi, Y. 1985. *J. Biol. Chem.* 260:14424–27
216. Stroop, S. D., Boyer, P. D. 1985. *Biochemistry* 24:2304–10
217. Gresser, M. J., Myers, J. A., Boyer, P. D. 1982. *J. Biol. Chem.* 257:12030–38
218. Cox, G. B., Jans, D. A., Fimmel, A. L., Gibson, F., Hatch, L. 1984. *Biochim. Biophys. Acta* 768:201–8
219. Hayashi, S., Oosawa, F. 1984. *Proc. Jpn. Acad.* 60:161–64
220. Kandpal, R. P., Boyer, P. D. 1987. *Biochim. Biophys. Acta* 890:97–105
221. Musier, K. M., Hammes, G. G. 1987. *Biochemistry* 26:5982–88
222. Rao, R., Senior, A. E. 1987. *J. Biol. Chem.* 262:17450–54
223. Wang, J. H. 1985. *J. Biol. Chem.* 260:1374–77

224. Wang, J. H., Joshi, V., Wu, J. C. 1986. *Biochemistry* 25:7996–8001
225. Di Pietro, A., Penin, F., Godinot, C., Gautheron, D. C. 1980. *Biochemistry* 19:5671–78
226. Noumi, T., Futai, M., Kanazawa, H. 1984. *J. Biol. Chem.* 259:10076–79
227. Stan-Lotter, H., Boyer, P. D. 1986. *Eur. J. Biochem.* 160:169–74
228. Kanazawa, H., Noumi, T., Matsuoka, I., Hirata, T., Futai, M. 1984. *Arch. Biochem. Biophys.* 228:258–69
229. Parsonage, D., Duncan, T. M., Wilke-Mounts, S., Kironde, F. A. S., Hatch, L., Senior, A. E. 1987. *J. Biol. Chem.* 262:6301–7
230. Noumi, T., Mosher, M. E., Natori, S., Futai, M., Kanazawa, H. 1984. *J. Biol. Chem.* 259:10071–75
231. Walker, J. E., Saraste, M., Runswick, M. J., Gay, N. J. 1982. *EMBO J.* 1:945–51
232. Fry, D. C., Kuby, S. A., Mildvan, A. S. 1985. *Biochemistry* 24:4680–94
233. Tagaya, M., Yagami, T., Fukui, T. 1987. *J. Biol. Chem.* 262:8257–61
234. Duncan, T. M., Parsonage, D., Senior, A. E. 1986. *FEBS Lett.* 208:1–6
235. Andrews, W. W., Hill, F. C., Allison, W. S. 1984. *J. Biol. Chem.* 259:14378–82
236. Tagaya, M., Noumi, T., Nakano, K., Futai, M., Fukui, T. 1988. *FEBS Lett.* 233:347–51
237. Wu, J. C., Wang, J. H. 1986. *Biochemistry* 25:7991–95
238. Hsu, S.-Y., Noumi, T., Takeyama, M., Maeda, M., Ishibashi, S., Futai, M. 1987. *FEBS Lett.* 218:222–26
239. Reddy, E. P., Reynolds, R. K., Santos, E., Barbacid, M. 1982. *Nature* 300:149–52
240. Parsonage, D., Wilke-Mounts, S., Senior, A. E. 1987. *J. Biol. Chem.* 262:8022–26
241. Parsonage, D., Al-Shawi, M. K., Senior, A. E. 1988. *J. Biol. Chem.* 263:4740–44
242. Yohda, M., Ohta, S., Hisabori, T., Kagawa, Y. 1988. *Biochim. Biophys. Acta* 933:156–64
243. Garboczi, D. N., Shenbagamurthi, P., Kirk, W., Hullihen, J., Pedersen, P. L. 1988. *J. Biol. Chem.* 263:812–16
244. Rao, R., Cunningham, D., Cross, R. L., Senior, A. E. 1988. *J. Biol. Chem.* 263:5640–45
245. Hollemans, M., Runswick, M. J., Fearnley, I. M., Walker, J. E. 1983. *J. Biol. Chem.* 258:9307–13
246. Esch, F. S., Böhlen, P., Otsuka, A. S.,
247. Yoshida, M., Allison, W. S. 1981. *J. Biol. Chem.* 256:9084–89
248. Yoshida, M., Allison, W. S., Esch, F. S., Futai, M. 1982. *J. Biol. Chem.* 257:10033–37
249. Yoshida, M., Poser, J. W., Allison, W. S., Esch, F. S. 1981. *J. Biol. Chem.* 256:148–53
249. Parsonage, D., Wilke-Mounts, S., Senior, A. E. 1988. *Arch. Biochem. Biophys.* 261:222–25
250. Ohtsubo, M., Yoshida, M., Ohta, S., Kagawa, Y., Yohda, M., Date, T. 1987. *Biochem. Biophys. Res. Commun.* 146:705–10
251. Esch, F. S., Allison, W. S. 1978. *J. Biol. Chem.* 253:6100–6
252. Andrews, W. W., Hill, F. C., Allison, W. S. 1984. *J. Biol. Chem.* 259:8219–25
253. Wu, J. C., Chuan, H., Wang, J. H. 1987. *J. Biol. Chem.* 262:5145–50
254. Engelbrecht, S., Lill, H., Junge, W. 1986. *Eur. J. Biochem.* 160:635–43
255. McCarty, R. E. 1982. In *Membranes and Transport*, ed. A. N. Martonosi, 2:599–603. New York: Plenum
256. Friedl, P., Hoppe, J., Schairer, H. U. 1984. *Biochim. Biophys. Res. Commun.* 120:527–33
257. Willson, T. A., Nagley, P. 1987. *Eur. J. Biochem.* 167:291–97
258. Cain, B. D., Simoni, R. D. 1986. *J. Biol. Chem.* 261:10043–50
259. Lightowlers, R. N., Howitt, S. M., Hatch, L., Gibson, F., Cox, G. B. 1987. *Biochim. Biophys. Acta* 894:399–406
260. Cain, B. D., Simoni, R. D. 1988. *J. Biol. Chem.* 263:6606–12
261. Lightowlers, R. N., Howitt, S. M., Hatch, L., Gibson, F., Cox, G. B. 1988. *Biochim. Biophys. Acta* 933:241–48
262. Vik, S. B., Cain, B. D., Chun, K. T., Simoni, R. D. 1988. *J. Biol. Chem.* 263:6599–605
263. Cox, G. B., Fimmel, A. L., Gibson, F., Hatch, L. 1986. *Biochim. Biophys. Acta* 849:62–69
264. Khorana, H. G., Braiman, M. S., Chao, B. H., Doi, T., Flitsch, S. L., et al. 1987. *Chem. Scr.* 27B:137–47
265. Viale, A. M., Vallejos, R. H. 1985. *J. Biol. Chem.* 260:4958–62
266. Mueller, D. M. 1988. *J. Biol. Chem.* 263:5634–39
267. Garin, J., Boulay, F., Issartel, J. P., Lunardi, J., Vignais, P. V. 1986. *Biochemistry* 25:4431–37
268. Bullough, D. A., Allison, W. S. 1986. *J. Biol. Chem.* 261:5722–30

Annu. Rev. Biochem. 1989. 58:137–71

THE BIOCHEMISTRY OF P-GLYCOPROTEIN-MEDIATED MULTIDRUG RESISTANCE[1]

Jane A. Endicott and Victor Ling

The Ontario Cancer Institute, The Princess Margaret Hospital, and The Department of Medical Biophysics, University of Toronto, Toronto, Ontario, M4X 1K9 Canada

CONTENTS

[1]Abbreviations used: MDR, multidrug resistance; PMA, 4β-phorbol 12β-myristate 13α-acetate; OAG, 1-oleoyl 2-acetylglycerol; P(BtO)$_2$, phorbol 12,13-dibutyrate; NASV, N-(p-azido-[3-^{125}I]salicyl)-N'-(β-aminoethyl)vindesine; NABV, N-(p-azido-[3,5-^{3}H]benzoyl)-N'-(β-aminoethyl)vindesine; DM, double minutes; HSR, homogeneously staining region.

137

0066-4154/89/0701-0137$02.00

PERSPECTIVES AND SUMMARY

Multidrug resistance (MDR) is a unique phenomenon in the study of cellular drug resistance. Cell lines exhibiting this phenotype have been selected for resistance to a single cytotoxic agent, yet they display a broad, unpredictable cross-resistance to a wide variety of unrelated cytotoxic drugs, many of which are used in cancer treatment. The drugs most often involved in MDR are alkaloids or antibiotics of plant or fungal origin, and they include the vinca alkaloids, anthracyclines, epipodophyllotoxins, and dactinomycin. Cross-resistance to alkylating agents such as melphalan, nitrogen mustard, and mitomycin C is occasionally observed. Collateral sensitivity (increased sensitivity) to membrane-active agents such as nonionic detergents, local anesthetics, steroid hormones, and calcium channel blockers often accompanies the development of MDR. The recognition that the emergence of a complex drug resistance phenotype of broad specificity in human tumors could limit successful chemotherapy has provided the impetus to study MDR cell lines as a model for clinical drug resistance.

Kinetic experiments reveal that MDR cells are able to maintain a lowered intracellular drug concentration to a large degree via the increased activity of an energy-dependent drug efflux mechanism. However, many factors may determine the extent of cellular drug accumulation [see (1) for a review]. This is not surprising, given the structural diversity of the compounds involved, their very different pharmacological properties, and the range of their intracellular targets. Drug transport studies indicate that MDR might be mediated by an alteration in the plasma membrane rather than a host of drug-specific enzymatic changes. Early genetic experiments suggested that a single genetic alteration mediates the decreased drug accumulation in MDR cell lines (2, 3). For example, MDR mutants can be derived by single-step selections with or without mutagenesis, and by DNA-mediated gene transfer. Revertants that have lost the complete MDR phenotype can also be selected in a single step. The gene product mediating MDR has now been isolated and alternatively called P-glycoprotein, P-170, or the MDR-associated glycoprotein (MDRG). Transfection experiments using either the human or mouse P-glycoprotein cDNA sequence show that increased P-glycoprotein expression is able to confer a complete MDR phenotype on drug-sensitive cells (4, 5).

Increased expression of P-glycoprotein in the plasma membrane is the most consistent change detected in MDR cells. DNA sequences isolated by their ability to cause MDR (termed *mdr* sequences) have been shown to encode P-glycoprotein (6). P-glycoprotein expression correlates with both the decrease in intracellular drug accumulation and the observed degree of drug resistance in many MDR cell lines. P-glycoprotein-specific monoclonal antibodies have confirmed that a large number of MDR cell lines have increased expression of P-glycoprotein and that the protein is conserved in size and immunological cross-reactivity across species. Indeed P-glycoprotein appears to be an example of a fundamental membrane transport protein that has been conserved through evolution from bacteria to man.

In higher eukaryotes, P-glycoprotein is in fact a small family of highly homologous proteins. Genetic studies have endeavored to determine the evolution, arrangement, localization, and control of this multigene family. The analysis of P-glycoprotein cDNA and genomic sequences has led to the proposal of a model for the structure of P-glycoprotein and its involvement in drug transport. The molecule has a tandemly duplicated structure, each half encoding six potential transmembrane domains and the consensus sequences for a nucleotide-binding site. Drugs are proposed to bind directly to P-glycoprotein and then to be actively effluxed from the cell via a pore or channel formed by the molecule's multiple transmembrane domains using energy derived from P-glycoprotein-mediated ATP hydrolysis.

A number of compounds, originally described as calcium channel blockers, calmodulin inhibitors, or lysosomotropic agents, have been found to "reverse" MDR. Many of these drugs are used to treat noncancer patients, and represent a basis upon which to design agents suitable as "chemosensitizers" for the treatment of clinical drug resistance. Expression of P-glycoprotein in human tissues and cancers is also being extensively explored.

In addition to P-glycoprotein-mediated MDR, there are increasing numbers of reports on the isolation of MDR cell lines that apparently do not overexpress P-glycoprotein. In general they exhibit a more limited range of cross-resistance, and are resistant to lower concentrations of drugs than cells that overexpress P-glycoprotein. Changes in topoisomerase II, protein kinase C, or specific glutathione transferase isozymes have been reported. These mechanisms of drug resistance will not be reviewed here, except as they bear on MDR mediated by P-glycoprotein.

Many previous reviews have ended with the conclusion that P-glycoprotein can be causative of MDR. This review will assume this as a tenet, and will instead concentrate on reviewing P-glycoprotein-mediated MDR by showing how the results of recent work may help us to gain further insights into the following important questions about P-glycoprotein. What mechanisms regulate P-glycoprotein expression? What structural features of the protein are important for function? What is the normal function of P-glycoprotein? What features of the protein are responsible for the diversity of the phenotype? What is the importance of the gene family to the normal function of P-glycoprotein and to the development of resistance? Does P-glycoprotein contribute to clinical drug resistance? A number of recent reviews describe the isolation and characterization of MDR cells, and complement this review (7–10).

CHARACTERIZATION OF THE P-GLYCOPROTEIN MULTIGENE FAMILY

P-glycoprotein has been unambiguously identified in many MDR cell lines through the extensive use of P-glycoprotein-specific monoclonal antibodies (11). However, the identification of the different P-glycoprotein molecules has progressed more rapidly using molecular biological techniques. Complete cDNA sequences have now been reported for two human (12, 12a) and two mouse P-glycoproteins (13, 14), and partial sequences have been published for two Chinese hamster sequences (15). Gene-specific DNA probes have been used to study the expression of P-glycoprotein. Transfection experiments have unambiguously shown that alterations in P-glycoprotein expression are sufficient to cause MDR. These experiments are also beginning to reveal how different members of the P-glycoprotein multigene family may mediate the MDR phenotype.

Monoclonal antibodies provide the most exact method for determining the expression of different P-glycoprotein genes in normal and tumor tissues and in MDR cell lines. The epitopes of three monoclonal antibodies, C219, C494, and C32 (11), have been mapped to the cytoplasmic domain of P-glycoprotein (E. Georges, et al, manuscript submitted). cDNA sequence analysis has confirmed that the epitope recognized by C219 is present in all P-glycoprotein molecules whose protein sequence is known, and that the antibody therefore represents a universal probe for the detection of P-glycoprotein. The monoclonal antibody C494 binds to the P-glycoprotein *pgp*1 product in human and hamster cells. A third monoclonal antibody in the series, C32, is hamster-specific and recognizes the P-glycoprotein products of hamster *pgp*1 and *pgp*2. These antibodies therefore are potentially useful for localizing the tissue-specific expression of different P-glycoprotein gene family members. A number of other P-glycoprotein-specific monoclonal antibodies have been reported (16–19), and these are described in Table 1. The monospecific antibody P7 (20) binds to a short sequence in the N-terminal region of the human *mdr*1 (*pgp*1) gene product, and is the first polyclonal antibody reported against a specific P-glycoprotein peptide. Antibody probing of P-glycoprotein expression will no doubt yield important advances in the future.

Cloning of the P-Glycoprotein Genes

P-glycoprotein-specific monoclonal antibodies have been used to isolate P-glycoprotein cDNA clones from libraries prepared in the expression vector, λgt11 (21, 22). A P-glycoprotein cDNA sequence was first isolated from the MDR CHO cell line CH^{R}B30 (21). The clone (pCHP1) was shown to encode

Table 1 P-glycoprotein-specific monoclonal antibodies

Antibody name	Cell line used in inoculation	Notes	Ref.
Hamster			
C219 C494 C32	CHO CH^{R}B30 and CEM/VLB$_{500}$	Each monoclonal antibody recognizes independent intracellular epitopes. Precipitates P-glycoprotein of 170 kd from CHO MDR CH^{R}C5 cells.	11
265/F4	CHO CH^{R}C5	Recognizes an external epitope. Precipitates P-glycoprotein of 170 kd from CHO MDR CH^{R}C5 cells.	16
Human			
MRK16 MRK17	K562/ADM	Recognize external epitopes. 170–180 kd P-glycoprotein from MDR K562/ADM and 2780^{AD} cells.	18
32G7 9A7 1F10	CEM/VLB$_{100}$	Precipitates a broad P-glycoprotein band of 180–210 kd from CEM/VLB$_{100}$ but not from CEM cells. 9A7 and 1F10 precipitate additional bands at 155 kd and 130 kd, respectively.	17
HYB-612	MC-IXC/VCR	Recognizes a 180 kd P-glycoprotein from MC-IXC/VCR, CEM/VLB$_{100}$, and hamster DC-3F MDR sublines.	19

P-glycoprotein by (*a*) its detection by three monoclonal antibodies (C219, C494, and C32) that bind to independent epitopes on the P-glycoprotein molecule (11), (*b*) its hybridization to an mRNA of 4.7 kb (a plausible length for P-glycoprotein mRNA given P-glycoprotein's molecular size) that is consistently expressed in increased amounts in MDR cells, and (*c*) its hybridization to amplified genomic DNA sequences in different MDR cell lines. This short probe of only 630 bp hybridized to 12 differentially amplified bands on Southern analysis of hamster genomic DNA following digestion with the restriction enzyme *EcoR1* (21). This result was the first to suggest that P-glycoprotein may be encoded by a multigene family. pCHP1 also cross-hybridized to human and mouse sequences, indicating that P-glycoprotein was conserved across species.

pCHP1 has also been used to isolate clones from a cDNA library (23) prepared from a drug-sensitive CHO cell line (15). Clones isolated from this library fell into two categories corresponding to transcripts from two different genes, which were termed P-glycoprotein genes *pgp*1 and *pgp*2. Four different P-glycoprotein cDNA transcripts were detected in this drug-sensitive cell line as a result of differential polyadenylation. A third Chinese hamster P-glycoprotein gene (*pgp*3) has now been identified by genomic and cDNA sequence analysis (24; J. A. Endicott, F. Sarangi, unpublished results).

P-glycoprotein genomic DNA and mRNA sequences are amplified in MDR cells. This has allowed P-glycoprotein sequences to be cloned by the techniques of either differential hybridization (25), or in-gel renaturation (26). These methods are advantageous because other genes coamplified with P-glycoprotein have also been isolated, and they may be found to contribute to the diversity of the MDR phenotype.

One study, using the hamster MDR cell line CH^{R}C5 and the technique of differential hybridization to labeled single-stranded cDNA from CH^{R}C5 and to cDNA from the parental drug-sensitive cell line AuxB1, isolated five different linked classes of transcripts (classes 1–5) on the basis of their hybridization to mRNAs of different lengths on Northern blot analysis (27, 28). Since this work was first described, a sixth amplified sequence (class 6) has been isolated (29). The class 2 cDNA clones were found to hybridize to an mRNA transcript of 4.7 kb, and to cross-hybridize to the P-glycoprotein cDNA clone pCHP1. This transcript was the only one consistently amplified in various MDR cell lines (28). Mapping experiments using pulse field gradient gel electrophoresis have shown that there are three class 2 genes in CH^{R}C5 (28). These have been designated class 2a, 2b, and 2c, and correspond to the three hamster *pgp* genes.

One of the hamster P-glycoprotein-specific cDNA clones, cp22 (1.3 kb), which maps to the highly conserved C-terminal domain of P-glycoprotein, was subsequently used to isolate clones from cDNA libraries prepared from human liver tissue and the drug-sensitive human liver cell line HepG2 (30). Sequences for two P-glycoprotein gene family members were isolated. One sequence corresponded to the previously reported human *mdr*1 sequence (12). The second was transcribed from another P-glycoprotein gene, which was termed *mdr*3 because it is apparently the human homologue of the hamster *pgp*3 gene (30, 12a).

Transcripts from human *mdr*3 are of special interest because they have been shown to undergo differential splicing, and because to date they have only been detected in human liver. Three different human *mdr*3 transcripts were isolated from both cDNA libraries. The predominant isolate aligned with the *mdr*1 sequence. The other two variant sequences have either an in-frame 21-base-pair insertion between the region encoding the highly conserved 'A' and 'B' nucleotide-binding consensus sequences in the C-terminal cytoplasmic domain of P-glycoprotein, or alternatively a deletion that removes sequences encoding the fifth potential transmembrane region from the C-terminal domain (see Figure 1). This was a deletion of 129 bp in cDNA clones isolated from the liver library. However, cDNAs isolated from the HepG2 cell line had extended this deletion by another 12 bp. It is not known whether these different transcripts each encode a functional P-glycoprotein molecule. Northern blot analysis shows that the human *mdr*3 mRNA transcript is consistently shorter than the human *mdr*1 sequence (4.1 kb versus 4.5 kb), partly as a result of differences in the lengths of the 3'-untranslated regions (30). Differences in 3'-untranslated region processing have also been reported for both hamster *pgp*1 and *pgp*2 transcripts (15). The biological significance of these observations is not known. Alternative splicing of mRNA transcripts has recently been reviewed (31).

The P-glycoprotein cDNA clone p5L-18 (502 bp) was isolated from the Chinese hamster MDR cell line DC-3F/VCRd-5L by differential hybridization to C_ot-fractioned genomic DNA from DC-3F/VCRd-5L cells and then to genomic DNA from the drug-sensitive parental cell line DC-3F (32, 33). p5L-18 hybridizes to two transcripts of 4.5 and 2.3 kb on Northern analysis of mRNA from hamster MDR cell lines (32). Similar short P-glycoprotein transcripts have been detected in the hamster MDR cell line, CH[R]C5, that are apparently aberrantly spliced and have undergone premature termination and polyadenylation (J. Endicott, P. F. Juranka, unpublished results). It is not known whether the truncated P-glycoprotein molecules potentially encoded by these transcripts are synthesized, and if they contribute at all to the MDR phenotype.

In-gel renaturation (26) has been used in an independent investigation to detect commonly amplified sequences in two independently selected MDR Chinese hamster cell lines (34). The clone pDR1.1 was isolated in this study and shown to contain DNA sequences that are amplified in both hamster MDR cell lines, but it does not detect any mRNA transcripts by Northern blot analysis. pDR1.1 was subsequently used to isolate 120 kb of contiguous

A.

B.

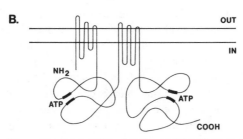

Figure 1 A. Schematic representation of P-glycoprotein. The proposed transmembrane regions are marked by numbered boxes. The two consensus sequences that form the potential ATP-binding fold common to P-glycoprotein, HlyB, and the bacterial periplasmic transport systems are marked by solid boxes lettered A and B (taken from Ref. 15). *B.* Schematic drawing of the orientation of P-glycoprotein in the cell plasma membrane (redrawn from Ref. 13).

genomic DNA amplified in hamster MDR LZ cells (35). A fragment of one cosmid clone, pDR4.7, was used to probe human genomic DNA from a number of MDR cell lines (36). *EcoR1* fragments of 13.5 kb and 4.5 kb and *HindIII* fragments of 4.4 kb and 10.5 kb were seen to hybridize to this P-glycoprotein-specific probe. Because of their consistent coamplification, the 13.5 kb *EcoR1* and the 4.4 kb *HindIII* fragments were called *mdr*1 sequences, and the 4.5 kb *EcoR1* and 10.5 kb *HindIII* fragments *mdr*2 sequences.

By means of a series of cloning steps, repeat-free, pDR4.7-positive sub-fragments were isolated from these amplified genomic sequences and were named pMDR1 (an 800 bp *Pvu*II fragment) and pMDR2 (a 1.0 kbp *Pst*I fragment). These two short gene-specific probes were then used to show that the *mdr*1 sequence is transcribed in a selection of MDR cell lines, but the *mdr*2 sequence is not expressed (36). The *mdr*2 sequence corresponds to the sequence of the *mdr*3 gene independently isolated from a human liver cDNA library (30). The genomic DNA probe pDR1.1 has also been used by a second group to isolate genomic clones from the vincristine-selected MDR CHO cell line, VCR 15 (37).

pMDR1 was used to isolate cDNA clones from the colchicine-selected human MDR cell line KB-C2.5 (12). The complete *mdr*1 P-glycoprotein coding sequence was determined from three overlapping cDNA clones. A second set of *mdr*1 cDNA clones were subsequently isolated from a vincris-tine-selected cell line, KB-V1 (38). The two *mdr*1 sequences are identical except for a cluster of nucleotide changes within a 16 bp region that results in an amino acid alteration at position 185. This residue is a valine in the *mdr*1 sequence isolated from the cell line KB-C2.5, and a glycine residue in the sequence isolated from KB-V1 cells. A glycine residue was also present at this position in the *mdr*1 sequence isolated from the cell line KB-A1 selected with adriamycin.

These findings suggested that a glycine-to-valine amino acid change oc-curred at position 185 in the P-glycoprotein sequence during the selection of the cell line KB-C2.5. Treatment with ethylmethane sulfonate (a mutagenic agent) was only used at the first steps of colchicine selection in this series of cell lines (39). The mutation was shown to appear first in the cell line KB-8-5-11-24, concurrently with the emergence of preferential resistance to colchicine (38). The mutated copy of the gene was preferentially amplified and the wild-type copies were lost. The amino acid change occurs within the N-terminal hydrophobic region of P-glycoprotein. cDNA sequence analysis predicts that this site is in the cytoplasmic domain, very close to the third potential transmembrane sequence. It was postulated that this residue forms part of a potential drug-binding site and that the change enhances P-glycoprotein's ability to bind and remove colchicine from the cell. This is the only known example of a specific amino acid change possibly being responsible for altering the drug-resistance profile of a P-glycoprotein-mediated MDR cell line.

Two P-glycoprotein sequences have been isolated from a mouse pre-B cell BALB/c line using cloned genomic *mdr* sequences (13, 14, 35). The λDR11 P-glycoprotein cDNA was isolated as a single full-length clone and has since been referred to as mouse *mdr*1 (13). This sequence is the mouse homologue of hamster *pgp*2 (15). The second mouse sequence to be determined (14) (called mouse *mdr*2) was constructed from two overlapping clones (λDR29 and λDR27), and is in fact the mouse homologue of hamster *pgp*3 and human *mdr*3 (24). Mouse *mdr*1 (*pgp*2) is the only P-glycoprotein cDNA clone that has been isolated as a full-length sequence and from a drug-sensitive cell line. Two different mouse P-glycoprotein cDNA sequences have also been isolated from MDR mouse J774.2 cell lines (40).

How might the P-glycoprotein multigene family contribute to the complexity of the MDR phenotype? P-glycoprotein molecules may differ in their efficiencies in transporting different drugs, and under stringent selection conditions more efficient P-glycoprotein molecules may be selected. Mouse J7.V3-1 cells selected for resistance to 40 nM vinblastine (40) have been found to preferentially express only one gene family member. In the human MDR KB-8-5-11-24 cell line, selected with 1000 ng/ml colchicine (39), mutation has altered a single amino acid residue in the *mdr*1 gene product (38). The human *mdr*3 (*pgp*3) gene can be differentially spliced (30). Human cells do not appear to have a gene equivalent to *pgp*2. Preliminary information using a P-glycoprotein-specific probe prepared from a highly conserved exon sequence (24) suggests that in some species the number of P-glycoprotein genes may be as high as five (L. Veinot-Drebot, W. Ng, unpublished results). The relative contributions of each of these events to determining the MDR phenotype could be envisaged to depend on both the number of P-glycoprotein genes available for transcription and the drug and stringency employed in the selection protocol. Complexity in the MDR phenotype may arise through the differential expression of a multigene family that may be further diversified by mutation and by differential splicing of P-glycoprotein mRNA molecules.

P-Glycoprotein Gene Transfection

Gene transfection is currently the experimental technique that provides the closest approximation to studying individually the contribution of each P-glycoprotein gene family member to the MDR phenotype. The first reported transfer of MDR to drug-sensitive recipient cells was by transfection of genomic DNA from the MDR CHO cell line CH^{R}C5 to mouse LTA and LMTK$^-$ cells (41). The P-glycoprotein-specific cDNA probe pCHP1 (21), and probes specific for genes coamplified with the P-glycoprotein genes (27), were subsequently used to show that the amplified P-glycoprotein sequences in the transfectants originate from the hamster donor DNA and that flanking sequences known to be coamplified with P-glycoprotein in the MDR donor cell line had not been transferred (42). The monoclonal antibody, C494 (11), which discriminates between mouse and hamster P-glycoprotein, confirmed that hamster P-glycoprotein was overexpressed. P-glycoprotein gene-specific probes have shown that the *pgp*1 gene is amplified and overexpressed in the transfectant cell lines (K. Deuchars, unpublished results). Thus increased hamster *pgp*1 expression can mediate a complete MDR phenotype in mouse cells.

The MDR phenotype has also been transferred to mouse cells using genomic DNA from human MDR cell lines. Primary and secondary transfectants of mouse NIH 3T3 cells transfected with high-molecular-weight DNA from the human MDR cell line KB-C1.5 have a pattern of cross-resistance to colchicine, vinblastine, and adriamycin similar to that of the donor cell line (43). P-glycoprotein-specific DNA probes were used to show that the human *mdr*1 gene and not the *mdr*3 gene (referred to as *mdr*2) is expressed in the transfectants. Genomic DNA from human MDR K562 cells has been transfected into drug-sensitive mouse Ltk⁻ cells, and human P-glycoprotein genomic sequences were cloned from the transfectant (44). Chromosome transfer has been used to transfect DNA from MDR LZ Chinese hamster cells to mouse drug-sensitive LTA cells (45). These two experiments did not determine which P-glycoprotein genes were overexpressed in the resulting transfectant cell lines.

cDNA transfer of P-glycoprotein provides convincing evidence that increased P-glycoprotein expression is sufficient to mediate MDR. Transfection of the mouse λDR11 (*pgp*2) cDNA sequence has been reported (4, 46). In one study, the sequence was transferred to hamster drug-sensitive LR73 cells and transfectants were subsequently selected with either adriamycin or colchicine (4). When compared to the parental cell line, the transfectants showed a 10–20-fold increase in cross-resistance to adriamycin, colchicine, and vincristine, and an 8–16-fold increase in P-glycoprotein mRNA expression. P-glycoprotein gene copy number varied from 10 to 60 copies/genome and the gene had undergone rearrangements upon integration. This mouse cDNA sequence can also confer the MDR phenotype on transfected cells without prior drug selection when expressed from a retroviral vector (47).

Human *mdr*1 (*pgp*1) cDNA sequences can also mediate MDR when used in transfection experiments (5, 38). Transfection of the sequence isolated from KB-C2.5 cells and coding for a valine residue at amino acid position 185 conferred MDR on mouse NIH 3T3 cells and human KB cells (5). The P-glycoprotein gene copy number in the transfected cell lines was variable, and did not correlate well with relative drug resistance. Cross-resistance to colchicine, vinblastine, and doxorubicin was tested, and the levels of relative resistance to these drugs ranged from 1.5 to 15-fold. This human *mdr*1 sequence has also been expressed in dog kidney MDCK cells from a retroviral vector (48). In another series of experiments, the two *mdr*1 sequences (see above) have been transfected separately into drug-sensitive KB-3-1 cells (38). In general the transfected sequences integrated only once into the host genome and were not amplified. The "mutant" P-glycoprotein isolated from a colchicine-selected cell line and coding for a valine residue at position 185 had an increased resistance to colchicine when compared to the resistance profile induced by expression of the other *mdr*1 sequence coding for a glycine residue at position 185.

In summary, cDNA and genomic transfection experiments have shown that increased expression of the human *mdr*1 (*pgp*1), hamster *pgp*1, and mouse *mdr*1 (*pgp*2) sequences can mediate the MDR phenotype. Attempts at repeating these experiments using the human *mdr*3 (*pgp*3) or mouse *mdr*2 (*pgp*3) sequences have not been reported. Significant expression of the human *mdr*3 gene has only been detected in liver (30). Thus there is a distinct possibility that in human cells, MDR may be mediated by increased expression of only the *mdr*1 P-glycoprotein gene, and that one of the P-glycoprotein sequences (termed '*mdr*' sequences), may not be associated with multiple drug resistance at all. The function of the *pgp*3 (*mdr*3) protein remains to be determined.

In rodent cells the situation is different. Three P-glycoprotein genes exist and either *pgp*1 or *pgp*2 sequences can mediate MDR. cDNA transfection of the mouse *pgp*2 (*mdr*1) sequence or increased expression of only the hamster *pgp*1 gene in genomic transfection experiments have been reported. At least two (*pgp*1 and *pgp*2) genes are expressed in the MDR cell line CH^{R}C5 (15; K. Deuchars, P. F. Juranka, J. Endicott, unpublished results). However, both sequences may not be required for MDR: either sequence may be fortuitously amplified and subsequently expressed as a result of its location close to the selected gene. In hamster cells, selection pressure has been proposed to act on *pgp*1 (29), as *pgp*3 coamplification was apparently unrelated to the MDR phenotype in a number of cell lines (28). Expression of the hamster *pgp*3 gene (like expression of the human and mouse *pgp*3 sequences) may also be unable to cause MDR.

The interesting observation remains that cDNA transfection of a single P-glycoprotein gene family member results in a variable cross-resistance phenotype being transferred to the recipient drug-sensitive cells. Cell lines that result from transfection of P-glycoprotein cDNA sequences apparently are less easily selected to high levels of drug resistance than cell lines transfected with genomic DNA from MDR cells. These observations suggest that other factors may contribute to the P-glycoprotein-mediated MDR pheno-type. If so, these factors must also be highly conserved because P-glycoprotein can mediate the phenotype across species.

P-Glycoprotein Expression in MDR Cell Lines

A number of MDR cell lines have been isolated that show a good correlation between the levels of DNA amplification, increased mRNA expression, and MDR. P-glycoprotein-specific DNA probes and monoclonal antibodies are now being used to study P-glycoprotein expression more accurately in MDR cell lines, especially those lines that have low levels of drug resistance and that may mimic the clinical situation more closely. The situation appears to be more complex than might first have been envisaged. There is sugges-tive evidence that the regulation of P-glycoprotein expression may occur at the levels of gene transcription, mRNA translation, and/or protein stability. A direct way to determine the relative importance of these events is to study changes in P-glycoprotein expression in serially selected cell lines. Only a limited number of cell lines have been analyzed systematically in this manner.

A series of MDR cell lines has been derived from human ovarian SKOV3 cells (G. Bradley, et al, manuscript submitted) by selection in vincristine or vinblastine. These cell lines have a rather complicated pattern of P-glycoprotein DNA, mRNA, and protein sequence amplification. At low levels of drug resistance (up to 64-fold relative resistance to vinblastine, and 16-fold relative resistance to vincristine in the cell line SKVLB0.03 selected with vinblastine), increase in P-glycoprotein mRNA and protein occurs with-out concomitant DNA amplification. P-glycoprotein gene amplification is only observed at subsequent steps. During the selection of the SKVCR0.25 and SKVCR2 cell lines to high levels of drug resistance, P-glycoprotein expression increased without any further alterations in the levels of DNA amplification or mRNA expression.

A human *mdr*1 probe and the 'slot-blot' technique have been used to quantitate P-glycoprotein DNA and mRNA levels in three series of human KB MDR cell lines selected with either colchicine, vinblastine, or adriamycin (39, 50). The colchicine-selected cell line, KB-8-5, was shown to have a threefold increase in both the level of P-glycoprotein mRNA and relative drug resistance without P-glycoprotein DNA amplification (50). DNA amplifica-tion was again only observed after further selection for resistance to higher colchicine concentrations. Gene copy number did not correlate with relative resistance in one derivative, KB-8-5-11-24. This cell line is 128-fold resistant to colchicine, but has only a ninefold amplification of the *mdr*1 and *mdr*2 sequences. As previously discussed, this increased preferential resistance to colchicine in KB-8-5-11-24 occurs concomitantly with a mutation in the P-glycoprotein sequence (38). Overexpression of human *mdr* sequences without P-glycoprotein DNA amplification has also been reported in MDR human sarcoma cells (51), and in MDR derivatives of the human breast cancer cell line, MDA-231 (22). A detailed study of DNA, mRNA, and drug resistance levels has been carried out on a series of human MDR cell lines (51a). All the cell lines had disproportionately high levels of P-glycoprotein mRNA expression when compared to the levels of DNA amplification. Discrepancies between the levels of P-glycoprotein DNA amplification, mRNA overexpression, and MDR have also been reported in several human revertant cell lines (36, 50, 52).

Fewer examples of differences in the degree of P-glycoprotein DNA amplification and mRNA and P-glycoprotein expression have been reported in rodent cells (28, 32). A detailed study of P-glycoprotein expression has been carried out in a series of vincristine-selected MDR mouse J774.2 cell lines (53, 54). The cell lines J7.V3–1, J7.V1–1, J7.V2–1 have equivalent levels of drug resistance. However the cell line J7.V3–1 has only one 30th the level of P-glycoprotein DNA amplification and one quarter the level of P-glycoprotein mRNA expression of the two cell lines J7.V1–1 and J7.V2–1. J7.V3–1 has been found to express one member of the P-glycoprotein gene family preferentially (40, 54).

In summary, it appears that MDR cell lines can be selected that show no P-glycoprotein DNA or RNA amplification, but only with low levels of drug resistance. Subsequent selection for high levels of resistance results in DNA amplification and overexpression of P-glycoprotein mRNA. In general, the levels of DNA, mRNA, and P-glycoprotein expression correlate reasonably well, but a number of exceptions have been reported. In addition to the most commonly observed phenomenon of increased P-glycoprotein mRNA and protein levels as a result of DNA amplification, increases in mRNA and protein expression can also occur without P-glycoprotein gene amplification. This suggests that P-glycoprotein may be transcriptionally and/or translationally regulated. Cell lines have also been described that express increased amounts of P-glycoprotein without concomitant DNA or RNA amplification. These experiments suggest translational control of P-glycoprotein mRNA. Changes in P-glycoprotein stability and/or activity may also affect the MDR phenotype.

The P-glycoprotein promoter sequence has been isolated from the human *mdr*1 gene (55, 56). The start sites for transcription have been mapped using the techniques of primer extension and S1 nuclease mapping. Transcription starts at two major start sites, 136 and 140 bp 5' to the first AUG sequence, in normal human kidney, liver, adrenal, and colon tissues and in the drug-sensitive HepG2 and KB-8-5 cell lines. In a number of MDR cell lines, additional minor transcripts are also initiated at other promoter sequences. In the adriamycin- and vinblastine-selected cell lines KB-A1 and KB-V1, minor transcripts (less than 20%) start at sites located 155 to 180 bases 5' to the first AUG codon. In the colchicine-selected lines KB-8-5-11, KB C2.5, and KB C4.5, mRNA initiation also occurs further 5' to the major start site at positions between nucleotides -480 and -630.

A genomic fragment has been isolated from a cosmid clone that contains the major human *mdr*1 P-glycoprotein transcription start site (55). This sequence can act as a promoter when linked to the CAT gene in monkey CV-1 and human KB cells (55). The putative promoter has a consensus CAAT box, two G/C boxes, and no TATA box. When P-glycoprotein transcription initiates at this start site, the cDNA encodes one noncoding exon. In KB-C2.5 cells, there are two additional noncoding exons when transcription begins at the upstream start site. The selective use of these different transcription start sites may explain the increased expression of P-glycoprotein mRNA without DNA amplification in these colchicine-selected cell lines (50).

Chromosomal Alterations in MDR Cells

The P-glycoprotein genes map to chromosome region 1q26 in hamster (57, 58), to chromosome 5 in mouse (49), and to chromosomal region 7q21–31 in man (59, 60). The P-glycoprotein amplicon has been elegantly studied in both hamster (CH^{R}C5) cells and a selection of human cell lines (27, 28, 51a). By use of pulse field gradient electrophoresis, it has been shown that 1100 kb of DNA can be coamplified with P-glycoprotein sequences (27, 29), and that this DNA codes for five additional genes (termed gene classes 1 and 3–6) (51a, 57). Within this domain, P-glycoprotein sequences must be quite extensive. In human and hamster cells, the *pgp*1 (*mdr*1) and *pgp*3 (*mdr*3) genes map within 500 kb (29, 30). An independent investigation had previously proposed a minimum size for a hamster P-glycoprotein gene of 75 kb (35). A study of the P-glycoprotein genes in CHO cells has shown that the exon arrangement of the P-glycoprotein gene is conserved both between the two halves of the gene and between gene family members (W. Ng, et al, 1989. *Mol. Cell. Biol.* In press).

Five gene classes that are coamplified with P-glycoprotein have been isolated, and they appear to map within the P-glycoprotein amplicon (27–29). However, only the product of the class 4 gene has been identified to date (61). This gene encodes V19 (alternatively known as sorcin), a cytosolic calcium-binding protein that is homologous to calpain (61). This protein had previously been identified in increased amounts in vincristine-resistant Chinese hamster and mouse cell lines (62). In one study (63), increases in sorcin expression paralleled changes in DNA amplification, suggesting that increased sorcin expression may result from gene amplification. The discovery of the sorcin protein and the subsequent cloning of the sorcin gene has provided an elegant example of two very different experimental approaches converging to the same result. Four other RNA transcripts are also overexpressed by being fortuitously coamplified with the P-glycoprotein genes. It is not known what role, if any, these additional proteins may play in modulating the MDR phenotype, but it is interesting to note that the overall organization of the genes identified within the P-glycoprotein amplicon is conserved among human, mouse, and hamster cells (51a, 63a).

Chromosomal mapping studies of the P-glycoprotein gene in MDR cells show that under certain selecting conditions the gene may be translocated to various positions in the genome and undergo extensive amplification. Double minutes (DMs) and homogeneously staining regions (HSRs) are both common karyotypic alterations found in many MDR cells. [For a review see (10).]

The CH^{R}C5 cell line is pseudodiploid with two abnormally banding chromosomal regions. Both these regions are strongly labeled in situ by a P-glycoprotein probe and by probes specific to the genes coamplified with P-glycoprotein (described in Ref. 29). By analyzing the series of cell lines that led to the isolation of the MDR cell line CH^{R}C5, an interesting history of DNA amplification, rearrangement, and translocation of the P-glycoprotein genes through the hamster genome can be traced (29). P-glycoprotein gene amplification has also been shown to be quite complex in a series of hamster MDR DC-3F-derived cell lines (64). However, P-glycoprotein gene amplification does not always result in such complex chromosomal aberrations. Three independently selected MDR CHO cell lines show 30–60-fold levels of amplification of the P-glycoprotein genes. Interestingly, these amplified copies appear to be located near the original locus of the P-glycoprotein genes in the long arm of chromosome 1 (65).

The structure of the amplified P-glycoprotein DNA sequences may affect P-glycoprotein gene expression, which in turn could cause differences in the MDR phenotype. An extensive karyotypic analysis has been carried out on the B16-BL6 murine melanoma cell line selected in doxorubicin (66). This cell line has extensive heteroploidy and abnormal banding regions, which show specific changes in staining intensity and in band length on increasing doxorubicin resistance. Doxorubicin resistance was also studied in two human cell lines, chosen because they have very different inherent sensitivities to the drug (67). HT1080 is a fibrosarcoma and LoVo a colon adenocarcinoma. Both were originally near diploid, with IC_{50} values for doxorubicin of 0.006 μg/ml and 0.10 μg/ml, respectively. On selection in increasing concentrations of doxorubicin, the MDR derivatives of the LoVo cell line displayed minimal chromosomal changes. However, cell lines derived from HT1080 cells exhibited marked aneuploidy, DMs, and HSRs. Chromosome 7q, which is the location of the human P-glycoprotein genes, was the most frequent site of genetic change. This experiment provides convincing evidence that different cell lines respond to drug differently and that the inherent drug sensitivity of a cell line may affect the subsequent selection and MDR phenotype of its drug-resistant derivatives.

A MODEL FOR THE FUNCTION OF P-GLYCOPROTEIN IN MDR CELLS

cDNA Sequence Analysis

Prior to DNA sequencing of P-glycoprotein cDNA clones, P-glycoprotein had been shown to be conserved across species by molecular weight determinations, identification of cross-reactivity with monoclonal antibodies (11), and DNA probing (21). cDNA sequence analysis has confirmed that there is a remarkable degree of homology between different P-glycoprotein gene family members within a species and across species.

An analysis of the hamster *pgp*1 and *pgp*2 protein sequences in the C-terminal cytoplasmic domain shows that they differ in only 13 amino acids (15). The homologue of each gene family member in different species can be identified by comparison of either the protein-coding or the 3'-untranslated sequences (15, 30). Multiple P-glycoprotein sequences have been reported in hamster and designated *pgp*1, *pgp*2, and *pgp*3 (15, 24). They are likely the result of gene duplication events. Sequence analysis suggests that the hamster *pgp*1 and *pgp*2 genes and the two human P-glycoprotein genes are undergoing concerted evolution (15, 30). However, the human *mdr*1 and *mdr*3 genes (30) are not as homologous as the hamster *pgp*1 and *pgp*2 genes (15). The hamster *pgp*3 sequence has also diverged from the hamster *pgp*1 and *pgp*2 sequences (24). These results again suggest that the P-glycoprotein product of *pgp*3 may have a different function from the products of the *pgp*1 and *pgp*2 genes that are implicated in MDR.

From an analysis of the human, mouse, and hamster P-glycoprotein cDNA sequences, a structure for P-glycoprotein and a model for its function have been proposed (12, 13, 68). The P-glycoprotein sequence consists of a tandem duplication of circa 590 amino acids, with an additional short sequence of circa 40 amino acids at the start and a longer (circa 60-amino-acid) linker region connecting the two halves of the protein. As P-glycoprotein is a plasma membrane protein, its sequences were analyzed for potential transmembrane domains. By using two different methods of calculation (69, 70), each half of the P-glycoprotein molecule can be divided into an N-terminal hydrophobic domain, and a more hydrophilic C-terminal region (see Figure 1). The hydrophobic domain can encode six potential transmembrane sequences. The orientation of P-glycoprotein across the plasma membrane is known from studies using monoclonal antibodies, which have localized the C-terminus to the cytoplasm (11, 68).

P-glycoprotein sequences were also compared to other protein sequences available in the GenBank, NBRF, and EMBL DNA Sequence Databases (12, 13, 68). The hydrophilic domains of P-glycoprotein showed sequence homology to a group of bacterial transport proteins involved in the uptake of various small molecules into the bacterial cell. These included HisP (histidine), MalK (maltose), and PstB (phosphate) transport proteins [for a review see (71)]. In particular, two short sequences that form the consensus sequences for the proposed nucleotide-binding domains in these proteins were identical to the P-glycoprotein sequences; but significant homology extended beyond this region. A remarkable degree of sequence homology was found between P-glycoprotein and the bacterial transport protein HlyB, which is responsible for the export of HlyA (a 107 kd protein) from strains of hemolytic bacteria [for a review see (72)]. An alignment of the C-terminal 232 amino acids of hamster P-glycoprotein (the *pgp2* gene product) with the C-terminal 230 amino acids of HlyB showed that only three breaks had to be introduced into the P-glycoprotein sequence to align it perfectly with the bacterial protein. Of the 228 possible amino acid matches, 107 (46.9%) were identical, and a further 63 (27.6%) were conserved substitutions (68). Such a high degree of sequence homology between a bacterial and a eukaryotic protein is remarkable and suggests that in addition to binding nucleotides, this region may contain domains for other common functions, possibly associated with energy-dependent transport.

Based on the sequence homology between P-glycoprotein and known transport proteins, a model for P-glycoprotein has been proposed in which P-glycoprotein actively effluxes drugs through a pore or channel in the membrane formed by the potential transmembrane domains of one or more P-glycoprotein molecules (12, 13, 68). Energy is presumably derived from ATP hydrolysis utilizing ATP bound to one or both proposed nucleotide-binding sites. By analogy with hemolysin transport, it has been proposed that an alternative to direct drug transport by P-glycoprotein might exist (68). HlyB exports HlyA, a 107-kd protein, from hemolytic *Escherichia coli*. P-glycoprotein might similarly transport a molecule from mammalian cells to which drugs could bind nonspecifically and "piggy-back" out of the cell. Such a model would explain P-glycoprotein's ability to transport a wide variety of hydrophobic compounds. However, the results of subsequent in vitro studies with photoaffinity-labeled compounds suggest that P-glycoprotein can bind drugs directly, though this binding has not been associated with active drug transport. Figure 1 shows a schematic of the P-glycoprotein structure and its proposed orientation through the plasma membrane. The number of P-glycoprotein molecules that are required to form the channel has not been determined.

Modification of P-Glycoprotein

A range of molecular sizes (130–200 kd) has been reported for P-glycoprotein based on its mobility on gel electrophoresis. A comparative study has shown that much of this heterogeneity can result from different experimental protocols (73). The mobility of P-glycoprotein in different gel systems is unusually sensitive to the conditions employed. The full-length P-glycoprotein cDNA sequences isolated from mouse and human cells predict a P-glycoprotein molecular size of circa 140 kd (12, 13). The higher molecular weights observed for the mature glycoprotein result from extensive P-glycoprotein posttranslational modification. This modification includes at least glycosylation and phosphorylation. The carbohydrate moiety can account for 30–40 kd of the mature P-glycoprotein.

GLYCOSYLATION   Recently it has been shown that the multiple forms of P-glycoprotein observed in a number of MDR mouse cell lines are likely the products of different P-glycoprotein genes that can undergo differential N-linked glycosylation (53, 54). The vinblastine-selected cell line J7.V1–1 and the colchicine-selected cell line J7.C1–100 both express a P-glycoprotein precursor of 125 kd that can undergo rapid ($T_{1/2}$ equals approximately 20 minutes), differential N-linked glycosylation to a mature P-glycoprotein of molecular size 135 kd or 140 kd respectively (53). The taxol-selected cell line J7.T1–50 expresses two distinct P-glycoprotein precursors with molecular sizes of 120 kd and 125 kd that have been shown to be the products of distinct P-glycoprotein genes (54, 74). These two precursors are endoH-sensitive, and treatment with this enzyme further decreases the observed molecular sizes by approximately 5 kd in each case (53).

P-glycoprotein glycosylation has also been studied in MDR KB cells (20). Only a single P-glycoprotein precursor of 140 kd is observed in these cell lines that undergoes maturation to a final molecular size of 170 kd through N-linked glycosylation. The P-glycoprotein half-life in KB-V1 cells was estimated to be between 48 and 72 hours (20). Earlier studies did not encourage the study of P-glycoprotein glycosylation. Both pronase treatment and cell growth in tunicamycin (75), and mutants with greatly reduced cell-surface glycosylation (76) had shown that P-glycoprotein glycosylation was not necessary for the MDR phenotype.

PHOSPHORYLATION   Early studies suggested that changes in P-glycoprotein phosphorylation can alter the MDR phenotype (77). The finding that P-glycoprotein is phosphorylated in vivo (20, 78, 79; E. Georges, unpublished results) and may be involved in the regulation of the drug efflux mechanism has led to renewed interest in P-glycoprotein phosphorylation.

Factors that may affect P-glycoprotein phosphorylation have been studied using partially purified membrane preparations from MDR mouse J774.2 sublines (79). Under certain conditions, P-glycoprotein was proposed to be a substrate for the cAMP-dependent protein kinase A. Phosphorylation occurs on both serine and threonine (but not tyrosine) residues in these cell lines (79,

80). Although it was suggested (79) that P-glycoprotein is not auto-phosphorylated, further evidence is required to rule out this possibility conclusively. These experiments were carried out in vitro, and there is a distinct possibility that inactivation of the P-glycoprotein enzymatic activity may have occurred.

P-glycoprotein appears to be phosphorylated in the basal state in MDR human K562/ADM cells and can be further phosphorylated, on different serine residues, by both known activators of protein kinase C (the phorbol esters PMA, and OAG) (81) and by agents that have been shown to reverse MDR (verapamil and trifluoperazine) (78). These results suggest that P-glycoprotein may be a substrate for different protein kinases.

Verapamil and trifluoperazine are able to inhibit the binding of drugs to P-glycoprotein and can reverse aspects of the MDR phenotype. In most cases these agents inhibit drug binding in a concentration-dependent manner, but evidence for direct competition at a drug-binding site is lacking. The results of the experiments described above suggest that verapamil and trifluoperazine may affect P-glycoprotein function by causing alterations in P-glycoprotein phosphorylation. The effect of P-glycoprotein phosphorylation on drug transport or on the MDR profile of MDR K562/ADM cells was not reported. Activation of protein kinase C by addition of the phorbol ester $P(BtO)_2$ has been reported to induce MDR in drug-sensitive MCF 7 cells and to increase the degree of drug resistance observed in MDR MCF 7 cells (82). MDR MCF 7 cells, however, have elevated levels of protein kinase C activity compared to the drug-sensitive parental cell line. This result, therefore, may not reflect a general property of MDR cells. However it does indicate that other protein changes in MDR cells that can affect protein kinase activity may cause alterations to the MDR phenotype.

The sorcin protein may be able to affect indirectly the MDR profile of cell lines in which it is overexpressed by altering protein kinase activity. As explained previously, sorcin is often overexpressed in MDR cells because its gene has been fortuitously coamplified with the P-glycoprotein genes (61). Antibodies against sorcin have been isolated (82a, 83). Sorcin can bind calcium (82a, 83), and its concentration in the cell has been reported to reach sufficiently high levels that it might be expected to interfere with cellular calcium metabolism (84).

In summary, the importance of P-glycoprotein phosphorylation to either its normal function or to its ability to transport drugs is not known. P-glycoprotein appears to be phosphorylated in the basal state, and a number of different treatments have been reported that increase P-glycoprotein phosphorylation. Verapamil and trifluoperazine can increase the level of P-glycoprotein phosphorylation, which does suggest that agents that can bind to P-glycoprotein can also alter its phosphorylation state. However, definitive evidence that P-glycoprotein phosphorylation affects the MDR phenotype is not yet available. Mouse P-glycoprotein encoded by a full-length P-glycoprotein cDNA clone, λDR11 (*pgp2*), expressed in drug-sensitive LR73 Chinese hamster cells after selection in adriamycin, was found to be phosphorylated on serine residues (E. Schurr, et al, manuscript submitted). The determination of which P-glycoprotein serine and threonine residues are phosphorylated followed by mutation of these residues and subsequent cDNA transfection should provide an exact method for studying the importance of P-glycoprotein phosphorylation to the MDR phenotype.

THE IDENTIFICATION OF AGENTS THAT BIND TO P-GLYCOPROTEIN

ATP binding to P-glycoprotein has been demonstrated. This confirms the predicted nucleotide-binding sites determined from cDNA sequence analysis. Recently it has been shown that some of the drugs involved in the MDR phenotype bind to P-glycoprotein in vitro, and this may be dependent on the presence of ATP. Moreover, a number of structurally diverse compounds, some of which are able to reverse MDR, can also bind to P-glycoprotein and can inhibit drug binding. Such in vitro studies on agents that bind to P-glycoprotein are the first to directly probe the mechanism of P-glycoprotein.

ATP Binding to P-Glycoprotein

P-glycoprotein has been shown to bind the ATP analogue 8-azido-ATP (85; E. Schurr, et al, manuscript submitted). In both reports, P-glycoprotein was identified by its subsequent immunoprecipitation with either the monoclonal antibody MRK16 (85), or C219 (E. Schurr, et al, manuscript submitted). The binding of 8-azido-ATP to P-glycoprotein in membrane vesicles prepared from the MDR human cell line KB-V1 can be inhibited by excess ATP and GTP, but not excess ADP or ribose-5-phosphate (85). Vinblastine at a concentration of 100 μM did not appear to affect the efficiency of the labeling reaction.

Membrane-enriched fractions prepared from Chinese hamster LR73 cells transfected with the cloned mouse *pgp2* (*mdr1*) cDNA contain mouse P-glycoprotein that can be specifically labeled with 8-azido-ATP (E. Schurr, et al, manuscript submitted). P-glycoprotein was immunopurified and digested with trypsin. Two discrete bands, labeled to differing intensity, were detected. This result suggests that both of the P-glycoprotein's potential nucleotide-binding sites that have been deduced from cDNA sequence analysis may be able to bind ATP.

P-glycoprotein has been purified by immunoaffinity chromatography from MDR human K562/ADM cells using the monoclonal antibody MRK16, and has been observed to possess ATPase activity (86). However, the activity of the purified P-glycoprotein preparation was quite low (only 1.2 nmol ATP/mg/min). Using a P-glycoprotein-antibody-Protein A-Sepharose complex, preliminary characterization of the P-glycoprotein ATPase activity has been carried out (86a). Vincristine or adriamycin were shown not to affect the enzyme activity. However, coincubation with verapamil or trifluoperazine (agents that have been shown to "reverse" MDR) did stimulate the P-glycoprotein ATPase.

The binding of vincristine to plasma membrane preparations from human MDR K562/ADM cells has been shown to depend on the presence of ATP (86b). In the presence of 3 mM ATP, vincristine bound with high affinity to the plasma membrane ($K_D = 2.4 \pm 0.4 \times 10^{-7}$ M). When ATP was omitted from the binding experiments, vincristine was still able to bind to the membrane, but the K_D value rose to $9.7 \pm 3.1 \times 10^{-6}$ M. Nonhydrolyzable ATP analogues have been shown to block vincristine high-affinity binding to the plasma membrane preparation. [^{3}H]vincristine binding to K562/ADM cell membranes could be inhibited by excess vincristine, vinblastine, actinomycinD, and adriamycin. Interestingly, verapamil and a selection of other agents known to reverse MDR are also able to block [^{3}H]vincristine binding. These data provide circumstantial evidence that ATP hydrolysis is required for drug binding and (presumably) subsequent drug efflux. Vinblastine efflux from plasma membrane vesicles prepared from MDR human KB-V1 cells has been shown to depend on the presence of ATP (87). In this system the drug transport reaction has an apparent K_m for ATP of 3.8×10^{-5} M and a K_m of approximately 2.0×10^{-6} M for vinblastine.

These experiments suggest that P-glycoprotein can bind ATP, and that ATP is required for active drug transport. P-glycoprotein has been immunopurified from human MDR K562/ADM cells and shown to be an ATPase. ATP appears to bind to two sites on the P-glycoprotein molecule (E. Schurr, et al, manuscript submitted), and in this regard it is interesting that members of the group of bacterial transport proteins that show significant homology to P-glycoprotein may also utilize two ATP-binding sites (88). This strengthens the argument that these bacterial proteins have not only sequence homology, but also functional homology to P-glycoprotein, and may provide an excellent model system in which to study the mechanism of the coupling of ATP hydrolysis to drug transport in P-glycoprotein.

Drug Binding to P-Glycoprotein

Vinblastine has been shown to bind in increased amounts to membrane vesicles prepared from human MDR KB-C4 cells as compared to membrane vesicles prepared from drug-sensitive KB-3-1 or revertant KB-R1 cell lines (89). The binding of [^{3}H]vinblastine could be inhibited by vinblastine, vincristine, and daunomycin, but not by excess of colchicine or actinomycinD.

The addition of 10 μg/ml verapamil to the labeling reaction decreased the vinblastine binding to the MDR KB-C4 cell membranes to the levels of binding observed to membranes prepared from drug-sensitive KB-3-1 cells.

A number of reports have shown that photoaffinity analogues of vinblastine, NABV and NASV, can bind to P-glycoprotein (90–93). Proteins with molecular sizes of 180 kd (the major-labeled band) and 150 kd were labeled with NABV or NASV in whole-cell particulate fractions prepared from the hamster MDR cell lines DC-3F/VCRd-5L, DC-3F/ADX, and a revertant cell line DC-3F/ADX-U (90). The labeled proteins could be subsequently immunoprecipitated with polyclonal antibody against P-glycoprotein, and up to 96% of the labeling could be competed with 200-fold excess vinblastine.

NABV and NASV have also been used to study vinblastine binding to membrane vesicles prepared from KB-3-1, KB-C4, and KB-R1 cell lines (91). The analogue did not bind to the drug-sensitive parental cell line KB-3-1 in these experiments, and it bound in decreased amounts to the revertant (KB-R1) cell membranes. The binding of these drug analogues to membrane vesicles prepared from the KB-C4 cells was inhibited by 1 μM vinblastine, and also by 100 μM vincristine and verapamil. 100 μM excess of colchicine or of dexamethasone (a drug that is not involved in the MDR phenotype) did not substantially affect the analogue binding to P-glycoprotein. This study was extended to include a number of other agents that reverse MDR (92). Agents such as chlorpromazine, trifluoperazine, thioridazine, and chloroquine were found to be poor inhibitors of NASV labeling of P-glycoprotein at concentrations at which they reverse MDR.

NASV has also been shown to bind to P-glycoprotein isolated from MDR CEM/VLB$_{1K}$ cells (93). Notably, both reserpine and verapamil (both agents that are able to reverse MDR) are able to compete with NASV for binding to P-glycoprotein. However, when other agents known to reverse MDR were studied, their potency as agents able to reverse MDR did not consistently correlate with their ability to inhibit NASV binding to P-glycoprotein.

NABV has also been used to identify vinblastine-binding proteins from drug-sensitive cells (94). These studies could isolate other proteins that may bind vinblastine and interact with P-glycoprotein, and provide an exciting method to probe other mechanisms that may cause vinblastine resistance and/or modulate the MDR phenotype.

Compounds that Reverse MDR

Agents that reverse MDR can be grouped into three categories: drug analogues, calcium antagonists and calmodulin inhibitors, and others. The mechanisms by which these agents reverse MDR is not fully understood. The relatively noncytotoxic daunorubicin analogue, N-acetyl daunorubicin, can inhibit both active efflux and unidirectional influx of daunorubicin in both drug-sensitive and drug-resistant cell lines (95). Essentially this compound acts as a decoy for its more cytotoxic relative. Drug analogues have not been widely reported in the literature (96, 97) since the discovery that certain calcium antagonists, and calmodulin inhibitors and other drugs currently being used for patient treatment, are effective in reversing MDR. However, these calcium antagonists and calmodulin inhibitors have not been widely applied as "chemosensitizing agents" for the treatment of cancer. There are a number of reports describing their use in cancer treatment in animal models (98–101), but the doses required to achieve effects on in vitro cell lines are frequently either toxic, or are unattainable in patients. Preliminary reports have appeared (102–106), and several trials are in progress.

These agents have been shown to reverse MDR by inhibiting drug efflux and thereby increasing drug accumulation in MDR cells. A correlation has been reported between calmodulin inhibitor potency and ability to restore adriamycin sensitivity to P388 cells (107), but no such correlation was observed with a series of calcium antagonists (108). The action of these agents as calcium antagonists and calmodulin inhibitors therefore appears to be distinct from their activity towards reversing the MDR phenotype.

Indeed, calcium channel blockers and calmodulin inhibitors have been reported to have a number of diverse and apparently unrelated activities. For example, verapamil is able to block the secretion of a number of hormones (109, 110), decrease sodium conductance in neuroblastoma and heart muscle cells (111), and inhibit certain receptor/ligand binding (112, 113). The detergents Tween 80 and LubrolWX can mimic verapamil's effects on drug transport, which suggests that verapamil's activity may result in part from its ability to generally perturb the cell membrane (114, 115). Prochlorperazine has been proposed to affect the function of integral membrane proteins by a similar general, nonspecific mechanism (116). Calmodulin inhibitors bind to hydrophobic regions of calmodulin that are exposed on the surface of the molecule following conformational changes induced by calmodulin calcium binding (117). By analogy to this system, these agents may modulate the MDR phenotype by binding to hydrophobic domains within the P-glycoprotein molecule. This binding may not define functionally important P-glycoprotein domains. The differing abilities of various drugs and compounds that can reverse MDR to inhibit other drug or drug analogue binding suggests that P-glycoprotein may have multiple drug-binding sites.

Lysosomotropic agents have also been shown to reverse MDR in different studies (118–123). Increases in both membrane fluidity (124–126) and nonspecific adsorptive endocytosis (127, 128) have been observed in MDR cells, suggesting an increase in membrane trafficking between the plasma membrane and membrane-bound intracellular organelles. Fluorescence studies show that MDR cells have proportionally less daunomycin associated with the nucleus than drug-sensitive cells (129). However, when the monoclonal antibody MRK16 was used to probe the location of P-glycoprotein in MDR human KB-C4 cells, it was found to be excluded from the coated pits on the plasma membrane, and from the endocytic vesicles and lysosomes (130). Lysosomotropic agents increase lysosomal and endosomal pH (131, 132). Their ability to reverse MDR may involve interference with a drug extrusion mechanism that involves drug uptake and transport through the lysosomal/ endosomal system. This mechanism may not involve P-glycoprotein directly.

There have been numerous reports in the past nine years describing the abilities of various calcium antagonists and calmodulin inhibitors to "chemosensitize" a wide variety of different MDR cell lines (108, 133, 99, 104, 129, 115, 92, 134–147). Though reversal of MDR is a common finding, the degree to which these agents achieve this varies. Almost complete reversal of the MDR phenotype has been reported in a number of cases (99, 104, 92, 129, 142–147), but only partial reversal in others (104, 115, 134, 137, 141, 145–147). Measurements of drug accumulation in these experiments show that complete reversal of the MDR phenotype is not always accompanied by an increased accumulation of drug to levels observed in the drug-sensitive parental cell lines (129).

The success with which agents can reverse MDR may depend on the agent used, and the degree and drug resistance to be overcome. Four calcium channel blockers (diltiazem, nicardipine, niludipine, and nimodipine) were found to circumvent the MDR phenotype completely in a vincristine-selected human MDR K562 cell line, but they could only reverse the adriamycin resistance of an adriamycin-selected P388 cell line 10–30-fold (99). Coincubation with diltiazem or nicardipine resulted in only a 50–70% increase in vincristine cytotoxicity in a P388 vincristine-selected MDR cell line. In a separate study, verapamil was able to reverse completely adriamycin resistance in human ovarian cancer cells with 3–6-fold increases in adriamycin resistance, but could only partially reverse resistance in cell lines with high (150-fold) resistance levels (104). In a CHO cell line, CH^RC5, taxol toxicity was enhanced to a greater degree than daunorubicin toxicity by several agents (138). Verapamil was reported to circumvent vincristine resistance in human K562/VCR cells, and yet to affect adriamycin cytotoxicity only marginally (147). In CEM/VLB_{100} cells, verapamil (at 10 μM) decreased the IC_{50} values for a number of vinca alkaloids 75–85-fold, but caused only slight (2–5-fold) decreases in IC_{50} values for resistance to certain anthracyclines, epipodophyllotoxins, and colchicine. The CEM/DOX cell line gave qualitatively similar results (145).

Calcium antagonists and calmodulin inhibitors have been reported not to chemosensitize drug-sensitive cells in a number of studies (92, 129, 133, 134, 136, 141, 143, 144). Other studies do report sensitization of drug-sensitive cells, though the effect is much less pronounced than that observed in MDR cells (108, 135, 138, 142). This effect may involve their ability to perturb the cell membrane. It may also depend on the method of assay. Phenothiazine and naphthalene-sulfonamide calmodulin inhibitors were not found to chemosensitize drug-sensitive P388 cells if the cytotoxicity measurements were made after a short drug exposure. However, when survival based on colony formation after long-term (in this report 120 hours) drug exposure was determined, cytotoxic effects were observed in both the drug-sensitive and drug-resistant cell lines (107).

Photoactivatable calcium channel blockers, azidopine, and calmodulin inhibitors have been used to provide evidence that these agents can bind directly to P-glycoprotein (148–150). Azidopine is a calcium antagonist that on uv irradiation forms a nitrene derivative capable of specifically cross-linking to protein (151). The specific half-maximal saturable photolabeling of [^{3}H]azidopine to P-glycoprotein prepared from hamster MDR DC-3F/VCRd-5L or DC-3F/ADX plasma membranes was reported to be at a [^{3}H]azidopine concentration of 1.07×10^{-7} M (148). This value is quite low when it is considered that this agent has been reported to bind to the voltage-dependent calcium channel isolated from guinea-pig skeletal muscle T-tubules with dissociation constants approaching 3×10^{-9} M (150a). The [^{3}H]azidopine binding to P-glycoprotein could be inhibited to varying extents with azidopine, nimodipine, nitrendipine, nifedipine, verapamil, and diltiazem, and could be stimulated by prenylamine and bepridil (148, 150). In 200-fold molar excess, vinblastine was most effective at inhibiting azidopine binding (92%); actinomycinD could inhibit 56% of the azidopine binding, adriamycin 19%, and colchicine only 7% (148). A low K_D for desmethoxyverapamil binding to KB-V1 membranes has also been reported (149).

Azidopine has also been used to label P-glycoprotein from colchicine-, vinblastine-, or taxol-selected MDR mouse J774.2 cell lines (150). The labeling patterns (including a P-glycoprotein doublet in the taxol-selected cell line) resembled that seen for immunoprecipitation or antibody staining of P-glycoprotein. Thus it appears that each P-glycoprotein molecule can bind azidopine. Verapamil, desmethoxyverapamil, diltiazem, and quinidine can increase drug accumulation, and can block vinblastine photoaffinity analogue binding to P-glycoprotein in membrane vesicles prepared from MDR KB-C4 or KB-V1 cells (149). These four drugs did not affect the uptake of vinblastine in drug-sensitive KB-3-1 cells. Colchicine consistently appears as a poor competitive inhibitor of both vinblastine and NASV binding to P-glycoprotein (89, 91, 92, 145). Prenylamine and bepridil have been reported to enhance [^{3}H]-azidopine binding to P-glycoprotein (148).

Verapamil increases ATP consumption in MDR cells to a greater extent than the addition of certain drugs involved in the MDR phenotype (152).

Verapamil has also been shown to stimulate P-glycoprotein's ATPase activity (86a). A number of MDR cells are collaterally sensitive to verapamil (115, 153), and yet maintain a lower intracellular verapamil concentration than their drug-sensitive counterparts. The ability of verapamil to increase ATP consumption reaches a maximum at a lower concentration (1 μM) than its ability to inhibit drug efflux in 2780^{AD} cells (>4 μM) (152). Similarly, calcium antagonists have been shown to be poor inhibitors of NASV binding to P-glycoprotein at concentrations at which they reverse MDR. These two effects may be related, and they may indicate that verapamil's ability to reverse MDR may not only depend on its ability to inhibit drug binding to P-glycoprotein. These agents may modulate the MDR phenotype by directly affecting P-glycoprotein function. This interaction has not been fully explored and may prove to be a useful model for probing P-glycoprotein's normal function.

EXPRESSION OF P-GLYCOPROTEIN IN NORMAL TISSUE AND TUMORS

The expression of P-glycoprotein in normal human tissues (154–160), and in a number of human tumor samples (128, 154, 156–165), has been determined using both P-glycoprotein-specific monoclonal antibodies and DNA probes. The tissue distribution of P-glycoprotein has also been studied in hamster (166, 167) and mouse (J. M. Croop et al, 1989, In press).

Studies on the tissue-dependent expression of P-glycoprotein are in the early stages. The detection of P-glycoprotein at very low levels using either protein-specific or DNA-specific probes is not yet reproducible. However, a number of studies have consistently shown that P-glycoprotein is expressed in normal human liver (154, 155), kidney (154–156, 160, 164), and adult adrenal tissue (154–157), and also in pancreas (155), colon (154, 155), and jejunum (154, 155). P-glycoprotein also appears to be expressed at relatively high levels during pregnancy on the luminal surface of the secretory epithelium of the mouse uterus (167a) and in the human placenta (156). There is a possibility that P-glycoprotein expression may be hormonally induced in these organs. Low detectable levels of P-glycoprotein have also been detected in a number of other human organs, notably the skin, stomach, skeletal muscle, and ovary (154). P-glycoprotein expression may not be equivalent in the same organs in different species (154–156, 166).

Within the organs that express P-glycoprotein at detectable levels, P-glycoprotein expression is quite localized. The monoclonal antibody MRK16 binds to human liver cells on both the biliary canalicular front of hepatocytes and the apical surface of epithelial cells lining the small biliary ductules (155). In the colon, jejunum, and rectum, expression is similarly localized to the apical surface of the superficial columnar epithelial cells (155). P-glycoprotein expression in the kidney is greatest in the brush border of the proximal tubules, again on the apical surface of the epithelial cells (155). Adrenal tissue shows greater expression in the cortex than the medulla (157). Expression of P-glycoprotein in the adrenal may be developmentally regulated, as fetal and neonatal adrenal tissues do not express P-glycoprotein (155–157).

The results of tissue distribution studies immediately suggest a possible normal role for P-glycoprotein in transport or secretion. P-glycoprotein may transport different compounds in different tissues, or, by analogy to its role in drug transport in MDR cells, it may be a common route for the export of toxic compounds. The high sequence conservation of P-glycoprotein across species suggests that its function is of fundamental importance. However, the fact that a point mutation within the P-glycoprotein sequence (38) can apparently change its drug resistance phenotype suggests that its specificity may be subtly modulated. It should also be considered that different members of the multigene family may have highly conserved but divergent functions.

Preneoplastic and neoplastic liver nodules and hepatocytes from regenerating liver have elevated levels of P-glycoprotein expression and are resistant to a large number of carcinogens (168, 169). A number of similarities between MDR cells and carcinogen-induced resistance to xenobiotics in hyperplastic liver nodules have been tabulated (170). P-glycoprotein expression may be specifically induced in these tissues as a mechanism for detoxifying the cells (123, 170). Alternatively, its activation may be part of a more complex, nonspecific response of the cell to metabolic or toxic stress, akin to the heat-shock or glucose-responsive genes (171).

The observed differences in the distribution of P-glycoprotein in normal tissue have to be considered in any attempts to determine whether expression of P-glycoprotein contributes to clinical drug resistance. Certain tumors originating from tissues with higher natural levels of P-glycoprotein expression may be intrinsically drug-resistant. In tissues with no detectable P-glycoprotein expression, P-glycoprotein-mediated MDR cells would have to be selected for during chemotherapy. There have been a few reports of P-glycoprotein expression increasing in patient tumor samples following chemotherapy (154, 161, 162). A systematic survey of the levels of P-glycoprotein expression in tumor samples from cancer patients undergoing chemotherapy is required, before the importance of P-glycoprotein expression to failure of chemotherapy can be critically evaluated.

P-GLYCOPROTEIN EXPRESSION IN LOWER ORGANISMS

As noted above, P-glycoprotein shows remarkable homology to a number of bacterial transport proteins. However, these are not the bacterial homologues of P-glycoprotein, since they possess narrowly defined specificities for the transport of molecules such as amino acids, phosphate, and small oligosaccharides (71). For example, the *chvA* gene product of *A. tumefaciens* is required for the transport of a small polysaccharide, β-1,2 glycan, from the bacterial cell, an early event in crown gall tumor formation [see (172) for a review]. The *chvA* gene has been cloned and the proposed open reading frame of 1.8 kb encodes a protein with a predicted molecular size of 64 kd (G. A. Cangelosi, et al, *J. Bacteriol.* 1989. In press). The protein has significant amino acid sequence conservation with both HlyB and P-glycoprotein. Based on this sequence analysis, the ChvA protein has also been proposed to have a C-terminal cytoplasmic domain that contains the consensus sequences for nucleotide binding, and an N-terminal sequence that encodes three pairs of hydrophobic sequences that may cross the inner cell membrane.

Other bacterial species contain functional equivalents of the *chvA* gene product, and they are all used to transport molecules from the bacterial cells to aid in colonizing plant tissue (173). They therefore appear (like HlyB transport) to be a mechanism of attack rather than defense.

Chloroquine-resistant and -sensitive *Plasmodium falciparum* strains accumulate chloroquine at equivalent rates; however, the chloroquine-resistant strains are able to efflux the drug 40–50 times more rapidly (174). These strains are also cross-resistant to drugs such as mefloquine and quinine. Drug efflux can be inhibited by verapamil, diltiazem, and TMB-8, and also by vinblastine and daunomycin (174). The transport system found in *P. falciparum* has many striking similarities to the drug efflux mediated by P-glycoprotein (175).

A number of mutations have been isolated from *Saccharomyces cerevisiae* that involve alterations in drug uptake and confer a pleiotropic drug resistance phenotype. The *S. cerevisiae PDR1* gene has been isolated (176), and has been shown by DNA sequence analysis to have a number of homologies to different proteins that regulate gene expression. Therefore the *PDR1* gene is not the yeast P-glycoprotein gene, but it may encode a protein that regulates its expression.

Thus, a number of lower organisms have now been shown to express P-glycoprotein-like molecules. A common finding is their involvement in various transport processes and this provides strong corroborative evidence that P-glycoprotein is an energy-dependent membrane transport pump. P-glycoprotein may have evolved in separate species to transport different molecules. Alternatively, it may represent a common transport function that is essential to all living cells. In higher eukaryotes, P-glycoprotein has been shown to be a family of homologous proteins, and with the generation of multiple P-glycoprotein molecules, the function of the protein may have diversified.

CONCLUDING REMARKS

Progress made in recent years has yielded definitive evidence that increased expression of P-glycoprotein causes multidrug resistance. These findings have provided a fundamental conceptual framework for further investigations of this complex phenotype. However, a full understanding of the mechanism(s) of MDR continues to be a challenge. Although it is clear that P-glycoprotein functions directly in MDR cells in the active transport of a variety of drugs, the details of how it is accomplished is not known. Basic questions, such as what defines the specificity of the pump, and how energy is transduced by P-glycoprotein for active efflux, remain to be answered. Undoubtedly, with the molecular genetic tools currently available, rapid progress will be made in defining the functional domains of P-glycoprotein. The availability of purified P-glycoprotein in sufficient quantities and the prospect of reconstituting function in a completely defined system should greatly increase our knowledge of the working mechanism of P-glycoprotein. The isolation of P-glycoprotein from different species will provide further insights into the conserved sequences that may be essential for function in this class of homologous transport proteins.

The mechanisms by means of which a wide variety of structural and metabolic changes are mediated in MDR cells will be much less easily delineated. Generalized changes such as increases (124–126) and decreases (177) in membrane fluidity, alterations in lipid composition and metabolism (178–180), decreased tumorigenic capability of MDR cells (181), and alterations in membrane potential (182), for example, have been described in different MDR cell lines. In addition, specific changes, such as increased expression of the epidermal growth factor receptor (183), a number of unidentified membrane proteins (184–187), and sorcin (62), have been reported. Alterations in the levels of expression of a number of unidentified proteins (188, 189), and in the levels of topoisomerase II (190) and glutathione transferase activities (191), have all been observed in particular cases.

Perturbation and reorganization of components in the plasma membrane of MDR cells must occur where a large amount of P-glycoprotein is present, and this has been revealed in the dramatic alteration of membrane ultrastructure in MDR cells (192). Such an alteration could account for a whole host of generalized and specific changes found in MDR cells. It is not known whether or not any of these changes play a role, directly or indirectly, in the MDR phenotype. Future studies with gene transfection will separate the important changes from the nonessential ones. However, the wide-ranging pleiotropic changes observed in the P-glycoprotein-mediated MDR cells do indicate that apparently unrelated biochemical reactions can be perturbed by alterations in P-glycoprotein expression. It is not known whether or not biological systems make use of P-glycoprotein for regulation of diverse metabolic activities. In this context, work already underway defining the regulatory sequences for expression, and investigating P-glycoprotein expression in normal tissues during development may provide additional insights.

A major impetus to the study of P-glycoprotein has come from the anticipation that MDR tumor cells may play a crucial role in determining the clinical response of cancer patients to chemotherapy. Findings to date suggest that increased expression of P-glycoprotein can be detected in some leukemias and solid tumors. Continued studies in this area will determine the generality of these initial observations. Should a correlation between the presence of P-glycoprotein tumor cells and nonresponse to chemotherapy be found, P-glycoprotein may prove to be an important diagnostic marker of tumor progression. Understanding the mechanisms through which such MDR cells arise may provide additional insights into malignant development.

Future efforts to improve therapeutic efficacy will entail targeting of drugs or antibodies to P-glycoprotein to sensitize resistant tumor cells. There is considerable optimism that knowledge of the biochemistry of P-glycoprotein can be exploited, and translated into clinical applications.

ACKNOWLEDGMENTS

The authors would like to thank all their colleagues who generously shared results before publication. We thank our colleagues at the Ontario Cancer Institute for helpful discussions, and especially Drs. L. Veinot-Drebot and E. Georges for critical reading of the manuscript. J. A. Endicott is the recipient of a Medical Research Council of Canada Graduate Studentship. The studies in the authors' laboratory were supported by the National Cancer Institute of Canada and by Public Health Service Grant CA37130 from the National Institutes of Health.

Literature Cited

1. Riordan, J. R., Ling, V. 1985. *Pharmacol. Ther.* 28:51–75
2. Baker, R. M., Ling, V. 1978. *Methods in Membrane Biology*, ed. E. D. Korn, Vol. 9, pp. 337–84. New York/London: Plenum
3. Ling, V. 1981. *Mitosis/Cytokinesis*, ed. A. M. Zimmerman, A. Fover, pp. 197–209. New York: Academic
4. Gros, P., Neriah, Y. B., Croop, J. M., Housman, D. E. 1986. *Nature* 323:728–31
5. Ueda, K., Cardarelli, C., Gottesman, M. M., Pastan, I. 1987. *Proc. Natl. Acad. Sci. USA* 84:3004–8
6. Ueda, K., Cornwell, M. M., Gottesman, M. M., Pastan, I., Roninson, I., et

al. 1986. *Biochem. Biophys. Res. Commun.* 141:956–62
7. Gerlach, J. H., Kartner, N., Bell, D. R., Ling, V. 1986. *Cancer Surv.* 5:25–46
8. Gottesman, M. M., Pastan, I. 1988. *Trends Pharmacol. Sci.* 9:54–58
9. Moscow, J. A., Cowan, K. H. 1988. *J. Natl. Cancer Inst.* 80:14–20
10. Bradley, G., Juranka, P. F., Ling, V. 1988. *Biochim. Biophys. Acta* 948:87–128
11. Kartner, N., Evernden-Porelle, D., Bradley, G., Ling, V. 1985. *Nature* 316:820–23
12. Chen, C.-j., Chin, J. E., Ueda, K., Clark, D. P., Pastan, I., et al. 1986. *Cell* 47:381–89

12a. van der Bliek, A. M., Kooiman, P.
M., Schneider, C., Borst, P. 1989.
Gene. In press
13. Gros, P., Croop, J., Housman, D. 1986.
Cell 47:371–80
14. Gros, P., Raymond, M., Bell, J., Housman, D. 1988. *Mol. Cell. Biol*. 8:2770–78
15. Endicott, J. A., Juranka, P. F., Sarangi, F., Gerlach, J. H., Deuchars, K. L., Ling, V. 1987. *Mol. Cell. Biol*. 7:4075–81
16. Lathan, B., Edwards, D. P., Dressler, L. G., Von Hoff, D. D., McGuire, W. L. 1985. *Cancer Res*. 45:5064–69
17. Danks, M. K., Metzger, D. W., Ashmun, R. A., Beck, W. T. 1985. *Cancer Res*. 45:3220–24
18. Hamada, H., Tsuruo, T. 1986. *Proc. Natl. Acad. Sci. USA* 83:7785–89
19. Meyers, M. B., Rittman-Grauer, L., Biedler, J. L. 1988. *Proc. Am. Assoc. Cancer Res*. 29:295 (Abstr.)
20. Richert, N. D., Aldwin, L., Nitecki, D., Gottesman, M. M., Pastan, I. 1988. *Biochemistry*. 27:7607–13
21. Riordan, J. R., Deuchars, K., Kartner, N., Alon, N., Trent, J., Ling, V. 1985. *Nature* 316:817–19
22. Fuqua, S. A. W., Moretti-Rojas, I. M., Schneider, S. L., McGuire, W. L. 1987. *Cancer Res*. 47:2103–6
23. Elliott, E. M., Okayama, H., Sarangi, F., Henderson, G., Ling, V. 1985. *Mol. Cell. Biol*. 5:236–41
24. Ng, W., Sarangi, F., Zastawny, R., Ling, V. 1988. *Proc. Am. Assoc. Cancer Res*. 29:308 (Abstr.)
25. Brison, O., Ardeshir, F., Stark, G. R. 1982. *Mol. Cell. Biol* 2:578–87
26. Roninson, I. B. 1983. *Nucleic Acids Res*. 11:5413–31
27. van der Bliek, A. M., Van der Velde-Koerts, T., Ling, V., Borst, P. 1986. *Mol. Cell. Biol*. 6:1671–78
28. de Bruijn, M. H. L., van der Bliek, A. M., Biedler, J. L., Borst, P. 1986. *Mol. Cell. Biol*. 6:4717–22
29. Borst, P., van der Bliek, A. M. 1988. *Molecular and Cellular Biology of Multidrug Resistance in Tumor Cells*, ed. I. B. Roninson, Ch. 12. Boca Raton, Fla: CRC
30. van der Bliek, A. M., Baas, F., Ten Houte de Lange, T., Kooiman, P. M., van der Velde-Koerts, T., Borst, P. 1987. *EMBO J*. 6:3325–31
31. Breitbart, R. E., Andreadis, A., Nadal-Ginard, B. 1987. *Annu. Rev. Biochem*. 56:467–95
32. Scotto, K. W., Biedler, J. L., Melera, P. W. 1986. *Science* 232:751–55
33. Fairchild, C. R., Ivy, S. P., Kao-Shan, C.-S., Whang-Peng, J., Rosen, N., et al. 1987. *Cancer Res*. 47:5141–48
34. Roninson, I. B., Abelson, H. T., Housman, D. E., Howell, N., Varshavsky, A. 1984. *Nature* 309:626–28
35. Gros, P., Croop, J., Roninson, I., Varshavsky, A., Housman, D. E. 1986. *Proc. Natl. Acad. Sci. USA* 83:337–41
36. Roninson, I. B., Chin, J. E., Choi, K., Gros, P., Housman, D. E., et al. 1986. *Proc. Natl. Acad. Sci. USA* 83:4538–42
37. Teeter, L. D., Atsumi, S., Sen, S., Kuo, T. 1986. *J. Cell Biol*. 103:1159–66
38. Choi, K., Chen, C., Kriegler, M., Roninson, I. B. 1988. *Cell* 53:519–29
39. Akiyama, S., Fojo, A., Hanover, J. A., Pastan, I., Gottesman, M. M. 1985. *Somatic Cell Mol. Genet*. 11:117–26
40. Hsu, S. I., Lothstein, L., Greenberger, L., Goei, S., Horwitz, S. B. 1988. *Proc. Am. Assoc. Cancer Res*. 29:296 (Abstr.)
41. Debenham, P. G., Kartner, N., Siminovitch, L., Riordan, J. R., Ling, V. 1982. *Mol. Cell. Biol*. 2:881–89
42. Deuchars, K. L., Du, R.-P., Naik, M., Evernden-Porelle, D., Kartner, N., et al. 1987. *Mol. Cell. Biol*. 7:718–24
43. Shen, D.-W., Fojo, A., Roninson, I. B., Chin, J. E., Soffir, R., et al. 1986. *Mol. Cell. Biol*. 6:4039–44
44. Sugimoto, Y., Tsuruo, T. 1987. *Cancer Res*. 47:2620–25
45. Gros, P., Fallows, D. A., Croop, J. M., Housman, D. E. 1986. *Mol. Cell. Biol*. 6:3785–90
46. Croop, J. M., Guild, B. C., Gros, P., Housman, D. E. 1987. *Cancer Res*. 47:5982–88
47. Guild, B. C., Mulligan, R. C., Gros, P., Housman, D. E. 1988. *Proc. Natl. Acad. Sci. USA* 85:1595–99
48. Pastan, I., Gottesman, M. M., Ueda, K., Lovelace, E., Rutherford, A. V., Willingham, M. C. 1988. *Proc. Natl. Acad. Sci. USA* 85:4486–90
49. Martinsson, T., Levan, G. 1987. *Cytogenet. Cell Genet*. 45:99–101
50. Shen, D. W., Fojo, A., Chin, J. E., Roninson, I. B., Richert, N., et al. 1986. *Science* 232:643–45
51. Scudder, S. A., Roninson, I. B., Davatelis, G., Fukumoto, M., Sikic, B. I. 1987. *Proc. Am. Assoc. Clin. Oncol*. 6:13 (Abstr.)
51a. van der Bliek, A. M., Baas, F., van der Velde-Koerts, T., Biedler, J. L., Meyers, M. B., et al. 1988. *Cancer Res*. 48:5927–32
52. Sugimoto, Y., Roninson, I. B., Tsuruo, T. 1987. *Mol. Cell. Biol* 7:4549–52
53. Greenberger, L. M., Williams, S. S.,

Horwitz, S. B. 1987. *J. Biol. Chem.* 262:13685–89

54. Greenberger, L. M., Lothstein, L., Williams, S. S., Horwitz, S. B. 1988. *Proc. Natl. Acad. Sci. USA* 85:3762–66

55. Ueda, K., Clark, D. P., Chen, C., Roninson, I. B., Gottesman, M. M., Pastan, I. 1987. *J. Biol. Chem.* 262:505–8

56. Ueda, K., Pastan, I., Gottesman, M. M. 1987. *J. Biol. Chem.* 262:17432–36

57. Jongsma, A. P. M., Spengler, B. A., van der Bliek, A. M., Borst, P., Biedler, J. L. 1987. *Cancer Res.* 47:2875–78

58. Teeter, L. D., Atsumi, S., Sen, S., Kuo, T. 1986. *J. Cell Biol.* 103:1159–66

59. Trent, J. M., Witkowski, C. M. 1987. *Cancer Genet. Cytogenet.* 26:187–90

60. Fojo, A., Lebo, R., Shimizu, N., Chin, J. E., Roninson, I. B., et al. 1986. *Somat. Cell Mol. Genet.* 12:415–20

61. van der Bliek, A. M., Meyers, M. B., Biedler, J. L., Hes, E., Borst, P. 1986. *EMBO J.* 5:3201–8

62. Meyers, M. B., Biedler, J. L. 1981. *Biochem. Biophys. Res. Commun.* 99:228–35

63. Meyers, M. B., Spengler, B. A., Chang, T. D., Melera, P. W., Biedler, J. L. 1985. *J. Cell Biol.* 100:588–97

63a. Stahl, F., Martinsson, T., Dahllof, B., Levan, G. 1988. *Hereditas* 108:251–58

64. Biedler, J. L., Chang, T., Scotto, K. W., Melera, P. W., Spengler, B. A. 1988. *Cancer Res.* 48:3179–87

65. Sen, S., Teeter, L. D., Kuo, T. 1987. *Chromosoma* 95:117–25

66. Slovak, M. L., Hoeltge, G. A., Ganapathi, R. 1986. *Cancer Res.* 46:4171–77

67. Slovak, M. L., Hoeltge, G. A., Trent, J. M. 1987. *Cancer Res.* 47:6646–52

68. Gerlach, J. H., Endicott, J. A., Juranka, P. F., Henderson, G., Sarangi, F., et al. 1986. *Nature* 324:485–89

69. Eisenberg, D., Schwarz, E., Komaromy, M., Wall, R. 1984. *J. Mol. Biol.* 179:125–42

70. Kyte, J., Doolittle, R. F. 1982. *J. Mol. Biol.* 157:105–32

71. Ames, G. F.-L. 1986. *Annu. Rev. Biochem.* 55:397–425

72. Mackman, N., Nicaud, J.-M., Gray, L., Holland, I. B. 1986. *Curr. Top Microbiol. Immunol.* 125:159–81

73. Greenberger, L. M., Williams, S. S., Georges, E., Ling, V., Horwitz, S. B. 1988. *J. Natl. Cancer Inst.* 80:506–10

74. Greenberger, L. M., Lothstein, L., Mellado, W., Yang, C.-P. H., Han, E., Horwitz, S. B. 1988. *Proc. Am. Assoc. Cancer Res.* 29:295 (Abstr.)

75. Beck, W. T., Cirtain, M. C. 1982. *Cancer Res.* 42:184–189

76. Ling, V., Kartner, N., Sudo, T., Siminovitch, L., Riordan, J. R. 1983. *Cancer Treat. Rep.* 67:869–74

77. Carlsen, S. A., Till, J. E., Ling, V. 1977. *Biochim. Biophys. Acta* 467:238–50

78. Hamada, H., Hagiwara, K.-I., Nakajima, T., Tsuruo, T. 1987. *Cancer Res.* 47:2860–65

79. Mellado, W., Horwitz, S. B. 1987. *Biochemistry* 26:6900–4

80. Roy, S. M., Horwitz, S. B. 1985. *Cancer Res.* 45:3856–63

81. Nishizuka, Y. 1984. *Nature* 308:693–98

82. Fine, R. L., Patel, J., Chabner, B. A. 1988. *Proc. Natl. Acad. Sci. USA* 85:582–86

82a. Hamada, H., Okochi, E., Oh-hara, T., Tsuruo, T. 1988. *Cancer Res.* 48:3173–78

83. Meyers, M. B., Schneider, K. A., Spengler, B. A., Chang, T.-D., Biedler, J. L. 1987. *Biochem. Pharmacol.* 36:2373–80

84. Koch, G., Smith, M., Twentyman, P., Wright, K. 1986. *FEBS Lett.* 195:275–79

85. Cornwell, M. M., Tsuruo, T., Gottesman, M. M., Pastan, I. 1987. *FASEB J.* 1:51–54

86. Hamada, H., Tsuruo, T. 1988. *J. Biol. Chem.* 263:1454–58

86a. Hamada, H., Tsuruo, T. 1988. *Cancer Res.* 48:4926–32

86b. Naito, M., Hamada, H., Tsuruo, T. 1988. *J. Biol. Chem.* 263:11887–91

87. Horio, M., Gottesman, M. M., Pastan, I. 1988. *Proc. Natl. Acad. Sci. USA* 85:3580–84

88. Higgins, C. F., Gallagher, M. P., Mimmack, M. L., Pearce, S. R. 1988. *BioEssays* 8:111–16

89. Cornwell, M. M., Gottesman, M. M., Pastan, I. 1986. *J. Biol. Chem.* 261:7921–28

90. Safa, A. R., Glover, C. J., Meyers, M. B., Biedler, J. L., Felsted, R. L. 1986. *J. Biol. Chem.* 261:6137–40

91. Cornwell, M. M., Safa, A. R., Felsted, R. L., Gottesman, M. M., Pastan, I. 1986. *Proc. Natl. Acad. Sci. USA* 83:3847–50

92. Akiyama, S.-I., Cornwell, M. M., Kuwano, M., Pastan, I., Gottesman, M. M. 1988. *Mol. Pharmacol.* 33:144–47

93. Beck, W. T., Cirtain, M. C., Glover, C. J., Felsted, R. L., Safa, A. R. 1988. *Biochem. Biophys. Res. Commun.* 153:959–66

94. Safa, A. R., Glover, C. J., Felsted, R. L. 1987. *Cancer Res.* 47:5149–54

95. Skovsgaard, T. 1980. *Cancer Res.* 40:1077–83
96. Inaba, M., Nagashima, K., Sakurai, Y., Fukui, M., Yanagi, Y. 1984. *Gann* 75:1049–52
97. Inaba, M., Nagashima, K. 1986. *Jpn. J. Cancer Res.* 77:197–204
98. Radel, S., Bankusli, I., Mayhew, E., Rustum, Y. M. 1988. *Cancer Chemother. Pharmacol.* 21:25–30
99. Tsuruo, T., Iida, H., Nojiri, M., Tsukagoshi, S., Sakurai, Y. 1983. *Cancer Res.* 43:2905–10
100. Formelli, F., Cleris, L., Carsana, R. 1988. *Cancer Chemother. Pharmacol.* 21:329–36
101. Tsuruo, T., Iida, H., Tsukagoshi, S., Sakurai, Y. 1985. *Cancer Treat. Rep.* 69:523–25
102. Dalton, W. S., Durie, B. G. M., Salmon, S. E., Grogin, T. M., Scheper, R. J., Meltzer, P. S. 1987. *Blood* 70: 245 (Abstr.)
103. Kerr, D. J., Graham, J., Cummings, J., Morrison, J. G., Thompson, G. G., et al. 1986. *Cancer Chemother. Pharmacol.* 18:239–42
104. Rogan, A. M., Hamilton, T. C., Young, R. C., Klecker, R. W., Ozols, R. F. 1984. *Science* 224:994–96
105. Bessho, F., Kinumaki, H., Kobayashi, M., Habu, H., Nakamara, K., et al. 1985. *Med. Pediatr. Oncol.* 13:199–202
106. Nooter, K., Oostrum, R., Janssen, A., Valerio, D., Bauman, J., et al. 1988. *Proc. Am. Assoc. Cancer Res.* 29:303 (Abstr.)
107. Ganapathi, R., Grabowski, D., Turinic, R., Valenzuela, R. 1984. *Eur. J. Cancer Clin. Oncol.* 20:799–806
108. Tsuruo, T., Iida, H., Tsukagoshi, S., Sakurai, Y. 1982. *Cancer Res.* 42: 4730–33
109. Devis, G., Somers, G., van Obberghen, E., Malaisse, W. J. 1975. *Diabetes* 24:547–51
110. Eto, S., Wood, J. M., Hutchins, M., Fleischer, N. 1974. *Am. J. Physiol.* 226:1315–20
111. Galper, J. B., Catterall, W. A. 1979. *Mol. Pharmacol.* 15:174–78
112. Gerry, R. H., Rauch, B., Colvin, R. A., Adler, P. N., Messineo, F. C. 1987. *Biochem. Pharmacol.* 36:2951–56
113. Waelbroeck, M., Robberecht, P., de Neef, P., Christophe, J. 1984. *Biochem. Biophys. Res. Commun.* 121:340–45
114. Fairhurst, A. S., Whittaker, M. L., Ehlert, F. J. 1980. *Biochem. Pharmacol.* 29:155–62
115. Cano-Gauci, D. F., Riordan, J. R. 1987. *Biochem. Pharmacol.* 36:2115–23
116. Dannenberg, A., Zakim, D. 1988. *Biochem. Pharmacol.* 37:1259–62
117. Zimmer, M., Hofmann, F. 1987. *Eur. J. Biochem.* 164:411–20
118. Zamora, J. M., Beck, W. T. 1986. *Biochem. Pharmacol.* 35:4303–10
119. Shiraishi, N., Akiyama, S., Kobayashi, M., Kuwano, M. 1986. *Cancer Lett.* 30:251–59
120. Inaba, M., Maruyama, E. 1988. *Cancer Res.* 48:2064–67
121. Goldberg, H., Ling, V., Wong, P. Y., Skorecki, K. 1988. *Biochem. Biophys. Res. Commun.* 152:552–58
122. Beck, W. T. 1987. *Biochem. Pharmacol.* 36:2879–87
123. Klohs, W. D., Steinkampf, R. W. 1988. *Cancer Res.* 48:3025–30
124. Rintoul, D. A., Center, M. S. 1984. *Cancer Res.* 44:4978–80
125. Siegfried, J. A., Kennedy, K. A., Sartorelli, A. C., Tritton, T. R. 1983. *J. Biol. Chem.* 258:339–43
126. Wheeler, C., Rader, R., Kessel, D. 1982. *Biochem. Pharmacol.* 31:2691–93
127. Sehested, M., Skovsgaard, T., van Deurs, B., Winther-Nielsen, H. 1987. *J. Natl. Cancer Inst.* 78:171–79
128. Basrur, V. S., Chitnis, M. P., Menon, R. S. 1985. *Oncology* 42:328–31
129. Willingham, M. C., Cornwell, M. M., Cardarelli, C. O., Gottesman, M. M., Pastan, I. 1986. *Cancer Res.* 46:5941–46
130. Willingham, M. C., Richert, N. D., Cornwell, M. M., Tsuruo, T., Hamada, H., et al. 1987. *J. Histochem. Cytochem.* 35:1451–56
131. Poole, B., Ohkuma, S. 1981. *J. Cell Biol.* 90:665–69
132. Maxfield, F. R. 1982. *J. Cell Biol.* 95:676–81
133. Ramu, A., Fuks, Z., Gatt, S., Glaubiger, D. 1984. *Cancer Res.* 44:144–48
134. Ganapathi, R., Grabowski, D. 1983. *Cancer Res.* 43:3696–99
135. Tsuruo, T., Iida, H., Tsukagoshi, S., Sakurai, Y. 1981. *Cancer Res.* 41: 1967–72
136. Merry, S., Fetherston, C. A., Kaye, S. B., Freshney, R. I., Plumb, J. A. 1986. *Br. J. Cancer* 53:129–35
137. Ozols, R. F. 1985. *Semin. Oncol.* 12:7–11
138. Racker, E., Wu, L.-T., Westcott, D. 1986. *Cancer Treat. Rep.* 70:275–78
139. Kessel, D., Wilberding, C. 1985. *Cancer Res.* 45:1687–91
140. Schuurhuis, G. J., Broxterman, H. J., van der Hoeven, J. J. M., Pinedo, H. M., Lankelma, J. 1987. *Cancer Chemother. Pharmacol.* 20:285–90

141. Klohs, W. D., Steinkampf, R. W., Havlick, M. J., Jackson, R. C. 1986. *Cancer Res.* 46:4352–56

142. Nakagawa, M., Akiyama, S., Yamaguchi, T., Shiraishi, N., Ogata, J., Kuwano, M. 1986. *Cancer Res.* 46:4453–57

143. Inaba, M., Fujikura, R., Tsukagoshi, S., Sakurai, Y. 1981. *Biochem. Pharmacol.* 30:2191–94

144. Slater, L. M., Murray, S. L., Wetzel, M. W., Wisdom, R. M., Duvall, E. M. 1982. *J. Clin. Invest.* 70:1131–34

145. Beck, W. T., Cirtain, M. C., Look, A. T., Ashmun, R. A. 1986. *Cancer Res.* 46:778–84

146. Ramu, A., Spanier, R., Rahamimoff, H., Fuks, Z. 1984. *Br. J. Cancer* 50:501–7

147. Tsuruo, T., Iida, H., Tsukagoshi, S., Sakurai, Y. 1983. *Cancer Res.* 43: 2267–72

148. Safa, A. R., Glover, C. J., Sewell, J. L., Meyers, M. B., Biedler, J. L., Felsted, R. L. 1987. *J. Biol. Chem.* 262:7884–88

149. Cornwell, M. M., Pastan, I., Gottesman, M. M. 1987. *J. Biol. Chem.* 262:2166–70

150. Yang, C.-P. H., Mellado, W., Horwitz, S. B. 1988. *Biochem. Pharmacol.* 37:1417–21

150a. Striessnig, J., Moosburger, K., Goll, A., Ferry, D. R., Glossmann, H. 1986. *Eur. J. Biochem.* 161:603–9

151. Glossmann, H., Ferry, D. R., Striessnig, J., Goll, A., Moosburger, K. 1987. *Trends Pharmacol. Sci.* 8:95–100

152. Broxterman, H. J., Pinedo, H. M., Kuiper, C. M., Kaptein, L. C. M., Schuurhuis, G. J., Lankelma, J. 1988. *FASEB J.* 2:2278–82

153. Biedler, J. L., Meyers, M. B., Spengler, B. A. 1988. *Proc. Am. Assoc. Cancer Res.* 29:295 (Abstr.)

154. Fojo, A. T., Ueda, K., Slamon, D. J., Poplack, D. G., Gottesman, M. M., Pastan, I. 1987. *Proc. Natl. Acad. Sci. USA* 84:265–69

155. Thiebaut, T., Tsuruo, T., Hamada, H., Gottesman, M. M., Pastan, I., Willingham, M. C. 1987. *Proc. Natl. Acad. Sci. USA.* 84:7735–38

156. Sugawara, I., Kataoka, I., Morishita, Y., Hamada, H., Tsuruo, T., et al. 1988. *Cancer Res.* 48:1926–29

157. Sugawara, I., Nakahama, M., Hamada, H., Tsuruo, T., Mori, S. 1988. *Cancer Res.* 48:4611–14

158. Mickley, L. A., Rothenberg, M. L., Hamilton, T. C., Ozols, R. F., Fojo, A. T. 1988. *Proc. Am. Assoc. Cancer Res.* 29:297 (Abstr.)

159. Shuin, T., Masuda, M., Yao, M., Kubota, Y., Sugimoto, Y., Tsuruo, T. 1988. *Proc. Am. Assoc. Cancer Res.* 29:307 (Abstr.)

160. Fojo, A. T., Shen, D.-W., Mickley, L. A., Pastan, I., Gottesman, M. M. 1987. *J. Clin. Oncol.* 5:1922–27

161. Bell, D. R., Gerlach, J. H., Kartner, N., Buick, R. N., Ling, V. 1985. *J. Clin. Oncol.* 3:311–15

162. Ma, D. D., Davey, R. A., Harman, D. H., Isbister, J. P., Scurr, R. D., et al. 1987. *Lancet* 1:135–37

163. Gerlach, J. H., Bell, D. R., Karakousis, C., Slocum, H. K., Kartner, N., et al. 1987. *J. Clin. Oncol.* 5:1452–60

164. Kakehi, Y., Kanamaru, H., Yoshida, O., Ohkubo, H., Nakanishi, S., et al. 1988. *J. Urology* 139:862–65

165. Mukaiyama, T., Mitsui, I., Shibata, H., Inoue, K., Ogawa, M. 1988. *Proc. Am. Assoc. Cancer Res.* 29:306 (Abstr.)

166. Baas, F., Borst, P., 1988. *FEBS Lett.* 229:329–32

167. Mukhopadhyay, T., Batsakis, J. G., Kuo, M. T. 1988. *J. Natl. Cancer Inst.* 80:269–75

167a. Arceci, R. J., Croop, J. M., Horwitz, S. B., Housman, D. 1988. *Proc. Natl. Acad. Sci. USA* 85:4350–54

168. Thorgeirsson, S. S., Huber, B. E., Sorrell, S., Fojo, A., Pastan, I., Gottesman, M. M. 1987. *Science* 236:1120–22

169. Fairchild, C. R., Ivy, S. P., Rushmore, T., Lee, G., Koo, P., et al. 1987. *Proc. Natl. Acad. Sci. USA* 84:7701–5

170. Cowan, K. H., Batist, G., Tulpule, A., Sinha, B. K., Myers, C. E. 1986. *Proc. Natl. Acad. Sci. USA* 83:9328–32

171. Watowich, S. S., Morimoto, R. I. 1988. *Mol. Cell. Biol.* 8:393–405

172. Stachel, S. E., Zambryski, P. C. 1986. *Cell* 47:155–57

173. Stanfield, S. W., Ielpi, L., O'Brochta, D., Helinski, D. R., Ditta, G. S. 1988. *J. Bacteriol.* 170:3523–30

174. Krogstad, D. J., Gluzman, I. Y., Kyle, D. E., Oduola, A. M. J., Martin, S. K., et al. 1987. *Science* 238:1283–85

175. Martin, S. K., Oduola, A. M. J., Milhous, W. K. 1987. *Science* 235:899–901

176. Balzi, E., Chen, W., Ulaszewski, S., Capieaux, E., Goffeau, A. 1987. *J. Biol. Chem.* 262:16871–79

177. Ramu, A., Glaubiger, D., Magrath, I. T., Joshi, A. 1983. *Cancer Res.* 43:5533–37

178. Ramu, A., Glaubiger, D., Weintraub, H. 1984. *Cancer Treat. Rep.* 68:637–41

179. Ramu, A., Glaubiger, D., Soprey, P., Reaman, G. H., Feuerstein, N. 1984. *Br. J. Cancer* 49:447–51

180. Ramu, A., Shan, T., Glaubiger, D. 1983. *Cancer Treat. Rep.* 67:895–99
181. Biedler, J. L., Chang, T., Meyers, M. B., Peterson, R. H. F., Spengler, B. A. 1983. *Cancer Treat. Rep.* 67:859–67
182. Vayuvegula, B., Slater, L., Meador, J., Gupta, S. 1988. *Cancer Chemother. Pharmacol.* 22:163–68
183. Meyers, M. B., Merluzzi, V. J., Spengler, B. A., Biedler, J. L. 1986. *Proc. Natl. Acad. Sci. USA* 83:5521–25
184. McGrath, T., Center, M. S. 1988. *Cancer Res.* 48:3959–63
185. Marsh, W., Center, M. S. 1986. *Biochem. Biophys. Res. Commun.* 138:9–16
186. McGrath, T., Center, M. S. 1987. *Biochem. Biophys. Res. Commun.* 145:1171–76

187. Marsh, W., Center, M. S. 1987. *Cancer Res.* 47:5080–86
188. Richert, N., Akiyama, S., Shen, D. W., Gottesman, M. M., Pastan, I. 1985. *Proc. Natl. Acad. Sci. USA* 82:2330–33
189. Shen, D.-W., Cardarelli, C., Hwang, J., Cornwell, M. M., Richert, N., et al. 1986. *J. Biol. Chem.* 261:7762–70
190. Yalowich, J. C., Roberts, D., Benton, S., Parganas, E. 1987. *Proc. Am. Assoc. Cancer Res.* 28:277 (Abstr.)
191. Batist, G., Tulpule, A., Sinha, B. K., Katki, A. G., Myers, C. E., Cowan, K. H. 1986. *J. Biol. Chem.* 261:15544–49
192. Arsenault, A. L., Ling, V., Kartner, N. 1988. *Biochim. Biophys. Acta* 938:315–21

Annu. Rev. Biochem. 1989. 58:173–94

STRUCTURE AND BIOSYNTHESIS OF PROKARYOTIC GLYCOPROTEINS[1,2]

Johann Lechner

Department of Biological Sciences, University of California S.B., Santa Barbara, California

Felix Wieland

Institut für Biochemie I, Im Neuenheimer Feld 328, 6900 Heidelberg, West Germany

CONTENTS

INTRODUCTION AND PERSPECTIVES

Biosynthesis of glycoproteins, as well as the structure of their glycoconjugates, has been the focus of investigation on complex carbohydrates for the

---

[1] Abbreviations used: CSG, cell-surface glycoprotein; PAS, periodic acid-Schiff; SDS-PAGE, sodium dodecyl sulfate-polyacrylamide gel electrophoresis; GlcNAc, N-acetylglucosamine, GalNAc, N-acetylgalactosamine.

[2] Dedicated to Dr. R. Purrmann on the occasion of his 75th birthday.

0000-4154/89/0701-0173$02.00

last decade. Most of our knowledge on this subject has been obtained by research on eukaryotic organisms, primarily on cultures of mammalian cells and on yeast. These studies have led to a detailed insight into the chemistry underlying protein-glycosylation, including the elucidation (1) of the coordinate action of various cell organelles during generation of the N-glycosyl bond, both in the construction of a variety of protein-linked carbohydrate structures, as well as in the transport of newly synthesized proteins. These results have overshadowed the observation that glycoproteins are also components of prokaryotic organisms, although reports on bacterial glycoproteins have existed for a few years (2–6). Prokaryotes do of course possess an extreme variety of carbohydrates as constituents of their cell surfaces, but these carbohydrates are either attached to lipids or are part of elongated glycan chains, and glycoconjugates that are linked to proteins or peptides generated by ribosomes had not been reported before.

After a short discussion of the experimental evidence for a covalent sugar-protein linkage in some of the prokaryotic glycoproteins, this review focuses on the cell-surface glycoprotein and the flagellins of halobacteria, since these are the bacterial glycoproteins studied in most detail.

The plasma membrane of halobacteria is not surrounded by a rigid sacculus (as is the case for typical eubacteria), but rather by a surface protein layer. This layer consists of the above-mentioned glycoprotein, hexagonally arranged in a two-dimensional crystal (17–19).

We first present a model of the CSG that summarizes the data available on its structure, showing the structure and binding sites of the three different types of saccharides along the primary structure of the halobacterial CSG. Next we give a more extensive description of the structural details, including some of the experimental methods used in their elucidation. A functional role for one of the saccharides in maintaining the cell shape is discussed in this context. We conclude the structural part by a short review of recent data on the three-dimensional structure of the CSG layer of halobacteria.

The chemistry of glycoprotein structure and biosynthesis is expected to be basically the same in eukaryotes and prokaryotes, but prokaryotes do not have the organelles shown to be involved in eukaryotic glycoprotein biosynthesis. This observation has raised the question as to how prokaryotes glycosylate proteins and transport glycoproteins to their cell surfaces? We address this question in the fifth section, and show that biosynthesis of the two N-glycosyl-linked saccharides involves dolichol monophosphate and lipid diphosphate-linked saccharide precursors. Sulfation and epimerization of the glycoconjugates are completed at the lipid-linked level, and the mature saccharides are transferred to the protein core on the cell surface. One type of N-linked saccharide is transiently methylated at its lipid-linked stage. The significance of this transient chemical modification for the biosynthesis of the corresponding N-glycosyl bond is discussed.

REPORTED PROKARYOTIC GLYCOPROTEINS

Following the first report (3) on a glycoprotein in halobacteria, the occurrence of glycoproteins has been claimed for several prokaryotes (see Table 1). For eubacteria the evidence of a covalent protein-carbohydrate linkage is still not absolutely convincing: lectin-binding and PAS-staining (for the method, see Ref. 20) of reported proteins, even after SDS-PAGE, might be caused by firmly associated but noncovalently linked saccharides. This is a constant danger in the field since carbohydrates are abundant in the bacterial cell wall, where most of the reported proteins were derived.

A recent report by Messner et al described the isolation of a N-glycosyl-linkage unit asparaginyl-rhamnose from the eubacterium *Bacillus stearother-mophilus* (21). The authors obtain this aminoacylsaccharide by treatment of a glycopeptide with hydrogen fluoride under mild conditions (0° C, 1 h). However, because this treatment is not expected to split peptide bonds to any substantial degree (22), the existence of a covalent bond between an eubacterial protein and a glycoconjugate still awaits confirmation.

Most of the prokaryotes suspected to possess glycoproteins belong to the archaebacteria, a third kingdom of life (2, 23–25). There is good evidence for a true glycoprotein in *Thermoplasma acidophilum*. Besides PAS-staining of this protein and of pronase-derived peptides, hydrogen fluoride treatment removed all carbohydrate but one glycosamine residue, implying a N-glycosyl bond (10) (see Table 1).

A couple of additional gram-negative archaebacteria may also contain glycoproteins as components of their cell walls. They show PAS-positive proteins and are also listed in Table 1. However, the bacterial glycoprotein studied in most detail so far is the CSG of the genus *Halobacterium* of archaebacteria (3–6). This cell-surface component and the halobacterial flagellins are the only prokaryotic glycoproteins for which the linkage units between carbohydrate molecules and protein have been isolated and chemically characterized (26, 27), thus providing chemical proof for the existence of a true glycoprotein.

STRUCTURE OF THE HALOBACTERIAL CELL-SURFACE GLYCOPROTEIN

Summary

The CSG of *Halobacterium salinarium*, according to DNA homology, values a halobacterial strain that belongs to the same species as *halobacterium halobium* (28). This CSG was originally reported to have a molecular mass of 200 kd and to contain 10–12% carbohydrate, attached to the protein via N- as well as O-glycosyl linkage units analogous to those found in eukaryotes (3, 4).

Table 1 Summary of various reports on bacterial glycoproteins[a]

Organism	Reported sugars	Proposed linkage	Evidence for covalent carbohydrate-protein linkage	Ref.
Eubacteria				
Streptococcus faecium	Glc, methylpentose	O	PAS-staining of CNBr-derived glycopeptides	7
Bacillus thuringiensis subsp. israelis	Neutral sugars GlcNAc, GalNAc	N	Lectin binding (WGA) after SDS-PAGE and transfer to nitrocellulose	8
Myxococcus xanthus	Neutral hexoses, aminosugars, uronic	—	SDS-PAGE, PAS-staining, binding to Dolichos lectin	9
Bacillus stearothermophilus	Rhamn, N-Acetylmannuronic acid, GlcNAc, Glc	N	Isolation of Asn-Rhamnose (?) (see text)	13, 21
Archaebacteria				
Halobacterium salinarium and *halobium*	Glc, Gal	O	β-Elimination products of glyco-peptide characterized, isolated Asn-Glc and Asn-GalNAc	3–6
	GlcA, GalA, IdA, 3-O-Me-GalA, GlcNAc, GalNAc	N		
Thermoplasma acidophilum	Man, Glc, Gal, GlcNAc	N	HF-treated GP shows GlcNAc only	10
Sulfolobus acidocaldarius	Hexosamine	—	SDS-PAGE, PAS-staining	11
Pyrodictum occultum	—	—	SDS-PAGE, PAS-staining	12
Walsby's square bacteria	Hexuronic acids, hexoses	—	Isolation of sulfated gly-copeptides	b
Methanogenium maris nigri	—	—	SDS-PAGE, PAS-staining	2
Methanoplanus limicola	—	—	SDS-PAGE, PAS-staining	14
Methanolobus tindarius	Glc, GalA	—	SDS-PAGE, PAS-staining	15
Methanothermus fervidus	Man, 3-O-Methylglucose, Gal, GlcNAc, GalNAc	—	SDS-PAGE, PAS-staining	16

[a] Abbreviations used: Gal, galactose; GalA, galacturonic acid; Glc, glucose; GlcA, glucuronic acid; GP, glycoprotein; HF, hydrogen fluoride; IdA, iduronic acid; 3-O-Me-GalA, 3-O-methylgalacturonic acid.
[b] F. Wieland, unpublished results.

More recent work on the detailed chemical structure of glycopeptides (26, 27, 29, 30) and the primary structure of the CSG of *H. halobium* (31) has corrected most of the original data. To summarize these data, a schematic representation of the CSG of *H. halobium* is given in Figure 1.

1. The entire polypeptide chain of the mature glycoprotein shows a single hydrophobic stretch of 21 amino acids, which is only 3 amino acids away from the C-terminus. Most probably this hydrophobic peptide serves as a membrane anchor (31).

2. Three different types of glycoconjugates, each involving a different carbo-hydrate protein linkage, are found.

   (*a*) At position 2 of the mature polypeptide, one glycosaminoglycan chain, constructed by a repeating sulfated pentasaccharide block, is linked to one protein molecule via the novel N-glycosyl linkage unit asparaginyl-GalNAc (27). The Asn involved is a constituent of the typical Asn-X-Ser/Thr acceptor sequence (see section on BIOSYNTH-ESIS).

   (*b*) Ten sulfated oligosaccharides that contain glucose, glucuronic acid, and iduronic acid are bound to the protein via the hitherto unknown N-glycosyl linkage unit asparaginylglucose (26). Again the Asn in-volved is a constituent of the Asn-X-Ser/Thr sequence.

   (*c*) About 15 glucosylgalactose disaccharides are O-glycosyl linked to a cluster of threonine residues close to the postulated transmembrane domain. In structure and composition, these disaccharides resemble a type of oligosaccharide found in animal collagens (3, 32–34).

With these features and the occurrence of a glycosaminoglycan, this pro-karyotic cell-surface glycoprotein shows striking structural parallels to the animal extracellular proteoglycan-collagen complexes. Similar glycoconju-gates seem also to occur in ascidia, where they are believed to serve a structural function resembling that of connective tissue in mammals (35, 36). In addition, in a series of recent reports (37–39), novel glycoproteins from transformed mammalian cells were described that exhibit similar struc-tural features (N-linked sulfated saccharides and glycosaminoglycanlike chains).

The CSG of halobacteria has a molecular mass of about 120 kd (core protein = 87 kd) and is extraordinarily acidic. Several factors contribute to this acidity: (*a*) more than 20% of its amino acids are aspartic and glutamic acids (3), (*b*) one mole of CSG contains about 40–50 moles of uronic acids within its glycoconjugates (29), and (*c*) about 40–50 moles of sulfate residues per one mole of glycoprotein are bound in ester linkage (40) to its carbohy-drates. This high degree of sulfation has allowed for an easy specific radioac-tive labeling of the glycoprotein: halobacteria do not reduce sulfate (41), but rather they incorporate $^{35}SO_4^{2-}$-label exclusively into their glycoconjugate-sulfate esters. This labeling has revealed that the CSG of halobacteria shares

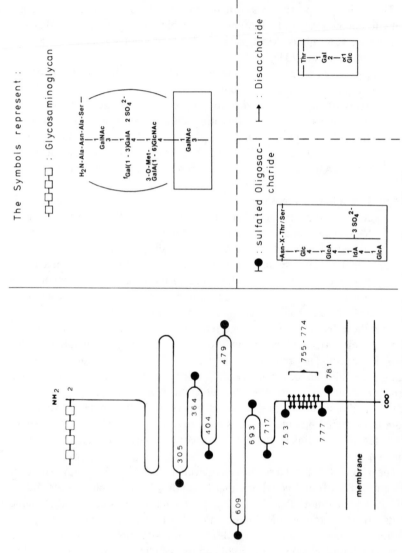

Figure 1 Schematic drawing of the CSG from halobacteria according to (30, 31).

with its flagellins one type of glycoconjugate, and has also greatly facilitated studies on the biosynthesis of the glycoproteins in this organism (42).

Glycosaminoglycan

Labeling with $^{35}SO_4^{2-}$ has been used to identify a glycoconjugate that may play a functional role in maintaining the physiological rod form of halobacterial cells: it was known that halobacteria grown in the presence of the antibiotic bacitracin are not able to maintain their rod shape, but grow as regular spheres. In addition, the lack of some carbohydrate in the CSG was reported (4). The antibiotic is known to inhibit glycosylation by binding to dolichol diphosphate, thereby preventing hydrolysis to dolichol monophosphate and re-entry of the monophosphate into the dolichol phosphate cycle (42a). Halobacteria were grown with $^{35}SO_4^{2-}$-in the absence and presence of bacitracin, and a radioactively labeled glycoprotein could be isolated from both the rod-shaped and spherical bacteria. These glycoproteins were analyzed by SDS-PAGE. The glycoprotein isolated from cells obtained after growth in the presence of bacitracin migrated slightly faster than the control, indicating a lower apparent molecular mass. To investigate the structural difference of the glycoproteins obtained from control cells and those grown in the presence of the antibiotic, pronase peptides from both $^{35}SO_4^{2-}$-labeled glycoproteins were prepared and separated by ion-exchange HPLC. Control cells yielded a series of $^{35}SO_4^{2-}$-labeled glycopeptides that eluted at a relatively low ionic strength, and in addition one peak of material eluted at much higher ionic strength. The array of peaks eluted at lower ionic strength was also present in the glycopeptides obtained from bacteria grown in the presence of bacitracin, but the material eluted at high ionic strength was missing (40). Bacitracin cannot diffuse into halobacteria (43), and therefore it was concluded that the missing sulfated saccharide might be involved in a form-giving or shape-maintaining function.

This finding has stimulated interest in the structure of the corresponding glycopeptide. It can be isolated in large amounts and turns out to contain a high-molecular-weight saccharide that consists of N-acetylglucosamine, N-acetylgalactosamine, galacturonic acid, galactose, and sulfate. In addition, it contains 3-O-methylgalacturonic acid. These five different monosaccharides together with two sulfate residues represent a sulfated pentasaccharide building block that is linearly repeated to give rise to a sulfated carbohydrate chain (30, 44). Two methods have been developed for specific fragmentation of the glycosaminoglycan peptide, each yielding a different subset of fragments. These fragments were isolated by gel filtration and ion-exchange chromatography. After quantitative and qualitative composition analyses of the resulting pure oligosaccharides, their structures were elucidated by permethylation analysis. From the data obtained, the resulting substructures could be combined to yield the complete structure of the glycosaminoglycan from halobacteria; as schematically depicted in Figure 1, upper right panel, the repeating pentasaccharide consists of a linear chain of GalNAc-GalA-GlcNAc with a branching galactofuranosyl residue linked to the GalA residue, and a methylated monosaccharide, 3-O-methylgalacturonic acid, branching from the GlcNAc residue. This pentasaccharide bears one sulfate ester in the 4-position of GalNAc; a second sulfated residue could not yet be localized, probably because it was hydrolyzed under the conditions of fragmentation of the carbohydrate chain. Ten such sulfated pentamers, on the average, make up one glycosaminoglycan chain, and one molecule of CSG contains one such chain (30).

The bond between this carbohydrate chain and the peptide is resistant to treatment with alkali, an indication for the N-glycosyl type. The linkage unit was isolated by HPLC after solvolysis of the glycopeptide with hydrogen fluoride under conditions that exclusively break O-glycosyl linkages, and was analyzed chemically and by FAB-mass spectroscopy. Surprisingly, the structure found is asparaginyl-GalNAc, a N-glycosyl linkage unit that has not been described before (27). Unlike the eukaryotic glycosaminoglycans, the halobacterial glycosaminoglycan does not contain a special carbohydrate linker region, but the reducing end GalNAc residue of the chain is linked directly to an asparagine residue. This asparagine residue is the second N-terminal amino acid of the mature CSG, and is a constituent of the typical N-glycosyl acceptor sequence Asn-X-Thr(Ser) (27) that has been established as the minimal peptide structure common to all eukaryotic N-glycosyl glycopeptides investigated so far (45–54).

Sulfated Oligosaccharides

In addition to the glycosaminoglycan, a family of sulfated glycopeptides have been isolated, which will be referred to as "sulfated oligosaccharide" peptides. In contrast to the glycosaminoglycan, incorporation of these sulfated oligosaccharides into the CSG is not inhibited by the presence of the antibiotic bacitracin (see above). Their carbohydrate composition has been determined by gas liquid chromatography and combined gas liquid chromatography–mass spectroscopy. The oligosaccharides contain exclusively glucose as neutral sugar, and in addition hexuronic acids and sulfate are present in stoichiometric amounts. Unexpectedly, in addition to glucuronic acid, iduronic acid is found as a stoichiometric constituent (29). Iduronic acid is a well-known constituent of animal glycosaminoglycans, but is not typically found [with a possible exception (55)] in prokaryotes. Permethylation analyses have shown that the oligosaccharides are linear chains of 1–4-linked monosaccharides, with glucose at the reducing end and a sulfate residue in ester linkage to the 3-position of each hexuronic acid (32, 56).

How are these unusual saccharides linked to the protein core? Attempts to eliminate the carbohydrates by treatment with alkali have failed, as in the case of the glycosaminoglycan. However, by high-voltage thin-layer electrophoresis, an aminoacylsaccharide can be isolated from the mixture of sulfated oligosaccharide peptides. After acid hydrolysis, this aminoacylsaccharide yields aspartic acid as the only amino acid. This finding strongly suggests that the sulfated oligosaccharides, like the glycosaminoglycan, are bound to the protein via an N-glycosyl linkage. Final proof has been obtained by isolation and characterization of this linkage unit: the homogeneous aminoacylsaccharide was subjected to solvolysis with hydrogen fluoride, and from the resulting mixture a compound was isolated that contained asparagine and glucose in a 1:1 ratio. The compound asparaginylglucose was confirmed by atom bombardment mass spectroscopy, and proton-NMR spectroscopy revealed that the glucose is linked to the amido nitrogen of asparagine in β-linkage (26). Thus, a second hitherto unknown N-glycosyl linkage unit occurs in the CSG of halobacteria.

According to quantitative analysis, about 10 sulfated oligosaccharides are linked to one copy of CSG. Sequence analysis of various sulfated oligosaccharide peptides show that, as for the glycosaminoglycan, the peptides surrounding the N-glycosyl linkage units are consistent with common eukaryotic "sequon" sequence Asn-X-Thr(Ser) (31, 56, 57). Figure 2 (bottom panel) gives the complete structures of typical sulfated oligosaccharide peptides. The saccharides range in their content of uronic acids and sulfate residues, and some contain an additional glucose residue at their nonreducing ends. Remarkably, the same type of glycoconjugate is present in the three flagellins of halobacteria (42, 57).

In addition, a possibility exists that a linkage unit of this composition occurs in animals as well. Shibata et al (58) have described a nephritogenic glycopeptide from rat kidney that contains only glucose and no GlcNAc. The carbohydrate moiety cannot be eliminated from the peptide by alkaline treatment, and therefore the authors conclude that it might be N-glycosyl linked. Final proof of this linkage unit in animals would be provided by isolation of the compound asparaginylglucose from a mammalian glycoprotein and chemical characterization of its structure.

Neutral Glycopeptides

Besides these two different types of sulfated saccharides, the glycosaminoglycan and the sulfated oligosaccharides, an additional type of conjugate is found in the CSG of halobacteria. From pronase digests of the glycoprotein, a fraction of neutral glycopeptides can be isolated (by gel filtration) whose size is between that of the glycosaminoglycan and that of the sulfated oligosaccharides (32). This fraction shows a high content of threonine, and the saccharides can be eliminated by alkaline treatment. They consist of glucose

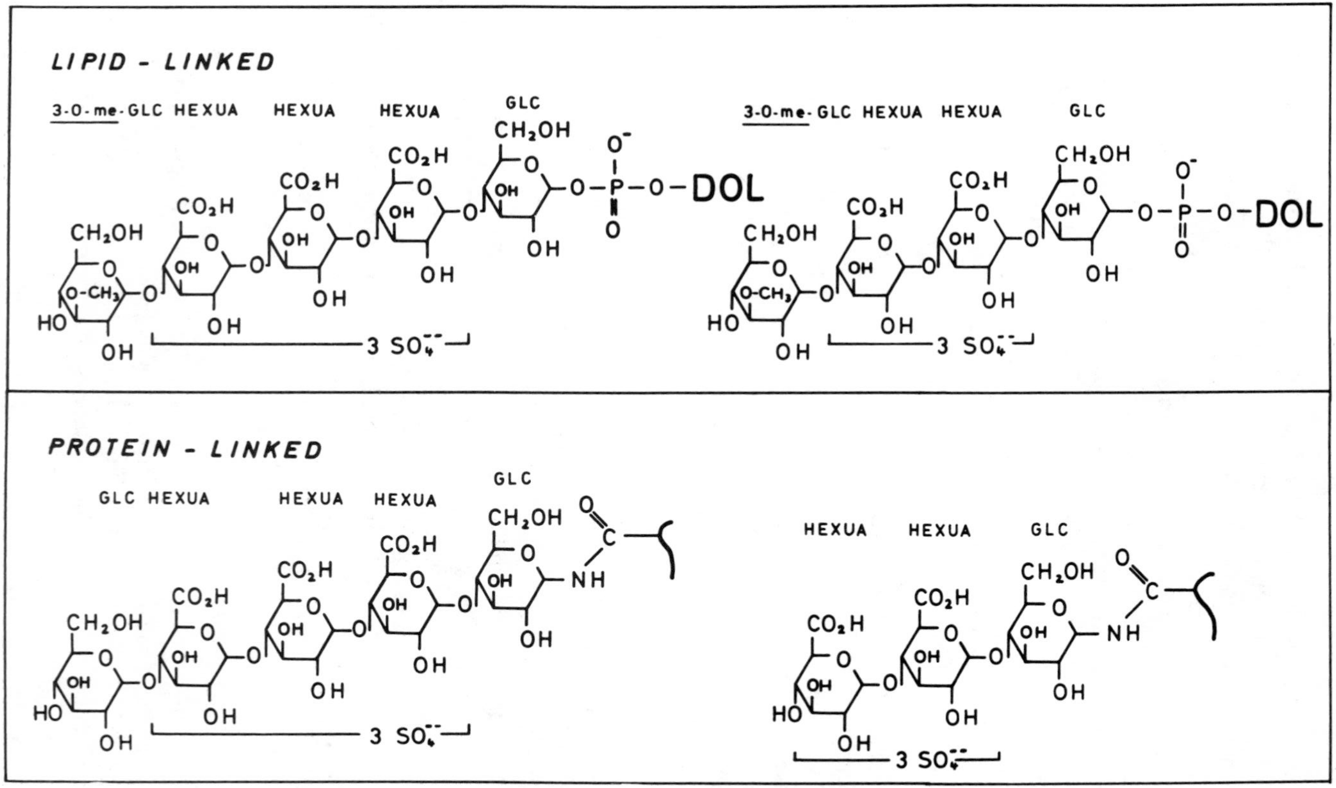

Figure 2 Typical structures of sulfated oligosaccharides from halobacteria in their lipid-linked stage (upper panel) and in their peptide-linked stage (lower panel).

and galactose in a 1:1 ratio, with a galactose residue O-glycosyl linked to the protein via a threonine residue (3, 57). Permethylation analyses have shown that these oligosaccharides resemble those found in collagen not only in composition, but also in being connected in 1–2 linkage (G. Mengele, G. Pual, F. Wieland, M. Sumper, unpublished results).

As all galactose is found in the glycopeptide fraction of intermediate molecular mass, it was concluded that about 15 Glc-1-2-Gal disaccharides are linked to a cluster of densely arranged threonine residues within the glycoprotein.

Protein Core

By use of antibodies against purified CSG that has been deglycosylated with hydrogen fluoride (G. Paul, F. Wieland, unpublished results), the gene coding for its protein core has been cloned by screening an expression vector [PIN-III (59)] library (31). A fragment of the gene has been obtained and used for hybridization screening of a 14–15 kb library of halobacterial DNA in the EMBL 4 phage. Eighteen positive clones obtained were again screened with an oligonucleotide probe designed to detect the coding region for the N-terminus of the mature CSG, using the N-terminal sequence of a 21-amino-acid peptide known from Edman degradation (G. Paul, F. Lottspeich, F. Wieland, unpublished results). The gene, recovered from the plaque that gave the strongest hybridization signal, revealed the complete amino acid sequence of the halobacterial CSG (31). The data obtained showed that the glycoprotein is synthesized with a signal sequence that resembles eukaryotic and prokaryotic signal peptides (60, 61): (a) It contains positively charged amino acids in the N-terminal region, (b) a stretch of at least eight hydrophobic amino acid residues is distant by six positions from the cleavage site, and (c) the sequence preceding the cleavage site is Ala-Ala-Ala, consistent with the proposed recognition sequence Ala-Ala for signal peptidases (62). The molecular mass of the mature polypeptide chain derived from its gene sequence is 87 kd. After addition of the molecular mass of all the glycoconjugates described above, a molecular mass of about 120 kd was calculated, which is in disagreement with the apparent molecular weight of 200,000, as estimated from SDS-PAGE.

Earlier data on partial peptide sequences and the stoichiometry of saccharides linked to the protein (27, 30, 32) have been confirmed and refined by these results. Twelve N-glycosyl recognition sites Asn-X-Thr(Ser) occur, one of which had been shown earlier to be unglycosylated (G. Paul, F. Lottspeich, F. Wieland, unpublished results). Therefore, all the remaining 11 sites are glycosylated, with the glycosaminoglycan at the asparagine in the second N-terminal position and the 10 sulfated oligosaccharides localized within the C-terminal two-thirds of the protein core. The above-mentioned cluster of O-linked disaccharides is localized at the far C-terminal part of the glycoprotein, close to its single transmembrane stretch of 21 hydrophobic amino acids, that makes up its C-terminus (see Figure 1).

Three-Dimensional Structure of the Surface Glycoprotein Layer of Halobacterium volcanii

Early electron microscopy studies on the *Halobacterium halobium* cell envelope (17) have shown the outer surface of the wall to be composed of a hexagonal array of morphological units with a 16.8 nm spacing.

Further progress in electron microscopy of halobacteria has been hampered by the high salt concentrations required to maintain the integrity of the regular surface array.

Recently, however, Kessel et al (63) were able to maintain intact cell envelopes of *Halobacterium volcanii*, a moderate halophilic species of halobacteria, under low salt concentrations (10 mM CaCl$_2$). This provided the authors with envelopes of sufficient quality to obtain electron micrographs after negative staining. From the data obtained, a three-dimensional reconstruction of the surface glycoprotein in a resolution of 2 nm was developed.

Correlation averages of untilted projections show a P6 symmetry. A ring structure with a 5 nm wide pore and a 5 nm wide ring is resolved into six distinct globular domains, each 3 nm in diameter. A radial arm extends from each of the ring domains in a pinwheel orientation when viewed in projection. The arms from adjacent rings seem to join near the threefold crystallographic axis. A second but smaller arm (the "interstitial domain"), directed toward the twofold axis, appears to end blindly.

Three-dimensional reconstruction from a tilt series of electron micrographs shows the glycoprotein to form a 4.5 nm high dome-shaped complex. Viewed from the outer surface, the complex shows a small opening in a depression just below the apex. The radial arms are seen at the base of the dome and adjacent to the interstitial domains. Viewed from the inner surface the narrow pore is opening into a "funnel" toward the cell membrane. The complex shows a ring comprising six domains from which the radial arms appear to emanate. The interstitial domains located midway between the radial arms project downward.

The polarity of the structure was derived by surface reconstruction from an area of the cell surface revealed by freeze etching and by examination of the folded-over edges of the wall.

X-ray studies (64) describe the cell envelope profile of *H. halobium* with a surface protein layer separated from a 2 nm thick "inner protein layer," which is directly opposed to the cell membrane by a 6.5 nm wide space of low electron density. The dome-shaped complexes, as revealed for the *H. volcanii* surface proteins, clearly constitute the (outer) surface layer of the envelope.

On the other hand, the amino acid sequence of the CSG of *H. halobium* suggests that the CSG has a membrane anchor (see above). Therefore a spacer element, separating the surface layer from the cell membrane, should be an integral element of the CSG. Accordingly, Kessel et al (63) proposed a model for the halobacterial cell envelope that combines their structural data obtained from *H. volcanii* with the chemical and sequence data obtained from *H. halobium* [see above and (31)] (Figure 3). They assume that the glycopeptide with the cluster of threonine-linked disaccharides provides a spacer element, emanating from the blindly ending interstitial domain toward the plasma membrane. The sequence of 20 amino acids left between this threonine cluster and the probable membrane anchor is supposed to create a small globular domain, which may be (part of) the "inner protein layer."

This combination of data from two halobacterial species that live under different salt conditions and display a different overall shape (a rod for *H. halobium* and a pleomorphic shape for *H. volcanii*) seems to be justified, based on the identity of the basic structural features of the cell envelopes of both species at a lower resolution level at which images can be obtained from *H. halobium* envelopes, as well. To establish this model, however, it will be necessary to confirm a strong homology between the *H. halobium* and *H. volcanii* CSGs amino acid sequences and to show that an equivalent of the cluster of threonine-linked disaccharides close to the membrane anchor as found in *H. halobium* does also exist in the *H. volcanii* protein.

In addition, it will be of interest whether the glycosaminoglycan, which seems to be important to maintain the rod shape of *H. halobium* (see above), is also present in the pleomorphic-shaped *H. volcanii*.

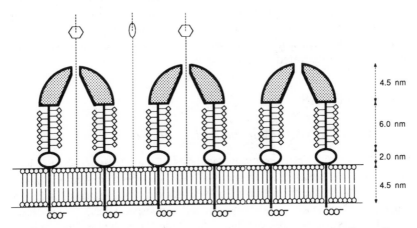

Figure 3 Schematic diagram of the cell-surface layer of *Halobacterium volcanii* according to (63).

HALOBACTERIAL FLAGELLINS

The halobacterial flagella have been described by Alam & Oesterhelt (65). Based on SDS-PAGE, they consist of a set of three related glycoproteins, and chemical analysis has shown that they contain sulfated oligosaccharides (42). Glycopeptides derived by pronase and subtilisin digestion of $^{35}SO_4^{2-}$-labeled flagellins turned out to be indistinguishable from the sulfated oligosaccharide-containing glycopeptides as derived from the CSG. Therefore, halobacterial flagellins and CSG have in common these sulfated oligosaccharides. No other glycoconjugates have been detected in the flagellins. By use of a polyclonal antibody (42), which cross-reacts with all flagellin subunits, a gene fragment of one of the flagellins was isolated from an expression library and sequenced (5).

This fragment has been used as a probe to identify and determine the nucleotide sequence of five different but highly homologous flagellin genes (65a). Two of these genes are arranged tandemly at one locus (flg A1 and A2), and the other three in a tandem arrangement at a different locus (flg B1, B2, and B3). With immunological methods, the products of flg A1 and A2 have been attributed to the flagellar glycoproteins of apparent molecular masses of 20 kd and 26 kd, respectively. A third flagellin subunit of 36 kd has not yet been attributed to one of the cloned genes. This protein, in addition to the N-linked sulfated oligosaccharides mentioned above, contains an alkaline labile modification of unknown nature (F. Wieland, unpublished observation). When in vivo glycosylation of flagellins is inhibited by removal of Mg^{2+}-ions from the growth medium (5, 41), or isolated flagella are deglycosylated by treatment with anhydrous hydrogen fluoride (42), individual flagellin species are obtained with apparent molecular masses higher than those expected for the molecular weight calculated from the corresponding cDNAs. Therefore, Gerl & Sumper (65a) suggest that (in addition to the alkaline labile modification) probably another postribosomal modification remains to be discovered within these unusual flagella.

BIOSYNTHESIS

Cell-Surface Glycoprotein and Flagellins of Halobacteria

First attempts to investigate the biosynthesis of the halobacterial CSG have been made by Mescher & Strominger, who described an activity in cell homogenates for the formation of polyisoprenyl phosphoglucose, polyisoprenyl phosphomannose, and polyisoprenyl diphospho-GlcNAc (43). From these findings, the authors concluded that lipid-linked intermediates are involved in the biosynthesis of the amino sugar–containing saccharide. Later investigations have not confirmed the presence of mannose in the amino sugar-containing saccharide, which turned out to be a glucosaminoglycan, as described above. It would be of interest, therefore, to determine if a lipid-linked mannose residue could serve as a precursor of mannose-containing membrane glycolipid sulfates in halobacteria (66). The activity of formation of lipid-linked N-acetyl-D-glucosamine is dependent on a high concentration of potassium ions, which predominate inside the halobacterial cells, and therefore transfer of GlcNAc has been suggested to occur inside the cell. In the meantime, the linkage unit of the amino sugar–containing glycosaminoglycan moiety has been shown to consist of asparaginyl GalNAc rather than GlcNAc (27), and therefore it will be of interest to learn about the role of this lipid diphosphate GlcNAc.

GLYCOSAMINOGLYCAN   According to recent results, the repeating unit saccharide strand is completed in a lipid-linked state including sulfation, and thereafter transferred "en bloc" to the nascent protein chain (44). After in vivo pulse labeling with $^{35}SO_4$ the lipid-linked intermediates appear on a SDS-PAGE as a regular pattern of 10–15 bands, which are thought to represent different chain lengths of the growing saccharide. Most likely the distance between adjacent bands is due to one repeating unit of the glycosaminoglycan. This implies that the glycosaminoglycan is assembled on the lipid carrier by polymerization of preformed pentasaccharides, a mechanism similar to the one described for the biosynthesis of the Salmonella O antigen (67). The above-mentioned inhibition by bacitracin of this transfer implies two conclusions: (a) The lipid involved is likely to be a diphosphate, as bacitracin complexes prenyl diphosphate compounds, thus inhibiting hydrolytic regeneration of the corresponding monophosphate (42a), and (b) transfer to protein of the glycosaminoglycan most likely takes place at the cell surface, as bacitracin does not penetrate the plasma membrane of halobacteria (43, 68). Thus, biosynthesis of the halobacterial glycosaminoglycan differs from the synthesis of animal glycosaminoglycans (69) in that an already sulfated lipid precursor occurs in halobacteria, whereas in animals the carbohydrate chain is established at the protein-linked level through step-by-step addition of monosaccharides (from their nucleotide-activated derivatives), and sulfation takes place only on the protein-linked saccharide strand. On the other hand, as mentioned before, sequence analysis has revealed a "sequon" acceptor peptide Asn-Ala-Ser for this transfer that results in the linkage unit Asn-GalNAc (27). Additional principal structural similarities with the animal glycosaminoglycans exist: both classes are linear strands composed of uronic acids, amino sugars, and sulfate, and although many animal glycosaminoglycans are O-glycosyl linked to protein via xylose, corneal keratan sulfate is bound to protein via a true N-glycosyl linkage (70).

SULFATED OLIGOSACCHARIDES   Studies on the biosynthesis of the second novel type of N-glycosyl linkage unit, Asn-Glc, have revealed some unexpected features: pulse-chase labeling experiments with $^{35}SO_4^{2-}$ demonstrated the existence of sulfated precursors of this type of glycoconjugate (56). The protein acceptors for this sulfated precursor material are the CSG and the set proteins of lower molecular weight that have been shown to represent the halobacterial flagellins (42). Thus, two different types of glycoproteins share the same pool of precursors.

The radioactive label has facilitated purification to homogeneity of a group of lipid-linked oligosaccharides that make up this pool (56). Detailed chemical analyses of these compounds has revealed that, as in the case of the glycosaminoglycan, completely sulfated lipid-linked precursors are established before transfer of the oligosaccharides to protein. The lipid is a C_{60}-polyprenol of the eukaryotic dolichyl- rather than of the bacterial undecaprenyl-type. The reducing-end glucose residue of the oligosaccharide part is linked to the dolichol via a monophosphate rather than a diphosphate bridge. Furthermore, the occurrence of iduronic acid in a lipid-activated saccharide has first been demonstrated in these precursors (29). This hexuronic acid is a typical constituent of eukaryotic extracellular glycoconjugates and is reported to result from epimerization of glucuronic acid residues within the completed, protein-linked saccharide chain (69).

To prove a precursor-product relationship of these lipid-linked sulfated oligosaccharides with the sulfated oligosaccharides from the CSG, their molecular structure has been determined by permethylation analysis. Typical structures are given in Figure 2, upper panel. Composition, linkage pattern, and sequence of the lipid-linked oligosaccharides turned out to be identical to those obtained from glycopeptides (Figure 2, lower panel) with one surprising exception: the lipid-linked saccharides contained a chemical modification that could not be found in the protein-linked saccharides. Specifically, position 3 of the oligosaccharides' peripheral, nonreducing-end glucose residue carries a methyl group. Except for this methylation, all structural details were identical in the lipid-linked and protein-linked sulfated oligosaccharides, and the pool of $^{35}SO_4^{2-}$-labeled lipid-linked percursors could quantitatively be chased into glycoprotein. Therefore, a true precursor-product relationship exists between the two species, and the methylation observed represents a transient modification of the lipid-linked oligosaccharides (71).

What might be the role of this transient methylation? To investigate this, sulfated glycoprotein biosynthesis has been analyzed under conditions where S-adenosylmethionine-dependent methylation is inhibited. Surprisingly, sulfated glycoprotein biosynthesis is depressed greatly, although (a) general protein biosyntheses (as assessed by incorporation of ^{35}S-methionine) is not altered, and (b) synthesis of an unmethylated pool of sulfated lipid-linked precursors is not inhibited, but rather this pool remains stable in pulse-chase experiments performed in the presence of the inhibitors of methylation.

Thus, a transient methylation is involved at some stage in the biosynthesis of the novel N-glycosyl linkage Asn-Glc.

Inhibition of this methylation could only be achieved by use of a combination of adenosine and homocysteine thiolactone, and not with the actual direct inhibitor S-adenosylhomocysteine. The two former substances do permeate cell membranes to a sufficient extent, whereas the latter does not (72). Therefore, it is assumed that construction of the dolichol monophosphate-linked oligosaccharides, including methylation of their peripheral glucose-residues, occurs at the cytosolic face of the plasma membrane (71, 73).

Where then does transfer to protein occur? To investigate this, advantage was taken of the finding that typical N-glycosyl "sequon" sequences are present around the N-glycosyl linkage units. Accordingly, a synthetic glycosyl acceptor peptide was employed in vivo that has successfully been used before in in vitro glycosylation studies (74). The hexapeptide Tyr-Asn-Leu-Thr-Ser-Val contains two ionic charges and cannot therefore permeate membranes. Interestingly, this acceptor-peptide becomes glycosylated with sulfated oligosaccharides when added to the medium of stirred suspensions of halobacteria (71). Thus, transfer to protein of the sulfated oligosaccharides occurs on the surface of the halobacterial cell. Partial characterization of the sulfated glycopeptides obtained after exogenous addition of acceptor peptide revealed that only the sulfated oligosaccharides are transferred to this acceptor, but no glycosaminoglycan. This finding indicates that, for the transfer of the glycosaminoglycan, structural information may be needed in addition to a general N-glycosyl acceptor sequence.

HYPOTHESIS FOR THE BIOSYNTHESIS OF N-LINKED SACCHARIDES IN HALOBACTERIA   A concluding scheme reflecting a present working hypothesis on the biosynthesis of the sulfated CSG of halobacteria is depicted in Figure 4. Two different N-glycosyl linkages are synthesized within this complex CSG, and the two biosynthetic pathways differ in the type of lipid phosphate involved. A dolichol monophosphate is used for the Asn-Glc-, and a lipid (most likely a dolichol) -diphosphate for the Asn-GalNAc-type. All three types of N-glycosyl linkages established so far have in common sequences that surround the binding asparagine residue and that fit the general formula Asn-X-Thr(Ser). This strongly supports the notion that the hydroxyl group of the amino acid next but one to Asn-carbohydrate is involved in the process of glycosyl transfer in a catalytic manner, and is not merely a structural recognition site for the transferase (75, 76). In addition to the hitherto unknown types of N-glycosyl linkages found in the CSG in halobacteria, differences exist in details of the biosynthetic pathways leading to sulfated glycoconjugates in halobacteria and eukaryotes. In eukaryotes, glycoconjugates are covalently modified at the protein-linked level in the Golgi apparatus, where trimming, further glycosylation, sulfation, or epimerization occurs. Halobacteria lack organelles like the Golgi apparatus and therefore must follow a different biosynthetic pathway leading to their glycoproteins. Thus, the oligosaccharides are completed and sulfated while still attached to dolichol on the cytosolic side of the cell membrane. Thereafter, they are translocated to the cell surface, possibly by a mechanism that involves a transient methylation of their peripheral glucose residues. Finally, transfer to the protein occurs at the cell surface, and with this generation of N-glycosyl linkages, the halobacterial cell surface is functionally equivalent to the luminal side of the endoplasmatic reticulum membrane in eukaryotic cells.

With the mechanism described, an unglycosylated protein core can be

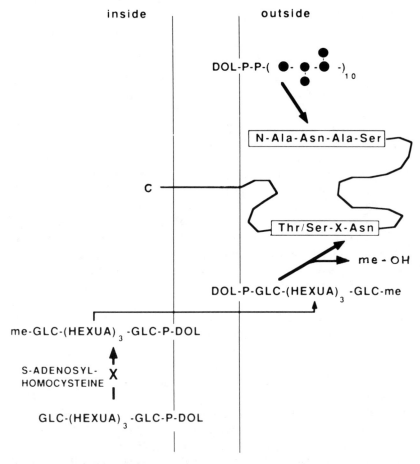

Figure 4 Schematic representation of the present hypotheses on the biosynthesis of the CSG of halobacteria. For details see text.

translocated through the cell membrane, and the corresponding glycoconjugates may pass this membrane in their lipid-linked stage. Thus, despite the lack of compartmentalization in prokaryotes, halobacterial CSG biosynthesis in essence resembles the mechanism of glycoprotein biosynthesis in eukaryotes.

More questions of interest remain to be answered, among them: How is the glycosaminoglycan translocated to the cell surface? Is it transported block by block, similar to the biosynthesis of the bacterial O-antigens (67), or is it completed inside the cell and then translocated through the membrane? What exactly is the role of the transient methylation of the sulfated dolichol monophosphate-oligosaccharides?

Further problems wait to be resolved: How are the O-glycosyl linkage units established, from nucleotide or from lipid-activated saccharide precursors, inside or outside the cell? What is the function of the sulfated oligosaccharides in the CSG and in the flagellins? How widespread in nature are the saccharide structures and linkage units described here, and may they be used for a taxonomic purpose?

The halobacterial system has turned out to be a far more complicated model for the investigation of structure, biosynthesis, and function of glycoproteins than had been anticipated at the onset. On the other hand, we hope that our more detailed biochemical knowledge, in combination with the help of a halobacterial transfection system (77), eventually will enable us to obtain insight into those problems. Specifically, the three-dimensional elucidation at the molecular level of carbohydrate structures presented here should teach us about the functions that glycoconjugates exert by protein-carbohydrate and carbohydrate-carbohydrate interactions.

Methanothermus fervidus *Glycoprotein*

Most recently, the composition of glycoconjugates from this extreme thermophilic archaebacterium was published [c. f. Table 1 (16)]. Although the structures of these saccharides have not yet been established, some unusual features of glycoprotein biosynthesis in this bacterium have been described (78):

1. Besides nucleosidediphosphate–activated monosaccharides, oligosaccharides with rather complex compositions have been found to occur as UDP-derivatives, as well. Disaccharides with this type of activation have exclusively been found to be involved in the biosynthesis of an archaebacterial pseudomurein, so far (79).

2. Processing (methylation) of oligosaccharides seems to occur at the UDP-linked level. Interestingly, as in halobacteria, 3-O-methylglucose is generated by methylation of the oligosaccharide at the precursor level, but

unlike in halobacteria, the modification is stable and found as a stoichiometric component of the glycoprotein.

3. Like in halobacteria, lipid-activated oligosaccharides occur with short-chain (C_{55}) dolichol—rather than undecaprenol. Unlike in halobacteria, exclusively dolicholdiphosphate oligosaccharides were found, with no dolicholmonophosphate oligosaccharide. Probably the oligosaccharides are transferred from UDP to dolicholphosphate.

4. Finally, as is the case in eukaryotes and halobacteria, a transient modification of the precursor oligosaccharides seems to take place. Rather than a transient methylation, here a transient glycosylation takes place with up to 6 or 8 glucose residues that were found in lipid-linked oligosaccharides, and are absent in the surface glycoprotein. This is reminiscent of the transient glycosylation with three glucose units of the unique dolichol-linked precursor saccharide for N-glycosylation in eukaryotes (80). Final insight into the mechanism of this transient modification depends on the knowledge of the molecular structures of the glycoconjugates in their precursor-linked as well as their protein-bonded state. Mysteriously, neither UDP- nor lipiddiphosphate-activated glucose or 3-O-methylglucose could be detected in *Methanothermus fervidus*. On the other hand, these monosaccharides are constituents of UDP-activated and dolichol-activated oligosaccharides. Therefore it will be of interest to elucidate the process of activation for transfer of glucose in this organism.

ACKNOWLEDGMENTS

We would like to thank Dr. H. König for discussions, Dr. W. Just for critical reading, and B. Schröter for assistance in the preparation of the manuscript.

Part of the work described here was supported by the Deutsche Forschungsgemeinschaft (SFB 43, Regensburg).

Literature Cited

1. Kornfeld, R., Kornfeld, S. 1985. *Annu. Rev. Biochem.* 54:631–64
2. Kandler, O., König, H. 1985. *The Bacteria*, Vol. VIII: *Archaebacteria*, pp. 413–57. New York: Academic
3. Mescher, M. F., Strominger, J. L. 1976. *J. Biol. Chem.* 251:2005–14
4. Mescher, M. F., Strominger, J. L. 1976. *Proc. Natl. Acad. Sci. USA* 73:2687–91
5. Sumper, M. 1987. *Biochim. Biophys. Acta* 906:69–79
6. Wieland, F. 1988. *Halophilic Bacteria*, Vol. II, ed. F. Rodriguez-Valera, pp. 55–65. Boca Raton, Fla: CRC
7. Kawamura, T., Shockman, G. D. 1983. *J. Biol. Chem.* 258:9514–21

8. Pfannenstiel, M. A., Muthukumar, G., Couche, G. A., Nickerson, K. W. 1987. *J. Bacteriol.* 169:796–801
9. Maeba, P. Y. 1988. *J. Bacteriol.* 166:644–50
10. Young, L. L., Haug, A. 1979. *Biochim. Biophys. Acta* 556:265–77
11. Michel, H., Neugebauer, D. C., Oesterhelt, D. 1980. *Electron Microscopy at Molecular Dimensions*, pp. 27–35. Berlin/Heidelberg/New York: Springer
12. Stetter, K. O., König, H., Stackebrandt, E. 1983. *Syst. Appl. Microbiol.* 4:535–51
13. Küpcü, Z., März, L., Messner, P., Sleytr, U. B. 1984. *FEBS Lett.* 173:185–90

14. Wildgruber, G., Thomm, M., König, H., Ober, K., Ricchiuto, T., Stetter, K. O. 1982. *Arch. Microbiol.* 132:31–36
15. König, H., Stetter, K. O. 1982. *Zentralb. Bakteriol. Microbiol. Hyg. Abt. Orig. C* 3:478–90
16. Nusser, E., Hartmann, E., Allmeier, H., König, H., Paul, G., Stetter, K. O. 1988. *Proc. 2nd Int. Workshop on S-Layers in Prokaryotes*, Vienna, ed. U. B. Sleytr, B. Messer, D. Pum, M. Sara, pp. 21–25. Berlin: Springer.
17. Houwink, H. L. 1956. *J. Gen. Microbiol.* 15:145–62
18. Robertson, H. D., Schreil, W., Reedy, M. 1982. *J. Ultrastruct. Res.* 80:148–57
19. Stoeckenius, W., Rowen, R. 1967. *J. Cell Biol.* 34:365–74
20. Segrest, J. P., Jackson, R. L. 1972. *Methods Enzymol.* 28:54–63
21. Messner, P., Sleytr, U. B. 1988. *FEBS Lett.* 228:317–20
22. Mort, A. J., Lamport, D. T. A. 1977. *Anal. Biochem.* 82:289–309
23. Balch, W. E., Magrum, L. J., Fox, G. E., Wolfe, R. S., Woese, C. R. 1977. *J. Mol. Evol.* 9:305–11
24. Woese, C. R. 1981. *Sci. Am.* 244(6):94–106
25. Woese, C. R., Wolfe, R. S. 1985. *The Bacteria*, Vol. VIII: *Archaebacteria.* New York: Academic
26. Wieland, F., Heitzer, R., Schaefer, W. 1983. *Proc. Natl. Acad. Sci. USA* 80:5470–74
27. Paul, G., Lottspeich, F., Wieland, F. 1986. *J. Biol. Chem.* 261:1020–24
28. Ross, H. N. M., Grant, W. D. 1985. *J. Gen. Microbiol.* 131:165–72
29. Wieland, F., Lechner, J., Sumper, M. 1986. *FEBS Lett.* 195:77–81
30. Paul, G., Wieland, F. 1987. *J. Biol. Chem.* 262:9587–93
31. Lechner, J., Sumper, M. 1987. *J. Biol. Chem.* 262:9724–29
32. Wieland, F., Lechner, J., Sumper, M. 1982. *Zentralb. Bakteriol. Mikrobiol. Abt. Orig. C* 3:161–70
33. Wieland, F. 1988. *Biochimie.* In press
34. Kornfeld, R., Kornfeld, S. 1980. *The Biochemistry of Glycoproteins and Proteoglycans*, ed. W. J. Lennarz, pp. 1–34. New York: Plenum
35. Albano, R. M., Mourao, P. A. S. 1986. *J. Biol. Chem.* 261:758–65
36. Mathews, M. B. 1975. *Connective Tissue: Macromolecular Structure and Evolution.* Berlin: Springer
37. Roux, L., Holojda, S., Sundblad, G., Freeze, H. H., Varki, A. 1988. *J. Biol. Chem.* 263:8879–89
38. Sundblad, G., Holojda, S., Roux, L.,

Varki, A., Freeze, H. H. 1988. *J. Biol. Chem.* 263:8890–96
39. Sundblad, G., Kajiji, S., Quaranta, V., Freeze, H. H., Varki, A. 1988. *J. Biol. Chem.* 263:8897–903
40. Wieland, F., Dompert, W., Bernhardt, G., Sumper, M. 1980. *FEBS Lett.* 120:120–24
41. Sumper, M., Herrmann, G. 1978. *Eur. J. Biochem.* 89:229–35
42. Wieland, F., Paul, G., Sumper, M. 1985. *J. Biol. Chem.* 260:15180–85
42a. Siewert, G., Strominger, J. L. 1967. *Proc. Natl. Acad. Sci. USA* 57:767–71
43. Mescher, M. F., Strominger, J. L. 1978. In *Energetics and Structure of Halophilic Microorganisms*, ed. S. R. Caplan, M. Ginzburg, pp. 503–12. Elsevier/North-Holland Biomedical
44. Wieland, F., Lechner, J., Bernhardt, G., Sumper, M. 1981. *FEBS Lett.* 132:319–23
45. Marshall, R. D. 1974. *Biochem. Soc. Symp.* 40:17–26
46. Sharma, C. B., Lehle, L., Tanner, W. 1981. *Eur. J. Biochem.* 116:101–8
47. Lehle, L., Bause, E. 1984. *Biochim. Biophys. Acta* 799:246–51
48. Struck, K. D., Lennarz, W. J., Brew, K. 1978. *J. Biol. Chem.* 253:5786–94
49. Hart, G. W., Brew, K., Grant, G. A., Bradshaw, R. A., Lennarz, W. J. 1979. *J. Biol. Chem.* 254:9747–53
50. Bause, E., Lehle, L. 1979. *Eur. J. Biochem.* 101:531–40
51. Ronin, C., Granier, C., Caseti, C., Bouchilloux, S., Van Rietschoten, J. 1981. *Eur. J. Biochem.* 118:159–64
52. Welply, J. K., Shenbagamurthi, P., Lennarz, W. J., Naider, F. 1983. *J. Biol. Chem.* 258:11856–63
53. Aubert, J. P., Helbeque, N., Loucheux-Lefebvre, M. H. 1981. *Arch. Biochem. Biophys.* 208:20–29
54. Ronin, C., Aubert, J. P. 1982. *Biochem. Biophys. Res. Commun.* 105:909–15
55. Lee, L., Cherniak, R. 1974. *Carbohydr. Res.* 33:387–91
56. Lechner, J., Wieland, F., Sumper, M. 1985. *J. Biol. Chem.* 260:860–66
57. Paul, G. 1986. *Characterization of sulfated glycoproteins from Halobacterium halobium.* PhD thesis. Univ. Regensburg. 126 pp.
58. Shibata, S., Saito, H., Nakanishi, H. 1982. *Biochim. Biophys. Acta* 714:456–70
59. Masui, Y., Mizumo, T., Inouye, M. 1984. *Biotechniques* 2:81–85
60. Perlman, D., Halvorson, H. O. 1983. *J. Mol. Biol.* 167:391–409
61. Michaelis, S., Beckwith, J. 1982. *Annu. Rev. Microbiol.* 36:435–65

62. Kyte, J., Doolittle, R. F. 1982. *J. Mol. Biol.* 167:391–409
63. Kessel, M., Wildhaber, I., Cohen, S., Baumeister, W. 1988. *EMBO J.* 7:1549–54
64. Blaurock, A. E., Stoeckenius, W., Oesterhelt, D., Scherphof, G. L. 1976. *J. Cell Biol.* 71:1–22
65. Alam, M., Oesterhelt, D. 1984. *J. Mol. Biol.* 176:459–75
65a. Gerl, L., Sumper, M. 1988. *J. Biol. Chem.* 263:13246–51
66. Kates, M., Kushwaha, S. C. 1976. *Lipids*, ed. G. Porcellati, G., Jacini, 1:276–94. New York: Raven
67. Robbins, P. W., Bray, D., Dankert, M., Wright, A. 1967. *Science* 158:1536–43
68. Mescher, M. F., Strominger, J. L. 1978. *FEBS Lett.* 89:37–41
69. Roden, L. 1980. *The Biochemistry of Glycoproteins and Proteoglycans*, ed. W. J. Lennarz, pp. 267–371. New York: Plenum
70. Stein, T., Keller, R., Stuhlsatz, H. W., Greiling, H., Ohst, E., et al. 1982.

Hoppe Seyler's Z. Physiol. Chem. 363:825–33
71. Lechner, J., Wieland, F., Sumper, M. 1985. *J. Biol. Chem.* 260:8984–89
72. Barber, J. R., Clarke, S. 1984. *J. Biol. Chem.* 259:7115–22
73. Lechner, J., Wieland, F., Sumper, M. 1986. *Syst. Appl. Microbiol.* 7:286–92
74. Lehle, L., Bause, E. 1984. *Biochim. Biophys. Acta* 799:246–51
75. Bause, E., Legler, G. 1981. *Biochem. J.* 195:639–44
76. Bause, E. 1984. *Biochem. Soc. Trans.* 12:514–17
77. Charlebois, R. L., Lam, W. L., Cline, S. W., Doolittle, W. F. 1987. *Proc. Natl. Acad. Sci. USA* 84:8530–34
78. Hartmann, E., König, H. 1989. *Arch. Microbiol.* In press
79. König, H., Kandler, O., Hammes, W. 1988. *Can. J. Microbiol.* In press
80. Turco, S. J., Stetron, B., Robbins, P. W. 1977. *Proc. Natl. Acad. Sci. USA* 74:4411–14

Annu. Rev. Biochem. 1989. 58:195–221

THE MECHANISM OF BIOTIN-DEPENDENT ENZYMES

Jeremy R. Knowles

Department of Chemistry, Harvard University, Cambridge, Massachusetts 02138

CONTENTS

TRANSFORMATIONS THAT REQUIRE BIOTIN

The enzymes that are involved in the metabolism of one-carbon units at the oxidation level of carbon dioxide have a problem. Considerations of chemical reactivity would favor the attack of a nucleophile on dissolved carbon dioxide to generate the new carboxylic acid. Yet at physiological pH and temperature the concentration of dissolved carbon dioxide in equilibrium with the atmosphere is only 10 μM, whereas that of bicarbonate ion is 200 μM (1). Is it preferable for an enzyme to use the small concentration of electrophilic carbon dioxide, or are there effective ways of fixing the more abundant bicarbonate ion? Nature has taken both routes. Carbon dioxide is the substrate for the rather sluggish and (therefore) abundant enzyme ribulose-1,5-bisphosphate carboxylase, as well as the vitamin K–dependent carboxylases, and phosphoenolpyruvate carboxykinase and carboxytransphosphorylase.

0066-4154/89/0701-0195$02.00

In contrast, bicarbonate ion is used by all the biotin-dependent carboxylases, by carbamoyl phosphate synthetase, and by phosphoenolpyruvate carboxylase (2). This review concerns the mechanism of action of those enzymes that use the cofactor biotin in bicarbonate-dependent carboxylation reactions. Other recent accounts of this field are listed (3–6).

Biotin, illustrated in Figure 1, is essential for a group of carboxylases that can be subdivided into three classes (7). In the reactions of Class I, biotin becomes carboxylated in a reaction that requires ATP, Mg(II), and bicarbonate, and the carboxyl group of carboxybiotin is transferred in a second step to such acceptors as pyruvate, propionyl-CoA, acetyl-CoA, β-methyl-crotonyl-CoA, geranyl-CoA, and urea. These reactions can be summarized:

$$\text{enz-biotin} + \text{ATP} + \text{HCO}_3^- \overset{\text{Mg(II)}}{\rightleftharpoons} \text{enz-biotin-CO}_2^- + \text{ADP} + \text{P}_i + \text{H}^+ \quad 1.$$

$$\text{enz-biotin-CO}_2^- + \text{R-H} \rightleftharpoons \text{enz-biotin} + \text{R-CO}_2^- \qquad 2.$$

The Class II enzymes mediate sodium transport in anaerobes, the transport being coupled to the decarboxylation of β-keto acids and their thioesters, such as oxalacetate, methylmalonyl-CoA, and glutaconyl-CoA. These systems are structurally and mechanistically related to the Class I enzymes, and their reactions can be summarized analogously:

$$\text{enz-biotin} + \text{R-CO}_2^- \rightleftharpoons \text{enz-biotin-CO}_2^- + \text{R-H} \qquad 3.$$

$$\text{enz-biotin-CO}_2^- + 2(\text{Na}^+)_{in} \rightleftharpoons \text{enz-biotin} + \text{HCO}_3^- + 2(\text{Na}^+)_{out} \qquad 4.$$

Finally, there is one Class III enzyme, transcarboxylase, that couples two carbon carboxylations:

$$\text{enz-biotin} + \text{oxalacetate} \rightleftharpoons \text{enz-biotin-CO}_2^- + \text{pyruvate} \qquad 5.$$

$$\text{enz-biotin-CO}_2^- + \text{propionyl-CoA} \rightleftharpoons \text{enz-biotin} + \text{methylmalonyl-CoA} \quad 6.$$

Structural Similarities

From a bewildering variety of protein subunit types and functions has now emerged a satisfyingly tidy pattern, as well as some striking amino acid sequence similarities that suggest strong evolutionary conservation among the carboxylases. Each class of reaction outlined above requires three functional elements: one active site for each of the two chemical transformations (that is, reactions 1 & 2, reactions 3 & 4, and reactions 5 & 6, and a translocation element that links the two active sites by allowing carboxybiotin to "visit" each in turn. In the simplest cases, the enzymes are made up of three different components. Thus acetyl-CoA carboxylase from *Escherichia coli* (an enzyme from Class I) is readily separated into an ATP-dependent biotin carboxylase that performs reaction 1, an acetyl-CoA carboxyltransferase that catalyzes

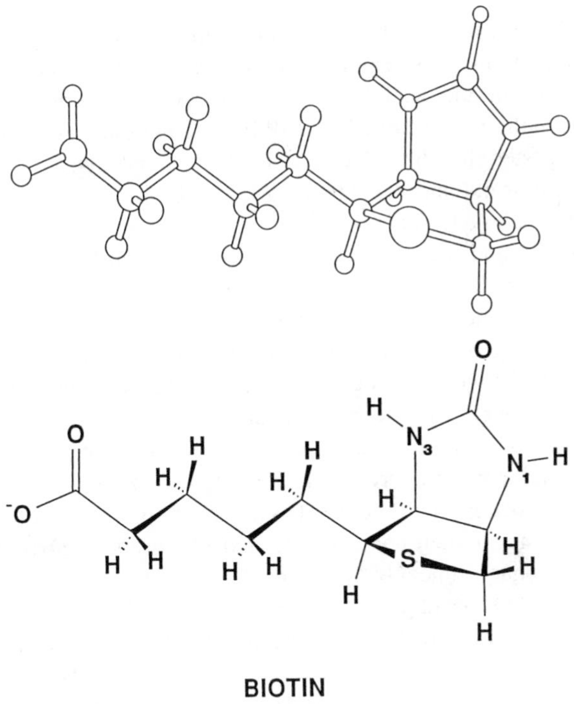

BIOTIN

Figure 1   The structure of biotin.

reaction 2 (R-H = acetyl-CoA), and a carboxyl carrier subunit that contains the covalently bound biotin that shuttles the carboxyl group between the two catalytic subunits (8–10). Analogously, transcarboxylase from *Propionibacterium shermanii* (the enzyme of Class III) is made up of three different kinds of subunit (the enzymes that catalyze reactions 5 and 6, and the biotin carrier protein) in an assembly having a central core of six propionyl-CoA carboxylases, each with two CoA ester substrate sites, linked through 12 biotinyl carrier subunits to six dimeric pyruvate carboxylases (11). In contrast, in higher organisms some or all of the three functions can lie on a single polypeptide chain, presumably as a consequence of gene fusion. For example, the acetyl-CoA carboxylase from chicken liver (12) and the pyruvate carboxylase from yeast (13) each carry both carboxylation active sites and the biotin attachment site on a single polypeptide. The sequences of the genes for these two enzymes have been determined (12a, 13), and have proved very informative.

The yeast pyruvate carboxylase gene encodes a polypeptide of 1178 amino acids, the sequence of which shows strong similarity to a number of other systems (see Figure 2) (13). First, and predictably, the lysine residue to which biotin is attached is readily identified from the -Ala-Met-Lys-Met- sequence that has been found at nine of the ten known biotin attachment sites (7).

Second, and of great interest evolutionarily, there is similarity between this 90-amino-acid biotinylated region of pyruvate carboxylase and the lipoic acid attachment site in the acetyltransferase component of the pyruvate de-hydrogenase multienzyme complex (14)! For each of these enzyme systems it has been suggested that the cofactor visits different catalytic sites on the end of a "swinging arm," and it is tempting to see in this sequence similarity a reflection of analogous functions for biotin and for lipoate as carriers of carboxyl groups and acetyl groups, respectively. [Further support for this view comes from the remarkable stretch of -Ala-Pro- residues that is seen upstream from the biotin attachment site in the biotin-containing α-subunit of oxalacetate decarboxylase (a Class II sodium pump) (15). This arrangement is very reminiscent of the hinge region rich in -Ala-Pro- sequences that has been found between the lipoate domains in the acetyltransferase from pyruvate dehydrogenase (14), and hints at an even closer relationship between lipoate and biotin than has been appreciated. These two cofactors are unusual: both contain sulfur, each is small, neither has any relation to nucleotides, both are covalently attached to their host proteins, and both probably arrived relatively late in evolutionary time (16).] Third, there is high sequence similarity between a 350-amino-acid portion near the middle of yeast pyruvate carboxy-lase, and the N-terminal half of the subunit of transcarboxylase that catalyzes the carboxylation of pyruvate (13). This similarity suggests that the central domain of pyruvate carboxylase binds pyruvate and an essential divalent metal cation, and catalyzes the carboxylation from carboxybiotin. Fourth, there is an upstream 120-amino-acid segment of pyruvate carboxylase that resembles parts of the α-subunit of human propionyl-CoA carboxylase and chicken acetyl-CoA carboxylase (13). The only common catalytic function amongst these three polypeptides is the ATP-dependent carboxylation of biotin by bicarbonate, so it seems likely that this segment represents part of the domain that recognizes ATP and catalyzes reaction 1. Finally, a 170-amino-acid long section of yeast pyruvate carboxylase is very similar to the N-terminal halves of the ATP-binding domains of many carbamoyl phosphate synthetases. As will become evident later, there are many similarities between the pathways followed by carbamoyl phosphate synthetase and by the biotin carboxylase activity of Class I enzymes, and the structural parallelism noted by Lim et al (13) increases our confidence that these two reactions have an equally close mechanistic relationship. The elegant sequence comparisons made by the Adelaide group (13), seductively suggesting the recruitment of functional domains in the assembly of a new catalytic entity, are illustrated in Figure 2.

Functional Similarities

The sequence comparisons exemplified above provide strong evidence for structural conservation among all the three classes of enzymes that use biotin

Pyruvate Carboxylase (yeast)

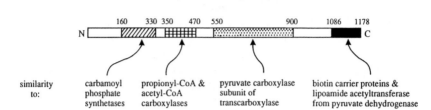

Figure 2   Similarities between regions of the amino acid sequence of yeast pyruvate carboxylase and various other proteins.

as a cofactor, and we may expect that this architectural similarity will be reflected in common mechanistic paths for enzyme-catalyzed transformations involving biotin. There are, indeed, only three types of process. First, the carboxylation of biotin that is effected by ATP and bicarbonate is shared (as reaction 1, above) by all Class I carboxylases. Second, the transfer of the carboxyl group from carboxybiotin to a carbon or nitrogen acceptor is common to all Class I carboxylases (as reaction 2), to all the transport systems of Class II (as reaction 3), and to each of the two chemical steps of the Class III enzyme, transcarboxylase (as reactions 5 and 6). The third type of process is the ion translocation step that is coupled to the decarboxylation of carboxybiotin as reaction 4 of the Class II pumps. In this review, we shall focus on the mechanistic aspects of the first two of these enzymic transformations. This necessarily precludes discussion of such facets of the subject as the biosynthesis of biotin, the transport and mobilization of the cofactor, the nature and specificity of the biotinylation process, and the assembly and regulation of the multienzyme complexes that are biotin-dependent. Many of these subjects have been reviewed elsewhere (3, 4, 17, 18).

THE CARBOXYLATION OF BIOTIN

The first reaction, 1, which is common to all Class I carboxylases, involves the carboxylation of biotin. For most carboxylases the cofactor is covalently attached to a lysine residue of the biotin carrier domain (for the multifunctional enzymes) or of the biotin carrier protein (for the multienzyme complexes). Early models of the carboxylases emphasized the fact that biotin attached to a lysine residue provides (in extenso) a "swinging arm" that is about 14 Å from C-2 of lysine to C-5 of biotin. Yet recent analogy with the lipoate domains of the acetyltransferase core of the pyruvate dehydrogenase complex, where it is believed that whole lipoate-carrying domains are hinged and mobile (19), would suggest a much longer "reach" for the biotin residue if this were needed.[1] In contrast to the enzymes that require biotin to be attached to its carrier subunit, two enzymes, the β-methylcrotonyl-CoA carboxylase from *Mycobacteria* (21) and the biotin carboxylase subunit of acetyl-CoA carboxylase from *E. coli* (9), catalyze the carboxylation of free biotin, which has allowed reaction 1 to be studied in isolation. Before any discussion of mechanism, however, two questions must be answered: what is the nature of the substrate (that is, is it bicarbonate or carbon dioxide), and what is the nature of the product, carboxybiotin?

Bicarbonate or Carbon Dioxide as Substrate?

The early ^{18}O tracer experiments of Kaziro et al (22) showed that when [^{18}O]bicarbonate is used as substrate for propionyl-CoA carboxylase in $H_2^{16}O$, two ^{18}O atoms end up in the new carboxyl group (of methylmalonyl-CoA) and the third is found in the P_i. These findings are most readily explained if bicarbonate is the substrate, and this conclusion was confirmed for pyruvate carboxylase by the direct observation of the initial reaction rates with bicarbonate and with carbon dioxide (23). Biotin-dependent enzymes are thus in a group with carbamoyl phosphate synthetase and phosphoenolpyruvate carboxylase, in using bicarbonate as the primary carbon source (2). Whether the bicarbonate subsequently dehydrates to carbon dioxide at the active site is discussed later.

N- or O-Carboxybiotin as Product?

The original isolation by Lynen et al (24) of the methyl ester of *N*-1 carboxybiotin from the biotin carboxylation reaction catalyzed by β-methylcrotonyl-CoA carboxylase led to the presumption that the enzymic intermediate is *N*-1 carboxybiotin. The relative inertness of this species generated some unease in the mechanistic community, and an ingenious alternative explanation was put forward (25), where *O*-carboxybiotin is the actual intermediate, and rearranges after methylation and during isolation to the more stable *N*-acyl material. This attractive proposal was ruled out, however, by the demonstration that authentic exogenous *N*-1 carboxybiotinol is a chemically and kinetically competent substrate[2] for the biotin carboxylase subunit of acetyl-CoA carboxylase (9). Biotin clearly carries the carboxyl group on *N*-1.

Some Suggested Mechanisms and Efforts to Discriminate Amongst Them

Of the rather large number of proposals that have been put forward for the mechanism of reaction 1, we summarize some of the mechanistically less

---

[1] In fact, the required arm length may in some cases be quite short, and the distance between the two active sites of transcarboxylase has been estimated to be only about 7 Å (20).

[2] This was the test, indeed, that Lynen (24) had originally said must be done.

egregious ones in Figure 3. Pathway **1** involves the activation of bicarbonate by ATP to produce carboxyphosphate, which then—either directly or after collapse to enzyme-bound carbon dioxide—carboxylates the N-1 of biotin. This pathway is close to (if chemically more reasonable than) the route originally presented by Kaziro et al (22) to accommodate the fact that one of the bicarbonate oxygens ends up in P_i. Pathway **2** involves the activation of biotin by phosphorylation on its urea oxygen, and was an early suggestion from Calvin & Pon (26) and from Lynen (27). The intermediate O-phosphobiotin then reacts with bicarbonate, either in a concerted fashion or by an "adjacent associative" route with pseudorotation at phosphorus (28, 29), to give the products. Pathway **3** also goes via O-phosphobiotin, but this intermediate is attacked "in line" by bicarbonate to give carboxyphosphate and the ureide anion of biotin, collapse of which (directly, or again, via carbon dioxide) gives N-1 carboxybiotin (29). The experimental evidence that bears upon these various mechanisms is enumerated below.

ISOTOPE EXCHANGE   If the carboxylation reaction, 1, were to follow a simple ping pong pathway, isotope exchange experiments would define whether bicarbonate (pathway **1**) or biotin (pathway **2** or **3**) is phosphorylated

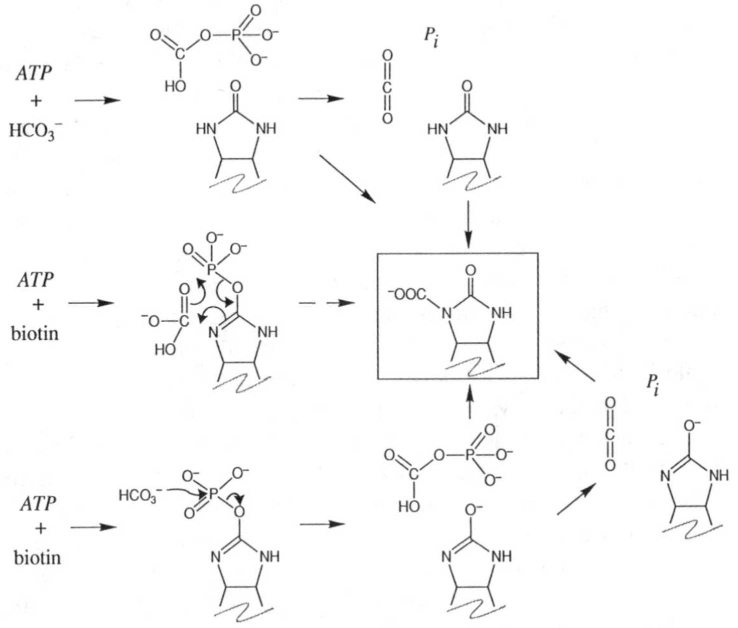

Figure 3   Some conceivable mechanistic pathways for the carboxylation of biotin. Pathway **1**, activation of bicarbonate to carboxyphosphate; pathways **2** and **3**, activation of biotin to O-phosphobiotin.

by ATP. Yet the consensus from a number of laboratories investigating several biotin-dependent carboxylases is that neither a rapid ATP/ADP exchange in the absence of biotin but dependent upon bicarbonate (which would support pathway **1**), nor a biotin-dependent ATP/ADP exchange in the absence of bicarbonate (which would indicate pathway **2** or **3**), can be observed (10, 22, 30–33). The failure to observe partial exchange reactions is quite common, of course, and may derive from the need to collect all the reaction components before the chemistry even of a partial reaction can occur (this is substrate synergism) (34), or from the fact that no product is released until all chemical processes at the active site are complete. Thus, while useful mechanistic information has been gleaned from the study of a number of different exchange reactions catalyzed by carboxylases (see, e.g., 33), the distinction at issue here cannot be made by this method.

When isotope exchange reactions fail, it has become natural to ask if the reason derives simply from a slow "product off" rate. Such behavior can often be uncovered by finding positional isotope exchange in appropriately labeled $[^{18}O]$-ATP by the method developed by Midelfort & Rose (35). The search for carboxylase-catalyzed exchange of ^{18}O from the peripheral β positions of ATP into the β,γ bridge position of reisolated material (see Figure 4A) has been conducted in three laboratories. Wimmer has investigated pyruvate carboxylase and propionyl-CoA carboxylase (M. Wimmer, private communication), and Tipton & Cleland (36) and Ogita & Knowles (37) have both studied the biotin carboxylase subunit of acetyl-CoA carboxylase. The latter enzyme system allows the positional isotope exchange process to be evaluated in the absence of biotin, and permits the study of possible synergistic effects of free biotin analogues such as N-1 methylbiotin [as had earlier been used for the carbamoyl phosphate reaction by Lane's group (10): this is discussed later]. In no case has any positional isotope exchange been observed. This is even true for the complete system (that is, including ATP, bicarbonate, biotin, P_i, and enzyme). In general, there are two reasons why the positional isotope exchange test can fail. First, the relevant intermediate state (in this case, enzyme-bound ADP and carboxyphosphate for pathway **1**, or enzyme-bound ADP and O-phosphobiotin for pathway **2**) may partition forward so strongly that the resynthesis of isotopically scrambled ATP is too rare an event to be detected. This explanation certainly cannot be ruled out, granting the expected high chemical reactivity of each of the putative intermediates, carboxyphosphate or O-phosphobiotin. The second reason why a positional isotope exchange reaction can fail is if the β-phospho group of enzyme-bound ADP is not torsionally free. Such suppression of phospho group rotation has been suggested before (e.g. 38), and there are, indeed, persuasive arguments why the positional isotope effect phenomenon may be the exception rather than the rule, even when the necessary intermediates form and partition appropriately (W. W. Cleland, private communication).

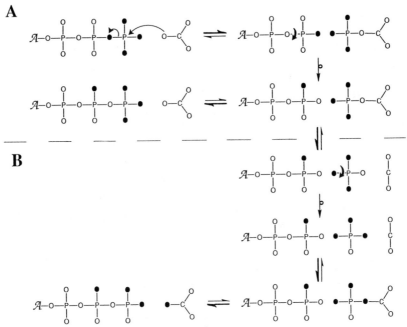

Figure 4 Pathways for conceivable isotope exchanges catalyzed by carboxylases: *A*, positional isotope exchange between the β, γ bridge position and the peripheral β positions of ATP; *B*, isotope exchange between bicarbonate and the peripheral γ positions of ATP.

INTERMEDIATE TRAPPING   The most direct way to determine the nature of the reaction intermediate in the carboxylation reaction would be to trap it from the steady state. This approach seems feasible if pathway **1** is followed, since Sauers et al (39) have estimated the half-life of carboxyphosphate in neutral aqueous solution to be on the order of 70 msec, and because Powers & Meister (40, 41) have reported the trapping of this species as its trimethyl ester from the steady-state reaction of carbamoyl phosphate synthetase. If pathway **2** or **3** is followed, however, the chances of trapping *O*-phospho-biotin seem slim, since the work of Blonski et al (42, 43) suggests that this species (even in esterified form) is extremely labile. In practice, efforts to trap carboxyphosphate from pyruvate carboxylase (33), from biotin carboxylase (37), and from phosphoenolpyruvate carboxylase (44), have all failed. It seems that if carboxyphosphate is formed, either it is in such low amount at the steady state, or it is so inaccessible to the diazomethane quencher (such sequestration would surely not be surprising for a labile reaction intermediate), that none can be detected in direct trapping experiments.

STEREOCHEMICAL TESTS   With the aim of narrowing the range of mechanistic possibility expressed in Figure 3, Hansen & Knowles determined the stereochemical fate of the γ-phospho group of ATP in the reaction catalyzed by pyruvate carboxylase (45). For this purpose, we made the $[\gamma\text{-}^{17}O, \gamma\text{-}^{18}O]$-$\gamma$-phosphorothioate of ATP (that is, ATPγS chiral at the γ phosphorus), used this as substrate for the carboxylase, and then determined the configuration of the chiral inorganic $[^{16}O, {}^{17}O, {}^{18}O]$thiophosphate product. On the basis that each act of enzyme-catalyzed phospho group transfer involves inversion of the configuration at phosphorus (46), we expected that pathways 1 and 2 would lead to overall inversion, and that pathway 3 would result in overall retention. Thus in pathway 1, bicarbonate would be phosphorylated with inversion, and carboxyphosphate would then react with biotin with no further change in the bonds to phosphorus. In pathway 2, O-phosphobiotin would be formed with inversion, and the subsequent pericyclic step, though chemically unprecedented, would be equivalent to a front-face displacement at phosphorus with retention. Pathway 3 would show one inversion in the formation of O-phosphobiotin, and a second inversion in the transfer to produce carboxyphosphate, giving retention overall. These stereochemical expectations are based upon the seemingly universal pattern that single enzyme-catalyzed displacements at phosphorus proceed with "in-line" geometry and invert the configuration (46). There could be exceptions to this pattern, of course, and it has been suggested that the generation of carboxyphosphate from O-phosphobiotin in pathway 3 could follow an "adjacent" attack (29), which, with the necessary pseudorotation at phosphorus (47), would give retention of the configuration for this step. Yet the lack of any enzymic precedent for pseudorotatory pathways makes one hesitant to make an exception to the corpus of knowledge on enzyme-catalyzed phospho group transfers, and the predicted stereochemical outcomes are therefore: inversion for pathways 1 and 2 and retention for pathway 3.

Of the three routes, pathway 3 is attractively close to the favored mechanism for an analogous enzyme, phosphoenolpyruvate carboxylase. This enzyme does not require biotin, but the ^{18}O labeling pattern (one of the bicarbonate oxygens ends in P_i) (48) and the stereochemical course (inversion at phosphorus) (49) has, with other results, led to the proposed pathway via carboxyphosphate shown in Figure 5. From this figure, the analogy with pathway 3 for biotin carboxylase is clear. So it was slightly surprising when biotin carboxylase was found to proceed with inversion at phosphorus, which (pace Ref. 29) effectively rules out pathway 3.

SUBSTRATE ANALOGUES   In 1972 Lane and his group (50) showed that incubation of biotin carboxylase with biotin, ADP, and carbamoyl phosphate led to the production of ATP. This result was taken to support pathway 1,

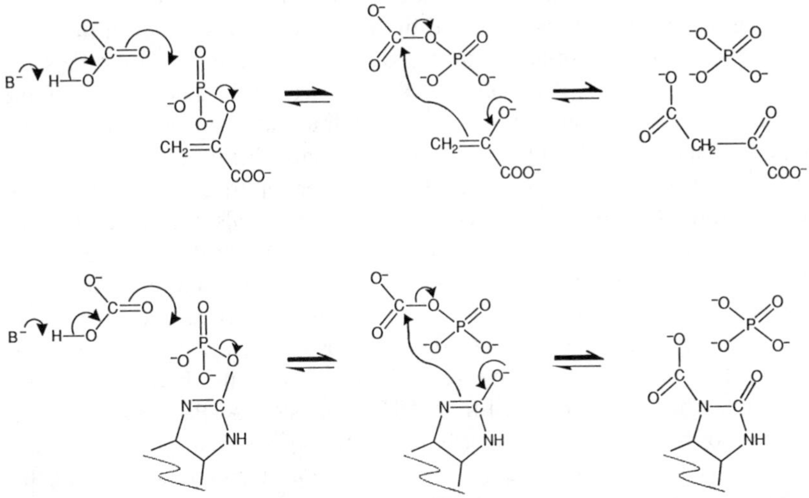

Figure 5   Mechanistic analogy between the favored pathway for phosphoenolpyruvate carboxylase (upper path) and pathway 3 (of Figure 3) for biotin carboxylation (lower path).

with carbamoyl phosphate acting simply as a chemically more stable isosteric analogue of carboxyphosphate. A concern about the possible covalent involvement of the required biotin was not fully resolved by the observation that N-1 methylbiotin can partly support the phospho group transfer reaction (10), since it is the urea oxygen that is phosphorylated in pathways 2 and 3. This possibility notwithstanding, the above conclusions were supported by Ashman & Keech (32), who made similar observations with pyruvate carboxylase. Additionally, these workers found that another analogue of carboxyphosphate, phosphonoacetate, inhibits the enzyme-catalyzed carboxylation reaction, though the binding is not very tight (K_i is 0.5 mM) (32). The presumption that carbamoyl phosphate acts as a surrogate for the carboxyphosphate intermediate of pathway 1 has been challenged by Kluger and his collaborators (28, 29), who suggested that carbamoyl phosphate could as well be an analogue of O-phosphobiotin, and that the observed synthesis of ATP could represent a reversal of the first step of pathway 2 or 3. While it is true that the interpretation of the behavior of substrate analogues always carries dangers, the requirement for biotin or a biotin derivative in the reaction of carbamoyl phosphate and ADP (10, 50), tilts the argument towards pathway **1**.

ATP HYDROLASE ACTIVITY   It has recently been shown by Climent & Rubio (51) that biotin carboxylase possesses a low ATPase activity in the absence of biotin, an activity that appears to be an intrinsic property of the carboxylase (51, 52). Thus the ATPase reaction is dependent on the presence of bicarbonate, its rate is increased by the presence of ethanol in the same proportion as in the normal biotin carboxylation reaction, it shows the same magnesium ion and potassium ion dependence, it is independent of the presence of avidin, and the thermal stabilities of the two activities run hand-in-hand (51). This hydrolysis of ATP seems, therefore, to be catalyzed by the carboxylase in what could well be a diversion from the normal fate of carboxyphosphate when biotin is present. Yet it can also be argued that the observed ATPase activity is adventitious, and that the dependence of this reaction on the presence of bicarbonate derives from some conformational activation of the carboxylase. To prove that the activity indeed comes from the hydrolytic diversion of a carboxyphosphate intermediate, it must be shown that bicarbonate is involved covalently. This has been done by establishing that the P_i produced from ATP in the slow ATPase reaction in the absence of biotin, derives its fourth oxygen from [^{18}O]bicarbonate rather than from the (unlabeled) water (37). These experiments therefore indicate that the ATPase reaction catalyzed by biotin carboxylase in the presence of bicarbonate and the absence of biotin occurs via a carboxyphosphate intermediate, and strengthen the view that the carboxylation of biotin itself proceeds analogously, following pathway 1.

The Mechanism of Biotin Carboxylation

It is clear from the above that the balance of evidence favors pathway 1. In the first step carboxyphosphate is formed, and in the second, this reactive species carboxylates biotin. Rubio (2) has stressed the mechanistic similarities between the biotin-dependent carboxylases and carbamoyl phosphate synthetase, and this relationship is made even closer by the amino acid sequence similarities between these enzymes discussed earlier (13). Mechanistically, the carboxylases and carbamoyl phosphate synthetase share the following features: (a) one of the three oxygens of bicarbonate ends up in product P_i; (b) the carboxylases show a bicarbonate-dependent ATPase activity in the absence of biotin, and the synthetase shows a bicarbonate-dependent ATPase activity in the absence of ammonia; (c) neither an ADP/ATP exchange nor an ATP/P_i exchange can be observed; (d) the steady-state kinetics of the reactions are consistent with no product being released until all the chemistry has occurred; and (e) the two groups of enzymes have very similar requirements for both monovalent and divalent metal cations. The carboxylases differ from carbamoyl phosphate synthetase in two respects: (f) trapping of the carboxyphosphate intermediate has been reported from the synthetase (40, 41), but analogous experiments with carboxylases have failed (33, 37, 44); and (g) positional isotope exchange in [^{18}O]-labeled ATP is observed with carbamoyl phosphate synthetase (53) but not with any of several carboxylases (36, 37).

These two differences do not seriously undermine the structural and mechanistic parallels between the two groups of enzymes, and it seems sensible to assume that the reaction pathways followed are analogous.

The second step of the biotin carboxylation reaction formally involves the attack by biotin's *N*-1 on the carboxyphosphate intermediate. As stated, this does not look like a facile process: the carbon being attacked is not especially electrophilic, and the nitrogen of biotin is a poor nucleophile. Considerable speculation and experimentation has gone into each aspect of this step.

THE ELECTROPHILE   First, how can carbon of carboxyphosphate become more electrophilic? Obvious possibilities would be to protonate the carboxylate or to bind a metal cation, yet this formulation creates several difficulties for the transfer of the carboxyl group from carboxybiotin, both in the reverse reaction (of ATP synthesis) and in the forward transfer to a carbon acceptor. The problem is one of avoiding the formation of the high-energy dianionic tetrahedral intermediate I that would derive from nucleophilic attack on a carboxylate. From recent model studies discussed below, such energetically unfavorable species can be avoided by providing a general acid catalyst as in II, or by following a "borderline" mechanism III such as has been proposed

for phospho group transfers (54). But an alternative to any such direct transfer is to create an electrophilic carbon center by prior dissociation of carboxyphosphate into carbon dioxide and P_i. The possible advantages of such a dissociation were first noted by Sauers et al (39), who pointed out that carbon dioxide is much more electrophilic than carboxyphosphate, and that the role of ATP in biotin carboxylation may be to collect a bicarbonate from solution and deliver a carbon dioxide molecule to biotin at the active site. Such a locally generated molecule of carbon dioxide could be appropriately positioned for reaction, thus providing a significant entropic advantage for the carboxylation step. ATP is thus used for dehydration and delivery, and carboxybiotin becomes a carbon dioxide carrier. Release of carbon dioxide is triggered at the active site, either from carboxyphosphate for the carboxylation of biotin, or, in subsequent steps of the reaction of carboxylases, from carboxybiotin for the carboxylation of carbon enolates.

Despite some effort, there is still no experimental evidence to support the view that carboxyphosphate dissociates to P_i and carbon dioxide at the active site of carboxylases. The experimental test that has been applied to carbamoyl phosphate synthetase (53, 55) and to the biotin carboxylase subunit of acetyl-CoA carboxylase (36, 37), looks for the enzyme-catalyzed exchange of ^{18}O between the solvent and the three peripheral γ oxygens of ATP. As illustrated in Figure 4B, if the P_i were to tumble at the active site thus rendering all its oxygens torsionally equivalent, resynthesis of ATP would effectively equilibrate the three oxygens of bicarbonate (which are in equilibrium with solvent, via free bicarbonate) with the γ phospho group of ATP. This exchange has not been observed in any of the cases examined. Yet the fact that positional isotope exchange is often not observed when we have every reason to expect it (37, 38), presumably because the ADP β-phospho group cannot rotate, would argue, a fortiori, that inorganic phosphate need not tumble, either. The failure to detect the ATP/solvent oxygen exchange, therefore, is not evidence against carbon dioxide as the reactive carboxylating entity. There is, however, in the related case of a phosphoenolpyruvate carboxylase, a suggestion that an enzyme-bound P_i can indeed tumble, and that carboxyphosphate does dissociate to P_i and carbon dioxide. Fujita et al (44) have shown that the methylated substrate analogue phosphoenol-α-ketobutyrate is cleaved into P_i and α-ketobutyrate in a bicarbonate-dependent reaction. Although no substrate carboxylation was found, when the reaction was done with $[^{18}O]$bicarbonate, multiple ^{18}O atoms were incorporated into the liberated P_i. This finding has been confirmed by M. O'Leary and J. O'Laughlin (private communication), who have further shown that phosphoenol-α-ketobutyrate contains multiple ^{18}O labels when reisolated after incubation in $H_2{}^{18}O$ with the enzyme. These results, albeit with a substrate analogue that does not complete the reaction, are consistent with the reversible cleavage of carboxyphosphate into P_i and carbon dioxide at the active site.

THE NUCLEOPHILE   The low nucleophilicity of biotin's N-1 has generated a great deal of bioorganic concern, and a number of proposals to circumnavigate the problem. Caplow's early suggestion that the decarboxylation of N-carboxyimidazolidone at low pH probably occurs by the unimolecular decomposition of the free acid, which would give carbon dioxide plus the isourea tautomer of imidazolidone (56, 57), led to the recognition by Bruice and others that the isourea is about 10^{10} times more nucleophilic than the urea form (58). This fact, with the knowledge that when urea itself acts nucleophilically it uses oxygen rather than nitrogen, resulted in the suggestion that carboxybiotin is really carboxylated on oxygen, and that Lynen's isolation of the N-1 carboxylated species (25) was a consequence of the isolation procedure. This seductive proposal neatly resolved both the nucleophilicity question and the acyl-transfer reactivity issue. Yet it was short-lived, since Lane's group proved in 1972 that an authentic N-1 carboxybiotin was both chemically and kinetically competent as an enzymic intermediate (50). Now, while this demonstration forced the abandonment of O-carboxybiotin and the facile acyl transfer from such an O-acylisourea, the greater nucleophilicity of an isourea could still be rescued. The groups of Kluger (28, 29) and of Kellogg (59), reviving an earlier suggestion of Calvin (26) and of Lynen (27), put forward O-phosphobiotin as an intermediate in the biotin carboxylase reaction (see, e.g., Figure 3, pathways 2 and 3) and produced ingenious model chemistry to support these ideas. Yet, as has become clear from earlier sections, the more banal reaction of N-1 of biotin with carboxyphosphate (or with carbon dioxide derived from it) now seems to be the most likely path, and we must ask whether our prejudices about the poor nucleophilicity of biotin's nitrogen are justified.

In 1985, Mildvan and his colleagues (60) published an NMR study of the exchange of the NH protons of biotin with water, and showed that under mild conditions the exchange rates were comparable to the rate of enzyme-catalyzed carboxylation of biotin. In principle, therefore, base-catalyzed N-deprotonation of biotin is fast enough without any further action by the enzyme, to accommodate the second step of pathway 1 (Figure 3). More recently, Perrin & Dwyer (61) have made some important corrections to the earlier work, and have discussed all the possible modes of proton exchange and the relevance of these possibilities to the mechanism of biotin-mediated carboxyl transfer. If carboxylation proceeds via carboxyphosphate and not carbon dioxide, then the preferred nucleophile will be the ureido anion IV. This species can be formed fast enough ($k \geqslant 12$ s^{-1} at pH 7.4, 25°C), and the deprotonation can be effected by an enzymic base. There seems to be no purpose in creating the isourea V, since O-protonation would surely reduce the nucleophilicity of N-1. To avoid dianionic tetrahedral intermediates, Perrin

& Dwyer (61) suggest the participation of a general acid (as is also indicated by the model experiments described below). If, however, carboxyphosphate collapses at the active site to produce carbon dioxide, then a concerted process (as in VI) is feasible. Indeed, from a study of the breakdown of carbamates

IV   V   VI

derived from amines of different basicities, Jencks' group concluded that the cleavage of carbamates of weakly basic amines occurs through a pre-associative pathway that may well be concerted (62). In the reverse (carboxylation) reaction, Rebek has pointed out (J. Rebek, private communication) that the collapse of the carboxyphosphate trianion at the active site would generate electrophilic carbon dioxide and a strong base (orthophosphate trianion) appropriate for the deprotonation of biotin at N-1. (Looking ahead to the question of carboxyl transfer to biotin from a β-keto acid, the analogous base would then be the carbon enolate formed after decarboxylation of the acid.) In short, this work on the proton exchange at N-1 of biotin has largely removed the concern of the past 20 years that biotin's nitrogen could not act directly as the nucleophile in the carboxylation process.

A MODEL SYSTEM   To select a single model reaction from a field that has received so much more than its share of attention from the bioorganic community may seem gratuitous, yet Blagoeva, Pojarlieff & Kirby (63) have reported a model reaction of particular appropriateness and simplicity. These workers investigated the cyclization of the hydantoic acid derivative VII (Figure 6) to the hydantoin VIII. At high pH, the reaction is specific base–general acid catalyzed, and avoids a dianionic tetrahedral intermediate, as shown in Figure 6. The authors point out that one may reasonably presume that N-1 of biotin is analogously deprotonated before carboxylation, while recognizing (for attack on a carboxyl species carrying a good leaving group, such as carboxyphosphate) the possibility of the "borderline" mechanism (as in III) discussed earlier.

This view of biotin carboxylation is consistent with what is known about the pH dependence of the reaction catalyzed by the biotin carboxylase subunit of acetyl-CoA carboxylase. The carboxylation reaction requires a catalytic

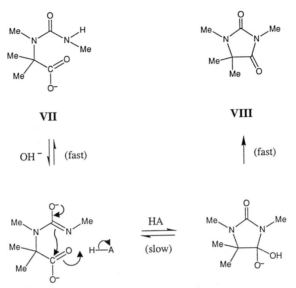

Figure 6 Mechanistic pathway for the cyclization of the hydantoic acid VII to the hydantoin VIII at high pH.

base of apparent pK_a of 6.6 in V_{max}/K_m (52). Interestingly, the pH dependence of the analogue of the reverse reaction, the phosphorylation of ADP by carbamoyl phosphate, is very similar, which may mean that the catalytic groups assume the same protonation states for catalysis of phospho transfer in either direction (52). The carboxylation reaction shows an unusual inverse solvent isotope effect on V_{max}/K_m (of 0.5 for biotin carboxylation at pH 8), which is most readily interpreted in terms of a thiol on the enzyme that has (essentially completely) transferred its proton to an oxygen or nitrogen base at the rate-limiting transition state (36). It seems possible that for the carboxylase to become active, an essential thiol has to relinquish its proton to a nitrogen or oxygen base on the enzyme. Before all the data are collected and a mechanism for biotin carobxylation is proposed, we discuss the second half-reaction, the carboxylation of carbon acceptors. By doing this, several parallels in mechanism between the two half-reactions become apparent, and a more economical pathway for each of them can then be put forward.

THE CARBOXYLATION OF CARBON ACCEPTORS

The second type of process catalyzed by biotin-dependent enzymes involves the transfer of the carboxyl group from *N*-1 carboxybiotin to a readily enolizable carbon center, as exemplified by reactions 2, 3, 5, and 6 listed at the beginning of this review. Two key facts about these transfers, concerning stereochemistry and proton exchange, were soon established, and drove the early mechanistic thinking. Stereochemically, all biotin-dependent carboxylases that have been investigated proceed with retention of the configuration at the carbon atom that undergoes carboxylation. Thus propionyl-CoA carboxylase (64, 65), pyruvate carboxylase (66), and each of the reactions catalyzed by transcarboxylase (67), all follow a retention path, and it seems reasonable to presume that this will prove to be the universal pattern for biotin-dependent carboxylases. The second mechanistic feature is that the enzymes do not catalyze the exchange of protons between solvent and substrate, without overall turnover. For example, propionyl-CoA carboxylase only incorporates solvent tritium into propionyl-CoA at a rate commensurate with the overall carboxylation-decarboxylation rate (64, 68), and enzymically synthesized [2(R)-^{3}H]propionyl-CoA does not lose the isotopic label unless all substrates are present (68). Analogous findings for pyruvate carboxylase (69) and for transcarboxylase (67) have been reported.

These two characteristics of the biotin-dependent carboxylases led to early proposals that the reactions were concerted, and to suggestions of the (chemically unprecedented) pericyclic process illustrated in Figure 7A (64, 68, 69). Yet neither the stereochemical course nor the absence of proton exchange requires a concerted process. Thus retention of stereochemistry is seen with the aldolases, for example, and this is quite compatible with proton abstraction to give an enol(ate) or enamine, which then reacts in a second step with a carbonyl electrophile. Analogously, the absence of proton exchange does not preclude a carbanionic intermediate: a high forward commitment, or an enzyme conjugate acid that is sequestered from solvent, readily accommodates the lack of an observable exchange. The stepwise pathways shown in Figure 7B and 7C must therefore be considered.

Is the Carboxyl Transfer Stepwise or Concerted?

To examine the concertedness of biotin-dependent carboxylases, the reaction of β-fluoropropionyl-CoA with propionyl-CoA carboxylase and with transcarboxylase was investigated (70, 71). Both enzymes catalyze the elimination of HF from the fluorinated substrate to form acryloyl-CoA, suggesting [by analogy with the great body of work on mechanism-based inactivation by fluorinated substrates of enzymes that require pyridoxal phosphate (72)] a stepwise reaction and the intermediacy of a substrate carbanion. Yet the elimination of HF catalyzed by propionyl-CoA carboxylase occurs at the same rate as ATP hydrolysis (71), which is just what would be expected if proton removal and carboxylation were coupled processes. Abeles and coworkers (71) argued, however, that the reaction does follow a stepwise path such as that shown in Figure 7B, in which the first-formed enolate attacks N-1 carboxybiotin in a second step. The pathway illustrated in Figure 7C is a

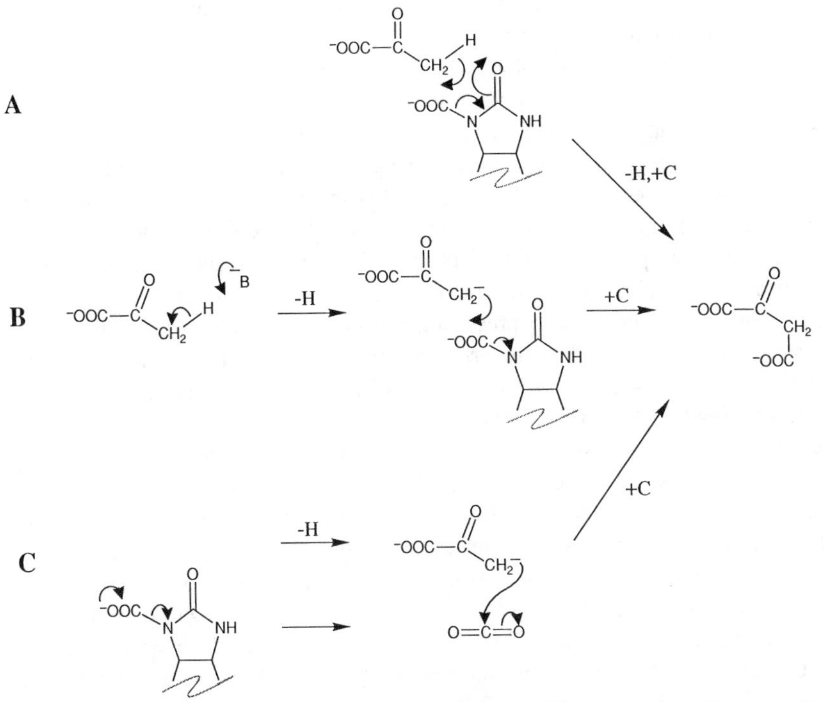

Figure 7 Concerted (*A*) and stepwise (*B*, *C*) pathways for the carboxylation of pyruvate by *N*-carboxybiotin, catalyzed by pyruvate carboxylase.

variant of that in Figure 7*B*, and follows the view that the electrophilic species is carbon dioxide rather than carboxybiotin.

In 1982, Kuo & Rose (73) sought to discover if exogenously generated enolpyruvate, which is a presumed intermediate in the reaction catalyzed by transcarboxylase, is handled as a substrate by this enzyme. They showed that transcarboxylase catalyzes the stereospecific ketonization of enolpyruvate, which is entirely consistent with the stepwise path. This C-protonation of enolpyruvate was only effected in the presence of the other substrate, methylmalonyl-CoA (which would drive biotin towards carboxybiotin), yet it transpired that a negligible fraction of the enolpyruvate was carboxylated. Thus although enolpyruvate appears to be a substrate for transcarboxylase in reprotonation, it is not carboxylated. This is a concern, since a true enolpyruvate intermediate should partition both ways: back to pyruvate and forward to oxalacetate.[3] Neither of the two experiments described, therefore, one using a

---

[3]Though it can be argued that the enolate of pyruvate is the natural intermediate, whereas the exogenously supplied material is the enol. This enol will bind to a form of the enzyme that may be in an inappropriate protonation state, and the resulting complex need not partition identically to that formed in the normal reaction.

substrate analogue and the other an exogenous intermediate, produces an entirely unambiguous answer.

The issue was settled in 1986, when two groups applied the double-isotope fractionation test to transcarboxylase (74) and to pyruvate carboxylase (75). The method can be summarized as follows. In a carboxylation reaction, a carbon-hydrogen bond is broken and a carbon-carbon bond is made. If the reaction were concerted (and providing that the rate-limiting transition state was that involving the covalency changes), then the reaction would show both a ^{2}H and a ^{13}C primary kinetic isotope effect (in pyruvate and in carboxy-biotin, respectively: Figure 7A). If, however, the reaction were stepwise, with the proton abstraction step preceding the carboxyl transfer step, then the reaction would show either one primary effect (i.e. a ^{2}H effect or a ^{13}C effect) if the proton abstraction step or the carboxyl transfer step were cleanly rate-limiting (Figure 7B), or both primary effects if the intermediate carbanion partitioned evenly between the forward and reverse reactions (in a "balanced" stepwise process). So the existence of one isotope effect but not the other, requires a stepwise reaction. However, if both a ^{2}H and a ^{13}C kinetic iso-tope effect are seen, the reaction may be concerted or balanced stepwise. These two situations can be distinguished by looking at the consequence of substrate deuteration on the ^{13}C isotope effect. In the concerted case, pyru-vate deuteriation (in Figure 7A) will not affect the discrimination between ^{12}C and ^{13}C carboxyl groups; the reaction will go slower, but the ^{12}C/^{13}C rate ratio will be unchanged. But in the stepwise case, pyruvate deuteri-ation (in Figure 7B) will slow the proton abstraction step, causing the car-boxylation step to be kinetically less significant, and resulting in a lower ob-served ^{13}C effect. [The above simple description sidesteps some fine points, such as the consequences of equilibrium effects and of slow "product off" rates: the original papers (74, 75) discuss these questions.] When this test was applied to transcarboxylase, the ^{13}C effect of 2.3% with protio-pyruvate was found to fall to 1.4% with deuterio-pyruvate (74). The results with pyruvate carboxylase showed that this carboxylation, too, is unequivocally stepwise.

These findings rule out the concerted pathway shown in Figure 7A (against which, it must be said, persuasive though nonexperimental arguments of an improbable geometry, unprecedented chemistry, and an unlikely tautomeriza-tion for the reverse reaction, had been leveled). Also ruled out are two concerted variants of Figure 7A (76, 77) that involve carboxylation by N-1 carboxybiotin in its isourea form with N-3. What remains, is the stepwise pathway shown in Figure 7B, and the corresponding version using carbon dioxide as the electrophile, in 7C. These pathways for transcarboxylation mirror the preferred route for the carboxylation of biotin, which is pathway **1** of Figure 3.

Tritium Transfer in Transcarboxylase

In 1976, Rose et al (78) investigated the possibility that in the transcarboxylase reaction, protons might be transferred between the noncarboxylated substrates, pyruvate and propionyl-CoA. One reason for doing this experiment was to test the concerted mechanism (Figure 7A). According to this formulation, biotin collects a proton from pyruvate, and, in its isourea form, migrates to the other active site where it becomes carboxylated by methylmalonyl-CoA and gives up this proton to produce propionyl-CoA. So, as long as the proton is not rapidly lost to solvent en route, one might hope to observe tritium transfer from pyruvate to propionyl-CoA, as illustrated in Figure 8. In practice, about 5% of the tritium labilized in the enzyme-catalyzed reaction of [^{3}H]pyruvate was found in the appropriate (2-pro-R) position of the propionyl-CoA product (78). This observation of tritium transfer was taken as evidence for the concerted mechanism in which the isourea form of biotin is the proton carrier between substrates and enzyme subunits. Yet we now know that transcarboxylase does not follow a concerted pathway. Moreover, the exchange rate of the hydroxyl group proton in the isourea form of biotin is likely to be very rapid [a rate constant of > 5000 s^{-1} has been estimated (75)], and it is arguable whether the migration of biotin from active site to active site in transcarboxylase would occur as fast as this (5, 79). The generation of biotin in its much less stable isourea form (for example, in the decarboxylation direction), also puts unnecessary strain on mechanistic credulity. So how can the tritium transfer result be accommodated, since we know that the concerted pathway is not followed?

Fortunately, the preferred mechanistic pathway for carboxyl transfer to and from active sites involves biotin in only two forms: the urea tautomer of biotin itself, and N-1 carboxybiotin. So it is the stable urea tautomer of biotin that can carry the tritium label without any significant concerns about the site-to-site migration rate, since the proton (presumably on N-1) exchanges quite slowly in neutral aqueous solution [the rate constant is about 5 s^{-1} (60, 61)]. Biotin carries the proton, and N-carboxybiotin carries the carboxyl group.

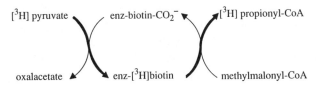

Figure 8 The putative pathway of a tritium label in pyruvate to propionyl-CoA, catalyzed by transcarboxylase.

THE MECHANISTIC PATHWAY FOR CARBOXYLASES

While no reaction mechanism is ever secure, and while we can only, at best, formulate minimalist proposals that are in reasonable accord with the facts, it would be improper not to attempt to summarize the foregoing in a suggested mechanistic pathway. This is done in Figure 9. The reader will notice at least one omission, namely the metal cations that are known to be essential for biotin carboxylation and for carboxyl transfer to carbon. Thus both Mg(II) and K(I) are required for the ATP-dependent activity of biotin carboxylase (51), and carboxyltransferases such as pyruvate carboxylase (80) and transcarboxylase (81) need a divalent cation for activity. But although a Mn(II) ion lies near the pyruvate binding site in pyruvate carboxylase (69), pyruvate is not an inner-sphere ligand (20, 82), and the role of the metal may be structural rather than participatory (83). Until these uncertainties have been fully resolved, therefore, we prefer not to invest the various cations with defined functions. In other respects, the pathway shown in Figure 9 is consistent with all the results that have been presented,[4] and aims to cast both the ATP-dependent carboxylation of biotin and the transfer of the carboxyl group from carboxybiotin to a carbon acceptor into the same mechanistic framework.

WHY BIOTIN?

In this final section, we ask why Nature goes to such biosynthetic lengths to assemble biotin. While such teleological questions are, of course, unanswerable, efforts to find answers have certainly sharpened our understanding of the function of many cofactors, and the question has a special piquancy here. The role of most cofactors seems to be to supplement and extend the range of chemistry that enzymes can mediate. Thus while the side chains of the 20 amino acids can handle much of the heterolytic chemistry required in metabolism, metal cations are used as superacids, thiamin provides an ylid, and pyridoxal is a particularly versatile electron sink. Very little redox chemistry is possible with amino acid side chains alone, and essentially all of these reactions need the help of cofactors (the nicotinamides, flavins, pterins, porphyrins, etc).

Looked at from this angle of chemical reactivity, biotin is a puzzle. Biotin is an unreactive molecule that seems to lack any interesting bioorganic properties that would extend the catalytic range or prowess of the enzymes that use it. What emerges from the mechanistic considerations of biotin-dependent systems is that the cofactor seems to be a vehicle for the carriage

---

[4]Though for pyruvate carboxylase, Attwood et al (75) prefer a slightly different sequence on the basis of the fit of predicted to experimental isotope effects.

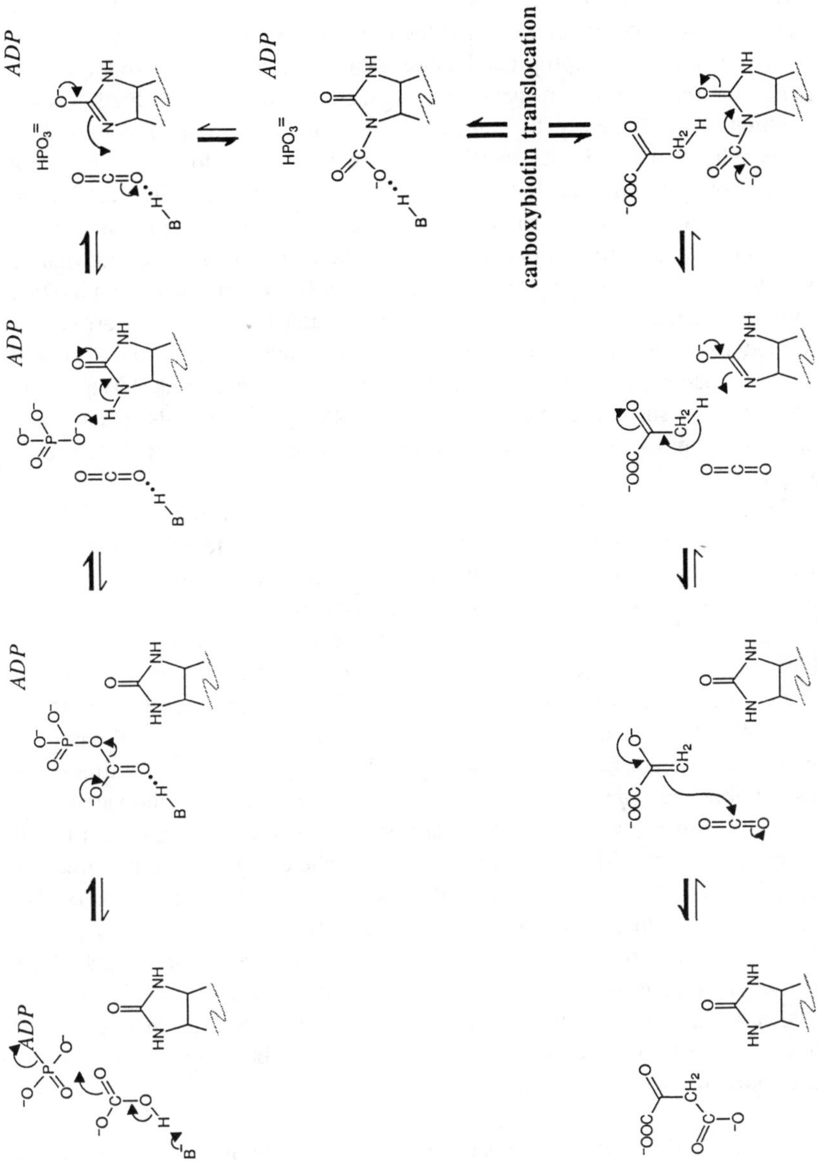

Figure 9 Proposed mechanistic pathway for the biotin-dependent carboxylases.

and delivery of carbon dioxide, as Sauers et al (39) originally suggested. How does biotin perform in these roles?

CARRIAGE   To be an effective carrier of carbon dioxide, and to avoid wasteful discharge before arrival at the transcarboxylation enzyme, carboxybiotin must have an appropriate kinetic stability without excessive thermodynamic inertness. In this respect, many cofactors activate their charges. Thus methionine is evidently too unreactive as a methyl group donor, and S-adenosylmethionine is the metabolic consequence. Analogously, P_i is not used in phosphorylation, but ATP is; acetate ion is not used for acetylation, rather acetyl-CoA is; and so on. In other cases, such as acetyl group transfer between CoA and dihydrolipoate, the chemical reactivity and thermodynamic stability are (very) roughly left unchanged. But for biotin, the carbon dioxide molecule actually becomes less reactive on attachment to its carrier. (A second example may be methylenetetrahydrofolate, which can be viewed as the transportable form of undesirably reactive formaldehyde.) We should not, then, be surprised or disappointed that carboxybiotin fails to show any special chemical reactivity: a kinetically stable species may be just what is required.

That carboxybiotin has a reasonably long half-life (>100 min at pH 8) is a consequence of two features of the system. First, as is clear from the crystal structure, the N-carboxyurea moiety of N-1 methoxycarbonylbiotin methyl ester is planar (84), and is therefore fully conjugated and (relative to carbon dioxide) immune from nucleophilic attack. Second, the bicyclic structure of biotin creates a steric crowding that decreases the lability of the molecule in decarboxylation, particularly by the acid-catalyzed route. Thus a recent study by Tipton & Cleland (85) of the mechanisms of decarboxylation of carboxybiotin at different pH values has shown that carboxybiotin is 30 times less susceptible than the monocyclic analogue (N-carboxyimidazolidone) to decarboxylation at low pH. These workers suggest that the second ring of biotin helps to prevent N-1 from achieving the tetrahedral geometry required for decarboxylation via a transition state analogous to VI. These workers also speculate (85) that this crowding has little effect on the mechanism that simply involves spontaneous C-N bond cleavage (such as is part of each of the carboxylation pathways outlined in Figure 9). In any event, there is no doubt that N-1 carboxybiotin is a much less labile carbamate than monocyclic or acyclic carboxyureas, and as such has the kinetic stability appropriate for the transport of the carboxyl group.

DELIVERY   On binding to the active site, carboxybiotin must relinquish the carboxyl group, either directly to an acceptor nucleophile, or by dissociation to carbon dioxide. In the search for unusual features of the biotin molecule that might bear on this problem, the possibility of a transannular interaction between sulfur and the urea carbonyl carbon has been rejected. This suggestion was originally triggered by the fact that these two atoms lie within 0.13 Å of their atomic radii in biotin crystals (60, 84). While such an interaction had been observed in thiacyclooctanone (86), no effects in the infrared, or in the ^{1}H or ^{13}C NMR, can be observed for biotin (87, 88). Even when BF_3 is coordinated to the urea oxygen, no significant transannular interaction between sulfur and the carbonyl group of biotin is detected (88). These findings have put the transannular effect to rest, but leave us with the unanswered question of why O-heterobiotin (in which oxygen replaces sulfur) can be incorporated into transcarboxylase and yet results in an inactive enzyme (89). [Though it should be mentioned that O-heterobiotin can support the growth of some microorganisms where it has been confirmed that the analogue is not converted to biotin itself in vivo (90).]

Currently the most persuasive suggestion to account for the delivery of the carboxyl group when carboxybiotin binds to the active site is due to Kluger and his group (91). These workers, recognizing that the decarboxylation of carboxybiotin is faster when bound to the active site of biotin-dependent enzymes than when free in solution (92), have calculated that rotation of the carboxyl group of carboxybiotin out of the plane of the urea ring may trigger the decarboxylation. The deconjugation that results from rotation of the carboxyl group both weakens the carbon-nitrogen bond, and leads to the migration of charge from the carboxylate oxygens towards the urea ring. This, then, possibly combined with desolvation into a hydrophobic environment (e.g. 93, 94), could constitute a neat way in which carbon dioxide could be released at the active site when carboxybiotin binds.

Whether the speculations that have been brought together in this section have any validity will only become clear when the structure of a biotin-dependent enzyme is determined to atomic resolution. Yet already we can see that biotin does solve, in the reasonably economic fashion that we expect from nature, the problems of the transfer of carboxyl groups in metabolism.

ACKNOWLEDGMENTS

This work was supported by the National Institutes of Health and the National Science Foundation. The author is especially grateful to a large number of colleagues for preprints of their unpublished work, and to a slightly smaller number for very helpful comments on this manuscript.

Literature Cited

1. Butler, J. N. 1982. *Carbon Dioxide Equilibria and their Applications*. Reading, Mass: Addison-Wesley. 259 pp.
2. Rubio, V. 1986. *Biosci. Rep.* 6(4):335–47
3. Dakshinamurti, K., Bhargava, H. N., eds. 1985. *Ann. NY Acad. Sci.*, Vol. 447. New York: NY Acad. Sci. 441 pp.
4. Wood, H. G., Barden, R. E. 1977. *Annu. Rev. Biochem.* 46:385–413
5. Mildvan, A. S., Fry, D. C., Serpersu, E. H. 1989. In *A Study of Enzymes*, Vol. 2, ed. S. A. Kirby. Boca Raton: CRC Press. In press
6. Attwood, P. V., Keech, D. B. 1984. *Curr. Top. Cell. Regul.* 23:1–55
7. Samols, D., Thornton, C. G., Murtif, V. L., Kumar, G. K., Haase, F. C., Wood, H. G. 1988. *J. Biol. Chem.* 263:6461–64
8. Guchhait, R. B., Polakis, S. E., Dimroth, P., Stoll, E., Moss, J., Lane, M. D. 1974. *J. Biol. Chem.* 249:6633–45
9. Guchhait, R. B., Polakis, S. E., Hollis, D., Fenselau, C., Lane, M. D. 1974. *J. Biol. Chem.* 249:6646–56
10. Polakis, S. E., Guchhait, R. B., Zwergel, E. E., Lane, M. D. 1974. *J. Biol. Chem.* 249:6657–67
11. Wood, H. G., Kumar, G. K. 1985. *Ann. NY Acad. Sci.* 447:1–22
12. Beaty, N. B., Lane, M. D. 1982. *J. Biol. Chem.* 257:924–29
12a. Takai, T., Yokoyama, C., Wada, K., Tanabe, T. 1988. *J. Biol. Chem.* 263: 2651–57
13. Lim, F., Morris, C. P., Occhiodoro, F., Wallace, J. C. 1988. *J. Biol. Chem.* 263:11493–97
14. Guest, J. R., Lewis, H. M., Graham, L. D., Packman, L. C., Perham, R. N. 1985. *J. Mol. Biol.* 185:743–54
15. Schwarz, E., Oesterhelt, D., Reinke, H., Beyreuther, K., Dimroth, P. 1988. *J. Biol. Chem.* 263:9640–45
16. Visser, C. M., Kellogg, R. M. 1978. *J. Mol. Evol.* 11:171–87
17. Shenoy, B. C., Wood, H. G. 1988. *FASEB J.* 2:2396–401
18. Shenoy, B. C., Paranjape, S., Murtif, V. L., Kumar, G. K., Samols, D., Wood, H. G. 1988. *FASEB J.* 2:2505–11
19. Graham, L. D., Guest, J. R., Lewis, H. M., Miles, J. S., Packman, L. C., et al. 1986. *Philos. Trans. R. Soc. London, Ser. A* 317:391–404
20. Fung, C. H., Mildvan, A. S., Allerhand, A., Komoroski, R., Scrutton, M. C. 1973. *Biochemistry* 12:620–29
21. Knappe, J., Schlegel, H-G., Lynen, F. 1961. *Biochem. Z.* 335:101–22
22. Kaziro, Y., Hass, L. F., Boyer, P. D., Ochoa, S. 1962. *J. Biol. Chem.* 237: 1460–68
23. Cooper, T. G., Wood, H. G. 1971. *J. Biol. Chem.* 246:5488–90
24. Lynen, F., Knappe, J., Lorch, E., Jütting, G., Ringelmann, E. 1959. *Angew. Chem.* 71:481–86
25. Bruice, T. C., Hegarty, A. F. 1970. *Proc. Natl. Acad. Sci. USA* 65:805–9
26. Calvin, M., Pon, N. G. 1959. *J. Cell. Comp. Physiol.* 54 (Suppl. 1): 51–74
27. Lynen, F. 1967. *Biochem. J.* 102:381–400
28. Kluger, R., Adawadkar, P. D. 1976. *J. Am. Chem. Soc.* 98:3741–42
29. Kluger, R., Davis, P. P., Adawadkar, P. D. 1979. *J. Am. Chem. Soc.* 101: 5995–6000
30. Scrutton, M. C., Utter, M. F. 1965. *J. Biol. Chem.* 240:3714–23
31. Barden, R. E., Fung, C. H., Utter, M. F., Scrutton, M. C. 1972. *J. Biol. Chem.* 247:1323–33
32. Ashman, L. K., Keech, D. B. 1975. *J. Biol. Chem.* 250:14–21
33. Wallace, J. C., Phillips, N. B., Snoswell, M. A., Goodall, G. J., Attwood, P. V., Keech, D. B. 1985. *Ann. NY Acad. Sci.* 447:169–88
34. Bridger, W. A., Millen, W. A., Boyer, P. D. 1968. *Biochemistry* 7:3608–16
35. Midelfort, C. F., Rose, I. A. 1976. *J. Biol. Chem.* 251:5881–87
36. Tipton, P. A., Cleland, W. W. 1988. *Biochemistry* 27:4325–31
37. Ogita, T., Knowles, J. R. 1988. *Biochemistry* 27:8028–33
38. Hilscher, L. W., Hanson, C. D., Russell, D. H., Raushel, F. M. 1985. *Biochemistry* 24:5888–93
39. Sauers, C. K., Jencks, W. P., Groh, S. 1975. *J. Am. Chem. Soc.* 97:5546–53
40. Powers, S. G., Meister, A. 1976. *Proc. Natl. Acad. Sci. USA* 73:3020–24
41. Powers, S. G., Meister, A. 1978. *J. Biol. Chem.* 253:1258–65
42. Blonski, C., Etemad-Moghadam, G., Gasc, M. B., Klaebe, A., Perie, J. J. 1983. *Phosphorus and Sulfur* 18:361–64
43. Blonski, C., Gasc, M. B., Hegarty, A. F., Klaebe, A., Perie, J. J. 1984. *J. Am. Chem. Soc.* 106:7523–29
44. Fujita, N., Izui, K., Nishino, T., Katsuki, H. 1984. *Biochemistry* 23:1774–79
45. Hansen, D. E., Knowles, J. R. 1985. *J. Am. Chem. Soc.* 107:8304–5
46. Knowles, J. R. 1980. *Annu. Rev. Biochem.* 49:877–919

47. Westheimer, F. H. 1968. *Acc. Chem. Res.* 1:70–78
48. Maruyama, H., Easterday, R. L., Chang, H.-C., Lane, M. D. 1966. *J. Biol. Chem.* 241:2405–12
49. Hansen, D. E., Knowles, J. R. 1982. *J. Biol. Chem.* 257:14795–98
50. Polakis, S. E., Guchhait, R. B., Lane, M. D. 1972. *J. Biol. Chem.* 247:1335–37
51. Climent, I., Rubio, V. 1986. *Arch. Biochem. Biophys.* 251:465–70
52. Tipton, P. A., Cleland, W. W. 1988. *Biochemistry* 27:4317–25
53. Wimmer, M. J., Rose, I. A., Powers, S. G., Meister, A. 1979. *J. Biol. Chem.* 254:1854–59
54. Jencks, W. P. 1981. *Chem. Soc. Rev.* 10:345–75
55. Rubio, V., Britton, H. G., Grisolia, S., Sproat, B. S., Lowe, G. 1981. *Biochemistry* 20:1969–74
56. Caplow, M. 1965. *J. Am. Chem. Soc.* 87:5774–85
57. Caplow, M., Yager, M. 1967. *J. Am. Chem. Soc.* 89:4513–21
58. Hegarty, A. F., Bruice, T. C., Benkovic, S. J. 1969. *Chem. Commun.*, pp. 1173–74
59. Visser, C. M., Kellogg, R. M. 1977. *Bioorg. Chem.* 6:79–88
60. Fry, D. C., Fox, T. L., Lane, M. D., Mildvan, A. S. 1985. *J. Am. Chem. Soc.* 107:7659–65
61. Perrin, C. L., Dwyer, T. J. 1987. *J. Am. Chem. Soc.* 109:5163–67
62. Ewing, S. P., Lockshon, D., Jencks, W. P. 1980. *J. Am. Chem. Soc.* 102:3072–84
63. Blagoeva, I. B., Pojarlieff, I. G., Kirby, A. J. 1984. *J. Chem. Soc. Perkin Trans. II:*745–51
64. Rétey, J., Lynen, F. 1965. *Biochem. Z.* 342:256–71
65. Arigoni, D., Lynen, F., Rétey, J. 1966. *Helv. Chim. Acta* 49:311–16
66. Rose, I. A. 1970. *J. Biol. Chem.* 245:6052–56
67. Cheung, Y. F., Chien, H. F., Walsh, C. T. 1975. *Biochemistry* 14:2981–86
68. Prescott, D. J., Rabinowitz, J. L. 1968. *J. Biol. Chem.* 243:1551–57
69. Mildvan, A. S., Scrutton, M. C., Utter, M. F. 1966. *J. Biol. Chem.* 241:3488–98
70. Stubbe, J., Abeles, R. H. 1977. J. Biol. Chem. 252:8338–40
71. Stubbe, J., Fish, S., Abeles, R. H. 1980. *J. Biol. Chem.* 255:236–42

72. Walsh, C. T. 1984. *Annu. Rev. Biochem.* 53:493–535
73. Kuo, D. J., Rose, I. A. 1982. *J. Am. Chem. Soc.* 104:3235–36
74. O'Keefe, S. J., Knowles, J. R. 1986. *Biochemistry* 25:6077–84
75. Attwood, P. V., Tipton, P. A., Cleland, W. W. 1986. *Biochemistry* 25:8197–205
76. Goodall, G. J., Prager, R., Wallace, J. C., Keech, D. B. 1983. *FEBS Lett.* 163:6–9
77. Kohn, H. 1976. *J. Am. Chem. Soc.* 98:3690–94
78. Rose, I. A., O'Connell, E. L., Solomon, F. 1976. *J. Biol. Chem.* 251:902–4
79. Fry, D. C., Fox, T., Lane, M. D., Mildvan, A. S. 1985. *Ann. NY Acad. Sci.* 447:140–51
80. Scrutton, M. C., Utter, M. F., Mildvan, A. S. 1966. *J. Biol. Chem.* 241:3480–87
81. Northrop, D. B., Wood, H. G. 1969. *J. Biol. Chem.* 244:5801–7
82. Reed, G. H., Scrutton, M. C. 1974. *J. Biol. Chem.* 249:6156–62
83. Carver, J. A., Baldwin, G. S., Keech, D. B., Bais, R., Wallace, J. C. 1988. *Biochem. J.* 252:501–7
84. Stallings, W. C., Monti, C. T., Lane, M. D., DeTitta, G. T. 1980. *Proc. Natl. Acad. Sci. USA* 77:1260–64
85. Tipton, P. A., Cleland, W. W. 1988. *J. Am. Chem. Soc.* 110:5866–69
86. Leonard, N. J., Milligan, T. W., Brown, T. L. 1960. *J. Am. Chem. Soc.* 82:4075–84
87. Bowen, C. E., Rausher, E., Ingraham, L. L. 1968. *Arch. Biochem. Biophys.* 125:865–72
88. Berkessel, A., Breslow, R. 1986. *Bioorg. Chem.* 14:249–61
89. Lane, M. D., Young, D. L., Lynen, F. 1964. *J. Biol. Chem.* 239:2858–64
90. Dyke, S. F. 1964. In *Chemistry of Natural Products,* ed. K. W. Bentley, 6:161–81. London: Interscience
91. Thatcher, G.R.J., Poirier, R., Kluger, R. 1986. *J. Am. Chem. Soc.* 108:2699–704
92. Moss, J., Lane, M. D. 1971. *Adv. Enzymol. Relat. Areas Mol. Biol.* 35:321–442
93. Kemp, D. S., Cox, D. D., Paul, K. G. 1975. *J. Am. Chem. Soc.* 97:7312–18
94. Crosby, J., Stone, R., Lienhard, G. 1970. *J. Am. Chem. Soc.* 92:2891–900

Annu. Rev. Biochem. 1989. 58:223–56

TWO-DIMENSIONAL NMR AND PROTEIN STRUCTURE[1,2]

Ad Bax

Laboratory of Chemical Physics, National Institute of Diabetes and Digestive and Kidney Diseases, National Institutes of Health, Bethesda, Maryland 20892

CONTENTS

[1]The US government has the right to retain a nonexclusive, royalty-free license in and to any copyright covering this paper.

[2]Abbreviations used: NMR, nuclear magnetic resonance; 1D, one-dimensional; 2D, two-dimensional; 3D, three-dimensional; BPTI, basic pancreatic trypsin inhibitor; TSP, trimethylsilylpropionate; ppm, parts per million; NOE, nuclear Overhauser enhancement; NOESY, 2D NOE spectroscopy; HOHAHA, homonuclear Hartmann-Hahn spectroscopy; TOCSY, total correlation spectroscopy; COSY, homonuclear J-correlated spectroscopy; E-COSY, exclusive COSY; PE-COSY, primitive E-COSY; RMSD, root-mean-square deviation; MD, molecular dynamics; RMD, restrained molecular dynamics; DG, distance geometry; CPI, carboxypeptidase inhibitor; BSPI-2, barley serine proteinase inhibitor 2; EGF, epidermal growth factor

PERSPECTIVES AND SUMMARY

Dramatic improvements in NMR methodology and instrumentation over the past 10 years have made it possible to determine the three-dimensional structures of small proteins in solution. The most important methodological development was the introduction of two-dimensional (2D) NMR, first proposed in 1971 (1) and finding widespread use about a decade later, after instrumental and computer requirements became available for this new class of experiments. 2D NMR spreads the severely overlapping one-dimensional NMR spectrum of a protein into two orthogonal frequency dimensions, giving the NMR spectrum an appearance that is somewhat similar to two-dimensional gel-electrophoresis maps. The resulting improvement in resolution has been a key factor in the detailed NMR structural studies conducted at present. Major instrumental developments concern the increase in magnetic field strength, leading to improved resolution of the ^1H NMR spectrum. This increase in field strength, combined with improved radiofrequency technology, has also led to a large increase in NMR sensitivity, needed for the study of proteins at millimolar concentrations.

Much of the early NMR structural work concentrated on BPTI, a small globular protein of 58 amino acids for which a high-resolution X-ray crystallographic structure was available. These initial studies set the ground rules for protein structure determination by NMR (2–9). A significant number of NMR structures have recently become available, although the resolution of many is poor in comparison with high-resolution crystal structures. Often, only the backbone conformation was given, and NMR spectra were too complex to derive complete side-chain definition. More recently, however, improvements in methodology, instrumentation, and data analysis have made it possible to determine NMR structures for small globular proteins with a precision that may be roughly comparable to an X-ray crystal structure determined at 2–2.5 Å resolution.

NMR enables for the first time the study of proteins in their "natural state," at physiological ionic strength and concentration, undistorted by crystal packing forces. This may provide new insights into the rules governing protein structure, function, and dynamics as well as in the protein folding problem. Because the number of new NMR protein structures appears to be growing at an exponential rate, this article is limited to surveying the strategy used for NMR structure determination, its power and limitations, and the prospects for structural studies of proteins significantly larger than 100 amino acids. Recently, a monograph (10) and several reviews have appeared dealing with the structure determination of proteins by NMR (11–13). In addition, a much larger body of review literature is available describing many of the modern NMR techniques used in this structure determination process (14–22).

NMR OF PROTEINS

In order to record NMR spectra of sufficient quality for detailed NMR structural studies, sample concentrations of at least 1 millimolar in 0.5 ml solution are required. The protein should be stable for at least 24 hours at or near room temperature, and the protein should be in the monomeric or at most dimeric form. This last restriction is imposed by the required tumbling rate of the protein that should be described by a correlation time, τ_c, shorter than about 10 ns (τ_c is approximately the time needed to change the protein orientation by one radian). For proteins, the higher the temperature and the lower the apparent molecular weight, the narrower the width of individual proton resonances. Narrow individual resonances are essential for the accurate measurement of the NMR parameters needed for detailed structural studies.

Figure 1 shows three one-dimensional (1D) NMR spectra of polypeptides of different molecular weights, recorded at a magnetic field strength that corresponds to a ^1H resonance frequency of 600 MHz, the highest field strength commercially available to date. For the small magainin-2 peptide (23 amino acids), many of the resonances, each originating from a particular proton in the peptide, are well resolved. For the larger BPTI (6.5 kd) and staphylococcal nuclease (18 kd), much smaller fractions of the protons yield nonoverlapping resonances. The resonance position in the spectrum reflects the shielding of the nucleus by the surrounding electrons, i.e. it reflects the electron density at the position of the nucleus (23a, 23b). Amide and aromatic protons are, on average, most deshielded and their resonance frequencies are about 6–10 ppm higher relative to a commonly used reference signal from the methyl protons in trimethylsilylpropionate (TSP). Note that according to NMR convention, frequency increases toward the left of the spectrum. CαH protons typically resonate between 3 and 5 ppm and methyl groups between 0 and 2 ppm. Differences in chemical shifts between two amide protons, for example, are caused by structural differences in their vicinity. However, to date, no clear correlation between local structure and chemical shift is available (24a, 24b).

Expansions of the 1D spectra in Figure 1 show small splittings for each of the resonances. These so-called J splittings are caused by a scalar interaction (J coupling) with neighboring protons, two or three chemical bonds removed. The absolute size of the couplings reflects the torsion angle between the C or N nuclei to which the hydrogens are attached.

A second, even more important source of structural information stems from the fact that the protons continuously exchange their nuclear magnetization with one another, at rates that depend on the sixth power of their interspatial distance, r^{-6}. By measuring these magnetization exchange rates, a set of

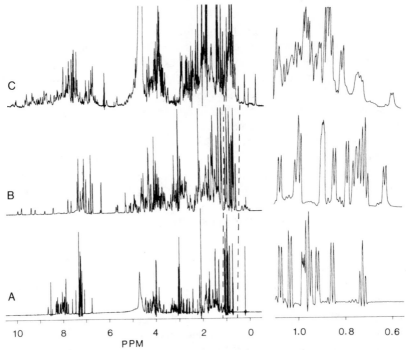

Figure 1   ¹H NMR spectra of (*A*) magainin 2 (23 amino acids) in 75% H_2O, 25% trifluoroetha-nol-d_3, (*B*) basic pancreatic trypsin inhibitor (58 amino acids) in ^2H_2O, and (*C*) staphylococcal nuclease (156 amino acids) in 90% H_2O/10% ^2H_2O. All spectra were recorded at 27°C at 600 MHz and identical digital filtering was used for all three spectra to sharpen the individual resonances. Note that because of hydrogen exchange most but not all amide protons are replaced by deuterons in the BPTI spectrum. In the expansions of the regions between the broken lines, shown on the right, J splittings of the resonances are observed.

interproton distances is obtained. This magnetization exchange is often referred to as the nuclear Overhauser effect (NOE) (25), and the most convenient method for measuring it is called the NOESY (NOE *s*pectroscopy) experiment (26, 27).

Other structural information can be derived from amide hydrogen exchange rates, which depend strongly on solvent accessibility and hydrogen bonding. It is well established that hydrogen-bonded amide protons are less labile than non–hydrogen-bonded ones (3, 28, 29). By dissolving the protein in D_2O solution, one can monitor which resonances disappear most slowly and thereby obtain information about the hydrogen bonding patterns in the protein. For example, the spectrum in Figure 1*b* has been recorded about one year after the protein was dissolved in D_2O, and all non–hydrogen-bonded plus many hydrogen-bonded protons were exchanged with solvent deuterons at the time this spectrum was recorded.

All structural information mentioned above becomes available only once firm individual resonance assignments have been established. The assignment procedure is one of the most difficult and tedious steps in protein structure determination. Below, various methods used for obtaining resonance assignments will be briefly discussed.

Resonance Assignment of Proteins

Wüthrich and coworkers have developed a standard approach for systematic resonance assignments in proteins (2, 6, 7, 30). However, it should be realized that in practice a pure systematic approach often will be insufficient for all but the smallest proteins and slightly modified procedures are often followed (31). Nevertheless, for conceptual reasons it may be useful to consider this standard assignment recipe.

The first step in the resonance assignment uses J-coupling (through-bond) information to classify resonances according to which type of amino acid they correspond. For example, glycine is the only residue where two protons interact with the amide proton, and alanine is the only residue where the $C\alpha H$ interacts directly with methyl protons. In contrast, His, Trp, Tyr, and Phe residues are difficult to distinguish because in each case, the $C\alpha H$ interacts with two $C\beta$ methylene protons that do not show any J coupling to any of the ring protons (more than three bonds removed). Therefore, these residues are also difficult to distinguish from Asp, Asn, Cys, and Ser residues.

The second step in the assignment procedure concerns the identification of the sequence-specific position of each amino acid, relying on through-space connectivity provided by the NOESY experiment. As will be discussed later, for any peptide backbone conformation, NOE interactions between adjacent amino acids are present, making it possible to search for unique dipeptide segments in the protein backbone (provided its sequence is known). Such unique dipeptides then present starting points for further sequential resonance assignments based on the NOESY experiment.

Through-Bond Correlation

A series of two-dimensional NMR experiments have been developed that permit identification of J-coupled protons (1, 32–45). The simplest such experiment is depicted in Figure 2a. In this COSY experiment, two radiofrequency pulses are applied, spaced by a variable time, t_1. Fourier transformation of the time domain data collected after the second pulse, during the time t_2, results in ^1H NMR spectra, in which the intensities of the individual resonances are sinusoidal functions of the time t_1. A particular resonance is modulated as a function of t_1 not only by its own chemical shift frequency, but also by the frequencies of protons that have a J coupling to the proton of interest. By repeating the pulse scheme of Figure 2 for a large number of different t_1 durations, it becomes possible to determine all modulation

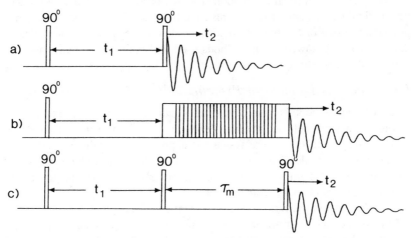

Figure 2 Pulse schemes of three of the most commonly used 2D NMR experiments: (*a*) the COSY experiment, (*b*) the HOHAHA experiment, and (*c*) the NOESY experiment. For each value of t_1, a spectrum is obtained by Fourier transformation of the data acquired during the time t_2. In consecutive experiments, t_1 is systematically incremented from 0 to about 100 ms, in steps of about 200 μs.

frequencies present for a particular resonance, i.e. the resonance frequencies of all J-coupled protons. This is done most easily using a two-dimensional Fourier transformation with respect to the time variables, t_1 and t_2, resulting in a frequency domain spectrum with frequency variables F_1 and F_2.

An example of such a spectrum, obtained for the antimicrobial peptide magainin 2 (46), is shown in Figure 3*a*. Information particularly important for resonance assignment is contained in the boxed region connecting the amide and the CαH protons. An expansion of this so-called fingerprint region (Figure 3*b*) shows a single correlation for each amino acid, with the exception of glycine residues where the NH proton shows a correlation to two Cα protons. Note that although all so-called cross peaks in the spectrum of Figure 3*b* have been labeled, the only residues that can immediately be identified from this part of the spectrum are the glycine residues. For type-specific assignments of the other amino acids, connectivity information between the CαH and side chain protons is needed. Thus, it is seen from Figure 3*a* that a CαH proton at 4.22 ppm correlates with a methyl group at 1.51 ppm, assigning these resonances to an Ala residue. Note that proline is the only amino acid that does not contain an amide hydrogen and this residue does not yield any resonances in the fingerprint region.

The CαH of the second Ala residue (A15) resonates at exactly the same position as CαH of E19, making identification of the NH proton of this second Ala residue ambiguous. However, a number of experimental techniques are available that can provide indirect or relayed connectivity. One such technique, referred to as homonuclear Hartmann-Hahn (HOHAHA) (38) or total correlation spectroscopy (TOCSY) (37), in principle permits the correlation of all protons within a given coupling network. A small region of such a HOHAHA spectrum, recorded for the protein hirudin (65 amino acids) displaying connectivities between NH protons and CαH and side chain protons, is shown in Figure 4. The spectrum has been recorded for a mixing time of 50 ms, sufficiently long for yielding a substantial number of HN-CβH and NH-CγH connectivities. Because of the absence of J coupling between the $CβH_2$ and the ring protons, connectivities between NH protons and ring protons of the Phe and His residues are usually not observed. This HOHAHA experiment, first applied to proteins by Clore, Gronenborn, and coworkers (47, 48), is rapidly gaining popularity.

The pulse scheme used in the HOHAHA experiment is quite complex, as depicted in Figure 2b. The second pulse in the COSY pulse scheme of Figure 2a now has been replaced by an integral number of repetitions of 49 pulses (38). These pulses are timed in such a way that during their application the effect of chemical shifts is temporarily removed. This then permits magnetization to flow freely from one proton to another, at a rate determined by their J coupling. As pointed out above, the HOHAHA experiment can remove ambiguities arising from coincident chemical shifts. In addition, this experiment offers relatively high sensitivity and resolution compared to the hitherto more commonly used COSY technique. For a detailed discussion of theoretical and experimental aspects of this class of experiments, the reader is referred to the literature (37–40).

A third class of experiments utilizes multiple-quantum transitions; these are transitions in which several nuclei participate simultaneously in a coherent manner (15, 41, 49–54). The multiple-quantum frequency itself can be measured (53, 54), which is always a linear combination of the resonance frequencies of the nuclei participating in such a transition. Alternatively, experiments exist that use special properties of multiple-quantum transitions to select certain types of amino acids that are capable of generating such a multiple-quantum transition (41, 49–52). For example, since glycine has only two coupled protons, assuming that the NH protons have been exchanged for deuterons, no triple-quantum transitions can be generated, and a spectrum free of glycine resonances can be obtained by using a so-called triple-quantum filter (41, 49). At first sight these multiple-quantum filtered spectra have the same appearance as regular COSY spectra. However, a closer inspection shows cross peak patterns that are characteristic for the type of amino acid (51, 52). The most popular of such filtered experiments are the double-quantum and triple-quantum filtered COSY experiments, although higher orders of filtering have also been demonstrated for small model compounds. In general, higher orders of multiple-quantum filtering result in simpler spectra but also in a lower signal-to-noise ratio.

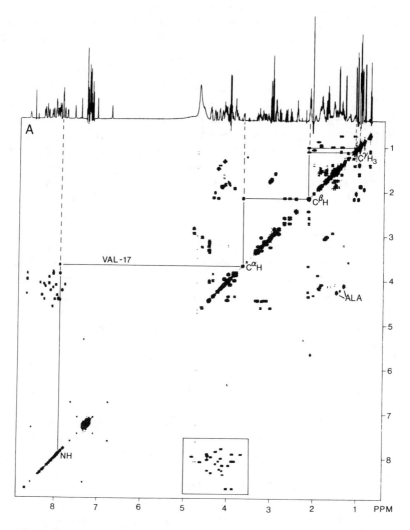

All COSY, multiple-quantum, and HOHAHA correlation techniques are limited to cases where the line width is not much larger than the size of the J coupling. For proteins larger than about 20 kd, this is no longer the case and these valuable sources of information disappear. Even for smaller proteins, it rapidly becomes more difficult to observe NH to side chain connectivities when the temperature is decreased (i.e. the viscosity and τ_c are increased).

Through-Space Correlation

After the J-correlated types of experiments have identified sets of J-coupled NH, CαH, and CβH resonances for the individual amino acids, it becomes

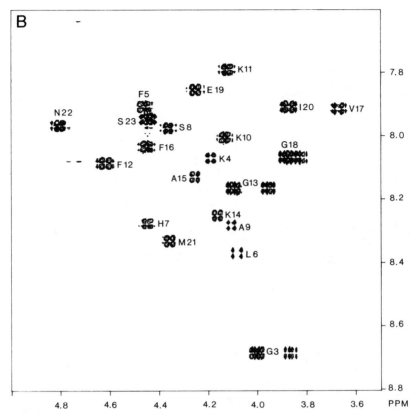

Figure 3 (*A*) 600 MHz COSY spectrum of 12 mg magainin 2 in 0.5 ml 75% H₂O/25% trifluoroethanol-d₃. The drawn lines in (*A*) identify the coupled protons of Val-17. CαH-CH₃ cross peaks of the two alanine residues are marked Ala. The boxed region (commonly known as the "fingerprint region") shows interactions between NH and CαH protons and is enlarged in (*B*). In this enlargement a single cross multiplet, consisting of a number of closely spaced multiplet components, is observed for each nonglycine amino acid with the exception of the two N-terminal residues, for which the NH protons exchange rapidly with the solvent. For glycine residues, two cross peaks are observed corresponding to the two nonequivalent CαH protons. For Gly-18, the chemical shifts of the two CαH protons differ by only 0.04 ppm, and the two multiplets nearly overlap. For both Gly-3 and Gly-13 the CαH chemical shifts differ by about 0.2 ppm, and the label is positioned in between the two cross peaks.

necessary to identify every amino acid not only by type, but also regarding its position in the polypeptide sequence. Most commonly this information is derived from NOESY experiments that yield 2D spectra that show correlations for pairs of protons that are in close proximity of each other. Most of such NOE correlations are short range, i.e. they correspond to protons that are less than five amino acids apart in the peptide sequence. However, a sub-

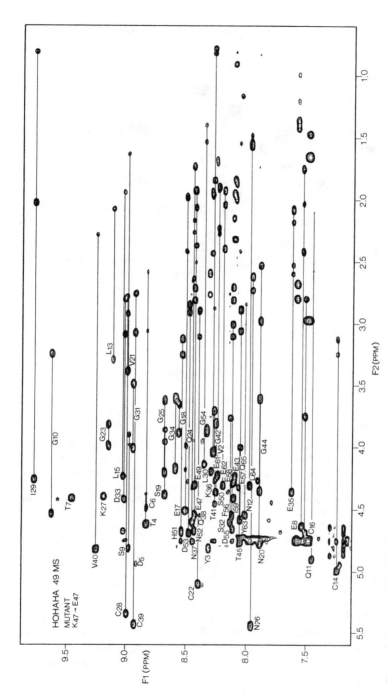

Figure 4 Part of a 600 MHz HOHAHA spectrum of a mutant of hirudin, recorded with a 49 ms mixing period and using a pulse scheme that avoids irradiation of the H$_2$O resonance (173). Reproduced from Folkers et al (78). Reprinted with permission from *Biochemistry*. Copyright 1989, American Chemical Society.

stantial number of long-range NOEs is also present, providing the crucial information about the backbone fold.

The pulse scheme of the NOESY experiment (Figure 2c) consists of three radiofrequency pulses (26, 27). The duration of the delay, t_1, between the first and the second pulse is again systematically incremented in successive experiments until a complete two-dimensional data set is obtained. The duration, τ_m, between the second and third pulse is kept constant and is generally referred to as the mixing period. For macromolecules, the rate of magnetization transfer between protons is proportional to $r^{-6}\tau_c$, where r is the interproton distance and τ_c is the molecular correlation time. For small proteins, τ_m is typically chosen between 50 and 200 ms. For the short mixing period, only relatively short interproton distances (shorter than about 3 to 3.5 Å) will give rise to observable correlations. For the longer mixing times, a much larger number of correlations can be observed, corresponding to distances up to 5 Å. In principle, because of the steep r^{-6} distance dependence of the magnetization transfer rate, one would expect that NMR permits very sensitive distance measurements. As will be discussed later, quantitation of the transfer rate and calculation of distances from such rates can be subject to serious errors. However, for sequence-specific assignments it suffices to use NOE cross peak intensities (NOEs) in a qualitative manner. Short distances (< 2.5 Å) always yield substantial NOEs, medium-range distances (2.5–3.5 Å) yield weaker NOEs, whereas distances larger than 3.5 Å are often too weak to be observed in the short mixing time NOESY spectra.

Distances between backbone (CαH, NH) protons, and between a backbone and CβH protons on adjacent amino acids, are termed *sequential* distances. These strongly depend on the protein structure, i.e. on the intervening ϕ, ψ, and χ torsion angles (5, 55). For example, for an α-helix, the sequential NH-NH distance, d_{NN}, is about 2.8 Å, and the distance between CαH of residue i and the NH proton of residue $i + 1$ ($d_{\alpha N}$) is about 3.5 Å. In contrast, in β-sheets d_{NN} is about 4.2 Å, and $d_{\alpha N}$ is 2.2 Å. Wüthrich and coworkers have tabulated sequential backbone distances for all secondary structural elements, including turns of types I and II, and the recently proposed half turn (56). They concluded that in the sterically allowed region of ϕ and ψ angles of the Ramachandran plot, at least one sequential distance is always less than 3 Å. The possible types of sequential NOE connectivities are schematically indicated in Figure 5a; the sequential backbone NOE intensities for various types of secondary structure are schematically depicted in Figure 5b. For each amino acid, at least one and usually two sequential connectivities can be observed in high-quality NOESY spectra. Since some amino acids have unique spin system topologies, the scalar connectivity networks obtained with the COSY and HOHAHA experiments have made it possible to determine what type of amino acid corresponds to the NH, CαH, and CβH resonances. Sequential connectivity between two residues of known type then can provide

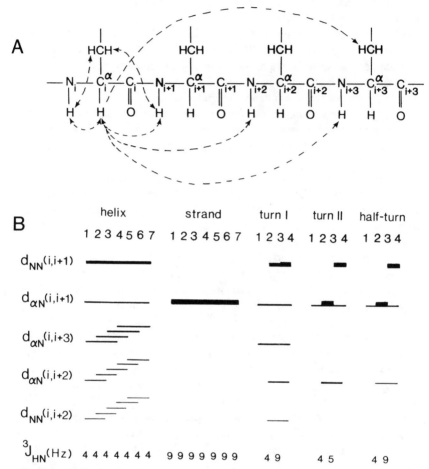

Figure 5 (A) Broken lines indicate some of the NOE interactions that may be observable in polypeptide chains. (B) NOE intensities and NH-CαH J couplings in several types of secondary structure. The thickness of the horizontal lines indicates the intensity of the NOEs.

an anchoring point in the sequence, provided that this dipeptide is unique in the known amino acid sequence of the protein. To avoid branching in a wrong direction, it is essential to have a large number of these unique dipeptides located along the protein backbone. If done carefully, the sequential assignment procedure can be applied to proteins of over 100 amino acids, as illustrated by a detailed example presented by Redfield & Dobson (57) for hen egg white lysozyme (129 residues).

At first sight, the procedure outlined above appears very straightforward. However, a requirement is that each NH and each CαH proton has its own unique chemical shift. On average, α-helixes show relatively little chemical shift dispersion, whereas β-sheeted domains typically show much better dispersion of both NH and CαH protons. In practice, the resolution with which a peak position can be determined reliably is at best ± 0.01 ppm, and following this criterion, even for small proteins, less than half the residues have unique chemical shifts for both NH and CαH protons (58). Therefore, even while there may be little overlap in the 2D NMR spectrum, it becomes difficult to determine which correlation peak corresponds to which particular pair of protons. For example, two overlapping NH resonances may each show intraresidual and sequential NH-CαH connectivities, and it is then unclear which of the four cross peaks corresponds to which of the amide protons. One commonly used procedure to alleviate this serious problem is to record a set of 2D NMR spectra under slightly different experimental conditions. Particularly the NH protons are often quite sensitive to pH and temperature, and different residues shift differently with temperature and pH. In addition, non–hydrogen-bonded NH protons often rapidly exchange when the protein is dissolved in D_2O, simplifying the NH-CαH region. As many as half a dozen or more NOE spectra may therefore be needed to remove most ambiguities in the sequential assignment procedure. More sophisticated approaches for solving this overlap problem rely on labeling with stable isotopes and on three-dimensional NMR experiments, to be discussed later.

The two major types of secondary structure in proteins, α-helixes and β-sheets, show characteristic, easily recognizable fingerprints in the NOESY spectrum. α-helixes are characterized by a sequence of short sequential NH-NH connectivities, corresponding to the 2.8 Å interproton distance. As an example, Figure 6 shows the amide region of the 500 MHz NOESY spectrum of hen egg white lysozyme, with the sequential series of d_{NN} connectivities corresponding to one of the α-helixes marked. Intraresidue NH-CαH distances in α-helixes are also short (2.4 Å) and give rise to intense correlations too. In addition, weak interresidue correlations often can be observed between NH and the CαH of the preceding residue ($d_{\alpha N}$ connectivity) and to the CαH of the residue three positions earlier in the sequence ($d_{\alpha N}$ ($i, i + 3$)). Correlations between NH resonances and the Cβ protons of the preceding residue (dβN) are also commonly observed, as well as weaker $d_{\alpha N}(i, i + 2)$ and $d_{\alpha N}(i, i + 4)$ connectivities. Antiparallel β-sheets are characterized by intense CαH-CαH cross peaks for residues on opposite strands, in addition to an intense sequential $d_{\alpha N}$ connectivity. Parallel β sheets show the strong sequential $d_{\alpha N}$ connectivity, but only weak CαH-CαH cross peaks are observed.

Measurement of J Couplings

The J coupling between protons three bonds apart has a characteristic dependence on the dihedral angle, θ. This dependence is described by a so-called Karplus equation (59):

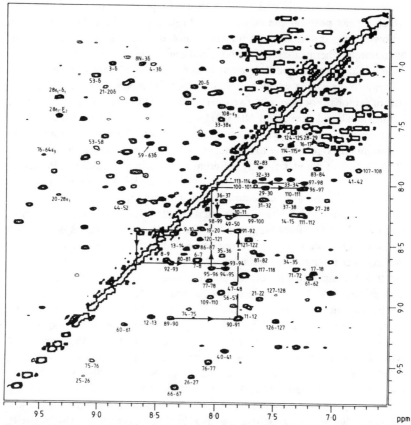

Figure 6   NH and aromatic region of the 500-MHz NOESY spectrum of hen lysozyme in H$_2$O, recorded with a 150-ms mixing time. Connectivities involving pairs of neighboring NH protons are labeled in the region below the diagonal. Long-range NH-NH connectivities and NH-aromatic connectivities are labeled in the region above the diagonal. The cross peaks corresponding to the α-helix from residues 89 to 101 are connected by drawn lines. From Redfield & Dobson (57). Reprinted with permission from *Biochemistry*. Copyright 1988, American Chemical Society.

$$J(\theta) = A\cos^2\theta - B\cos\theta + C \qquad\qquad 1.$$

where the constants A, B, and C have been determined empirically (60–65) and θ angles of 0 and 180° correspond respectively to *cis* and *trans* arrangements of the hydrogens. For the CαH-NH coupling, A, B, and C values of 6.4, 1.4, and 1.9 appear to give best agreement between the measured couplings and the X-ray structure of the basic pancreatic trypsin inhibitor (65). For CαH-CβH couplings, A = 9.5, B = 1.6, and C = 1.8 give good results (64).

Accurate measurement of J couplings in the overlapping one-dimensional spectrum of a protein can be very difficult. If a particular proton is coupled to only one other proton (e.g. NH in all but Gly residues), a measure for the size of the J coupling can be obtained from the antiphase multiplet structure of the cross peak in a COSY spectrum (43). It should be noted however, that the measured coupling can be a serious overestimate of the actual coupling if the line width is larger than the J coupling (66, 67). Therefore, it is important to record the 2D COSY spectrum with very high resolution in the F_2 dimension where one wishes to measure the coupling. Careful use of this procedure, including line shape analysis of the antiphase multiplet structure, showed excellent agreement between the measured J values and X-ray ϕ backbone angles in hen egg white lysozyme (68).

For measurement of χ_1 angles in residues that contain a single β hydrogen (Thr, Val, Ile), a similar procedure as described above for the ϕ angle can be employed, provided that the amide proton is exchanged for a deuteron. Couplings between $C\alpha H$ and β-methylene protons are more difficult to measure. In principle, measurement from a high-resolution COSY spectrum may be possible by comparison with simulated cross multiplet patterns (69). A more elegant approach uses a modified COSY experiment, such as the E-COSY (42) or PE-COSY (44, 68) method, which simplifies the cross peak multiplet structure. The measured coupling represents the time average (on the 100-ms time scale) of the actual J coupling. Therefore, interpretation of the measured coupling may be difficult because it is not always possible to distinguish whether a J value corresponds to a single dihedral angle or whether it represents a rapid averaging of J values from quite different dihedral angles. In cases where the protein does not exist in a single conformation, but rather as an equilibrium of rapidly interconverting conformers, the NOE and J coupling contain information about the time average of the distance to the power negative six, r^{-6}, and the time average of the J coupling, respectively. Both averaging processes are strongly nonlinearly dependent on geometry, and the presence of multiple conformations therefore often results in NOEs and J couplings that could not simultaneously be consistent with a single conformation (70).

Stereospecific Assignments of Methylene and Methyl Resonances

Until recently, usually no stereospecific assignments of β-methylene protons or of the methyl groups in leucine and valine residues were made. To keep the structure calculations simple, a distance measured to a methylene proton is therefore entered into the computer program as a distance to the geometric center between the two methylene protons; a 1 Å correction is added to the upper bound of the measured distance to correct for the difference in position of the so-called "pseudo atom" relative to the true proton (71). Similarly, a distance to a methyl group of valine or leucine usually is entered as a distance to the geometric center of the methyl protons, after increasing the upper bound for the measured distance 2.4 Å. For the nondistinguishable C2H/C6H and C3H/C5H protons of rapidly reorienting Phe or Tyr rings, a 2 Å correction is used. Intraresidue corrections and corrections on short range distances in regions of known secondary structure can be chosen smaller (71–73).

The pseudo-atom description of magnetically indistinguishable nuclei (such as the three protons of a rapidly rotating methyl group) provides a simple means for entering distance information concerning these nuclei. Because the pseudo-atom approach involves the use of looser distance constraints, the local structure is less well defined compared to the case where distances could be measured to individual protons. Recently, a very interesting alternative method has been proposed that avoids the pseudo-atom approach. By relaxing the chirality constraints on prochiral centers in the structure determination program, it is possible to input the distances as measured and have the program figure out which of the prochiral protons is which (74, 75). This new approach appears very promising, and at present it is not clear whether any hidden pitfalls could be associated with this method.

Whenever stereospecific assignments of prochiral resonances can be made, this should give the most reliable and accurate distance information. As demonstrated for valyl methyl groups by Zuiderweg et al (76a) and for β-methylene sites by Hyberts et al (69), stereospecific assignments of these resonances may be possible. For example, for valine residues in an α-helix, for steric reasons the side chains have to be in the g^+ conformation, positioning the $C\gamma^1$ methyl group much closer to the valine NH proton than the $C\gamma^2H_3$ protons (76a). Provided that the χ_1 torsion angle is close to one of its standard conformers, stereospecific assignment for non–α-helical valines and for leucine and β-methylene protons often can also be determined by using a combination of J coupling and intraresidual NOE information (76a, 55). Unambiguous stereospecific assignments for all residues, including glycine, can be obtained by biosynthetic incorporation of stereospecifically deuterated amino acids in the protein (58, 76b).

Preliminary results obtained by a number of research groups indicate large improvements in defining the local protein conformation when stereospecific assignments are used (77, 78). At present it is still unclear what percentage of the possible stereospecific assignments can be determined in an average small protein. So far, it appears that in practice only about 50% of the possible stereospecific assignments can be made in an unambiguous manner. However, it is expected that further improvements in methodology could increase this number significantly.

Accuracy of NMR Distance Determination

The NOESY experiment measures the rate at which protons exchange their nuclear magnetization. The off-diagonal resonance at coordinates $(F_1, F_2) = (\Omega_A, \Omega_B)$ corresponds to magnetization transferred from proton A to proton B during the mixing period, τ_m, of the NOESY experiment (Figure 2a). The intensity of this off-diagonal resonance can be measured with an accuracy of about 10%. As mentioned earlier, the NOE buildup rate depends on r^{-6}, where r is the distance between the two protons; therefore, one might hope to determine very accurate distances ($\pm 1.5\%$) from the NMR measurement. There are two major reasons why such high accuracy cannot be obtained in practice: local internal motion within the protein and indirect or relayed NOE effects.

The rate, k, at which the NOE effect transfers magnetization from spin A to spin B is given by:

$$k = [34.2\tau_c/(1 + 4\omega^2\tau_c^2) - 5.7\tau_c] \times 10^{10}r^{-6} \qquad 2.$$

where ω is 2π times the ^{1}H resonance frequency (usually about 5×10^8). As can be seen from this equation, for $\omega\tau_c \gg 1$, k increases linearly with τ_c. For τ_c values smaller than $1.1/\omega$, k changes sign, and consequently, cross peaks in NOESY spectra of small peptides are opposite in sign relative to the diagonal resonances. A protein usually cannot be described by a single correlation time, however. In particular, the long amino acid side chains are subject to rapid motions of considerable amplitude. This increased local motion reduces the NOE buildup rate k in macromolecules. Moreover, the interproton distance, r, is modulated by the local motion and the measured NOE buildup rate reflects $<r^{-3}>^2$ (fast local motion) or $<r^{-6}>$ (slow local motion) (70). A recent molecular dynamics study of the effect of picosecond motions in proteins indicates that the effects of distance fluctuations are largely compensated by angular fluctuations, and that NOE cross peak intensities in the presence of fast local motion in practice may be analyzed using a $<r^{-6}>$ model (79a).

A second problem in quantitating NOE buildup rates is caused by indirect NOE effects. Consider three protons, A, B, and C, arranged in a linear fashion, with interatomic distances $r_{AB} = 2$ Å, $r_{AC} = 4$ Å, and $r_{BC} = 2$ Å. Direct magnetization transfer from A to B is 64 times faster than from A to C. However, in the experimental scheme, as soon as magnetization has been transferred from A to B, it can be transferred on to C because of the strong BC NOE interaction. To avoid this indirect effect, it is necessary to work at very short mixing times, such that only a small fraction of A magnetization has transferred to B by the end of the mixing time. Under such conditions, only very short distances yield cross peaks with sufficient intensity for reliable measurement. If the approximate macromolecular structure is known (as it may be for DNA oligomers), one may be able to calculate the severity of the indirect effects and correct the analysis of the NOE intensities (79, 80). For proteins, such procedures are less straightforward. However, as shown by Marion et al (81), if a global fold structure is available from the inaccurate NMR data, it is possible to refine such a structure by taking the indirect NOE contributions into account in a similar manner as proposed by Keepers & James for DNA (79b). In contrast to DNA, for proteins it does not appear essential to measure distances with very high accuracy for obtaining a high-resolution structure. Model calculations demonstrate that even with very loose distant constraints, it is possible to define the protein structure with high precision (75, 82–90). This is attributed to the fact that the polypeptide folds back on itself, providing long-range distant constraints that are absent in DNA.

Most commonly, the NOE "distances" are classified in three categories: 1.8–2.7, 1.8–3.3, and 1.8–5 Å, for example. This classification is somewhat arbitrary but functions quite well in practice. Intense NOE cross peaks in a NOESY spectrum recorded with a short mixing time (about 50 ms) are classified as 1.8–2.7 Å; the weak resonances in such a spectrum fall in the second category. Additional NOEs that show up at longer mixing times (e.g. 200 ms) may be due to indirect effects and are classified in the 1.8–5 Å group. The lower boundary is kept the same in each of these groups because local motion could severely attenuate the NOE, even for short distances. Computer simulations suggest that the use of much tighter constraints does not lead to a dramatic improvement in determination of the protein structure (86–88).

Determination of Tertiary Structure

Determination of the relative orientations of segments of secondary structure requires the use of so-called long-range NOEs between residues that are more than five residues apart in the sequence. In the first step of this process, long-range NOE cross peaks between two protons can only be used if both protons have unique chemical shifts, i.e. are at least 0.01 ppm removed from any other proton. Usually, relatively few of such NOEs can be found in a typical NOESY spectrum of a protein. However, even with few long-range NOE contacts, it is possible to obtain some idea about the relative orientations of individual domains. From the low-resolution structure thus obtained, it then becomes possible to assign a larger number of long-range NOEs, permitting the generation of a more accurate structure, which may permit additional long-range NOEs to be identified.

The crude structure mentioned above may be obtained by manual model building or by using molecular graphics. However, this type of structure is very qualitative and more work is needed to find the structure that best fits the

NMR data, without violating any of the standard geometrical constraints, such as bond angles, bond lengths, and van der Waals radii. Most commonly used are two approaches, the so-called distance geometry approach and restrained molecular dynamics. The merits of each of these methods for NMR structure determination have been discussed in a number of places (13, 73, 75, 82–90), and only a very brief discussion of this essential step of the structure determination process is presented here.

Distance geometry is a mathematical tool that can be used to obtain protein structures that satisfy internuclear distances determined with NMR. This type of mathematical problem has been addressed long before interproton distances in proteins could be measured (91), but recent developments in protein NMR have provided a new impetus for further perfection of such procedures, in particular with respect to speed and the necessity to handle large arrays of data. Also, efficient algorithms had to be developed to overcome one of the major difficulties in using distance geometry: the problem of local minima. Of the large variety of algorithms suitable for determination of protein structure from NMR data (86, 88, 92–97), two quite different distance geometry programs now appear to be most widely used: DISGEO (86) and DISMAN (88). The DISMAN program does not work in distance space and therefore it is not a true distance geometry technique; instead, it relies on minimization in torsion angle space rather than Cartesian space. DISMAN uses a so-called variable target function. First, only NMR constraints between residues that are close in the polypeptide sequence determine the local folding; distances further and further apart in the sequence are then gradually incorporated in the target function. The DISGEO program uses n-dimensional space to overcome the local minimum problem. A detailed comparison of the two programs demonstrates that both are capable of faithfully reproducing a protein structure from NMR data (89). The root-mean-square deviation (RMSD) of a set of NMR structures, calculated from the same set of data but using randomly selected starting conditions, is often used to assess the quality of the NMR structure. In this respect, it was found that the RMSDs for the DISGEO tend to be smaller than for DISMAN. However, comparison of the DISMAN and DISGEO structures with the original crystal structure from which the artificial NMR constraints were derived, showed that the difference in RMSD was largely caused by the fact that in regions of the protein that are relatively poorly constrained by NMR data, the DISGEO program tends to produce slightly expanded structures, insufficiently reflecting the lack of NMR constraints. Better agreement between the calculated structure and NMR data and a lowering of the computed energy of the protein can be obtained by subjecting the distance geometry structure to a restrained molecular dynamics simulation (48), to be discussed below.

Restrained molecular dynamics (RMD) is a conceptually relatively simple alternative to distance geometry algorithms. It uses a modification of the regular molecular dynamics simulation programs (98–101). Restrained molecular dynamics solves Newton's equations of motion, with the potential energy, V, defined by:

$$V = V_{bond} + V_{vdW} + V_{angle} + V_{dihedr} + V_{coulomb} + V_{NMR} \qquad 3.$$

where V_{bond}, V_{angle}, and V_{dihedr} keep bond lengths, angles, and chirality at their equilibrium values. The van der Waals and electrostatic interactions are described by V_{vdW} and $V_{coulomb}$. V_{NMR} contains the NMR constraints; it has the effect of pulling the protons that show an NOE interaction closer to the measured distance, r_{ij}. Similarly, V_{NMR} may also contain J-coupling information by including a torsion term. Restrained molecular dynamics was first applied to the NMR structure determination problem by Kaptein et al (102) and Clore et al (103). Although, in principle, an arbitrary starting structure may be used for the restrained molecular dynamics calculation (82), in practice a starting structure obtained by means of distance geometry algorithms (13, 104, 105) or by model building is often used (102, 106). Because of the kinetic energy present in the protein during the dynamics simulation, this procedure can be efficient at overcoming the problem of local minima. The RMD approach requires a relatively large amount of computational time compared to the distance geometry methods. This problem can be overcome by using a simplified potential energy function, where all nonbonded contact interactions are described by a single term. By using a lower cutoff distance, the number of nonbonded interactions is also decreased significantly. This process, referred to as simulated annealing (107), is computationally more efficient than RMD and yields structures of similar quality (75, 85). A combined use of the first stages of the DISGEO structure calculation procedure followed by simulated annealing appears to be an extremely fast structure calculation method, capable of providing high-quality structures (108).

For experimental data, one typically uses a large number (20 to 100) of different starting structures (or starting distances for DISGEO) for the DG or MD algorithms. If a significant fraction of the thus calculated structures satisfies all NMR constraints, and shows small root-mean-square (RMS) deviations from one another (<2 Å for all backbone atoms), this indicates that the calculated structures must be close to the actual solution structure. If serious violations of NMR constraints remain, or if the RMS deviation between the various structures is too large, reanalysis of the NMR spectra is necessary. Another useful indication of the agreement between the NMR data and the molecular structure can be obtained if an NOE spectrum is calculated from the obtained protein structure. Comparison of the real and the calculated

NOE spectra gives a visual impression of the agreement between the experimental NMR spectrum and the protein structure (55, 81).

Quality of NMR Structures

Until recently, NMR protein structures were determined at a low level of
detail, showing the global backbone fold typically with RMS deviations of at
least 1.5 to 2 Å. The definition of side chains, if presented at all, was
significantly poorer. More recently, with the use of stereospecific assignments and the availability of stronger magnetic fields, this situation appears to
be changing, although to date very few "high-resolution" NMR protein
structures have actually been published (77, 78, 89, 130). Nevertheless, even
at relatively low resolution, a comparison of structural features in solution and
in the crystalline state is possible, as briefly discussed below.

NMR structures have been determined for a number of proteins for which a
high-resolution crystal structure was already available. Proteins for which
good agreement between the two types of structure is found include BPTI,
potato carboxypeptidase inhibitor (CPI) (105), and barley serine proteinase
inhibitor 2 (BSPI-2) (109). For BPTI, minor differences are found at the
surface of the protein (89). For CPI, small deviations from the crystal
structure were reported for two regions of the backbone (105). For BSPI-2,
the small RMS difference (1.9 Å) for the backbone atom positions between
crystal and solution structure appears not to be localized in particular regions
of the protein (110).

For epidermal growth factor (EGF) no crystal structure was available, but
three groups independently determined very similar structures for the two
domains of this polypeptide (111–115) despite significant differences in the
quality of the NMR data available to them. The Japanese group (112, 113)
found a different relative orientation of the two domains, but this may be due
to different conditions under which the spectra were recorded, and it is not
certain the two domains have a fixed relative geometry under all conditions.

A few significant differences between NMR and X-ray structures have also
surfaced recently. Most interesting, two-dimensional ^{1}H-^{113}Cd correlation
experiments (116–118) revealed that the cadmium coordination in the solution
and crystal structure (119, 120) of metallothionein-2a is quite different (116).
At present, it is unclear whether this is due to a problem in the X-ray
refinement procedure, or to crystallization of a minor component of the
protein. Differences between the crystal structure and the solution data have
been reported by Nettesheim et al (121) for the human complement protein
C3A. These authors attributed the substantial differences in helical structure,
observed both at the C and N termini, to crystal packing forces.

Discrepancies between the X-ray crystal structure and NMR data of α-
bungarotoxin were reported by Inagaki et al (122). A recent detailed NMR

structural study by Basus et al (123) confirmed a different orientation of Trp-28, and showed the presence of a more extended β-sheet structure in solution, making it more similar to the homologous cobratoxin crystal structure. This NMR study also identified four errors in the primary sequence of the protein. These errors had remained undetected in the 2.5 Å crystal structure (124). Other incorrect primary sequences were detected for metallothionein-2a from rabbit liver (56), protease inhibitor IIA from bull seminal plasma (125a, 125b), and for toxin II from *Radianthus paumotensis* (126).

The structure of the α-amylase inhibitor, Tendamistat (74 amino acids) was determined simultaneously and independently by the crystallographic group of Huber (127) and by Kline, Braun, and Wüthrich (129, 130). The backbone folds of the solution and crystal structures were virtually identical, although small differences in some of the side chain conformations were observed between the two structures. The NMR structure had been obtained using a very large number (842) of NOEs and many stereospecific assignments and dihedral constraints from J couplings, and the solution structure is of much higher quality than most NMR structures. RMS deviations between the NMR structures were 0.85 Å for the backbone, 1.04 Å for the backbone plus the interior amino acid side chains, and 1.53 Å for all heavy atoms. The differences between the solution and crystal structures were 1.0, 1.3, and 1.8 Å, respectively.

Solution structures of two other small proteins with similarly high resolution, also using stereospecific assignments and dihedral constraints, have recently been derived by Driscoll et al (77) and by Folkers et al (78). As an example, Figure 7 shows a superposition of 42 structures obtained for the antihypertensive and antiviral protein BDS-I from the sea anemone *Anemonia sulcata*. The high degree of similarity of the 42 structures reflects the large number (> 500) of NOE and dihedral constraints used for this small protein (46 amino acids).

PROSPECTS FOR THE STRUCTURE DETERMINATION OF LARGER PROTEINS

The NMR study of proteins significantly larger than 10 kd is hampered by a number of factors. Most importantly, the molecular correlation time increases nearly linearly with molecular weight and leads to a significant increase in line width. This not only increases crowding in the 2D NMR spectra, it also reduces sensitivity, particularly for the essential J correlation methods (32–45). The molar concentrations of nonaggregating protein, on average, decrease with increasing molecular weight, lowering NMR sensitivity even further.

One important approach, to be discussed below, for obtaining detailed

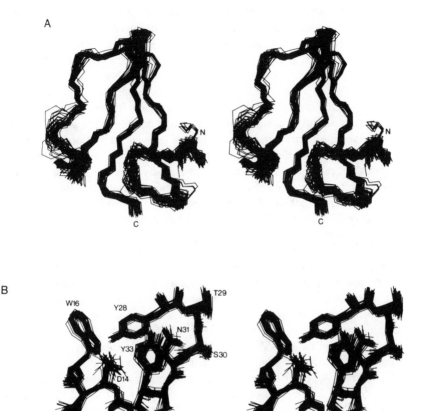

Figure 7   Stereoviews of the solution structure of the protein BDS-I from the sea anemone *Anemonia sulcata.* (*A*) superposition of 42 backbone structures and (*B*) all atoms of residues 13–16 and 28–33. The structures have been calculated with the simulated annealing procedure; 42 out of 50 calculated structures showed correct stereochemistry and no distance violations larger than 0.5 Å. From Driscoll et al (77). Reprinted with permission from *Biochemistry.* Copyright 1989, American Chemical Society.

NMR structural information relies on the incorporation of stable isotopes (^{15}N, ^{13}C, ^{2}H) in the protein. However, it should be noted that for a number of proteins larger than 10 kd virtually complete resonance assignments have been reported without the use of isotopic labeling. These include plastocyanin (131a–c), *Escherichia coli* thioredoxin (132), hen egg white lysozyme (57), and acylphosphatase (133). These proteins all are relatively "NMR friendly," showing good resonance dispersion and permitting the use of relatively high protein concentrations without significant aggregation.

For proteins that are genetically expressed in microorganisms, incorporation of stable isotopes can be relatively easy, and permits use of alternative methods for obtaining structural and assignment information. Random labeling with ^{15}N, ^{13}C, or ^{2}H can be achieved by feeding the microorganism with labeled precursors, such as $^{15}NH_4Cl$, ^{13}C-succinate, $^{2}H_2O$, and ^{2}H-glucose. For some experiments, to be discussed below, it is beneficial to label only certain types of amino acids. This can be accomplished by feeding with suitably labeled amino acid precursors. For efficient incorporation, the use of auxotrophic strains of the microorganism may be required (134). In the case of ^{15}N labeling, transaminase activity must be reduced by flooding the cells with excess of unlabeled amino acids (135) or by genetic modification (136). In cases where the expression occurs very rapidly (less than one hour), use of auxotrophic strains may not be essential, although scrambling among certain types of amino acids can then be expected (137). Two-dimensional isotope labeled NMR has recently been reviewed by Griffey & Redfield (138).

Double Labeling with ^{13}C and ^{15}N

One particularly powerful method for obtaining sequence-specific assignments relies on the use of double labeling; one type of amino acid (for example, Met) is labeled with ^{15}N and another type of amino acid (for example, Leu) is labeled with ^{13}C in its carbonyl position. As first demonstrated by Kainoshi & Tsuji (139), this approach permits immediate identification of the Leu-Met dipeptide because the carbonyl resonance of the Leu prior to the Met residue shows a J splitting caused by interaction with the ^{15}N nucleus. A powerful and very sensitive extension of this technique does not detect the ^{13}C signal, but uses the more intense ^{1}H signal to detect the attached ^{15}N nucleus in a two-dimensional experiment (140–143). ^{15}N nuclei adjacent to ^{13}C will show the J splitting in the ^{15}N dimension of the 2D spectrum, identifying the dipeptide (144). Figure 8 shows two such correlation spectra, obtained for staphylococcal nuclease (18 kd) doubly labeled with ^{15}N Met and $^{13}C'$ Leu or $^{13}C'$ Lys.

^{2}H Labeling

The ^{1}H line width problem can be overcome by incorporating deuterium, which dilutes the ^{1}H density in the protein and therefore reduces dipolar broadening (145, 146). Deuterium has a seven times smaller magnetogyric ratio compared to protons, and consequently its dipolar broadening effect on remaining protons is quite small. This effect has been demonstrated most dramatically by LeMaster & Richards (58) in a study of thioredoxin with random deuteration of the nonexchangeable protons. As they pointed out, the loss in concentration of hydrogens is compensated in part by the resulting narrower line widths. A substantial gain in sensitivity is observed in the NH-NH region of the NOE spectrum because no dilution of these exchangeable resonances occurs, and because during the mixing period of the NOESY

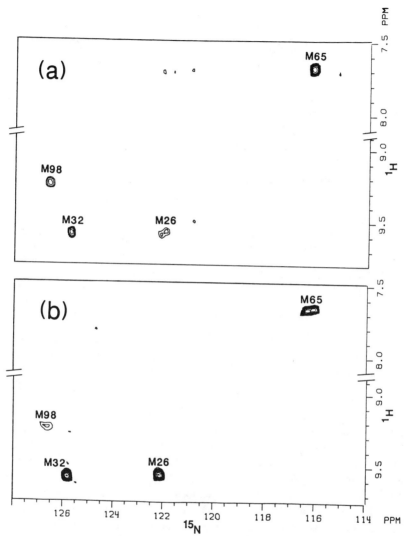

Figure 8 Comparison of ¹H-¹⁵N shift correlation spectra of 1.5 mM staphylococcal nuclease in 90% H₂O/10% D₂O, labeled with (*a*) [1-¹³C]Leu and [¹⁵N]Met and (*b*) [1-¹³C]Lys and [¹⁵N]Met. Met-26 is preceded by a leucine residue and shows a splitting caused by the carbonyl-nitrogen J coupling. Similarly, Met-65 and Met-98 are preceded by lysine residues and show J splittings. From Torchia et al (137).

experiment these resonances lose less magnetization to (deuterated) aliphatic side chains (58, 147). An example in the increase in multiplet resolution observed upon 75% deuteration is shown in Figure 9. The possibility to deuterate stereospecifically (58) is also expected to be important for the study of medium-size proteins. Furthermore, the old idea (145) of incorporating

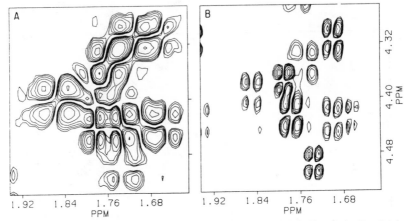

Figure 9 Comparison of small regions of the COSY spectra of *E. coli* thioredoxin. Panel A is recorded for a natural abundance sample, and panel B is the corresponding region for the 75% uniformly ^{2}H-labeled sample. The improved resolution observed for the fractionally deuterated sample is partly due to the longer T_2 relaxation time and hence narrower line width, but mainly is a result of the fact that coupling from other protons to the two protons involved in the coherence transfer is largely eliminated by the isotopic dilution. From LeMaster & Richards (58). Reprinted with permission from *Biochemistry*. Copyright 1988, American Chemical Society.

certain protonated amino acids into an otherwise fully deuterated protein offers interesting possibilities for spectral simplification and resonance assignment (137).

Isotope-Edited Experiments

It is relatively straightforward to exclusively observe hydrogens that are directly attached to either ^{15}N or ^{13}C (148). As demonstrated by a number of research groups (149–153), it is then relatively straightforward to incorporate this editing procedure into two-dimensional experiments, greatly simplifying the corresponding spectra. These edited 2D spectra thus can show a great reduction in spectral crowding, as demonstrated in Figure 10. However, it should be noted that for experiments involving incorporation of ^{13}C, the strong heteronuclear dipolar ^{13}C-^{1}H interaction causes significant extra broadening of the ^{1}H resonances, making this type of editing approach less suitable for J-correlated 2D experiments. Complete ^{15}N labeling of proteins offers another possibility for spectral simplification of the amide region: by labeling in the F_1 dimension with the ^{15}N chemical shift instead of the amide ^{1}H shift, many degeneracies present in the regular NOESY spectrum can be removed, facilitating sequential assignment (154).

^{13}C and ^{15}N Experiments

The sensitivity of ^{13}C and ^{15}N NMR spectra of proteins at natural abundance is too low for permitting advanced 2D NMR experiments of proteins of a

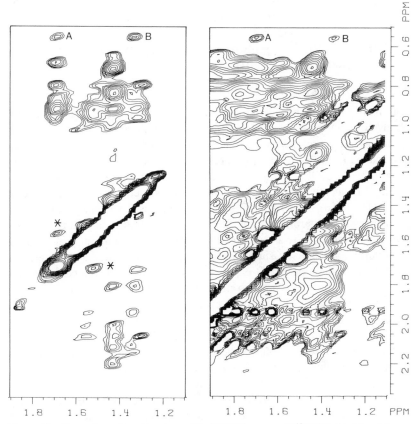

Figure 10 Comparison of small regions of the NOESY spectra of [^{13}Cβ]Ala lambda repressor. The left panel shows the result of NOE editing and exhibits interactions only to ^{13}C-labeled protons, i.e. to the alanine methyl groups. The right panel shows the identical region of the conventional NOESY spectrum. Both spectra were recorded with a 300 ms NOE mixing time. From Bax & Weiss (149). Reprinted with permission from the *Journal of Magnetic Resonance*. Copyright 1987, Academic.

significant size. However, as convincingly demonstrated by Markley and coworkers (155, 156), when isotopic enrichment is used in combination with large sample volumes, a number of very powerful experiments become feasible. For example, using uniform labeling with ^{13}C permits detection of J connectivities between adjacent carbons (156). Because ^{13}C has a significantly larger spread in chemical shifts than ^{1}H, spectral crowding may be reduced. Moreover, ^{13}C chemical shifts for a certain type of carbon in a particular amino acid are quite predictable, facilitating spectral assignment. Sequential connectivity determination becomes possible when the protein is also labeled with ^{15}N, and ^{13}C–^{15}N 2D correlation experiments are performed (157). Once complete ^{13}C assignments have been obtained, other 2D NMR experiments (140–143) can be used for correlating the ^{13}C resonances with chemical shifts of their directly attached protons (158, 159). J correlation through multiple bond ^{1}H-^{13}C or ^{1}H-^{15}N is often also possible, providing assignment and structural information (159–161).

Three-Dimensional NMR

For small proteins, the overlap present in one-dimensional NMR is largely removed in 2D NMR spectra. In a similar fashion, overlap present in the 2D spectra of larger proteins can be removed by extending the experiment to three dimensions (162–164). The first protein experiments have concentrated on combining the NOESY and HOHAHA pulse schemes, yielding spectra that display J connectivity in one plane and NOE connectivity in a perpendicular plane (165, 166). The final 3D spectrum contains a very large number of resonances, and interpretation of the entire information content at present is a formidable task. Software, currently under development in a number of laboratories, will greatly simplify the interpretation process and permit full utilization of this type of 3D experiment. A drawback of this particular 3D experiment is that there are two magnetization transfer steps involved: one NOE step and one HOHAHA mixing period. The efficiency of the HOHAHA step, which depends on the J coupling and is attenuated by increased line width, rapidly decreases for increasing molecular weight. Therefore, it is not certain at present whether this particular 3D experiment will be useful for the study of proteins significantly larger than 10 kd.

A more promising experiment, in this respect, combines heteronuclear chemical shift correlation with the regular NOESY experiment (167). By using uniform ^{15}N labeling of the protein, it is possible to separate the part of the regular NOE spectrum that shows interactions with amide protons according to the ^{15}N chemical shift. The number of resonances in such a spectrum is identical to the number of resonances in the regular NOE spectrum and sensitivity is very similar. However, spectral overlap is almost completely removed in the 3D spectrum. An example of a slice taken through a 3D spectrum of the protein staphylococcal nuclease is shown in Figure 11. Measurement time for 3D spectra is necessarily longer than for 2D spectra and can require up to several weeks of precious instrument time. However, the large amount of information present in such spectra may make this approach the method of choice for problems too complicated to be handled by conventional 2D NMR experiments.

DISCUSSION

NMR structural studies are not limited to aqueous solutions. Studies of the membrane-active peptides, melittin and glucagon, have been conducted in

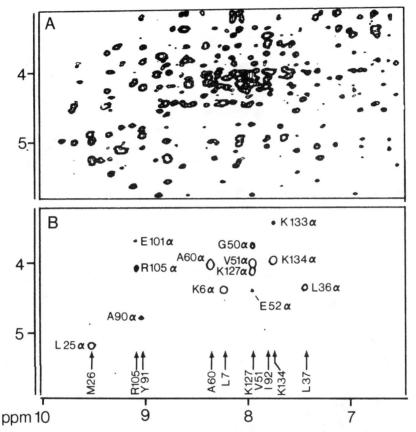

Figure 11   NH-CαH NOE interactions observed for a 1.5 mM sample of ¹⁵N-labeled staphylococcal nuclease in 90% H₂O, using a 125 ms NOE mixing time. Panel A corresponds to the
regular 2D NOESY spectrum; panel B is the corresponding region taken from a three-dimensional
NMR spectrum where the NOESY spectra are separated according to the ¹⁵N chemical shift of the
amide nitrogen. The section shown displays NH-CαH correlations only for those amides that
have a ¹⁵N chemical shift of 122.3 ± 0.3 ppm. Adapted from Marion et al (172).

micelles, indicating conformations significantly different from either solution
or crystal structures (168, 169). A study of filamentous bacteriophage coat
protein in micelles yielded interesting information about both structure and
dynamics of this polypeptide (170).

Even among crystallographers skepticism seems to be disappearing about
the possibility of using NMR to determine solution structures of small proteins. At present, solution structures of nearly 50 small proteins or protein
domains have been published or are in the process of being published.
Continuing rapid developments in NMR methodology and equipment are
extending the size limits of proteins under NMR investigation. Already, a

low-resolution NMR structure has been obtained for the *lac* repressor headpiece interacting with its operator (171). Similar studies for the *cro* and *lambda* repressor-operator complexes are under way. For suitable small proteins, the structure can sometimes be determined very fast, in less than a few months. However, the time and the amount of work needed for larger molecules rapidly increases. Most likely, therefore, even in the distant future, detailed NMR structural studies will largely remain limited to relatively small proteins or protein domains (< 40 kd).

NMR structures, even at relatively low resolution, offer valuable complementary information to X-ray crystallography. NMR may become particularly useful for characterization of protein surfaces. In addition, NMR can probe motions over a time scale spanning 10 orders of magnitude and can provide significant insights in problems regarding protein folding and dynamics.

ACKNOWLEDGMENTS

I thank Ted Becker, Marius Clore, Angela Gronenborn, Lewis Kay, David LeMaster, Dominique Marion, Dennis Torchia, and Attila Szabo for stimulating discussions and useful comments and Ingrid Pufahl for assistance during the preparation of this manuscript.

Literature Cited

1. Jeener, J. 1971. Ampere International Summer School. Basko Polje, Yugoslavia
2. Wagner, G., Wüthrich, K. 1982. *J. Mol. Biol.* 155:347–66
3. Wagner, G., Wüthrich, K. 1982. *J. Mol. Biol.* 160:343–61
4. Kumar, A., Wagner, G., Ernst, R. R., Wüthrich, K. 1980. *Biochem. Biophys. Res. Commun.* 95:1–6
5. Billeter, M., Braun, W., Wüthrich, K. 1982. *J. Mol. Biol.* 155:321–46
6. Wüthrich, K., Wider, G., Wagner, G., Braun, W. 1982. *J. Mol. Biol.* 155:311–20
7. Wider, G., Lee, K. H., Wüthrich, K. 1982. *J. Mol. Biol.* 155:367–88
8. Wüthrich, K., Billeter, M., Braun, W. 1984. *J. Mol. Biol.* 180:715–40
9. Wagner, G. 1983. *Q. Rev. Biophys.* 16:1–57
10. Wüthrich, K. 1986. *NMR of Proteins and Nucleic Acids.* New York: Wiley
11. Wemmer, D. E., Reid, B. R. 1985. *Annu. Rev. Phys. Chem.* 36:105–37
12. Clore, G. M., Gronenborn, A. M. 1987. *Protein Eng.* 1:275–88
13. Kaptein, R., Boelens, R., Scheek, R. M., van Gunsteren, W. F. 1988. *Biochemistry* 27:5389–95

14. Bax, A. 1982. *Two-dimensional Nuclear Magnetic Resonance.* Boston: Reidel
15. Ernst, R. R., Bodenhausen, G., Wokaun, A. 1987. *Principles of Nuclear Magnetic Resonance in One and Two Dimensions.* Oxford: Clarendon
16. Chandrakumar, N., Subramanian, S. 1987. *Modern Techniques in High Resolution FT NMR.* New York: Springer
17. Croasmun, W. R., Carlson, R. M. K., eds. 1987. *Two-dimensional NMR Spectroscopy.* New York: VCH
18. Sanders, J. K. M., Hunter, B. K. 1987. *Modern NMR Spectroscopy.* Oxford: Univ. Press
19. Derome, A. E. 1987. *Modern NMR Techniques for Chemistry Research.* Oxford: Pergamon
20. Morris, G. A. 1986. *Magn. Reson. Chem.* 24:371–403
21. Benn, R., Günther, H. 1983. *Angew. Chem. Int. Ed. Engl.* 22:350–80
22. Kessler, H., Gehrke, M., Griesinger, C. 1988. *Angew. Chem. Int. Ed. Engl.* 27:490–536
23a. Ramsey, N. F. 1952. *Phys. Rev.* 86:243–57
23b. Emsley, J. W., Feeney, J., Sutcliffe, L. H. 1965. *High Resolution Nuclear*

Magnetic Resonance Spectroscopy, Vol.
1, pp. 65–103. Oxford: Pergamon
24a. Pardi, A., Wagner, G., Wüthrich, K.
1983. *Eur. J. Biochem.* 137:445–54
24b. Gross, K. H., Kalbitzer, H. R. 1988.
J. Magn. Reson. 76:87–99
25. Noggle, J. H., Schirmer, R. H. 1971.
The Nuclear Overhauser Effect, Chemical Applications. New York: Academic
26. Jeener, J., Meier, B. H., Bachmann, P.,
Ernst, R. R. 1979. *J. Chem. Phys.*
71:4546–53
27. Macura, S., Ernst, R. R. 1980. *Mol.
Phys.* 41:95–117
28. Englander, S. W., Kallenbach, N. R.
1984. *Q. Rev. Biophys.* 16:521–655
29. Tuchsen, E., Woodward, C. 1985. *J.
Mol. Biol.* 185:405–19
30. Billeter, M., Braun, W., Wüthrich, K.
1982. *J. Mol. Biol.* 155:321–46
31. Englander, S. W., Wand, A. J. 1987.
Biochemistry 26:5953–58
32. Aue, W. P., Bartholdi, E., Ernst, R.
R. 1976. *J. Chem. Phys.* 64:2229–
46
33. Bax, A., Freeman, R. 1981. *J. Magn.
Reson.* 44:542–61
34. Nagayama, K., Kumar, A., Wüthrich,
K., Ernst, R. R. 1980. *J. Magn. Reson.*
40:321–34
35. Eich, G., Bodenhausen, G., Ernst, R.
R. 1982. *J. Am. Chem. Soc.* 104:3731–
32
36. Bax, A., Drobny, G. 1985. *J. Magn.
Reson.* 61:306–20
37. Braunschweiler, L., Ernst, R. R. 1983.
J. Magn. Reson. 53:521–28
38. Bax, A., Davis, D. G. 1985. *J. Magn.
Reson.* 65:355–60
39. Rance, M. 1987. *J. Magn. Reson.*
74:557–64
40. Griesinger, C., Otting, G., Wüthrich,
K., Ernst, R. R. 1988. *J. Am. Chem.
Soc.* In press
41. Piantini, U., Sørensen, O. W., Ernst,
R. R. 1982. *J. Am. Chem. Soc.* 104:
6800–1
42. Griesinger, C., Sørensen, O. W., Ernst,
R. R. 1985. *J. Am. Chem. Soc.*
107:6394–96
43. Marion, D., Wüthrich, K. 1983.
Biochem. Biophys. Res. Commun.
113:967–74
44. Mueller, L. 1987. *J. Magn. Reson.*
72:191–96
45. Marion, D., Bax, A. 1988. *J. Magn.
Reson.* In press
46. Zasloff, M. 1987. *Proc. Natl. Acad.
Sci. USA* 874:5449–53
47. Clore, G. M., Martin, S. R.,
Gronenborn, A. M. 1986. *J. Mol. Biol.*
191:553–61
48. Clore, G. M., Sukumaran, D. K.,

Nilges, M., Gronenborn, A. M. 1987.
Biochemistry 26:1732–45
49. Shaka, A. J., Freeman, R. 1983. *J.
Magn. Reson.* 51:169–73
50. Boyd, J., Dobson, C. M., Redfield, C.
1985. *FEBS Lett.* 186:35–40
51. Müller, N., Ernst, R. R., Wüthrich, K.
1986. *J. Am. Chem. Soc.* 108:6482–92
52. Boyd, J., Redfield, C. 1986. *J. Magn.
Reson.* 68:67–84
53. Rance, M., Wright, P. E. 1986. *J.
Magn. Reson.* 66:372–78
54. Zuiderweg, E. R. P., Mollison, K. W.,
Henkin, J., Carter, G. W. 1988. *Biochemistry* 27:3568–80
55. Arseniev, A., Schultze, P., Wörgötter,
E., Braun, W., Wagner, G., et al. 1988.
J. Mol. Biol. 201:637–57
56. Wagner, G., Neuhaus, D., Wörgötter,
E., Vasak, M., Kägi, J. H. R.,
Wüthrich, K. 1986. *J. Mol. Biol.*
187:131–35
57. Redfield, C., Dobson, C. M. 1988.
Biochemistry 27:122–36
58. LeMaster, D. M., Richards, F. M.
1988. *Biochemistry* 27:142–50
59. Karplus, M. 1963. *J. Am. Chem. Soc.*
85:2870–71
60. Ramachandran, G. N., Chandrasekaran,
R., Kopple, K. D. 1971. *Biopolymers*
10:2113–31
61. Cung, M. T., Marrand, M., Neel, J.
1974. *Macromolecules* 7:606–13
62. Bystrov, V. F. 1976. *Prog. Nucl. Magn.
Reson. Spectrosc.* 10:41–81
63. Feeney, J. 1975. *Proc. R. Soc. London
Ser. A* 345:61–72
64. DeMarco, A., Llinas, M., Wüthrich, K.
1978. *Biopolymers* 17:637–50
65. Pardi, A., Billeter, M., Wüthrich, K.
1984. *J. Mol. Biol.* 180:741–51
66. Neuhaus, D., Wagner, G., Vasak, M.,
Kägi, J. H. R., Wüthrich, K. 1985. *Eur.
J. Biochem.* 151:257–73
67. Bax, A., Lerner, L. 1988. *J. Magn.
Reson.* 79:429–38
68. Blazer, J., Dobson, C. M., Redfield, C.
1988. Abstr. #P15-24. *XIII ICMRBS
Conf., Madison, Wis.*
69. Hyberts, S. G., Märki, W., Wagner, G.
1987. *Eur. J. Biochem.* 164:625–35
70. Kessler, H., Griesinger, C., Lautz, J.,
Muller, A., van Gunsteren, W. F., Berendsen, H. J. C. 1988. *J. Am. Chem.
Soc.* 110:3393–96
71. Wüthrich, K., Billeter, M., Braun, W.
1983. *J. Mol. Biol.* 169:949–61
72. Clore, G. M., Gronenborn, A. M.,
Brünger, A. T., Karplus, M. 1985. *J.
Mol. Biol.* 186:435–55
73. Holak, T. A., Prestegard, J. H., Forman, J. D. 1987. *Biochemistry* 4652–60
74. Pardi, A., Hare, D. R., Selsted, M. E.,
Morrison, R. D., Bassolino, D. A.,

Bach, A. C. II. 1988. *J. Mol. Biol.* 201:625–36

75. Nilges, M., Clore, G. M., Gronenborn, A. M. 1988. *FEBS Lett.* 239:129–36

76a. Zuiderweg, E. R. P., Boelens, R., Kaptein, R. 1985. *Biopolymers* 24:601–11

76b. LeMaster, D. M. 1987. *FEBS Lett.* 223:191–96

77. Driscoll, P. C., Clore, G. M., Beress, L., Gronenborn, A. M. 1989. *Biochemistry.* In press

78. Folkers, P. J. M., Clore, G. M., Driscoll, P. C., Dodt, J., Kohler, S., Gronenborn, A. M. 1989. *Biochemistry.* In press

79a. LeMaster, D. M., Kay, L. E., Brünger, A. T., Prestegard, J. H. 1988. *FEBS Lett.* 236:71–76

79b. Keepers, J. W., James, T. L. 1984. *J. Magn. Reson.* 57:404–26

80. Zhou, N. Z., Bianucci, A. M., Pattabiraman, N., James, T. L. 1987. *Biochemistry* 26:7905–13

81. Marion, D., Genest, M., Ptak, M. 1987. *Biophys. Chem.* 28:235–44

82. Brünger, A. T., Clore, G. M., Gronenborn, A. M., Karplus, M. 1986. *Proc. Natl. Acad. Sci. USA* 83:3801–5

83. Clore, G. M., Brünger, A. T., Karplus, M., Gronenborn, A. M. 1986. *J. Mol. Biol.* 191:523–51

84. Clore, G. M., Nilges, M., Brünger, A. T., Karplus, M., Gronenborn, A. M. 1987. *FEBS Lett.* 213:269–77

85. Nilges, M., Gronenborn, A. M., Brünger, A. T., Clore, G. M. 1988. *Protein Eng.* 2:27–38

86. Havel, T. F., Wüthrich, K. 1984. *Bull. Math. Biol.* 46:673–98

87. Havel, T. F., Wüthrich, K. 1985. *J. Mol. Biol.* 182:281–94

88. Braun, W., Go, N. 1985. *J. Mol. Biol.* 186:611–26

89. Wagner, G., Braun, W., Havel, T. F., Schaumann, T., Go, N., Wüthrich, K. 1987. *J. Mol. Biol.* 196:611–39

90. Kaptein, R., Zuiderweg, E. R. P., Scheek, R., Boelens, R., van Gunsteren, W. F. 1985. *J. Mol. Biol.* 182:179–82

91. Blumenthal, L. M. 1970. *Theory and Applications of Distance Geometry.* New York: Chelsea.

92. Kuntz, I. D., Crippen, G. M., Kollman, P. A., Kimelman, D. 1976. *J. Mol. Biol.* 106:983–94

93. Kuntz, I. D., Crippen, G. M., Kollman, P. A. 1979. *Biopolymers* 18:939–57

94. Crippen, G. M. 1977. *J. Comp. Phys.* 24:96–107

95. Crippen, G. M., Havel, T. F. 1978. *Acta Crystallogr. A* 34:282–84

96. Wako, H., Scheraga, H. A. 1982. *J. Protein Chem.* 1:85–117

97. Sippl, M. J., Scheraga, H. A. 1986. *Proc. Natl. Acad. Sci. USA* 83:2283–87

98. McCammon, J. A., Gelin, B. R., Karplus, M. 1977. *Nature* 267:585–90

99. McCammon, J. A., Wolyness, P. G., Karplus, M. 1979. *Biochemistry* 18:927–42

100. Karplus, M., McCammon, J. A. 1983. *Annu. Rev. Biochem.* 52:263–300

101. Levitt, M. 1983. *J. Mol. Biol.* 170:723–64

102. Kaptein, R., Zuiderweg, E. R. P., Scheek, R. M., Boelens, R., van Gunsteren, W. F. 1985. *J. Mol. Biol.* 182:179–82

103. Clore, G. M., Gronenborn, A. M., Brünger, A. T., Karplus, M. 1985. *J. Mol. Biol.* 186:435–55

104. Clore, G. M., Nilges, M., Sukumaran, D. K., Brünger, A. T., Karplus, M., Gronenborn, A. M. 1986. *EMBO J.* 5:2729–35

105. Clore, G. M., Gronenborn, A. M., Nilges, M., Ryan, C. A. 1987. *Biochemistry* 26:8012–23

106. Zuiderweg, E. R. P., Scheek, R. M., Boelens, R., van Gunsteren, W. F., Kaptein, R. 1985. *Biochimie* 67:707–15

107. Kirkpatrick, S., Gelatt, C. D., Vecchi, M. P. 1983. *Science* 220:671–80

108. Nilges, M., Clore, G. M., Gronenborn, A. M. 1988. *FEBS Lett.* 229:317–24

109. Clore, G. M., Gronenborn, A. M., Kjaer, M., Poulsen, F. M. 1987. *Protein Eng.* 1:305–11

110. Clore, G. M., Gronenborn, A. M., James, M. N. G., Kjaer, M., McPhalen, C. A., Poulsen, F. M. 1987. *Protein Eng.* 1:313–18

111. Montelione, G. T., Wüthrich, K., Nice, E. C., Burgess, A. W., Scheraga, H. A. 1987. *Proc. Natl. Acad. Sci. USA* 84:5226–30

112. Kohda, D., Go, N., Hayashi, K., Inagaki, F. 1988. *J. Biochem.* 103:741–43

113. Kohda, D., Inagaki, F. 1988. *J. Biochem.* 103:554–71

114. Carver, J. A., Cooke, R. M., Esposito, G., Campbell, I. D., Gregory, H., Sheard, B. 1986. *FEBS Lett.* 205:77–81

115. Cooke, R. M., Wilkinson, A. J., Baron, M., Pastore, A., Tappin, M. J., et al. 1987. *Nature* 327:339–41

116. Frey, M. H., Wagner, G., Vasak, M., Sørensen, O. W., Neuhaus, D., et al. 1985. *J. Am. Chem. Soc.* 107:6847–51

117. Otvos, J. D., Engeseth, H. R., Wehrli, S. 1985. *J. Magn. Reson.* 61:579–84

118. Live, D., Armitage, I. M., Dalgarno, D. C., Cowburn, D. J. 1985. *J. Am. Chem. Soc.* 107:1775–77

119. Furey, W. F., Robbins, A. H., Clancy, L. L., Winge, D. R., Wang, B. C., Stout, C. D. 1986. *Science* 231:704–10

120. Furey, W. F., Robbins, A. H., Clancy, L. L., Winge, D. R., Wang, B. C., Stout, C. D. 1987. *Metallothionein II,* ed. J. H. R. Kägi, Y. Kojima, pp. 139–48. Basel:Birkhäuser

121. Nettesheim, D. G., Edalji, R. P., Mollison, K. W., Greer, J., Zuiderweg, E. R. P. 1988. *Proc. Natl. Acad. Sci. USA* 85:5036–40

122. Inagaki, F., Hider, R. C., Hodges, S. J., Drake, A. F. 1985. *J. Mol. Biol.* 183:575–90

123. Basus, V. J., Billeter, M., Love, R. A., Stroud, R. M., Kuntz, I. D. 1988. *Biochemistry* 27:2763–71

124. Kosen, P. A., Finer-Moore, J., McCarthy, M. P., Basus, V. J. 1988. *Biochemistry* 27:2775–81

125a. Strop, P., Wider, G., Wüthrich, K. 1983. *J. Mol. Biol.* 166:641–65

125b. Frank, G. 1983. *J. Mol. Biol.* 166:665–68

126. Wemmer, D. E., Kumar, N. V., Metrione, R. M., Lazdunski, M., Drobny, G., Kallenbach, N. R. 1986. *Biochemistry* 25:6842–49

127. Pflugrath, J., Wiegand, E., Huber, R., Vertesy, L. 1986. *J. Mol. Biol.* 189:383–86

128. Deleted in proof

129. Kline, A. D., Braun, W., Wüthrich, K. 1986. *J. Mol. Biol.* 189:377–82

130. Kline, A. D., Braun, W., Wüthrich, K. 1988. *J. Mol. Biol.* 204:675–724

131a. Driscoll, P. C., Hill, A. O., Redfield, C. 1987. *Eur. J. Biochem.* 170:279–92

131b. Chazin, W. J., Rance, M., Wright, P. E. 1988. *J. Mol. Biol.* 202:603–22

131c. Chazin, W. J., Wright, P. E. 1988. *J. Mol. Biol.* 202:623–36

132. Dyson, H. J., Holmgren, A., Wright, P. E. 1989. *Biochemistry.* In press

133. Saudek, V., Williams, R. J. P., Ramponi, G. 1988. *J. Mol. Biol.* 199:233–37

134. Weiss, M. A., Jeitler-Nilsson, A., Fischbein, N. J., Karplus, M., Sauer, R. T. 1986. In *NMR in the Life Sciences,* ed. E. M. Bradbury, C. Nicolini, 107:37–48. *NATO ASI Ser.* London: Plenum

135. Griffey, R. H., Redfield, A. G., Loomis, R. E., Dahlquist, F. W. 1985. *Biochemistry* 24:817–22

136. LeMaster, D. M., Richards, F. M. 1985. *Biochemistry* 24:7263–68

137. Torchia, D. A., Sparks, S. W., Bax, A. 1988. *Biochemistry* 27:5135–41

138. Griffey, R. H., Redfield, A. G. 1987. *Q. Rev. Biophys.* 19:51–82

139. Kainoshi, M., Tsuji, T. 1982. *Biochemistry* 21:6273–79

140. Bendall, M. R., Pegg, D. T., Doddrell, D. M. 1983. *J. Magn. Reson.* 52:81–117

141. Bax, A., Griffey, R. H., Hawkins, B. L. 1983. *J. Magn. Reson.* 55:301–15

142. Griffey, R. H., Poulter, C. D., Bax, A., Hawkins, B. L., Yamaizumi, Z., Nishimura, S. 1983. *Proc. Natl. Acad. Sci. USA* 80:5895–97

143. Roy, S., Redfield, A. G., Papastavros, M. Z., Sanchez, V. 1984. *Biochemistry* 23:4395–400

144. Griffey, R. H., Redfield, A. G., McIntosh, L. P., Oas, T. G., Dahlquist, F. W. 1986. *J. Am. Chem. Soc.* 108:6816–17

145. Markley, J. L., Putter, I., Jardetzky, O. 1968. *Science* 161:1249–51

146. Kalbitzer, H. R., Leberman, R., Wittinghofer, A. 1985. *FEBS Lett.* 180:40–42

147. Torchia, D. A., Sparks, S. W., Bax, A. 1988. *J. Am. Chem. Soc.* 110:2320–21

148. Freeman, R., Mareci, T. H., Morris, G. A. 1981. *J. Magn. Reson.* 42:341–45

149. Bax, A., Weiss, M. A. 1987. *J. Magn. Reson.* 71:571–75

150. Senn, H., Otting, G., Wüthrich, K. 1987. *J. Am. Chem. Soc.* 109:1090–92

151. Rance, M., Wright, P. E., Messerle, B. A., Field, L. D. 1987. *J. Am. Chem. Soc.* 109:1591–93

152. Fesik, S. W., Gampe, R. T., Rockway, T. W. 1987. *J. Magn. Reson.* 74:366–71

153. McIntosh, L. P., Dahlquist, F. W., Redfield, A. G. 1987. *J. Biomol. Struct. Dyn.* 5:21–34

154. Gronenborn, A. M., Bax, A., Wingfield, P. T., Clore, G. M. 1989. *FEBS Lett.* In press

155. Markley, J. L. 1989. In *Methods in Enzymology,* ed. T. L. James, N. Oppenheimer. In press

156. Oh, B. H., Westler, W. M., Darba, P., Markley, J. L. 1988. *Science* 240:908–11

157. Westler, W. M., Stockman, B. J., Hosoya, Y., Miyake, Y., Kainosho, M., Markley, J. L. 1988. *J. Am. Chem. Soc.* 110:6256–58

158. Stockman, B. J., Reily, M. D., Westler, W. M., Ulrich, E. L., Markley, J. L. 1989. *Biochemistry* 28:230–36

159. Westler, W. M., Kainosho, M., Nagao, H., Tomonaga, N., Markley, J. L. 1988. *J. Am. Chem. Soc.* 110:4093–95

160. Bax, A., Sparks, S. W., Torchia, D. A. 1988. *J. Am. Chem. Soc.* 110:7926–27

161. Clore, G. M., Bax, A., Wingfield, P., Gronenborn, A. M. 1988. *FEBS Lett.* 238:17–21

162. Vuister, G. W., Boelens, R. 1987. *J. Magn. Reson.* 73:328–33

163. Griesinger, C., Sørensen, O. W., Ernst, R. R. 1988. *J. Magn. Reson.* 73:574–79

164. Griesinger, C., Sørensen, O. W., Ernst, R. R. 1987. *J. Am. Chem. Soc.* 109:7227–29

165. Oschkinat, H., Griesinger, C., Kraulis, P. J., Sørensen, O. W., Ernst, R. R., et al. 1988. *Nature* 332:374–76

166. Vuister, G. W., Boelens, R., Kaptein, R. 1988. *J. Magn. Reson.* 80:176–85

167. Fesik, S. W., Zuiderweg, E. R. P. 1988. *J. Magn. Reson.* 78:588–93

168. Braun, W., Wider, G., Lee, K. H., Wüthrich, K. 1983. *J. Mol. Biol.* 169:921–48

169. Brown, L. R., Braun, W., Kumar, A., Wüthrich, K. 1982. *Biophys. J.* 37:19–32

170. Schiksnis, R. A., Bogusky, M. J., Tsang, P., Opella, S. J. 1987. *Biochemistry* 26:1373–81

171. Boelens, R., Scheek, R. M., Lamerichs, R. M. J. N., de Vlieg, J., van Boom, J. H., Kaptein, R. 1987. *J. Mol. Biol.* 193:213–16

172. Marion, D., Kay, L. E., Sparks, S. W., Torchia, D. A., Bax, A. 1989. *J. Am. Chem. Soc.* In press

173. Bax, A., Sklenar, V., Gronenborn, A. M., Clore, G. M. 1987. *J. Am. Chem. Soc.* 109:6511–13

Annu. Rev. Biochem. 1989. 58:257-85

PROTEIN RADICAL INVOLVEMENT IN BIOLOGICAL CATALYSIS?

Jo Anne Stubbe

Department of Chemistry, Massachusetts Institute of Technology, Cambridge, Massachusetts 02139

CONTENTS

PERSPECTIVES AND SUMMARY

Initially this review was intended to summarize the broad topic of radical involvement in biological transformations (1, 2), but because of the tremendous breadth of that topic it is now focused on the potential involvement of protein radicals derived by oxidation of amino acid residues, in mediating catalytic transformations. A misunderstanding exists that radical intermediates are highly reactive and uncontrollable with respect to their chemistry, and therefore, are not generally involved in biological transformations. It is clear from the organic chemistry literature that this is indeed a misconception and that radicals can be involved in both regioselective and stereoselective reactions that are subject to the steric and stereoelectronic control of their

257

0066-4154/89/0701-0257$02.00

environments (3). At present it is too early to make generalizations about what type of transformations may involve protein radicals. However, thus far, in the majority of cases in which protein radicals have been implicated, cleavage of nonacidic C-H bonds is an essential part of the overall transformation.

Amino acid residues most likely to be involved in catalysis involving radical intermediates are those most easily oxidized: tryptophan, tyrosine, cysteine, histidine, and, perhaps, methionine. The oxidation potentials for these residues are well within the realm of those achievable by biological oxidants (4). The oxidized amino acid residues, of course, are not stable and, therefore, must be stabilized by the resting protein or generated transiently prior to catalysis. For the systems that have been described, the modes of generation of the putative protein radicals are amazingly diverse. However, in all cases except the flavoprotein mitochondrial amine oxidase, the enzymes have metal ion requirements for radical production (iron, cobalt, copper, or manganese).

While the presence of radicals derived by oxidation of amino acids has clearly been demonstrated in a number of proteins that catalyze intriguing transformations, in no case has the chemical and kinetic competence of these protein radicals been unambiguously demonstrated. This review focuses, therefore, on a few of the better-characterized protein radical–requiring systems in order to demonstrate the feasibility of their involvement and the methods presently available to study such systems. In addition, it is hoped that in the future, protein radical involvement in the other enzymatic systems will be considered as a viable mechanistic option.

RIBONUCLEOTIDE REDUCTASES

The elegant studies of Reichard, Ehrenberg, Thelander, Sjöberg, and their collaborators have presented us with the first definitively characterized stable protein radical to date: a radical derived from tyrosine residue 122 that is an integral part of one of the subunits of the *Escherichia coli* ribonucleotide reductase (RDPR) (5–7). Ribonucleotide reductases play a central role in DNA biosynthesis by catalyzing the conversion of ribonucleotides to deoxynucleotides. The state of phosphorylation of the nucleotide substrate, i.e. di or triphosphate, is a function of the organism (19). In all reductases examined thus far, the reduction of substrates is accompanied by oxidation of protein dithiols (Eq. 1), the resulting cystine is then reduced by an electron transfer chain in vivo composed of thioredoxin and thioredoxin reductase or glutaredoxin and glutathione reductase. The *E. coli* reductase has served as a prototype for most eukaryotic reductases (yeast, mouse, calf thymus, herpes, vaccinia, and pseudorabies viral), while the *Lactobacillus leichmannii* reductase is a prototype for most prokaryotic reductases. Discussion will be limited to these two enzymes.

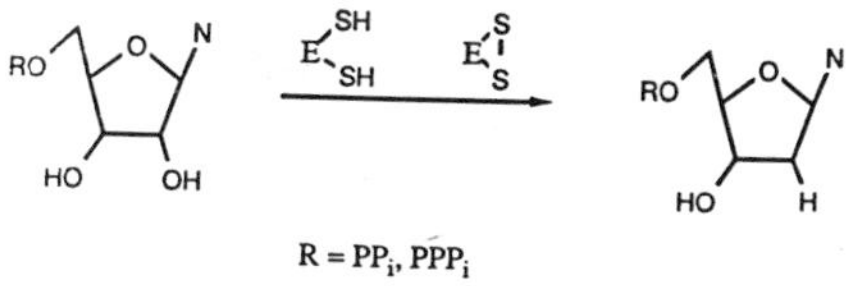

R = PP$_i$, PPP$_i$ 1.

I. E. Coli *Ribonucleotide Reductase*

The *E. coli* reductase is composed of two types of subunits. The large subunit, B$_1$, contains two equivalent protomers of 86 kd based on the gene sequence (8, 88). This subunit binds the NDP substrates, dNTP and ATP allosteric effectors, and contains the redox-active thiols. The smaller subunit, B$_2$, also consists of two equivalent protomers of 43.3 kd, and contains a unique cofactor: a binuclear iron center composed of two nonequivalent high-spin irons antiferromagnetically coupled through a μ oxo bridge and a single organic tyrosyl radical located at residue 122 (Figure 1). The tyrosyl radical, which is essential for catalytic activity, has an absorbance at 410 nm (ϵ = 3250 cm^{-1}M^{-1}) and a unique EPR doublet signal at g = 2.0042. Recent ENDOR studies of G. Babcock and coworkers (unpublished results) are consistent with a tyrosyl radical rather than a tyrosyl radical cation as initially proposed. Reduction of this radical with hydroxyurea or removal of the iron center from the B$_2$ subunit with imidazole and 8-hydroxyquinoline 5-sulfonate destroys the tyrosyl radical as well as the catalytic activity. The tyrosyl radical can be regenerated in the former case by enzymatic reduction of the iron center under aerobic conditions (61, 62) or in the latter case by addition of Fe(II) and O$_2$ (6). In both cases the amount of tyrosyl radical restored is proportional to the catalytic activity regenerated. The active site of

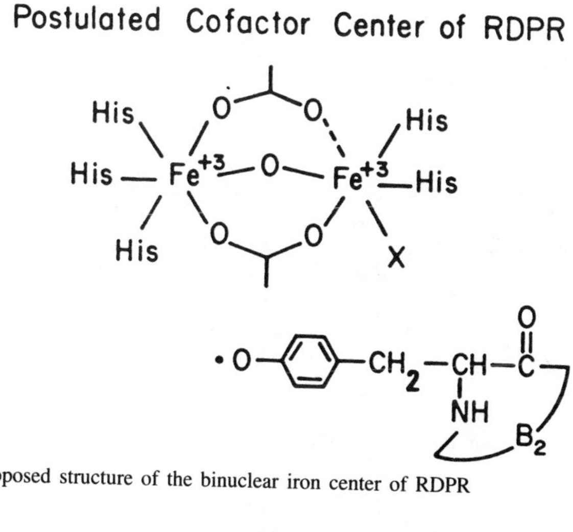

Postulated Cofactor Center of RDPR

Figure 1   Proposed structure of the binuclear iron center of RDPR

this enzyme is at the interface between the two subunits (6, 7). The requirement for the tyrosyl radical led to the proposal of a radical-mediated reduction reaction. However, it was not until the studies of Stubbe and coworkers that a reasonable hypothesis for this process was proposed (9).

PROPOSED MECHANISM   A mechanism consistent with all of the available experimental data and based on a chemical model system is indicated in Figure 2 (9). This hypothesis predicts that a radical (derived from tyrosine or another amino acid residue) abstracts a hydrogen atom from the 3'-position of the ribonucleotide to form a 3'-ribonucleotide radical. Subsequent to protonation of the 2'-hydroxyl, catalyzed by one of the redox-active thiols, and loss of H_2O, the cation radical intermediate is generated. This intermediate would then be reduced by dithiol oxidation via two one-electron transfers and protonation to produce the 3'-deoxynucleotide radical. This product radical would then be reduced to the dNDP product by oxidation of the tyrosine to regenerate the tyrosyl radical.

EVIDENCE IN SUPPORT OF THE PROPOSED MECHANISM   The postulated mechanism makes predictions that have been experimentally tested. The use of isotopically labeled [3'-^{3}H, U-^{14}C] or [3'-^{2}H]NDPs as substrate and measurement of the isotope effect on dNDP production unambiguously es-

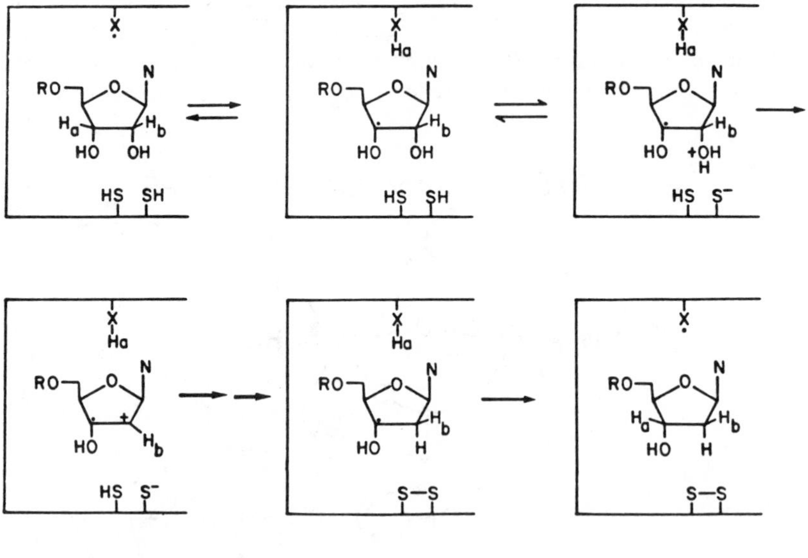

X = protein

Figure 2 Postulated mechanism of reduction of nucleotides to deoxynucleotides catalyzed by ribonucleotide reductase

tablished that RDPR is capable of mediating 3' carbon-hydrogen bond cleavage. ^{T}(V/K) isotope effects ranging from 1.5 to 4.7 were observed. In addition, the use of [3'-^{2}H]NDPs and NMR spectroscopy established that the hydrogen abstracted from the 3' position in the substrate is returned to the 3' position in the product. While this information is consistent with the mechanism proposed in Figure 2, it is also consistent with a heterolytic mechanism of bond cleavage. Attempts to demonstrate the homolytic nature of the bond cleavage took advantage of the fact that the tyrosyl radical (X·) has an absorbance at 410 nm and that during turnover, the radical is transiently reduced to tyrosine, with loss of absorbance at 410 nm, and ultimately reoxidized. The use of stopped-flow kinetics methods under a variety of conditions involving changes of pH, temperature, and substrate has thus far failed to observe any change in absorption at 410 nm during the course of the reduction reaction (M. Ator, J. Stubbe, D. Ballou, unpublished results). This may not be surprising given that recently reported rate constants for cation radical production and for α-ketone radical reduction are $\sim 10^6$ sec and 10^8 $M^{-1}s^{-1}$ (with thiolate) in similar nonenzymatic systems (10, 11). These values are several orders of magnitude faster than the turnover number of the reductase, making it probable that the concentration of reduced radical species determined by disappearance of the 410 nm absorption would be too low to detect experimentally.

Having failed to provide evidence for radical intermediates in the normal reduction process, efforts have focused on two alternative approaches: 1. the use of modified substrates whose transformations might be accompanied by a change in rate-determining step or alternative chemistry diagnostic of radical intermediates; and 2. the use of site-directed mutagenesis to alter the active site and hence rate-determining steps of the normal reduction process. The former approach has been applied with moderate success as outlined subsequently and as recently reviewed (9). Recent cloning and overproduction of each of the subunits of reductase in conjunction with recent active site labeling experiments make the latter approach very feasible (12–15).

USE OF ALTERNATE SUBSTRATES AS POTENTIAL RADICAL TRAPS: INTERACTION OF 2'-N$_3$NDPS WITH RDPR   The interaction of the alternative substrates 2'-halo-2'-deoxynucleotides, 2'XNDPs where X=F, Cl, Br, I, with RDPR have been studied in detail, and a common theme emerges. In all cases the initial step is enzyme-mediated cleavage of the 3'-carbon hydrogen bond. With 2'-halo deoxynucleotides (e.g. 2'ClUDP), which are potent irreversible inhibitors of RDPR, the results of studies using a variety of isotopically labeled derivatives are consistent with the mechanism proposed in Figure 3. Evidence firmly supports enzyme-mediated conversion of [3'-^{3}H]-2'-ClNDPs to 3'-keto (2'-^{3}H) dNDPs and subsequently production of PPi, uracil, and the 2-methylene-3(2-^{3}H)-furanone responsible for enzyme in-activation. This mechanism, in analogy with that for normal substrate reduc-tion, postulates a tyrosyl radical–mediated hydrogen atom abstraction to produce a 3'-nucleotide radical and reduced tyrosine and ultimately reoxida-tion of the latter. Stopped-flow kinetics utilizing a variety of 2'-halo-2'-dNDPs (halo = Cl, Br, F, and N =cytosine, adenine, and uracil) under a variety of conditions (pH 5 to 9, temperature 0 to 25°) have failed, however, to detect any change in tyrosyl radical absorption during turnover (M. Ator, J. Stubbe, D. Ballou, unpublished results).

A detailed study of the interaction of a second type of mechanism-based inhibitor, 2'-azido-2'-deoxynucleotide (N$_3$NDP), first reported by Thelander et al (16), has provided the first support for correlation of the cleavage of the 3' carbon-hydrogen bond of substrate and reduction of the tyrosyl radical (17). N$_3$UDP inactivates the protein in a single turnover, making analysis of the products challenging. The use of isotopically labeled N$_3$NDPs [3'-^{3}H, 5'-^{3}H, β-^{32}P and ^{15}N] has recently allowed identification of most of the

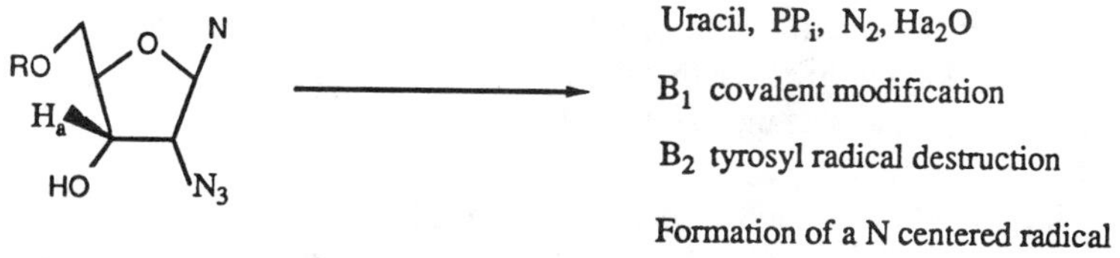

Uracil, PP$_i$, N$_2$, Ha$_2$O

B$_1$ covalent modification

B$_2$ tyrosyl radical destruction

Formation of a N centered radical   2.

products accompanying enzyme inactivation (Eq. 2). The production of N$_2$, PPi, and uracil and cleavage of the 3'-carbon hydrogen bond, have been established. In addition, labeling of the 5' position of the sugar moiety resulted in stoichiometric covalent modification of the B$_1$ subunit (17). This latter point provides strong support for the active site being located at the interface between the B$_1$ and B$_2$ subunits. The most intriguing observation, however, is that reduction of the tyrosyl radical (monitored by loss in the absorbance at 410 nm or by EPR spectroscopy) is accompanied by formation of a new radical species (18, 87). The observed EPR spectrum was identical regardless of the base moiety of N$_3$NDP (cytosine, uracil, adenine). Isotopic labeling studies utilizing [N$_3$-^{15}N]UDP have unequivocally demonstrated that the EPR signal is due to an unpaired electron residing on a single nitrogen derived from the N$_3$ moiety of N$_3$NDP. Studies utilizing [1'-^{2}H], [2'-^{2}H],

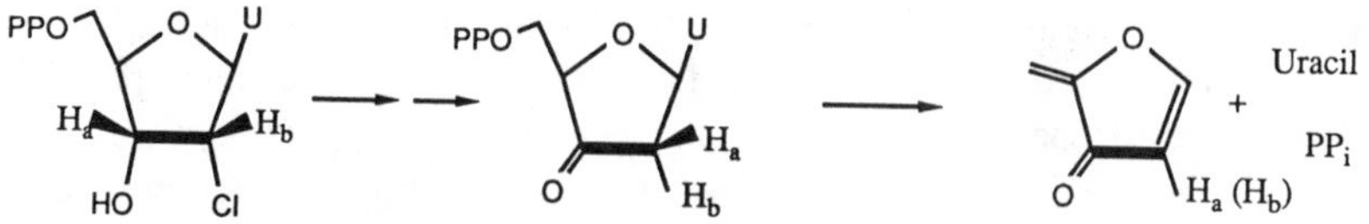

Figure 3 Minimal mechanism for inactivation of RDPR by ClUDP

[3'-^{2}H], and [4'-^{2}H] N$_3$UDPs, however, leave still unresolved the nature of the spin 1/2 nucleus responsible for what is thought to be a hyperfine interaction (6.5 gauss splitting) with the unpaired electron on nitrogen (9). Studies by Sjöberg et al have ruled out the possible involvement of a coupling interaction derived from protons on the B$_2$ subunit (18). Interactions with a proton derived from the B$_1$ subunit presently cannot be eliminated. Given that covalent modification by the sugar moiety occurs uniquely on B$_1$, this possibility must be seriously entertained. Efforts are under way to unravel the source of the hyperfine coupling by preparing deuterium-labeled B$_1$. In addition, studies with [2'-^{13}C]N$_3$UDP should determine if the nitrogen-based radical is still attached to the carbohydrate moiety.

A working model that accounts for the experimental information, with the exception of this spin 1/2 hyperfine interaction, is shown in Figure 4. Several important conclusions can be drawn from these studies. The first is that RDPR can produce a substrate analogue–derived radical, consistent with the hypothesis in Figure 4 that the azide moiety of 2'N$_3$NDP is acting as an intramolecular radical trap. The second point is that RDPR is capable of mediating cleavage of the 3' carbon hydrogen bond. If cleavage of the bond could be correlated with reduction of the tyrosyl radical, then strong support for the proposed mechanisms (Figures 2–4) will have been provided. To investigate this prediction, [3'-^{2}H]N$_3$UDP was prepared and the kinetics of enzyme inactivation and loss of the tyrosyl radical absorbance at 410 nm were compared with studies on the analogous protonated material. The results shown in Figures 5a and 5b indicate that there is an isotope effect on the rate of inactivation, and an isotope effect on the loss of the tyrosyl radical (S. Salowe, J. Stubbe, unpublished results). In contrast to the prediction, however, the rate of loss of the tyrosyl radical is less than the rate of inactivation. Unfortunately, complex kinetics are observed in all three cases (Figures 5a, b, c). While the mechanism shown in Figure 4 has been constructed to accommodate these complex kinetic observations, i.e. tyrosyl radical can be regenerated if 3 is in equilibrium with 3a, followed by slow conversion to products, the direct correlation between carbon-hydrogen bond cleavage and tyrosyl radical reduction is not observed and is much more complex than anticipated. Thus, N$_3$UDP has provided the useful piece of information that RDPR can generate substrate-derived radicals, but whether this observation can be applied to the studies on normal substrates and 2' halonucleotides, thereby supporting the proposed mechanism in Figures 2 and 3, awaits further experimentation.

II. Adenosylcobalamin-Dependent Ribonucleotide Reductase

Ribonucleotide reductases have also been isolated from other prokaryotes, (*Anabaena, Corynebacterium nephridiis, Sphaerophorus varius, Lactobacillus acidophilus, Clostridium tetanomorphum, Clostridium stricklandii, Rhi-*

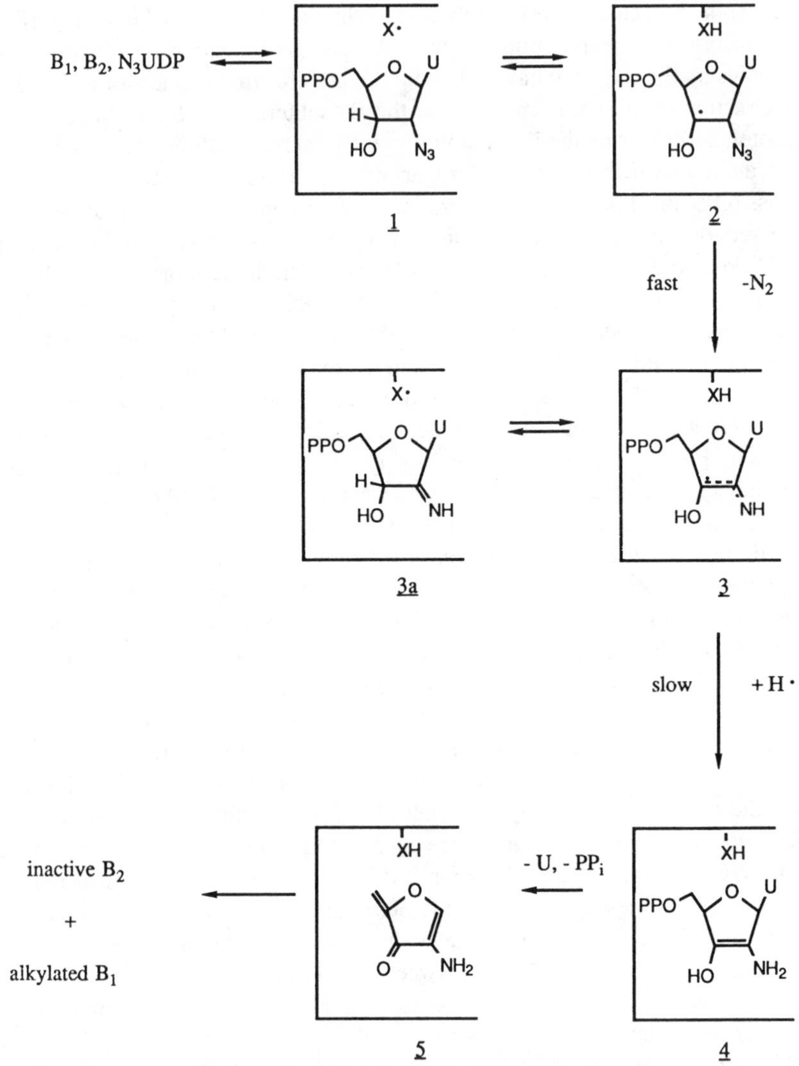

Figure 4 Proposed mechanism for inactivation of RDPR by N_3UDP

zobium lequminusarum, Pseudomonas stutzeri, Streptomyces aureofaciens),
and shown to be quite different in terms of cofactor requirement and structure
from the RDPR isolated from *E. coli* (19). The enzyme from *L. leichmannii*
(RTPR) that catalyzes the conversion of NTPs to dNTPs has been studied
most extensively. It is composed of a single polypeptide of 76 kd and utilizes
adenosylcobalamin (AdoCbl) as a cofactor (Figure 6). This protein has been
thought to be unique among AdoCbl-requiring proteins, in that it does not

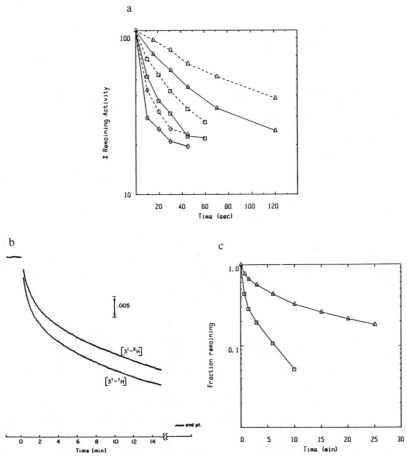

Figure 5   Kinetics of the interaction of [3'-²H] and [3'-¹H]N₃UDP with RDPR.

(*a*) Time-dependent inactivation. Symbols used: - - - - [²H]N₃UDP; ——— [¹H]N₃UDP; Δ, □, ◇ different concentrations of N₃UDP

(*b*) Time-dependent loss of the tyrosyl radical

(*c*) Comparison of rate of loss of the Tyrosyl Radical and rate of inactivation. Symbols used: Δ, tyrosyl radical; □, loss of activity.

catalyze the prototypical rearrangement reaction (Eq. 3) (20). However, as will be discussed, recent data supports the postulate that RTPR may serve as a

$$R_1R_2XCCHR_3R_4 \rightleftharpoons R_1R_2HCCXR_3R_4$$

$$RCHXH_aHOH \longrightarrow \boxed{RCHH_aCHOH\underline{X}} \longrightarrow RCHH_aCHO \quad 3.$$

prototype for all AdoCbl-dependent enzymes and that protein radicals may be directly involved in catalysis.

RTPR catalyzes two reactions, the reduction of NTPs to dNTP accom-

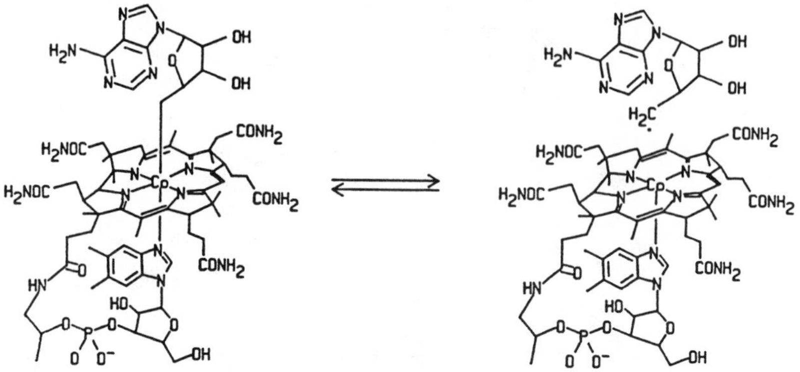

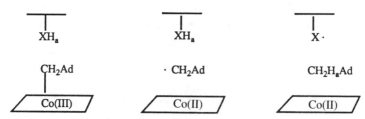

Figure 6 Homolytic cleavage of the carbon-cobalt bond of adenosylcobalamin

panied by thiol oxidation in analogy with *E. coli* RDPR and the exchange of [^{3}H] from the 5'-position of AdoCbl with solvent in the presence of a reductant and an allosteric effector (20, 21). Postulated mechanisms for both of these reactions are indicated in Figures 2 and 7 (21).

EVIDENCE IN SUPPORT OF THE POSTULATED MECHANISM OF THE REDUC-TION REACTION   The early studies by Blakley et al using stopped-flow kinetics and rapid quench EPR indicated that cobalamin (II) and presumably 5'deoxyadenosyl radical are produced in a kinetically competent fashion upon incubation of RTPR, substrate, and AdoCbl (22, 23). This observation initially led to the proposal that the 5'-deoxyadenosyl radical is the X· in Figure 2. Studies of Harris et al, analogous to those described above for the *E. coli* reductase, have established that the enzyme is involved in cleavage of the 3'-carbon-hydrogen bond of the substrate and that the hydrogen abstracted from the 3' position in the NTP is returned to the 3' position in the product (24).

However, results from efforts to establish the involvement of the 5'-deoxyadenosyl radical as the species responsible for 3'-hydrogen atom abstraction have required reconsideration of the proposed mechanism. The

Figure 7 Postulated mechanism of exchange of [5'-^{3}H]AdoCbl with the media catalyzed by RTPR

role of the 5'-deoxyadenosyl radical was investigated in a single turnover experiment in which prereduced RTPR in the absence of external reductant was incubated with either [3'-^{3}H]UTP or [3'-^{3}H]ATP. After 5 min, 0.9 equivalent of dUTP (or dATP) per equivalent of RTPR was isolated, and the AdoCbl was isolated by HPLC and analyzed for radioactivity. No radioactivity (<0.01 equivalent) was detected. These results indicate that even though the 3' carbon hydrogen bond of the nucleotide was cleaved, the X· (Figure 2) is not the 5'-deoxyadenosyl radical. Multiple turnover experiments resulted in a similar conclusion (25). No [^{3}H] from [3'-^{3}H]NTP was found in the cofactor, and all of the [^{3}H] could be accounted for in substrates and products. The explanation put forth to account for these results is indicated in Figure 7, in which the 5'-deoxyadenosyl radical abstracts a hydrogen atom from a protein residue, which then mediates the chemistry (Figure 2).

A similar mechanism was proposed by Hogenkamp and coworkers to account for the second reaction catalyzed by RTPR, washout of ^{3}H from [5'-^{3}H]AdoCbl to solvent in the presence of reductant and an allosteric effector (Figure 7) (21). The rate of washout from AdoCbl is slow, however, compared to the rate of normal turnover, and since it is not limited by the rate of dissociation of the cofactor from the enzyme and ^{3}H is not transferred from the cofactor to product, these observations indicate that this exchange reaction is not on the normal catalytic pathway. However, the observation of exchange is consistent with the presence of a protein radical intermediate. Recent active-site labeling experiments have fueled speculation that X· is a thiyl radical (26).

RTPR was thought to be unique among AdoCbl-dependent enzymes because it does not catalyze a rearrangement reaction. However, recent studies with ClUTP and RTPR indicate that its mechanism might not differ from that of the extensively investigated dioldehydrase and ethanolamine ammonia lyase (Eq. 3). As in the case of *E. coli* RDPR, the RTPR has been shown recently to catalyze a 1,2-hydrogen migration, conversion of [3'-^{3}H]ClNTP to [2'proS-^{3}H]3'-keto-dUTP (27). The unique role of AdoCbl postulated for this protein, as well as the ability of the protein to mediate a 1,2-hydrogen shift, has suggested the possibility that RTPR might serve as a prototype to explain extremely large isotope effects ($k_H/k_T = 150$) for the transfer of [^{3}H] from [5'-^{3}H]AdoCbl to product in the dioldehydrase and ethanolamine ammonia lyase–catalyzed reactions (20) (Section III).

OTHER PROTEINS THAT MAY INVOLVE AMINO ACID RADICALS

III. Adenosylcobalamin-Dependent Rearrangement Reactions

Dioldehydrase and ethanolamine ammonia lyase (Eq. 3) are two of the few enzymatic systems where substrate-derived radicals have been shown to be produced in a chemically and kinetically competent fashion. Elegant studies from numerous laboratories are consistent with the general mechanism for these rearrangements summarized in Figure 8 (20, 28, 29). The first step, as in the case of all AdoCbl-requiring enzymes, has been shown to be homolytic cleavage of the carbon cobalt bond to produce cobalamin(II) and the 5'-deoxyadenosyl radical. This latter species has been convincingly shown to be involved in mediation of a hydrogen atom transfer between substrate and product. However, one aspect of this overall scheme has not been satisfactorily explained: the observation of an k_H/k_T isotope effect of 125 for dioldehydrase (28) and 160 for ethanolamine ammonia lyase (29) for the transfer of ^{3}H from cofactor to the product. The isotope effect was calculated based on the assumption that every hydrogen that is transferred from substrate to product goes through 5'-deoxyadenosine as an obligate intermediate. In 1982 Cleland put forth an alternative mechanism to account for these observed isotope effects (30). He postulated that the 5'-deoxyadenosyl radical serves as a radical chain initiator to produce an alternative substrate hydrogen atom abstractor. This alternative radical may, in fact, be a protein radical. If conversion of substrate to product occurs 9 out of 10 times through this alternative pathway, then the observed (^{3}H) selection effect is reduced to the range of the normally observed ^{3}H selection effects. Recent EXEEM studies and isotopic labeling studies form Babior's group support the contention that an alternative hydrogen atom abstractor exists (31, 32). This proposal, i.e. involvement of a protein radical as a hydrogen atom abstractor, is analogous to the one we were forced to consider for the RTPR system. Given that both ethanolamine ammonia lyase and dioldehydrase have a thiol residue(s) that when modified results in loss in activity, it is reasonable to postulate that X· in Figure 8 is a thiyl radical.

A comment on the role of the thiyl radical as a potential hydrogen atom

Proposed Mechanism of Diol Dehydratase

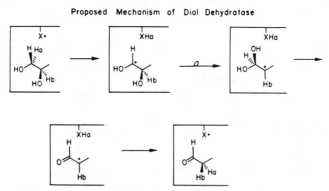

Figure 8   General mechanism of AdoCbl-dependent rearrangement reactions

abstractor is also warranted. While it is generally accepted that thiols are capable of transferring a hydrogen atom to carbon-centered radicals because of the relative weakness of their bond dissociation energies (91 + 1.5 kcal/mol) (33, 89), their role as hydrogen atom abstractors is contraindicated by most commonly held beliefs. Hydrogen atom abstraction of the 3'-H of a nucleoside by a thiyl radical is allowed on thermodynamic grounds because the C-H homolytic bond dissociation energy is ~91 kcal/mole. Akhlaq et al recently devised a method to detect thiyl radical–mediated hydrogen atom abstractions (34), which are usually masked in the presence of high RSH concentrations due to thiol-mediated repair. Akhlaq et al took either *cis*-2,5 Me$_2$THF or *trans*-2,5 Me$_2$THF and measured the production of a mixture of *cis* and *trans* 2,5 Me$_2$THF (Eq. 4). Thiols in this case, as in the earlier studies

by Huyser & Kellogg (35), do in fact induce a chain reaction. Thus, the feasibility of thiyl radicals mediating hydrogen atom abstractions has reasonable chemical precedent. The final step in the postulated mechanism, thiyl radical abstraction of a hydrogen atom from 5' deoxyadenosine, is thermodynamically unfavorable by 18 kcal. The driving force for this reaction would have to be provided by reformation of the C-Co bond to produce AdoCbl. This process has recently been shown to be exothermic by 25 to 30 kcal/mole (91). The establishment of the role of any postulated thiol in an enzymatic reaction will be facilitated by the cloning, sequencing, and expression of any of these genes and use of site-directed mutagenesis methods to replace the cysteine residues. Efforts are presently under way with both RTPR and ethanolamine ammonia lyase to investigate the putative role of thiyl radicals in the chemistry.

An interesting paper of Hartmanis & Stadtman provides further credibility to the postulated role of protein radicals (36). Specifically, a dioldehydrase that does not require AdoCbl has been partially purified from *Chlostridium glycolicum* and shown to possess a protein radical that is destroyed by reduction with hydroxyurea concomitant with loss of dioldehydrase activity. The properties of this radical are similar to those observed in *E. coli* RDPR.

In addition to the binuclear iron center, tyrosyl radical, and AdoCbl-dependent systems discussed above, recent studies from Follmann's laboratory indicate that ribonucleotide reductase isolated from *Brevibacterium ammoniagenes* requires manganese for activity (37). Based on recent chemical model systems from Lippard's and Wieghardt's laboratories, it is reasonable to postulate the existence of a novel binuclear Mn(III)-dependent system that will also allow protein radical–mediated catalysis (38, 39). (See Section IV.)

IV. Photosystem II: $H_2O \rightarrow O_2$ Evolving System

In plants and algae, photosystem II couples light-induced charge separation to the oxidation of H_2O to O_2 and reduction of plastoquinone. A summary of the details of this complex system composed of a minimum of seven polypeptides and multiple cofactors including chlorophyll, quinones, manganese, and two organic radical species designated $Z^{\cdot +}$ and $D^{\cdot +}$ is beyond the scope of this review (40, 41). However, evidence will be briefly summarized that has defined a role for a protein tyrosyl radical designated $Z^{\cdot +}$ generated by the reduction of $P^{\cdot +}680$, the oxidized reaction center's primary donor, which is a specialized monomer or dimer of chlorophyll. This residue ($Z^{\cdot +}$) is directly involved in the reoxidation of the reduced manganese center, which plays a central role in the oxidation of H_2O to oxygen.

The reaction center core contains a heterodimer of homologous polypeptides designated D_1 and D_2. In the dark, D_2 displays a very stable EPR signal associated with a radical species $D^{\cdot +}$, and D_1 contains a residue, $Z^{\cdot +}$, which is directly involved in reduction of chlorophyll $P^{\cdot +}680$ and has visible and EPR spectral properties almot identical to those of $D^{\cdot +}$. Recent efforts from Babcock's laboratory using auxotrophic photosynthetic cyanobacteria strains have developed in vivo methods to isotopically label tyrosine residues with deuterium (42). Results from EPR experiments allowed assignment of the $D^{\cdot +}$ signal to an unprotonated tyrosyl radical. The observed EPR signal differs from the well-characterized tyrosyl radical EPR signal of RDPR (43). The differences are presumably due to the sensitivity of the β-methylene hydrogens of tyrosine to the dihedral angle, which measures their orientation to the aromatic ring and causes alterations in the observed hyperfine interactions. Recently Dreyfus et al have used the *psbD* gene of cyanobacterium *Synechocystis* (44) corresponding to D_2, and site-directed mutagenesis methods to further probe the identity of $D^{\cdot +}$. After examining D_1 and D_2 for homologous sequences containing a tyrosine residue, they replaced Tyr 160 in D_2 by site-directed mutagenesis with a Phe. While this mutant grows photosynthetically and evolves O_2, the dark-adapted EPR signal of $D^{\cdot +}$ is missing. While the function of $D^{\cdot +}$ has not been unambiguously established (45), its $Z^{\cdot +}$ counterpart is directly involved in electron-transfer reactions in photosystem II. By symmetry arguments, spectroscopic properties, and the homologous sequence of the D_1- and D_2- containing tyrosine peptides, $Z^{\cdot +}$ is also proposed to be a tyrosine radical.

The interaction of a tyrosyl radical with a four-manganese cluster is an intriguing result, given the recently characterized RDPR from *B. ammoniagenes* reported by Follmann's laboratory (37). This protein contains an uncharacterized manganese center and activity, which is destroyed by hydroxyurea. This latter trait is reminiscent of the hydroxyurea-dependent, one-electron reduction and destruction of the tyrosyl radical in the *E. coli* reductase. Given recent model studies of Sheats et al (38) and Wieghardt et al (39), which describe a binuclear manganese complex isomorphous to the binuclear iron model complex for *E. coli* reductase, one could easily postulate existence of a novel binuclear manganese (III) center-tyrosyl radical complex in this reductase. The photosystem II center and the reductase center may possess some functional similarities. Both the $D^{\cdot+}$ and $Z^{\cdot+}$ centers of photosystem II can mediate redox chemistry on the manganese cluster involved in oxidation of H_2O to O_2. It is also possible that the tyrosyl radical of reductase can mediate redox reductase chemistry on the manganese or iron center in the *B. ammoniagenes* and *E. coli* respectively. In *E. coli*, electron transfer between the tyrosyl radical on one protomer of B_1 and the unoxidized tyrosine on the second protomer of B_1 through the binuclear metal center could make both protomers of the B_1 subunit active and negate the need for the postulated half site's reactivity (Diagram 1).

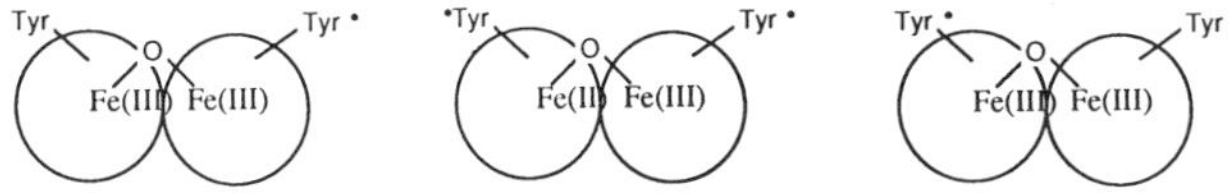

Diagram 1

Future studies will be directed toward elucidation of the structures of the novel manganese centers and their interactions with tyrosyl radicals in their defined biochemical system.

$$CH_3COCO_2H + SCoA \rightleftharpoons CH_3COSCoA + HCO_2H$$

5.

V. Pyruvate Formate Lyase

Pyruvate formate lyase (PFL), which plays a central role in anaerobic glucose metabolism in *E. coli* and catalyzes the conversion of pyruvate and coenzyme A to acetyl CoA and formate (Eq. 5), has been studied by Knappe and coworkers and more recently by Kozarich and coworkers (46, 47). This protein also possesses a putative radical essential for catalysis, but it is not yet well characterized.

Pyruvate formate lyase is a homodimer (170 kd) that can be converted from an inactive to an active form by posttranslational modification by an "activating enzyme," a monomer of 30 kd. Both proteins have been purified to homogeneity (46, 48). The activating enzyme requires Fe^{2+}, S-adenosylmethionine, dihydroflavodoxin, and inactive PFL, and produces methionine, 5'-deoxyadenosine, oxidized flavodoxin, and active PFL. Intriguingly, the activated form of PFL displays an EPR signal at $g = 2.0037$ with a doublet splitting of 15 gauss that has been attributed to a radical located on an amino acid residue. Quantification of the EPR signal gives one electron per enzyme molecule (J. Knappe, personal communication). Isotopic labeling experiments undertaken to establish the identity of the oxidized amino acid residue indicate that when the cells are grown in D_2O, the doublet EPR signal collapses to a singlet; this observation allows the conclusion that the observed large coupling is due to the interaction of the unpaired electron with a hydrogen nucleus. In addition, results from ENDOR spectroscopy are consistent with coupling of the electron with two sets of nonexchangeable hydrogen nucleii with much smaller coupling constants of 3.6 and 5.7 gauss (49). Using PFL isolated from cysteine autotrophic mutant strains grown in the presence of $[3\text{-}^2H_2]$ cysteine or prototrophic grown bacteria on ^{33}S-sulfate in D_2O (H_2O), Knappe and coworkers have unambiguously established that the $g = 2.0037$ signal is not due to a sulfur-centered radical. Knappe (personal communication) has also ruled out the possibility, via as-yet-unpublished isotopic labeling experiments with auxotrophic mutants, that the radical resides on the common amino acids except for histidine, aspartate, and asparagine. The identity of this novel radical is eagerly awaited. The possibility must be considered that some unique cofactor instead of a protein residue might be the source of this radical.

While the details of the activation process remain obscure, Knappe has postulated that an Fe-adenosyl complex might be involved (Eq. 6) (46).

This proposal is supported by model chemical systems. Formation of an organometallic derivative of iron (III) has precedence in the work of Floriani & Calderazzo (50), who demonstrated that N,N'-ethylenebis(salicylideneimidato)iron can be converted to a benzyl derivative of N,N'-ethylenebis (salicylideneiminato)iron (III) [Fe(saen)R, R= Bz] by reaction with reductant Na-Hg and benzylchloride. Under a variety of conditions this molecule can decompose to produce R-R (Eq. 7), which could arise via homo-

$$2Fe(saen)R \longrightarrow 2Fe(saen) + R\text{-}R \qquad 7.$$

lytic cleavage of the C-Fe bond. The detailed mechanism of product production, however, remains to be established. Additional support for homolytic cleavage of Fe-C bonds is provided by the result of early studies by Yamamoto et al (51), who showed that incubation of diethylbis(dipyridyl)iron ($FeEt_2(dipyridyl)_2$) under anaerobic conditions in the presence of alcohol or H_2O resulted in the quantitative production of ethane and Fe^{2+} (dipy)$_2$ (51). Finally, the postulated mechanism for activating enzyme (Eq. 6) predicts that 5'-deoxyadenosyl radical can generate an organic radical, a prediction with ample precedent in the AdoCbl literature. This activating system, although the details of the mechanism are at present speculative, may represent another method in addition to Fe^{2+} oxidation to form a binuclear iron center and C-Co bond homolysis, for generation of a protein radical.

Since the production of the protein radical is required for PFL to be active, it is reasonable to postulate that it is involved in catalysis. Recently Kozarich et al (47) and J. Knappe et al (personal communication) have proposed the following general mechanism for transformation of pyruvate into acetyl CoA and formate (Figure 9).

The best support for this proposal is provided by model chemistry (Eq. 8) (52), discussed in detail by Kozarich et al (47).

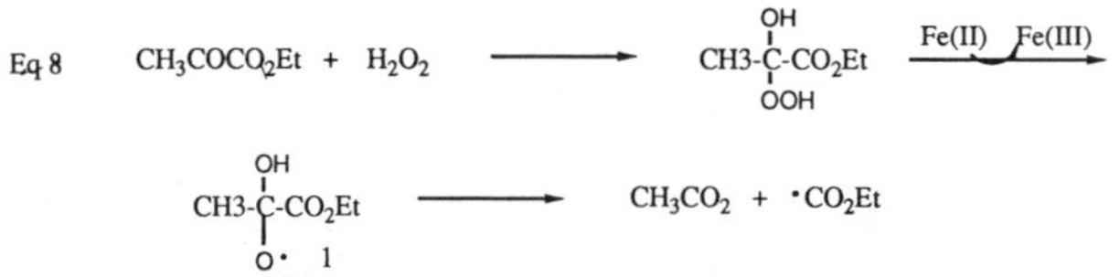

$$Eq\ 8 \quad CH_3COCO_2Et + H_2O_2 \longrightarrow CH3\text{-}\overset{OH}{\underset{OOH}{C}}\text{-}CO_2Et \xrightarrow{Fe(II)\quad Fe(III)}$$

$$CH3\text{-}\overset{OH}{\underset{O\cdot\ 1}{C}}\text{-}CO_2Et \longrightarrow CH_3CO_2 + \ ^\bullet CO_2Et \qquad\qquad 8.$$

Subsequent to nucleophilic attack of peroxide on the ketone of ethyl pyruvate and homolytic cleavage of the peroxide bond in a Fenton-type process, an oxy radical is produced, which decomposes to ethyl formate radical and acetate.

Intermediate 1 (Eq. 8) is analogous to the intermediate proposed in the PFL reaction produced either by addition of thiyl radical to the ketone of pyruvate

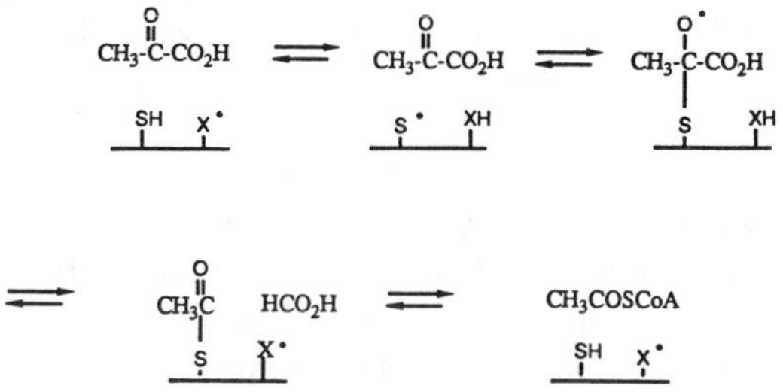

Figure 9 Proposed mechanism for pyruvate formate lyase

or by an addition of thiolate to ketone of pyruvate followed by oxidation of the thiohemiacetal of pyruvate (Figure 9). Collapse of this intermediate yields the well-established acetyl thioenzyme and the putative "formate radical." The latter would reabstract a hydrogen atom from the protein residue to regenerate the protein radical and produce formate.

Recent results of Knappe et al indicate that incubation of PFL with [^{32}P] hypophosphite ($H_2PO_2^-$), a formate analogue, results in destruction of the protein radical, loss of enzymatic activity, and covalent modification of cysteine 418 at the active site. The detailed reaction mechanism of normal substrate turnover and hypophosphite inactivation are under intensive study.

VI. Prostaglandin H Synthetase

Prostaglandin H synthetase (PGHS) catalyzes the first two steps in the biosynthesis of prostaglandins, prostacyclins, and thromboxanes: oxidation of arachidonic acid (AA) to produce 9,11 endoperoxide C-15 peroxide, PGG$_2$, and its subsequent reduction to the C-15 alcohol (Eq. 9).

The enzyme is a dimer of 72 kd that contains a single protoporphyrin IX prosthetic group required for both the cyclooxygenase and peroxidase activities (53).

The mechanism for the cyclooxygenase reaction was postulated by Hamberg & Samuelsson to involve radical intermediates, the first step being hydrogen atom abstraction from the allylic C-13 carbon of AA (54). An alternative first step involves general base-catalyzed production of the C-13 carbanion of AA with subsequent electron transfer reducing protoporphyrin Fe(III) to Fe(II) and producing the C-13 radical. However, this latter mechanistic option has been eliminated because under anaerobic conditions incubation of AA with PGHS failed to convert the EPR-active Fe(III) signal to an EPR-silent Fe(II) (55). Therefore, the first mechanistic option raises a number of interesting questions: 1. how is the hydrogen atom removed? and 2. how can evidence be obtained to support the existence of a radical intermediate subsequent to hydrogen atom removal?

Recent studies by Karthein et al have addressed the first question and formulated the following hypothesis (Figure 10), which is consistent with the observations of Kulmacz et al that lipid peroxides are obligatory feedback activators of the cyclooxygenase activity and, in addition, provide a link between the enzymatic cyclooxygenase and peroxidase activities (55, 57, 58). In this proposed mechanism, a peroxide is required to generate an iron IV porphyrin π cation radical (Compound I), which is then reduced by a protein residue to form Fe(IV) porphyrin and an amino acid radical (Compound II). The amino acid radical is proposed to abstract a hydrogen atom from AA. This proposal, based on the postulated mechanism for ribonucleotide reductases and the extensive electronic and EPR literature of intermediates produced in peroxidase reactions (cytochrome c, horseradish, chloroperoxidases), has stimulated a series of spectroscopic investigations of the interaction of PGHS with peroxides (56, 59, 60).

Stopped-flow kinetic studies and low-temperature EPR studies have indicated that incubation of PGHS with PGG$_2$ results in rapid production (4 × 10^7 M^{-1}s^{-1} at 1°C) of compound I (55, 56). The spectral properties of I are

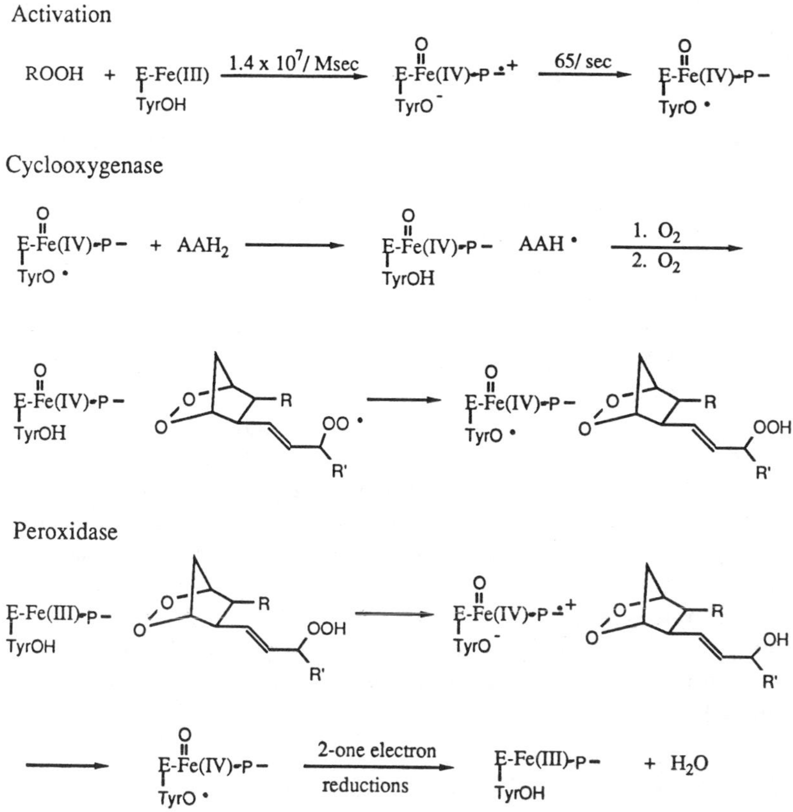

Figure 10   Proposed mechanism of the cyclooxygenase activity of PGHS.   –P– is protoporphyrin IX

consistent with the Fe(IV) = O. Spectroscopically the rapid conversion of compound I to II (65 sec^{-1} at 1°C) is detected. Compound II has visible spectral properties similar to the ES complex of cytochrome C peroxidase and consistent with an iron (IV) porphyrin (90). In addition, compound II has a doublet EPR signal at g = 2.005 (22 gauss splitting). This signal has been assigned to a tyrosyl radical cation[1] since it is very similar to that reported by Sjöberg et al for the *E. coli* reductase (43). No absorption at 410 nm is observed as the Soret band of the heme has a 30-fold higher extinction coefficient than phenoxy radicals. The putative tyrosyl radical is produced transiently and in a kinetically competent fashion (k = 65 sec $^{-1}$ at 1°C, and the turnover number of PGHS is 260 sec^{-1} at 30°C).

This mechanistic proposal (Figure 10) presents an alternative method of generating protein radicals, which on closer inspection is in many ways analogous to a proposal made for the generation (in vivo) of the tyrosyl radical and binuclear nonheme iron center in ribonucleotide reductase (9, 61–63). In both cases (Figure 11), subsequent to ROOH or O_2-mediated activation, a ferryl species is produced. In the reductase case an additional reducing equivalent is required to generate the binuclear iron center and remove the potential reactive ferryl moiety. In the case of PGHS, the "activated" ferryl species is present during multiple turnovers. This aspect of the mechanism is unappealing; however, it also provides a reasonable explanation for how PGHS might mediate its own self destruction.

The second prediction made by the postulated mechanism in Figure 10 is that carbon-centered radicals are produced in both the cyclooxygenase and peroxidase reactions. Recent spin-trapping experiments using 2-methyl-2-nitrosopropane (NtB) and PGHS reconstituted with Mn^{2+} protoporphyrin IX[2] have shown that a carbon-centered radical is produced, similar to that reported in crude extracts of rat seminal vesicles (64, 65). A second radical species present in much lower concentrations has not been further characterized. The loss of β-hydrogen coupling when octadeuterated (5, 6, 8, 9, 11, 12, 14, 15) arachidonic acid replaced protonated-AA in the experiment is indicative of a carbon-centered AA radical. Indomethacin, a potent inhibitor of cyclooxygenase activity, also inhibits production of the EPR species. Unfortunately, the appearance of this EPR signal is observed only after many cycles of the normal cyclooxygenase reaction are completed. While a hypothesis has been put forth by Mason et al to account for the observed lag time for EPR signal production, the kinetic competence of the radical species remains to be satisfactorily demonstrated.

[1]Phenol radical cations have pka's ~5 and therefore it is more reasonable that the observed signal is due to a phenoxy radical.

[2]Mn protoporphyrin supports cyclooxygenase activity and not peroxidase activity.

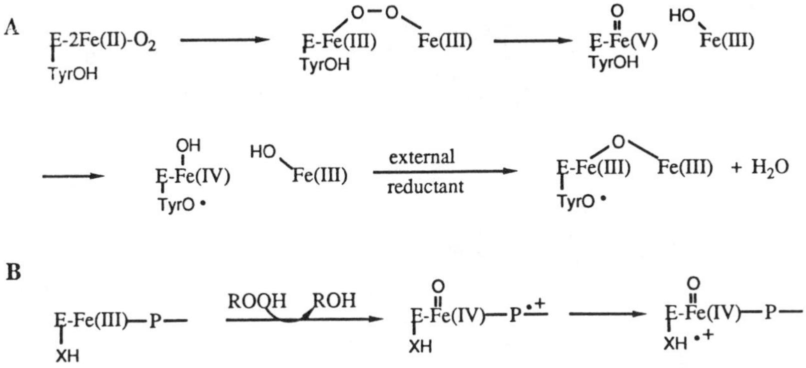

A

$$F\text{-}2Fe(II)\text{-}O_2 \longrightarrow F\text{-}Fe(III) \overset{O-O}{\frown} Fe(III) \longrightarrow F\text{-}Fe(V) \overset{O}{\underset{\parallel}{}} \overset{HO}{\searrow} Fe(III)$$
$$\text{TyrOH} \qquad\qquad \text{TyrOH} \qquad\qquad\qquad \text{TyrOH}$$

$$\longrightarrow F\text{-}Fe(IV) \overset{OH}{\underset{}{}} \overset{HO}{\searrow} Fe(III) \xrightarrow[\text{reductant}]{\text{external}} F\text{-}Fe(III) \overset{O}{\frown} Fe(III) + H_2O$$
$$\text{TyrO}\bullet \qquad\qquad\qquad\qquad\qquad \text{TyrO}\bullet$$

B

$$F\text{-}Fe(III)\text{—}P\text{—} \xrightarrow{\text{ROOH} \quad \text{ROH}} F\text{-}\overset{O}{\underset{\parallel}{Fe}}(IV)\text{—}P\overset{\bullet+}{\underline{}} \longrightarrow F\text{-}\overset{O}{\underset{\parallel}{Fe}}(IV)\text{—}P\text{—}$$
$$\text{XH} \qquad\qquad\qquad\qquad \text{XH} \qquad\qquad\qquad \text{XH}\,\bullet+$$

Figure 11   Postulated mechanism for generation of the tyrosyl radical: *(a)* in RDPR and *(b)* in PGHS

VII. Lipoxygenases

Lipoxygenases catalyze the allylic dioxygenation of fatty acids. For recent reviews concerning the chemical mechanism of lipoxygenation, see Veldink et al and Corey (66–68). Soybean lipoxygenase-1 (SBLO), the prototype for this class of proteins, catalyzes the conversion of linoleic acid to 13-S-hydroperoxy-*cis*-9-*trans*-11-octadecadienoic acid (Eq. 10). SBLO is 100 kd in size and as isolated contains one atom of nonheme high-spin iron (II) per molecule and is catalytically inactive. Addition of one equivalent of ROOH generates an active protein and produces a high-spin FeIII center that pos-

$$H_{11}C_5 \diagup\!\!\diagdown\!\!\diagup\!\!\diagdown C_7H_{14}CO_2H \longrightarrow \overset{HOO}{H_{11}C_5\diagup}\!\!\diagdown\!\!\diagup\!\!\diagdown C_7H_{14}CO_2H \qquad 10.$$

sesses an EPR $g = 6.0$ signal and a weak electronic absorption spectrum at 320 nm. Because of the importance of lipoxygenation to the biosynthesis of the physiologically important prostaglandins and leukotrienes, the mechanism of this class of enzyme has been studied extensively; a number of possibilities have been considered as outlined in recent papers by Corey and coworkers (68, 69) (Figure 12).

Case 1 (Figure 12), suggested by de Groot and coworkers, includes a self-activation in which the FeIII protein is converted to an FeII X·, where X· is a protein radical capable of mediating a hydrogen atom abstraction (71). Case 2 involves proton abstraction by a general base on the protein, which occurs concomitantly with electron transfer to the Fe(III) center on the protein. Case 3, recently championed by Corey and coworkers, postulates deprotonation and electrophilic addition to FeIII to produce an organo iron intermediate (69, 70). Subsequent homolytic Fe-C bond cleavage or direct

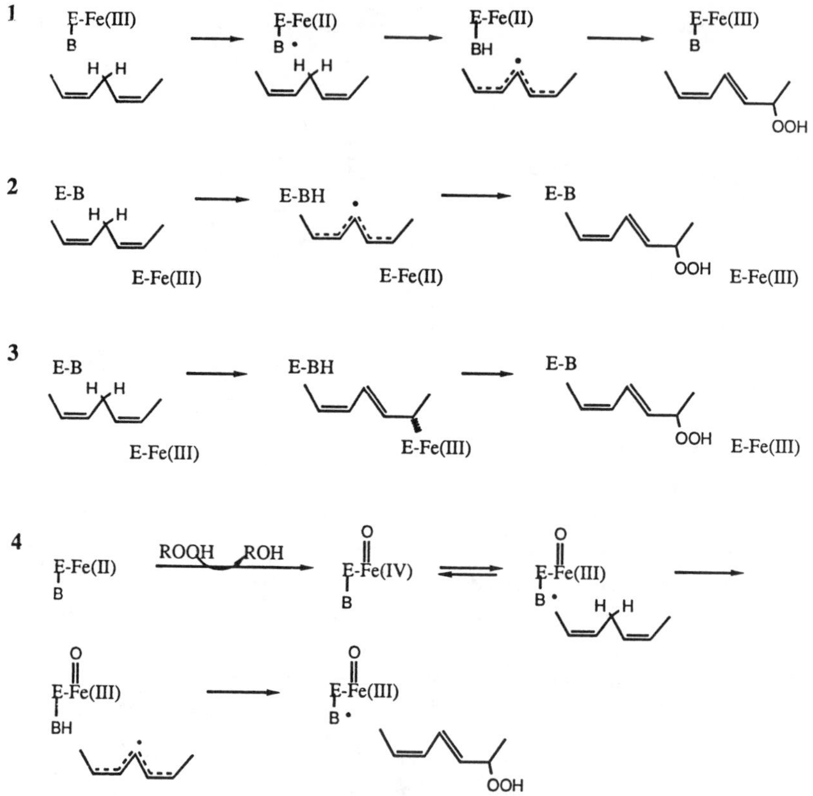

Figure 12   Proposed mechanisms for soybean lipoxygenase

insertion of O_2 would result in product production. The subtle mechanistic distinctions between options 2 and 3 (Figure 12) are reminiscent of those appearing in the literature for the last 20 years in AdoCbl-dependent chemistry. Case 4 might also be considered based on the recent postulate of Dietz et al (56) concerning the mechanism of PGHS outlined above. In this case heterolytic cleavage of the O-O bond of the activating peroxide, rather than the homolytic cleavage in Cases 1, 2, and 3, would result in production of an Fe(IV) species, which would be rapidly reduced by an amino acid residue to form the Fe(III) X· (protein radical lipoxygenase).

 This proposal (Case 4) is testable by identifying the products produced when one equivalent of ROOH is used to generate active lipoxygenase and by careful redox titration of the activated protein to determine if one or two reducing equivalents are required to produce the Fe(II) SBLO. Case 4 may be inconsistent with several reports in the literature. The first is the absence of an organic radical at $g=2.000$ in the EPR spectrum of the activated protein.

Recent EPR and magnetic susceptibility data (72) are consistent with the active "yellow" form of lipoxygenase possessing a high-spin ferric center. At the high concentrations of enzyme required to do the EPR spectroscopy, however, a problem with quantitation arises, since only 45% of the protein is in the FE(III) form. The presence of an antiferromagnetically coupled radical with this iron center may in part account for the poor quantitation, although the magnetic susceptibility measurements at 10°K do not appear to show the required deviation from Curie behavior.

The second apparent inconsistency results from studies of the inactivation of the "active" yellow enzyme with linoleic acid under anaerobic conditions, which resulted in disappearance of the EPR signal and production of high-spin ferrous iron. These results would be difficult to explain by Case 4 (Figure 12), as Fe(III) XH R· should be EPR active. However, in the absence of O_2, it is conceivable that R· could be oxidized to R+, resulting in the production of Fe(II) inactivated protein. Identification of the products produced during inactivation, that is, whether they are consistent with carbon radical or cation intermediates, would define the feasibility of Case 4 (Figure 12). Unfortunately, the vast majority of product identification experiments have been carried out in the presence of both linoleic acid and the peroxide of linoleic acid—allowing cycling of the metal center, and making quantitative association of products with a defined pathway difficult. In general, the bulk of evidence now favors options 2 and 3 of Figure 12. However, while there is no evidence for protein radical involvement, Case 4 cannot be unambiguously ruled out at present.

Mechanisms 1, 2, and 4 in Figure 12 propose the intermediacy of radicals in these reactions. Spin-trapping studies by De Groot et al using anaerobic conditions, soybean lipoxygenase, linoleic acid, and the spin-trap 2-methyl-2-nitrosopropane, resulted in trapping of a linoleic acid carbon-centered radical (71). However, the intermediacy of the trapped species in the SBLO reaction has not been unambiguously demonstrated. Subsequent to these initial studies, numerous reports of spin-trapping peroxy radicals have appeared in the literature. The pitfalls of these spin-trapping methods have been clearly pointed out by recent studies of Connor et al (73). They have shown using (4-pyridyl-2-oxide)-N-t-butylnitrone 4-POBN and $^{17}O_2$ that previous reports claiming to have trapped peroxyradicals are in error and that in reality these workers had trapped a carbon-centered radical. The necessity for isolating and identifying the structure of the adducts trapped and using appropriate isotopic labeling methods is clearly demonstrated by these studies (73).

More recently, radical clock substrate analogues have been utilized in a number of systems to provide evidence for potential radical intermediates along a reaction pathway (92). Corey prepared a cyclopropyl derivative of arachidonic acid 1 (Diagram 2). Incubation of 1 with SBLO at atmospheric pressure resulted in time-dependent inactivation of **SBLO**. However, as the $[O_2]$ was increased, compound 1 was converted to numerous products: 15-lipoxygenation without ring opening (50%) of time, 11, 8 and 5-lipoxygenation without ring opening (28%), cyclopropyl ring cleavage product (14%), and 14-formyl-5,8,11,13-tetradecatetraenolic acid (7%) (68, 69). These

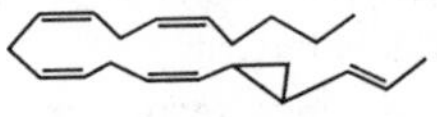

Diagram 2

results, in addition to recent model studies, have been interpreted to support mechanism 3, Figure 12. Additional studies are required to unravel the subtleties of this reaction mechanism.

VIII. Mitochondrial Amine Oxidases

Extensive reviews on flavin-dependent dehydrogenases, oxidases, and monooxygenases (74, 75) summarize a richness in chemistry, diversities in mechanism, and difficulty encountered experimentally in distinguishing among subtle alternative mechanistic pathways. Recent studies from Silverman's laboratory have provided compelling evidence for the involvement of substrate radical intermediates in the mitochondrial amine oxidase (MAO)–catalyzed reaction and have postulated a role for a protein thiyl radical in the oxidation process (76, 77).

MAO catalyzes the oxidation of biogenic amines and amines with nonacidic hydrogens. Based on extensive electrochemical literature, the mechanism shown in Figure 13 has been postulated for amine oxidation. The first step involves electron transfer to the flavin to form a flavin semiquinone radical. To accommodate the large isotope effects observed on amine oxidation and the lack of observation of a flavin semiquinone intermediate by Husain et al (78), a rapid equilibrium must exist between amine and oxidized flavin and the amine cation radical and flavin semiquinone, which lies in favor of the former. Subsequently, rate-determining proton removal followed by a second electron transfer is postulated to result in the production of imine and reduced

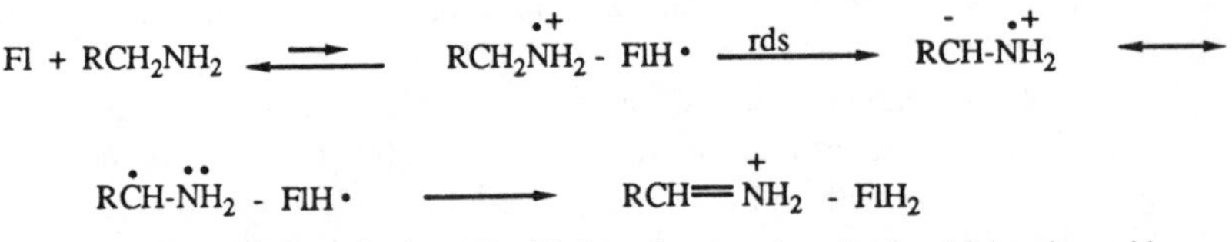

Figure 13   Postulated mechanism of oxidation of amines by mitochondrial amine oxidase

flavin. The support for this proposed scheme is based on a variety of alternate substrates that contain radical traps and studies with a number of mechanism-based inhibitors.

Compelling evidence for substrate-derived radical intermediates is provided by Silverman & Zieske's studies (79) on the interaction of MAO with 1-phenylcyclobutylamine (PCBA). PCBA is both a substrate and an inactivator with a partition rate of 325:1. The first product produced has been shown to be 2-phenyl-1-pyrroline (4) (Figure 14), and the inactivation has been shown to result from covalent modification of the flavin. The postulated sequence of events that can account for these results is shown in Figure 14. Subsequent to one-electron transfer to the flavin to form the amine cation radical 1, ring opening to 2 is postulated. This step has ample chemical precedent in the work of Maeda & Ingold (80), who have shown that N-propyl-N-cyclobutylamine radical ring can open at an extrapolated rate constant of $1.2 \times 10^5 \sec^{-1}$ (25°C). Ample chemical precedent for the cyclization of 2 to 3 also exists (81). A second one-electron oxidation would then result in production of the first observed metabolite 4. A variety of other amine analogues, including cyclopropylamines, allylamines (76, 82), (aminoalkyl)trimethylsilanes (83), and (aminoalkyl)trimethylgermines (84), in addition to the cyclobutylamines, have produced strong support for the involvement of radical intermediates in amine oxidations.

Results with a number of phenylcyclopropyl amine, mechanism-based inhibitors, have allowed Silverman to expand the mechanism shown in Figure 13 to involve a thiyl radical (Figure 15) (77).

This postulate has arisen from extensive studies of the inactivation of

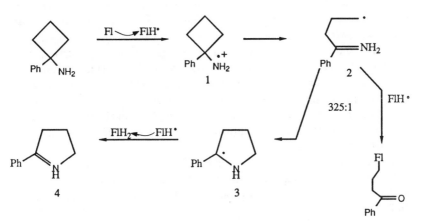

Figure 14 Reaction of PCBA with mitochondrial amine oxidase

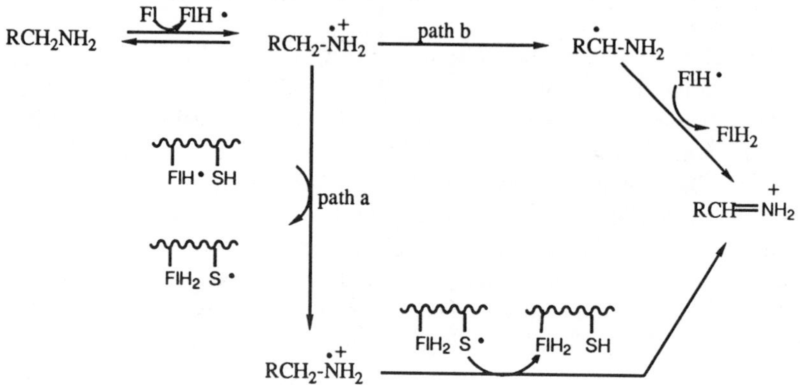

Figure 15   Expansion of the mechanism of oxidation of amines by mitochondrial amine oxidases to include an active-site thiol residue

MAO by 1-phenylcyclopropylamine (PCPA) (76). Silverman and coworkers showed that eight molecules of PCPA are required to inactivate MAO and that this inactivation is the result of two pathways: an irreversible pathway in which the N-5 position of the covalently bound flavin is modified and a reversible covalent pathway in which an amino acid residue is modified. The reversible adducts can reform active enzyme with a $t_{1/2}$ =65 min, approximately seven times faster than the rate of irreversible inactivation. Thus, the stoichiometry of eight moles PCPA per inactivation event can be rationalized. Extensive chemical modification and degradation studies revealed that the reversible adduct is the result of covalent modification of a cysteine (Figure 15). The proximity of the flavin and cysteine within the active site is demonstrated by results with a number of additional phenylcylcopropyl amine analogues, which are irreversible inhibitors of MAO. With *trans* 2-phenylcyclopropylamine, only an amino acid residue is modified, while with N-(1-methylcyclopropyl)amine, only the flavin is modified. This proximity has suggested to Silverman as indicated in Figure 15 that the flavin anion radical could be in equilibrium with a cysteine radical. The mechanism of amine oxidation (in some cases) might involve hydrogen atom abstraction (pathway a) rather than proton transfer followed by electron transfer mechanism (pathway b) (Figure 15).

   These studies and a variety of additional results from Silverman's laboratory provide compelling evidence for MAO-mediated production of substrate radicals and for the intriguing hypothesis that thiyl radicals may participate in amine oxidation.

IX. Galactose Oxidase

Galactose oxidase from *Dactylium dendroites* is a mononuclear copper protein composed of a single polypeptide of 68 kd that catalyzes the transformation shown in Eq. 11.

$$RCH_2OH + O_2 \longrightarrow RCHO + H_2O_2 \qquad\qquad 11.$$

While this enzyme has been studied intensively for years, the detailed mechanism of this oxidation and the role of the metal center have remained elusive (85). Recent studies of Whittaker & Whittaker (86) have suggested that in part, the problems associated with mechanistic elucidation are related to the heterogeneity of the native protein as isolated. They have shown that one-electron reduction of the native protein results in catalytic inactivation and an increase in the Cu^{2+} fraction of the protein that is EPR accessible (0.7 Cu^{2+}/mole protein.)[3]

On the other hand, oxidation of the native protein by one electron results in highest specific activity observed for this protein and EPR inaccessibility (<0.14 Cu^{2+}). These results, and additional spectroscopic analyses, have allowed these workers to postulate that the active form of the protein is a Cu^{2+} protein radical. The protein radical could be tyrosine, cysteine, or tryptophan. The EPR inaccessibility of the active species is proposed to be related to antiferromagnetic exchange coupling between copper and the amino acid radical. A unique mechanism (Figure 16) has been proposed for galactose oxidase–mediated oxidation.

Future studies should unravel the details of the elusive mechanism.

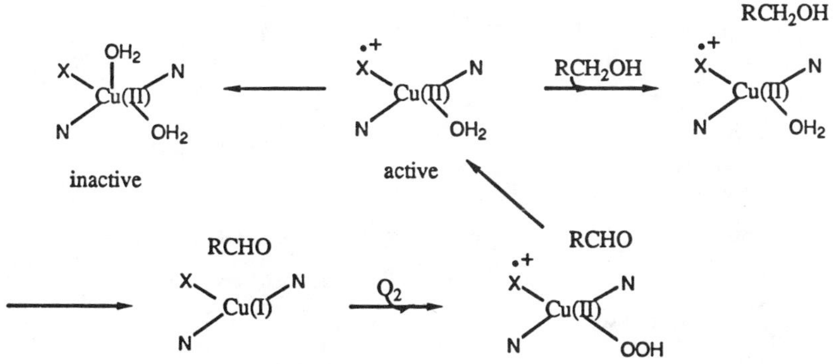

Figure 16   Postulated mechanism of galactose oxidase

[3]The native protein as isolated only contains 0.76 Cu^{2+}/mole.

284   STUBBE

Literature Cited

1. Stubbe, J. 1988. *Biochemistry* 27:3893–900
2. Williams, R. J. P. 1985. *Philos. Trans. R. Soc. London Ser. B* 311:593–603
3. Giese, B. 1986. In *Radicals in Organic Synthesis: Formation of Carbon-Carbon Bonds,* ed. J. G. Baldwin, Vol. 5. Elmsford, NY: Pergamon
4. Harriman, A. 1987. *J. Phys. Chem.* 91:6102–104
5. Reichard, P., Ehrenberg, A. 1983. *Science* 221:514–19
6. Thelander, L., Reichard, P. 1979. *Annu. Rev. Biochem.* 48:133–58
7. Gräslund, A., Sahlin, M., Sjöberg, B. M. 1985. *Environ. Health Perspect.* 64:139–49
8. Carlson, J., Fuchs, J. A., Messing, J. 1984. *Proc. Natl. Acad. Sci. USA* 81:4294–97
9. Ashley, G. W., Stubbe, J. 1985. *Pharmacol. Ther.* 30:301–29
10. Behrens, G., Koltzenburg, G., Schultz-Frohlinde, D. 1982. *Z. Naturforsch. Teil C* 37:1205–27
11. Akhlaq, M. S., von Sonntag, C. 1986. *J. Am. Chem. Soc.* 108:3542–44
12. Salowe, S. P., Stubbe, J. 1986. *J. Bacteriol.* 165:363–66
13. Sjöberg, B. M., Hahne, S., Karlsson, M., Jörnvall, H., Goransson, M., Uhlin, B. E. 1986. *J. Biol. Chem.* 261:5658–62
14. Nilsson, O., Lundquist, T., Hahne, S., Sjöberg, B. M. 1988. *Biochem. Soc. Trans.* In press
15. Larsson, A., Sjöberg, B. M. 1986. *EMBO J.* 5:2037–40
16. Thelander, L., Larsson, B., Hobbs, J., Eckstein, F. 1976. *J. Biol. Chem.* 251:1398–405
17. Salowe, S. P., Ator, M., Stubbe, J. 1987. *Biochemistry* 26:3408–12
18. Sjöberg, B. M., Gräslund, A., Eckstein, F. 1983. *J. Biol. Chem.* 258:8060–67
19. Lammers, M., Follmann, H. 1983. *Struct. Bonding (Berlin)* 54:27–91
20. Babior, B. M., Krouwer, J. S. 1979. *CRC Crit. Rev. Biochem.* 6:35–102
21. Hogenkamp, H. P. C., Ghambeer, R. K., Brownson, C., Blakley, R. L., Vitols, E. 1968. *J. Biol. Chem.* 243:799–808
22. Orme-Johnson, W. H., Beinert, H., Blakley, R. L. 1974. *J. Biol. Chem.* 249:2238–43
23. Tamao, Y., Blakley, R. L. 1973. *Biochemistry* 12:24–34
24. Stubbe, J., Ackles, P., Segal, R., Blakley, R. L. 1981. *J. Biol. Chem.* 256:4843–46
25. Ashley, G. W., Harris, G., Stubbe, J. 1986. *J. Biol. Chem.* 261:3958–64
26. Lin, A.-N. I., Ashley, G. W., Stubbe, J. 1987. *Biochemistry* 26:6905–909
27. Ashley, G. W., Harris, G., Stubbe, J. 1988. *Biochemistry.* 27:4305–10
28. Frey, P. A., Essenberg, M. K., Abeles, R. H. 1971. *J. Am. Chem. Soc.* 93:1242–51
29. Zagalak, B., Friedrich, W. eds. 1979. *Vitamin B_{12}.* Berlin/New York: de Gruyter
30. Cleland, W. W. 1982. *Crit. Rev. Biochem.* 13:385–428
31. Tan, S. L., Kopczynski, M. G., Bachovchin, W. W., Orme-Johnson, W. H., Babior, B. M. 1986. *J. Biol. Chem.* 261:3483–85
32. O'Brien, R. J., Fox, J. A., Kopczynski, M. G., Babior, B. M. 1985. *J. Biol. Chem.* 260:16131–36
33. Akhlaq, M. S., Schumann, H. P., von Sonntag, C. 1987. *Int. J. Radiat. Biol.* 51:91–102
34. Akhlaq, M. S., Al-Baghdadi, S., von Sonntag, C. 1987. *Carbohydr. Res.* 164:71–83
35. Huyser, E. S., Kellogg, R. M. 1966. *J. Org. Chem.* 31:3366–69
36. Hartmanis, M. G. N., Stadtman, T. C. 1987. *Proc. Natl. Acad. Sci. USA* 84:76–79
37. Willing, A., Follmann, H., Auling, G. 1988. *Eur. J. Biochem.* 170:603–11
38. Sheats, J. E., Czernuszewicz, R. S., Dismukes, G. C., Reingold, A., Petrouleas, V., et al. 1986. *J. Am. Chem. Soc.* 109:1435–44
39. Wieghardt, K., Pohl, K., Gebert, W. 1983. *Angew. Chem. Int. Ed.* 22:727
40. Babcock, G. T. 1987. *New Comprehensive Biochemistry: Photosynthesis,* ed. J. Amese, pp. 125–58. Amsterdam: Elsevier
41. Dismukes, G. C. 1986. *Photochem. Photobiol.* 43:99–115
42. Barry, B. A., Babcock, G. T. 1987. *Proc. Natl. Acad. Sci. USA* 84:7099–103
43. Sjöberg, B. M., Reichard, P., Gräslund, A., Ehrenberg, A. 1978. *J. Biol. Chem.* 253:6863–65
44. Dreyfus, R. J., Barry, B. A., Babcock, G. T., McIntosh, L. 1988. *Proc. Natl. Acad. Sci. USA* 85:427–30
45. Styring, S., Rutherford, A. W. 1987. *Biochemistry* 26:2401–05
46. Knappe, J., Neugebauer, F. A., Blaschkowski, H. P., Ganzler, M. 1984. *Proc. Natl. Acad. Sci. USA* 81:1332–35
47. Brush, E. J., Lipsett, K. A., Kozarich,

J. W. 1988. *Biochemistry* 27:2217–22
48. Conradt, H., Blaschkowski, H. P., Knappe, J., Hohmann-Berger, M., Hohmann, H. P. 1984. *Arch. Biochem. Biophys.* 228:133–42
49. Knappe, J. 1986. FEBS Meet. Aug. Abstr. 06.01.50
50. Floriani, C., Calderazzo, F. 1971. *J. Chem. Soc. A* 1971:3665–68
51. Yamamoto, A., Morifujii, K., Ikeda, S., Saito, T., Uchida, Y. et al. 1968. *J. Am. Chem. Soc.* 90:1878–83
52a. Heinisch, G., Lotsch, G. 1985. *Angew. Chem. Int. Ed. Eng.* 24:692–93
52b. Heinisch, G., Lotsch, G. 1985. *Tetrahedron* 41:1199–205
53. Marnett, L. J., Eling, T. E. 1983. In *Reviews in Biochemistry Toxicology*, ed. E. Hodgson, J. R. Bend, R. M. Philpot. 5:135–72. New York: Elsevier Biomedical
54. Hamberg, M., Samuelsson, B. 1967. *J. Biol. Chem.* 242:5329–35
55. Dietz, R., Karthein, R., Nastaimczyk, W., Ruf, H. H. 1988. *Eur. J. Biochem.* 171:313–20
56. Dietz, R., Nastaimczyk, W., Ruf, H. H. 1988. *Eur. J. Biochem.* 171:321–28
57. Kulmacz, R. J. 1986. *Arch. Biochem. Biophys.* 249:273–85
58. Kulmacz, R. J., Miller, J. F., Lands, W. E. 1980. *Biochem. Biophys. Res. Commun.* 918–22
59. Lambeir, A. M., Markey, C. M., Dunford, H. B., Marnett, L. J. 1985. *J. Biol. Chem.* 260:14894–96
60. Kulmacz, R. J., Tsai, A. L., Palmer, G. 1986. *J. Biol. Chem.* 262:10524–31
61. Eliasson, R., Jornvall, H., Reichard, P. 1986. *Proc. Natl. Acad. Sci. USA* 83:2373–77
62a. Fontecave, M., Gräslund, A., Reichard, P. 1987. *J. Biol. Chem.* 262:12332–36
62b. Fontecave, M., Eliasson, R., Reichard, P. 1987. *J. Biol. Chem.* 262:12325–31
63. Sanders-Loehr, J. 1988. *Oxidases & Related Redox Systems*, ed. H. S. Mason. New York: Liss
64. Mason, R. P., Kalyanaraman, B., Tainer, B. E., Eling, T. E. 1980. *J. Biol. Chem.* 255:5019–22
65. Schreiber, J., Eling, T. E., Mason, R. P. 1986. *Arch. Biochem. Biophys.* 249:126–36
66. Veldink, G. A., Vliegenthart, J. F. G. 1984. *Adv. Inorg. Biochem.* 6:139–61
67. Vliegenthart, J. F. G., Veldink, G. A.

1982. *Free Radicals Biol.* 5:29–64
68. Corey, E. J. 1986. In *Stereochemistry of Organic and Bioorganic Transformations*, W. Bartmann, K. B. Sharpless, pp. 1–12. New York: VCH
69. Corey, E. J., Nagata, R. 1987. *J. Am. Chem. Soc.* 109:8107–108
70. Corey, E. J., Walker, J. C. 1987. *J. Am. Chem. Soc.* 109:8108–109
71. deGroot, J. J. M. C., Veldink, G. A., Vliegenhardt, J. F. G., Boldingh, J., Weaver, R. 1975. *Biochem. Biophys. Acta* 377:71–79
72. Petersson, L., Slappendel, S., Feiters, M. C., Vliegenhardt, J. F. G. 1987. *Biochem. Biophys. Acta* 913:228–37
73. Connor, H. D., Fischer, V., Mason, R. P. 1986. *Biochem. Biophys. Res. Commun.* 141:614–21
74. Walsh, C. T. 1980. *Acc. Chem. Res.* 13:148–55
75. Bruice, T. C. 1980. *Acc. Chem. Res.* 13:256–62
76. Silverman, R. B., Yanasaki, R. B. 1984. *Biochemistry* 23:1322–31
77. Silverman, R. B., Zieske, P. 1986. *Biochem. Biophys. Res. Commun.* 135:154–59
78. Husain, M., Edmondson, D. E., Singer, T. P. 1982. *Biochemistry* 21:595–600
79. Silverman, R. B., Zieske, P. A. 1986. *Biochemistry* 25:341–46
80. Maeda, Y., Ingold, K. U. 1980. *J. Am. Chem. Soc.* 102:328–31
81. Tanner, D. D., Rahimi, R. M. 1979. *J. Org. Chem.* 44:1674–72
82. Vazquez, M. L., Silverman, R. B. 1985. *Biochemistry* 24:6538–43
83. Silverman, R. B., Banik, G. M. 1987. *J. Am. Chem. Soc.* 109:2219–20
84. Silverman, R. B., Vadnere, M. K. 1987. *Bioorg. Chem.* 15:328–32
85. Kosman, D. J. 1985. In *Copper Proteins and Copper Enzymes*, ed. R. Lanne, 2:1–26. Boca Raton, Fla: CRC Press
86. Whittaker, M. M., Whittaker, J. W. 1987. *J. Biol. Chem.* 263:6074–80
87. Ator, M., Salowe, S. P., Stubbe, J., Emptage, M. H., Robins, M. J. 1984. *J. Am. Chem. Soc.* 106:1886–87
88. Nilsson, O., Aberg, A., Sjöberg, B. M. 1988. *Nucleic Acids Res.* 16:4174
89. Griller, D., Kanabus-Kaminska, J. M., Maccoll, A. 1988. *J. Mol. Struct. Theochem.* 163:125–31
90. Fishel, L. A., Villafranca, J. E., Mauro, J. M., Kraut, J. 1987. *Biochemistry* 26:351–60
91. Halpern, J. 1985. *Science* 227:869–73
92. Griller, P., Ingold, K. U. 1980. *Acc. Chem. Res.* 13:317–23

Annu. Rev. Biochem. 1989. 58:287–307

MOLECULAR BIOLOGY OF ALZHEIMER'S DISEASE

Benno Müller-Hill

Institute for Genetics, University of Cologne, Weyertal 121, D-5000 Koeln 41, Federal Republic of Germany

Konrad Beyreuther

Center for Molecular Biology, University of Heidelberg, Im Neuenheimer Feld 282, D-6900 Heidelberg 1, Federal Republic of Germany

CONTENTS

SUMMARY AND PERSPECTIVES

The amyloid first seen by Alois Alzheimer in the brain of a deceased demented patient (1) turned out to be an extraordinary challenge for biochemists and molecular biologists alike. The extreme insolubility of this polypeptide first challenged the protein chemists attempting to sequence it (2, 3). Then the questions arose: Was the amyloid the product of a slow virus (4) or was it the product of a human gene? If the latter, did the structure of its

287

0066-4154/0701-0287$02.00

gene in an Alzheimer patient differ significantly from its structure in a normal person? Was the amyloid the direct cause of Alzheimer's disease?

The techniques of molecular biology have apparently produced unambiguous answers to these questions within a very few years. Amyloid is the breakdown product of a large protein ubiquitously produced by a human gene, and there is no significant difference between the gene of an Alzheimer patient and the gene of a normal person. This moves the PAD gene (*p*recursor of *A*lzheimer's *d*isease A4 amyloid gene) out of the limelight into the rank and file. Yet, there may be surprises in store, if the hint that it is rather close to the primary cause of the disease turns out to be right.

MOLECULAR PATHOLOGY AND BIOCHEMISTRY OF AMYLOID DEPOSITS

Alzheimer's disease results in a progressive loss of intellectual function and appears to occur both sporadically and in an autosomal dominant (familial) form (5–8). Alzheimer's disease affects more than 5% of the population above the age of 65 years. In the United States, more than two million people suffer from this degenerative disorder of the central nervous system (6). The brains of Down's patients who grow to adulthood degenerate in much the same way as those of Alzheimer's patients (9–11).

One of the key features and the major diagnostic lesion of Alzheimer's disease is the deposits of abnormal amyloid proteins in intracellular and extracellular locations (1, 12, 13). The cellular dysfunction and death that eventually result from these deposits are common consequences of diseases termed amyloidosis, which are characterized by the deposition of abnormal fibrillar proteins in the extracellular and intracellular spaces (20, 23; see below).

In Alzheimer's disease and adult Down's syndrome, proteinaceous deposits accumulate in the form of abnormal fibers within the perikaryon of neurons and in degenerating nerve cell processes. Their large aggregates found in neuronal cell bodies are called neurofibrillary tangles (1). Extracellular protein filaments are found in two locations (1, 12, 13). The first includes the neuritic plaque or senile plaque, and the amyloid plaque. Plaques are spherical lesions of 10–200μm in diameter that have in common extracellular deposits of 6–10 nm protein filaments. The typical neuritic or senile plaque has a central amyloid core of abnormal protein filaments surrounded by degenerating and frequently swollen nerve terminals. The number of plaques and neurofibrillary tangle–bearing neurons in the cerebral cortex shows a correlation with appearance and severity of dementia (14). Thus a detailed knowledge of the proteins in the amyloid deposits, the genes that code for them, and the tissues of amyloid origin will enable the pathogenic mechanism to be determined and should help to understand the causes of neuronal death in Alzheimer's disease. The second extracellular site for deposition of protein filaments in Alzheimer's disease and adult Down's syndrome is the walls of intracortical and meningeal blood vessels, although this is a less frequent deposition than plaques and neurofibrillary tangles (12, 13).

The intracellular and extracellular protein deposits in Alzheimer's disease have the tinctorial property of green birefringence under polarized light following Congo red staining and are therefore all called "amyloid" (15). This generic term, which has the meaning of starchlike or celluloselike, was coined in 1854 by Virchow, who believed in a polysaccharide composition of amyloid on the basis of its iodine staining (16, 17). But the proteinaceous nature of amyloid became immediately apparent in 1859 when Friedreich & Kekulé published the result of their chemical analysis of amyloid (18). The term "amyloid" is now generally used to define aggregates of 6–10 nm protein filaments that share the following properties: Congophilia and green birefringence under polarized light, β-pleated sheet secondary structure and cross-β X-ray diffraction pattern, and a fibrillar quaternary structure that is seen in the electron microscope to consist of 6–10 nm protein filaments (15, 19, 20). In addition, all amyloids are highly insoluble under physiological condition, and amyloid fibrils can therefore be separated from soluble proteins by differential centrifugation.

Proteins that can aggregate into amyloid fibers are termed "amyloidogenic." Amyloid can be deposited in a variety of organs as the result of a number of different disease processes, and is known to cause cellular dysfunction and death (2, 3, 20–32). The surprising biochemical diversity of amyloidogenic proteins includes at present nine different biochemical classes that can be formed into amyloid fibers as a consequence of increased local concentration, as variants or as proteolytic fragments (2, 3, 20–32). Protein AL, which is of immuno-globulin light chain origin, was identified as the amyloid fibril protein of myeloma-associated primary amyloidosis (20, 21). Serum amyloid A (SAA) is an apoprotein of high-density lipoprotein (HDL) and presumed to be the precursor for amyloid A protein, the main protein component of amyloid fibrils in reactive amyloidosis (22). Prealbumin variants constitute AF amyloid fibrils of familial amyloidotic polyneuropathy and one type of senile cardiac amyloidosis (23). Precalcitonin is the amyloid subunit of medullar carcinoma of the thyroid (24). Gamma trace, a variant of cystatin C that is an inhibitor of cysteine proteases, forms amyloid deposits in hereditary cerebral hemorrhage with amyloidosis of Icelandic type (HCHWA-I) (25), and β_2-microglobulin is the monomer in hemodialysis-associated amyloid (26). Insulinoma amyloid proteins are homologous to the calcitonin gene–related peptide (27). The amyloid subunit associated with unconventional virus infections is the Protease-resistant Protein (PrP), which is found in

Creutzfeldt-Jakob Disease, Gerstmann-Sträussler syndrome, kuru, and scrapie (28, 29). In Alzheimer's disease and adult Down's syndrome, amyloid A4 protein (3, 30, 31), also termed β protein (2, 32), is the amyloid subunit. Although there is no obvious homology between the known amyloidogenic proteins. it appears evident that common principles govern the mechanisms of amyloid formation. Creation of amyloid fibrils in vitro from the known amyloidogenic protein precursor was successful for amyloid AL by proteolysis (33) and for β_2-microglobulin by concentration (34). This suggests the amyloid fibrils formed are homopolymers of the characteristic protein, with no requirement for other subunits. Furthermore, amyloidogenic primary structures that can adopt a β-pleated sheet structure and aggregate to fibrils can be generated in vitro from the precursors in two different ways. These hold up as an example to the two basic mechanisms discussed for amyloidosis: increasing the concentration if the precursor molecule itself forms the amyloid fibril, or proteolysis if only a proteolytic fragment of the precursor assembles into a fibrillar quaternary structure. The latter mechanism is also relevant for Alzheimer's disease, as we discuss in detail in subsequent sections. In addition, as has become apparent over the last years, several amyloid-associated substances may also be relevant for the process of amyloid formation in Alzheimer's disease: amyloid P component (an elastase inhibitor in vitro) and other protease inhibitors (35, 36), glycosaminoglycans (37), aluminum silicates or other inorganic components (30, 38), ubiquitin (39, 40), tau (41–49), and possibly as-yet-unidentified factors, such as proteases. The mechanism by which amyloid-related factors function, as well as whether they have a function in amyloidogenesis at all, is unclear and remains largely unexplored. The identification of the known components of the three types of proteinaceous lesions in Alzheimer's disease is discussed in detail below.

Neurofibrillary Tangles

Neurofibrillary tangles are intraneuronal deposits found mainly within cell bodies of pyramidal neurons. It has been pointed out that tangles in the hippocampus are localized to neurons that connect the hippocampus to other areas of the cortex, basal forebrain, thalamus, and hypothalamus (50), thus implicating tangles in a local pathology, which may account for the memory disorder of Alzheimer's disease (1, 5, 6). Therefore the analysis of the molecular nature and structure of neurofibrillary tangles may reveal how they are produced and may contribute to the understanding of the degenerating processes that underlie Alzheimer's disease.

Neurofibrillary tangles can be visualized and quantified in the perikarya of affected neurons, in neurites surrounding a plaque, and as argyrophilic threads in the neuropil with argentic impregnation techniques, Congo red staining showing the typical green birefringence, or fluorescence microscopy with thioflavin-S (1, 15, 51, 52). The fine structure of tangles was first described by Kidd (53) and Terry (54); electron microscopy showed them to consist of pairs of approximately 10 nm protein fibers helically twisted about each other with a cross-over roughly every 80 nm. These paired helical filaments (PHF) are wound in a left-handed manner and apparently consist of transversely oriented subunits arranged in a double-helical stack (55, 56). The subunits have a molecular mass of approximately 100 kd and a three-domain core substructure. X-ray diffraction patterns from human brain fractions enriched in PHF indicated a cross-β conformation, since the predominant X-ray scatter was a sharp reflection at 4.76-Å spacing and a diffuse one at about 10.6-Å spacing (57). These features are typical of β-pleated sheet structures and were also found for other classes of amyloid proteins formed in vivo or in vitro (19, 20, 23, 33). The proposed cross-β conformation of PHF and the aforementioned transversely oriented three-domain core subunit structure of PHF make it most unlikely that the filaments are formed by a simple collapse of normal cytoskeletal elements, such as neurofilaments.

Two additional approaches have been used for the characterization of the composition of PHF by several groups. One approach to studying the molecular nature of PHF was based on immunologic cross-reactivity, since the insolubility of neurofibrillary tangles has created great difficulty in purifying PHF and isolating their polypeptides. The other approach has been to develop methods for the solubilization and purification of PHFs, the major tangle constituent, a prerequisite to their biochemical analyses. Immunological studies have identified candidate proteins that were also found in tangle isolates by chemical methods, such as amyloid A4 protein (β protein) (30, 31, 77), ubiquitin (39, 40), and the microtubule-associated protein phosphoprotein tau (41–49). Other tangle constituents identified on the basis of cross-reactivity were not reported to be present in PHF isolates analyzed by biochemical methods, such as the 160– and 200–kd neurofilament subunits (58–60) and the microtubule-associated protein 2 (MAP-2) (61–63). An apparent explanation relating these findings came from a re-examination of some of these monoclonal antibodies that had been reported to bind to neurofibrillary tangles. All of the antibodies included in the study were found to cross-react with phosphorylated microtubule-associated protein tau but no longer with isolated tangles after dephosphorylation treatment (64, 65).

The biochemical composition of neurofibrillary tangles has been approached by several groups. Two groups used rigorous purification procedures including protease treatments to remove molecules that are not an integral part of the tangle and only associated with or decorating the paired helical filaments (30, 31, 47, 66). Electron microscopy of negatively stained tangle isolates showed indeed the association of PHF with an amorphous outer covering, which could be removed by protease treatment, leaving the three-domain structure of PHFs intact (47). This amorphous material was shown to contain ubiquitin sequences and tau epitopes, which disappeared upon pronase digestion (47). Although the protease-resistant core was reported to contain sequences originating from the three repeating C-terminal segments of tau, which apparently remained intimately associated with PHF, the identity of the major repeating subunit of PHF was suggested not to be derived from tau (47–49).

The issue whether PHF is made up of amyloid A4 protein (β protein) also remains controversial (30, 31, 47–49, 66). The isolation of amyloid A4 protein (β protein) after formic acid extractions has been reported for tangle preparations, including a pepsin digestion step in the purification protocols (30, 31, 66). When the N-terminal sequences of the purified A4 polypeptides were determined, they were shown to be identical to that of A4 polypeptides extracted from purified amyloid plaques of Alzheimer's and Down's patients apart from the extent of N-terminal heterogeneity (3, 30, 31). The sequence of these A4 proteins from the intracellular and extracellular amyloid source is the same as that determined by Glenner & Wong for β protein isolated from meningeal blood vessels of both Alzheimer's and Down's patients (2, 32). The N-terminal heterogeneity encouraged Masters et al to speculate on a possible chronological sequence of plaque and tangle formation (30). The greater degree of heterogeneity was suggested to reflect a greater age of the A4 protein (β protein) monomer in neurofibrillary tangles.

PHF could thus be among the first structures to be formed from A4, which was presumed to be derived from a larger precursor protein of neuronal origin (30). Later, newly produced amyloid A4 precursor proteins or fragments thereof would spill into the extracellular space to form amyloid plaque cores. Finally the excess A4 precursor would reach the blood vessels, giving rise to the youngest amyloid (30). The claim that amyloid A4 protein (β protein) is a structural component of PHF has been challenged with the argument of the usual cross-contamination between neurofibrillary tangles, amyloid cores, and vascular amyloid and insolubility of PHF (67). However, several groups have reported amino acid compositions of neurofibrillary tangles and molecular weight determinations that are similar to the 4kd-amyloid A4 protein (β protein) (30, 31, 66, 68–70). Nevertheless, the ultrastructural difference between paired helical filaments and the typical 6–10 nm straight amyloid fibril (see below), and the difficulty in staining extracellular amyloid deposits with antibodies raised against PHF and vice versa, remain largely unexplored (77).

It is still tempting to consider the possibility that PHF and plaque amyloid are derived from the same amyloidogenic precursor protein as originally postulated by Divry (71) and recently by Masters et al and Guiroy et al (3, 30, 31, 66). Isolation of amyloid A4 protein (β protein) from the neurofibrillary tangles present in Alzheimer's disease (30), adult Down's syndrome (31), and Parkinson-dementia of Guam (66) support this concept and are indirect evidence that amyloid A4 protein (β-protein) is a structural component of PHF. Such a concept is compatible with part of tau protein constituting a component of the core of PHF, and tau playing a role in the de novo polymerization of PHF, since fibrils with a similar morphology of PHF have been obtained from purified tau protein (72). Accordingly, the difficulty in demonstrating cross-reactivity between PHF and extracellular amyloid deposits in Alzheimer's disease could reflect masked epitopes in intracellular amyloid deposits due to tau binding (47, 77).

Amyloid Plaque Cores

The fully developed neuritic plaques are spherical lesions, classically consisting of an extracellular central core of closely packed, radiating 6–10 nm amyloid fibrils, surrounded by a rim of swollen degenerating neuritic processes filled with paired helical filaments, dense bodies, multilamellar bodies, and vesicles (73–76). This collection of degenerating neurites is frequently intertwined with glial processes. Plaques made up almost exclusively of amyloid material are called "amyloid plaques." There is some evidence that the gradual process of infiltration, condensation, and finally crystallization of amyloidogenic protein subunits is the basic process underlying the pathogenesis of the plaque (77). Thus the "amorphous" plaque cores first defined by von Braunmühl (78) may represent the phase of condensation of amyloid subunits. These amorphous amyloid plaque cores seem to be more prevalent in relatively younger patients with Down's syndrome than in those with Alzheimer's disease (3, 79).

Plaque formation is not limited to the presence of neurofibrillary tangles. The remarkable observations on the plaques in the cortex of aged mammals, such as monkey, orangutan, polar bear, and dog, show clearly for the first time that "neuritic" plaques can develop in the absence of neurofibrillary tangles (80, 81).

Biochemical studies have revealed that the plaque core protein in both Alzheimer's disease and Down's syndrome is formed from a 4500-dalton protein. This protein was designated as amyloid A4, which refers to the molecular mass in kd (3). The sequence of amyloid A4 protein was identical to the sequence of the β protein isolated by Glenner & Wong from meningeal amyloid of postmortem brains from patients with Down's syndrome (32) and Alzheimer's disease (2) with one difference: amyloid A4 proteins and β protein from Down's individuals had glutamic acid at position 11 (3, 32), whereas glutamine was reported at this position for the β protein from Alzheimer cases (2). The cDNA sequence has the codon for glutamic acid at this position (82, 96–98). As has been pointed out before, it is likely that the paired helical filaments of neurofibrillary tangles are also a different form of this same amyloid A4 protein (β protein) (30, 31, 66, 68, 70).

The basic features of the amyloid A4 protein from amyloid plaque cores may be summarized as follows. The full-length protein consists of only 42 to 43 residues (82). The A4 monomer readily aggregates into dimers (9 kd), tetramers (18 kd), and higher oligomeric species. The solubility of the A4 fibrils in both amyloid plaque cores and neurofibrillary tangle preparations is altered dramatically after treatment of the native amyloid fibrils with concentrated formic acid. The aggregational properties of both the native A4 monomer and the synthetic peptides corresponding to the 42/43 residues in the A4 sequence show a dependence on pH, concentration, and ionic strength. So far, no evidence has been obtained for the presence of carbohydrate or the effect of any other form of posttranslational modification other than proteolysis. Sequencing of amyloid A4 protein extracted with formic acid from amyloid plaque core and tangle preparations has proved difficult because of the presence of ragged N-termini. It was shown that this raggedness is not the result of the pepsin digestion step included in the original purification procedures of the tangles and amyloid plaque cores (3, 30), since ragged N-termini were also found for fibrils isolated by protease-free methods (77). The latter notion and some of the basic features of the A4 protein (β protein) have been confirmed in other laboratories, mainly on the basis of amino acid composition and molecular weight determinations (67–70, 83–85).

Derivation of the protein sequence of amyloid A4 protein has allowed comparison of antisera to the native and synthetic A4 molecules, the mapping of relevant epitopes, and the development of immunochemical assays and purification procedures. A major epitope for amyloid plaque cores was identified between residues 10 and 23 on the A4 molecule (3, 83, 85), and a minor epitope for neurofibrillary tangles was found in the N-terminal region (30). The latter epitope is exposed only on a subpopulation of all tangles and found only in tangles in the cell soma in neocortical areas of the brain (77). Antibodies raised against the amino-terminal 28 residues of amyloid A4 protein (β protein) recognized the cortical and microvascular amyloid of aged mammals with brain amyloid, suggesting conservation of amyloid A4 protein (81).

Biophysical approaches to the structure of the A4 filaments have been performed using synthetic peptides of different lengths homologous to the A4 protein (β protein). Synthetic fibrils displaying the typical fibrillar electron microscopic appearance and Congo red staining with green birefringence were reported to be spontaneously formed by synthetic peptides corresponding to residues 1–28, 12–28, and 15–28 of the A4 protein (β protein) sequence (23, 86). X-ray diffraction from fibrils formed by the synthetic peptide homologous to residues 1–28 of the A4 protein (β protein) indicates a cross-β-conformation (87), which means that the fibrils consist of β-sheets with peptide chains running approximately perpendicular to the fibril axis. The same conformation is also present in in vivo amyloid of plaque cores (57).

How many years are required for the formation of an amyloid plaque? The age of amyloid A4 protein in amyloid plaque cores isolated from postmortem brains of patients with Alzheimer's disease was recently judged by the rate of racemization of the amino acids aspartic acid and asparagine (88). On the assumption that approximately 0.15% D-aspartate are formed per year (89), the 4.57% D-aspartate of amyloid of plaque core correspond to approximately 30 years, the time it takes to fully develop these lesions. Immunocytochemistry on population-based surveys of amyloid A4 protein deposition also estimated 30 years for the age of the amyloid (90).

What do we know of the other constitutents of amyloid plaques? The biochemical data on the composition of the major protein species (3) suggest that with the exception of the inorganic components (30, 38), other contenders are at best minor components or the result of the postmortem leakage of serum proteins into the brain parenchyma. However, as discussed before, some of the known amyloid-associated constituents may be of relevance for amyloid fibril and plaque formation. They may act as accelerators or inhibitors of amyloidosis, such as amyloid P component, protease inhibitors, glycosaminoglycans, inorganic components, and proteases (35–38).

Cerebrovascular Amyloid

In more than 90% of the cases of Alzheimer's disease, extracellular amyloid fibrils similar to those of neuritic plaques also occur within the vessel walls of the leptomeninges and cerebral cortex (91). This amyloid Congophilic angiopathy affects mainly the media and adventitita of small veins and arteries. No significant correlation was found between the amount of vascular amyloid and the clinical severity of the dementia in Alzheimer's disease (92).

Among the seven known pathological conditions in which amyloid depositions consistently occur in vascular walls, four are localized only to the cerebral vessels: Alzheimer's disease (12, 13), adult Down's syndrome (9–11), hereditary cerebral hemorrhage with amyloidosis of Dutch type (HCHWA-D) (93, 94), and hereditary cerebral hemorrhage with amyloidosis of Icelandic type (HCHWA-I) (25). The first cerebral amyloid protein to be biochemically characterized was gamma-trace, a variant of cystatin C that is an inhibitor of cysteine proteases. This amyloid is found in HCHWA-I, an autosomal dominant form of cerebral amyloidosis identified within families in Iceland (25). HCHWA-I patients usually die before the age of 40 years due to massive intracerebral hemorrhages. An amino acid substitution of glutamine for leucine at position 68 of cystatin C, a 13-kd protein related to kininogens and stefins that is present in human brain, thyroid, adrenal gland, and pancreas, may represent a point mutation in these families (25) and may lead to an abnormal, amyloidogenic protein that cannot be normally processed. But the possibility that the substitution is harmless cannot be excluded for this and other variant proteins identified as major components in familial amyloidosis (23, 25).

The second cerebrovascular amyloid protein biochemically identified was the major protein of the amyloid of Congophilic angiopathy in Alzheimer's disease and aged Down's syndrome. Glenner & Wong (2, 32) decided to purify this vascular amyloid from amyloid-laden (amyloidotic) leptomeningeal tissues, which can be efficiently separated from parenchymal brain tissues, and thus avoided cross-contamination between vascular amyloid, amyloid of plaque cores, and neurofibrillary tangles. Using collagenase digestion, the major contaminant in crude homogenates of meninges could be removed (2). Amyloid fibers are resistant to collagenase treatment and maintain their tinctorial properties and folding such as Congo red staining with green birefringence and β-pleated sheet configuration as determined by X-ray diffraction (2). Amyloid was extracted with 6 M guanidine-HCl and fractionated by gel permeation chromatography. Extracts from amyloidotic vessels from Alzheimer's disease cases resulted in a peak fraction corresponding to a protein of 4.2 kd. No peak at the same position was found when age-matched control cases were examined (2). The 4.2 kd protein was further fractionated into three peaks designated $\beta1-\beta3$ by high-performance liquid chromotography. The same protocol was repeated on amyloidotic leptomeningial tissues from two cases of adult Down's syndrome (ages 61 and 62), and the same proteins were isolated (32). The three β proteins were subsequently shown to have similar amino acid compositions. The $\beta1$ and $\beta2$ proteins were also found to be homologous by N-terminal amino sequence analysis (2, 32). The initial amino acid sequence yielded 22 (2) and 24 (32) amino acids that showed no homology to known sequences. β-Protein and amyloid A4 protein extracted with formic acid from plaque core and neurofibrillary tangle preparations (3, 30, 31, 66) are similar but not identical fragments of the same amyloidogenic protein. In addition to showing that the amyloid A4 protein extracted from plaque cores and neurofibrillary tangle isolates of patients with Alzheimer's disease is the same as found in aged Down's syndrome, it was also demonstrated that the A4 protein in Down's syndrome has less ragged N-termini than in Alzheimer's disease (3, 31). Recently the N-terminal state of the β protein has also been determined and shown to be heterogeneous (95). The N-terminal amino acid sequence of amyloid protein isolated from cortical microvessels (95) of patients with Alzheimer's disease is identical to the N-terminal sequence reported for meningeal vascular amyloid (2), but exists in two forms with regard to the N-terminus. Approximately 60% of the protein has an N-terminal aspartic acid residue, and the remaining 40% of the protein an N-terminal alanine residue, the latter corresponding to residue 2 of amyloid A4 protein (β protein). Both cortical vascular amyloid proteins were eluted together from reversed-phase columns (95), and are therefore not equivalents of two of the three β proteins, $\beta 1$–$\beta 3$, described by Glenner & Wong (2, 32). The $\beta 1$–$\beta 3$ proteins may represent protein species with different C–terminals.

The amyloid subunit in HCHWA-D is similar to the β protein (A4 protein) of Alzheimer's disease and Down's syndrome (93, 94). However, the HCHWA-D amyloid protein was shown to be three residues shorter than the A4 protein and consists of 38 or 39 instead of 42 residues. In addition to N-terminal heterogeneity, with N-terminal aspartic acid and alanine (position 2) in a 3:1 ratio, the C-terminal residues 40–42 were missing (93). The C-terminal shortening appears to alter the physical properties of the fibrils and makes the amyloid protein from leptomeningal vessel walls more soluble. Whether a genetic variant of the HCHWA-D amyloid protein precursor accounts for the clinical feature of predominant vascular amyloid deposition in HCHWA-D, leading to recurrent intracerebral hemorrhages and early death in the fifth to sixth decades of life, remains to be determined.

The biochemical findings described for cerebrovascular amyloid indicate that of the four disease processes characterized by cerebrovascular amyloidosis, Alzheimer's disease, Down's syndrome, and HCHWA-D are characterized by a homologous amyloid protein, the A4 protein/β protein. The chemical differences at the N- and C-terminal of the A4 protein/β protein isolated from the different amyloid deposits may simply reflect different proteolytic processing of the amyloid precursor proteins in the vessel wall and brain parenchyma.

cDNAs CODING FOR AMYLOID A4 PRECURSORS

The availability of the N-terminal sequence of the amyloid A4 protein (β protein) (2, 3) made possible a search for the corresponding cDNA. Four groups succeeded at almost the same time (82, 96–98). The decision as to which kind of library to use seemed a gamble. Was amyloid A4 protein synthesized in brain or was it synthesized elsewhere and only deposited there? Normal postmortem procedures (taking 12–24 hours) lead to drastic degradation of mRNA. Two groups screened cDNA libraries constructed from human fetal brain mRNA (82, 98). The two others used the same commercial brain cDNA library from an adult cancer patient (96, 97). One of the fetal libraries gave a full-length clone coding for an open reading frame of 695 codons following an AUG (82). The other clones were substantially shorter (96–98). Degradation of mRNA and insufficient methylation at potential internal *Eco* RI sites during cDNA synthesis were the major reasons for this discrepancy.

The cDNA could now be used to search by RNA-blot hybridization for the presence or absence of mRNA in various tissues. It came as a surprise that the cDNA probe detected mRNA in brain, spleen, thymus, pancreas, muscle, kidney, liver, lung, small intestine, heart, and adrenal gland alike (98). This made it likely that the protein may be expressed in all tissues and that the gene could be a housekeeping gene.

The cDNA probe was also used to look for the presence of homologous mRNA or DNA in the tissues of other organisms. All four groups used human, mouse, or hamster hybrid cell lines to determine the human chromosome harboring the gene coding for the amyloid precursor. These Southern blot analyses show clear hybridization, with the mouse or hamster genes indicating close similarity in sequence (82, 96–98). Recently, the mouse and rat genes have been sequenced (99, 100). A comparison of human and mouse sequences revealed 22, of human and rat sequences 19 and of mouse and rat sequences 7 amino acid exchanges indicating strong structural constraints. Three amino acid substitutions, R to G (position 601), Y to F (position 606), and H to R (position 609) distinguish the human sequence of the amyloid A4 protein region from those of both rat and mouse. Human *PAD* (amyloid A4 precursor gene) cDNA has also been used successfully as probe for in situ analysis of the mouse brain (101).

The presence of two bands in Northern blots with mRNA from human fetal brain (82) suggested the existence of different messages, possibly arising as different splice products. Further analysis of the cDNA library constructed from mRNA from human fetal brain produced only cDNAs that were similar to the one already described: eight out of eight full-length clones were identical with the one published before (A. Unterbeck, personal communication). The situation in the rat brain cDNA library was similar. Again eight out of eight full-length clones were identical (100). Three groups, two of them not previously involved in the isolation of amyloid A4 precursor cDNA, searched various cDNA libraries and found a second (102, 103) and a third (104) additional splice product. All three groups reported that the additional splice products carried internally 56 additional residues, which code for an analogue of the Kunitz protease inhibitor. The homology between the protein sequence of this and other inhibitors of the same protein family varies between 40 and 47%. One group demonstrated that this putative protease inhibitor did indeed act in vitro as a trypsin inhibitor (104). All groups reported that the product with the extra exon was found in all tissues, but in lower amounts in some brain tissues. One group found a slightly larger insert that carries some additional codons downstream from the trypsin inhibitorlike region (104). The corresponding 19-residue sequence shows 47% homology to residues 18–36 of OX-2 antigen, a protein common to neurons and thymocytes (105).

The existence of the three different cDNAs raises the question as to whether they are really different splice products. The analysis of genomic DNA is consistent with this plausible assumption (104). A second question is whether differential splicing is disturbed in brain tissue from patients with Alzheimer's or Down's syndrome. One case of fetal Down's brain suggests that this may be so (103). Whereas RNA blot hybridization of mRNA from normal fetal brain gives only a weak signal with the trypsin inhibitor probe, it gives a strong signal with a probe lacking the inhibitor region. A corresponding Northern blot with mRNA from the brain of a Down's fetus gives a different picture. Here Tanzi et al (103) find less of the mRNA lacking the trypsin inhibitor like region but more of the species which carries it.

The question has been asked as to how much of each mRNA variant is produced by the various cell types in the various regions of human brains from patients with Alzheimer's disease and control individuals (106–111). This raises the question of differential stability of such mRNA species in the human brain. In animals like the rat (100), the mouse (101), or the monkey (108, 109), brain slices can be obtained without postmortem delay. This is prohibited by law or taboo in humans. Ideally the postmortem times have to be as short as possible. One hour (109) has already been reported. One of the authors of this review (B. Müller-Hill) has a strong aversion to this development in which postmortem times become shorter and shorter. This is clearly not a question of the best possible science but of values which have their origin outside of science. We will not consider this aspect further here, although a general discussion of this question seems necessary.

The fact that brains of adult individuals with Down's syndrome contain amyloid deposits comparable to those of Alzheimer patients (3, 9–11, 32, 79) has led some researchers to suspect that Alzheimer's disease is the effect of overproduction of the precursor of amyloid A4 protein. In situ hybridization with *PAD* cDNA probes showed indeed that the amyloid precursor mRNA is found in different amounts in various regions of the brain (106–109). Amyloid deposits, however, are not found only in those regions where most of the mRNA is found (100). Two studies seemed to indicate an excess of *PAD* mRNA in brain sections from Alzheimer patients when compared to normal individuals (107, 109). The results were discussed to implicate a role for gene regulation in Alzheimer's disease, but could also be due to an unpredictable differential decay of mRNA in human brains obtained after different postmortem times.

The mechanisms leading to differential splicing are in general not understood. Thus we can only state here that different amounts of the two main forms (with or without RNA coding for the protease inhibitor) of *PAD* mRNA are found in various human and animal tissues (111). Whether different amounts of these mRNAs are involved in the etiology of Alzheimer's disease is another question. It has, however, been reported that in brains of Alzheimer patients the relative amount of the mRNA form lacking the protease inhibitor coding region was relatively increased in some brain regions but not in others (111).

If differential splicing is involved at all in the etiology of Alzheimer's disease, it must be brought about by one or several trans-acting factors. The dominance of the inheritance of familial Alzheimer's disease requires, moreover, that an neomorphic splicing factor is present and not absent in the brain cells of Alzheimer patients. It is difficult to see how in the case of Down's patients a simple duplication of one gene could bring about the drastic change of splice pattern observed (103). Cooperativity of many small multigenic effects could of course be responsible, a rather unpleasant possibility, since it is almost impossible to analyze such a situation.

AMYLOID A4 PRECURSOR PROTEINS

Cloning and cDNA sequencing have indicated that the self-aggregating amyloid A4 protein of Alzheimer's disease is encoded as part of three larger proteins by the *PAD* gene (82, 102–104). The corresponding proteins have 695 residues ($PreA4_{695}$), 751 residues ($PreA4_{751}$), and 770 residues ($PreA4_{770}$). These three forms of human PreA4 proteins are identical for 694 of the 695 residues encoded by human $PreA4_{695}$-cDNA (82), and diverge only upstream of codon 288, where additional 57 or 76 codons are expressed in $PreA4_{751}$ and $PreA4_{770}$ cDNA, respectively (102–104). As mentioned in the preceding chapter, the insert in $PreA4_{770}$ encodes in addition to a 57-residue trypsin inhibitor domain, which is also found in $PreA4_{751}$, a 19-residue sequence of unknown function, which is, however, homologous to an internal sequence of a cell-surface protein expressed on neurons and thymocytes (105). The common 694 residues of all three PreA4 proteins display the typical structural features of cell-surface receptors: an amino-terminal signal sequence followed by three extra-cellular domains, a transmembrane region of 24-residues, and a cytoplasmic domain of 47 residues (82). The amyloid subunit, which was shown to be maximally either 42 or 43 residues long, corresponds to residues 597–639 of the sequence of $PreA4_{695}$. About one third of the amyloid A4 protein is derived from the putative transmembrane region that includes residues 626–648 of $PreA4_{695}$. The amino-terminus of amyloid A4 protein is derived from the third extracellular N-glycan domain, which spans residues 289–624 of $PreA4_{695}$. The additional domain of the $PreA4_{751}$ and $PreA4_{770}$ proteins is inserted between the second and third domain of $PreA4_{695}$ (102–104). The hundred residues of the second domain include a stretch of seven uninterrupted threonine residues and are extremely rich in acidic residues, but contain only two basic residues. The N-terminal domain consists of approximately 175 residues, is rich in cysteines, and may be stabilized by S-S bridges. Furthermore, the precursor sequences include two functional sites for N-glycosylation (112), and several potential sites for O-glycosylation (113) and tyrosine-sulfation (114).

To determine membrane insertion, transmembrane orientation, and post-translational processing, the biogenesis of PreA4 proteins was studied in a cell-free system programmed with SP6 transcripts of the full-length cDNAs (112). Here the precursors were shown to be integral, glycosylated membrane proteins that span the lipid bilayer once. The carboxy-terminal domain was located at the cytoplasmic site of the membrane, and the domains following the transient signal sequence of 17 residues were shown to face the opposite side of the membrane.

Expression of the carboxy-terminal 100 residues of the precursor (termed A4CT) comprising the amyloid A4 part and the cytoplasmic domain produced protein products with a high tendency to aggregate. Proteinase K treatment of these A4CT aggregates produced peptides of the size of amyloid A4 (112). This finding supports the concept that there is a precursor-product relationship between precursor and amyloid A4 protein.

The experimentally determined relative molecular mass of 91.5 kd for $PreA4_{695}$ and 103 kd for $PreA4_{770}$ synthesized in vitro in the presence of membranes is substantially higher than the theoretical values of 78.6 kd and 88.6 kd, respectively (82, 104, 112; T. Dyrks, K. Beyreuther, unpublished). An explanation for this unexpected electrophoretic mobility on SDS gels would be reduced SDS binding of the very acidic second extracellular domain that includes 45 negatively charged residues. Expression of the 303 amino-terminal residues of PreA4 that contain this anionic domain resulted indeed in a protein with the expected abnormal migration in the gel (112).

Immunochemical studies revealed a patchy and punctate appearance of preA4 proteins on neurons in rat brain (100). This location and the suggested proteoglycan (or O-glycosylation) nature of the amyloid precursor secreted from the rat pheochromocytoma cell line PC12 (115) led to the proposal of a function of the A4 precursor proteins in cell-cell or cell-matrix interaction, or both (100), and of its synaptic localization and metabolism (100, 112, 115).

AMYLOID A4 PRECURSOR GENE

In a report dealing with a partial sequence of the genomic locus, the amyloid A4 protein precursor gene has been given the name *PAD* for *Precursor of Alzheimer's Disease* A4 amyloid protein gene (126). This gene is also termed AD-AP gene (96), amyloid β-protein gene (98), APP gene (97), A4 amyloid gene (123), APP gene (103), or CVAP gene (101) on the basis of cDNA cloning.

The isolation of cDNA clones coding for part or all of the precursor of the amyloid A4 protein (82, 96–98) made it easy to pinpoint the human chromosome on which the *PAD* gene was located. All the groups involved used the same method: Southern analysis of DNAs originating from various hybrid animal/human cell lines that carried just one or very few different human chromosomes. The analysis was unambiguous. All the groups reported that human chromosome 21 carries the *PAD* gene (82, 96–98). This result was almost expected. The fact that Down's patients above the age of 30 regularly show amyloid plaques in their brains in an amount comparable to Alzheimer patients had led to the prediction that the amyloid precursor might be placed on chromosome 21, the chromosome in excess in Down's syndrome (116). So it seemed straightforward to look for duplications of the *PAD* gene in Alzheimer patients. One group promptly reported such duplications (117), and claimed erroneously that another one had done it too (118). But the other group had in fact used a SOD probe. Four other groups also looking for duplications could not find a single case of them (119–122). These latter results were in line with the reports of two groups that there is no linkage between familial Alzheimer's disease *(FAD)* locus and the restriction polymorphism of the *PAD* gene (123, 124). While this analysis removes the possibility that the amyloid is the direct cause of Alzheimer's disease, it does not exclude that it may be close to the direct cause. It may be added here that the exact location of the *FAD* gene that causes Alzheimer's disease has not been found yet. The *lod score* of 3 reported is too low to allow a confident assignment (125).

The 5' end of the *PAD* gene coding for the precursor of amyloid A4 protein has been sequenced (126). Inspection and analysis of this sequence reveals the typical features of the promoter of a housekeeping gene and putative transcription factor binding sites and sites for DNA methylation. It has been suggested that at least four mechanisms could act to control the *PAD* gene and could be of relevance for progression of amyloid deposition in Alzheimer's disease: a stress-related heat shock control element (HSE), an oncogene-related AP-1/FOS binding site, a potential binding at a GC-rich element, and the possible methylation of a CpG region (126). The analysis of the structural part of the gene is in progress in the laboratories of the authors of this review. In addition to the four intron-exon junctions already reported, 28 more have so far been found.

AMYLOID A4 PRECURSOR PROTEIN AND ALZHEIMER's DISEASE

What can we learn from the recent biochemical and genetic work on the cause of Alzheimer's disease? The nonlinkage between the *FAD* and the *PAD* genes (123, 124) excludes any difference in structure of the *PAD* genes as an explanation for the occurrence of familial Alzheimer's disease.

It must therefore be another factor or other factors that may directly or indirectly use the *PAD* gene or its products as targets. We first look at the structure of the protein product of the *PAD* gene. There is general agreement that its sequence is most likely compatible with that of a receptor anchored by a membrane-spanning region in membranes of the cell. Both in vitro and in vivo experiments support this notion (100, 112; A. Weidemann, J. Behr, C. L. Masters, K. Beyreuther, unpublished). The possibility that the amyloid precursor may be degraded to a particular active peptide (127) seems unlikely in view of the finding that a nerve cell line secretes a stable, high-molecular-weight form of the precursor (115). The latter result is not in conflict with the existence of transmembrane species of the amyloid precursors (112; A. Weidemann, K. Beyreuther, unpublished). In nature, several strategies have been discovered whereby the subcellular localization of a protein is converted from integral transmembrane to secretory. One of these strategies is selective proteolytic release of an extracytoplasmic domain of an integral membrane protein as is the case in conversion of the IgA receptor into secretory component (128) and of rat liver Golgi sialyltransferase into sialyltransferase of body fluids (129). The same mechanism applies for the generation of secretory PreA4 proteins (A. Weidemann, K. Beyreuther, unpublished). This processing should yield two fragments, secretory PreA4 corresponding to the extracellular part of PreA4 proteins, and the C-terminal part residing in the membrane. The size of secretory PreA4 proteins is consistent with the assumption that the specific proteolytic processing step occurs upstream of the N-terminal amino acid of the amyloid sequence. Thus a physiological process such as the proteolytic release of secretory PreA4 proteins from the integral transmembrane PreA4 species could concomitantly produce a C-terminal membrane-bound fragment the N-terminus of which would be that of amyloid A4.

In view of the finding of Dyrks et al (112) that such carboxy-terminal precursor fragments (A4CT) have a high tendency to aggregate when not inserted into membranes, we suggest amyloid formation may occur in three steps. The first step could be the conversion of the PreA4 proteins from integral transmembrane to secretory PreA4 species of the extracytoplasmic domains and a putative residual transmembrane fragment comprising the C-terminal 100 residues of the precursor (A4CT). The second hypothetical step is membrane destruction and aggregation of the putative A4CT subunits. Finally, proteolytic cleavage could lead to the removal of those sequences, probably hindering further aggregation. Thus the amyloid A4 subunits could continue aggregation into fibrils, fibril bundles, and amyloid. It is noteworthy in this context that the C-terminus of amyloid A4 protein is situated just in the middle of the transmembrane region of the amyloid precursors (82). This is a site of a membrane protein that is only accessible to proteases under pathological conditions such as disruption of membranes. We therefore suggest that if amyloid deposition occurs at clinical target sites, such as in neurons or between synapses, amyloid formation as an irreversible process could be the crucial event for the clinical expression of dementia. The cellular location of the amyloid precursor proteins on or near the plasma membranes of neurons (100) and its postulated function in cell-cell interaction or cell-matrix interaction or both (100, 115) are consistent with this hypothesis.

The presence of two amyloid precursor proteins with protease inhibitor function (102–104) may provide a clue to the elucidation of their function. Proteases play important roles for cells extending processes to facilitate growth cone-elongation through the tissue. The corresponding target cells secrete protease inhibitors to stop process extensions and to promote stable cell-cell interactions, such as formation of stable synapses (130). A function of the amyloid precursor protein in neurogenesis or axon and dendrite sprouting or more generally in tissue regeneration is particularly intriguing in conjunction with the finding that pathological axon and/or dendrite sprouting may occur in Alzheimer's disease (131, 132) and contribute to membrane depletion, which in turn could give rise to amyloid pathology.

ACKNOWLEDGMENTS

We thank Colin L. Masters, Gerd Multhaup, J. Michael Salbaum, Andreas Weidemann, Wilfred Goldmann, Peter Fischer, Caroline Hilbich, Ursula Mönning, Thomas Dyrks, Justin Beer, Nicoletta Catteruccia, Dirk Bunke, Gerhard König, Jie Kang, and Hans-Georg Lemaire for helpful comments, Eveline Schlichtmann and Santosh Pinto for technical assistance, and Beulah Pinto and Ramzija Suljic for their invaluable secretarial assistance. Work in the authors' laboratories was made possible by grants from the Deutsche Forschungsgemeinschaft through SFBs 74 and 317, the Bundesministerium für Forschung und Technologie, the Thyssen Foundation, the Boehringer Ingelheim Fonds, the Cusanus Werke, and the Fonds der Chemischen Industrie.

Literature Cited

1. Alzheimer, A. 1907. *Allg. Z. Psychiatr. Psych. Gerichtl. Med.* 64:146–48
2. Glenner, G. G., Wong, C. W. 1984. *Biochem. Biophys. Res. Commun.* 120:1131–35
3. Masters, C. L., Simms, G., Weinman, N. A., Multhaup, G., McDonald, B. L., Beyreuther, K. 1985. *Proc. Natl. Acad. Sci. USA* 82:4245–49
4. Gajdusek, D. C. 1977. *Science* 197: 943–60
5. Terry, R. D., Katzman, R. 1983. In *The Neurology of Aging*, ed. R. Katzman, R. D. Terry, pp. 51–84. Philadelphia: Davis
6. Katzman, R. 1986. *New Engl. J. Med.* 314:964–73
7. Price, D. L. 1986. *Annu. Rev. Neurosci.* 9:489–512
8. Heston, L. L., Mastri, A. R., Anderson, V. E., White, J. 1981. *Arch. Gen Psychiatry* 38:1085–90
9. Malamud, N. 1972. In *Aging and the Brain*, ed. C. M. Gaitz, pp. 63–87. New York: Plenum
10. Owens, D., Dawson, J. C., Losin, S. 1971. *Am. J. Ment. Defic.* 75:606–12
11. Wisniewski, K. E., Wisniewski, H. M., Wen, G. Y. 1985. *Ann. Neurol.* 17:278–82

12. Scholz, W. 1938. *Z. Gesamte Neurol. Psychiatr.* 162:694–715
13. Pantelakis, S. 1954. *Monatsschr. Psychiatr. Neurol.* 128:219–56
14. Roth, M., Tomlinson, B. E., Blessed, G. 1966. *Nature* 209:109–10
15. Divry, P., Florkin, M. 1927. *C. R. Soc. Biol.* (Paris) 97:1808–10
16. Virchow, R. 1854. *Virchows Arch.* 6:135–38
17. Virchow, R. 1859. *Cellularpathologie,* pp. 334–57. Berlin:Hirschwald
18. Friedreich, N., Kekulé, A. 1859. *Virchows Arch.* 16:50–56
19. Glenner, G. G., Eanes, E. D., Bladen, H. A., Linke, R. P., Termine, J. D. 1974. *J. Histochem. Cytochem.* 22:1141–58
20. Glenner, G. G. 1980. *New Engl. J. Med.* 302:1283–92; 1330–45
21. Glenner, G. G., Terry, W., Harada, M., Isersky, C., Page, D. 1971. *Science* 172:1150–51
22. Meek, R. L., Hoffmann, J. S., Benditt, E. P. 1986. *J. Exp. Med.* 163:499–510
23. Castano, E. M., Fragione, B. 1988. *Lab. Invest.* 58:122–32
24. Sletten, K., Westermark, P., Natvig, J. B. 1976. *J. Exp. Med.* 143:993–98
25. Ghiso, J., Jensson, O., Frangione, B. 1986. *Proc. Natl. Acad. Sci. USA* 83:2974–78
26. Gejyo, F., Yamada, T., Odani, S., Nakagawa, Y., Arakawa, M., et al. 1985. *Biochem. Biophys. Res. Commun.* 129:701–6
27. Westermark, P., Wernstedt, C., O'Brien, T. D., Hayden, D. W., Johnson, K. H. 1987. *Annu. J. Pathol.* 127:414–19
28. Oesch, B., Westaway, D., Walchli, M., McKinley, M. P., Kent, S. B. H., et al. 1985. *Cell* 40:735–46
29. Masters, C. L., Beyreuther, K. 1988. In *Novel infectious Agents and the Central Nervous System.* Ciba Found. Symp. 35:24–36. Chicester: Wiley
30. Masters, C. L., Multhaup, G., Simms, G., Pottgiesser, J., Martins, R. N., Beyreuther, K. 1985. *EMBO J.* 4:2757–63
31. Beyreuther, K., Multhaup, G., Simms, G., Pottgiesser, J., Schröder, W., et al. 1986. In *Discussions in Neuroscience,* ed. A. Bignami, L. Bolis, C. L. Gajdusek, 3:68–79. Geneva: FESN
32. Glenner, G. G., Wong, C. W. 1984. *Biochem. Biophys. Res. Commun.* 122:1131–35
33. Glenner, G. G., Ein, D., Eanes, E. D., Bladen, H. A., Terry, W., Page, D. L. 1971. *Science* 174:712–14
34. Connors, L. H., Shirahama, T., Skinner, M., Fenves, A., Cohen, A. S.

1985. *Biochem. Biophys. Res. Commun.* 131:1063–68
35. Coria, F., Castano, E., Prelli, F., Larrondo-Lillo, M., van Duinen, S., et al. 1988. *Lab. Invest.* 58:454–58
36. Abraham, C. R., Selkoe, D. J., Potter, H. 1988. *Cell* 52:487–501
37. Snow, A. D., Willmer, J., Kisilevsky, R. 1987. *Hum. Pathol.* 18:506–10
38. Candy, J. M., Klinowski, J., Perry, R. H., Perry, E. K., Fairbairn, A., et al. 1986. *Lancet* 1:354–56
39. Mori, H., Kondo, J., Ihara, Y. 1987. *Science* 235:1641–44
40. Perry, G., Friedman, R., Shaw, G., Chan, V. 1987. *Proc. Natl. Acad. Sci. USA* 84:3033–36
41. Brion, J. P., Passareiro, H., Nunez, J., Flament-Durand, J. 1985. *Arch. Biol.* 95:229–35
42. Ihara, Y., Nukina, N., Miura, R., Ogawara, M. 1986. *J. Biochem.* 99:1807–10
43. Ihara, Y. 1987. *J. Neurochem.* 48(Suppl.):14
44. Wood, J. G., Mirra, S. S., Pollock, N. J., Binder, L. I. 1986. *Proc. Natl. Acad. Sci. USA* 83:4040–43
45. Grundke-Iqbal, I., Iqbal, K., Quinlan, M., Tung, Y. C., Zaidi, M. S., Wisniewski, H. M. 1986. *J. Biol. Chem.* 261:6084–89
46. Kosik, K. S., Joachim, C. L., Selkoe, D. 1986. *Proc. Natl. Acad. Sci. USA* 83:4044–48
47. Wischik, C. M., Novak, M., Thogersen, H. C., Edwards, P. C., Runswick, M. J., et al. 1988. *Proc. Natl. Acad. Sci. USA* 85:4506–10
48. Goedert, M., Wischik, C. M., Crowther, R. A., Walker, J. E., Klug, A. 1988. *Proc. Natl. Acad. Sci. USA* 85:4051–55
49. Wischik, C. M., Novak, M., Edwards, P. C., Klug, A., Tichelaar, W., Crowther, R. A. 1988. *Proc. Natl. Acad. Sci. USA* 85:4884–88
50. Hyman, B. T., van Hoesen, G., Damasio, A. R., Barnes, C. L. 1984. *Science* 225:1168–70
51. Braak, H., Braak, E., Grundke-Iqbal, I., Iqbal, K. 1986. *Neurosci. Lett.* 65:351–55
52. Terry, R. D., Hansen, L. A., De Teresa, R., Davies, P., Tobias, H., Katzman, R. 1987. *J. Neuropathol. Exp. Neurol.* 46:262–68
53. Kidd, M. 1963. *Nature* 197:192–93
54. Terry, R. D. 1963. *Exp. Neurol.* 22:629–42
55. Wischik, C. M., Crowther, R. A., Stewart, M., Roth, M. 1985. *J. Cell Biol.* 100:1905–12
56. Crowther, R. A., Wischik, C. M. 1985. *EMBO J.* 5:3661–65

57. Kirschner, D. A., Arbraham, C., Selkoe, D. J. 1986. *Proc. Natl. Acad. Sci. USA* 83:503–7
58. Miller, C. C. J., Brion, J. P., Calvert, R., Chin, T. K., Eagles, P. A. M., et al. 1986. *EMBO J.* 5:269–76
59. Perry, G., Rizzuto, N., Antilio-Gambetti, L., Gambetti, P. 1985. *Proc. Natl. Acad. Sci. USA* 82:3916–20
60. Sternberger, N. H., Sternberger, L. A., Ulrich, J. 1985. *Proc. Natl. Acad. Sci. USA* 82:4274–76
61. Nukina, N., Ihara, Y. 1983. *Proc. Jpn. Acad.* 59:284–92
62. Kosik, K. S., Duffy, L. K., Dowling, M. M., Abraham, C., McCluskey, A., Selkoe, D. 1984. *Proc. Natl. Acad. Sci. USA* 81:7941–45
63. Yen, S. H., Dickson, D. W., Crowe, A., Butler, M., Shelanski, M. L. 1987. *Annu. J. Pathol.* 126:81–87
64. Ksiezak-Reding, H., Dickson, D. W., Davies, P., Yen, S. H. 1987. *Proc. Natl. Acad. Sci. USA* 84:3410–14
65. Nukina, N., Kosik, K. S., Selkoe, D. J. 1987. *Proc. Natl. Acad. Sci. USA* 84:3415–19
66. Guiroy, D. C., Miyazaki, M., Multhaup, G., Fischer, P., Garruto, R. M., et al. 1987. *Proc. Natl. Acad. Sci. USA* 84:2073–77
67. Selkoe, D. J., Ihara, Y., Salazar, F. J. 1982. *Science* 215:1243–45
68. Gorevic, P. D., Goni, F., Pons-Estel, B., Alvarez, F., Peress, N. S., Frangione, B. 1986. *J. Neuropathol. Exp. Neurol.* 45:647–64
69. Selkoe, D. J., Abraham, C. R., Podlisny, M. B., Duffy, L. K. 1986. *J. Neurochem.* 46:1820–34
70. Kidd, M., Allsop, D., Landon, M. 1985. *Lancet* 1:278
71. Divry, P. J. 1927. *J. Belg. Neurol. Psychiatr.* 27:643–57
72. Montejo de Garcini, E., Avila, J. 1987. *J. Biochem.* 102:1415–21
73. Kidd, M. 1964. *Brain* 87:307–27
74. Terry, R. D., Gonatas, N. K., Weiss, M. 1964. *Annu. J. Pathol.* 44:269–97
75. Wisniewski, H. M., Terry, R. D. 1973. In *Progress in Neuropathology*, ed. H. M. Zimmermann, 2:1–26. New York: Gruen & Stratton
76. Miyakawa, T., Katsuragi, S., Watanabe, K., Shimoji, A., Ikeuchi, Y. 1986. *Acta Neuropathol.* 70:202–8
77. Masters, C. L., Beyreuther, K. 1988. In *Aging of the Brain*, ed. R. D. Terry. 38:183–204. New York: Raven
78. von Braunmühl, A. 1937. In *Handbuch der speziellen pathologischen Anatomie und Histologie*, ed. W. Scholz, 13:337–539. Berlin:Springer-Verlag
79. Allsop, D., Kidd, M., Landon, M., Tomlinson, A. 1986. *J. Neurol. Neurosurg. Psychiatry* 49:886–92
80. Struble, R. G., Price, D. L. Jr., Cork, L. C., Price, D. L. 1985. *Brain Res.* 361:267–75
81. Selkoe, D. J., Bell, D. S., Podlisny, M. B., Price, D. L., Cork, L. C. 1987. *Science* 235:873–77
82. Kang, J., Lemaire, H. G., Unterbeck, A., Salbaum, J. M., Masters, C. L., et al. 1987. *Nature* 325:733–36
83. Wong, C. W., Quaranta, V., Glenner, G. G. 1985. *Proc. Natl. Acad. Sci. USA* 82:8729–32
84. Roher, A., Wolfe, D., Palutke, M., KuKuruga, D. 1986. *Proc. Natl. Acad. Sci. USA* 83:2662–66
85. Allsop, D., Landon, M., Kidd, M., Lowe, J. S., Reynolds, G. P., Gardner, A. 1986. *Neurosci. Lett.* 68:252–56
86. Castano, E. M., Ghiso, J., Prelli, F., Gorevic, P. D., Migheli, A., Frangione, B. 1986. *Biochem. Biophys. Res. Commun.* 141:782–89
87. Kirschner, D. A., Inouye, H., Duffy, L. K., Sinclair, A., Lind, M., Selkoe, D. 1987. *Proc. Natl. Acad. Sci. USA* 84:6953–57
88. Shapira, R., Austin, G. E., Mirra, S. S. 1988. *J. Neurochem.* 50:69–74
89. Masters, P. M. 1983. *J. Am. Geriatr. Soc.* 31:426–34
90. Davies, L., Wolska, B., Hilbich, C., Multhaup, G., Martins, R., et al. 1988. *Neurology* 38:1688–93
91. Glenner, G. G., Henry, J. H., Fujihara, S. 1981. *Ann. Pathol.* 1:120–29
92. Mountjoy, C. Q., Tomlinson, B. E., Gibson, P. H. 1982. *J. Neurol. Sci.* 57:89–103
93. Prelli, F., Castano, E. M., van Duinen, S. G., Bots, G. T. A. B., Luyendijk, W., Frangione, B. 1988. *Biochem. Biophys. Res. Commun.* 151:1150–55
94. van Duinen, S. G., Castano, E. M., Prelli, F., Bots, G. T. A. B., Luyendrijk, W., Frangione, B. 1987. *Proc. Natl. Acad. Sci. USA* 84:5991–94
95. Pardridge, W. M., Vinters, H. V., Yang, J., Eisenberg, J., Choi, T. B., et al. 1987. *J. Neurochem.* 49:1394–401
96. Goldgaber, D., Lerman, M. I., McBride, O. W., Saffiotti, U., Gajdusek, D. C. 1987. *Science* 235:877–80
97. Robakis, N. K., Ramakrishna, N., Wolfe, G., Wisniewski, H. M. 1987. *Proc. Natl. Acad. Sci. USA* 84:4190–94
98. Tanzi, R. E., Gusella, J. F., Wakins, P. C., Bruns, G. A. P., St. George-Hyslop, P., et al. 1987. *Science* 235:880–84
99. Yamada, T., Sasaki, H., Furuya, H.,

Miyata, T., Goto, I., Sakaki, Y. 1987. *Biochem. Biophys. Res. Commun.* 149:665–71
100. Shivers, B., Hilbich, C., Multhaup, G., Salbaum, J. M., Beyreuther, K., Seeburg, P. H. 1988. *EMBO J.* 7:1365–70
101. Bendotti, C., Forloni, G. L., Morgan, R. A., O'Hara, B. F., Oster-Granite, M. L., et al. 1988. *Proc. Natl. Acad. Sci. USA* 85:3628–32
102. Ponte, P., Gonzales-DeWhitt, P., Schilling, J., Miller, J., Hsu, D., et al. 1988. *Nature* 331:525–27
103. Tanzi, R. E., McClatchey, A. I., Lamperti, E. D., Villa-Komaroff, L., Gusella, J. F., Neve, R. L. 1988. *Nature* 331:528–30
104. Kitaguchi, N., Takahashi, Y., Tokushima, Y., Shiojiri, S., Ito, H. 1988. *Nature* 331:530–32
105. Clark, M. J., Gagnon, J., Williams, A. F., Barclay, A. N. 1985. *EMBO J.* 4:113–18
106. Goedert, M. 1987. *EMBO J.* 6:3627–32
107. Cohen, M. L., Golde, T. E., Usiak, M. F., Younkin, L. H., Younkin, S. G. 1988. *Proc. Natl. Acad. Sci. USA* 85:1227–31
108. Bahmanyar, S., Higgins, G. A., Goldgaber, D., Lewis, D. A., Morrison, J. H., et al. 1987. *Science* 237:77–80
109. Higgins, G. A., Lewis, D. A., Bahmanyar, S., Goldgaber, D., Gajdusek, D. C., et al. 1988. *Proc. Natl. Acad. Sci. USA* 85:1297–301
110. Lewis, D. A., Higgins, G. A., Young, W. G., Goldgaber, D., Gajdusek, D. C., et al. 1988. *Proc. Natl. Acad. Sci. USA* 85:1691–95
111. Palmert, M. R., Golde, T. E., Cohen, M. L., Kovacs, D., Tanzi, R. E., et al. 1988. *Science* 241:1080–84
112. Dyrks, T., Weidemann, A., Multhaup, G., Salbaum, J. M., Lemaire, H.-G., et al. 1988. *EMBO J.* 7:949–57
113. Fransson, L. A. 1987. *Trends Biochem. Sci.* 12:406–8
114. Huttner, W. B. 1987. *Trends Biochem. Sci.* 12:361–63
115. Schubert, D., Schroeder, R., La Cor-

biere, M., Saitoh, T., Cole, G. 1988. *Science* 241:223–26
116. Lejeune, J., Gautier, M., Turpin, R. 1959. *C. R. Acad. Sci. (Paris)* 248:602–3
117. Delabar, J. M., Goldgaber, D., Lamour, A., Nicole, A., Huret, J. L., et al. 1987. *Science* 235:1390–92
118. Schweber, M., Tuson, C. 1987. *Neurology* 37 (Suppl. 1):222
119. Tanzi, R. E., Bird, E. D., Latt, S. A., Neve, R. L. 1987. *Science* 238:666–69
120. Podlisny, M. B., Lee, G., Selkoe, D. J. 1987. *Science* 238:669–71
121. St. George-Hyslop, P. H., Tanzi, R. E., Polinsky, R. J., Neve, R. L., Pellen, D., et al. 1987. *Science* 238:668–69
122. Murdoch, G. H., Manuelidis, L., Kim, J. H., Manuelidis, E. E. 1988. *Nucleic Acids Res.* 16:357
123. Van Broeckhoven, C., Genthe, A. M., Vandenbergh, A., Horsthemke, B., Backhovens, H., et al. 1987. *Nature* 329:153–55
124. Tanzi, R. E., St. George-Hyslop, P. H., Haines, J. L., Polinsky, R. J., Nee, L., et al. 1987. *Nature* 329:156–57
125. St. George-Hyslop, P. H., Tanzi, R. E., Polinsky, R. J., Haines, J. L., Nee, L., et al. 1987. *Science* 235:885–89
126. Salbaum, J. M., Weidemann, A., Lemaire, H.-G., Masters, C. L., Beyreuther, K. 1988. *EMBO J.* 7:2807–13
127. Allsop, D., Wong, C. W., Ikeda, S. I., Landon, M., Kidd, M., Glenner, G. G. 1988. *Proc. Natl. Acad. Sci. USA* 85:2790–94
128. Mostov, K. E., Kraehenbuhl, J. P., Blobel, G. 1980. *Proc. Natl. Acad. Sci. USA* 77:7257–61
129. Paulson, J. C., Weinstein, J., Ujita, E. L., Riggs, K. J., Lai, P. H. 1987. *Biochem. Soc. Trans.* 15:618–20
130. Patterson, P. H. 1985. *J. Physiol.* 80:207–11
131. Scheibel, A. B. 1983. In *Alzheimer's Disease*, ed. B. Reisberg, pp. 69–73. New York: The Free Press
132. Geddes, J. W., Monaghan, D. T., Cotman, C. W., Lott, I. T., Kim, R. C., Chui, H. C. 1985. *Science* 230:1179–81

Annu. Rev. Biochem. 1989. 58:309–50

ANIMAL GLYCOSPHINGOLIPIDS AS MEMBRANE ATTACHMENT SITES FOR BACTERIA[1]

Karl-Anders Karlsson

Department of Medical Biochemistry, University of Göteborg, P.O. Box 33031, S-400 33 Göteborg, Sweden

CONTENTS

[1]The glycolipid nomenclature used follows IUB-IUPAC CBN recommendations for 1976 (1), and the abbreviated names and condensed representation of sugar chains follows the IUB-IUPAC JCBN recommendations from 1980 (2) and 1985 (3). The designation *isoreceptors* is used to mean isomeric receptor substances having in common the specifically recognized sequence but located in different environments, i.e. having separate neighboring groups. Primary binding sequence or epitope means the common part of isoreceptors that is specifically recognized by the microbial protein, and secondary binding sequence or epitope means a neighboring part to the primary sequence that may influence binding (see Figure 1). Receptor drift means a change in the microbial protein resulting in a slight shift of the primary and/or secondary binding epitope.

309

0066-4154/89/0701-0309$02.00

PERSPECTIVES AND SUMMARY

Interest is growing rapidly in the molecular aspects of microbe association to animal cell surfaces. This is in part due to recent improvements in technology that make the subject accessible for experiments, especially concerning glycolipids. In addition, detailed receptor knowledge is essential to meet the need for alternate ways of diagnosing, treating, and preventing important infections.

Attachment to the host, most important to mucous membranes, is the initial event of the infectious process and considered an important virulence factor. This may be followed by host membrane penetration or tissue invasion, which is required for viruses, bacterial toxins, and several bacterial cells, where membrane-close lipid-linked oligosaccharides are of particular interest.

The association may be less specific, e.g. based on hydrophobic or electrostatic interactions, but in the case of a pronounced tropism of an infection for animal species or tissue (selection of target), it is logical to assume a specific molecular recognition. The substances involved are in the majority of cases proteins on the surface of the virus, bacterium (or bacterial toxin), yeast or parasite (hemagglutinin, agglutinin, lectin, adhesin), and receptor proteins or carbohydrates of the animal host. The dominating part of receptor substances are glycoconjugates, possibly due to their abundance on epithelial cells. A generalization is developing that receptors on intact surfaces are mainly carbohydrates, while association to damaged tissue is more favorable to subepithelial connective tissue (e.g. collagen) or typical wound proteins like fibrin, laminin, and fibronectin. One should realize that our knowledge is still extremely limited in this field. Firstly, the molecular topology and dynamics of animal cell surfaces are mostly unknown. Probably, only a small fraction of surface glycoconjugates, the most complex and diverse of biological substances, has been structurally defined up to now, if all human and animal tissues and cells are considered. Varying accessibility of potential receptors for binding (masking, crypticity) is an almost neglected area. The establishment of the biological relevance of receptor specificities of microbes requires sophisticated knowledge not only of the molecular details of the specificity but also the balance in vivo of strictly membrane-bound and secreted receptor substances. Today we cannot even adequately assay the binding of a microbe to various candidate receptors in vitro. However, characterization of glycolipid receptors, which may be considered strictly membrane-bound and therefore of special relevance, is at present especially rewarding due to improved assay and structural methods. The situation is analogous to the rapidly expanding field of tumor antigens, where not only are the great majority of chemically characterized antigens glycolipids, but they may also be the most relevant biologically.

The intention of this review, on a rather novel subject with limited information, is to present the current technical data. Emphasis is on characteristics and properties in common for the few glycolipids known to be selected as receptors for bacterial systems, which should be of primary interest for future work. Part of the presentation is therefore more suggestions than experimentally proven facts. Most of the more microbiological aspects, including biological relevance and attachment or adhesion as a virulence factor, are not covered.

Although this is the first paper on this topic in *Annual Review of Biochemistry,* there are already a number of monographs and reviews on the general subject of microbial attachment, adhesion, or invasion. Specific topics include viruses (4, 5), bacteria (6–14), bacterial toxins (15–18), lectins (19–21), and carbohydrate-binding proteins (22). See (23, 24) for proceedings from recent meetings.

CELL-SURFACE GLYCOSPHINGOLIPIDS

The membrane-linked glycoconjugates are of three kinds: glycoproteins (25–29) proteoglycans (30, 31), and glycolipids. Also of relevance for microbe interactions are extracellular, surface-associated but not membrane-anchored secreted forms of glycoproteins, including mucins (25) and proteoglycans (31). Therefore, carbohydrate chains make up the major part of cell surfaces, especially on mucous membranes, and are primary collision partners for microbes. The different groups of glycoconjugates have many oligosaccharide structures in common. On the other hand, there are important differences of structural specificity. The proteoglycans have their own core saccharides with repetitive, often highly sulfated and carboxylated, sequences. O-linked chains of glycoproteins have much in common with glycolipids, except that the peptide-linked monosaccharide is GalNAcα compared with Glcβ linked to ceramide. This makes lactose, Galβ4Glc, specific for glycolipids, which may explain why many bacteria have selected this specificity (see below). In contrast, mannose, which frequently is present in N-linked saccharides of glycoproteins, has not yet been detected in glycolipids of higher animals, although it exists in invertebrates (see below). Therefore, the large number of bacteria known to bind mannose selectively recognize glycoproteins, although they also may carry glycolipid-specific adhesins. Such differences must be kept in mind when assaying for microbe interactions and using for technical reasons potential receptors only within one class of substances.

The physiological meaning of the great diversity of surface carbohydrates is still largely unknown (32). They are considered to play a recognition role in growth, differentiation, and cell-cell interactions (33–35) as well as in malignancy (35–40). Recent data indicate that glycolipids may have important modulation roles in receptor-mediated membrane processes (41, 42).

One technical advantage in the handling of lipid-linked oligosaccharides is that each molecule has only one saccharide, compared to glycoproteins and proteoglycans, which have several, often different oligosacchrides linked to the same peptide. Also, the lipid tail may anchor the glycolipid in various assay media for analysis of binding properties (see below). This may explain the relatively rapid expansion at present of our knowledge of glycolipid antigens and receptors.

Structural Diversity and Tissue Specificity

There are several hundred known glycolipid sequences (43–46). Most have 10 or fewer sugars, but some have been proven by FAB mass spectrometry and NMR spectroscopy to have 20–30 sugars (47), and even larger glycolipids have been proposed (44). Glycolipids are divided into separate series based on the nature of the core saccharide (Table 1). These cores may be repeated or extended and finally ended with blood group and other antigenic determinants, including sialic acids and sulfate groups. Hybrids between basic sequences of the separate series have been reported, and with improved technology many novel structures have been identified (see examples of Table 2). One may note the diversity and unusual sequences of invertebrates, like methylated monosaccharides, internal Fuc, or sialic acid linked to another sialic acid through the N-glycoloyl group. When looking into minor substances, similar structures may eventually be found also in mammalian tissues.

The lipophilic part, ceramide, varies in both the fatty acid and long-chain base in level of hydroxylation, chain length, unsaturation, and branching (43–46, 78). The ceramide hydroxylation appears to affect the sugar conformation and is of relevance for bacterial binding (see below).

There is a distinct expression of separate glycolipids in different animals and tissues, which is of relevance for the tropism of infections. Unfortunately, there is no covering review on this important subject, but there are examples given in general presentations on glycolipids (39, 40, 43–46, 78). Classical examples of species differences are blood group glycolipids of erythrocytes (44), and more recent examples are kidney (44, 51, 82) and intestine (78, 79, 83). Concerning tissue specificity, nervous and non-nervous tissues were

Table 1   Different glycosphingolipid series based on type of core saccharide

Series and sequence	Trivial name of oligosaccharide	Binding bacteria[a]
Simple members		
GalβCer		
GlcβCer		
XylβCer		
FucαCer		
Galβ4GlcβCer	Lactose	Many bacteria
Manβ4GlcβCer		
Globo series		
Galα4Galβ4GlcβCer	Globotriaose	*E. coli*
		Shiga toxin
GalNAcβ3Galα4Galβ4GlcβCer	Globotetraose	*Actinomyces*
Isoglobo series		
Galα3Galβ4GlcβCer	Isoglobotriaose	
GalNAcβ3Galα3Galβ4GlcβCer	Isoglobotetraose	*Actinomyces*
Lacto and lactoneo series		
GlcNAcβ3Galβ4GlcβCer	Lactotriaose	*Streptococcus*
Galβ3GlcNAcβ3Galβ4GlcβCer	Lactotetraose	
Galβ4GlcNAcβ3Galβ4GlcβCer	Lactoneotetraose	*Staphylococcus*
Ganglio series		
GalNAcβ4Galβ4GlcβCer	Gangliotriaose	*Pseudomonas*
Galβ3GalNAcβ4Galβ4GlcβCer	Gangliotetraose	
Isoganglio series		
Galβ3GalNAcβ3Galβ4GlcβCer	Isogangliotetraose	
Gala series		
Galα4GalβCer	Galabiose	*E. coli*
		Shiga toxin
Muco series		
Galβ4Galβ4GlcβCer	Mucotriaose	
Gala-6 series[b]		
Galβ6GalβCer		
Mollus and isomollus series[c]		
Manα3Manβ4GlcβCer	Mollustriaose	
Manα4Manβ4GlcβCer	Isomollustriaose	
GlcNAcβ2Manα3Manβ4GlcβCer	Mollustetraose	
Arthro series[d]		
GlcNAcβ3Manβ4GlcβCer	Arthrotriaose	

[a] Compare Tables 3, 4, and 6. Only terminal primary epitope is exemplified for binding.
[b] Proposed in Ref. 48 for invertebrate glycolipids.
[c] Proposed in Ref. 49 for invertebrate glycolipids.
[d] Proposed in Ref. 50 for insect glycolipids.

Table 2 Examples of novel glycosphingolipid structures

Sequences that are hybrids of the basic series of Table 1	Ref.
Lactoneo and globo series	
Galβ4(Fucα3)GlcNAcβ6(Galβ3)GalNAcβ3Galα4Galβ4GlcβCer	51
Lacto and ganglio series	
GlcNAcβ3(GalNAcβ4)Galβ4GlcβCer	52
Ganglio and lactoneo series	
GalNAcβ4(NeuAcα3)Galβ4GlcNAcβ3Galβ4GlcβCer	53, 54
GalNAcβ4(NeuAcα3)Galβ4GlcNAcβ3(GalNAcβ4)Galβ4GlcβCer	55
Lacto and lactoneo series	
Galβ3(Fucα4)GlcNAcβ3Galβ4GlcNAcβ3Galβ4GlcβCer	56
Globo and ganglio series	
GalNAcα3GalNAcβ3Galβ3GalNAcβ4Galβ4GlcβCer	57
Isoglobo, lactoneo, and ganglio series	
Fucα3GalNAcβ3Galα3Galβ4GlcNAcβ3(GalNAcβ4)Galβ4GlcβCer	58

Other novel structures	
NeuAcα3Galβ3(NeuAcα6)GalNAcβ4Galβ4GlcβCer	59–61
NeuAcα3Galβ3(NeuAcα6)GlcNAcβ3Galβ4GlcβCer	62, 63
NeuAcα6Galβ4GlcNAcβ3(Galβ4GlcNAcβ6)Galβ4GlcβCer	64
SO3GlcAβ3Galβ4GlcNAcβ3Galβ4GlcβCer	65, 66
GalNAcα3(Fucα2)Galβ3GalNAcα3(Fucα2)Galβ4GlcNAcβ3Galβ4GlcNAcβ3Galβ4GlcβCer	67
NeuAcα3Galβ3GalNAcα3(Fucα2)Galβ4GlcNAcβ3Galβ4GlcβCer	68
Fucα2Galβ4(Fucα3)GlcNAcβ3Galβ4(Fucα3)GlcNAcβ3Galβ4GlcβCer	69
NeuAcα3Galβ4GlcNAcβ3Galβ4(Fucα3)GlcNAcβ3Galβ4GlcβCer	70
NeuNH2α3Galβ4GlcβCer	41
Lactone of: Galβ3GalNAcβ4(NeuAcα8NeuAcα3)Galβ4GlcβCer	71
Lactone of: NeuAcα3Galβ4GlcβCer	72
NeuAcα3Galβ3[NeuAcα6(Fucα4)GlcNAcβ3]Galβ4GlcβCer	73
Galβ3GalNAcβ4(NeuGcα3)Galβ3GalNAcβ4Galβ4GlcβCer	74
MeGlcNAcα4GalNAcα3[6-(2-aminoethylphosphonyl)Galα2](2-aminoethyl-phosphonyl6)Galβ4(2-aminoethyl-phosphonyl6)GlcβCer[a,b]	75
MeFucα2MeXylβ4(MeGalNAcα3)Fucα4GlcNAcβ2Manα3-(Xylβ2)Manβ4GlcβCer[a,b]	49
MeNeuGcα5MeNeuGcα3Galβ4GlcβCer[a,b]	76
Galβ6Galβ6Galβ6GalβCer[a]	48
GlcNAcβ3Galβ3GalNAcα4GalNAcβ4GlcNAcβ3Manβ4GlcβCer[a]	77

[a] Invertebrate origin.
[b] Me as prefix means O-methylated (position known but not given here).

classical comparisons (44, 46), and more recent studies have analyzed rabbit (80) and rat (81) tissues and also separate compartments of the same tissue like the small (78, 79, 83, 84) and large (85, 86) intestine, also including the ceramide variation. In addition to the long known phenotypic and very strict expression of blood group glycolipids in various tissues (44, 78, 84), an

interesting investigation is analyzing the genetic control of globoseries glycolipids of kidney in inbred strains of mice (51, 82, 87) and the genetic
polymorphism of ganglioside expression in mouse organs (88). Altogether
these examples document the possibility of distinct tropisms for microbes
based on receptor glycolipids. One example is given below for *Vibrio
cholerae*.

Location to the Surface Monolayer

Glycolipids in final form are generally considered surface located with intracellular occupation during processing (44, 78). It is of interest that most, if
not all, sphingolipids, including the phosphosphingolipids, are exclusively
placed in the outer monolayer of the surface membrane bilayer (78). This may
contribute to the stability of the cells' passive passage of material, since
sphingolipids are especially suited to form laterally oriented intermolecular
hydrogen bonds within this monolayer (78, 89). A striking relation exists in
the tissues between the number of hydroxyl groups (one to four) of the
ceramide and the apparent functional demands on the membrane (78),
suggesting an essential role in this part of the membrane. The first one or two
sugars may add to this function (89–91). Thus erythrocytes have the least
hydroxylated ceramide species of glycolipids, while epithelial cells of the
intestine have the most hydroxylated species (78). This molecular barrier is of
relevance for fusion or invasion with pathogens. In this regard we have
identified a binding specificity common for several viruses, apparently restricted to the first sugar and part of the epithelial type of ceramide (K.-A.
Karlsson, et al, unpublished results), thus interfering with the noted interactions as a possible prerequisite for fusion. The recognition by many
bacteria of lactose, the first two sugars of glycolipids, may be used similarly
(see below).

In some cases glycolipids may not be confined to the plasma membrane. In
polymorphonuclear neutrophils, lactosylceramide is mainly intracellular in
lysosomal granules (92). Of interest for bacterial binding, there are significant
levels of glycolipids in the intestinal lumen, both shown for the rat (93) and
human (94), derived from the rapid renewal of epithelial cells. Bacterial
enzymes appear to degrade more complex glycolipids to mostly lactosylceramide, which is a potential receptor substance for major normal anaerobes
and several pathogens (see below).

The largely unknown molecular topology on the cell surface will be
discussed below.

Novel Assaying of Binding

The traditional methods for revealing carbohydrate-based binding specificities
for microbes (6, 12) have been either direct assaying of inhibition of binding

with commercially available monosaccharides or simple oligosaccharides, or indirect approaches based on selective chemical or enzymatic modification of target cell carbohydrates. For technical reasons, complex oligosaccharides from relevant tissues have not been available for analysis. Examples of the few exceptions are Man-containing oligosaccharides collected from several sources and used to screen for inhibition of Man-binding bacteria (95, 96), N-acetyllactosamine oligosaccharides used for inhibition of binding of *Staphylococcus saprophyticus* (97), and a series of saccharides used to define the specificity of toxin A of *Clostridium difficile* (98). Specially synthetized saccharides have been used in the cases of *Actinomyces* (99) and *Streptococcus pneumoniae* (100). An elegant technique of sialidase treatment and resialylation with specific transferases has been developed both for viruses (5) and bacteria (101), revealing inportant receptor variants of influenza virus (102).

The recently published solid-phase methods for binding of viruses (103, 104) and bacteria (105, 106) based on separated glycolipids on thin-layer chromatograms, have opened new possibilities for the detection and characterization of receptor carbohydrates for microbes (104, 107–111). Such a method was first used for overlay with cholera toxin (112) and antibodies (113, 114), and has also been applied for eukaryotic cells (115). The plastic treatment (114) allows a successful anchoring of lipid tails, but free oligosaccharides may be covalently coupled in various ways after development (116, 117). Bound microbe can be detected either through direct radiolabeling or with antibodies using different techniques (106, 117). For rough estimation of binding strength, detection limits from dilution series on the thin-layer plate or curves from glycolipids coated in microtiter wells may be used (106). Total cell or tissue extracts may be run directly, but for a successful detection of minor receptor species among major membrane phospholipids, a preparation of total glycolipids free of nonglycolipid substances is preferable, and is now possible (118). More about glycolipid receptor characterization is given below. One may also detect a microbe interaction by overlay on separated glycoproteins (119, 120). However, as noted above, the logical working up for identification (see below) is complicated by the presence of several different oligosaccharide chains in the same glycoprotein, requiring saccharide release and further fractionation for a final answer.

The advantages and drawbacks of this rational assay technology have been discussed in detail elsewhere (106). The limitation is the devotion to glycolipids, which may lack some relevant receptor specificities like Man. There is, however, the potential of effective screening for receptor substances, since one may include a series of reference mixtures on the plate covering receptor candidates from many animals and tissues. One additional advantage is the multivalent presentation of receptors capable of picking up also low-affinity specificities, which escape detection with the traditional inhibition assays using free oligosaccharides (see also below for Shiga toxin). In this sense the overlay methods are more relevant for the situation at the cell surface. One may also note the threshold phenomenon demonstrated before for cell binding to substituted polyacrylamide gels (121). Although a particular oligosaccharide sequence may well be recognized by a ligand, there is no binding below a certain level of receptor density, in part due to shear forces induced by the washing procedure. In this way the solid-phase methods are less sensitive than solution and suspension inhibition methods (e.g. inhibition of hemagglutination by bacteria with saccharides). Again, this threshold phenomenon may be adequate for comparison with the cell-surface interactions. In practice the binding data from the solid-phase binding therefore appear very simple: either there is binding or not, showing up as for instance autoradiographic spots or no spots against the control background. The differentiation is thus made between binders and nonbinders, defining a detection level under standard conditions and estimating relative avidities for positive binders. Without access to good soluble inhibitors, no more precise binding parameters can be defined. However, as already illustrated in many cases, this information is sufficient at the present stage of development.

Glycolipid Recognition by Bacterial Toxins and Cells

It is not appropriate here to discuss the relevance of receptor specificities to the virulence of microbes. There is a common assumption that the specific attachment step is required for the colonization or infection with bacteria (6–13). However, experimentally this is far from proven using adequate infection models. The situation at the colonization and potential binding site is very complex, including many factors. In the case of bacterial toxins or viruses, the requirement for a receptor may be more obvious, since the ligand has to get into the host cell to express an effect or reproduce. Concerning toxins, glycolipid receptor coating of insensitive cells has induced sensitivity (15). For viruses, cleavage of sialic acid from host cells has prevented both the binding and infection, and enzymatic resialylation has restored the activity (5). Binding sites for sialic acid on influenza virus were recently proven by X-ray crystallography of the complex of sialyllactose and nicked virus hemagglutinin (122). No convincing corresponding data for bacteria have yet been presented, although the very likely necessity for receptors in many cases will probably soon be proven by use of synthetic receptor analogues (see below). One obvious complication is that most bacteria appear to have conserved not only one but several binding specificities, indicating multistep mechanisms (see also below). It has long been known that lymphocytes may agglutinate with bacteria, suggested to be based on protein-carbohydrate binding specificities (123, 124). Recently, evidence has been given for carbohydrate receptor involvement in the phagocytosis of bacteria by animal cells, named lectinophagocytosis (14).

An impressive list of saccharides proposed as receptors for bacterial and other systems (7) was recently published. However, both the detailed specificities and the nature of the natural receptors are largely unknown. Most of these reported saccharides exist in glycolipid form. Therefore, a number of systems will be accessible for the informative approach reviewed here. Although only a few toxins and bacteria have so far been analyzed in this way, some general conclusions will be allowed.

BACTERIAL TOXINS   Glycolipid-binding toxins are gathered in Table 3. In the classical case of cholera toxin, there is no convincing evidence yet that receptor activity is present in glycoprotein form (138). This may be surprising in view of a recently published glycolipid sequence with an extended ganglio core (see Table 2), where the terminal pentasaccharide is expected to exist in peptide-linked form, although not yet isolated. In the case of heat-labile *Escherichia coli* toxin, with very similar specificity to that of cholera toxin, there was a binding to glycoprotein in addition to GM1 ganglioside, when the two toxins were tested in parallel (139). This may be due to slightly different structures of their binding sites, allowing a minor shift in binding epitope on the saccharide (see also below). The following sequence may hypothetically be present in glycoproteins and active for heat-labile toxin: Galβ3GalNAcβ4(NeuAcα3)Galβ4GlcNAc (compare CAD-active glycoproteins, which are lacking the terminal Gal, see Table 2). This may be the basis for the binding of heat-labile toxin. Cholera toxin, however, may not accept GlcNAc instead of Glc (Table 3).

It would be of interest to know if the glycoprotein is able to mediate the effect for heat-labile toxin, or if the novel cholera toxin receptor on an extended glycolipid is active when coated on resistant cells. One argument favoring the penetration mechanism is the proximity of the binding epitope to the bilayer, making possible the insertion of the A subunit into the membrane (16). This effect may not be produced with the binding epitope at some distance from the membrane. Therefore there may be functional and nonfunctional receptors depending on ability to mediate the toxin action. In case of Shiga toxin, separate cell lines of HeLa cells are known to carry the same or similar number of binding sites for the toxin, although only some of the cells are sensitive to toxin action (140). Glycolipid analysis and overlay with toxin indicated that sensitive cells contained Galα4Galβ Cer, which was practically lacking in insensitive cells (132). The proposal was therefore made that the receptor sequence directly linked to lipid mediated penetration (functional receptor due to bilayer proximity of epitope), while Galα4Gal-containing glycoproteins were positive binders without mediating the effect (nonfunc-

Table 3 Glycosphingolipid receptors for bacterial toxins

Microorganism	Toxin	Target tissue	Proposed receptor sequence	Ref.
Vibrio cholerae	Cholera toxin	Small intestine	Galβ3GalNAcβ4(NeuAcα3)Galβ4GlcβCer	15, 16
E. coli	Heat-labile toxin	Intestine	Galβ3GalNAcβ4(NeuAcα3)Galβ4GlcβCer and "Galactoproteins"	15, 16
Clostridium tetani	Tetanus toxin	Nerve membrane	Galβ3GalNAcβ4(NeuAcα8NeuAcα3)Galβ4GlcβCer	15, 16
Clostridium botulinum	Botulinum toxin A and E	Nerve membrane	NeuAcα8NeuAcα3Galβ3GalNAcβ4(NeuAcα8NeuAcα3)Galβ4GlcβCer and Free fatty acids	15, 16, 125, 126
Clostridium botulinum	Botulinum toxin B, C, and F	Nerve membrane	NeuAcα3Galβ3GalNAcβ4(NeuAcα8NeuAcα3)Galβ4GlcβCer	15, 16, 127, 128
Clostridium botulinum	Botulinum toxin B	Nerve membrane	GalβCer, Free fatty acids	128
Clostridium perfringens	Delta toxin	Cell lytic	GalNAcβ4(NeuAcα3)Galβ4GlcβCer	129
Clostridium difficile	Toxin A	Large intestine	Galα3Galβ4GlcNAcβ3Galβ4GlcβCer	98, 130, K.-A. Karlsson, I. Lönnroth, et al, unpublished results
Shigella dysenteriae	Shiga toxin	Large intestine	Galα4Galβ4GlcβCer Galα4Galβ4GlcβCer GlcNAcβ4GlcNAc	15–17, 131–134 144
E. coli	Vero toxin or Shiga-like toxin	Intestine	Galα4Galβ4GlcβCer Galα4Galβ4GlcβCer	135–137, K.-A. Karlsson, A. A. Lindberg et al, unpublished results

tional receptors). Attempts to induce sensitivity by coating with the rather nonpolar digalactosylceramide have not yet been successful. There is a notable finding for botulinum toxins of apparently two binding specificities (125, 126, 128). The heavy chain but not the light chain of toxin B was shown to bind ganglioside. The intact neurotoxin and a chymotrypsin-induced fragment consisting of part of the heavy and part of the light chain linked by a disulfide bridge bound to GalβCer and fatty acids (128). This fragment did not bind ganglioside. The possible importance of this is discussed below.

Several of the sialic acid–recognizing toxins (cholera, heat-labile, tetanus, botulinum A, B, C, F, and Delta toxins) have similarities in their binding specificities. The requirement for membrane proximity for toxin penetration may explain the selection of bilayer-close gangliosides. To achieve tissue specificity, e.g. botulinum toxins for separate neural membranes (15, 16, 128), slightly different binding epitopes on particular gangliosides may have been chosen. Using the approach with molecular modeling described below, the binding epitope on the cholera toxin receptor is proposed (Figure 2). More binding data are, however, required for a good comparison with the other toxins. When more information on the sequences and conformation (141–143) of the toxins becomes available, the binding sites may be located for a more precise modeling of toxin-receptor interactions. It is also clear that the binding epitope on Galα4Gal must differ between E. coli and the Shiga toxin (see below and Table 8). On the other hand, the binding patterns of Shiga toxin and Vero toxin appear very similar (K.-A. Karlsson, A. A. Lindberg, et al, unpublished results). Inhibition studies with soluble receptor analogues have established the biological relevance of this specificity for Shiga toxin (132), and direct binding studies on rabbit intestine support this (133). We were not able to confirm the proposed binding of Shiga toxin to chitobiose (144).

An interesting recent finding is the binding specificity for toxin A of C. difficile (Table 3). Apparently there is a strict requirement for the proposed trisaccharide, which is abundant on some animal cells as shown by antibody (145, 146). However, glycolipid-based receptor has not yet been detected by overlay of toxin or antibody on preparations of human large intestine, the target tissue for toxin action (U. Galili, K.-A. Karlsson, I. Lönnroth, S. Teneberg, et al, unpublished results). Only small amounts of a receptor pentaglycosylceramide were detected in human erythrocytes (U. Galili, K.-A. Karlsson, I. Lönnroth, S. Teneberg, et al, unpublished results), possibly explaining a binding of specific antibodies (agglutination) to aged erythrocytes (145). Furthermore, human sera contain antibodies specific for this trisaccharide and making up as much as one percent of the immunoglobulin fraction (146). The natural antigens for these antibodies are probably located on parasites (147, 148) and bacteria (U. Galili, et al, unpublished results).

This indicates that this saccharide sequence is not present to any appreciable extent on human cells and the biological relevance of the toxin A specificity in human large intestine is questionable. On the other hand, a proposal has been made (U. Galili, personal communication) that pathogenic bacteria of the intestine may induce toxin receptors by catalyzing the transfer of Galα3 residues to endogenous acceptors. Further work is required to test this fascinating hypothesis.

BACTERIAL CELLS   Glycolipid-binding bacteria are gathered in Table 4. Of these only a few have been characterized in some detail for binding specificity using a sufficient number of receptor candidates to allow epitope approximation (see below, e.g. Nos. 1, 3, 10, and 12 of Table 4). Urinary tract infection with *E. coli* and Galα4Gal is by far the most studied of bacterial infections, and a covering review of adherence and pathogenesis was recently published (168). Concerning inhibition with receptor analogues of Galα4Gal, epitope characteristics, and receptor variants, aspects will appear in the sections below. For *Actinomyces naeslundii* it has not yet been definitely settled which of Galβ and GalNAcβ is the essential receptor sequence, although in glycolipids the latter seems to be preferred (164). Although the biological relevance is far from proven in the case of lactosylceramide binders, this large group of bacteria will be treated separately below.

CHARACTERISTICS OF THE RECOGNITION OF GLYCOLIPIDS

Although only a few systems have yet been analyzed, there are some reappearing features in the binding and in the glycolipids having been selected for binding. They are of importance for further experimental processing and are therefore discussed in some detail.

Microbes are Able to Bind Sequences Placed Nonterminally in the Saccharide Chain

Table 5 gathers examples of bacterial toxins and cells that are capable of recognizing internal sequences. We have evidence for this property for several viruses also (K.-A. Karlsson, et al, to be published). One consequence is that there may be a number of separate glycolipids of different sizes having the receptor sequence in common but with varying neighboring sequences. I have designated such receptor substances as *isoreceptors* in analogy with isoenzymes and other names, regardless of receptor activity (108). Compared to the receptor sequence placed terminally in a chain, a neighboring sequence may improve, unaffect, diminish, or eliminate a binding. Relative affinities

Table 4 Glycosphingolipid receptors for bacterial cells

Microorganism	Target tissue	Proposed specificity	Ref.
E. coli	Urinary	Galα4Galβ	149–151
E. coli[a]	Urinary	GlcNAcβ	152, 153
Propionibacterium	Skin, intestine	Galβ4Glcβ	104, 154–156
Several bacteria[b]	Diverse	Galβ4Glcβ	104, 154–156
Streptococcus pneumoniae	Respiratory	GlcNAcβ3Gal	100
E. coli CFA/I	Intestine	NeuAcα8	157
E. coli[a]	Urinary	NeuAcα3Gal	119, 158
E. coli[c]	Intestine	NeuGcα3Galβ4GlcβCer GalNAcβ4(NeuAcα3)Galβ4GlcβCer	159–162
Staphylococcus saprophyticus[a]	Urinary	Galβ4GlcNAc	97
Actinomyces naeslundii[d]	Mouth	Galβ, GalNAcβ, Galβ3GalNAcβ, GalNAcβ3Galβ	110, 163, 164
Pseudomonas	Respiratory	GalNAcβ4Gal	165
Neisseria gonorrhoeae[e]	Genital	Galβ4Glcβ NeuAcα3Galβ4GlcNAc	166, 167

[a] Not yet shown to bind glycolipids, although this is very likely.

[b] See Table 6.

[c] F. K. de Graaf, K.-A. Karlsson, et al, in preparation. Results presented at the 14th International Carbohydrate Symposium in Stockholm, August 14–19, 1988.

[d] Although both Gal and GalNAc, as well as several saccharides, are soluble inhibitors of binding, GalNAcβ has been proposed as the primary binding epitope in glycolipids (164).

[e] Results on ganglioside binding were presented at the 14th International Carbohydrate Symposium in Stockholm, August 14–19, 1988.

Table 5 Glycosphingolipid structures (isoreceptors) carrying primary binding sequences (underlined) being recognized in nonterminal positions by bacterial cells and toxins*

Ligand	Isoreceptor	Ref.
Cholera toxin	NeuAcα3Galβ3GalNAcβ4(NeuAcα3)Galβ4GlcβCer	15
	Galβ3GalNAcβ4(NeuAcα8NeuAcα3)Galβ4GlcβCer	15, 74
	GalNAcβ4Galβ3GalNAcβ4(NeuAcα3)Galβ4GlcβCer	107
	Fucα2Galβ3GalNAcβ4(NeuAcα3)Galβ4GlcβCer	169
	Galβ3GalNAcβ4(R-NeuAcα3)Galβ4GlcβCer	170
	R = Lucifer yellow CH, Rhodamine, or DNP	
Tetanus toxin	NeuAcα3Galβ3GalNAcβ4(NeuAcα8NeuAcα3)Galβ4GlcβCer	15
	NeuAcα8NeuAcα3Galβ3GalNAcβ4(NeuAcα8NeuAcα3)Galβ4GlcβCer	15
Shiga toxin	Galα3Galα4Galβ4GlcβCer	131, 132
	GalNAcβ3Galα4Galα4Galβ4GlcβCer	
E. coli	GalNAcβ3Galα4Galα4Galβ4GlcβCer	151
	GalNAcα3GalNAcβ3Galα4Galα4Galβ4GlcβCer	
	Fucα2Galβ3GalNAcβ3Galα4Galα4Galβ4GlcβCer	
	NeuAcα3(NeuAcα6)Galβ3GalNAcβ3Galα4Galα4Galβ4GlcβCer	
E. coli K99[a]	GalNAcβ4(NeuGcα3)Galβ4GlcβCer	
Pseudomonas[b]	Galβ3GalNAcβ4Galβ4GlcβCer	165
Actinomyces naeslundii[c]	Galβ3GalNAcβ4Galβ4GlcβCer	163, 164
Propionibacterium granulosum[d]	Galβ3(NeuAcα6)GalNAcβ4Galβ4GlcβCer	156
	GalNAcβ3Galβ4GlcβCer	
	GlcNAcβ3Galβ4GlcβCer	
	Galα3Galβ4GlcβCer	
	GalNAcβ3Galβ4GlcβCer	
	Galβ3(Fucα4)GlcNAcβ3Galβ4GlcβCer	

*The list shows selected examples of isoreceptors only. Compare with Tables 3 and 4.
[a] Results were presented at the 14th International Carbohydrate Symposium in Stockholm, August 14–19, 1988 (F. K. de Graaf, K.-A. Karlsson et al, in preparation).
[b] This species is most probably a lactosylceramide binder (K.-A. Karlsson, et al, unpublished) with binding properties very similar to that of Neisseria gonorrhoeae ([166), see also Table 8].
[c] See footnote d of Table 4.
[d] Most bacteria of Table 6 bind in this way. See also Table 8.

for isoreceptors may be used technically to approximate the binding epitope on a receptor sequence (107–111, 171). This is outlined below.

The internal binding probably requires a certain design of the receptor-binding proteins, which should have a biological meaning (see below). One may note that this internal binding apparently has resulted in some confusion in the interpretation of binding specificities, notably in the field of toxins. Before it was clearly realized (104, 107, 111, 131, 151, 154, 171), the assumption was a terminal binding as in the case of most antibodies. This in combination with unsafe binding data gave conclusions of "unspecificity," since several different terminal sequences were interpreted to bind. Therefore the internal binding may initially complicate the processing compared with antibody binding, and a number of both positive and negative isoreceptors have to be isolated and characterized for proper epitope analysis (see below).

Many Bacteria Carry the Lactosylceramide Specificity

A surprisingly large number of bacteria have been shown to express the lactosylceramide specificity (Table 6). This is comparable only with the many Man-binders (6–14). Most of the bacteria bind identical or very similar to *Propionibacterium granulosum,* which has been characterized in detail (156). However, partly due to the inability to inhibit with free lactose based on a low affinity, the biological relevance of the binding has not yet been proven. The adhesin for *Neisseria gonorrhoeae* is at present being genetically cloned (166; M. So et al, in preparation), making relevant testing studies soon accessible.

Both gram-negative and gram-positive bacteria with various colonization tissues are binders, including both the normal flora and pathogens. Thus, if this specificity has a meaning for adhesion, evidently a number of bacteria may compete for the same receptor. Interestingly, the dominating anaerobes of the normal flora of human large intestine, including *Bacteroides, Clostridium, Fusobacterium, and Lactobacillus,* may all use this specificity. Lactosylceramide is present in the colon epithelium (86) but not in the small intestinal epithelium (84), coinciding with the normal colonization. Aspects on the membrane-close lactosylceramide as a site to establish a secondary binding are given below.

In some cases, after antibiotic treatment the serious pseudomembraneous colitis may develop. This is based on overgrowth with *C. difficile* and production of toxin (Table 3), on behalf of *Bacteroides* and others that are sensitive to common antibiotics (172). These bacteria thus share one binding specificity, and one potential therapy is recolonization with normal flora and competition at the receptor level. Recent experiments with *Lactobacillus* to treat colitis (173) and with *Lactobacillus* to treat recurrent urinary tract infection (174) show promising results. Using the normal microflora to

Table 6   Bacteria shown to express lactosylceramide binding*

Bacterium	Colonization or target tissue	Comment
Bacteroides fragilis[a]	Large intestine	Several strains tested
Bacteroides ovatus[a]	Large intestine	Several strains tested
Bacteroides vulgatus[a]	Large intestine	Several strains tested
Bacteroides distasonis[a]	Large intestine	Several strains tested
Bacteroides thetaiotamicron[a]	Large intestine	Several strains tested
Lactobacillus fermentum[a]	Large intestine	ATCC 9338
Lactobacillus acidophilus[a]	Various places	ATCC 4356
Fusobacterium necrophorus[a]	Large intestine	ATCC 25286
Fusobacterium varium[a]	Large intestine	ATCC 8501
Clostridium difficile[a]	Large intestine	Several strains tested
Clostridium botulinum[a]	Large intestine	ATCC 25763 and 25764
Propionibacterium granulosum[b]	Skin, large intestine	
Propionibacterium acne[a]	Skin	
Propionibacterium freudenreichii[c]	Milk products	This is a receptor variant, see Table 8
Actinomyces viscosus[a,d]	Mouth	
Actinomyces naeslundii[a,d]	Mouth	
Shigella dysenteriae[a]	Large intestine	Several strains tested
Shigella flexnerii[a]	Large intestine	Several strains tested
Shigella sonnei[a]	Large intestine	
Salmonella typhimurium[a]	Large intestine	Inconsistent results from several strains
E. coli[a]	Intestine	A number of intestinal strains gave inconsistent results
Vibrio cholerae[a,e]	Small intestine Large intestine	Separate biotypes (classical, El Tor) and serotypes (Inaba, Ogawa) gave consistent results
Campylobacter jejunii[a]	Intestine	
Hemophilus influenzae[a]	Respiratory tract	
Yersinia pseudotuberculosis[f]	Intestine	Several strains tested
Yersinia pestis[f]	Intestine	Attenuated strain
Neisseria gonorrhoeae[g]	Genital tract	
Pseudomonas aeruginosa[h]	Respiratory tract	

*If not otherwise stated, the specificity is similar to that reported in detail for *Propionibacterium granulosum* (156) and *Actinomyces* (164). The binding was characterized with the thin-layer overlay technique (106), using radiolabeled bacteria and a number of well-characterized glycolipids.

[a] K.-A. Karlsson, A. A. Lindberg, et al, in preparation.
[b] Refs. 154, 156.
[c] Refs. 154, 155.
[d] Ref. 164.
[e] See also Table 7.
[f] K.-A. Karlsson, H. Wolf Watz, et al, in preparation.
[g] Ref. 166.
[h] See footnote b of Table 5.

promote health in various parts of the human body was recently discussed more generally (175, 176). As is clear from Table 6, most potential ecological niches (intestine, oral cavity, respiratory tract, genitalia, skin) are represented by lactosylceramide binders.

V. cholerae may become an interesting object in the future for the analysis of several aspects of receptor-mediated phenomena. Some comments are therefore given from the lactosylceramide-binding property (Table 6) on the tissue tropism in the intestine and potential relation to the blood group phenotype of susceptible people (Table 7). Several both carbohydrate and noncarbohydrate specificities have been described for *V. cholerae* (177). The former vary between different serotypes and biotypes and are not related to virulence. The latter is based on a soluble hemagglutinin (protease) apparently present in all pathogenic variants. The lactosylceramide binding is also present in all major serotypes and biotypes (Table 6).

Cholera is characterized by hypersecretion (diarrhea) in the small intestine, induced by cholera toxin. As briefly summarized in Table 7, isoreceptor glycolipids for bacterial cells of *V. cholerae* exist both in the small and large intestine. However, only the large intestine has blood group–independent free lactosylceramide. The only positive isoreceptor in the small intestine is Lewis a-active glycolipid, which is a dominating species of Lewis-positive nonsecretors, but only a minor glycolipid of other Lewis-positive phenotypes. This suggests that Lewis-positive nonsecretors may have a higher risk of contracting cholera. In fact, although the Lewis phenotype has not yet been considered, a few limited investigations show a relation with blood group O and nonsecretion (178–181). It has also been known for a long time that the blood group distribution in areas with cholera disease is shifted compared with noncholera areas (182). Therefore a reanalysis of this question is highly desirable using typing reagents also for Lewis phenotypes. One may note the possible colonization of *V. cholerae* in the large intestine of any individual, where the toxin is inactive, explaining why symptomless individuals may spread the disease.

If this relation of disease to blood group phenotype is valid, one would expect similar outcomes for other lactosylceramide-binding pathogens (Table 6), provided that the balance of expression in the actual tissue between free blood group–independent lactosylceramide and Lewis glycolipid makes the latter decisive. Interestingly, nonsecretion has recently been shown to predispose for infection by *Hemophilus influenzae* (183), and this may be expected also for infections with other members of Table 6, e.g. *N. gonorrhoeae* and *Pseudomonas aeruginosa*.

Thus this sophisticated receptor-mediated selection is based on two combined facts or properties. One is the exclusive presence of the receptor specificity (lactose) only in lipid-bound form. The other is the internal binding

Table 7 Receptor expression parameters in human intestine in relation to cholera disease and Lewis blood group phenotypes[a]

Isoreceptor glycolipid	Binder of Vibrio cholerae	Lewis positive				Lewis negative			
		Small intestine[b]		Large intestine[b]		Small intestine[b]		Large intestine[b]	
		Se+	Se−	Se+	Se−	Se+	Se−	Se+	Se−
Lactosylceramide — Galβ4GlcβCer	+	−	−	+	+	−	−	+	+
Lewis a glycolipid — Galβ3(Fucα4)GlcNAcβ3Galβ4GlcβCer	+	(+)	+++	+	+	−	−	+	−
Blood group O glycolipid[c] — Fucα2Galβ3GlcNAcβ3Galβ4GlcβCer	−	+	−	+	+	+++	−	+	−
Lewis b glycolipid[d] — Fucα2Galβ3(Fucα4)GlcNAcβ3Galβ4GlcβCer	−	+++	−	+	−	−	−	−	−
Blood group A glycolipid[e] — GalNAcα3(Fucα2)Galβ3GlcNAcβ3Galβ4GlcβCer	−	(+)	−	+	−	+++	−	+	−
Blood group B glycolipid[f] — Galα3(Fucα2)Galβ3GlcNAcβ3Galβ4GlcβCer	−	(+)	−	+	−	+++	−	+	−
Blood group A and Lewis glycolipid[e] — GalNAcα3(Fucα2)Galβ3(Fucα4)GlcNAcβ3Galβ4GlcβCer	−	+++	−	+	−	−	−	−	−
Blood group B and Lewis glycolipid[f] — Galα3(Fucα2)Galβ3(Fucα4)GlcNAcβ3Galβ4GlcβCer	−	+++	−	+	−	−	−	−	−

[a] Data were taken from recently published results on the expression of blood group glycolipids in the small (84) and large (86) intestine. Overlay analysis has been performed with Vibrio cholerae for binding to glycolipid preparations from epithelial cells of single individuals and from a bank of glycolipids. Only case 3 from Ref. 84 was shown to be a positive binder, which was a Lewis positive nonsecretor. The detailed specificity for Vibrio cholerae is very similar to that of Propionibacterium granulosum (156), see also Tables 4 and 8. The details of Vibrio cholerae binding will be reported in detail (K.-A Karlsson, et al, in preparation). The primary receptor sequence, lactose, is underlined in the isoreceptors presented.

[b] There is a rough evaluation of relative amounts in case of small intestine. For large intestine only presence or absence is indicated.

[c] Provided the individual is blood group O.

[d] The relative amounts given are based on blood group A and B negative individual.

[e] Provided the individual is blood group A positive.

[f] Provided the individual is blood group B positive.

of an isoreceptor, where only one particular blood group glycolipid is receptor-active among many others in the target tissue for infection. The potential importance of membrane proximity for bacterial binding is discussed below.

Glycolipid Crypticity and Receptor Accessibility

One important question concerning lactosylceramide and other shorter glycolipids in relation to bacterial adhesion is the accessibility or availability of the epitope for binding close to the membrane bilayer. Classical studies have demonstrated (mainly for erythrocytes of various animal species and some cultivated cell lines) that one- and two-sugar glycolipids on normal cells are inaccessible for galactose oxidase–mediated radiolabeling. NIL cells transformed with polyoma virus showed, however, a much better labeling of lactosylceramide due to a less complex surface of transformed cells (184). More recent studies have shown that even longer-chain glycolipids may be poorly labeled in erythrocytes of some species, suggesting a tight interaction laterally with proteins and other membrane components (185). An interesting comparison has been done of glycolipids inserted in vesicles and dissolved in tetrahydrofuran (186). The best-labeled glycolipid in both cases was the four-sugar globoside ending with GalNAcβ. The five-sugar Forssman glycolipid ending with GalNAcα was difficult to label both on vesicles and in solution, while lactosylceramide was not labeled on vesicles but rather good in solution.

Such results may argue against lactosylceramide being available for bacterial binding on cell surfaces, especially on target epithelial cells, which are still more complex than the cells so far analyzed with galactose oxidase. However, the two situations may differ in several respects. The fact that the Forssman glycolipid is not even labeled in solution may indicate a conformation effect in the molecule. I therefore compared conformations calculated by the Hard Sphere Exo Anomeric (HSEA) method (see further below) for globoside and Forssman glycolipid (the latter differing only by an additional GalNAcα). Interestingly, the terminal HO-6 groups, being the oxidation sites, pointed in opposite directions in the two molecules, toward the ceramide side facing the bilayer in the Forssman glycolipid and toward the bulk phase in the case of globoside. This may explain the different results with the two glycolipids, although both were probably presenting their terminal sugars to the bulk phase. The more bilayer-close lactosylceramide was labeled in solution but not on the membrane, indicating a sterical effect from the surface layer. The calculated conformation of lactosylceramide with the epithelial type of ceramide (see below and Figure 3) has the HO-6 group of Gal pointing toward the membrane and therefore not easily accessible for oxidation. (The nonepithelial type, which was used by these authors, may on the other hand be more exposed, see below.) The lactosylceramide-binding bacteria have, however, been proposed to bind on the side being directed toward the bulk phase (see details below). Therefore the data from studies with galactose oxidase may not be directly relevant for bacterial binding. In the case of Forssman glycolipid we have found *E. coli* strains that readily agglutinate sheep erythrocytes based on the binding to this glycolipid (187). Labeling with galactose oxidase of sheep erythrocytes showed, however, that Forssman glycolipid was mainly in the cryptic state (185). The accessibility of a certain epitope is thus probably dependent both on the molecular conformation itself and on masking from other cell-surface substance. Modeling of a complex glycolipid with several fucose branches being specifically recognized by monoclonal antibody indicated binding to a nonpolar side of the chain (188). It was proposed that this side was exposed to the outside, while the polar side was lying down on the bilayer surface. Cholera toxin (see above and also below) requires a bilayer-close receptor epitope, evidently accessible at the surface of target small intestinal epithelial cells, which are probably much more crowded than erythrocytes.

Recent theoretical models for particle-particle and cell-cell interactions include specific bonding and nonspecific repulsion from freely mobile repellers (189). A substantial redistribution of repellers from the contact area may appear depending on the balance of forces involved. Therefore, a primary collision between a bacterium and an animal cell (possibly based on a first-step receptor, see below) may induce exposition of lactosylceramide not accessible for galactose oxidase. Work on intricate model systems is required to prove this.

Closely Related Receptor-Binding Variants

The overlay method for binding to a number of isoreceptor glycolipids separated on a thin-layer plate is ideally suited for revealing receptor-binding variants of microbes. Slight differences in amino acid sequence in or at the binding site of the receptor-binding protein are expected to produce separate binding preferences for isoreceptors based on the internal binding. Thus, different binding patterns are shown by autoradiography, although the minimum recognized receptor sequence may be identical. Table 8 illustrates some examples for lactosylceramide- and Galα4Gal-binders. Similar variants are cholera toxin and heat-labile toxin (Table 3, see also below and Figure 2), and several other toxins of Table 3. Slightly different specificities for Man binding were shown by inhibition of yeast agglutination by several bacteria using Man oligosaccharides (95), and variants of *E. coli* of Table 8 were in fact picked up by hemagglutination tests (187).

The two lactosylceramide-binding *Propionibacterium* species have separate preferences for ceramide structures (see below). The binding for *Neisseria* is probably identical to that for *Pseudomonas* (Tables 4 and 6), although the

Table 8   Examples of receptor-binding variants*

Lactosylceramide-binding bacteria

Isoreceptor	Propionibacterium freudenreichii[c]	Propionibacterium granulosum[d]	Neisseria gonorrhoeae[e]
Galβ4GlcβCer[a]	+	−	−
Galβ4GlcβCer[b]	−	+	(+)
GalNAcβ4Galβ4GlcβCer	−	+	+
Galβ3GalNAcβ4Galβ4GlcβCer	−	+	+
GlcNAcβ3Galβ4GlcβCer	−	+	−
Galα3Galβ4GlcβCer	−	+	−

Shiga toxin and *E. coli*

	Shiga toxin[f]	E. coli[g] pPIL	Pap5	Prs
Galα4Galβ4GlcNAcβ3Galβ4GlcβCer	+	+++	+	−
Galα4Galβ4GlcβCer	+	+++	+++	−
GalNAcβ3Galα4Galβ4GlcβCer	(+)	+++	+++	+
GalNAcβ3GalNAcβ3Galα4Galβ4GlcβCer	−	+++	+++	+++
GalNAcα3GalNAcβ3Galα4Galβ4GlcβCer	−	+++	+++	+++

*The primary receptor sequence has been underlined.
[a] Ceramide with nonhydroxy fatty acid and dihydroxy long-chain base, see Table 9.
[b] Ceramide with 2-hydroxy fatty acid and/or trihydroxy long-chain base, see Table 9.
[c] Ref. 155.
[d] Ref. 156.
[e] Ref. 166.
[f] Ref. 132.
[g] The general approach and some of the data have been published (187). Cloned genes for variant adhesins were expressed in *E. coli* HB 101, which were labeled metabolically with ^{35}S-methionine and analyzed for binding to a large series of glycolipids by the overlay method based on a thin-layer chromatogram (106). These data will be described in detail (K.-A. Karlsson, S. Normark et al, in preparation).

weak binding to free lactosylceramide was not observed for *Pseudomonas* (165). Our preliminary conclusion is that all three *E. coli* variants have Galα4Gal as a basic requirement (primary epitope, see definition below). However, the Prs adhesin has lost binding to terminal Galα4Gal and compensated with improved affinity to neighboring groups (secondary epitope, see definition below), where the terminal sequences for the two last binders differ (GalNAcβ and GalNAcα, respectively).

For influenza virus, two sialic acid–binding variants (α3 and α6, respectively) were elegantly shown to be based on one single amino acid substitution (Gln and Ile, respectively) of the binding site of the hemagglutinin (5, 102, 122). In the present examples, molecular cloning techniques presently applied on *Neisseria* (166; M. So, et al, to be published) and *E. coli* (187; S Normark, et al, to be published) will help in elucidating if similar mutations make up the basis for the binding specificities.

Influence of Ceramide Structure on Epitope Presentation

The large variation in ceramide structures, including in the level of hydroxylation of the fatty acid (none to two) and in the long-chain base (two or three), is not unexpected to influence the conformation and presentation of ceramide-close epitopes. The majority of lactosylceramide binders (Table 6) show a selective binding to molecular species with the higher level of hydroxylation (two or three free hydroxyl groups), but do not bind to lactosylceramide with only one free hydroxyl of the ceramide (156). Conversely, two bacterial species, including *Propionibacterium freudenreichii* (155), show a selective binding to lactosylceramide with less hydroxylated ceramide but no binding to other species (Table 9). Interestingly, this preference coincides with a distinct tissue localization of these molecular species (Table 9). In the human, and several animals, the epithelial cells of mucous surfaces (e.g. intestine), where many of the bacteria of Table 6 colonize, contain glycolipids with exclusively hydroxy fatty acids and a large amount of trihydroxy base, which are absent from the subepithelial tissue (78, 84). Free lactosylceramide of human colon epithelium is of this type (86).

As the molecular species with an extra hydroxyl group on the base were equally good binders as those with the hydroxyl group on the fatty acid (Table 9), it is unlikely that this part is directly involved in the binding to the protein. Rather, the interpretation was that the adhesins differed in their selecting of two separate epitopes on lactose, which were differently directed from the bilayer depending on ceramide structure. Support for this exists in X-ray (91) and NMR (190) data. The crystal conformation of GalβCer with hydroxy fatty acid had the hexose in a bent, L-like conformation in relation to ceramide. In contrast, NMR studies of deuterium-labeled GlcβCer with nonhydroxy fatty acid in multilamellar dispersions in water showed all Glc headgroup orientations projecting straight up from the bilayer region into the aqueous phase. Although pure species of lactosylceramide have to be analyzed in similar ways before final interpretations, it is likely that such separate conformers may explain the selected bacterial specificities.

The binding results were obtained from overlay on thin-layer plates (plastic treatment) and from glycolipids coated in microtiter wells. Evidence for a ceramide influence on antibody binding to glycolipids has been obtained both from living cells (191) and from model vesicles (192). The precise data from sulfated GalβCer (ceramide-close epitope) and a population of high- and low-affinity polyclonal antibodies showed that lengthening the fatty acid distinctly increased surface exposure, while hydroxylation decreased exposure of the actual epitope (192).

Table 9 Influence of ceramide structure on the binding of two variants of *Propionibacterium* to lactosylceramide

Isoreceptor	Ceramide components[a]	Localization[b]		Propionibacterium freudenreichii[c]	Propionibacterium granulosum[d,e]
		Epithelial	Subepithelial		
Galβ4Glcβ<u>Cer</u>	Nonhydroxy fatty acid Dihydroxy base	−	+	+	−
	2-Hydroxy fatty acid Dihydroxy base	+	−	−	+
	Nonhydroxy fatty acid Trihydroxy base	+	−	−	+
	2-Hydroxy fatty acid Trihydroxy base	+	−	−	+
Galβ3GalNAcβ4Galβ4Glcβ<u>Cer</u>	Nonhydroxy fatty acid Dihydroxy base	−	+	−[f]	+[g]
	2-Hydroxy fatty acid Trihydroxy base	+	−	−	+

[a] Only overall composition is given. For details, see References 78, 84, 86, 155, and 156.
[b] In human tissues the distinction is clear, as shown for the small (84) and large (86) intestine.
[c] From Ref. 155.
[d] From Refs. 155 and 156.
[e] This bacterium represents most of the species of Table 6 concerning binding preference.
[f] The disaccharide added to lactose blocks the binding, see Table 8.
[g] Apparently, the importance of ceramide hydroxylation, as is evident for free lactosylceramide, disappears when lactose is extended with additional sugars, indicating an effect on presentation also from the saccharide (155, 156).

Biological Aspects on Some Receptor Binding Characteristics

LOW-AFFINITY INTERACTIONS   The binding of Shiga toxin to Vero cells may be used as a working definition of a low-affinity interaction (132). Free receptor oligosaccharide at fairly high concentration showed no detectable inhibition, while the same saccharide multivalently linked to albumin was a good inhibitor (successive improvement with increasing number of linked saccharides). Similarly, free lactose is unable to inhibit lactosylceramide-binding bacteria (156, 164). However, in this case, albumin-linked lactose is ineffective, interpreted as a nonoptimal presentation of the relevant epitope (see also ceramide influence above). Low-affinity interactions also exist for many viruses (5) and eukaryotic lectins (38). Classical inhibition assays based on free saccharides fail to detect a binding in such cases, but binding is easy to show with glycolipids in solid-phase assays or saccharides multivalently linked as neoglycoproteins (38, 132). Many bacterial specificities were, however, picked up by inhibition with free saccharides and are therefore of higher affinity, like Man-binders (95), Galα4Gal-binders, GalNAc- and Gal-binders, and several toxins (Tables 3 and 4).

The reason why the two most common bacterial specificities known at present, to Man and lactose, have different levels of affinity is still unclear. Certain habitats and microenvironments for binding may require a low-affinity interaction. The receptor saccharide for cholera toxin, active in the small intestine, is not expected to appear free at the membrane since the degradation pathway from glycolipid may not permit this. On the other hand, the Shiga toxin working in the large intestine may be exposed for Galα4Gal sequences enzymatically released from glycoproteins, since in erythrocytes the major part of these sequences have been shown to exist in glycoproteins and not in glycolipids (193). The low affinity of lactosylceramide binding may balance this specificity as a second-step receptor (see below).

INTERNAL BINDING   The property of binding to sequences located within a saccharide chain is a novel feature picked up mainly from binding to glycolipids. It differs from antibodies recognizing comparable series of glycolipids, which almost all bind in the nonreducing end (39, 40, 44). Possibly, eukaryotic lectins may share the same property (194). The design of the lectin may therefore be adapted to accommodate the continuation of the chain, and this design should have a biological purpose. Although no bacterial lectin structure has yet been solved, the influenza virus hemagglutinin has a pocketlike site apparently accepting only a terminal NeuAc (122). On the other hand, crystal structures of microbial transport proteins in complex with sugars have shown the binding site between two major domains of the protein, and in the case of maltose binding, a cyclic maltose oligosaccharide was as good a binder as free maltose (22; F. A. Quiocho, unpublished).

Antibodies may differ, since antigenic determinants, including the basis of blood groups (see Table 7), which differ between individuals, have mostly a terminal location. Microbes may thus avoid such differences by selecting core sequences, which may be more tissue specific (see glycolipid series of Table 1). However, this internal binding may be more important for an efficient receptor drift and potental shift of target cell. The variants of Table 8 illustrate mutations for changes in preference for neighboring groups with a retained requirement for the minimum receptor sequence (in these examples lactose and Galα4Gal, respectively). One may imagine fewer point mutations required for shifting between isoreceptors than between two separate minimum sequences like lactose and Galα4Gal. Many examples may be given where separate isoreceptors have tissue specificities to make a change of target cell possible. Thus, if free lactosylceramide is absent in the cell (see small intestinal epithelium of Table 7), a shift of preference from R-GlcNAcβ3Galβ4Glc to R-GalNAcβ4Galβ4Glc may be sufficient to change host cell (see *Neisseria* of Table 8). For *E. coli* of Table 8, an almost loss of binding to globoside, which is dominating in human kidney, and preference for Forssman glycolipid (added GalNAcα3) may shift to dog kidney, where this glycolipid is a major species. In fact many isolates from dog urine carry this specificity (K.-A. Karlsson, T. Korhonen, et al, unpublished results).

It may be practical for this novel binding phenomenon to define the minimum receptor sequence (e.g. lactose or Galα4Gal) as the *primary binding sequence or epitope* and the decisive additions in isoreceptors as *secondary binding sequence or epitope*. The corresponding designations for the protein are *primary and secondary binding sites*. A *receptor drift* is a shift in binding specificity assumed to be based on a mutation in either the primary or secondary binding site or both, and affect either the primary or secondary binding epitopes or both. It is of interest that a variant specificity may be rather stable. Thus the binding type for *Neisseria* (Table 8) was conserved through isogenic variants with or without pilus expression (166).

Therefore the general postulate is that the internal binding has been developed mainly as an effective means of receptor drift to select target cell, where a change in secondary specificity is less expensive than a change in primary specificity.

SPECIFICITY AND PROXIMITY FOR MEMBRANE BILAYER   The lactosylceramide binding is interesting since lactose in bound form is known only in glycolipids, making it membrane-anchored and not appearing in glycoproteins in secretions. Free lactose in fairly large amounts as existing in milk and milk products may be unable to inhibit (e.g. in throat and gastrointestinal tract or excreted in urine) due to the low-affinity binding as discussed above. Therefore the selected recognition of lactosylceramide may ensure a membrane association.

Lactosylceramide is, however, present in most epithelial cells and cannot mediate the distinct tissue tropism for many of the bacteria of Table 6. Therefore, lactosylceramide may be used as a *second-step receptor* to establish the adhesion. Most bacteria carry more than one binding specificity. The purpose of the common Man binding to glycoproteins existing on all cells may be to associate peripherally to a potential target cell to find the receptor mediating the tropism. After this a lateral redistribution of surface components (see discussion on accessibility above) may expose lactosylceramide to establish the adhesion. The low-affinity binding may be important to avoid a direct binding to lactosylceramide, without a selection of cell through the first-step receptor (compare cholera toxin and its direct high-affinity binding to the GM1 glycolipid receptor).

Many pathogenic bacteria of Table 6 are invasive, and binding to lactosylceramide may mediate the essential proximity for a certain protein factor to work, which has been cloned from *Yersinia pseudotuberculosis* (195). Recent invasion data for *N. gonorrhoeae* (196), *Shigella* (197), and *E. coli* and *Salmonella* (198) may later be used in conjunction with the cloned lactosylceramide-binding adhesin (166; M. So, et al, in preparation) to find out a possible relation.

An importance for membrane proximity in the case of toxin action is also of potential interest. There is evidence that adherent *V. cholerae* may contribute more to the pathogenesis of cholera (199), probably through a more effective delivery of toxin molecules to receptor sites, than the production of toxin in the intestinal lumen. It is of interest that several toxin-producing bacteria (Table 3) are lactosylceramide binders (Table 6), including *V. cholerae, C. difficile, C. botulinum,* and *Shigella dysenteriae*. It is possible that a second-step establishment of bacterial cell adhesion close to the membrane bilayer is an efficient means of reaching laterally diffusible partially cryptic glycolipid receptors by in situ–produced toxin. In the case of the very potent botulinum toxin, which is often produced in food or intestine and is taken up and exerts its effect on nerve membranes at myoneural junctions (16), the interesting second binding specificity (Table 3) to GalβCer, a major nerve membrane component (43, 44), may be essential for penetration. Thus, a primary attachment to sialic acid–containing receptor, mediated by the heavy subunit of the toxin, is followed by association to GalβCer (the binding epitope probably composed of both polar and nonpolar parts at the membrane border), mediated by a region of both subunits, to penetrate the membrane.

INHIBITION OF BINDING

If attachment is an essential primary event for colonization and infection, the interference with binding seems a logical goal for medicine. However, with the lack of more precise data on receptor specificities and due to the difficulties of synthesis of natural and modified carbohydrates, this field is just beginning. There are several ways of interfering with binding or adhesion. As the molecular basis is protein-carbohydrate interaction, one may use one of these components in natural or synthetic form. A prophylactic treatment with possibly life-long effect is vaccination against the protein component. This may be only a minor protein located at the tip of the pilus (200, 201), maybe only a few hundred copies per bacterial cell, requiring cloning techniques to produce enough material. Synthetic peptides corresponding to several regions of the Galα4Gal-binding pilus have been used to immunize and induce protection in a mouse pyelonephritic model (202). An interesting treatment, as already noted above, is to deliver normal flora, which may compete with pathogens, possibly using the same receptors, where vaccination therapy is impossible (for instance the lactosylceramide specificity).

One argument against using optimized receptor analogues is the potential interference with assumed physiological carbohydrate functions (32–42, 203). However, if the detailed binding epitopes on the same receptor sequence differ in the two cases, which is likely, then sophisticated drug design may produce selective effects. Also, a potential treatment may in most cases possibly be only acute and not for long periods.

A recent pioneering report on the crystal structure of the complex between the influenza virus hemagglutinin and its receptor, sialyllactose, had an optimistic note on drug design based on the interactions revealed (122). No doubt such an approach may prove to be an important supplement to traditional therapy, e.g. in serious cases with antibiotic resistance or rapid antigenic drift.

Natural Inhibitors

Selection of receptor variants of influenza virus may be based on the presence of macromolecules probably carrying the receptor epitopes (5). It is an old assumption that receptor sequences of secreted macromolecules, notably mucins, may compete with the cell association of microbes (6–18). We urgently need to investigate the balance between membrane-bound and secreted receptors, but we still lack relevant methods. However, a recent study (J. Parkkinen, et al, in preparation) focused on soluble compounds in human urine and found inhibitory substances for less frequent *E. coli* colonizers like S-fimbriated strains, but was not able to detect substances, neither of high- nor low-molecular weight, that inhibited the very frequent Galα4Gal specificity. Thus natural inhibitors based on receptor specificities may be important. Attempts are being made to prepare natural substances to prevent infection, e.g. oligomannoside glycopeptides of plant origin being potent inhibitors of *E. coli* with type 1 pili (96).

Synthetic or Modified Compounds

In the case of *E. coli* K88, it was possible to inhibit the adhesive activity by peptides derived from the adhesin and by synthetic peptides (204). Synthetic efforts in the bacterial adhesion field have, however, mostly been concerned with carbohydrate receptors. The following examples are not comprehensive, but rather illustrate the approach for analyzing protein-carbohydrate interactions.

A most impressive series of papers has been concerned with synthesis of systematically modified blood group oligosaccharides to test the influence on binding to antibodies and plant lectins (205, 206, and references therein), of considerable interest for work on bacterial receptors. A common feature for several systems was formulated as the hydrated polar gate principle (205, 207). According to this, there are key polar groupings for hydrogen bonding adjacent to a relatively large lipophilic surface making up the binding epitope on the sugar. Crystallographic studies of complexes, especially the very high-affinity complex of L-arabinose and its transport protein in *E. coli* (22), documented that hydrogen bonds and van der Waals contacts are the dominant forces that stabilize the interactions.

In a few cases synthetic oligosaccharides have been used to retrieve information on the binding site of bacterial adhesins. *A. viscosus* and *A. naeslundii* were compared in detail using 16 synthetic substances and inhibition of coaggregation with *Streptococcus sanguis* (99). A systematic study is being performed on Galα4Gal and binding of *E. coli* (208), using inhibition of hemagglutination as assay (G. Magnusson, et al, unpublished results). Earlier proposals described the binding epitope on the nonpolar side of the disaccharide (151), see also Figure 1 below. Of the seven hydroxyl groups of the disaccharide, five were concluded to be engaged in hydrogen bonding with the protein, one was intramolecularly bonded, and one in contact with the bulk phase. NMR and HSEA calculations showed very similar conformations for the natural disaccharide and the synthetic analogues. To test these analogues on the binding of Shiga toxin (Table 3), they first have to be linked to protein for soluble multivalency or to lipid for overlay assay, due to the low-affinity interaction (see above).

Two notable recent examples of receptor analogues have interesting features for potential applications. One is the methylumbelliferylαMan, which was 10^3 times more efficient than MeαMan in inhibiting type 1–piliated *E. coli* binding to intestinal epithelial cells (209). The aglycon itself or the αGlc derivative were inactive, documenting the specificity for αMan. The second example is the inhibition of Shiga toxin having Galα4Gal specificity (Table 3) with the N-trifluoroacetyl analogue of globotetraose, GalNAcβ3Galα4Galβ4Glc (210). This analogue showed a 50% inhibition at 1.2 mg/ml of the binding of toxin to Vero cells, while globotetraose at much

338   KARLSSON

higher concentration had no inhibitory effect (132, 210). For comparison, the multivalently linked Galα4Gal (25 moles of saccharide per mole of bovine serum albumin) showed 50% inhibition at 0.07 mg/ml (132). The improvement in binding for the analogue is difficult to evaluate but may be several hundred times. The conclusion from these examples is that a modification outside the primary epitope (being specifically recognized by the protein) may exert a dramatic improvement in binding. It is also likely that changes in this secondary epitope have a better chance of improving a binding than modifications in the primary epitope, which has to keep enough character to retain the specificity of binding. Therefore a general principle of the design of receptor analogues may prove to be modulations of the secondary epitopes of isoreceptors.

A calculated model of the glycolipid corresponding to the N-trifluoroacetyl tetrasaccharide is reproduced in Figure 1. It is of interest that solid-phase binding data for the Shiga toxin to the natural glycolipid with N-acetyl group showed a weaker binding than to the triglycosylceramide with terminal

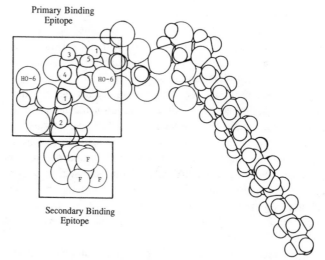

Figure 1 Calculated conformation of the N-trifluoroacetyl analogue of globoside, GalNAcβ3Galα4Galβ4GlcβCer. The free tetrasaccharide has been shown to be a potent inhibitor of Shiga toxin binding to Vero cells (210). The projection shows the side proposed to bind the toxin. The primary binding sequence is Galα4Gal, identical with *E. coli* (Table 4), but with slightly shifted epitopes (Table 8). There is a continuous nonpolar surface made up of ring hydrogens (H-1, H-3, H-4, and H-5 of Galβ and H-1 and H-2 of Galα are marked) surrounded by key polar groups (tested so far only in case of *E. coli*). The trifluoro group is designated secondary binding epitope and is exposed on the same side as the primary binding epitope. The N-acetyl and N-trifluoroacetyl groups have opposite effects on the binding, indicating an improved binding through hydrogen bonding.

Galα4Gal (132). Therefore, the N-acetyl and N-trifluoroacetyl groups have opposite effects on binding, possibly implying a strengthening hydrogen bonding. In the case of the αMan derivative, however, the effect must be mediated by strong van der Waals interactions.

An important aspect is the conversion of the requirement for multivalency of natural receptor sequences to univalent active analogues, since this makes synthetic preparations more realistic. Also, if there is need for resorption from the intestine, the molecule in question probably must have a limit in size.

ANALYSIS OF GLYCOLIPID RECEPTORS

The technology for receptor characterization based on glycolipids is at present very informative. A particularly useful characteristic is the microbial property of binding to internal portions of a saccharide chain.

Detection of a Receptor Glycolipid

Detection of microbial binding to specific glycolipids is highly facilitated by overlay on a thin-layer plate as discussed above. This was the basis for most of the systems gathered in Tables 3, 4, and 6. A large number of receptor candidates may be assayed together with positive and negative controls and test of background level, and the multivalency of presentation picks up also low-affinity binders, which escape traditional inhibition assays based on free oligosaccharides. In this respect the plate mimics the situation on the target membrane.

One vital methodological supplement for other glycoconjugates may be enzymatic or chemical release of the heterogeneous oligosaccharides from the peptide and linking them up to a paraffin chain at the reducing end. This would yield both single molecules with homogeneous saccharide for classical glycolipid isolation, and hydrophobic tails for efficient solid-phase assaying. The important balance between various membrane-bound and secreted glycoconjugates could then be investigated more specifically. Today overlay to intact glycoproteins separated by electrophoresis is accessible (119, 120).

Isolation and Structural Analysis. Identification of Isoreceptors

It is logical to choose the target tissue for infection as the source for analysis and preparation of a receptor. Methods have been adapted for the isolation of total glycolipid mixtures free of nonglycolipid contaminants (118, 211), suited for overlay analysis and also direct inlet mass spectrometry. However, the target tissue may have levels of receptor material that are too low for a realistic isolation (108). Therefore we routinely use glycolipid mixtures of many animal species and tissues to detect isoreceptors and define preparative sources. The levels of glycolipid species differ greatly between various cells, shown very clearly for erythrocytes (44) and intestine (83–86). Also, some of the sources may already have been worked up for glycolipid structures, making it possible to pick isoreceptors from a bank.

The isolation of glycolipid species is the most laborious step, and is being successively improved by advanced chromatographic techniques. Any isolated substance can now be completely analyzed for primary structure, including use of mass spectrometry (212) and NMR spectroscopy (212, 213). More development is, however, required for the direct analysis of mixtures. We have used optimized derivatives and electron ionization from fractionally evaporated mixtures to compare mass spectral scans directly with thin-layer chromatograms of the same mixture (83, 214), providing good information on sequence and ceramide composition. One interesting possibility is the direct desorption of glycolipids from the thin-layer plate (215, 216), to identify a particular band located by overlay with microbe on a mixture. With escalation of the instrumentation to the informative tandem analysis (217), one may envisage a successive desorption over the plate using fast atom bombardment followed by a selected transfer of molecular and fragment ions for collision and analysis in the second instrument.

Conclusion about the primary epitope is given when the minimum sequence is recognized and is the common denominator of all positive binders. The situation may be complicated by several specificities being present in the same microbe, and by the fact that some isoreceptors are negative although carrying the primary epitope (sterical hindrance from neighboring groups). A sufficient number of both positive and negative isoreceptors is required for epitope dissection (see below).

Pure glycolipid species are estimated for relative binding avidities using dilutions on a thin-layer plate or coated in microtiter wells (106). This is simplified by the threshold phenomena referred to above. More precise binding parameters may not yet be defined due to low-affinity interactions and to lack of purified binding proteins.

Approximation of Binding Epitopes on Isoreceptors Using a Comparison of Binding Preferences with Conformation Analysis

The internal binding (Table 5) is the basis for isoreceptors with varying neighboring groups to the primary receptor sequence affecting the binding differently. This is being used for epitope dissection (107–111, 171), which is based on a comparison of binding preferences to isoreceptors with their calculated conformations, using HSEA or similar analysis (218). The sterical shapes of neighboring groups are decisive for access to the primary sequence, and information on the conformation may therefore help defining which part (side) of the primary sequence is involved in the interaction with the binding protein. Also, positive and negative effects from neighboring groups may suggest enzymatic and chemical modifications to further elucidate the binding epitope as a rational preinformation for the more laborious organic synthesis. This is at present being elaborated for several specificities. As this is a novel approach, two brief illustrations will be given to support the statements. One is taken from the classical cholera toxin and one from the large number of lactosylceramide-binding bacteria.

The minimum receptor glycolipid with optimal binding for cholera toxin is the GM1 ganglioside (Table 3). Equally effective are GM1 containing N-glycoloylneuraminic acid and GM1 oxidized with galactose oxidase (219). Derivatives of GM1 after mild periodate oxidation of the NeuAc tail appear to have activities similar to those of intact GM1 (170). Also, the extended GM1 sequence (Table 2) had identical activity with GM1 (74). Adding NeuAcα8 to NeuAc of GM1 (GD1b) reduces activity about 50 times (74), while removing Galβ (GM2) reduces activity 2000 times (74). Fucα2 (169) or GalNAcβ4 (107) added on Galβ3 of GM1 reduces but does not block the binding. These data indicate the essential parts of GM1 and also that substitutions on GM1 are tolerated, meaning an internal binding (Table 5).

Figure 2 shows two separate projections exposing opposite sides of GM1 as calculated by the HSEA method (218). The GM1 pentasaccharide in solution was shown by NMR to possess an identical conformation with the calculated one (220). Using the binding data above gathered from various sources, it seems clear that the binding epitope for the toxin is on the side projected to the left, including part of the NeuAc with the carboxyl group (cannot be reduced or blocked), the acetyl group of GalNAc, the nonpolar side of Galβ3 (ring hydrogens H-1, H-3, H-4, and H-5 marked), and possibly HO-2 of Galβ4. However, the methyl group of NeuAc is outside the epitope, since it may be replaced by hydroxymethyl as in NeuGc without affecting the binding. Similarly, the back side of NeuAc (right projection) exposes HO-9 and HO-8, which may be periodate oxidized and linked to large groups with retained activity. Also, another NeuAc may be limited to HO-8 with only some loss of activity. For the terminal Galβ, HO-6 (top) may be oxidized to aldehyde with no loss of activity. Substitutions at HO-2, HO-3, and HO-4 of Galβ3 are pointing mostly away from the binding side (left), explaining why these do not completely block binding. 4Glcβ may not be essential for the binding since it may be replaced with 3GalNAc (74; Table 2) without reducing binding. However, 4G1cNAcβ in this place may block binding, since such sequence should exist in glycoproteins, although it has not yet been shown. As speculated above, this 4GlcNAcβ may be tolerated by the heat-labile toxin of E. coli (Table 3), explaining why this toxin may bind both glycolipid and glycoprotein. In the left projection the HO-2 of Glcβ is indicated. In case of

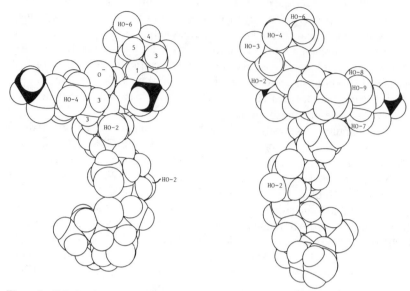

Figure 2   Calculated conformation of the receptor glycolipid for cholera toxin, the ganglioside GM1 (see Table 3). The two projections differ by 180° with the ceramide down (only first six carbon atoms of each chain) and Galβ3 on top. The left projection exposes the side proposed to carry the binding epitope, based on binding data from various papers. The nonpolar (unspecific ?) side of Galβ3 (ring hydrogens H-1, H-3, H-4, and H-5 are marked), the acetamido group of GalNAcβ (methyl carbon in black), the carboxyl of NeuAc (marked O⁻), and possibly HO-2 of Galβ4 are involved in the interaction with the toxin. Glcβ, the acetyl of NeuAc (methyl carbon in black), C-7 to C-9 of NeuAc with modifications and substitutions, and HO-2, HO-3, HO-4, and HO-6 of Galβ3 are probably not essential for the binding. These data illustrate the epitope dissection approach based on internal binding (see text).

GlcNAc the acetamido group is located in this place and may give a sterical hindrance for cholera but not for heat-labile toxin and explain the binding data.

Deleting the terminal Galβ (GM2) reduces binding considerably, which may be due to loss of the nonpolar interaction with this side of Galβ. This secondary epitope is therefore a candidate for replacement with other nonpolar groups to test possible effects on binding (see the example above with methylumbelliferylαMan as a potent inhibitor of bacterial binding). A second obvious modulation site is the acetamido group of GalNAcβ.

The conclusion from this illustration of comparative epitope dissection is that the binding epitope on GM1 ganglioside for cholera toxin may be rather restricted in size, in a crowded region specifically involving three or four sugars but with a total size corresponding to only one or two sugars.

The majority of lactosylceramide binders of Table 6, including *V. cholerae*

(Table 7), show a detailed specificity identical or very similar to that of *P. granulosum* (156), see also Table 5. Figure 3 shows two isoreceptors with five sugars: one is Le^a-5, a positive binder, the other is Le^x-5, a positional isomer without binding activity. HSEA modeling of a number of isoreceptors with lactose as the primary sequence (110, 111) suggests that the side of lactose being recognized is that projected toward the viewer of Figure 3, including the nonpolar side of Glcβ (ring hydrogens H-1, H-3, and H-5 are marked) and a polar part of Galβ. HO-2 of Glcβ is important since N-acetyllactosamine is negative. The major difference between the positive (top) and negative (bottom) binder as seen on this side is that the terminal Galβ (top) is exposing its nonpolar side, while Fucα (bottom) is exposing its more polar side. Modeling the interaction with water molecules (111) makes this difference still more clear. As shown in Table 7 for *V. cholerae,* additional substitutions on the active Le^a-5 sequence (top) block binding, as Fucα2 on Galβ, which substituent is common for all other blood group antigens (Table 7). Therefore only Le^a-5 glycolipid among all blood group antigens is a positive binder for *V. cholerae* and several other bacteria, suggesting a prevalence for certain blood group phenotypes to acquire disease (see above).

Connecting Experiments

A defined receptor glycolipid may be used for assaying cloning of genes for bacterial adhesins (166, 200), which may be crystallized for binding site localization and classical drug design (221), or may be used for vaccination. Even without this information on the protein, the approach described above for binding epitope localization may be valuable as a rational basis for synthetic work. Of importance is the design of univalent receptor analogues with improved binding, which may replace the multivalency required for activity of several important natural receptors. Such tools will be decisive for proving the biological relevance of many of the specificities discussed in this paper.

CONCLUSIONS

Partly based on novel assay methods and improved techniques for isolation and structural analysis, interesting information on glycolipid receptors for microbes is being gathered. Some properties of glycolipids, like strict membrane association and potential proximity of a binding epitope to the membrane bilayer, may make glycolipids especially suited as receptors. Of special interest is the property of microbes in general to recognize sequences placed internally in the saccharide chain, which may facilitate mutational shifts in receptor specificity. It is also of technical value for assigning the binding epitope based on a comparison of binding preferences to isoreceptors with

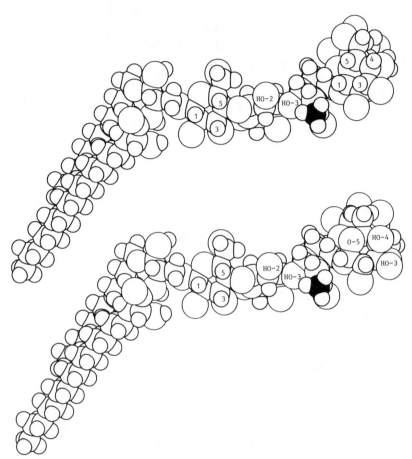

Figure 3 Calculated conformations of two isoreceptors for several lactosylceramide-binding bacteria, including *Vibrio cholerae* (Tables 5–7). The top sequence is a positive binder and Le^a-5, Galβ3(Fucα4)GlcNAcβ3Galβ4GlcβCer, and the bottom sequence is a negative binder and Le^x-5, Galβ4(Fucα3)GlcNAcβ3Galβ4GlcβCer. The two substances are thus positional isomers with similar overall shapes. The projections show the side of lactose concluded to bind the bacteria (110, 111), including the nonpolar side of Glcβ (ring hydrogens H-1, H-3, and H-5 are marked) and a polar side of Galβ (HO-2 and OH-3 marked). One difference between the two isoreceptors on this side is the nonpolar appearance of the terminal Galβ3 of the active Le^a-5 *(top)*, with the ring hydrogens H-1, H-3, H-4, and H-5 marked, while the inactive Le^x-5 *(bottom)* has the polar side of Fucα exposed (HO-3, HO-4, and O-5 marked). This difference in polarity of the secondary epitope may explain the difference in binding. The methyl carbon of GlcNAcβ is black.

calculated conformations. The common lactosylceramide specificity may create a second-step bilayer-close establishment of bacterial adhesion and may also, together with a protein factor, mediate cell invasion. The interesting binding preferences to lactosylceramide isoreceptors may be the basis for a prevalency for cholera and other diseases among certain blood group phenotypes, primarily Lewis positive nonsecretors. The need for proximity to membrane is satisfied for several bacterial toxins that bind glycolipids. Several of the toxin-producing bacteria bind lactosylceramide, adding to this proximity.

Many data on glycolipid receptors are recent and have to be confirmed. Especially the biological relevance for adhesion to target cells has to be proven in most cases. However, several binding characteristics are developed enough to allow a rational synthesis of soluble receptor analogues, and these in combination with genetically manipulated bacteria may produce valuable information in colonization and infection models.

ACKNOWLEDGMENTS

My own work included in this review was supported by grants from The Swedish Medical Research Council (Grant No. 3967) and Symbicom Ltd. I am indebted to Klaus Bock for help with Figure 2 and to Nils Calander, Per-Georg Nyholm, and Irmin Pascher for help with Figures 1 and 3. This paper is dedicated to my mother, Maja Karlsson, on the occasion of her 85th birthday, September 20, 1988.

Literature Cited

1. The Nomenclature of Lipids. 1977. *Eur. J. Biochem.* 79:11–21
2. Abbreviated Terminology of Oligosaccharide Chains. 1982. *J. Biol. Chem.* 257:3347–51
3. Nomenclature of Glycoproteins, Glycopeptides and Peptidoglycans. 1987. *J. Biol. Chem.* 262:13–18
4. Lonberg-Holm, K., Philipson, L., eds. 1981. *Virus Receptors,* Part 2: *Animal Viruses, Receptors and Recognition,* Ser. B, Vol. 8. London: Chapman & Hall. 217 pp.
5. Paulson, J. C. 1985. In *The Receptors,* ed. P. M. Conn, pp. 131–219. London: Academic
6. Beachey, E. H., ed. 1980. *Bacterial Adherence, Receptors and Recognition,* Ser. B, Vol. 6. London: Chapman & Hall. 466 pp.
7. Mirelman, D., ed. 1986. *Microbial Lectins and Agglutinins.* New York: Wiley. 443 pp.

8. Savage, D. C., Fletcher, M., eds. 1985. *Bacterial Adhesion.* New York: Plenum. 476 pp.
9. Marshall, K. C., ed. 1984. *Microbial Adhesion and Aggregation.* Berlin: Springer-Verlag. 424 pp.
10. Horwitz, M. A., ed. 1988. *Bacteria-Host Cell Interaction.* New York: Liss. 393 pp.
11. Berkeley, R. C. W., Lynch, J. M., Melling, J., Rutter, P. R., Vincent, B., eds. 1980. *Microbial Adhesion to Surfaces.* Chichester, England: Horwood. 559 pp.
12. Jones, G. W., Isaacson, R. E. 1983. *CRC Crit. Rev. Microbiol.* 10:229–60
13. Sharon, N. 1987. *FEBS Lett.* 217:145–57
14. Ofek, I., Sharon, N. 1988. *Infect. Immun.* 56:539–47
15. Eidels, L., Proia, R. L., Hart, D. A. 1983. *Microbiol. Rev.* 47:596–620
16. Middlebrook, J. L., Dorland, R. B. 1984. *Microbiol. Rev.* 48:199–221

17. O'Brien, A. D., Holmes, R. K. 1987. *Microbiol. Rev.* 51:206–20
18. Fishman, P. H. 1982. *J. Membr. Biol.* 69:85–97
19. Lis, H., Sharon, N. 1986. *Annu. Rev. Biochem.* 55:35–67
20. Liener, I. E., Sharon, N., Goldstein, I. J., eds. 1986. *The Lectins: Properties, Functions, and Applications in Biology and Medicine.* London: Academic. 600 pp.
21. Olden, K., Parent, J. B., eds. 1987. *Vertebrate Lectins.* New York: Van Nostrold Reinhold. 255 pp.
22. Quiocho, F. A. 1986. *Annu. Rev. Biochem.* 55:287–315
23. Lark, D. L., ed. 1986. *Protein-Carbohydrate Interactions in Biological Systems.* London: Academic. 464 pp.
24. Switalski, L., Höök, M., Beachey, E. H., eds. 1988. *Molecular Mechanisms of Microbial Adhesion.* Berlin: Springer-Verlag. 226 pp.
25. Berger, E. G., Buddecke, E., Kamerling, J. P., Kobata, A., Paulson, J. C., Vliegenthart, J. F. G. 1982. *Experientia* 38:1129–62
26. Schachter, H. 1984. *Clin. Biochem.* 17:3–14.
27. Kornfeld, R., Kornfeld, S. 1984. *Annu. Rev. Biochem.* 54:631–64
28. Ivatt, R. J., ed. 1984. *The Biology of Glycoproteins.* New York: Plenum. 449 pp.
29. Iwase, H. 1988. *Int. J. Biochem.* 20:479–91
30. Höök, M., Kjellén, L., Johansson, S., Robinson, J. 1984. *Annu. Rev. Biochem.* 53:847–69
31. Hassell, J. R., Kimura, J. H., Hascall, V. C. 1986. *Annu. Rev. Biochem.* 55:539–67
32. Cook, G. M. W. 1986. *J. Cell Sci.* Suppl. 4:45–70
33. Brandley, B. K., Schnaar, R. L. 1986. *J. Leukocyte Biol.* 40:97–111
34. Roseman, S. 1985. *J. Biochem.* 97:709–18
35. Hakomori, S.-i. 1986. *Sci. Am.* 254:44–53
36. Prokazora, N. V., Dyatlovitskaya, E. V., Bergelson, L. D. 1988. *Eur. J. Biochem.* 172:1–6
37. Olden, K., Bernard, B. A., Humphries, M. J., Yeo, T.-K., Yeo, K.-T., et al. 1985. *Trends Biochem. Sci.* 10:78–82
38. Monsigny, M., Kieda, C., Roche, A.-C. 1983. *Biol. Cell* 47:95–110
39. Hakomori, S.-i. 1984. *Annu. Rev. Immunol.* 2:103–26
40. Yogeeswaran, G. 1983. *Adv. Cancer Res.* 38:289–350
41. Hanai, N., Dohi, T., Nores, G. A.,

Hakomori, S.-i. 1988. *J. Biol. Chem.* 263:6296–301
42. Usuki, S., Hoops, P., Sweeley, C. C. 1988. *J. Biol. Chem.* 263:10595–99
43. Sweeley, C. C., Siddiqui, B. 1977. In *The Glycoconjugates,* ed. M. I. Horowitz, W. Pigman, 1:459–540. New York: Academic
44. Kanfer, J. N., Hakomori, S.-i., eds. 1983. *Sphingolipid Biochemistry.* New York: Plenum. 485 pp.
45. Hawthorne, J. N., Ansell, G. B., eds. 1982. *Phospholipids.* Amsterdam: Elsevier. 484 pp.
46. Wiegandt, H., ed. 1985. *Glycolipids.* Amsterdam: Elsevier. 320 pp.
47. Egge, H., Kordowicz, M., Peter-Katalinic, J., Hanfland, P. 1985. *J. Biol. Chem.* 260:4927–35
48. Matsubara, T., Hayashi, A. 1986. *J. Biochem.* 99:1401–8
49. Sugita, M., Nakano, Y., Nose, T., Itasaka, O., Hori, T. 1984. *J. Biochem.* 95:47–55
50. Dennis, R. D., Geyer, R., Egge, H., Menges, H., Stirm, S., Wiegandt, H. 1985. *Eur. J. Biochem.* 146:51–58
51. Sekine, M., Suzuki, M., Inagaki, F., Suzuki, A., Yamakawa, T. 1987. *J. Biochem.* 101:553–62
52. Shigeta, K., Ito, Y., Ogawa, T., Kirihata, Y., Hakomori, S.-i. 1987. *J. Biol. Chem.* 262:1358–62
53. Blanchard, D., Piller, F., Gillard, B., Marcus, D., Cartron, J.-P. 1985. *J. Biol. Chem.* 260:7813–16
54. Gillard, B. K., Blanchard, D., Bouhours, J.-F., Cartron, J.-P., van Kuik, J. A., et al. 1988. *Biochemistry* 27:4601–6
55. DeGasperi, R., Koerner, T. A. W., Quarles, R. H., Ilyas, A. A., Ishikawa, Y., et al. 1987. *J. Biol. Chem.* 262:17149–55
56. Kannagi, R., Levery, S. B., Hakomori, S.-i. 1985. *J. Biol. Chem.* 260:6410–15
57. Ostrander, G. K., Levery, S. B., Hakomori, S.-i., Holmes, E. H. 1988. *J. Biol. Chem.* 263:3103–10
58. Levery, S. B., Eaton, H. L., Ostrander, G. K., Holmes, E. H., Salyan, E. K., Hakomori, S.-i. 1989. *J. Biol. Chem.* In press
59. Ohashi, M. 1981. In *Glycoconjugates,* ed. T. Yamakawa, T. Osawa, S. Handa, pp. 33–38. Tokyo: Japan Sci. Soc. Press
60. Taki, T., Hirabayashi, Y., Ishikawa, H., Ando, K., Kon, K., et al. 1986. *J. Biol. Chem.* 261:3075–78
61. Murayama, K., Levery, S. B., Schirrmacher, V., Hakomori, S.-i. 1986. *Cancer Res.* 46:1395–1402
62. Fukushi, Y., Nudelman, E., Levery, S.

B., Higuchi, T., Hakomori, S.-i. 1986.
Biochemistry 25:2859–66
63. Månsson, J.-E., Fredman, P., Bigner,
D. D., Molin, K., Rosengren, B., et al.
1986. *FEBS Lett.* 201:109–13
64. Takamizawa, K., Iwamori, M., Mutai,
M., Nagai, Y. 1986. *J. Biol. Chem.*
261:5625–30
65. Chou, D. K. H., Schwarting, G. A.,
Evans, J. E., Jungwala, F. B. 1987. *J.
Neurochem.* 49:865–73
66. Ariga, T., Kohriyama, T., Freddo, L.,
Latov, N., Saito, M., et al. 1987. *J.
Biol. Chem.* 262:848–53
67. Clausen, H., Levery, S. B., Nudelman,
E., Baldwin, M., Hakomori, S.-i. 1986.
Biochemistry 25:7075–85
68. Clausen, H., Levery, S. B., Nudelman,
E. D., Stroud, M., Salyan, M. E. K.,
Hakomori, S.-i. 1987. *J. Biol. Chem.*
262:14228–34
69. Nudelman, E., Levery, S. B., Kaizu,
T., Hakomori, S.-i. 1986. *J. Biol.
Chem.* 261:11247–53
70. Fukuda, M. N., Dell, A., Tiller, P.
R., Varki, A., Klock, J. C., Kukuda,
M. 1986. *J. Biol. Chem.* 261:2376–
83
71. Riboni, L., Sonnini, S., Acquotti, D.,
Malesci, A., Ghidoni, R., et al. 1986. *J.
Biol. Chem.* 261:8514–19
72. Nores, G. A., Dohi, T., Taniguchi, M.,
Hakomori, S.-i. 1987. *J. Immunol.*
139:3171–76
73. Nudelman, E., Fukushi, Y., Levery, S.
B., Higuchi, T., Hakomori, S.-i. 1986.
J. Biol. Chem. 261:5487–95
74. Nakamura, K., Suzuki, M., Inagaki, F.,
Yamakawa, T., Suzuki, A. 1987. *J.
Biochem.* 101:825–35
75. Araki, S., Abe, S., Odani, S., Ando,
S., Fujii, N., Satake, M. 1987. *J. Biol.
Chem.* 262:14141–45
76. Smirnova, G. P., Kochetkov, N. K.,
Sadovskaya, V. L. 1987. *Biochim. Biophys. Acta* 920:47–55
77. Dennis, R. D., Geyer, R., Egge, H.,
Peter-Katalinic, J., Li, S.-C., et al.
1985. *J. Biol. Chem.* 260:5370–77
78. Karlsson, K.-A. 1982. In *Biological
Membranes*, ed. D. Chapman, 4:1–74.
London: Academic
79. Breimer, M. E., Hansson, G. C., Karlsson, K.-A., Leffler, H. 1982. *J. Biol.
Chem.* 257:557–68
80. Iwamori, M., Nagai, Y. 1981. *Biochim.
Biophys. Acta* 665:214–20
81. Iwamori, M., Shimomura, J.,
Tsuyuhara, S., Nagai, Y. 1984. *J.
Biochem.* 95:761–70
82. Sekine, M., Yamakawa, T., Suzuki, A.
1987. *J. Biochem.* 101:563–68
83. Breimer, M. E., Hansson, G. C., Karls-

son, K.-A., Leffler, H. 1981. *J.
Biochem.* 90:589–609
84. Björk, S., Breimer, M. E., Hansson, G.
C., Karlsson, K.-A., Leffler, H. 1987.
J. Biol. Chem. 262:6758–65
85. Hansson, G. C., Karlsson, K.-A., Thurin, J. 1984. *Biochim. Biophys. Acta*
792:281–92
86. Holgersson, J., Strömberg, N., Breimer, M. E. 1988. *Biochimie.* In press
87. Sekine, M., Nakamura, K., Suzuki, M.,
Inagaki, F., Yamakawa, T., Suzuki, A.
1988. *J. Biochem.* 103:722–29
88. Nakamura, K., Hashimoto, Y., Yamakawa, T., Suzuki, A. 1988. *J. Biochem.*
103:201–8
89. Boggs, J. M. 1987. *Biochim. Biophys.
Acta* 906:353–404
90. Abrahamsson, S., Dahlén, B., Löfgren,
H., Pascher, I., Sundell, S. 1977. In
Structure of Biological Membranes, ed.
S. Abrahamsson, I. Pascher, pp. 1–23.
New York: Plenum
91. Pascher, I., Sundell, S. 1977. *Chem.
Phys. Lipids* 20:175–91
92. Symington, F. W., Murray, W. A.,
Bearman, S. I., Hakomori, S.-i. 1987.
J. Biol. Chem. 262:11356–63
93. Gustafsson, B. E., Karlsson, K.-A.,
Larson, G., Midtvedt, T., Strömberg,
N., et al. 1986. *J. Biol. Chem.* 261:
15294–300
94. Larson, G., Watsfeldt, P., Falk, P., Leffler, H., Koprowski, H. 1987. *FEBS
Lett.* 214:41–44
95. Firon, N., Ofek, I., Sharon, N. 1984.
Infect. Immun. 43:1088–90
96. Neeser, J.-R., Koellreutter, B.,
Wuersch, P. 1986. *Infect. Immun.* 52:
428–36
97. Gunnarsson, A., Mårdh, P.-A., Lundblad, A., Svensson, S. 1984. *Infect. Immun.* 45:41–46
98. Krivan, H. C., Clark, G. F., Smith, D.
F., Wilkins, T. D. 1986. *Infect. Immun.*
53:573–81
99. McIntire, F. C., Crosby, L. K., Barlow,
J. J., Matta, K. L. 1983. *Infect. Immun.*
41:484–50
100. Andersson, B., Dahmén, J., Frejd, T.,
Leffler, H., Magnusson, G., et al. 1983.
J. Exp. Med. 158:559–70
101. Loomes, L. M., Uemura, K.-I., Childs,
R. A., Paulson, J. C., Rogers, G. N., et
al. 1984. *Nature* 307:560–63
102. Rogers, G. N., Paulson, J. C., Daniels,
R. S., Skehel, J. J., Wilson, I. A.,
Wiley, D. C. 1983. *Nature* 304:76–79
103. Hansson, G. C., Karlsson, K.-A., Larson, G., Strömberg, N., Thurin, J., et
al. 1984. *FEBS Lett.* 170:15–18
104. Holgersson, J., Karlsson, K.-A., Karlsson, P., Norrby, E., Örvell, C., Ström-

berg, N. 1985. In *World's Debt to Pasteur*, ed. H. Koprowski, S. A. Plotkin, pp. 273–301. New York: Liss
105. Hansson, G. C., Karlsson, K.-A., Larson, G., Strömberg, N., Thurin, J. 1985. *Anal. Biochem.* 146:158–63
106. Karlsson, K.-A., Strömberg, N. 1987. *Methods Enzymol.* 138:220–32
107. Bock, K., Karlsson, K.-A., Strömberg, N., Teneberg, S. 1988. In *Immunology of Complex Carbohydrates*, ed. A. Wu, pp. 153–86. New York: Plenum.
108. Karlsson, K.-A. 1986. *Chem. Phys. Lipids* 42:153–72
109. Karlsson, K.-A. 1987. *Pure Appl. Chem.* 59:1477–88
110. Karlsson, K.-A. 1988. In *Molecular Mechanisms of Microbial Adhesion*, ed. L. Switalski, M. Höök, E. H. Beachey, pp. 77–96. Berlin: Springer-Verlag
111. Calander, N., Karlsson, K.-A., Nyholm, P.-G., Pascher, I. 1988. *Biochimie*. In press
112. Magnani, J. L., Smith, D. F., Ginsburg, V. 1980. *Anal. Biochem.* 109:399–402
113. Magnani, J. L., Brockhaus, M., Smith, D. F., Ginsburg, V., Blaszczyk, M., et al. 1981. *Science* 212:55–56
114. Brockhaus, M., Magnani, J. L., Blaszczyk, M., Steplewski, Z., Koprowski, H., et al. 1981. *J. Biol. Chem.* 256:13223–25
115. Swank-Hill, P., Needham, L. K., Schnaar, R. L. 1987. *Anal. Biochem.* 163:27–35
116. Magnani, J. L. 1985. *Anal. Biochem.* 150:13–17
117. Higashi, H., Hirabayashi, Y., Ito, M., Yamagata, T., Matsumoto, M., et al. 1987. *J. Biochem.* 102:291–96
118. Karlsson, K.-A. 1987. *Methods Enzymol.* 138:212–20
119. Parkkinen, J., Rogers, G. N., Korhonen, T., Dahr, W., Finne, J. 1986. *Infect. Immun.* 54:37–42
120. Prakobphol, A., Murray, P. A., Fischer, S. J. 1987. *Anal. Biochem.* 164:5–11
121. Weigel, P. H., Schnaar, R. L., Kuhlenschmidt, M. S., Schnell, E., Lee, R. T., et al. 1979. *J. Biol. Chem.* 254:10830–38
122. Weis, W., Brown, J. H., Cusack, S., Paulson, J. C., Skehel, J. J., Wiley, D. C. 1988. *Nature* 333:426–31
123. Pistole, T. G. 1981. *Annu. Rev. Microbiol.* 35:85–112
124. Teodorescu, M., Mayer, E. P. 1982. *Adv. Immunol.* 33:307–51
125. Takamizawa, K., Iwamori, M., Kozaki, S., Sakaguchi, G., Tanaka, R., et al. 1986. *FEBS Lett.* 201:229–32
126. Kamata, Y., Kozaki, S., Sakaguchi, G.,

Iwamori, M., Nagai, Y. 1986. *Biochem. Biophys. Res. Commun.* 140:1015–19
127. Ochanda, J. O., Syuto, B., Ohishi, I., Naiki, M., Kubo, S. 1986. *J. Biochem.* 100:27–33
128. Kozaki, S., Ogasawara, J., Shimote, Y., Kamata, Y., Sakaguchi, G. 1987. *Infect. Immun.* 55:3051–56
129. Jolivet-Reynand, C., Alouf, J. E. 1983. *J. Biol. Chem.* 258:1871–77
130. Clark, G. F., Krivan, H. C., Wilkins, T. D., Smith, D. F. 1987. *Arch. Biochem. Biophys.* 257:217–29
131. Brown, J. E., Karlsson, K.-A., Lindberg, A. A., Strömberg, N., Thurin, J. 1983. In *Glycoconjugates*, ed. M. A. Chester, D. Heinegård, A. Lundblad, S. Svensson, pp. 678–79. Lund, Sweden: Rahms
132. Lindberg, A. A., Brown, J. E., Strömberg, N., Westling-Ryd, M., Schultz, J. E., Karlsson, K.-A. 1987. *J. Biol. Chem.* 262:1779–85
133. Jacewicz, M., Clausen, H., Nudelman, E., Donohue-Rolfe, A., Keusch, G. T. 1986. *J. Exp. Med.* 163:1391–404
134. Mobasaleh, M., Donohue-Rolfe, A., Jacewicz, M., Grand, R. J., Keusch, G. T. 1988. *J. Infect. Dis.* 157:1023–31
135. Lingwood, C. A., Law, H., Richardson, S., Petric, M., Brunton, J. L., et al. 1987. *J. Biol. Chem.* 262:8834–39
136. Cohen, A., Hannigan, G. E., Williams, B. R. G., Lingwood, C. A. 1987. *J. Biol. Chem.* 262:17088–91
137. Waddell, T., Head, S., Petric, M., Cohen, A., Lingwood, C. 1988. *Biochem. Biophys. Res. Commun.* 152:674–79
138. Critchley, D. R., Strenli, C. H., Kellie, S., Ansell, S., Patel, B. 1982. *Biochem. J.* 204:209–19
139. Holmgren, J., Fredman, P., Lindblad, M., Svennerholm, A.-M., Svennerholm, L. 1982. *Infect. Immun.* 38:424–33
140. Eiklid, K., Olsnes, S. 1980. *J. Receptor Res.* 1:199–213
141. Pronk, S. E., Hofstra, H., Groendijk, H., Kingma, J., Swarte, M. B. A., et al. 1985. *J. Biol. Chem.* 260:13580–84
142. Reed, R. A., Mattai, J., Shipley, G. G. 1987. *Biochemistry* 26:824–32
143. Ribi, H. O., Ludwig, D. S., Mercer, K. L., Schoolnik, G. K., Kornberg, R. D. 1988. *Science* 239:1272–76
144. Keusch, G. T., Jacewicz, M., Donohue-Rolfe, A. 1986. *J. Infect. Dis.* 153:238–48
145. Galili, U., Basbaum, C. B., Shohet, S. B., Buehler, J., Macher, B. 1987. *J. Biol. Chem.* 262:4683–88

146. Galili, U., Clark, M. R., Shohet, S. B., Buehler, J., Macher, B. 1987. *Proc. Natl. Acad. Sci. USA* 84:1369–73
147. Towbin, H., Rosenfelder, G., Wieslander, J., Avila, J. L., Rojas, M., et al. 1987. *J. Exp. Med.* 166:419–32
148. Avila, J. L., Rojas, M., Towbin, H. 1988. *J. Clin. Microbiol.* 26:126–32
149. Källenius, G., Möllby, R., Svensson, S. B., Winberg, J., Lundblad, A., et al. 1980. *FEMS Microbiol. Lett.* 7:297–302
150. Leffler, H., Svanborg Edén, C. 1980. *FEMS Microbiol. Lett.* 8:127–34
151. Bock, K., Breimer, M. E., Brignole, A., Hansson, G. C., Karlsson, K.-A., et al. 1985. *J. Biol. Chem.* 260:8545–51
152. Väisänen-Rehn, V., Korhonen, T., Finne, J. 1983. *FEBS Lett.* 159:233–36
153. Rehn, M., Klemm, P., Korhonen, T. 1986. *J. Bacteriol.* 168:1234–42
154. Hansson, G. C., Karlsson, K.-A., Larson, G., Lindberg, A. A., Strömberg, N., Thurin, J. 1983. In *Glycoconjugates,* ed. M. A. Chester, D. Heinegård, A. Lundblad, S. Svensson, pp. 631–32. Lund, Sweden: Rahms
155. Strömberg, N., Ryd, M., Lindberg, A. A., Karlsson, K.-A. 1988. *FEBS Lett.* 232:193–98
156. Strömberg, N., Karlsson, K.-A. 1989. *J. Biol. Chem.* In press
157. Lindahl, M., Faris, A., Wadström, T. 1982. *Lancet* 2:280
158. Parkkinen, J., Finne, J., Achtman, M., Väisänen, V., Korhonen, T. 1983. *Biochem. Biophys. Res. Commun.* 111:456–61
159. Faris, A., Lindahl, A., Wadström, T. 1980. *FEMS Microbiol. Lett.* 7:265–69
160. Lindahl, M., Wadström, T. 1984. *Vet. Microbiol.* 9:249–57
161. Smit, H., Gaastra, W., Kamerling, J. P., Vliegenthart, J. F. G., de Graaf, F. K. 1984. *Infect. Immun.* 46:578–84
162. Lindahl, M., Brossmer, R., Wadström, T. 1987. *Glycoconjugate J.* 4:51–58
163. Brennan, M. J., Joralmon, R. A., Cisar, J. O., Sandberg, A. 1986. *Infect. Immun.* 55:487–89
164. Strömberg, N., Karlsson, K.-A. 1989. *J. Biol. Chem.* In press
165. Krivan, H. C., Ginsburg, V., Roberts, D. D. 1988. *Arch. Biochem. Biophys.* 260:493–96
166. Strömberg, N., Deal, C., Nyberg, G., Normark, S., So, M., Karlsson, K.-A. 1988. *Proc. Natl. Acad. Sci. USA* 85: 4902–6
167. Deal, C. D., Strömberg, N., Nyberg, G., Normark, S., Karlsson, K.-A., So, M. 1987. *Antonie van Leeuwenhoek* 53:425–30
168. Reid, G., Sobel, J. D. 1987. *Rev. Infect. Dis.* 9:470–87
169. Iwamore, M., Shimomura, J., Nagai, Y. 1985. *J. Biochem.* 97:729–35
170. Spiegel, S. 1985. *Biochemistry* 24: 5947–52
171. Karlsson, K.-A., Bock, K., Strömberg, N., Teneberg, S. 1986. In *Protein-Carbohydrate Interactions in Biological Systems,* ed. D. L. Lark, pp. 207–13. London: Academic
172. Finegold, S. M. 1986. *Scand. J. Infect. Dis.* Suppl. 49:160–64
173. Gorbach, S. L., Chang, T.-W., Goldin, B. 1987. *Lancet* 2:1519
174. Bruce, A. W., Reid, G. 1988. *Can. J. Microbiol.* 34:339–43
175. Lee, A. 1985. *Adv. Microb. Ecol.* 8:115–62
176. Tannock, G. W. 1988. *Microbiol. Sci.* 5:4–8
177. Booth, B. A., Sciortino, C. V., Finkelstein, R. A. 1986. In *Microbial Lectins and Agglutinins,* ed. D. Mirelman, pp. 169–82. New York: Wiley
178. Barna, D., Pagnio, A. S. 1977. *Ann. Human Biol.* 4:489–92
179. Chaudhuri, A., De, S. 1977. *Lancet* 2:404
180. Chaudhuri, A., Das Adhikary, C. R. 1978. *Trans. R. Soc. Trop. Med. Hyg.* 72:664–65
181. Levine, M. M., Nalin, D. R., Rennels, M. B., Hornick, R. B., Sotman, S., et al. 1979. *Ann. Human Biol.* 6:369–74
182. Bodmer, W., Cavalli-Sforza, L. L. 1976. *Genetics, Evolution and Man.* San Francisco: Freeman
183. Blackwell, C. C., Jonsdottir, K., Hanson, M. F., Seir, D. M. 1986. *Lancet* 2:687
184. Gahmberg, C. G., Hakomori, S.-i. 1975. *J. Biol. Chem.* 250:2438–46
185. Lampio, A., Rauvala, H., Gahmberg, C. G. 1986. *Eur. J. Biochem.* 157:611–16
186. Lampio, A., Siissalo, I., Gahmberg, C. G. 1988. *Eur. J. Biochem.* 178:87–91
187. Lund, B., Marklund, B.-I., Strömberg, N., Karlsson, K.-A., Normark, S. 1988. *Mol. Microbiol.* 2:255–63
188. Kaizu, T., Levery, S. B., Nudelman, E., Stenkamp, R. E., Hakomori, S.-i. 1986. *J. Biol. Chem.* 261:11254–58
189. Torney, D. C., Dembo, M., Bell, G. I. 1986. *Biophys. J.* 49:501–7
190. Skarjune, R., Oldfield, E. 1982. *Biochemistry* 21:3154–60
191. Kannagi, R., Nudelman, E., Hakomori, S.-i. 1982. *Proc. Natl. Acad. Sci. USA* 79:3470–74
192. Crook, S. J., Boggs, J. M., Vistnes, A.

I., Koshy, K. M. 1986. *Biochemistry* 25:7488–94

193. Haselberger, C. G., Schenkel-Brunner, H. 1982. *FEBS Lett.* 149:126–28

194. Maynard, Y., Baenziger, J. U. 1982. *J. Biol. Chem.* 257:3788–94

195. Isberg, R. R., Falkow, S. 1985. *Nature* 317:262–64

196. Shaw, J. H., Falkow, S. 1988. *Infect. Immun.* 56:1625–32

197. Watanabe, H., Nakamura, A. 1985. *Infect. Immun.* 48:260–62

198. Small, P. L., Isberg, R. R., Falkow, S. 1987. *Infect. Immun.* 55:1674–79

199. Chitnis, D. S., Sharma, K. D., Kamat, R. S. 1982. *J. Med. Microbiol.* 15:43–51

200. Lindberg, F., Lund, B., Johansson, L., Normark, S. 1987. *Nature* 328:84–87

201. Hanson, M. S., Brinton, C. Jr. 1988. *Nature* 332:265–68

202. Schmidt, M. A., O'Hanley, P., Lark, D., Schoolnik, G. K. 1988. *Proc. Natl. Acad. Sci. USA* 85:1247–51

203. Rademacher, T. W., Parekh, R. B., Dwek, R. A. 1988. *Annu. Rev. Biochem.* 57:785–838

204. Jacobs, A. A. C., Venema, J., Leeven, R., van Pelt-Heerschap, H., de Graaf, F. K. 1987. *J. Bacteriol.* 169:735–41

205. Lemieux, R. U., Venot, A. P., Spohr, U., Bird, P., Mendal, G., et al. 1985. *Can. J. Chem.* 63:2664–68

206. Spohr, U., Lemieux, R. U. 1988. *Carbohydr. Res.* 174:211–37

207. Lemieux, R. U. 1984. *Proc. 7th Int. Symp. Med. Chem.* Vol. 1, pp. 329–51. Uppsala, Sweden

208. Kihlberg, J., Frejd, T., Jansson, K., Sundin, A., Magnusson, G. 1988. *Carbohydr. Res.* 176:271–86

209. Firon, N., Ashkenazi, S., Mirelman, D., Ofek, I., Sharon, N. 1987. *Infect. Immun.* 55:472–76

210. Lindberg, A. A., Svensson, S. 1986. *Int. Patent Appl. N. PCT/SE85/00539.* Int. Publ. N. WO 86/04064

211. Breimer, M. E., Hansson, G. C., Karlsson, K.-A., Leffler, H. 1983. *J. Biochem.* 93:1473–85

212. Sweeley, C. C., Nunez, H. A. 1985. *Annu. Rev. Biochem.* 54:765–801

213. Koerner, T. A. W., Scarsdale, J. N., Prestegard, J. H., Yu, R. K. 1984. *J. Carbohydr. Chem.* 3:565–80

214. Breimer, M. E., Hansson, G. C., Karlsson, K.-A., Leffler, H., Pimlott, W., Samuelsson, B. E. 1979. *Biomed. Mass Spectrom.* 6:231–41

215. Kushi, Y., Handa, S. 1985. *J. Biochem.* 98:265–68

216. Tamura, J., Sakamoto, S., Kubota, E. 1987. *JEOL News* 23:20–25

217. Domon, B., Costello, C. E. 1988. *Biochemistry* 27:1534–43

218. Thøgersen, H., Lemieux, R. U., Bock, K., Meyer, B. 1982. *Can. J. Chem.* 60:44–57

219. Fishman, P. H., Pacuszka, T., Hom, B., Moss, J. 1980. *J. Biol. Chem.* 255:7657–64

220. Sabesan, S., Bock, K., Lemieux, R. U. 1984. *Can. J. Chem.* 62:1034–45

221. Hol, W. G. J. 1986. *Angew. Chem. Int. Ed.* 25:767–78

Annu. Rev. Biochem. 1989. 58:351–75

DNA TOPOISOMERASE POISONS AS ANTITUMOR DRUGS

Leroy F. Liu

Department of Biological Chemistry, Johns Hopkins University School of Medicine, 725 N. Wolfe Street, Baltimore, Maryland 21205

CONTENTS

PERSPECTIVES AND SUMMARY

DNA undergoes conformational and topological changes during many cellular processes such as replication and transcription. Topoisomerases have probably evolved to solve these conformational and topological changes in DNA. Two types of topoisomerases have been isolated from both prokaryotic and eukaryotic cells. They perform their functions by introducing transient protein-bridged DNA breaks on one (type I) or both DNA strands (type II)

351

0066-4154/89/0701-0351$02.00

(reviewed in 1–5). Both types of topoisomerases have been isolated from mammalian cells. While either one of the two topoisomerases can relax supercoiled DNA, only mammalian DNA topoisomerase II can unlink two intertwined DNA circles via its strand-passing activity. The relaxation activities of mammalian DNA topoisomerases are probably involved in the relaxation of supercoils generated during various DNA transactions such as replication and transcription (6–8). Their highly polarized and specific distribution on cellular chromatin may reflect DNA conformational changes due to active transcription as suggested by the twin-supercoiled-domain model of transcription, in which negative and positive supercoiling waves are generated behind and ahead of the transcription complex, respectively. The unique unlinking activity of DNA topoisomerase II is essential for segregating replicated daughter chromosomes. The essential role of topoisomerase II in cell proliferation is also reflected in its abundance in proliferating cells (9).

Mammalian DNA topoisomerase II is both functionally and structurally related to bacterial DNA topoisomerase II (DNA gyrase). Like bacterial DNA topoisomerase II, which is an important therapeutic target of quinolone antibiotics, mammalian DNA topoisomerase II is also the cellular target of many potent antitumor drugs from diverse chemical classes (reviewed in 10–15). These antitumor drugs, referred to as topoisomerase II poisons, affect mammalian DNA topoisomerase II in a specific manner. They interfere with the breakage-rejoining reaction of mammalian DNA topoisomerase II by trapping a key covalent reaction intermediate, termed the cleavable complex. Similar to topoisomerase II poisoning, mammalian DNA topoisomerase I has recently been shown to be the target of camptothecin, a plant alkaloid with strong antitumor activity. Camptothecin specifically interferes with the breakage-reunion reaction of topoisomerase I, also by trapping the enzyme in a putative covalent reaction intermediate, the cleavable complex (16). These drug-stabilized cleavable complexes, in contrast to other protein-DNA complexes, can be converted to protein-linked DNA breaks by treatment with a strong protein denaturant such as SDS. However, in the absence of protein denaturants, cleavable complexes can be reversibly converted to noncleavable complexes. This type of reversible DNA damage represents a new type of DNA lesion in mammalian cells. While very little is known about the "repair" of this type of DNA damage, drug-stabilized topoisomerase cleavable complexes are lethal to proliferating cells and are responsible for the antitumor activity of topoisomerase poisons.

Interference with DNA metabolism appears to be the principal mechanism of tumor cell killing by the topoisomerase poisons as with many other antitumor drugs. The transient nature of cleavable complexes suggests that the effect of these drugs may be due to their interaction with the machinery of DNA metabolism. While the cell-killing mechanism of topoisomerase poisons remains unclear, specific arrest of replication forks by drug-stabilized topoisomerase–DNA cleavable complexes appears to be one possible mechanism of cell killing at least for DNA topoisomerase I poisons (17). Cell killing by mammalian topoisomerase II poisons appears more complex. Studies of bacterial topoisomerase II poisons, the quinolone antibiotics, have shown that the induction of cell division inhibitors due to replication arrest and SOS induction can lead to cell death. Whether treatment of mammalian cells with topoisomerase poisons leads to lethality by analogous mechanisms remains to be tested. The recent rapid advances in the area of cell division regulation in eukaryotic cells may shed light on possible cell-killing mechanisms of topoisomerase poisons.

Research on DNA topoisomerases has progressed from DNA enzymology into developmental therapeutics. Basic understanding of the biochemistry and molecular biology of topoisomerases can now be rapidly applied to clinical pharmacology. Topoisomerase poisons also offer a unique opportunity for probing the mechanisms of cell killing and DNA repair processes in mammalian cells. Such studies are likely to lead to a more rational drug design and better strategies in cancer chemotherapy.

MAMMALIAN DNA TOPOISOMERASES

DNA topoisomerases are a unique class of enzymes that change the topological state of DNA by breaking and rejoining the phosphodiester backbone of DNA (reviewed in 1–5). Based on fundamental differences in their reaction mechanisms, DNA topoisomerases are classified into two types. Type I DNA topoisomerases change the topological state of DNA by transiently breaking one strand of the DNA double helix and therefore characteristically change the linking number of DNA in multiples of unity. Type II DNA topoisomerases catalyze the strand-passing reaction by making transient, enzyme-bridged, double-strand breaks and consequently change the linking number of DNA in multiples of two. The importance of topoisomerases in various DNA transactions, such as replication, transcription, and recombination, in bacteria, yeast, and *Drosophila,* has been discussed in a number of reviews (1–5). This review focuses on mammalian DNA topoisomerases, especially human DNA topoisomerases, and a class of unusual inhibitors, topoisomerase poisons.

Mammalian DNA Topoisomerase I

Human DNA topoisomerase I (100 kd), the archetype of mammalian DNA topoisomerase I, is a monomeric protein, encoded by a single-copy gene located on human chromosome 20q12–13.2 (18–20). Like other eukaryotic type I DNA topoisomerases, human topoisomerase I catalyzes the relaxation of both positive and negative supercoils. No energy cofactor is required for catalysis. Its high content (43%) of charged amino acids (19) might explain why charged molecules can modulate its catalytic activity. Mg(II) ion and polycations stimulate the reaction, while polyanions inhibit the reaction (18, 21). HMG proteins and histone H1, which are highly charged proteins, bind tightly to mammalian topoisomerase I and stimulate its catalytic activity (22, 23). The association of topoisomerase I with nucleosomes and other macromolecules has also been shown (22, 24).

A putative reaction intermediate of mammalian DNA topoisomerase I has been characterized. This intermediate can be detected as protein-linked, single-strand DNA breaks, albeit at a low level, when the topoisomerase I reaction is terminated with a strong detergent such as SDS (25, 26). Topoisomerase I is found covalently linked to the 3'-phosphoryl end of the broken DNA strand via a tyrosyl phosphate bond (25, 26). A simple model for topoisomerase I–catalyzed relaxation reaction is shown in Figure 1. The transient covalent intermediate, termed the cleavable complex (C), is at rapid equilibrium with at least one other complex, the noncleavable complex (B). Exposure of the cleavable complexes to a strong protein denaturant, such as SDS or alkali, results in single-strand DNA breaks and the covalent linking of topoisomerase I to the 3'-phosphoryl end of the broken DNA strand. The relative rotation of the two broken ends in the cleavable complex leads to relaxation of supercoiled DNA. Sequencing studies of topoisomerase I cleavage sites have revealed a preference for a loosely defined 5 bp sequence comprised of 4 bp to the 3' side of the cleavage site and one bp to the 5' side (27–29). The DNA structural feature recognized by topoisomerase I is unclear. However, recent studies have shown that the local twist angle variation forms a pattern common to all cleavage sites (C. C. Shen, C. K. J. Shen, personal communication).

Spontaneous cleavage can also occur on single-stranded DNA (presumably in the duplex region of a single-stranded DNA), and has been demonstrated by intra- and intermolecular strand transfers (30, 31). Spontaneous cleavage at the site of a nick on a duplex DNA has also been demonstrated (32). The linearization of the nicked DNA circle by topoisomerase I followed by cyclization in the presence of another circular DNA is presumably responsible for the catenation reaction of topoisomerase I (32). These results suggest that the 5' end of the transient break is not tightly held by topoisomerase I at all times during the breakage-reunion reaction.

The roles of topoisomerase I have not been clearly defined. However, its swivel-like activity is likely to be important for DNA replication, RNA transcription, and other DNA functions. Its abortive reaction intermediate in which the enzyme is covalently linked to the 3' terminus of the broken DNA strand has also been implicated in illegitimate recombination (30–33). Purified mammalian DNA topoisomerase I, in the absence of topoisomerase II, can support the complete replication of SV40 DNA in a cell-free replication system (7, 8). However, the product of DNA replication is a multiply interlocked molecule consisting of two replicated daughter SV40 DNA molecules (7, 8). Since the function of DNA topoisomerase I can be fully substituted for by DNA topoisomerase II in this cell-free system, it remains uncertain whether the DNA topoisomerase I activity is normally involved in DNA replication in vivo.

A role for mammalian DNA topoisomerase I in RNA transcription has been suggested (reviewed in 34, 35). Using the specific topoisomerase I poison, camptothecin, topoisomerase I cleavage sites have been determined and shown to be restricted to the transcribed region of both human ribosomal gene repeats and the tyrosine aminotransferase gene. The presence of topoisomerase I cleavage sites also paralleled transcription activation (34, 35). Zhang et al (35) have suggested that topoisomerase I is involved in modulating the torsional waves (supercoiling waves) generated by transcription (36, 37). However, it is not clear whether topoisomerase I prefers one of the two supercoiled domains. Two other reports have also presented evidence

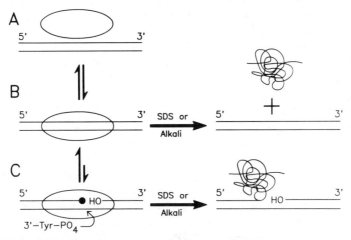

Figure 1 A proposed reaction mechanism for mammalian DNA topoisomerase I.

Mammalian DNA topoisomerase I is proposed to form at least two different complexes with DNA that are in rapid equilibrium: the noncleavable complex (B) and the cleavable complex (C). The cleavable complex is presumably the transient covalent reaction intermediate. Relative rotation of the two broken DNA ends in the cleavable complex leads to DNA relaxation. Exposure of the cleavable complex to a strong protein denaturant, such as SDS or alkali, results in single-strand DNA breaks and the covalent linking of the topoisomerase I polypeptide chain to the 3'-phosphoryl end of the broken DNA strand through a phosphotyrosyl bond. Camptothecin stabilizes the cleavable complex and converts it into a nonproductive ternary complex.

that topoisomerase I is involved in RNA transcription. Studies using partially purified rat RNA polymerase I showed the involvement of topoisomerase I in RNA transcription (38). Microinjection of anti-topoisomerase I antibodies into nuclei of *Chironomous tentans* salivary gland cells also led to blockage of transcription elongation (39).

Mammalian DNA Topoisomerase II

Human DNA topoisomerase II (170 kd), the archetype of mammalian DNA topoisomerase II, is a homodimeric protein, encoded by a single copy gene on human chromosome 17q21-22 (40, 41). Like other eukaryotic type II DNA topoisomerases and phage T4 DNA topoisomerase, it catalyzes the strand-passing of two DNA segments in a reaction coupled to ATP hydrolysis (40). Different from bacterial DNA topoisomerase II (DNA gyrase), no supercoiling activity has been detected using the purified mammalian enzyme (40, 42). Sequence analysis of type II DNA topoisomerases suggests that human DNA topoisomerase II, like other eukaryotic type II DNA topoisomerases, evolved from bacterial DNA topoisomerase II by fusion of the two gyrase subunits into a single polypeptide (41). It remains unexplained why the supercoiling activity cannot be detected using the purified mammalian enzyme. However, recent studies have shown that the major gyrase function, as far as transcription is concerned, may be mediated by its relaxation activity rather than its supercoiling activity (37).

A partial reaction of mammalian DNA topoisomerase II has been characterized. The addition of a strong detergent to a topoisomerase II reaction results in the cleavage of a small population of the DNA molecules and the covalent linking of a topoisomerase polypeptide to the 5' phosphoryl end of each broken DNA strand. Both single- and double-strand DNA breaks are produced (43). The cleavage reaction is rapid and can be reversed by the addition of 0.5 M NaCl prior to SDS addition (43). DNA remains superhelical after salt reversal, suggesting that no free ends are generated in this partial reaction (43). Based on these results, a simple two-state model has been proposed for this partial reaction (Figure 2) (44). A topoisomerase II–DNA cleavable complex, which is presumably the key covalent intermediate in the strand-passing reaction or is related to it, is proposed to be in rapid equilibrium with at least one other topoisomerase II–DNA complex, the non-cleavable complex (44). Differing from other protein-DNA complexes and the noncleavable complex, exposure of the cleavable complex to a strong protein denaturant such as SDS results in DNA cleavage and the covalent linking of a topoisomerase II subunit to the 5'-phosphoryl end of the broken DNA strand. The 3'-hydroxyl end is recessed by four bases (43). Interaction of a second DNA segment with the cleavable complex presumably triggers the strand-passing reaction. The low level of cleavable complexes in a normal topoisomerase II reaction is consistent with its proposed transient nature.

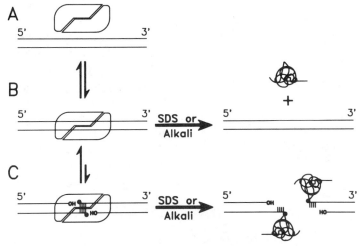

Figure 2   A partial reaction of mammalian DNA topoisomerase II.

Mammalian DNA topoisomerase II, a homodimer, is proposed to form at least two different complexes with DNA that are in rapid equilibrium: the noncleavable complex (B) and the cleavable complex (C). The cleavable complex is presumably the transient covalent intermediate or one related to it. The interaction of the second DNA (incoming DNA) with the cleavable complex (C) may trigger the DNA-dependent ATPase activity and the translocation of the second DNA through the cleavable complex (not shown). Additional cleavable complexes may also exist after passing the second DNA during the reaction (not shown). Intercalative antitumor drugs and epipodophyllotoxins (VP-16 and VM-26) stabilize the cleavable complexes and block the strand-passing reaction.

The biological functions of mammalian DNA topoisomerase II are not completely understood. Studies in the SV40 cell-free replication system have shown that topoisomerase II is essential for segregating completely replicated daughter molecules (7, 8). Its strand-passing activity is also important for the fork movement. However, either topoisomerase I or II can serve to remove the topological intertwines of the two parental DNA strands during the elongation step of SV40 DNA replication in vitro (7, 8). Studies, using a topoisomerase II–specific inhibitor, VM-26, in SV40-infected monkey cells suggested similar roles (9). It seems likely that topoisomerase II is involved in the replication and segregation of chromosomal DNA. A possible role for topoisomerase II in RNA transcription is much less clear. Studies in *Escherichia coli* have suggested that DNA topoisomerase II is involved in the removal of positive supercoiling (torsional) waves generated by transcription (37). Eukaryotic DNA topoisomerase II may have a similar role. Different from DNA topoisomerase I, DNA topoisomerase II appears to be uniformly distributed along the chromosomes, and is not enriched in actively transcribed regions (45, 46). On the other hand, mapping of topoisomerase II cleavage sites using the topoisomerase II poison, VM-26 (teniposide), has revealed strong cleavage sites on both 3' and 5' ends of the *Drosophila* heat-shock genes *hsp70* (46). The significant increase of topoisomerase II cleavage sites on the 3' end of the *hsp*70 genes upon induction of heat-shock transcription suggests a possible role for topoisomerase II in this region during active transcription (46). The enhanced 3' end cleavage may be due to the high degree of positive supercoiling in this region during active transcription, as suggested by the twin-supercoiled-domain model of RNA transcription (36, 37). Strong topoisomerase II cleavage sites in the SV40 enhancer region have also been demonstrated in SV40-infected monkey cells (47, 48). Further experiments are necessary to clarify the possible role of topoisomerase II in transcription. Like other type II DNA topoisomerases, mammalian DNA topoisomerase II can also promote illegitimate recombination in vitro (49).

MAMMALIAN DNA TOPOISOMERASE II POISONS

Intercalative Antitumor Drugs

Many DNA intercalators have been shown to have antitumor activity (reviewed in 50–53). Among them, acridines and anthracyclines have been studied most extensively (54–59). Despite extensive effort in analogue synthesis, the antitumor activity of these intercalators was not understood. No single known parameter of these intercalators (e.g. DNA-binding strength, drug hydrophobicity, and ability to inhibit DNA synthesis) correlated with drug cytotoxicity or antitumor activity (54–60). However, structural specificity of these drugs was clearly noted. This is best exemplified by a pair of acridines, m-AMSA and o-AMSA. Both compounds intercalate DNA equally well, but only m-AMSA has strong antitumor activity (61). It also appeared that a minimal binding affinity is necessary for antitumor activity (58). These studies led to the suggestion that the antitumor activity of acridines was due to the drugs' intercalative mode of DNA binding and a specific interaction of acridines with a nuclear receptor (58, 59, 61). The clue that DNA damage might be responsible for the antitumor activity of acridines came from studies of m-AMSA, which induced limited fragmentation of chromosomal DNA in cultured mammalian cells (62, 63). This result led to the speculation that the nuclear receptor might be a nuclease (58, 59, 61).

Studies using the alkaline elution technique also showed that adriamycin, ellipticine, ethidium bromide, and actinomycin D induced DNA strand breaks in cultured mammalian cells (64–69). In addition, protein-DNA crosslinks were also observed (64–69). Strand breaks and protein-DNA crosslinks seemed to derive from the same DNA damage since they were produced synchronously at a molar ratio of close to unity (65). The disappearance of the strand breaks and protein-DNA crosslinks upon removal of drugs from the culture media followed a similar pattern (65). These results led to the suggestion that one terminus of the broken strand might be linked to a protein molecule (65). However, this type of DNA damage, protein-associated breaks, was initially considered not to be cytotoxic (65, 67, 68, 70, 71). This conclusion was partly due to the lack of correlation between drug cytotoxicity and the levels of protein-associated breaks produced by a number of intercalators from different chemical classes, and partly due to the misconception that all intercalators induced protein-associated breaks (65). It was rationalized that protein-associated breaks were produced by a nuclease, or by a topoisomerase activated to counteract the torsional change induced by drug intercalation (66). This interpretation was later shown to be incorrect, and contrary to the early report, ethidium bromide, a strong intercalator with no antitumor activity, was shown to be inactive in inducing protein-linked breaks (72, 73).

The acridine derivative, m-AMSA, induces large amounts of protein-associated DNA breaks in cultured cells and has been particularly useful for biochemical studies of protein-associated DNA breaks. Using exonucleases, Marshall et al demonstrated that the 5' termini of DNA fragments produced in cells exposed to m-AMSA were blocked by protein (74). The similarity between m-AMSA-induced DNA damage and topoisomerase II–linked DNA breaks in a topoisomerase II partial reaction led to the initial testing of m-AMSA using purified mammalian DNA topoisomerase II (44). m-AMSA, but not its inactive isomer, o-AMSA, was shown to stimulate topoisomerase II–mediated DNA cleavage (44). ATP stimulated the reaction severalfold. Both single- and double-strand breaks were observed (44, 71, 75). Topoisomerase II was covalently linked to the 5' phosphoryl end of each broken DNA strand via a phosphotyrosyl linkage (76). m-AMSA-induced DNA cleavage was rapid and reversible: if prior to SDS addition, the reaction mixture was diluted or NaCl (0.5 M) was added, DNA cleavage was greatly reduced (71, 75). Since linear DNA is just as sensitive to m-AMSA-induced cleavage as supercoiled DNA, altered template supercoiling cannot be the cause of topoisomerase II–mediated DNA cleavage. Nelson et al (44) proposed that m-AMSA interfered with the breakage-reunion reaction of DNA topoisomerase II by trapping the putative key reaction intermediate, the cleavable complex (see Figure 2). m-AMSA presumably stabilizes the cleavable complex by forming a nonproductive drug-enzyme-DNA ternary complex.

Studies with other intercalative antitumor drugs produced similar results. Ellipticine and its derivative, 2-methyl-9-hydroxyl-ellipticinium acetate, were shown to stimulate topoisomerase II–mediated DNA cleavage using purified mammalian DNA topoisomerase II (75). Furthermore, like m-AMSA and o-AMSA, 2-methyl-9-hydroxyl-ellipticinium, a much more potent antitumor drug than ellipticine, was also more effective in stimulating topoisomerase

II–mediated DNA cleavage in the purified system (74). Other intercalative antitumor drugs such as adriamycin, daunomycin, actinomycin D, mitoxantrone, and bisantrene also stimulated topoisomerase II–mediated DNA cleavage (71). However, ethidium bromide, a strong intercalator with no antitumor activity, did not stimulate topoisomerase II–mediated DNA cleavage (71). Two rather unusual features of the cleavage reaction were observed: (a) Intercalators from different chemical classes stimulated topoisomerase II–mediated DNA cleavage at different sites, whereas intercalators from the same chemical class stimulated cleavage at similar sites (71). (b) High concentrations of intercalators inhibited topoisomerase II–mediated DNA cleavage (71). The inhibition is presumably due to template blockage by intercalators. This template blockage effect can also explain the nonspecific inhibition of the strand-passing activity of topoisomerase II by DNA intercalators, including ethidium bromide (71). The different cleavage patterns are more difficult to explain. In the absence of drugs, topoisomerase II produces a cleavage pattern of its own, referred to as background cleavage. Many of the drug-stimulated cleavage sites overlap with these background cleavage sites (71). It seems possible that each drug selectively stimulates a subset of the background cleavage sites. This selectivity may reflect both specific drug-DNA interactions and enzyme-DNA interactions, consistent with the ternary complex hypothesis. This interpretation also implies that drug-DNA interaction is necessary for the stimulation of topoisomerase II–mediated DNA cleavage. The appearance of both single- and double-strand breaks may reflect the presence of one drug binding site for each topoisomerase II subunit. Indeed, at low drug concentrations, single-strand breaks predominate (44).

The stimulatory effect of intercalative antitumor drugs on topoisomerase II–mediated DNA cleavage in the purified system immediately suggested that DNA damage induced by intercalative antitumor drugs in cultured mammalian cells was mediated by topoisomerase II. A number of experiments verified this conclusion. Rowe et al (76) were able to demonstrate m-AMSA-dependent DNA cleavage in a nuclear extract. The protein factor(s) responsible for m-AMSA-dependent DNA cleavage was purified using the cleavage assay. The purified protein factor was identified as DNA topoisomerase II by a number of criteria (76). Yang et al (47) have also studied the effect of m-AMSA on cellular SV40 chromatin. Both form II (nicked form) and form III (linear form) SV40 DNA were recovered in cells treated with m-AMSA. Both forms of SV40 DNA were covalently bound to protein and were selectively precipitated by antibodies against purified mammalian DNA topoisomerase II (47). The sites of double-strand DNA breaks were mapped to similar positions as those produced in the purified system using purified SV40 DNA and calf thymus DNA topoisomerase II (47). Studies in isolated nuclei also led to the same conclusion (77, 78). These results provide strong evidence that topoisomerase II is solely responsible for the unusual cellular DNA damage induced by intercalative antitumor drugs.

Studies using purified topoisomerase II showed that drug stimulation of topoisomerase II–mediated DNA cleavage was a reversible reaction and was insensitive to DNA conformation (44, 72, 75). These results do not support the previous notion that DNA damage induced by intercalators reflects the repair of DNA conformational distortion by a topoisomerase or a nuclease (66). It seems more likely that a drug-DNA-topoisomerase ternary complex is responsible for the DNA damage induced by intercalative antitumor drugs. Consistent with this interpretation, X-ray irradiation of cells treated with intercalative antitumor drugs did not alter the amounts of protein-linked DNA breaks (79). To test whether this type of DNA damage is responsible for drug cytotoxicity, the amounts of protein-linked breaks produced by a number of acridines derivatives were measured both in cells and in reactions using purified topoisomerase II. The correlation between protein-linked breaks and drug cytotoxicity was excellent, strongly suggesting that protein-linked DNA breaks represent a form of lethal DNA damage (76, 80). This result is in conflict with previous reports that showed no correlation between drug cytotoxicity and protein-associated DNA breaks (65, 67, 68, 70, 71). However, previous experiments compared intercalators from many different chemical classes. It seems possible that drugs from different chemical classes may produce cleavable complexes with different properties (e.g. sites of cleavage, and half-lives of cleavable complexes). In addition, other parameters, such as membrane interaction and drug transport, may also complicate the comparison.

Nonintercalative Antitumor Epipodophyllotoxins

In addition to intercalative antitumor drugs, two nonintercalative glycosidic derivatives of podophyllotoxins, VP-16 (etoposide) and VM-26 (teniposide), were also shown to be topoisomerase II poisons. These podophyllotoxin derivatives have significant antitumor activity (reviewed in 81–83). Despite the structural similarity between podophyllotoxin and its derivatives, their physiological effects are very different. Podophyllotoxin, an inhibitor of tubulin polymerization, blocks cells at mitosis, while the epipodophyllotoxins, VP-16 and VM-26, block cells in late S and G_2 phases of the cell cycle (84, 85). VP-16 and VM-26 inhibit both DNA synthesis and RNA synthesis, and induce extensive chromosomal DNA fragmentation in cultured mammalian cells (86–90). Protein-DNA crosslinks were also detected in cells treated with these drugs (90). VM-26 is about 5–10-fold more potent than VP-16 in inducing all these physiological effects (91). The lack of any chemical reactivity of VP-16 and VM-26 with purified DNA suggested that there might be a cellular target for these drugs. Long et al (92) suggested that the target of VP-16 and VM-26 might be topoisomerase II and showed that both drugs inhibited the decatenation activity of topoisomerase II. The successful demonstration that cellular topoisomerase II can mediate intercalator-induced DNA damage prompted investigation into the possible role of topoisomerase II in the action of these epipodophyllotoxins. Using purified calf thymus topoisomerase II, both VP-16 and VM-26 were found to cause extensive DNA fragmentation (93, 94). VM-26 was about 5–10-fold more active than VP-16 in this purified system (93, 94). DNA topoisomerase II was covalently linked to each 5' phosphoryl end of the broken DNA strand (93). Again, the cleavage reaction was reversible. Dilution of the reaction or addition of 0.5 M NaCl to the reaction prior to SDS treatment resulted in reduced cleavage (93). VP-16 and VM-26 also inhibited the strand-passing activity of topoisomerase II (93). These results suggest that, like intercalative antitumor drugs, VP-16 and VM-26 interfere with the breakage-reunion reaction of mammalian DNA topoisomerase II by trapping the key putative reaction intermediate, the cleavable complex (93). However, different from intercalative antitumor drugs, neither VP-16 nor VM-26 intercalates DNA (93).

To further establish that topoisomerase II mediates epipodophyllotoxin-induced DNA damage in cells, a number of experiments have been performed. Yang et al (48) studied the effects of VM-26 on cellular SV40 chromatin in SV40-infected monkey cells. They showed that, similar to the intercalative topoisomerase II poisons, protein-linked single- and double-strand breaks were produced on cellular SV40 DNA in cells treated with VM-26. At low concentrations of VM-26, most breaks were single-strand breaks. At higher concentrations, double-strand breaks were observed. Both form II (nicked form) and form III (linear form) SV40 DNA were selectively immunoprecipitated by antisera against mammalian topoisomerase II (48). Furthermore, cleavage sites on cellular SV40 DNA occurred at sites nearly identical to those produced on SV40 DNA using purified calf thymus DNA topoisomerase II and VM-26 (48). Studies in cultured L1210 cells also showed that more than 80% of cellular topoisomerase II molecules were specifically linked to chromosomal DNA in cells treated with 100 μM, VM-26 (95). Together, these results provide convincing evidence that topoisomerase II is the cellular protein mediating epipodophyllotoxin-induced DNA damage.

The nature of the topoisomerase II–mediated cellular DNA damage has also been investigated. One of the unusual properties of topoisomerase II–DNA cleavable complexes is their instability at high temperatures (e.g. 65°C). Brief exposure of cleavable complexes to 65°C causes rapid reversal of the cleavage reaction. To test whether VM-26-induced protein-linked DNA breaks were in fact cleavable complexes in cultured cells, VM-26-treated cells were briefly heated to 65°C. The rapid reduction of protein-linked DNA breaks paralleled the release of topoisomerase II molecules from cellular DNA as measured by immunoblotting with topoisomerase II–specific antibodies (Y.-H Hsiang, L. F. Liu, unpublished results). These results strongly suggest that the majority of protein-linked DNA breaks reflect the reversible formation of cleavable complexes in cells treated with VM-26. The conversion of cleavable complexes to protein-linked breaks occurs when cells are lysed with a strong protein denaturant such as SDS or alkali.

Altered Regulation of Topoisomerase II in Tumor Cells

It has been established that the cellular level of mammalian topoisomerase II is highly sensitive to the growth state of cells (96–101). Cellular topoisomerase II levels rapidly increase when quiescent cells are stimulated to proliferate, and decrease when cells are induced to differentiate (6, 96–102). Proliferating cells are in general more sensitive to topoisomerase II poisons than quiescent cells (103–112). The higher cellular level of topoisomerase II in proliferating cells can partly explain their greater sensitivity to topoisomerase II poisons. Studies have shown that both proliferating normal and tumor cells contain high levels of topoisomerase II (6). However, topoisomerase II is apparently regulated very differently in tumor cells versus "normal" cells. Hsiang et al (6) showed that the level of topoisomerase II in cultured "normal" primary human skin fibroblasts is tightly controlled by cell density and serum growth factors (6). At higher cell densities or lower serum concentrations, the cellular level of topoisomerase II is reduced in primary human skin fibroblasts and in NIH 3T3 cells (6, 14). Serum replenishment in serum-starved primary human skin fibroblasts or NIH 3T3 cells rapidly restored the high cellular level of topoisomerase II (6, 14). However, these culture conditions had much less effect on topoisomerase II levels in tumor cells (6). Furthermore, the cellular level of topoisomerase II in tumor cells remained relatively constant throughout the early G_1, S, G_2, and M phases of the cell cycle (6, 14). These results suggest that the level of topoisomerase II is primarily controlled during the G_0/G_1 phase of the cell cycle and that this regulation is altered in tumor cells.

In tumor cells, the parallel reduction of sensitivity of both topoisomerase II levels and proliferation inhibition in response to serum deprivation and contact inhibition suggests that a high topoisomerase II level is important for cell proliferation. The altered regulation of topoisomerase II in tumor cells probably occurs at the level of RNA transcription as well as protein degradation (6, 113). However, other changes of topoisomerase II regulation in tumor cells are also possible but have not been investigated. The high level of topoisomerase II in proliferating tumor cells and its altered regulation may partly explain why many potent antitumor drugs are topoisomerase II poisons.

It is uncertain whether topoisomerase II is structurally altered in certain tumor cells.

MAMMALIAN DNA TOPOISOMERASE I POISONS—CAMPTOTHECINS

Camptothecin, a cytotoxic plant alkaloid isolated from *Camptotheca acuminata* of the *Nyssaceae* family, has a broad spectrum of antitumor activity in experimental animals (reviewed in 114). Camptothecin is a highly phase-specific cytotoxic drug; S-phase cells are selectively killed (17). Like topoisomerase II poisons, camptothecin also induces sister chromatid exchanges and chromosomal aberrations (115–117). In cultured mammalian cells, it strongly inhibits both DNA and RNA synthesis, and induces fragmentation of chromosomal DNA (118–124). Both inhibition of RNA synthesis and fragmentation of chromosomal DNA are rapidly "reversible" (118). Inhibition of DNA synthesis, on the other hand, is only partially reversible (121, 123). Camptothecin does not bind to or react with purified DNA. Neither does camptothecin inhibit purified RNA or DNA polymerases (123). These results led to the suggestion that a cellular target might mediate the action of camptothecin. Camptothecin was initially tested for its possible effect on mammalian DNA topoisomerase II (16). Surprisingly, camptothecin had no inhibitory effect on purified mammalian DNA topoisomerase II (16). However, it produced large amounts of single-strand DNA breaks in the presence of mammalian DNA topoisomerase I (16). Topoisomerase I was covalently linked to the 3' phosphoryl end of the broken DNA strand (16). Topoisomerase I–mediated DNA cleavage in the presence of camptothecin is rapid, and prolonged incubation does not lead to more cleavage (16). In fact, less DNA cleavage is normally observed with prolonged incubation when the lactone form of camptothecin is used, presumably due to the hydrolysis of lactone to the inactive sodium form. Treatment of the reaction mixture with a strong protein denaturant, such as SDS or alkali, is required to induce topoisomerase I–mediated DNA strand breaks. If prior to treatment with a strong protein denaturant, the reaction is diluted, challenged with excess second DNA, or adjusted to 0.5 M NaCl, DNA cleavage is greatly reduced (16). Together, these results suggest that camptothecin interferes with the breakage-reunion reaction of mammalian DNA topoisomerase I by trapping the key covalent reaction intermediate, the cleavable complex (see Figure 1). The cleavable complex, which is a productive reaction intermediate in the relaxation reaction and is present in small amounts, is in rapid equilibrium with the noncleavable complex. In the presence of camptothecin, the cleavable complex is stabilized and becomes nonproductive in the relaxation reaction. However, camptothecin-stabilized cleavable complexes are still in rapid equilibrium with noncleavable complexes. The exposure of drug-stabilized cleavable complexes to a strong protein denaturant leads to topoisomerase I–linked, single-strand breaks. Based on this proposed mechanism, camptothecin can therefore be considered as a topoisomerase I poison. The majority of camptothecin-stimulated DNA cleavage sites occur at positions similar to the background cleavage sites of topoisomerase I in the absence of camptothecin (16). While a few new sites were induced, some of the background cleavage sites were induced, some of the background cleavage sites also seemed suppressed (125, 126). The basis for site specificity of this reaction is not clear.

In cultured mammalian cells, camptothecin induces both DNA strand breaks and protein-DNA crosslinks (95, 127). These protein-DNA crosslinks most likely reflect protein-linked DNA breaks, as camptothecin-induced single-strand breaks on intracellular SV40 DNA are covalently linked by protein (95). The identity of the protein that was crosslinked to cellular DNA in camptothecin-treated cells was shown to be topoisomerase I by immunoblotting with topoisomerase I–specific antibodies (95). The reversibility of camptothecin-induced DNA damage in cells has also been investigated (H.-Y. Hsiang, L. F. Liu, unpublished results). As for topoisomerase II poisons, brief heating of camptothecin-treated cells to 65°C rapidly abolished DNA strand breaks and protein-DNA crosslinks. Furthermore, topoisomerase I is also rapidly released from chromosomal DNA as evidenced by immunoblot analysis using antisera against purified DNA topoisomerase I. In an in vitro system with purified components, brief heating of a pre-equilibrated (at 37°C) reaction mixture containing purified mammalian DNA topoisomerase I, camptothecin, and DNA, to 65°C also rapidly reduced cleavable complexes. Together, these results strongly suggest that camptothecin-induced DNA damage in cells is due to the stabilization of camptothecin-topoisomerase I–DNA cleavable complexes.

A number of recent experiments have suggested that camptothecin-induced, topoisomerase I–DNA cleavable complexes are responsible for most if not all the biological effects of camptothecin. Studies of a large number of camptothecin derivatives have shown a quantitative correlation between drug-induced cytotoxicity (also antitumor activity) and the potency of drugs in stimulating topoisomerase I–DNA cleavable complex formation (H.-Y. Hsiang, L.F. Liu, M. Wall, M. C. Wani, A. W. Nicholas, G. Manikumar, S. Kirschenbaum, R. Silber, M. Potmesil, unpublished results). A camptothecin-resistant topoisomerase I was also purified from a camptothecin-resistant cell line, suggesting that topoisomerase I is the cytotoxic target of camptothecin (128). A yeast *top1* deletion strain has been shown to be resistant to camptothecin, while its isogenic wild type is sensitive (129). Induction of *din3*, a damage-inducible gene, following treatment with camptothecin, has also been demonstrated in yeast. Another yeast DNA repair mutant, *rad52*, is shown to be hypersensitive to camptothecin and topoisomerase II poisons (129). These results strongly suggest camptothecin damages DNA via DNA topoisomerase I.

BACTERIAL DNA TOPOISOMERASE II POISONS—QUINOLONE ANTIBIOTICS

The mechanism of action of mammalian DNA topoisomerase II poisons is very similar to that of nalidixic acid, a bacterial DNA topoisomerase II poison (reviewed in 2, 130, 131). Nalidixic acid and other quinolone antibiotics are potent bactericidal agents. The target of nalidixic acid is GyrA, the breakage-reunion subunit of bacterial DNA topoisomerase II. Nalidixic acid interferes with the breakage-reunion reaction of bacterial DNA topoisomerase II by trapping a covalent protein-DNA intermediate. SDS treatment of the trapped intermediate results in single- and double-strand DNA breaks and the covalent linking of the GyrA subunit to the 5'-phosphoryl end of the broken DNA strand. In the case of double-strand breaks, which are generated at higher nalidixic acid concentrations, the 3' hydroxyl end is recessed by four bases. The properties of the trapped intermediate are very similar to those of topoisomerase II–DNA cleavable complexes trapped by mammalian DNA topoisomerase II poisons. Interestingly, the epipodophyllotoxin, VP-16, which is a known mammalian topoisomerase II poison, is also a bacterial topoisomerase II poison. On the other hand, m-AMSA, which is a mammalian topoisomerase II poison and a T4 DNA topoisomerase poison, is not a bacterial topoisomerase II poison (L. F. Liu, unpublished results).

The hypothesis that the target of nalidixic acid is *GyrA* has recently been challenged by Shen & Pernet (132). In a series of experiments, they demonstrated that nalidixic acid binds DNA but not topoisomerase II. Furthermore, the K_i values of various quinolone antibiotics parallel their binding affinity for DNA. Consistent with these results, Tornaletti & Pedrini (133) found that nalidixic acid and norfloxacin, a more potent quinolone antibiotic, can unwind the DNA double helix in the presence of Mg^{2+}. Similarly, studies of a large number of daunorubicin congeners have shown that the potency of intercalative anthracyclines in stimulating mammalian DNA topoisomerase II–DNA cleavable complexes paralleled their strength of intercalation (A. L. Bodley, L. F. Liu, M. Israel, R. Seshadri, Y. Koseki, F. C. Giuliani, R. Silber, S. Kirschenbaum, M. Potmesil, unpublished results). These results are consistent with the hypothesis that drug-DNA interaction is an important parameter in the formation of the enzyme-drug-DNA ternary complex.

The various biological effects of bacterial topoisomerase II poisons have been well documented (reviewed in 2, 130, 131, 134). Nalidixic acid strongly inhibits DNA replication and is a potent inducer of SOS repair (130, 135–138). The rapid and strong bactericidal action of nalidixic acid is most likely due to its trapping of the abortive topoisomerase II–DNA complex on chromosomal DNA rather than its inhibition of the catalytic function of topoisomerase II, since the *nal*^{s} allele is completely dominant in a partial bacterial diploid (139). Why drug-trapping of abortive topoisomerase II–DNA complexes can lead to rapid cell killing and low incidence of drug resistance is not completely clear (140). However, studies of coupling between cell division and DNA replication suggest that nalidixic acid can kill bacteria by multiple mechanisms (141–146). These multiple killing mechanisms may explain the potency as well as low incidence of drug resistance of quinolone antibiotics.

Cell division in bacteria is tightly coupled to DNA replication. Inhibition of DNA replication leads to inhibition of cell division by a number of mechanisms. Coupling between DNA replication and cell division through SOS induction is mediated through two septum formation inhibitors, SfiA (SulA) and SfiC (141–146). SfiA, but not SfiC, is negatively regulated by LexA. SfiC is positively regulated by RecA (141). Both inhibitors prevent cell division by interacting with the cell division protein FtsZ (SulB), and can lead to lethality (141). SfiA is a short-lived protein with a half-life of 10 min and is normally proteolyzed by Lon protease. In the absence of Lon, the half-life of SfiA increases to 170 min, and cell killing through SfiA increases 20,000 fold (143, 147). In the absence of Lon, cell killing through SfiC increases only 13 fold (147). In *lon*$^-$ strains, SfiA-dependent inhibition of cell division is primarily responsible for the observed rapid killing by nalidixic acid (148). Another mechanism of SOS-independent division inhibition is also known to operate when DNA replication is inhibited either at the initiation or elongation step (142). This SOS-independent division inhibition is also associated with cell death (142). The efficiency of coupling is modulated by the cAMP-cAMP receptor protein complex (142). Similar to SfiA and SfiC-dependent division inhibition, the target of the SOS-independent division inhibition is also the cell division protein, FtsZ (142).

Whether DNA degradation is also a direct cause of cell killing is unclear. Nalidixic acid treatment is known to induce DNA degradation (149, 150). However, studies using different repair mutants of *E. coli* B strains (naturally *lon*$^-$) show that DNA degradation, as measured by the release of acid-soluble counts, correlates with survival rather than cell kill (150). Extensive DNA degradation was observed only in mutant strains unable to respond to SOS induction (150). This apparent paradox can be explained by the known inhibitory effect of SOS functions on DNA degradation. In strains where SOS repair is induced, DNA degradation is inhibited. The induction of SOS repair in *E. coli* B strains (*lon*$^-$) is lethal because of the induction of the cell division inhibitor, SfiA (145). The relationship between DNA degradation and cell killing therefore seems complex even in the bacterial model.

POSSIBLE MECHANISM(S) OF CELL KILLING BY MAMMALIAN TOPOISOMERASE POISONS

The mechanism of tumor cell killing by topoisomerase poisons is still not understood. It seems certain, however, that the unusual DNA damage, topoisomerase-DNA cleavable complexes, is responsible for cell killing as well as other cellular responses such as elevated levels of sister chromatid exchanges and chromosomal aberrations. The reversibility of the DNA damage also raises the question of how the cell killing signal is triggered (see Figure 3). The possibility that other cellular processes must interact with the cleavable complexes to trigger cell killing is suggested from a number of experiments. The development of a thermotolerant state in CHO cells, following heat-shock treatment, protected cells from topoisomerase II poisons (151). However, the level of formation of cleavable complexes induced by VP-16 as well as the cellular level of topoisomerase II did not change (151). Dinitrophenol, an uncoupler of oxidative phosphorylation, also protected L1210 cells against topoisomerase II poisons (152). Again, the amounts of drug-induced cleavable complexes did not change (152). Although the mechanism of protection is unclear, these experiments suggest that the level of cleavable complexes alone is not sufficient to predict cytotoxicity. Cellular processing of the cleavable complexes was suggested to be an important event mediating the biological effects of cleavable complexes (151, 152). Interaction of the cleavable complexes with replication forks, helicases, proteases, or nucleases has been suggested (44). It is interesting to note that except for tubulin binders and perhaps a few other antitumor drugs, almost all other types of antitumor drugs, including antimetabolites, bifunctional alkylating agents, and topoisomerase poisons, affect DNA metabolism. Analogous agents in bacteria are potent inducers of the SOS response. This suggests that proliferating mammalian cells may be particularly sensitive to certain types of DNA damage, similar to lon^- strains of E. coli. However, the mechanism of cell death induced by DNA damage is still unclear. Topoisomerase poisons should serve as useful tools for studies on cell-killing mechanisms by antitumor drugs.

Inhibition of DNA Synthesis

Both topoisomerase I and topoisomerase II poisons inhibit chromosomal DNA replication in mammalian cells and viral DNA replication. Studies in the cell-free SV40 replication system showed that the formation of cleavable complexes inhibited DNA replication and that the inhibition of DNA replica-

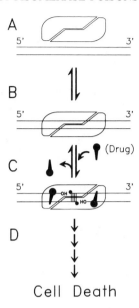

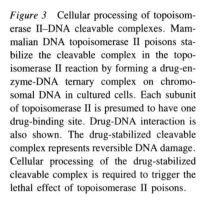

Figure 3 Cellular processing of topoisomerase II–DNA cleavable complexes. Mammalian DNA topoisomerase II poisons stabilize the cleavable complex in the topoisomerase II reaction by forming a drug-enzyme-DNA ternary complex on chromosomal DNA in cultured cells. Each subunit of topoisomerase II is presumed to have one drug-binding site. Drug-DNA interaction is also shown. The drug-stabilized cleavable complex represents reversible DNA damage. Cellular processing of the drug-stabilized cleavable complex is required to trigger the lethal effect of topoisomerase II poisons.

tion is most likely due to the collision betwen the replication forks and the cleavable complexes. In the case of the topoisomerase I poison, camptothecin, the arrest of the DNA replication fork by cleavable complexes may be primarily responsible for drug cytotoxicity, since aphidicolin completely blocks camptothecin-induced cytotoxicity (H.-Y. Hsiang, M. Lihou, L. F. Liu, unpublished results). This interpretation is also consistent with the known S-phase specificity of camptothecin (17). In the case of topoisomerase II poisons, arrest of the replication forks may only contribute partly to drug cytotoxicity, since aphidicolin only partially reduced drug cytotoxicity (M. Lihou, personal communication). This interpretation is also consistent with the fact that topoisomerase II poisons at high concentrations are not highly specific for S-phase cells. The difference between mechanisms of cell killing by topoisomerase I and II poisons is also reflected by studies with dinitrophenol (DNP). DNP is known to protect bacteria from the rapid bactericidal action of nalidixic acid (153). Studies in cultured L1210 cells showed similar protection by DNP against topoisomerase II poisons (152). However, camptothecin cytotoxicity is not alleviated by DNP treatment (152).

Inhibition of Cell Division

Many cytotoxic drugs, including topoisomerase poisons, induce G_2 delay at lower doses (17, 84). This delay is analogous to transient cell division inhibition (filamentation) in bacteria treated with bactericidal agents. Prolonged cell division inhibition is lethal to bacteria (141–146). Whether prolonged G_2 delay due to cell division inhibition is responsible for the cytotoxicity of topoisomerase poisons and other cytotoxic drugs in mammalian cells is unclear. As in *E. coli,* cell division in yeast is controlled by a regulatory network (154). A protein kinase designated p34^{cdc2}, which controls entry into mitosis, is controlled by a number of positive (*cdc25$^+$, niml$^+$*) and negative (*weel$^+$*) regulatory factors (154–158). In *Xenopus* oocytes and other cells, maturation-promoting factor (MPF) appears in late G_2 phase and is able to promote entry into mitosis (159–161). More recently, two groups have shown that yeast *cdc2* kinase is immunologically and functionally related to a component of *Xenopus* MPF (162, 163). It remains to be tested whether certain DNA damage and replication arrest by cytotoxic drugs causes cell division inhibition through inhibition of MPF activation in mammalian cells.

DNA Degradation

Recent studies in conA-stimulated mouse splenocytes showed that camptothecin or VM-26 treatment caused fragmentation of chromosomal DNA into small pieces that were not protein-linked (164). In a separate study, chromatin was degraded into oligonucleosomes in HL60 cells treated with either VP-16 or camptothecin. Within 6 hrs, the majority of the chromosomal DNA was converted into small DNA pieces (S. Kaufman, personal communication). DNA degradation is presumably due to the activation of a nuclease as a consequence of cleavable complex formation. It is not clear whether DNA degradation is just the result of cell lysis or the cause of cell killing. There are two major morphologically and biochemically distinct modes of death in nucleated eukaryotic cells, necrosis and apoptosis (165, 166). In the case of apoptosis, DNA degradation into oligonucleosomes by a Ca^{2+}- and Mg^{2+}-dependent nuclease is well documented (165, 166). Apoptosis is often dependent upon active metabolism and protein synthesis by the dying cells (165, 166). Recent studies of topoisomerase II poisons also showed that cells were partially protected by treatment with the protein synthesis inhibitor, cycloheximide (167, 168). These results suggest that cell killing by topoisomerase poisons might involve the induction of an active cell-killing mechanism, and might share some common steps with apoptosis.

OTHER CELLULAR RESPONSES AND DRUG RESISTANCE

Like many other cytotoxic drugs, the cellular responses to topoisomerase poisons are complex. Induction of the transcription of certain heat-shock genes (169, 170), stimulation of differentiation of murine erythroleukemia cells (MELC) (171), G_2 delay (17, 84), elevated levels of sister chromatid exchanges, and chromosomal aberrations (115–117) have been observed. Some of these cellular responses are analogous to responses in bacteria induced by the bacterial DNA topoisomerase II poison, nalidixic acid. Nalidixic acid is known to stimulate the transcription of heat-shock genes in *E. coli* (169, 170). Nalidixic acid also causes delays in cell division due to replication arrest and SOS induction (141–146). Induction of *RecA* as part of the SOS response may be analogous to the induction of sister chromatid exchanges in mammalian cells (130, 136). Stable DNA replication (sdr), which is also induced by nalidixic acid as part of the SOS response (130, 136), may be analogous to re-replication during S-phase interruption by cytotoxic drugs (172, 173). The onion-skin type of DNA replication of lambda DNA in SOS-induced *E. coli* cells also indicates the loss of reinitiation control (174, 175). No analogous processes in *E. coli* can be related to stimulation of MELC differentiation by mammalian topoisomerase II poisons. However, Nomura et al (176) showed that fusion of MELC with cells treated with mitomycin C or UV (typical SOS inducers) led to stimulation of MELC differentiation. Newly synthesized proteins, responsible for stimulation of MELC differentiation, have also been partially purified (176, 177).

The identification of topoisomerase II as the target for a number of clinically important antitumor drugs has led to re-investigation of drug resistance mechanisms. Selection of resistant tumor cells in culture by stepwise increase of certain cytotoxic drugs has been shown to lead to the amplification of the MDR1 gene and cross resistance to a number of structurally unrelated drugs, including topoisomerase II poisons (178, 179). This multiple drug resistance phenotype is apparently due to the overexpression of the MDR1 gene, which encodes a 170-kd membrane P-glycoprotein involved in drug efflux (178, 179). Recent studies using topoisomerase II poisons have unveiled new drug resistance mechanisms unrelated to the overexpression of the MDR1 gene (180–189). More studies are necessary to establish the mechanisms of drug resistance to topoisomerase poisons. It is expected, however, that multiple resistance mechanisms may exist. Studies of the drug-resistance mechanisms to topoisomerase poisons will not only provide useful information for clinical intervention in cancer chemotherapy but also lead to better understanding of the cell-killing mechanism of topoisomerase-targeting antitumor drugs.

ACKNOWLEDGMENTS

I am grateful to Drs. Peter D'Arpa and Annette L. Bodley, and Mr. Hui Zhang for their critical reading of the manuscript. This work was supported by NIH Grants CA39662 and GM27731. LFL is a recipient of the ACS Faculty Research Awards.

Literature Cited

1. Cozzarelli, N. R. 1980. *Science* 207:953–60
2. Gellert, M. 1981. *Annu. Rev. Biochem.* 50:879–910
3. Liu, L. F. 1983. *CRC Crit. Rev. Biochem.* 15:1–24
4. Wang, J. C. 1985. *Annu. Rev. Biochem.* 54:665–97
5. Wang, J. C. 1987. *Biochim. Biophys. Acta* 909:1–9
6. Hsiang, Y.-H., Wu, H.-Y., Liu, L. F. 1988. *Cancer Res.* 48:3230–35
7. Yang, L., Liu, L. F., Li, J. J., Wold, M. S., Kelly, T. J. 1986. *UCLA Symp. Mol. Cell. Biol.* 47:315–26
8. Yang, L., Wold, M. S., Li, J. J., Kelly, T. J., Liu, L. F. 1987. *Proc. Natl. Acad. Sci. USA* 84:950–54
9. Snapka, R. M. 1988. *Mol. Cell. Biol.* 8:515–21
10. Ross, W. E. 1985. *Biochem. Pharmacol.* 34:4191–95
11. Chen, G. L., Liu, L. F. 1986. *Annu. Rep. Med. Chem.* 21:257–62
12. Bodley, A. L., Liu, L. F. 1987. *Bristol-Myers Cancer Symp.* Vol. 9, *Mechanisms of Drug Resistance in Neoplastic Cells*, ed. P. V. Woolley, K. D. Tew, pp. 277–86. San Diego: Academic
13. Glisson, B. S., Ross, W. E. 1987. *Pharmacol. Ther.* 32:89–106
14. Hsiang, Y.-H., Wu, H.-Y., Liu, L. F. 1988. *Biochem. Pharmacol.* 17:1801–2
15. Bodley, A. L., Liu, L. F. 1988. *Biotechnology.* 6:1315–19
16. Hsiang, Y-H., Hertzberg, R., Hecht, S., Liu, L. F. 1985. *J. Biol. Chem.* 260:14873–78
17. Li, L. H., Fraser, T. J., Olin, E. J., Bhuyan, B. K. 1972. *Cancer Res.* 32:2563–50
18. Liu, L. F., Miller, K. G. 1981. *Proc. Natl. Acad. Sci. USA* 76:3487–91
19. D'Arpa, P., Machlin, P. S., Ratrie, H. III, Rothfield, N. F., Cleveland, D. W., Earnshaw, W. C. 1988. *Proc. Natl. Acad. Sci. USA* 85:2543–47
20. Juan, C-C., Hwang, J., Liu, A. A., Whang-Peng, J., Knutsen, T., et al. 1988. *Proc. Natl. Acad. Sci. USA* 85:8910–13
21. Ishii, K., Futaki, S., Uchiyama, H., Nagasawa, K., Andoh, T. 1987. *Biochem. J.* 241:111–19
22. Javaherian, K., Liu, L. F. 1982. *Nucleic Acids Res.* 11:461–72
23. Javaherian, K., Liu, L. F., Wang, J. C. 1978. *Science* 199:1345–46
24. Rose, K. M., Szopa, J., Han, F-S., Cheng, Y-C., Richter, A., Scheer, U. 1988. *Chromosoma* 92:411–16

25. Champoux, J. J. 1977. *Proc. Natl. Acad. Sci. USA* 74:3800–4
26. Champoux, J. J. 1981. *J. Biol. Chem.* 256:4805–9
27. Edwards, K. A., Halligan, B. D., Davis, J. L., Nevera, N. L., Liu, L. F. 1982. *Nucleic Acids Res.* 10:2565–76
28. Been, M. D., Burgess, R. R., Champoux, J. J. 1984. *Nucleic Acids Res.* 12:3097–4014
29. Anderson, A. H., Gocke, E., Bonven, B. J., Nielsen, O. F., Westergaard, O. 1985. *Nucleic Acids Res.* 13:1543–57
30. Halligan, B. D., Davis, J. L., Edwards, K. A., Liu, L. F. 1982. *J. Biol. Chem.* 257:3995–4000
31. Been, M. D., Champoux, J. J. 1981. *Proc. Natl. Acad. Sci. USA* 78:2883–87
32. McCoubreg, W. K. Jr., Champoux, J. J. 1986. *J. Biol. Chem.* 261:5130–37
33. Bullock, P., Champoux, J. J., Botchan, M. 1985. *Science* 230:954–58
34. Stewart, A. F., Schutz, G. 1987. *Cell* 50:1109–17
35. Zhang, H., Wang, J. C., Liu, L. F. 1988. *Proc. Natl. Acad. Sci. USA* 85:1060–64
36. Liu, L. F., Wang, J. C. 1987. *Proc. Natl. Acad. Sci. USA* 84:7024–27
37. Wu, H.-Y., Shyy, S. H., Wang, J. C., Liu, L. F. 1988. *Cell* 53:433–40
38. Garg, L. C., DiAngelo, S., Jacob, S. T. 1987. *Proc. Natl. Acad. Sci. USA* 84:3185–88
39. Egyhazi, E., Durban, E. 1987. *Mol. Cell. Biol.* 12:4308–16
40. Miller, K. G., Liu, L. F., Englund, P. T. 1981. *J. Biol. Chem.* 256:9334–39
41. Pflugfelder, M. T., Liu, L. F., Liu, A. A., Tewey, K. M., Whang-Peng, J., et al. 1988. *Proc. Natl. Acad. Sci. USA* 85:7177–81
42. Halligan, B. D., Edwards, K. A., Liu, L. F. 1985. *J. Biol. Chem.* 260:2475–82
43. Liu, L. F., Rowe, T. C., Yang, L., Tewey, K. M., Chen, G. L. 1983. *J. Biol. Chem.* 258:15365–70
44. Nelson, E. M., Tewey, K. M., Liu, L. F. 1984. *Proc. Natl. Acad. Sci. USA* 81:1361–65
45. Earnshaw, W. C., Halligan, B. D., Cooke, C. A., Heck, M. M. S., Liu, L. F. 1985. *J. Cell Biol.* 100:1706–15
46. Rowe, T. C., Wang, J. C., Liu, L. F. 1986. *Mol. Cell. Biol.* 6:985–92
47. Yang, L., Rowe, T. C., Nelson, E. M., Liu, L. F. 1985. *Cell* 41:127–32
48. Yang, L., Rowe, T. C., Liu, L. F. 1985. *Cancer Res.* 45:5872–76
49. Bae, Y.-S., Kawasaki, I., Ikeda, H.,

Liu, L. F. 1988. *Proc. Natl. Acad. Sci. USA* 85:2076–80

50. Marshall, B., Darkin, S., Ralph, R. K. 1983. *Trends Biochem. Sci.* 8:212–14
51. Denny, W. A., Baguley, B. C., Cain, B. F. 1983. In *Molecular Aspects of Anticancer Drug Action*, ed. S. Neidle, M. J. Waring. New York: Macmillan
52. Waring, M. J. 1981. *Annu. Rev. Biochem.* 50:159–92
53. Marshall, B., Ralph, R. K. 1985. *Adv. Cancer Res.* 44:267–93
54. Cain, B. F., Atwell, G. J., Seelye, R. N. 1971. *J. Med. Chem.* 14:311–15
55. Cain, B. F., Atwell, G. J. 1974. *Eur. J. Cancer* 10:539–49
56. Cain, B. F., Atwell, G. J. 1976. *J. Med. Chem.* 19:1124–29
57. Baguley, B. C., Cain, B. F. 1982. *Mol. Pharmacol.* 22:486–92
58. Baguley, B. C., Nash, R. 1981. *Eur. J. Cancer* 17:671–79
59. Baguley, B. C., Denny, W. A., Atwell, G. J., Cain, B. F. 1981. *J. Med. Chem.* 24:520–25
60. Ferguson, L. R., Denny, W. A. 1979. *J. Med. Chem.* 22:251–55
61. Wilson, W. R., Baguley, B. C., Wabelin, L.P.G., Waring, M. J. 1981. *J. Mol. Pharmacol.* 20:404–14
62. Furlong, N. B., Sato, J., Brown, T., Chavez, F., Hurlbert, R. B. 1978. *Cancer Res.* 38:1329–35
63. Ralph, R. K. 1980. *Eur. J. Cancer* 16:595–600
64. Ross, W. E., Glaubiger, D. L., Kohn, K. W. 1978. *Biochim. Biophys. Acta* 519:23–30
65. Ross, W. E., Glaubiger, D. L., Kohn, K. W. 1979. *Biochim. Biophys. Acta* 562:41–50
66. Ross, W. E., Bradley, M. O. 1981. *Biochim. Biophys. Acta* 654:129–34
67. Zwelling, L. A., Michaels, S., Erickson, L. C., Ungerleider, R. S., Nichols, M., Kohn, K. W. 1981. *Biochemistry* 20:6553–63
68. Zwelling, L. A., Michaels, S., Kerrigan, D., Prommier, Y., Kohn, K. W. 1982. *Biochem. Pharmacol.* 31:3261–67
69. Zwelling, L. A., Michaels, S., Erickson, L. C., Ungerleider, R. S., Nichols, M., Kohn, K. W. 1981. *Biochemistry* 20:6553–63
70. Zwelling, L. A., Kerrigan, D., Michaels, S. 1982. *Cancer Res.* 42:2687–91
71. Pommier, Y., Zwelling, L. A., Mattern, M. R., Erickson, L. C., Kerrigan, D., et al. 1983. *Cancer Res.* 43:5718–24
72. Tewey, K. M., Rowe, T. C., Yang, L.,

Halligan, B. D., Liu, L. F. 1984. *Science* 226:466–68
73. Rowe, T. C., Kupfer, G., Ross, W. E. 1985. *Biochem. Pharmacol.* 34:2483–87
74. Marshall, B., Ralph, R. K., Hancock, R. 1983. *Nucleic Acids Res.* 11:4251–56
75. Tewey, K. M., Chen, G. L., Nelson, E. M., Liu, L. F. 1984. *J. Biol. Chem.* 259:9182–87
76. Rowe, T. C., Chen, G. L., Hsiang, Y-H., Liu, L. F. 1986. *Cancer Res.* 46:2021–26
77. Pommier, Y., Kerrigan, D., Schwartz, R., Zwelling, L. A. 1982. *Biochim. Biophys. Acta* 107:576–83
78. Pommier, Y., Schwartz, R. E., Kohn, K. W., Zwelling, L. A. 1984. *Biochemistry* 23:3194–4001
79. Pommier, Y., Zwelling, L. A., Schwartz, R. E., Mattern, M. R., Kohn, K. W. 1984. *Biochem. Pharmacol.* 33:3909–12
80. Covey, J. M., Kohn, K. W., Kerrigan, D., Tilchen, E. J., Pommier, Y. 1988. *Cancer Res.* 48:860–65
81. Goldsmith, M. A., Carter, S. K. 1973. *Eur. J. Cancer* 9:477–82
82. Schmoll, H. 1982. *Cancer Treatment Rev.* 9:21–30
83. Long, B. H., Brattain, M. G. 1984. In *Etoposide(VP-16): Current Status and Developments*, ed. B. F. Issell, F. M. Muggia, S. K. Carter, D. Schaefer, J. Schurig, pp. 63–86. New York: Academic
84. Krishan, A., Paika, K., Frei, E. III. 1985. *J. Cell Biol.* 66:521–30
85. Misra, N. C., Roberts, D. W. 1975. *Cancer Res.* 35:99–105
86. Grieder, A., Maurer, R., Stahelin, H. 1974. *Cancer Res.* 34:1788–93
87. Loike, J. D., Brewer, C. F., Sternlight, H., Gensler, W. J., Horowitz, S. B. 1978. *Cancer Res.* 38:2688–93
88. Loike, J. D., Horowitz, S. B. 1976. *Biochemistry* 15:5443–48
89. Roberts, D. W., Hilliard, S., Peck, C. 1980. *Cancer Res.* 40:4225–23
90. Wozniak, A. J., Ross, W. E. 1983. *Cancer Res.* 43:120–24
91. Long, B. H., Musial, S. T., Brattain, M. G. 1984. *Biochemistry* 23:1183–88
92. Minocha, A., Long, B. H. 1984. *Biochem. Biophys. Res. Commun.* 122:165–70
93. Chen, G. L., Yang, L., Rowe, T. C., Halligan, B. D., Tewey, K. M., Liu, L. F. 1984. *J. Biol. Chem.* 259:13560–66
94. Ross, W., Rowe, T., Glisson, B., Yalowich, J., Liu, L. F. 1984. *Cancer Res.* 44:5857–60

95. Hsiang, Y.-H., Liu, L. F. 1988. *Cancer Res.* 48:1722–26
96. Tricoli, J. V., Sahai, B. M., Mccormick, P. J., Jarlinski, S. J., Bertram, J. S., Kowalski, D. 1985. *Exp. Cell Res.* 158:1–14
97. Diguet, M., Lavenot, C., Harper, F., Mirambeau, G., DeRecondo, A-M. 1983. *Nucleic Acids Res.* 11:1059–75
98. Miskimins, R., Miskimins, W. K., Bernstein, H., Shimizu, N. 1983. *Cell Res.* 146:53–62
99. Taudou, G., Mirambeau, G., Lavenot, C., der Garabedian, A., Vermeersch, J., Duguet, M. 1984. *FEBS Lett.* 176:431–35
100. Nelson, W. G., Cho, K. R., Hsiang, Y-H., Liu, L. F., Coffey, D. S. 1987. *Cancer Res.* 47:3246–50
101. Heck, M. M., Earnshaw, W. C. 1987. *J. Cell Biol.* 103:2569–81
102. Bodley, A., Wu, H-Y., Liu, L. F. 1987. *NCI Monogr.* 4:31–35
103. Chow, K. C., Ross, W. E. 1987. *Mol. Cell. Biol.* 9: 3119–23
104. Sullivan, D. M., Latham, M. D., Ross, W. E. 1987. *Cancer Res.* 47:3973–79
105. Sullivan, D. M., Glisson, B. S., Hodges, P. K., Smallwood-Kentro, S., Ross, W. E. 1986. *Biochemistry* 25: 2248–56
106. Sullivan, D. M., Chow, K-C., Glisson, B. S., Ross, W. E. 1987. *NCI Monogr.* 4:73–78
107. Epstein, R. J., Smith, P. J. 1988. *Cancer Res.* 48:297–303
108. Estey, E., Adlakha, R. C., Hittelman, W. N., Zwelling, L. A. 1987. *Biochemistry* 26:4338–44
109. Markovits, J., Pommier, Y., Kerrigan, D., Covey, J. M., Tilchen, E. J., Kohn, K. W. 1987. *Cancer Res.* 47:2050–55
110. Potmesil, M., Hsiang, Y.-H., Liu, L. F., Bank, B., Grossberg, H., et al. 1988. *Cancer Res.* 48:3537–43
111. Schneider, E., Darkin, S. J., Robbie, M. A., Wilson, W. R., Ralph, R. K. 1988. *Biochim. Biophys. Acta* 949:264–72
112. Zwelling, L. A., Estey, E., Silberman, L., Doyle, S., Hittelman, W. 1987. *Cancer Res.* 47:251–57
113. Heck, M. M. S., Hittelman, W. N., Earnshaw, W. C. 1988. *Proc. Natl. Acad. Sci. USA* 85:1086–90
114. Horwitz, S. B. 1975. *Antibiotics*, Vol. III, *Mechanism of Action of Antimicrobial and Antitumor Agents*, ed. J. W. Corcoran, F. E. Hahn, pp. 48–57. New York: Springer-Verlag
115. Lim, M., Liu, L. F., Jacobson-Kram, D., Williams, J. R. 1986. *Cell Biol. Toxicol.* 2:485–94
116. Pommier, Y., Kerrigan, D., Covey, J. M., Kao-Shan, C. S., Whang-Peng, J. 1988. *Cancer Res.* 48:512–16
117. Pommier, Y., Zwelling, L. A., Kao-Shan, C. S., Whang-Peng, J., Bradley, M. O. 1985. *Cancer Res.* 45:143–49
118. Abelson, H. T., Penman, S. 1972. *Nature New Biol.* 237:144–46
119. Kessel, D. 1971. *Biochim. Biophys. Acta* 246:225–32
120. Wu, R. S., Kuman, A., Warner, J. R. 1971. *Proc. Natl. Acad. Sci. USA* 68:3009–14
121. Kessel, D., Bosmann, H. B., Lohr, K. 1972. *Biochim. Biophys. Acta* 269:210–16
122. Bosmann, H. B. 1970. *Biochem. Biophys. Res. Commun.* 41:1412–20
123. Horwitz, S. B., Chang, C-K., Grollman, A. P. 1971. *Mol. Pharmacol.* 7:632–44
124. Horowitz, M. S., Horwitz, S. B. 1971. *Biochem. Biophys. Res. Commun.* 45:723–27
125. Thomsen, B., Mollerup, S., Bonven, B. J., Frank, R., Blocker, H., et al. 1987. *EMBO J.* 6:1817–23
126. Perez-Stable, C., Shen, C. C., Shen, C-K. J. 1988. *Nucleic Acids Res.* 16:7975–92
127. Mattern, M. R., Mong, S. M., Bartus, H. F., Mirabelli, C. K., Crooke, S. T., Johnson, R. K. 1987. *Cancer Res.* 47:1793–98
128. Andoh, T., Ishii, K., Suzuki, Y., Ikegami, Y., Kusunoki, Y., et al. 1987. *Proc. Natl. Acad. Sci. USA* 84:5565–69
129. Nitiss, J., Wang, J. C. 1988. *Proc. Natl. Acad. Sci. USA* 85:7501–5
130. Drlica, K. 1984. *Microbiol. Rev.* 48:273–89
131. Pedrini, A. M. 1979. In *Antibiotics, Vol. 5*, pp. 154–75. Berlin: Springer-Verlag
132. Shen, L. L., Pernet, A. G. 1985. *Proc. Natl. Acad. Sci. USA* 82:307–11
133. Tornaletti, S., Pedrini, A. M. 1988. *Biochem. Biophys. Acta* 949:279–87
134. Goss, W. A., Cook, T. M. 1975. In *Antibiotics*, Vol. 3, pp. 174–96. Berlin: Springer-Verlag
135. Little, J. W., Mount, D. W. 1982. *Cell* 29:11–12
136. Walker, G. C. 1985. *Annu. Rev. Biochem.* 54:425–57
137. Walker, G. C. 1984. *Microbiol. Rev.* 48:60–93
138. Walker, G. C., Marsh, L., Dodson, L. A. 1985. *Annu. Rev. Genet.* 19:103–26
139. Hane, M. W., Wood, T. H. 1969. *J. Bacteriol.* 99:238–41
140. Wolfson, J. S., Hooper, D. C. 1985.

Antimicrob. Agents Chemother. 28:581–86

141. D'Ari, R., Huisman, O. 1983. *J. Bacteriol.* 156:243–50
142. Jaffe, A., D'Ari, R., Norris, V. 1986. *J. Bacteriol.* 165:66–71
143. Maguin, E., Lutkenhaus, J., D'Ari, R. 1986. *J. Bacteriol.* 166:733–38
144. Tormo, A., Dopazo, A., De La Campa, A., Aldea, M. 1985. *J. Bacteriol.* 164:950–53
145. Gottesman, S., Halpern, E., Trisler, P. 1981. *J. Bacteriol.* 148:265–73
146. Huisman, O., D'Ari, R. 1981. *Nature* 290:797–99
147. Huisman, O., D'Ari, R. 1983. *J. Bacteriol.* 153:169–75
148. Kantor, G. J., Deering, R. A. 1968. *J. Bacteriol.* 95:520–30
149. Cook, T. M., Brown, K. G., Boyle, J. V., Goss, W. A. 1966. *J. Bacteriol.* 92:1510
150. Grigg, G. W. 1970. *J. Gen. Microbiol.* 61:21–25
151. Li, G. C. 1987. *NCI Monogr.* 4:99–103
152. Kupfer, G., Bodley, A., Liu, L. F. 1987. *Proc. 1st Conf. DNA Topoisomerases in Cancer Chemotherapy. NCI Monogr.* 4:37–40
153. Deitz, W. H., Cook, T. M., Goss, W. A. 1966. *J. Bacteriol.* 91:768–73
154. Nurse, P. 1985. *Trends Genet.* 1:51–55
155. Nurse, P. 1975. *Nature* 256:547–51
156. Nurse, P., Thuriaux, P. 1980. *Genetics* 96:627–37
157. Russell, P., Nurse, P. 1986. *Cell* 45:145–53
158. Simanis, V., Nurse, P. 1986. *Cell* 45:261–68
159. Lohka, M. J., Maller, J. L. 1985. *J. Cell Biol.* 101:518–23
160. Miake-Lye, R., Kirschner, M. W. 1985. *Cell* 41:165–75
161. Dunphy, W. G., Newport, J. W. 1988. *J. Cell Biol.* 106:2047–56
162. Dunphy, W. G., Brizuela, L., Beach, D., Newport, J. 1988. *Cell* 54:423–31
163. Gautier, J., Norbury, C., Lohka, M., Nurse, P., Maller, J. 1988. *Cell* 54:433–39
164. Jaxel, C., Taudou, G., Portemer, C., Mirambeau, G., Panijel, J., Duguet, M. 1988. *Biochemistry* 7:95–99
165. Duvall, E., Wyllie, A. H. 1986. *Immunol. Today* 7:115–19
166. Wyllie, A. H. 1987. *J. Pathol.* 153:313–16
167. Schneider, E., Lawson, P. A., Ralph, R. K. 1988. *Biochem. Pharmacol.* In press

168. Chow, K. C., King, C. K., Ross, W. E. 1988. *Biochem. Pharmacol.* 37:1117–22
169. Krueger, J. H., Walker, G. C. 1984. *Proc. Natl. Acad. Sci. USA* 81:1499–1503
170. Morgan, R. W., Christman, M. F., Jacobson, F. S., Storz, G. 1986. *Proc. Natl. Acad. Sci. USA* 83:8059–63
171. Nomura, S., Oishi, M. 1983. *Proc. Natl. Acad. Sci. USA* 80:210–14
172. Johnston, R. N., Feder, J., Hill, A. B., Sherwood, S. W., Schimke, R. T. 1986. *Mol. Cell Biol.* 6:3373–81
173. Woodcock, D. M., Cooper, I. A. 1981. *Cancer Res.* 41:2483–90
174. Schnos, M., Inman, R. B. 1985. *Virology* 145:304–12
175. Schnos, M., Inman, R. B. 1982. *J. Mol. Biol.* 159:457–65
176. Nomura, S., Yamagoe, S., Kamiya, T., Oishi, M. 1986. *Cell* 44:663–69
177. Watanabe, T., Oishi, M. 1987. *Proc. Natl. Acad. Sci. USA* 84:6481–85
178. Riordan, J. R., Ling, V. 1985. *Pharmacol. Ther.* 28:51–75
179. Roninson, I. B., ed. 1988. In *Molecular and Cellular Biology of Multidrug Resistance in Tumor Cells.* New York: Plenum.
180. Danks, M. K., Yalowich, J. C., Beck, W. T. 1987. *Cancer Res.* 47:1297–1301
181. Beck, W. T., Cirtain, M. C., Danks, M. K., Felsted, R. L., Safa, A. R., et al. 1987. *Cancer Res.* 47:5455–60
182. Charcosset, J. Y., Saucier, J. M., Jacquemin-Sablon, A. 1988. *Biochem. Pharmacol.* 37:2145–49
183. Bakic, M., Chan, D., Andersson, B. S., Beran, M., Silberman, L., et al. 1987. *Biochem. Pharmacol.* 36:4067–77
184. Drake, F. H., Zimmerman, J. P., McCabe, F. L., Bartus, H. F., Per, S. R., et al. 1987. *J. Biol. Chem.* 262:16739–47
185. Glisson, B., Gupta, R., Smallwood-Kentro, S., Ross, W. 1986. *Cancer Res.* 46:1934–38
186. Glisson, B., Gupta, R., Hodges, P., Ross, W. 1986. *Cancer Res.* 46:1939–42
187. Per, S. R., Mattern, M. R., Mirabelli, C. K., Drake, F. H., Johnson, R. K., Crooke, S. T. 1987. *Mol. Pharmacol.* 32:17–25
188. Pommier, Y., Schwartz, R. E., Zwelling, L. A., Kerrigan, D., Mattern, M. R., et al. 1987. *Cancer Res.* 46:611–16
189. Pommier, Y., Kerrigan, D., Kohn, K. W. 1987. *NCI Monogr.* 4:83–87

Annu. Rev. Biochem. 1989. 58:377–401

MULTIPLE ISOTOPE EFFECTS ON ENZYME-CATALYZED REACTIONS

Marion H. O'Leary

Department of Biochemistry, University of Nebraska-Lincoln, Lincoln, Nebraska 68583

CONTENTS

PERSPECTIVES AND SUMMARY

Isotope effects have long been popular as a method for studying mechanisms of chemical reactions (1, 2). Application of this method to enzymatic reactions has come slowly, in part because of the difficulty of making measure-

377

ments of the necessary precision, and in part because of the difficulty of interpreting the variations in rate that occur in multistep reactions. The theory and practice of isotope effects has now reached the stage where a variety of interesting mechanistic studies are possible (3–9), including use of heavy-atom isotope effects (5), application to multireactant enzymes (9a, 10), pH dependence of isotope effects (11, 12), proton inventory studies (9), relative deuterium and tritium isotope effects (13), the remote label method (14), and the use of deuterium and tritium isotope effects for the study of tunneling (J. Klinman, personal communication).

Multiple isotope effects are a recent addition to this battery of techniques. They represent another example of the idea that isotopes are a minimal perturbation on a reaction system, and thus are particularly amenable to interpretation. In this case, an isotope effect for a substrate containing two isotopic substitutions is compared with the isotope effects for the corresponding singly substituted compounds. In various cases the two effects may be multiplicative (the simplest case), more than multiplicative, or less than multiplicative. The theory of this approach was developed by Cleland et al (15) and independently by Knowles et al (16, 17). The same technique has been used in studies of mechanisms of organic reactions (18, 19).

The multiple isotope effect technique can be used at two levels. First, it can be used qualitatively to distinguish between stepwise and concerted mechanisms, and in the case of stepwise mechanisms, it can be used to distinguish whether the step that gives rise to the hydrogen isotope effect precedes or follows the step that gives rise to the heavy-atom isotope effect. At a more quantitative level, the technique can be used to calculate isotope effects on individual steps and relative rates of various steps in an enzyme-catalyzed reaction. As we will see below, the latter approach requires a number of assumptions and cannot always be applied convincingly.

We first describe the nomenclature and theory of the multiple isotope effect method. Then we apply the method to reactions of gradually increasing complexity. In the last section, we consider limitations and prospects.

THEORY

The basic theory of the multiple isotope effect method has been described by Cleland et al (15).

Definitions

The nomenclature used for describing isotope effects is that of Northrop (20), in which isotope effects are designated by a leading superscript and the heavier isotope appears in the denominator of the rate or equilibrium ratio. Thus, $^{13}k = k^{12}/k^{13}$; $^D(V/K) = (V/K)^H/(V/K)^D$; $^{13}K_{eq} = K_{eq}^{12}/K_{eq}^{13}$.

An *intrinsic isotope effect* is the isotope effect on the rate constant for a single step in a multistep reaction sequence. Observed isotope effects are often smaller than intrinsic isotope effects because other, non–isotope-sensitive steps are partially rate-determining. An isotope effect on an enzyme-catalyzed reaction is a function of the intrinsic isotope effect(s) for the isotope-sensitive step(s) in the overall process and of the rate of that step(s) relative to rates of other steps. Thus, intrinsic isotope effects reflect differences between ground-state and transition-state structure; observed isotope effects reflect intrinsic isotope effects and relative rates of various steps within the mechanism.

Rates of enzyme-catalyzed reactions are sometimes limited by rates of chemical transformations within the enzyme-substrate complex and sometimes by rates of dissociation of substrates or products (21). In the latter case, the substrate or product involved is said to be *sticky*. Operationally, this means that the rate of chemical transformation of the enzyme-substrate complex is similar to (or faster than) the rate of substrate or product dissociation.

Isotope Effects and Enzyme Kinetics

Most applications of the multiple isotope effect method involve isotope effects for elements heavier than hydrogen. Such heavy atom isotope effects are usually measured by the competitive method, in which the isotopic content (e.g. $^{13}C/^{12}C$) of a substrate or product is measured over the course of the reaction by means of isotope-ratio mass spectrometry (22). In careful experiments, the precision of such measurements is $\pm$ 0.001 or better in the rate ratio. The isotope effects so obtained are those on V/K. Heavy-atom isotope effects on V can be obtained only by separate measurement of V for labeled and unlabeled substrates. This is seldom practical for heavy-atom isotope effects. Isotope effects measured by use of radioactive labels are also isotope effects on V/K. Isotope effects on V/K reflect steps in the reaction mechanism beginning with the binding of the isotopic substrate and ending with the first irreversible step, usually the release of the first product (8).

The effect of order of addition of substrates on the observed isotope effect has been described by Klinman et al (9a), and by Cook & Cleland (10). In a multisubstrate mechanism, if the isotopic substrate is the last substrate to bind, then the binding and dissociation of the preceding substrate(s) do not affect the isotope effect. On the other hand, when the isotopic substrate is not the last to bind, the isotope effect must be extrapolated to zero concentrations of the other substrates in order to obtain the full isotope effect.

The term *rate-determining step* is widely used in kinetics and has often found use in enzyme kinetics as well. Unfortunately, the term has no generally agreed-upon meaning, being used for the individual convenience of the individual investigator. The problem is exacerbated by the fact that the rate-determining step may change with reaction conditions, and the step that limits the overall rate may be after the first irreversible step and thus may not be reflected in the isotope effect. Several writers have tried to clarify this issue (8, 21, 23). The term *rate-determining step* should be reserved for the step whose transition state has the highest free energy; where several transition states have similar free energies, the steps can be referred to as *partially rate-determining*.

Quantitative Relationships

The relationship between a number of specific mechanisms and the corresponding isotope effects has been described by Cook & Cleland (10). For a multistep mechanism of the type:

$$E + S \rightleftharpoons \ldots \rightleftharpoons ES_n \underset{k_{-n}}{\overset{k_n}{\rightleftharpoons}} EP_1 \rightleftharpoons \ldots \rightleftharpoons E + P \qquad 1.$$

in which the only isotope-sensitive step is the step with rate constant k_n, the observed isotope effect is given by:

$$^{13}(V/K) = [^{13}k_n + c_f + c_r {}^{13}K_{eq}]/[1 + c_f + c_r] \qquad 2.$$

where $^{13}k_n$ is the isotope effect on k_n, and $^{13}K_{eq}$ is the isotope effect on the equilibrium constant for the overall reaction. The terms c_f and c_r contain only terms for carbon-12. The term c_f is called the "forward commitment" and reflects the rate of the isotope-sensitive step relative to the rates of preceding steps in the sequence. More specifically:

$$c_f = k_n/k_{substrate} \qquad 3.$$

where $k_{substrate}$ is the "net rate constant" (see Cleland, 24) for the release of substrate from intermediate ES_n. Factor c_r, the "reverse commitment," reflects the rate of reverse reaction through the isotope-sensitive step relative to release of product from complex EP_1 and is defined by:

$$c_r = k_{-n}/k_{product} \qquad 4.$$

where $k_{product}$ is the "net rate constant" for the release of the first product from intermediate EP_1.

When the isotope-sensitive step is irreversible (e.g. in many decarboxylations), c_r goes to zero, and the equation can be simplified.

The Multiple Isotope Effect Method

The multiple isotope effect method is applied by measuring a ^{13}C or other isotope effect, then deuterating the substrate and measuring the ^{13}C isotope effect again under the same conditions. Provided that the ^{13}C and deuterium isotope effects are significantly different from unity, deuteration may change the ^{13}C isotope effect. Deuteration is assumed to affect commitments, but not intrinsic isotope effects (a consequence of the rule of the geometric mean; see below).

It should be noted that in the analysis that follows we assume that the deuterium isotope effect is normal (i.e. >1). This will ordinarily be true for primary isotope effects, but it is not necessarily true for secondary isotope effects or for solvent isotope effects. If the deuterium isotope effect in question is inverse, then the signs of the inequalities that follow are reversed.

Deuteration of the substrate can have several effects on the ^{13}C isotope effect. The first possibility is:

$$^{13}(V/K)_H = {}^{13}(V/K)_D \qquad\qquad 5.$$

(The following subscript indicates that the substrate contains H or D at the second isotopic site.) This requires that $c_f = c_r = 0$. In this case, k_n is the step sensitive to both ^{13}C and deuterium, and all other steps are fast compared to this step. Further, the observed carbon and deuterium isotope effects are equal to the intrinsic isotope effects.[1]

The second possibility is:

$$^{13}(V/K)_H < {}^{13}(V/K)_D \qquad\qquad 6.$$

Thus, either c_f or c_r (or both) is decreased by deuteration. This requires that k_n (and k_{-n} if the reaction is reversible) decreases on deuteration. Thus, the deuterium-sensitive step is the same as the ^{13}C-sensitive step. This case differs from the first case in that other non-isotope-sensitive steps are also slow in this case.

The third and fourth possibilities begin with:

$$^{13}(V/K)_H > {}^{13}(V/K)_D \qquad\qquad 7.$$

and deuteration has increased c_f and/or c_r.

Because we are assuming that deuteration affects only a single step in the mechanism, the increase must be either in c_f (in which case, the deuterium-sensitive step precedes the ^{13}C-sensitive step) or in c_r (in which case, the

---

[1]Equality of the two isotope effects can also arise from a cancellation of effects in a more complex mechanism. See the case of malate synthase below.

deuterium-sensitive step follows the ^{13}C-sensitive step) but not both. In the former case:

$$[^{13}(V/K)_H - 1]/[^{13}(V/K)_D - 1] = {^D(V/K)}/{^DK_{eq}} \qquad 8.$$

Alternatively, if the ^{13}C-sensitive step is first:

$$[^{13}(V/K)_H - {^{13}K_{eq}}]/[^{13}(V/K)_D - {^{13}K_{eq}}] = {^D(V/K)} \qquad 9.$$

If the equilibrium isotope effects are known [see Cleland (25) for a compilation of equilibrium isotope effects] and are not all equal to unity, Equations 5, 6, 8, and 9 can be used to determine the order of isotope-sensitive steps.

Numerical Solutions

The second purpose of the multiple isotope effect method is to determine values for intrinsic isotope effects and relative rate constants for various steps in the mechanism, which are combined to give rise to the two commitments. Analysis of relative rates is more difficult than distinction among mechanism types. It is most successful in cases where the isotope-sensitive step is irreversible ($k_{-n} = 0$, and thus $c_r = 0$), but even in that case, it is not invariably successful.

We expect to have experimental values for $^D(V/K)$, $^{13}(V/K)_H$, and $^{13}(V/K)_D$. The problem comes in using these three values (and whatever other information is available) to extract intrinsic isotope effects and commitments. In case 1 above, all commitments are zero, and the intrinsic isotope effects are equal to the observed isotope effects. In case 2 (isotope-sensitive steps the same), the value of c_f for $^{13}(V/K)$ is the same as the value of c_f for $^D(V/K)$, and the same is true for c_r. If c_r is zero, the carbon isotope effect with deuterated substrate is given by:

$$^{13}(V/K)_D = [^{13}k_n + c_f/^Dk_n]/[1 + c_f/^Dk_n] \qquad 10.$$

Thus, there are three measured isotope effects and three unknowns ($^{13}k_n$, ^Dk_n, and c_f, and the system can be solved in closed form.

In the third and fourth cases the situation is more complex. Even if c_r is zero, the problem is that c_f for $^D(V/K)$ is not the same as c_f for $^{13}(V/K)$. Additional information or additional assumptions are needed if the system is to be solved for a unique set of intrinsic isotope effects and commitments.

The Rule of the Geometric Mean

It is implicit in the equations derived above that isotope effects are independent; that is, substitution of deuterium for hydrogen in a molecule does not change the intrinsic carbon isotope effect, and vice versa. This is a manifestation of the rule of the geometric mean, which was originally derived experimentally and was then justified theoretically by Bigeleisen (26).

The validity of the rule of the geometric mean has been the subject of a variety of theoretical treatments (26–29). Although the rule generally works quite well, deviations are expected when vibrations involving the first isotopic site are coupled to vibrations involving the second isotopic site (27), particularly when two hydrogen isotope effects are used. Large deviations are expected when tunneling is significant (28, 29).

Experimentally, deviations from the rule of the geometric mean have been shown in elimination reactions involving substrates such as $R-CL_2-CH_2-X$ (L = H, D, T) (30) and in proton rearrangements of porphyrins (31). Deviations also occur when both isotopic substitutions are hydrogen in the enzymatic dehydrogenation of formic acid and in glucose-6-phosphate dehydrogenase (see below).

ISOTOPE EFFECTS IN IRREVERSIBLE REACTIONS

Irreversible reactions are the simplest to treat because c_r drops out of the isotope effect equation. In the following section we will consider a number of cases of this type.

Single-Step Reactions

The simplest enzyme mechanism is one involving the binding of a single substrate (or, if several substrates are involved, binding of the isotopic substrate as the last substrate), followed by a single, irreversible chemical step, followed by product dissociation. Such a reaction can be represented by:

$$E + S \underset{k_2}{\overset{k_1}{\rightleftharpoons}} ES \overset{k_3}{\longrightarrow} EP \rightleftharpoons E + P \qquad 11.$$

where k_1 and k_2 are the rate constants for substrate binding and dissociation, and k_3 is that for the (irreversible) isotope-sensitive step. The isotope effect is given by:

$$^{13}(V/K) = [^{13}k_3 + k_3/k_2]/[1 + k_3/k_2] \qquad 12.$$

The forward commitment in this case is simply k_3/k_2. The rate constant for substrate binding (k_1) does not enter this equation.

If substrate dissociation from the ES complex is rapid compared to the chemical step (i.e. $k_2 >> k_3$) then:

$$^{13}(V/K) = {}^{13}k_3 \qquad\qquad 13.$$

$$^{D}(V/K) = {}^{D}k_3 \qquad\qquad 14.$$

$$^{13}(V/K)_H = {}^{13}(V/K)_D \qquad\qquad 15.$$

If rapid steps intervene prior to k_3, Equations 13–15 will still be true. Equation 15, in particular, is diagnostic for such cases.

FORMATE DEHYDROGENASE   A clear case of this type is formate dehydrogenase, which catalyzes the reaction:

$$HCO_2^- + NAD^+ \rightarrow CO_2 + NADH \qquad\qquad 16.$$

Kinetic studies reveal that substrate binding is ordered, with NAD^+ binding first (32). Substrate binding and dissociation are rapid compared to catalysis.

Carbon isotope effects were measured by isotope-ratio analysis of the CO_2 formed in the reaction. Hydrogen isotope effects were measured by comparison of steady-state kinetic parameters for HCO_2^- and DCO_2^-. At pH 7.8, $^{13}(V/K)_H = 1.042 \pm 0.001$, $^{13}(V/K)_D = 1.043 \pm 0.003$ (32), and $^{D}(V/K) = 2.2$ (33). These results are consistent with the predictions of Equations 13–15 and indicate that the hydride transfer step is the only kinetically significant step (note that this eliminates mechanisms in which electron transfer is temporally separated from hydrogen-atom transfer). The isotope effects observed are the full intrinsic isotope effects. The intrinsic hydrogen isotope effect is surprisingly small and may indicate that the transition state is quite asymmetric.

Similar experiments were conducted with other nucleotides. In every case, carbon isotope effects are the same for deuterated and undeuterated formate (33).

A second application of the multiple isotope effect technique is possible in this case. Carbon-4 of the pyridine ring of NAD^+ undergoes a hybridization change as a result of the hydride transfer, and $^{D}(V/K) = 1.23$ at this position. This value indicates that motion of this hydrogen is taking place during the hydride transfer. For DCO_2^-, the value is 1.07. According to Equation 15, we expected a value of 1.23. The discrepancy represents a breakdown of the rule of the geometric mean and occurs because of the tunneling associated with simultaneous, coupled motion of the two hydrogens (33). Thus, caution is required when the multiple isotope effect method is applied to cases involving two hydrogens.

AMP NUCLEOSIDASE   The hydrolysis of the glycosidic bond in AMP is catalyzed by the enzyme AMP nucleosidase as well as by aqueous acid:

$$\text{adenosine-5'-P} + H_2O \rightarrow \text{adenine} + \text{ribose-5-P} \qquad\qquad 17.$$

Isotope effects were measured for nitrogen, carbon, and secondary hydrogen in the glycosidic bond (34). In this case, isotope effects were measured by scintillation counting of $^3H/^{14}C$ ratios. The precision that can be obtained in such studies is lower than that obtained by isotope-ratio measurements. An estimate of the precision of the data can be obtained by looking at the interaction of ^{15}N and ^{14}C isotope effects. Because of the small sizes of heavy-atom isotope effects, the two isotope effects should be independent. For the acid-catalyzed reaction, $^{15}k = 1.030 \pm 0.002$, $^{14}k = 1.044 \pm 0.003$ for the nitrogen-14-containing substrate, and 1.033 ± 0.005 for the nitrogen-15-containing substrate. These last two numbers should be equal. Thus, it is likely that the uncertainties in the measurements are somewhat larger than the stated experimental errors.

The principal question in this study is whether the measured isotope effects in the enzymatic reaction are totally intrinsic; that is, whether substrate and products dissociate rapidly and there is only one reaction step. $^D(V/K)$ is 1.045 $\pm$ 0.002 and $^{14}(V/K)$ is 1.035 $\pm$ 0.002. In a single-step reaction, we expect to see $^{14}(V/K)_D = 1.035$, whereas the observed value is 1.037 ± 0.006.

However, the problem with this study is that the deuterium isotope effect is so small that the extent to which it might perturb the ^{14}C isotope effect is within experimental error for all possible mechanisms. If the reaction were stepwise, the ^{14}C isotope effect would be expected to decrease from 1.035 to about 1.034, and this change would not be seen, given the precision of the measurements. This illustrates a limitation of the multiple isotope effect method: If the change in commitment caused by the perturbing isotope is too small, unattainably high precision is required to distinguish among competing mechanisms.

PREPHENATE DEHYDROGENASE   Deuterium substitution does not change a ^{13}C isotope effect if the deuterium-sensitive and ^{13}C-sensitive steps are the same and all preceding steps are fast (cf Equation 13). If the reaction takes place in a single chemical step, but substrate dissociation is not fast (cf Equations 6, 11, and 12; k_3/k_2 decreases on deuteration), then deuteration of the substrate will increase the ^{13}C isotope effect.

An example of this type is prephenate dehydrogenase, which catalyzes the reaction:

The alternate substrate deoxoprephenate ($R=H_2$) reacts at 78% (V) or 18% (V/K) of the rate of prephenate ($R=O$). Overall, the reaction involves hydride transfer, loss of CO_2, and aromatization of the ring. Three mechanisms can be envisioned, depending on whether hydride transfer precedes, follows, or is simultaneous with decarboxylation. These three can be clearly distinguished by the multiple isotope effect technique.

For synthetic reasons, multiple isotope effect studies were carried out with deoxoprephenate (35). For deuteration at the position undergoing hydride transfer, $^D(V/K) = 2.3$, $^{13}(V/K)_H = 1.0033$, and $^{13}(V/K)_D = 1.0103$. This result is uniquely consistent with a mechanism in which the ^{13}C- and D-sensitive steps are the same and some other slow step precedes the chemical step (cf Equation 6). That is, decarboxylation and hydride transfer must be concerted. Because of the large thermodynamic driving force accompanying decarboxylation and aromatization, it is safe to assume that $c_r = 0$. However, the fact that deuteration increases the carbon isotope effect indicates that c_f is not zero.

With no further assumptions, it is possible to calculate $^Dk_3 = 7.3$, $^{13}k_3 = 1.015$, and $k_3/k_2 = 3.7$. The intrinsic carbon isotope effect is surprisingly small compared to other intrinsic isotope effects on decarboxylations, perhaps because the transition state is early in this case. It is also interesting that the reaction is concerted, since all other oxidative decarboxylations studied to date (e.g. malic enzyme, isocitrate dehydrogenase, 6-phosphogluconate dehydrogenase) are stepwise, with hydride transfer preceding decarboxylation (see below).

The isotope effects tell us that some step preceding the chemical step is partially limiting, but they do not tell us whether this step is substrate binding and dissociation or a conformation change that prepares the enzyme for catalysis. Hermes et al (35) argue for the latter, since kinetic evidence suggests that the substrates are not sticky.

Stepwise Mechanisms With Both Isotope-Sensitive Steps the Same

Most enzymatic reactions require more than a single chemical step. Many cases can be summarized as:

$$E + S \underset{k_2}{\overset{k_1}{\rightleftharpoons}} ES_1 \underset{k_4}{\overset{k_3}{\rightleftharpoons}} ES_2 \overset{k_5}{\longrightarrow} EP \rightleftharpoons E + P \qquad 19.$$

Where k_1 and k_2 are substrate binding and dissociation, k_3 is the first chemical step (or perhaps in the more general case, an isomerization of the ES complex), k_4 is the reverse of this step, and k_5 is the ^{13}C-sensitive step. For the present we assume that k_5 is irreversible. If no step except k_5 is sensitive to ^{13}C substitution:

$$^{13}(V/K) = [^{13}k_5 + k_5/k_4(1 + k_3/k_2)]/[1 + k_5/k_4(1 + k_3/k_2)] \qquad 20.$$

If substrate binding and dissociation are fast, this reduces to:

$$^{13}(V/K) = [^{13}k_5 + k_5/k_4]/[1 + k_5/k_4] \qquad 21.$$

which is essentially the same as Equation 12.

If deuteration affects k_5, then the situation is like prephenate dehydrogenase—deuteration will increase the observed ^{13}C isotope effect. However, if deuteration affects k_3 and k_4, then the ^{13}C isotope effect will decrease on deuteration and the isotope effects will obey Equation 8.

HISTIDINE DECARBOXYLASE   This enzyme catalyzes the reaction:

$$\text{histidine} \rightarrow CO_2 + \text{histamine} \qquad 22.$$

For the pyruvate-dependent enzyme from *Lactobacillus* 30a, $^{13}(V/K)_H = 1.0334 \pm 0.0005$. Deuteration of the substrate at the α-position gives a small but poorly known deuterium isotope effect of perhaps 1.2 because of the change in hybridization of the α-carbon during the decarboxylation step. For the deuterated substrate, $^{13}(V/K)_D = 1.0346 \pm 0.0002$ (36). The increase in carbon isotope effect on deuteration requires that the deuterium- and ^{13}C-sensitive steps are the same.

Because $^D(V/K)$ is not accurately known, the intrinsic isotope effects cannot be approached in the same way they were above. Instead, the analysis started with a range of estimated values for the intrinsic ^{13}C isotope effect (this procedure is probably satisfactory because of the number of analogous cases that have been studied) and used these to calculate the intrinsic deuterium isotope effect and the ratio k_5/k_4 (in this case, $k_3/k_2 \ll 1$).

The use of a secondary isotope effect to change a ^{13}C isotope effect obeys the same equations as used above for a primary isotope effect, but the application is difficult because the secondary isotope effect is so small.

The histidine decarboxylase from *Morganella morganii* requires pyridoxal 5'-phosphate, rather than pyruvate, for activity. A similar analysis of this enzyme revealed a very similar set of kinetic parameters (37).

Stepwise Mechanisms with Deuterium-Sensitive Step First

If the mechanism can be expressed by Equation 19 with k_5 the ^{13}C-sensitive step and k_3 and k_4 the deuterium-sensitive steps, then according to Equation 8 the ^{13}C isotope effect decreases when the substrate is deuterated. The ^{13}C isotope effect with deuterated substrate is given by:

$$^{13}(V/K)_D = \frac{^{13}k_5 + \ ^Dk_4(k_5/k_4)[1 + (k_3/k_2)/(^DK_{eq}{}^Dk_4)]}{1 + \ ^Dk_4(k_5/k_4)[1 + (k_3/k_2)/(^DK_{eq}{}^Dk_4)]} \qquad 23.$$

In this case, the kinetic isotope effect on k_4 and the equilibrium deuterium isotope effect are needed. A number of deuterium isotope effects have been summarized by Cleland (25). If substrate binding and dissociation are fast, then Equation 23 simplifies to:

$$^{13}(V/K)_D = [^{13}k_5 + \ ^Dk_4 \ (k_5/k_4)]/[1 + \ ^Dk_4(k_5/k_4)] \qquad 24.$$

In this case, measurement of $^{13}(V/K)_H$, $^{13}(V/K)_D$, and $^D(V/K)$ permits calculation of $^{13}k_5$, $^Dk_4$, and k_5/k_4. However, when substrate dissociation is slow, additional measurements or assumptions are necessary to obtain a complete solution in closed form.

MALIC ENZYME   Malic enzyme catalyzes the divalent metal-dependent oxidative decarboxylation of malic acid:

$$malate + NADP^+ \rightleftharpoons CO_2 + pyruvate + NADPH \qquad 25.$$

The reaction is reversible, but if the pyruvate produced is removed as fast as it is formed, the reaction can be treated as being irreversible.

The reaction is believed to proceed in two steps: Oxidation of malate to form enzyme-bound oxalacetate, followed by decarboxylation. According to Equation 8, deuteration of carbon-2 of malate should decrease the observed isotope effect. On the other hand, if the reaction is concerted, the carbon isotope effect will stay the same or increase on deuteration.

The measured isotope effects are $^{13}(V/K)_H = 1.031$, $^{13}(V/K)_D = 1.025$, and $^D(V/K) = 1.47$ (15). This is consistent with the expectation that hydride transfer precedes decarboxylation and that enzyme-bound oxalacetate is an intermediate. The fit of the isotope effects to Equation 8 is excellent.

A complete solution for the intrinsic isotope effects and rate ratios requires additional information. Grissom & Cleland (38) generated the enzyme-NADPH-oxalacetate complex and measured its partitioning between product formation and release of starting materials. This procedure was repeated with deuterated nucleotide. A complete set of kinetic parameters can then be derived: $^{13}k_5 = 1.044$; $^D(V/K) = 5.7$; $k_5/k_4 = 10$; $k_3/k_2 = 3.3$. The latter ratio may also include a contribution from an enzyme conformation change in k_1 and k_2. Interestingly, the intrinsic carbon isotope effect is only slightly different from the value of 1.049 for the Mg^{2+}-catalyzed decarboxylation of oxalacetate (39).

The enzyme from the plant *Crassula argenta* gives similar isotope effects (40).

ISOCITRATE DEHYDROGENASE   The oxidative decarboxylation of isocitric acid is chemically similar to the oxidative decarboxylation of malic acid in that an enzyme-bound beta-ketoacid is believed to be an intermediate. The overall reaction is:

$$\text{Isocitrate} + \text{NADP} \rightleftharpoons CO_2 + \text{2-ketoglutarate} + \text{NADPH} \qquad 26.$$

The catalytic mechanism is random sequential, and catalysis seems to be more rapid than product release (41, 42). As expected for such a mechanism, isotope effects under optimum conditions are small: $^{13}(V/K) = 0.9989$ (43) and $^{D}(V/K) = 1.00$ (10). Isotope effects of this magnitude cannot be used safely in the multiple isotope method. However, as the pH is lowered, the carbon isotope effect increases. The largest value obtained was $^{13}(V/K)_H = 1.028$ at pH 4.1, and $^{13}(V/K)_D = 1.005$ (44). Because of instability of enzyme and nucleotide, these values could be obtained only with difficulty, and values at still lower pH could not be measured. The carbon isotope effects were fitted to limiting isotope effects at low pH of $^{13}(V/K)_H = 1.04$ and $^{13}(V/K)_D = 1.016$. The deuterium isotope effect increases only slightly with decreasing pH. Qualitatively, it appears that hydride transfer and decarboxylation are separate steps, but these data do not fit Equation 8, and the reason is not known.

6-PHOSPHOGLUCONATE DEHYDROGENASE   This enzyme catalyzes a reaction that is formally similar to the previous two:

$$\text{6-phosphogluconate} + \text{NADP}^+ \rightleftharpoons CO_2 + \text{ribulose 5-P} + \text{NADPH} \qquad 27.$$

The mechanism may proceed in a stepwise fashion, with hydride transfer first, followed by decarboxylation. Consistent with this expectation, $^{D}(V/K) = 1.61$, $^{13}(V/K)_H = 1.0096$, and $^{13}(V/K)_D = 1.0081$ (45). Sufficient data are not available to permit calculation of unique intrinsic isotope effects.

DIHYDROOROTATE OXIDASE   Two deuterium isotope effects can also be used in multiple isotope effect studies. The enzyme catalyzes the reaction:

28.

The elimination of the two hydrogens is from the *anti* positions (46). Pascal & Walsh (46) synthesized various deuterated forms of the substrate and measured the rates. For deuteration at H_5, $^D(V/K) = 2.93$; at H_4, $^D(V/K) = 2.99$. When substrate deuterated at H_5 is used, the isotope effect at H_4 is reduced to 2.09. Qualitatively, this is consistent with a stepwise mechanism for hydrogen removal (Equations 8 and 9) but not with a concerted mechanism (Equations 5 and 6). Unfortunately, it is not possible to use the isotope effects to determine which hydrogen is removed first. Isotope exchange studies suggest that the hydrogen at C-5 is removed first (46).

MALATE SYNTHASE   This enzyme catalyzes the reaction:

acetyl CoA + glyoxylate $\rightleftharpoons$ malate + CoA   29.

Qualitatively, the mechanism involves hydrogen abstraction from acetyl CoA followed by (or concomitant with) nucleophilic attack on carbon-2 of glyoxylate. The question is whether the mechanism is synchronous or stepwise. The enzyme does not catalyze hydrogen exchange into the methyl group of acetyl CoA (47), but that is an unreliable indicator of mechanism, since the abstracted proton might be sequestered in the active site.

Trideutero acetyl CoA gives $^D(V/K) = 1.3$ (47). This value represents the product of the primary isotope effect for the hydrogen being transferred and the secondary isotope effects for the two hydrogens remaining behind. In spite of this complication, this is the proper isotope effect to use in the multiple isotope effect study. The intramolecular kinetic isotope effect using CH_2D acetyl CoA is 3.8 (48), and this value gives a lower limit on the intrinsic hydrogen isotope effect for the hydrogen being transferred.

Carbon isotope effects for carbon-2 of glyoxylate are $^{13}(V/K)_H = 1.0037 \pm 0.0004$ and $^{13}(V/K)_D = 1.0037 \pm 0.0007$ with trideuterated acetyl CoA. These values are so small that they are unlikely to be intrinsic isotope effects.

The equality of carbon isotope effects is at first glance most consistent with a single-step mechanism in which hydrogen transfer and carbon-carbon bond formation are synchronous and all other steps are fast. However, if that mechanism were correct, then the observed hydrogen isotope effect should be equal to the intrinsic isotope effect, and that is not the case.

Clark et al (47) argue that if the mechanism were concerted, the carbon isotope effect should increase to about 1.011 on deuteration of acetyl CoA; instead, the mechanism must be stepwise. The lack of a deuterium effect on the ^{13}C isotope effect may be due to the occurrence of an early rate-determining step that is not isotope sensitive.

Interpretation of the carbon isotope effects is made more difficult by the fact that glyoxylate is at least 99% hydrated in aqueous solution, and there may be an equilibrium carbon isotope effect on that hydration. The unhydrated aldehyde is presumably the substrate for the condensation reaction. If the carbon isotope effect on the hydration were approximately 1.003, then the corrected carbon isotope effects would be approximately 1.000, and the multiple isotope effect method could not be applied. However, even in this case, the conclusion that the reaction is stepwise would still stand.

RIBULOSE BISPHOSPHATE CARBOXYLASE   The first step in the C_3 pathway of photosynthesis is catalyzed by the enzyme ribulose bisphosphate carboxylase:

$$\text{Ribulose-1,5-bisphosphate} + CO_2 \rightarrow 2 \text{ 3-phosphoglyceric acid} \qquad 30.$$

A variety of mechanistic information suggests that the reaction is stepwise (49): A hydrogen is first abstracted from carbon-3 of the substrate, and at this position $^D(V/K) = 2.1$ at pH 8 (50). Following a proton transfer within the intermediate, reaction with CO_2 occurs. Carbon isotope effects are $^{13}(V/K)_H = 1.029 \pm 0.001$ and $^{13}(V/K)_D = 1.021 \pm 0.003$ (51). These values are qualitatively consistent with the stepwise mechanism, but they do not fit Equation 8, probably because of the occurrence of a second deuterium-sensitive step following carboxylation.

Stepwise Mechanisms with the ^{13}C-Sensitive Step First

Such cases are expected to fit Equation 9. Although specific examples have not been fitted to this mechanism, note that reactions such as malic enzyme, which fit Equation 8 in the forward direction, will fit Equation 9 in the reverse direction.

Stepwise Mechanisms with Two Isotope-Sensitive Steps

If two steps are deuterium sensitive or if two steps are heavy-atom sensitive, then Equations 5–9 can no longer be used. If the relative sizes of the isotope effects on the two steps can be predicted (if they are expected to be equal, or if one should clearly be larger than the other), then a set of predictions analogous to those used above can be made. This situation is likely to be common for nitrogen isotope effects and when solvent isotope effects are used to perturb the heavy-atom isotope effect.

PHENYLALANINE AMMONIA-LYASE   This enzyme catalyzes the reaction:

$$\text{Phenylalanine} \rightleftharpoons NH_4^+ + \text{cinnamate} \qquad\qquad 31.$$

The enzyme contains a dehydroalanine cofactor (52) that is presumed to activate the nitrogen atom of the substrate by formation of a covalent bond. Cinnamate is released from the enzyme before ammonia, possibly because hydrolysis of the covalent dehydroalanine-ammonia complex is slow (52).

The nitrogen-15 isotope effect was measured by isolation of released ammonia and conversion to molecular nitrogen for isotope-ratio analysis. Studies with phenylalanine were not very satisfactory. Isotope effects were small ($^D(V/K) = 1.15$ at carbon-3), presumably because dissociation of substrates and/or products is slow. Instead, isotope effect studies were carried out with 2,5-dihydrophenylalanine, which reacts at about 7% the rate of phenylalanine; thus, we expect that in this case, substrate and product dissociation are fast compared to catalysis. For this compound, $^D(V/K) = 2.0$, and this value is pH independent. The value of $^{15}(V/K)$ increases with decreasing pH because of the acid-base equilibrium involving the substrate amino group ($^{15}K_a = 1.016$). When the isotope effect is calculated for unprotonated substrate, $^{15}(V/K)_H = 1.0047$, and $^{15}(V/K)_D = 0.9921$ (53). These results indicate that the reaction is stepwise.

The quantitative analysis is more complex in this case than in previous cases for a couple of reasons. First, the reaction is reversible and it may not be safe to assume that the first C-N bond-breaking step is irreversible. Second, at least two steps have significant nitrogen isotope effects: the step in which the covalent complex between substrate and dehydroalanine is formed, and the step in which the substrate C-N bond is broken. Hermes et al (53) argue that the mechanism is stepwise, with C-H bond breaking preceding C-N bond breaking.

REVERSIBLE REACTIONS

The qualitative distinction among the three classes of mechanisms (Equations 5–9) still holds, whether or not the second isotope-sensitive step is reversible, provided the experiments are conducted in such a way that the release of the first product from the enzyme is irreversible. In fact, it is possible that some of the cases above should have been considered reversible within this context (e.g. malic enzyme and isocitrate dehydrogenase), but this change would not affect the conclusion about concerted vs stepwise mechanisms.

Quantitative analysis of reversible reactions is difficult. For most reversible reactions, the number of experimental observables is inadequate to provide a unique set of intrinsic isotope effects and rate constant ratios. The cases below illustrate a variety of approaches to this problem.

GLUCOSE-6-PHOSPHATE DEHYDROGENASE   The reaction catalyzed is:

Glucose-6-phosphate + NADP$^+$ $\rightleftharpoons$ 6-phosphogluconolactone + NADPH   32.

Although the equilibrium lies on the right, hydride transfer is reversible. The isotope effects are $^D(V/K) = 2.97$ for the hydrogen being transferred, $^{13}(V/K)_H = 1.0165$, and $^{13}(V/K)_D = 1.0316$ (15) for carbon-1. The increase in carbon isotope effect on deuteration indicates that the reaction is concerted, with the ^{13}C- and deuterium-sensitive steps the same. The hydrogen isotope effect was also measured for the 4-position of the pyridine ring of NADP$^+$. Narrow limits could be calculated for the various intrinsic isotope effects and rate ratios.

In a subsequent study, the same isotope effects were measured in D_2O (54). In this solvent, $^{13}(V/K)_H = 1.0110$, $^{13}(V/K)_D = 1.0242$, and $^D(V/K) = 1.81$. The intrinsic hydrogen isotope effect (for the hydrogen being transferred) appears to be smaller in D_2O than in H_2O. As in the case of formate dehydrogenase (see above), coupled motion of more than a single hydrogen in the transition state can lead to a breakdown of the rule of the mean. The analysis is somewhat complicated because the effects of D_2O are primarily on steps other than the chemical step (binding and dissociation or conformation change steps), but the fundamental analysis is probably correct.

TRANSCARBOXYLASE   O'Keefe & Knowles (55, 56) studied the biotin-dependent carboxylation of pyruvate by malonyl CoA:

pyruvate + malonyl CoA $\rightleftharpoons$ oxalacetate + acetyl CoA   33.

The carbon isotope effect was obtained by reduction of oxalacetate to malate, and then isotopic analysis of carbon-4 of malate. For trideuteropyruvate, $^D(V/K) = 1.39$. For the carbon being transferred, $^{13}(V/K)_H = 1.0227 \pm 0.0008$ and $^{13}(V/K)_D = 1.0141 \pm 0.001$.

The isotope effects fit Equation 8. Presumably, enolization of pyruvate is the first step, and this is followed by carboxylation. However, the isotope effects do not distinguish between a reaction with direct carboxyl transfer between carboxybiotin and the enolate and a transfer involving enzyme-bound CO_2.

PROLINE RACEMASE   This case is complex because the reaction is, of course, reversible. Extensive kinetic and isotope effect studies have been reported by Knowles and collaborators (16, 17, 57–60). The reaction is:

$$\text{D-proline} \rightleftharpoons \text{L-proline} \qquad\qquad 34.$$

The enzyme operates by a two-base mechanism: one base abstracts the hydrogen from one face of the substrate, and the other base donates a hydrogen to the other face. In the presence of substrate, the hydrogen bound to one enzyme site cannot exchange with the solvent and cannot be transferred to the other site. At high substrate concentrations, hydrogen transfer between these two sites in the free enzyme may become rate-determining.

The multiple isotope effect approach can be used to determine whether the two hydrogen transfers are simultaneous or stepwise. However, because one of the hydrogens involved is ultimately derived from the solvent, the experimental design is more complex. The isotope effect was measured by running the reaction in an H_2O/D_2O mixture and then comparing the deuterium content of the product with that of the solvent. The same measurement was then conducted with the deuterated substrate. The reaction was made nearly irreversible by running to only about 15% completion.

With proline, the isotope effect for hydrogen incorporation from solvent was 2.78. With deuterated proline the value was 2.70 (16, 17). The mechanism of this reaction involves two partially rate-determining steps, and both steps have significant hydrogen isotope effects at both sites. This apparently occurs because the t es at both sites are thiols. Thiols show an equilibrium isotope effect of about 2X for the exchange reaction:

$$\text{HOD} + \text{R-SH} \rightleftharpoons \text{HOH} + \text{R-SD} \qquad\qquad 35.$$

Thus, both steps show hydrogen isotope effects at both sites. The first step will show a primary hydrogen isotope effect at one site and an equilibrium isotope effect at the other. In the second step, these roles will be reversed. The equilibrium fractionation into the enzyme site was confirmed in separate experiments (58).

APPLICATIONS USING SOLVENT ISOTOPE EFFECTS

As we noted earlier, the multiple isotope effect method works because the second isotopic substitution represents a minimal perturbation of the chemical and kinetic properties of the system. Changing the solvent from H_2O to D_2O can be used in the same way, but some significant problems arise.

First, changing from H_2O to D_2O may change the rates of several steps in a multistep mechanism. Further, D_2O can affect substrate binding, so D_2O

effects on V/K cannot safely be used in the same way as substrate deuterium effects on V/K. One is tempted to use solvent isotope effects on V, which are independent of substrate binding, but V may include steps subsequent to the first irreversible step, and thus may not provide a proper comparison.

Second, D_2O can affect pK_a values for catalytic groups on the enzyme. Procedures for including this effect are well understood (9), but measurements at a single pH (pD) must be interpreted with caution.

Third, the validity of the rule of the geometric mean in H_2O/D_2O comparisons must be considered. A consequence of the rule of the mean is that geometry must be invariant with isotopic substitution. Protein conformation might be altered by substitution of D_2O for H_2O, with a corresponding change in intrinsic isotope effects, but several lines of evidence suggest that this is not so. First, solvent isotope effects on enzymatic reactions are generally of the magnitude expected from consideration of organic reactions; they are not significantly larger, as would have been expected if these conformational issues were important. Second, proton-inventory studies of a variety of enzymes are consistent with the concept that solvent isotope effects on enzymatic reactions are usually derived from one or two significant hydrogens, rather than from a large number (9). Thus, we assume for the present that intrinsic isotope effects are the same in H_2O as in D_2O [except for cases in which the rule of the mean breaks down due to multiple coupled hydrogen motion (54)].

Theory

Solvent isotope effects on enzymatic reactions have been discussed by Venkatasubban & Schowen (9). For multiple isotope effect studies, the best approach is to write equations analogous to Equation 2 for both solvent species and then try to make reasonable assumptions about the solvent isotope effect on each reaction step.

Several different experimental approaches have been used for studying solvent isotope effects on other isotope effects. In the case of proline racemase described above, deuteration of a primary site was used to change the primary isotope effect obtained in an H_2O/D_2O mixture. In the case of alcohol dehydrogenase described below, deuteration of the substrate was used to perturb the solvent isotope effect. In other cases, heavy-atom isotope effects have been compared in H_2O and in D_2O.

YEAST ALCOHOL DEHYDROGENASE   Yeast alcohol dehydrogenase catalyzes the reduction of a number of aromatic aldehydes:

$$Ar\text{-}CHO + NADH \rightleftharpoons Ar\text{-}CH_2OH + NAD^+ \qquad 36.$$

For acetaldehyde and similar substrates, catalysis is faster than substrate and product dissociation (61). For aromatic aldehydes, catalysis is limiting. For example, p-methoxybenzaldehyde gives a primary hydrogen isotope effect for reduction by NADD of 3.6 (62). The corresponding primary tritium effect is 6.3, which obeys the Swain-Schaad relationship (63) and suggests that the hydride transfer step is fully rate-limiting (13). Consistent with this is the observation that the solvent isotope effect on the reduction is 0.50 ± 0.05 for reduction by NADH and 0.58 ± 0.06 for reduction by NADD (62).

Thus, the hydride transfer step must be fully rate-limiting, and this reaction corresponds to case 1 above. Interestingly, these results also require that NADH dissociation is rapid compared to catalysis.

GLUTAMATE DECARBOXYLASE   The pyridoxal phosphate-dependent gluta-mate decarboxylase from *E. coli* catalyzes the irreversible reaction:

$$\text{glutamate} \rightarrow CO_2 + \text{GABA} \qquad\qquad 37.$$

In H_2O, $^{13}(V/K) = 1.018$, whereas in D_2O, $^{13}(V/K) = 1.009$ (64). Solvent isotope effects are large: $^{D_2O}(V/K) = 2.6$, and $^{D_2O}V = 5.0$. A proton-inventory analysis shows that the isotope effect on V reflects contributions from several hydrogens.

Substrate dissociation is rapid and decarboxylation is irreversible, so the reaction can be treated as having two kinetically significant steps (cf Equations 19 and 21). Models indicate that the intrinsic carbon isotope effect should be about 1.06 (65). With this value in hand, it is possible without further assumptions to calculate a ratio of solvent isotope effects on k_5 (the decarboxylation step) and k_4 (the Schiff-base interchange step). Using the steady-state solvent isotope effects, we estimate that the solvent isotope effect is about 3 on decarboxylation and about 7 on Schiff-base interchange. The latter isotope effect may be due to the fact that a large number of protons change places in this step and/or to the fact that Schiff-base interchange involves a conformation change. Nitrogen isotope effects on this reaction are also consistent with this mechanism (66).

Similar studies have been carried out for histidine decarboxylase (36).

BENZOYLFORMATE DECARBOXYLASE   This enzyme is similar to pyruvate decarboxylase and catalyzes the thiamine pyrophosphate-dependent reaction:

$$\text{Ar-CO-CO}_2^- \rightarrow \text{Ar-CHO} + CO_2 \qquad\qquad 38.$$

The substrate first reacts with the enzyme-bound cofactor to form a tetrahedral adduct, which then undergoes decarboxylation.

Carbon isotope effects were measured in H_2O and in D_2O for five sub-strates with ring substituents covering a range of electron-withdrawing capacities. The values in H_2O range from 1.002 to 1.019. Values in D_2O were about half the corresponding values in H_2O. Solvent isotope effects on V/K ranged from 1.4 to 3.1 (67).

Using an intrinsic carbon isotope effect of 1.051 (68), we can calculate that the ratio of the solvent isotope effect on the decarboxylation step to that on the preceding steps is about 0.5, probably because there is a relatively large solvent isotope effect on formation of the tetrahedral adduct. This is not surprising, since formation of the adduct involves at least one proton transfer from the solvent. Attempts to fit the carbon isotope effects to Equation 8 were not successful, presumably because several steps show significant solvent isotope effects (67).

ADENOSINE DEAMINASE   The reaction catalyzed is:

$$\text{adenosine} \rightarrow \text{NH}_3 + \text{inosine} \tag{39.}$$

The mechanism appears to involve an addition-elimination sequence at the site of substitution on the adenine ring. Chemical modification and pH studies have been interpreted to indicate that an enzyme sulfhydryl group is involved in the reaction (69).

For adenosine, $^{15}(V/K) = 1.0048$ in H_2O and 1.0023 in D_2O. The values are pH independent, even though V/K is pH dependent. The solvent isotope effect on V/K is 0.77.

Larger isotope effects were obtained with 7,8-dihydro-8-oxoadenosine, which reacts at 10% of the rate of adenosine. For this compound, $^{15}(V/K) = 1.0150$ in H_2O and 1.0131 in D_2O. The solvent isotope effect is 0.45 on V/K. The inverse solvent isotope effect probably indicates that a sulfhydryl group acts as a proton donor at the active site of the enzyme.

The fact that the nitrogen isotope effect decreases in D_2O rules out mechanisms in which there is only a single kinetically significant step. From isotope effects and other evidence, the reaction appears to have at least two kinetically significant steps: first, addition of some nucleophile to form a tetrahedral adduct, then loss of ammonia from this adduct. The fact that the nitrogen isotope effect is smaller in D_2O suggests that the solvent isotope effect on the C-N bond cleavage step is smaller than that on some other step. At some point (probably prior to the first step) the nitrogen must be protonated, and this protonation can lead to a nitrogen isotope effect of about 1.016. A similar analysis has been conducted for alanine dehydrogenase (70).

CARBONIC ANHYDRASE   The reversible dehydration of bicarbonate is catalyzed by carbonic anhydrase:

$$\text{HCO}_3^- + \text{H}^+ \rightleftharpoons \text{CO}_2 + \text{H}_2\text{O} \tag{40.}$$

The enzyme requires Zn^{2+} for activity. The kinetic mechanism appears to be ping-pong, with the dehydration half reaction separated from the proton transfer between enzyme and solvent. The solvent isotope effect on V is about 5 because the proton transfer is rate-determining (71).

The carbon isotope effect on the dehydration of HCO_3^- is 1.010, and this value is not changed if D_2O is substituted for H_2O, if Co^{2+} is substituted for Zn^{2+}, or if the viscosity of the solvent is changed (72). The measured isotope effect is only slightly larger than the equilibrium isotope effect on dehydration.

The lack of a solvent isotope effect on the carbon isotope effect is consistent with the ping-pong mechanism; otherwise, the carbon isotope effect should change in D_2O. The isotope effect is approximately equal to that seen in the nonenzymatic loss of OH^- from HCO_3^- (P. Paneth, M. H. O'Leary, unpublished), and this, together with the lack of a D_2O effect on the carbon isotope effect, suggests that the dehydration step is rate-limiting in the dehydration half-reaction.

PYRUVATE CARBOXYLASE   The biotin-dependent carboxylation of pyruvic acid occurs in two stages: The first is the ATP-dependent carboxylation of biotin, and the second is carboxylation of pyruvate (presumably as the enol or enolate) to form oxalacetate. The reaction is reversible. Attwood et al (73) studied the second half-reaction in the reverse direction:

$$\text{oxalacetate} \rightarrow CO_2 + \text{pyruvate} \qquad\qquad 41.$$

which occurs by transfer of carbon-4 to biotin, and then decomposition of carboxybiotin.

In H_2O, $^{13}(V/K)$ is 1.032; the value in D_2O is 1.025. The solvent isotope effect on V/K is 2.1. When the reaction is run in the opposite direction, trideuteropyruvate gives an isotope effect of 2.8. Dissociation of oxalacetate from the enzyme-substrate complex is fast (74). It appears that the carboxylation is stepwise, as suggested above for the case of transcarboxylase.

CONCLUSIONS AND PROSPECTS

The change in an observed isotope effect as a result of a second isotopic substitution can often be used to learn about the order of the isotope-sensitive steps. In favorable cases, intrinsic isotope effects and rate ratios can be extracted from the data. This is most practical in reactions in which one of the isotope-sensitive steps is irreversible. The ideal cases are ones in which a primary deuterium isotope effect is used to perturb a ^{13}C isotope effect.

The assumption that only one step in the reaction mechanism is affected by each isotopic substitution is an important one. When this is not so, Equations

5–9 do not apply and it is more difficult to solve for intrinsic isotope effects and rate ratios.

Precision and Uncertainty

Heavy-atom isotope effects are always small; the uninitiated accuse practitioners of making "much ado about nothing." Although heavy-atom isotope effects can be measured with a reproducibility of ± 0.001 or better, not all reported data are of that quality. Careful attention to experimental detail is essential. Separate experiments must be statistically independent (22). Systematic errors have always been a serious problem in heavy-atom isotope effect studies, and it is not always clear that these have been eliminated.

Problems are even more severe when the perturbing isotope has only a small isotope effect and thus can cause only a small change in the original isotope effect. Problems also arise when the original isotope effect is much smaller than the expected intrinsic value. In that case, the change in isotope effect due to the second isotopic substitution becomes quite small and the precision of measurement may not be adequate to permit a clear distinction among mechanisms. The magnitudes of isotope effects can often be increased by working on the limbs of the pH-rate profile.

Reversibility

Perhaps the most serious increase in complexity in the multiple isotope effect method occurs when we move from reactions in which one of the isotope-sensitive steps is irreversible to reactions in which all steps up to the release of the first product are reversible. This change does not affect distinctions between concerted and stepwise mechanisms based on Equations 5–9, but it does affect our ability to calculate individual isotope effects and rate ratios.

The most serious problem is more subtle than this. The mathematical treatment given above assumes that the release of the first product is irreversible and that subsequent steps need not be considered. The history of this assumption is that it is derived from steady-state kinetics, where initial rates are measured, and the concentrations of products are effectively zero in the experiments. However, the same conditions do not necessarily hold in the measurement of heavy-atom isotope effects. When isotope effects are measured by the analysis of products, the initial measurement is usually taken at 10–15% reaction, and occasionally as high as 25% reaction. Even at 10% reaction, an appreciable amount of product has accumulated, and if the equilibrium constant for the reaction is not too far from unity, a significant amount of reverse reaction may occur. This will cause the observed isotope effect to be smaller than the actual isotope effect.

Even in the case of malic enzyme, where the equilibrium favors decarboxylation, this can be a problem. In early measurements of the carbon isotope effect, values near 1.02 were obtained. When lactate dehydrogenase and NADH were added, so that pyruvate release became irreversible, the isotope effect increased to 1.03 (15). We are left to wonder whether decarboxylation and CO_2 release might be reversible in other cases. If so, then the measured isotope effect will be smaller than the actual effect.

Theoretical Limitations

The principal theoretical limitation on the use of the multiple isotope effect method is the limitation on the applicability of the rule of the geometric mean. When this rule fails, the observed isotope effect will change upon isotopic substitution, even if the reaction in question involves only a single step.

Failure of the rule of the geometric mean may be fairly common in cases in which the effect of hydrogen/deuterium substitution on a hydrogen isotope effect is considered. If the sites are sufficiently close together that both give rise to isotope effects, then the sites are also sufficiently close together that vibrational coupling will occur, and the rule of the geometric mean will break down.

It is becoming clear that a number of enzymatic reactions that involve hydrogen transfer also involve a significant degree of hydrogen tunneling in the hydrogen transfer transition state. When tunneling occurs, then the rule of the geometric mean may break down, particularly when the second isotopic substitution also involves hydrogen. The importance of the analogous phenomenon in multiple isotope effect studies using heavy-atom isotope effects is not currently known.

Literature Cited

1. Bigeleisen, J., Wolfsberg, M. 1958. Adv. Chem. Phys. 1:15–76
2. Melander, L., Saunders, W. H. Jr. 1980. Reaction Rates of Isotopic Molecules. New York: Wiley-Interscience
3. Cleland, W. W., O'Leary, M. H., Northrop, D. B. 1977. Isotope Effects on Enzyme-Catalyzed Reactions. Baltimore: Univ. Park Press
4. Klinman, J. P. 1978. Adv. Enzymol. 46:415–94
5. O'Leary, M. H. 1978. Transition States of Biochemical Processes, ed. R. D. Gandour, R. L. Schowen, pp. 285–316. New York: Plenum
6. Cleland, W. W. 1982. CRC Crit. Rev. Biochem. 13:385–428
7. Cleland, W. W. 1987. Bioorg. Chem. 15:283–302
8. Northrop, D. B. 1981. Annu. Rev. Biochem. 50:103–31
9. Venkatasubban, K. S., Schowen, R. L. 1984. CRC Crit. Rev. Biochem. 17:1–44
9a. Klinman, J. P., Humphries, H., Voet, J. G. 1980. J. Biol. Chem. 255:11648–51
10. Cook, P. F., Cleland, W. W. 1981. Biochemistry 20:1790–96
11. Cook, P. F., Cleland, W. W. 1981. Biochemistry 20:1797–805
12. Cook, P. F., Cleland, W. W. 1981. Biochemistry 20:1805–16
13. Northrop, D. B. 1975. Biochemistry 14:2644–51
14. O'Leary, M. H., Marlier, J. F. 1979. J. Am. Chem. Soc. 101:3300–6
15. Hermes, J. D., Roeske, C. A., O'Leary, M. H., Cleland, W. W. 1982. Biochemistry 21:5106–14
16. Belasco, J. G., Albery, W. J., Knowles, J. R. 1983. J. Am. Chem. Soc. 105:2475–77
17. Belasco, J. G., Albery, W. J., Knowles, J. R. 1986. Biochemistry 25:2552–58
18. Koch, H. F., Dahlberg, D. B. 1980. J. Am. Chem. Soc. 102:6102–7
19. Koch, H. F., McLennan, D. J., Koch, J. G., Tumas, W., Dobson, B., Koch, N. H. 1983. J. Am. Chem. Soc. 105:1930–37
20. Northrop, D. B. 1977. See Ref. 3, pp. 122–57

21. Cleland, W. W. 1975. *Acc. Chem. Res.* 8:145–51
22. O'Leary, M. H. 1980. *Methods Enzymol.* 64:83–104
23. Ray, W. J. Jr. 1983. *Biochemistry* 22:4625–37
24. Cleland, W. W. 1975. *Biochemistry* 14:3220–24
25. Cleland, W. W. 1980. *Methods Enzymol.* 64:104–25
26. Bigeleisen, J. 1955. *J. Chem. Phys.* 23:2264–67
27. Ishida, T., Bigeleisen, J. 1976. *J. Chem. Phys.* 64:4775–89
28. Saunders, W. H. Jr. 1985. *J. Am. Chem. Soc.* 107:164–69
29. Huskey, W. P., Schowen, R. L. 1983. *J. Am. Chem. Soc.* 105:5704–6
30. Amin, M., Price, R. C., Saunders, W. H. Jr. 1988. *J. Am. Chem. Soc.* 110:4085
31. Limbach, H.-H., Hennig, J., Gerritzen, D., Rumpel, H. 1982. *Faraday Discuss. Chem. Soc.* 74:229–43
32. Blanchard, J. S., Cleland, W. W. 1980. *Biochemistry* 19:3543–50
33. Hermes, J. D., Morrical, S. W., O'Leary, M. H., Cleland, W. W. 1984. *Biochemistry* 23:5479–88
34. Parkin, D. W., Schramm, V. L. 1987. *Biochemistry* 26:913–20
35. Hermes, J. D., Tipton, P. A., Fisher, M. A., O'Leary, M. H., Morrison, J. F., Cleland, W. W. 1984. *Biochemistry* 23:6263–75
36. Abell, L. M., O'Leary, M. H. 1988. *Biochemistry* 27:5933–39
37. Abell, L. M., O'Leary, M. H. 1988. *Biochemistry* 27:5927–33
38. Grissom, C. B., Cleland, W. W. 1985. *Biochemistry* 24:944–48
39. Grissom, C. B., Cleland, W. W. 1986. *J. Am. Chem. Soc.* 108:5582–83
40. Grissom, C. B., Willeford, K. O., Wedding, R. T. 1987. *Biochemistry* 26:2594–96
41. Uhr, M. L., Thompson, V. W., Cleland, W. W. 1974. *J. Biol. Chem.* 249:2920–27
42. Northrop, D. B., Cleland, W. W. 1974. *J. Biol. Chem.* 249:2928–31
43. O'Leary, M. H., Limburg, J. A. 1977. *Biochemistry* 16:1129–35
44. Grissom, C. B., Cleland, W. W. 1988. *Biochemistry* 27:2934–43
45. Rendina, A. R., Hermes, J. D., Cleland, W. W. 1984. *Biochemistry* 23:6257–62
46. Pascal, R. A. Jr., Walsh, C. T. 1984. *Biochemistry* 23:2745–52
47. Clark, J. D., O'Keefe, S. J., Knowles, J. R. 1988. *Biochemistry* 27:5961–71
48. Lenz, H., Eggerer, H. 1976. *Eur. J. Biochem.* 65:237–46
49. Miziorko, H. M., Lorimer, G. H. 1983. *Annu. Rev. Biochem.* 52:507–35
50. Van Dyk, D. E., Schloss, J. V. 1986. *Biochemistry* 25:5145–56
51. Roeske, C. A., O'Leary, M. H. 1984. *Biochemistry* 23:6275–84
52. Havir, E. A., Hanson, K. R. 1975. *Biochemistry* 14:1620–29
53. Hermes, J. D., Weiss, P. M., Cleland, W. W. 1985. *Biochemistry* 24:2959–67
54. Hermes, J. D., Cleland, W. W. 1984. *J. Am. Chem. Soc.* 106:7263–64
55. O'Keefe, S. J., Knowles, J. R. 1986. *J. Am. Chem. Soc.* 108:328–29
56. O'Keefe, S. J., Knowles, J. R. 1986. *Biochemistry* 25:6077–84
57. Albery, W. J., Knowles, J. R. 1986. *Biochemistry* 25:2572–77, and preceding papers in *Biochemistry* 25
58. Belasco, J. G., Bruice, T. W., Albery, W. J., Knowles, J. R. 1986. *Biochemistry* 25:2558–64
59. Albery, W. J., Knowles, J. R. 1987. *J. Theor. Biol.* 124:137–71
60. Albery, W. J., Knowles, J. R. 1987. *J. Theor. Biol.* 124:173–89
61. Klinman, J. P. 1981. *CRC Crit. Rev. Biochem.* 10:39–80
62. Welsh, K. M., Creighton, D. J., Klinman, J. P. 1980. *Biochemistry* 19:2005–16
63. Swain, C. G., Stivers, E. C., Reuwer, J. F. Jr., Schaad, L. J. 1958. *J. Am. Chem. Soc.* 80:5885–93
64. O'Leary, M. H., Yamada, H., Yapp, C. J. 1981. *Biochemistry* 20:1476–81
65. Marlier, J. F., O'Leary, M. H. 1986. *J. Am. Chem. Soc.* 108:4896–99
66. Abell, L. M., O'Leary, M. H. 1988. *Biochemistry* 27:3325–30
67. Weiss, P. M., Garcia, G. A., Kenyon, G. L., Cleland, W. W., Cook, P. F. 1988. *Biochemistry* 27:2197–205
68. Jordan, F., Kuo, D. J., Monse, E. U. 1978. *J. Am. Chem. Soc.* 100:2872–78
69. Weiss, P. M., Cook, P. F., Hermes, J. D., Cleland, W. W. 1987. *Biochemistry* 26:7378–84
70. Weiss, P. M., Chen, C.-Y., Cleland, W. W., Cook, P. F. 1988. *Biochemistry* 27:4814–22
71. Silverman, D. N., Vincent, S. H. 1984. *CRC Crit. Rev. Biochem.* 14:207–55
72. Paneth, P., O'Leary, M. H. 1987. *Biochemistry* 26:1728–31
73. Attwood, P. V., Tipton, P. A., Cleland, W. W. 1986. *Biochemistry* 25:8197–205
74. Cheung, Y.-F., Walsh, C. T. 1976. *Biochemistry* 15:3749–55

Annu. Rev. Biochem. 1989. 58:403–26

QUINOPROTEINS, ENZYMES WITH PYRROLO-QUINOLINE QUINONE AS COFACTOR

Johannis A. Duine and Jacob A. Jongejan

Department of Microbiology and Enzymology, Delft University of Technology, Julianalaan 67, 2628 BC Delft, The Netherlands

CONTENTS

PERSPECTIVES AND SUMMARY

The structure elucidation of the cofactor of methanol dehydrogenase was an important landmark in the history of pyrrolo-quinoline quinone (PQQ). In addition, as was already apparent at the announcement of the structure, and has been confirmed in the intervening 10 years, several other PQQ-containing bacterial dehydrogenases, so-called quinoproteins, exist. Most (if not all) of these dehydrogenases are situated in the periplasm of Gram-negative bacteria. They are involved in the primary oxidation step of substrates like alcohols (besides methanol dehydrogenase, different quinoprotein dehydrogenases exist for oxidation of ethanol, glycerol, quinate, polyvinyl alcohol, and polyethylene glycol), amines (methylamine dehydrogenase), and aldose sugars (glucose dehydrogenase). Depending on the nature of the dehydrogenase, electron transfer occurs to a special cytochrome c, cytochrome

403

0066-4154/89/0701-0403$02.00

b, copper protein (amicyanin), or the bacterial ubiquinones. Certain quinoprotein dehydrogenases are involved in incomplete microbial oxidations (e.g.
vinegar and gluconic acid production), while others function in a special
respiratory chain, providing the organism with an auxiliary energy system. In
other bacterial species, possessing a set of dehydrogenases catalyzing the
same reaction but with different cofactors, the role of quinoproteins is still
unclear. Not all bacteria are able to provide their quinoprotein apoenzyme
with PQQ, but nevertheless they produce the apoenzyme constitutively.
Therefore, depending on the growth substrate and the occurrence of other
metabolic pathways in the organism, supplementation of the medium with
PQQ is stimulatory or indispensable for growth in these cases. Excretion of
PQQ occurs with many bacterial species growing on alcohols, and mixed
cultures have been described consisting of a nonproducer (though harboring a
quinoprotein apoenzyme) and a PQQ producer.

The role of PQQ is not restricted to bacteria or to dehydrogenases. Several
copper-containing oxidases from a variety of organisms appear to be
quinoproteins: bovine plasma amine oxidase; hog kidney diamine oxidase;
human placental lysyl oxidase; fungal galactose oxidase; bacterial methylamine oxidase; pea seedling amine oxidase. Examples are also found in other
subclasses of enzymes and in the group of iron-containing enzymes: bovine
adrenal medulla dopamine β-hydroxylase; soybean lipoxygenase-1; bacterial
nitrile hydratase. Moreover, strong arguments exist to propose a role for PQQ
as cofactor in several pyridoxal phosphate (PLP)–dependent enzymes, e.g. in
certain ω-amino transferases and in decarboxylases. In fact, the presence of
PQQ has recently been established in dopa decarboxylase. From the published
spectroscopic and mechanistic data of some of these enzymes, a function of
PQQ as cofactor can be deduced. This means that the prevailing view on the
mechanism of these enzymes, once thought to be well understood, has to be
reconsidered.

The current list of quinoproteins shows that many are involved in the
degradation or biosynthesis of mammalian bioregulators. Therefore, design of
inhibitors blocking the action or the biosynthesis of PQQ can be expected to
shed light on several physiological phenomena. Studies on the role of vitamin
B6 have been carried out frequently with unspecific carbonyl-group reactive
reagents. Since PQQ is also sensitive to these compounds and at least one of
the PLP-dependent enzymes contains PQQ, the design of specific inhibitors is
necessary to redefine the significance of PLP and to establish the role of PQQ.
With respect to the latter, the development of analytical procedures is highly
relevant to establishing the distribution, occurrence, and production of PQQ.
Although it is known that tyrosine and a glutamic acid moiety are the
precursors of PQQ biosynthesis in methylotrophic bacteria, how either the
synthesis or the covalent attachment to apoquinoproteins occurs is still unknown for mammals. The available information indicates that the process

might proceed on a protein matrix in situ, resulting in PQQ that is covalently bound to a lysine residue via an amide bond. Although administration of PQQ to certain organisms has physiological effects, the reason for this is not clear, since the reactivity of PQQ with nucleophiles almost certainly leads to conversion into a variety of compounds, as indicated by model studies of PQQ and amino acids, giving oxazoles and related condensation products.

The present review indicates that PQQ has come of age, as it is widespread and versatile, comparable to the common cofactors. In this respect, the question can be posed why this cofactor has been overlooked in so many well-studied enzymes. Related to this: could there exist other still undetected cofactors? The history of PQQ and quinoproteins emphasizes that cofactor enzymology, especially that of systematic cofactor identification, could be rewarding. At the moment, the specific features of quinoproteins cannot be indicated. Since several quinoproteins have their counterpart in groups of enzymes with other cofactors, sometimes occurring in one and the same organism, comparative enzymology of these examples could reveal the unique properties of quinoproteins.

IN SEARCH OF QUINOPROTEINS

Indications of the existence of PQQ can be traced back to the 1950s. Several research groups interested in the nonphosphorylative degradation of glucose via gluconic acid discovered a so-called glucose dehydrogenase showing unusual behavior with respect to a stimulating factor (see for instance Ref. 1). In 1964, Hauge showed that this enzyme had a cofactor with unprecedented properties (2). Curiously, this landmark remained unnoticed until progress was made in a different field. Growing interest in the 1970s in production of single cell protein and biogas stimulated development of the microbial enzymology of methylotrophic bacteria. Several enzymes in these bacteria appeared to have unusual cofactors, one of these being methanol dehydrogenase (3). Since the cofactor of this enzyme was claimed to be a pteridine derivative for many years, the discovery that it was an o-quinone with two nitrogens was a real breakthrough (4, 5). Soon after (1979), structure elucidation was achieved by applying X-ray diffraction to a crystalline derivative (6) and by applying mass spectroscopy and NMR to the compound itself (see Figure 1 for the structure) and to a reduced form (7).

Since ESR spectroscopy had already revealed a unique structure for the cofactor (4, 5), even before the complete structure was elucidated, we attempted to determine whether PQQ might occur in other enzymes and bacteria as well. It soon appeared that the answer was affirmative (8, 9), so that PQQ's versatility as cofactor and occurrence in a variety of bacteria was indicated in the report on structure elucidation (7). In the intervening 10 years, several other bacterial quinoprotein dehydrogenases have been discovered in our

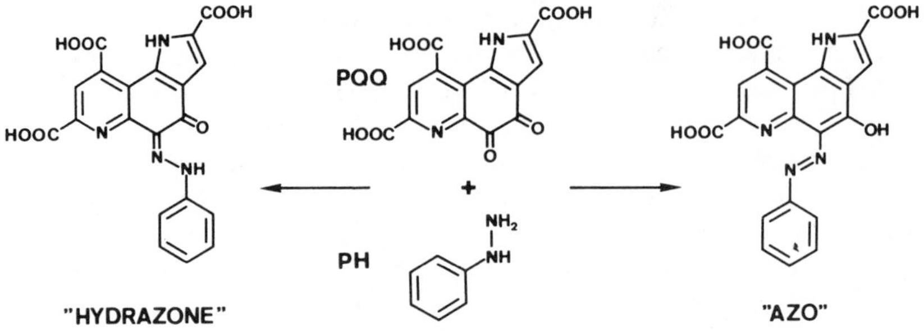

Figure 1   Adducts of PQQ formed with phenylhydrazine (PH).

laboratory, as well in those of Ameyama and coworkers and other Japanese groups (see Refs. 10 and 11 for citations).

To determine whether PQQ is also a cofactor in mammalian enzymes appeared more difficult. We chose copper-containing amine oxidases as the target for the following reasons. Although pyridoxal phosphate was claimed to be the organic cofactor in these enzymes [see for instance the latest edition of the enzyme nomenclature recommendations (12)], real evidence for the assumption had never been provided. On the other hand, general agreement existed with respect to the presence of a carbonyl group in the cofactor. From the similarities with methylamine dehydrogenase and the fact that PQQ has two carbonyl groups, these enzymes were considered attractive candidates to explore for the presence of PQQ. However, it was known that the organic cofactor could not be removed from the enzyme unless the protein was hydrolyzed, and reaction of PQQ with nucleophilic amino acids had already been noted (13). Therefore, to circumvent this difficulty, it was reasoned that analysis of PQQ could not be performed in the enzyme hydrolysate [a view later proven essentially correct (14)]. Instead, the cofactor had to be derivatized in situ to a stable adduct prior to enzyme hydrolysis. Using dinitrophenylhydrazine as the derivatizing reagent, the presence of PQQ could be unequivocally established in bovine plasma amine oxidase by comparing the absorption spectra and chromatographic properties of the adduct isolated from the enzyme with a model compound, this despite the fact that only 6% of the expected amount could be detected in these initial studies (15).

Some doubt has been cast on the method's suitability in providing evidence for PQQ in these enzymes (16). Raman spectroscopy of hydrazine-treated enzymes has been advocated by Dooley and coworkers as more suitable in this respect (17–19). In this method, spectra of derivatized enzyme and model compound are compared. However, the comparison relies completely on a spectroscopic technique used in a nonquantitative way. Moreover, the authors were apparently unaware of the reason for the low yield of PQQ-hydrazone, since they compared the spectra of their derivatized enzymes with the incorrect model compound. For earlier research on the reasons for the low yields had shown that upon addition of a hydrazine to these enzymes, immediate formation of the so-called azo-compound is observed (Figure 1). Quantitative hydrazone formation requires long incubation times under an oxygen atmosphere (19a). The elucidation of the conditions led to the "hydrazine method," which enabled the quantitative analysis of PQQ in several amine oxidases, ranging in distribution from microbes to man.

After the presence of PQQ in these oxidases had been established, it was reasoned that the cofactor might also have been overlooked in other copper-containing subclasses of oxidoreductases. Dopamine β-hydroxylase seemed an attractive target for this search, since great difficulties exist in explaining its mechanism from the presence of only copper ions (20). After the presumption appeared to be correct for this hydroxylase (21), it seemed interesting to investigate other metalloenzymes. Also in the case of soybean lipoxygenase-1, the presence of only a single Fe ion can hardly explain the mechanism and the stereospecificity of the enzyme. A satisfactory explanation can, however, be given now from the detection of PQQ and by proposing the existence of an iron-PQQ complex in the enzyme (22). The lesson to be learned from all these examples is that, although some metalloenzymes have been studied for years, the presence of PQQ has remained unnoticed. Therefore, could this also be the case for well-known enzymes with an organic cofactor? Screening the literature on pyridoxoproteins [enzymes requiring pyridoxal phosphate (PLP) as cofactor] revealed that several of them contain an additional carbonyl-group cofactor, the identity of which has been claimed to be a PLP-derivative, although evidence for this is lacking. A novel analysis procedure, based on derivatizing and extracting the formed adduct with hexanol, established the presence of PQQ in dopa decarboxylase from pig kidney (23).

The search for quinoproteins resulted in the discovery of a variety of examples, but probably not the whole catalogue. However, from the present status it can already be concluded that PQQ has come of age as a cofactor. The role of quinoproteins in microbial oxidations (10), and the enzymology of bacterial quinoproteins (11) and of PQQ (24), have been reviewed. The present review summarizes the progress with microbial quinoproteins and highlights the recently discovered quinoproteins in plants and animals. Structural and mechanistic data are examined in an attempt to reveal the role of PQQ in the catalytic process.

ENZYMOLOGY OF QUINOPROTEINS

Dehydrogenases

All quinoprotein dehydrogenases hitherto detected are situated in the periplasmic space of Gram-negative bacteria. Although this could indicate a special suitability of this type of dehydrogenase for that location, the reason is unknown. The enzymes are involved in the primary degradation steps of nonphosphorylated substrates, converting them to either aldehydes or acids, after which the products are channelled into common metabolic pathways. Since aldehydes are very toxic for the cell, delicate mechanisms should exist to prevent their accumulation and to provide fine tuning to the capacity for their use by metabolic pathways inside the cell. Another interesting but uninvestigated aspect is the transport of PQQ and the assemblage to holoenzyme in the periplasmic space (asssuming transport of protein through the membrane in unfolded form). Most of these enzymes have been described in earlier reviews (10, 11), except quinate dehydrogenase, which was recently detected (25). Therefore, only enzymes for which substantial progress has been reported during the past few years with respect to mechanism, structure, or insight into the biological role, will be discussed.

METHANOL DEHYDROGENASE (MDH) (EC 1.1.99.8)   This enzyme occurs in Gram-negative bacteria that convert methanol into formaldehyde [N.B. the Gram-positive bacterium *Nocardia* sp. 239 uses a quite different, NAD-dependent, PQQ-containing enzyme (26)]. Depending on the organism, substrate specificity varies, some enzymes oxidizing not only a large number of primary but also secondary aliphatic alcohols (27); others oxidize aromatic alcohols (28). In addition, formaldehyde is an excellent substrate for all MDHs, while higher aldehydes are not. It has been known for a long time that ammonium salts are necessary in the in vitro assay as activators (29). However, certain amines (e.g. 2-bromoethylamine) or amino acid esters (e.g. ethylglycine) are even better activators (J. A. Duine, unpublished results). On the other hand, certain MDHs have a substrate specificity extended to higher aliphatic aldehydes, while they are activated by aliphatic primary amines (27). This indicates that substrate specificities for alcohols and aldehydes do not run parallel, and specificities for activators have similar restrictions as those for aldehydes. The latter suggests that activation occurs in the active site.

Insight has recently been obtained into which step becomes activated in the catalytic cycle of MDH (30). It appears that the oxidation of substrate is the rate-limiting step, which is enhanced enormously by addition of activator. The intermediates have been characterized with respect to absorption spectrum and redox state of the two PQQs. Even the enzyme-substrate complex could be isolated by making use of the large deuterium effect on the substrate oxidation step and by performing the reaction in the absence of activator (the activating effect concerns the decomposition of the complex). Cyclopropanol seems a mechanism-based inhibitor, only attacking the oxidized form of MDH. The proposed structure of the inhibition product (31), namely the C(5)-ω-propionaldehyde adduct of PQQ, has meanwhile been proven (32).

The large deuterium effect on substrate oxidation and the fact that cyclopropylmethanol is a substrate and not an inhibitor make a sequential one-electron transfer mechanism very unlikely.

Structural data on MDH are still scarce. The enzyme occurs in monomeric, but mostly in (homo)dimeric form (33). Production probably proceeds in the usual way, the protein having a leader sequence that is removed after transfer through the cytoplasmic membrane (34). The genes of MDH have been cloned and sequenced from a number of bacteria (35, 36, 36a). A mutant of *Methylobacterium organophilum* has been obtained, unable to provide the protein with PQQ (37). Although supplementation of the medium with PQQ restored growth on methanol (as well as ethanol, indicating that MDH is also functional for the metabolism of higher alcohols), a delicate group is probably involved in the binding of PQQ, since in vitro reconstitution of PQQ-free MDH to activity has never been reported.

Progress has also been made with respect to the mechanism in vivo. Indirect evidence suggested already that cytochrome c_L is the natural electron acceptor for MDH (38). However, using this compound in an in vitro assay, very low activities were obtained and the mechanism involved was proposed to be similar to that responsible for the autoreduction phenomenon observed for ferricytochrome c_L (39). Using the components from *Hyphomicrobium* X (40), it has been shown (41) that ferricytochrome c_L is an excellent electron acceptor for all reduced forms of MDH at pH 7 but not at the common pH of the in vitro assay (pH 9). However, substrate oxidation is very slow at pH 7 and ammonium salts show no activation at this pH. Although all these observations explain why cytochrome c_L behaves overall as a very poor electron acceptor in the in vitro assay, the question remains how substrate oxidation proceeds in vivo. A compound has been found in anaerobically prepared cell extracts that stimulates the reaction at pH 7 and that is certainly not an ammonium salt (42). This suggests that methanol oxidation in vivo probably also requires an activator. Since the compound is destroyed by oxygen in the presence of cytochromes, this explains why in vitro assays require addition of ammonium salts as artificial activators as well as a buffer of pH 9.

METHYLAMINE DEHYDROGENASE (MADH) (EC 1.4.99.3)   The enzyme occurs in several Gram-negative bacteria able to grow on methylated amines. Primary aliphatic amines are converted into the corresponding aldehydes and ammonia. A variant with a preference for aromatic amines has also been described (43). Since the cofactor is covalently bound to the protein, the identification of MADH as a quinoprotein relied on indirect evidence so that the possibility was also mentioned that the cofactor might turn out to be a PQQ-like compound (9). Recently it has been claimed indeed that the cofactor is PQQ without carboxylic acid groups, attached to the protein by way of ether bridges (44). However, conclusive evidence for the proposed structure was not provided, and application of the hydrazine method on the enzyme from *Thiobacillus versutus* revealed the usual adduct, that is the hydrazone of PQQ (19a). Therefore the cofactor in MADH is likely to be genuine PQQ, most probably bound to the protein via an amide bond. In this respect, the progress that has been made in resolving the three-dimensional structure of MADH (45, 46) may help elucidate the precise mode of binding of the cofactor by identifying the residues used for attachment.

Redox titrations and kinetic measurements have revealed many details of the catalytic cycle of MADH (47, 48). For this enzyme, ammonia is an activator at low but an inhibitor at high concentrations, changing the absorption spectra of several intermediates of the enzyme in the catalytic cycle (49). It is still uncertain whether transfer of the two reduction equivalents, originating from substrate oxidation, occurs to one subunit (resulting in $PQQH_2$) or to the two small subunits by partitioning them (resulting in the formation of two PQQH·s that could eventually disproportionate). Some controversy exists regarding the natural electron acceptor. Amicyanin, a type I blue copper protein, is an excellent electron acceptor for MADH. On the other hand, it has been proposed that a cytochrome c could also function as electron acceptor (50, 51). Using the rate constants obtained by stopped-flow kinetic measurements and estimated concentrations of the components in the periplasm in a simulation, it was shown that the oxidation rates of whole cells can be adequately accounted for by amicyanin-mediated electron transfer (52).

GLUCOSE DEHYDROGENASE (GDH) (EC 1.1.99.17)   This enzyme has been found in many bacterial species, either as holoenzyme or as apoenzyme [e.g. all *Escherichia coli* laboratory strains harbor the apo-form and the enzyme becomes functional in vitro as well as in vivo upon PQQ addition (53)]. The apoenzyme becomes easily reconstituted so that many biological assays are based on it (24). Most aldose sugars (hexoses, pentoses) are substrates. However, some warning should be made concerning comparisons of substrate specificities in vivo and in vitro. Although it was suggested that large changes in substrate specificities occur on disruption of *Acinetobacter calcoaceticus* cells (54), recent findings indicate that this organism produces a soluble as well as a membrane-bound GDH with quite different structural properties as well as substrate specificities (55, 56). Since soluble GDH activity is only detected after disruption of the cells, its function in the cell is unclear. Electron transfer occurs to the soluble cytochrome b562, isolated from the same organism (57). In contrast, the membrane-bound GDH transfers its electrons to ubiquinone (58).

Soluble GDH (59, 60) and membrane-bound GDH (61, 62) have been characterized from this organism, and the genes have been cloned and sequenced for both enzymes (55, 63) and PQQ (64). Other bacteria like *E. coli* (61) or *Pseudomonas* species (65, 66) only contain membrane-bound GDH. The soluble GDH consists of two subunits with M_r 48,000 and two PQQs (59). In contrast, the membrane-bound GDH is monomeric (M_r 87,000) (61). Addition of substrate to the soluble GDH shifts the absorption maximum of the oxidized (350 nm) to the reduced form (338 nm). Binding of cofactor requires the presence of Ca or Mg ions, which become very firmly bound after reconstitution [except for membrane-bound GDH, a phenomenon that has also been observed in the case of other quinoprotein dehydrogenases (67)]. Modification of the enzyme with chemical reagents has revealed several interesting details of the active site (68).

The significance of the distribution of apo-GDHs and holo-GDHs is not understood (69). The precursors for biosynthesis of PQQ in methylotrophic bacteria are tyrosine and a glutamic acid moiety (70, 71), and arguments have been put forward to assume that the process occurs on a protein matrix (24, 70). It seems very likely that biosynthesis of polyamines, compounds indispensable for regulation of cell division, also occurs in these bacteria via pyridoxo-quinoprotein decarboxylases (see section on dopa decarboxylase). Therefore, those species unable to produce holoenzyme might be deficient in one or more steps leading to production of free PQQ. Note added in proof: the presence of PQQ in glutamate decarboxylase from *E. coli* has been established (B. W. Groen, J. A. Duine, unpublished results).

Oxidases

Quinoprotein oxidases occur in a wide variety of organisms and are involved in oxidation of amine and hydroxyl groups. In all cases, the enzymes contain PQQ as well as Cu ions. If the suggestion (72) is correct that nitroalkane oxidase contains PQQ as well as flavin mononucleotide (FMN), then other groups can also be oxidized and a combination of PQQ with an organic cofactor is also possible.

AMINE OXIDASES (AOs) (EC 1.4.3.6)   These so-called copper-containing amine oxidases have many properties in common and occur under different names [plasma amine oxidase, diamine oxidase, lysyl oxidase (EC 1.4.3.13)] in mammals, plants, and fungi. A recent member of this group is methylamine oxidase, found in the Gram-positive bacterium *Arthrobacter* P1 (73). The eukaryotic AOs have been investigated already for many years and the results discussed in several reviews. The discovery that not pyridoxal phosphate but PQQ is the organic cofactor has introduced new avenues of research in this field, as will be discussed in the following.

The enzymes consist of two subunits with closely similar primary structure and two well-separated Cu ions situated at chemically distinct sites (74). Titration with hydrazines reveals one reactive group per enzyme molecule, in accordance with the product formed with the hydrazine method, either the azo-compound or the hydrazone, isolated in the same stoichiometry for several of these enzymes (75). Since PQQ is covalently bound to the protein but the subunits can be easily dissociated, these data suggest that there is an asymmetric cofactor distribution. A peptide has been isolated from pig kidney diamine oxidase containing the PQQ-hydrazone, most probably bound to the peptide via an amide bond consisting of one of the carboxylic acid groups of PQQ and a lysyl residue (76).

A curious point in view of the afore-mentioned structural data is that a few publications report the conversion of two molecules of substrate per enzyme molecule under anaerobic conditions (77, 78). Since it is generally agreed that reduction of the Cu(II) ions does not take place under these conditions, the explanation for this phenomenon is that there exists still another electron-accepting group in these enzymes. A reason that this has remained undetected could be that AOs show half-of-the-sites reactivity toward hydrazine inhibition, i.e. each subunit contains one PQQ, but after one of these has reacted, the other becomes inactive. This hypothesis could be checked by applying another analysis procedure for PQQ to these enzymes, e.g. the hexanol extraction procedure (23). Another explanation, however, might be the presence of a different electron-accepting group. It has been reported that after treatment of enzyme with substrate under anaerobic conditions, one extra SH group can be detected in the enzyme (79, 80). Thus if it is assumed that $PQQH_2$ can transfer its reduction equivalents to an S-X group, this could adequately explain why two substrate molecules are converted.

Besides the amide bond, perhaps there exists another interaction between PQQ and the protein, responsible for the atypical quinoprotein absorption spectrum in view of its rather structureless appearance with a broad band at 490 nm. Even after anaerobic treatment with substrate, most of the enzymes show an absorption spectrum that does not reflect the presence of $PQQH_2$, only a decrease in absorbance of the 490 nm band occurring. On the other hand, there are a few exceptions where substrate addition leads to the formation of a set of pronounced absorption maxima (77, 81, 82) and the formation of an organic free radical, as apparent from ESR spectroscopy (82). A tentative interpretation of the absorption spectra suggests the presence of PQQ (maxima at 430 and 403 nm, similar to those of oxidized methanol dehydrogenase), PQQH· [maxima at 460 and 360 nm, comparable to those of free PQQH· (83) and $PQQH_2$ (330 nm, comparable to that of the reduced forms of methylamine and methanol dehydrogenases). The variety of PQQ redox forms suggests interplay with another electron acceptor of the enzyme, for instance the presumed S-X group or the Cu(II) ion(s). The fact that a decrease of the Cu(II) ESR signal occurs (82), concomitant with formation of the organic free radical, suggests that the latter possibility is the most likely. This means that in certain AOs (77, 81) or under certain conditions (82), transfer of electrons occurs between PQQ and the Cu ions, suggesting the existence of interaction, and the removal of a structural constraint responsible for the atypical absorption spectrum.

Titration of PQQ in acetate buffer with $CuSO_4$ solutions shows that there is a strong binding, leading to a complex in which PQQ occurs in hydrated form. Similar experiments with $PQQH_2$ showed that it becomes immediately oxidized on addition of the Cu(II) ions (84). However, model systems have been reported (85) in which this oxidation does not occur. From the structure of PQQ, at least three sites may provide suitable ligands for Cu(II) binding. Especially the site occupied by Cu in galactose oxidase (Figure 2) seems well suited. It is, however, very unlikely that such a complex exists in AOs. Using fluoridated hydrazines to derivatize the PQQ in the enzyme, NMR relaxation measurements indicated that the distance between the F atoms and the Cu ions is too large for such a complex (86). On the other hand, removal of Cu from the enzymes has effects on the absorption spectrum and addition of substrate or hydrazine have slight effects on the ESR signal of Cu(II). Cu-depleted enzymes are able to oxidize the substrate but are not catalytically active since the reoxidation step does not proceed. Since all these phenomena suggest some kind of interaction, perhaps this could occur at the upper side of the PQQ molecule, which is a redox-inactive moiety for Cu and a location compatible with the distance calculated (86). On the other hand, the possibility that Cu has merely an allosteric effect without close interaction, cannot be excluded at the moment.

GALACTOSE OXIDASE (GAO) (EC 1.1.3.9)   GAO is produced by several fungi and can be isolated from their spent culture media. The enzyme oxidizes the 6-hydroxyl group of galactose to the aldehyde and H_2O_2. Most of the

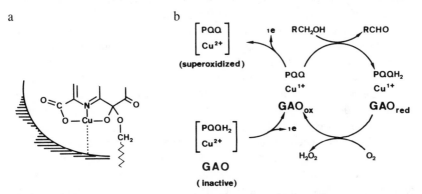

Figure 2   Complex of PQQ and Cu (*a*) in galactose oxidase (GAO), and the catalytic cycle of GAO (*b*).

research has been performed on the enzyme from *Dactylium dendroides*. GAO is a monomeric copper protein containing a single Cu ion, which until recently was considered the only cofactor. Ingenious mechanisms have been invoked to explain accommodation of the two reduction equivalents derived from the substrate, e.g. by assuming a Cu(III) redox state of the cofactor (87, 88). Since it has now been found that GAO contains one covalently bound PQQ (89), these mechanistic proposals must be revised.

In retrospect, the reason the mechanisms have not been challenged for many years is that heterogeneity with respect to enzyme redox form and activity was diagnosed only recently. Preparations isolated by common procedures appear now to consist largely of an inactive enzyme form (GAO$_{in}$), implying that interpretations of spectroscopic data have been based largely on the contributions from the inactive enzyme form, while e.g. interpretations based on kinetic measurements refer to the active enzyme form. Although the activating effect of ferricyanide treatment was known for some time (88), the preparation and characteristics of the different enzyme forms were only recently described (90). By increasing the redox potential of the solution, an oxidized form was obtained (GAO$_{ox}$) that was fully active. Treatment of this form with substrate under anaerobic conditions gave a reduced form (GAO$_{red}$). By including the presence of PQQ in the considerations, the spectroscopic features can be accounted for. A scheme of the catalytic cycle is given in Figure 2.

GAO$_{ox}$ is an ESR-silent enzyme form with respect to copper signals. On the other hand, a tiny amount of an organic free radical seems to be present (90) with characteristics reminiscent of those of PQQ (4, 5). The absorption spectrum (in the 300–500 nm region) is similar to that of methylamine dehydrogenase in its oxidized form. Therefore, GAO$_{ox}$ seems to contain Cu(I) and PQQ.

Reducing conditions transform GAO$_{ox}$ into GAO$_{in}$. ESR measurements of GAO$_{in}$ show the presence of Cu(II), with characteristics (91) as have been found for plasma amine oxidase, but no organic free radical (although this may be largely obscured by the Cu(II) signal in the $g = 2$ region). The absorption spectrum shows a significant shoulder at 340 nm, a value that is more or less similar to that of quinoprotein dehydrogenases containing PQQH$_2$. This indicates that GAO$_{in}$ contains Cu(II) and PQQH$_2$. In view of the redox potentials of free PQQ and Cu(II), either they occur at different locations in the enzyme or the ligands of Cu(II) induce a lowering of its redox potential below that of PQQ (85).

GAO$_{red}$ is produced from GAO$_{ox}$ on addition of substrate. Since two reduction equivalents are now transferred to the enzyme, this explains why the properties of GAO$_{red}$ differ from those of GAO$_{in}$. Based on the same reasoning, GAO$_{red}$ would contain Cu(I) and PQQH$_2$. The absorption spectrum of this enzyme form is structureless. Although free $PQQH_2$ has an absorption maximum at 302 nm (92), a maximum at higher wavelengths is found for reduced quinoprotein dehydrogenases. Different ligandation could be responsible for this discrepancy.

Theoretically, a superoxidized form of the enzyme could exist (GAO_{sup}), containing Cu(II) and PQQ, a couple having the same redox state as found in the quinoprotein amine oxidases. Perhaps the redox potential applied so far was not high enough to convert GAO into this form. On the other hand, it cannot be excluded that GAO_{sup} occurs in vivo and in untreated enzyme preparations. If so, reaction with substrate could lead to GAO_{in}, so that the presence of substantial amounts of GAO_{in} in isolated preparations might indeed reflect the occurrence of GAO_{sup}. Other indirect evidence can be derived from the observation that cyanide addition is seen to displace a phenolate ion, as judged by ESR spectroscopy (93). Since H_2O and cyanide add to the C(5) carbonyl group of PQQ (but not to PQQH· and $PQQH_2$) and the hydrated group becomes deprotonated above pH 7, such effects could be explained from a complex of Cu(II) and PQQ in GAO_{sup}. If this appears to be correct, it is questionable whether GAO functions as an oxidase in vivo since GAO_{ox} might be an artifact. Since the one-electron acceptor ferricyanide is able to change the redox state of GAO, the conversion of GAO_{in} into GAO_{sup} could be catalyzed by a component from the respiratory chain and the occurrence of GAO in the culture medium resulting from cell lysis.

Based on the enzyme forms discussed, a catalytic cycle can be proposed for GAO (Figure 2). Since PQQ in so many quinoprotein dehydrogenases is able to oxidize a hydroxyl group, it seems logical to propose that the same occurs in GAO. Oxidation will be preceded by binding of the alcohol group to C(5) (as mentioned already in the case of methanol dehydrogenase), since it was found that 6-deoxygalactose does not bind to GAO (93). The role of Cu could be to assist in the oxidation of $PQQH_2$ to PQQ by molecular oxygen, suggesting a close interaction between the cofactors. Since kinetic experiments reveal that cyanide competes with galactose for the same binding site, the effects of cyanide on Cu(II) as judged from ESR spectroscopy, can best be explained by assuming a complex of PQQ and Cu as depicted in Figure 2.

Oxygenases

Reactions involving the enzyme-catalyzed insertion of oxygen normally require the presence of cofactors that are capable of activating dioxygen (94). Although such properties are not immediately apparent from the structure of PQQ itself, combinations of PQQ and suitable metal ions could well be effective in this respect. A complex involving PQQ and Fe has been proposed to function in the lipoxygenase-catalyzed insertion of dioxygen (22).

DOPAMINE   β-HYDROXYLASE   (MONOOXYGENASE)   (EC   1.14.17.1)
(DβH)   DβH catalyzes the conversion of dopamine to noradrenaline according to the following reaction:

$$\text{dopamine} + O_2 + 2(H) \rightarrow \text{noradrenaline} + H_2O$$

The enzyme has been the subject of intensive study [for a recent review see
(20)]. Both its central role in the catecholamine biosynthetic pathway, as well
as the mechanistic details of the hydroxylation reaction, have drawn much
attention. However, despite extensive investigations, the presence of PQQ in
this enzyme has been completely overlooked in the past. Using phenylhydrazine as an active site–directed inhibitor, the hydrazone of PQQ and phenylhydrazine could be extracted in near stoichiometric amounts (approximately one
PQQ per subunit of tetrameric enzyme) from (commercial) preparations of
DβH, purified by established procedures (19a, 21). The possibility that PQQ
is attached to these samples as a result of nonspecific adsorption [tight binding
of a variety of anions to DβH has recently been reported (95)], can be refuted
on the basis of concomitant formation of adduct and inactivation of DβH as
well as the purification procedure, from which covalent binding is deduced
(21).

The functional presence of PQQ in DβH warrants a reappraisal of a
long-standing debate regarding the role and the amount of copper present in
the enzyme (96). On the basis of existing information, PQQ has been proposed as an anchoring site for the docking of dopamine and structurally
related substrates (21). Although the well-documented electrophilicity of the
C5-carbonyl group of PQQ could well account for such a function, recent
results concerning the presence and individual binding characteristics of two
copper sites per DβH subunit (97), open the possibility for a more directly
involved role for PQQ. In Figure 3, a tentative configuration is depicted, in

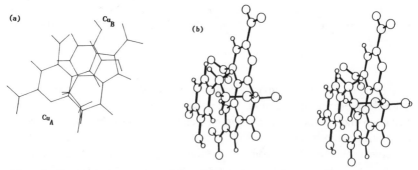

Figure 3 Hypothetical structure of PQQ, substrate and Cu ions in dopamine β-hydroxylase
active site (*a*) top view, (*b*) stereoscopic view.

which a molecule of PQQ bridges the two copper centers. The attached dopamine substrate, arising from addition to the C5-carbonyl group of PQQ, is accommodated in a sandwich arrangement (binding of substrates that lack the amino group, e.g. p-cresol, might thus be more difficult).

On the basis of this model, puzzling observations regarding the mechanism of action of DβH can be accounted for. Several groups have deduced a distance of more than 6–7 Å for the two copper centers from the fact that no paramagnetic coupling can be detected in the resting Cu(II)/Cu(II) enzyme (98, 99). Nevertheless, both copper ions become reduced with equal rates (100). Reconstitution of copper-depleted enzyme shows a high degree of similarity in the binding characteristics of the copper sites (101, 102, but see 97), yet, addition of cyanide reveals distinct differences in the ESR signals of the individual centers (91).

Although ascorbate has been established as the functional reductant in vivo (103), ascorbate, ferrous cyanide, and substrates like dopamine and tyramine can act as reducing agents in vitro. Whether these reductants require separate sites is not clear at the moment. However, based on geometric considerations, a similar site for ascorbate and ferrous cyanide has been suggested (104).

Considering PQQ as a bridging ligand, fairly equal affinities for the Cu_A and Cu_B positions can be expected (J. Reedijk, personal communication). Differential behavior observed upon addition of cyanide can be explained by assuming one cyanide ion to become liganded to Cu_B, while the second molecule adds to the C5-carbonyl group of PQQ, thus invoking only a minor perturbation of the Cu_A ESR parameters. In the Cu(II)/Cu(II) state, PQQ can hardly sustain paramagnetic coupling of the copper centers as the overall conjugation is impaired by the sp^3 character of carbon-5. However, upon reduction of one of the copper sites, PQQ may adopt a more planar configuration, thus facilitating rapid equilibration of reducing equivalents over both coppers. In agreement with this view, a large ligand reorganization of the copper sites upon reduction has been noted (105). The tentative model (Figure 3) allows the reduction of Cu(II) by substrate, in the absence of additional reductants, to occur via electron transfer involving the phenolic hydroxyls. In this view, the existence of separate sites is not required. To investigate the spatial relationship between the copper sites and PQQ in DβH, the approach used to estimate the distance between PQQ and copper in plasma amine oxidases (86) could be worthwhile. Studies on the inhibition of DβH with trifluoromethyl-substituted phenylhydrazines are under way.

Peptidyl-α-amidating monooxygenase is an enzyme that has been found recently in mammalian tissues (106). Although the enzyme still awaits characterization, the properties that are already known suggest that it could be similar to DβH with respect to cofactor identity.

LIPOXYGENASE (LPO) (EC 1.13.11.12)   LPO catalyzes the stereospecific and regiospecific insertion of molecular oxygen into unsaturated fatty acids containing a pentadiene system. It is generally accepted that nonheme iron is the only cofactor in LPO, present in a ratio of one Fe per enzyme molecule. From the large amount of data available for soybean lipoxygenase-1, a mechanism has been proposed wherein Fe has a dualistic role, namely removal of hydrogen from the substrate followed by insertion of molecular oxygen into the free radical substrate (107, 108). A different view has been proposed, the so-called organo-iron mechanism, based on results obtained with mechanism-based inhibitors (109). Proton abstraction, catalyzed with a very strong basic group, is supposed to occur with concurrent electrophilic attack of Fe on the substrate.

The discovery of PQQ in soybean lipoxygenase-1 (22) sheds light on this long-standing problem. Strong arguments have been put forward to propose an interaction of PQQ with Fe, giving a complex well suited to function as an electron relay system, explaining how the regioselective and enantio-selective steps could proceed in the enzyme (Figure 4). In addition, the redox states of the different enzyme forms were indicated and from this, a catalytic cycle for the enzyme proposed (Figure 4).

Mammalian organisms contain a variety of lipoxygenases that have not been characterized so far. However, the properties that are already known suggest similarity with soybean lipoxygenase-1 so that they might also be quinoproteins.

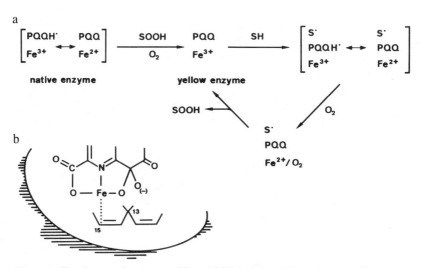

Figure 4   The electron relay system of Fe and PQQ in the catalytic cycle (*a*) and lipoxygenase (*b*).

Hydratases and Decarboxylases

The common characteristic of these subclasses of enzymes is group transfer. Furthermore, although the overall reaction catalyzed by these enzymes is formally redox-neutral, in fact for several enzymes the separate steps in the cycle are redox reactions. In a number of cases, PLP has been shown to play this role (shuttling between pyridoxal phosphate and pyridoxamine phosphate). In principle, however, PQQ is also suited for this task since it shows redox behavior and is an acceptor for several transferable groups [addition occurs at its C(5) position of water, ammonia, cyanide, and several other nucleophilic compounds (13)]. The following examples show or strongly suggest the involvement of PQQ in these enzyme classes.

NITRILE HYDRATASE   The enzyme has been discovered in several bacteria and catalyzes the conversion of aliphatic nitriles into the corresponding amides [this in contrast to nitrilase (EC 3.5.5.1), which converts these substrates into the corresponding carboxylic acids plus ammonia]. It has been proposed that besides Fe(III)-ions, the enzyme contains PQQ (110). Although no quantitative data have been reported, the activity in a biological PQQ assay and the spectral change induced by phenylhydrazine treatment of the enzyme, suggest that this could indeed be the case. Since it is already known that hydration occurs of PQQ in an aqueous solution (13), this phenomenon was used to explain the mechanism of the enzyme (111). Chemical hydration of nitriles occurs in the presence of peroxides (112). Since H_2O_2 is formed on oxidation of $PQQH_2$ (113), it could be imagined that the enzymatic catalysis consists of a cyclic process in which the redox function of PQQ is used instead of its ability to form hydrates.

DIHYDROXYPHENYLALANINE (DOPA) DECARBOXYLASE (DDC) OR AROMATIC AMINO ACID DECARBOXYLASE (EC 4.1.1.28)   DDC from pig kidney has been intensively studied (114). It contains PLP, and the absorption spectrum shows maxima at 335 and 420 nm. After removal of PLP, a shoulder at 335 nm is still observed in the spectrum of the apoenzyme. Although it has been suggested that this belongs to a vitamin B6–like compound, the structure has never been elucidated. Since the properties of this compound were reminiscent of covalently bound PQQ in enzymes previously thought to contain PLP, we attempted its identification. The hydrazine method failed in this case, but using the hexanol extraction procedure, the presence of one PLP and one PQQ per enzyme molecule could be established (115).

The finding of cofactor other than PLP in a decarboxylase is not unique, since several decarboxylases have a pyruvyl group as cofactor. The functionality of this group has been demonstrated in the case of histidine decarboxylase from a *Lactobacillus* strain (116). In the mechanism, a pyruvamide moiety in the protein chain acts as an electron sink (117), and a similar function could be devised for the o-quinone moiety in PQQ. However, such a mechanism does not account for a role for PLP in DDC. Therefore, in the following, the existing data in the literature are reviewed to provide arguments for a role for the cofactors.

Starting with the absorption spectrum of the isolated enzyme, we propose that the absorption band at 420 nm belongs to PQQ, occurring in the form of an adduct [e.g. by reaction of the C(5) carbonyl group of PQQ with an ϵ-amino group of a lysine residue]. A similar band has been found in the oxidized forms of quinoprotein dehydrogenases: a double band in the 390–412 nm region (118) of methanol dehydrogenase (the variation related to the type of external ligand involved) and the maximum at 440 nm of methylamine dehydrogenase (49). We further propose that this adduct reacts with the substrate, that is the internal amino group is exchanged for the incoming one of the substrate or inhibitor. This is supported by the observation (119) that addition of D-substrate analogues (which are not decarboxylated) give an increase of the absorbance and shifts of the maximum (from 390 nm to 440 nm) depending on the nature of the analogue and the pH. The proposal is also in accordance with the observation that the first transient induced by addition of genuine substrates is in the 420 nm region (120). D-tryptophan analogues give rise to time-dependent inhibition and deamination. However, only part of the PLP is transformed into PMP, and activity of the inactivated enzyme is only partly restored on PLP addition (121). Therefore, this could indicate that part of the adducts of PQQ and this type of D-substrate analogue slowly decompose under deamination (PLP addition restoring activity), while the decomposition of the other part leads to derivatization of PQQ (e.g. to an oxazole), resulting in inactivation as it cannot be reversed with PLP. Model studies (122) have revealed that PQQ is indeed very suited to catalyze decarboxylation, and a cyclic scheme that explains this reaction as well as dead end product (oxazole) formation has been derived.

Even better support for the proposed mechanism is provided by the behavior of α-methyl-L-DOPA, which is decarboxylated under concomitant deamination to dihydroxyphenylacetone (123), accompanied by the nearly complete disappearance of the 420 nm band and a strong increase in the 330 nm region (maximum at 323 nm) (124). Curiously, the conversion of PLP into PMP was negligible and scarcely any free ammonia was found. If it is accepted, however, that PQQ is responsible for both reactions, the amino groups should be transferred to this cofactor, and the reaction stops after all the PQQ is converted into the amino form of $PQQH_2$. Evidence for this is provided by the absorption spectra, the rise of the absorption at 323 nm being precisely what occurs in methylamine dehydrogenase on transfer of the amino group from the substrate to PQQ (49). A hitherto incomprehensible observation (123) was that about 10 times more product was found than could be expected from the amount of enzyme used. Since the latter was based on the amount of PLP, an explanation could be that these enzyme preparations had only 10% of the stoichiometric amount of PLP so that in fact the amount of PQQ was 10 times higher [in this context it should be noted that PLP determinations based on dissolving these enzymes in 0.1 M NaOH are incorrect, since PQQ in this solvent also has substantial absorbance in this region (92)].

The phenomena related to preparation of the so-called apoenzyme should also be discussed, as they indicate the role of the cofactors. Addition of hydroxylamine to DDC induces a strong absorption band at 390 nm (124). This maximum is also observed upon hydroxylamine treatment of the oxidized form of methanol dehydrogenase (7). The band of inhibited DDC at 390 nm slowly converts into a band at 340 nm. Presumably, this is a process in which PQQ becomes reduced by the hydroxylamine, since $PQQH_2$ in some reduced quinoproteins (e.g. methanol and glucose dehydrogenase) has a similar maximum. After dialysis, the apoenzyme does not contain PLP [using a method in which PLP is determined as the phenylhydrazone in the supernatant (125)], is not active but activity can be partly restored by addition of PLP, and the absorption spectrum shows a shoulder in the 330–340 nm region. Thus it appears that this treatment removes PLP and transforms the covalently bound PQQ into a reduced form. The fact that PLP addition results in nearly complete restoration of the original spectrum, suggests that PLP binding makes possible the oxidation of reduced PQQ so that the adducts related with the 420 nm adsortion band can be formed and activity is restored. Since there is also a large increase at 335 nm, PLP most probably forms a bond with the protein (an absorption band with this maximum has been found in several pyridoxoproteins and ascribed to a substituted aldimine adduct or to a Schiff base in apolar environment). Since it is necessary to add PLP to the buffers used in certain purification steps and addition to the assay gives higher activities, binding is rather loose.

Since PLP addition to DDC prevents the deamination reaction to occur of the afore-mentioned inhibitors, it is conceivable that its protective effect concerns the rapid acceptance of the decarboxylated substrate from PQQ, after which release of the product can occur (a tentative scheme for the mechanism with genuine substrate is given in Figure 5). Obviously, the latter step is enantio-selective and stereo-selective, since decarboxylated products from D-amino acids or α-methylamino acids are not accepted, and reduced PLP-amino acid adducts in the L-form are strong inhibitors, while such adducts in the D-form are not.

Characteristic features similar to those mentioned here for DDC have been

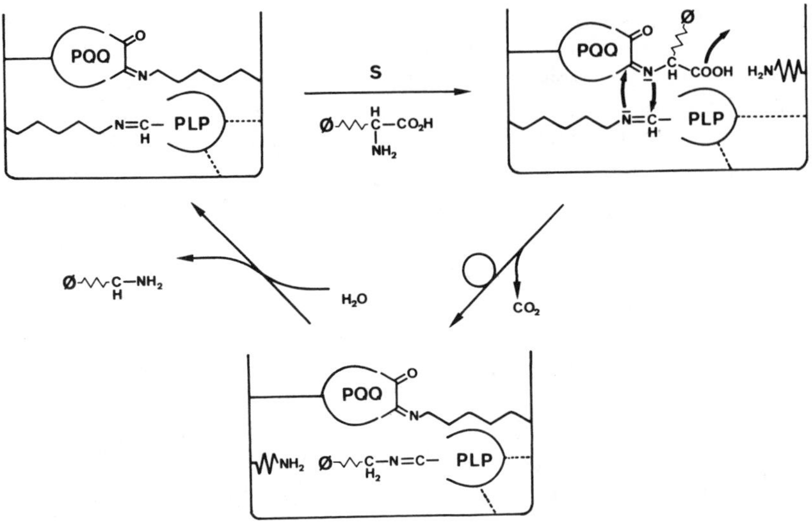

Figure 5 The interplay between PQQ and PLP in the mechanism of dopa decarboxylase.

reported for some other amino acid decarboxylases. Moreover, examination of the current literature on pyridoxoproteins shows that there are strong arguments to suggest the presence of PQQ in other groups, as discussed below.

The group of transaminases is well established. It can be subdivided into enzymes transferring the α-amino group and those transferring a terminal amino group of the amino acids, the so-called ω-amino acid transaminases. The subgroups appear quite different from each other with respect to protein structure as well as cofactor composition. To illustrate the latter, among the ω-amino acid transaminases several examples have been described where the same situation exists as in dopa decarboxylase, that is besides PLP another nonproteinaceous compound is present with properties very similar to those of covalently bound PQQ in DDC [e.g. L-lysine 6-aminotransferase (EC 2.6.1.36) (126), arnithine-oxo-acid aminotransferase (EC 2.6.1.13) (127), ω-amino acid:pyruvate aminotransferase (EC 2.6.1.18) (128), and 4-aminobutyrate aminotransferase (EC 2.6.1.19) (129, 130)]. In view of this striking similarity, it is proposed that these enzymes are also pyridoxoquinoproteins.

What role could PQQ play in transaminations? In model studies (122), it was observed that rapid conversion occurs of the terminal amino group of several α,ω-diamino acids into an aldehyde group. Since it is known that PLP forms stable adducts with the ϵ-amino group of lysine, in a pyridoxoquinoprotein enzyme the reactivity of PQQ toward terminal amino groups could be used for their removal, followed by transfer to PLP which in turn transfers it to an α-ketoacid acceptor. Although the nature of the covalently bound cofactor still awaits identification, it is obvious that the mechanism for this subgroup can no longer be indicated in the usual way of a pyridoxoprotein transaminase.

ACKNOWLEDGMENTS

For making unpublished data available to us we thank our colleagues, especially Hans Frank, John van Wielink, Mario van Kleef, and Rob van der Meer. We acknowledge Dr. H. Peters, Unilever Research Laboratories, for assistance in preparing Figure 3.

Literature Cited

1. Szymona, M., Doudoroff, M. 1960. *J. Gen. Microbiol.* 22:167–83
2. Hauge, J. G. 1964. *J. Biol. Chem.* 239:3630–39
3. Anthony, C., Zatman, L. J. 1967. *Biochem. J.* 104:960–69
4. Duine, J. A., Frank Jzn, J., Westerling, J. 1978. *Biochim. Biophys. Acta* 524:277–87
5. Westerling, J., Frank Jzn, J., Duine, J. A. 1979. *Biochem. Biophys. Res. Commun.* 87:719–24
6. Salisbury, S. A., Forrest, H. S., Cruse, W. B. T., Kennard, O. 1979. *Nature* 280:843–44
7. Duine, J. A., Frank Jzn, J., Verwiel, P. E. J. 1980. *Eur. J. Biochem.* 108:187–92
8. Duine, J. A., Frank Jzn, J., van Zeeland, J. K. 1979. *FEBS Lett.* 108:443–46
9. de Beer, R., Duine, J. A., Frank Jzn, J., Large, P. G. 1980. *Biochim. Biophys. Acta* 622:370–74
10. Duine, J. A., Frank Jzn, J., Jongejan, J. A. 1986. *FEMS Microbiol. Rev.* 32:165–78
11. Duine, J. A., Frank Jzn, J., Jongejan, J. A. 1987. *Adv. Enzymol.* 59:169–212
12. Enzyme Nomenclature. 1984. *Recommendations (1984) of the Nomenclature Committee of the I.U.B.* Academic
13. Dekker, R. H., Duine, J. A., Frank Jzn, J., Verwiel, P. E. J., Westerling, J. 1982. *Eur. J. Biochem.* 125:69–73
14. van Kleef, M. A. G., Dokter, P., Mulder, A. C., Duine, J. A. 1987. *Anal. Biochem.* 162:143–49
15. Lobenstein-Verbeek, C. L., Jongejan, J. A., Frank Jzn, J., Duine, J. A. 1984. *FEBS Lett.* 170:305–9
16. Hartmann, C., Klinman, J. P. 1988. *Biofactors* 1:41–49
17. Moog, R. S., McGuirl, M. A., Cote, C. E., Dooley, D. M. 1981. *Proc. Natl. Acad. Sci. USA* 83:8435–39
18. Williamson, P. R., Moog, R. S., Dooley, D. M. M., Kagan, H. M. 1986. *J. Biol. Chem.* 261:13602–5
19. Knowles, P. F., Pandeya, K. B., Rius, F. X., Spencer, C. M., Moog, R. S., et al. 1987. *Biochem. J.* 241:603–8
19a. van der Meer, R. A., Jongejan, J. A., Duine, J. A. 1987. *FEBS Lett.* 221:299–304
20. Stewart, L. C., Klinman, J. P. 1988. *Annu. Rev. Biochem.* 57:551–92
21. van der Meer, R. A., Jongejan, J. A., Duine, J. A. 1988. *FEBS Lett.* 231:303–7
22. van der Meer, R. A., Duine, J. A. 1988. *FEBS Lett.* 235:194–200
23. Groen, B. W., van der Meer, R. A., Duine, J. A. 1988. *FEBS Lett.* 237:98–102
24. Duine, J. A., Jongejan, J. A. 1989. *Vitamins and Hormones*, Vol. 46. In press
25. van Kleef, M. A. G., Duine, J. A. 1988. *Arch. Microbiol.* 150:32–36
26. Duine, J. A., Frank Jzn, J., Berkhout, M. P. J. 1984. *FEBS Lett.* 168:217–21
27. Bamforth, C. W., Quayle, J. R. 1978. *Biochem. J.* 169:677–86
28. Yamanaka, K., Tsuyuki, Y. 1983. *Agric. Biol. Chem.* 47:2173–78
29. Sperl, G. T., Forrest, H. S., Gibson, D. T. 1974. *J. Bacteriol.* 118:541–50
30. Frank Jzn, J., Dijkstra, M., Duine, J. A., Balny, C. 1988. *Eur. J. Biochem.* 174:331–38
31. Dijkstra, M., Frank Jzn, J., Jongejan, J. A., Duine, J. A. 1984. *Eur. J. Biochem.* 140:369–73
32. Frank Jzn, J., Duine, J. A. 1989. *Eur. J. Biochem.* In press
33. Dijkstra, M., van den Tweel, W. J. J., de Bont, J. A. M., Frank Jzn, J., Duine, J. A. 1985. *J. Gen. Microbiol.* 131:3163–69
34. Davidson, V. L., Neher, J. W., Cec-

chini, G. 1985. *J. Biol. Chem.* 260:9642–47
35. Anderson, D. J., Lidstrom, M. E. 1988. *J. Bacteriol.* 170:2254–62
36. Harms, N., de Vries, G. E., Maurer, K., Hoogendijk, J., Stouthamer, A. H. 1987. *J. Bacteriol.* 169:3969–75
36a. Machlin, S. M., Tam, P. E., Bastien, C. A., Hanson, R. S. 1988. *J. Bacteriol.* 170:141–48
37. Biville, F., Mazodier, P., Gasser, F., van Kleef, M. A. G., Duine, J. A. 1988. *FEMS Microbiol. Lett.* 52:53–58
38. Anthony, C. 1982. *The Biochemistry of Methylotrophs.* London: Academic
39. Beardsmore-Gray, M., O'Keeffe, D. T., Anthony, C. 1983. *J. Gen. Microbiol.* 129:923–33
40. Dijkstra, M., Frank Jzn, J., van Wielink, J. E., Duine, J. A. 1988. *Biochem. J.* 251:467–74
41. Dijkstra, M., Frank Jzn, J., Duine, J. A. 1988. *Biochem. J.* 257:87–94
42. Dijkstra, M., Frank Jzn, J., Duine, J. A. 1988. *FEBS Lett.* 227:198–202
43. Nozaki, M. 1987. *Methods Enzymol.* 142:650–55
44. McIntire, W. S., Stults, J. T. 1986. *Biochem. Biophys. Res. Commun.* 141:562–68
45. Vellieux, F. M. D., Frank Jzn, J., Swarte, M. B. A., Groendijk, H., Duine, J. A., et al. 1986. *Eur. J. Biochem.* 154:383–80
46. Vellieux, F. M. D., Swarte, M. B. A., Groendijk, H., Huitema, F., Kalk, K. H., et al. 1989. In *1st Symp. on PQQ and Quinoproteins.* Dordrecht: Kluwer Academic. In press
47. Husain, M., Davidson, V. L., Gray, K. A., Knaff, D. B. 1987. *Biochemistry* 26:4139–43
48. McIntire, W. S. 1987. *J. Biol. Chem.* 262:11012–19
49. Kenney, W. C., McIntire, W. 1983. *Biochemistry* 22:3858–68
50. Chandrasekar, R., Klapper, M. H. 1986. *J. Biol. Chem.* 261:3616–19
51. Fukomori, Y., Yamanaka, T. 1987. *J. Biochem.* 101:441–45
52. van Wielink, J. E., Frank Jzn, J., Duine, J. A. 1989. See Ref. 46, pp. 269–78
53. Hommes, R. W. J., Postma, P. W., Neijssel, O. M., Tempest, D. W., Dokter, P., Duine, J. A. 1984. *FEMS Microbiol. Lett.* 24:329–33
54. Dokter, P., Pronk, J. T., van Schie, B. J., van Dijken, J. P., Duine, J. A. 1987. *FEMS Microbiol. Lett.* 43:195–200
55. Cleton-Jansen, A.-M., Goossen, N.,

Wenzel, T. J., van de Putte, P. 1988. *J. Bacteriol.* 170:2121–25
56. Matsushita, K., Shinagawa, E., Adachi, O., Ameyama, M. 1989. See Ref. 46, pp. 69–78
57. Dokter, P., van Wielink, J. E., van Kleef, M. A. G., Duine, J. A. 1988. *Biochem. J.* 254:131–38
58. Matsushita, K., Nonobe, M., Shinagawa, E., Adachi, O., Ameyama, M. 1987. *J. Bacteriol.* 169:205–9
59. Dokter, P., Frank Jzn, J., Duine, J. A. 1986. *Biochem. J.* 239:163–67
60. Geiger, O., Goerisch, H. 1986. *Biochemistry.* 25:6043–48
61. Ameyama, M., Nonobe, M., Shinagawa, E., Matsushita, K., Takimoto, K., Adachi, O. 1986. *Agric. Biol. Chem.* 50:49–57
62. Matsushita, K., Shinagawa, E., Inoue, T., Adachi, O., Ameyama, M. 1986. *FEMS Microbiol. Lett.* 37:141–44
63. Cleton-Jansen, A. M., Goossen, N., Vink, K., van de Putte, P. 1989. See Ref. 46, pp. 79–86
64. Goossen, N., Vermaas, D. A. M., van de Putte, P. 1987. *J. Bacteriol.* 169:303–7
65. Matsushita, K., Ameyama, M. 1982. *Methods Enzymol.* 89:149–54
66. Duine, J. A., Frank Jzn, J., Jongejan, J. A. 1983. *Anal. Biochem.* 133:239–43
67. Groen, B. W., van Kleef, M. A. G., Duine, J. A. 1986. *Biochem. J.* 234:611–15
68. Imanaga, Y. 1989. See Ref. 46, pp. 87–96
69. Neijssel, O. M. 1987. *Microbiol. Sci.* 4:87–90
70. van Kleef, M. A. G., Duine, J. A. 1988. *FEBS Lett.* 237:91–97
71. Houck, D. R., Hanners, J. L., Unkefer, C. J. 1988. *J. Am. Chem. Soc.* 110:6920–21
72. Tanizawa, K., Moriya, T., Kido, T., Tanaka, H., Soda, K. 1989. See Ref. 46, pp. 43–45
73. van Iersel, J., van der Meer, R. A., Duine, J. A. 1986. *Eur. J. Biochem.* 161:415–19
74. Barker, R., Boden, N., Cayley, G., Charlton, S. C., Henson, R., et al. 1979. *Biochem. J.* 177:289–302
75. Duine, J. A., Jongejan, J. A., van der Meer, R. A. 1987. *Biochemistry of Vitamin B6,* ed. T. Korpela. P. Christen, pp. 243–54. Basel: Birkhauser
76. van der Meer, R. A., van Wassenaar, P. D., van Brouwershaven, J. H., Duine, J. A. 1989. *Biochem. Biophys. Res. Commun.* In press

77. Rinaldi, A., Giartosio, A., Floris, G., Medda, R., Finazzi-Agro, A. 1984. *Biochem. Biophys. Res. Commun.* 120: 242–49
78. Klinman, J. P., Hartmann, C. H., Janes, S. 1989. See Ref. 46, pp. 297–306
79. Suva, R. H., Abeles, R. H. 1978. *Biochemistry* 17:3538–46
80. Zeidan, H., Watanabe, K., Piette, L. H., Yasunobu, K. T. 1980. *J. Biol. Chem.* 255:7621–26
81. Hill, J. M., Mann, P. J. G. 1964. *Biochem. J.* 91:171–82
82. Dooley, D. M., McGuirl, M. A., Peisach, J., McCracken, J. 1987. *FEBS Lett.* 214:274–78
83. Faraggi, M., Chandrasekar, R., McWhirter, R. B., Klapper, M. H. 1986. *Biochem. Biophys. Res. Commun.* 139:955–66
84. Jongejan, J. A., van der Meer, R. A., Van Zuylen, G. A., Duine, J. A. 1987. *Recl. Trav. Chim. Pays-Bas* 106:365
85. Suzuki, S., Sauray, T., Itoh, S., Ohshiro, Y. 1988. *Inorg. Chem.* 27: 591–92
86. Williams, T. J., Falk, M. C. 1986. *J. Biol. Chem.* 261:15949–54
87. Hamilton, G. A. 1981. In *Copper Proteins*, ed. T. G. Spiro, pp. 193–218. New York: Wiley-Interscience
88. Ettinger, M. J., Kosman, D. J. 1981. See Ref. 87, pp. 219–62
89. van der Meer, R. A., Jongejan, J. A., Duine, J. A. 1988. Submitted for publication
90. Whittaker, M. M., Whittaker, J. W. 1988. *J. Biol. Chem.* 263:6074–80
91. Blackburn, N. J., Collison, D., Sutton, J., Mabbs, F. E. 1984. *Biochem. J.* 220:447–54
92. Duine, J. A., Frank Jzn, J., Verwiel, P. E. J. 1981. *Eur. J. Biochem.* 118:395–99
93. Kosman, D. J. 1984. In *Copper Proteins and Copper Enzymes*, ed. R. Lontie, 2:1–26. Boca Raton: CRC
94. Hayaishi, O. 1974. In *Molecular Mechanisms of Oxygen Activation*, ed. O. Hayaishi, pp. 1–28. New York: Academic
95. Colombo, G., Papadopoulos, N. J., Ash, D. E., Villafranca, J. J. 1987. *Arch. Biochem. Biophys.* 252:71–80
96. Ljones, T., Skotland, T. 1984. See Ref. 93, pp. 131–57
97. Blackburn, N. J., Concannon, M., Shahiyan, S. K., Mabbs, F. E., Collison, D. 1988. *Biochemistry* 27:6001–8
98. Walker, G. A., Kon, H., Lovenberg, W. 1977. *Biochim. Biophys. Acta* 482: 309–15
99. Ljones, T., Flatmark, T., Skotland, T., Petersson, L., Bäckstroem, D., Ehrenberg, A. 1978. *FEBS Lett.* 92:81–89
100. Klinman, J. P., Brenner, M. C. 1988. In *4th Int. Symp. on Oxidases and Related Redox Systems.* In press
101. Klinman, J. P., Krueger, M., Brenner, M., Edmondson, D. E. 1984. *J. Biol. Chem.* 259:3399–402
102. Ash, D. E., Papdopoulos, N. J., Colombo, G., Villafranca, J. J. 1984. *J. Biol. Chem.* 259:3395–98
103. Diliberto, E. J. Jr., Allen, P. L. 1980. *J. Biol. Chem.* 256:3385–93
104. Stewart, L. C., Klinman, J. P. 1987. *Biochemistry* 26:5302–9
105. Scott, R. A., Sullivan, R. J., DeWolf, W. E. Jr., Dolle, R. E., Kruse, L. I. 1988. *Biochemistry* 27:5411–17
106. Glembotski, C., Eipper, B. A., Manis, R. E. 1984. *J. Biol. Chem.* 259:6385–92
107. Vliegenthart, J. F. G., Veldink, G. A. 1982. In *Free Radicals in Biology*, ed. W. A. Prior, 5:29–64. New York: Academic
108. Schewe, T., Rapoport, S. M., Kuhn, H. 1986. *Adv. Enzymol.* 58:191–272
109. Corey, E. J., Nagata, R. 1987. *J. Am. Chem. Soc.* 109:8107–8
110. Nagasawa, T., Yamada, H. 1987. *Biochem. Biophys. Res. Commun.* 147: 701–9
111. Sugiura, Y., Kuwahara, J., Nagasawa, T., Yamada, H. 1988. *Biochem. Biophys. Res. Commun.* 154:522–28
112. Wiberg, K. B. 1955. *J. Am. Chem. Soc.* 77:2519–22
113. Itoh, S., Kato, N., Mure, M., Ohshiro, Y. 1987. *Bull. Chem. Soc. Jpn.* 60:420–22
114. Voltattorni, C. B., Giartosio, A., Turano, C. 1987. *Methods Enzymol.* 142: 179–87
115. Groen, B. W., van der Meer, R. A., Duine, J. A. 1988. See Ref. 23
116. Lane, R., Manning, J., Snell, E. E. 1976. *Biochemistry* 15:4180–84
117. Walsh, C. 1979. *Enzymatic Reaction Mechanisms*, p. 807. San Francisco:
118. Duine, J. A., Frank Jzn, J. 1980. *Biochem. J.* 187:213–19
119. Voltattorni, C. B., Minelli, A., Dominici, P. 1983. *Biochemistry* 22:2249–54
120. Minelli, A., Charteris, A. T., Voltattorni, C. B., John, R. A. 1979. *Biochem. J.* 183:361–68
121. O'Leary, M. H., Baugh, R. L. 1977. *J. Biol. Chem.* 252:7168–73
122. van Kleef, M. A. G., Jongejan, J. A.,

Duine, J. A. 1989. See Ref. 46, pp. 217–26

123. Barboni, E., Voltattorni, C. B., D'Erme, M., Fiori, A., Minelli, A., et al. 1981. *Biochem. Biophys. Res. Commun.* 99:576–83

124. Voltattorni, C. B., Minelli, A., Turano, C. 1971. *FEBS Lett.* 17:231–35

125. Wada, H., Snell, E. E. 1961. *J. Biol. Chem.* 236:2089–95

126. Soda, K., Misono, H., Yamamoto, T. 1969. *Biochim. Biophys. Acta* 177:364–67

127. Sanada, Y., Shiotani, T., Okuno, E., Katunuma, N. 1976. *Eur. J. Biochem.* 69:507–15

128. Yonaha, K., Toyama, S., Kagiyama, H. 1983. *J. Biol. Chem.* 258:2260–65

129. John, R. A., Fowler, L. J. 1976. *Biochem. J.* 155:645–51

130. Tamaki, N., Aoyama, H., Kubo, K., Ikeda, T., Hama, T. 1982. *J. Biochem.* 92:1009–17

Annu. Rev. Biochem. 1989. 58:427–52

DNA CONFORMATION AND PROTEIN BINDING[1]

Andrew A. Travers

MRC Laboratory of Molecular Biology, Hills Road, Cambridge, England

CONTENTS

PERSPECTIVES AND SUMMARY

The selective binding of a protein to a particular DNA sequence requires the recognition by the protein of an ensemble of steric and chemical features that

---

[1]Terminology used: This review uses specific terms to describe the geometry of the DNA double helix. The definitions of the particular terms used are summarized below. For a more comprehensive discussion and illustrative diagrams the reader should consult Dickerson et al (113a).

Relative roll (θ_R) is the rotation of the long axis of a base pair relative to that of its neighbor in the direction of the major or minor groove. This quantity is positive if the base pairs open to the minor groove, negative if they open to the major groove.

Relative tilt (θ_t) is the rotation around the short axis of a base pair relative to that of its neighbor toward either sugar-phosphate backbone.

Propeller twist is the dihedral angle between the planes of the individual bases in a single base pair. This quantity is normally in the range of 5–25°.

Slide is the displacement of one base pair relative to that of its neighbor as measured along the long axis of each base pair.

0066-4154/89/0701-0427$02.00

in total delineate the binding site. This process of *selective recognition* (1) is essentially comparable to the binding of any ligand, be it enzymatic substrate or allosteric effector, to a protein molecule. In the particular case of protein-DNA interactions, original models for selective recognition invoked principally hydrogen bonding interactions between the proteins and the individual bases (2–6), the remainder of the DNA molecule being considered to lack sufficient information for selectivity. However, during the past decade, it has become apparent that both the local conformation and the local configuration of a DNA molecule can have a profound influence on protein-DNA interactions and in some cases, notably for the histone octamer (7) and the *Escherichia coli trp* repressor (8), can by themselves act as the major, and possibly the sole, determinants of selectivity. This type of recognition has been termed *indirect readout* (8) or *analogue recognition* (9) in contradistinction to the *direct* or *digital recognition* of individual bases or base pairs.

The basis for this latter mode of recognition stems from the realization that the structure of the DNA double helix is not monotonous but instead exhibits marked sequence-dependent variation. This variation is apparent both from the detailed structure of crystals of short sequences of A and B type DNA (10) and the specificity of cleavage of nucleases such as DNase I (11). The conclusion from such studies is that individual dinucleotide steps or slightly longer short sequences can assume preferred conformation(s) that differ for different sequences. This variation in structure is reflected in such parameters as groove width, local twist, displacement of the average base pair plane from the helical axis, and the inclination of a base pair plane to its neighbor. The details of this local variation, which have been recently reviewed elsewhere (12), are a major element in the recognition of binding sites by proteins.

A further aspect of conformational variation is the flexibility of DNA structures. In solution the local structure of DNA should not be considered static. Instead many measurements of DNA structure, particularly those involving imino proton exchange monitored by nuclear magnetic resonance techniques (13, 14), indicate that the structure of a given DNA sequence is an equilibrium mixture of many different conformational states. It is the sum of these average properties that determines the gross flexibility of long DNA molecules, a topic rigorously reviewed by Hagerman (15). It follows from these arguments that the recognition of a particular conformation by a protein may involve the selection of a subset of molecules at any instant. Similarly, the ability of any particular sequence to assume a range of conformations allows the possibility of deformation of the DNA on protein binding. This deformation can result from the tight bending of the DNA, as in the case of the nucleosome core particle (16), or from a change in the local twist either as an overwinding (17, 18) or as an underwinding (19–21). It is these latter deformations that are crucial elements of the enzymatic manipulation of DNA in such processes as the initiation of transcription and DNA replication and also site-specific recombination.

CONFORMATIONAL FLEXIBILITY OF DNA

In any discussion of the conformational flexibility of DNA and its role in recognition by proteins, it is important to appreciate the nature of the information provided by various experimental techniques. In general, it is possible to distinguish between those techniques, such as X-ray crystallography, that provide a detailed description of a particular static structure, and those that measure the time-averaged properties of a particular DNA sequence. Included in this latter category are the chemical and enzymatic probing of DNA, the investigation of DNA bending by gel electrophoresis, circular dichroism measurements, and nuclear magnetic resonance studies. Two important considerations arise from the various methodologies. First, to what extent are the reported crystal structures of short DNA sequences representative of the average conformation in solution? Second, certain techniques, notably the use of OH radical (22), often make use of solution conditions (low ionic strength, no magnesium ion), which may destabilize the DNA structure to a certain degree (23). Put another way, in a dynamic system the investigative technique may shift the position of any equilibrium between conformational isomers.

The relation of the crystal structures of defined DNA sequences to their average solution conformation is crucial to understanding the detailed geometry of the DNA on the protein surface. In general, for any reasonably flexible molecule, such as DNA, it must be assumed that the process of crystallization will select a particular conformation, that is determined, at least in part, by the crystal packing forces. This conformation need not necessarily be the predominant form in solution. Nevertheless, there is considerable experimental evidence that the conformation of certain sequences in the crystalline state correlates with that in solution. For example, DNase I in solution cleaves the dodecamer d(CGCGAATTCGCG) most rapidly at base steps that in the crystal structure of the same sequence have the highest helical twist (10, 11). In addition, the CD spectra of various DNA sequences (24) correlate well with the average properties of the same sequences in crystalline form. In particular the CD spectra can be related to the average rise per base step in the crystals, which in turn is a measure of the A- or B-like character of the double helix (25). The major apparent discrepancy between a crystal structure and the inferred solution properties of a DNA sequence arises in the case of oligo (dA).(dT) tracts. Such tracts are unusual in that the formation of the highly propeller twisted conformation unique to such tracts (27, 28) is a cooperative process that, in general, requires the interaction of four or more consecutive AT base pairs (14, 30–32). This interaction is strongly dependent on the length of the (dA).(dT) tract (33) and thus can be highly sensitive to the environmental conditions (30). In this instance it seems probable that the structures of the (dA).(dT) tracts in crystals (27–29) may represent one extreme of a spectrum of conformations. Conversely, the inferred solution structure depends on experimental measurements made in the presence of EDTA, an agent known to favor the A-like character in DNA (24).

The structural rules for the sequence-dependent variation in crystals of different DNA sequences form a consistent pattern. One fundamental sequence-dependent property is the determination of groove width (34) within the B-DNA family of helixes. Mixed sequences and GC-rich regions are associated with a wider minor groove, whereas AT tracts, whether homopolymer (dA).(dT) or alternating (dAT).(dAT), are invariably associated with a narrow minor groove (34). In the A-like family of helixes, of which the studied examples are predominantly GC rich, the minor groove is again wide (12). Two other generalizations about the conformational properties of particular base steps may be made from known crystal structures. (A base step is defined here as a DNA structural unit that contains two adjacent base pairs and their connecting sugar-phosphate backbones.) First, there is a strong tendency for the conformation of pyridimine-purine steps to depend on the local sequence context. For example, in the dodecamer d(CGCATA-TATGCG), the efficient stacking at ApT steps is optimized by a decrease in twist resulting in a decrease in stacking and a consequent increase in twist to an average 40° at the neighboring TpA steps (34) as originally postulated by Klug et al (35). Conversely, in the sequence d(CTCTAGAG), the TpA step is underwound with a twist angle of 21° (36). This bistable character of pyrimidine-purine steps arises from the ability of purines to engage in stacking interactions either with the neighboring base on the same strand or with the corresponding base on the opposite strand (37). Second, in general at pyrimidine-purine steps the average base pair planes are inclined to each so that the roll angle opens to the minor groove, whereas the converse is usually observed at purine-pyrimidine steps (10, 38). The exception to this general statement is the GC step in the sequence GGC, which in the crystal structure G_4C_4 rolls to open the minor groove (39).

In the context of protein binding, it is important to clarify what is meant by the conformational flexibility or rigidity of DNA. Both crystal and solution studies indicate that DNA sequences can adopt a range of conformations. The crucial point is that the ranges of form available to different short sequences may not necessarily overlap to a substantial degree and may also differ in extent of variation. Thus the homopolymer poly(dA).(dT) is exceptional in its failure to undergo the B→A transition in fibers (40). By contrast, alternating copolymers, such as poly (dAT).(dAT) and poly (dAG).(dCT), which can be reconstituted into nucleosomes, must necessarily be able to assume the wide range of conformational variants required for the tight wrapping of DNA on a protein surface (41). Such sequences represent one extreme of conformational flexibility. This property correlates both with a low melting temperature (42, 43) and with a rapid exchange rate of hydrogen-bonded imino protons (14).

A further important aspect of conformational flexibility is that the range of conformational variation available to particular short sequences is dependent not only on the nature of the sequence but also on the local torsion. In general it seems probable that the greater the torsional constraint on a DNA molecule (as expressed in a right-handed sense), the greater will be the conformational constraints (44). For DNA on protein surfaces, the local torsion can vary substantially, as observed in the crystal structures of the 434 repressor headpiece (45, 46) and the nucleosome core particle (48). This variation implies that the stringency of selection of sequences on the basis of conformation in a binding site will vary according to the local constraints. Similarly, when the average torsion of DNA free in solution is altered by variation of ionic conditions or by the application of positive or negative torsional strain, the energetically preferred range of conformations may differ substantially from that of "relaxed" or "unconstrained" DNA in a normal environment (47). Such strain-induced variability is an important factor in the maintenance of the superhelical density in vivo (48) and in the function of promoter DNA (23, 49).

THE ROLE OF DNA STRUCTURE IN PROTEIN-DNA INTERACTIONS

(a) Local Conformation and Protein Binding

Many proteins, of which nucleases are the best example, interact with a relatively short region of DNA with low selectivity. In two such cases there is now substantial evidence that it is a particular local conformation of the DNA that is necessary for a productive interaction.

One of the best-studied examples of this type of interaction involves bovine pancreatic deoxyribonuclease I (DNase I). The cleavage rates of this enzyme vary along a given DNA sequence, indicating that the enzyme recognizes sequence-dependent structural variations of the DNA double helix (11). In particular, DNase I cleaves DNA at a low rate at homopolymeric tracts of (dA).(dT) or (dG).(dC) (Ref. 30) relative to "random" sequence DNA. Under normal digestion conditions the enzyme cleaves the DNA sugar-phosphate backbone so that the separation of cleavage sites on complementary sites is 2–4 bp, a value consistent with recognition of the minor, but not the major, groove of DNA. Consequently, it was suggested that the width of the minor groove was a major determinant of the DNase I cleavage rate, the enzyme cutting most frequently where the minor groove was of average width, rather than narrow as in homopolymeric tracts of (dA).(dT) or expanded as in tracts of oligo (dG).(dC). Although this proposal was consistent with the cleavage patterns of naked DNA, it did not adequately explain the enhanced cleavage at certain sites on protein-bound DNA.

This importance of the minor groove in the mechanism of action of DNase I was recently confirmed by the solution to 2Å resolution of the crystal structure of DNase I bound to an octanucleotide (50). In this structure, an exposed loop of the enzyme binds in the minor groove of DNA, such that contacts are made with both flanking sugar-phosphate backbones while a tyrosine and an arginine residue penetrate into the groove itself. The most striking feature of the complex is that the DNA is deformed, showing a 21.5° bend toward the major groove and away from the bound enzyme. This bend results in a widening of the minor groove to 15 Å from an average width of 11–12 Å. The principal contribution to the bend comes from the base step that is immediately adjacent to the step that is cleaved. This structure is consistent with the notion that the failure of DNase I to cleave oligo (dA).(dT) tracts is a consequence of the narrow minor groove hindering access of the tyrosine and arginine residues. However, the width of the minor groove in the complex is similar to that in GC-rich DNA sequences (e.g. Ref. 39), and consequently the expectation would be that oligo (dG).(dC) tracts would be more sensitive than random DNA sequences. An alternative explanation is that the cleavage selectivity of DNase I is determined, at least in part, by the flexibility or bendability of DNA (50). Thus any sequence that cannot be readily deformed to the appropriate conformation would show a reduced rate of cleavage. Conversely, any DNA segment already stabilized in the preferred conformation for cleavage would exhibit an enhanced susceptibility. In the former category are sequences such as oligo (dA).(dT) and to a lesser extent oligo (dG).(dC), which by virtue of their base stacking interactions are believed to be conformationally rigid (27, 29). Similarly, the minor groove on the inside of small DNA circles (< 300 bp) would, by this explanation, be less sensitive to DNase I because the local curvature of the DNA is constrained to the opposite direction to that required for cleavage. By contrast, when DNA is wrapped on the surface of a protein, as in the nucleosome core particle, the direction of curvature is such that an outward-facing minor groove could be in the correct orientation for rapid cleavage. For DNase I, comparison of the crystal structures with and without bound DNA shows that the protein itself is not significantly deformed by the interaction with DNA (50). The probability of cleavage of a particular base step is thus likely to be a measure of the ability of that base step to assume the appropriate conformation.

A second example of the influence of local DNA conformation on protein binding and activity involves the eukaryotic topoisomerase I. This enzyme relaxes positively or negatively supercoiled DNA and interacts with specific sites which, in general, contain a TpA base step (51). The rate of relaxation is dependent on the superhelical density (52), and therefore it has been suggested that the enzyme requires a particular conformation of DNA for activity. One possibility is that the probability of transient unwinding at the TpA step, which is the most conformationally sensitive of all base steps under the influence of negative superhelicity (23), would be strongly correlated with superhelical density.

In the two examples discussed so far, the selectivity of the protein for a productive interaction is low relative to proteins, such as the λ C_I repressor whose interactions with DNA are dominated by direct contacts with the base pairs in the binding site (53). Nevertheless, substantial selectivity can be established in the absence of simple complementarity between the amino acid side chains of a protein and the functional groups on the bases. In the crystallized complex of the *E. coli trp* repressor with its 19 bp operator, there are no direct hydrogen-bonded or nonpolar contacts between the protein and the DNA that can account for the selectivity of the interaction (8). Instead, direct hydrogen-bonded contacts are principally to the phosphate groups in the operator, although well-ordered water molecules can make polar contacts with the functional groups of mutationally sensitive base pairs (54). These detailed interactions suggest that a major determinant of specificity is the ability of the operator DNA to assume a particular overall conformation that allows interaction with a repressor dimer (53). A necessary condition for the validity of this suggestion is that the energy required to deform other DNA sequences to the appropriate conformation be too high to permit the formation of a stable complex.

The sequence of the *trp* operator in the crystallized complex is d(TGTAC-TAGTTAACTAGTAC), in which mutational studies show that the CTAG sequences centered 4 bp from the dyad are the most crucial for favorable binding (54). In the complex this sequence has A type character with high slide and low twist at the CpT and TpA steps and significant positive roll at the TpA and ApG steps, resulting in a widening of the minor groove directing curvature toward the protein (8). By contrast, the helical axis at the TpA step at the dyad is bent away from the protein, again a consequence of positive roll. These deviations from uniform B-DNA allow the contacts between the protein and the sugar-phosphate backbones of the DNA strands to be optimized.

The essential element of the *trp* operator, ACTAG, is unusual in that it contains CTAG, one of the rarest tetranucleotide sequences in *E. coli* (8) and has the characteristic that it is preferentially cut by micrococcal nuclease (55), an enzyme that preferentially cleaves at unwound base steps (56). The structure of CTAG in the bound operator may be related to that in the crystal structure of the self-complementary DNA oligomer d(CTCTAGAG) (Ref. 36). In this structure the dominant feature is a cross-strand purine-purine stacking at TpA step that directs the phosphate backbone into an unusual conformation. In this respect the structures of the crystalline DNA and the operator site are comparable, although the positive roll at TpA is greater in the latter case. Nevertheless, in this context the thermal lability of the TpA step could, in principle, contribute to the formation of a stable DNA complex by providing a low energetic barrier to any necessary structural rearrangement on binding.

The *trp* operator-repressor complex, as crystallized, appears to be a good example of selectivity of protein binding that is determined in large part by the potential conformation of the DNA-binding site. Nevertheless, there are two caveats to this conclusion. First, although there are no direct contacts between the amino acid side chains and mutationally sensitive base pairs, three well-ordered water molecules mediate hydrogen-bonded interactions between each half repressor and base pairs in the half operator. Significantly, mutations in the amino acids that form hydrogen bonds with these water molecules result both in reduced affinity, and in some examples, altered sequence selectivity of the repressor for the operator (56a). This result raises the possibility that the water molecules are themselves an essential component of the recognition surface of the repressor and may mediate, at least in part, direct readout of the operator sequence acting, as it were, by proxy for the protein. In this situation, amino acid substitution might destabilize the water structure and permit direct contacts between the amino acid side chains and the bases, thus creating the potential for an altered selectivity. Second, the N-terminal arms of the repressor do not, in the crystal structure, form defined contacts with operator DNA, although removal of these arms reduces the in vitro affinity of the repressor for the operator by 30–100-fold (J. Carey, quoted in Ref. 8).

Whatever the relative contributions of direct and indirect readout to the selectivity of the *trp* repressor–operator interaction, this selectivity is somewhat lower than that achieved in some complexes in which direct contacts between the protein and base pairs contribute to the binding energy. Thus in vitro measurements of *trp* repressor specificity show a 10^4-fold preference for the operator over a random nonoperator site (57), in contrast to a value of 10^8 for the *lac* repressor (58).

(b) DNA Bendability and Protein Binding

A particular aspect of conformational constraints on protein binding is the bendability of DNA. The bending of DNA around a bound protein or protein complex is a ubiquitous phenomenon that has been recognized by such diverse techniques as X-ray crystallography (16), electron microscopy (59), the anomalous retardation of protein-DNA complexes on polyacrylamide gels

Table 1 Proteins that bend DNA—a representative selection

Protein	Methods for detection of bending	Refs.
Catabolite activator protein	Gel retardation, electron microscopy, facilitation of ligation of small circles	59–61
Tn3 resolvase	Gel retardation	115
E. coli integration host factor	Gel retardation	116, 117
LexA repressor	Gel retardation	118
Phage 434 repressor	Gel retardation, X-ray crystallography	17, 45, 46
Drosophila heat-shock transcription factor	Gel retardation	119
Bovine papillomavirus E2 protein	Gel retardation	120
E. coli RNA polymerase	Gel retardation	121, 121a
DnaA protein	Electron microscopy	113
Phage lambda O protein	Electron microscopy	122
DNA gyrase	Electron microscopy	122a
Phage lambda intasome	Electron microscopy	25
Core nucleosome	X-ray crystallography	16
Pancreatic DNase	X-ray crystallography	50
cer binding protein	Sequence analysis	122b

(60), and the facilitation of closure of small DNA circles containing an appropriate protein-binding site (61). Table 1 lists the examples of such complexes so far reported.

The property of DNA that is relevant to complexes in which it does not follow a straight path is the bendability of the molecule. DNA does not behave as an isotropic rod (62–64); it may bend more easily in one plane than another and thus possesses anisotropic flexibility. In a long molecule this will depend on the overall effect of the distribution of short sequences within it that are differentially flexible in different directions. In many large DNA-protein complexes, including the nucleosome and those involved in the enzymatic manipulation of DNA in replication and recombination, the DNA is wrapped tightly around a central core of protein (20). In such structures all the grooves (both major and minor) on the inside of the curve must narrow somewhat because of the compression associated with bending, while those on the outside of the curve become correspondingly wider. It is the ability of particular short sequences to assume these required conformations that enables the DNA structure to accommodate the deformation associated with bending and thus determines the bendability of the binding site.

A paradigm for the role of DNA bendability in selective binding is the nucleosome core particle. In this structure ~145 bp are wrapped in about 1.8 left-handed superhelical turns of diameter 86 Å about an octamer of the histone proteins H3, H4, H2A, and H2B (16). Although in the chromosome the histone octamer is associated with a great variety of DNA sequences, studies of nucleosome positioning in both reconstituted and naturally occurring systems have shown that these proteins can adopt well-defined, often precise, locations with respect to the primary DNA sequence. To resolve this apparent paradox, Trifonov & Sussman (62) proposed that nucleosome positioning was determined not by base-specified protein-DNA contacts but by the anisotropic flexibility of the DNA double helix. To test this hypothesis experimentally, it must be established that the bending preference of a particular DNA molecule is the same when reconstituted into a nucleosome core particle as it is when constrained in the absence of any bound proteins. To approach this question, a 169 bp DNA fragment of bacterial origin was covalently closed into a small relaxed circle (7). Such a molecule, if bent uniformly, would have an average diameter of ~170 Å, that is, about twice that of the superhelix on the nucleosome core. The preferred direction of curvature, or the rotational orientation of the DNA sequence, was established using DNase I as a probe. This enzyme was assumed to cleave the minor groove preferentially on the outside of the circle, a property consistent with the geometry of a DNase I–octanucleotide complex (50). The resulting cleavage pattern demonstrated that the DNA in the circle adopted a highly preferred rotational orientation. When this same DNA fragment, now in linear form, was reconstituted into a nucleosome core particle, the angular setting of the DNA remained largely conserved in going from circle to nucleosome. This result is fully consistent with the view that the bendability of DNA can be a major determinant of nucleosome positioning.

If the rotational positioning of DNA on a nucleosome is influenced by particular sequences, it would be expected that, in a population of nucelosome core particles, the occurrence of such sequences would exhibit a periodic modulation that would reflect the structural periodicity of the DNA molecule lying on the surface of the histone octamer. By the use of two techniques, *statistical sequencing* and *direct sequencing* of nucleosome core DNA from chicken erythrocytes, the natures of the short DNA sequences associated with the bending of nucleosomal DNA were established (7, 65). The former technique determines the general sequence distribution in the population as a whole, whereas the latter determines the precise sequence of a small sample of molecules. Both techniques show that in general (A + T)-rich sequences are preferentially placed where the minor groove faces approximately inward toward the histone octamer and the center of curvature, whereas (G + C)-rich sequences prefer to occupy positions where the minor groove points outward. Thus the periodic modulations of (A + T)- and (G + C)-rich sequences have essentially identical frequencies but opposite phases. By Fourier analysis of the sequences of 177 different core DNA molecules, it is possible to assign a quantitative preference for each and every dinucleotide and trinucleotide preference. This analysis clearly showed that the positioning of some sequences on the nucleosome core was determined by a trinucleotide or longer sequence, rather than by dinucleotide components acting independently. Thus of the seven trinucleotides containing the dinucleotide ApA/TpT, only three, ApApA/TpTpT, ApApT/ApTpT, and TpApA/TpTpA, show significant relative preferences for an inward-facing minor groove, with respective amplitudes of $\pm$ 36%, 30%, and 20% about their mean values. Indeed the dinucleotide ApA/TpT in isolation (that is, the sequence ApA not flanked by A on either side and the sequence TpT not flanked by T) does not exhibit any significant periodic modulation. By contrast, periodic modulations of the GpC-containing trinucleotides are all in same phase but differ substantially in amplitude, the trinucleotide GpCpC/GpGpC showing the largest modulation of $\pm$45%. From this data Calladine & Drew (64) have constructed a simplified description that expresses the anisotropic flexibility of DNA solely in terms of the rotational preferences of dinucleotide sequences, with the exception that it was necessary to distinguish between the GC-containing trinucleotides. With one exception (66), an algorithm based on these preferences accurately predicts the positions of all nucleosomes reconstituted in vitro.

Several points should be emphasized with respect to these conclusions. First, the data clearly show that rotational positioning depends on sequence preferences and not absolute sequence requirements. In any piece of mixed sequence DNA, it is unlikely that all the bending preferences in that sequence can be simultaneously satisfied by the configuration assumed by the whole molecule, and the overall setting will be determined by the balance of local preferences. Consequently, there will usually be a few helix segments for which rotational position is imposed not by local constraints but by the preferred configuration of the whole molecule. Second, rotational positioning by itself is insufficient, in principle, to specify a precise translational position for a histone octamer on a defined DNA sequence. Such specification requires sequence markers that occur nonperiodically and would be expected to be found either at unique locations within the nucleosome, such as the dyad, or at positions asymmetrically related about the dyad. More detailed analysis of nucleosome DNA sequences reveals precisely these features. The sequence preferences at the dyad differ from those at any other outward-facing minor groove (65, 67). In addition, the average sequence organization of the four double-helical turns spanning the dyad is asymmetrically arranged with respect to the dyad (43). Finally, the nucleosome core particles used for the isolation of DNA clones for sequencing were prepared under conditions that would allow the sliding of octamers along the DNA. Under these conditions we would expect a histone octamer to position itself at optimal binding sites that are not necessarily identical to those occupied in vivo. The experiment thus selects binding sites that satisfy the requirements of the isolated histone octamer as for in vitro nucleosome reconstitution.

A crucial question is to what extent the sequence preferences observed on the nucleosome also apply to other nucleoprotein complexes containing bent DNA. In two cases, that of the 434 repressor headpiece–operator complex (45, 46) and that of the catabolite activator protein interaction with the *lac* regulatory region (68), there is sufficient information available to answer this question. Both these complexes are examples of interactions where it has been shown that the conformational flexibility of DNA complements identified or assumed direct base-specific recognition.

The complex of CAP with its binding site in the *lac* control region, a site for which the protein has the highest affinity of all characterized naturally occurring targets, is of particular interest. The existence of significant DNA bending has been inferred from the anomalous mobility of the complex in polyacrylamide gels (60), from the effect of CAP on the rate of ligation of small circular DNA molecules containing the *lac* control region (61), and also from electron micrographs of the complex (59). In the complex the strong binding interactions span 28–30 bp of DNA as indicated by ethylation interference experiments (69). By relating the known crystal structure of free CAP protein complexed with cAMP at 2.5 Å resolution, Warwicker et al (70) have proposed models in which a "ramp" of positive electrostatic potential running around three sides of the protein provides a surface for interaction with 30–44 bp of DNA in a tightly bent configuration. These calculations suggest that the DNA is bent through $\sim 150°$, or, on average 35–50° per double helical turn. This approach does not yield a unique solution, the two most favored models differing principally in the planarity of bending; the first suggesting a planar bend, the second a left-handed supercoil. The second is more consistent with the observed change of -0.18 in linking number on the binding of CAP to DNA (71).

Since sequence conservation in CAP-binding sites is essentially confined to the central double helical turn, bending of this magnitude implies that there should be a marked dependence of binding on the bending preferences of DNA sequences with the binding site. This site contains an interrupted inverted repeat with potential major groove contacts through a helix-turn-helix motif on each side of the center of symmetry. Thus the dyad at the center must have its minor groove pointing in toward the protein. At this position and at 10 bp on either side are AT-rich sequences, which on the nucleosome occur preferentially where the minor groove is narrowed. Similarly, exactly out of phase are sequences that preferentially occupy an outward-facing minor groove. This precise correspondence between the *lac* CAP-binding site and nucleosomal DNA is consistent with both the known direction of bending in the CAP-DNA complex (72) and its inferred magnitude (70, 73). Possibly significantly, this correspondence is not as apparent in other weaker naturally occurring CAP-binding sites.

To test the inference (74) that these sequences are related to the bending of

DNA, Gartenberg & Crothers (68) exhaustively mutagenized the CAP-binding site, concentrating on two particular regions (-11 and -16) where the minor groove of DNA points respectively toward and away from the protein surface. They found that many of the mutations in these regions, but not mutations situated two double helical turns from the dyad, affected both the affinity of CAP for the binding site and the mobility of CAP-DNA complexes in polyacrylamide gels. The results strongly suggest that the magnitude of DNA bending in the complex, as indicated by gel mobility, correlated with the binding affinity of CAP. From the effects of the sequence changes on gel mobility, Gartenberg & Crothers (68) concluded that the inferred bending was more consistent with a dependence on the properties of dinucleotides than on those of mononucleotides, but there was insufficient data to test a dependence on trinucleotides. In general, the dinucleotide dependence of bending in the CAP-DNA complex correlates well with that observed on the nucleosome, AT-rich sequences being favored where the minor groove points in and disfavored where the minor groove points out, with the opposite pattern for GC-rich sequences (68). Conversely, the sequence data from the nucleosomal DNA can be used to calculate the likelihood of flexure of the mutated CAP-binding sites (64). With one exception, there is an excellent correlation between the observed mobilities of the mutant complexes and the calculated probability of bending (Figure 1). This result establishes that the sequence-dependent bending preferences in the nucleosome are also applicable to other nucleoprotein complexes.

A similar correspondence can also be shown to exist between bending preferences in the nucleosome and the binding of 434 repressor to its operator site. Recently, Aggarwal et al (46) have determined, to 2.5 Å resolution, the X-ray structure of a complex between the 434 repressor DNA-binding domain and a 20 bp operator fragment TATACAAGAAAGTTTGTACT. This struc-
                                                        TATGTTCTTTCAAACATGAA
ture is closely related to an earlier crystal structure of the 434 DNA-binding domain with a 14 bp DNA fragment (45). In both complexes, the DNA structure is distorted. There is an overall bend with an average radius of curvature of 65 Å. As in the nucleosome, bending is not smooth but is concentrated at two sites symmetrically disposed 2–3 base steps from the central region, which is itself relatively straight. The most striking feature of the structure is a narrowing of the minor groove at the center, which is accompanied by an overwinding of the base steps in this region by several degrees. The crucial importance of the flexibility of this central region for the interaction with the protein is demonstrated by the observation that the introduction of a nick on one strand at its central phosphodiester bond increased the affinity of the operator for the repressor by approximately five-fold (17). By contrast, a similar nick 4 bp outside the operator had no

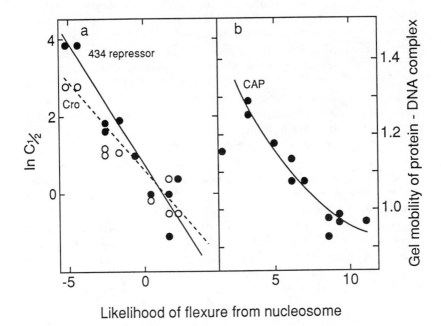

Figure 1 Comparison of the position-dependent sequence preferences in nucleosomal DNA with those occurring (*a*) at the midpoint of the 434 operator and (*b*) in the CAP binding site in the *lac* regulatory region. The likelihood of flexure (probability of bending) is calculated from the matrix of rotation in Ref. 64. For the 434 operator the experimental data is taken from Koudelka et al (123) and is plotted as the logarithm of the concentration of 434 repressor (●) and of Cro protein (○) necessary to half fill its binding site on DNA. For the CAP binding site, data is taken from Gartenberg & Crothers (68) and is plotted as the rate of migration of the complex through polyacrylamide gels relative to the rate for the wild-type complex. Plots are taken from Ref. 75 and are reproduced by kind permission of Dr. H. R. Drew.

effect on this affinity. In a complementary experiment, a mutation, introduced into the repressor near the interface of the dimer, resulted in a failure to discriminate between nicked and intact operators. It was assumed, but not directly demonstrated, that this mutation increases the flexibility of the dimer interaction.

The demonstrable requirement for conformational flexibility of the central region is wholly consistent with the effect of mutations in this region on the affinity for the repressor. Thus, replacement of the central d(ATAT) by d(ACGT) weakens the binding affinity by a factor of 50, while replacement by d(AAAA) increases it by a factor of 5 (17). Significantly, the mutant repressor is less sensitive to these substitutions. Again, however, the clear pattern is that the required conformation is favored by AT-rich sequences and disfavored by GC-rich sequences. When the probability of deformation for the central region is calculated from the observed sequence-dependent bending preferences in the nucleosome, assuming that the structure of the complex does not change as the central sequence is altered, there is a strong correlation between the observed affinity of the operator for the repressor and the predicted probability of deformation (75). This result again points to the generality of the nucleosomal sequence preferences and indicates that the structural deformations at the center of the complexed 434 operator must be similar to those encountered at an inward-facing minor groove on the nucleosome.

The question at issue is whether these sequence preferences are related solely to bending or also to other properties of DNA such as torsional flexibility (or rigidity) (46). The structural feature common to inward-facing minor grooves on the nucleosome and the central region of the complexed 434 operator is an unusually narrow minor groove. The establishment of this structure is favored by a particular base-stacking conformation that concomitantly increases the twist. Any relaxation of the requirement for a narrow minor groove in the complexed operator would be expected to expand the minor groove at the center and at the same time slightly alter the angular register of the two bends.

A second phage-coded protein that binds the 434 operator is the Cro repressor. The conformation of operator DNA complexed differs in significant respects from that of the same DNA complexed with the C_I repressor (75a). In the Cro-operator complex, the DNA is bent only slightly in comparison with the tight bend of <100 Å radius in the repressor-operator complex. Similarly, in the Cro-operator complex the DNA is uniformly overwound with an average helical repeat of 10.1 bp/turn, in contrast to the 10.5 bp/turn in the repressor-operator complex (46, 75a). In each of these cases, the binding of the protein determines the precise conformation of the DNA. Both complexes are similar in that there are no direct contacts between the base pairs and the protein in the vicinity of the operator dyad, yet the local conformation of the DNA in this region differs subtly with respect to the minor groove width in the two complexes (75a). Significantly, the slightly wider minor groove in the Cro-operator complex is reflected in a reduced degree of sequence dependence when compared with that for the repressor-operator complex. In other words, the proportional change in binding affinity for a given sequence change at the operator dyad is less for Cro than for the C_I repressor (Figure 1).

The influence of DNA conformation on the binding of the 434 repressor to its operator differs from the case of the structurally similar λ C_I repressor. In the latter case, there are no dramatic kinks or distortions in the bound DNA, whose helical axis is essentially straight (53). The selectivity of the binding is thus dominated by base-specific interactions as inferred from genetic studies (76). Similarly, it cannot be assumed that the tight bending of DNA in a nucleoprotein complex is reflected in the sequence organization of the binding site. For example, the binding sites of nucleosomes containing histone H5 lack the strong rotational sequence signals characteristic of core nucleosome particles (43). This observation suggests that the additional binding energy provided by the clamping of the two superhelical turns by histone H5 may be sufficient to override the bending preferences of some DNA sequences.

(c) Intrinsic Curvature and Facilitated Wrapping

In DNA there are certain sequences that, even in the absence of external forces imposed by proteins, impart a preferred direction of curvature on a DNA molecule, i.e. give an intrinsic bend (61, 77–80). Both for naturally and artificially designed sequences, any short homopolymeric runs of A or T nucleotides longer than 3 bp (such as AAAA or TTTTT), when periodically repeated in approximate phase with the helical repeat, can confer a detectable amount of curvature on an isolated DNA molecule (61, 79–82). The molecular basis for this curvature is believed to be the ability of homopolymeric runs of (dA).(dT) to assume a structure in which the base pairs are highly propeller twisted (27). This arrangement maximizes the stacking interactions and permits the formation of a spine of bifurcated hydrogen bonds connecting adjacent base pairs (27, 28). The resultant double helical structure is thus both more stable (or conformationally rigid) and also structurally distinct from adjacent DNA sequences.

The ability of short runs of (dA).(dT) to confer intrinsic curvature on a DNA molecule in the absence of external constraints suggests that appropriate placement of such runs within a binding site for a particular protein might allow the DNA to adopt a preferred configuration, which could then facilitate the wrapping of DNA around that protein (41, 72, 81). Three separate lines of evidence support this view. First, Hsieh & Griffith (82) have examined the positioning of nucleosomes on two DNA segments containing regions of sequence-directed curvature. For both segments, a 223 bp sequence of kinetoplast DNA from *Crithidia fasiculata* and a ~200 bp sequence from SV40 containing transcription and replication terminators nucleosomes reconstituted with a 2–7-fold higher probability in the curved regions relative to adjacent DNA. Second, E. Bertrand-Burgraff and J. F. Lefevre (personal communication) have observed that certain mutations in the *E. coli ada* promoter upstream of the −35 region increase the apparent intrinsic curvature of the DNA and also increase the association constant of RNA polymerase with the promoter. A similar result has been obtained for two phage SP82 promoters (83). In both these examples there is a correlation between curvature and the binding of particular protein, but in neither case has it been established that curvature per se is causal.

A necessary condition for the facilitation of DNA wrapping by intrinsic curvature is that the configuration of DNA in solution be very similar, or identical, to that on the protein surface, that is, the rotational orientation of the oligo (dA).(dT) tracts, relative to the direction of curvature, should be the same in both situations. In three different nucleoprotein complexes containing bent DNA, the core nucleosome (65), the *E. coli* RNA polymerase-promoter complex (85), and the $\gamma\delta$ resolvase-site II (86) complex, short tracts of $(dA)_n.(dT)_n$ 5–6 bp in extent have a strong tendency to adopt an orientation in which the homopolymer run crosses from an outward-facing to an inward-facing minor groove or vice versa. For longer runs of $(dA)_n.(dT)_n$ where $n \geq$ 7, the homopolymer tracts either avoid an outward-facing minor groove in regions of bent DNA or preferentially occupy positions such as the outer arms of a histone octamer binding site where the magnitude of curvature is least (65). It immediately follows from the geometry of curvature that on the protein surface the junctions of $(dA)_{5-6}.(dT)_{5-6}$ tracts with flanking sequences cannot be structurally equivalent. All structural studies of the local conformation of such tracts in solution in the absence of protein are also consistent with this conclusion. Thus, in two crystal structures of short DNA sequences containing (dA).(dT) tracts, $d(CGCA_6GCG)$ and $d(CGCA_5TGCG)$ (27, 29), the direction of bending is consistent with that observed on protein surfaces. The crucial point is that in both cases the (dA).(dT) tract itself is essentially straight, and that the most abrupt changes in the direction of the helical axis occur for $d(CGCA_6GCG)$ at the CA step and at the GC step 3' to the dA tract. At both these steps the normal purine-purine clash (38) is exaggerated by a difference in propeller twist, resulting in a large positive roll angle for the pyrimidine-purine step and a corresponding negative roll angle for the purine-pyrimidine step. A similar pattern is observed in the crystals of $d(CGCA_5TGCG)$ (Ref. 27) and $d(CGCA_3T_3GCG)$ (Ref. 28). However, in the former case the asymmetric DNA sequence is packed in both possible orientations in approximately equal proportions in the same crystal lattice. This observation would be consistent with the possibility that the orientation of this sequence within the crystal is entirely determined by crystal packing forces and is independent of the sequence (29). While there is clear evidence that crystal packing forces can influence the conformation of the base pairs at the extremities of such short DNA sequences (87), it seems less likely that these forces could qualitatively affect the geometry of the junctions of the conformationally rigid (dA).(dT) tracts with flanking base pairs. An alternative interpretation of the result of DiGabriele et al (29) is that the oligo (dA).(dT) tract has the inherent potential to adopt two possible configurations as a consequence of being flanked on either side by the bistable base step TG/CA.

Indeed, in small 169 bp circles of *Leishmania* kDNA, sequences of the general type CA_nT_mG where $n + m = 5$–6 show precisely this behavior (A. Travers, S. C. Satchwell, quoted in Ref. 41). If it is argued that the (dA).(dT) tracts are the principal determinants of the direction of curvature and that any strain resulting from the constraints imposed by circularization are compensated for by conformational changes in the more flexible regions of kDNA, i.e. those lacking (dA).(dT) tracts, then this latter observation is also consistent with the view that the rotational orientation of (dA).(dT) tracts is, on average, the same when free in solution as on the protein surface.

These conclusions from direct studies of local conformation are at variance with attempts to deduce the local conformation of curved DNA containing (dA).(dT) tracts from the global configuration of such molecules as inferred from their relative electrophoretic mobilities in polyacrylamide gels. From essentially the same data set, three different models for the geometry of curved DNA have been proposed (80, 89, 90), all of which differ, to a greater or lesser degree, from the determined crystal structures. These models propose that in solution the oligo (dA).(dT) tract be oriented such that the minor groove at the center of the tract points approximately toward the center of curvature. There are three inherent difficulties in this approach. First, there is insufficient information in the relative mobilities themselves to deduce local conformations without additional ad hoc assumptions. Second, it is necessarily assumed that the conformation and configuration of DNA is the same under gel running conditions as it is when in free solution (89, 90). At present there is no experimental evidence to support this assumption. If passage of a curved DNA molecule through a polyacrylamide gel involved deformation of the three-dimensional configuration, such deformation would inevitably be reflected in the deformation of local conformation, a process that could, in principle, involve the (dA).(dT) tracts themselves. Third, all models are essentially static and assume that the structure of a $(dA)_n.(dT)_n$ tract is invariant with length if $n \geqslant 3$. However, the length of the tract that confers optimum curvature is dependent on the precise conditions of gel running. Thus at 20°C the optimum length is $n = 6$–7 bp, while at 4°C the optimum length is 5–6 bp (32). This change in optimum length is probably a consequence of the lower thermal stability of short (dA).(dT) tracts (30), suggesting that the implicit assumption of stasis is unjustified.

Other evidence also indicates that, in solution, an oligo (dA).(dT) tract is not a static structure but instead is an equilibrium mixture of different structures. First, the sensitivity of (dA).(dT) tracts to cleavage by DNase I is strongly temperature dependent, significant differences being observed between 4 and 37°C (30). These differences correlate with the temperature dependence of anomalous mobility. Second, Burkhoff & Tullius (91, 92) have demonstrated for both *Crithidia* kDNA and synthetic constructs of defined sequence that hydroxyl radical attack is more rapid at the 5' end than at the 3' end of (dA) tracts. Third, the rate of imino proton exchange for AT base pairs at the 5' and 3' extremities of (dA).(dT) tracts is some 10-fold faster than that of the central base pairs, the effect being most pronounced at the 5' end (14). Both these latter observations are consistent with a structural instability or "fraying" in the (dA) tract that propagates to a greater or lesser extent from the 5' end. Third, for certain sequences that induce intrinsic bending, there is a direct correlation between this thermal stability and the magnitude of curvature. From these observations it can be inferred that the observed anomalous mobility is a measure of the time-averaged curvature of an equilibrium mixture and does not necessarily reflect the magnitude and direction of curvature of a single defined structure. It remains to be established that the position of this equilibrium is the same in the gel as it is for DNA in free solution.

All structural models for (dA).(dT) tracts depend crucially on the observation that the gel mobility of sequences of the type $(A_6XXXXA_6XXXX)_n$ is, in general, very similar to that of sequences in which every alternate $(dA)_6(dT)_6$ tract is inverted (60, 90). The inference from this result is that the direction of bending is largely independent of the orientation of the (dA).(dT) tract. However, this observation is only true when the running buffer contains EDTA. When A tract–containing oligomers are electrophoresed in the absence of EDTA or in the presence of Mg^{2+}, the mobilities of the oligomers depend crucially on whether the A tracts are arranged all in the same orientation, or with one or more tracts inverted with respect to the others (92a). It follows that under these conditions the inference of symmetry is no longer valid. Since Mg^{2+} is an essential component of all living cells, it seems probable that the structure assumed by (dA).(dT) tracts in the presence of this ion, of which the crystal structure of $d(CGCA_6GCG)$ is a notable example (27), is the one that is biologically relevant. Conversely, the structure deduced from gel mobilities may represent a "relaxed" but biologically irrelevant structure.

A second possible functional role of intrinsic bending has been documented on studies on the enzymatic activity of EcoR1 restriction endonuclease. The recognition sequence, d(GAATTC), for this enzyme is bent both in solution (93, 94) and in the crystalline state (10). When complexed with the enzyme, this sequence is significantly distorted (95), with one "kink" at its center and others at both termini. By modifying base-pair substituents within the recognition sequence, Diekmann & McLaughlin (93) have shown that certain modified sequences with different curvature also significantly decrease relative specificity (approximate k_{cat}/K_m). Since the decrease in specificity cannot be adequately explained in terms of the known contacts between the protein and DNA, the authors propose that the curvature of the helical axis is an essential factor for optimal catalysis.

(d) The Local Geometry of Protein-Bound DNA

In nucleoprotein complexes the helical axis of the DNA is often bent. Such a change in direction of the helical axis can, in principle, result at the local level from a roll deformation (θ_R) in which the short axes of the adjacent base pair planes are inclined relative to each other (10, 38, 96), or a tilt deformation (θ_T) in which the corresponding long axes are inclined relative each or from a combination of both. In those cases where the nature of the deformation has been directly determined (8, 46), the "ironclad principle" of Dickerson (97) that the dominant determinant of bending is a roll deformation remains, so far, inviolate.

Nevertheless, it is apparent from the sequence analysis of nucleosomal and *E. coli* promoter DNA that the local geometry of certain sequences may vary (85). For example, whereas, in general, in core nucleosomal DNA the frequencies of occurrence of the complementary trinucleotides ApApA and TpTpT vary in phase with each other, in promoter DNA the preferred position of TpTpT is, on average, shifted by two nucleotides 3' to that of ApApA on the same strand. This displacement or sequence stagger is also characteristic of nucleosomal DNA situated immediately adjacent to the dyad. Mutational analysis of the CAP binding site reveals a similar pattern (68). At a minor groove facing inward to the protein surface, certain purine-purine dinucleotides when placed on the 5' side of this point enhance bendability to a greater extent than the complementary pyrimidine-pyrimidine dinucleotides, while the converse pattern is characteristic of the 3' side (68).

What is the structural explanation for this nonequivalence of complementary di- or trinucleotides? In the case of nucleosomal DNA, the large sequence stagger occurs in a region of high torsion (18), suggesting that the variation in sequence stagger reflects changes in either the local torsion or the magnitude of curvature. If this were so, one possible explanation for the observed structural variation would be that tighter bending would require more extreme deformation, perhaps involving additional "roll" points. The preferred explanation of Gartenberg & Crothers (68) for the nonequivalence is that both roll and tilt components contribute to the dinucleotide wedge. An alternative structural interpretation would be that the major deformation is a roll or wedge 3' to a purine-purine dinucleotide or 5' to a pyrimidine-pyrimidine dinucleotide. The only example of this type of asymmetry in crystal structures occurs at an outward-facing minor groove in the *trp* repressor-operator complex (8). For this sequence, CTAG, the sense of the asymmetry is reversed as expected (68). The major deformation here is a roll wedge between the complementary dinucleotides, and there is no systematic tilt component apparent. Nevertheless it should be emphasized that this single example may not necessarily be representative.

A further aspect of the geometry of protein-bound DNA is the conformation adopted by the preferred trinucleotides ApApA/TpTpT at inward-facing minor grooves. This trinucleotide, unlike homopolymeric runs of $(dA)_n(dT)_n$ where $n \geqslant 4$, confers only weak intrinsic curvature on unconstrained DNA (31, 32). An important question is therefore whether the structural principles that determine the preferred conformation of ApApA in nucleoprotein complexes are related to those that direct the intrinsic curvature dependent on longer runs of (dA).(dT). A simple explanation for the apparent difference in the sequence dependence of protein-induced and intrinsic curvature is that in unconstrained DNA a run of at least four consecutive AT base pairs is required to stabilize the conformation with high propeller twist, whereas on the protein surface this conformation could be stabilized by local or distant interactions with the polypeptide chain and hence could be adopted by the shorter $(dA)_3$ run under most conditions (85). Precisely this structure is observed in the central region of the protein 434 operator (46), although significantly a second run of $(dA)_3$ in the same structure has a less lightly wound conformation. Since the sequence preferences at the center of the 434 operator correlate quantitatively with those in nucleosomal DNA (75), it seems probable that the basis for conformational selection at an inward-facing minor groove on the nucleosome is the ability to adopt a narrow minor groove consequent upon a high propeller twist and high helical twist.

DNA CONFIGURATION AND PROTEIN BINDING

The formation of nucleosomelike structures is a common feature of processes such as transcription, site-specific recombination, and initiation of DNA replication that involve the enzymatic manipulation of DNA (20). In all these processes the wrapping of DNA about a protein core precedes the nucleation of unstacking that is an essential prelude to strand separation. The binding sites for such complexes should therefore contain sequences that would facilitate wrapping and also additional sites at which unstacking could be initiated. Typically these latter sites contain the thermally labile TpA base step, which is preferentially unwound by negative torsional strain (23, 98).

A relatively simple example of a process in which DNA wrapping precedes unwinding is the initiation of transcription by bacterial RNA polymerase. In this reaction the initial binding of the enzyme to the promoter site is followed by the unstacking of DNA base pairs within the highly conserved -10 region prior to the strand separation of about one turn of DNA around the transcription startpoint (recently reviewed in Ref. 99). In general, the functional promoter extends for 40–60 bp upstream of the transcription startpoint. Within this region occur periodic modulations of the short sequences correlated with DNA bending in nucleosomal DNA (85). In certain promoters the DNA upstream of the -35 region exhibits significant intrinsic curvature (101,

102), a property that is correlated in certain cases with promoter activity (83, 100). What is the functional significance of the bendability of promoter DNA? For the *lac*UV5 promoter Amouyal & Buc (100) have shown that the initial binding of RNA polymerase is accompanied by a change in linking number of approximately -1 turn, yet in this initial complex there is no evidence for extensive unstacking in promoter DNA. This is consistent with their interpretation that in this complex the DNA is tightly wrapped as a left-handed supercoil. The subsequent strand separation at the transcription start point during open complex formation is accompanied by a relatively small further decrease in linking number, suggesting that the topological unwinding, which in this case would be predominantly "writhe," is, at least in part converted to a decrease in double-helical twist. Such a scenario implies that the configuration of the protein-DNA complex must change substantially during this process. Evidence for such a change has been obtained both from thermodynamic measurements, which have been interpreted in terms of a substantial conformational shift of the polymerase protein (103), and also from footprinting studies, which show that the polymerase distorts the spacer DNA between the -35 and -10 regions to different extents in different promoters (104). This latter observation is consistent with the proposal (105) that during the formation of the "open" complex, realignment of the relative spatial position of the -10 and -35 regions occurs.

This simple model for the interaction of RNA polymerase with DNA implies that the energy expended in bending the DNA around the protein can be converted into a torsional stress that unwinds the double helix. It follows that if the DNA is intrinsically bent in a configuration that facilitates wrapping, the initial association of the enzyme with the DNA will be favored at the expense of the subsequent unwinding step. This is indeed observed for the *ada* promoter in which an increase in the intrinsic curvature of the upstream region increases the association constant for the promoter under negative superhelical strain but decreases the rate of the subsequent step (E. Bertrand-Burgraff, J. F. Lefevre, personal communication). Similarly, negative torsional strain can affect promoter activity either by increasing the association constant or by increasing the rate of open complex formation but not, in general, by increasing both constants (99, 106). This effect depends on both the type of promoter and also the precise reaction conditions and presumably reflects the most energetically favorable modes of partition of the available torsional strain.

A second process in which a localized unwinding of the DNA double helix is necessary is the initiation of DNA replication. The origins of replication on both phage lambda DNA and *E. coli* DNA contain tandemly repeated sequences that bind respectively the lambda O protein and the *E. coli* DnaA protein (107–110). In addition, the lambda origin is intrinsically bent (110, 111), a feature that is also observed for certain eukaryotic replication origins

(112). The initial recognition of replication origins by either DnaA protein or O protein involves cooperative binding to form a complex in which the DNA is wrapped in a negative supercoil around a protein core (113). In the case of the *E. coli oriC* complex, DnaA protein can by itself induce strand separation within a 13 bp repeated sequence (21), whereas for the lambda complex additional proteins appear to be necessary for localized unwinding (114). Although the precise mechanism by which the local unwinding of the DNA is accomplished remains to be established, it has been suggested, in both cases, that the wrapping of the DNA in the initial complex destabilizes the AT-rich regions at which unstacking occurs.

CONCLUSIONS

All recent studies point to the conclusion that two aspects of DNA structure are crucial determinants of its specific interaction with DNA-binding proteins. First, the sequence-dependent polymorphism of DNA allows the protein to recognize structural features, both global and local, of the double helix in addition to direct interactions with individual base pairs. The relative contributions of these two components is highly variable, ranging from the core nucleosome particle in which structural recognition is dominant (7) to the lambda C_I repressor where selectivity is determined principally by base-specific interactions (53). Between these extremes are proteins such as CAP (68) and the phage 434 repressor (46), in which both direct and indirect readout are significant. The second crucial property of DNA is its conformational flexibility. This is utilized in interactions with proteins both to promote enzymatic manipulation and also to allow a snug fit between the protein surface and the double helix. For interactions of this type in which the DNA structure is deformed, the DNA is not always a passive partner. In the case of both the histone octamer (43) and the *trp* repressor (8), it has been suggested that the sequence of the DNA binding site can also influence the conformation of the bound protein, although for DNase I it is clear that the structure of the enzyme when bound to DNA is essentially the same as when free (50). The conformational flexibility of DNA is also an important determinant of its response to negative superhelical strain, allowing changes both in the local conformation of particular base steps (23) and also in the global configuration of the molecule.

Acknowledgments

I would particularly like to thank the many scientists who have generously sent me reprints and preprints of their most recent work. Without such cooperation this review would have been outdated almost before it was written.

Literature Cited

1. Helene, C., Lancelot, G. 1983. *Prog. Biophys. Mol. Biol.* 39:1–68
2. von Hippel, P. H., McGhee, J. D. 1972. *Annu. Rev. Biochem.* 41:231–300
3. Seeman, W. C., Rosenberg, J. M., Rich, A. 1976. *Proc. Natl. Acad. Sci. USA* 73:804–8
4. von Hippel, P. H., Berg, O. G. 1986. *Proc. Natl. Acad. Sci. USA* 83:1608–12
5. Ebright, R. H., Cossart, P., Gicquel-Sanzey, B., Beckwith, J. 1984. *Proc. Natl. Acad. Sci. USA* 81:7274–78
6. Pabo, C. O., Sauer, R. T. 1984. *Annu. Rev. Biochem.* 53:293–321
7. Drew, H. R., Travers, A. A. 1985. *J. Mol. Biol.* 186:773–90
8. Otwinowski, Z., Schevitz, R. W., Zhang, R.-G., Lawson, C. L., Joachimiak, A., et al. 1988. *Nature* 355:321–29
9. Drew, H. R., Travers, A. A. 1985. *Nucleic Acids Res.* 13:4445–67
10. Dickerson, R. E., Drew, H. R. 1981. *J. Mol. Biol.* 149:761–86
11. Lomonossoff, G. P., Butler, P. J. G., Klug, A. 1981. *J. Mol. Biol.* 149:745–60
12. Drew, H. R., McCall, M. J., Calladine, C. R. 1988. *Annu. Rev. Cell Biol.* 4:1–20
13. Patel, D. J., Kozlowski, S. A., Ikuta, S., Itakura, K., Bhatt, R., Hare, D. R. 1982. *Cold Spring Harbor Symp. Quant. Biol.* 47:197–206
14. Leroy, J. L., Kochoyan, M., Huyrh-Dinh, T., Gueron, M. 1988. *J. Mol. Biol.* 200:223–38
15. Hagerman, P. J. 1988. *Annu. Rev. Biophys. Biophys. Chem.* 17:265–86
16. Richmond, T. J., Finch, J. T., Rushton, B., Rhodes, D., Klug, A. 1984. *Nature* 311:532–37
17. Koudelka, G. B., Harbury, P., Harrison, S. C., Ptashne, M. 1988. *Proc. Natl. Acad. Sci. USA* 85:4633–37
18. Richmond, T. J., Searles, M. A., Simpson, R. T. 1988. *J. Mol. Biol.* 199:161–70
19. Spassky, A., Kirkegaard, K., Buc, H. 1985. *Biochemistry* 24:2723–31
20. Echols, H. 1986. *Science* 233:1050–56
21. Bramhill, D., Kornberg, A. 1988. *Cell* 52:743–55
22. Tullius, T. D., Dombroski, B. A. 1985. *Science* 230:679–81
23. Drew, H. R., Weeks, J. R., Travers, A. A. 1985. *EMBO J.* 4:1025–32
24. Fairall, L., Martin, S., Rhodes, D. 1989. *EMBO J.* 8:In press
25. Heinemann, U., Lauble, H., Frank, R., Blocker, H. 1987. *Nucleic Acids Res.* 15:9531–50
26. Deleted in proof
27. Nelson, H. C. M., Finch, J. T., Luisi, B. F., Klug, A. 1987. *Nature* 330:221–26
28. Coll, M., Frederick, C. A., Wang, A. H.-J., Rich, A. 1987. *Proc. Natl. Acad. Sci. USA* 84:8385–89
29. DiGabriele, A., Sanderson, M. S., Steitz, T. A. 1989. *Proc. Natl. Acad. Sci. USA* 86:In press
30. Drew, H. R., Travers, A. A. 1984. *Cell* 37:491–502
31. Diekmann, S. 1986. *FEBS Lett.* 195:53–56
32. Koo, H. S., Wu, H.-M., Crothers, D. M. 1986. *Nature* 320:501–6
33. Diekmann, S., von Kitzing, E. 1988. In *Structure and Expression, Vol. 3, DNA Bending and Curvature*, ed. W. K., Olson, M. H. Sarma, R. H. Sarma, M. Sundaralingam, pp. 57–67. New York: Adenine
34. Yoon, C., Prive, G. G., Goodsell, D. S., Dickerson, R. E. 1988. *Proc. Natl. Acad. Sci. USA* 85:6332–36
35. Klug, A., Jack, A., Viswamitra, M. A., Kennard, O., Shakked, Z., Steitz, T. A. 1979. *J. Mol. Biol.* 131:669–80
36. Hunter, W. N., D'Estaintot, B. L., Kennard, O. 1989. *Biochemistry* 28: In press
37. Calladine, C. R., Drew, H. R. 1984. *J. Mol. Biol.* 178:773–82
38. Calladine, C. R. 1982. *J. Mol. Biol.* 161:343–52
39. McCall, M., Brown, T., Kennard, O. 1985. *J. Mol. Biol.* 183:385–96
40. Arnott, S., Hukins, D. W. L. 1974. *J. Mol. Biol.* 88:551–52
41. Travers, A. A., Klug, A. 1987. *Philos. Trans. R. Soc. London, Ser. B* 317:537–61
42. Gotoh, O., Tagashira, Y. 1981. *Biopolymers* 20:1033–42
43. Satchwell, S. C., Travers, A. A. 1989. *EMBO J.* 8:229–38
44. Kramer, H., Amouyal, M., Nordheim, A., Muller-Hill, B. 1988. *EMBO J.* 7:547–56
45. Anderson, J. E., Ptashne, M., Harrison, S. C. 1987. *Nature* 326:846–52
46. Aggarwal, A. K., Rodgers, D., Drottar, M., Ptashne, M., Harrison, S. C. 1988. *Science* 242:899–907
47. Levitt, M. 1978. *Proc. Natl. Acad. Sci. USA* 75:640–44
48. Wang, J. C. 1985. *Annu. Rev. Biochem.* 54:665–97
49. Spassky, A., Rimsky, S., Buc, H., Busby, S. 1988. *EMBO J.* 7:1871–79
50. Suck, D., Lahm, A., Oefner, C. 1988. *Nature* 332:464–68

51. Anderson, A. H., Gocke, E., Bonven, B. J., Nielsen, D. F., Westergaard, O. 1985. *Nucleic Acids Res.* 13:1543–57
52. Camilloni, G., Di Martino, E., Caserta, M., Di Mauro, E. 1988. *Nucleic Acids Res.* 16:7071–85
53. Jordan, S. R., Pabo, C. O. 1988. *Science* 242:893–99
54. Bass, S., Sugiono, P., Arvidson, D. N., Gunsalus, R. P., Youderian, P. 1987. *Genes Dev.* 1:565–72
55. Flick, J. T., Eissenberg, J. C., Elgin, S. C. R. 1987. *J. Mol. Biol.* 190:619–33
56. Drew, H. R. 1984. *J. Mol. Biol.* 176:535–57
56a. Bass, S., Sorrells, V., Youderian, P. 1988. *Science* 242:240–45
57. Carey, J. 1988. *Proc. Natl. Acad. Sci. USA* 85:975–79
58. Linn, S.-Y., Riggs, A. D. 1975. *Cell* 4:107–11
59. Gronenborn, A. M., Nermut, M. V., Eason, P., Clore, G. M. 1984. *J. Mol. Biol.* 179:751–57
60. Wu, H.-M., Crothers, D. M. 1984. *Nature* 308:509–13
61. Kotlarz, D., Fritsch, A., Buc, H. 1986. *EMBO J.* 5:799–803
62. Trifonov, E. N., Sussman, J. L. 1980. *Proc. Natl. Acad. Sci. USA* 77:3816–20
63. Widom, J. 1985. *Bioessays* 2:11–14
64. Drew, H. R., Calladine, C. R. 1987. *J. Mol. Biol.* 195:143–73
65. Satchwell, S. C., Drew, H. R., Travers, A. A. 1986. *J. Mol. Biol.* 191:659–75
66. Kefalas, P., Gray, F. C., Allan, J. 1988. *Nucleic Acids Res.* 16:501–7
67. Turnell, W. G., Satchwell, S. C., Travers, A. A. 1988. *FEBS Lett.* 232:263–68
68. Gartenberg, M. R., Crothers, D. M. 1988. *Nature* 333:824–29
69. Majors, J. 1977. PhD thesis. Harvard Univ.
70. Warwicker, J., Engelman, B. P., Steitz, T. A. 1987. *Proteins* 2:283–89
71. Buc, H., Amouyal, M., Buckle, M., Herbert, M., Kolb, A., et al. 1987. In *RNA Polymerase and the Regulation of Transcription*, ed. W. S. Reznikoff, R. R. Burgess, J. E. Dahlberg, C. A. Gross, M. T. Record, Jr., M. P. Wickens, pp. 115–25. New York: Elsevier
72. Zinkel, S. S., Crothers, D. M. 1987. *Nature* 328:178–81
73. Liu-Johnson, H.-N., Gartenberg, M. R., Crothers, D. M. 1986. *Cell* 47:995–1005
74. Travers, A. A., Klug, A. 1987. *Nature* 327:280–81
75. Drew, H. R., McCall, M. J., Calladine, C. R. 1989. In *Biological Aspects of DNA Topology*, ed. N. Cozzarelli, J. C.

Wang. New York: Cold Spring Harbor Lab. In press
75a. Wolberger, C., Dong, Y., Ptashne, M., Harrison, S. C. 1988. *Nature* 335: 789–95
76. Nelson, H. C. M., Sauer, R. T. 1985. *Cell* 42:549–58
77. Marini, J. C., Levene, S. D., Crothers, D. M., Englund, P. T. 1982. *Proc. Natl. Acad. Sci. USA* 79:7664–68
78. Hagerman, P. J. 1984. *Proc. Natl. Acad. Sci. USA* 81:4632–36
79. Hagerman, P. J. 1985. *Biochemistry* 24:7033–37
80. Ulanovsky, L. E., Trifonov, E. N. 1987. *Nature* 326:720–22
81. Trifonov, E. N. 1985. *CRC Crit. Rev. Biochem.* 19:89–106
82. Hsieh, C.-H., Griffith, J. D. 1988. *Cell* 52:535–44
83. McAllister, C. F., Achberger, E. C. 1988. *J. Biol. Chem.* 263:11743–49
84. Deleted in proof
85. Travers, A. A. 1988. *Nucleic Acids Mol. Biol.* 2:136–48
86. Falvey, E., Grindley, N. D. F. 1987. *EMBO J.* 6:815–21
87. Dickerson, R. E., Goodsell, D. S., Kopka, M. L., Pjura, P. E. 1987. *J. Biomol. Struct. Dyn.* 5:557–80
88. Deleted in proof
89. Calladine, C. R., Drew, H. R., McCall, M. J. 1988. *J. Mol. Biol.* 201:127–37
90. Koo, H.-S., Rice, J. A., Crothers, D. M. 1988. *Proc. Natl. Acad. Sci. USA* 85:1763–67
91. Burkhoff, A. M., Tullius, T. D. 1987. *Cell* 48:935–43
92. Burkhoff, A. M., Tullius, T. D. 1988. *Nature* 331:455–57
92a. Diekmann, S. 1987. *Nucleic Acids Res.* 15:247–65
93. Diekmann, S., McLaughlin, L. W. 1988. *J. Mol. Biol.* 202:823–34
94. Hagerman, P. J. 1988. In *Unusual DNA Structures*, ed. R. D. Wells, S. C. Harvey, pp. 196–207. Springer-Verlag
95. McClarin, J. A., Frederick, C. A., Wang, B.-C., Greene, P., Boyer, H. W., et al. 1986. *Science* 234:1526–41
96. Fratini, A. V., Kopka, M. L., Drew, H. R., Dickerson, R. E. 1982. *J. Biol. Chem.* 257:14686–14707
97. Dickerson, R. E., Kopka, M. L., Pjura, P. 1985. In *Biological Macromolecules and Assemblies*, ed. F. A. Jurnak, A. McPherson, 2:37–126. New York: Wiley
98. McClellan, J. A., Palecek, E., Lilley, D. M. J. 1986. *Nucleic Acids Res.* 14:9291–309
99. Travers, A. A. 1987. *CRC Crit. Rev. Biochem.* 22:181–219

100. Amouyal, M., Buc, H. 1987. *J. Mol. Biol.* 195:795–808
101. Bossi, L., Smith, D. M. 1984. *Cell* 39:643–52
102. Gourse, R. L., de Boer, H. A., Nomura, M. 1986. *Cell* 44:197–205
103. Roe, J. H., Burgess, R. R., Record, M. T. 1985. *J. Mol. Biol.* 184:441–53
104. Auble, D. T., de Haseth, P. L. 1988. *J. Mol. Biol.* 202:471–82
105. Stefano, J. E., Gralla, J. D. 1982. *Proc. Natl. Acad. Sci. USA* 79:1069–72
106. Bertrand-Burggraf, E., Schnarr, M., Lefevre, J. F., Daune, M. 1984. *Nucleic Acids Res.* 12:7741–52
107. Furth, M. E., Blattner, F. R., McLeester, C., Dove, W. F. 1977. *Science* 198:1046–51
108. Scherer, G. 1978. *Nucleic Acids Res.* 5:3141–55
109. Fuller, R. S., Kaguni, J. M., Kornberg, A. 1981. *Proc. Natl. Acad. Sci. USA* 78:7370–74
110. Zahn, K., Blattner, F. R. 1985. *EMBO J.* 4:3605–16
111. Zahn, K., Blattner, F. R. 1987. *Science* 236:416–22
112. Snyder, M., Buchman, A. R., Davis, R. W. 1986. *Nature* 324:87–89
113. Fuller, R. S., Funnell, B. E., Kornberg, A. 1984. *Cell* 38:889–900
113a. Dickerson, R., Bansal, M., Calladine, C. R., Diekmann, S., Hunter, W. N., et al. 1989. *EMBO J.* 8:1–4
114. Dodson, M., Echols, H., Wickner, S., Alfano, C., Mensa-Wilmot, K., et al. 1986. *Proc. Natl. Acad. Sci. USA* 83:7638–42
115. Hatfull, G. F., Noble, S. M., Grindley, N. D. F. 1987. *Cell* 49:103–10
116. Prentki, P., Chandler, M., Galas, D. J. 1987. *EMBO J.* 6:2479–87
117. Robertson, C. A., Nash, H. A. 1988. *J. Biol. Chem.* 263:3554–57
118. Lloubes, R., Granger-Schnarr, M., Lazdunski, C., Schnarr, M. 1988. *J. Mol. Biol.* 204:1049–54
119. Shuey, D. J., Parker, C. S. 1986. *Nature* 323:459–61
120. Moskaluk, C., Bastia, D. 1988. *Proc. Natl. Acad. Sci. USA.* 85:1826–30
121. Heumann, H., Metzger, W., Niehörster, M. 1986. *Eur. J. Biochem.* 158:575–79
121a. Kuhnke, G., Fritz, H. J., Ehring, R. 1987. *EMBO J.* 6:507–13
122. Dodson, M., Roberts, J., McMacken, R., Echols, H. 1985. *Proc. Natl. Acad. Sci. USA* 82:4678–82
122a. Kirchausen, T., Wang, J. C., Harrison, S. C. 1985. *Cell* 41:933–43
122b. Summers, D. K., Sherratt, D. J. 1988. *EMBO J.* 7:851–58
123. Koudelka, G. B., Harrison, S. C., Ptashne, M. 1987. *Nature* 326:886–88

Annu. Rev. Biochem. 1989. 58:453–508

THE STRUCTURE AND REGULATION OF PROTEIN PHOSPHATASES

Philip Cohen

Department of Biochemistry, University of Dundee, Scotland

CONTENTS

0066-4154/89/0701-0453$02.00

INTRODUCTION AND SCOPE OF THE REVIEW

The reversible phosphorylation of proteins is now recognized to be a major mechanism for the control of intracellular events in eukaryotic cells. Processes as diverse as metabolism, contractility, membrane transport and secretion, the transcription and translation of genes, cell division, fertilization, and even memory, are all regulated by this versatile posttranslational modification. The phosphorylation (or dephosphorylation) of serine, threonine, and tyrosine residues triggers conformational changes in regulated proteins that alter their biological properties, the level of phosphorylation at any instant reflecting the relative activities of the protein kinases and phosphatases that catalyze the interconversion processes. The structures and functions of many protein kinases and their mechanisms of regulation have been defined. Some are specific for a particular substrate, while others phosphorylate many intracellular proteins. Protein kinase activities are controlled in a variety of ways, but most frequently by second messengers and allosteric effectors, and by phosphorylation (1). In this article, progress in elucidating the structure, substrate specificity, and regulation of the major serine- and threonine-specific protein phosphatases in eukaryotic cells is reviewed. Their structures are presented first, since this information has turned out to be critical to comprehending their physiological roles and regulation. Mitochondrial protein phosphatases, which have been the subject of recent reviews (2, 3), are not considered, nor are the protein tyrosine phosphatases, whose intracellular targets include the receptors for certain growth factors (4, 5).

CLASSIFICATION OF PROTEIN PHOSPHATASES

For many years, the nature of the serine/threonine-specific protein phosphatases was controversial. A large number of phosphatase preparations had been described, but little attempt had been made to establish whether they were related. In 1983, it was suggested that most, if not all, of the protein phosphatases in the literature could be explained by four principal catalytic subunits with broad and overlapping substrate specificities, and simple criteria were introduced that could be used to distinguish them, even in tissue extracts (6–8). The enzymes were subdivided into two groups (Type 1 and Type 2) depending on whether they dephosphorylated the β subunit of phosphorylase kinase specifically and were inhibited by nanomolar concentrations of two small heat- and acid-stable proteins, termed inhibitor-1 (I-1) and inhibitor-2 (I-2) (Type 1, protein phosphatase-1), or whether they dephosphorylated the α subunit of phosphorylase kinase preferentially and were insensitive to I-1 and I-2 (Type 2). Type 2 phosphatases could, in turn, be subclassified into three distinct enzymes, 2A, 2B, and 2C (PP-2A, PP-2B, and PP-2C) in a number of ways (Table 1), but most simply by their dependence on divalent cations. PP-2B and PP-2C had absolute requirements for Ca^{2+} and Mg^{2+}, respectively (6, 7), while PP-2A, like protein phosphatase-1 (PP-1), was active toward most substrates in the absence of divalent cations. Subsequently other useful criteria were introduced for distinguishing PP-1 and PP-2A (Table 1).

The procedures for classifying protein phosphatases are applicable to invertebrate as well as mammalian enzymes. Fruit flies (29, 30) and starfish (31) contain Type 1 and Type 2A protein phosphatases, indistinguishable from the mammalian enzymes by the criteria in Table 1. PP-2A has also been identified in the brine shrimp (32, 33) and soil amoeba (34, 35), and PP-2B in the fruit fly, sea urchin, and squid [reviewed in (36)].

In this review only phosphatases that have been purified to homogeneity or exhibit novel regulatory features are considered in detail. The identities of partially purified enzymes were discussed in (7, 37, 38). The protein phosphatases present in skeletal muscle are usually considered first, because information about their structure and regulation is most complete, and this tissue provides the model with which others are compared.

THE STRUCTURES OF PROTEIN PHOSPHATASES

Protein Phosphatase-1

CATALYTIC SUBUNIT   Isolation of the catalytic subunit of PP-1 (PP-1C) involves treatment with ethanol at ambient temperature, which dissociates PP-1C from higher-molecular-mass complexes containing other subunits (39, 40). Undegraded PP-1C is a monomeric 37-kd protein (41), but in the absence of proteinase inhibitors is converted during purification to a 33-kd species (41, 42) through the loss of about 30 residues from the C-terminus (P. Cohen, unpublished; 42).

Two full-length clones encoding Type 1 catalytic subunits have been isolated from a rabbit skeletal muscle cDNA library. One, termed PP-1α, codes for a 330-residue 37.5-kd protein (43), and the other, PP-1β, for a 311-residue 35.4-kd protein (44). The nucleotide and amino acid sequences of PP-1α and PP-1β are identical after amino acid 34 of PP-1α, including the 3'-untranslated region. However, the nucleotide sequences of the 5'-untranslated regions and coding regions up to amino acid 33 of PP-1α are completely different. PP-1α is encoded in a 28-kb EcoR1 fragment of rabbit genomic DNA. The central and C-terminal regions of PP-1β are also encoded in this 28-kb fragment, while the N-terminal 14 amino acids are encoded in a 15-kb EcoR1 fragment (43). These results suggest that PP-1α and PP-1β may be derived from the same gene, either by transcription initiation at different promoters followed by differential splicing, or by alternative splicing at the 5'-end of a common precursor message. The major mRNA species hybridizing to the cDNA clones coding for either PP-1α or PP-1β is 1.6 kb in rabbit

Table 1 Useful methods for distinguishing protein phosphatases

Method	PP-1	PP-2A	PP-2B	PP-2C	References
1. Preference for the α or β subunit of phosphorylase kinase	β subunit	α subunit	α subunit	α subunit	6, 7
2. Inhibition by I-1 and I-2	yes[a]	no[b]	no	no	6, 7
3. Absolute requirement for divalent cations	no[c]	no[c]	yes(Ca^{2+})	yes(Mg^{2+})	6, 7
4. Stimulation by calmodulin	no	no	yes	no	9, 10
5. Inhibition of trifluoperazine	no	no	yes	no	10
6. Inhibition by okadaic acid	yes[d]	yes[d]	yes[d] (weak)	no	11, 11a
7. Phosphorylase phosphatase activity	high	high	very low	very low	6
8. Activity toward histone H1 Phosphorylated by protein kinase C	very low	very high			12
9. Activity toward casein phosphorylated by A-kinase	very low	high			13
10. Effect of highly basic proteins on phosphorylase phosphatase activity	inhibition[e]	activation[f]			14–18
11. Effect of heparin on phosphorylase phosphatase activity	inhibition[e,g]	no effect or[h] activation			18–21
12. Effect of p-nitrophenyl phosphate on phosphorylase phosphatase activity	activation	inhibition			21a
13. Binding to heparin-Sepharose at 0.1 M NaCl	retained	excluded			22

[a] PP-1c is inhibited instantaneously by I-1 or I-2, but inhibition of some high-molecular-mass forms of PP-1 is time dependent (23–25), while other forms may be resistant to I-2 (26).

[b] PP-2Ac was reported to be inhibited by I-2 at extremely high (micromolar) concentrations (27), but other investigators were unable to confirm this result (210).

[c] Dephosphorylation of some substrates (e.g. I-1) is strongly stimulated by Mn^{2+} (8)

[d] PP-1 and PP-2A are inhibited with I$_{50}$ values of 20 nM and 0.2 nM, respectively, in the standard assay (P. Cohen, unpublished). The I$_{50}$ for PP-2B is about 5 μM (11).

[e] Inhibition by highly basic proteins (histone H1, protamine, polylysine) and heparin is only observed with phosphorylase a as substrate. With most other substrates (e.g. glycogen synthase, pyruvate kinase, acetyl CoA carboxylase, hormone-sensitive lipase), basic proteins and heparin are activators rather than inhibitors of PP-1 (18, 28).

[f] PP-2A is activated at low (submicromolar) concentrations of basic proteins, but inhibited at higher concentrations (16, 17). Optimal differentiation between PP-1 and PP-2A is observed at about 30 μg/ml protamine (18).

[g] Heparin binds to many proteins and the concentration required for 50% inhibition depends on the protein concentration in the assay. 90% inhibition is observed with 150 μg/ml heparin at 1 mg/ml serum albumin (18).

[h] Whether any activation is observed depends on the form of PP-2A, and the presence or absence of Mn^{2+} (18, 21).

skeletal muscle, liver, brain, heart, kidney, adrenal, lung, and intestine (43, 44, 44a; E. F. da Cruz e Silva, P. T. W. Cohen, unpublished).

Partial amino acid sequencing of the 33-kd catalytic subunit from rabbit skeletal muscle, which is probably derived from the glycogen-bound form of the enzyme (see below), showed that this preparation contains PP-1α (45). PP-1C isolated from glycogen-bound PP-1 had five amino acid differences (145 residues sequenced) from the primary structure deduced from the cDNA in the region where PP-1α and PP-1β are identical (44). These differences are probably explained by the presence of allelic genes in the population, as found for other rabbit proteins (46–48).

Immunoblotting of skeletal muscle extracts using antibody to PP-1C revealed a major cross-reacting species of 70 kd, and it was suggested that the native form of PP-1C was much larger than 37 kd (49). This possibility has been excluded by cDNA cloning and Northern blot analyses (43, 44). The identity of the 70-kd protein is unclear.

Two protein phosphatase catalytic subunits (34- and 30-kd) were isolated from rabbit liver (50) before PP-1C and PP-2AC were recognized as distinct entities. Nevertheless, they are probably Type 1 phosphatases since they have been used to assay for I-1 and I-2 (51). PP-1C from rat liver was reported to show 37-, 35-, and 34-kd bands on SDS-PAGE (52). The latter two species may be proteolysis products.

PP-1C isolated from rabbit liver in the absence of proteinase inhibitors comigrated with the 33-kd form of PP-1C from rabbit skeletal muscle on SDS-PAGE and both enzymes yielded identical peptide maps upon digestion with *S. aureus* V8 proteinase or papain (52a). The sequences of cDNA clones encoding rabbit liver PP-1C showed that it was the same gene product as skeletal muscle PP-1α (52a).

PP-1C was purified from fruit flies as a 33-kd enzyme (52b). A cDNA clone encoding PP-1α was isolated from a *Drosophila* head library and the deduced amino acid sequence (302 residues) showed 92% identity with PP-1α from skeletal muscle, demonstrating that the structure of PP-1 is remarkably conserved (V. Dombradi, P. T. W. Cohen, unpublished).

In *Aspergillus*, a mutant (BIM G) was isolated whose cell division was blocked at a late stage of mitosis. The gene affected was isolated and the deduced amino acid sequence showed about 80% identity with PP-1α from mammalian skeletal muscle. Most of the changes are near the N- and C-termini, and identity with mammalian PP-1α is 92% between residues 42 and 298 (97% homology, if conservative substitutions are included—J. H. Doonan, N. R. Morris, personal communication). This finding further emphasizes the remarkable structural conservation of PP-1C during evolution.

THE GLYCOGEN-BOUND FORM OF PP-1 (PP-1G)   The major active forms of PP-1 in skeletal muscle are associated with glycogen (8, 53, 54) and myofibrils (25). The former, termed PP-1G, is a 1 : 1 complex between PP-1C and a much larger, and highly asymmetric, glycogen-binding (G) subunit (23). Although originally isolated as a 103-kd protein (23), the native G subunit has an apparent molecular mass of 161 kd (55, 56), cleavage to the 103-kd G' subunit, and even smaller 40–80 kd fragments occurring during purification (23, 57). Fragments of the G subunit as small as 40 kd retain the ability to bind to PP-1C and glycogen, and the sites phosphorylated by A-kinase (56, 57). The G subunit or its fragments bind to glycogen with a K_{app} of about 4 nM (56).

Khatra (58) reported that the glycogen-bound form of PP-1 contained one major protein-staining band corresponding to PP-1C, plus a number of much fainter bands of higher molecular mass. Based on a value of 83 kd determined by sedimentation/diffusion analysis, it was concluded that the native enzyme was a dimer composed of two catalytic subunits. As discussed in (57), this preparation is almost certainly a complex between PP-1C and ~40-kd fragments of the G subunit that stain very poorly with Coomassie blue. Fragmentation of the G subunit also explains the wide variation in gel filtration behavior of PP-1 reported over many years [discussed in (23, 37)]. PP-1 activity associated with glycogen is immunoprecipitated quantitatively by antibody to the G subunit (57).

Villar-Moruzzi (59) isolated PP-1 from the glycogen fraction and also reported that it was simply PP-1C. It seems likely that the G subunit was dissociated or denatured during the precipitation with 50% acetone that was employed at one stage of the purification.

SARCOPLASMIC RETICULUM—ASSOCIATED PP-1G   In skeletal muscle, large amounts of PP-1 activity are associated with sarcoplasmic reticulum (SR) membranes (60), which can be released with detergents [(25), M. J. Hubbard, P. Cohen, unpublished]. This enzyme is PP-1G, as judged by its glycogen-binding properties, immunoblotting and immunotitration experiments using antibody to the G subunit, and phosphopeptide mapping (25; M. Hubbard, unpublished). SR-associated PP-1G could interact directly with membranes or with an SR-associated protein. It does not appear to be associated with glycogen particles that are attached to SR membranes, since incubation with α-amylase fails to release the enzyme (M. Hubbard, P. Cohen, unpublished).

Villar-Moruzzi & Heilmeyer (61) reported that SR-associated phosphatase was simply PP-1C. It seems likely that the G subunit was dissociated or denatured during precipitation with 50% acetone, which was used at one stage of purification.

THE MYOFIBRILLAR FORM OF PP-1 (PP-1M)   As much PP-1 activity is associated with the myofibrils of skeletal muscle as is bound to glycogen (25).

The myofibrillar enzyme PP-1M, when separated from the thick filaments, will rebind to purified actomyosin (25). PP-1M does not bind to glycogen and is not immunoprecipitated by antibody to the G subunit. Conversely, PP-1G does not bind to actomyosin (25). The apparent molecular mass of PP-1M estimated by gel-filtration (110 kd) is lower than that of undegraded PP-1G, but much higher than that of PP-1C. PP-1C derived from either PP-1G or PP-1M does not bind to either glycogen or actomyosin (25).

PP-1M has a severalfold higher myosin phosphatase/phosphorylase phosphatase activity ratio than PP-1G, but its enhanced myosin phosphatase activity is lost upon dissociation to PP-1C (62). Although not yet purified to homogeneity, PP-1M is likely to be composed of PP-1C complexed to a myofibrillar-binding M subunit, which also enhances the myosin phosphatase activity of the catalytic subunit, as discussed later. A myofibrillar form of PP-1 with similar properties is also present in cardiac myofibrils (62).

Two protein phosphatases have been identified in the cytosol of turkey gizzard, termed smooth muscle phosphatases III and IV, which have high activity toward native myosin (63). Phosphatase IV binds tightly to myosin and can be purified by chromatography on myosin-Sepharose (26). It shows 58-kd and 40-kd protein-staining bands on SDS-PAGE, which comigrate during PAGE under nondenaturing conditions. The native enzyme has an apparent molecular mass of 150 kd by gel-filtration, shows a high degree of specificity for the β subunit of phosphorylase kinase, but is unaffected by I-2. These results suggest that the 40-kd component may be PP-1C and that interaction with the 58-kd component prevents inhibition by I-2. It will be important to establish whether the 40-kd component becomes sensitive to I-1 and I-2 after dissociation from the 58-kd species. A myosin phosphatase isolated from chicken gizzard (64) may be the same enzyme, but was insufficiently characterized for definitive conclusions to be drawn.

THE ACTIVE FORMS OF PP-1 ARE LARGELY PARTICULATE   The active forms of PP-1 in striated muscles are largely bound to glycogen, SR, and myofibrils. In liver, brain, or kidney homogenates, 50% of Type 1 protein phosphatase activity is sedimented at 10,000 × g, and up to 70% at 100,000 × g. By contrast, little Type 2 protein phosphatase activity is sedimented at 100,000 × g (A. Chisholm, P. Cohen, unpublished). About 70% of hepatic PP-1 activity sedimenting between 10,000 × g and 100,000 × g is associated with microsomes and 30% with glycogen, as judged by measurements of I-2-sensitive phosphorylase phosphatase activity (8, 24, 65). Centrifugation of rabbit reticulocyte lysates at 150,000 × g sediments 20–25% of the PP-1 activity, but only 3% of the PP-2A (66). PP-1 is the major protein phosphatase associated with the microsomes of the parotid gland (66a) and with the postsynaptic densities of the brain (67).

PP-1 is also largely particulate in invertebrates. In starfish, PP-1 is the major protein phosphatase in the 100,000 × g pellet, and PP-2A the major phosphatase in the supernatant (31). 35% of the PP-1 activity in homogenates of fruit fly heads (but almost no PP-2A activity) is sedimented at 100,000 × g (29).

High levels of PP-1 are present in rat liver nuclei (68, 69). After correction for DNA recovery, PP-1 activity in nuclei was 40% of that present in a liver homogenate (69). Since nuclei occupy about 6% of the volume of an hepatocyte, the nuclear activity of PP-1 is at least five times higher than the average extranuclear activity. Nuclear PP-1 appears to be associated with chromatin, since it can be released by incubation with DNAase (68) or 0.5 M NaCl (69). PP-1 activity released by either procedure is PP-1C (68, 69).

An active form of PP-1 has been purified to apparent homogeneity from dog liver (21a). The preparation contained the 37-kd PP-1C and a 75-kd protein present in approximately equimolar amounts. Several properties of this enzyme resemble those of the hepatic microsomal enzyme (210).

THE INACTIVE CYTOSOLIC FORM OF PP-1 (PP-1I)   Although active PP-1 in skeletal muscle is largely particulate, an inactive cytosolic form (PP-1I), not considered in the previous discussion, is also present. This species was first identified in bovine adrenal cortex (70) and termed the MgATP-dependent protein phosphatase, because preincubation with MgATP is required to generate activity. Subsequently, the skeletal muscle enzyme was shown to consist of an inactive phosphatase (Fc), the activation of which required MgATP and another protein (Fa) (71). Factor Fa was shown to be associated with a protein kinase having activity toward glycogen synthase (72), and purification to homogeneity established that it was the enzyme glycogen synthase kinase-3 (GSK3) (73, 74).

The substrate specificity and characteristics of PP-1I were found to be similar to those of PP-1, suggesting that it might contain PP-1C (75, 76). Following reports that I-2 was required for interconversion of PP-1I between its active and inactive forms (77, 78), an enzyme with identical properties was produced by incubating purified I-2 with PP-1C (79). This enzyme consists of a 1:1 complex between I-2 and PP-1C, and its activation by GSK3 and MgATP is triggered by the phosphorylation of a threonine residue on I-2 (79, 80). PP-1I can be formed by incubating I-2 with either the undegraded 37-kd PP-1C or the proteolysed 33-kd species. However, activation of the complex formed with 33-kd enzyme is accompanied by dissociation of I-2 (79), whereas activation of PP-1I formed with undegraded enzyme is not (81). Thus the C-terminal fragment removed from PP-1C during conversion to the 33-kd species, strengthens association between PP-1C and the phosphorylated form of I-2, although it is not required to bind the dephosphorylated form of I-2.

PP-1I was subsequently purified from skeletal muscle (81, 82) and shown to have the same structure and properties as enzyme reconstituted from I-2 and PP-1C (81).

Other investigators reported that PP-1I was composed of a single 70-kd protein (83, 84), but subsequently confirmed that it was a complex of PP-1C and I-2 (85, 86). The 70-kd protein is probably an impurity.

PP-1I, being cytosolic, is extracted more easily from skeletal muscle than the active forms (PP-1G and PP-1M), and can be released by stirring minced tissue in low–ionic strength buffer. By contrast, homogenization is required to extract PP-1G, and high salt to extract PP-1M (A. Chisholm, P. Cohen, unpublished). The claim that PP-1I is the major form of PP-1 in muscle (82, 87) is explained by the use of extraction conditions that do not release PP-1G and PP-1M.

Protein phosphatase 1I has been identified in adrenal cortex (70), liver (71, 88), heart (71, 88–90), smooth muscle (91), brain (92), fruit flies (29), and starfish (31). PP-1I from bovine heart was shown to consist of two major proteins that correspond to PP-1C and I-2 (90).

It was reported that PP-1I can be activated by Mn^{2+} in the absence of MgATP and GSK3 (78, 93), but later work showed that this only happens if I-2 is partially proteolysed (81, 90).

INACTIVE PP-1 IN ERYTHROCYTE CYTOSOL   A Type 1 protein phosphatase was purified from human erythrocyte cytosol that was inactive in the absence of Mn^{2+} (94). The purified enzyme showed two major protein-staining bands, 36 kd and 62 kd, on SDS-PAGE. Based on a molecular mass of 135 kd for the native enzyme and relative staining intensities of the two bands, it was suggested that the enzyme contained two 36-kd subunits and one 62-kd subunit. Freezing and thawing denatured the 62-kd component and decreased the molecular mass of the native enzyme to 36 kd. The 62- and 36-kd proteins could also be separated by hydrophobic chromatography. These experiments demonstrated that the 36-kd protein was PP-1C. However no evidence was presented that the separated 62-kd protein could recombine with PP-1C. Further information is therefore needed to establish that the 62-kd protein is a subunit and not an impurity. In view of its Mn^{2+} dependence, this enzyme could be a proteolysed form of PP-1I, fragmentation of I-2 preventing detection of this component by SDS-PAGE. The suggestion (94) that this enzyme is the "peak I phosphatase" identified in (95) is incorrect, since the latter is a Type 2A phosphatase (21).

Inhibitor-1 and Inhibitor-2

Following a report that liver and muscle extracts contained heat-stable proteins that inhibit phosphorylase phosphatase (96), Huang & Glinsmann (97) identified two such proteins, I-1 and I-2. They discovered that I-1 was only effective when phosphorylated by cyclic AMP–dependent protein kinase (A-kinase), whereas I-2 did not require phosphorylation to be an inhibitor.

I-1 was purified to homogeneity by Nimmo & Cohen (98), who showed it to be a specific inhibitor of PP-1. The molecular mass calculated from the sequence [18.7 kd (99, 100)] is in close agreement with that determined by sedimentation equilibrium [19.2 kd (98)]. This demonstrates that I-1 is monomeric in solution. Abnormally low binding of detergent may explain the high apparent molecular mass of 26 kd determined by SDS-PAGE (97, 98), and its unusual gel-filtration behavior is accounted for by its asymmetry (98). The circular dichroism spectrum suggests that a high proportion of the molecule is random coil and that there is almost no α helix, although some β-structure may be present (101). The lack of ordered structure may explain its extreme stability to heat, low pH, organic solvents, and detergents. I-1 is not an artifact produced by heating muscle extracts at 100°C, since a protein possessing identical inhibitory properties, and with the same chromatographic and gel-filtration behavior, can be partially purified from muscle extracts that have not been heat treated (102).

I-1 comprises 166 residues (100). The N-terminal methionine is acetylated, and about half the molecules isolated have lost residues 165 and 166 (99). The protein has a very low content of hydrophobic amino acids, tyrosine, tryptophan, and cysteine are absent, and 13% of the residues are proline. The residue phosphorylated by A-kinase is Thr-35 (99). The peptides 2–66 and 9–54 retain the full inhibitory potency of I-1, provided that Thr-35 is phosphorylated. However, other peptides, such as 22–54 and 13–41, are inactive (103). These observations suggest that residues 10–20, the most hydrophobic region of I-1, may be required for inhibition, in addition to the phosphorylation of Thr-35. Ser-67 is phosphorylated in about half the I-1 molecules isolated (99), indicating that this residue is phosphorylated in vivo. However, the role of this modification is unknown, because fragments of I-1 that lack Ser-67 (residues 2–66, 9–54) are fully active.

I-1 is the only physiological substrate of A-kinase phosphorylated on threonine rather than serine. I-1 is phosphorylated in vitro at a comparable rate to the best substrates of A-kinase (104), yet the decapeptide IRRRRPTPA surrounding Thr-35 is not phosphorylated at a significant rate (105). This differs from other peptides of the sequence Arg-Arg-X-Ser-Y, corresponding to the phosphorylation sites of other physiological substrates, which are phosphorylated at comparable rates to the parent protein (106). A region of I-1, not in the immediate vicinity of the phosphorylation site, must be required for phosphorylation.

I-1 has been partially purified from dog liver (51) and purified to homogeneity from rabbit liver (100). The amount of I-1/g tissue is similar in rabbit liver and rabbit skeletal muscle, and protein sequencing has established that the rabbit liver and muscle proteins are almost certainly the same gene product. Immunoblotting experiments show that I-1 in rabbit brain, heart, kidney, uterus, and adipose tissue comigrates with the skeletal muscle and liver proteins on SDS-PAGE (100). I-1 has an identical electrophoretic mobility in rat skeletal muscle, kidney, brain, heart, uterus, and adipose tissue, although its apparent molecular mass (30 kd) is slightly higher than that of the rabbit protein (26 kd) (100, 107).

Surprisingly, no I-1 could be detected in rat liver either by immunoblotting or activity measurements, irrespective of the age, sex, or strain of the animals (100). These results disagree with those of Shenolikar & Steiner (108), who reported that antibodies to rabbit I-1 recognized a 26-kd protein in rat liver extracts of identical mobility to the rabbit protein. Although I-1 is undetectable in rat and mouse liver, it is present in guinea pig, pig, and sheep liver (100).

Brain not only contains I-1 (100, 109), but an additional isoform, termed "dopamine and cyclic AMP–regulated phosphoprotein" (DARPP). Its distribution correlates with dopamine-innervated neurones that possess D-1 dopamine receptors, being highly concentrated in the basal ganglia (110, 111). DARPP purified from bovine caudate nucleus shares many properties with I-1, including heat and acid stability, asymmetry, and phosphorylation by A-kinase on a threonine residue (112). It inhibits PP-1 with identical potency to I-1, inhibition only being observed after phosphorylation by A-kinase (113). DARPP comprises 202 residues (114), and like I-1, the true molecular mass (22.6 kd) is lower than that estimated by SDS-PAGE (32 kd). DARPP shows 60% identity with I-1 over the region comprising residues 9–50, consistent with the finding that the peptide 9–54 of I-1 is fully active. After residue 50 there is very little homology, apart from a short stretch between residues 77 and 92. I-1 lacks the 16 consecutive acidic residues found in DARPP (residues 120–135). DARPP is larger than I-1 because it has a 36-residue C-terminal extension. DARPP represents about 0.25% of the total protein in extracts of bovine caudate nuclei (112), indicating that its concentration in dopaminoceptive neurones is about 10-fold higher than in rabbit skeletal muscle (98, 115). DARPP has also been identified in bovine adrenal chromaffin cells, parathyroid cells, and choroid plexus (115a).

I-2 was first purified from skeletal muscle by Foulkes & Cohen (115), although improved isolation procedures have been published subsequently (116–118). It has also been isolated as a 1:1 complex with PP-1C (i.e. as PP-1I). The molecular mass calculated from the amino acid sequence is 22.8 kd (119), similar to the value of 25.5 kd determined by sedimentation/diffusion analysis (115). This finding demonstrates that I-2 is a monomer.

Abnormally low binding of detergent may explain the high apparent molecular mass (31 kd) determined by SDS-PAGE (115), and its anomalous gel-filtration behavior is accounted for by its asymmetry (115). Like I-1, the activity of I-2 is stable to heating at 100°C and exposure to low pH and detergents.

I-2 comprises 204 residues. The N-terminal alanine residue is acetylated and about 40% of I-2 molecules isolated have lost residues 203 and 204 (119). The residue phosphorylated by GSK3 is Thr-72 (119). Like I-1, I-2 has a low content of hydrophobic amino acids, there is a single tryptophan, and cysteine is absent. The amino acid sequence has recently been confirmed by cDNA cloning, Ser-204 being followed by a stop codon (120). This demonstrates that I-2 is synthesized as a 22.8-kd protein. The 60-kd protein recognized by antibodies to I-2 in immunoblotting experiments (121) cannot therefore be I-2. Isolation of cDNA clones coding for I-2 and PP-1C also exclude the proposal made in (87) that these proteins are synthesized as a single protein.

There is no sequence homology between I-1 and I-2. Residues 128–137 of I-2 (LSPEEREKKR) show eight identities and one conservative replacement (Arg for Lys) with residues 131–140 (LSPEEEEKRR) of c-fos (119), which corresponds to the first 10 residues of the third exon of c-fos (122). The significance of this observation is unknown, although other proteins are known that are homologous to the third exon of c-fos (123).

I-2 partially purified from either rabbit liver (124) or rabbit reticulocytes (125) has a molecular mass on SDS-PAGE (31 kd) identical to that of the skeletal muscle protein. Indeed, the protein comigrates on SDS-PAGE in all rabbit tissues tested (skeletal muscle, diaphragm, liver, heart, kidney, lung, and brain), as judged by immunoblotting (100, 126).

Strålfors (127) purified I-2 to homogeneity from rat adipose tissue. Its properties and two-dimensional peptide maps were very similar to the rabbit skeletal muscle protein, although its apparent molecular mass on SDS-PAGE was slightly greater (33 kd vs 31 kd). The apparent molecular mass in rat skeletal muscle, heart, liver, kidney, brain, uterus, and lung) is also 33 kd (100), suggesting that I-2 may be identical in all tissues.

Strålfors (127) detected a larger 40-kd protein that inhibited PP-1 after its elution from SDS/polyacrylamide gels. Unlike I-2, this inhibitor was lost when the extracts were heated at 100°C, suggesting that it may coprecipitate with another protein during heat treatment. Further information about the 40-kd protein is awaited with interest.

I-2 and PP-1I have been detected in fruit fly heads (29) and starfish oocytes (31). Invertebrate I-2 comigrated with rabbit I-2 on SDS-PAGE, and recombined with rabbit PP-1C to form a "hybrid" PP-1I that could be activated by mammalian GSK3 and MgATP. This implies that Thr-72, or an analogous residue, is conserved in invertebrates.

Protein Phosphatase-2A

CATALYTIC SUBUNIT   Isolation of the catalytic subunit of protein phosphatase-2A (PP-2AC), like PP-1C, involves an ethanol treatment to dissociate it from higher-molecular-mass complexes containing other subunits. The original procedures (39, 40) copurified PP-1C and PP-2AC, and many early preparations were mixtures (128). Pure preparations of PP-2AC have been isolated from rabbit skeletal muscle (41, 42), bovine heart (129–131), porcine kidney (132), rat liver (133), and soil amoeba (34). When purified in the presence of proteinase inhibitors, the apparent molecular mass of PP-2AC from rabbit skeletal muscle is 36 kd (41). A slightly smaller 33.5-kd species isolated in the absence of proteinase inhibitors (42, 134) may be proteolytically "nicked" [discussed in (135)]. A monoclonal antibody directed against PP-2AC from bovine cardiac muscle identified a single immunoreactive protein of the same molecular mass in extracts of bovine heart, brain, liver, smooth muscle, and skeletal muscle, rat cardiac muscle, brain, liver, spleen, and intestine, as well as rabbit liver, chicken gizzard, and human HT-29 cells (136). Peptide mapping (41, 42) and immunological studies (27, 41, 137) indicate that PP-1C and PP-2AC from rabbit skeletal muscle are the products of distinct genes.

Two full-length clones encoding Type 2A catalytic subunits, termed 2Aα and 2Aβ, have been isolated from a rabbit skeletal muscle cDNA library (138, 139). Both clones encode 309-residue, 35.6-kd proteins, values in close agreement with those determined by SDS-PAGE. The amino acid sequences of PP-2Aα and PP-2Aβ are 97% identical, but at the nucleotide level there is only 82% identity in the coding region and the 130 differences are spread throughout the cDNA sequences. Furthermore, the 3' noncoding regions are completely different. PP-2Aα and PP-2Aβ are therefore the products of distinct genes. Seven amino acid differences lie in the first 30 residues from the N-terminus, at positions 3, 5, 14, 24, 26, 29, and 30. Four of these are conservative, Asp for Glu, Val for Ile, Arg for Lys, and Thr for Ser. The eighth difference is at residue 108.

cDNA clones encoding two different Type 2A protein phosphatase catalytic subunits were isolated from a porcine kidney epithelial cell line cDNA library, and also termed 2Aα and 2Aβ (140). Overlapping clones yielded the complete sequence of 2Aα and an almost complete sequence of 2Aβ (lacking the first 16 amino acid residues). The amino acid sequences of 2Aα from rabbit skeletal muscle and porcine kidney epithelial cells were identical; 2Aβ only differed at residue 250.

A cDNA clone coding for PP-2Aα was isolated from a bovine adrenal cDNA library (141). After correction of three sequencing errors [discussed in (139, 140)], the deduced structure showed only one change from the rabbit muscle and pig kidney epithelial enzymes (at position 55).

Subsequently, cDNA clones coding for PP-2Aα were isolated from human liver (142) and rat liver (143), and for PP-2Aβ from human liver (142) and rabbit liver (144). Human liver PP-2Aα had the same amino acid sequence as did the rabbit muscle enzyme, while rat liver PP-2Aα showed a single conservative replacement at position 5. Human liver PP-2Aβ differed from rabbit muscle PP-2Aβ at one position, with Ala at position 5 instead of Thr. PP-2Aβ was the same gene product in rabbit skeletal muscle and liver (144).

The cDNA cloning suggests that PP-2Aα is the same gene product in each mammalian tissue examined, as is PP-2Aβ. The structural conservation between different mammalian species and recent finding that the amino acid sequence of PP-2AC from fruit fly brain is 93% identical to those of the mammalian enzymes, with nearly all the differences near the N-terminus (S. Orgad, Y. Dudai, P. T. W. Cohen, unpublished), demonstrates that PP-2AC is a very highly conserved protein.

The major mRNA species hybridizing to cDNA clones specific for either PP-2Aα or PP-2Aβ is 2 kb in a variety of mammalian tissues (139, 141–143). PP-2Aα mRNA was reported to be about 10 times more abundant than PP-2Aβ mRNA in each mammalian tissue examined, while mRNA for PP-2Aα and PP-2Aβ was about 10-fold more abundant in brain and heart than skeletal muscle, liver, kidney, and ovary (44a). A minor 2.8 kb mRNA species hybridizes to PP-2A mRNA even at high stringency (139, 143) and appears to be related to PP-2Aα (44a).

The primary structures of PP-1C and PP-2AC are strikingly homologous (44). Homology extends from residues 23 to 292 of the coding region of PP-2AC, and only a single deletion in PP-1C and one in PP-2AC are necessary to maximize homology. The overall sequence identity is 43% (50% between residues 23 and 292), and homology rises to 59% (67% between residues 23 and 292) if conservative substitutions are included. These findings provide a structural basis for previous observations that some monoclonal antibodies to PP-2AC cross-react with PP-1C (27) and vice versa (137).

PP-1C and PP-2AC show homology to an open reading frame in bacteriophage λ (orf 221). The identity is 35% over the region comprising residues 38–158 of PP-2AC (and the corresponding region of PP-1C), and homology is 49% if conservative substitutions are included (145). This suggests that orf 221 in λ may encode a protein phosphatase. The homologous gene in bacteriophage ϕ80 shows similar identity (145). Residues 55–95 of PP-2AC (and the corresponding region of PP-1C) show some homology to a region of mammalian alkaline phosphatases (44, 139). However, this is not the region where the phosphate-binding site or identified metal ligand sites of alkaline phosphatases are located.

Partial amino acid sequencing of PP-2AC from rabbit skeletal muscle (140) has demonstrated that it contains PP-2Aα. Whether PP-2Aβ is also present in this preparation is unknown.

Table 2 High-molecular mass Type-2A protein phosphatases that have been purified to near homogeneity from eukaryotic cells[a]

Reference	Enzyme termed	Tissue	Classification	Suggested subunit structure
147	protein phosphatase-H	rat liver	PP-2A1+PP-2A2	ND
148	protein phosphatase-1B	rat liver	PP-2A1	ABC2
149	protein phosphatase-II	rat liver	PP-2A2	AC2
150	eIF-2 phosphatase	rabbit reticulocyte	PP-2A2	AC
151a,b	phosphorylase phosphatase	rabbit liver	PP-2A2	AC
152	protein phosphatase E	bovine heart	PP-2A2	AC
153	protein phosphatase	pig heart	PP-2A1	ABC
154	myosin phosphatase	chicken gizzard	PP-2A2	AC
146	PT-1	bovine heart	PP-2A1	ABC
146	PT-2	bovine heart	PP-2A2	AC
63	smooth muscle phosphatase-1	turkey gizzard	PP-2A1	ABC
155	myosin light chain phosphatase	bovine aorta	PP-2A1	ND
156	peak-IV	rabbit skeletal muscle	PP-2A2	ND
134	protein phosphatase-H-II	rabbit skeletal muscle	PP-2A2	AC or AC2
135	protein phosphatase-2A0	rabbit skeletal muscle	PP-2A0	AB'C2
135	protein phosphatase-2A1	rabbit skeletal muscle	PP-2A1	ABC2
135	protein phosphatase-2A2	rabbit skeletal muscle	PP-2A2	AC
157	PCS H1	rabbit skeletal muscle	PP-2A1	ND
157	PCS H2	rabbit skeletal muscle	PP-2A2	ND
157	PCS L	rabbit skeletal muscle	PP-2A2	ND
157	PCS M	rabbit skeletal muscle	PP-2A2 or novel PP-2A	ND
158	protein phosphatase	hen oviduct	PP-2A2	AC
159	MAP-2 phosphatase	bovine brain	PP-2A1	ABC
160	LP-1	pig brain	PP-2A0 or PP-2A1	ND
33	mRNP phosphatase	brine shrimp	PP-2A2	(AC)2
21	phosphatase-I	human erythrocyte	novel PP-2A	AB''C
21	phosphatase-II	human erythrocyte	PP-2A1	ABC
21	phosphatase-IV	human erythrocyte	PP-2A2	ABC

[a]The classification of these enzymes is discussed in (7, 37, 135, 161), and in the text. The classification of less highly purified enzymes was discussed in (7). ND, not determined.

THE NATIVE FORMS OF PP-2A   Chromatography of skeletal muscle, reticulocyte, liver, or heart extracts on DEAE-cellulose resolves two major forms of PP-2A, termed PP-2A1 (0.2 M NaCl) and PP-2A2 (0.3 M NaCl) (18, 37, 66, 135, 146). Freezing and thawing the fractions in the presence of 0.25 M mercaptoethanol enhances the activity of PP-2A1 severalfold and PP-2A2 slightly, and reveals a new form eluting at 0.12 M NaCl, termed PP-2A0 (37). PP-2A1 and PP-2A2 have been purified to homogeneity from many tissues and species, and PP-2A0 from rabbit skeletal muscle (Table 2). PP-2A0 and

PP-2A1 are each composed of three subunits, termed A, B', and C (2A0) and A, B, and C (2A1), while PP-2A2 contains only two subunits, A and C. The A and C subunits of each species comigrate on SDS-PAGE. The apparent molecular mass of the C subunits is 36 kd, while that of the A subunit is 60 kd or 70 kd depending on the electrophoretic system employed [discussed in (21, 135)].

A polyclonal antibody to PP-2AC from rabbit skeletal muscle inactivates the catalytic subunits released from rabbit liver PP-2A0, PP-2A1, and PP-2A2 in a manner similar to that of PP-2AC isolated from skeletal muscle (65). Peptide maps of the catalytic subunits of PP-2A0, PP-2A1, and PP-2A2 from rabbit skeletal muscle are identical, and indistinguishable from those of PP-2AC isolated from the same tissue (135). It will be important to establish whether the two different subunits identified by cDNA cloning, 2Aα and 2Aβ (which would not be distinguished by peptide mapping), are associated with different forms of PP-2A.

Peptide maps of the A subunits of PP-2A0, PP-2A1, and PP-2A2 are identical. The B' and B subunits have apparent molecular masses of 54 kd and 55 kd, respectively, and their peptide maps are distinct (135). These findings indicate that PP-2A0 and PP-2A1 differ from PP-2A2 in possessing additional B' or B subunits. There is general agreement that the subunit structure of PP-2A2 is AC, but the subunit structure of PP-2A1 has been reported as ABC or ABC2 (Table 2).

A new form of PP-2A has recently been identified in human erythrocytes (21). This enzyme elutes from anion exchange columns before PP-2A1, but its subunit composition is distinct from that of PP-2A0. In addition to the A and C subunits, it contains a 72-kd subunit, termed B'' (Table 2).

A form of PP-2A was identified in skeletal muscle that eluted after PP-2A1 and slightly before PP-2A2 (157). Its activity was seen most clearly after cleavage with an endogenous Ca^{2+}-dependent proteinase, which enhanced its ability to dephosphorylate I-1 about 10-fold, without changing its activity toward phosphorylase a. The phosphorylase phosphatase activity of this species was stimulated to a greater extent by protamine and polylysine, and activation required lower concentrations of these basic polypeptides than with PP-2A1 or PP-2A2. The purified enzyme showed three protein-staining bands on SDS-PAGE, two of which could be equated with the A and C subunits. The third component (72–75 kd) may be a third subunit, but it appeared to be present in substoichiometric amounts, as judged by staining intensity with Coomassie blue or silver. This enzyme's composition is similar to that of the novel form of PP-2A recently isolated from human erythrocytes (21), but its elution position relative to PP-2A1 is quite different. It may represent a further form of PP-2A, or (if the 72–75-kd protein is an impurity) a form of PP-2A2 in which the A and/or C subunits have been modified in some unknown way.

Several lines of evidence suggest that PP-2A2 may be derived during purification from PP-2A0, PP-2A1, or other forms of PP-2A, and hence not exist in vivo. Tamura & Tsuiki [148] noted the tendency of hepatic PP-2A1 to convert to PP-2A2 during purification, a phenomenon that probably explains why the B subunit was substoichiometric in an earlier preparation (147). Rechromatography of rabbit liver PP-2A1 on DEAE-cellulose (37) or rabbit skeletal muscle PP-2A1 on Mono Q (157) causes partial conversion to PP-2A2. The generation of rabbit skeletal muscle PP-2A2 can be avoided by altering the purification protocol (135). PP-2A2 may be formed by either the degradation or dissociation of the B and/or B' subunits of PP-2A0 and PP-2A1. The B subunit is degraded far more rapidly by trypsin or chymotrypsin than are the A and C subunits (162). The use of phenyl-Sepharose in several isolation procedures (33, 134, 158), which involves exposure to high concentrations of NaCl or ammonium sulfate, may have promoted dissociation of the B and/or B' subunits, explaining why PP-2A2 was isolated by these investigators.

Takeda and coworkers (153) dissociated and reassociated the subunits of PP-2A1 from porcine heart. The A and C subunits could reassociate to form a complex. The isolated B subunit could bind to this complex, but not to the isolated C subunit. This suggests that B (and presumably B' and B'' also) interact with A rather than C.

A nuclear, chromatin-associated form of PP-2A has been identified (68). Its apparent molecular mass by gel-filtration is 70 kd, but it could be dissociated to an active 35-kd species by treatment with 4 M urea or freeze-thawing in the presence of 2-mercaptoethanol.

A protein phosphatase was purified from extracts of bovine kidney mitochondria that closely resembled PP-2A2 (163). This enzyme is distinct from the mitochondrial enzymes that dephosphorylate the pyruvate dehydrogenase and branched-chain ketoacid dehydrogenase complexes. The purified enzyme is composed of two subunits with apparent molecular masses of 60 kd and 34 kd. The subunits can be dissociated by gel-filtration in 6 M urea–1.4 M NaCl, demonstrating that the 34-kd species is the catalytic subunit. Antibody to PP-2AC from rabbit skeletal muscle, but not preimmune serum, inactivates the mitochondrial enzyme. Like PP-2A, the phosphorylase phosphatase activity of this enzyme is stimulated by basic proteins and spermine.

Protein Phosphatase-2B

This enzyme has recently been reviewed in detail (36, 164) and will therefore be discussed relatively briefly. PP-2B was first identified as an activity that dephosphorylated the α subunit of phosphorylase kinase (165), and shown to be a Ca^{2+}-dependent calmodulin-stimulated enzyme subsequently (9). When purified to homogeneity from skeletal muscle (10), SDS-PAGE revealed that its structure bore a striking resemblance to that of calcineurin, one of the two major calmodulin-binding proteins in brain (166). Further experiments confirmed that calcineurin was a Ca^{2+}/calmodulin-regulated protein phosphatase (9).

PP-2B comprises up to 1% of total brain protein, but it is not a major calmodulin-binding protein in peripheral tissues, representing only 0.03% of the soluble protein in skeletal muscle. Nevertheless, PP-2B activity in skeletal muscle and brain extracts is similar (8), suggesting that the brain enzyme may have a lower specific activity, under standard assay conditions (36).

The brain enzyme is a heterodimer composed of a 61-kd (A) subunit and a 19-kd (B) subunit (166–169). The A subunit is the calmodulin-binding component (167, 170, 171) and contains the catalytic site (172, 173), while the B subunit is the Ca^{2+}-binding component (169, 170, 174). Its amino acid sequence (169) revealed the presence of four "EF hand" Ca^{2+}-binding domains (175), indicating that it contains the four high-affinity Ca^{2+}-binding sites of the native protein (170). The B subunit shows a high degree of homology with other members of the "EF hand" Ca^{2+}-binding proteins at the regions of the four Ca^{2+}-binding loops. There are marked differences between the B subunit and either troponin C or calmodulin at the N- and C-termini of domains II and IV and the peptides connecting Ca^{2+}-binding domains I and II and III and IV. The overall sequence identity is only 35% with calmodulin and 29% with troponin C.

An important feature of the B subunit from bovine brain is the myristylation of its amino-terminal glycine residue (176), which might explain why it is associated with particulate fractions of brain as well as cytosol (36). It would be of interest to know if the B subunit is also myristylated in other tissues, such as skeletal muscle, where PP-2B is almost entirely soluble (177).

Calmodulin-regulated protein phosphatases isolated from skeletal muscle (10), platelets (178, 179), placenta (180), heart (178), and sea urchin eggs (181) are also composed of two subunits with molecular masses similar to those of the A and B subunits of brain PP-2B. The A subunit of the purified skeletal muscle enzyme migrates as a doublet (61-kd/58-kd) on SDS-PAGE (9, 10). However, immunoblotting of muscle extracts only revealed the 61-kd species (177), suggesting that the 58-kd band is formed during purification, presumably by proteolysis. The low molecular mass of 55-kd reported for the pancreatic A subunit may also result from proteolysis, and its apparent lack of a B subunit is most likely due to the poor staining intensity of this component when low amounts of enzyme are analyzed (182).

PP-2B has been identified in all mammalian tissues examined as a Ca^{2+}/calmodulin-dependent, trifluoperazine-sensitive activity (8) or by immunoblotting with antibody to the brain enzyme (177, 180, 183, 184).

Whereas the B subunit is easily detected by immunoblotting, the A subunit is often barely seen, suggesting that isoforms of the A subunit may exist that cross-react poorly with antibody to the brain enzyme (177). This may explain an earlier report that the protein was barely detectable in non-neural tissues (185). The conserved nature of the B subunit has facilitated detection of PP-2B-like proteins in several lower eukaryotes (177, 181, 186, 187).

A cDNA clone encoding the C-terminal half of the catalytic domain of the A subunit of PP-2B has been isolated from a mouse cDNA library (188). Residues 1–111 of this clone show 32% identity (51% homology) with residues 189–300 of PP-1α and 35% identity (45% homology) with residues 183–294 of PP-2A, although two insertions of seven and six amino acids in the sequence of PP-2B are necessary to maximize homology. The putative calmodulin-binding domain is located between amino acid residues 176 and 200 of the PP-2B clone.

Protein Phosphatase-2C

PP-2C has been partially purified from canine heart as a casein phosphatase (189), and to homogeneity from turkey gizzard as a myosin P-light chain phosphatase (190, 191), from rat liver as a glycogen synthase phosphatase (192) or phosphofructokinase phosphatase (193, 194), and from rabbit liver and rabbit skeletal muscle (195). Each enzyme is a monomeric protein with a molecular mass of about 43 kd on SDS-PAGE. A higher value of 48 kd (192) may be explained by the different electrophoretic system employed by these investigators. The claim that PP-2C from rat liver is a dimer (193, 194, 196) is an error, for reasons discussed in (195).

Two forms of PP-2C have been isolated from rabbit skeletal muscle or rabbit liver, termed PP-2C1 and PP-2C2 (195). PP-2C1 migrates as a 44-kd protein and PP-2C2 as a 42-kd protein on SDS-PAGE. Peptide mapping indicates that PP-2C1 and PP-2C2 are isozymes (195), a suggestion confirmed by partial sequence analysis. There were 49 identities over the 62 residues where the two proteins could be compared directly (197), establishing that PP-2C1 and PP-2C2 are the products of different genes. Surprisingly, amino acid sequences comprising over 250 residues of PP-2C2 have failed to reveal any homology with PP-1C and PP-2AC (C. MacGowan, P. Cohen, unpublished).

Peptide maps of PP-2C1 from skeletal muscle are identical to those of hepatic PP-2C1. Similarly, peptide maps of PP-2C2 from rabbit muscle and liver are indistinguishable (195). This indicates that the same two isozymes of PP-2C are present in both tissues.

PP-2C has also been identified in rabbit brain and adipose tissue (8) and human erythrocyte cytosol (95). Higher levels are found in liver and brain than in skeletal muscle (8, 195).

Other Protein Phosphatases

Five years after the original suggestion that PP-1, PP-2A, PP-2B, and PP-2C are the principal Ser/Thr-specific protein phosphatases in the cytoplasmic compartment of mammalian cells, there is still little evidence for the existence of additional enzymes.

A Mn^{2+}-dependent ribosomal protein S6 phosphatase was isolated from rabbit reticulocyte lysates using 40S ribosomal subunits phosphorylated by A-kinase as substrate (198). The final preparation contained a major 56-kd protein, that was claimed to be the phosphatase, but the specific activity was extremely low. The major S6 phosphatase in mammalian cells is PP-1 (as discussed later), and these investigators subsequently reported that the enzyme was inhibited by I-2 (199). It therefore seems likely that the dephosphorylation of S6 was catalyzed by PP-1, this enzyme being a minor component of their final preparation.

A rat liver enzyme, called protein phosphatase T because of its activity toward casein phosphorylated on a threonine residue by casein kinase-2 and apparent preference for phosphothreonine-containing peptides (200), has subsequently been classified as a Type 2A protein phosphatase, possibly PP-2A0 (201). Similarly, an enzyme reported to dephosphorylate histone H1 phosphorylated by protein kinase C (202) has also now been classified as a Type 2A enzyme (133).

A "specific" myosin light chain phosphatase was isolated from the cytosol of rabbit skeletal muscle (203). The enzyme was retained by phosphorylated myosin light chain–Sepharose in the presence of Mg^{2+} and displaced by EDTA. Its activity was dependent on Mg^{2+} and unaffected by Ca^{2+}, and it had low phosphorylase phosphatase activity. These properties indicate that this enzyme is PP-2C (195), although its reported molecular mass (70 kd) and apparent failure to dephosphorylate phosphorylase kinase and glycogen synthase are at variance with this suggestion.

A protein phosphatase was isolated from porcine brain, termed latent phosphatase (LP)-2 (160). LP-2 had very low phosphorylase phosphatase activity, but activity could be increased up to 40-fold by freezing and thawing in the presence of 0.2 M mercaptoethanol. The elution position of LP-2 from DEAE-cellulose, its activation by freeze-thawing, high phosphorylase phosphatase activity (after activation), and insensitivity to I-2 suggest it is similar to PP-2A1. However, a single 49-kd band was detected by SDS-PAGE. If the 49-kd band is not an impurity, LP-2 may represent a novel phosphatase catalytic subunit, and detailed sequence analysis is awaited with interest. LP-2 was reported to dephosphorylate the β-adrenergic receptor much more effectively than PP-1 or PP-2A (203a).

Recently a cDNA clone was isolated from a rabbit liver library that encodes part of a protein phosphatase distinct from PP-1, PP-2A, PP-2B, and

PP-2C. This enzyme, termed PP-X, shows 45% and 65% amino acid sequence identity, respectively, to PP-1C and PP-2AC from rabbit liver (144). Further work is required to establish whether PP-X represents the catalytic subunit of PP-2A-like enzymes isolated from rabbit skeletal muscle (157), rat liver nuclei (68), or mitochondria (163), or whether it is a novel enzyme.

SUBSTRATE SPECIFICITIES AND PHYSIOLOGICAL ROLES OF PROTEIN PHOSPHATASES 1, 2A, and 2C

These enzymes have very broad and overlapping specificities in vitro. As discussed below, they account for virtually all detectable phosphatase activity in muscle and liver extracts toward some 20 phosphoproteins that control glycogen metabolism, muscle contractility, protein synthesis, cholesterol synthesis, fatty acid synthesis, glycolysis/gluconeogenesis, and aromatic amino acid breakdown. However, their specificities are likely to be even wider in vivo. They are present at similar concentrations in all mammalian tissues examined (8, 21), and dephosphorylate hormone-sensitive lipase from adipose tissue (28), synapsin-1, protein III, protein KF (A. Nairn, unpublished), and microtubule-associated protein-2 (159) from brain, tyrosine hydroxylase from adrenal medulla (P. Cohen, J. Haavik, unpublished), and spectrin from erythrocyte membranes (21, 95) at high rates. However, the mere observation that a protein is dephosphorylated by PP-1, PP-2A, and PP-2C in vitro does not prove that these enzymes act on that protein in vivo. In this section the physiological roles of these enzymes are discussed, with emphasis on the problems posed by their broad substrate specificities in vitro.

Glycogen Metabolism

The neuronal and hormonal control of glycogen metabolism is mediated via changes in the phosphorylation states of glycogen phosphorylase, phosphorylase kinase, and glycogen synthase [for recent reviews see (204–207)]. The glycogenolytic enzymes are activated, and glycogen synthase inhibited, by phosphorylation.

   PP-1 and PP-2A are the only enzymes in skeletal muscle with significant activity toward glycogen phosphorylase and glycogen synthase (8, 65), and further fractionation by ion exchange chromatography and gel-filtration has failed to resolve any other phosphatases that are active toward these substrates (37). Two lines of evidence suggest that PP-1 plays an important role in the regulation of muscle glycogen metabolism. Firstly, at pH 7 and physiological ($\sim$1 mM) Mg^{2+}, PP-1 accounts for 85–90% of the phosphorylase phosphatase activity and for 60–75% of the glycogen synthase phosphatase activity (depending on which serines are phosphorylated) (8, 65) in muscle extracts. PP-1 also accounts for >95% of the phosphorylase kinase phosphatase activity (toward the β-subunit serine that plays the dominant role in activation) (8, 65). Secondly, PP-1 is bound specifically to protein-glycogen particles (as PP-1G), whereas PP-2A is not (53, 54, 65).

However, a role for PP-2A in muscle glycogen metabolism is not excluded. For example, if glycogen is depleted, the glycogen-metabolizing enzymes will be released into the cytosol where they may be dephosphorylated by PP-2A, as well as PP-1. Furthermore, in the absence of Ca^{2+} (i.e. in resting muscle), PP-2A is the major enzyme acting on the α subunit of phosphorylase kinase (8), the phosphorylation of which enhances activity up to twofold [reviewed in (206)]. PP-2A is also the only significant I-1 phosphatase activity in the absence of Ca^{2+} (8, 37), and may control glycogen metabolism indirectly by stimulating PP-1 through this mechanism, as discussed later.

The activity of PP-2A toward the glycogen-metabolizing enzymes (and other substrates) is also enhanced by certain basic proteins and polyamines (Table 1). By contrast, PP-1 is inhibited or stimulated by these substances depending on the substrate (18, 28). At optimal levels of basic proteins, PP-2A can account for ~50% of the phosphorylase phosphatase activity in muscle extracts (135), and at optimal spermine for most of the glycogen synthase phosphatase activity (208). Although there is not yet any evidence that basic proteins and polyamines are activators of PP-2A in vivo, they could be mimicking another substance that stimulates PP-2A in vivo.

The amount of PP-1 activity relative to PP-2A is somewhat lower in rabbit liver than in rabbit skeletal muscle, and PP-1 accounts for only 50% of the phosphorylase phosphatase activity, 75% of the phosphorylase kinase (β subunit) phosphatase activity, and 20–40% of the glycogen synthase phosphatase activity, when liver extracts are assayed under standard conditions with glycogen-metabolizing enzymes from muscle as substrates (8). As in skeletal muscle, PP-1 is the only phosphatase associated with glycogen, but the proportion bound to hepatic glycogen is lower, as judged by measurements of phosphorylase phosphatase activity (8, 24, 65). Nevertheless, several lines of evidence indicate that the glycogen-bound form of hepatic PP-1 (PP-1G) plays a major role in the regulation of glycogen synthesis.

1. When assayed using muscle glycogen synthase as substrate, hepatic PP-1G has a 50-fold higher glycogen synthase phosphatase/phosphorylase phosphatase activity ratio than its skeletal muscle counterpart and a 5-fold higher ratio than PP-1 associated with hepatic microsomes (209). The selectivity of hepatic PP-1G is even greater toward highly phosphorylated preparations of liver glycogen synthase (a different isozyme), since this substrate is poorly activated by hepatic microsomal PP-1, cytosolic PP-1, or hepatic PP-2A (210). Most of the glycogen synthase phosphatase activity is therefore accounted for by hepatic PP-1G, when the "natural

substrate" is employed (210). The critical importance of the source and state of phosphorylation of glycogen synthase, extent of dilution of the extracts, and precise incubation conditions when measuring hepatic glycogen synthase phosphatase activity have been reviewed (211).

2. The glycogen-bound form of hepatic PP-1 is lost selectively in severely diabetic rats, or rats that have been adrenalectomized and starved for 48 h, treatments that are known to prevent glycogen synthesis in vivo (212, 213, 213a).

3. Phosphorylase a is the key in vivo regulator of glycogen synthase phosphatase activity in fed animals, and as discussed later, hepatic PP-1G is potently inhibited by this protein, whereas PP-2A and PP-2C are not.

Hepatic PP-1G is also likely to inactivate hepatic phosphorylase a and phosphorylase kinase in vivo, although a role for PP-2A is not excluded for the reasons discussed above.

PP-2C activity is higher in liver than skeletal muscle, yet still accounts for only a small proportion of hepatic glycogen synthase phosphatase in liver extracts when assays are performed with muscle glycogen synthase as substrate (8). When assayed by the activation of liver glycogen synthase, PP-2C was reported to be the major glycogen synthase phosphatase detected after chromatography of rat liver cytosol on DEAE-cellulose (192, 214). This observation reemphasizes the importance of the type of substrate, and confirms that PP-2A dephosphorylates hepatic glycogen synthase very poorly. However, in these studies glycogen (containing hepatic PP-1G) was removed prior to chromatography (192, 214). In addition, these results conflict with the observation that glycogen synthase phosphatase activity in liver cytosol can be inhibited almost totally by I-2, even in the presence of Mg^{2+} (210). Incubation of hepatocytes with 0.1–1.0 μM okadaic acid (a potent inhibitor of PP-1 and PP-2A, Table 1) rapidly produces marked increases in the phosphorylation states of glycogen synthase and glycogen phosphorylase, confirming that PP-1 and PP-2A, rather than PP-2C, are the dominant phosphatases acting on the regulatory enzymes of hepatic glycogen metabolism in vivo (11a).

Muscle Contractility

The myosin P-light chain is phosphorylated by Ca^{2+}/calmodulin-dependent myosin light chain kinases in striated and smooth muscles. In smooth muscle, phosphorylation initiates contraction, while in fast twitch skeletal muscle P-light chain phosphorylation correlates with the potentiation of isometric twitch tension that occurs during repetitive stimulation of the fibers. P-light chain phosphorylation also occurs in cardiac muscle in vivo, but its physiological role is less clear [reviewed in (215)]. Four lines of evidence strongly suggest that the myofibril-associated form of PP-1 (PP-1M) is the enzyme that dephosphorylates the myosin P-light chain in skeletal and cardiac muscle (25, 62).

1. PP-1M is the only phosphatase that is bound specifically to actomyosin.
2. PP-1M has severalfold enhanced myosin phosphatase activity compared to other forms of PP-1.
3. PP-1M dephosphorylates the myosin P-light chain more rapidly when it is complexed to the heavy chain in "native myosin." By contrast, PP-2A is very active toward isolated myosin P-light chains, but ineffective toward this substrate when it is bound to the heavy chain.
4. As a consequence of 2 and 3, PP-1M accounts for ~60% of the phosphatase activity toward native myosin in skeletal muscle (PP-1G accounting for the remainder), and for 90% of the myosin phosphatase activity in bovine cardiac muscle (where PP-1G is almost absent).

PP-2A was reported to be the only enzyme in bovine cardiac cytosol with significant activity toward the isolated P-light chain or "native" myosin (146), although activity toward the latter substrate was much lower. These observations are in agreement with Refs. 25, 62. However, the former investigators (146) concluded that PP-2A was the major myosin phosphatase in cardiac muscle because significant myosin phosphatase activity was not detected in the myofibrillar fraction in their study. The reason for this is unknown.

The conclusion that PP-1M is the major myosin phosphatase activity in striated muscles is consistent with observations made in avian smooth muscle. In turkey gizzard, smooth muscle phosphatases III and IV are the principal activities that dephosphorylate the myosin P-light chain when "native myosin" is employed as substrate, while smooth muscle phosphatases I and II, classified as PP-2A1 and PP-2C respectively in (161), are very active toward isolated myosin P-light chains, but ineffective in dephosphorylating native myosin (63, 191). Smooth muscle phosphatase IV appears to be a Type 1 phosphatase as discussed earlier, and although isolated from cytosol, it binds to myosin (26, 216) and may be the smooth muscle equivalent of PP-1M.

Okadaic acid, which does not affect myosin light chain kinase activity, enhances the development of tension when added to chemically skinned smooth muscle fibers from guinea pig intestine (217). This indicates that PP-2C, which is unaffected by okadaic acid (11), is not a myosin phosphatase in vivo. Smooth muscle phosphatase IV dephosphorylates the myosin P-light chain when added to skinned avian smooth muscle fibers (218). However, PP-2A2 also relaxed chemically skinned uterine smooth muscle fibers (219). It would therefore seem important to evaluate the relative effectiveness of different phosphatase preparations in such experiments.

Although PP-1M is likely to be the major myosin phosphatase in vivo, a role for PP-2A in the regulation of muscle contractility is not excluded. For example, smooth muscle myosin light chain kinase is phosphorylated in vitro by A-kinase, which reduces its affinity for calmodulin. This could represent one of the mechanisms by which cyclic AMP relaxes smooth muscle (220), although its physiological significance is still controversial (215). PP-2A1 is very effective in dephosphorylating myosin light chain kinase in vitro (63).

At least four other proteins in cardiac muscle (troponin I, C-protein, phospholamban, and the calcium channel) are phosphorylated by A-kinase in vivo and appear to be involved in the regulation of cardiac muscle contractility by adrenalin (221–223). PP-2A1 and PP-2A2 account for virtually all the troponin I phosphatase activity in bovine cardiac cytosol (146). PP-1C and PP-2AC rapidly dephosphorylate a threonine residue on C-protein, which accounts for 30–40% of the phosphate introduced by A-kinase, but on prolonged incubation only PP-2AC can remove the rest of the phosphate, which is mainly attached to serine (223). Although troponin I and C-protein may be dephosphorylated by PP-2A in vivo, these studies were carried out using isolated cardiac troponin and C-protein. It will be important to repeat these experiments using C-protein complexed to myosin, and troponin I bound to the thin filaments, in view of the critical importance of using "native myosin" for identification of the physiologically relevant myosin phosphatase. In this connection, some 20% of the troponin I phosphatase activity could not be extracted from the cardiac myofibrillar pellet (146). The identity of this phosphatase was not investigated.

Infusion of PP-1C and PP-2AC into guinea pig myocytes abolished the increase in the voltage-dependent calcium current induced by the β-adrenergic agonist isoprenaline (224). Infusion of I-2 into the myocytes caused a 23% increase in the calcium current in the absence of isoprenaline, increased the amplitude of the calcium current induced by isoprenaline, and increased twofold to threefold the time required to reverse the effect of isoprenaline following its removal from the system. Okadaic acid also markedly increased the calcium current when added to isolated myocytes (225). These observations suggest that PP-1 plays a role in regulating the cardiac calcium current in vivo. Whether another phosphatase, such as PP-2A, is also involved is unclear. A protein phosphatase associated with canine SR that dephosphorylated phospholamban was partially inhibited by I-1, suggesting that it was a Type 1 phosphatase (226). However, a phospholamban phosphatase subsequently purified from canine cardiac cytosol was hardly affected by I-2, and its subunit composition suggested that it might be PP-2A1 (226a).

In soil amoebae the actin-activated myosin ATPase is inhibited by the phosphorylation of three serines near the C-terminus of the myosin heavy chain. These residues can be dephosphorylated, and activity recovered, by incubation with PP-2AC isolated from soil amoeba (34). Whether this enzyme, or another phosphatase, is responsible for dephosphorylating the myosin heavy chain in vivo is unknown.

Protein Synthesis

S6, the major phosphoprotein of the 40S ribosomal subunit, is multiply phosphorylated in a variety of cells in response to insulin and other growth factors that stimulate protein synthesis. Phosphorylation of S6 is catalyzed by a mitogen-stimulated S6 kinase, which is itself activated by phosphorylation (227, 228). S6 can also be phosphorylated in vitro by A-kinase and in vivo in response to cyclic AMP–elevating hormones, such as glucagon (229).

A significant percentage of PP-1 activity in rabbit reticulocyte lysates sediments with ribosomes, whereas PP-2A does not (66). 40S ribosomal subunits from *Xenopus* oocytes in which S6 had been phosphorylated by a mitogen-stimulated S6 kinase isolated from the same source, was dephosphorylated much more efficiently by rabbit skeletal muscle PP-1C than by PP-2AC (230). 80% of the S6 phosphatase activity in oocyte extracts could be blocked by the addition of I-2, while microinjection of I-2 into intact oocytes caused a threefold increase in the phosphate content of S6 within 15 min. Similarly, PP-1 is the major S6 phosphatase in mouse 3T3 cells (231). PP-1 is also the major phosphatase in rat liver extracts toward S6 phosphorylated by A-kinase (209). However, a role for PP-2A in regulating S6 phosphorylation is not excluded, because it is the major activity in mouse 3T3 cells that inactivates the mitogen-stimulated S6 kinase (227, 228).

Phosphorylation of a single serine on the α-subunit of initiation factor eIF-2 inhibits protein synthesis in reticulocytes. Phosphorylation is catalyzed by a protein kinase that is inhibited by heme, and by a double-stranded RNA-dependent protein kinase that is activated by autophosphorylation [reviewed in (232)]. The α subunit of eIF-2 is dephosphorylated efficiently in vitro by PP-1 and PP-2A, but not by PP-2C (6, 233), and PP-2A2 has been purified as an eIF-2 phosphatase (150). However, eIF-2 forms a preinitiation complex with methionyl tRNA, GTP, and the 40S ribosome, and is therefore bound to ribosomes in vivo. PP-1 is specifically associated with ribosomes (66), and addition of I-2 to rabbit reticulocyte lysates increases the phosphorylation of eIF-2 and inhibits peptide chain initiation (234). PP-1 may therefore be the most important eIF-2 phosphatase in vivo. However, a role for PP-2A is not excluded. For example, dephosphorylation and inactivation of the double-stranded RNA–dependent protein kinase is catalyzed in vitro by PP-2A, but not by PP-1, and I-2 does not inhibit the dephosphorylation of this kinase in reticulocyte lysates (235).

The β subunit of eIF-2 is phosphorylated by casein kinase-II. The α and β subunits of eIF-2 are dephosphorylated at similar rates in vitro by PP-2A2, but only the α subunit is dephosphorylated in reticulocyte lysates, even if the lysates are supplemented with purified PP-2A2 (236). Interaction of eIF-2 with GTP and methionyl tRNA in the lysates appears to prevent access to the phosphorylation site on the β subunit, explaining why phosphorylation of the β subunit is a relatively stable modification in vivo. This example further emphasizes the importance of using the "native" substrate, when studying protein phosphatases.

In the brine shrimp, PP-2A dephosphorylates messenger ribonucleoproteins phosphorylated by a casein kinase-II activity (32, 33), another level at which protein synthesis could be controlled.

Oocyte Maturation

PP-1 has been identified in *Xenopus* oocytes, and its ability to dephosphorylate phosphorylase a (237) or the β subunit of phosphorylase kinase (230) is blocked by microinjection of I-2 into the oocytes. Microinjection of I-1 (238) or I-2 (237) delays the entry into meiosis of oocytes treated with progesterone, suggesting that PP-1-catalyzed dephosphorylation of one or more proteins controls maturation.

Microinjection of PP-1C or PP-2AC into starfish oocytes prevents the formation of maturation-promoting factor and breakdown of the nuclear envelope in response to 1-methyl adenine (239), establishing that protein phosphorylation is required for oocyte maturation by this hormone. The protein phosphatase involved in regulating this process in vivo is unknown, but PP-1 might not be involved, because microinjection of I-1 or I-2 did not promote maturation (240).

Hepatic Metabolism

Hydroxymethylglutaryl CoA (HMG-CoA) reductase, the rate-limiting enzyme in cholesterol synthesis, is inactivated by an AMP-activated protein kinase, which is itself activated by phosphorylation catalyzed by a "kinase kinase" (241, 242). PP-1, PP-2A, and PP-2C account for all the HMG-CoA reductase phosphatase and AMP-activated kinase phosphatase activity in rat liver extracts (8, 38), PP-2C being particularly active. However, HMG-CoA reductase is bound to the endoplasmic reticulum, and the major HMG-CoA reductase phosphatase associated with liver microsomes is PP-1 (209).

The inhibition of hepatic glycolysis and fatty acid synthesis and stimulation of gluconeogenesis and aromatic amino acid breakdown by glucagon are mediated by the A-kinase–catalyzed phosphorylation of serine residues on the regulatory enzymes of these pathways. PP-1, PP-2A, and PP-2C account for virtually all the phosphatase activity in rat and rabbit liver extracts toward these enzymes, namely 6-phosphofructo-2-kinase/fructose 2, 6 bisphosphatase (PF2K/F2,6Pase), pyruvate kinase, 6-phosphofructo-1-kinase (PF1K), and fructose 1,6 bisphosphatase (F1,6Pase) (glycolysis/gluconeogenesis), acetyl CoA carboxylase and ATP citrate lyase (fatty acid synthesis), and phenylalanine hydroxylase (amino acid breakdown) (8, 38, 65, 243). PP-2A accounts for most of the activity toward each substrate, even when basic proteins or spermine are excluded from the assays. For example, antibody to PP-2AC immunoprecipitates 65% of the activity toward PF1K, 70% of the activity toward PF2K/F2,6Pase or phenylalanine hydroxylase, 75% of the pyruvate kinase phosphatase activity, and >90% of activity toward F1,6Pase in dilute liver extracts (65). I-2 only inhibits pyruvate kinase phosphatase activity by 15–20% (65, 243) and phosphatase activity toward PF1K, PF2K/F2,6Pase, F1,6Pase, phenylalanine hydroxylase, acetyl CoA carboxylase, and ATP citrate lyase by <10% (8, 38, 65, 243). Fractionation of the extracts by anion exchange chromatography and gel-filtration confirms that PP-2A is the major phosphatase acting on these proteins (38, 243), although PP-2C has significant activity toward PF1K, pyruvate kinase, and acetyl CoA carboxylase (8, 38, 193).

Acetyl CoA carboxylase is also phosphorylated by casein kinase-I and casein kinase-II at sites distinct from those phosphorylated by A-kinase. PP-2A is the major activity that dephosphorylates these sites (244). The protein kinase cascade that inactivates HMG-CoA reductase also inactivates acetyl CoA carboxylase (242) by phosphorylating a distinct serine. The major phosphatase that dephosphorylates this critical residue is also PP-2A (P. Cohen, D. G. Hardie, unpublished).

The high activity of PP-2A in vitro toward the regulatory enzymes of glycolysis/gluconeogenesis, fatty acid synthesis, and amino acid breakdown and its cytosolic location suggests that it may be the major phosphatase acting on these substrates in vivo. The finding that addition of okadaic acid to intact cells rapidly stimulates gluconeogenesis (hepatocytes) and lipolysis (adipocytes), inhibits fatty acid synthesis from acetate (adipocytes), and stimulates phosphorylation of the regulatory enzymes of these pathways (11a), supports the view that PP-2A, rather than PP-2C, is the dominant phosphatase in vivo.

In summary, while progress has been made in identifying the physiological targets of protein phosphatases, the preceding discussion has highlighted the difficulties and pitfalls involved in such analyses. An enzyme that may appear to be the major dephosphorylating activity under certain assay conditions may not be under others, and use of the "native" form of the substrate may be critical to avoid erroneous conclusions.

The realization that the active forms of PP-1 are largely particulate, and the concept of targetting subunits that direct this enzyme to particular locations and modify its substrate specificity, have provided major clues to some of the physiological roles of this enzyme. There is strong evidence that PP-1 plays a major role in the regulation of glycogen metabolism in skeletal muscle and liver and in the dephosphorylation of myosin in striated and smooth muscles. It is implicated in the control of protein synthesis and the meiotic maturation of cells, and is likely to dephosphorylate many membrane-bound proteins, such as the calcium channel in cardiac muscle and HMG-CoA reductase in hepatic microsomes.

PP-2A is likely to dephosphorylate many proteins in vivo, particularly in cytosol where it is usually the principal activity. Nevertheless, definitive evidence for its involvement in any process is still lacking, although the further exploitation of okadaic acid and its analogues may prove useful in defining the physiological roles of PP-1 and PP-2A (11a). The role of PP-2C in vivo is even less clear. Although purified as a myosin light chain phosphatase and a glycogen synthase phosphatase, it does not seem to dephosphorylate these substrates in vivo. When compared to PP-1, PP-2A, and PP-2B, PP-2C has a very low specific activity toward all substrates tested so far, even at optimal Mg^{2+} (10–20 mM). At physiological Mg^{2+} ($\sim$1 mM), the activity would be even lower (8).

Microinjection of protein phosphatases and/or their inhibitors into cells may become a powerful approach for probing their physiological roles, but mutant cells defective in a particular protein phosphatase would be even more valuable. Only one is available at present, namely the BIM G mutant of *Aspergillus,* which has altered cell polarity, is blocked at a late stage of mitosis, and encodes a protein that is strikingly homologous to mammalian PP-1α, as discussed earlier (J. Doonan, R. Morris, personal communication). This indicates that PP-1α plays a role in cell cycle control in *Aspergillus.*

The ppd1 mutant of the yeast *Saccharomyces cerevisae* was reported to be defective in a particular protein phosphatase activity (245), but subsequent biochemical studies failed to confirm these observations and structural analysis of the PPD1 gene now suggests it is not a protein phosphatase (246).

SPECIFICITY AND REGULATION OF PROTEIN PHOSPHATASE-2B

PP-2B from skeletal muscle is completely dependent on Ca^{2+} for activity in the presence or absence of calmodulin, the $A_{0.5}$ for this cation being close to 1 μM. In the presence of saturating concentrations of calmodulin (0.03 μM) V_{max} is increased about 10-fold without any effect on the K_m for substrates (10). The $A_{0.5}$ for calmodulin is 6 nM (10), and there are no obvious differences in calmodulin stimulation among the enzymes from various sources that have been tested (36).

Calmodulin appears to activate PP-2B by neutralizing the inhibitory effect of a 4-kd domain on the A subunit, distinct from the calmodulin-binding domain. This inhibitory domain is extremely susceptible to proteolysis, and its removal yields a Ca^{2+}-dependent enzyme that cannot be further stimulated by calmodulin (36, 247). This suggests that the Ca^{2+}-dependent activity observed in the absence of calmodulin, which is presumably mediated by the binding of Ca^{2+} to the B subunit, might be catalyzed by proteolytically "nicked" enzyme that has lost the inhibitory domain. In this case, the native enzyme would be completely dependent on calmodulin for activity. However, PP-2B from brain only binds to phosphorylated myosin P-light chain–Sepharose in the presence of Ca^{2+} from which it can be displaced with EGTA (248). The binding of Ca^{2+} to the B subunit may therefore be essential for interaction of the A subunit with substrates.

PP-2B from brain binds one mol of calmodulin per mol with very high affinity ($K_d = 0.1$ nM). Activation is reversible, does not involve dissociation of the A and B subunits, but is accompanied by changes in their conformation. Stimulation by Ca^{2+} is cooperative, indicating that more than one mol of Ca^{2+} per mol is required for activation. However, exactly how many Ca^{2+} sites on the B subunit, or on calmodulin, must be occupied for activation is unknown. There is no evidence that PP-2B is regulated in vivo by any mechanism other than Ca^{2+} and calmodulin (36).

The most effective substrates for PP-2B so far identified are proteins that regulate the activities of protein kinases and phosphatases, namely inhibitor-1 (102) and its isoform DARPP (113), the regulatory RII subunit of A-kinase (249), the α subunit of phosphorylase kinase (10), and the calmodulin-dependent cyclic AMP phosphodiesterase (250). Although it is not yet established that these proteins are physiological substrates for PP-2B, they are all substrates for A-kinase. This suggests that PP-2B may allow signals that act via Ca^{2+} to attenuate the actions of cyclic AMP in certain cells, a concept discussed later [see also (36)].

The high levels of PP-2B in the brain suggest that it may have a number of substrates in this tissue. Neuronal substrates dephosphorylated at similar rates to the proteins mentioned above are a particulate protein of unknown function termed protein KF (251), and the microtubule-associated proteins MAP-2 and tau-factor (252). G-substrate, a protein found in the cerebellar Purkinje cells, which is phosphorylated specifically by cyclic GMP–dependent protein kinase, is dephosphorylated by PP-2B somewhat more slowly than DARPP (251).

In contrast to PP-1, PP-2A, and PP-2C, PP-2B does not dephosphorylate, at significant rates, the rate-limiting enzymes of metabolic pathways. These include the enzymes of glycogen metabolism, glycolysis/gluconeogenesis, fatty acid synthesis, cholesterol synthesis, and aromatic amino acid breakdown (6–8, 37, 38, 243). PP-2B is ineffective against protein synthesis initiation factor eIF-2 (6), although ribosomal protein S6 from *Xenopus* ovary is dephosphorylated, albeit at a slower rate than the α subunit of phosphorylase kinase (230).

A peptide corresponding to residues 81–99 of the RII subunit of A-kinase was dephosphorylated by PP-2B with K_m and V_{max} similar to those of the intact protein (249). The structural determinants for substrate recognition must therefore be contained within this sequence. Removal of the N-terminal four residues increased K_m fivefold and decreased V_{max} sevenfold, and removal of additional residues from the N-terminal end continued to worsen its effectiveness as a substrate. However, definitive conclusions about the structural requirements for specificity cannot be drawn from a single example, since the sequence surrounding the phosphoserine on the α subunit of phosphorylase kinase does not show any structural analogy to the RII-subunit peptide (253).

The introduction of PP-2B or its antibodies into cells has been used to obtain preliminary evidence for its involvement in several processes, including regulation of the calcium current in snail neurones (254), dephosphorylation of proteins involved in sperm motility (184), and exocytosis in paramecium (186). Further details are given in (36).

THE CONTROL OF PROTEIN PHOSPHATASE ACTIVITY

Regulation of Protein Phosphatase-1 by Cyclic AMP

CONTROL BY INHIBITOR-1   I-1 is phosphorylated in vitro at a rate comparable to that of the best substrates for A-kinase (104), and no other protein kinase has been found that phosphorylates Thr-35 of I-1 at a significant rate. The phosphorylated form of I-1 inhibits PP-1C with an I_{50} of about 1 nM [e.g. (23, 113)], and kinetic analyses suggest that it inhibits PP-1C by binding to a region distinct from the substrate-binding site (255). The average concentration of I-1 in skeletal muscle is 1.5–1.8 μM (102, 115), about twofold higher than that of all forms of PP-1 combined [estimated from data in (23, 25, 81, 115)]. The concentration of I-1 in rabbit liver is similar (100), and PP-1 activity slightly lower (8), than in skeletal muscle.

Phosphorylase phosphatase activity decreases in perfused rat hind limb muscle, and the amount of a heat-stable phosphatase inhibitor increases, within 2 min of infusion of adrenalin (256). In rabbit skeletal muscle, the heat-stable phosphatase inhibitor generated in response to adrenalin is inactivated by preincubation with protein phosphatase, and copurifies with I-1 (257). The percentage of I-1 in the active (phosphorylated) form increases from 31 $\pm$ 7% in control animals to 70 $\pm$ 12%, after injection of adrenalin in vivo (257). In perfused rabbit hind limb, activated I-1 is <10% that in control animals, increasing to 60% when adrenalin (10–100 nM) is added to the perfusate. Half-maximal effects are observed at about 1 nM adrenalin (258). Similar results are obtained in the perfused rat hemicorpus, where the β-

adrenergic agonist isoproterenol increases the activation state of I-1 from <10% to 70% within a min, without affecting the total amount of I-1 (259).

The amount of active I-1 in perfused rat hind limb doubles from 15 to 30% after 60 min perfusion with 0.5 nM isoproterenol (259). This effect is prevented by 3.3 nM insulin, which also suppresses the 40% rise in cyclic AMP produced by 0.5 nM isoproterenol. Insulin has no effect on I-1 activity or cyclic AMP at ≥10 nM isoproterenol, or in the presence of propranolol, a β-adrenergic antagonist (259). The effect of insulin can therefore be explained by its ability to suppress the rise in cyclic AMP produced by low concentrations of isoproterenol.

The activation state of I-1 in rat epidydamal fat pads increases from 47 ± 5% to 61 ± 5% after exposure to adrenalin and decreases to 27 ± 3% after exposure to insulin (107). The ^{32}P content of I-1 in fat pads in vivo increases by 35% in response to adrenalin and decreases by 30% in response to insulin (107). The high level of I-1 phosphorylation under basal conditions is surprising, since substrates for A-kinase in adipose tissue, such as hormone-sensitive lipase, should be essentially dephosphorylated under these conditions. I-1 may be dephosphorylated very slowly under basal conditions so that phosphorylation is substantial even when A-kinase activity is very low. Alternatively, adipose tissue might contain an I-1 kinase, distinct from A-kinase, that phosphorylates Thr-35. It will be important to establish that alterations in the in vivo ^{32}P-content of adipose I-1 result from changes in the phosphorylation of Thr-35, and not some other residue, since rabbit skeletal muscle I-1 is phosphorylated in vivo on at least one serine residue (Ser-67) (99).

Injection of glucagon (0.1 mg or 1.0 mg) in vivo increases the activation state of I-1 in rabbit liver from 14% to 42% (100). Incubation of rat caudate slices with either 0.1 mM dopamine (which elevates cyclic AMP in these neurones) or cyclic AMP analogues, increases the phosphorylation state of the I-1 isoform DARPP considerably (260). By contrast, depolarizing agents that increase the flux of calcium into neurones do not (110).

In summary, the concentration of I-1 in vivo is higher than that of PP-1 in several tissues, and its activation state is modulated by hormones that increase or decrease the intracellular concentration of cyclic AMP. Changes in the level of I-1 phosphorylation are therefore likely to be accompanied by alterations in the activity of PP-1 in vivo. The physiological relevance of I-1 has been questioned (261, 262) because the native form of PP-1 in skeletal muscle (i.e. PP-1G) is inhibited less effectively [less rapidly (23)] than PP-1C in vitro. However, these arguments are invalid since PP-1C dissociates from PP-1G in response to adrenalin, as discussed below. Nevertheless, some forms of PP-1 may be resistant to I-1 in vivo.

I-1 can be dephosphorylated by PP-1 in vitro, but in contrast to other substrates, the presence of Mn^{2+} is essential (102, 255). The $A_{0.5}$ for Mn^{2+} is

0.2 mM (263), much higher than the free concentration of Mn^{2+} in vivo [<1 μM (264)]. PP-1 may not therefore be an I-1 phosphatase in vivo. The only enzymes with significant I-1 phosphatase activity in the absence of Mn^{2+} are PP-2A and PP-2B (8, 37). PP-2A is likely to be the major I-1 phosphatase at basal levels of Ca^{2+}, while PP-2B accounts for 70–90% of the I-1 phosphatase activity in muscle and brain extracts in the presence of Ca^{2+} (8, 177).

The dephosphorylation of I-1 and certain other proteins by PP-2B may allow signals that act via Ca^{2+} to attenuate the action of cyclic AMP in certain cells (Figure 1). The dephosphorylation of I-1 should activate PP-1, promoting the dephosphorylation of substrates for A-kinase. Similarly, dephosphorylation of the RII subunit of A-kinase by PP-2B (249) facilitates its reassociation with the catalytic subunit to reform the inactive holoenzyme, further antagonizing the action of cyclic AMP. An isoform of calmodulin-dependent cyclic AMP phosphodiesterase is phosphorylated by A-kinase in vitro, increasing the $A_{0.5}$ for calmodulin 15-fold (250, 265, 266). Dephosphorylation (reactivation) is catalyzed by PP-2B at a rate similar to that of the α subunit of phosphorylase kinase (250).

Although the physiological significance of the hypothesis illustrated in Figure 1 has still to be established, one situation where it might operate is in the dopaminoceptive neurones of the brain, where I-1 is replaced by DARPP. In these neurones, dopamine acting through cyclic AMP increases the phosphorylation states of several proteins, including DARPP, and prevents glutamate (acting through Ca^{2+}) from depolarizing certain neurones (113, 267). The ability of dopamine to decrease the firing rate of these neurones

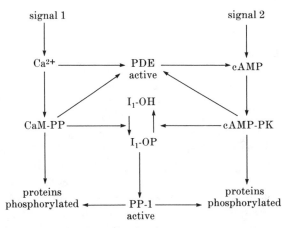

Figure 1   Mechanisms by which signals that act through Ca^{2+} may attenuate signals that act via cyclic AMP (cAMP). Abbreviations: phosphodiesterase, PDE; calmodulin-dependent protein phosphatase (i.e. PP-2B), CaM-PP; dephosphorylated and phosphorylated forms of inhibitor-1, I_1-OH, I_1-OP; protein phosphatase-1, PP-1; cyclic AMP–dependent protein kinase, cAMP-PK.

relatively slowly may involve the phosphorylation of DARPP. Conversely, termination of the action of dopamine by glutamate may be achieved through the activation of PP-2B. In support of this hypothesis, PP-2B and DARPP have a strikingly similar localization in basal ganglia and neurones of the caudate projecting to the globus pallidus and substantia nigra (111), see also (36), and are present at very high concentrations in these regions.

In skeletal muscle, the dephosphorylation of I-1 and the α subunit of phosphorylase kinase by PP-2B may represent devices for deemphasizing the effects of adrenalin during muscle contraction, ensuring that glycogen is not depleted too rapidly during nerve stimulation at low frequencies. This concept is discussed in greater detail in (36).

In summary, I-1 appears to have four potential roles in vivo.

1. It provides a mechanism for amplifying the effects of cyclic AMP, since many substrates for PP-1 are themselves phosphorylated by A-kinase.
2. Inactivation of PP-1 by I-1 minimizes the loss of ATP inherent in any phosphorylation-dephosphorylation cycle, which may be critical in some situations, such as the adrenergic control of glycogenolysis in skeletal muscle (see below).
3. I-1 allows cyclic AMP to increase the phosphorylation of proteins that are not substrates for A-kinase, since PP-1 is active toward the substrates of other protein kinases (see below).
4. The dephosphorylation of I-1 by PP-2A and PP-2B allows PP-1 activity to be controlled by substances that regulate these other protein phosphatases. This protein phosphatase cascade may allow calcium to attenuate the effects of cyclic AMP in certain cells.

PHOSPHORYLATION OF THE GLYCOGEN-BINDING SUBUNIT   The G subunit of PP-1G is phosphorylated very rapidly by A-kinase (23, 56) and the sequences surrounding each of the two major phosphoserine residues (site-1 and site-2) have the structure Arg-Arg-X-Ser-Y (268), frequently found at phosphorylation sites of physiological substrates for A-kinase (1). Site-2 is located only 19 residues C-terminal to site-1 in the primary structure (268a).

Phosphorylation of the G subunit does not alter the phosphorylase phosphatase activity of PP-1G under standard assay conditions, but causes dissociation from PP-1C in vitro (56, 268a), accounting for earlier observations that phosphorylation increases the rate of inactivation of PP-1G by I-1 (23) and decreases the proportion of PP-1 activity that sediments with glycogen (54). Identical effects are observed in vivo in response to adrenalin. Injection of the hormone into rabbits decreases PP-1 activity associated with glycogen by 50%, with a corresponding increase in the cytosolic content, while the G subunit remains attached to glycogen (54). Phosphorylation of site-1 increases from 0.56 to 0.83 mol/mol in vivo after injection of adrenalin (55). The high level of site-1 phosphorylation under basal conditions may be explained by the finding that this phosphoserine is resistant to PP-1 in vitro (56). Alternatively, site-1 may be phosphorylated by a further enzyme distinct from A-kinase. Site-2 is dephosphorylated by PP-1 in vitro at a low rate (56). Its in vivo phosphorylation state is unknown.

In summary, adrenalin promotes the release of PP-1C from the G subunit in vivo, causing its translocation from the glycogen-protein particles to the cytosol. This should prevent PP-1C from dephosphorylating phosphorylase a and glycogen synthase, which remain bound to glycogen (54). PP-1C released into the cytosol is presumably inactivated by I-1, preventing the dephosphorylation of cytosolic proteins (Figure 2). Inhibition of PP-1 can explain how high levels of phosphorylase a are formed in resting muscle in response to adrenalin, despite the very low activity of phosphorylase kinase activity under these conditions [reviewed in (206)]. It can also explain how adrenalin increases the phosphorylation of glycogen synthase on serine residues that are not phosphorylated by A-kinase (205, 269).

The concentration of phosphorylase in mammalian skeletal muscle is almost 0.1 mM. The cyclic phosphorylation and dephosphorylation of this enzyme could therefore exhaust ATP levels very quickly, unless it were regenerated via glycolysis. It may therefore be critical to inactivate PP-1 during adrenergic stimulation of resting muscle, since ATP formation via glycolysis is very slow under these conditions [discussed in (206)].

INHIBITION BY THE REGULATORY SUBUNIT OF A-KINASE   The regulatory (R) subunit of A-kinase was reported to inhibit the dephosphorylation of phosphorylase a by a phosphatase preparation that could have been PP-1C, PP-2AC, or a mixture (270). The catalytic (C) subunit of A-kinase or the RC complex were about fivefold less effective as inhibitors. Since interaction of cyclic AMP with the R subunit promotes dissociation from the C subunit, it was suggested that the phosphatase might be inhibited in vivo by R subunit in response to cyclic AMP. However the I_{50} for R subunit was 5 μM (270), 10-fold higher than its intracellular concentration in any mammalian tissue (271). The A-kinase used in these studies was isolated from rabbit skeletal muscle, which contains predominantly the RI subtype, but similar results (I_{50} = 10 μM) were subsequently obtained using RII subunit from bovine heart and PP-1C from rabbit skeletal muscle (272).

More recently, the RII subunit was found to inhibit skeletal muscle PP-1G (273). Figure 1 of this report shows ~20% inhibition by 1.2 μM RII subunit from bovine heart, similar to earlier work (270, 272). However, RII subunit from horse heart appears to be a stronger inhibitor (I_{50} = 0.4 μM). Other

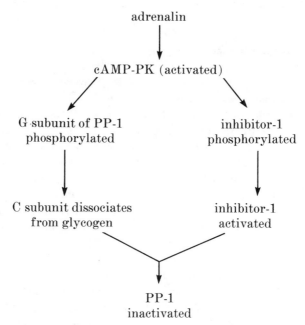

Figure 2   Molecular mechanisms by which adrenalin inactivates the glycogen-bound form of protein phosphatase-1(PP-1) in skeletal muscle.

investigators found that 1 μM RII subunit from bovine heart inhibited PP-1C by 30%, but also reported that 0.1 μM RII inhibited by 50% the rate of activation of PP-1I (4 nM) by GSK3 (274). It was suggested that interaction of the RII subunit with PP-1I prevents activation by GSK3 (274). However, a simpler interpretation is that the RII subunit inhibits I-2 phosphorylation by a competitive mechanism, since it is also a substrate for GSK3 (275). Whatever the explanation, these observations are unlikely to have physiological relevance, as the molar concentration RII:PP-1I in skeletal muscle is 1:7, not 25:1.

The RII subunit is a more potent inhibitor of PP-1 (273, 276) and PP-2A (276) if it is phosphorylated (at Ser-95) by the C subunit of A-kinase. In these experiments, the RII subunit was thiophosphorylated to produce a derivative resistant to dephosphorylation, and the reported I_{50} values were in the range 0.05–0.15 μM. However, phosphorylation of any substrate will inhibit the dephosphorylation of others by simple competition. No evidence was presented that thiophosphorylation of the RII subunit increases its effectiveness as an inhibitor to a greater extent than thiophosphorylation of any other substrate. Furthermore, the extent of phosphorylation of Ser-95 in vivo is unclear.

In summary, inhibition of PP-1 by the R subunit of A-kinase is an attractive idea, analogous to the regulation of PP-1 by I-1, but convincing evidence for its operation in vivo is lacking.

Regulation of Protein Phosphatase-1I

Activation of the cytosolic form of PP-1 (PP-1I) is complex and not yet fully understood, but the following events appear to take place during the activation-deactivation cycle (85, 89, 277, 278, 280, 281).

1. Phosphorylation of Thr-72 on I-2 triggers a change in the conformation of PP-1C, converting it from an inactive to an active conformation.
2. Activated PP-1C dephosphorylates Thr-72 by an intramolecular reaction. Until this has happened PP-1C cannot dephosphorylate exogenous substrates, presumably because phosphorylated I-2 occupies the active site. Thus dephosphorylation of I-2, as well as phosphorylation, is necessary for activation.
3. Following dephosphorylation, I-2 may transfer to an inhibitory site (80, 117) where it induces a slow ($t \approx 30$ min) conversion of PP-1C to the inactive conformation.

I-2 from rabbit skeletal muscle is phosphorylated by casein kinase-II (52) on serines 86, 120, and 121 (282), at a rate comparable to those of other physiological substrates (282). This does not activate PP-1I directly, but enhances the rate of phosphorylation of Thr-72 (52, 263) and rate of activation (52) by GSK3. I-2 is also phosphorylated in vitro by A-kinase without any effect on PP-1I activity (52, 277). Phosphorylation occurs in the N-terminal cyanogen bromide fragment (282).

Two forms of PP-1I were resolved on phenyl-Sepharose or blue-Sepharose. The species eluting later on phenyl-Sepharose and not retained by blue-Sepharose was activated by MgATP and GSK3, while the other was not (86). The reason why the latter cannot be activated is unknown, although prior phosphorylation of PP-1I by casein kinase-I in vitro has been reported to prevent activation by MgATP and GSK3 (283).

Analysis of the in vivo phosphorylation state of rabbit skeletal muscle I-2 by fast atom bombardment mass spectrometry revealed that the only residues phosphorylated in normally fed animals are Ser-86 (0.7 mol/mol), Ser-120 (0.3 mol/mol), and Ser-121 (0.3 mol/mol) (284), indicating that phosphorylation by casein kinase-II occurs in vivo. Injection of adrenalin or insulin in vivo under conditions that cause maximal inactivation and activation of glycogen synthase, respectively, does not alter the phosphorylation states of these residues. No residue in the N-terminal cyanogen bromide peptide is phosphorylated in response to adrenalin (284), demonstrating that I-2 is not phosphorylated by A-kinase in vivo. Phosphorylation of Thr-72 is not observed under any condition (284), although very low levels of phosphorylation ($<5\%$) might have escaped detection.

The phosphorylation state of I-2 has also been measured in mouse diaphragm (285) and rat adipocytes (286) following incubation with [^{32}P]Pi and immunoprecipitation of I-2 from the extracts. About 90% (285) or $>95\%$ (286) of the [^{32}P]-radioactivity is present as phosphoserine and the remainder as phosphothreonine. The large amounts of phosphoserine are consistent with the studies of rabbit skeletal muscle I-2 (284), but more work is needed to establish whether the small amounts of phosphothreonine detected are located at Thr-72 or some other site, or derived from traces of contaminating proteins. In contrast to the results obtained with rabbit skeletal muscle I-2, insulin was reported to increase the content of ^{32}P-phosphoserine of mouse diaphragm I-2 by 18 $\pm$ 6% after five min (286).

PP-1I is of considerable interest as the first example of a protein phosphatase that is activated by a protein kinase in vitro, but its physiological role is unclear. In skeletal muscle, PP-1G is likely to be the form of PP-1 involved in the regulation of glycogen metabolism, and PP-1M the form that dephosphorylates myosin. In other tissues the active forms of PP-1 are also largely particulate. Thr-72 of I-2 is barely phosphorylated in vivo, if at all, nor is there any evidence that hormones regulate the phosphorylation of this residue or the activity of PP-1I. The unique mechanism of activation of PP-1I means that in vivo activity cannot be deduced from the in vivo phosphorylation state of I-2. Furthermore, the proportion of active and inactive PP-1I in tissue extracts has not yet been determined, although the recent report that low concentrations (0.3 mM) of NaF and phenylmethanesulfonylfluoride prevent the inactivation of PP-1I (286a) may facilitate such measurements. On the other hand, it is difficult to believe that PP-1I has no physiological role, because this enzyme and its activation mechanism are even present in invertebrates (29, 31) and have therefore been conserved over a wide span of evolution.

It is possible that the function of I-2 is simply to act as a storage site, or a buffer, for PP-1C when it is not bound to targetting subunits. Additionally, I-2 might inactivate PP-1C released adventitiously into the cytosol from particulate fractions, as a result of the proteolysis of targetting subunits, such as the G subunit. Alternatively, newly synthesized PP-1C might be in an inactive conformation, activation requiring interaction with I-2 and the reversible phosphorylation of this protein, prior to the delivery of PP-1C to a targetting protein and insertion at the correct subcellular location. The properties of this intriguing system seem more appropriate to storage, protection, and repair, rather than hormonal regulation.

Thr-72 is dephosphorylated efficiently in vitro by PP-1, PP-2A, and PP-2B

(117). However, if all the I-2 in cells is complexed to PP-1, as seems likely (86, 287), Thr-72 (if it is phosphorylated in vivo) is probably dephosphorylated intramolecularly by PP-1. PP-2A appears to be the major phosphatase acting on the serine residues phosphorylated by casein kinase-II (288). PP-1 does not dephosphorylate these residues at a significant rate, except in the presence of protamine (288).

Regulation of Hepatic Protein Phosphatase-1

CONTROL BY PHOSPHORYLASE a   The activation (dephosphorylation) of glycogen synthase observed in liver extracts only occurs after a lag period, corresponding to the time required for reconversion of phosphorylase a to b (289). The key factor preventing reactivation of glycogen synthase is phosphorylase a, since the lag can be shortened by addition of substances that accelerate the conversion of phosphorylase a to b (glucose, caffeine), abolished if phosphorylase a is removed by immunoprecipitation, or prolonged by addition of more phosphorylase a (but not phosphorylase b) [reviewed in (290, 291)].

The glycogen synthase phosphatase sensitive to inhibition by phosphorylase a is associated with hepatic glycogen (292) and is a Type 1 protein phosphatase (24, 65, 209, 210), termed here hepatic PP-1G. The apparent resolution of glycogen synthase phosphatase from phosphorylase phosphatase by ion exchange chromatography (292) or gel-filtration (293) can be explained by the separation of proteolytically modified and unmodified forms of PP-1, which have very different phosphorylase phosphatase/glycogen synthase phosphatase activity ratios (24).

Since hepatic PP-1G dephosphorylates phosphorylase a as well as glycogen synthase, inhibition by the former protein could occur through simple competition. Four lines of evidence demonstrate that the effect is much too powerful to be explained in this way (294).

1. The I_{50} for phosphorylase a is in the nanomolar range, more than 1000-fold lower than its K_m as a substrate.
2. Tryptic digestion increases the I_{50} for phosphorylase a 1000-fold, without affecting the K_m for phosphorylase a as a substrate.
3. The dephosphorylation of glycogen synthase by PP-1G from skeletal muscle is only inhibited by phosphorylase a at concentrations 1000-fold higher than those that inhibit hepatic PP-1G.
4. Phosphorylase a does not inhibit the dephosphorylation of glycogen synthase by hepatic PP-2A.

These observations suggest that inhibition of glycogen synthase dephosphorylation by phosphorylase a is an allosteric effect. By analogy with

PP-1G from skeletal muscle, hepatic PP-1G might also be composed of PP-1C complexed to a G subunit. However, either the hepatic G subunit is different or an additional component is present, to explain the remarkable sensitivity of hepatic PP-1G to phosphorylase *a* (294) and its very high glycogen synthase phosphatase activity (209). The inhibition of hepatic PP-1G by phosphorylase *a* is a mechanism for inhibiting glycogen synthesis as glycogenolysis is activated and vice versa. The level of phosphorylase *a* in vivo is increased by signals that elevate cyclic AMP or Ca^{2+} (Figure 3) and

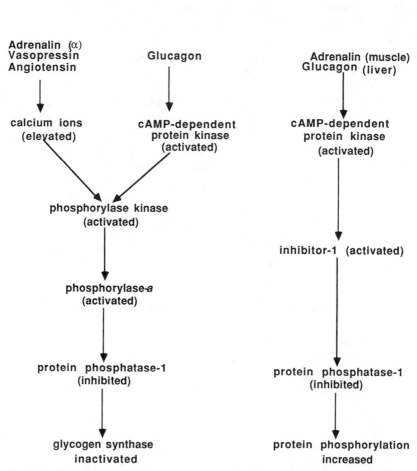

Figure 3   Allosteric inhibition of hepatic protein phosphatase-1G promotes the phosphorylation (inactivation) of glycogen synthase in response to calcium or cyclic AMP, and is analogous to inhibition by inhibitor-1.

decreased by insulin and glucose. Inhibition by phosphorylase *a* is analogous to inhibition by I-1 in that both mechanisms allow a protein kinase to inhibit PP-1. The major difference is that phosphorylase *a* inhibits PP-1 in response to signals that elevate cyclic AMP or Ca^{2+}, whereas I-1 only inhibits PP-1 in response to cyclic AMP–elevating agents (Figure 3).

The phosphorylation of rat liver glycogen synthase increases in vivo in response to glucagon, phosphorylation occurring primarily in a peptide distinct from that labeled by A-kinase in vitro (295). Since rat liver lacks I-1 (100), inhibition of hepatic PP-1G by phosphorylase *a* is the simplest mechanism that can explain these findings (Figure 3).

Many reports of the activation of hepatic glycogen synthase phosphatase by insulin are explained by the ability of this hormone to decrease phosphorylase *a* (296–299). Insulin exerts its effect by lowering cyclic AMP levels that have been increased by cyclic AMP–elevating hormones, and by antagonizing the actions of hormones that mobilize intracellular calcium [reviewed in (299)]. The binding of glucose to phosphorylase *a* stimulates conversion to phosphorylase *b* by PP-1, both inhibiting glycogenolysis and activating glycogen synthesis in response to elevated blood glucose (290, 291).

Inhibition of hepatic PP-1G by phosphorylase *a* only operates in the fed state. In fasted animals, glycogen synthase is largely dephosphorylated, despite a high level of phosphorylase *a* (300, 301). These observations are explained by the very low levels of glycogen in fasted animals. The dephosphorylation of glycogen synthase in extracts prepared from glycogen-depleted livers is not inhibited by phosphorylase *a*, but inhibition is restored when extracts are prepared in a glycogen-containing buffer (301). The I_{50} of hepatic PP-1G for phosphorylase *a* increases 20-fold following digestion of the glycogen fraction with α-amylase, and sensitivity to phosphorylase *a* can be restored by readdition of glycogen (209). The phosphorylase *a*–glycogen complex is the physiologically relevant inhibitor of hepatic PP-1G (209).

Although I-1 is absent in rat liver, it is present at high levels in rabbit and other mammalian livers (100). Hepatic PP-1G is as sensitive to phosphorylase *a* in rabbits as in rats (D. Schelling, P. Cohen, unpublished), demonstrating that both mechanisms for inhibiting PP-1 coexist in rabbit liver. The major role of I-1 in liver may be to inhibit other forms of PP-1, or to inhibit the dephosphorylation of phosphorylase *a* by hepatic PP-1G.

The major protein phosphatase associated with hepatic microsomes is also a Type 1 enzyme (24, 209, 210). It has a fivefold lower glycogen synthase phosphatase/phosphorylase phosphatase activity ratio than does hepatic PP-1G, greater sensitivity to I-1 and I-2, and does not bind to glycogen (209). The dephosphorylation of glycogen synthase by the hepatic microsomal enzyme is inhibited by the phosphorylase *a*–glycogen complex in a manner similar to that of hepatic PP-1G (24, 209). However, there is currently no evidence that this mechanism is involved in regulating the phosphorylation states of microsomal proteins. The dephosphorylation of two microsomal phosphoproteins, HMG-CoA reductase and ribosomal protein S6, by microsomal PP-1 is not inhibited by phosphorylase a (209). How glucagon stimulates the phosphorylation (inactivation) of HMG-CoA reductase in rat liver (241) is unknown. The finding that glucagon decreases the phosphorylase phosphatase activity of rat hepatic PP-1 by 20% within five min (301a) suggests that yet a further regulatory mechanism may be operating.

INHIBITION OF HEPATIC PP-1G BY CALCIUM   The dephosphorylation of glycogen synthase by rat hepatic PP-1G is inhibited by calcium ions (302, 303). Inhibition is maximal at 0.3 μM calcium, and greatly enhanced by increasing glycogen in the assays from 1.5 to 20 mg/ml. The effects are greater using hepatic glycogen synthase, rather than muscle glycogen synthase. The dephosphorylation of phosphorylase a is unaffected by calcium. At 0.3 μM calcium and 20 mg/ml glycogen, reactivation of liver glycogen synthase is inhibited by 85%. Inhibition of hepatic PP-1G by calcium could contribute to the increased phosphorylation of hepatic glycogen synthase observed in response to calcium-mobilizing hormones. Inhibition by calcium is unaffected by calmodulin antagonists or by addition of calmodulin to the assays.

EFFECTS OF DIABETES, INSULIN, AND STARVATION   Treatments that cause long-term deprivation of insulin (severe alloxan-induced diabetes, 48 h starvation after adrenalectomy) prevent the activation of glycogen synthase (212, 213, 213a, 304–306) by decreasing hepatic PP-1G. Restoration of glycogen synthase phosphatase activity can be achieved by giving insulin to diabetic animals for 68 h (recovery is negligible after 20 h and only partial after 44 h), or by administration of glucose or glucocorticoids to starved, adrenalectomized animals. The effects of glucocorticoids are prevented by actinomycin D, suggesting that mRNA synthesis is required. Synthesis of at least one component of hepatic PP-1G is therefore controlled by insulin and glucocorticoids, perhaps that responsible enhancing its glycogen synthase phosphatase activity (209). Full restoration of synthase phosphatase activity in hepatocytes isolated from diabetic animals requires cortisol and thyroid hormone, in addition to insulin (307). Synthase phosphatase activity is elevated 60% in the livers of hyperthyroid rats and decreased by 40% in hypothyroid animals (307a).

Small decreases in the phosphorylase phosphatase activity of rat liver extracts prepared from diabetic animals that were reversible by insulin administration (304, 308), or small increases by administration of insulin to normal animals (309), were observed by some investigators, but not others (310, 311). Subsequently, a twofold decrease in the phosphorylase phosphatase activity of PP-1 was observed in liver extracts prepared from alloxan diabetic rats (312). There was no change in PP-1 activity in skeletal muscle and no change in hepatic or muscle PP-2A. The effect on hepatic PP-1 is not explained by increased levels of I-1 or I-2. A larger (62–79%) decrease in the phosphorylase phosphatase activity of rat hepatic PP-1 in alloxan diabetic animals, with no change in PP-2A, was reported by others (313). However, in these studies assays were only made after the cytosol had been subjected to anion exchange chromatography. Activatable PP-1I was also decreased to some extent. Administration of insulin to the diabetic animals for only five min was reported to increase active PP-1 by 45% and activatable PP-1I by 36% (313).

A 35% increase in the phosphorylase phosphatase activity of hepatic PP-1 was also observed by other investigators within five min of an intravenous injection of insulin plus glucose. The effects were preserved after gel-filtration, but lost when the extracts were diluted (301a).

Incubation of Swiss mouse 3T3-D1 cells with insulin, EGF, or PDGF was reported to increase the phosphorylase phosphatase activity of PP-1 by 40% within five min, provided that glycogen (2 mg/ml) was included in the extraction buffer (314). Other investigators reported that insulin (but not EGF) increased the S6 phosphatase activity of PP-1 by 60% in Swiss mouse 3T3 cells, but the effects were only observed after exposure to the hormone for 2 h (231).

Taken together, these observations suggest that the activity of PP-1 can be altered acutely by insulin, but further work is needed to elucidate the molecular basis for these effects.

THE DEINHIBITOR   A protein isolated from the glycogen fraction of dog liver prevents inactivation of PP-1 by I-1 or I-2 (315). This "deinhibitor" can be largely resolved from PP-1 by anion exchange chromatography, but only in the presence of thiols. Like I-1 and I-2, the deinhibitor is not destroyed by heat, acid, or SDS. Its apparent molecular mass on SDS-PAGE is 9 kd (316).

The deinhibitor does not activate PP-1I, but in its presence PP-1I attains a twofold higher activity following incubation with GSK3 and MgATP (316, 317). The deinhibitor protein also stimulates the phosphorylase phosphatase activity of microsomal PP-1 (fivefold) and cytosolic PP-1 (sevenfold) (210). Since hepatic microsomes do not contain significant amounts of I-1 or I-2, the deinhibitor must activate microsomal PP-1 by neutralizing a different inhibitory subunit. The deinhibitor hardly affects the phosphorylase phosphatase activity of hepatic PP-1G, perhaps because it is already bound to this enzyme. Surprisingly, the deinhibitor does not stimulate the glycogen synthase phosphatase activity of PP-1 in rat liver cytosol, microsomes, or glycogen-particles (210). At concentrations 100-fold higher than those required to prevent inhibition by I-1 and I-2, the deinhibitor allows PP-1 to de-phosphorylate I-1 in the absence of Mn^{2+} (317, 318). No effect of the deinhibitor on PP-2A activity has been detected (319).

The deinhibitor is inactivated by incubation with A-kinase and MgATP (320) and reactivated by PP-2A (320–322). Complete inactivation of purified deinhibitor was accompanied by the incorporation of only 0.02 mol phosphate/9 kd protein (320). To account for this finding, it was suggested that the deinhibitor preparation might be 98% phosphorylated as isolated, but the effects of PP-2A on activity (320) seem to exclude this possibility. It is more likely that the deinhibitor is only a minor component of the final preparation.

In summary, the deinhibitor may be a subunit of hepatic PP-1G, and responsible, at least in part, for its relative insensitivity to I-1 and I-2. It does not appear to be responsible for the high synthase phosphatase activity of hepatic PP-1G. The effects of A-kinase and MgATP suggest that the de-inhibitor might modulate the response of hepatic PP-1G to I-1. However, its regulation by phosphorylation has not yet been demonstrated in vivo. Further-more, the deinhibitor may merely be a proteolytic fragment of another protein, such as the G subunit [see (23)].

Phosphorylation of Protein Phosphatase-1 by pp60^{v-src}

PP-1C is phosphorylated in vitro by the protein tyrosine kinase pp60^{v-src} (323). Under optimal conditions phosphorylation reached 0.35 mol/mol protein and phosphorylase phosphatase activity decreased by 50%. Subsequent experiments showed that phosphorylation to 0.22 mol/mol increased the K_m for phosphorylase a, phosphorylase kinase, and glycogen synthase by 1.5–2-fold (324). Phosphorylation did not prevent inactivation by I-1 or the recombination of PP-1C with I-2 to form PP-1I. However, pp60^{v-src} could not phosphorylate PP-1I, before or after activation by GSK3 and MgATP, suggesting that steric hindrance by I-2, rather than an altered conformation of PP-1C, prevented phosphorylation. These observations suggest that the tyro-sine phosphorylation site may be close to a binding site for I-2. Conversion of PP-1C from the 37-kd to the 33-kd form by limited proteolysis released all the phosphate, and the 33-kd species was not a substrate for pp60^{v-src}. The phosphotyrosine must therefore be near the C-terminus, perhaps Tyr-304 of PP-1α.

Phosphorylation of PP-1C is slow, even at high concentrations of pp60^{v-src}. PP-1I is not a substrate, nor is it known whether other forms of PP-1 that occur in vivo (PP-1G, PP-1M, etc) can be phosphorylated. Furthermore, no evidence that PP-1 is phosphorylated on a tyrosine residue in vivo has been presented. Cell transformation by Rous sarcoma virus is accompanied by large increases in serine and threonine phosphorylation of cellular proteins (325, 326), but evidence that this is mediated by pp60^{v-src}-catalyzed inhibition of PP-1 is lacking.

Regulation of Protein Phosphatase-2A

ROLE OF THE A AND B SUBUNITS   Interaction of the A subunit with PP-2AC suppresses activity toward some substrates, such as phosphorylase a (21), without affecting activity toward others, such as spectrin (21). Further addition of the B subunit suppresses activity toward a variety of substrates (e.g. phosphorylase a, glycogen synthase, and spectrin) to a greater extent (21, 153). Consistent with these findings, PP-2A1 has a much lower activity toward myosin and phosphorylase a than PP-2A2 (146, 148), while selective removal of the B subunit from PP-2A1 by limited proteolysis greatly enhances activity toward myosin and phosphorylase a (162). By contrast, the troponin I phosphatase activity of PP-2A1 is slightly higher than that of PP-2A2 (146). PP-2A0, in its native state, has extremely low activity (37, 243), implying that the B' subunit suppresses activity more strongly than does the B subunit. The B'' subunit (Table 2) also suppresses the activity toward phosphorylase a, spectrin, and histone H1 more strongly than does the B subunit (21).

Certain procedures used in the purification of PP-2A, such as precipitation with ammonium sulfate, may weaken the inhibitory effects of the A and B/B'/B'' subunits on the C subunit, yielding preparations whose basal activity is much higher than those of the native enzymes (discussed in (135)]. Activation without dissociation to the C subunit can also result from other harsh treatments, such as freezing-thawing (18).

ACTIVATION BY BASIC PROTEINS AND POLYAMINES   A heat-stable protein in rabbit kidney cortex (327), subsequently identified as histone H1 (328), was found to stimulate kidney phosphorylase phosphatase activity. These observations led to the discovery of "latent" phosphorylase phosphatases in skeletal (17) and smooth muscle (16) that were markedly activated by histone H1 and other highly basic polypeptides, such as protamine and polylysine. These enzymes were insensitive to I-1 and I-2 (329), dephosphorylated the α subunit of phosphorylase kinase preferentially (330), and were subsequently identified as forms of PP-2A (18, 135). In one study, histone H1, protamine, and polylysine stimulated the dephosphorylation of all nine substrates tested, and activation with some substrates, such as phosphorylase kinase (10–30-fold), glycogen synthase labeled in sites-3 (9–50-fold), and phenylalanine hydroxylase (13–230-fold), was much greater than with phosphorylase a (2.5–8-fold) (18).

In general, the effects of basic proteins are mediated by interaction with the phosphatase and not its substrates (135, 157). PP-2A1 is stimulated to a greater extent than PP-2A2 with most, but not all, substrates (18). Activation is not accompanied by dissociation to PP-2AC (18), and the latter is stimulated by basic proteins to a much smaller extent than PP-2A1 and PP-2A2 (135). Basic proteins usually elicit a more striking activation than treatments that promote dissociation to PP-2AC (18). These observations suggest that basic proteins exert their effects by interaction with the A and B subunits.

Polylysine and protamine are not constituents of mammalian cells, and histone H1 is located in the nucleus. A nuclear form of PP-2A has been identified (68), but whether it would be affected by histone H1 complexed to DNA in chromatin is unknown. Most probably, basic polypeptides are not activators in vivo, although they might mimic another substance that regulates PP-2A in vivo.

The polyamine, spermine, stimulates PP-2A in vitro (208). Half-maximal activation is observed at 0.2 mM, with optimal effects at 2 mM. Above 2 mM spermine is inhibitory. Spermine is a more potent activator than spermidine, while putrescine is ineffective. In general, substrates whose dephosphorylation is stimulated to the greatest extent by spermine are those whose dephosphorylation is increased most markedly by basic proteins. The dephosphorylation of glycogen synthase (phosphorylated at sites-3) by PP-2A0 or PP-2A1 is stimulated 8–15-fold and the dephosphorylation of phenylalanine hydroxylase 6–7-fold. With five other substrates, activation of PP-2A1 ranges from 1.3 to 3-fold. Activation of PP-2A0, 2A1, and 2A2 by spermine is greater than that of PP-2AC.

The activity of PP-1G or PP-1C toward glycogen synthase (phosphorylated at sites-3) is stimulated two to three-fold by 1–2 mM spermine, with half-maximal effects at 0.1–0.2 mM. By contrast, the dephosphorylation of some other substrates, including glycogen synthase phosphorylated at site-2, is inhibited (208). The sites-3 region of skeletal muscle glycogen synthase is dephosphorylated relatively specifically and within minutes in response to insulin (331), and in isolated rat diaphragm activation can be mimicked by addition of putrescine or spermidine (the precursors of spermine) to the perfusate (332). It has therefore been speculated that activation of PP-1 and/or PP-2A by spermine might play a role in the stimulation of glycogen synthase by insulin (208). However, evidence to support (or refute) this idea is lacking. Polyamines are present in large amounts in cells, and the concentration of spermine would be in the millimolar range, if it were distributed uniformally. However, polyamines are mostly bound to nucleic acids and proteins (333, 334), the free concentration of spermine in cells is unknown, nor is it known whether the free concentration fluctuates in response to extracellular signals.

In contrast to other substrates, spermine inhibits the dephosphorylation of phosphorylase a by PP-2A. This is explained by the interaction of spermine with phosphorylase (208, 335).

Purified PP-2A2 and PP-2AC were reported to be activated up to four-fold by submicromolar concentrations of insulin in vitro (335a). However, other investigators observed only a twofold activation of PP-2AC by insulin, and failed to see any stimulation of purified PP-2A0, PP-2A1, and PP-2A2 (135).

Insulin binds very tightly to PP-1C, but without affecting its activity (137). It is unlikely that these direct effects of insulin on purified phosphatases have physiological significance.

In summary, it appears that the native forms of PP-2A are present in cells as low-activity forms, which have a tremendous potential for activation. However, whether the full activity of this enzyme is ever expressed in vivo is unclear, and with the possible exception of spermine, no candidates for physiological activators have yet emerged.

Regulation of Protein Phosphatase-2C

PP-2C is dependent on Mg^{2+}, but this cation is unlikely to regulate activity, since its free concentration does not fluctuate significantly in vivo. Furthermore, the $A_{0.5}$ for Mg^{2+} in tissue extracts is 1 mM (8), similar to the free concentration in vivo, indicating that PP-2C may only be operating at 50% of V_{max}.

Incubation of rat hepatocytes with insulin was reported to cause a transient rise in a Mg^{2+}-dependent pyruvate kinase phosphatase activity, which was maximal about two min after addition of hormone (336). Activation persisted after gel-filtration and appeared to result from a decrease in the $A_{0.5}$ for Mg^{2+}, resulting in 70% activation at 1 mM Mg^{2+}. Further work is required to establish whether this phosphatase is PP-2C and the molecular mechanism of activation.

Other Regulatory Mechanisms

CHANGES IN SUBSTRATE CONFORMATION   Substances that interact with PP-1, PP-2A, and PP-2C directly will alter the rates of dephosphorylation of many proteins. However, selective effects can be achieved by the interaction of metabolites with substrates, and several examples have already been discussed. The binding of glucose to hepatic phosphorylase a stimulates its dephosphorylation by PP-1 or PP-2A (290, 291), while AMP, an allosteric activator of phosphorylase, inhibits the dephosphorylation of this protein selectively (243, 337). Interaction of cyclic AMP with the RII subunit of A-kinase promotes dissociation from the C subunit, and increases 5–10-fold the rate of dephosphorylation of RII by PP-2AC, without affecting the dephosphorylation of other substrates (129).

SUBSTRATE COMPETITION   Inhibition of the dephosphorylation of one protein by the phosphorylation of another, when both phosphoproteins compete for the active site of the same phosphatase, could be a mechanism for increasing the phosphorylation states of some proteins in vivo. A possible example of this phenomenon is the inhibition by phosphorylase kinase of the dephosphorylation of phosphorylase a by muscle PP-1 (338–340).

SUMMARY

Four major serine/threonine-specific protein phosphatase catalytic subunits are present in the cytoplasm of animal cells. Three of these enzymes, PP-1, PP-2A, and PP-2B, are members of the same gene family, while PP-2C appears to be distinct. PP-1, PP-2A, and PP-2B are complexed to other subunits in vivo, whereas PP-2C has only been isolated as a monomeric protein.

PP-1, PP-2A, and PP-2C have broad and overlapping specificities in vitro, and account for virtually all measurable activity in tissue extracts toward a variety of phosphoproteins that regulate metabolism, muscle contractility, and other processes. Their precise functions in vivo are unknown, although important clues to the physiological roles of PP-1 and PP-2A are provided by the effects of okadaic acid and by the subcellular localization of PP-1. The active forms of PP-1 are largely particulate, and their association with subcellular structures is mediated by "targetting subunits" that direct PP-1 to particular locations, enhance its activity toward certain substrates, and confer important regulatory properties upon it. This concept is best established for the glycogen-bound enzymes in skeletal muscle and liver (PP-1G) and the myofibrillar form (PP-1M) in skeletal muscle.

The activities of PP-1 and PP-2B are controlled by the second messengers cyclic AMP and calcium. The activity of PP-2B is dependent on calcium and calmodulin, while PP-1 is controlled in a variety of ways that depend on the form of the enzyme and the tissue. PP-1 can be inhibited by cyclic AMP in a variety of cells through the A-kinase-catalyzed phosphorylation of inhibitor-1 and its isoforms. Phosphorylation of the glycogen-binding subunit of PP-1G by A-kinase promotes translocation of the catalytic subunit from glycogen particles to cytosol in skeletal muscle, inhibiting the dephosphorylation of glycogen-metabolizing enzymes. Allosteric inhibition of hepatic PP-1G by phosphorylase a occurs in response to signals that elevate cyclic AMP or calcium, and prevents the activation of glycogen synthase in liver. PP-1 can also be activated indirectly by calcium through the ability of PP-2B to dephosphorylate inhibitor-1. This control mechanism may operate in dopaminoceptive neurones of the brain and other cells. The inactive cytosolic form of PP-1 (PP-1I) can be activated in vitro through the glycogen synthase kinase-3–catalyzed phosphorylation of its inhibitory subunit (inhibitor-2), but the physiological significance is unclear. PP-2A exists in cells as low-activity forms that have a tremendous potential for activation, but the mechanisms that regulate this enzyme and PP-2C in vivo are unknown.

Literature Cited

1. Cohen, P. 1988. *Proc. R. Soc. London Ser. B* 234:115–44
2. Bradford, A. P., Yeaman, S. J. 1986. *Advances in Protein Phosphatases*, ed. W. Merlevede, J. Di Salvo, 3:73–105. Belgium: Leuven Univ. Press
3. Reed, L. J., Damuni, Z. 1987. *Advances in Protein Phosphatases*, ed. W. Merlevede, J. DiSalvo, 4:59–76 Belgium: Leuven Univ.
4. Tonks, N. K., Diltz, C. D., Fischer, E. H. 1988. *J. Biol. Chem.* 263:6722–30
5. Tonks, N. K., Diltz, C. D., Fischer, E. H. 1988. *J. Biol. Chem.* 263:6731–37
6. Ingebritsen, T. S., Cohen, P. 1983. *Eur. J. Biochem.* 132:255–61
7. Ingebritsen, T. S., Cohen, P. 1983. *Science* 221:331–38
8. Ingebritsen, T. S., Stewart, A. A., Cohen, P. 1983. *Eur. J. Biochem.* 132:297–307
9. Stewart, A. A., Ingebritsen, T. S., Manalan, A., Klee, C. B., Cohen, P. 1982. *FEBS Lett.* 137:80–84
10. Stewart, A. A., Ingebritsen, T. S., Cohen, P. 1983. *Eur. J. Biochem.* 132:289–95
11. Bialojan, C., Takai, A. 1988. *Biochem. J.* 256:283–90
11a. Haystead, T. A. J., Sim, A. T. R., Carling, D., Honnor, R. C., Tsukitani, Y., et al. 1989. *Nature* 337:78–81
12. Jakes, S., Schlender, K. K. 1988. *Biochim. Biophys. Acta.* 967:11–16
13. Agostinis, P., Goris, J., Waelkens, E., Pinna, L. A., Marchiori, F., Merlevede, W. 1987. *J. Biol. Chem.* 262:1060–64
14. Burchell, A., Cohen, P. 1977. *Abstr. 11th FEBS Meet.* A1–4, 003
15. Gratecos, D., Detwiler, T. C., Hurd, S., Fischer, E. H. 1977. *Biochemistry* 16, 4812–17
16. Di Salvo, J., Waelkens, E., Gifford, D., Goris, J., Merlevede, W. 1983. *Biochem. Biophys. Res. Commun.* 117:493–500
17. Mellgren, R. L., Schlender, K. K. 1983. *Biochem. Biophys. Res. Commun.* 117:501–8
18. Pelech, S., Cohen, P. 1985. *Eur. J. Biochem.* 148:245–51
19. Gergely, P., Erdodi, F., Bot, G. 1984. *FEBS Lett.* 169:45–48
20. Erdodi, F., Csortos, C., Bot, G., Gergely, P. 1985. *Biochim. Biophys. Acta* 827:23–29
21. Usui, H., Imazu, M., Maeta, K., Tsukamoto, H., Azume, K., Takeda, M. 1988. *J. Biol. Chem.* 263:3752–61
21a. Goris, J., Merlevede, W. 1988. *Biochem. J.* 254:501–7
22. Erdodi, F., Csortos, C., Bot, G., Gergely, P. 1985. *Biochem. Biophys. Res. Commun.* 128:705–12
23. Strålfors, P., Hiraga, A., Cohen, P. 1985. *Eur. J. Biochem.* 149:295–303
24. Alemany, S., Pelech, S., Brierley, C. H., Cohen, P. 1986. *Eur. J. Biochem* 156:101–10
25. Chisholm, A. A. K., Cohen, P. 1988. *Biochim. Biophys. Acta* 968:392–400
26. Pato, M. D., Kerc, E. 1985. *J. Biol. Chem.* 260:12359–66
27. Brautigan, D. L., Gruppuso, P. A., Mumby, M. 1986. *J. Biol. Chem.* 251:14924–28
28. Olssen, H., Belfrage, P. 1987. *Eur. J. Biochem.* 168:399–405
29. Orgad, S., Dudai, Y., Cohen, P. 1987. *Eur. J. Biochem.* 164:31–38
30. Dombradi, V., Friedrich, P., Bot, G. 1987. *Comp. Biochem. Physiol. B* 87: 857–61
31. Pondaven, P., Cohen, P. 1987. *Eur. J. Biochem.* 167:135–40
32. Thoen, C., Van Hove, L., Cohen, P., Slegers, H. 1985. *Biochem. Biophys. Res. Commun.* 131:84–90
33. Thoen, C., Van Hove, L., Slegers, H. 1987. *Eur. J. Biochem.* 163:503–11
34. McClure, J. A., Korn, E. D. 1983. *J. Biol. Chem.* 258:14570–75
35. Shacter, E., McClure, J. A., Korn, E. D., Chock, P. B. 1985. *Arch. Biochem. Biophys.* 242:523–31
36. Klee, C. B., Cohen, P. 1988. *Molecular Aspects of Cellular Regulation*, ed. P. Cohen, C. B. Klee, 5:225–48. Amsterdam: Elsevier Biomed. Press
37. Ingebritsen, T. S., Foulkes, J. G., Cohen, P. 1983. *Eur. J. Biochem.* 132: 263–74
38. Ingebritsen, T. S., Blair, J., Guy, P. S., Witters, L., Hardie, D. G. 1983. *Eur. J. Biochem.* 132:275–81
39. Brandt, H., Killilea, S. D., Lee, E. Y. C. 1974. *Biochem. Biophys. Res. Commun.* 61:598–604
40. Brandt, H., Capulong, Z. L., Lee, E. Y. C. 1975. *J. Biol. Chem.* 250:8038–44
41. Tung, H. Y. L., Resink, T. J., Hemmings, B. A., Shenolikar, S., Cohen, P. 1984. *Eur. J. Biochem.* 138:635–41
42. Silberman, S. R., Speth, M., Nemani, R., Ganapathi, M. K., Paris, H., et al. 1984. *J. Biol. Chem.* 259:2913–22
43. Cohen, P. T. W. 1988. *FEBS Lett.* 232:17–23
44. Berndt, N., Campbell, D. G., Caudwell, F. B., Cohen, P., da Cruz e Silva, E. F., et al. 1987. *FEBS Lett.* 223:340–46

44a. Khew-Goodall, Y., Hemmings, B. A. 1988. *FEBS Lett.* 238:265–68
45. Johnson, G. L., Brautigan, D. L., Shriner, C., Jaspers, S., Arino, J., et al. 1987. *Mol. Endocrinol.* 1:745–48
46. Hunter, T., Munro, A. 1969. *Nature* 223:1270–72
47. Galizzi, A. 1970. *Eur. J. Biochem.* 17: 49–55
48. Cohen, P. T. W., Burchell, A., Cohen, P. 1976. *Eur. J. Biochem.* 66:347–56
49. Brautigan, D. L., Shriner, C. L., Gruppuso, P. A. 1985. *J. Biol. Chem.* 260:4295–302
50. Khandelwal, R., Vandenheede, J. R., Krebs, E. G. 1976. *J. Biol. Chem.* 251:4850–58
51. Goris, J., Defreyn, G., Vandenheede, J. R., Merlevede, W. 1978. *Eur. J. Biochem.* 91:457–64
52. DePaoli-Roach, A. A. 1984. *J. Biol. Chem.* 259:12144–52
52a. Cohen, P. T. W., Schelling, D. L., da Cruz e Silva, O. B., Barker, H., Cohen, P. 1989. Submitted
52b. Dombradi, V., Gergely, P., Bot, G., Friedrich, P. 1987. *Biochem. Biophys. Res. Commun.* 144:1175–81
53. Cohen, P. 1978. *Curr. Top. Cell. Regul.* 14:117–96
54. Hiraga, A., Cohen, P. 1986. *Eur. J. Biochem.* 161:763–69
55. MacKintosh, C., Campbell, D. G., Hiraga, A., Cohen, P. 1988. *FEBS Lett.* 234:189–94
56. Hubbard, M. J., Cohen, P. 1989. *Eur. J. Biochem.* In press
57. Hiraga, A., Kemp, B. E., Cohen, P. 1987. *Eur. J. Biochem.* 163:253–58
58. Khatra, B. S. 1986. *J. Biol. Chem.* 261:8944–52
59. Villar-Moruzzi, E. 1986. *Arch. Biochem. Biophys.* 247:155–64
60. Villar-Moruzzi, E. 1986. See Ref. 2, pp. 225–36
61. Villar-Moruzzi, E., Heilmeyer, L. M. G. 1987. *Eur. J. Biochem.* 169:659–67
62. Chisholm, A. A. K., Cohen, P. 1988. *Biochim. Biophys. Acta* 971:163–69
63. Pato, M. D., Adelstein, R. S. 1983. *J. Biol. Chem.* 258:7047–54
64. Yoshida, M., Yagi, K. 1986. *J. Biochem.* 99:1027–36
65. Alemany, S., Tung, H. Y. L., Shenolikar, S., Pilkis, S. J., Cohen, P. 1984. *Eur. J. Biochem.* 145:51–56
66. Foulkes, J. G., Ernst, V., Levin, D. H. 1983. *J. Biol. Chem.* 258:1439–43
66a. Mieskes, G., Soling, H. D. 1987. *Eur. J. Biochem.* 167:377–82
67. Shields, S. M., Ingebritsen, T. S., Kelly, P. T. 1985. *J. Neurosci.* 5:3414–22

68. Jakes, S., Mellgren, R. L., Schlender, K. K. 1986. *Biochim. Biophys. Acta* 888:138–42
69. Kuret, J., Bell, H., Cohen, P. 1986. *FEBS Lett.* 203:197–202
70. Merlevede, W., Riley, G. A. 1966. *J. Biol. Chem.* 241:3517–24
71. Goris, J., Defreyn, G., Merlevede, W. 1979. *FEBS Lett.* 99:279–82
72. Vandenheede, J. R., Yang, S. D., Goris, J., Merlevede, W. 1980. *J. Biol. Chem.* 255:11768–74
73. Hemmings, B. A., Yellowlees, D., Kernohan, J. C., Cohen, P. 1981. *Eur. J. Biochem.* 119:443–51
74. Woodgett, J. R., Cohen, P. 1984. *Biochim. Biophys. Acta* 788:339–47
75. Cohen, P. 1980. *Molecular Aspects of Cellular Regulation*, ed. P. Cohen, 1:255–68. Amsterdam: Elsevier Biomed.
76. Stewart, A. A., Hemmings, B. A., Cohen, P., Goris, J., Merlevede, W. 1981. *Eur. J. Biochem.* 115:197–205
77. Vandenheede, J. R., Goris, J., Yang, S. D., Camps, T., Merlevede, W. 1981. *FEBS Lett.* 127:1–3
78. Yang, S. D., Vandenheede, J. R., Merlevede, W. 1981. *J. Biol. Chem.* 256:10231–34
79. Hemmings, B. A., Resink, T. J., Cohen, P. 1982. *FEBS Lett.* 150:319–24
80. Resink, T. J., Hemmings, B. A., Tung, H. Y. L., Cohen, P. 1983. *Eur. J. Biochem.* 133:455–61
81. Tung, H. Y. L., Cohen, P. 1984. 145:57–64
82. Ballou, L. M., Brautigan, D. L., Fischer, E. H. 1983. *Biochemistry* 22:3393–99
83. Yang, S. D., Vandenheede, J. R., Goris, J., Merlevede, W. 1980. *J. Biol. Chem.* 255:11759–67
84. Vandenheede, J. R., Yang, S. D., Merlevede, W. 1981. *J. Biol. Chem.* 256:5894–900
85. Jurgensen, S., Shacter, E., Huang, C. Y., Chock, P. B., Yang, S. D., et al. 1984. *J. Biol. Chem.* 259:5864–70
86. Vandenheede, J. R., Vanden-Abele, C., Merlevede, W. 1986. *Biochem. Biophys. Res. Commun.* 135:367–73
87. Brautigan, D. L., Ballou, L. M., Fischer, E. H. 1982. *Biochemistry* 21:1977–82
88. Yang, S. D., Vandenheede, J. R., Goris, J., Merlevede, W. 1980. *FEBS Lett.* 111:201–4
89. Price, D. J., Li, H. C. 1985. *Biochem. Biophys. Res. Commun.* 128:1203–10
90. Price, D. J., Tabarini, D., Li, H. C. 1986. *Eur. J. Biochem.* 158:635–45
91. Di Salvo, J., Jiang, M. J. 1982.

Biochem. Biophys. Res. Commun. 108:534–40
92. Yang, S. D., Fong, Y. L. 1985. *J. Biol. Chem.* 260:13464–70
93. Yang, S. D., Vandenheede, J. R., Merlevede, W. 1981. *FEBS Lett.* 126:57–60
94. Kiener, P. A., Carroll, D., Roth, B. J., Westhead, E. W. 1987. *J. Biol. Chem.* 262:2016–24
95. Usui, H., Kinohara, N., Yoshikawa, K., Imazu, M., Imaoka, T., Takeda, M. 1983. *J. Biol. Chem.* 258:10455–63
96. Brandt, H., Lee, E. Y. C., Killilea, S. D. 1975. *Biochem. Biophys. Res. Commun.* 63:950–56
97. Huang, F. L., Glinsmann, W. H. 1976. *Eur. J. Biochem.* 70:419–26
98. Nimmo, G. A., Cohen, P. 1978. *Eur. J. Biochem.* 87:341–51
99. Aitken, A., Bilham, T., Cohen, P. 1982. *Eur. J. Biochem.* 126:235–46
100. MacDougall, L., Campbell, D. G., Hubbard, M. J., Cohen, P. 1988. *Biochim. Biophys. Acta* 1010:218–26
101. Cohen, P., Nimmo, G. A., Shenolikar, S., Foulkes, J. G. 1978. *FEBS Symp.* 54:161–69
102. Nimmo, G. A., Cohen, P. 1978. *Eur. J. Biochem.* 87:353–65
103. Aitken, A., Cohen, P. 1982. *FEBS Lett.* 147:54–58
104. Cohen, P., Rylatt, D. B., Nimmo, G. A. 1977. *FEBS Lett.* 76:182–86
105. Chessa, G., Borin, G., Marchiori, F., Meggio, F., Bruna, A. M., Pinna, L. A. 1983. *Eur. J. Biochem.* 135:609–14
106. Glass, D. B., Krebs, E. G. 1980. *Annu. Rev. Pharmacol. Toxicol.* 20:363–88
107. Nemenoff, R. A., Blackshear, P. J., Avruch, J. 1983. *J. Biol. Chem.* 258:9437–43
108. Shenolikar, S., Steiner, A. 1984. *Adv. Cyclic Nucleotide Protein Phosphorylation Res.* 17:405–15
109. Detre, J. A., Nairn, A. C., Aswad, D. W., Greengard, P. 1984. *J. Neurosci.* 4:2843–49
110. Walaas, S. I., Greengard, P. 1984. *J. Neurosci.* 4:84–98
111. Ouimet, C. C., Miller, P., Hemmings, H. C., Walaas, S. I., Greengard, P. 1984. *J. Neurosci.* 4:111–24
112. Hemmings, H. C., Nairn, A. C., Aswad, D. W., Greengard, P. 1984. *J. Neurosci.* 4:99–110
113. Hemmings, H. C., Greengard, P., Tung, H. Y. L., Cohen, P. 1984. *Nature* 310:503–8
114. Williams, K. R., Hemmings, H. C., LoPresti, M. B., Konigsberg, W. H., Greengard, P. 1986. *J. Biol. Chem.* 261:1890–903
115. Foulkes, J. G., Cohen, P. 1980. *Eur. J. Biochem.* 105:195–203
115a. Hemmings, H. C., Greengard, P. 1986. *J. Neurosci.* 6:1469–81
116. Yang, S. D., Vandenheede, J. R., Merlevede, W. 1981. *FEBS Lett.* 132:293–95
117. Tonks, N. K., Cohen, P. 1984. *Eur. J. Biochem.* 145:65–70
118. Cohen, P., Foulkes, J. G., Holmes, C. F. B., Nimmo, G. A., Tonks, N. K. 1988. *Methods Enzymol.* 159:427–37
119. Holmes, C. F. B., Campbell, D. G., Caudwell, F. B., Aitken, A., Cohen, P. 1986. *Eur. J. Biochem.* 155:173–82
120. Wang, A., De Paoli-Roach, A. A. 1987. *Fed. Proc. Abstr.* 899:207–9
121. Gruppuso, P. A., Johnson, G. L., Constantinides, M., Brautigan, D. L. 1985. *J. Biol. Chem.* 260:4288–94
122. Van Straaten, F., Muller, R., Curran, T., Van Beveren, C., Verma, I. M. 1983. *Proc. Natl. Acad. Sci. USA* 80:3183–87
123. Cochran, B. H., Zullo, J., Verma, I. M., Stiles, C. D. 1984. *Science* 226:1080–82
124. Chisholm, A. A. K., Cohen, P. 1985. *Biochim. Biophys. Acta* 847:155–58
125. Reddy, P., Ernst, V. 1983. *Biochem. Biophys. Res. Commun.* 117:1089–96
126. Roach, P., Roach, P. J., De Paoli-Roach, A. A. 1985. *J. Biol. Chem.* 260:6314–17
127. Strålfors, P. 1988. *Eur. J. Biochem.* 171:199–207
128. Ingebritsen, T. S., Foulkes, J. G., Cohen, P. 1980. *FEBS Lett.* 119:9–15
129. Chou, C. K., Alfano, J., Rosen, O. M. 1977. *J. Biol. Chem.* 252:2855–59
130. Killilea, S. D., Aylward, J. H., Mellgren, R. L., Lee, E. Y. C. 1978. *Arch. Biochem. Biophys.* 191:638–46
131. Shacter-Noiman, E., Chock, P. B. 1983. *J. Biol. Chem.* 258:4214–19
132. Schlender, K. K., Wilson, S. E., Mellgren, R. L. 1986. *Biochim. Biophys. Acta* 872:1–10
133. LeVine, H., Sahyoun, N., McConnell, R., Bronson, D., Cuatrecasas, P. 1984. *Biochem. Biophys. Res. Commun.* 118:278–83
134. Paris, H., Ganapathi, M. K., Silberman, S. R., Aylward, J. H., Lee, E. Y. C. 1984. *J. Biol. Chem.* 259:7510–18
135. Tung, H. Y. L., Alemany, S., Cohen, P. 1985. *Eur. J. Biochem.* 148:253–63
136. Mumby, M. C., Green, D. D., Russell, K. L. 1985. *J. Biol. Chem.* 260:13763–70
137. Speth, M., Lee, E. Y. C. 1984. *J. Biol. Chem.* 259:4027–30

138. da Cruz e Silva, O. B., Cohen, P. T. W. 1987. *FEBS Lett.* 266:176–78
139. da Cruz e Silva, O. B., Alemany, S., Campbell, D. G., Cohen, P. T. W. 1987. *FEBS Lett.* 221:415–22
140. Stone, S. R., Hofsteenge, J., Hemmings, B. A. 1987. *Biochemistry* 26:7215–20
141. Green, D. D., Yang, S. I., Mumby, M. C. 1987. *Proc. Natl. Acad. Sci. USA* 84:4880–84
142. Arino, J., Woon, C. W., Brautigan, D. L., Miller, T. B., Johnson, G. L. 1988. *Proc. Natl. Acad. Sci. USA* 85:4252–56
143. Kitagawa, Y., Tahira, T., Ikeda, I., Kikuchi, K., Tsuiki, S., Sugimura, T., Nagao, M. 1988. *Biochim. Biophys. Acta* 951:123–29
144. da Cruz e Silva, O. B., da Cruz e Silva, E. F., Cohen, P. T. W. 1988. *FEBS Lett.* 242:106–10
145. Cohen, P. T. W., Collins, J., Coulson, A. F., Berndt, N., da Cruz e Silva, O. B. 1988. *Gene* 69:131–34
146. Mumby, M. C., Russell, K. L., Garrard, L. J., Green, D. D. 1987. *J. Biol. Chem.* 262:6257–65
147. Lee, E. Y. C., Aylward, J. H., Mellgren, R. L., Killilea, S. D. 1979. *Miami Winter Symp.* 16:483–99
148. Tamura, S., Tsuiki, S. 1980. *Eur. J. Biochem.* 111:217–24
149. Tamura, S., Kikuchi, H., Kikuchi, K., Hiraga, A., Tsuiki, S. 1980. *Eur. J. Biochem.* 104:347–55
150. Crouch, D., Safer, B. 1980. *J. Biol. Chem.* 255:7918–24
151a. Khandelwal, R. L., Enno, T. L. 1985. *J. Biol. Chem.* 260:14335–43
151b. Nestler, E. J., Walaas, S. I., Greengard, P. 1984. *Science* 225:1357–64
152. Li, H. C. 1981. *Cold Spring Harbor Conf. Cell Prolif.* 8:441–57
153. Imaoka, T., Imazu, M., Usui, H., Kinohara, N., Takeda, M. 1983. *J. Biol. Chem.* 245:6642–49
154. Onishi, H., Umeda, J., Uchiwa, H., Watanabe, S. 1982. *J. Biochem.* 91:265–71
155. Werth, D. K., Haeberle, J. R., Hathaway, D. R. 1982. *J. Biol. Chem.* 257:7306–9
156. Lee, E. Y. C., Silberman, S. R., Ganapathi, M. K., Paris, H., Petrovic, S. 1981. *Cold Spring Harbor Conf. Cell Prolif.* 8:425–39
157. Waelkens, E., Goris, J., Merlevede, W. 1987. *J. Biol. Chem.* 262:1049–59
158. Kanayama, K., Wada, K., Negami, A., Yamamura, H., Tanabe, T. 1985. *FEBS Lett.* 184:78–81
159. Patterson, C. L., Flavin, M. 1986. *J. Biol. Chem.* 261:7791–96
160. Yang, S. D., Yu, J. S., Fong, Y. L. 1986. *J. Biol. Chem.* 261:5590–96
161. Pato, M. D., Adelstein, R. S., Crouch, D., Safer, B., Ingebritsen, T. S., Cohen, P. 1983. *Eur. J. Biochem.* 132:283–87
162. Pato, M. D., Kerc, E. 1986. *J. Biol. Chem.* 261:3770–74
163. Damuni, Z., Reed, L. J. 1987. *J. Biol. Chem.* 262:5133–38
164. Klee, C. B., Draetta, G. F., Hubbard, M. J. 1988. *Adv. Enzymol.* 61:149–200
165. Antoniw, J. F., Cohen, P. 1976. *Eur. J. Biochem.* 68:45–54
166. Klee, C. B., Krinks, M. H. 1978. *Biochemistry* 17:120–26
167. Sharma, R. K., Desai, R., Waisman, D. M., Wang, J. H. 1979. *J. Biol. Chem.* 254:4276–80
168. Wallace, R. W., Lynch, T. J., Tallant, E. A., Cheung, W. Y. 1979. *J. Biol. Chem.* 254:377–82
169. Aitken, A., Klee, C. B., Cohen, P. 1984. *Eur. J. Biochem.* 139:663–71
170. Klee, C. B., Crouch, T. H., Krinks, M. H. 1979. *Proc. Natl. Acad. Sci. USA* 76:6270–73
171. Carlin, R. K., Grab, D. J., Siekevitz, P. 1981. *J. Cell. Biol.* 89:449–55
172. Merat, D. L., Hu, Z. Y., Carter, T. C., Cheung, W. Y. 1985. *J. Biol. Chem.* 260:11053–59
173. Gupta, R. C., Khandelwal, R. L., Sulakhe, P. V. 1985. *FEBS Lett.* 190:104–8
174. Klee, C. B., Krinks, M. H., Manalan, A. J., Draetta, G. F., Newton, D. L. 1985. *Advances in Protein Phosphatases*, ed. W. Merlevede, J. Di Salvo, 1:135–46. Belgium: Leuven Univ. Press
175. Kretsinger, R. H. 1980. *CRC Crit. Rev. Biochem.* 8:119–74
176. Aitken, A., Cohen, P., Santikarn, S., Williams, D. H., Calder, A. G., Klee, C. B. 1982. *FEBS Lett.* 150:314–18
177. Krinks, M. H., Manalan, A. S., Klee, C. B. 1985. *Fed. Proc.* 44:707a
178. Wolf, H., Hofmann, F. 1980. *Proc. Natl. Acad. Sci. USA* 77:5852–55
179. Tallant, E. A., Wallace, R. W. 1985. *J. Biol. Chem.* 260:7744–51
180. Pallen, C. J., Valentine, K. A., Wang, J. H., Hollenberg, M. O. 1985. *Biochemistry* 24:4727–30
181. Isawa, F., Ishiguro, K. 1986. *J. Biochem.* 99:1353–58
182. Burnham, D. B. 1985. *Biochem. J.* 231:335–41
183. Chantler, P. D. 1985. *J. Cell. Biol.* 101:207–16
184. Tash, J. S., Klee, C. B., Means, R. L., Tran, A., Krinks, M. H., Means, A. R. 1988. *J. Cell. Biol.* 106:1625–33

185. Wallace, R. W., Tallant, E. A., Cheung, W. Y. 1980. *Biochemistry* 19:1831–37
186. Klumpp, S., Steiner, A. S., Schultz, J. E. 1983. *Eur. J. Cell. Biol.* 32:164–70
187. Momayezi, M., Lumpert, C. J., Kersken, H., Gras, U., Plattner, H., et al. 1987. *J. Cell. Biol.* 105:181–89
188. Kincaid, R., Nightingale, M. S., Martin, B. H. 1988. *Proc. Natl. Acad. Sci. USA* 85:8983–87
189. Li, H. C., Hsiao, K. Y., Chan, W. W. S. 1978. *Eur. J. Biochem.* 84:215–25
190. Pato, M. D., Adelstein, R. S. 1980. *J. Biol. Chem.* 255:6535–38
191. Pato, M. D., Adelstein, R. S. 1983. *J. Biol. Chem.* 258:7055–58
192. Hiraga, A., Kikuchi, K., Tamura, S., Tsuiki, S. 1981. *Eur. J. Biochem.* 119:503–10
193. Mieskes, G., Brand, I. A., Soling, H. D. 1984. *Eur. J. Biochem.* 140:375–83
194. Mieskes, G., Soling, H. D. 1985. *Biochem. J.* 225:665–70
195. McGowan, C., Cohen, P. 1987. *Eur. J. Biochem.* 166:713–22
196. Mieskes, G., Soling, H. D. 1985. *FEBS Lett.* 181:7–11
197. McGowan, C., Campbell, D. G., Cohen, P. 1987. *Biochim. Biophys. Acta* 930:279–82
198. Wollny, E., Watkins, K., Kramer, G., Hardesty, B. 1984. *J. Biol. Chem.* 259:2484–92
199. Tipper, J., Wollny, E., Fullilove, S., Kramer, G., Hardesty, B. 1986. *J. Biol. Chem.* 261:7144–50
200. Donella-Deana, A., Marchiori, F., Meggio, F., Pinna, L. A. 1982. *J. Biol. Chem.* 257:8565–68
201. Donella-Deana, A., Pinna, L. A. 1988. *Biochim. Biophys. Acta* 968:179–85
202. Sahyoun, N., Le Vine, H., McConnell, R., Bronson, D., Cuatrecasas, P. 1983. *Proc. Natl. Acad. Sci. USA* 80:6760–64
203. Morgan, M., Perry, S. V., Ottaway, J. 1976. *Biochem. J.* 157:687–97
203a. Yang, S. D., Fong, Y. L., Benovic, J. L., Sibley, D. R., Caron, M. G., Lefcowitz, R. J. 1988. *J. Biol. Chem.* 263:8856–58
204. Cohen, P. 1983. *Philos. Trans. R. Soc. London Ser. B* 302:13–25
205. Cohen, P. 1986. *Enzymes* 17:361–97
206. Cohen, P. 1988. See Ref. 36, pp. 123–44
207. Cohen, P. 1988. See Ref. 36, pp. 145–94
208. Tung, H. Y. L., Pelech, S., Fisher, M. J., Pogson, C. I., Cohen, P. 1985. *Eur. J. Biochem.* 149:305–13

209. Schelling, D. L., Leader, D. P., Zammit, V. A., Cohen, P. 1988. *Biochim. Biophys. Acta* 927:221–31
210. Bollen, M., Vandenheede, J. R., Goris, J., Stalmans, W. 1988. *Biochim. Biophys. Acta* 969:66–77
211. Stalmans, W., Bollen, M. 1987. See Ref. 3, pp. 391–408
212. Bollen, M., Stalmans, W. 1984. *Biochem. J.* 217:427–34
213. Bollen, M., Dopere, F., Goris, J., Merlevede, W., Stalmans, W. 1984. *Eur. J. Biochem.* 144:57–63
213a. Bollen, M., Keppens, S., Stalmans, W. 1988. *Diabetologia* 31:711–13
214. Hiraga, A., Tamura, S., Kikuchi, K., Tsuiki, S. 1987. *J. Biochem.* 101:1161–68
215. Stull, J. T. 1988. See Ref. 36, pp. 91–122
216. Sellers, J. R., Pato, M. D. 1984. *J. Biol. Chem.* 259:11383–90
217. Takai, A., Bialojan, C., Troschka, M., Ruegg, J. C. 1987. *FEBS Lett.* 217:81–84
218. Hoar, P. E., Pato, M. D., Kerrick, W. G. L. 1985. *J. Biol. Chem.* 260:8760–64
219. Haeberle, J. R., Hathaway, D. R., DePaoli-Roach, A. A. 1985. *J. Biol. Chem.* 260:9965–68
220. Adelstein, R. S., Eisenberg, E. A. 1980. *Annu. Rev. Biochem.* 49:921–56
221. Jeacocke, S. A., England, P. J. 1980. *FEBS Lett.* 122:129–32
222. Solaro, R. J. 1986. In *Protein Phosphorylation in Heart Muscle,* ed. R. J. Solaro, pp. 129–56. Boca Raton: CRC Press
223. Schlender, K. K., Hegazy, M. G., Thysseril, T. J. 1987. *Biochim. Biophys. Acta* 928:312–19
224. Hescheler, J., Kameyama, M., Trautwein, W., Mieskes, G., Soling, H. D. 1987. *Eur. J. Biochem.* 165:261–66
225. Hescheler, J., Mieskes, G., Ruegg, J. C., Takai, A., Trautwein, W. 1988. *Pfluegers Arch.* 412:248–52
226. Kranias, E. G., Di Salvo, J. 1986. *J. Biol. Chem.* 261:10029–32
226a. Kranias, E. G., Steenaart, N. A. E., Di Salvo, J. 1988. *J. Biol. Chem.* 263:15681–87
227. Ballou, L. M., Jeno, P., Thomas, G. A. 1988. *J. Biol. Chem.* 263:1188–94
228. Jeno, P., Ballou, L. M., Novak-Hofer, I., Thomas, G. 1988. *Proc. Natl. Acad. Sci. USA* 85:406–10
229. Wettenhall, R. E. H., Cohen, P., Caudwell, B., Holland, R. 1982. *FEBS Lett.* 148:207–13
230. Andres, J. L., Johansen, J. W., Maller,

J. L. 1987. *J. Biol. Chem.* 262:14389–93

231. Olivier, A. R., Ballou, L. M., Thomas, G. 1988. *Proc. Natl. Acad. Sci. USA* 85:4720–24

232. Pain, V. M. 1986. *Biochem. J.* 235:625–37

233. Stewart, A. A., Crouch, D., Cohen, P., Safer, B. 1980. *FEBS Lett.* 119:16–19

234. Ernst, V., Levin, D. H., Foulkes, J. G., London, I. M. 1982. *Proc. Natl. Acad. Sci. USA* 79:7092–96

235. Petryshyn, R., Levin, D. H., London, I. M. 1982. *Proc. Natl. Acad. Sci. USA* 79:6512–16

236. Crouch, D., Safer, B. 1984. *J. Biol. Chem.* 259:10363–68

237. Foulkes, J. G., Maller, J. L. 1982. *FEBS Lett.* 150:155–60

238. Huchon, D., Ozon, R., Demaille, J. G. 1981. *Nature* 294:358–59

239. Meijer, L., Pondaven, P., Tung, H. Y. L., Cohen, P., Wallace, R. W. 1986. *Exp. Cell Res.* 163:483–99

240. Pondaven, P., Meijer, L. 1986. *Exp. Cell Res.* 163:477–88

241. Ingebritsen, T. S., Gibson, D. M. 1980. See Ref. 75, pp. 63–93

242. Carling, D., Zammit, V., Hardie, D. G. 1987. *FEBS Lett.* 223:217–22

243. Pelech, S., Cohen, P., Fisher, M. J., Pogson, C. I., El-Maghrabi, M. R., Pilkis, S. J. 1984. *Eur. J. Biochem.* 145:39–49

244. Witters, L. A., Bacon, G. W. 1985. *Biochem. Biophys. Res. Commun.* 130:1132–38

245. Matsumoto, K., Uno, I., Kato, K., Ishikawa, T. 1985. *Yeast* 1:25–38

246. Tanaka, K., Matsumoto, K., Toh-e, A. 1989. *Mol. Cell. Biol.* 9:757–68

247. Hubbard, M. J., Klee, C. B. 1988. *Biochemistry.* In press

248. Tonks, N. K., Cohen, P. 1983. *Biochim. Biophys. Acta* 747:191–93

249. Blumenthal, D. K., Takio, K., Hansen, R. S., Krebs, E. G. 1986. *J. Biol. Chem.* 261:8140–45

250. Sharma, R. K., Mooibroek, M., Wang, J. H. 1988. See Ref. 36, pp. 267–97

251. King, M. M., Huang, C. Y., Chock, P. B., Nairn, A. C., Hemmings, H. C., et al. 1984. *J. Biol. Chem.* 259:8080–83

252. Goto, S., Yamamoto, H., Fukunaga, K., Iwasa, T., Matsukado, Y., Miyamoto, E. 1985. *J. Neurochem.* 45:276–83

253. Yeaman, S. J., Watson, D. C., Dixon, G. H., Cohen, P. 1977. *Biochem. J.* 162:411–21

254. Chad, J. E., Eckert, R. 1986. *J. Physiol.* 378:31–51

255. Foulkes, J. G., Strada, S. J., Henderson, P. J. F., Cohen, P. 1983. *Eur. J. Biochem.* 132:309–13

256. Tao, S. H., Huang, F. L., Lynch, A., Glinsmann, W. H. 1978. *Biochem. J.* 176:347–50

257. Foulkes, J. G., Cohen, P. 1979. *Eur. J. Biochem.* 97:251–56

258. Khatra, B. S., Chiasson, J., Shikama, H., Exton, J. H., Soderling, T. R. 1980. *FEBS Lett.* 114:253–56

259. Foulkes, J. G., Cohen, P., Strada, S. J., Everson, W. V., Jefferson, L. S. 1982. *J. Biol. Chem.* 257:12493–96

260. Walaas, S. I., Aswad, D. W., Greengard, P. 1983. *Nature* 301:69–71

261. Laloux, M., Hers, H. G. 1979. *FEBS Lett.* 105:239–43

262. Khatra, B. S., Soderling, T. R. 1983. *Arch. Biochem. Biophys.* 227:39–51

263. Tonks, N. K. 1985. *The structure and regulation of protein phosphatases.* PhD thesis. Univ. Dundee

264. Ash, D. E., Schramm, V. L. 1982. *J. Biol. Chem.* 257:9261–64

265. Sharma, R. K., Wang, J. H. 1985. *Proc. Natl. Acad. Sci. USA* 82:2603–7

266. Sharma, R. K., Wang, J. H. 1986. *J. Biol. Chem.* 261:1322–28

267. Nestler, E. J., Walaas, S. I., Greengard, P. 1984. *Science* 225:1357–64

268. Caudwell, F. B., Hiraga, A., Cohen, P. 1986. *FEBS Lett.* 194:85–90

268a. Hubbard, M. J., Cohen, P. 1989. Submitted

269. Poulter, L., Ang, S. G., Gibson, B., Holmes, C. F. B., Caudwell, F. B., et al. 1988. *Eur. J. Biochem.* 175:497–510

270. Gergely, P., Bot, G. A. 1977. *FEBS Lett.* 82:269–72

271. Walsh, D. A., Cooper, R. H. 1979. In *Biochemical Actions of Hormones,* ed. G. Litwack, 6:1–75. New York: Academic

272. Stewart, A. A. 1982. *Regulation of protein phosphatases involved in the control of glycogen metabolism in skeletal muscle.* PhD thesis. Univ. Dundee

273. Khatra, B. S., Printz, R., Cobb, C. E., Corbin, J. D. 1985. *Biochem. Biophys. Res. Commun.* 130:567–73

274. Jurgensen, S. R., Chock, P. B., Taylor, S., Vandenheede, J. R., Merlevede, W. 1985. *Proc. Natl. Acad. Sci. USA* 82:7565–69

275. Hemmings, B. A., Aitken, A., Cohen, P., Rymond, M., Hofmann, F. 1982. *Eur. J. Biochem.* 127:473–81

276. Vereb, G., Erdodi, F., Toth, B., Bot, G. 1986. *FEBS Lett.* 197:139–42

277. Villar-Moruzzi, E., Ballou, L. M., Fischer, E. H. 1984. *J. Biol. Chem.* 259:5857–63
278. Vandenheede, J. R., Yang, S. D., Merlevede, W., Jurgensen, S., Chock, P. B. 1985. *J. Biol. Chem.* 260:10512–16
279. Deleted in proof
280. Li, H. C., Price, D. J., Tabarini, D. 1985. *J. Biol. Chem.* 260:6416–26
281. Ballou, L. M., Fischer, E. H. 1986. *Enzymes* 17:311–61
282. Holmes, C. F. B., Kuret, J., Chisholm, A. A. K., Cohen, P. 1986. *Biochim. Biophys. Acta* 870:408–16
283. Agostinis, P., Vandenheede, J. R., Goris, J., Meggio, F., Pinna, L. A., Merlevede, W. 1987. *FEBS Lett.* 224:385–90
284. Holmes, C. F. B., Tonks, N. K., Major, H., Cohen, P. 1987. *Biochim. Biophys. Acta* 929:208–19
285. DePaoli-Roach, A. A., Lee, F. T. 1985. *FEBS Lett.* 183:423–29
286. Lawrence, J. C., Hiken, J., Burnette, B., DePaoli-Roach, A. A. 1988. *Biochem. Biophys. Res. Commun.* 150:197–203
286a. Bollen, M., Stalmans, W. 1988. *Biochem. J.* 255:327–33
287. Yang, S. D., Vandenheede, J. R., Merlevede, W. 1983. *Biochem. Biophys. Res. Commun.* 113:439–45
288. Agostinis, P., Goris, J., Vandenheede, J. R., Waelkens, E., Pinna, L. A., Merlevede, W. 1986. *FEBS Lett.* 207:167–72
289. Stalmans, W., deWulf, H., Hers, H. G. 1971. *Eur. J. Biochem.* 18:582–87
290. Hers, H. G. 1976. *Annu. Rev. Biochem.* 45:167–89
291. Stalmans, W. 1976. *Curr. Top. Cell. Regul.* 11:51–97
292. Mvumbi, L., Dopere, F., Stalmans, W. 1983. *Biochem. J.* 212:407–16
293. Tan, A. W. H., Nuttall, F. Q. 1978. *Biochim. Biophys. Acta* 522:139–50
294. Alemany, S., Cohen, P. 1986. *FEBS Lett.* 198:194–202
295. Ciudad, C., Camici, M., Ahmad, Z., Wang, Y., DePaoli-Roach, A. A., Roach, P. J. 1984. *Eur. J. Biochem.* 142:511–20
296. Witters, L. A., Avruch, J. 1978. *Biochemistry* 17:406–10
297. Bishop, J. M. 1970. *Biochem. Biophys. Acta* 208:208–18
298. Miller, T. B., Garnache, A., Vicalvi, J. J. 1981. *J. Biol. Chem.* 256:2851–55
299. Stalmans, W., Van de Werve, G. 1981. *Short Term Regulation of Liver Metabolism*, ed. L. Hue, G. Van de Werve, pp. 119–38. Amsterdam: Elsevier Biomed.
300. Hue, L., Bontemps, F., Hers, H. G. 1975. *Biochem. J.* 152:105–14
301. Mvumbi, L., Stalmans, W. 1987. *Biochem. J.* 246:367–74
301a. Toth, B., Bollen, M., Stalmans, W. 1988. *J. Biol. Chem.* 263:14061–66
302. Van de Werve, G. 1981. *Biochem. Biophys. Res. Commun.* 102:1323–29
303. Mvumbi, L., Bollen, M., Stalmans, W. 1985. *Biochem. J.* 232:697–704
304. Tan, A. W. H., Nuttall, F. Q. 1976. *Biochim. Biophys. Acta* 445:118–30
305. Miller, T. B., Hazen, R., Larner, J. 1973. *Biochem. Biophys. Res. Commun.* 53:466–77
306. Whitton, P. D., Hems, D. A. 1975. *Biochem. J.* 150:153–65
307. Miller, T. B., Garnache, A., Crux, J. 1984. *J. Biol. Chem.* 259:12470–74
307a. Bollen, M., Stalmans, W. 1988. *Endocrinology* 122:2915–19
308. Khandelwal, R. J., Zinman, S. M., Zebrowski, E. J. 1977. *Biochem. J.* 168:541–48
309. Shahed, A. R., Mehta, P. P., Chalker, D., Allman, D. W., Gibson, D. M., Harper, E. T. 1980. *Biochem. Int.* 1:486–92
310. Gold, A. H., Dickemper, D., Haverstick, D. M. 1979. *Mol. Cell. Biochem.* 25:47–59
311. Dopere, F., Stalmans, W. 1980. *Arch. Int. Physiol. Biochem.* 88:B267–68
312. Foulkes, J. G., Jefferson, L. S. 1984. *Diabetes* 33:576–79
313. Dragland-Meserve, C. J., Webster, D. K., Botelho, L. H. P. 1985. *Eur. J. Biochem.* 146:699–704
314. Chan, C. P., McNall, S. J., Krebs, E. G., Fischer, E. H. 1988. *Proc. Natl. Acad. Sci. USA* 85:6257–61
315. Defreyn, G., Goris, J., Merlevede, W. 1977. *FEBS Lett.* 79:125–28
316. Goris, J., Waelkens, E., Merlevede, W. 1983. *Biochem. Biophys. Res. Commun.* 116:349–54
317. Goris, J., Waelkens, E., Camps, T., Merlevede, W. 1984. *Adv. Enzyme Regul.* 22:467–84
318. Goris, J., Camps, T., Defreyn, G., Merlevede, W. 1981. *FEBS Lett.* 134:189–93
319. Goris, J., Walsh, D. A., Merlevede, W. 1984. *Biochem. Biophys. Res. Commun.* 125:293–98
320. Goris, J., Parker, P. J., Waelkens, E., Merlevede, W. 1984. *Biochem. Biophys. Res. Commun.* 120:405–10
321. Goris, J., Waelkens, E., Merlevede, W. 1985. *FEBS Lett.* 188:262–66

322. Goris, J., Waelkens, E., Merlevede, W. 1986. *Biochem. J.* 239:109–14
323. Johansen, J. W., Ingebritsen, T. S. 1986. *Proc. Natl. Acad. Sci. USA* 83: 207–11
324. Johansen, J. W., Ingebritsen, T. S. 1987. *Biochem. Biophys. Acta* 928:63–75
325. Cooper, J. A., Hunter, J. 1981. *Mol. Cell. Biol.* 1:165–78
326. Decker, S. 1981. *Proc. Natl. Acad. Sci. USA* 78:4112–15
327. Wilson, S. E., Mellgren, R. L., Schlender, K. K. 1982. *FEBS Lett.* 146:331–34
328. Wilson, S. E., Mellgren, R. L., Schlender, K. K. 1983. *Biochem. Biophys. Res. Commun.* 116:581–86
329. Yang, S. D., Vandenheede, J. R., Merlevede, W. 1984. *Biochem. Biophys. Res. Commun.* 118:923–28
330. Mellgren, R. L., Wilson, S. E., Schlender, K. K. 1984. *FEBS Lett.* 176:291–93
331. Parker, P. J., Caudwell, F. B., Cohen, P. 1983. *Eur. J. Biochem.* 130:227–34
332. Huang, L. C., Chang, L. Y. 1980. *Biochim. Biophys. Acta* 613:106–15
333. Tabor, C. W., Tabor, H. 1976. *Annu. Rev. Biochem.* 45:285–306
334. Tabor, C. W., Tabor, H. 1984. *Annu. Rev. Biochem.* 53:749–90
335. Killilea, S. D., Mellgren, R. L., Aylward, J. H., Lee, E. Y. C. 1978. *Biochem. Biophys. Res. Commun.* 81: 1040–46
335a. Speth, M., Alejandro, R., Lee, E. Y. C. 1984. *J. Biol. Chem.* 259:3475–81
336. Lopez-Alarcon, C., Mojena, M., Monge, L., Feliu, J. E. 1986. *Biochem. Biophys. Res. Commun* 134:292–98
337. Gratecos, D., Detwiler, J., Fischer, E. H. 1974. In *Metabolic Interconversions of Enzymes* 1973, ed. E. H. Fischer, E. G. Krebs, H. Neurath, E. R. Stadtman, pp. 43–52. Heidelberg: Springer-Verlag
338. Gergely, P., Bot, G. 1981. *Acta Biochim Biophys. Acad. Sci. Hung.* 16:163–78
339. Bot, G., Varsanyi, M., Gergely, P. 1975. *FEBS Lett.* 50:351–54
340. Gergely, P., Vereb, G., Bot, G. 1976. *Biochem. Biophys. Acta* 429:809–16

Annu. Rev. Biochem. 1989. 58:509–31

GENE CONVERSION AND THE GENERATION OF ANTIBODY DIVERSITY

Lawrence J. Wysocki

Division of Basic Sciences, Department of Pediatrics, National Jewish Center for Immunology and Respiratory Medicine, Denver, Colorado 80206

Malcolm L. Gefter

Department of Biology, Massachusetts Institute of Technology, Cambridge, Massachusetts 02139

PERSPECTIVES AND SUMMARY

Maintaining an arsenal against the unexpected requires diversity. Vertebrates have achieved antibody diversity by several strategies. One is the maintenance of a large repertoire of germline V (variable) genes. The others are somatic processes that shuffle germline V genes and possibly introduce de

509

novo sequence diversity. Gene conversion has been suggested as a process that maintains sequence homogeneity in multigene families. Paradoxically, it has also been implicated in the creation and maintenance of antibody diversity. In this review we summarize the evidence for the role played by gene conversion or, more generally, of information transfer, in generating germline and somatic forms of antibody diversity. Gene conversion plays a major and critical role in the somatic construction of the chicken light chain V region repertoire. It has probably also played an important role in the evolution and diversification of Ig (immunoglobulin) V and C (constant) gene segments, although other recombinational processes probably have also contributed. While studies in vitro have suggested that gene conversion may contribute to the somatic diversification of mammalian V genes, in vivo evidence in support of this conclusion has not been obtained.

THE PHENOMENON OF GENE CONVERSION

Gene conversion was originally observed in lower eukaryotic organisms in which all of the products of a single meiotic event can be captured and subjected to phenotypic analyses (reviewed in 1). For a given pair of distinguishable alleles, A and a, half of the products of a meiotic event should contain allele A and half should contain allele a. Aberrancies were often noted, however, in which genetic information was apparently not conserved, as revealed by unequal segregation ratios (3:1, 5:3, 6:2). Such instances, in which one allele apparently increases in copy number while another allele decreases, have been termed "gene conversions."

With the advent of DNA transformation technology, intra- and interchromosomal gene conversions were revealed in yeast (2–4). In these experiments, cloned genes determining selectable phenotypes were introduced into the yeast genome in nonallelic chromosomal locations with respect to a homologous gene. Potential conversion events between the two genes were identified by various selection strategies, and the nature of a given intergenic interaction was evaluated by tetrad analysis following sporulation. Nonreciprocal events were revealed by aberrant segregation ratios upon sporulation.

The mechanism(s) of gene conversion has been the subject of much experimental attention (1). Most models posit single strand invasions between interacting homologous genes, resulting in strand displacement and the formation of heteroduplex DNA. Synthesis of DNA that is associated with heteroduplex formation or with mismatch repair of heteroduplex DNA is believed to result in conversions (1, 5). The term "gene conversion" has been applied liberally to genetic exchanges in higher eukaryotic organisms in which the formation of heteroduplex DNA is a postulated intermediate. In the strictest sense, a conversion should result in a nonreciprocal exchange of homologous genetic information. To our knowledge, nonreciprocality has never been formally demonstrated in any mammalian system. In mammalian systems the term conversion is thus often used loosely to refer to incompletely defined genetic exchanges in which all of the genetic participants cannot be isolated and characterized to confirm or exclude nonreciprocality.

Evidence that gene conversion occurs in mammals was first reported in a classic paper by Slightom et al (6), who cloned and sequenced fetal globin genes from the homologous chromosomes of a single individual. The premise for their work was the paradox that fetal globin genes are duplicated in widely divergent species, such as mouse and man, yet products of these nonallelic duplicated genes often are more similar in amino acid sequence to one another within a given species than are the products of comparable alleles in different species.

Slightom and coworkers found evidence of intergenic gene conversion between nonallelic G_γ and A_γ genes as revealed by the fact that one continuous stretch of an intervening sequence was nearly identical in the two genes on one chromosome and very different from the corresponding sequence in the A_γ allele on the homologous chromosome. They argued convincingly that at some point in evolution sequence information from the G_γ gene had converted a portion of the nonallelic A_γ gene. Their data support a role for gene conversion in the maintenance of homogeneity within nonallelic members of a multigene family and in the generation of polymorphisms within a species by diversification of alleles. Their interpretation is attractive because it explains the paradox described above without the need to invoke parallel and independent duplications of fetal globin genes following speciation.

Since the report of Slightom et al (6), many examples of information transfer between members of mammalian multigene families have been documented, including class I and class II genes of the major histocompatibility locus, tubulin genes, histone genes, amyloid A genes, urinary protein genes, and immunoglobulin constant and variable region genes (7–33). Experimental models for mammalian gene conversion have been generated using transfected genes determining selectable phenotypes that either integrate or remain extrachromosomal (34–38). In both cases results have been obtained that support convincing but not airtight arguments for genuine nonreciprocal transfer of genetic information. In one well-characterized system, gene conversions were observed to involve contiguous blocks of DNA sequences, were more frequent when homologous interacting loci were of greater length, and were dependent upon proximity of interacting genes (36, 37).

SOMATIC GENE CONVERSION IN AVIAN IMMUNOGLOBULIN VARIABLE GENES

Reynaud et al (33) and Thompson & Neiman (39) have provided direct experimental evidence that the somatic diversification of the chicken light chain V region repertoire is achieved by a segmental gene conversion process. Unlike mammalian B cells, in which a somatic assembly process involving large multigene families of V gene segments contributes significantly to V region diversity, chicken B cells apparently rearrange only one light chain V gene segment and one joining segment (40). Assembly of this single Vλ1 segment to a single 3' Jλ joining segment is revealed in Southern blotting experiments by a uniform rearranged fragment that hybridizes with Vλ1 DNA probes. Despite apparent usage of only one light chain V gene, bursal B cells display functional diversification during antigen-independent stages of B cell development (41, 42), and light chain proteins isolated from chicken serum are structurally diverse by isoelectric focusing criteria (43). Clearly the light chain repertoire is diversified during B cell development in the chicken. Although only a single light chain Vλ gene is rearranged in chicken B cells, many other unrearranged Vλ gene segments were apparently present in the genome, as evidenced by extensive cross hybridizations in Southern blots performed using a Vλ1 probe (40).

Reynaud and coworkers cloned and sequenced all of the germline Vλ genes, and found, in addition to the Vλ1 gene, an unrearranged set consisting of 25 closely linked members, which were all apparently nonfunctional (pseudogenes) by several criteria, including coding sequence truncations, absence of consensus signals for rearrangement, missing leader sequences, and translation termination codons. The pseudogenes were tightly linked: no two adjacent members were more than 1.6 kilobases distant, and the closest pair was separated by only 22 bases.

To determine the structural basis of the apparent diversification of the single expressed Vλ1Jλ gene, Reynaud et al (33) cloned rearranged Vλ1Jλ genes from bursal B cells at two stages of development: day 18 of embryonic development and 21 days post-hatching. Sequences of these rearranged Vλ1Jλ genes revealed extensive somatic alterations. Surprisingly, nearly all of the alterations were identical to sequences located somewhere within the 25 pseudogenes (Figure 1): only 8 of 214 somatic nucleotide substitutions in rearranged Vλ1Jλ genes were unaccounted for by sequences within the pseudogene pool, and most of these exceptions were near border regions for postulated conversion events. This observation strongly implicates the pseudogenes as donors of the somatic alterations seen within the rearranged Vλ1Jλ genes. The lengths of the transferred information varied from approximately 10 to 100 nucleotides. In the sample of 12 rearranged Vλ1Jλ genes examined, the most frequent donors were the most homologous pseudogenes, but these were also the most proximal to the rearranged VλlJλ recipient. Proximity and homology have both been implicated as important variables in mammalian models of gene conversion (34, 37).

The regions of the rearranged VλlJλ genes most extensively altered by this somatic process were those encoding the complementarity determining regions. Extensive divergences in these regions are evident among the members of the pseudogene pool, in contrast to the relative conservation among them of framework sequences. These sequence relationships among the pseudogene donors nearly guarantee conservation in the Vλ1 target of those regions most important for V region structure (frameworks) and diversification of those regions most important in antigen binding (complementarity determining regions) (44).

All 12 of the cloned rearranged VλlJλ genes were somatically altered to some degree, indicating that somatic alteration was widespread among bursal B cells. This conclusion was supported by Southern blotting experiments

```
                              FR  1
                         6                                              17
Germline  Vλ1            TCC TCG GTG TCA GCG AAC CCG GGA GGA ACC GTC AAG

3W2  (rearranged)        G-- --- --- --- --A --- -T- --- --- --- --- G--

Ψ V#2&7                  G-- --- --- --- --A --- -T- --- --- --- --- G--


                             CDR  2
                         44                      51
Germline  Vλ1            TAT GAC AAC ACC AAC AGA CCC TCG

3W4  (rearranged)        --- AG- --- -A- G-G --- --- ---

Ψ v#18                   --- AG- --- -A- G-G --- --- ---


                             FR  3
                         74                  79
Germline  Vλ1            CGA GCC GAC GAC AAT GCT

18D-6  (rearranged)      -A- -T- --G --- G-G ---

Ψv#4,8,20,24             -A- -T- --G --- G-G ---
```

Figure 1 Somatic gene conversion in chicken Vλ1 genes. Sequences of somatically altered rearranged Vλ1 genes are compared to the germline *(top)* version, in three different regions. Lines indicate identity with the germline sequence. Sequences of potential pseudogene donors for the somatic alterations are shown at the bottom. Data taken from Reynaud et al (33).

performed on DNA from whole bursal B cell populations. In such experiments, the progressive loss of particular germline restriction enzyme sites within the rearranged VλlJλ gene segment could be seen as development progressed (33, 39). This observation, together with the fact that the rearranged VλlJλ genes from day 21 birds had more alterations than did those from 18-day embryos, suggested that somatic alterations were sustained sequentially in single B cell lineages. Alternatively the somatic alterations might be sustained as a single burst, followed by the preferential selection of more extensively altered variants.

To distinguish between these alternatives, Thompson & Neiman (39) transformed bursal B cells with avian leukosis virus and obtained clonal cell lines, as revealed by the genomic site of integration of the viral genome and the rearrangement of only one of the two possible Vλl alleles that could be distinguished on the basis of a restriction fragment length polymorphism. (They used an F1 chicken.) One transformed B cell line thus obtained contained rearranged VλlJλ genes that had uniformly (in all cells of the clone) lost a particular restriction site within VλlJλ, indicating that a common conversion event was apparently sustained by all members of the clone. With respect to other restriction site losses within VλlJλ however, the line was heterogeneous, indicating that these somatic alterations were punctuated by cell division cycles following the viral integration event. This result suggests that the information transfer process proceeds sequentially through cycles of cell division.

Southern blotting analyses of DNAs derived from lines of transformed bursal B cells showed no obvious alteration of the restriction map defining the pseudogene pool, suggesting that the transfer mechanism may indeed be nonreciprocal and thus of the conversion type (33). The possibility that a reciprocal event is followed by the segregation of participating chromatids, however, remains to be formally excluded.

Weill & Reynaud (44) estimate, on the basis of cell division time and the average number of conversion events sustained in VλlJλ genes obtained from the 21-day-old chicken (4-7 events), that one successfully propagated conversion event occurs every 10–15 cell generations. Whereas division without cell death should generate 10^{20} B cells at 21 days post-hatching, the chicken contains, in fact, only 6×10^9 B cells at this stage, suggesting that perhaps many of the somatic alterations are detrimental to the survival of the B cell.

Despite extensive modification of the rearranged VλlJλ allele, no modifications are evident in the nonrearranged Vλl allele, which is only 1.7 kilobases removed from the Jλ gene segment. Rearrangement thus apparently activates a cis-acting transfer process. The basis for the cis-specificity of the transfer process could be transcriptional activation of the rearranged allele. A significant body of data suggests that transcription is necessary for immunoglobulin gene rearrangements in mammals, including both V gene assembly and isotype switching (45–50). Yancopoulos & Alt suggest that the basis of this effect may be related to the accessibility of transcribed DNA sequences to recombination-mediating enzymes (46). Of interest in this regard is the observation by Thompson & Neiman of DNase hypersensitive sites unique to the rearranged Vλ1 allele (39).

CONVERSIONS IN MAMMALIAN IMMUNOGLOBULIN GENES

Mammalian Antibody Diversity

Diversity in mammalian antibodies originates from large multigene germline V gene segment families and from three types of somatic processes that operate on this gene pool (reviewed in 51). In addition, due to extensive polymorphisms, germline diversity within a species exceeds that of the individual (27, 52–54). Germline V genes are segregated into five qualitatively different kinds of segments, and during the ontogeny of any one B lymphocyte one combination of these segments assembles to encode complete and contiguous V genes for the heavy and the light chains (51). A V_H, D, and J_H segment combination assemble to encode the complete heavy chain V region, and a V_L, J_L pair assemble to encode the complete light chain V region. This somatic assembly process operating on the V gene segment pools, each with multiple members, probably generates more than 10^7 different V regions, distinguishable by the particular combination of segments encoding them. It thus differs markedly from the process that generates avian diversity, in which only single Vλ and Jλ gene segments are utilized for the light chain and in which a segmental gene conversion apparently replaces the mammalian assembly mechanism as a means of creating diversity.

For any given pair of assembling V gene segments, inconsistent sequence information has been observed at the segment junctures (55–59). This junctional diversity is the consequence of a variation in the precise point of ligation of the segments. In the case of the heavy chain V genes, the addition of sequence information that is not derived from either of the assembling segments also contributes to junctional diversity (56, 57).

Sequences of expressed hybridoma V genes have shown that they frequently differ from those of the corresponding germline V gene segments. These somatic alterations are generated by the third and poorly understood somatic diversification process termed "somatic mutation," despite a dearth of information regarding the actual mechanism (reviewed in 60). Studies of model murine immune responses to defined antigens have revealed that somatic alterations are apparently introduced exclusively during the antigen-driven stages of B cell clonal expansion and differentiation (61–65). Somatic alterations are sustained over multiple cell division cycles (66–70). Finally, selection of B cells with receptor antibodies of high affinity during the immune response results in a concomitant selection of somatically altered antibodies of increased affinity (reviewed in 60).

The possible role of gene conversion in generating, over an evolutionary time scale, the diverse pool of germline V gene segments, and over a physiological time scale in creating somatic alterations in mammalian V genes, is discussed below.

The Minigene Hypothesis

Comparisons of the amino acid sequences of a large collection of myeloma variable regions showed that some subregions were more conserved than others. Four conserved areas on each chain, called framework regions (FR), are interrupted by three hypervariable regions, termed complementarity determining regions (CDR) because of their apparent role in contacting antigen (71, 72). Kabat and coworkers (73–75) observed that many myeloma antibodies had V regions in which the sequence of one or more of these FR or CDR segments were identical. Yet for any given pair of V regions under comparison, identity in one subregion was not always correlated with identity in another subregion (e.g. FR1 vs FR2): CDR and FR subregions appeared to be segregating independently. A major thrust of research at the time of this observation was to elucidate the genetic basis of antibody diversity. All such theories were bound by two extreme hypothetical conditions: at one extreme, one germline gene was somatically modified to create the entire repertoire of V genes, and at the other extreme, the entire repertoire of V genes was directly encoded in the germline and expressed somatically unmodified (76–78). Kabat and coworkers' observation of independently segregating V subregions led them to propose the minigene hypothesis, an attractive compromise to these two extremes (73–75). They suggested that the millions of V region structures expressed by an individual were the expressed products of a more limited number of minigenes encoding individual subregions (CDRs and FRs). For any given B cell one of the many possible combinations of the FR and CDR minigenes was proposed to assemble somatically.

Subsequent cloning and sequencing analyses revealed that to a limited extent, the hypothesis was correct, inasmuch as the J segments encode FR4, and CDR3 in the heavy chain is encoded by a composite of D segment and junctional sequences (79). But in contrast to the expectations of the minigene hypothesis, FR1, CDR1, FR2, CDR2, and FR3 regions of both heavy and light chains were encoded as a unit by gene segments termed V_H and V_L, respectively (80–82). What is the genetic explanation for this discrepancy between the apparent independent assortment of genes for FR1–FR3 regions as revealed by protein sequences and their association as a single genetic unit in the form of a V_H or V_L gene segment? Comparisons of germline sequences of related V genes suggests that recombinational mechanisms have shuffled segments of V gene sequence information over an evolutionary time scale.

Germline Conversions

VARIABLE GENES   A number of groups have now reported sequence data for immunoglobulin germline V_H and V_L genes of humans, mice, and rabbits, suggesting that information transfer among V genes has played an important evolutionary role in the generation and diversification of these multigene families (24–31).

Two kinds of observations have been noted and used as evidence in support of intergenic transfer of information between Ig V genes in the germline.

First, V genes that have recently duplicated, as evidenced by near sequence identity over long stretches, may differ significantly in sequence in highly localized segments. Bentley & Rabbitts (24) compared two nonallelic human V_K genes (HK101 and HK137) that exhibited homology over 13.5 kilobases of sequence as indicated by heteroduplex analysis. In 940 bases of contiguous sequence, including the V_K coding regions, the two genes differed in only 10 positions, but 7 of the 10 were clustered in a 39-base-pair stretch encoding the first CDR. The clustering of differences in such a short segment suggests that the CDR1 region of one of the genes was recently altered by a conversion event involving a related gene. Events of this sort that apparently involve the segmental transfer of V gene information in the germline over evolutionary time could explain the apparent independent assortment of FR and CDR segments that were noted by Kabat and his colleagues, as proposed by Baltimore (83). Results of Jaenichen et al (25) further support this suggestion.

Jaenichen and coworkers performed sequence comparisons of a number of human germline V_K gene segments. Comparison of the translated amino acid sequences of these genes revealed apparent independent assortment of CDR and FR regions. When the nucleotide sequences were compared, however, the boundaries of the subregions of shared homology often did not coincide with CDR or FR boundaries. This result implies that segmental information transfers among ancestral members of this V_K gene family were not limited by CDR and FR boundaries. Analogous results in support of a germline conversion mechanism that diversifies rabbit V_H genes were reported by McCormack et al and are illustrated in Figure 2.

The second kind of observation used in support of germline intergenic transfer of V gene information is illustrated by data from Bentley & Rabbitts (24). They identified two more distantly related V_K genes, HK102 and HK122, that differ from each other in five of the same seven locations where HK101 and HK137 differ. HK102 is identical to HK101 in six of seven of the diverged positions and HK122 is identical to HK137 in five of the same seven

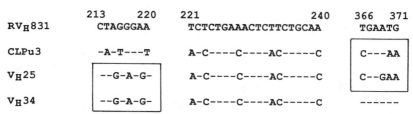

Rabbit V_H genes

	213 220	221 240	366 371
RV_H831	CTAGGGAA	TCTCTGAAACTCTTCTGCAA	TGAATG
CLPu3	-A-T---T	A-C----C----AC-----C	C---AA
V_H25	--G-A-G-	A-C----C----AC-----C	C--GAA
V_H34	--G-A-G-	A-C----C----AC-----C	------

Figure 2 Evidence of germline transfer of information among rabbit V_H genes. Lines indicate identity with top (RVH831) sequence. Assuming evolution occurred by duplication and point mutation exclusively, on the basis of sequences between nucleotides 221 and 240, VH25, VH34, and CLpμ3 shared a common ancestor not shared by RVH831. Similarly, sequences between positions 213 and 220 suggest that VH25 and VH34 have diverged most recently from a common ancestor not shared with CLpμ3. Sequences between positions 366 and 371, however, contradict these inferred evolutionary relationships and suggest instead that VH25 and CLpμ3 have recently diverged from a common ancestor not shared with VH34 and RVH831, which similarly diverged from a common ancestor not shared by VH25 and VH34. Data taken from McCormack et al (31).

positions. HK102 and HK122 thus embody most of the sequence expected for postulated ancestral donors in a transfer event that resulted in the segmental divergence of HK101 and HK137. To exclude the possibility of information transfer events while attempting to generate an evolutionary tree relating the four human V_K sequences would require the parallel and independent acquisition and fixation of five mutations in two different ancestral genes. The need to otherwise invoke parallel independent mutational events to construct evolutionary lineages of V genes is the second argument used to support a role for recombination in their germline evolution. This argument has been made by several authors (26–29), and is clearly illustrated by Clarke & Rudikoff (26) using sequences of the murine T15 V_H gene subfamily.

Four genes belonging to the T15 V_H subfamily were previously cloned and sequenced from inbred strain Balb/c mice. No two members of this highly homologous set differ in nucleotide sequence by more than 15% in their coding regions. Clarke & Rudikoff obtained the sequence of an expressed hybridoma antibody V_H gene called 6G6 from a CBA mouse (26). Comparison of the sequence of this gene with the other T15 members suggested that it was an allele of the V11 gene of Balb/c mice, differing in the coding region by only 3%, or in nine nucleotide positions. At five of these nine positions a second T15 Balb/c gene called V13 shared sequence identity with 6G6 even though this pair of genes was apparently less related (95% identical) than were V11 and 6G6. At two other of the nine discrepancies between V11 and 6G6, a third even less related gene (90% homologous to 6G6) was identical to 6G6. To account for these sequence relationships purely on the basis of gene duplication and point mutation would require minimally the parallel independent acquisition in two genes of four point mutations, two of which are silent. A persuasive argument can be put forth favoring informational transfer among V genes, as opposed to parallel independent point mutations, as the source of shared homologies like those described above. In some cases [e.g. the T15 subfamily (26)] parallel independent mutations would have to be silent and hence would likely not be selected. Furthermore, individual V genes are apparently not under strong evolutionary selection pressure, which presumably would be needed to fix parallel independent point mutations. Lack of a strong selection pressure is suggested from the large size of the V gene family (53, 84), from which loss of any single member might be expected to only minimally affect survival (85). Evidence supporting this contention is derived from three observations. First, a large proportion of V genes are nonfunctional [Rechavi et al (30) calculate that approximately 40% are pseudogenes]. Second, expansions and contractions in V gene number are frequent and seen even within species (27, 53, 54, 86, 87). Third, V genes are highly polymorphic, suggesting that they are rapidly diverging (27, 52–54).

The mechanisms by which transfer of genetic information between V genes occurred in the germline histories of mammalian species can only be inferred, since all of the participants in a given transfer event cannot be isolated. Most authors have favored the mechanism of gene conversion, generally because short segments of information seem to be involved and because multiple recombinational events of reciprocal kind would be needed to account for the segmental nature of the homologies. Nevertheless, the precise nature of the mechanism(s) that recombine V gene information in the germline are unknown. Unequal recombination, for example (88, 89), is also a likely contributor to the germline shuffling of V gene information.

The hallmarks of unequal recombination are gene duplication and deletion. Duplications and deletions of V gene segments are apparent both between and within species (27, 52–54, 86, 87, 90–92). The human genome probably contains greater than 50 Vλ gene segments (90–92), while inbred mice have only 2 closely related Vλ gene segments (86) and a third one that is more distantly related (93, 94). Hybridization patterns between and within species support this contention of rapid changes in V gene copy number. Perlmutter et al (27) performed hybridization analyses with a DNA probe from a member of the T15 V_H subfamily on rat and mouse DNA. While rats have at least 14 related genes, mice have fewer than 6. Differences between inbred strains of mice were also evident: the number of germline DNA fragments hybridizing to a T15 V_H probe varied from 2 to 6. The predominating V_H gene utilized in the immune response of strain A mice to p-azophenylarsonate is lacking from the genome of strain Balb/c as revealed by hybridization (54) and cloning experiments (L. Wysocki, unpublished). In a more global study, Brodeur et al (53) and D'Hoostelaere et al (52) used DNA probes representing most of the known murine V_H and V_K subfamilies in hybridization analyses under nonstringent conditions to identify related genes within various inbred strains of mice. With most of the probes, differences in the number of hybridizing fragments among strains were obvious. These observations of apparent expansions and contractions of V genes support a role for unequal crossing over in the evolution of V genes.

One caveat regarding inferred changes in V gene copy number on the basis of hybridization data is the possibility that multiple segmental exchanges between V genes so extensively alters their sequences that they are no longer detectable with a given DNA probe. A drastically modified V gene may, in fact, exist in the same surrounding DNA context as before, and by such a criteria be allelic to a gene with a very different sequence located at the same position on the homologous chromosome. Construction of linkage relationships by overlapping cloning methods should be very useful in clarifying allelic relationships. Such relationships based on chromosomal address as well as on conventional sequence homologies would be very useful in understanding the evolutionary scenarios by which V gene families were generated.

CONSTANT GENES   Several reports document evidence for conversions between Ig constant region genes during their evolution (18–23). Particularly convincing are combined data from the work of Schrier et al (95) and Ollo & Rougeon (18), who cloned and sequenced the a and b alleles of the γ2a and γ2b genes. The a and the b alleles of γ2a were found to differ in 146 out of 1912 positions, making them two of the most diverged alleles known. Surprisingly, when sequences of the γ2a and γ2b nonallelic genes were compared, most of the sequence differences between the two γ2a alleles were found to be identical to sequences at the corresponding location in one or both of the γ2b alleles (the γ2b alleles only differ in 12 positions from one another). In 70 of 134 places where γ2a[a] differed from γ2a[b], the γ2a[a] nucleotide could be found in the same position in the closely linked γ2b[a] gene. Similarly, in 54 of 134 places where the γ2a alleles differed, the γ2a[b] nucleotide could be found in the corresponding position of the linked γ2b[b] gene. In only ten positions did γ2a allelic differences not correspond to γ2b sequences. The data strongly implicate information transfer(s) from an ancestral γ2b gene to an ancestral γ2a gene.

Ollo & Rougeon (18) suggest that intergenic exchanges of the sort they describe simultaneously homogenize and diversify related genes. Homogenization has apparently occurred in the sense that the γ2a alleles more closely resemble the γ2b alleles than did the ancestral versions of these genes. On the other hand, the two γ2a alles differ dramatically, precisely because of the intergenic transfers of information. Understanding how gene conversion can create or maintain diversity within a multigene family is difficult in view of the fact that it has been invoked to maintain sequence homogeneity in gene families such as the ribosomal RNA genes (96–98). A major implication of this work is that intergenic diversity is sacrificed at the expense of interallelic diversity. This conclusion is not inconsistent with the observation that V genes are extensively polymorphic and many of them are organized into subfamilies of closely (≥80%) related members (53).

Ollo & Rougeon favor the interpretation of gene conversion rather than unequal crossing over as the transfer mechanism involved in the γ2a-γ2b exchanges. In the latter case, a convoluted series of recombinational events would be needed to account for the sequence data and would result in gene duplications and deletions. As is the case with V genes, however, evidence for duplications and deletions in Ig constant genes is abundant (23, 99, 100). Most relevantly, the Japanese wild mouse has duplicated γ2a genes (99). Thus, reciprocal and nonreciprocal exchange mechanisms probably both have played a role in IgC gene evolution.

The extent of information transfer in Ig constant region genes is much more dramatic than the subtle differences thus far catalogued for mammalian germline V region genes. In part, this may be due to the larger size of the V gene pool, and the near impossibility of identifying true alleles (28). With more possible donors and recipients, sequential exchange events sustained over a long period of time may tend to scramble V gene sequence information, obscuring individual events. Extensive polymorphisms in V genes further aggravate this problem, since donor and recipient sequences may segregate within the species, or more specifically between two different inbred strains in the case of mice. The two are obviously related, since scrambling of information creates polymorphisms which can then segregate. The complexity of the germline V gene pool is not as great in chickens as in mammals. Not surprisingly, one of the clearest cases of germline V gene information exchange is seen in chickens.

The germline Vλ1 gene of CB chickens (inbred) differs by nine nucleotides in the coding region from the cloned germline Vλ1 gene of an outbred White Leghorn chicken (33). (These are the single Vλ1 genes that rearrange and are expressed in the respective strains.) In seven of these nine positions both of the corresponding nucleotides are found within members of the pseudogene pool of the CB chicken, and in five of those seven positions, the two nucleotides found in the Vλ1 genes are the only nucleotides found in the corresponding positions in the pseudogene pool (33).

Somatic Conversions

The mammalian antibody V region repertoire is generated by a somatic assembly process that links germline gene segments in a continuum to encode the complete variable genes that are poised for expression (51). Further diversity is generated at the segment boundaries during assembly. Whether additional diversity is generated via a gene conversion mechanism like the one that operates in the avian immune system is addressed below.

IN VITRO   Several examples of apparent conversions between IgV genes in hybridoma cell lines have now been obtained (32, 101–103). Dildrop et al (32) and Krawinkel et al (101) reported such an event between an expressed hybridoma V_H gene and a related V_H germline gene. The variant was selected from a parental hybridoma, B1.8, that expressed a surface IgD molecule with a well-defined idiotype and antigen specificity (103). To preclude the isolation of variants that had simply lost surface IgD expression (the largest category), selection was implemented using a pair of anti-idiotypic antibodies. Variants that had lost the property of binding to one but not the other of these two anti-idiotypes were selected by fluorescence-activated cell sorting (103).

Sequence analysis of the expressed V_H gene of one such variant indicated that it was altered by 15 nucleotide substitutions in a segment between codons 11 and 70 (102). Eight of the encoded ten amino acid replacements were clustered in CDR2. The contiguous stretch of sequence defining the altered segment of DNA (from codon 11 to codon 70) was found to be identical to that of a neighboring V_H gene belonging to the same subfamily. This neighboring V_H gene had been previously cloned and sequenced by Bothwell et al (29), and its identity with the somatic variant, precisely in the region of somatic alteration, strongly implicates it as a donor in a recombinational event that introduced the somatic alterations. The event was likely either a gene conversion or a double reciprocal recombination because sequence on either side of the somatically altered segment was identical to that of the V_H gene expressed by the parental B1.8 hybridoma. The left breakpoint in the potential recombinational event was bordered by palindromic sequences, which were also correlated with conversion events in other genes (102). Unfortunately, in the absence of selection, to definitively recover the copy of the donor gene that actually participated in the recombinational event was not possible. Normal chromosomal segregation events and possible nondisjunctional events, as well, may partition into separate cell lineages the selected and the unselected genetic participants of a recombinational event. Thus, whether the event was genuinely nonreciprocal, as expected of a gene conversion, could not be determined.

D. Panka and M. Margolies (personal communication) have recently obtained similar results implicating the somatic transfer of information between an expressed hybridoma V_H gene and unrearranged germline donors. Variants of a parental hybridoma (26-10) producing an antibody with specificity for digoxin were selected for reduced binding to digoxin by fluorescence-activated cell sorting (104) and subjected to sequencing analyses. Of seven isolates, four revealed only single amino acid replacements in the V_H region, while three contained multiple amino acid replacements. The mRNA encoding the V_H regions of two of the three with multiple replacements have been completely sequenced. One of the variants, R9, has a total of seven nucleotide substitutions clustered between codons 24 and 38 in the V_H region. A V_H nucleotide sequence reported by Givol et al (105) matches this variant stretch of sequence with the exception of a single nucleotide difference. As in the case discussed above, these data strongly implicate the somatic transfer of information from an unexpressed V_H gene related to the one reported by Givol et al to the V_H gene expressed by hybridoma 26-10. As in the case discussed above, the exact nature of the recombinational event could not be determined due to the possible segregation of selected and unselected genetic participants.

Whether information transfer in either of the cases described above does not simply involve the complete replacement of a rearranged V gene by the rearrangement of another unidentified upstream V_H gene has not been demonstrated (Figure 3). Events of this kind, which are apparently mediated by consensus sequences for immunoglobulin rearrangement, have been reported by two laboratories (106, 107). Ruling them out is difficult because sequences are known for only a minority of the many germline V genes.

IN VIVO

The preimmune repertoire   The finding that, in vitro, somatic variants of hybridoma antibodies apparently result from information transfer from other unexpressed V genes lends support to the contention that such transfers of information play a role in the evolution of germline V genes and suggests that similar mechanisms may operate in vivo during the somatic construction of the expressed repertoire. Although this question has not been systematically addressed, several observations indicate that it probably does not occur frequently.

In many cases, the V_H and V_L segments of expressed hybridoma variable regions have been cloned and used as DNA probes to isolate the corresponding germline V_H or V_L segment (29, 52, 61–65, 108–118). Often the germline V gene segments were found to be identical to the expressed V gene segments. If segmental exchanges of information contributed significantly to the generation of the V genes expressed in the preimmune repertoire, use of expressed V_H or V_L gene segments in genomic searches for their germline counterparts should frequently yield ambiguous germline candidates.

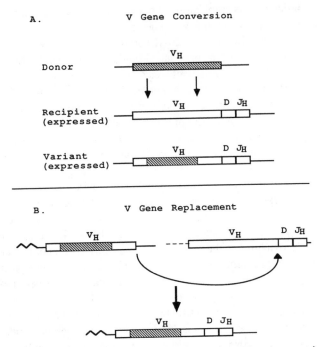

Figure 3 Alternative mechanisms to explain blocks of somatic alterations sustained by expressed hybridoma V_H genes in vitro. V gene replacement *(bottom)* is apparently mediated by signal sequences involved in the somatic assembly process (105, 106).

Data from Manser and Gefter (62, 63) can be used to address this issue more directly. They generated hybridomas from the preimmune murine B cell compartment and screened the hybridomas for the expression of a single V_H gene by using only a segment of the corresponding germline V_H gene as a probe. The probe spanned a portion of the coding region of the V_H gene from nucleotides 45 to 178. When the V_H genes expressed by the selected hybridomas were sequenced, the sequences were always identical to that of the complete germline V_H gene segment, even in the region not spanned by the probe (nucleotides 179–294). Had gene conversion altered preimmune somatic V_H structure, evidence of this should have been seen in the 3' segments of the expressed genes (nucleotides 179–294), which should have differed from the same segment of the original germline V_H gene from which the probe was derived.

The immune repertoire Sequences of expressed V genes from myelomas and hybridomas often differ from those of their germline counterparts. The somatic alterations have been operationally termed "somatic mutations." A wide body of data now suggests that somatic alterations are introduced during the antigen-driven stages of B cell differentiation (61–65). Sampling of immune B cells through the production of hybridomas has revealed that a high proportion of those participating in the secondary immune response express somatically altered V genes (reviewed in 60). Informatively, most of the altered V regions have a higher affinity for the eliciting antigen than do germline V regions that are encoded by the same combination of V gene segments (61). Since the somatic alteration process is not instructive (117, 119, 120), the B cells are apparently subjected to a selection pressure in which the affinity of the expressed antibody product for antigen is a critical variable determining their degree of clonal expansion and continued participation in the response. Because of its importance in creating the high-affinity antibody that characterizes the state of immunity, the somatic alteration process has been the focus of much attention.

One of the most clearly characterized examples of somatic alteration is seen in the immune response to phosphorylcholine, in which a large proportion of the elicited antibodies have V regions encoded by somatically altered versions of a common germline V_H gene segment, called T15. Crews et al (113) cloned and sequenced four germline V_H genes belonging to the T15 family and found them to be >85% identical in sequence. When the sequences of the mutated derivatives of the T15 V_H gene (expressed in myelomas and hybridomas) were compared to the sequences of the three closely related family members, only one of 24 somatic alterations could have been derived from the related V_H genes by a recombinational event. The authors suggest that gene conversion or recombinational processes are an unlikely origin for the somatic alterations.

Similar arguments have been made for somatically altered V genes in other immune responses. The two germline V_H genes that are most homologous to one that predominates in encoding anti-p-azophenylarsonate antibodies cannot account for the vast majority of the sustained somatic alterations (108). Most somatic alterations in murine $V\lambda$ genes cannot be accounted for as being derived from the other germline $V\lambda$ gene (120, 121) (inbred mice have only two closely related $V\lambda$ genes). Somatic alterations in the J_H and J_K segments, whose germline families have only four members each, most often have no counterpart in any of the remaining J segments (60). Together, these observations indicate that somatic alterations in expressed V genes are not apparently derived by information transfer from donor genes that are most homologous to them.

This interpretation has a serious caveat, however. Somatic alterations that improve the affinity of a V region structure for an immunogen result in the enhanced clonal selection of the B cell expressing the somatic variant (61). Thus, B cells that are sampled by hybridoma production inevitably express V genes with somatic alterations that confer to the V region enhanced affinity for the immunogen. Alterations that are detrimental or even neutral with respect to conferring affinity changes are likely to be underrepresented in most hybridoma samplings. Therefore, even if a given expressed V gene is undergoing somatic conversions with its most homologous relatives, hybridomas expressing these alterations probably would not be isolated unless they happened to specifically confer an improvement in affinity for the antigen used to elicit the immune response.

De novo addition of somatic alterations might be expected to generate nonfunctional V regions with a very high frequency. Nonconservative replacements of conserved framework amino acids that contribute to V region superstructure might be one category of unfavorable alteration. Translation termination codons introduced anywhere in the V gene would be another. One possible strategy to introduce somatic alterations while minimizing V region destruction would be to use sequences located within the genome as donor information for the somatic alterations. If the donors were homologous to segments encoding CDR regions, framework disruptions could be minimized. Furthermore, evolution could play a role in eliminating the introduction of translation termination codons during the somatic alteration process by eliminating such sequences from the germline donor population.

It is generally assumed that the probability of genetic recombination is related to the degree of sequence identity between participants. Given the intensity of the clonal selection process, however, less frequent exchange events involving less identical participants could be favorably selected if they resulted in affinity improvements. Somatic alterations in expressed V genes could then, in theory, be derived from more distant genetic relatives. Of relevance in this regard is the finding of a third Vλ-like gene (93, 94) and several reports of members of the Ig supergene family with regions homologous to V and J segments (122–126).

Some recent results from this laboratory address this issue of weakly homologous donors for somatic alterations. Two independently derived hybridomas expressing somatically altered versions of a V_H gene of common germline origin share four identical and consecutive amino acid replacements in CDR2, and exhibit the hallmarks expected of a region encoded by a Kabat minigene (Figure 4) (M. Margolies, personal communication). (The two sequences have sustained different somatic alterations elsewhere.) Nucleotide sequences of the two V_H genes in the region of shared somatic alterations reveal, however, that they differ at two positions. These differences indicate that a minimum of two different donors would be needed to account for the common amino acid replacements, assuming gene conversion as the mechanism of their introduction.

Oligonucleotide probes spanning the regions of shared alterations were

CDR-2

```
                51                                              60
                I    N    P    G    N    G    Y    T    K    Y
GERMLINE:      ATT  AAT  CCT  GGA  AAT  GGT  TAT  ACT  AAG  TAC

   Variant #1:          H              K                   I    H
   Variant #2:          H              K                   I    H
```

Figure 4   Consecutive shared somatic alterations in two expressed V_H genes derived from a common germline V_H gene. The sequence of the germline copy is shown at the top. Amino acid replacements in each of the somatic variants are indicated below.

used to isolate from the genome the most identical fragments of DNA that could have served as donors for these alterations. Multiple and independent clonings of the same fragments of DNA demonstrated that the most identical germline segments encoded only 18 of 24 nucleotides defining the region of shared somatic alterations. These data, taken together with the sequence differences between the somatic variants at the nucleotide level, indicate that a minimum of three independent recombinational events would be needed to account for the shared set of somatic alterations, if indeed this were the mechanism of their introduction. In addition, the cloned germline segments of DNA differed from the V_H gene immediately beyond the boundaries defined by the probes used for their isolation and had no substantial homology to any other sequenced V gene. On the basis of these results, and similar results from Chien & Scharff (127), the transfer of segments of genomic information into expressed V genes is an unlikely explanation for the origins of somatic alterations.

The mechanism of the somatic alteration of mammalian V genes thus remains a major enigma in molecular immunology, and although several models have been proposed (128–130), hard data in support of them have proved to be elusive.

CONCLUSIONS

Sequence analyses of germline Ig gene segments indicate that recombinational mechanisms have contributed greatly to their evolutionary construction. The precise natures of the recombinational mechanisms are beyond unambiguous identification, but several clues suggest that gene conversion probably has played a major role. Conversion has been invoked as a mechanism to homogenize members of a gene family. Several reported cases of intergenic conversions show, however, that in the course of homogenizing nonallelic genes, new alleles are created (6–12 18, 20). Gene conversion could then be responsible for some of the extensive polymorphisms seen in IgV gene families.

A segmental gene conversion process apparently plays a major role in the diversification of the chicken light chain V region repertoire during B cell development (33). The predictable quality of this somatic process should lend it to further investigation. Rearrangement of the Vλ1 gene apparently activates the somatic conversion process and may also activate transcription (39). The strong correlation between transcription of Ig genes and their rearrangement in mammalian systems suggests that transcription may indirectly mediate recombinational processes by providing accessibility to recombination-mediating enzymes (45–48). Whether transcription of the chicken Vλ1 gene is involved in the process of its conversion, however, remains to be determined.

Somatic conversion processes apparently can alter the structures of expressed mammalian V genes in vitro (32, 101). These events are rare, however, and strong selection strategies are required to isolate the resulting somatic variants. Events like this could operate in vivo, but if so, they are probably infrequent. No evidence in support of a role for gene conversion in the somatic construction of either the preimmune or the postimmune mammalian V gene repertoire has yet been reported.

ACKNOWLEDGMENTS

Supported by American Cancer Society grants to Malcolm Gefter (#NP-6R) and to Lawrence Wysocki (#IM-525) and by a grant from the National Institutes of Health (5 R37 AI13357-13) awarded to Malcolm Gefter.

Literature Cited

1. Radding, C. M. 1978. *Annu. Rev. Biochem.* 47:847–80
2. Klein, H. L., Petes, T. D. 1981. *Nature* 289:144–48
3. Jackson, J. A., Fink, G. R. 1981. *Nature* 292:306–11
4. Scherer, S., Davis. R. W. 1980. *Science* 209:1380–84
5. Egel, R. 1981. *Nature* 290:191–92
6. Slightom, J. L., Blechl, A. E., Smithies, O. 1980. *Cell* 21:627–38
7. Mellor, A. L., Weiss, E. H., Ramachandran, R. A., Flavell, R. A. 1983. *Nature* 306:792–95
8. Pease, L. R., Schulze, D. H., Pfaffenbach, G. M., Nathenson, S. G. 1983. *Proc. Natl. Acad. Sci. USA* 80:242–46
9. Weiss, E. H., Golden, L., Zakut, R., Mellor, A., Fahrner, K., et al. 1983. *EMBO J.* 2:453–62
10. Widera, G., Flavell, R. A. 1984. *EMBO J.* 3:1221–25
11. Geliebter, J., Zeff, R. A., Schulze, D. H., Pease, L. R., Weiss, E. H., et al. 1986. *Mol. Cell. Biol.* 6:645–52
12. Gorski, J., Mach, B. 1986. *Nature* 322:67–70
13. David, C. S., McCormick, J. F., Lafuse, W. P., Hirose, S. 1986. *Transplantation* 42:429–-33
14. Sullivan, K. F., Lau, J. T., Cleveland, D. W. 1985. *Mol. Cell. Biol.* 5:2454–65
15. Lowell, C. A., Potter, D. A., Stearman, R. S., Morrow, J. F. 1986. *J. Biol. Chem.* 261:8442–52
16. Clark, A. J., Chave-Cox, A., Ma, X., Bishop, J. O. 1985. *EMBO J.* 4:3167–71
17. Liu, T. J., Liu, L., Marzluff, W. F. 1987. *Nucleic Acids Res.* 15:3023–39

18. Ollo, R., Rougeon, F. 1983. *Cell* 32:515–23
19. Akimenko, M.-A., Heidmann, O., Rougeon, F. 1984. *Nucleic Acids Res.* 12:4691–701
20. Marame, B., Akimenko, M.-A., Rougeon, F. 1987. *Nucleic Acids Res.* 15:6171–79
21. Udey, A. J., Blomberg, B. 1988. *Nucleic Acids Res.* 16:2959–69
22. Lefranc, M.-P., Helal, A.-N., de Lange, G., Chaabani, H., van Loghem, E., Lefranc, G. 1986. *FEBS Lett.* 196:96–102
23. Flanagan, J. G., Lefranc, M.-P., Rabbitts, T. H. 1984. *Cell* 36:681–88
24. Bentley, D. L., Rabbitts, T. H. 1983. *Cell* 32:181–89
25. Jaenichen, H. R., Pech, M., Lindenmaier, W., Wildgruber, N., Zachau, H. G. 1984. *Nucleic Acids Res.* 12:5249–63
26. Clarke, S, H., Rudikoff, S. 1984. *J. Exp. Med.* 159:773–82
27. Perlmutter, R. M., Berson, B., Griffin, J. A., Hood, L. 1985. *J. Exp. Med.* 162:1998–2016
28. Loh, D. Y., Bothwell, A. L. M., White-Scharf, M. E., Imanishi-Kari, T., Baltimore, D. 1983. *Cell* 33:85–93
29. Bothwell, A. L. M., Paskind, M., Reth, M., Imanishi-Kari, T., Rajewsky, K., Baltimore, D. 1981. *Cell* 24:625–37
30. Rechavi, G., Ram, D., Glazer, L., Zakut, R., Givol, D. 1983. *Proc. Natl. Acad. Sci. USA* 80:855–59
31. McCormack, W. T., Laster, S. M., Marzluff, W. F., Roux, K. H. 1985. *Nucleic Acids Res.* 13:7041–55
32. Dildrop, R., Bruggemann, M., Radbruch, A., Rajewsky, K., Beyreuther, K. 1982. *EMBO J.* 5:635–40
33. Reynaud, C.-A., Anquez, V., Grimal, H., Weill, J.-C. 1987. *Cell* 48:379–88
34. Liskay, R. M., Stachelek, J. L., Letsou, A. 1984. *Cold Spring Harbor Symp. Quant. Biol.* 49:183–89
35. Subramani, S., Rubnitz, J. 1985. *Mol. Cell. Biol.* 5:659–66
36. Liskay, R. M., Stachelek, J. L. 1986. *Proc. Natl. Acad. Sci. USA* 83:1802–6
37. Liskay, R. M., Letsou, A., Stachelek, J. L. 1987. *Genetics* 115:161–67
38. Rubnitz, J., Subramani, S. 1987. *Somatic Cell Mol. Genet.* 13:183–87
39. Thompson, C. B., Neiman, P. E. 1987. *Cell* 48:369–78
40. Reynaud, C.-A., Anquez, V., Dahan, A., Weill, J.-C. 1985. *Cell* 40:283–91
41. Lydyard, P. M., Grossi, C. E., Cooper, M. D. 1976. *J. Exp. Med.* 144:79–97
42. Huang, H. V., Dreyer, W. J. 1978. *J. Immunol.* 121:1738–47
43. Jalkanen, S., Jalkanen, M., Granfors,

K., Toivanen, P. 1984. *Nature* 311:69–71
44. Weill, J.-C., Reynaud, C.-A. 1987. *Science* 238:1094–98
45. Yancopoulos, G. D., Alt, F. W. 1985. *Cell* 40:271–81
46. Yancopoulos, G. D., Alt, F. W. 1986. *Annu. Rev. Immunol.* 4:339–68
47. Blackwell, K. T., Moore, M. W., Yancopoulos, G. D., Suh, H., Lutzker, S., et al. 1986. *Nature* 324:585–89
48. Lutzker, S., Alt, F. W. 1988. *Mol. Cell. Biol.* 8:1849–52
49. Lutzker, S., Rothman, P., Pollock, R., Coffman, R., Alt, F. W. 1988. *Cell* 53:177–84
50. Yancopoulos, G. D., DePinho, R. A., Zimmerman, K. A., Lutzker, S. G., Rosenberg, N., Alt, F. W. 1986. *EMBO J.* 5:3259–66
51. Honjo, T. 1983. *Annu. Rev. Immunol.* 1:499–528
52. D'Hoostelaere, L. A., Huppi, K., Mock, B., Mallett, C., Potter, M. 1988. *J. Immunol.* 141:652–61
53. Brodeur, P. H., Riblet, R. 1984. *Eur. J. Immunol.* 14:922–30
54. Siekevitz, M., Gefter, M. L., Brodeur, P., Riblet, R., Marshak-Rothstein, A. 1982. *Eur. J. Immunol.* 12:1023–32
55. Tonegawa, S. 1983. *Nature* 302:575
56. Alt, F. W., Baltimore, D. 1982. *Proc. Natl. Acad. Sci. USA* 79:4118–22
57. Desiderio, S. V., Yancopoulos, G. D., Paskind, M., Thomas, E., Boss, M. A., et al. 1984. *Nature* 311:752–55
58. Sakano, H., Huppi, K., Heinrich, G., Tonegawa, S. 1979. *Nature* 280:288–94
59. Max, E. E., Seidman, J. G., Leder, P. 1979. *Proc. Natl. Acad. Sci. USA* 76:3450–54
60. Moller, G., ed. 1987. *Immunol. Rev.* Vol. 96. 162 pp.
61. Wysocki, L., Manser, T., Gefter, M. L. 1986. *Proc. Natl. Acad. Sci. USA* 83:1847–51
62. Manser, T., Huang, S., Gefter, M. L. 1984. *Science* 226:1283–88
63. Manser, T. 1986. *J. Immunol.* 139:234–38
64. Manser, T., Gefter, M. L. 1986. *Eur. J. Immunol.* 16:1439–47
65. Kaartinen, M., Griffiths, G. M., Markham, A. F., Milstein, C. 1983. *Nature* 304:320–24
66. McKean, D., Huppi, K., Bell, M., Staudt, L., Gerhard, W., Weigert, M. 1984. *Proc. Natl. Acad. Sci. USA* 80:3180–84
67. Clarke, S. H., Huppi, K., Ruezinsky, D., Staudt, L., Gerhard, W., Weigert, M. 1985. *J. Exp. Med.* 161:687–704
68. Cleary, M. L., Meker, T. C., Levy, S.,

Lee, E., Trela, M., et al. 1986. *Cell* 44:97–106
69. Claflin, J. L., Berry, J., Flaherty, D., Dunnick, W. 1987. *J. Immunol.* 138:3060–68
70. Blier, P. R., Bothwell, A. 1987. *J. Immunol.* 139:3996–4006
71. Wu, T. T., Kabat, E. A. 1970. *J. Exp. Med.* 132:211–50
72. Kabat, E. A., Wu, T. T. 1971. *Ann. NY Acad. Sci.* 190:382–97
73. Kabat, E. A., Wu, T. T., Bilofsky, H. 1978. *Proc. Natl. Acad. Sci. USA* 75:2429–33
74. Kabat, E. A., Wu, T. T., Bilofsky, H. 1980. *J. Exp. Med.* 152:72–84
75. Kabat, E. A., Wu, T. T., Bilofsky, H. 1980. *J. Exp. Med.* 149:1299–313
76. Dreyer, W. J., Bennett, J. C. 1965. *Proc. Natl. Acad. Sci. USA* 54:864–68
77. Hood, L., Talmage, D. W. 1970. *Science* 168:325–34
78. Gally, J. A., Edelman, G. M. 1970. *Nature* 227:341–48
79. Sakano, H., Kurosawa, Y., Weigert, M., Tonegawa, S. 1981. *Nature* 290:562–70
80. Hozumi, N., Tonegawa, S. 1976. *Proc. Natl. Acad. Sci. USA* 73:3628–32
81. Lenhard-Schuller, R., Hohn, B., Brack, C., Hirama, M., Tonegawa, S. 1978. *Proc. Natl. Acad. Sci. USA* 74:4709–13
82. Sakano, H., Maki, R., Kurosawa, Y., Roeder, W., Tonegawa, S. 1980. *Nature* 286:676–83
83. Baltimore, D. 1981. *Cell* 24:592–94
84. Livant, D., Blatt, C., Hood, L. 1986. *Cell* 47:461–70
85. Kelsoe, G., Farina, D. 1987. *Evolution and Vertebrate Immunity*, pp. 163–74. Austin: Univ. Tex. Press
86. Scott, C. L., Potter, M. 1984. *J. Immunol.* 132:2638–43
87. Rechavi, G., Bienz, B., Ram, D., Ben-Neriah, Y., Cohen, J. B., et al. 1982. *Proc. Natl. Acad. Sci. USA* 79:4405–9
88. Tartof, K. D. 1975. *Annu. Rev. Genet.* 9:355–85
89. Smith, G. P. 1976. *Science* 191:528–35
90. Anderson, M. L. M., Brown, L., McKenzie, E., Kellow, J. E., Young, B. D. 1985. *Nucleic Acids Res.* 13:2931–41
91. Tsujimoto, Y., Croce, C. M. 1984. *Nucleic Acids Res.* 12:8407–14
92. Anderson, M. L. M., Szajnert, M. F., Kaplan, J. C., McColl, L., Young, B. D. 1984. *Nucleic Acids Res.* 12:6647–61
93. Sanchez, P., Cazenave, P. A. 1987. *J. Exp. Med.* 166:265–70
94. Dildrop, R., Gause, A., Muller, W., Rajewsky, K. 1987. *Eur. J. Immunol.* 17:731–34

95. Schrier, P. H., Bothwell, A. L. M., Mueller-Hill, B., Baltimore, D. 1981. *Proc. Natl. Acad. Sci. USA* 78:4495–99
96. Nagylaki, T., Petes, T. D. 1982. *Genetics* 100:315–37
97. Hood, L., Campbell, J. H., Elgin, S. C. R. 1975. *Annu. Rev. Genet.* 9:305–53
98. Dvorak, J., Jue, D., Lassner, M. 1987. *Genetics* 116:487–98
99. Shimizu, A., Hamaguchi, Y., Yaoita, Y., Moriwaki, K., Kondo, K., Honjo, T. 1982. *Nature* 298:82–84
100. Scott, C. L., Mushinski, F. J., Huppi, K., Weigert, M., Potter, M. 1982. *Nature* 300:757–60
101. Krawinkel, U., Zoebelein, G., Bruggemann, M., Radbruch, A., Rajewsky, K. 1983. *Proc. Natl. Acad. Sci. USA* 80:4997–5001
102. Krawinkel, U., Zoebelein, G., Bothwell, A. L. M. 1986. *Nucleic Acids Res.* 14:3873–80
103. Bruggemann, M., Radbruch, A., Rajewsky, K. 1982. *EMBO J.* 5:629–34
104. Herzenberg, L. A., Kipps, T. J., Peterson, L., Parks, D. R. 1985. In *Biotechnology and Diagnostics*, ed. H. Koprowski, S. Ferrone, A. Albertini, 1:3–16. Amsterdam: Elsevier
105. Givol, D., Zakut, R., Effron, K., Rechavi, G., Ram, D., Cohen, J. B. 1981. *Nature* 292:426–30
106. Kleinfield, R., Hardy, R. R., Tarlinton, D., Dangl, J., Herzenberg, L. A., Weigert, M. 1986. *Nature* 322:843–46
107. Reth, M., Gehrmann, P., Petrac, E., Wiese, P. 1986. *Nature* 322:840–42
108. Siekevitz, M., Huang, S.-Y., Gefter, M. L. 1983. *Eur. J. Immunol.* 13:123–32
109. Seidman, J. G., Max, E. E., Leder, P. 1979. *Nature* 280:370–75
110. Max, E. E., Seidman, J. G., Leder, P. 1979. *Cell* 21:793–99
111. Selsing, E., Storb, U. 1981. *Cell* 25:47–58
112. Bernard, O., Hozumi, N., Tonegawa, S. 1978. *Cell* 15:1133–44
113. Crews, S., Griffin, J., Huang, H., Calame, K., Hood, L. 1981. *Cell* 25:59–66
114. Even, J., Griffiths, G. M., Berek, C., Milstein, C. 1985. *EMBO J.* 4:3439–45
115. Wysocki, L. J., Gridley, T., Huang, S., Grandea, A. G., Gefter, M. L. 1987. *J. Exp. Med.* 166:1–11
116. Sanz, I., Capra, J. D. 1987. *Proc. Natl. Acad. Sci. USA* 84:1085–89
117. Pech, M., Hochtl, J., Schnell, H., Zachau, H. G. 1981. *Nature* 291:668–70
118. Schiff, C., Milili, M., Hue, I., Rudi-

koff, S., Fougereau, M. 1986. *J. Exp. Med.* 163:573–87
119. Gearhart, F. J., Bogenhagen, D. F. 1983. *Proc. Natl. Acad. Sci. USA* 80:3439–43
120. Manser, T., Parhami-Seren, B., Margolies, M. N., Gefter, M. L. 1987. *J. Exp. Med.* 166:1456–63
121. Cumano, A., Rajewsky, K. 1986. *EMBO J.* 5:2459–68
122. Maddon, P. J., Littman, D. R., Godfrey, M., Maddon, D., Chess, L., Axel, R. 1985. *Cell* 42:93–104
123. Hedrick, S. M., Nielsen, E. A., Kavaler, J., Cohen, D. I., Davis, M. M. 1984. *Nature* 308:153–58
124. Sukhatme, V. P., Sizer, K. C., Voll-

mer, A. C., Hunkapiller, T., Parnes, J. R. 1985. *Cell* 40:591–97
125. Mostov, K. E., Friedlander, M., Blobel, G. 1984. *Nature* 308:37–43
126. Williams, A. F., Gagnon, J. 1982. *Science* 216:696–703
127. Chien, N., Pollock, R. R., Desaymard, C., Scharff, M. D. 1988. *J. Exp. Med.* 167:954–73
128. Golding, B. G., Gearhart, P. J., Glickman, B. W. 1987. *Genetics* 115:169–76
129. Steele, E. J., Pollard, J. W. 1987. *Mol. Immunol.* 24:667–73
130. Kolchanov, N. A., Solovyov, V. V., Rogozin, I. B. 1987. *FEBS Lett.* 214:87–91

Annu. Rev. Biochem. 1989. 58:533–73

ICOSAHEDRAL RNA VIRUS STRUCTURE[1]

Michael G. Rossmann and John E. Johnson

Department of Biological Sciences, Purdue University, West Lafayette, Indiana 47907

CONTENTS

PERSPECTIVES AND SUMMARY

Viruses provide well-defined systems for the study of structural biology at near atomic resolution. They are nucleoprotein particles designed to transport specific genes between cells of a host and between hosts. Depending on the

[1]Abbreviations used: AMV, alfalfa mosaic virus; BBV, black beetle virus; BMV, brome mosaic virus; BPMV, beanpod mottle virus; CCMV, cowpea chlorotic mottle virus; CPMV, cowpea mosaic virus; FMDV, foot-and-mouth disease virus; HIV, human immunodeficiency virus; HRV, human rhinovirus; NβV, *Nudaurelia capensis β* virus; NIm, neutralizing immunogenic; SBMV, southern bean mosaic virus; STNV, satellite tobacco necrosis virus; TBSV, tomato bushy stunt virus; TCV, turnip crinkle virus; TMV, tobacco mosaic virus; TYMV, turnip yellow mosaic virus.

0066-4154/89/0701-0533$02.00

virus type, the genetic information may be coded on either RNA or DNA, which may be single-stranded or double-stranded. The nucleic acid is packaged in a capsid composed of either protein or a lipid membrane and protein. Figure 1 schematically illustrates the diversity found in viral capsids and the type and organization of the encapsulated nucleic acid (1).

It has been more than a decade since the 2.9 Å structures of tomato bushy stunt virus (TBSV; 2) and tobacco mosaic virus (TMV) disk protein (3) initiated high-resolution structural virology. Since then, at least 14 other virus

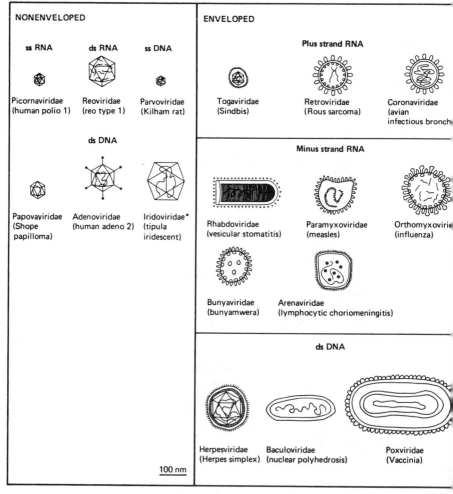

Figure 1 Diversity found in viral capsids and the type and organization of the encapsulated nucleic [Reprinted with permission from Matthews (1). Copyright by CRC Press, Inc.]

structures and a number of viral surface antigens have been determined at atomic resolution (Table 1). New virus structure determinations are being made at an ever-increasing rate as a consequence of advances in technology. In particular, the availability of intense synchrotron X-radiation has made it possible to collect diffraction data at very fast rates before the crystals are excessively damaged. Supercomputers have made it feasible to process and analyze the vast quantities of data. Finally, experience in the utilization of noncrystallographic symmetry (13) (symmetry that pertains to a virion, but is not propagated throughout the crystal) has not only simplified the structure solution of viruses, but also increased the precision of the results compared to smaller protein structures (14). The major remaining obstacles in the determination of atomic-resolution virus structures are the ability to produce

Table 1 Three-dimensional high-resolution structures of spherical viruses

		Capsid[a] symmetry	Reference
Plant RNA viruses:			
Tombus group	TBSV	$T = 3$	2
	TCV	$T = 3$	4
Sobeamo group	SBMV	$T = 3$	5
Como group	CPMV	$P = 3$	6
	BPMV	$P = 3$	73
	STNV	$T = 1$	7
Animal RNA viruses:			
Rhino	HRV14	$P = 3$	8
	HRV1A	$P = 3$	94
Entero	Polio Mahoney 1	$P = 3$	9
	Polio Sabin 3	$P = 3$	J. M. Hogle, private communication
Cardio	Mengo	$P = 3$	10
Aphtho	FMDV 01K	$P = 3$	10a
Insect RNA viruses:			
	BBV	$T = 3$	11

[a] $T = n$ ($n = 1, 3, 4, 7, \ldots$) relates to the triangulation number (12). $P = 3$ implies a pseudo $T = 3$ surface lattice arrangement where there are three nonidentical but similarly folded subunits arranged as in a $T = 3$ lattice.

sufficient quantity of purified and stable virus and (perhaps even more difficult) to produce suitable crystals that diffract well. Nevertheless, there is now a promise of a very rapid expansion of structural information.

Perhaps the most surprising result that has emerged is that small, spherical RNA viruses all have similar tertiary and quaternary structures. For instance, each of the three larger capsid proteins of picornaviruses has the same fold and they are similar to the single protein that is used in the capsid of many plant RNA viruses. Thus, it would seem rather probable that these and other viruses have evolved from a common primordial virus, which presumably had little host specificity. In spite of these similarities, the properties of these viruses are diverse, in part as a result of mutations, insertions, and deletions in the capsid proteins. The advent of rapid RNA and DNA sequencing techniques has provided the chemical decoration to the available structures, which permits a more rational analysis of capsid properties such as cell attachment, penetration, uncoating, assembly, and mechanisms of neutralization.

This review concentrates primarily on the structures of spherical viruses that have been determined to high resolution. Thus, no attempt is made to review seriously the results of electron microscopy or other techniques (e.g. chemical cross-linking and NMR) that can be employed to analyze structure. A necessary complement to three-dimensional structure determination is knowledge of nucleic acid or capsid protein sequences. Such information in itself can give considerable insight into function, but a detailed analysis of the very large viral sequence information base will be omitted.

In light of the rapid expansion of structural results (15, 16), earlier reviews of the subject were comprehensive (for their time) but dated. Liljas (17) has written an excellent and relatively recent account that mostly covers the earlier plant virus work but predates the high-resolution results on spherical animal viruses. Similarly, Harrison's review (18) is even more dated but gives a thorough conceptual basis, which is unlikely to be repeated in future reviews. There are other authors that cover special topics, such as chemical stability (19), assembly (20), or evolution (21), but these accounts can now be updated in light of new structural information.

The first part of this chapter reviews the virus structures determined by X-ray crystallography and mentions some observations derived from image-processed electron micrographs. Emphasis is given to a discussion of the common features in the subunit tertiary structures and to a comparison of the quaternary organization in the viral particles. The second part of this chapter considers the evolution of viruses and the functional role of structural proteins. However, excluded from this review is work on the structural components of viruses, such as the surface antigens of influenza virus (22, 23), which have been reviewed elsewhere (24).

INTRODUCTION

The size and complexity of virus capsids and the availability of substantial quantities of virus have largely dictated the early course of high-resolution studies. Small, nonenveloped, single-stranded, RNA plant viruses were the first spherical viruses to be studied because it is relatively easy to obtain gram quantities of many plant viruses. Mammalian and insect virus structures, which can be isolated only in mg quantities, have been elucidated more recently. Larger, more complex, virus structures have been determined at moderate resolution by image analysis of electron micrographs and X-ray diffraction. Virus samples analyzed in this way have usually been negatively stained, but the use of frozen hydrated samples (25) has allowed the direct visualization of virus particles. Table 1 lists spherical virus structures reported at high resolution. In addition, the structure of TMV, a cylindrical virus, is known at near atomic resolution (3, 25a).

Enveloped viruses have been difficult to study at moderate or high resolution because, in general, their flexible, nonsymmetric, capsids prevent crystallization or image analysis of electron micrographs. Two exceptions are the closely related sindbis and semliki forest viruses where the glycoprotein surface antigens have regular $T = 4$ symmetry. High-resolution studies of components of an enveloped virus (influenza virus) have been successful. The hemagglutinin (22) and neuraminidase (23) protein structures have each been determined at high resolution by first removing them from the membrane by proteolytic cleavage, then crystallizing the isolated proteins. Similarly, the hexon unit of adenovirus, a virus without a lipid coat but with spikes that inhibit crystallization, has been studied at high resolution (26). Structural studies of retroviruses, such as human immunodeficiency virus (HIV), probably can be performed only by analyzing individual components or aggregates of components of the capsid. Some of the enveloped viruses, such as herpes simplex virus or alphaviruses, have a symmetric nucleoprotein core. Other viruses (e.g. reo- and rotaviruses) also have cores, but these are enveloped by an outer protein rather than lipid shell. It may be possible to crystallize isolated cores for study by X-ray diffraction or electron microscopy image processing techniques.

QUATERNARY STRUCTURE

Crick & Watson (27) proposed that the protein coat structures of most simple spherical viruses, whether animal, plant, or bacterial, were based on the design of regular polyhedra. They recognized (28) that the information in the viral nucleic acid would be sufficient to code for a protein whose molecular weight was only a fraction of that of the intact virus capsid. Thus, they inferred that the nucleic acid is encapsulated by multiple copies of identical protein subunits, necessitating identical environments and, hence, limiting the possible assembly to the Platonic solids.

It soon became apparent that simple viruses invariably had an icosahedral structure. For instance, by using X-ray diffraction techniques, Caspar (29) demonstrated that TBSV was icosahedral; later, Huxley & Zubay (30) and Nixon & Gibbs (31) observed that turnip yellow mosaic virus (TYMV) was also icosahedral. However, the principles laid down by Crick & Watson required modification because the number of subunits per virus was usually larger than the 60 permitted with icosahedral symmetry. This observation led Caspar & Klug (12) to develop the concepts of quasi-symmetry that have been the foundation for the structural classification of viruses for a quarter century. High-resolution structures show that quasi-equivalence of subunit contacts is frequently violated. Nevertheless, the quaternary structure of viral particles follows the anticipated organizational patterns predicted by Caspar & Klug.

The surface lattices described by Caspar & Klug are identified by a T (triangulation) number, equal to the number of quasi-equivalent units in an icosahedral asymmetric unit. Thus, there would be $60T$ subunits in the whole capsid. The value of T must be restricted, such that $T = (h^2 + hk + k^2)f^2$ where h, k, and f are integers, if the quasi-symmetry is to be retained. Although Caspar & Klug considered quasi-equivalent environments for identical objects, it is now apparent that these objects need not be identical to form the predicted lattice. This was first found in the structure determination of human rhinovirus 14 (HRV14; 8) and was described as pseudo-symmetry with a P rather than a T number.

Figure 2a depicts the quaternary structures of the small RNA spherical viruses that have been determined at high resolution. In each case, the individual protein subunits are represented by a trapezoidal unit. All these virions have icosahedral symmetry and, thus, contain 60 identical asymmetric units. Within each of these units there may be a single protein ($T = 1$ capsid), three copies of a single protein ($T = 3$ capsid) each with a slightly different environment, or three different protein domains each with a unique primary structure but similar tertiary structure as is the case for the $P = 3$ picornaviruses or plant comoviruses.

A $T = 1$ capsid with only one protein subunit is the least frequently observed quaternary structure in native viruses. Only small satellite viruses, such as tobacco necrosis virus (STNV; see Table 1), are known to have this capsid type. These are dependent on the coinfection of the host by a helper virus for genes that code for nonstructural proteins that are required for replication (34). The genome for STNV codes only for the capsid protein. The small size of the STNV RNA is accommodated by a shell with an interior

(a)

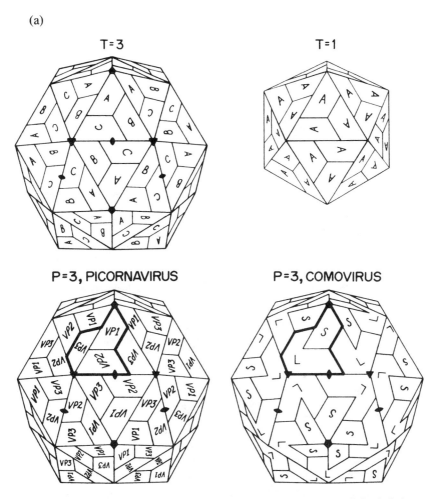

Figure 2 (*a*) Different capsids observed in high-resolution virus structures. All the shells have exact icosahedral symmetry. The $T = 1$ shell contains 60 subunits related by icosahedral symmetry. Each subunit is represented by a trapezoid that is the approximate shape of the β-barrel when viewed from the top. All subunits in the $T = 1$ capsid are identical and labeled A for comparison with the $T = 3$ capsid. The icosahedral asymmetric unit of $T = 1$ viruses is one subunit and the threefold symmetry in the central triangle is exact. The asymmetric unit of the $T = 3$ capsid is the central triangle containing subunits A, B, and C. The subunits labeled A, B, and C have the same amino acid sequence but are in slightly different environments. While similar to the triangle in the $T = 1$ structure, the threefold axis relating A, B, and C is not exact. This quasi-threefold axis relates the quasi-sixfold axes (left and right vertexes of the triangle) to a fivefold axis (top vertex). Like the $T = 1$ structure, the $T = 3$ structures are formed by identical subunits with the β-barrel fold. The $P = 3$ picornavirus shell, technically a $T = 1$ particle, is closely related to the $T = 3$ shell, being formed by 180 β-barrel domains. The three subunits in the central triangle (labeled VP1, VP2, and VP3) are, however, distinct proteins. The comovirus

(b)

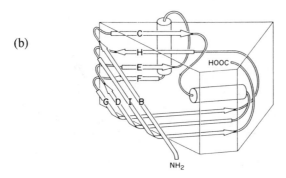

shell is very similar to the picornavirus capsid, with 180 β-barrels forming the shell. However, there are only two protein types. The large protein (labeled L in the figure) is composed of two β-barrel domains (the equivalent of VP2 and VP3) covalently linked together. The small subunit (S) is a single β-barrel domain. [Reprinted with permission from Sehnke et al (32). Copyright by Elsevier Science Publishing.]

(*b*) Diagrammatic representation of the polypeptide fold of one subunit of poliovirus found also in the shell-forming portion of all viral subunit structures determined to date. Shown also is the nomenclature for the secondary structural elements βB, βC, . . . , βI. [Reprinted with permission from Hogle et al (9). Copyright by the American Association for the Advancement of Science.]

(*c*) See next page. Main-chain hydrogen bonding scheme of the SBMV A subunit. Hydrogen bonds were postulated whenever the N-O distances were less than 3.4 Å and the angle C-O-N was less than 110°. [Reprinted with permission from Silva & Rossmann (33). Copyright by Academic Press Inc. (London) Ltd.]

radius of 60 Å. Other $T = 1$ capsids are commonly observed as an in vitro reassembly product of capsid proteins modified by removal of the amino terminus; examples are southern bean mosaic virus (SBMV; 35–37), turnip crinkle virus (TCV; 38, 39), alfalfa mosaic virus (AMV; 40), and brome mosaic virus (BMV; 41). The structures of reassembly products of AMV (normally forms a bacillus-shaped particle) and SBMV (normally forms $T = 3$ icosahedral particles) have been determined at low resolution (42, 43).

RNA viruses competent to infect a host without a helper virus must at least contain genes for an RNA-directed RNA polymerase and a capsid protein. The minimum size RNA for these viruses is roughly 1.5×10^6 daltons. Packaging of such an RNA requires a shell with a radius larger than that found in $T = 1$ shells. Capsids with $T = 3$ surface lattices are of adequate size to package genomes of up to about 3.0×10^6 daltons. Each subunit has a core structure of eight antiparallel β-strands arranged as in a barrel. Structures solved with $T = 3$ symmetry include TBSV, SBMV, black beetle virus (BBV), and TCV (see Table 1). The $T = 3$ capsid is most common among isometric plant viruses.

Animal viruses (BBV and other nodaviruses being exceptions) rarely have $T = 3$ capsids. Instead, they have pseudo $P = 3$ symmetry, where the three

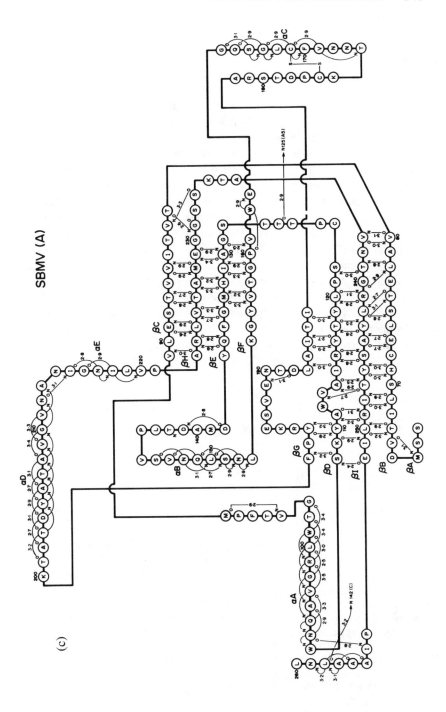

covalently identical subunits are replaced by subunits of similar fold but quite different amino acid sequence. Picornaviruses have $P = 3$ symmetry and are the most common simple RNA viruses infecting mammalian systems. Structures of two rhinovirus serotypes, two poliovirus serotypes, Mengo virus, and foot-and-mouth disease virus (FMDV) have been determined to high resolution (see Table 1). Each of the three larger proteins have the eight-stranded antiparallel β-barrel motif (Figure 2*b*) described below. The protein subunits result from proteolytic cleavage of a polyprotein (44). If all three β-barrels were covalently linked together [as they are prior to assembly (45)], they would form the equivalent of a single large subunit. That may be the case in nepoviruses (46). Cowpea mosaic virus (CPMV) and beanpod mottle virus (BPMV) are examples where two of the domains remain covalently linked together in the mature virion. These comoviruses have two capsid proteins, one of 43 kd and the other of 24 kd. The domains of the large protein occupy the positions of VP2 and VP3 in the picornavirus capsid and the small subunit occupies the position of VP1 (Figures 2*a* and 3).

Viruses in Figure 4 are larger or more complex than the small RNA viruses described above. Some of these viruses have been crystallized and are currently under study by X-ray diffraction, others have been studied only by electron microscopy. Polyoma virus is a double-stranded DNA virus whose structure has been determined at 22 Å resolution using single crystal X-ray diffraction (49) and recently extended to 8 Å resolution (S. C. Harrison, R.

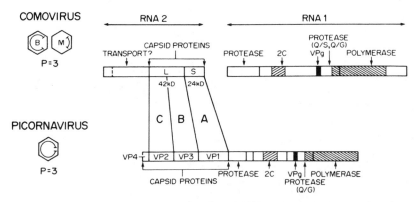

Figure 3 Comparison of the genome organization of CPMV and picornaviruses. RNA2 *(left)* and RNA1 *(right)* of CPMV are shown aligned with the RNA of picornaviruses. The molecular weight and function are marked for each gene product. Regions in the two genomes are shaded where the amino acid sequences had been recognized as homologous (47, 48). The 42 kd structural protein in CPMV (L) contains β-barrel domains that correspond in location to VP2 and VP3 in picornaviruses. The 24 kd protein in CPMV (S) corresponds to protein VP1 in picornaviruses. The letters C, B, and A indicate the positions occupied by each of these β-barrels in the $T = 3$ quasi-equivalent surface lattice. [Adapted from Franssen et al (47). Copyright by IRL Press Ltd.]

Liddington, et al, private communication). The quaternary structure of the capsid did not display the expected quasi-symmetry of the $T = 7$ capsid. The quasi-equivalence concept (12) predicted that the $T = 7$ capsid surface lattice should consist of 72 capsomeres (60 hexameric units and 12 pentameric units) resulting in 420 protein subunits. The observed structure displays pentamers at all the predicted capsomere positions, resulting in (60 + 12) pentamers or 360 subunits. The low-resolution X-ray results have been supported by electron microscopy studies of closely related SV40 (50), by tubular structures found in polyoma virus preparations composed entirely of pentamers (51) and recently a 5 Å resolution electron density map of SV40 (S. C. Harrison, R. Liddington, et al, private communication).

Nudaurelia capensis β virus (NβV) has been studied by electron microscopy using samples prepared by both the negative stain (52) and frozen

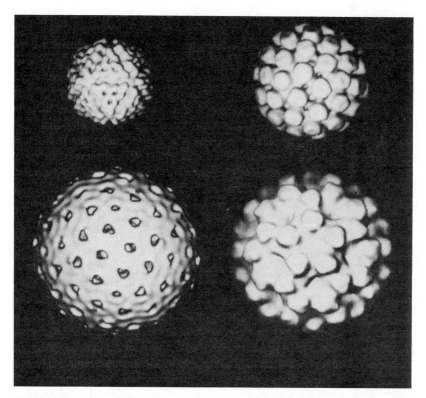

Figure 4 Three-dimensional reconstructions of virus particles derived from electron micrographs of frozen hydrated particles shown on an equivalent scale. *Nudaurelia capensis β* virus is in the top left corner, simian virus 40 in the top right, rotavirus in the lower left, and sindbis virus in the lower right. (By courtesy of N. Olsen, T. S. Baker, and S. D. Fuller.)

hydrated method (53). Particles are roughly 400 Å in diameter. The subunits form trimeric clusters at all the predicted threefold sites on the $T = 4$ surface lattice. The trimeric cylinders protrude 40 Å above the surface of the capsid. The virus has been crystallized, and diffraction patterns extend to 2.8 Å resolution (32).

Parvoviruses are small icosahedral ss-DNA viruses. The particles are not well characterized; however, they appear to be $T = 1$ capsids composed primarily of essentially one protein type whose molecular weight is 61 kd. The particle diameter is roughly 260 Å and the molecular weight is around 6×10^6. It would seem possible that the single coat protein has three tandem domains arranged with pseudo $P = 3$ symmetry. Crystals of canine parvovirus that diffract to about 2.8 Å resolution have been grown (54).

Reovirus capsids are composed of two shells encapsulating 10 double-stranded RNA molecules. The outer shell contains three protein types and has a diameter of roughly 750 Å. The inner shell contains one major protein type and three minor protein types. It has a diameter of roughly 500 Å. Reoviruses and the related rotaviruses have been the subject of numerous electron microscopy studies. Electron micrographs of frozen hydrated samples that show that the outer coat has $T = 13$ symmetry have recently been analyzed (55).

Togaviruses contain a lipid membrane. The outer shell, which is partially embedded in the membrane, has $T = 4$ icosahedral symmetry (56). The particles have a maximum diameter of 640 Å. Crystals of two togaviruses, sindbis and semliki forest virus, have been grown in a number of laboratories, although to date diffraction patterns do not extend beyond 40 Å resolution. Sindbis and semliki forest virus have been studied at moderate resolution using frozen hydrated samples (56, 57). In the outer shell, the subunits are clustered at the trimer sites of the $T = 4$ lattice. Each trimer has a molecular weight of approximately 100,000. The core lying within the lipid membrane can be isolated from the native virions but is much less stable than the intact virions (58). These cores have $T = 3$ symmetry and a diameter of 400 Å, and contain a single-stranded RNA molecule. The core protein has been crystallized, and these crystals diffract to high resolution (59). Sequence homology suggests that the core protein of sindbis virus may have a similar fold and organization as in the $T = 3$ viruses (60).

TERTIARY STRUCTURE

Figure 2b depicts the protein fold in the shell-forming portion of all RNA isometric viral subunits determined to date. There are two back-to-back four-stranded β-sheets that follow a "jelly roll" topology (61) (Figure 5). This is shown in a hydrogen-bonding diagram for SBMV (Figure 2c). The minimum number of residues required to form a β-barrel domain is approximately

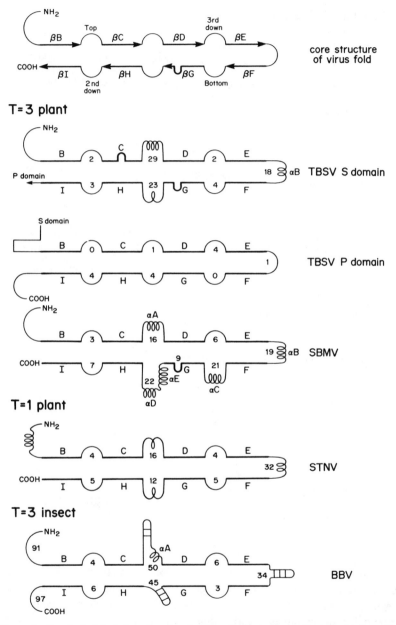

Figure 5 Topological depiction of various known subunit structures. Compare the secondary structural elements to those in Figures 2*b* or 15. Animal viruses tend to be more elaborate in their insertions between β-strands than the plant viruses. Figure continues over next two pages.

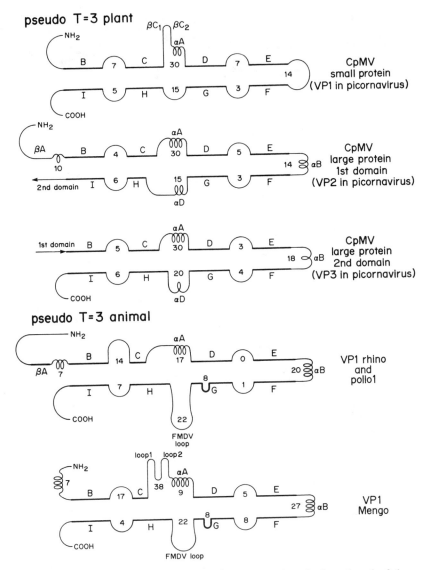

150. Insertions can occur between strands connected at the broad end of the trapezoid (i.e. between the β-strands βC and βD, βE and βF, βG and βH; see Figures 2b and 2c for nomenclature of the secondary structure elements). The topological depictions in Figure 5 illustrate that animal viruses are frequently more elaborate in their insertions than plant viruses, possibly because animal viruses have additional functions such as receptor binding and the need to

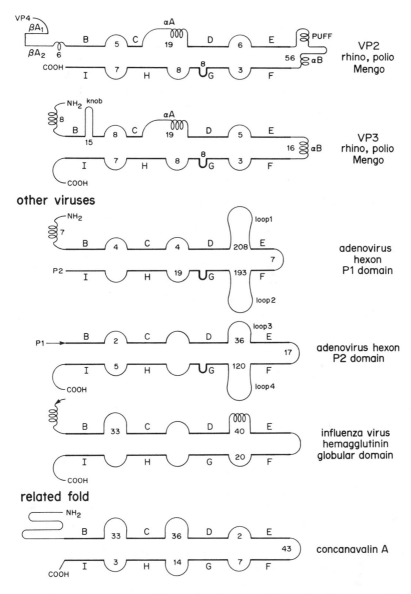

escape from immune surveillance by the host. The subunits have widely differing amino acid sequences that show no obvious homology. The thickness of the protein shell is roughly 30 Å, while the narrow and long dimensions of the trapezoidal subunit are 30 Å and 60 Å.

In many plant and insect virus capsid proteins, as well as in sindbis core protein, the first 20 to 100 amino-terminal residues contain a large number of basic amino acids associated with the negatively charged RNA. Hence, this amino-terminal domain does not obey the icosahedral symmetry and appears to be "disordered." It is not known whether or not these basic amino-terminal structures are totally different within each virus or whether there is order that might be visualized if the virus were oriented, as for instance by a magnetic field (62). NMR studies of AMV (63, 64), cowpea chlorotic mottle virus (CCMV; 65), TBSV (66), and SBMV (67, 68) have shown that the basic amino-terminal residues are more mobile on swelling the virus or in the disassembled protein subunits than they are in the intact virus. This suggests that without RNA most of the amino-terminal residues do not fold into a regular structure.

The portion of the amino terminus just preceding the eight-stranded β-barrel is "ordered" in the C subunits of TBSV and SBMV, but is disordered in the A and B subunits (see Figure 2b for definition of subunit nomenclature). This β-strand participates in the interaction of the CC_2 dimers but not of the AB_5 dimers. The insect virus BBV, although possessing a basic, disordered, amino terminus, uses a different part of the protein chain to differentiate the two types of dimers. Picornaviruses, although not possessing a basic random amino-terminal domain, nevertheless use the antiparallel β-strands βA_1-βA_2 of VP2 (corresponding to subunit C in plant viruses) to form the equivalent CC_2 dimer contact.

The carboxy termini of small spherical viral subunits are usually on the outside of the virus particle. The TBSV subunit contains a second carboxy-terminal β-barrel domain of similar topology as the shell domain. It is outside the shell and forms, with a twofold-related subunit, a large protrusion. Some of the picornavirus subunits have up to 25 residues as a carboxy-terminal extension on the surface of the particle. Insect BBV is an exception in which the carboxy terminus of the subunit makes an excursion within the shell.

PROTEIN-RNA INTERACTIONS

Packing of RNA into virus capsids requires mechanisms for neutralizing of charge. Viruses that do not have basic amino termini (tymoviruses, como-viruses, picornaviruses) contain polyamines (69). The latter presumably play the role of the basic residues in the other viruses in neutralizing the charge of the RNA for packaging.

Packing of RNA also requires recognition of the correct nucleic acid strands. Specific sequences of AMV RNA (70) and TMV RNA (71) have a preference of binding to corresponding coat protein aggregates. Protein-nucleic acid interactions in spherical viruses have been proposed on the basis of the distribution of basic residues on the internal surface of the protein coat subunits (72). However, with the exception of the plant comoviruses, it has not been possible to associate electron density with nucleic acid in any high-resolution structure of a spherical virus.

Although both picornaviruses and comoviruses contain polyamines and lack high concentrations of basic residues in the protein subunit, only the comoviruses contain significant portions of RNA that conform to the icosahedral symmetry of the coat protein. Nearly 20% of the BPMV genome can be seen in the electron density circulating around the particle threefold axes (73). Sixty icosahedrally related sets of seven ribonucleotides were easily recognized and showed the polarity and are conformationally similar to one strand of an A-type RNA helix. The bases are stacked, but they must correspond to an average of many primary sequences. Four other ribonucleotides could be fitted with much less certanty to a region of lower density that appeared to connect the ends of the threefold-related well-ordered units. The lower density corresponds to regions where the RNA chain might enter or leave the trimeric "rings." The ordered nucleotides lie in a pocket on the interior surface of the large subunit in the BPMV structure. The 5' to 3' direction of the RNA is adjacent to and antiparallel with the N to C polypeptide direction, which connects the amino and carboxy domains of the large subunit. The pocket is rich in hydrophilic amino acids, and the side chains of these residues extend toward the RNA molecule. However, there are only about two interactions of less than 3.2 Å (Figure 6).

Why is a portion of the RNA icosahedrally ordered in BPMV while it is not

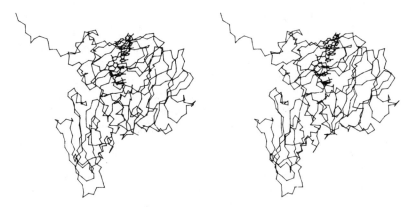

Figure 6 A stereographic view of the C_α backbone of the large and small subunits of BPMV shown with the ordered RNA (bold lines), which is located below the carboxy domain of the large subunit. The icosahedral asymmetric unit illustrated corresponds to the protomer (VP4, VP2, VP3, VP1) of a picornavirus. The ordered RNA in BPMV is located where the VP4-VP2 autocatalytic cleavage takes place in rhinovirus.

visible in the similar picornaviruses? These are two obvious differences between comoviruses and all other viruses that have been studied structurally to date: (a) two different comovirus RNA molecules that display no sequence similarity are encapsulated into separate particles, and (b) comoviruses readily form empty capsids in vivo. Both these observations suggest weak interactions of the RNA with the coat protein and possible packaging of the RNA at a stage when much of the capsid is already formed. These properties might permit the RNA to recognize more easily an icosahedral environment. It follows that nepoviruses and tymoviruses may also have some icosahedrally ordered RNA.

EVOLUTION

High-resolution structure determination of viruses has vividly demonstrated the common structural motif, the eight-stranded antiparallel β-barrel, that underlies the architecture of many, if not most, spherical viruses. It is, however, not only the conserved capsid protein topologies but also sequence homology with other genes as well as the conserved gene order that strongly suggest a single primordial origin for a wide variety of viruses and certainly for small spherical RNA viruses (74, 75).

Franssen et al (47) and Argos et al (48) showed that the virally coded polymerase, the genomic protein VPg, and the protease of polioviruses are homologous in sequence and gene order with that of CPMV. [Later it was shown that the gene order of the structural proteins was also the same (6; Figure 3).] This was perhaps the first formal recognition of an evolutionary link between animal and plant viruses. These comparisons have been extended even further, namely to the tripartite viruses AMV, BMV, and cucumber mosaic virus, the single genome of the cylindrical TMV, and the animal sindbis virus (76). Attempts at finding homology between picornaviruses and plant virus structural proteins had failed when viewed only from the primary sequence of amino acids. Among animal viruses, Fuller & Argos have proposed homology between sindbis virus core protein and FMDV VP3 (60) and of hepatitis B core protein to Mengo virus VP3 (77). Also, Argos (78) has suggested that there is homology between the plant viruses TYMV coat protein and SBMV coat protein.

Goldbach (74) suggests that there are two superfamilies of RNA viruses. Viruses within each group have similarities of translational strategies, gene order, and gene sequence. Examination of some more conserved sequences, such as that of RNA polymerase, show that both families have probably diverged from a common ancestral virus (79; Figure 7). In addition, there must have occurred gene fusion with host genes as well as gene rearrangements. For instance, the isometric viruses mostly use eight-stranded β-barrel

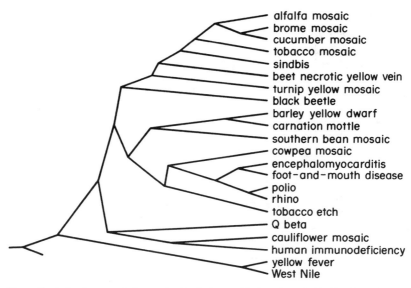

Figure 7 A dendogram of plant and animal viruses (79) based on similarities found in viral nucleic acid polymerase sequences. The sequences were aligned using the Gly-Asp-Asp motif (154) found in many polymerases. Divergence was measured by the method of Needleman & Wunsch (155). (By courtesy of A. Gibbs.)

domains as capsid proteins, whereas some cylindrical viruses use motifs based on α-helixes [e.g. TMV (3, 25a) and pf 2 phage (80)]. These capsid proteins will have had a different evolutionary history and have been separately fused with a similar replicative portion of the genome (75).

Three-dimensional biological structures are generally conserved over far larger time spans than primary protein sequences. Therefore, they can be used for alignment where amino acid sequences have diverged greatly (47, 81). The major differences between homologous structures usually correspond to deletions and insertions, while the essential polypeptide folding motif is maintained. The extent of these changes can be used as a rough measure of evolutionary divergence. If these differences are particularly great, then the probability of divergence, as opposed to convergence to a functionally suitable fold, is low. In comparisons of viral genomes, Palmenberg (82) and others have shown that viral coat protein genes are usually the least conserved. However, the degree of similarity of the tertiary structure of icosahedral viral capsid proteins is considerable. Table 2 compares the degree of similarity of viral capsid proteins with such benchmarks as a comparison of the α and β chains of hemoglobin, a comparison of the NAD-binding domains in two different dehydrogenases, and a comparison of hen egg white lyso-

Table 2 Comparison of proteins

Comparison	Protein 1[a]	Protein 2	Number of equivalenced residues	Percentage of equivalences		rms[b] (Å)	MBC/C[c]	κ[d] (°)	Δ[d] (Å)
				Protein 1	Protein 2				
Benchmarks	Hb (β)	Hb (α)	139	95	99	1.9	—		
	GAPDH (NAD)	LDH (NAD)	96	65	67	2.9	1.24		
	T4L	HEWL	78	48	60	4.1	1.53		
Plant viruses	SBMV (C)	TBSV (C)	176	84	93	2.2	1.13	3.1	0.4
HRV14 vs Mengo	VP1 (HRV14)	VP1 (Mengo)	181 (150)[e]	63 (64)	65 (61)	2.6	1.22	1.7	1.4
	VP2 (HRV14)	VP2 (Mengo)	215 (173)	82 (83)	84 (81)	1.9	0.88	1.7	0.3
	VP3 (HRV14)	VP3 (Mengo)	219 (178)	93 (91)	95 (94)	1.6	1.10	1.5	0.5
HRV14	VP1 (HRV14)	VP3 (HRV14)	168	58	71	3.7	1.34		
	VP2 (HRV14)	VP3 (HRV14)	152	58	64	2.8	1.23		
	VP1 (HRV14)	VP2 (HRV14)	124	43	47	3.2	1.27		
Mengo	VP1 (Mengo)	VP3 (Mengo)	156	56	68	3.1	1.27		
	VP2 (Mengo)	VP3 (Mengo)	146	57	63	2.8	1.34		
	VP1 (Mengo)	VP2 (Mengo)	125	45	49	3.1	1.19		
HRV14 vs SBMV	VP1 (HRV14)	SBMV (A)	123	43	59	3.3	1.42	3.5	2.5
	VP2 (HRV14)	SBMV (C)	134	51	64	2.8	1.40	19.1	1.1
	VP3 (HRV14)	SBMV (B)	136	58	65	2.6	1.34	15.3	0.6

[a] Abbreviations: Hb (α) and Hb (β) are the α and β chains of horse hemoglobin, GAPDH (NAD) and LDH (NAD) are the NAD-binding domains of glyceraldehyde-3-phosphate and lactate dehydrogenases, T4L and HEWL are T4 phage and hen egg white lysozymes, and SBMV and TBSV are southern bean mosaic and tomato bushy stunt viruses. Letters in parentheses refer to the A, B, and C subunits.

[b] rms (Å) is the root-mean-square deviation between equivalenced C_α atoms.

[c] MBC/C is the minimum base change per codon between equivalenced residues.

[d] κ (°) and Δ (Å) are the rotation and translation of the center of gravity of the protein subunit relative to the icosahedral axes to obtain the best superposition.

[e] Numbers in parentheses refer to the superposition of the β-barrel domain only.

[Reprinted with permission of Luo et al (10).]

zyme with T4 phage lysozyme. The latter two comparisons have about the same or less similarity than a comparison of plant and animal RNA viral capsid proteins. The similarity of structure and function of the above-mentioned benchmarks have usually been accepted as evidence for divergent evolution (81, 83–85). Hence, the various simple icosahedral RNA viruses are also likely to have evolved from a common precursor.

In general, there is greater structural similarity between equivalent viral capsid proteins of Mengo virus and HRV14 than there is between the different viral capsid proteins of either Mengo virus or HRV14 (Table 2). Thus, VP1, VP2, and VP3 most likely diverged from each other (the primordial picornaviruses) before the polyprotein of Mengo virus diverged from that of HRV14. Argos et al (86) suggest that a precursor protein to the viral capsid proteins may have been related to an ancient receptor binding protein, such as the lectin concanavalin A, which not only has a similar fold, but also can compete with poliovirus for HeLa cell receptors. The observation by Roberts et al (26) that the hexon unit of the DNA-containing adenovirus includes two distinct β-barrels similar to the viral protein of the plant and animal RNA viruses may also have a bearing on the origin of these and other viruses, as perhaps also the crude similarity of the globular component of the hemagglutinin spike of influenza virus (22).

Superpositions of various virus coat protein structures gave not only the amino acid alignments (Table 3) but also the relative orientation and position of the viral proteins in the capsid (Table 2). The alterations between the quaternary structure of Mengo virus and HRV14 are negligible, as are also those between the plant viruses TBSV and SBMV. However, there are significant changes between the $T = 3$ and $P = 3$ shells in the relative arrangement of their capsid proteins. Differences are as great as a 19° rotation and 2.5 Å translation in a comparison of the C subunit of SBMV and VP1 of HRV14.

Virally coded genes frequently show a low level of similarity to genes in host organisms. For instance, picornavirus proteases show homology to serine proteases of both eukaryotes and prokaryotes, except that the essential serine becomes a cysteine (87). This suggests the coevolution of host and parasite with occasional transfer of genetic information.

PROCESSING

Controlled, limited proteolysis plays a major role in the regulation of many biological processes. The BBV particles undergo a postassembly maturation in which a peptide of 44 amino acids is autocatalytically cleaved from the carboxy terminus (11). This transforms the particle from an unstable provirion to a mature virion of extraordinary stability (88).

Table 3 Alignment of Mengo virus and HRV14 sequences for VP1, VP2, and VP3[a]

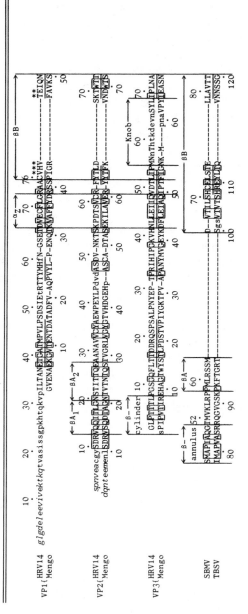

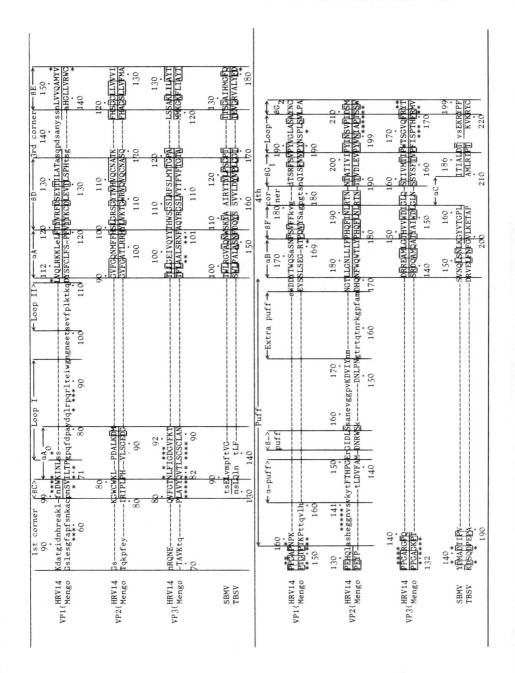

Table 3 (Continued)

	FMDV loop	βH	2nd corner	βI

```
                  202    210    221     222
VP1[ HRV14  FYDGYSHD-DAETQYCITVLNHM-SSMAFRI---vnehdehktLVKIRVYHRAKH-VEAWIPRAPRa--LPvtsigrtnypkntepvikkrkgdiksy
     Mengo  WYNGHKRfDNTGDLIEIAPNSDF-GTLFEAG---tkp------DIKFTKYLRKN-MRVFCPRPTVfipWtl sgdkiamtpragvlmle
                  202    210                             239       250        260        270        284
                  **.*  ****    **.**   **.* *           .*.***              ***  ***  ****:**

VP2[ HRV14       TRH---NN-VSLMVIPIAPLTNPTGAPTSLPITVTIAPMC-TEFSGIRSks------ivpq
     Mengo       EDH---AS--TLVIAVWIPLTNAGASTDLITHSLRQPVR-PVFNGLRNievlsrq
                  210    220    230    240       250        260

VP3[ HRV14       D------PDTVHSA GFLSCAYQTSLIPETTGQVYLISFISACPFKLRLMKDTQTIs-----
     Mengo       G------tcQ4VITMDGWTVSDUPLYPFCPTSAKILTNWSAGKDSLKxPISPAPWspq
                  *     ****:******   **** **  **** ****    **  ****:**
                         190    200    210    220    230        234
                                                                 ****:**
                                                                 qtvalte

            αD    αE
            202    210    220    230
SBMV  KTatdyat-avgvnanignNILVP-ARLVTAMEGS AWNTRLYASYT-I-RDTEPTAAAL
TBSV  NDsatvdq-----------KLIL-GQLGIAHYEGA WMWELFLARS-V-TLFFPPTNT
            230    240    250    260    270
```

---

a Shown also is the alignment of major structural segments in the shell domains of SBMV and TBSV with Mengo virus and HRV14. Horizontal bars indicate regions of structural alignments among VP1, VP2, VP3, and the two plant virus coat proteins. Sequences given in lower case letters have no structural equivalences. Deletions relative to other sequences are indicated by dashes. A blank in the plant viruses indicates that a portion of the sequence has been omitted from the alignments. An asterisk (*) indicates residues in the pit (Mengo) or canyon (HRV14) lining or ligands for Ca^{2+} binding (SBMV and TBSV). Sequences given in italic letters are disordered in the X-ray structures. Sequences that are identical between HRV14 and Mengo virus and between SBMV and TBSV have been boxed. Alignment of sequences that have no structural equivalence is somewhat arbitrary.

Picornavirus RNA is translated as a single polyprotein, which is subsequently cleaved into its functional proteins by two virally coded proteases (Figure 8). Similarly, the two RNA strands of CPMV and BPMV are each translated as a polyprotein and then are cleaved by virally coded proteases. These proteases are highly specific for amino acid pairs such as Gln-Gly and Tyr-Gly in polioviruses. The cleavage sites are positioned in regions between the β-barrel domains. After cleavage, the new ends could easily reposition themselves into the orientation observed in crystallized virions without disrupting the contacts between β-barrels. The probable manner in which the carboxyl ends of VP0 and VP3 would be associated with the amino ends of VP3 and VP1, respectively, before proteolysis can readily be visualized (Figure 9). Furthermore, the proposed connectivity shown in Figure 9 is consistent with the polypeptide connection between the two β-barrels in the larger of the two viral capsid proteins of CPMV (6) and BPMV (73). These observations and the pervasiveness and stability of the β-barrel arrangement strongly suggest that they fold and organize themselves within a protomer even before they are cleaved from the nascent polyprotein.

The potential cleavage sites are not all recognized by picornavirus proteases. Contrary to expectation, some of these are on the protomer surface, where they occur within the capsid proteins. However, unlike the actual cleavage sites, these pseudo-sites invariably occur in rigid portions of the

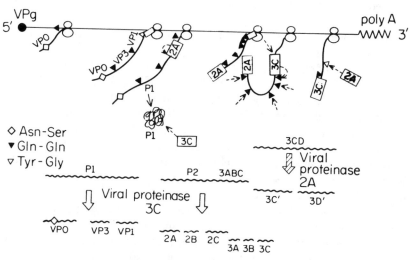

Figure 8   Steps in the proteolytic processing of the initial translation product of poliovirus RNA by two virus-encoded proteinases. [Reprinted with permission from Toyoda et al (89). Copyright by Cell Press.]

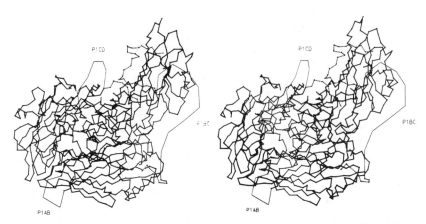

Figure 9 Stereographic view of the C_α backbone of one HRV14 protomer. Hypothetical linkages of VP1, VP2, VP3, and VP4 are indicated by thinner lines labeled P1AB, P1BC, and P1CD as they might occur in the precursor polyprotein prior to posttranslational cleavage. Dashed portions correspond to terminal segments of the native structure that might have rearranged subsequent to processing. [Reprinted with permission from Arnold et al (45).]

β-barrel. Presumably, flexibility is needed to permit the protein substrate to accommodate itself into the active center.

The processing of picornaviral protomers into VP0, VP3, and VP1 is a necessary prerequisite to pentamer formation (90). The amino termini of VP3, released by cleavage, associate with each other in the formation of the five-stranded β-cylinder (Figure 10) and stabilize the formation of 14S pentamers. Thus, the posttranslational processing directs, in part, the subsequent assembly of the virion.

The last step of picornaviral maturation (formation of an infectious particle) is cleavage of VP0 into VP4 and VP2. This apparently occurs early after the RNA has been packaged. The structures of HRV14, poliovirus, and Mengo virus each show the close proximity of a serine residue to the carboxy end of VP4. In HRV1A, this portion of VP4 is disordered. The site of the cleavage is deeply buried in the viral capsid, inaccessible to proteases. Hence, VP0 cleavage may be an autocatalytic process with serine acting as a nucleophile as in other serine proteases, such as trypsin. However, the virus structures show no base such as histidine in the vicinity of the catalytic serine. It has, therefore, been suggested (8, 45) that the RNA provides the essential base and thereby also initiates the catalytic step during assembly. Interestingly, the ordered RNA in BPMV (73) would be in precisely the right position and orientation to perform this catalysis (Figure 6). Also, Arnold et al (45) have shown that EMC VP0 remains unprocessed in a rabbit reticulocyte system unless a strong bivalent base is present. The folding of the protomer suggests that the function of this cleavage is to switch from the requirements of

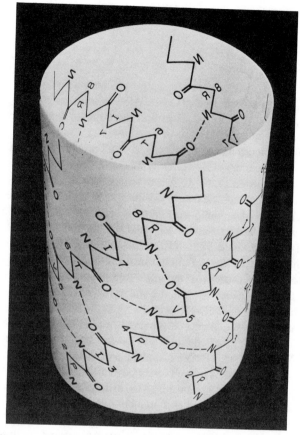

Figure 10 Hydrogen bonding arrangement in five-stranded parallel β-cylinder made by the amino ends of VP3 about each icosahedral fivefold axis in Mengo virus. [Reprinted with permission from Luo et al (10). Copyright by the American Association for the Advancement of Science.]

assembly (in which VP0 is required to maintain the integrity of the 6S and 12S assembly units) to an infectious particle, which requires the separate disassembly of VP4 and VP2 during membrane attachment and penetration.

Some emphasis has been placed here on picornavirus processing, as this is probably best understood in terms of structure. Polyprotein processing is used by numerous viruses, as for example in picornaviruses, comoviruses, nodaviruses, and retroviruses. However, other strategies should be noted. For instance, quite a few viruses make subgenomic copies of RNA components that contain only a single gene (e.g. SBMV, TMV, and AMV require subgenomic RNA copies for the translation of the coat protein). Other forms of processing, such as phosphorylation (e.g. polyoma virus), are also common phenomena in guiding the assembly of virus particles.

ASSEMBLY

The assembly of viruses occurs via discrete intermediate units sometimes aided by a temporary scaffold (91, 92). Among simple spherical viruses, the relationship between assembly units and three-dimensional structure is best documented for picornaviruses.

Assembly of picornaviruses proceeds from 6S protomers of VP0, VP3, and VP1, via 14S pentamers of five 6S protomers to mature virions (93). The final step involves inclusion of the RNA into empty capsids or partially assembled shells with simultaneous cleavage of VP0 into VP2 and VP4 (see section on processing). Both the amino and carboxy ends of VP1 and VP3 are intertwined with each other, and if VP4 and VP2 are considered as VP0, then VP0 is also intertwined with VP1 and VP3. This strongly suggests the protomer has the structure heavily outlined in Figure 2a and shown in Figure 9. These 6S protomers are woven into 12–14S pentamers and are stabilized by a fivefold β-cylinder formed from the amino ends of the VP3s (Figure 10) as well as interactions of the VP4 (94).

A similar assembly sequence is observed in plant viruses, in particular for SBMV (95), TBSV (96), and TCV (97), where the building blocks correspond to VP1-VP3 or VP2-VP2 dimers and where the formation of intermediates with fivefold symmetry may be an important step in the formation of $T = 1$ (43) and $T = 3$ (95) capsids. Two different types of dimers (AB$_5$ and CC$_2$) occur in assembled virions, which may have a functional role in assembly. The two types of dimers differ by an approximately 35–40° rotation, permitting the hydrophobic divided contact between CC$_2$ dimers (Figure 11; 33, 96) that is absent in AB$_5$ dimers. The divided contact is occupied by the βA arm, ordered only in the C subunits (see section on tertiary structure). In the insect BBV, the "arm" in the icosahedral twofold groove is not connected to the βB strand of the C subunit, but probably corresponds to residues near the carboxy terminus. An autocatalytic cleavage, which occurs during the particle maturation (see section on processing), releases the carboxy-terminal portion of the protein from the BBV subunit, thereby increasing the particle stability.

The differing subunit interactions within the AB$_5$ and CC$_2$ dimers represent a breakdown of the quasi-equivalence principle of Caspar & Klug (12). While Caspar & Klug considered two-dimensional shapes, in reality capsid proteins are three-dimensional, which makes quasi-equivalence impossible. Relevant here is that $P = 3$ viruses, such as picornaviruses, have totally different nonequivalent contacts and yet display the same surface lattice. More blatant

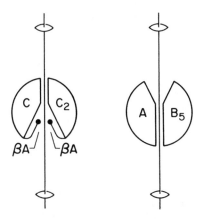

Figure 11 Diagrammatic representation of the two possible states of dimers in $T = 3$ viruses. The CC_2 dimer occurs about the icosahedral twofold axes and has the ordered βA arm between subunits. The AB_5 dimer occurs about the quasi-twofold axes. [Reprinted with permission from Rossmann (98). Copyright by Academic Press, Inc.]

examples of a breakdown in quasi-equivalence are polyoma virus (49, 51) and SV40 (50; S. C. Harrison, R. Liddington, et al, private communication), which are aggregated from pentamers rather than from the pentameric and hexameric units.

The study of disassembly has been particularly fruitful for plant viruses, some of which swell on decreasing pH (99) or removal of metal ions (100–102). In a study of swollen TBSV, Robinson & Harrison (103) show that the quasi-threefold contacts come apart and the protruding domains rotate relative to the shell domains. The amount of rotation is 103° for the icosahedral CC_2 dimers, but only 30° for the quasi-twofold AB_5 dimers.

Disassembly for most animal viruses is mediated by receptor attachment, membrane fusion (for enveloped viruses), and endocytosis. Chow et al (104) have shown that the amino end of VP4 is myristylated. The myristate might act as an anchor in the cell membrane during assembly or disassembly. In either case, it is not immediately obvious how the myristate group is internalized (assembly) or emerges from the virus (disassembly). This process is probably related to the loss of RNA and VP4 in the first steps of cell attachment (44, 105).

RECEPTOR RECOGNITION

Most animal viruses initiate infection and entry into host cells by attaching to receptors on the host cell membrane. The normal function of these receptors is not generally known but, for instance, reovirus cell entry is mediated by the β-adrenergic receptor (106). The receptor attachment site must remain sufficiently conserved in order for the viruses to continue to bind to the same receptors. Yet the viral surface, in viruses that infect higher vertebrates, undergoes rapid change to escape the neutralizing antibodies of the immune

Table 4 Rhino- and enterovirus comparisons (minimum base change)[a]

	1	2	3	4	5	6	7
Surface—non-canyon:							
1. Polio 1 Mahoney	0.0	0.94	1.54	1.46	1.33	1.33	1.60
2. Polio 3 Sabin		0.0	1.46	1.43	1.51	1.31	1.72
3. Coxsackie B3			0.0	1.78	1.45	1.73	1.79
4. HRV14				0.0	1.28	1.48	1.82
5. HRV89					0.0	1.25	1.21
6. HRV2						0.0	1.18
7. HRV1A							0.0
Surface—canyon:							
1. Polio 1 Mahoney	0.0	0.52	1.01	1.16	0.86	1.10	0.70
2. Polio 3 Sabin		0.0	0.94	1.15	0.81	1.09	0.82
3. Coxsackie B3			0.0	1.00	1.16	1.05	1.03
4. HRV14				0.0	0.85	1.03	0.52
5. HRV89					0.0	0.83	0.51
6. HRV2						0.0	1.15
7. HRV1A							0.0
Internal:							
1. Polio 1 Mahoney	0.0	0.15	0.59	0.62	0.69	0.68	0.91
2. Polio 3 Sabin		0.0	0.61	0.65	0.70	0.70	0.94
3. Coxsackie B3			0.0	0.63	0.64	0.63	0.82
4. HRV14				0.0	0.61	0.62	0.87
5. HRV89					0.0	0.27	0.31
6. HRV2						0.0	0.18
7. HRV1A							0.0

[a] Comparisons with HRV1A apply only to VP1.
[Reprinted with permission of Rossmann & Palmenberg (107).]

system. Thus, there is an apparent conflict between the requirements for cell attachment and avoidance of antibody binding. The surfaces of rhino-, polio- and Mengo viruses have deep clefts. In rhino- and enteroviruses, a canyon encircles each fivefold axis. The amino acids that line the canyon (or pit in Mengo virus) are far more conserved than the residues elsewhere on the surface (Table 4) (107). The epitopes that bind neutralizing antibodies are hypervariable. Hence, it has been hypothesized that these deep depressions on picornaviral surfaces are the site of host cell attachment (Figure 12), where the residues on the floor of the canyon would be inaccessible to antibodies and, hence, not under pressure to change.

A number of circumstantial pieces of evidence support the "canyon hypothesis":

1. Site-specific mutations of some residues lining the canyon of HRV14 cause alteration in the binding affinity of the virus to HeLa cell membranes (108).

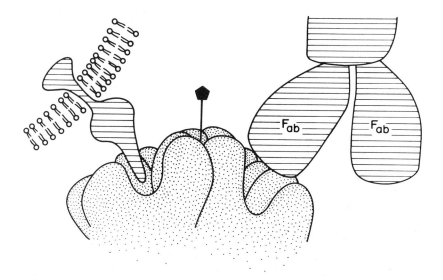

THE CANYON HYPOTHESIS

Figure 12   The presence of depressions on the picornavirus surface suggests a strategy for the evasion of immune surveillance. The dimensions of the putative receptor binding sites (the canyon in HRV14, the pit in Mengo virus) sterically hinder an antibody's *(top right)* recognition of residues at the base of the site, while still allowing recognition and binding by a smaller cellular receptor *(top left)*. This would allow for receptor specificity, while at the same time permit evolution of new serotypes by mutating residues about the rim of the canyon or pit. [Reprinted with permission from Luo et al (10). Copyright by the American Association for the Advancement of Science.]

2. Large conformational changes that occur in Mengo virus on altering pH are entirely concentrated in the pit area predicted to be the attachment site (108a). In the conditions used for producing the native Mengo virus crystals, the virus is noninfectious and does not attach to membranes. However, the conditions used in the crystal for altering pH give greater infectivity and membrane-binding ability. Thus, the "pit" area is very likely the site of receptor attachment.
3. Certain antiviral drugs that bind to HRV14 cause a conformational change in the canyon and also inhibit attachment of viruses to membranes (109).
4. The site of receptor attachment to influenza virus hemagglutinin spikes has been mapped both by use of mutations (110) and binding of appropriate polysaccharides (111). Here, also, the site is in a depression and is highly conserved. A similar situation may pertain in the glycoprotein surface of HIV (112).
5. Anti-idiotype antibodies to a monoclonal that binds to the major rhinovirus

receptor does not bind to the virus (112a). Presumably, the shape of antibodies is such that they cannot enter the canyon even if they in part mimic the viral surface.

6. In BBV, an insect virus that enters the host cell by means of attachment to receptors but is not subject to immune pressure, there is no surface depression (canyon or pit) but a surface protrusion (11).

Colonno et al (113) have shown that rhinoviruses are of two types (the major and minor groups) according to which kind of receptor they require. Within a group, viruses of different serotype compete for attachment. Crowell et al have shown that most different types of picornaviruses use different receptors on HeLa cells (105) although, for instance, coxsackie A21 competes with rhinoviruses of the major group. Colonno et al have been able to produce a hybridoma that produces antibody that binds to the major receptor for rhinoviruses (113), thus blocking viral attachment to cells in vitro. This monoclonal antibody has been used also to identify the receptor that is a glycoprotein whose subunit molecular weight is around 90,000 (114). All-away & Burness have shown that encephalomyocarditis virus uses glycophorin A as receptor in the process of hemagglutination (115, 116), although the receptor on cells infected by these viruses could be a different molecule.

NEUTRALIZATION BY ANTIBODIES

The immune system is the major defense against viral invasion in higher animals. Its stimulation by vaccines remains, by far, the most common defense that can be mustered. While antibodies can bind, in general, to all parts of a viral surface, only certain limited surface areas are useful targets for neutralization.

There are approximately four, nonoverlapping, epitopes on the surface of hemagglutinin spikes of influenza virus to which neutralizing antibodies can bind (117). Escape mutations to each of these sites are necessary to cause a shift in the antigenic properties of the virus. The subsequent absence of neutralizing antibodies of any of these sites appears to give rise to influenza pendamics.

The neutralizing epitopes (neutralizing immunogenic sites or NIm sites) of HRV14 (118, 119), poliovirus (9, 120), and FMDV (see 121) have been investigated by techniques similar to those used in the study of influenza virus (117). It was found for HRV14 that there were 60 sets of four types of distinctly different, non-overlapping, epitopes. Possibly, therefore, a new rhinovirus serotype occurs whenever there has been a significant change in each of the four epitopes. On the other hand, the limited number of poliovirus serotypes and the close resemblance of poliovirus to rhinovirus suggest that some other factors are at work. Hogle et al (9, 120) have shown that poliovirus has four NIm sites similar in position to those found on HRV14. However, these are distributed among the three different serotypes, with as yet no clear evidence that one serotype possesses all four NIm sites. It is reported that the neutralizing sites of Mengo (122) and FMDV (121, 123–131; A. A. M. Thomas, private communication) have similar positions.

Many other techniques, such as the use of synthetic peptides as antigens to mimic parts of each of the capsid protein, have been used to examine poliovirus and FMDV. It is clear, however, that such techniques do not necessarily map the viral surface. For instance, Chow et al (132) found a series of peptides, which map to the internal amino portion of VP1 in polio 1 Mahoney, that elicit neutralizing antibody.

Neutralizing antibodies can cause viral cross-linking (133) and may interfere with viral functions such as cell attachment, membrane penetration, or uncoating. As few as one antibody can neutralize polioviruses (134, 135), bivalent antibody attachment is sometimes necessary for neutralization (134), and such attachment may be accompanied by an isoelectric change of the virus (136).

The NIm sites of HRV14, NIm-IA, NIm-IB (both on VP1), NIm-II (primarily on VP2), and NIm-III (primarily on VP3), are distant from the nearest twofold-related immunogenic sites by 120, 120, 50, and 60 Å, respectively. Lower and upper limits of the distance of the two antibody-binding sites on an immunoglobulin molecule probably lie in the range 50 to 180 Å. As an antibody itself has a twofold axis, the organization of antigenic sites on the virus surface could be consistent with binding across twofold axes (137). However, model-building experiments suggest that divalent attachment between symmetry-related NIm-IB sites in HRV14 would be difficult.

It is difficult, or perhaps even impossible, to crystallize complexes of spherical viruses with antibodies, as the molecular complexes may differ in terms of the number of occupied sites. Instead, crystals can be formed of the Fab components of some of the neutralizing monoclonal antibodies (137), and the resultant structure can then be "docked" against the known virus epitope structure. Colman et al (138) have studied the structure of an Fab fragment complexed with the influenza virus neuraminidase spike. The structure of the antigen was not significantly perturbed from the native conformation. However, formation of the complex on the virus inhibits the neuraminidase action.

NEUTRALIZATION WITH ANTIVIRAL AGENTS

There are, as yet, few antiviral agents that have been found acceptable for clinical application. Nevertheless, the study of antiviral agents is likely to have the greatest impact on future attempts at curbing viral disease. Among the best functionally characterized agents are a series of compounds that have

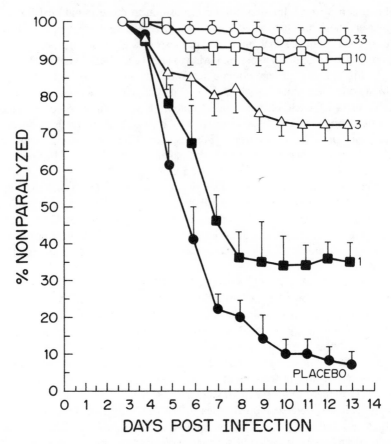

Figure 13 Therapy with a WIN compound against coxsackie A9 virus in suckling mice. The latter were treated once a day for 5 days. The curves show the effect of different relative dosages in mg/kg of mouse/day. [Reprinted with permission from Diana et al (149).]

been synthesized by the Sterling-Winthrop Research Institute and that inhibit uncoating of many entero- and rhinoviruses (139). Similar compounds have also been examined by Nippon La Roche (140, 141) and Wellcome Research Laboratories (142) as well as other investigators (143). Some of these compounds may also inhibit attachment (109). Early studies were primarily on arildone (144–147), a compound produced by the Sterling-Winthrop Research Institute. While most tests of these compounds have been made only in tissue culture, the compounds can inhibit paralysis in mice infected with coxsackie A9 virus (Figure 13) (148, 149).

A number of these compounds (Figure 14) have been studied structurally when complexed to HRV14 (150, 151) and HRV1A (94). They bind into a hydrophobic pocket within the VP1 β-barrel (Figures 15 and 16) associated in

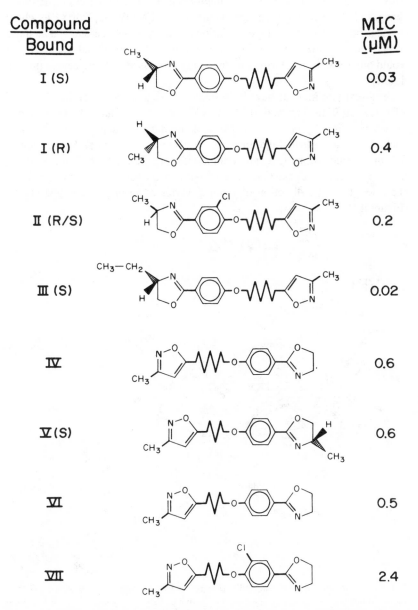

Compound Bound — **MIC (μM)**

I (S) — 0.03

I (R) — 0.4

II (R/S) — 0.2

III (S) — 0.02

IV — 0.6

V (S) — 0.6

VI — 0.5

VII — 2.4

Figure 14 Formulae of some antiviral compounds that have been examined on binding to HRV14. Compound VII has been examined when binding to HRV14 as well as HRV1A. Shown also are the in vitro activities against HRV14 measured in terms of the concentration (μM) required to reduce the plaque counts by a factor of two [minimal inhibitory concentration (MIC)]. The direction of binding within the WIN pocket is indicated. [Reprinted with permission from Badger et al (150).]

HRV14 with large (up to 4 Å in main-chain C_α positions) conformational changes (151). The major change occurs in the "FMDV loop" of VP1, which crosses the base of the canyon. Various lines of evidence show that these compounds stabilize the virions by inhibiting conformational changes that would otherwise occur during the uncoating process and by increasing protein rigidity.

Drug-resistant mutants have been found and sequenced (152). Mutations of VP1 Cys 199 to Trp and of VP1 Val 188 to Leu have been analyzed three-dimensionally (150, 152). Most of the mutants mapped into the WIN pocket, causing steric hindrance to drug entry or binding. Some shorter five-membered aliphatic chain compounds are still effective for mutants found by screening against compounds with longer seven-membered aliphatic chain lengths. The direction of binding varies according to compound size (Figure 14).

HRV14

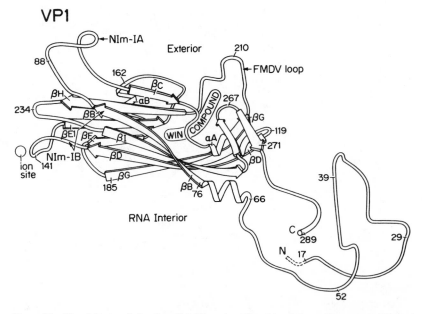

Figure 15 The eight-stranded antiparallel β-barrel, as found in viral capsid protein 1 (VP1) of human rhinovirus, is the site of attachment of antiviral WIN compounds. They bind into the hydrophobic internal pocket as shown. NIm-IA and NIm-IB are hypervariable, external neutralizing immunogenic sites on HRV14. Secondary structural elements (βB, βC, . . ., αA, αB, FMDV loop) and approximate sequence numbers are shown. [Reprinted with permission from Luo et al (10). Copyright by the American Association for the Advancement of Science.]

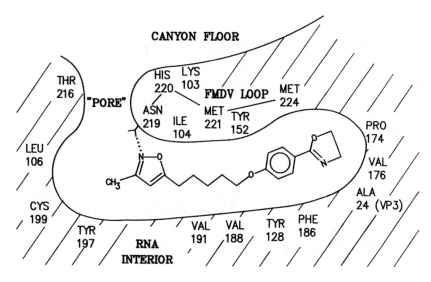

Figure 16 Diagrammatic representation of compound VI bound in the WIN pocket. The orientation is opposite to that of compound I (151). Residues lining the pocket were selected as having atoms within 3.6 Å of any atom in compound I(S). [Reprinted with permission from Badger et al (150).]

The presence of a hydrophobic pocket in the β-barrels of picornaviruses, which, when filled by a suitably fitting compound, inhibits viral disassembly, suggests that the pocket is functionally important. It may be that this hole is required for protein flexibility to permit assembly or disassembly. As many viruses are based on the same structural motif, it is reasonable to anticipate that most icosahedral viruses may have a functional requirement for a pocket of this kind. Thus, this pocket may be a suitable antiviral target in many viruses (153). If a compound can be found that binds into the pocket of a particular group of viruses, there should result an inhibition to uncoating. As the pocket, at least in picornaviruses, is particularly conserved, the probability of natural escape mutations is likely to be low, certainly much lower than the occurrence of escape mutations to neutralizing antibodies. It would seem improbable that a WIN pocket could be entirely filled by large amino acid side chains, for these would have the same effect of inhibiting assembly and disassembly as a WIN compound and would make it impossible for the virus to propagate. Thus, while escape mutations may occur to a particular compound, other compounds would still be able to act as uncoating inhibitors. The compounds that have been found to be efficacious are mostly low in their toxicity, possibly because there are no protein structures that resemble the viral capsid protein required in cell metabolism.

CONCLUSION

Various technical advances have made it possible to determine three-dimensional structures of simple viruses at atomic resolution. It is probable that the number of such determinations will continue to accelerate and the types of viruses that will be studied will increase in diversity and complexity. The functional information on attachment, fusion, assembly and disassembly, RNA recognition, processing of viral components, and neutralization by antibodies and antiviral agents that has already been derived from such studies has exceeded all anticipations.

ACKNOWLEDGMENTS

We are grateful to N. Olsen, T. S. Baker, and S. D. Fuller for preparation of Figure 4, to Kathy Shuster for producing Figures 2a, 2c, 5, 11, 12, 14, 15, and 16, and to Sharon Wilder and Sharon Fateley for help in the preparation of this manuscript. M. G. R. is supported by grants from NIH, NSF, and the Sterling-Winthrop Research Institute. J. E. J. is supported by grants from NIH and NSF.

Literature Cited

1. Matthews, R. E. F. 1983. *A Critical Appraisal of Viral Taxonomy.* Boca Raton: CRC Press
2. Harrison, S. C., Olson, A. J., Schutt, C. E., Winkler, F. K., Bricogne, G. 1978. *Nature* 276:368–73
3. Bloomer, A. C., Champness, J. N., Bricogne, G., Staden, R., Klug, A. 1978. *Nature* 276:362–68
4. Hogle, J. M., Maeda, A., Harrison, S. C. 1986. *J. Mol. Biol.* 191:625–38
5. Abad-Zapatero, C., Abdel-Meguid, S. S., Johnson, J. E., Leslie, A. G. W., Rayment, I., et al. 1980. *Nature* 286:33–39
6. Stauffacher, C. V., Usha, R., Harrington, M., Schmidt, T., Hosur, M. V., Johnson, J. E. 1987. In *Crystallography in Molecular Biology,* ed. D. Moras, J. Drenth, B. Strandberg, D. Suck, K. Wilson, pp. 293–308. New York/London: Plenum
7. Liljas, L., Unge, T., Jones, T. A., Fridborg, K., Lövgren, S., et al. 1982. *J. Mol. Biol.* 159:93–108
8. Rossmann, M. G., Arnold, E., Erickson, J. W., Frankenberger, E. A., Griffith, J. P., et al. 1985. *Nature* 317:145–53
9. Hogle, J. M., Chow, M., Filman, D. J. 1985. *Science* 229:1358–65
10. Luo, M., Vriend, G., Kamer, G., Minor, I., Arnold, E., Rossmann, M. G., et al. 1987. *Science* 235:182–91
10a. Acharya, R., Fry, E., Stuart, D., Fox, G., Rowlands, D., Brown, F. 1989. *Nature.* In press
11. Hosur, M. V., Schmidt, T., Tucker, R. C., Johnson, J. E., Gallagher, T. M., et al. 1987. *Proteins* 2:167–76
12. Caspar, D. L. D., Klug, A. 1962. *Cold Spring Harbor Symp. Quant. Biol.* 27:1–24
13. Rossmann, M. G. 1972. *The Molecular Replacement Method.* New York: Gordon & Breach
14. Arnold, E., Vriend, G., Luo, M., Griffith, J. P., Kamer, G., et al. 1987. *Acta Crystallogr. A* 43:346–61
15. Rossmann, M. G., Rueckert, R. R. 1987. *Microbiol. Sci.* 4:206–14
16. Rossmann, M. G. 1987. *BioEssays* 7:99–103
17. Liljas, L. 1986. *Prog. Biophys. Mol. Biol.* 48:1–36
18. Harrison, S. C. 1983. *Adv. Virus Res.* 28:175–240
19. Argos, P., Johnson, J. E. 1984. In *Biological Macromolecules and Assemblies,* ed. F. A. Jurnak, A. McPherson, 1:1–43. New York: Wiley
20. Casjens, S., King, J. 1975. *Annu. Rev. Biochem.* 44:555–611

21. Baltimore, D. 1980. *Ann. NY Acad. Sci.* 354:492–97
22. Wilson, I. A., Skehel, J. J., Wiley, D. C. 1981. *Nature* 289:366–73
23. Colman, P. M., Varghese, J. N., Laver, W. G. 1983. *Nature* 303:41–44
24. Wiley, D. C., Skehel, J. J. 1987. *Annu. Rev. Biochem.* 56:365–94
25. Chiu, W. 1986. *Annu Rev. Biophys. Biophys. Chem.* 15:237–57
25a. Namba, K., Stubbs, G. 1986. *Science* 231:1401–6
26. Roberts, M. M., White, J. L., Grütter, M. G., Burnett, R. M. 1986. *Science* 232:1148–51
27. Crick, F. H. C., Watson, J. D. 1956. *Nature* 177:473–75
28. Crick, F. H. C., Watson, J. D. 1957. In *Ciba Foundation Symposium on the Nature of Viruses*, ed. G. E. W. Wolstenholme, E. C. P. Millar, pp. 5–13. Boston: Little Brown
29. Caspar, D. L. D. 1956. *Nature* 177:475–76
30. Huxley, H. E., Zubay, G. 1960. *J. Mol. Biol.* 2:189–96
31. Nixon, H. L., Gibbs, A. J. 1960. *J. Mol. Biol.* 2:197–200
32. Sehnke, P. C., Harrington, M., Hosur, M. V., Li, Y., Usha, R., et al. 1988. *J. Crystal Growth* 90:222–30
33. Silva, A. M., Rossmann, M. G. 1987. *J. Mol. Biol.* 197:69–87
34. Jones, T. A., Liljas, L. 1984. *J. Mol. Biol.* 177:735–67
35. Sehgal, O. P., Hsu, C. H., White, J. A., Van, M. 1979. *Phytopathol. Z.* 95:167–77
36. Tremaine, J. H., Ronald, W. P. 1978. *Virology* 91:164–72
37. Erickson, J. W., Rossmann, M. G. 1982. *Virology* 116:128–36
38. Leberman, R., Finch, J. T. 1970. *J. Mol. Biol.* 50:209–13
39. Crowther, R. A., Amos, L. A. 1972. *Cold Spring Harbor Symp. Quant. Biol.* 36:489–94
40. Bol, J. F., Kraal, B., Brederode, F. Th. 1974. *Virology* 58:101–10
41. Pfeiffer, P., Hirth, L. 1975. *FEBS Lett.* 57:144–48
42. Fukuyama, K., Abdel-Meguid, S. S., Johnson, J. E., Rossmann, M. G. 1983. *J. Mol. Biol.* 167:873–94
43. Erickson, J. W., Silva, A. M., Murthy, M. R. N., Fita, I., Rossmann, M. G. 1985. *Science* 229:625–29
44. Rueckert, R. R. 1986. In *Fundamental Virology*, ed. B. N. Fields, D. M. Knipe. pp. 357–90. New York: Raven
45. Arnold, E., Luo, M., Vriend, G., Rossmann, M. G., Palmenberg, A. C., et al. 1987. *Proc. Natl. Acad. Sci. USA* 84:21–25
46. Francki, R. I. B., Milne, R. G., Hatta, T. 1985. *Atlas of Plant Viruses*, Vol. II, pp. 23–38. Boca Raton: CRC
47. Franssen, H., Leunissen, J., Goldbach, R., Lomonossoff, G., Zimmern, D. 1984. *Embo J.* 3:855–61
48. Argos, P., Kamer, G., Nicklin, M. J. H., Wimmer, E. 1984. *Nucleic Acids Res.* 12:7251–67
49. Rayment, I., Baker, T. S., Caspar, D. L. D., Murakami, W. T. 1982. *Nature* 295:110–15
50. Baker, T. S., Drak, J., Bina, M. 1988. *Proc. Natl. Acad. Sci. USA* 85:422–26
51. Baker, T. S., Caspar, D. L. D., Murakami, W. T. 1983. *Nature* 303:446–48
52. Finch, J. T., Crowther, R. A., Hendry, D. A., Struthers, J. K. 1974. *J. Gen. Virol.* 24:191–200
53. Olson, N. H., Baker, T. S., Mu, W. B., Johnson, J. E., Hendry, D. A. 1987. *Proc. 45th Ann. Meet. Elec. Microsc. Soc. Am.*, ed. G. W. Bailey, pp. 650–51. San Francisco: San Francisco Press
54. Luo, M., Tsao, J., Rossmann, M. G., Basak, S., Compans, R. W. 1988. *J. Mol. Biol.* 200:209–11
55. Prasad, B. V. V., Wang, G. J., Clerx, J. P. M., Chiu, W. 1988. *J. Mol. Biol.* 199:269–75
56. Fuller, S. D. 1987. *Cell* 48:923–34
57. Vogel, R. H., Provencher, S. W., von Bonsdorff, C. H., Adrian, M., Dubochet, J. 1986. *Nature* 320:533–35
58. Wengler, G., Boege, U., Bischoff, H., Wahn, K. 1982. *Virology* 118:401–10
59. Boege, U., Cygler, M., Wengler, G., Dumas, P., Tsao, J., et al. 1989. Submitted
60. Fuller, S. D., Argos, P. 1987. *EMBO J.* 6:1099–105
61. Richardson, J. S. 1979. *Adv. Protein Chem.* 34:167–339
62. Torbet, J. 1987. *Trends Biochem. Sci.* 12:327–30
63. Andree, P. J., Kan, J. H., Mellema, J. E. 1981. *FEBS Lett.* 130:265–68
64. Kan, J. H., Andree, P. J., Kouijzer, L. C., Mellema, J. E. 1982. *Eur. J. Biochem.* 126:29–33
65. Vriend, G., Hemminga, M. A., Verduin, B. J. M., De Wit, J. L., Schaafsma, T. J. 1981. *FEBS Lett.* 134:167–71
66. Munowitz, M. G., Dobson, C. M., Griffin, R. G., Harrison, S. C. 1980. *J. Mol. Biol.* 141:327–33
67. McCain, D. C., Virudachalam, R., Markley, J. L., Abdel-Meguid, S. S.,

Rossmann, M. G. 1982. *Virology* 117:501–3
68. McCain, D. C., Virudachalam, R., Santini, R. E., Abdel-Meguid, S. S., Markley, J. L. 1982. *Biochemistry* 21:5390–97
69. Cohen, S. S., McCormick, F. P. 1979. *Adv. Virus Res.* 24:331–87
70. Jaspars, E. M. J. 1985. In *Molecular Plant Virology*, ed. J. W. Davies, 1:155–221. Boca Raton: CRC
71. Holmes, K. C. 1980. *Trends Biochem. Sci.* 5:4–7
72. Rossmann, M. G., Chandrasekaran, R., Abad-Zapatero, C., Erickson, J. W., Arnott, S. 1983. *J. Mol. Biol.* 166:73–80
73. Chen, Z., Stauffacher, C., Li, Y., Schmidt, T., Bomu, W., et al. 1989. Submitted
74. Goldbach, R. 1987. *Microbiol. Sci.* 4:197–202
75. Ahlquist, P., Strauss, E. G., Rice, C. M., Strauss, J. H., Haseloff, J., Zimmern, D. 1985. *J. Virol.* 53:536–42
76. Haseloff, J., Goelet, P., Zimmern, D., Ahlquist, P., Dasgupta, R., Kaesberg, P. 1984. *Proc. Natl. Acad. Sci. USA* 81:4358–62
77. Argos, P., Fuller, S. D. 1988. *EMBO J.* 7:819–24
78. Argos, P. 1981. *Virology* 110:55–62
79. Gibbs, A. 1987. Private communication
80. Makowski, L. 1984. In *Biological Molecules and Assemblies*, Vol. 1, ed. F. A. Jurnak, A. McPherson, pp. 203–53. New York: Wiley
81. Matthews, B. W., Rossmann, M. G. 1985. *Methods Enzymol.* 115:397–420
82. Palmenberg, A. C. 1989. *Proc. ICN-UCI Internatl. Conf. Virol.* In press
83. Rossmann, M. G., Liljas, A., Brändén, C. I., Banaszak, L. J. 1975. *The Enzymes* 11:61–102
84. Rossmann, M. G., Argos, P. 1976. *J. Mol. Biol.* 105:75–95
85. Remington, S. J., Matthews, B. W. 1978. *Proc. Natl. Acad. Sci. USA* 75:2180–84
86. Argos, P., Tsukihara, T., Rossmann, M. G. 1980. *J. Mol. Evol.* 15:169–79
87. Bazan, J. F., Fletterick, R. J. 1988. *Proc. Natl. Acad. Sci. USA* 85:7857–61
88. Gallagher, T. M., Rueckert, R. R. 1988. *J. Virol.* 62:3399–406
89. Toyoda, H., Nicklin, M. J. H., Murray, M. V., Anderson, C. W., Dunn, J. J., et al. 1986. *Cell* 45:761–70
90. Palmenberg, A. C. 1982. *J. Virol.* 44:900–6
91. King, J., Hall, C., Casjens, S. 1978. *Cell* 15:551–60

92. Casjens, S., Adams, M. B. 1985. *J. Virol.* 53:185–91
93. Rueckert, R. R. 1976. In *Comprehensive Virology*, ed. H. Fraenkel-Conrat, R. R. Wagner, 6:131–213. New York: Plenum
94. Kim, S., Smith, T. J., Chapman, M. S., Rossmann, M. G., Pevear, D., et al. 1989. Manuscript in preparation
95. Rossmann, M. G., Abad-Zapatero, C., Hermodson, M. A., Erickson, J. W. 1983. *J. Mol. Biol.* 166:37–83
96. Harrison, S. C. 1984. *Trends Biochem. Sci.* 9:345–51
97. Golden, J. S., Harrison, S. C. 1982. *Biochemistry* 21:3862–66
98. Rossmann, M. G. 1984. *Virology* 134:1–11
99. Bancroft, J. B. 1970. *Adv. Virus Res.* 16:99–134
100. Hsu, C. H., Sehgal, O. P., Pickett, E. E. 1976. *Virology* 69:587–95
101. Hull, R. 1978. *Virology* 89:418–22
102. Bancroft, J. B., Hills, G. J., Markham, R. 1967. *Virology* 31:354–79
103. Robinson, I. K., Harrison, S. C. 1982. *Nature* 297:563–68
104. Chow, M., Newman, J. F. E., Filman, D., Hogle, J. M., Rowlands, D. J., Brown, F. 1987. *Nature* 327:482–86
105. Crowell, R. L., Hsu, K. H. L., Schultz, M., Landau, B. J. 1987. In *Positive Strand RNA Viruses*, ed. M. A. Brinton, R. R. Rueckert, pp. 453–66. New York: Liss
106. Co, M. S., Gaulton, G. N., Tominaga, A., Homcy, C. J., Fields, B. N., Greene, M. I. 1985. *Proc. Natl. Acad. Sci. USA* 82:5315–18
107. Rossmann, M. G., Palmenberg, A. C. 1988. *Virology* 164:373–82
108. Colonno, R. J., Condra, J. H., Mizutani, S., Callahan, P. L., Davies, M. E., Murcko, M. A. 1988. *Proc. Natl. Acad. Sci. USA* 85:5449–53
108a. Kim, S., Boege, U., Krishnaswamy, S., Minor, I., Smith, T. J., et al. 1989. Manuscript in preparation
109. Pevear, D. C., Fancher, M. J., Felock, P. J., Rossmann, M. G., Miller, M. S., et al. 1989. *J. Virol.* In press
110. Rogers, G. N., Paulson, J. C., Daniels, R. S., Skehel, J. J., Wilson, I. A., Wiley, D. C. 1983. *Nature* 304:76–78
111. Weis, W., Brown, J. H., Cusack, S., Paulson, J. C., Skehel, J. J., Wiley, D. C. 1988. *Nature* 333:426–31
112. Matthews, T. J., Weinhold, K. J., Lyerly, H. K., Langlois, A. J., Wigzell, H., Bolognesi, D. P. 1987. *Proc. Natl. Acad. Sci. USA* 84:5424–28

112a. Colonno, R. J. 1988. *BioEssays* 5:270–74
113. Colonno, R. J., Callahan, P. L., Long, W. J. 1986. *J. Virol.* 57:7–12
114. Tomassini, J. E., Colonno, R. J. 1986. *J. Virol.* 58:290–95
115. Allaway, G. P., Burness, A. T. H. 1986. *J. Virol.* 59:768–70
116. Allaway, G. P., Burness, A. T. H. 1987. *J. Gen. Virol.* 68:1849–56
117. Wiley, D. C., Wilson, I. A., Skehel, J. J. 1981. *Nature* 289:373–78
118. Sherry, B., Rueckert, R. 1985. *J. Virol.* 53:137–43
119. Sherry, B., Mosser, A. G., Colonno, R. J., Rueckert, R. R. 1986. *J. Virol.* 57:246–57
120. Page, G. S., Mosser, A. G., Hogle, J. M., Filman, D. J., Rueckert, R. R., Chow, M. 1988. *J. Virol.* 62:1781–94
121. Baxt, B., Vakharia, V., Moore, D. M., Franke, A. J., Morgan, D. O. 1989. *J. Virol.* In press
122. Luo, M., Rossmann, M. G., Palmenberg, A. C. 1988. *Virology.* 166:503–14
123. Strohmaier, K., Franze, R., Adam, K. H. 1982. *J. Gen. Virol.* 59:295–306
124. Baxt, B., Morgan, D. O., Robertson, B. H., Timpone, C. A. 1984. *J. Virol.* 51:298–305
125. Bittle, J. L., Houghten, R. A., Alexander, H., Shinnick, T. M., Sutcliffe, J. G., et al. 1982. *Nature* 298:30–33
126. Robertson, B. H., Morgan, D. O., Moore, D. M. 1984. *Virus Res.* 1:489–500
127. Pfaff, E., Mussgay, M., Böhm, H. O., Schulz, G. E., Schaller, H. 1982. *EMBO J.* 1:869–74
128. Meloen, R. H., Puyk, W. C., Meijer, D. J. A., Lankhof, H., Posthumus, W. P. A., Schaaper, W. M. M. 1987. *J. Gen. Virol.* 68:305–14
129. McCullough, K. C., Crowther, J. R., Carpenter, W. C., Brocchi, E., Capucci, L., et al. 1987. *Virology* 157:516–25
130. Grubman, M. J., Zellner, M., Wagner, J. 1987. *Virology* 158:133–40
131. Xie, Q. C., McCahon, D., Crowther, J. R., Belsham, G. J., McCullough, K. C. 1987. *J. Gen. Virol.* 68:1637–47
132. Chow, M., Yabrov, R., Bittle, J., Hogle, J., Baltimore, D. 1985. *Proc. Natl. Acad. Sci. USA* 82:910–14
133. Thomas, A. A. M., Vrijsen, R., Boeyé, A. 1986. *J. Virol.* 59:479–85
134. Icenogle, J., Shiwen, H., Duke, G.,

135. Gilbert, S., Rueckert, R., Anderegg, J. 1983. *Virology* 127:412–25
Wetz, K., Willingman, P., Zeichhardt, H., Habermehl, K. O. 1986. *Arch. Virol.* 91:207–20
136. Emini, E. A., Jameson, B. A., Wimmer, E. 1983. *Nature* 304:699–703
137. Smith, T. J., Rossmann, M. G., Rueckert, R. R. 1989. Manuscript in preparation
138. Colman, P. M., Laver, W. G., Varghese, J. N., Baker, A. T., Tulloch, P. A., et al. 1987. *Nature* 326:358–63
139. Diana, G. D., McKinlay, M. A., Otto, M. J., Akullian, V., Oglesby, C. 1985. *J. Med. Chem.* 28:1906–10
140. Ninomiya, Y., Ohsawa, C., Aoyama, M., Umeda, I., Suhara, Y., Ishitsuka, H. 1984. *Virology* 134:269–76
141. Ninomiya, Y., Aoyama, M., Umeda, I., Suhara, Y., Ishitsuka, H. 1985. *Antimicrob. Agents Chemother.* 27:595–99
142. Tisdale, M., Selway, J. W. T. 1984. *J. Antimicrob. Chemother.* 14 (Suppl. A):97–105
143. Eggers, H. J., Rosenwirth, B. 1988. *Antiviral Res.* 9:23–36
144. Caliguiri, L. A., McSharry, J. J., Lawrence, G. W. 1980. *Virology* 105:86–93
145. McSharry, J. J., Caliguiri, L. A., Eggers, H. J. 1979. *Virology* 97:307–15
146. Diana, G. D., Salvador, U. J., Zalay, E. S., Johnson, R. E., Collins, J. C., et al. 1977. *J. Med. Chem.* 20:750–56
147. Diana, G. D., Salvador, U. J., Zalay, E. S., Carabateas, P. M., Williams, G. L., et al. 1977. *J. Med. Chem.* 20:757–61
148. McKinlay, M. A., Steinberg, B. A. 1986. *Antimicrob. Agents Chemother.* 29:30–32
149. Diana, G. D., Pevear, D. C., Otto, M. J., Rossmann, M. G. 1989. *Pharmacol. Therapeut.* In press
150. Badger, J., Minor, I., Kremer, M. J., Oliveira, M. A., Smith, T. J., et al. 1988. *Proc. Natl. Acad. Sci. USA* 85:3304–8
151. Smith, T. J., Kremer, M. J., Luo, M., Vriend, G., Arnold, E., et al. 1986. *Science* 233:1286–93
152. Badger, J., Krishnaswamy, S., Kremer, M. J., Oliveira, M. A., Rossmann, M. G., et al. 1989. *J. Mol. Biol.* In press
153. Rossmann, M. G. 1988. *Proc. Natl. Acad. Sci. USA* 85:4625–27
154. Kamer, G., Argos, P. 1984. *Nucleic Acids Res.* 12:7269–82
155. Needleman, S. D., Wunsch, C. D. 1970. *J. Mol. Biol.* 48:443–53

Annu. Rev. Biochem. 1989. 58:575–606

THE HEPARIN-BINDING (FIBROBLAST) GROWTH FACTOR FAMILY OF PROTEINS

Wilson H. Burgess and Thomas Maciag

Laboratory of Molecular Biology, Jerome H. Holland Laboratory for the Biomedical Sciences, American Red Cross, 15601 Crabbs Branch Way, Rockville, Maryland 20855

CONTENTS

0066-4154/89/0701-0575$02.00

PERSPECTIVES AND SUMMARY

Polypeptide growth factors are hormone-like modulators of cell proliferation and differentiation in vitro and in vivo. These functions are mediated, in part, by interaction of the growth factors with relatively high-affinity cell-surface receptors and subsequent alterations in gene expression within responsive cells. The complete cascade of signals initiated by growth factor occupancy of cell-surface receptors that are responsible for the mitogenic or differentiating responses are not known. However, activation of tyrosine kinases, changes in cyclic nucleotide metabolism, inositol-3-phosphate levels, and intracellular free calcium levels are thought to be the mediators of polypeptide growth factor action. Abnormal regulation of one or more of these signal pathways may explain the mechanism of action of cellular oncogenes and resulting phenotypes associated with cellular transformation.

The previous review in this series by James & Bradshaw (1) is likely to be the last comprehensive article on polypeptide growth factors. We focus this review on the structures and functions associated with the fibroblast or heparin-binding growth factor (HBGF) family of polypeptides. We refer the reader to the following references for recent reviews of other polypeptide growth factors: epidermal growth factor (2), nerve growth factor (1), platelet-derived growth factors (3, 4), insulin-like growth factors (5), transforming growth factors (6–8), hematopoietic growth factors (9), and their receptors (10–12).

The HBGF family presently consists of five structurally related polypeptides. The genes for each have been cloned and sequenced. Two of the members, HBGF-1 and HBGF-2, have been characterized under many different names (see below), but most often as acidic and basic fibroblast growth factor, respectively. Three recent additions to the family are oncogene products of related sequence. The normal gene products influence the general proliferation capacity of the majority of mesoderm- and neuroectoderm-derived cells. They are capable of inducing angiogenesis in vivo and may play important roles in early development. Although little is known about the true physiological functions of these proteins or their mechanism of action, the cloning of their genes, the identification of their receptors, and the identification of related oncogenes should provide the experimental basis to increase our understanding of these important areas.

NOMENCLATURE, PURIFICATION, AND PRELIMINARY CHARACTERIZATIONS

General History

The impetus for the identification of the HBGFs originated with the premise that soluble factors are responsible for the regulation of a variety of physiological processes, including development, regeneration, and wound repair. As early as 1939, extracts of brain were cited as a rich source of factors that promoted fibroblast proliferation in vitro (13, 14). These observations continued into the late 1960s and early 1970s, when preparations of pituitary-derived thyroid-stimulating hormone and leutenizing hormone (15–17) provided by the NIAMD Hormone Distribution Program at NIH were found to contain a potent polypeptide mitogen for murine 3T3 cells and chondrocytes. Hugo Armelin later described the effects of pituitary extracts on the growth of 3T3 cells (18), and Gospodarowicz and colleagues characterized the polypeptide from acid extracts of bovine pituitary in 1974 (19). The protein was named ovarian cell growth factor, and it was shown to be distinct biochemically from other known hormones or growth factors (19). At this time, it seemed that extraction conditions were important in maintaining biological activities, suggesting the presence of additional mitogens in the pituitary extracts (20). In 1975, an extraction and purification procedure, similar to the ovarian cell growth factor protocol, was used to purify a polypeptide mitogen for 3T3 cells from bovine pituitary (21) and later from brain (22). The polypeptide was named fibroblast growth factor (FGF), and the purification procedure would become the standard for its preparation. It was subsequently characterized as a basic polypeptide possessing mitogenic activity for mesoderm- and neuroectoderm-derived cells (23). Attempts to characterize the structure of the growth factor further were hampered by relatively high levels of contaminating fragments of myelin basic protein (24).

During this period, the activity of an additional polypeptide mitogen for human endothelial cells was characterized from extracts of neural tissue (25). The relationship of this and other biological activities described from other sources (20, 26–30) to FGF was not clear. Thomas and coworkers were independently able to detect the presence of a polypeptide with an acidic isoelectric point from acid extracts of bovine brain (31) and this polypeptide was named acidic FGF. Acidic FGF did not interact with antibodies against myelin basic protein (31), and purified fragments of myelin basic protein were not biologically active in cell culture (32). Interestingly, acidic FGF possessed mitogenic activity for endothelial cells (33), a criteria that appeared to distinguish it from basic FGF (25, 33, 34). Ultimately, the endothelial cell growth factor would be described as an acidic polypeptide (35) with a precursor-product relationship to acidic FGF (36, 37). The biological activity associated with basic FGF would be separated from contaminating fragments of myelin basic protein (38), and acidic FGF (39) and basic FGF (40) would be purified and both characterized as potent heparin-binding polypeptide mitogens for endothelial cells (41, 42). Structural characterization of acidic FGF (43) and basic FGF (44) would further demonstrate that these growth factors are related polypeptides (37) and serve to establish the basis for a larger family of polypeptide growth factors.

The Role of Source Tissue and Biological Assays in Establishing Nomenclature

Prior to the characterization of the structures for acidic FGF and basic FGF, a number of laboratories had identified polypeptide mitogens from tissue, organ, and cell culture sources that influenced the growth of FGF-responsive target cells. These individual efforts were dedicated to the identification of polypeptide growth factors that modulate growth and differentiation of neurotropic (reviewed in 44, 45) and angiogenic (reviewed in 114) responsive cells and included eye-derived growth factor-1 and -2 (46), endothelial cell growth factor (47), heparin-binding growth factor-α and -β (48), anionic and basic brain-derived growth factor (49), hypothalamus-derived growth factor (50), retinal-derived growth factor (51), astroglial growth factor-1 and -2 (52), cartilage-derived growth factor (53), myogenic growth factor (54), pituitary growth factor (55), bone-derived growth factor (56), anionic endothelial growth factor (57), prostatic growth factor (58, 59), and prostatropin (60). Although these growth factors would ultimately be characterized as structural equivalents of either acidic FGF or basic FGF, the variety of names that describe their source or target cells demonstrate the potential biological significance of the FGF family. Further, the diversity of the nomenclature would also be resolved by the ability of the individual biological activities from numerous tissue sources to bind immobilized heparin.

Heparin Interactions, Physical Properties, and Nomenclature Clarification

In 1983, Thornton and her colleagues demonstrated that the sulfated glycosaminoglycan, heparin, could potentiate the biological activity of crude preparations of acidic FGF (34). This observation provided the basis for the discoveries that heparin was able to bind basic FGF (41) and acidic FGF (61). Indeed, Shing and his coworkers were the first to describe that the interaction between heparin and basic FGF could be used as an efficient affinity chromatography method for the purification of basic FGF (62). Subsequently, the principle of heparin adsorption was applied to acidic FGF (61). These studies also demonstrated that binding of acidic FGF and basic FGF to heparin were different, and could be used to distinguish between acidic FGF and basic FGF (reviewed in 63). While heparin potentiated the biological activity of acidic FGF (34, 64), and acidic FGF eluted from immobilized heparin near 1.0 M NaCl, basic FGF required at least 1.6 M NaCl for elution (41). In addition, heparin did not augment the mitogenic activity of basic FGF (37), another criterion that discriminates between the two polypeptides. These descriptive criteria led Lobb and his colleagues (63) to suggest that the nomenclature for the FGFs could be separated into two classes: the Class I heparin-binding growth factors (HBGF-1) describing the acidic polypeptides, and the Class II heparin-binding growth factors (HBGF-2) describing the basic polypeptides. This classification system did not anticipate the identification of additional gene products related structurally to acidic and basic FGF. However, the fact that two of the three recently identified polypeptides interact with heparin (reviewed in 65 and below) indicates we should take advantage of the flexibility and simplicity of this nomenclature (i.e. HBGF-1, -2, -3, . . . etc), as has been done for the interleukin family of hematopoietic growth factors.

The interaction of HBGF-1 and HBGF-2 with heparin has provided more than a simplified affinity-based purification procedure and convenient nomenclature system. Heparin has been shown to potentiate the biological activities of HBGF-1 (reviewed in 63), but not HBGF-2. The enhanced activity of the HBGF-1 : heparin complex varies from several to one hundred fold (63). Although the precise mechanism responsible for the potentiation of the biological activity of HBGF-1 has not been deduced, experimental evidence suggests that the mechanism may involve two synergistic pathways: the stabilization of tertiary structure and the prevention of proteolytic modification. It is known that heparin can induce immunologically sensitive epitope exposure in HBGF-1 (64), restore biological activity to preparations of native (64) and recombinant HBGF-1 (66), and increase the affinity of HBGF-1 for its receptor (64). Further, heparin protects HBGF-1 and HBGF-2 from heat (67, 68) and acid (68) inactivation and prevents proteolytic modification of HBGF-1 by trypsin (67), plasmin (67), and thrombin (69). Sommer & Rifkin (70) have shown that heparin also protects HBGF-2 from trypsin degradation and show complete protection at HBGF-2 to heparin ratios of ~10 : 1 (w/w). In contrast to the extensive digestion of HBGF-1 by trypsin and plasmin, thrombin inactivation at low concentrations of protease involves the selective cleavage of HBGF-1 in the carboxy-terminal domain at arginine 122–threonine 123 (69). Interestingly, higher concentrations of thrombin that favor further fragmentation of HBGF-1, do not modify HBGF-2 (69). Thus, the interaction between heparin and HBGF-1 may involve the formation of a stable glycosaminoglycan-polypeptide complex that is resistant to inactivation by cell culture–derived proteases.

STRUCTURES OF HBGF-1 and HBGF-2

Primary Structures from Amino Acid Sequences

The complete structure of bovine HBGF-1 (residues 15–154) was first described by Gimenez-Gallego et al (43). The amino acid sequence of full-length bovine HBGF-1 (36) is shown in Figure 1, and the amino-terminal acetylation of alanine is indicated. Amino acid sequence analysis of truncated forms of human HBGF-1 has been described (71–73). Crabb et al (74) used methods similar to those described in the elucidation of the bovine HBGF-1 sequence (36) (i.e. fast atom bombardment mass spectrometry) to determine the sequence of full-length human HBGF-1 (1–154) and the identity of the blocking group.

The structures described for bovine acidic FGF (HBGF-1; 15–154) and acidic FGF-II or α-ECGF (HBGF-1; 21–154) result from proteolytic cleavage

```
                                            ▼▼▼      ▼
hHBGF-1      Ac-AEGEITTFTA--LTE----KFNLPPGNYKKPKL--LYCSNG-G      (34)
bHBGF-1      Ac-     T                L              LYCSNG-G    (34)
hHBGF-2      GTMAAGSITTLPA--LPED-GGSGAFPPGHFKDPKR--LYCKNG-G      (37)
bHBGF-2                                                          (37)
mHBGF-3      SSLEPSWPTTGPGTRLRRDAGGRGGVYEHLGGAPRRRKLYCATK--      (53)
hHBGF-4      LSLARLPVAAQPK--EAAVQSGAGDYLLG-IKRLRR--LYCNVGIG      (93)
hHBGF-5      MSSSSASSSPAASLGSQGSGLEQSSFQWSLGARTGS--LYCRVGIG      (97)

             ▼   ▼ ▼    ▼           ▼        ▼ ▼ ▼          ▼
hHBGF-1      HFLRILPDGTVDGTRDRSDQHIQLQLSAESVGEVYIKSTETGQYLA      (80)
bHBGF-1      Y          K              C  I              F       (80)
hHBGF-2      FFLRIHPDGRVDGVREKSDPHIKLQLQAEERGVVSIKGVCANRYLA      (83)
bHBGF-2                                                          (83)
mHBGF-3      YHLQLHPSGRVNGSLENSAYSI-LEITAVEVGVVAIKGLFSGRYLA      (98)
hHBGF-4      FHLQALPDGRIGGAHADTRDSL-LELSPVERGVVSIFGVASRFFVA      (138)
hHBGF-5      FHLQIYPDGKYNGSHEANMLSV-LEIFAVSQGIVGIRGVFSNKFLA      (142)

             ▼   ▼ ▼    ▼   ▼ ▼ ▼
hHBGF-1      MDTDGLLYGSQTPNEECLFLERLEENHYNTYISKKHA---------      (117)
bHBGF-1                                                          (117)
hHBGF-2      MKEDGRLLASKCVTDECFFFERLESNNYNTYRSRKYT---------      (120)
bHBGF-2                                                    S      (154)
mHBHF-3      MNKRGRLYASDHYNAECEFVERIHELGYNTYASRLYRTGSSGPGAQ      (144)
hHBGF-4      MSSKGKLYGSPFFTDECTFKEILLPNNYNAYESYKYP---------      (175)
hHBGF-5      MSKKGKLHASAKFTDDCKFRERFQENSYNTYASAIHRTEKTGRE--      (186)

                                            ▼▼▼
hHBGF-1      -----EKNWFVGLKKNGSCKRGPRTHYGQKAILFLPLPVSSD         (154)
bHBGF-1           H       RS L    F                           (154)
hHBGF-2      -------SWYVALKRTGQYKLGSKTGPGQKAILFLPMSAKS          (154)
bHBGF-2                       P                               (154)
mHBGF-3      RQPGAQRPWYVSVNGKGRPRRGFKTRRTQKSSLFLPRVLGHK        (186)
hHBGF-4      -------GMFIALSKNGKTKKGNRVSPTMKVTHFLPRL             (206)
hHBGF-5      ------WYVALNKRGKAKRGCSPRVKPQHISTHFLPRFKQ           (220)
```

Figure 1 Protein sequence of human (h) and bovine (b) HBGF-1, HBGF-2, murine int-2 (mHBGF-3), human hst/KS3 (hHBGF-4), and human FGF-5 (hHBGF-5). The sequences are aligned to maximize homologies in the regions of overlap with HBGF-1 and HBGF-2. The numbers on the right indicate the number of the last amino acid residue on the line, where numbering begins at residue 1 of HBGF-1, -3, -4, and -5. The numbering for HBGF-2 refers to the fourth residue shown (A) as residue 1 (see text). The arrows indicate residues that are absolutely conserved in the sequences shown. The complete amino acid sequences of HBGF-3, -4, and -5 are not shown.

of intact HBGF-1 at lysine-14:phenylalanine-15 or glycine-20:asparagine-21, respectively (36, 43, 75–77). Sequence analysis of various forms of human HBGF-1 (71–74) demonstrate the generation of similar amino-terminal truncations, an approximate 92% sequence identity with the bovine protein, conservation of the positions of two of the three cysteine residues, and the presence of a potential N-linked glycosylation site. The implications of these changes are discussed later in this review.

HBGF-2 has also been purified from bovine brain as a single-chain polypeptide of approximately 15–16 kd by a number of laboratories (78). Alternative bovine tissue sources that have yielded HBGF-2 include corpus luteum (79), prostate (74), kidney (80), adrenal gland (81), macrophage (82), retina (83), pituitary (84, 85), liver (86), and testes (87). The structure of HBGF-2 was first described by Esch and colleagues (88), and the sequence of HBGF-2 is shown in Figure 1. Larger polypeptide forms of bovine (89, 90) and human (91–93) HBGF-2 have also been purified, and limited NH_2 terminal sequence analysis of the polypeptides (89, 91, 92) suggest that they represent NH_2-terminal truncations of a larger precursor form. Although the significance of these precursor polypeptides is not known, Moscatelli et al (95) have described the presence of an immunoreactive and receptor competent 25 kd form of HBGF-2 in guinea pig brain that can be converted to an 18 kd form of HBGF-2 by selective digestion with trypsin.

Analysis of the structures for bovine HBGF-1 and HBGF-2 demonstrate a 55% identity (96) between the two polypeptides (Figure 1). This sequence similarity together with similar organ sources and biological activities suggests that HBGF-1 and HBGF-2 may have originated from a common ancestral gene. Although the structures of HBGF-1 and HBGF-2 weakly resemble the structures for human IL-1-α and IL-1-β and for the neuropeptides of the tachykinin and bombesin classes (43), the biological significance of this similarity is not known.

Primary Structures from cDNA Clones and Organization of the Genes for HBGF-1 and HBGF-2

Complementary DNA clones encoding the human HBGF-1 (97) and human (98, 99) and bovine (100) HBGF-2 have been isolated and sequenced. The predicted amino acid sequences derived from the cDNAs agree with the structures determined by protein sequence analysis. The HBGF-1 and HBGF-2 polypeptides are encoded by separate genes, each of which appear to be present in a single copy (97, 98, 100). Chromosomal localization analysis demonstrates the gene for human HBGF-1 is positioned on human chromosome 5 between bands 5q 31.3 and 5q 33.2 (97), whereas the gene for human HBGF-2 is located on chromosome 4 (101). Both genes are similar in their overall organization, containing three exons separated by two relatively large introns. The relative positions of the intron-exon boundaries are similar (102, 103). This feature is also shared by the *Xenopus* HBGF-2 gene (104). Other examples where related growth factor genes map to different chromosomes include insulin-like growth factor-I and -II (105) and the A and B chains of human platelet-derived growth factor (106). The genes encoding human c-fms (the receptor for colony stimulating factor-1) and human platelet-derived growth factor receptor are also localized on the short arm of chromosome 5 (97, 107). Similarly, the genes for epidermal growth factor and interleukin-2 share with HBGF-2 localization to chromosome 4 (105).

The major difference that stands out after analysis of the cDNAs for HBGF-1 and HBGF-2 involves the precise location of the amino terminus of the two proteins. Nucleotide sequence analysis of the human HBGF-1 cDNA demonstrates that the open reading frame, which corresponds to the protein sequence of HBGF-1 (1–154) (36, 74), is flanked by termination codons, a feature not found in the HBGF-2 cDNA (97, 100). Further, the sequence flanking the translation initiation codon (GACCAAUGG) is in agreement with the consensus sequence for initiation of translation (GACCCAUGG) in eukaryotes (108). The fact that a similar sequence (GAGCCAUGG) is observed flanking a potential translation initiation codon in the cDNA for HBGF-2, which would lead to the translation of HBGF-2 (1–154), has led several investigators to propose that this is indeed the mature form of HBGF-2 (103). The isolation and primary sequence determination of forms of HBGF-2 containing amino-terminal sequence that extends 5' to this proposed initiator methionine (92) suggests alternative mechanisms of HBGF-2 translation are responsible for some of the amino-terminal diversity of the HBGF-2 protein.

Both HBGF-1 and HBGF-2 lack classical consensus signal peptide sequences (97, 100); a feature also characteristic of the precursors of interleukin-1α and interleukin-1β (109, 110). This observation raises the same questions as to mechanisms of secretion for the HBGFs as have been raised for the interleukins. Although release of the growth factors can certainly occur following cell lysis, alternative mechanisms of secretion may involve the formation of complexes between the HBGFs and carrier/binding proteins.

Posttranslational Modifications

The evidence of acetylation of the amino-terminal alanine of full-length human and bovine HBGF-1 (1–154) and a form of HBGF-2 (1–154) has been described above. This modification is, to date, the only defined posttranslational addition to HBGF-1 or HBGF-2. The presence of potential N-linked glycosylation sites in human HBGF-1 and HBGF-2 was noted above. There is, however, no direct evidence for glycosylation of these proteins. The ambiguities surrounding the actual size of HBGF-2 and the fact that very little protein structural data exists for these amino terminally extended forms, makes it impossible to rule out any form of posttranslational addition for the HBGF-2 polypeptides.

The major observations relating to potential posttranslational cleavage of various lengths of the amino-terminal domains of HBGF-1 and HBGF-2 have come from the amino-terminal sequencing of HBGFs isolated by different investigators from various sources using different purification protocols. Whereas it is not yet known whether proteolytic processing occurs in vivo to generate the different forms of the HBGFs described above, increasing evidence indicates that at least for HBGF-1, these truncated forms are artifacts of purification procedures. McKeehan & Crabb (111) reported a detailed study of different molecular and chromatographic forms of HBGF-1 from bovine brain and provided convincing evidence that the heterogeneity is due to a combination of proteolytic cleavage and disulfide rearrangements. They conclude, as we did (36), that a unique, specific protease may be responsible for the generation of HBGF-1 (21–154). We have also shown by Western blot analysis that cleavage of a full-length HBGF-1 (1–154) occurs during extraction of the protein from bovine brain (A. M. Shaheen, W. H. Burgess unpublished). Further, it appears that all of the immunoreactive HBGF-1 in fresh tissue is full-length (Figure 2). Together these data indicate that the multiple truncated forms of HBGF-1 isolated and sequenced are not physiologically relevant and thus, do not represent any type of processing in vivo. Although similar results relating to proteolytic modifications of HBGF-2 have been described (112, 113), the uncertainty as to the actual size of HBGF-2 in vivo makes it difficult to speculate about the physiological significance, if any, of the differences in size of the various forms.

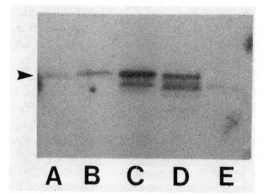

Figure 2   Western blot analysis using affinity-purified antibodies against HBGF-1 and ^{125}I-labeled Protein A. The figure shows the autoradiogram visualizing proteins that were separated on a 15% acrylamide SDS gel, transferred to nitrocellulose, and probed with HBGF-1 antibodies. Lane A contains HPLC-purified HBGF-1 (1–154); lane B contains bovine brain homogenized in SDS sample buffer; lane C contains a neutral extract of bovine brain; lane D contains an acidified extract of bovine brain; and lane E contains HPLC-purified HBGF-1 (21–154).

BIOLOGICAL PROPERTIES

Induction of DNA Synthesis and Cellular Proliferation In Vitro

HBGF-1 and HBGF-2 are extremely potent inducers of DNA synthesis in a variety of normal diploid mammalian cell types from mesoderm and neuroectoderm lineages. These include endothelial cells (114–116), smooth muscle cells (117a,b, 118), adrenal cortex cells (119, 120), prostatic (121–123), and retina epithelial cells (28, 46, 83, 124–127), oligodendrocytes (128), astrocytes (129), chondrocytes (130–133), myoblasts (54, 56, 128, 134), and osteoblasts (135–137). Although human melanocytes respond to the mitogenic influence of HBGF-2 but not HBGF-1 (138), most avian and mammalian cell types respond to both polypeptides (reviewed in 139). In most cases, the mitogenic effect of the HBGFs on normal diploid mammalian cells significantly delays their premature senescence in vitro (116, 140). HBGF-1 and HBGF-2 are also potent mitogens for a large number of established cell lines including BALB/c and Swiss 3T3 cells, LEII cells, BHK21, CHO cells, and rhabdomyosarcoma cells, among others (reviewed in 65, 78, 96, 139, 141–143). Indeed, HBGF-1 and HBGF-2 induce early G_0 to G_1 cell cycle events (134 and reviewed in 23, 144) and join the PDGFs as polypeptide inducers of cellular competence (145). HBGF-2 also induces cell division in soft agar (146, 147).

Induction of Chemotaxis In Vitro

The first report of stimulation of chemotaxis of endothelial cells by angiogenic preparations was that of Mullins & Rifkin (148). Although the polypeptide responsible for the chemotactic activity was not identified, these studies provided a basis for two of the necessary functions of a true angiogenic factor, i.e. stimulation of endothelial cell migration and proliferation. The abilities of HBGF-1 and HBGF-2 to stimulate the proliferation of a variety of cell types in vitro, including endothelial cells, have been summarized above. In addition, preparations of HBGF-1 have been shown to be chemotactic for endothelial cells, fibroblasts, and astroglial cells (149, 150). These results provide a basis for a role of HBGFs in wound repair, angiogenesis, and perhaps neurite outgrowth and tumor metastasis (see below).

The fact that cell division and cell migration in response to HBGF-1 require mutually exclusive cytoskeletal arrangements was described by Linemeyer et al (151). They also reported that high concentrations of recombinant HBGF-1 (15–154), which inhibited thymidine incorporation into BALB/c 3T3 cells, stimulated chemokinesis and chemotaxis of these cells. These results indicate that the chemotactic and mitogenic activities of HBGF-1 may be distinct.

Thus, localized and transient high concentrations of HBGF-1 could stimulate the migration of endothelial or fibroblast cells to the site of a wound or tumor, and diffusion-mediated decreases in the concentration of growth factor would result in stimulation of proliferation of the recruited cells.

It is presently unclear whether the chemotactic activities of the HBGFs are mediated through the same receptor systems involved in the induction of cellular proliferation. A better understanding of the mechanisms of induction of these two phenomena will be possible if the mitogenic and chemotactic activities of the HBGFs can be dissociated at the structural level as was shown for PDGF (152).

HBGF-Induced Protein Expression

The transmembrane signaling pathway responsible for HBGF-mediated cellular proliferation is not known. Evidence has been provided that the mitogenic signal is not mediated by polyphosphoinositide hydrolysis and protein kinase C activation (153–155). Indeed, it has been demonstrated that protein kinase C activators antagonize the mitogenic activities of HBGFs on endothelial cells (156) by reduction of the number of HBGF receptors on the cell surface (157). These results are consistent with studies of the mechanism of EGF or EGF/insulin mitogenic activities (158, 159). There are, however, several reports of activation of gene expression by cells in response to HBGFs that may provide insight into their mechanisms of action.

Not surprisingly, the HBGFs induce rapid and transient expression of c-fos mRNA (160) as do other growth factors or serum. Similar results are observed for c-myc mRNA (161, 162), although its increase appears to follow that of c-fos (161, 162). These data together with the fact that c-myc induction can occur in the presence of cycloheximide argue that the increase in c-myc mRNA is not secondary to growth (160).

It is known that HBGF-2 increases the activity of ornithine decarboxylase in normal and protein kinase C–deficient BALB/c 3T3 cells (163). In addition, the HBGF prototypes activate the phosphorylation of a synthetic peptide on serine residues based upon the sequence of one of the phosphorylation sites of 40S ribosomal subunit S6, and this phosphorylation proceeds by a protein kinase C–independent pathway (164). Further, the induction of DNA synthesis occurs in HBGF-2-responsive, protein kinase C–deficient cells (165). HBGF-2 also induces a G_0 to G_1 transition in quiescent cells in vitro, a feature it has in common with the mitogenic activity of PDGF (145). While HBGF-1 and HBGF-2 both induce actin mRNA expression in vitro, the levels of actin mRNA induced by HBGF-2 are sustained for a much longer period of time (166). In addition, analytical two-dimensional gel analysis of BALB/c 3T3 cell polypeptide expression independently induced by the HBGFs demonstrates that HBGF-1 and HBGF-2 induce identical changes in intracellular polypeptide expression (166). In contrast, HBGF-2 induced the selective expression of an extracellular M_r 43,000 polypeptide (166). Further, the changes in polypeptide expression observed with the HBGFs resemble the effects of PDGF on polypeptide expression in vitro.

HBGF-2 also induces the transcription of ribosomal genes after its addition to quiescent bovine endothelial cells (167). This event occurs during the G_0 to G_1 transition period after the increase in the nonhistone nucleolar protein, nucleolin (167). Interestingly, immunoreactive HBGF-2 could be detected by indirect immunofluorescence in association with the nucleolus, and chloroquine inhibits its nucleolar association (167).

HBGFs and Differentiation

In addition to the mitogenic and chemotactic activites of HBGF-1 and HBGF-2, both growth factors are known to promote cellular differentiation in vitro. HBGF-2 induces adipocyte differentiation in vitro (168) and the NGF-like biological activities of HBGF-1 and HBGF-2 include the promotion of neurite extension in PC12 cells (169–171), a rat pheochromocytoma cell line.

The ability of HBGF-1 and HBGF-2 to promote the proliferation of cells derived from embryologically diverse germ layers predicts the importance of the HBGFs as physiological regulators of mesoderm and neuroectoderm development (172–177). The presence of HBGF-2 in *Xenopus* eggs suggests that the HBGFs may regulate the induction of mesoderm (173, 174, 176). Further, HBGF-2 is present in the unfertilized chicken egg and chick embryo (178). The ability of the HBGFs to influence development also emphasizes their importance as neurotropic and angiogenic modulators in the adult, consistent with early reports of the influence of HBGF-2 on amphibian limb regeneration (179, 180). Indeed, the angiogenic activity of the HBGFs are well established (88, 181–187), and both polypeptides at very low concentrations promote and coordinate the site-specific formation of neovessels in situ (188). Moreover, HBGF-1 and HBGF-2 may play important roles in corneal (189) and epidermal (190, 191) ulceration, chondrofication (135, 136, 137, 192, 193), myogenesis (194), steroidgenesis (195), and wound repair (184, 196–198), in general (114). These polypeptides also promote the survival and differentiation of a variety of cells that originate from the neural crest, including hippocampal and cortex-derived cells in vitro (128, 129, 171, 199–203). In addition, the HBGFs are present in amphibian mesoderm, and HBGF-1 has been isolated from embryonic chick brain (177) and kidney (204). Although the precise function of the HBGFs during embryogenesis is not clear, these data argue that they may play a major role during the development of the nervous, skeletal, and vascular systems (reviewed in 143).

Distribution of HBGFs

While the HBGFs have been isolated from a variety of tissues (reviewed in 44, 48, 114), usually as a result of detection by heparin affinity, it has proven difficult to define biological sites of synthesis and storage (205–207). Limited immunofluorescence staining of neural tissue suggests that HBGF-1 is associated with the cytosol of neurons in bovine cerebral cortex (208) and rat neocortex (209). In the rat, HBGF-2 was immunologically localized to the neuronal somata, axons, and the proximal area of dendrites (209). In both studies, the HBGFs were not found as extracellular polypeptides (208, 209). In contrast with the immunocytochemical data, HBGF-1 and HBGF-2 are present in most tissues in significant concentrations, usually 40–120 pM per gram. These concentrations differ from the low levels of the HBGF mRNA transcripts observed in situ, an observation that emphasizes the degree of technical difficulty in obtaining their cDNA clones from human and bovine tissue cDNA libraries (97, 100). However, expression of the mRNA transcripts for the HBGFs has been observed in a variety of normal and transformed cell types in vitro. Expression of the HBGF-2 mRNA transcript has been observed in bovine retinal pigment epithelial cells (210), bovine adrenal cortex cells (211), and bovine endothelial cells from the brain (212), aorta (213–215), adrenal cortex (212), and capillary bed (212–216). In addition, human umbilical vein endothelial cells (214), bovine pituitary follicular cells (217), human retinoblastoma cells (218), and human rhabdomyosarcoma cells (219) also express the HBGF-2 mRNA transcript. Likewise, the mRNA transcript for HBGF-1 is expressed by human glioma (220), medullablastoma (221), rhabdomyosarcoma (221), foreskin fibroblasts (117a,b), and vascular smooth muscle cells (117a,b) in vitro. Interestingly, the HBGF-1 mRNA transcript was not detected by Northern blot analysis of RNA derived from cultured human umbilical vein endothelial cells in vitro (117a,b). The presence of the HBGF-2 polypeptide has been documented not only in those cells expressing the HBGF-2 mRNA transcript in vitro, but also in other cell culture systems including cervical carcinoma (214), embryo lung fibroblasts (214), melanoma (214), retinoblastoma (222), human hepatoma (214, 223, 224), Ewing's sarcoma (225), bovine follicular cells (226), bovine and chicken embryo fibroblasts (214), and cells derived from murine peritoneal exudate (227) or bladder tumors (222). The broad diversity of cell types capable of expressing the HBGFs suggests that these polypeptide mitogens may play a fundamental role in the genesis of a variety of pathological situations in the nervous and vascular systems. Thus, it is particularly important to understand the extracellular fate of the HBGFs.

It has been demonstrated that HBGF-1 and HBGF-2 bind to basement membranes in the eye (228, 229). Pretreatments of the tissue sites with heparitinase significantly diminish this interaction; collagenase or chondroitinase ABC pretreatment had little effect (228). Similarly, Folkman and coworkers have demonstrated that Descement's membrane, the extracellular matrix of the corneal endothelial cell in situ, is a rich source of HBGF-2 presumably bound to heparin-like structures (229). These data confirm earlier in vitro observations using bovine corneal and aorta endothelial cells that HBGF-2 is present within the subendothelial cell extracellular matrix (223). Indeed, HBGF-2 could be solubilized by treatment of the extracellular matrix with high salt concentrations, heparin, heparitinase, and heparinase, but not with hyaluronidase, chondroitinase, chondro-4-sulfatase, or chondro-6-sulfatase (229). It has also been reported that the inhibition of the mitogenic activities associated with the extracellular matrix by protamine sulfate can be attributed to the inhibition of HBGF-2 (230). Further, HBGF-1 binds to gelatin and collagen IV and pretreatment of the collagens with heparin can prevent the association of HBGF-1 with these extracellular matrix proteins (188). These data suggest that the HBGFs are associated with the extracellular matrix in a manner that may permit the release of growth factor through the activities of hydrolytic enzymes, or perhaps, the expression of appropriate proteoglycans. In addition, these studies present the extracellular matrix as a dynamic component ultimately capable of modulating gene expression.

THE HBGF RECEPTORS

Distribution and High- vs Low-Affinity Sites

The mitogenic response initiated by the HBGFs is mediated by a plasma membrane–bound high-affinity binding domain. Radioreceptor binding assays established for HBGF-1 demonstrate the presence of saturable high-affinity binding sites on a variety of cell types, including endothelial cells, smooth muscle cells, fibroblasts, gliomas, chondrocytes, hepatocytes, and epithelial cells (44, 64, 220, 226, 231). The high-affinity HBGF-1 receptor has a K_d of 50–500 pM with $0.5–5 \times 10^4$ sites per cell (44, 64, 220, 226, 231). High-affinity binding sites for HBGF-1 have also been demonstrated on intact endothelium in vivo (232). Low-affinity HBGF-1 binding sites have been characterized on hepatomas, and have been shown to be cell-associated heparin-like molecules (231).

Radioreceptor binding studies with HBGF-2 have demonstrated both high- and low-affinity binding sites on fibroblasts, endothelial cells, myoblasts, and tumor cells from various species (226, 233–237). In addition, high-affinity HBGF-2 receptors have been demonstrated in bovine and rat brain (238, 239). The K_d for the high-affinity HBGF-2 receptor varies from 10 to 200 pM with $0.2–10 \times 10^4$ sites per cell, while the K_d for the low-affinity sites varies from 2 to 10 nM with $0.5–2 \times 10^6$ sites per cell (226, 233–237). As with HBGF-1, the low-affinity binding sites for HBGF-2 have been shown to be cell-associated heparin-like molecules (236, 240). The low-affinity HBGF-2 bind-ing site was characterized as heparin-like due to its sensitivity to heparinase treatment but not other glycolytic enzymes (236). In addition, low-affinity binding could be abolished by incubating the cells with heparin during the binding assay, or by washing the cells with 2.0 M NaCl (236). The low-affinity HBGF-2 binding sites appear not to participate in HBGF-2-mediated signal transduction as measured by induction of plasminogen activator activ-ity by endothelial cells (236).

Characterization by Cross-Linking Studies

Radiolabeled HBGF-1 has been cross-linked to high-affinity receptors with apparent molecular weights of 130,000–165,000 (117a,b, 220, 226, 231, 234, 241–243). The electrophoretic mobilities of these cross-linked receptor species are not altered by the presence of reducing agents, suggesting that the high-affinity HBGF-1 binding domain is a single-chain polypeptide (231, 234, 241). Further evidence for the polypeptide nature of the HBGF-1 receptor is its susceptibility to cleavage with cyanogen bromide and trypsin (241). The variability in the molecular weight assignment for the HBGF-1 receptor, as deduced from cross-linking studies, may be the result of pro-teolysis during sample preparation, variable glycosylation patterns between different cell types, or variations in the method of analysis. Alternatively, the two major HBGF-1 binding domains of M_r 150,000 and M_r 130,000 identified on the majority of cell types examined may represent two separate receptors that are cross-reactive with HBGF-1 and HBGF-2, in a manner analogous to the cross-reactivity of IGF-I and insulin with their respective receptors (10).

Radiolabeled HBGF-2 has been covalently cross-linked to high-affinity receptors with apparent molecular weights of 125,000–150,000 (219, 225, 226, 233, 235). As with HBGF-1, often two cross-linked bands of M_r 125,000 and 145,000 are detected. Unlabeled HBGF-2 will compete with radiolabeled HBGF-2, resulting in the loss of both of these bands in cross-linking experiments (219, 233, 235). In addition, it has been demonstrated that HBGF-1 and HBGF-2 both compete for the 145,000 and 125,000 recep-tors in cross-linking assays, with HBGF-2 appearing to have a higher apparent affinity for the M_r 145,000 receptor species (244). Thus, the M_r 145,000 and M_r 125,000 receptors detected with radiolabeled HBGF-2 are similar to those detected with radiolabeled HBGF-1 and may be the same receptor(s). Addi-tional data are required to determine whether there is a single class or a family of receptors for the HBGFs.

Several compounds have been shown to inhibit the HBGFs from interacting with their receptors. The polyanionic compound suramin has been shown to inhibit the binding of both HBGF-1 and HBGF-2 to their receptors (49, 245). Another polyanionic compound, protamine sulfate, has been shown to inhibit the binding of HBGF-1 to its receptor and block its mitogenic effects (49). Similar compounds may be of clinical importance in pathological situations that involve abnormal expression of the HBGFs.

The receptor for HBGF-2 has been shown to be a glycoprotein (246). The lectin, wheat-germ agglutinin (WGA) was shown to effectively inhibit both radiolabeled HBGF-2 binding to its receptor and the mitogenic effects of HBGF-2 (246). Enzymatic deglycosylation with glycopeptidase-F demonstrated that the two HBGF-2 receptors have core proteins of relative molecular weights of 100,000 and 125,000 (246). Use of inhibitors of glycosylation demonstrated that these two receptor core proteins are glycosylated to high mannose–containing forms of M_r 115,000 and M_r 140,000 and finally, processed to mature forms of M_r 130,000 and M_r 150,000 (246). Additional data demonstrated that glycosylation was required for binding of radiolabeled HBGF-2. Similar results have been obtained for the HBGF-1 receptor, including adsorption of cross-linked radiolabeled HBGF-1 : receptor complexes to immobilized WGA (R. Friesel, T. Maciag, unpublished data). These results support the notion that the HBGF-1 and HBGF-2 receptors are closely related cell-surface glycoproteins.

Fate of the Ligand

Binding and cross-linking experiments have demonstrated that HBGF-1 induces rapid down-regulation of its receptor with subsequent internalization of the ligand (49, 64, 237, 241, 247). Intracellular HBGF-1, internalized as a result of receptor-mediated endocytosis, has been shown to be remarkably resistant to degradation (231, 247). In contrast, several other polypeptide growth factors are rapidly degraded upon internalization (1, 248–251). The appearance of HBGF-1 degradation products does not occur until approximately 2–3 hours postinternalization (247). At this time, two major intracellular HBGF-1 fragments appear with M_r of 15,000 and 10,000 (247). Although intact HBGF-1 may persist intracellularly for up to 6 hours (231, 247), the M_r 15,000 and M_r 10,000 fragments persist for up to 24 hours (247). The degradation of internalized HBGF-1 can be inhibited by the lysosomaltropic agent, chloroquine, an observation consistent with internalized HBGF-1 being targeted to the lysosomal compartment (248).

The intracellular fate of radiolabeled HBGF-2 after internalization by receptor-mediated endocytosis appears similar to that of HBGF-1. Internalized radiolabeled HBGF-2 remains intact for the first half hour after internalization (240). After 2 hours, intact M_r 18,000 HBGF-2 was converted to a M_r 16,000 form that persisted for up to 24 hours (240). Cells exposed to chloroquine did not degrade internalized radiolabeled HBGF-2 (240). These data suggest that as with HBGF-1, degradation of HBGF-2 occurs in the lysosomal compartment with the generation of relatively large fragments of the polypeptide.

Tyrosine Kinase Activities

The binding of HBGFs to their receptors rapidly induces intracellular phosphorylation. HBGF-1 induces the rapid tyrosine phosphorylation of a M_r 135,000 protein in the plasma membranes of Swiss 3T3 cells (252). This phosphorylation event could be inhibited by protamine, a compound that prevents HBGF-1 from binding to its receptor (252). In addition, both HBGF-1 and HBGF-2 have been shown to induce the tyrosine phosphorylation of an M_r 90,000 protein in intact Swiss 3T3 fibroblasts (243). More recently, HBGF-1 has been shown to induce the tyrosine phosphorylation of M_r 150,000 and M_r 130,000 proteins in a concentration-dependent manner consistent with its mitogenic activity. Kinetic data indicate that phosphorylation of the M_r 150,000 and M_r 130,000 proteins occurs within 30 seconds of exposure to HBGF-1, indicating that this is one of the earliest events in HBGF-1–mediated signal transduction (242). The tyrosine phosphorylation of a M_r 90,000 protein occurs three minutes after exposure to HBGF-1, suggesting that it is a substrate for an HBGF-1–induced tyrosine kinase (242, 243). Finally, cross-linked radiolabeled HBGF-1:receptor complexes of M_r 170,000 and 150,000 could be immunoprecipitated with antiphosphotyrosine antibodies (242). These data suggest that HBGF-1 and HBGF-2 stimulate proliferation of a similar spectrum of target cells by interacting through similar receptor systems. It is not clear at present whether the M_r 150,000 and M_r 130,000 receptors represent different gene products or are related through a precursor-product relationship. Determination of the primary structure of both of these proteins will resolve this issue, and establish the relationship of these receptors to the tyrosine kinase family of polypeptide receptors (reviewed in 10).

The induction of tyrosine phosphorylation by HBGFs is consistent with the observations that the early and late mitogenic events initiated by the HBGFs on epithelial cells (153, 154) and fibroblasts (155) are not mediated by the activation of protein kinase C or the intracellular hydrolysis of polyphosphoinositides (153–155). Further, it has been demonstrated that the protein kinase C activator β-phorbol 12,13-dibutyrate antagonizes the mitogenic activity of HBGF-2 on endothelial cells (156) by a mechanism that reversibly reduces the number of HBGF receptors present on the cell surface (157). A similar mechanism has been demonstrated for the antagonism of HBGF-1 receptor function by γ-IFN (253). In contrast, terminal differentiation of skeletal muscle myoblasts has been shown to correlate with irreversible loss of HBGF receptors from the cell surface (254).

STRUCTURE FUNCTION RELATIONSHIPS OF HBGF-1 AND HBGF-2

Implications of Conserved Structures

The fact that HBGF-1 and HBGF-2 are nearly identical in their spectrum of biological activities and may indeed act via the same population of cell-surface receptors indicates similar structures are utilized by these growth factors in their various functions. As described earlier, HBGF-1 and HBGF-2 share greater than 50% amino acid sequence identity. In addition, the conservation is, to date, nearly independent of the species source of the protein, there being perhaps as few as two amino acid differences between human and bovine HBGF-2 and 13 differences between human and bovine HBGF-1. There may be even fewer differences between avian and human HBGF-1 (168; T. Mehlman, W. H. Burgess, unpublished data). The highly conserved nature of these structures indicates a point mutation rate of 6.5 mutations or less per 100 residues per 100 million years (255). As noted by Harper et al (255), the mutation rate is similar to that of the cytochrome c's and may indicate that nearly the entire polypeptide structure has functional significance. It is presently unclear why significant sequence variation between HBGF-1 and HBGF-2 has occurred, although the fact that it has indicates additional functions will be described for which the two proteins differ significantly in their activities. However, to date the amino acid sequences of HBGF-1 or HBGF-2 have provided little insight into the structural requirements of HBGF function. The argument that conserved structure implies important functions may extend to the amino-terminal domains of these proteins. As discussed earlier, the amino-terminal truncations that result in the different forms of HBGF-1 and HBGF-2, described by many laboratories, do not affect significantly their apparent affinities for immobilized heparin or their potencies as mitogens. The fact that these regions are as highly conserved as the rest of the molecules argues that functional significance will ultimately be assigned to these regions of the proteins as well.

Thus, the overall sequence identity between HBGF-1 and HBGF-2 from different species has limited the ability to identify "nonessential" regions of the proteins in an attempt to limit the number of site-directed mutations or synthetic peptides required to begin rigorous structure-function studies.

Functional Activities of Synthetic Peptides

The problems associated with the identification of peptides of functional significance in highly conserved structures described above has been overcome, in part, by Baird et al (256) in their studies of HBGF-2. These investigators reported the synthesis of 25 peptides, which together encompass and overlap the entire sequence of HBGF-2 as described by Ueno et al (112).

They report the identification of two functional domains in the primary structure of HBGF-2 based on the ability of the synthetic peptides to interact with HBGF receptor, bind radiolabeled heparin in a solid phase assay, and inhibit HBGF-2 stimulation of thymidine incorporation into 3T3 fibroblasts. Using the numbering system of the authors (residues 1–5 = P-A-L-P-E; see Figure 1), statistically significant functional activities could be assigned to peptides corresponding to residues (24–68)-NH$_2$ and (106–115)-NH$_2$ of HBGF-2.

Similarly, Schubert et al (257) demonstrated that one of these peptides, (24–68)-NH$_2$, is able to stimulate cell-substratum adhesion of PC12 cells, a feature shared by peptides based on the sequence of residues 93–120 and residues 1–24 of HBGF-2. The authors conclude from these studies that neither net charge, amphiphilic secondary structure, nor the combination of the two is sufficient to promote cell adhesion, suggesting that specific sequences are involved, and the phenomenon is receptor mediated. To date, these studies are the best attempts at demonstrating specific functions associated with specific regions of HBGF-2. Clearly, specificity has been demonstrated with respect to inhibition or mimicry of HBGF-2 function. However, the approach still suffers from the relatively high (generally 10–100 μM) concentrations of peptide needed to observe the effects. Future studies and additional synthesis should improve the potencies of these peptides in bioassays.

Effects of Chemical Modification

Attempts to elucidate the importance of specific residues in the HBGFs that are important to their functions have been limited primarily to modifications of cysteines or cystines and to lysines. Although these approaches suffer from the extensive characterization of the proteins required to be reasonably certain that the chemical modifications are specific, they have provided useful information as to the importance of certain residues. The information obtained can then be used as a rational basis for the design of site-directed mutagenesis experiments utilizing the expression vectors for HBGF-1 and HBGF-2 described below.

Despite the fact that the complete, or nearly complete, amino acid sequences for human and bovine HBGF-1 and HBGF-2 from a variety of tissue and cellular sources have been reported, the assignment of intrachain disulfide bonds, if any, have not been made for any of the reported structures. The relative positions of the cysteine residues in human and bovine HBGF-1 and HBGF-2 have been discussed. Using the numbering system described in Figure 1, the fact that only cysteines 30 and 97 are conserved in all species of HBGF-1 and HBGF-2 has led several investigators to assume that these residues are involved in the formation of intramolecular disulfide bonds.

Whereas this is a reasonable assumption, there is little direct evidence to support it. It is equally reasonable to assume that free cysteines are required at these positions for some of the biological activities displayed by HBGF-1 and HBGF-2.

Harper et al (255) provided indirect evidence for a disulfide bond between cysteine-30 and cysteine-97 in their determination of the structure of human HBGF-1 (15–154). Esch et al (76) concluded following alkylation experiments that human HBGF-2 contains all free cysteines or the disulfides scramble at such a rate as to give the appearance that all are free. Similarly, McKeehan & Crabb (60) provided evidence that the multiple fractions of HBGF-1 eluted from reversed-phase columns are due, in part, to scrambled disulfide bonds and that treatment with reducing agents can result in single peaks of HBGF-1. Others (258) have proposed that the cysteines in HBGF-2 are involved in disulfide linkage to small molecules unrelated to the growth factor.

Whatever the native states of the cysteine residues in HBGF-1 or HBGF-2, their role in biological function remains unclear. It has been reported that alkylation of the free cysteines of human HBGF-1 without prior reduction has no effect on its mitogenic activity and that quantitative alkylation following reduction leaves the protein with reduced but significant activity (74). In contrast, it has been reported that quantitative alkylation of otherwise fully active recombinant HBGF-1 (21–154) abolishes all detectable receptor binding activity (66). Seno et al (259) have addressed the role of cysteines in HBGF-2 through systematic site-directed mutagenesis, and these results will be described later in this chapter.

To date, the most informative studies documenting the effects of chemical modification of HBGF-1 function are those of Harper & Lobb (260). Briefly, they were able to show that limited reductive methylation of bovine HBGF-1 (15–154) with formaldehyde and sodium cyanoborohydride resulted in significant methylation only of lysine-132 (90% modified, 60% dimethyllysine). The modified protein exhibits a significantly reduced apparent affinity for immobilized heparin (eluted at 0.7 M NaCl vs 1.2 M NaCl for unmodified HBGF-1), a fourfold reduction in its ability to stimulate DNA synthesis in 3T3 cells and a reduced ability to compete with labeled ligand in a radioreceptor assay. The position of this lysine residue is conserved in all forms of HBGF-1 or HBGF-2 characterized to date. It is clear from these studies that modification of lysine-132 has a significant effect on heparin-binding activity; however, the effects on receptor binding and their mitogenic activity may reflect the dependence of human HBGF-1 on heparin-binding for full biological activity (as discussed earlier in this review). Although the caveats discussed earlier regarding the caution required in interpretation of chemical modification data apply to these studies, site-directed mutagenesis data described below also supports the conclusions regarding the importance of lysine-132 to the heparin-binding properties of HBGF-1.

HBGF cDNA Expression in Prokaryotic and Eukaryotic Cells and Site-Directed Mutagenesis

The cloning of genomic DNA and cDNAs for HBGF-1 and HBGF-2 have been described earlier in this review. These accomplishments provide the most powerful tools for direct assessment of structure-function relationships of this family of growth factors. The ability to express the cDNAs that code for these proteins in prokaryotic and eukaryotic expression systems provides the capabilities to (a) produce practically unlimited quantities of growth factor, (b) produce chimeric proteins, and (c) introduce defined changes in the primary structures via site-directed mutagenesis.

High-level expression in *Escherichia coli* of functional human HBGF-1 (15–154) and (21–154) have been described (66, 151). To date, only low-level expression of full-length HBGF-1 (1–154) has been reported (66). High-level expression of truncated forms of human HBGF-2 (9–154) have also been successful (261). Although the definition of full-length HBGF-2 is still in question, Squires et al (261) reported expression of HBGF-2 (3–154) in *E. coli*, which has bioactivity that is indistinguishable from that of human placental HBGF-2 of the same sequence.

Barr et al (263) reported the expression of biologically active HBGF-1 and HBGF-2 in the yeast *Saccharomyces cerevisiae*. More importantly, they were able to obtain expression of full-length HBGF-1 (1–154) and a form of HBGF-2 of the same size (1–154). Each of the expressed proteins was processed in vivo as judged by quantitative removal of the initiation codon–derived methionine and acetylation of the amino-terminal alanine of HBGF-1 (100%) and HBGF-2 (50%). Although the definition of authentic HBGF-2 has not yet been established, the expression system described appears to be well suited for large-scale production of authentic HBGF-1 for therapeutic applications.

The identification of the HBGF-like oncogenes int-2, hst/KS3, and FGF 5 (referred to in Figure 1 and below as HBGFs 3, 4, and 5 respectively), suggests that HBGF-1 and HBGF-2 may contain transformation potential. Indeed, HBGF-1 and HBGF-2 induce the expression of fibrinolytic enzymes (264–267), which strengthens the correlation between enhanced plasminogen activator expression and mammalian cell transfomation in vitro (268, 269). The construction of cDNA expression vectors of HBGF-1 and HBGF-2 that are suitable for transfection of eukaryotic cells has made it possible to examine the effects of overexpression of these proteins or related chimerics in vitro.

To date, all reports of HBGF transfection experiments of mammalian cells have utilized fibroblast cell lines of nonhuman origin, none of which absolutely require HBGFs for growth (as do human endothelial cells). Neufeld et al (262) reported that transfection of a human HBGF-2 cDNA expression vector into BHK-21 fibroblasts resulted in the accumulation of high concentrations of biologically active protein. This phenomena correlated with the ability of the cells to grow in serum-free media without added growth factors and to proliferate in soft agar, indicating a loss of anchorage-dependent growth. In addition, antibodies to HBGF-2 could not inhibit the proliferation of these cells, indicating a possible intracellular mechanism of growth factor action. Sasada et al (270) reported similar results for mouse BALB/c 3T3 cells transfected with HBGF-2 cDNA. The major difference between the two results is Sasada et al (270) report HBGF-2 antibodies do reverse the proliferative effect, indicating that although HBGF-2 lacks a classical signal peptide sequence, it may be released from the cell. Rogelj et al (271) reported that transfection of HBGF-2 cDNA into NIH 3T3 cells did not result in foci formation, unusual morphologies, or tumorigenicity, whereas cells transfected with an immunoglobulin signal peptide:HBGF-2 chimeric plasmid did. A similar phenotype is achieved following the expression of a growth hormone signal peptide HBGF-2 chimeric but not HBGF-2 in BALB/c 3T3 cells (272). The growth hormone signal peptide chimeric induced the transformed phenotype at a more rapid frequency, and the chimeric protein was found to be secreted and modified posttranslationally (272), whereas the IgG signal peptide chimeric was not (271). Immunological characterization of the extracellular HBGF-2 demonstrated the presence of HBGF-2 of larger apparent molecular weight than expected, presumably the result of glycosylation (272). In addition, the extracellular forms of HBGF-2 displayed poor mitogenic activity.

These results are consistent with the findings of Jaye et al (273) using 3T3 cells transfected with an HBGF-1 (21–154) expression vector. They report transfected cells grow at a faster rate and to a higher density, but that growth in soft agar requires the addition of exogenous growth factor. Together, these results indicate that transformation by HBGF-1 and HBGF-2 may require both high levels of intracellular expression and an extracytoplasmic action of the growth factor. Thus, future structure-function studies may include analysis of the interaction of the HBGF polypeptides with intracellular regulatory or structural proteins in addition to studies on their interactions with heparin and the extracellular domains of cell-surface receptors.

The introduction of defined and specific changes in the primary structure of HBGF-1 or HBGF-2 by site-directed mutagenesis of cDNA expression vectors is the most powerful tool for dissecting structure-function relationships. Results of such experiments are not yet generally available and can be complicated by potential secondary effects of the changed residue on secondary or tertiary structure. Two examples will be discussed here. One is a report by Seno et al (259) on the results of systematic site-directed mutagenesis of each of the four cysteines in human HBGF-2 to serine residues, the other unpublished results related to the relative importance of lysine 132 to the interaction of HBGF-1 with immobilized heparin.

As discussed above, the native states of the cysteine residues of HBGF-1 or HBGF-2 are presently unclear as are the importance of these residues to the functions of the proteins. Seno et al (259) changed each cysteine codon in HBGF-2 cDNA to serine, both one at a time and in combinations, to produce 15 different modified versions of HBGF-2 in *E. coli*. They report that substitution of serine for cysteine at positions 77, 95, or 100 has no effect on the ability of the protein to stimulate DNA synthesis, whereas modification of cysteine 33 results in a 60% decrease in specific activity. Similarly, substitution of cysteines 77, 95, and 100 together yields HBGF-2 with a specific activity 90% that of unmodified protein, whereas modification of all four cysteines results in a 90% decrease in specific activity. These data indicate that intrachain disulfide bonds are not required for biological activity of HBGF-2 and that a free cysteine at position 33 may indeed be necessary.

We have begun to explore the effects of site-directed mutagenesis of the HBGF-1 (21–154) cDNA described by Jaye et al (66). The results described above by Harper et al (255) on the significant reduction in the apparent affinity of HBGF-1 for immobilized heparin by selective modification of lysine 132 led us to examine substitution of a glutamic acid residue for lysine at this position (W. H. Burgess, M. Jaye, J. A. Winkles, unpublished results). Our results are consistent with those of Harper et al (255) in that the mutant protein can be eluted from heparin-Sepharose with as little as 0.5 M NaCl as compared to the 1.0–1.1 M concentrations required to elute wild-type recombinant HBGF-1 (21–154). It can be anticipated that many additional reports of site-directed mutagenesis of the cDNAs for HBGF-1 and HBGF-2 will be forthcoming and that these reports will add significantly to our understanding of the structure-function relationships of these proteins.

ADDITIONAL MEMBERS OF THE HBGF FAMILY OF POLYPEPTIDES

The transforming potential of high-level expression of HBGF-1 or HBGF-2 cDNAs or chimeric proteins in eukaryotic cells has been described. These observations take on added significance with the identification of new oncogenes that have significant (30–50%) amino acid sequence identity with HBGF-1 and HBGF-2. To date, three different oncogenes with such similarity have been described and thus, represent three new members of the HBGF family of proteins.

Int-2—HBGF-3

The product of the int-2 gene was the third member of the HBGF family identified (274–277). The int-2 gene was originally isolated as a cellular gene that becomes active transcriptionally after integration (int) of the mouse mammary tumor virus with the cellular DNA of murine cells in vitro (274). The int-2 gene contains an open reading frame coding for 231 amino acids. The region of the predicted protein sequence with homology to HBGF-1 and HBGF-2 of murine int-2 is shown in Figure 1 (depicted as HBGF-3). The protein also contains a relatively short (16-residue) amino-terminal extension and a longer (45-residue) carboxy-terminal extension (not shown). Because the int-2 (HBGF-3) polypeptide has not been characterized, it is not clear whether these extensions affect putative heparin-binding or HBGF-like biological activities. The amino-terminal domain contains a relatively short hydrophobic sequence, which could function as an atypical signal peptide structure.

No transcripts of int-2 (HBGF-3) have yet been identified in normal adult tissue (278). Transcripts have been detected by Northern blot analysis of 7.5-day-egg cylinder and embryonal carcinoma cells, which led Jakobovits et al (279) to propose that expression may be confined to extraembryonic endoderm. In contrast, Wilkinson et al (277) used in situ hybridization and Northern blot analysis to study the expression of the int-2 (HBGF-3) mRNA in the developing mouse from early gestation through mid-somite stages and conclude that expression is limited to parietal rather than visceral endoderm, and that int-2 (HBGF-3) mRNA accumulates in newly formed migrating mesoderm, neuroepithelial cells of the hindbrain, and the pharyngeal pouches.

HST/KS3—HBGF-4

The fourth member of the HBGF family was identified as a genomic DNA fragment derived from a human stomach tumor (hst) capable of inducing murine 3T3 cell transformation in vitro (280). The transforming potential associated with the hst gene (HBGF-4) has also been detected in human DNA derived from chronic myelogenous leukemic leukocytes, gastric tumors, colon tumors, and colon mucosa (281, 282). Two potential open reading frames were identified and the putative translational product of the first open reading frame encodes a polypeptide containing 206 residues (Figure 1). The hst (HBGF-4) polypeptide contains a long NH_2-terminal extension including a long hydrophobic leader sequence, and possesses a shorter COOH-terminal extension relative to HBGF-1 and HBGF-2. Independently, an identical gene (KS3) was identified by transfection of 3T3 cells with DNA derived from a human Kaposi sarcoma (KS) lesion (283, 284), an angiosarcoma of presumed endothelial cell-mesenchymal cell origin that occurs at a high frequency in immunosuppressed individuals, including patients with acquired immuno-deficiency syndrome.

Unlike the int-2 (HBGF-3) oncogene, the mechanism of activation of HST/KS3 (HBGF-4) has not yet been established (284), although it may involve overexpression (285). Similarly, whereas the growth-promoting activities of int-2 (HBGF-3) are not known, the HST/KS3 (HBGF-4) protein is capable of promoting the growth of NIH 3T3 cells (284) and vascular endothelial cells in vitro (285).

FGF-5—HBGF-5

The most recent addition to the HBGF family of growth factors was identified using methods similar to those used to identify HBGF-4 (286, 287). DNA derived from established human tumor cell lines was transfected into 3T3 cells, which were selected under growth factor–deficient conditions, and led to the identification of the FGF-5 (HBGF-5) gene (287). The FGF-5 (HBGF-5) protein consists of 267 residues and contains extended amino- and carboxy-terminal domains relative to HBGF-1 or HBGF-2. A classical signal peptide sequence is present in the amino-terminal domain of FGF-5 (HBGF-5), a feature shared by HST/KS3 (HBGF-4). Both the HBGF-4 and HBGF-5 gene products are expressed as extracellular heparin-binding growth factors (285, 287).

In summary, several features are shared by all or most of the members of the HBGF family of polypeptides. The genomic structures for all five members contain a similarly positioned three exon: two intron format. The HBGF-2, -3, and -4 genes contain several GC boxes in the 5' noncoding region, a feature termed the HpaII tiny fragment feature (288). The established or derived primary structures are 35% to 55% identical and the relative positions of several amino acid residues are absolutely conserved (Figure 1). All five HBGF mRNA transcripts contain relatively long 5' untranslated regions, and HST/KS3 (HBGF-4) and FGF-5 (HBGF-5) contain a second open reading frame in this region (280, 287). A similar feature is found in the PDGF-B chain mRNA and is involved in its translational regulation (289).

FUTURE DIRECTIONS

The structural definition of HBGF-1 and HBGF-2 as neurotropic and an-giogenic modulators, should yield insight into the mechanisms that control early development, embryogenesis, and organogenesis. The abilities of the HBGFs to regulate wound repair and neovessel formation in vivo and the characterization of additional HBGFs with oncogenic potential support a fundamental role for HBGFs in physiology and pathology. These advances have begun to provide insight into the roles of HBGFs in human disease involving mesoderm- and neuroectoderm-derived cell types.

New and exciting questions have been generated; among the more intriguing remains the mechanism of secretion of HBGF-1 and HBGF-2 and subsequent deposition as latent extracellular matrix signals. In addition, much work is still needed to define the HBGF-receptor systems, to determine whether the structural similarities shared by the ligands are reflected at the receptor level, and to identify the signal transduction mechanisms utilized by the different HBGFs.

ACKNOWLEDGMENTS

We would like to thank our colleagues who shared data prior to its publication. We thank Linda Peterson and Sally Young for preparation of the manuscript. This work was supported, in part, by National Institutes of Health Grants HL35762 to W.B. and HL32348 to T.M.

Literature Cited

1. James, R., Bradshaw, R. A. 1984. *Annu. Rev. Biochem.* 53:259–92
2. Gill, G. N., Bertics, P. J., Santon, J. B. 1987. *Mol. Cell. Endocrinol.* 51:169–86
3. Ross, R., Raines, E. W., Bowen-Pope, D. F. 1986. *Cell* 46:155–69
4. Deuel, T. F. 1987. *Annu. Rev. Cell Biol.* 3:443–92
5. Froesch, E. R., Schmid, C., Schwander, J., Zapf, J. 1985. *Annu. Rev. Physiol.* 47:443–67
6. Massague, J. 1987. *Cell* 49:437–38
7. Keski-Oja, J., Moses, H. L. 1987. *Med. Biol.* 65:13–20
8. Roberts, A. B., Sporn, M. B. 1988. *Adv. Cancer Res.* 51:107–41
9. Webb, D. R., Pierce, C. W., Cohen, S., eds. 1985. *Molecular Basis of Lymphokine Action.* Clifton, NJ: Humana
10. Yarden, Y., Ulrich, A. 1988. *Annu. Rev. Biochem.* 57:443–78
11. Schlessinger, J. 1988. *Biochemistry* 27:3119–23
12. Carpenter, G. 1987. *Annu. Rev. Biochem.* 56:881–914
13. Trowell, O. A., Chir, B., Willmer, E. N. 1939. *J. Exp. Biol.* 16:60–70
14. Hoffman, R. S. 1940. *Growth* 4:361–76
15. Holley, R. W., Kiernan, J. A. 1968. *Proc. Natl. Acad. Sci. USA* 68:300–4
16. Clark, J. L., Jones, K. L., Gospodarowicz, D., Sato, G. 1972. *Nature* 236:180–81
17. Corvol, M. T., Mulemud, C. J., Sokoloff, L. 1972. *Endocrinology* 90:262–71
18. Armelin, H. A. 1973. *Proc. Natl. Acad. Sci. USA* 70:2702–6
19. Gospodarowicz, D., Jones, K. L., Sato, G. 1974. *Proc. Natl. Acad. Sci. USA* 71:2295–99
20. Gospodarowicz, D. 1974. *Nature* 249:123–27
21. Gospodarowicz, D. 1975. *J. Biol. Chem.* 250:2515–19
22. Gospodarowicz, D., Bialecki, H., Greenberg, G. 1978. *J. Biol. Chem.* 253:3736–43
23. Gospodarowicz, D. 1983. *J. Pathol.* 14:201–23
24. Westfall, F. C., Lennon, V. A., Gospodarowicz, D. 1978. *Proc. Natl. Acad. Sci. USA* 75:4675–78
25. Maciag, T., Cerundolo, J., Isley, S., Kelley, P. R., Forand, R. 1979. *Proc. Natl. Acad. Sci. USA* 76:5674–78
26. Kiss, F. A., Krompecher, I. 1962. *Acta Biol. Acad. Sci. Hung. Suppl.* 4:39–40
27. Polverini, P. J., Cotran, R. S., Gimbrone, M. A., Unanue, E. R. 1977. *Nature* 269:804–6
28. Arruti, C., Courtois, Y. 1978. *Exp. Cell Res.* 117:283–92
29. Glaser, B. M., D'Amore, P. A., Michels, R. G., Patz, A., Fenselau, A. J. 1980. *J. Cell Biol.* 84:298–304
30. Kellett, J. G., Tanaka, T., Rowe, J. M., Shin, R. P. C., Friesen, H. G. 1981. *J. Biol. Chem.* 256:54–58
31. Thomas, K. A., Riley, M. C., Lemmon, S. K., Baglan, N. C., Bradshaw, R. A. 1980. *J. Biol. Chem.* 255:5517–20
32. Addison, G. M., Jackson, P., Shape, B. M., Thompson, R. J. 1984. *Horm. Metab. Res.* 16:311–14

33. Lemmon, S. K., Riley, M. C., Thomas, K. A., Hoover, G. A., Maciag, T., Bradshaw, R. A. 1982. *J. Cell Biol.* 95:162–69
34. Thornton, S. C., Mueller, S. N., Levine, E. M. 1983. *Science* 222:623–25
35. Maciag, T., Hoover, G. A., Weinstein, R. 1982. *J. Biol. Chem.* 257:5333–36
36. Burgess, W. H., Mehlman, T., Marshak, D. R., Fraser, B. A., Maciag, T. 1986. *Proc. Natl. Acad. Sci. USA* 83: 7216–20
37. Thomas, K. A., Gimenez-Gallego, G. 1986. *Trends Biochem. Sci.* 11:81–84
38. Lemmon, S. K., Bradshaw, R. A. 1983. *J. Cell. Biochem.* 21:195–208
39. Thomas, K. A., Rios-Candelore, M., Fitzpatrick, S. 1984. *Proc. Natl. Acad. Sci. USA* 81:357–61
40. Bohlen, P., Baird, A., Esch, F., Ling, N., Gospodarowicz, D. 1984. *Proc. Natl. Acad. Sci. USA* 81:5364–68
41. Shing, Y., Folkman, J., Sullivan, R., Butterfield, C., Murray, J., Klagsbrun, M. 1984. *Science* 223:1296–99
42. D'Amore, P. A., Klagsbrun, M. 1984. *J. Cell Biol.* 99:1545–49
43. Gimenez-Gallego, G., Rodkey, J., Bennett, C., Rios-Candelore, M., DiSalvo, J., Thomas, K. A. 1985. *Science* 230:1385–88
44. Baird, A., Esch, F., Mormede, P., Ueno, N., Ling, N., et al. 1986. *Recent Prog. Horm. Res.* 42:143–205
45. Gospodarowicz, D., Ferrara, N., Schweigerer, L., Neufeld, G. 1987. *Endocrinol. Rev.* 8:96–113
46. Courty, J., Chevallier, B., Moenner, M., Loret, C., Lagente, O., et al. 1986. *Biochem. Biophys. Res. Commun.* 136:102–8
47. Burgess, W. H., Mehlman, T., Friesel, R., Johnson, W. V., Maciag, T. 1986. *J. Biol. Chem.* 260:11389–92
48. Lobb, R. R., Fett, J. W. 1984. *Biochemistry* 23:6295–99
49. Huang, J. S., Huang, S. S., Kuo, M.-D. 1986. *J. Biol. Chem.* 261:11600–11607
50. Klagsbrun, M., Shing, Y. 1985. *Proc. Natl. Acad. Sci. USA* 82:805–9
51. Lobb, R., Sasse, J., Sullivan, R., Shing, Y., D'Amore, P., et al. 1986. *J. Biol. Chem.* 261:1924–28
52. Pettmann, B., Weibel, M., Sensenbrenner, M., Labourdette, G. 1985. *FEBS Lett.* 189:102–8
53. Sullivan, R., Klagsbrun, M. 1985. *J. Biol. Chem.* 260:2399–403
54. Kardami, E., Spector, D., Strohman, R. C. 1985. *Proc. Natl. Acad. Sci. USA* 82:8044–47
55. Rowe, J. M., Henry, S. F., Friesen, H. G. 1986. *Biochemistry* 25:6421–25
56. Hauschka, P. V., Mavrakos, A. E., Iafrati, M. D., Doleman, S. E., Klagsbrun, M. 1986. *J. Biol. Chem.* 261:12665–74
57. Conn, G., Hatcher, V. B. 1984. *Biochem. Biophys. Res. Commun.* 124: 262–68
58. Story, M. T., Sasse, J., Jacobs, S. C., Lawson, R. K. 1987. *Biochemistry* 26: 3843–49
59. Matuo, Y., Nishi, N., Matsui, S., Sandberg, A. A., Issacs, J. T., Wada, F. 1987. *Cancer Res.* 47:188–92
60. McKeehan, W. L., Crabb, J. W. 1987. *Anal. Biochem.* 164:563–69
61. Maciag, T., Mehlman, T., Friesel, R., Schreiber, A. B. 1984. *Science* 225: 932–35
62. Shing, Y., Folkman, J., Murray, M., Klagsbrun, M. 1983. *J. Cell Biol.* 97:395a
63. Lobb, R. R., Harper, J. W., Fett, J. W. 1986. *Anal. Biochem.* 154:1–14
64. Schreiber, A. B., Kenney, J., Kowalski, W. J., Friesel, R., Mehlman, T., Maciag, T. 1985. *Proc. Natl. Acad. Sci. USA* 82:6138–42
65. Thomas, K. A. 1988. *Trends Biochem. Sci.* 13:327–28
66. Jaye, M., Burgess, W. H., Shaw, A. B., Drohan, W. N. 1987. *J. Biol. Chem.* 262:16612–17
67. Rosengart, T. K., Johnson, W., Friesel, R., Clark, R., Maciag, T. 1988. *Biochem. Biophys. Res. Commun.* 152: 432–40
68. Gospodarowicz, D., Cheng, J. 1986. *J. Cell. Physiol.* 128:475–84
69. Lobb, R. R. 1988. *Biochemistry* 27:2572–78
70. Sommer, A., Rifkin, D. B. 1989. *J. Cell Physiol.* In press
71. Harper, J. W., Strydom, D. J., Lobb, R. R. 1986. *Biochemistry* 25:4097–103
72. Gautschi-Sova, P., Muller, T., Bohlen, P. 1986. *Biochem. Biophys. Res. Commun.* 140:874–80
73. Gimenez-Gallego, G., Conn, G., Hatcher, V. B., Thomas, K. A. 1986. *Biochem. Biophys. Res. Commun.* 135: 541–48
74. Crabb, J. W., Armes, L. G., Carr, S. A., Johnson, C. M., Roberts, G. D., et al. 1986. *Biochemistry* 25:4988–93
75. Bohlen, P., Esch, F., Baird, A., Gospodarowicz, D. 1985. *EMBO J.* 4:1951–56
76. Esch, F., Ueno, N., Baird, A., Hill, F., Denoroy, L., et al. 1985. *Biochem. Biophys. Res. Commun.* 133:554–62
77. Strydom, D. J., Harper, J. W., Lobb, R. R. 1986. *Biochemistry* 25:945–51

78. Gospodarowicz, D., Ferrara, N., Schweigerer, L., Neufeld, G. 1987. *Endocrinol. Rev.* 8:95–114

79. Gospodarowicz, D., Cheng, J., Lui, G. M., Baird, A., Esch, F., Bohlen, P. 1985. *Endocrinology* 117:2383–91

80. Baird, A., Esch, F., Bohlen, P., Ling, N., Gospodarowicz, D. 1985. *Regul. Peptides* 12:201–13

81. Gospodarowicz, D., Baird, A., Cheng, J., Lui, G. M., Esch, F., Bohlen, P. 1986. *Endocrinology* 118:82–90

82. Baird, A., Mormede, P., Bohlen, P. 1985. *Biochem. Biophys. Res. Commun.* 126:359–64

83. Baird, A., Esch, F., Gospodarowicz, D., Guillemin, R. 1985. *Biochemistry* 24:7860–65

84. Smith, J. A., Winslow, D. P., O'Hare, M. J., Rudland, P. S. 1984. *Biochem. Biophys. Res. Commun.* 119:311–18

85. Bohlen, P., Baird, A., Esch, F., Ling, N., Gospodarowicz, D. 1984. *Proc. Natl. Acad. Sci. USA* 81:5364–68

86. Ueno, N., Baird, A., Esch, F., Shimasaki, S., Ling, N., Guillemin, R. 1986. *Regul. Peptides* 16:135–45

87. Ueno, N., Baird, A., Esch, F., Ling, N., Guillemin, R. 1987. *Mol. Cell. Endocrinol.* 49:189–94

88. Esch, F., Baird, A., Ling, N., Ueno, N., Hill, F., et al. 1985. *Proc. Natl. Acad. Sci. USA* 82:6507–11

89. Ueno, N., Baird, A., Esch, F., Ling, N., Guillemin, R. 1986. *Biochem. Biophys. Res. Commun.* 138:580–88

90. Klagsbrun, M., Smith, S., Sullivan, R., Shing, Y., Davidson, S., et al. 1987. *Proc. Natl. Acad. Sci. USA* 84:1839–43

91. Story, M. T., Esch, F., Shimasaki, S., Sasse, J., Jacobs, S. C., Lawson, R. K. 1987. *Biochem. Biophys. Res. Commun.* 142:702–9

92. Sommer, A., Brewer, M. T., Thompson, R. C., Moscatelli, D., Presta, M., Rifkin, D. B. 1987. *Biochem. Biophys. Res. Commun.* 144:543–50

93. Too, C. K. L., Murphy, P. R., Hamel, A.-M., Friesen, H. G. 1987. *Biochem. Biophys. Res. Commun.* 144:1128–34

94. Deleted in proof

95. Moscatelli, D., Joseph-Silverstein, J., Manejias, R., Rifkin, D. B. 1987. *Proc. Natl. Acad. Sci. USA* 84:5778–82

96. Thomas, K. A. 1987. *FASEB J.* 1:434–40

97. Jaye, M., Howk, R., Burgess, W. H., Ricca, G. A., Chiu, I.-M., et al. 1986. *Science* 233:543–45

98. Abraham, J. A., Whang, J. L., Tumolo, A., Mergia, A., Friedman, J., et al. 1986. *EMBO J.* 5:2523–28

99. Kurokawa, T., Sasada, R., Iwane, M., Igarashi, K. 1987. *FEBS Lett.* 213:186–94

100. Abraham, J. A., Mergia, A., Whang, J. L., Tumolo, A., Friedman, J., et al. 1986. *Science* 233:545–48

101. Mergia, A., Eddy, R., Abraham, J. A., Fiddes, J. C., Shows, T. B. 1986. *Biochem. Biophys. Res. Commun.* 138:644–51

102. Abraham, J. A., Whang, J. L., Tumolo, A., Mergia, A., Fiddes, J. C. 1986. *Cold Spring Harbor Symp. Quant. Biol.* 51:657–68

103. Gospodarowicz, D., Neufeld, G., Schweigerer, L. 1987. *J. Cell. Physiol.* 5:15–26

104. Kimelman, D., Abraham, J. A., Haaparanta, T., Palisi, T. M., Kirschner, M. W. 1988. *Science* 242:1053–56

105. McKusick, V. A. 1984. *Clin. Genet.* 27:207–39

106. Betsholtz, C., Johnsson, A., Heldin, C.-H., Westermark, B., Lind, P., et al. 1986. *Nature* 320:695–99

107. Sherr, C. J., Rettenmier, C. W., Sacca, R., Roussel, M. F., Look, A. T., Stanley, E. R. 1985. *Cell* 41:665–76

108. Kosak, M. 1984. *Nucleic Acids Res.* 12:857–72

109. March, C. J. 1985. *Nature* 315:641–47

110. Auron, P. E., Webb, A. C., Rosenwasser, L. J., Mucci, S. V., Rich, A., et al. 1984. *Proc. Natl. Acad. Sci. USA* 81:7907–11

111. McKeehan, W. L., Crabb, J. W. 1987. *Anal. Biochem.* 164:563–69

112. Ueno, N., Baird, A., Esch, F., Ling, N., Guillemin, R. 1986. *Biochem. Biophys. Res. Commun.* 138:580–88

113. Klagsbrun, M., Smith, S., Sullivan, R., Shing, Y., Davidson, S., et al. 1987. *Proc. Natl. Acad. Sci. USA* 84:1839–43

114. Folkman, J., Klagsbrun, M. 1987. *Science* 235:442–47

115. Gospodarowicz, D., Greenburg, G., Bialecki, H., Zetter, B. 1978. *In Vitro* 14:85–118

116. Maciag, T., Hoover, G. A., Stemerman, M. B., Weinstein, R. 1981. *J. Cell Biol.* 91:420–26

117a. Winkles, J. A., Friesel, R., Burgess, W. H., Howk, R., Mehlman, T., et al. 1987. *Proc. Natl. Acad. Sci. USA* 84:7124–28

117b. Gospodarowicz, D., Bohlen, P., Guillemin, R. 1985. *Proc. Natl. Acad. Sci. USA* 82:6507–10

118. Chen, J.-K., Hoshi, H., McKeehan, W. L. 1988. *In Vitro Cell. Mol. Biol.* 24:199–204

119. Schweigerer, L., Neufeld, G., Friedman, J., Abraham, J. A., Fiddes, J. C., Gospodarowicz, D. 1987. *Endocrinology* 120:796–800
120. Gospodarowicz, D., Ill, C. R., Hornsby, P. J., Gill, G. N. 1977. *Endocrinology* 100:1080–89
121. McKeehan, W. L., Adams, P. S., Rosser, M. P. 1984. *Cancer Res.* 44:1998–2010
122. Hoshi, H., McKeehan, W. L. 1984. *Proc. Natl. Acad. Sci. USA* 81:6413–17
123. Chaproniere, D. M., McKeehan, W. L. 1986. *Cancer Res.* 46:819–24
124. Barritault, D., Arruti, C., Courtois, Y. 1981. *Differentiation* 18:29–42
125. Courty, J., Loret, C., Moenner, M., Chevallier, B., Lagente, O., et al. 1985. *Biochemie* 67:265–69
126. Plouet, J., Courty, J., Olivie, M., Courtois, Y., Barritault, D. 1984. *Cell. Mol. Biol.* 30:105–10
127. Tarsio, J. F., Rubin, N. A., Russell, P., Gregerson, D. S., Reid, T. W. 1983. *Exp. Cell Res.* 146:71–78
128. Eccleston, P. A., Silberberg, D. H. 1985. *Dev. Brain Res.* 21:315–18
129. Wu, D. K., Maciag, T., de Villis, J. 1988. *J. Cell. Physiol.* 136:367–71
130. Katoh, Y., Takayama, S. 1984. *Exp. Cell Res.* 150:131–40
131. Kato, Y., Iwamoto, M., Koike, T. 1987. *J. Cell Physiol.* 133:491–98
132. Kato, Y., Gospodarowicz, D. 1985. *J. Cell Biol.* 100:477–85
133. Kato, Y., Gospodarowicz, D. 1985. *J. Cell Biol.* 100:486–95
134. Wice, R., Milbrandt, J., Glaser, L. 1987. *J. Biol. Chem.* 262:1810–17
135. Canalis, E., Lorenzo, J., Burgess, W. H., Maciag, T. 1987. *J. Clin. Invest.* 79:52–58
136. Rodan, S. B., Wesolowski, G., Thomas, K., Rodan, G. A. 1987. *Endocrinology* 121:1917–23
137. Canalis, E., Lian, J. B. 1988. *Bone* 9:243–46
138. Halaban, R., Ghosh, S., Baird, A. 1987. *In Vitro Cell Dev. Biol.* 23:47–52
139. Lobb, R. R. 1988. *Eur. J. Clin. Invest.* 18:321–36
140. Gospodarowicz, D., Bialecki, H. 1978. *Endocrinology* 103:854–58
141. Thomas, K. A., Gimenez-Gallego, G. 1986. *Trends Biochem. Sci.* 11:1–4
142. Gospodarowicz, D., Neufeld, G., Schweigerer, L. 1987. *J. Cell. Physiol. Suppl.* 5:15–26
143. Gospodarowicz, D., Neufeld, G., Schweigerer, L. 1986. *Cell Differentiation* 19:1–17
144. Maciag, T. 1984. *Prog. Hemostasis Thromb.* 9:33–58
145. Stiles, C. D., Pledger, W. J., VanWyk, J. J., Antoniades, H. N., Scher, C. D. 1979. *Proc. Natl. Acad. Sci. USA* 76:1279–83
146. Huang, S. S., Kuo, M.-D., Huang, J. S. 1986. *Biochem. Biophys. Res. Commun.* 39:619–25
147. Rizzino, A., Ruff, E. 1986. *In Vitro Cell Dev. Biol.* 22:749–55
148. Mullins, D. E., Rifkin, D. B. 1984. *J. Cell. Physiol.* 119:247–54
149. Terranova, V. P., DiFlorio, R., Lyall, R. M., Hic, S., Friesel, R., Maciag, T. 1985. *J. Cell Biol.* 101:2330–34
150. Senior, R. M., Huang, S. S., Griffin, G. L., Huang, J. S. 1986. *Biochem. Biophys. Res. Commun.* 141:67–72
151. Linemeyer, D. L., Kelly, L. J., Menke, J. G., Gimenez-Gallego, G., DiSalvo, J., Thomas, K. A. 1987. *Biotechnology* 5:960–65
152. Senior, R. M., Huang, J. S., Griffin, G. L., Deuel, T. F. 1985. *J. Cell Biol.* 100:351–56
153. Magnaldo, I., L'Allemain, G., Chambard, J. C., Moenner, M., Barritault, D., Pouyssegur, J. 1986. *J. Biol. Chem.* 261:16916–22
154. Chambard, J. C., Paris, S., L'Allemain, G., Pouyssegur, J. 1987. *Nature* 326:800–3
155. Moenner, M., Magnaldo, I., L'Allemain, G., Barritault, D., Pouyssegur, J. 1987. *Biochem. Biophys. Res. Commun.* 146:32–40
156. Doctrow, S. R., Folkman, J. 1987. *J. Cell Biol.* 104:679–87
157. Hoshi, H., Kan, M., Mioh, H., Chen, J.-K., McKeehan, W. L. 1988. *FASEB J.* 2:2797–800
158. L'Allemain, G., Pouyssegur, J. 1986. *FEBS Lett.* 197:344–48
159. Besterman, J. M., Watson, S. P., Cuatrecasas, P. 1986. *J. Biol. Chem.* 261:723–27
160. Stumpo, D. J., Blackshear, P. J. 1986. *Proc. Natl. Acad. Sci. USA* 83:9453–57
161. Muller, R., Bravo, R., Burckhardt, J., Curran, T. 1984. *Nature* 312:716–20
162. Kruijer, W., Cooper, J. A., Hunter, T., Verma, I. M. 1984. *Nature* 312:711–16
163. Hovis, J. G., Stumpo, D. J., Halsey, D. L., Blackshear, P. J. 1986. *J. Biol. Chem.* 261:10380–86
164. Pelech, S. L., Olwin, B. B., Krebs, E. G. 1986. *Proc. Natl. Acad. Sci. USA* 83:5968–72
165. Herschman, H. R. 1985. *Mol. Cell. Biol.* 5:1130–35

166. Rybak, S. M., Lobb, R. R., Fett, J. W. 1988. *J. Cell. Physiol.* 136:312–18
167. Bouche, G., Gas, N., Prats, H., Baldin, V., Tauber, J. P., et al. 1987. *Proc. Natl. Acad. Sci. USA* 84:6770–74
168. Serrero, G., Khoo, J. C. 1982. *Anal. Biochem.* 120:351–59
169. Togari, A., Baker, D., Dickens, G., Guroff, G. 1983. *Biochem. Biophys. Res. Commun.* 114:1189–93
170. Togari, A., Dickens, G., Kuzuya, H., Guroff, G. 1985. *J. Neurosci.* 5:307–16
171. Neufeld, G., Gospodarowicz, D., Dodge, L., Fujii, D. K. 1987. *J. Cell. Physiol.* 131:131–40
172. Risau, W. 1986. *Proc. Natl. Acad. Sci. USA* 83:3855–59
173. Slack, J. M. W., Darlington, B. G., Heath, J. K., Godsave, S. F. 1987. *Nature* 326:197–200
174. Knochel, W., Born, J., Hoppe, P., Loppnow-Blinde, B., Tiedemann, H., et al. 1987. *Naturwissenschaften* 74:604–6
175. Kimelman, D., Kirschner, M. 1987. *Cell* 51:869–77
176. Grunz, H., McKeehan, W. L., Knochel, W., Born, J., Tiedemann, H., Tiedemann, H. 1988. *Cell Differ.* 22:183–90
177. Risau, W., Gautschi-Sova, P., Bohlen, P. 1988. *EMBO J.* 7:959–62
178. Seed, J., Olwin, B. B., Hauschka, S. D. 1988. *Dev. Biol.* 128:50–57
179. Gospodarowicz, D., Mescher, A. L. 1980. *Ann. NY Acad. Sci.* 339:151–74
180. Mescher, A. L., Munaim, S. I. 1986. *Anat. Rec.* 214:424–31
181. Thomas, K. A., Rios-Candelore, M., Gimenez-Gallego, G., DiSalvo, J., Bennett, C., et al. 1985. *Proc. Natl. Acad. Sci. USA* 82:6409–13
182. Gospodarowicz, D., Cheng, J., Lui, G.-M., Baird, A., Bohlen, P. 1984. *Proc. Natl. Acad. Sci. USA* 81:6963–67
183. Lobb, R. R., Alderman, E. M., Fett, J. W. 1985. *Biochemistry* 24:4969–73
184. Davidson, J. M., Klagsbrun, M., Hill, K. E., Buckley, A., Sullivan, R., et al. 1985. *J. Cell Biol.* 100:1219–27
185. Shing, Y., Folkman, J., Haudenschild, C., Lund, D., Crum, R., Klagsbrun, M. 1985. *J. Cell. Biochem.* 29:275–87
186. Fett, J. W., Bethune, J. L., Vallee, B. L. 1987. *Biochem. Biophys. Res. Commun.* 146:1122–31
187. Hayek, A., Culler, F. L., Beattie, G. M., Lopez, A. D., Cuevas, P., Baird, A. 1987. *Biochem. Biophys. Res. Commun.* 147:876–80
188. Thompson, J. A., Anderson, K. A., DiPietro, J. M., Zweibel, J. A., Zametta, M., et al. 1988. *Science* 241:1352
189. Thompson, P., Desbordes, J. M.,
190. Buntrock, P., Buntrock, M., Marx, I., Kranz, D., Jentzsch, K. D., Heder, G. 1984. *Exp. Pathol.* 26:247–54
191. Fourtanier, A. Y., Courty, J., Muller, E., Courtois, Y., Prunieras, M., Barritault, D. 1986. *J. Invest. Dermatol.* 87:76–80
192. Phadke, K. 1987. *Biochem. Biophys. Res. Commun.* 142:448–53
193. Hamerman, D., Taylor, S., Kirschenbaum, I., Klagsbrun, M., Raines, E. W., et al. 1987. *Proc. Soc. Exp. Biol. Med.* 186:384–89
194. Seed, J., Hauschka, S. D. 1988. *Dev. Biol.* 128:40–49
195. Baird, A., Hsueh, A. J. W. 1986. *Regul. Peptides* 16:243–50
196. Katoh, Y., Kodama, K., Ishikawa, T. 1985. *Exp. Cell Res.* 161:111–16
197. Koschinsky, T., Bunting, C. E., Rutter, R., Gries, F. A. 1987. *Diabete Metab.* 13:318–25
198. Folkman, J. 1987. *Eur. J. Cancer Clin. Oncol.* 23:361–63
199. Walicke, P., Cowan, W. M., Ueno, N., Baird, A., Guillemin, R. 1986. *Proc. Natl. Acad. Sci. USA* 83:3012–16
200. Eccleston, P. A., Jessen, K. R., Mirsky, R. 1987. *Dev. Biol.* 124:409–17
201. Unsicker, K., Reichert-Preibsch, H., Schmidt, R., Pettmann, B., Labourdette, G., Sensenbrenner, M. 1987. *Proc. Natl. Acad. Sci. USA* 84:5459–63
202. Lipton, S. A., Wagner, J. A., Madison, R. D., D'Amore, P. A. 1988. *Proc. Natl. Acad. Sci. USA* 85:2388–92
203. Schubert, D., Ling, N., Baird, A. 1988. *J. Cell Biol.* 104:635–43
204. Risau, W., Ekblom, P. 1986. *J. Cell Biol.* 103:1101–7
205. Baird, A., Ueno, N., Esch, F., Ling, N. 1987. *J. Cell. Physiol. Suppl.* 5:101–6
206. Wadzinski, M. G., Folkman, J., Sasse, J., Devey, K., Ingber, D., Klagsbrun, M. 1987. *Clin. Physiol. Biochem.* 5:200–9
207. Logan, A., Logan, S. D. 1986. *Neurosci. Lett.* 69:162–65
208. Huang, S. S., Tsai, C. C., Adams, S. P., Huang, J. S. 1987. *Biochem. Biophys. Res. Commun.* 144:81–87
209. Pettmann, B., Labourdette, G., Weibel, M., Sensenbrenner, M. 1986. *Neurosci. Lett.* 68:175–80
210. Schweigerer, L., Malerstein, B., Neufeld, G., Gospodarowicz, D. 1987. *Biochem. Biophys. Res. Commun.* 143:934–40
211. Schweigerer, L., Neufeld, G., Fried-

man, J., Abraham, J. A., Fiddes, J. C., Gospodarowicz, D. 1987. *Endocrinology* 420:796–800

212. Schweigerer, L., Neufeld, G., Friedman, J., Abraham, J. A., Fiddes, J. C., Gospodarowicz, D. 1987. *Nature* 352:257–59

213. Vlodavsky, I., Folkman, J., Sullivan, R., Fridman, R., Ishai-Michaeli, R., et al. 1987. *Proc. Natl. Acad. Sci. USA* 84:2292–96

214. Moscatelli, D., Presta, M., Joseph-Silverstein, J., Rifkin, D. B. 1986. *J. Cell. Physiol.* 129:273–76

215. Vlodavsky, I., Fridman, R., Sullivan, R., Sasse, J., Klagsbrun, M. 1987. *J. Cell. Physiol.* 131:402–8

216. Baird, A., Ling, N. 1987. *Biochem. Biophys. Res. Commun.* 142:428–35

217. Ferrara, N., Schweigerer, L., Neufeld, G., Mitchell, R., Gospodarowicz, D. 1987. *Proc. Natl. Acad. Sci. USA* 84:5773–77

218. Schweigerer, L., Neufeld, G., Gospodarowicz, D. 1987. *Invest. Ophthalmol. Vis. Sci.* 28:1838–43

219. Schweigerer, L., Neufeld, G., Mergia, A., Abraham, J. A., Fiddes, J. C., Gospodarowicz, D. 1987. *Proc. Natl. Acad. Sci. USA* 84:842–46

220. Libermann, T. A., Friesel, R., Jaye, M., Lyall, R., Westermark, B., et al. 1987. *EMBO J.* 6:1627–32

221. Lobb, R. R., Rybak, S. M., St. Clair, D. K., Kett, J. W. 1986. *Biochem. Biophys. Res. Commun.* 139:861–67

222. Chodak, G. W., Shing, Y., Borge, M., Judge, S. M., Klagsbrun, M. 1986. *Cancer Res.* 46:5507–10

223. Klagsbrun, M., Sasse, J., Sullivan, R., Smith, J. A. 1986. *Proc. Natl. Acad. Sci. USA* 83:2448–52

224. Presta, M., Moscatelli, D., Joseph-Silverstein, J., Rifkin, D. B. 1986. *Mol. Cell. Biol.* 6:4060–66

225. Schweigerer, L., Neufeld, G., Gospodarowicz, D. 1987. *J. Clin. Invest.* 80:1516–20

226. Moenner, M., Badet, J., Chevallier, B., Tardieu, M., Courty, J., Barritault, D. 1987. *Current Communications in Molecular Biology: Angiogenesis Mechanisms and Pathobiology*, pp. 52–57. New York: Cold Spring Harbor Lab.

227. Baird, A., Mormede, P., Bohlen, P. 1985. *Biochem. Biophys. Res. Commun.* 126:358–64

228. Jeanny, J. C., Fayein, N., Moenner, M., Chevallier, B., Barritault, D., Courtois, Y. 1987. *Exp. Cell Res.* 171:63–75

229. Folkman, J., Klagsbrun, M., Sasse, J.,

Wadzinski, M., Ingber, D., Vlodavsky, I. 1988. *Am. J. Pathol.* 130:393–400

230. Neufeld, G., Gospodarowicz, D. 1987. *J. Cell. Physiol.* 132:287–94

231. Kan, M., DiSorbo, D., Hou, J., Hoshi, H., Mansson, P. E., McKeehan, W. L. 1988. *J. Biol. Chem.* 263:11306–13

232. Rosengart, T. K., Ferrans, V. J., Casscells, W., Kupferschmid, J. P., Maciag, T., Clark, R. E. 1988. *J. Vasc. Surg.* 7:311–17

233. Neufeld, G., Gospodarowicz, D. 1985. *J. Biol. Chem.* 260:13860–66

234. Olwin, B. B., Hauschka, S. D. 1986. *Biochemistry* 25:3487–92

235. Moenner, M., Chevallier, B., Badet, J., Barritault, D. 1986. *Proc. Natl. Acad. Sci. USA* 83:5024–28

236. Moscatelli, D. 1987. *J. Cell. Physiol.* 131:123–30

237. Clegg, C. H., Linkhart, T. A., Olwin, B. B., Hauschka, S. D. 1987. *J. Cell Biol.* 105:949–56

238. Courty, J., Dauchel, M. C., Mereau, A., Badet, J., Barritault, D. 1988. *J. Biol. Chem.* 263:11217–20

239. Imamura, T., Tokita, Y., Mitsui, Y. 1988. *Biochem. Biophys. Res. Commun.* 155:583–94

240. Moscatelli, D. 1988. *J. Cell Biol.* 107:753–59

241. Friesel, R., Burgess, W. H., Mehlman, T., Maciag, T. 1986. *J. Biol. Chem.* 261:7581–84

242. Friesel, R., Burgess, W. H., Maciag, T. 1989. *Mol. Cell. Biol.* In press

243. Coughlin, S. R., Barr, P. J., Cousens, L. S., Fretto, L. J., Williams, L. T. 1988. *J. Biol. Chem.* 263:988–93

244. Neufeld, G., Gospodarowicz, D. 1986. *J. Biol. Chem.* 261:5631–37

245. Coffey, R. J., Leof, E. B., Shipley, G. D., Moses, H. L. 1987. *J. Cell. Physiol.* 132:143–48

246. Feige, J. J., Baird, A. 1988. *J. Biol. Chem.* 263:14023–27

247. Friesel, R., Maciag, T. 1988. *Biochem. Biophys. Res. Commun.* 151:957–64

248. Carpenter, G., Cohen, S. 1979. *Annu. Rev. Biochem.* 48:193–216

249. King, A. C., Cuatrecasas, P. 1981. *New Engl. J. Med.* 305:77–88

250. Nilsson, J., Thyberg, J., Heldin, C. H., Wasteson, A., Westermark, B. 1983. *Proc. Natl. Acad. Sci. USA* 80:5592–96

251. Walker, F., Burgess, A. W. 1987. *J. Cell. Physiol.* 130:255–61

252. Huang, S. S., Huang, J. S. 1985. *J. Biol. Chem.* 261:9568–71

253. Friesel, R., Komoriya, A., Maciag, T. 1987. *J. Cell Biol.* 104:689–96

254. Olwin, B. B., Hauschka, S. D. 1988. *J. Cell Biol.* 107:761–69
255. Harper, J. W., Strydom, D. J., Lobb, R. R. 1986. *Biochemistry* 25:4097–103
256. Baird, A., Schubert, D., Ling, N., Guillemin, R. 1988. *Proc. Natl. Acad. Sci. USA* 85:2324–28
257. Schubert, D., Ling, N., Baird, A. 1987. *J. Cell Biol.* 104:635–43
258. Esch, F., Baird, A., Ling, N., Veno, N., Hill, F., et al. 1985. *Proc. Natl. Acad. Sci. USA* 82:6507–11
259. Seno, M., Sasada, R., Iwane, M., Sudo, K., Kurokawa, T., et al. 1988. *Biochem. Biophys. Res. Commun.* 151: 701–8
260. Harper, J. W., Lobb, R. R. 1988. *Biochemistry* 27:671–78
261. Squires, C. H., Childs, J., Eisenberg, S. P., Polverini, P. J., Sommer, A. 1988. *J. Biol. Chem.* 263:16297–302
262. Neufeld, G., Mitchell, R., Ponte, P., Gospodarowicz, D. 1988. *J. Cell. Biol.* 106:1385–94
263. Barr, P. J., Cousens, L. S., Lee-Ng, C. T., Medina-Selby, A., Masiarz, F. R., et al. 1988. *J. Biol. Chem.* 263:16471–78
264. Gross, J. L., Moscatelli, D., Jaffe, E. A., Rifkin, D. B. 1982. *J. Cell Biol.* 94:974–81
265. Gross, J. L., Moscatelli, D., Rifkin, D. B. 1983. *Proc. Natl. Acad. Sci. USA* 80:2623–27
266. Presta, M., Mignatti, P., Mullins, D. E., Moscatelli, D. 1985. *Biosci. Rep.* 5:783–88
267. Moscatelli, D., Presta, M., Rifkin, D. B. 1986. *Proc. Natl. Acad. Sci. USA* 83:2091–95
268. Kwaan, H.-C., Astrup, T. 1964. *J. Pathol. Bacteriol.* 87:409–14
269. Strickland, A., Beers, W. 1976. *J. Biol. Chem.* 251:5694–702
270. Sasada, R., Kurokawa, T., Iwane, M., Igarashi, K. 1988. *Mol. Cell. Biol.* 8:588–94
271. Rogelj, S., Weinberg, R. A., Fanning, P., Klagsbrun, M. 1988. *Nature* 331: 173–75
272. Blam, S. G., Mitchell, R., Tischer, E., Rubin, J. S., Silva, M., et al. 1989. *Oncogene.* In press
273. Jaye, M., Lyall, R. M., Mudd, R., Schlessinger, J., Sarva, N. 1988. *EMBO J.* 7:963–69
274. Moore, R., Casey, G., Brookes, S., Dixon, M., Peters, G., Dickson, C. 1986. *EMBO J.* 5:919–24
275. Dickson, C., Peters, G. 1987. *Nature* 326:833
276. Smith, R., Peters, G., Dickson, C. 1988. *EMBO J.* 7:1013–22
277. Wilkinson, D. G., Peters, G., Dickson, C., McMahon, A. P. 1988. *EMBO J.* 7:691–95
278. Dickson, C., Smith, R., Brookes, S., Peters, G. 1984. *Cell* 37:529–36
279. Jakobovits, A., Shackleford, G. M., Varmus, H. E., Martin, G. R. 1986. *Proc. Natl. Acad. Sci. USA* 83:7806–10
280. Taira, M., Yoshida, T., Miyagawa, K., Sakamoto, H., Terada, M., Sugimura, T. 1987. *Proc. Natl. Acad. Sci. USA* 84:2980–84
281. Yoshida, T., Sakamoto, H., Miyagawa, K., Little, P. F. R., Terada, M., Sugimura, T. 1987. *Biochem. Biophys. Res. Commun.* 142:1019–24
282. Yoshida, T., Miyagawa, K., Odagiri, H., Sakamoto, H., Little, P. F. R., et al. 1987. *Proc. Natl. Acad. Sci. USA* 84:7305–9
283. DelliBovi, P., Basilico, C. 1987. *Proc. Natl. Acad. Sci. USA* 84:5660–64
284. DelliBovi, P., Curatola, A. M., Kern, F. G., Greco, A., Ittmann, M., Basilico, C. 1987. *Cell* 50:729–37
285. DelliBovi, P., Curatola, A. M., Newman, K. M., Sato, Y., Moscatelli, D., et al. 1988. *Mol. Cell. Biol.* 8:2933–41
286. Zhan, X., Goldfarb, M. 1986. *Mol. Cell. Biol.* 6:3541–44
287. Zhan, X., Bates, B., Hu, X., Goldfarb, M. 1988. *Mol. Cell. Biol.* 8:3487–95
288. Bird, A. P. 1986. *Nature* 321:209–13
289. Collins, T., Ginsberg, D., Boss, J., Orkin, S. H., Pober, J. S. 1985. *Nature* 316:748–50

Annu. Rev. Biochem. 1989. 58:607–33

THE BACTERIAL PHOTOSYNTHETIC REACTION CENTER AS A MODEL FOR MEMBRANE PROTEINS[1]

D. C. Rees, H. Komiya, and T. O. Yeates

Department of Chemistry and Biochemistry, Molecular Biology Institute, University of California, Los Angeles, California 90024

J. P. Allen and G. Feher

Department of Physics, University of California, San Diego, La Jolla, California 92093

CONTENTS

PERSPECTIVES AND SUMMARY

Membrane proteins participate in many fundamental cellular processes. Until recently, an understanding of the function and properties of membrane pro-

---

[1]Abbreviations used: *Rb. sphaeroides, Rhodobacter sphaeroides; Rps. viridis, Rhodopseudomonas viridis;* RC, reaction center

607

teins was hampered by an absence of structural information at the atomic level. A landmark achievement toward understanding the structure of membrane proteins was the crystallization (1) and structure determination (2–5) of the photosynthetic reaction center (RC) from the purple bacteria *Rhodopseudomonas viridis*, followed by that of the RC from *Rhodobacter sphaeroides* (6–17). The RC is an integral membrane protein–pigment complex, which carries out the initial steps of photosynthesis (reviewed in 18). RCs from the purple bacteria *Rps. viridis* and *Rb. sphaeroides* are composed of three membrane-associated protein subunits (designated L, M, and H), and the following cofactors: four bacteriochlorophylls (Bchl or B), two bacteriopheophytins (Bphe or ϕ), two quinones, and a nonheme iron. The cofactors are organized into two symmetrical branches that are approximately related by a twofold rotation axis (2, 8). A central feature of the structural organization of the RC is the presence of 11 hydrophobic α-helixes, approximately 20–30 residues long, which are believed to represent the membrane-spanning portion of the RC (3, 9). Five membrane-spanning helixes are present in both the L and M subunits, while a single helix is in the H subunit. The folding of the L and M subunits is similar, consistent with significant sequence similarity between the two chains (19–25). The L and M subunits are approximately related by the same twofold rotation axis that relates the two cofactor branches.

RCs are the first membrane proteins to be described at atomic resolution; consequently they provide an important model for discussing the folding of membrane proteins. The structure demonstrates that α-helical structures may be adopted by integral membrane proteins, and provides confirmation of the utility of hydropathy plots in identifying nonpolar membrane-spanning regions from sequence data. An important distinction between the folding environments of water-soluble proteins and membrane proteins is the large difference in water concentration surrounding the proteins. As a result, hydrophobic interactions (26) play very different roles in stabilizing the tertiary structures of these two classes of proteins; this has important structural consequences. There is a striking difference in surface polarity of membrane and water-soluble proteins. However, the characteristic atomic packing and surface area appear quite similar.

A computational method is described for defining the position of the RC in the membrane (10). After localization of the RC structure in the membrane, surface residues in contact with the lipid bilayer were identified. As has been found for soluble globular proteins, surface residues are less well conserved in homologous membrane proteins than the buried, interior residues. Methods based on the variability of residues between homologous proteins are described (13); they are useful (*a*) in defining surface helical regions of both membrane and water-soluble proteins and (*b*) in assigning the side of these helixes that are exposed to the solvent. A unifying view of protein structure suggests that water-soluble proteins may be considered as modified membrane proteins with covalently attached polar groups that solubilize the proteins in aqueous solution.

INTRODUCTION

Despite the central role of membrane proteins in cellular processes, comparatively little structural information has been available concerning the atomic details of their structural organization. This is in striking contrast to the situation prevailing for water-soluble proteins, for which more than 200 atomic structures have been determined. The α-helix is believed to be an important structural motif in membrane proteins, as suggested by the formation of α-helical structures by polypeptides in nonaqueous solvents (27). The first glimpse of a membrane protein structure was provided for bacteriorhodopsin from the electron microscopy studies of Henderson & Unwin (28). A striking feature of this structure was the arrangement of the membrane-spanning region of bacteriorhodopsin into seven rodlike features identified as α-helices. Although a description of the structure at atomic resolution was not possible, this work motivated extensive efforts in understanding the folding of membrane proteins. A prominent area of this research was the development of methods for the identification and characterization of membrane-spanning helixes from sequence data (reviewed in 29, 30, 35).

This situation has changed recently with the structure determination of photosynthetic reaction centers (RCs) from purple bacteria by X-ray diffraction. RCs are integral membrane protein complexes, which carry out the initial photosynthetic steps of light-induced electron transfer from a donor to a series of acceptor species (reviewed in 18). The kinetics of these processes have been extensively studied spectroscopically, and a general picture of the structural organization of the RC was provided by a variety of biophysical, biochemical, and molecular biology studies. A critical development in obtaining detailed structural information was the crystallization (1) and structure determination (2–5) of the RC from *Rhodopseudomonas viridis* by Michel, Deisenhofer, Huber and coworkers. For the first time, X-ray crystallography revealed the atomic structure of a membrane protein, and significantly, one of great biological interest. Subsequently, the structure of the RC from *Rhodobacter sphaeroides* has been described (6–17). Knowledge of these structures has opened a window into the molecular architecture of membrane protein systems.

Since the RC structure determinations, a number of conference proceedings and reviews have appeared that discuss various aspects of the photosynthetic function of the RC (31–34). Many important studies related to electron transfer and spectroscopic studies of RCs are described in these collections. This review focuses only on aspects of the RC structure relevant to the folding of membrane proteins. Emphasis is placed on the surface properties of the membrane-spanning region of the RC, and comparison between folding characteristics of membrane and water-soluble proteins. The discussion is based on structural and sequence information obtained for RCs from purple bacteria; these are to date the best-characterized membrane protein system.

Notation

The following conventions are used throughout this review: the three RC protein subunits are designated L, M, and H. Residue numbers are based on the *Rb. sphaeroides* protein sequences (19–21). Cofactors are abbreviated as follows: bacteriochlorophyll (Bchl or B), bacteriopheophytin (Bphe or ϕ), quinone (Q), iron (Fe), and carotenoid (C). The special pair of bacteriochlorophyll is represented as $(Bchl)_2$ or D. The two cofactor branches are designated A and B (8), which correspond to the L and M branches, respectively, in the description of the RC from *Rps. viridis* (2–4). Branch affiliations of the different cofactors are indicated by a subscript A or B. LDAO and BOG represent the detergent molecules lauryl dimethylamine-N-oxide and β-octyl glucoside, respectively. The term "soluble protein" specifically refers to globular proteins soluble in aqueous solutions in the absence of detergents or membranes.

STRUCTURE OF THE RC

RCs from purple bacteria such as *Rps. viridis* and *Rb. sphaeroides* typically contain one copy each of three membrane-associated subunits designated L, M, and H (reviewed in 18). These labels (Light, Medium, and Heavy) were assigned on the basis of apparent mobilities on SDS gels, although subsequent sequence analyses indicated that the actual molecule weights increase in the order H < L < M. Certain RCs, including the one from *Rps. viridis* (but not from *Rb. sphaeroides*), contain a tightly associated cytochrome molecule. Amino acid sequences have been determined for RCs from four purple bacteria, *Rb. sphaeroides* (19–21), *Rhodobacter capsulatus* (22), *Rps. viridis* (23, 24), and *Rhodospirillum rubrum* (25). Determination of the primary sequences for these proteins provided important structural information. Sequence homology between the L and M subunits was evident, thus suggesting that they originally were derived from a common precursor subunit. Hydropathy plots (29, 30, 35) of the amino acid sequences of the L, M, and H subunits indicate a total of 11 membrane-spanning helices in the RC; the analysis was based on the occurrence of stretches of 20–30 predominantly hydrophobic residues. The L and M subunits each exhibit five such regions, while the H subunit contains only one region.

Associated with these proteins are a group of cofactors, including four Bchl, two Bphe, two quinones, and a nonheme iron. Two of the Bchl are organized into the special pair or dimer, $(Bchl)_2$, which is the primary electron donor for the photosynthetic reactions. The electron transfer reaction proceeds from the dimer to an intermediate acceptor (ϕ_A), a primary quinone (Q_A), and a secondary quinone (Q_B). Also associated with RCs is a single carotenoid molecule, except in certain carotenoidless mutants such as *Rb. sphaeroides* strain R-26.

Detailed structural determinations of membrane proteins, including RCs, require high-resolution crystallographic analyses, which in turn require the availability of suitable crystals. Ordered two-dimensional arrays of membrane proteins, in particular bacteriorhodopsin (28), have provided valuable structural information, but these analyses have not progressed to the approximately 3 Å resolution required for atomic model building. Garavito & Rosenbusch (36), and Michel (1) succeeded in obtaining well-ordered three-dimensional crystals of two membrane proteins: porin from *Escherichia coli* and the RC from *Rps. viridis,* respectively. These accomplishments opened the field of X-ray structural analysis of membrane proteins. Approaches to the crystallization of membrane proteins have been reviewed (37, 38, 38a).

The structure of the RC from *Rps. viridis* was initially determined to 3 Å resolution (2), using X-ray diffraction techniques and the method of multiple isomorphous replacement. The resolution and refinement of this structure has been subsequently extended to 2.3 Å (5). Crystallization of the carotenoidless strain R-26 (6, 6a, 15), wild-type strain 2.4.1 (14, 39), and strain Y (40) of the RC from *Rb. sphaeroides* have been reported. We have determined structures of the *Rb. sphaeroides* RCs from the carotenoidless strain R-26 and the wild-type strain 2.4.1 of *Rb. sphaeroides* at 2.8 Å and 3.0 Å resolution, respectively (8–14). This model of the RC from *Rb. sphaeroides* R-26 is used for the figures and analyses described in this review. A second, independent structure determination of the RC from *Rb. sphaeroides* R-26 has also been presented at 3.2 Å resolution (16, 17). The structure of the RC from *Rb. sphaeroides* is similar to that of the RC from *Rps. viridis;* this allowed the use of the method of molecular replacement to solve the initial crystallographic phase problem (7, 16).

The overall folding of the polypeptide chains and the cofactor arrangement in the *Rb. sphaeroides* R-26 RC is illustrated in Figure 1. A striking feature of the RC is the approximate symmetry evident in the structure, with both the two cofactor branches and the L and M subunits related by a twofold rotation axis. In view of the complexity of the interactions, it is convenient to describe initially the structures of the protein and cofactors separately. It must be kept in mind, however, that the cofactors and the protein subunits are intimately associated with each other, and it is unlikely that structures of any individual component could be maintained in isolation.

Protein Structure

A total of 11 hydrophobic α-helixes are observed in the L, M, and H subunits; they create a framework that organizes the cofactors (Figures 1 and 2). The general positions of the helixes in the protein sequences had been predicted from hydropathy analyses. Since these helixes create an apolar region approximately 35 Å wide, it is assumed that they represent the membrane-spanning region of the RC (2, 3, 9). The L and M subunits each have five apolar helixes designated A, B, C, D, and E, preceded by the appropriate subunit letter (i.e. LA, LB, etc). The H subunit has one membrane-spanning helix designated HA. The LD, LE, MD, and ME helixes form a core structure that interacts extensively with the cofactors. In contrast, the A, B, and C helixes of the L and M subunits, and the HA helix are located on the periphery of the RC, away from the cofactor rings. The tilts and curvatures of these helixes (3, 9) influence the number of residues necessary to span the membrane. The average tilt angle of the 11 membrane-spanning helixes is 22° from the twofold axis that relates the L and M subunits; the axes of individual helixes may differ by up to 65° (the angle between the LD and MD helixes). In addition, the C and E helixes exhibit substantial kinking, and have effective radii of curvature of 30–70 Å.

More than half of the residues in the RC, including most of the H subunit, are located outside the membrane-spanning region. The orientation of the RC in the cell membrane has been established by chemical modification, and by studies of cytochrome c binding (reviewed in 18, 43). The studies permit assignment of the "top" surface of the RC (as indicated in Figure 1) to the periplasmic side of the membrane, while the "bottom" surface faces the cytoplasm. Helical segments in these outer regions are designated in relation

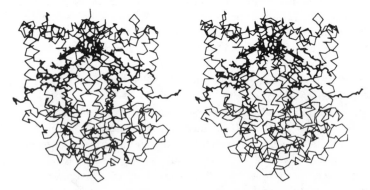

Figure 1   Stereoview of the cofactors and Cα backbone of the protein subunits of the RC from the carotenoidless mutant strain R-26 of *Rb. sphaeroides*. The twofold axis is aligned vertically in the paper, with the cytoplasmic side of the RC at the bottom of the figure. The figure was prepared with the FRODO graphics program (41). Modified from Ref. 9.

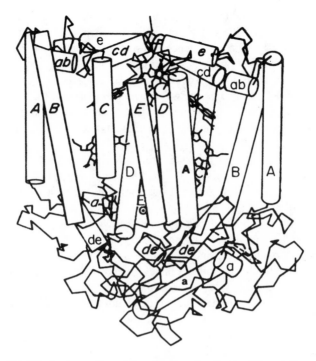

Figure 2   The RC structure with protein subunits and cofactors. The α-helices have been approximated by straight cylinders. Helixes of the L subunit are lettered in plain type, while helixes of the M subunit are lettered in italic type. H subunit helixes (**A** and **a**) are in bold font. The phytyl and isoprenoid tails of the cofactors have been truncated. The view is in the same orientation as Figure 1. The figure was modified from Ref. 9 and was prepared with the aid of a program described in Ref. 42.

to the transmembrane helixes that they connect. An interesting arrangement on the surface of the periplasmic side of both the L and M subunits are two "interrupted" helixes [designated as the I helixes (9)] composed of three helical segments connecting helixes A and B, C and D, and one following helix E (Figure 2). These helical segments are designated ab, cd, and e, respectively. The bulk of the H subunit lies on the cytoplasmic side, and consists of several β-sheets organized into a globular domain.

Cofactor Structure

The cofactors are arranged into two branches (2, 8), designated A and B, illustrated in Figure 3. For clarity, the phytyl and isoprenoid tails have been removed from the cofactors. Along either branch, the sequence of cofactors is $(Bchl)_2$, Bchl, Bphe, Q, and Fe. The ring centers of adjacent cofactors are separated by approximately 10–13 Å (2, 8). Individual distances between cofactor centers are listed in Table 1 of Ref. 8. The RC models have firmly established the structural basis for two important points that had been deduced spectroscopically: (*a*) the dimer interaction of two Bchl molecules in the special pair (Bchl)$_2$, whose existence had been proposed on the basis of EPR (44, 45) and ENDOR (46, 47) experiments, and (*b*) the sequence of the electron transfer steps in the order (Bchl)$_2$ to Bphe to Q. An unexpected aspect of the RC structures, however, is the existence of two cofactor branches, although the spectroscopic evidence indicates the existence of only one photochemically active branch. The active branch has been identified as the A branch (2, 8, 48).

The cofactor rings in each branch are approximately related by the same twofold rotation axis that relates the L and M subunits (3, 9). The (Bchl)$_2$ and the Fe are positioned close to this axis. Atoms in the cofactor rings can be superimposed to within 0.8 Å–1.5 Å by a twofold rotation of one branch about this axis. Significant departures from this symmetrical arrangement are observed for the phytyl and isoprenoid tails.

In addition to the cofactors associated with the primary photochemical events, the RC binds two other classes of molecules: carotenoid and detergents.

1. carotenoid: RCs bind a single carotenoid molecule, except in the carotenoidless mutant strain R-26 of *Rb. sphaeroides*. Carotenoids are long, conjugated polyenes that have been located in the structures of the RCs from both *Rps. viridis* (5) and *Rb. sphaeroides* (11). The carotenoid spher-

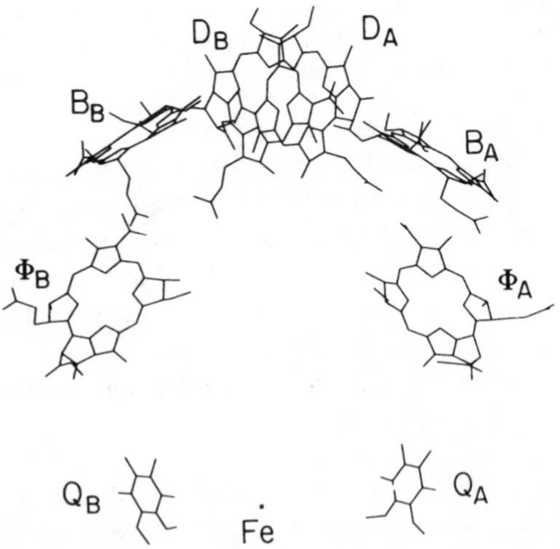

Figure 3 Cofactor structure of the RC from *Rb. sphaeroides* R-26. Phytyl and isoprenoid tails of the cofactors have been omitted for clarity. The cofactors are displayed in the same orientation as shown in Figure 1.

oidene in the RC from *Rb. sphaeroides* strain 2.4.1 adopts a boomerang-shaped structure that curves around the MA and MB helixes, passing near the B_B ring.

2. detergent: Most of the detergent molecules present in the crystal lattice are too disordered to observe crystallographically. A few exceptions have been noted, however. In the RC from *Rps. viridis,* a molecule of LDAO has been identified between the MD and HA helixes (J. Deisenhofer, personal communication). Two molecules of BOG have been identified in the structure of the RC from the carotenoidless strain R-26 of *Rb. sphaeroides* (11). One BOG molecule binds to the site occupied by the carotenoid in the RC from *Rb. sphaeroides* 2.4.1. This provides an illustration of how detergent molecules may influence the properties of membrane proteins by occupying specific ligand-binding sites. A second, less well defined, BOG is bound to the L subunit, near B_A and ϕ_A. Spectral changes observed in RCs solubilized in different detergents (49) may be due to direct interaction between detergent molecules and cofactors.

RC AS A MODEL FOR THE FOLDING OF MEMBRANE PROTEINS

With the exception of the RC, our knowledge of the tertiary structure of membrane proteins is primitive relative to that of water-soluble, globular proteins. A general picture of soluble proteins has emerged over the past 30 years, which emphasizes (*a*) the efficiently packed, relatively nonpolar interior, and (*b*) the polar surface, which minimizes the surface energy in an aqueous environment. These characteristics are a direct consequence of the influence of hydrophobic interactions (26). An important distinction between the folding environment of soluble proteins and membrane proteins is the relative absence of water in the bilayer region. Since this results in a much different role for hydrophobic interactions in stabilizing the structure of membrane proteins, a comparative analysis of the structures of proteins in these two classes is central to understanding the role that the solvent plays in protein folding. Although the RC may not prove to belong to the most prevalent class of membrane proteins (given the large cofactor component), it currently represents the best-defined structure and consequently commands our interest as a model to investigate the structural organization of membrane proteins.

Surface Area and Volume of the RC

A view of membrane protein structure in terms of a collection of hydrophobic, membrane transversing α-helixes was satisfyingly confirmed by the three-dimensional structures of the RC. As in the case of soluble proteins, it would, however, be surprising if there were only one structural motif that characterized the folding of transmembrane proteins. Given the observed (or antic-ipated) diversity in the detailed structure of the polypeptide chains of soluble and membrane proteins, how can a general comparison of structures between these two classes be achieved? Clearly, a characterization of protein tertiary structures that is independent of the details of the local folding pattern of the polypeptide is required.

One approach to this problem utilizes the concepts of molecular surface area and volume introduced and implemented by Richards (50). These ideas have provided a quantitative basis for characterizing soluble proteins with efficiently packed, apolar interiors and a polar surface for favorable solvent interactions. The RC structure provides an opportunity to perform comparable analyses on a membrane protein.

SURFACE AREA   The energy required to increase the surface area of a liquid is given by the product of the surface tension of the liquid and the change in surface area. The decreased surface tension of hydrocarbon liquids ($\sim$30 cal Å^{-2}) compared to water (105 cal Å^{-2}) (Ref. 51) suggests that an increase in surface area of the solvent surrounding a protein would require less energy in a membrane than in water. Consequently, membrane proteins might be expected to have a larger surface area than soluble proteins of the same molecular weight (if, for example, more favorable packing contacts could be achieved at the expense of a higher surface energy). Following Richards, surfaces of proteins may be defined by rolling a spherical probe around the van der Waals surface of a protein. The accessible surface area, A_s, is determined from the area of the surface generated by the center of the probe (52). For this calculation, van der Waals radii for the protein atoms were taken from Richmond & Richards (53), while all cofactor atoms were assigned radii of 1.8 Å. With a solvent probe radius of 1.4 Å, A_s for the *Rb. sphaeroides* RC was calculated to be 34,800 Å^2. The corresponding value for soluble proteins was estimated by an empirical relationship between A_s and molecular weight M established by Miller et al (54) for soluble, oligomeric proteins:

$$A_s = 5.3 \, M^{0.76} \qquad\qquad 1.$$

For a protein with $M = 10^5$ (the molecular weight of the RC from *Rb. sphaeroides*), Equation 1 predicts $A_s = 33,400 \, \text{Å}^2$. The agreement between the calculated surface area (based on observations on soluble proteins) and the observed value for RC indicates that there is no significant difference in surface area between membrane and soluble proteins of comparable size. The surface energies of soluble and membrane proteins must be similar, despite the differences in surface tensions between hydrocarbon liquids and water.

Surface energies are influenced by both the solvent surface tension, and the protein-solvent interaction energy (51). Comparable surface areas for membrane and soluble proteins could reflect weaker (van der Waals) interactions between proteins and hydrocarbons in the membrane, compared to stronger (hydrogen bond) interactions possible between proteins and water in aqueous solutions.

Unlike for a smooth sphere, the surface areas of objects with irregular or rough surfaces (including macromolecules) are not uniquely defined. Hence, it is essential that surface area comparisons of proteins be performed with identical van der Waals and probe radii. A measure of the roughness of protein surfaces can be derived from the dependence of the surface area on the parameters of the calculation. The accessible surface area is unsuitable for this analysis, however, and it is necessary to adopt a second type of surface, the molecular surface. As described by Richards (50), the molecular surface is defined as a continuous envelope stretched over the van der Waals surface of a protein; it describes the position of the inner surface of the probe sphere as the probe moves in contact with the van der Waals surface. In contrast, the accessible surface is defined by the position of the center of the probe sphere as the probe moves in contact with the van der Waals surface; hence the accessible surface is always displaced from the van der Waals surface of the protein. Since the displacement varies with the probe radius, the value of the calculated surface area will be affected by both the surface roughness and the displacement. The variation in molecular surface area with probe radius, however, directly provides information on surface roughness. For a perfectly smooth object such as a sphere, the molecular surface will be independent of the probe radius. In contrast, the molecular surface area of an irregular object, such as a sponge, will depend significantly on the probe radius. A larger surface area will be calculated for small probes that can penetrate into the pores of the sponge, as compared to larger probes which are excluded from the pores.

The surface roughness of a protein may be characterized by a parameter, D, which is calculated from the variation in molecular surface area, A, with probe radius, r, through the relationship (55, 56):

$$D = 2 - d(\log A)/d(\log r) \qquad\qquad 2.$$

As the surface becomes more irregular, D increases from the value 2 for a smooth surface, to a value $D \le 3$. D has properties of a (noninteger) dimension known as a "fractal dimension," which has found application in the description of a wide range of physical and mathematical phenomena (55). D is defined only for a certain range of probe radii; in the limit of both small and large probe radii, D will equal 2 for macromolecular surfaces. For small probe radii, the probe interacts predominantly with the spherical van der Waals spheres describing the protein atoms, whereas large probes are sensitive only to the overall shape of the molecule. D will be maximal for probe radii in the size range of the irregular surface features. For a protein, this corresponds to the approximate size of water molecules and side chains, with $1.5\ \text{Å} < r < 3\ \text{Å}$.

The variation in A with r is illustrated in Figure 4 for a representative sample of both monomeric and oligomeric soluble proteins, and the RC from *Rb. sphaeroides*. Curves are presented for both the entire RC molecule, as well as atoms located in the membrane-spanning region (defined more pre-

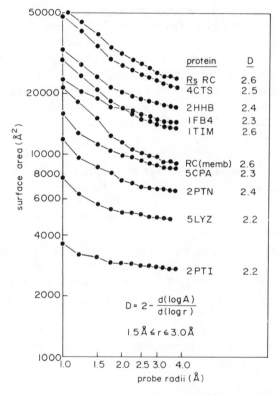

Figure 4 Variation in molecular surface area with probe radii plotted on a log-log scale for selected monomeric and oligomeric proteins. Molecular surface areas were calculated with Connelly's MS program (57). The D value (Eq. 2) provides a measure of the overall surface roughness of each protein. D was calculated with probe radii in the range $1.5\ \text{Å} < r < 3\ \text{Å}$. The following coordinate sets for soluble proteins were used from the Brookhaven Protein Data Bank (58): 2PTI (pancreatic trypsin inhibitor), 5LYZ (lysozyme), 2PTN (trypsin), 5CPA (carboxypeptidase A), 1TIM (triose phosphate isomerase), 1FB4 (immunoglobulin Kol), 2HHB (hemoglobin), 4CTS (citrate synthase). Surface areas for the RC from *Rb. sphaeroides* were calculated for both the entire RC (Rs RC), and for atoms in the membrane-spanning region [RC (memb)].

cisely below). Programs developed by Connelly (57) were used for the area calculations. Van der Waals radii for the protein atoms were assigned the values 1.8 Å, 1.7 Å, 1.4 Å, 1.8 Å, and 0.8 Å for carbon, nitrogen, oxygen, sulfur, and metal atoms, respectively. (Differences in these curves and those reported in Ref. 56 are due to the use of smaller van der Waals radii in the earlier work.) The trend apparent from this figure is that larger (oligomeric) proteins have larger values of D, i.e. they are more irregular than smaller proteins. Thus, large proteins are not simply smaller proteins scaled up in size. There is no significant difference in overall surface roughness of the RC relative to soluble, oligomeric proteins of comparable size. Previous studies (56) indicated that the surfaces of soluble proteins are not uniformly irregular, but rather exhibit variations in roughness between different regions of the protein surface. Comparable studies examining the local variations in surface roughness of the RC structure have not yet been performed.

ATOMIC VOLUMES AND PACKING   The volumes of buried atoms in a protein may be calculated with the Voronoi construction (50, 59, 60). In this method, planes are drawn that are perpendicular bisectors to all the vectors between pairs of atoms in the structure. These planes intersect to define a unique polyhedron around each atom. Only buried atoms (with zero accessible surface area) are included in the calculation, to ensure that a closed polyhedron with a defined volume is constructed. The atomic volume is defined by the volume of the polyhedron surrounding the atom. An important conclusion from volume calculations is that the packing density of buried atoms in soluble proteins is the same as that observed in crystals of small organic molecules; i.e. interior atoms in soluble proteins are efficiently packed (50). Volumes of buried atoms in the membrane-spanning region of the RC have been calculated, and are similar to those observed for interior atoms in soluble proteins such as carboxypeptidase A (Table 1) and ribonuclease S (60). Consequently, the same efficient packing that characterizes soluble proteins is also maintained in the RC structure.

Stabilization of the Tertiary Structure of Membrane Proteins

The RC maintains a well-defined tertiary structure in the membrane-spanning region, despite the decrease in significance of hydrophobic interactions relative to soluble proteins. Based on the RC structure, the following types of interactions appear to impart the necessary structural specificity in the transmembrane region (10):

1. Atomic packing in the transmembrane region. The observed efficient packing of atoms in the RC structure stabilizes the tertiary structure by maximizing van der Waals contacts between atoms, and minimizing the adverse consequences of cavities (61).

Table 1 Volumes (Vol) with standard deviations (SD) of buried atoms in the membrane-spanning region of the RC from *Rb. sphaeroides* and the water soluble globular protein carboxypeptidase A (adapted from Ref. 10)

Atom type	RC		Carboxypeptidase A	
	Vol, Å^3	SD, Å^3	Vol, Å^3	SD, Å^3
Main-chain atoms				
N	13	2	14	2
Cα	12	3	12	2
C	8	1	8	1
O	21	4	22	3
Pro N	10	1	10	1
Side-chain atoms				
CβH	13	3	13	1
CβH$_2$	21	6	23	8
CH	21	2	21	3
CH$_2$	14	3	14	2
CH$_3$	31	6	34	5
Aromatic C	19	7	18	5
His ring	16	4	15	4
OH	25	4	24	5
O/N	21	5	24	4
Trp	16	6	17	5

2. Polar interactions between transmembrane helixes. A major polar interaction that will stabilize the transmembrane helical arrangement is provided by the four histidine ligands on the D and E helixes, which coordinate the iron atom. More general types of electrostatic effects, such as helix dipole interactions, may be involved in stabilizing the dominantly antiparallel arrangement of the transmembrane helixes (62). On average, less than one interhelical hydrogen bond is present between the polar side chains of residues on different helixes. No salt bridges between membrane helixes are observed.

3. Protein structures outside the membrane-spanning region. Several types of organized protein structures are observed in regions of the RC exposed to the aqueous environment (Figure 2). The two periplasmic I helixes on both the L and M subunits may serve as a strap that holds the transmembrane helixes together on the periplasmic sides. Structures such as the β-sheet region, as well as contacts between the L and M subunits and the H subunit, may also stabilize the membrane-spanning structure on the cytoplasmic side. However, the H subunit does not seem to be essential for maintaining the RC structure, since its removal does not significantly change the kinetics of electron transfer up to (and including) the reduction of the primary quinone (49). Furthermore, the RC structure in the green bacterium *Chloroflexus aurantiacus* is stable despite the absence of an H subunit (63).

Membrane vs Soluble Proteins: Analogy to Crystal Morphology

Water-soluble and membrane (RC) proteins seem similar in terms of the geometrical criteria of surface area and volume. The most striking difference between these two classes of proteins is the chemical nature of the exposed surface groups. To minimize surface energies, soluble proteins fold to generate a polar surface, while membrane proteins require an apolar surface. This behavior is similar to the effect of solvent conditions on the morphology of small molecule crystals. Gibbs demonstrated that the equilibrium morphology of a crystal will have the minimum surface free energy (64). Since different crystal faces have different exposed chemical groups, changing solvent conditions will alter the crystal morphology so as to maintain the state with lowest surface energy. For example, polar crystal faces with exposed carboxyl groups dominate the morphology of succinic acid crystals grown from water, whereas more apolar crystal faces with exposed methylene carbons are prominent in crystals grown from apolar solvents or by sublimation (65). The interior packing of succinic acid molecules remains, however, unchanged under these different solvent conditions. Thus, crystal morphology may be viewed as being analogous to the "morphologies" of water-soluble and membrane proteins; i.e. the surface composition of proteins is sensitive to solvent (i.e. water or bilayer) conditions, but the same type of efficient package is maintained. This behavior suggests that water-soluble proteins may be considered as modified membrane proteins with covalently attached polar groups that make the proteins soluble in aqueous solutions.

Position of the RC in the Membrane

An important aspect of the characterization of membrane proteins is to define the region of interaction between the protein and the membrane. For the RC, this region is composed of contiguous stretches of 20–30 apolar residues, which were identified from an analysis of the sequence data. The three-dimensional structures of the RC strongly support these assignments by demonstrating that these apolar regions are organized into 11 α-helices, that create a hydrophobic band approximately 35 Å wide. This band is essentially devoid of charged residues. Satisfying as this picture is, a direct demonstration of the membrane-spanning region of the RC is, however, difficult to achieve. In both the *Rps. viridis* and *Rb. sphaeroides* crystals, the phospholipids have been replaced by detergent molecules, which, with one or two exceptions, are disordered and therefore not observable by X-ray diffraction. A general location of the disordered detergents has been obtained by low-resolution neutron diffraction studies of the RC from *Rps. viridis* (65a). These studies localized the binding region of the detergent on the surface of the RC that surrounds the 11 α-helices. The precise location of the boundaries of the detergent-binding region (assuming the boundary is sharply defined), and by inference, the membrane-spanning region of the RC, could not be determined at the 15 Å resolution of this neutron diffraction study.

The position of the membrane-spanning region of the RC was determined indirectly by an analysis of the energetics of the RC-membrane interaction (10). It is based on the decrease in hydrophobic free energy when nonpolar regions of a membrane protein are placed into a lipid bilayer. Various potential functions have been developed to estimate the free energy of transfer, ΔG_H, between apolar and aqueous solvents. In general, ΔG_H is expressed as a product of two terms: (a) the surface area of the region involved in the transfer between solvents and (b) a surface free energy term. Following Eisenberg & McLachlan (66), ΔG_H may be expressed as a sum involving the solvent accessible surface area of an atom i, A_{si}, and the surface free energy $\Delta \sigma_i$ for each atom type:

$$\Delta G_H = \sum_i \Delta \sigma_i A_{si} \qquad \qquad 3.$$

where the sum is over all atoms i. The surface free energies of transfer between a nonpolar and an aqueous phase for different atom types are (in cal Å^{-2}) $\Delta \sigma(C) = 16$; $\Delta \sigma(N, O) = -6$; $\Delta \sigma(O^-) = -24$; $\Delta \sigma(N^+) = -50$; $\Delta \sigma(S) = 21$. These values were determined empirically (66) by fitting experimentally obtained values for the free energy of transfer of amino acids to an energy function similar to Equation 3.

With Expression 3 for ΔG_H, the equilibrium position of the RC in a bilayer can be established by determining the position of minimum energy (subject to the assumptions and limitations described below). An initial estimate of the location of the membrane-spanning region of the RC was determined by evaluating ΔG_H for sections of the RC that were 5 Å thick (Figure 5). For these calculations, the RC was sectioned normal to the local twofold axis (defined as the z axis), with the Fe atom at the origin ($z = 0$). The values of ΔG_H provide an estimate of the free energy of transfer from the membrane to water of the surface atoms in a particular 5 Å thick section. A region of the RC approximately 40 Å thick exhibits a large hydrophobic energy ΔG_H; this presumably represents the membrane-spanning region. Integration of the area under the curve of Figure 5 (correcting for the 5 Å slab width) yields a value of about 20 kcal/mole for ΔG_H per helix. This is consistent with an estimate of 30 kcal/mole for a single transmembrane helix (30). More detailed calculations support the near coincidence of the twofold axis with the membrane normal (10). Accordingly, the membrane-spanning region of the RC from Rb. sphaeroides is approximately 40 Å wide, and extends from the Fe atom on one side (cytoplasmic), to a position approximately 10–15 Å beyond the center of the dimer on the opposite (periplasmic) side.

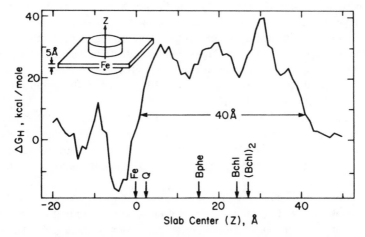

Figure 5 The energy, ΔG_H (Eq. 3), required to transfer a 5 Å thick section of the RC from the membrane to water for different positions (in 1 Å increments) of the RC. The normal of the section is parallel to the twofold symmetry axis z of the RC, as indicated schematically (inset). The projected locations of the centers of the cofactors onto the z axis are indicated. The position of the Fe was arbitrarily chosen as zero. Arrow (40 Å) indicates the membrane-spanning region. The energy calculations were performed on the experimentally determined three-dimensional structure of the RC from *Rb. sphaeroides*. Figure modified from Ref. 10.

What exactly does this 40 Å wide region represent? The Eisenberg & McLachlan expression for ΔG_H estimates the free energy of transfer from a nonaqueous to a water environment (66). Hence, the 40 Å wide slab should represent the total extent of the RC surface shielded from water. This would include both the region of the RC in contact with the fatty acid tails, as well as the region in contact with the polar head groups. It might seem surprising that the region of the RC in contact with the polar head groups would exhibit a positive ΔG_H for transfer to water, since the head groups themselves might be expected to resemble water more closely than an apolar solvent. As a relevant model, it is instructive to consider the interactions between carbohydrate and protein in glycoproteins. X-ray structures of glycoproteins (67, 68) have shown that the carbohydrate chains interact directly with parts of the protein surface, shielding those regions from exposure to water. Although carbohydrate groups are highly polar (69), they cover regions of the protein surface that have significant numbers of hydrophobic residues (67, 68). Apparently, even these polar molecules have nonpolar surface regions that interact preferentially with hydrophobic residues of the protein (69a). Consequently, a positive ΔG_H for the region of the RC in contact with the polar head groups seems plausible. Since the head group layer has a thickness of 5 Å (68), this region of the RC surface would consist of surface groups with z values of approximately 0–5 Å and 35–40 Å, with the convention illustrated in Figure

5. The region of the RC interacting with the nonpolar, fatty acid tail part of the bilayer of the RC will then extend over $z = 5$–35 Å.

The accuracy with which the membrane-spanning region of the RC can be identified depends critically on the assumptions made in the analysis. These assumptions fall into three general categories (10):

1. The membrane may be approximated by a planar slab with uniform thickness on all sides of the RC.

2. There is a sharp interface between the membrane and the aqueous solution, as well as between the RC and the membrane. Interactions between the RC and other proteins (such as the antennae complex) are neglected in this model. Similarly, the possible penetration of water into the bilayer region has been neglected.

3. Only hydrophobic energies are explicitly considered in this model. Other contributions to membrane–protein energetics, including electrostatic image charges arising at the protein–membrane–water boundaries, have been neglected.

A measure of the validity of these assumptions can be obtained by comparing the calculated width of the membrane with the experimentally determined values. Small angle X-ray scattering studies of *Rb. sphaeroides* vesicles yield a membrane thickness of 45 ± 5 Å (70). Likewise, small angle scattering studies of vesicles containing vaccenic acid [the major fatty acid component in the membrane of *Rb. sphaeroides* (71)] indicate that the total membrane thickness of these vesicles is 38 Å (72). The agreement between the calculated and observed values for membrane thickness indicates that the assumptions made in this analysis are reasonable approximations.

Identification of Residues Exposed to the Membrane by Sequence Analysis

In addition to the position of the transmembrane helixes in the membrane discussed in the previous section, the characterization of residues in contact with the lipid bilayer is also important for structural analyses of membrane proteins (10). Figure 6 depicts the approximate position of Cα atoms of residues of the 11 transmembrane helixes in the membrane. As described above, the nonpolar region of the bilayer extends over $z = 5$–35 Å, while the head groups are located in the regions $z = 0$–5 Å and $z = 35$–40 Å. Most of the charged groups found within the membrane-spanning region are contained in the head group zones.

Residues on each helix that are in contact with the membrane bilayer were identified by tabulating the accessible surface area for each residue (10). Residues with more than half of their accessible surface area exposed to the membrane are circled in Figure 6, while residues with 20–50% of their surface area exposed to the membrane are capped with a semicircular arc. The

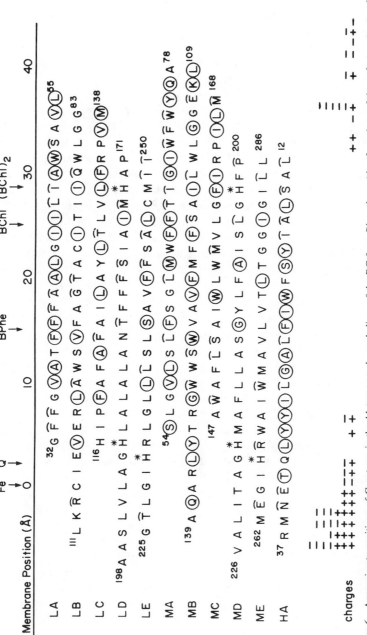

Figure 6 Approximate positions of Cα atoms in the 11 transmembrane helices of the RC from *Rb. sphaeroides*. Locations of the cofactors (*top*) and all charged residues (*bottom*) of the RC are indicated. Residues in circles have >50% of their surface area exposed to the membrane, while capped residues have 20–50% of their surface area exposed to the membrane. The remaining residues are primarily buried inside the protein. Histidine residues marked with an asterisk are ligands to Fe or Bchl. Amino acids are designated by the single-letter code. Figure modified from Ref. 10.

remaining residues are primarily buried inside the RC. Helixes on the periphery of the RC (A, B, and C) have more residues exposed to the membrane than the core helixes (D and E). In many instances, residues exposed to the membrane are spaced at multiples of three to four residues, which corresponds to the repeat distance of the α-helix. This periodicity will be examined more quantitatively in a later section.

The average amino acid composition may be determined for the membrane-spanning α-helixes for the RCs from *Rb. sphaeroides, Rb. capsulatus, R. rubrum*, and *Rps. viridis*. A total of 808 residues are in the membrane-spanning regions of the A, B, C, D, and E helixes of the L and M subunits of these four RCs (as aligned in Ref. 13). This analysis is restricted to residues whose $C\alpha$ atoms are in the nonpolar region of the bilayer (the z values of the $C\alpha$ atoms are in the range 5 Å $< z <$ 35 Å in the convention of Figures 5 and 6). Assignment of residue location in the membrane is based on the RC structure from *Rb. sphaeroides*, and is assumed to remain valid for the other three purple bacteria. The amino acid composition, in order of decreasing abundance (in percent of total residues in the indicated region) is: Leu(15%), Ala(14%), Phe(12%), Ile(10%), Gly(9%), Val(7%), Trp(6%), Ser(6%), Thr(5%), Met(4%), Pro(3%), Arg(2%), Cys(2%), Tyr(1%), His(1%), Asn(1%), Gln(1%), Glu(1%), Asp(0%), Lys(0%). As expected, there are a large number of apolar residues, and only very few charged residues, in the membrane-spanning region of the RC.

The distribution of residues between different environments present within the membrane may also be analyzed from the sequence and structural alignments. Different amino acids exhibit different preferences between the exposed surface positions and the buried interior sites. Of the most abundant amino acids in the membrane, the apolar residues Leu, Ile, Phe, and Val tend to be located on the side of the helix exposed to the membrane, whereas Trp, Thr, and Ser, show no particular preference between the interior and surface sides. Ala and Gly prefer to be located on the helix side facing the protein interior. Surface-facing residues are defined as having >20% of their surface area exposed to the membrane in the RC structure of *Rb. sphaeroides*.

Comparison of aligned sequences from *Rb. sphaeroides, Rb. capsulatus, Rps. viridis*, and *R. rubrum* indicates that 35% (71/202) of residues in the transmembrane helixes of the L and M subunits are identical in all four sequences (10, 13). Again, this analysis considers only residues in the nonpolar region of the bilayer. Significant variation in the pattern of sequence conservation between buried and membrane-exposed residues of the transmembrane helixes (Figure 6) is observed (10). 46% (52/112) of all buried residues are identical in all four sequences, whereas only 10% (5/50) of residues with more than half of their area exposed to the membrane are conserved. This suggests that fewer restrictions are placed on residues that are

j	1 2 3	4 5 6	7 8 9	10 11 12	13 14 15	16 17 18	19 20 21
Rb. sphaeroides	SLG	VLS	LFS	GLM	WFF	TIG	IWF
Rb. capsulatus	IAG	TVS	LAF	GAA	WFF	TIG	VWY
R. rubrum	TTG	VLS	LVF	GFF	AIE	IIG	FNL
Rps. viridis	ASG	IAA	FAF	GST	AIL	IIL	FNM
Surface Exp	+	+ +	+	+	+ +	+	+

Vj	4 4 1	3 3 2	2 3 2	1 4 4	2 2 3	2 1 2	3 2 4

Figure 7 Calculation of the variability profile for the region of the MA helix in the nonpolar region of the bilayer. The sequence alignment was taken from Ref. 13. The relative residue position in this alignment is indicated by j; V_j is the variability index (number of different amino acids) for position j; and a surface exposure of "+" indicates that more than 50% of the surface area of the residue is exposed to the membrane in the RC from *Rb. sphaeroides*.

exposed to the membrane, indicating that there are few specific interactions between protein and lipid. The high tolerance to substitution of residues exposed to the membrane is analogous to the situation in globular proteins, whose surface residues also have a higher tolerance to substitutions than the buried residues (73, 74).

The periodicity of residues in a surface α-helical structure that are exposed to the membrane, coupled with the increased sequence variability of exposed residues, suggests the possibility of identifying exposed residues by analyzing the sequence alignments of homologous proteins. Assuming (*a*) that the sequence represents a transmembrane helix and (*b*) that the helix is positioned at the protein surface of the membrane-spanning region, then residues in contact with the bilayer may be identified from the pattern of hypervariable positions occurring with a periodicity of about 3.6 residues in a family of sequence alignments.

Fourier transform methods provide a quantitative approach for characterizing the periodicity of conserved and variable residues in a family of aligned sequences (13). The first step in this process is to construct a variability profile, V, for a particular family of sequences. The V_j element of this profile is defined by the number of different types of amino acid residues that are observed at a given position j in a family of aligned sequences. Construction of V for the MA helix of the RC is illustrated in Figure 7. Qualitative inspection of this profile suggests that more variable positions are associated with membrane exposed positions. To search for periodicities in V, the Fourier transform power spectra, $P(\omega)$, of V is calculated:

$$P(\omega) = \left[\sum_{j=1}^{N} (V_j - \bar{V}_j) \cos(j\omega) \right]^2 + \left[\sum_{j=1}^{N} (V_j - \bar{V}_j) \sin(j\omega) \right]^2 \qquad 4.$$

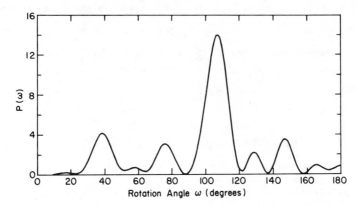

Figure 8 The Fourier transform power spectrum, P(ω) (Eq. 4), calculated for the variability profile of the transmembrane MA helix (Figure 7). The peak at $\omega = 105°$ corresponds to a periodicity of 3.4 residues/turn, and is consistent with an α-helical conformation for this sequence.

where N is the number of residues in the sequence; ω is the angular rotation angle between residues around a helical axis (it equals 100° for an ideal helix); and $\overline{V}_j$ is the mean value of V_j for the entire sequence. Similar expressions have been used by Eisenberg et al (75) and Cornette et al (76) to describe periodicity in the hydrophobicity profiles of proteins.

The P(ω) curve calculated from the variability profile for the MA helix is illustrated in Figure 8. The prominent peak near $\omega = 105°$ corresponds approximately to the periodicity of an ideal α-helix. Most of the P(ω) curves for the RC transmembrane helixes exhibit maxima at values of ω somewhat larger than the 100° expected for an ideal α-helix. In contrast, the periodicity in the hydrophobicity index of residues in helixes from soluble proteins corresponds to a value $\omega = 97.5°$ (76). The larger value of ω observed for membrane helixes may represent a slight overwinding of the helix, or more plausibly, a systematic shift in exposed residues due to interactions with adjacent helixes. For example, $\omega = 103°$ describes the periodicity of exposed residues in a coiled coil pair of α-helixes (77).

The α-helical character of the P(ω) curve may be described by the parameter ψ (13), which is defined by the average value of P(ω) in the α-helical range ($90° < \omega < 120°$), relative to the average value of P(ω) over the entire range:

$$\psi = \left[(1/30) \int_{90°}^{120°} P(\omega)d\omega \right] \bigg/ \left[(1/180) \int_{0°}^{180°} P(\omega)d\omega \right] \qquad 5.$$

Larger values of ψ correspond to a greater fraction of the $P(\omega)$ curve in the α-helical region. The following values of ψ were found for the A, B, C, D, and E helixes, respectively: 2.3, 2.9, 1.9, 1.6, and 0.9. These values of ψ were calculated using sequence alignments for the RCs from *Rb. sphaeroides, Rb. capsulatus, Rps. viridis,* and *R. rubrum,* and combining both the L and M subunit sequences. The more peripheral helixes (A and B) have larger values of ψ than the core helixes (D and E). This is consistent with the analysis that membrane-exposed residues are more poorly conserved than buried residues. These results show that ψ provides a measure of the surface exposure of a helix. This might prove useful in deriving additional information about the three-dimensional structure of membrane proteins from sequence data.

The enhanced variability of residues on the surface of an α-helix that is exposed to solvent, compared to those that face the interior side, provides an approach for identifying the topology of membrane-spanning helixes. First, the variability profile is constructed from aligned sequences of the helical regions. Next, the residue positions with greatest variability consistent with an α-helical periodicity are determined by fitting a cosine curve with $\omega = 100°$ to the variability profile. The residue positions for which this Fourier series has the greatest amplitude correspond to the most variable positions. Calculation of these positions for the 11 RC transmembrane helixes shows a strong correlation between the most variable positions and the exposed positions illustrated in Figure 6. Consequently, it is possible to assign surface-exposed sides of helixes on the basis of sequence conservation alone, without consideration of the chemical nature of the different amino acids.

The variability profile may also be used to predict the presence of α-helical segments, which are usually identified from hydropathy plots or hydrophobic moment analyses (29, 30, 35). The procedure is as follows: ψ values are calculated for a sequence contained within a window of defined size (typically 11–19 residues long). This window is moved along the sequence one residue at a time; at each position the value of ψ is determined. Regions of high ψ values (greater than ~2) correspond to a sequence that shows a strong helical periodicity and hence can be associated with a surface helix. A plot of ψ vs residue number for an 11 sequence alignment of homologous RC proteins (21) is illustrated in Figure 9. Sequence regions corresponding to the A and B helixes are evident as regions of high ψ. Local peaks in ψ are also associated with the C and D helixes, as well as the helical ab and cd segments of the interrupted helix on the periplasmic surface of the RC. Thus, this method can be used to make surface helical assignments, without any consideration of the chemical nature of the different amino acids. The requirement for surface helixes implies that this method is not applicable to α-helixes that are either completely buried inside protein or that are completely surrounded by lipid. Although we discussed only α-helixes in membrane proteins, these methods may also be applicable to soluble proteins, and to the characterization of surface-exposed β-sheets.

An experimental approach for determining the variability profile of a sequence that does not require a number of homologous sequences has recently been described (78). Techniques of site-directed mutagenesis provide experimental procedures to determine the number of different amino acids that are tolerated at a given sequence position. Positions that accept only a small number of different amino acids are classified as having a low variability index, while positions that accept a large number of different amino acids have a high variability index. In studies on the λ repressor (a water-soluble protein), the number of substitutions that were allowed at a specific sequence position was approximately proportional to the surface exposure of that residue. This conclusion, as well as the observation that "α-helical and β-strand regions might be recognized by characteristic patterns" of high and low variability (78), is consistent with the above observations derived from sequence alignments of homologous proteins.

CONCLUDING REMARKS

A knowledge of the structures of bacterial RCs represents only the first stage in understanding the folding and properties of membrane proteins. Important questions remain concerning the actual folding and assembly mechanism for the RC in the bacterial membrane and the generality of the conclusions

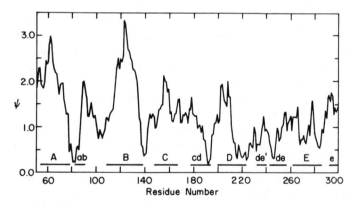

Figure 9 Calculation of ψ (Eq. 5) for a sliding window of 19 residues moving along an alignment of 11 homologous RC sequences (from Ref. 21). The sequences include both L and M subunits from bacterial RCs, and the related D1 and D2 proteins from plant photosystem II. Residue numbers correspond to the sequence of the M subunit from *Rb. sphaeroides*. Location in the sequence of α-helixes are indicated by the labeled horizontal bars. ψ is a measure of the preferential conservation of residues on one side of a (surface) helix. Large peaks for the A and B helixes are consistent with their location on the periphery of the transmembrane region of the RC.

described here for the folding and structure of other membrane proteins. The next best characterized membrane protein, bacteriorhodopsin, is believed to have a polar interior (79), in contrast to the apolar interior of the membrane-spanning region of the RCs. Whether RCs and bacteriorhodopsin represent two distinct structural motifs, or are simply limiting cases of a variety of intermediate cases, can only be decided as other high-resolution structures of membrane proteins become available. As with soluble proteins, a variety of folding patterns for membrane proteins are anticipated. The structure of the outer membrane protein porin, for which diffraction quality crystals are available, should be of special interest since this protein is believed to have a predominantly β-sheet organization (80), in contrast to the helical structures that have so far been described. Recent progress in structural and molecular biology techniques holds great promise for rapid progress in the characterization of other membrane proteins.

ACKNOWLEDGMENTS

The authors thank D. Eisenberg, L. DeAntonio, A. Chirino, J. Deisenhofer, W. DeGrado, and D. C. Wiley for discussions and comments. Research in the laboratories of the authors has been supported by grants from NIH and NSF. DCR is an A. P. Sloan research fellow.

Literature Cited

1. Michel, H. 1982. *J. Mol. Biol.* 158:567–72
2. Deisenhofer, J., Epp, O., Miki, K., Huber, R., Michel, H. 1984. *J. Mol. Biol.* 180:385–98
3. Deisenhofer, J., Epp, O., Miki, K., Huber, R., Michel, H. 1985. *Nature* 318:618–24
4. Michel, H., Epp, O., Deisenhofer, J. 1986. *EMBO J.* 5:2445–51
5. Deisenhofer, J., Michel, H. 1988. See Ref. 33, pp. 1–3
6. Allen, J. P., Feher, G. 1984. *Proc. Natl. Acad. Sci. USA* 81:4795–99
6a. Feher, G. 1983. *Meet. Biophys. Soc. (natl. lecture), Feb. 13–16, 1983, San Diego, Calif*
7. Allen, J. P., Feher, G., Yeates, T. O., Rees, D. C., Deisenhofer, J., et al. 1986. *Proc. Natl. Acad. Sci. USA* 83:8589–93
8. Allen, J. P., Feher, G., Yeates, T. O., Komiya, H., Rees, D. C. 1987. *Proc. Natl. Acad. Sci. USA* 84:5730–34
9. Allen, J. P., Feher, G., Yeates, T. O., Komiya, H., Rees, D. C. 1987. *Proc. Natl. Acad. Sci. USA* 84:6162–66
10. Yeates, T. O., Komiya, H., Rees, D. C., Allen, J. P., Feher, G. 1987. *Proc. Natl. Acad. Sci. USA* 84:6438–42
11. Yeates, T. O., Komiya, H., Chirino, A., Rees, D. C., Allen, J. P., Feher, G. 1988. *Proc. Natl. Acad. Sci. USA* 85:7993–97
12. Allen, J. P., Feher, G., Yeates, T. O., Komiya, H., Rees, D. C. 1988. *Proc. Natl. Acad. Sci. USA* 85:8487–91
13. Komiya, H., Yeates, T. O., Rees, D. C., Allen, J. P., Feher, G. 1988. *Proc. Natl. Acad. Sci. USA* 85:9012–16
14. Allen, J. P., Feher, G., Yeates, T. O., Komiya, H., Rees, D. C. 1988. See Ref. 33, pp. 5–11
15. Chang, C.-H., Schiffer, M., Tiede, D. M., Smith, U., Norris, J. 1985. *J. Mol. Biol.* 186:201–3
16. Chang, C.-H., Tiede, D., Tang, J., Smith, U., Norris, J. R., Schiffer, M. 1986. *FEBS Lett.* 205:82–86
17. Tiede, D. M., Budil, D. E., Tang, J., El-Kabbani, O., Norris, J. R., et al. 1988. See Ref. 33, pp. 13–20
18. Okamura, M. Y., Feher, G., Nelson, N. 1982. In *Photosynthesis,* ed. Govindjee, pp. 195–272. New York: Academic
19. Williams, J. C., Steiner, L. A., Ogden,

R. C., Simon, M. I., Feher, G. 1983. *Proc. Natl. Acad. Sci. USA* 80:6505–9
20. Williams, J. C., Steiner, L. A., Feher, G., Simon, M. I. 1984. *Proc. Natl. Acad. Sci. USA* 81:7303–7
21. Williams, J. C., Steiner, L. A., Feher, G. 1986. *Proteins* 1:312–25
22. Youvan, D. C., Bylina, E. J., Alberti, M., Begusch, H., Hearst, J. E. 1984. *Cell* 37:949–57
23. Michel, H., Weyer, K. A., Gruenberg, H., Lottspeich, F. 1985. *EMBO J.* 4:1667–72
24. Michel, H., Weyer, K. A., Gruenberg, H., Dunger, I., Oesterhelt, D., Lottspeich, F. 1986. *EMBO J.* 5:1149–58
25. Bélanger, G., Bérard, J., Corriveav, P., Gingras, G. 1988. *J. Biol. Chem.* 263:7632–38
26. Kauzmann, W. 1959. *Adv. Protein Chem.* 14:1–63
27. Singer, S. J. 1962. *Adv. Protein Chem.* 17:1–68
28. Henderson, R., Unwin, P. N. T. 1975. *Nature* 257:28–32
29. Eisenberg, D. 1984. *Annu. Rev. Biochem.* 53:595–623
30. Engelman, D. M., Steitz, T. A., Goldman, A. 1986. *Annu. Rev. Biophys. Biophys. Chem.* 15:321–53
31. Michel-Beyerle, M. E., ed. 1985. *Antennas and Reaction Centers of Photosynthetic Bacteria.* Berlin: Springer-Verlag
32. Biggens, J., ed. 1987. *Progress in Photosynthesis Research.* Dordrecht: Martinus Nijhoff
33. Breton, J., Vermeglio, A., eds. 1988. *The Photosynthetic Bacterial Reaction Center.* New York: Plenum
34. Kirmaier, C., Holten, D. 1987. *Photosynthesis Res.* 13:225–60
35. Kyte, J., Doolittle, R. F. 1982. *J. Mol. Biol.* 157:105–32
36. Garavito, R. M., Rosenbusch, J. P. 1980. *J. Cell Biol.* 86:327–29
37. Michel, H. 1983. *Trends Biochem. Sci.* 8:56–59
38. Garavito, R. M., Rosenbusch, J. P. 1986. *Methods Enzymol.* 125:309–28
38a. Allen, J. P., Feher, G. 1989. In *Crystallization of Membrane Proteins,* ed. H. Michel. Boca Raton: CRC. In press
39. Frank, H. A., Taremi, S. S., Knox, J. R. 1987. *J. Mol. Biol.* 198:139–41
40. Ducruix, A., Reiss-Husson, F. 1987. *J. Mol. Biol.* 193:419–21
41. Jones, T. A. 1985. *Methods Enzymol.* 115:157–71
42. Lesk, A. M., Hardman, K. D. 1982. *Science* 216:539–40
43. Bachofen, R., Wiemken, V. 1986. In *Photosynthesis III,* ed. L. A. Stahelin, C. J. Artnzen, pp. 620–31. Berlin: Springer-Verlag
44. McElroy, J. D., Feher, G., Mauzerall, D. C. 1969. *Biochim. Biophys. Acta* 172:180–83
45. Norris, J. R., Uphaus, R. A., Crespi, H. L., Katz, J. J. 1971. *Proc. Natl. Acad. Sci. USA* 68:625–28
46. Feher, G., Hoff, A. J., Isaacson, R. A., Ackerson, L. C. 1975. *Ann. NY Acad. Sci.* 244:239–59
47. Norris, J. R., Scheer, H., Katz, J. J. 1975. *Ann. NY Acad. Sci.* 244:260–80
48. Knapp, E. W., Fischer, S. F., Zinth, W., Sander, M., Kaiser, W., et al. 1985. *Proc. Natl. Acad. Sci. USA* 82:8463–67
49. Debus, R. J., Feher, G., Okamura, M. Y. 1985. *Biochemistry* 24:2488–500
50. Richards, F. M. 1977. *Annu. Rev. Biophys. Bioeng.* 6:151–76
51. Israelachvili, J. N. 1985. *Intermolecular and Surface Forces.* New York: Academic
52. Lee, B., Richards, F. M. 1971. *J. Mol. Biol.* 55:379–400
53. Richmond, T. J., Richards, F. M. 1978. *J. Mol. Biol.* 119:537–55
54. Miller, S., Lesk, A. M., Janin, J., Chothia, C. 1987. *Nature* 328:834–36
55. Mandelbrot, B. B. 1983. *The Fractal Geometry of Nature.* San Francisco: Freeman
56. Lewis, M., Rees, D. C. 1985. *Science* 230:1163–65
57. Connolly, M. 1983. *J. Appl. Crystallogr.* 16:548–58
58. Bernstein, F., Koetzle, T. F., Williams, G. J. B., Meyer, E. F., Brice, M. D., et al. 1977. *J. Mol. Biol.* 112:535–42
59. Richards, F. M. 1974. *J. Mol. Biol.* 82:1–14
60. Finney, J. L. 1975. *J. Mol. Biol.* 96:721–32
61. Rashin, A. A., Iofin, M., Honig, B. 1986. *Biochemistry* 25:3619–25
62. Hol, W. G. J. 1985. *Prog. Biophys. Mol. Biol.* 45:149–95
63. Pierson, B. K., Thornber, J. P., Seftor, R. E. B. 1983. *Biochim. Biophys. Acta* 723:322–26
64. Gibbs, J. W. 1928. *Collected Works of J. W. Gibbs.* New York: Longmans
65. Berkovitch-Yellin, Z. 1985. *J. Am. Chem. Soc.* 107:8239–53
65a. Roth, M., Lewit-Bentley, A., Michel, H., Diesenhofer, J., Huber, R. 1988. *Am. Crystallogr. Assoc. Meet., Philadelphia, Pa.* Abstr. PJ34
66. Eisenberg, D., McLachlan, A. D. 1986. *Nature* 319:199–203

67. Deisenhofer, J. 1981. *Biochemistry* 20:2361–70
68. Wilson, I. A., Skehel, J. J., Wiley, D. C. 1981. *Nature* 289:366–73
69. Wolfenden, R., Liang, Y-L. 1988. *J. Biol. Chem.* 263:8022–26
69a. Vyas, N. K., Vyas, M. N., Quiocho, F. A. 1988. *Science* 242:1290–95
70. Pape, E. H., Menke, W., Weick, D., Hosemann, R. 1974. *Biophys. J.* 14:221–32
71. Marinetti, G. V., Cattieu, K. 1981. *Chem. Phys. Lipids* 28:241–51
72. Lewis, B. A., Engelman, D. M. 1983. *J. Mol. Biol.* 166:211–17
73. Smith, E. L. 1968. *Harvey Lect.* 62:231–56
74. Chothia, C., Lesk, A. M. 1986. *EMBO J.* 5:823–26
75. Eisenberg, D., Weiss, R. M., Terwilliger, T. 1984. *Proc. Natl. Acad. Sci. USA* 81:140–44
76. Cornette, J. L., Cease, K. B., Margalit, H., Spouge, J. L., Berzofsky, J. A., DeLisi, C. 1987. *J. Mol. Biol.* 195:659–85
77. Crick, F. H. C. 1953. *Acta Crystallogr.* 6:689–97
78. Reidhaar-Olson, J. F., Sauer, R. T. 1988. *Science* 241:53–57
79. Engelman, D. M., Zaccai, G. 1980. *Proc. Natl. Acad. Sci. USA* 77:5894–98
80. Rosenbusch, J. P. 1974. *J. Biol. Chem.* 249:8019–29

Annu. Rev. Biochem. 1989. 58:635–69

PHOSPHOLIPID BIOSYNTHESIS IN YEAST

George M. Carman

Department of Food Science, Cook College, New Jersey Agricultural Experiment Station, Rutgers University, New Brunswick, New Jersey 08903

Susan A. Henry

Department of Biological Sciences, Carnegie Mellon University, Pittsburgh, Pennsylvania 15213

CONTENTS

0066-4154/89/0701-0635$02.00

INTRODUCTION

The yeast *Saccharomyces cerevisiae* has become an organism of choice for biochemical, molecular, and genetic dissections of the regulation of phospholipid biosynthesis. Yeast is as easily grown as most bacteria, and *S. cerevisiae* has been used in classical genetic studies for decades. In recent years, a powerful molecular genetics has been developed, making it possible to isolate at will virtually any gene that has previously been identified by mutation. These techniques, coupled with technologies that enable the cloning of virtually any gene for which the gene product has been purified, have lead to an explosive increase in the rate of isolation of structural and regulatory genes in this organism in the last few years. *S. cerevisiae* has another powerful advantage for such molecular studies. In *S. cerevisiae,* targeted integrative transformation is accomplished with ease, allowing the investigator to manipulate cloned DNA in vitro and then transform it back into the organism at a desired location in the genome (1). In virtually all other eukaryotic organisms, the process of integrative transformation is much more random, giving the investigator little control. Thus, the potential for genetic engineering in *S. cerevisiae* is unmatched in the eukaryotic world.

From the point of view of an analysis of the biosynthesis and regulation of phospholipids, *S. cerevisiae* has additional attractions. As is discussed in this review, the basic pathways of phospholipid metabolism in *S. cerevisiae* are, with few exceptions, similar to those found in higher eukaryotes, and these pathways are increasingly well characterized. Yeast also displays the intracellular compartmentalization and subcellular membranes typical of eukaryotes, important features because of the increasing evidence for specificity of localization of the phospholipid biosynthetic enzymes. Finally, *S. cerevisiae* exhibits an unusual plasticity in its tolerance for alteration in its phospholipid composition (2), facilitating the isolation of nonconditional as well as conditional mutants with lesions in the phospholipid biosynthetic pathways.

PHOSPHOLIPID BIOSYNTHETIC PATHWAYS IN YEAST

The major phospholipids found in *S. cerevisiae* membranes are phosphatidylcholine (PC), phosphatidylethanolamine (PE), phosphatidylinositol (PI), and phosphatidylserine (PS) (3, 4). Mitochondrial membranes also contain phosphatidylglycerol (PG) and cardiolipin (CL) (3, 4). The synthesis of phospholipids in yeast is carried out by enzymes common to higher eukaryotes, with the exception of the synthesis of PS. In yeast, PS is synthesized from CDP-diacylglycerol (CDP-DG) and serine (5), whereas in higher eukaryotes (6, 7), PS is synthesized by an exchange reaction between PE and serine. The biosynthetic pathways for the synthesis of phospholipids in *S. cerevisiae* were identified primarily by Lester and coworkers (8–11), and are shown in Figure 1. As in higher eukaryotes (6, 7), PE and PC are synthesized in yeast by two alternative pathways (3). In both pathways PE and PC are derived from phosphatidate (PA). In the primary pathway of phospholipid synthesis, PE and PC are derived from PA by the reaction sequence PA → CDP-DG → PS → PE → phosphatidylmonomethylethanolamine (PMME) → phosphatidyl-dimethylethanolamine (PDME) → PC. An auxillary pathway exists in yeast for PE and PC synthesis, which is used by the ethanolamine- and choline-requiring mutants *(chol/pss)* defective in the synthesis of PS (12–15). These mutants synthesize PE and PC by the CDP-ethanolamine- and CDP-choline-based pathways (16) by the reaction sequences PA → diacylglycerol → PE and PA → diacylglycerol → PC. PI and the mitochondrial phospholipids PG and CL are also derived from CDP-DG (Figure 1). As in animal cells (17), PI in yeast is the precursor of the polyphosphoinositides PI 4-phosphate (PIP) and PI 4,5-bisphosphate (PIP_2).

The phospholipid biosynthetic enzymes in *S. cerevisiae* are localized in the cytosolic, mitochondrial, and microsomal fractions of the cell. Glycerophos-phate acyltransferase, CDP-DG synthase, PS synthase, PI synthase, and cholinephosphotransferase (Figure 1, reactions 1–3, 16, 10, respectively) are associated with the outer mitochondrial and microsomal membranes (18). PA phosphatase (Figure 1, reaction 14) is associated with the cytosolic, mitochondrial, and microsomal fractions of the cell (19, 20). PS de-carboxylase and phosphatidylglycerophosphate (PGP) synthase (Figure 1, reactions 4 and 17, respectively) are strictly inner mitochondrial membrane-associated enzymes (18). The PE methyltransferases (Figure 1, reactions 5–7) are strictly microsomal-associated enzymes (18). Inositol 1-phosphate (I-1-P) synthase is associated with the cytosolic fraction of the cell (21). The loca-tions of the phospholipid biosynthetic enzymes not mentioned in this section have not been determined.

PHOSPHOLIPID BIOSYNTHETIC ENZYMES PURIFIED FROM YEAST

The phospholipid biosynthetic enzymes must be purified to homogeneity to fully understand the mechanisms responsible for the synthesis of phospholip-ids and their function in yeast membranes. The kinetic analysis and reaction mechanism, an understanding of substrate specificity, and modulation of activity of pure enzymes are important to understanding the mode of action and control of these enzymes in vivo. In addition, pure enzymes are needed to prepare antibodies to be used as probes to study the regulation of enzyme

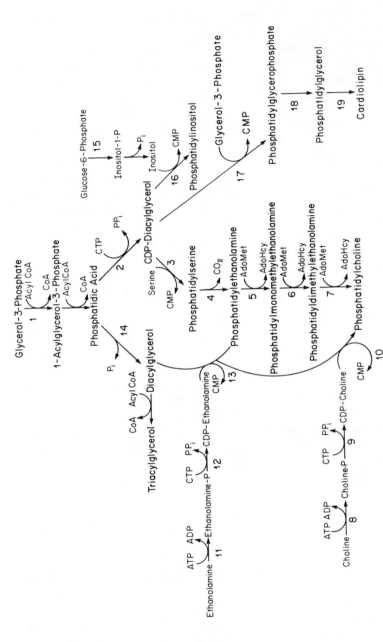

Figure 1   Phospholipid biosynthetic pathways in *S. cerevisiae*. The indicated reactions are catalyzed by the following enzymes: 1. glycerol-3-phosphate acyltransferase; 2. CDP-DG synthase; 3. PS synthase; 4. PS decarboxylase; 5. PE methyltransferase; 6 and 7. phospholipid methyltransferase; 8. choline kinase; 9. cholinephosphate cytidylyltransferase; 10. cholinephosphotransferase; 11. ethanolamine kinase; 12. ethanolaminephosphate cytidylyltransferase; 13. ethanolaminephosphotransferase; 14. PA phosphatase; 15. I-1-P synthase; 16. PI synthase; 17. PGP synthase; 18. PGP phosphatase; 19. CL synthase.

formation in vivo. The purification and properties of the phospholipid biosynthetic enzymes purified from yeast are discussed in the next sections.

Purification and Properties of CDP-DG Synthase

CDP-DG synthase (CTP:phosphatidate cytidylyltransferase, EC 2.7.7.41) catalyzes the formation of CDP-DG and PP_i from CTP and PA (22). The study of this enzyme is important because the product of the reaction, CDP-DG, is the source of the phosphatidyl moiety in the primary route of synthesis of the major phospholipids PC, PE, PI, and PS. Kelley & Carman (23) have purified CDP-DG synthase to apparent homogeneity from the mitochondrial fraction of *S. cerevisiae*. The purification scheme includes Triton X-100 solubilization of mitochondrial membranes, affinity chromatography with CDP-DG-Sepharose, and hydroxylapatite chromatography. A summary of the purification scheme is presented in Table 1. The overall purification of CDP-DG synthase over the cell extract using this scheme is 2309-fold, with an activity yield of 32%. Successful purification of CDP-DG synthase is absolutely dependent on the stabilization of enzyme activity during solubilization with Triton X-100 and following chromatography with hydroxylapatite. Enzyme activity is stabilized during solubilization with Triton X-100 by including potassium and magnesium ions in the solubilization buffer. The hydroxylapatite-purified enzyme is stabilized by the addition of the substrate CTP to the purified enzyme preparation. The major purification (162-fold relative to the Triton X-100 extract) of the enzyme is achieved by the CDP-DG-Sepharose chromatography step. The CDP-DG-Sepharose affinity resin was originally developed by Dowhan and coworkers (24, 25) for the purification of PGP synthase from bacteria. The CDP-DG-Sepharose resin used for CDP-DG synthase purification is the high-capacity affinity resin prepared as described by Fischl & Carman (26). Effective binding of the enzyme to the affinity resin occurs because the CDP-DG synthase reaction is favored in the reverse direction (23, 27). Binding of the Triton X-100–solubilized enzyme to the affinity resin is dependent on the presence of the cofactor magnesium and Triton X-100 in the chromatography buffer. Elution of the enzyme from the resin is dependent on both the substrate CTP and NaCl in the elution buffer. The purification procedure yields a nearly homogeneous protein preparation as shown by native and SDS-polyacrylamide gel electrophoresis. In the presence of 1 mM CTP, the purified enzyme is 90–100% stable for approximately six months and 50% stable for one year of storage at −80°C.

A radiation inactivation analysis (28, 29) of the native mitochondrial-bound enzyme and purified enzyme suggests that a functional CDP-DG synthase has a target size (or native molecular weight) of 114,000 (23). Based on the results of SDS-polyacrylamide gel electrophoresis, the native enzyme appears

Table 1 Purification of CDP-DG synthase from *S. cerevisiae*[a]

Purification step	Total units (nmol/min)	Protein (mg)	Specific activity (units/mg)	Purification (-fold)	Yield (%)
1. Cell extract	2700	4427	0.61	1.0	100
2. Mitochondria	2106	744	2.83	4.64	78
3. Triton X-100 extract	1602	300	5.33	8.74	59
4. CDP-DG-Sepharose	1362	1.57	868	1422	50
5. Hydroxylapatite	866	0.61	1409	2309	32

[a] The data are based on starting with 50 g (wet weight) of cells. Data are from Kelley & Carman (23).

to be a dimer composed of identical subunits with a molecular weight of 56,000 (23). Maximum CDP-DG synthase activity is dependent on $MgCl_2$ and Triton X-100 at pH 6.5. The energy of activation is 9 kcal/mol, the enzyme is labile above 30°C, and thioreactive agents inhibit activity. CDP-DG synthase follows typical saturation kinetics toward CTP and PA when activity is measured with a mixed micelle substrate of Triton X-100 and PA (23). The true K_m values for CTP and PA are 1 mM and 0.5 mM, respectively, and the V_{max} is 4.7 μmol/min/mg. The results of kinetic experiments and the ability of the enzyme to catalyze a variety of isotopic exchange reactions suggest that CDP-DG synthase catalyzes a sequential Bi Bi reaction. CDP-DG synthase binds to CTP prior to PA, and PPi is released prior to CDP-DG in the reaction sequence (23). dCTP is both a substrate (23) and a competitive inhibitor of the enzyme (23, 27).

Purification and Properties of PS Synthase

PS synthase (CDP-diacylglycerol:L-serine *O*-phosphatidyltransferase, EC 2.7.8.8) catalyzes the formation of PS by displacing CMP from CDP-DG with serine (30). The product of the reaction, PS, is the precursor to the major phospholipids PE and PC in the primary biosynthetic pathway. Mutants defective in PS synthase show abnormal patterns of phospholipid synthesis and physiological properties (12, 13, 31). PS synthase has been purified to apparent homogeneity from the microsomal fraction of wild-type *S. cerevisiae* (5). Successful purification of PS synthase requires the solubilization of the enzyme from the microsomal fraction with Triton X-100, CDP-DG-Sepharose affinity chromatography, and ion-exchange chromatography with DEAE-cellulose. A summary of the purification of PS synthase is presented in Table 2. Solubilization of PS synthase from microsomes with 1% Triton X-100 is dependent on the presence of $MnCl_2$ in the solubilization buffer (26, 32). The major purification of the enzyme is attained by affinity chromatography with CDP-DG-Sepharose. Binding of the enzyme to the affinity resin is dependent on Triton X-100 and $MnCl_2$ in the chromatography buffer. Elution of the enzyme from the resin is dependent on the substrate CDP-DG and NaCl in the elution buffer. CDP-DG-Sepharose affinity chromatography affords a 558-fold purification of PS synthase over the Triton X-100 extract. The CDP-DG-Sepharose resin used for PS synthase purification is a low-capacity affinity resin (26). PI synthase does not bind to this resin under the conditions used to bind PS synthase (5). The purification procedure results in the isolation of a nearly homogeneous protein preparation as shown by native and SDS-polyacrylamide gel electrophoresis. The subunit molecular weight of the purified PS synthase preparation is 23,000 as determined by SDS-polyacrylamide gel electrophoresis (5), and the electroblotting of PS synthase activity on nitrocellulose paper (33). The enzyme has also been purified from strain VAL2C(YEpCHO1) (5). VAL2C(YEpCHO1) bears the structural gene for PS synthase on a hybrid plasmid that directs the overproduction of the enzyme (34). The availability of the PS synthase overproducing strain facilitates the acquisition of larger amounts of pure enzyme. Microsomal-associated PS synthase requires a 370-fold purification from strain VAL-2C(YEpCHO1) compared to approximately a 2000-fold purification from the wild-type strain to attain similar final specific activities of pure enzyme with a subunit molecular weight of 23,000. Overall, PS synthase is purified about 5000-fold relative to the cell extract of wild-type cells. The purified enzyme is 100% stable for about six months at $-80°C$.

When PS synthase is partially purified in the presence of the protease inhibitor PMSF, a protein with a subunit molecular weight of 30,000 as well as the 23,000-dalton subunit of PS synthase is isolated (35). The 23,000-dalton subunit of PS synthase is a proteolytic cleavage product of the 30,000-dalton subunit of the enzyme (35).

PS synthase is dependent on either $MnCl_2$ or $MgCl_2$ and Triton X-100 for maximum activity at the pH optimum of 8.0 (5). The enzyme follows normal saturation kinetics with respect to CDP-DG and serine when activity is

Table 2 Purification of PS synthase from *S. cerevisiae*[a]

Purification step	Total units (nmol/min)	Protein (mg)	Specific activity (units/mg)	Purification (-fold)	Yield (%)
1. Microsomes	591	493	1.2	1	100
2. Triton X-100 extract	579	241	2.4	2	98
3. CDP-DG-Sepharose	457	0.34	1340	1120	77
4. DE-53 chromatography	421	0.18	2300	1920	71

[a] The data are based on starting with 75 g (wet weight) of cells. Data are from Bae-Lee & Carman (5).

measured with a mixed micelle substrate of Triton X-100 and CDP-DG. The K_m values for CDP-DG and serine are 83 μM and 0.83 mM, respectively, and the V_{max} is 4 μmol/min/mg (5, 36). PS synthase uses dCDP-DG as well as CDP-DG as a substrate (5). The enzyme is labile above 40°C and is sensitive to thioreactive agents (5). The reaction mechanism for PS synthase is sequential, where the enzyme binds to CDP-DG before serine and PS is the first product released in the reaction sequence. This mechanism is based on the results of enzyme kinetic experiments (36), isotopic exchange reactions between substrates and products (5), and a stereochemical analysis of the reaction using ^{31}P NMR (37).

Purification and Properties of PI Synthase

PI synthase (CDP-diacylglycerol : *myo*-inositol 3-phosphatidyltransferase, EC 2.7.8.11) catalyzes the formation of PI and CMP from CDP-DG and inositol (38). The product of the reaction, PI, is essential to the growth and metabolism of *S. cerevisiae* (39–42). Fischl & Carman (26) have purified PI synthase from the microsomal fraction of *S. cerevisiae*. The purification procedure includes the solubilization of microsomal membranes with Triton X-100, affinity chromatography with CDP-DG-Sepharose, and negative chromatography with the chromatofocusing resin PBE 94. PI synthase can be solubilized with 1% Triton X-100 in solubilization buffer containing either MgCl$_2$ or MnCl$_2$ (26, 32). To eliminate PS synthase activity from the solubilized enzyme preparation, PI synthase is solubilized with buffer containing MgCl$_2$ (26). The major purification of PI synthase is achieved by affinity chromatography with CDP-DG-Sepharose. The affinity resin used for purification is a high-capacity resin (26). Binding of PI synthase to the resin is dependent on the presence of Triton X-100 and enzyme cofactor (MgCl$_2$ or MnCl$_2$). Elution of the enzyme is dependent on both the substrate CDP-DG and hydroxylamine-HCl in the elution buffer. The affinity chromatography step results in a 400-fold purification of the enzyme over the Triton X-100 solubilization step. The chromatofocusing run-through step removes minor protein contaminants. A summary of the purification scheme for PI synthase is presented in Table 3. PI synthase was purified 1000-fold over the microsomal fraction and 3300-fold relative to the activity in cell extracts. The purification scheme leads to the isolation of a nearly homogeneous protein preparation as shown by polyacrylamide gel electrophoresis under nondenaturing conditions (43) and in the presence of SDS (26). The subunit molecular weight of the purified enzyme preparation is 34,000 as determined by SDS-polyacrylamide gel electrophoresis (26) and the electroblotting of PI synthase activity on nitrocellulose paper (33). The purified enzyme is 100% stable for at least six months and about 50% stable for four years when stored at −80°C.

Table 3   Purification of PI synthase from *S. cerevisiae*[a]

Purification step	Total units (nmol/min)	Protein (mg)	Specific activity (units/mg)	Purification (-fold)	Yield (%)
1. Microsomes	2000	2500	0.8	1	100
2. Triton X-100 extract	1500	1250	1.2	1.5	75
3. CDP-DG-Sepharose	1200	2.5	480	600	60
4. Chromatofocusing run-through	1200	1.5	800	1000	60

[a] The data are based on starting with 100 g (wet weight) of cells. Data are from Fischl & Carman (26).

Maximum PI synthase activity is dependent on either $MnCl_2$ or $MgCl_2$ and Triton X-100 at the pH optimum of 8.0 (26). The K_m values for CDP-DG and inositol are 66 μM and 0.21 mM, respectively, as determined with a mixed micelle substrate of Triton X-100 and CDP-DG (36). Based on the results of kinetic experiments (36) and the ability of the enzyme to catalyze isotopic exchange reactions between substrates and products (44), PI synthase catalyzes a Bi Bi sequential reaction mechanism. The enzyme binds to CDP-DG before inositol, and PI is released prior to CMP in the reaction sequence (44). PI synthase is labile above 60°C and is inactivated by thioreactive agents (26). The activation energy for the reaction is 35 kcal/mol (26).

Purification and Properties of PI Kinase

PI kinase (ATP:phosphatidylinositol 4-phosphotransferase, EC 2.7.1.67) catalyzes the formation of PIP and ADP from PI and ATP (45). PI kinase is the first enzyme in the phosphorylation sequence of PI $\rightarrow$ PIP $\rightarrow$ PIP_2. The regulation of PI kinase should play a major role in phosphoinositide turnover and cell growth in *S. cerevisiae*. PI kinase has been purified 8000-fold from *S. cerevisiae* by Belunis et al (46). The purification procedure includes Triton X-100 solubilization of microsomal membranes, DE-52 chromatography, hydroxylapatite chromatography, octyl Sepharose chromatography, and two consecutive Mono Q chromatographies. A summary of the purification scheme is presented in Table 4. The procedure results in the isolation of a protein with a subunit molecular weight of 35,000 that is 96% homogeneous as based on native and SDS-polyacrylamide gel electrophoresis. The pure enzyme is about 95% stable for at least three months of storage at -80°C.

Maximum PI kinase activity is dependent on magnesium ions and Triton X-100 at the pH optimum of 8.0 (46). The true K_m values for PI and MgATP are 70 μM and 0.3 mM, respectively, and the true V_{max} is 4.7 μmol/min/mg (46). The turnover number for the enzyme is 166 min^{-1}. Results of kinetic and isotopic exchange reactions indicate that PI kinase catalyzes a sequential

Table 4 Purification of PI kinase from *S. cerevisiae*[a]

Purification step	Total units (nmol/min)	Protein (mg)	Specific activity (units/mg)	Purification (-fold)	Yield (%)
1. Cell extract	5970	10,119	0.59	1	100
2. Microsomes	4418	2,559	1.72	2.9	74
3. Triton X-100 extract	3104	794	3.9	6.6	52
4. DE-52	1086	175	6.2	10.5	18
5. Hydroxylapatite	1032	57	6.2	10.5	18
6. Octyl Sepharose	743	5.64	131.7	223.2	12.4
7. Mono Q I	558	0.24	2325	3940.6	9.3
8. Mono Q II	380	0.08	4750	8050.8	6.3

[a] Data are based on starting with 200 g (wet weight) of cells. Data are from Belunis et al (46).

Bi Bi reaction mechanism (46). The enzyme binds to PI prior to ATP, and PIP is the first product released in the reaction. The enzyme catalyzes the reverse reaction, and the equilibrium constant for the reaction indicates that the reverse reaction is favored in vitro (46). It is likely that in vivo the reaction is driven in the forward direction, since PIP is converted to PIP_2, which is then hydrolyzed to diacylglycerol and inositol trisphosphate (17). The activation energy for the reaction is 31.5 kcal/mol, and the enzyme is thermally labile above 30°C (46). PI kinase activity is inhibited by calcium ions and thioreactive agents (46). Various nucleotides, including adenosine and AdoHcy, do not affect PI kinase activity.

Purification and Properties of I-1-P Synthase

I-1-P synthase (EC 5.5.1.4) catalyzes the formation of L-inositol 1-phosphate from D-glucose 6-phosphate (21). Mutants defective in I-1-P synthase (47) lose cell viability upon inositol starvation (39, 41). The cytosolic-associated I-1-P synthase has been purified to apparent homogeneity from *S. cerevisiae* by Donahue & Henry (21). Purification of the enzyme requires streptomycin sulfate and ammonium sulfate fractionation followed by chromatography with DEAE-cellulose, hexyl-agarose, Bio-Gel A-0.5 m, and DEAE-cellulose. Table 5 summarizes the purification of the enzyme. The major purification of the enzyme is attained with the first DEAE-cellulose column with an 18-fold increase in specific activity over the ammonium sulfate fractionation step. Inositol 1-phosphate synthase is purified by this procedure 500-fold over the activity in cell extracts with a yield of 21%. The purification scheme results in the isolation of a nearly homogeneous protein preparation as evidenced by native and SDS-polyacrylamide gel electrophoresis. The apparent subunit molecular weight of the enzyme is 62,000 as determined by SDS-polyacrylamide gel electrophoresis. Gel filtration analysis of the purified

Table 5 Purification of I-1-P synthase from *S. cerevisiae*[a]

Purification step	Total units (nmol/h × 10^{-3})	Protein (mg)	Specific activity (units/mg × 10^{-3})	Purification (-fold)	Yield (%)
1. Crude extract	466	9338	0.05	1	
2. Ammonium sulfate	637	3350	0.19	3.8	
3. DEAE-cellulose I	990	293	3.4	67.6	100
4. Hexyl agarose	524	37	14.2	283	53
5. Bio-Gel A-0.5 m	247	11.6	22.8	455	25
6. DEAE-cellulose II	202	8.3	24.3	486	21

[a] The data are based on starting with 250 g (wet weight) of cells. Data are from Donahue & Henry (21).

enzyme indicates a native molecular weight of 240,000, suggesting that the native enzyme is composed of four identical subunits. The purified enzyme is completely stable for six months when stored at $-80°C$.

The pH optimum for inositol 1-phosphate synthase is 7.0 and maximum activity is dependent on NAD and ammonium ions (21). The K_m values for glucose 6-phosphate and NAD are 1.18 mM and 8 μM, respectively (21). 2-Deoxyglucose 6-phosphate and glucitol 6-phosphate are inhibitors of the enzyme (21).

REGULATION OF PHOSPHOLIPID BIOSYNTHETIC ENZYMES IN YEAST

Phospholipid biosynthetic enzymes in yeast are regulated by the availability of water-soluble phospholipid precursors (11, 20, 48–56), the growth phase of cells (20, 57, 58), the membrane phospholipid composition (44, 59), and by phosphorylation by cAMP-dependent protein kinase (60).

Regulation of Expression of Phospholipid Biosynthetic Enzymes by Water-Soluble Phospholipid Precursors

The enzyme activities catalyzing the biosynthesis of PC by the reaction sequence PA → CDP-DG → PS → PE → PMME → PDME → PC are subject to a form of coordinate regulation. These enzyme activities, which include CDP-DG synthase (48, 61), PS synthase (49, 51, 53, 61), PS decarboxylase (52), and the phospholipid N-methyltransferases (11, 49, 52, 54–56) are repressed when wild-type exponential-phase cells are grown in medium containing inositol plus choline. CDP-DG synthase (48, 61) and PS synthase (53, 61) activities are also repressed when wild-type exponential-phase cells are grown in medium containing inositol plus ethanolamine and inositol plus serine. Water-soluble phospholipid precursors have no effect on

CDP-DG synthase (48, 61), PS synthase (49, 53, 61), and the phospholipid N-methyltransferases (49, 55) in the absence of inositol. The addition of inositol alone to the growth medium causes a partial repression of CDP-DG synthase (48, 61), PS synthase (49, 53, 61), and the phospholipid N-methyltransferases (55). Cytosolic-associated I-1-P synthase activity is also repressed by the availability of inositol in the growth medium (62). Greenberg et al (63) have shown that PGP synthase activity is also repressed in wild-type cells by inositol plus choline.

The availability of antibodies to CDP-DG synthase (23), PS synthase (53), and I-1-P synthase (21) have permitted experiments to determine if the regulation of these enzymes by phospholipid precursors occurs at the level of enzyme formation. Under the growth conditions where a reduction in enzyme activity occurs, there is a corresponding reduction of the CDP-DG synthase $M_r = 56,000$ (61), the PS synthase $M_r = 23,000$ (53, 61), and the I-1-P synthase $M_r = 62,000$ (21, 61) subunits as determined by an immunological analysis.

Unlike the enzymes leading to the formation of PC in the primary biosynthetic route and I-1-P synthase, PI synthase activity (49, 59, 61) and its $M_r = 34,000$ subunit (59, 61) levels are not regulated in exponential-phase wild-type cells in response to water-soluble phospholipid precursors.

The regulation of the enzymes in the primary biosynthetic route of PC synthesis by water-soluble phospholipid precursors is absolutely dependent upon inositol. The mechanism by which water-soluble precursors affect the enzymes is not yet clear. The fact that inositol is required for the coordinate regulation of these enzymes suggests that PC biosynthesis is coordinated with inositol synthesis (64). Furthermore, mutants with lesions in genes whose wild-type products exert a positive (i.e. the *ino2* and *ino4* strains) or a negative (i.e. the *opi1* strain) effect upon I-1-P synthase expression (21, 62, 65, 66) exert pleiotropic effects upon the coordinately regulated enzymes CDP-DG synthase (48, 61), PS synthase (49, 50, 53, 61), and the phospholipid N-methyltransferases (49, 67).

When wild-type cells are grown in inositol-containing medium with exogenous ethanolamine or choline, the enzyme activities in the primary biosynthetic route for PC synthesis are repressed, presumably in response to the utilization of the CDP-ethanolamine- and CDP-choline-based auxillary pathways for PE and PC synthesis. PA phosphatase catalyzes the formation of the diacylglycerol needed for these auxillary reactions. The addition of inositol to the growth medium of wild-type cells results in an elevation in PA phosphatase activity (20). The addition of ethanolamine and choline to medium with or without inositol has no effect on PA phosphatase activity (20). Like the enzymes in the primary route of PC biosynthesis, PA phosphatase activity is affected in the *ino2* and *opi1* regulatory mutants (20). These findings suggest that PA phosphatase regulation is also coupled to inositol biosynthesis. Little is known about the regulation of the CDP-ethanolamine- and CDP-choline-based pathway enzymes in response to water-soluble phospholipid precursors. However, it is known that the CDP-choline-based pathway is regulated by the rate of choline transport in polyamine-stimulated cells (68).

Regulation of Expression of Phospholipid Biosynthetic Enzymes by the Growth Phase

The primary PC biosynthetic pathway enzymes are also coordinately regulated with the growth phase of cells. Maximum CDP-DG synthase, PS synthase, and the phospholipid N-methyltransferase activities are found in the exponential phase of growth (58). When wild-type cells enter the stationary phase of growth, CDP-DG synthase, PS synthase, and the phospholipid N-methyltransferase activities decrease 2.5–5-fold (58). The decrease in CDP-DG synthase (69) and PS synthase (58) activities in the stationary phase is not due to a reduction in enzyme synthesis or an increase in enzyme turnover. Therefore another mechanism must be involved with the regulation of these enzymes during the growth phase. It is worth noting that the major change in phospholipid composition of stationary-phase cells as compared to exponential-phase cells is a doubling of the PI-to-PS ratio (58). Studies with purified PS synthase reconstituted into unilamellar PC-PE-PI-PS vesicles have shown that increases in the PI-to-PS ratio in vesicles cause a 2–3-fold decrease in enzyme activity (44). It is possible that the alterations in the PI-to-PS ratio of stationary-phase cells compared with exponential-phase cells may be responsible in part for the reduction in PS synthase activity as well as the other enzyme activities in the primary route of PC biosynthesis. This correlation, however, may be an oversimplification of the regulation that occurs in the stationary-phase cells.

PI synthase activity and subunit levels are not significantly affected by the growth phase of wild-type cells (57, 58). However, PI kinase activity is affected by the growth phase (57). PI kinase activity increases 2–2.5-fold in the stationary phase as compared to the exponential phase of growth (57). Preliminary studies of PI kinase activity in cell extracts suggest that the enzyme is down regulated in exponential-phase cells by phosphorylation by cAMP-dependent protein kinase (57).

PA phosphatase activity increases 2–3-fold as wild-type cells enter the stationary phase of growth (19). In addition to being used for PE and PC biosynthesis in the auxillary CDP-ethanolamine- and CDP-choline-based pathways (14), the product of the PA phosphatase reaction, diacylglycerol, is used for triacylglycerol synthesis. The increase in PA phosphatase activity in the stationary phase of wild-type cells correlates with an increase in triacylglycerol biosynthesis (19).

Regulation of PS Synthase Activity

Purified PS synthase is the most extensively studied enzyme of yeast phospholipid metabolism. PS synthase activity is regulated by a number of factors, which are discussed below.

REGULATION BY INOSITOL   As indicated above, inositol by itself and in concert with other water-soluble precursors including serine, ethanolamine, and choline controls the expression of PS synthase by an enzyme repression mechanism (53, 61). Studies with purified PS synthase have also shown that inositol inhibits PS synthase directly. Inositol is a noncompetitive inhibitor of PS synthase with respect to CDP-DG and serine with a K_i of 65 μM (36). The inhibition of PS synthase activity by inositol may contribute to the rapid rate of PI synthesis at the expense of PS synthesis in wild-type cells grown in the presence of 75 μM inositol (36). However, increases in the rate of PI synthesis in response to inositol supplementation are primarily attributed to the availability of inositol for the PI synthase reaction (36).

REGULATION BY PHOSPHOLIPIDS   To gain insight into the modulation of PS synthase activity by a changing phospholipid environment, the purified enzyme has been reconstituted into unilamellar phospholipid vesicles containing its substrate CDP-DG (44). Since PS synthase is purified in the presence of Triton X-100 (5), it is necessary to reconstitute the enzyme into vesicles directly from mixed micelles containing pure enzyme, phospholipid, and Triton X-100. Reconstitution of the enzyme is achieved by removing detergent from an octyl glucoside/phospholipid/Triton X-100/enzyme mixed micelle by Sephadex G-50 superfine chromatography (44). In this procedure, the vesicles containing PS synthase elute near the void volume of the column and are well separated from the detergent molecules. The average size of the vesicles containing reconstituted PS synthase is 90 nm in diameter. Each vesicle contains about 1 molecule of PS synthase, which is reconstituted asymmetrically with 80–90% of its active site facing outward. PS synthase has been reconstituted into PC/PE/PI/PS vesicles in which the ratios of PI to PS approximate in vivo conditions of both wild-type and mutant cells. Increases in the ratios of PI to PS in vesicles result in a 2–3-fold decrease in PS synthase activity (44). Changes in the ratios of PC to PE do not significantly affect reconstituted PS synthase activity (44). This modulation in PS synthase activity by changes in the PI-to-PS ratio is enough to account for some of the in vivo fluctuations in the PS content. For example, the addition of inositol to wild-type cells results in an approximate 2-fold increase in the cellular ratio of PI to PS (49, 61, 70). The cellular ratio of PI to PS also increases about 2-fold when wild-type cells grown in the absence of inositol enter the stationary phase of growth (61). Under both of these growth conditions, the cellular ratio of PC to PE is not significantly affected (49, 58, 61, 70). The *ino2* and *ino4* mutants have reduced PS synthase activity (50, 61), which is due in part to reduced PS synthase mRNA levels (50). However, the reduced PS synthase activity in the *ino2* and *ino4* mutants is not fully due to the reduction of the message (50). It is possible that some of the reduced PS synthase activity in the *ino2* and *ino4* mutants is due to the elevated ratios of PI to PS found in the membranes of these cells (50).

REGULATION BY PHOSPHORYLATION   PS synthase has been shown to be phosphorylated in vivo and in vitro by cAMP-dependent protein kinase (60). PS synthase activity in cell extracts is reduced in the *bcy1* mutant (which has high cAMP-dependent protein kinase activity) and elevated in the *cyr1* mutant (which has low cAMP-dependent protein kinase activity) when compared with wild-type cells. The reduced PS synthase activity in the *bcy1* mutant correlates with elevated levels of the phosphorylated form of the PS synthase $M_r = 23,000$ subunit. The elevated PS synthase activity in the *cyr1* mutant correlates with reduced levels of the phosphorylated form of the enzyme. There is negligible phosphorylation of the PS synthase $M_r = 23,000$ subunit from stationary-phase cells, further indicating the role of cAMP-dependent protein kinase regulation of the enzyme.

Pure PS synthase is phosphorylated by the cAMP-dependent protein kinase catalytic subunit. This phosphorylation results in a 60–70% reduction in PS synthase activity. The cAMP-dependent protein kinase catalytic subunit catalyzes the incorporation of 0.7 mol of phosphate per mol of PS synthase $M_r = 23,000$ subunit. The specific cAMP-dependent protein kinase inhibitor prevents the phosphorylation of PS synthase and the inhibition of its activity by the catalytic subunit. Analysis of peptides derived from protease-treated labeled-PS synthase has shown that only one peptide is labeled. Phosphoamino acid analysis of the labeled PS synthase indicates that the enzyme is phosphorylated at a serine residue.

The PS synthase $M_r = 30,000$ subunit, which is a precursor of the $M_r = 23,000$ subunit (35), is not phosphorylated either in vivo or in vitro by cAMP-dependent protein kinase (60).

GENETICS OF PHOSPHOLIPID SYNTHESIS IN YEAST

Mutants defining both structural and regulatory genes involved in phospholipid synthesis have been identified in *S. cerevisiae*. The unusual tolerance of *S. cerevisiae*, under laboratory conditions, for perturbations in its phospholipid content makes the prediction of mutant phenotypes difficult and results in some unusual and unexpected phenotypes. While this has caused difficulty in the identification and analysis of some mutants, it has also made it possible to isolate and analyze mutants with tight and unconditional lesions in important enzymes. For some yeast mutants with defects in phospholipid synthesis, it has not yet been possible to establish, even tentatively, whether the lesion is in a regulatory gene or in a structural gene. These cases will be clearly identified in the following discussion. In other cases, as will be discussed, the evidence for identification of a structural gene may be suggestive but less than conclusive.

Structural Gene Mutants

The *chol* mutants were isolated as choline auxotrophs (71), and in the original study, no biochemical analysis was performed, although the gene was genetically mapped to the V chromosome. The work of Kovac et al (15) and Atkinson et al (12, 13) demonstrated that *chol* mutants exhibit diminished PS biosynthesis. The ethanolamine auxotroph with reduced PS synthase activity described by Nikawa & Yamashita (14) appears to be another *chol* isolate. The mutant studied by Kovac et al (15) expresses PS synthase with an apparently altered K_m for serine. This mutant synthesizes PS if exposed to a high exogenous concentration of serine. The *chol* mutant studied in detail by Atkinson and colleagues (12) has little or no residual PS synthase activity. When the *CHOl* gene was isolated by complementation of a *chol* mutant phenotype (34), transformed yeast cells containing the gene at high copy number express PS synthase at elevated specific activity.

The data discussed above are consistent with, and supportive of, the hypothesis that the *CHOl* gene encodes PS synthase, but do not provide a conclusive level of evidence for the identification of a structural gene. For example, a mutant defect leading to an altered K_m of an enzyme could be produced by an alteration in a gene required for posttranslational modification, as could mutant phenotypes resulting in no activity. Mutations in regulatory genes required for expression of a particular gene can also result in loss of activity. Overexpression from a plasmid is very suggestive of the gene dosage effects most commonly associated with structural genes, but regulatory effects can also be responsible. Unambiguous evidence that PS synthase is encoded by the *CHOl* gene was provided by Kiyono et al (35), who demonstrated that the N-terminal sequence of purified PS synthase matches the sequence obtained by DNA sequencing of the cloned *CHOl* gene. Subsequently, Kohlwein et al (72) demonstrated that a portion of the *CHOl* gene expressed as a fusion protein in *Escherichia coli* produces a protein cross-reactive to antibody raised in response to PS synthase, which had been purified to homogeneity from yeast (5). Furthermore, antibody raised in response to the *CHOl* fusion protein purified from *E. coli* cross-reacts with PS synthase purified individually from the mitochondria and from the microsomal membrane fractions of yeast. These results confirm that mitochondrial and microsomal forms of the enzyme are both encoded by the *CHOl* gene.

The identification of the *CHO1* gene as the structural gene for PS synthase is discussed in detail to set a standard for the identification of structural genes. Present technology dictates that the minimum acceptable level of evidence provide a direct and unambiguous connection between the protein in question and the gene purported to encode it. This is particularly important in the case of genes encoding the membrane-associated enzymes of phospholipid biosynthesis, because so few of the gene products have been purified to homogeneity and studied at the protein level.

Only two structural genes encoding phospholipid biosynthetic enzymes of yeast have met these strict criteria. PS synthase, as described above, is one, and the other is the cytoplasmic enzyme, I-1-P synthase, which is the gene product of the *INO1* locus (73). The *ino1* mutants were isolated as strict inositol auxotrophs (47) and were subsequently demonstrated to lack the cytoplasmic activity of I-1-P synthase (62). However, mutations at a number of loci were shown to confer inositol auxotrophy and to result in a lack of I-1-P synthase activity (62). The first substantial evidence that the *INO1* gene might encode I-1-P synthase came with the purification of the enzyme and the production of specific antibody (21). It was then demonstrated that some *ino1* mutants that lack the catalytic activity of I-1-P synthase express a protein cross-reactive to the antibody raised in response to the purified I-1-P synthase subunit. Indeed, in such mutants, the inactive subunit is the same molecular weight (M_r = 62,000) as the active subunit (21, 74). Thus, the mutants express a mutationally altered protein, a result that supported the hypothesis that the *INO1* is the structural gene for I-1-P synthase (21). However, final proof of the identity of the *INO1* gene came with its isolation and sequencing and subsequent comparison of the DNA sequence with the chemically determined amino acid composition and N-terminal sequence of the purified I-1-P synthase subunit (73).

The isolation of a mutant with an altered K_m for PI synthase was reported by Nikawa & Yamashita (75). This mutant *(pis)* requires high exogenous levels of inositol for growth and, in crude extracts of mutant cells, the activity of PI synthase shows an apparently altered K_m for inositol. A DNA fragment capable of complementing the mutant phenotype in vivo was isolated, and yeast cells transformed with it have elevated PI synthase activity in crude extracts as compared to extracts of nontransformed wild-type cells (76). Disruption of the *PIS* sequence in vivo in yeast is a lethal event. These results support the hypothesis that the *PIS* gene encodes PI synthase. Nevertheless, the evidence accumulated so far does not include any direct comparison of the DNA sequence with PI synthase purified from yeast cells.

A mutant *(cdg1)* defective in CDP-DG synthase activity has been identified in yeast (77). In this case, it is not at all clear whether the mutation has occurred in a regulatory or structural gene. The *cdg1* mutant (77) has about 25% of the wild-type level of CDP-DG synthase activity. Studies on heterozygous (i.e. *cdg1/CDG1*) diploid cells showed levels of enzyme intermediate between wild-type (i.e. *CDG1/CDG1*) and homozygous mutant *(cdg1/cdg1)* cells, suggestive of a gene dosage effect. However, the enzyme purified from this mutant was similar in every regard to the enzyme purified from wild-type cells. The *cdg1* mutant is also pleiotropic and exhibits overproduction of many phospholipid biosynthetic enzymes.

Mutants with defects in the phospholipid N-methyltransferases that catalyze the synthesis of PC via PE (Figure 1) have been described by several laboratories. Yamashita and coworkers (54, 55) reported the isolation of mutants that were auxotrophic for choline and appeared to have specific defects in phospholipid methylation. These mutants fell into two categories: those *(pem1* mutants) unable to methylate PE (i.e. PE → PMME), and those *(pem2* mutants) unable to carry out the two final methylations (i.e. PMME → PDME → PC). Yamashita and coworkers (78) speculated that the mutants define two genes encoding two phospholipid methyltransferases, which between them were responsible for the three-step conversion of PE to PC. However, Greenberg et al (65, 70) described a mutant *(opi3)* that had been isolated on the basis of an inositol overproduction phenotype. This mutant also exhibited a lesion in the final two phospholipid methylations and synthesized virtually no PC. In terms of its phospholipid defect, this mutant was similar, if not identical, to the *pem2* mutant described by Yamashita et al (55). However, the *opi3* mutant is not a choline auxotroph despite its rather bizarre phospholipid composition (Table 6). Also, the inositol excretion phenotype exhibited by the *opi3* mutant was suggestive of an inability to regulate inositol metabolism, a phenotype previously associated with mutants defective in regulation of phospholipid biosynthesis (65, 66).

Additional attempts to isolate *S. cerevisiae* phospholipid methylation mutants by screening for choline auxotrophs resulted only in the isolation of five new *chol* mutants (31). However, another class of mutants, designated *cho2* (64, 79, 80), with an inositol excretion phenotype were found to have a defect in the first phospholipid methylation (i.e. PE → PMME). Again, the phospholipid methylation defect appeared similar to that reported by Yamashita & Oshima for the *pem1* mutants (54). The *cho2* mutants, however, are not choline auxotrophs (79). Kodaki & Yamashita (78) reported the isolation of two DNA fragments encoding sequences *PEM1* and *PEM2* which, in vivo, complement the choline auxotrophies and the phospholipid methylation defects of the *pem1* and *pem2* mutants, respectively. Summers et al (79) reported the isolation of the *CHO2* gene, and McGraw & Henry (81) reported the isolation of the *OPI3* gene. The restriction maps of the *PEM1* and *CHO2* sequences are virtually identical, as are the *OPI3* and *PEM2* genes, suggesting that they are, in fact, resolations of the same genes.

Given the above results, what then is the explanation for the discrepancy in

Table 6 Phospholipid composition of *S. cerevisiae* strains[a]

Strain	Medium[b]	Relative amount of label incorporated (%)							Ref.
		PI	PS	PE	PMME	PDME	PC	Other	
wild-type	I+	21.8	6.6	17.7	0.8	2.9	42.2	8.0	79
cho2-1	I+	24.8	4.1	51.0	—	—	9.1	11.0	79
cho2-δ	I+	27.3	5.2	49.6	—	—	6.8	11.1	79
opi3-3	I+	30.6	2.8	5.3	34.9	12.5	3.0	10.3	70
opi3-δ	I+	35.0	3.0	5.6	44.0	2.0	—	10.4	81
opi3-δ	I+D+	21.1	4.6	11.6	12.0	35.8	—	14.9	81
opi3-δ	I+C+	24.7	5.1	12.2	8.9	1.8	39.4	7.9	81
ino1-13	I+	28.6	4.9	13.2	1.1	1.9	37.8	12.5	81
ino4-38	I+	36.3	4.1	23.8	2.0	7.8	11.1	14.9	67
ino4-39	I+	37.0	3.8	19.9	1.5	9.6	15.3	12.9	67
ino4-40	I+	32.3	5.2	21.1	2.2	10.8	15.7	12.7	67
ino2-2	I+	36.5	4.6	28.3	1.3	9.5	11.9	7.9	67
ino2-21	I+	38.3	4.9	27.9	1.3	8.2	11.3	8.1	67
cho1-1, ino1-13	I+E+	32.8	—	15.0	1.0	1.1	41.7	8.4	67
cho1-1, ino1-13	I+C+	29.4	—	4.8	—	—	57.8	8.0	13
cho1-3	I+	39.1	1.0	6.4	0.8[c]		34.3	8.4	13
cho1-4	I+	36.1	1.4	18.6	1.7[c]		36.4	5.8	31

[a] Numbers in the body of the table represent the proportion of total lipid-soluble phosphorous associated with each lipid. Category "Other" is the sum of other phospholipids not specifically listed.
[b] The abbreviations used are: I+, plus inositol; I+D+, plus inositol and DMME; I+C+, plus inositol and choline; I+E+, plus inositol and ethanolamine.
[c] Values represent the sum of PMME and PDME.

the reported phenotypes of the *pem1*, *pem2* mutants versus the *cho2*, *opi3* mutants? Gene disruptions that eliminate major portions of the coding sequences of the *CHO2* and *OPI3* genes were used to construct null mutants in vivo. The null mutants, like the original *cho2* and *opi3* mutants, are not choline auxotrophs, exhibit inositol excretion phenotypes and have substantial alterations in their phospholipid compositions (Table 6). However, several combinations of mutations in a single haploid strain result in choline auxotrophy. Among the genotypes associated with choline auxotrophy is the double mutant, *cho2,opi3*. Other combinations of mutations involving regulatory gene defects also exhibit choline auxotrophy (79). It seems most probable, therefore, that the original *pem1* and *pem2* mutant isolates of Yamashita and coworkers (54, 55) are composites containing several mutations.

A number of questions remain. Most importantly, do any of the mutations discussed above and/or the DNA sequences that complement them define structural genes for the phospholipid N-methyltransferases? Kodaki & Yamashita (78) believe that *PEM1* and *PEM2* sequences encode structural genes, and the *CHO2* (79) and *OPI3* (81) clones appear to be identical to *PEM1* and *PEM2*, respectively. However, the evidence for identification of structural genes of the phospholipid methyltransferases is not yet compelling. The phospholipid N-methyltransferases of yeast have not yet been successfully purified and characterized at the protein level, so it is not possible to compare the molecular weights of the proteins predicted from DNA sequences with the molecular weights of the native proteins. No evidence linking the DNA sequences to the protein encoded exists, other than the lack of enzymatic activity in mutant crude extracts and its restoration after transformation with the cloned DNA. Hopefully, the existence of the clones for these genes will facilitate identification and purification of their gene products from yeast.

Remaining problems are the curious inositol excretion phenotypes of *cho2* and *opi3* mutants and the question of why these mutants are not choline auxotrophs. The question of the inositol excretion phenotype is addressed following the discussion of transcription regulation. A complete understanding of the lack of choline auxotrophy in *cho2* and *opi3* strains requires discussion of the overall tolerance of *S. cerevisiae* strains for alterations in phospholipid content, a topic that is taken up after the discussion of regulatory mutants. However, part of the explanation for the lack of choline auxotrophy in *cho2* mutants may be the fact that *cho2* cells grown in the absence of monomethylethanolamine (MME) or choline supplement do retain some PC synthesis. The total methyltransferase activity measured in vitro suggests that the cells retain about 10% of the wild-type capacity to methylate PE (79). This residual activity cannot be ascribed to the "leakiness" of partially defective enzyme, since the *cho2* null mutant has the same phenotype. However, the *cho2,opi3* double mutant exhibits no residual phospholipid methylation (79), and the cloned *OPI3* gene is capable of partial restoration of the *cho2* methylation defect in vivo (81). These two results taken in combination suggest that the *OPI3 (PEM2)* gene product is capable, at least to a limited degree, of carrying out the function of the *CHO2 (PEM1)* gene product. The *opi3* null mutant synthesizes no detectable PC (Table 6) and is not a choline auxotroph. However, all *opi3* mutants when grown in absence of supplement accumulate detectable levels of PDME. Since the *opi3 (pem2)* mutants are believed to be defective in the two final methylations (i.e. PMME → PDME → PC), the presence of PDME needs to be explained. Since PDME is not detected in the *cho2,opi3* double mutant (79), the explanation would again appear to be overlapping functions for the two gene products. It seems probable that the *CHO2/PEM1* gene product is capable, to a limited degree, of carrying out the second methylation (i.e. PMME → PDME) but not the third methylation (i.e. PDME → PC).

Recently, there have also been a number of reports of isolation of mutants with defects in the synthesis of PC via the auxillary Kennedy pathway (16). Using colony autoradiographic methods, Hjelmstad & Bell (82) isolated mutants defective in cholinephosphotransferase (*cpt* mutants). The 22 *cpt* mutants isolated are recessive and fall into three genetic complementation groups. The largest number of mutants fall into the *cpt1* complementation group and exhibit reductions in choline phosphotransferase ranging from 2- to 10-fold. The activities of two of the *cpt1* isolates differ from wild-type on the basis of their apparent K_m for CDP-choline. The *CPT1* gene was cloned by complementation of the *cpt1* phenotype, and a gene disruption was performed. The null mutant is viable and falls into the *cpt1* complementation group, confirming the identity of the cloned sequence. The null mutant exhibits a 5-fold reduction in cholinephosphotransferase activity. The residual cholinephosphotransferase activity in all of the mutants, including the mutant produced by gene disruption, exhibits CMP sensitivity. The presence of the cloned *CPT1* gene on a multicopy plasmid leads to elevated activity of choline phosphotransferase in wild-type cells. Hosaka & Yamashita (83) also reported the isolation of a mutant defective in cholinephosphotransferase activity. This mutant was isolated by second site suppression of a choline-sensitive strain. The residual activity in the mutant shows an altered apparent K_m for CDP-choline. The relationship of this mutant to the mutants of Hjelmstad & Bell (82) is unknown. The genetic and biochemical evidence presented by Hjelmstad & Bell (82) is quite supportive of the identification of *CPT1* as the structural gene for cholinephosphotransferase but, as yet, there is no direct evidence linking the *CPT1* sequence to cholinephosphotransferase.

The isolation of a mutant *(cct)* exhibiting thermolabile synthesis of CDP-choline was also described by Nikawa et al (84). The mutant appears to have a lesion in cholinephosphate cytidylyltransferase. Subsequently, the *CCT* gene was cloned by complementation of the *cct* mutant (85). The authors believe that the cloned gene, which has been sequenced, represents the structural gene for cholinephosphate cytidylytransferase, but the evidence is as yet indirect.

Hjelmstad & Bell (86) have isolated mutants *(ept)* defective in ethanolaminephosphotransferase. These mutants were isolated using a colony autoradiographic technique similar to the method employed in the isolation of cholinephosphotransferase mutants. Nine mutants, falling to five complementation groups, were identified. The *ept1* mutants exhibit reductions in ethanolaminephosphotransferase activity ranging from 30- to 90-fold, while *ept2* mutants exhibit only 2–3-fold reduction in activity. The *EPT1* gene was cloned by complementation of the *ept1* mutant phenotype, and transformants carrying the gene on a multicopy plasmid exhibit extreme (20–30-fold) overproduction of ethanolaminephosphotransferase activity. A disruption of the *EPT1* gene in vivo is viable and produces a mutant with *ept1* phenotype. Interestingly, *ept1* mutants exhibited reductions (3.5–7-fold) in cholinephosphotransferase. Transformants carrying *EPT1* in multicopy overproduced this same activity. However, in contrast to *cpt1* mutants, the residual choline phosphotransferase activity in *ept1* mutants is CMP insensitive. Hjelmstad &

Bell (86) believe that the *EPT1* gene product possesses both ethanol-aminephosphotransferase and cholinephosphotransferase activities. This idea is particularly intriguing in light of evidence concerning overlapping substrate specificities in the phospholipid methylation pathway.

Mutants with Defects in the Regulation of Phospholipid Synthesis

Most of the mutants with defects in the regulation of phospholipid synthesis were isolated on the basis of defects in regulation of I-1-P synthase in response to inositol. These mutants fell into two general categories: (*a*) those unable to derepress I-1-P synthase and other coregulated activities of phospholipid biosynthesis and, (*b*) those unable to repress I-1-P synthase and other coregulated activities.

The mutants (*ino2* and *ino4*) that are unable to derepress I-1-P synthase were first identified as inositol auxotrophs (47) and were shown to be unlinked genetically to the *INO1* structural gene. These mutants were shown to lack I-1-P synthase activity in crude extracts (62). They were found to lack material cross-reactive to antibody produced in response to I-1-P synthase subunit (21) even when grown in the presence of low levels of inositol, which permitted partial derepression of I-1-P synthase in wild-type cells. Thus, the *ino2* and *ino4* mutants were judged to be defective in derepression of I-1-P synthase, and since they were recessive mutations, it was postulated that the wild-type *INO2* and *INO4* gene products were required for expression of I-1-P synthase. In other words, the *ino2* and *ino4* lesions appear to define loci encoding positive regulatory factors (21). The *INO4* gene has recently been cloned by complementation of an *ino4* mutant (87).

Subsequently, it was discovered that *ino2* and *ino4* mutant strains exhibit other alterations in their patterns of phospholipid synthesis. For example, PC biosynthesis via the methylation pathway is much reduced in *ino2* and *ino4* mutants (67), and the mutants express low but constitutive levels of PS synthase (50). Thus, *ino2* and *ino4* strains have altered expression of a number of phospholipid biosynthetic activities which, in wild-type strains, are regulated in response to inositol and choline (11, 49, 50, 54, 64, 88). Yamashita et al (54) also reported the isolation of a mutant that exhibits altered regulation of the phospholipid N-methyltransferases. In this case, the mutant showed full repression of these activities when inositol alone was present, whereas full repression of wild-type cells requires choline in addition to inositol.

The *opi1* mutants are the best studied of a class of mutants (Opi⁻) isolated on the basis of an inositol overproduction and excretion assay (65). The *opi1* mutants (65, 66) exhibit constitutive expression of the same subset of activities that are affected in *ino2* and *ino4* strains. In other words, *opi1* mutants express constitutive levels of I-1-P synthase (66), PS synthase (49, 50, 53), and the phospholipid N-methyltransferases (49). The *opi1* mutants also express CDP-DG synthase constitutively (48). Since the *opi1* mutants are genetically recessive and are not linked to any known structural gene involved in phospholipid biosynthesis, it is presumed that the *OPI1* gene product is a *trans*-acting negative regulatory factor.

Various double mutant combinations using *ino2* or *ino4* in combination with Opi⁻ mutations have been constructed. In all cases, the double mutants are Ino⁻ in phenotype (65, 66, 89). Thus, the *ino2* and *ino4* mutations in genetic terms are said to be epistatic to the Opi⁻ mutations. This result implies that the gene products encoded by the *OPI* genes have a less direct role in regulating the structural genes than do the gene products of the *INO2* or *INO4* loci. Alternatively, the action of the *OPI* gene products may require the presence of the *INO2* and *INO4* gene products. By either interpretation, the genetic data suggest that the regulatory system controlling phospholipid synthesis is under positive regulation involving, directly or indirectly, the *INO2* and *INO4* gene products. Secondarily, the system is modulated by negative factors encoded by the *OPI1* gene and presumably by other *OPI* loci. The regulation in response to these regulatory genes is known to occur at the level of abundance of transcripts of the structural genes, at least in the case of the *INO1* and *CHO1* genes (50, 80). This topic is further explored in the section on transcription regulation.

Phospholipid Requirements of Yeast Cells

The studies of mutants defective in phospholipid biosynthesis and regulation have produced insights into the physiological requirements of yeast cells for phospholipids. It has been clear for some time that *S. cerevisiae* strains retain viability and grow with extreme alterations in phospholipid content (2). For example, *cho1* mutants are ethanolamine/choline auxotrophs, and the lesion in PS biosynthesis in the most extreme *cho1* mutant is absolute and unconditional (12, 34). Essentially no detectable PS synthase activity remains, yet *cho1* cells grow and divide, producing cellular membranes devoid of PS (12, 13, 34, 62) (Table 6). Furthermore, when *cho1* cells are supplemented with choline, they synthesize very little PE (Table 6). Some residual PE synthesis always remains in *cho1* cells, however, even when there is no exogenous ethanolamine and no PS to serve as precursor for PE biosynthesis. The residual PE synthesis is due to turnover of ethanolamine from sphingolipid metabolism (90, 91). Indeed, the Eam⁻ mutants, which are defective in sphingolipid metabolism, turn over ethanolamine at an elevated rate and thus suppress the ethanolamine auxotrophy of *cho1* mutants (90). Under conditions in which *cho1* cells are growing in the presence of choline, they produce phospholipid compositions composed largely of PI and PC (Table 6).

The PI content of *cho1* cells is elevated under all growth conditions, and it appears that PI can substitute for PS under these circumstances (12, 13). Since both lipids are anionic, carrying a net charge of minus 1, charge balance in the phospholipid composition is preserved. However, if PI is capable of substituting for PS, the reverse is clearly not true. In inositol auxotrophs (i.e. *ino1* mutants) deprived of inositol, PI content drops rapidly, growth ceases, and death ensues, despite somewhat elevated rates of PS synthesis (39, 41).

The phospholipid compositions of *ino2, ino4, opi3,* and *cho2* mutants grown in the absence of supplementation (Table 6) suggest that yeast cells tolerate major changes in the relative proportions of the zwitterionic phospholipids (i.e. PE, PMME, PDME, PC). The *ino2* and *ino4* mutants, for example, exhibit elevated PE levels, accumulate PMME and PDME above wild-type levels, and have reduced PC content. The *cho2* mutants have elevated PE and decreased PC content when they are grown in the absence of choline or MME. The *opi3* mutants on the other hand, when grown in the absence of exogenous choline, have reduced PE levels and synthesize no detectable PC. However, *opi3* mutants accumulate PMME to more than 30% of their phospholipid content. Growth studies with the *opi3* mutants suggest that there are detectable physiological consequences to such extreme variations in phospholipid content. For example, *opi3* mutants are also temperature sensitive for growth (i.e. they will not grow at 37°C) if MME is present in their growth medium. Furthermore, while *opi3* cells will grow continuously in the absence of exogenous choline, they lose viability under these conditions if permitted to enter stationary-growth phase. Loss of viability in stationary *OPI3* cells can be prevented by the addition of choline or, to a lesser extent, dimethylethanolamine (DME). When *opi3* cells are grown in the presence of DME or choline, PE content returns to normal and PMME accumulation is much reduced. However, only the presence of choline leads to PC synthesis. The presence of DME in the growth medium results in accumulation of elevated PDME levels, but no PC is synthesized [Table 6, (81)].

Taken overall, the growth and phospholipid composition studies using *ino1, ino2, ino4, cho1, cho2, opi3,* and various double mutant combinations support some surprising conclusions about the requirements for phospholipids in *S. cerevisiae*. First, yeast cells can grow and produce membranes devoid of PS and PC. PI can largely substitute for PS, while the reverse is not true. PDME appears to be able to substitute almost entirely for PC. The same cannot be said of PMME. Finally, the *cho2* mutants appear capable of substituting PE for a major portion of their PC content. It is clear, however, that PE cannot totally substitute for PC. The *cho2,opi3* double mutant, which accumulates PE and cannot synthesize any PC in the absence of exogenous choline, is a choline auxotroph. The *cho1* mutants tolerate very reduced PE and elevated PC levels, but it is not possible to eliminate PE entirely because of ongoing turnover of sphingolipids. It is, therefore, not clear whether PE can be eliminated entirely. It can be concluded, however, that while the zwitterionic lipids (i.e. PE, PMME, PDME, and PC) can substitute for one another to a large degree, some amount of PDME or PC is required for growth, particularly at higher temperatures.

MOLECULAR BIOLOGY OF PHOSPHOLIPID SYNTHESIS IN YEAST

The isolation and molecular analysis of genes encoding the enzymes of phospholipid biosynthesis provides the foundation for future detailed analyses of protein structure, studies on the targeting of these proteins into membrane compartments, as well as the discussion of the mechanisms of transcriptional regulation. Not all of the isolated sequences that are discussed have been proven unambiguously to be structural genes, but the molecular analysis, including DNA sequencing, should rapidly lead to such identification. In the following discussion, only those genes whose DNA sequence has been reported are considered.

Isolation and Analysis of Structural Genes

The *CHO1* gene is the most extensively studied of all genes encoding phospholipid biosynthetic enzymes. The *CHO1* gene was first isolated from a yeast genomic library on a high-copy-number plasmid by complementation of a *cho1* mutation (34). Subsequently, the complementing DNA fragment was shown to encode a transcript of 1.2 kb, the abundance of which is regulated in the same fashion as the level of PS synthase activity (50). The regulation of this transcript will be discussed further in the section on transcriptional regulation. The *CHO1* gene was sequenced by Kiyono et al (35) and by Nikawa et al (92). Nikawa et al have referred to the gene as *PSS*. The designation *CHO1* has precedence, however, because of its first use by Lindegren et al (71). The sequences reported by Nikawa et al (92) and Kiyono et al (35) are identical. The *CHO1* gene contains an open reading frame capable of encoding a protein of 30,804 daltons. The predicted protein contains extensive hydrophobic stretches. The protein predicted from the *CHO1* sequence has a molecular weight of 30,804, whereas the molecular weight determined by electrophoresis of PS synthase purified from microsomal membranes is a subunit molecular weight of 23,000 (5). Kiyono et al (35) showed that when PS synthase is partially purified in the presence of protease inhibitors, a protein of 30,000 daltons can be detected in addition to the 23,000-dalton form. They postulated that proteolysis may be involved in activation of PS synthase. Kohlwein et al (72) showed that PS synthase partially purified independently from mitochondria and from microsomal membranes contains both 30,000- and 23,000-dalton forms. Thus, the conversion of the enzyme to the lower-molecular-weight form is not involved in targeting to specific membrane compartments.

The *INO1* gene, which encodes I-1-P synthase, has also been sequenced (73). An open reading frame capable of encoding a protein of 62,842 daltons is contained in the sequence, providing excellent correspondence to the molecular weight of 62,000 for the subunit of I-1-P synthase determined by SDS-polyacrylamide gel electrophoresis (21). The amino acid sequence predicted from the DNA sequence of the *INO1* gene did not contain any long hydrophobic stretches, consistent with identification of I-1-P synthase as a cytoplasmic protein (21).

Nikawa & Yamashita (76) reported the cloning of a DNA fragment that complemented a mutation *(pis)* confirming a defect in PI synthase. The *PIS* clone was subsequently sequenced (42). The PIS coding region contains an open reading frame capable of encoding a protein of 220 amino acids with a predicted molecular weight of 24,823. The predicted protein contains hydrophobic stretches consistent with its being an integral membrane protein. It also contains a conserved region homologous to a sequence in the *CHO1* gene encoding PS synthase and PGP synthase gene from *E. coli* (42). Since these enzymes and PI synthase all use CDP-DG as substrate, Nikawa & Yamashita speculated that the conserved region may constitute a binding site for CDP-DG (42). One significant problem is the discrepancy between the 24,823 molecular weight predicted from the DNA sequence and the experimentally determined 34,000 molecular weight of purified PI synthase (26, 33). While it is possible that posttranslational modifications explain the discrepancy, no data to this effect is yet available.

Kodaki & Yamashita (78) reported the DNA sequence of the *PEM1* and *PEM2* genes, which they believe to be the structural genes for the phospholipid N-methyltransferases. The *PEM1* sequence, which is believed to be the structural gene for the enzyme that carries out the first phospholipid methylation (i.e. PE → PMME), contains an open reading frame capable of encoding a protein of 869 amino acids with a predicted molecular weight of 101,202. The *PEM2* gene, which is thought to encode the bifunctional enzyme responsible for the two final methylations (i.e. PMME → PDME → PC), contains an open reading frame capable of encoding a protein of 206 amino acids with a predicted molecular weight of 23,150. Both sequences predict proteins with long hydrophobic stretches. In addition, there are regions within the two sequences that showed homology at the amino acid sequence level. Furthermore, homology was detected between these two sequences and several other enzymes from other organisms that are known to catalyze methyl transfers using AdoMet as the methyl donor.

Transcriptional Regulation

PS synthase, I-1-P synthase, and the phospholipid N-methyltransferases are subject to coordinate regulation by inositol and choline (11, 49, 50, 53–56, 64). PI synthase, on the other hand, is constitutively expressed (49, 59). Genes thought or proven to encode these enzymes have been isolated and sequenced and, in several cases, analyzed with respect to regulation of transcript abundance. A growing body of evidence suggests that coordinate regulation in response to choline and inositol occurs at the level of mRNA abundance.

Hirsch & Henry (80) showed that the *INO1* transcript is regulated in response to inositol and choline. The *INO1* transcript is most abundant in cells grown in the absence of inositol and choline. When inositol is present in the growth medium, the abundance of the *INO1* transcript is reduced 10-fold. The addition of choline with inositol leads to a further threefold reduction (i.e. 30-fold below fully derepressed level) in the level of the *INO1* transcript. However, the addition of choline to the growth medium in the absence of inositol has no effect. A similar pattern of regulation was detected when the expression of the *CHO1* transcript was examined (50). The *CHO1* transcript is fully derepressed in cells grown in the absence of inositol and choline. Addition of inositol to the growth medium leads to an approximate 40% reduction in the abundance of *CHO1* transcript. Addition of inositol plus choline leads to an 85% reduction in the level of *CHO1* transcript. The relative level of *CHO1* transcript correlates very well with the relative levels of PS synthase activity in crude extracts (50).

The levels of the *INO1* and *CHO1* transcripts are also affected by the *ino2, ino4,* and *opi1* regulatory mutations (50, 80). The *INO1* transcript is expressed at repressed levels in *ino2* and *ino4* cells even when the cells are grown in low levels of inositol that permit derepression of the *INO1* transcript in wild-type cells. The *INO1* transcript is expressed constitutively at fully derepressed levels in *opi1* cells (80). The *CHO1* transcript is expressed at a low, but constitutive level in *ino2* and *ino4* cells and at a high constitutive level in *opi1* cells (50). Thus, it appears that the *INO2, INO4,* and *OPI1* gene products exert their effects by controlling expression of transcription of the *INO1* and *CHO1* structural genes.

These same regulatory genes also affect expression of the phospholipid N-methyltransferases and CDP-DG synthase (48, 49, 67, 66). Indeed, as previously discussed, these same enzymes are subject to a common pattern of regulation: full derepression when cells are grown in the absence of soluble precursors of phospholipid synthesis, partial repression in the presence of inositol, and full repression when inositol and choline are present. Accumulating genetic and molecular evidence suggests that this pattern of regulation occurs at a transcriptional level and is mediated in *trans* by the gene products of the *INO2, INO4,* and *OPI1* genes and possibly other *OPI* genes. By analogy with other transcriptionally controlled structural genes in *S. cerevisiae* (93) and higher eukaryotes in general, it is likely that this regulation is mediated by interaction of the regulatory proteins with repeated elements located in the 5' ends of the structural genes.

Since a number of the structural genes encoding coregulated phospholipid biosynthetic enzymes have been sequenced, it is possible to search the 5' noncoding regions for conserved repeated sequences that might be involved in the regulation. Tables 7 and 8 reveal that there are two classes of conserved sequences in the 5' ends of the *INO1, CHO1, PEM1/CHO2, PEM2/OPI3,* and *PIS* genes; a 9-bp repeat and an 11-bp repeat. Table 9 shows the precise positioning of these two classes of sequences relative to the 5' end of the *INO1* gene. A deletion analysis of the 5' end of the *INO1* gene demonstrated that elimination of all six copies of the 9-bp repeat leads to constitutive expression of the *INO1* gene (94). An *INO1* promoter was constructed, from which all but two nucleotides 5' to the TATA box had been deleted, eliminating all copies of the 9-bp repeat, but retaining two copies of the 11-bp element (Table 9). This truncated *INO1* promoter supports expression of full derepressed levels of transcript and is insensitive to the presence of inositol or inositol plus choline. For this reason, J. Hirsch speculated that the 9-bp repeat might be involved in repression of the *INO1* gene in response to inositol (94). However, the 11-bp element of the *INO1* gene may be responsible for the unregulated expression of the promoter lacking the 9-bp repeat. Clearly, further analysis of the cloned genes will be necessary to establish the roles of these sequences in the coordinate regulation of phospholipid synthesis.

What Metabolic Signals are Involved in the Coordinate Regulation of Phospholipid Synthesis?

In the preceding section we have discussed a general transcription control mechanism that is involved in the coordinate regulation of phospholipid synthesis. Genetic and molecular studies have defined regulatory genes that encode regulatory factors that interact in *trans,* either directly or indirectly, with sequences adjacent to the structural genes. It is clear that this control mechanism is sensitive to the presence of soluble precursors of the phospholipids. Recent studies have suggested that it is not the presence of the soluble precursors themselves that activate the regulation, but rather their participation in phospholipid synthesis. This insight grows out of analysis of the curious secondary phenotypes of some of the structural gene mutants, in particular the *cho1, cho2,* and *opi3* mutations. As discussed previously, *CHO1* has been demonstrated to be the structural gene for PS synthase, and a growing body of evidence suggests that the *cho2* and *opi3* mutants represent lesions in the structural genes for the phospholipid N-methyltransferases.

Table 7 Occurrence of a 9-bp conserved sequence in the 5' ends of structural genes encoding phospholipid biosynthetic enzymes[a]

Gene/Orientation of sequence	Sequence detected									Homology to consensus (OF 9)
INO1 5' → 3'	T	T	A	T	G	A	A	A	T	7
	A	T	G	C	G	G	A	A	T	7
	A	T	G	T	G	A	A	A	T	9
	A	T	G	T	G	A	A	A	A	8
	A	T	G	T	T	A	A	T	T	7
INO1 3' → 5'	A	T	G	T	G	A	A	T	T	8
CHO1/PSS 3' → 5'	A	T	G	T	G	A	A	A	G	8
PEM1/CHO2 5' → 3'	T	T	G	T	T	A	A	A	T	7
	C	T	G	T	G	A	A	A	A	7
	A	T	G	T	T	A	G	A	T	7
	A	T	G	T	G	A	A	T	T	8
	A	T	A	T	G	A	A	A	A	7
	A	T	G	T	G	A	A	G	A	7
PEM1/CHO2 3' → 5'	G	T	G	T	G	A	A	C	T	7
	A	T	G	A	G	A	A	A	A	7
PEM2/OPI3 5' → 3'	A	T	G	T	G	G	A	A	A	7
	T	T	G	T	G	A	A	A	T	8
PIS 5' → 3'	A	T	G	T	G	A	C	T	T	7
	T	T	G	T	A	A	A	A	T	7
	A	T	A	T	G	A	A	C	T	7
PIS 3' → 5'	T	T	G	A	G	A	A	A	T	7
	A	T	G	T	T	A	A	A	A	7
	A	T	G	G	G	A	A	A	A	7
Consensus	A	T[b]	G	T	G	A	A	A	T	
Summary of substitutions: A	6	–	3	2	1	21	21	16	8	
T	5	23	–	9	4	–	–	4	14	
C	1	–	–	1	–	–	1	2	–	
G	1	–	20	1	8	2	1	1	1	

[a] The occurrence of the repeated sequence with variations permitted in seven of the nine bases was detected by searching published DNA sequences using the "DNA Inspector II" program on a Macintosh SE computer. Data on *INO1* gene obtained from J. Hirsch (94). Search of other published sequences conducted by J. Lopes (personal communication).
[b] entirely conserved

Table 8 Occurrence of an 11-bp conserved sequence in the 5' ends of structural genes encoding phospholipid biosynthetic enzymes[a]

Gene / Orientation of sequence	Sequence detected											Homology to consensus (of 11)
INO1 5' → 3'	C	C	T	T	T	T	T	C	T	T	C	10
	C	C	T	T	T	T	T	T	T	G	G	10
	C	C	T	T	T	T	T	T	T	T	C	11
INO1 3' → 5'	A	C	T	T	C	T	T	T	T	T	C	9
CHO1/PSS 5' → 3'	C	C	T	A	T	T	T	T	T	T	T	9
PEM1/CHO2 5' → 3'	C	C	T	T	T	T	A	T	T	C	C	9
	C	C	T	T	T	T	T	T	T	T	T	10
	G	C	T	T	T	T	T	C	T	T	C	9
PEM1/OPI3 5' → 3'	C	C	T	T	G	T	T	G	T	T	C	9
PIS	not detected											
Consensus	C	C[b]	T[b]	T	T	T[b]	T	T	T[b]	T	C	
Summary of substitutions: A	1	—	—	1	—	—	1	—	—	—	—	
T	—	—	9	8	7	9	8	6	9	7	2	
C	7	9	—	—	1	—	—	2	—	1	6	
G	1	—	—	—	1	—	—	1	—	1	1	

[a] The occurrence of the repeated sequence with variations permitted in 9 of the 11 bases was detected by searching published DNA sequences using the "DNA Inspector II" program on a Macintosh SE computer. Search of published sequences conducted by J. Lopes (personal communication).
[b] entirely conserved

Mutations in the *cho1* gene eliminate PS biosynthesis, making *cho1* mutants dependent upon exogenous ethanolamine or choline. In the absence of exogenous ethanolamine or choline, PC biosynthesis is rapidly interrupted in *cho1* mutants (31). Curiously, *cho1* cells also exhibit an inositol excretion phenotype when ethanolamine or choline is absent from the growth medium (31). The inositol excretion phenotype suggests that I-1-P synthase is improperly regulated in *cho1* cells in which PC biosynthesis is interrupted. Indeed, I-1-P synthase is not repressed by inositol in *cho1* cells deprived of ethanolamine or choline (31).

A similar conditional Opi⁻ phenotype was found to be associated with *cho2* and *opi3* mutations. The *cho2* mutants excrete inositol unless MME, DME, or choline is present in the growth medium. Ethanolamine is ineffective in relieving the inositol excretion phenotype of *cho2* cells. In the case of *opi3* mutants, the inositol excretion phenotype is alleviated by the presence of

Table 9 Sequence of the 5' end of the *INO1* gene[a]

```
          10         20         30         40         50         60         70         80         90        100
          |          |          |          |          |          |          |          |          |          |

    GCCTTTTTCT TCGTTCCTTT TGTTCTTCAC GTCCTTTTTA TGAAATACGT GCCGGTGTTC CGGGTTGGAT GCCGAATCGA AAGTGTTGAA TGTGAAATAT
  1 CGGAAAAAGA AGCAAGGAAA ACAAGAAGTG CAGGAAAAAT ACTTTATGCA CGGCCACAAG GCCCAACCTA CGCCTTAGCT TTCACAACTT ACACTTTATA

    GCGGAGGCCA AGTATGCGCT TCGGCGGCTA AATGCGGCAT GTGAAAAGTA TTGTCTATTT TATCTTCATC CTTCTTTCCC AGAATATTGA ACTTATTTAA
101 CGCCTCCGGT TCATACGCGA AGCCGCCGAT TTACGCCGTA CACTTTTCAT AACAGATAAA ATAGAAGTAG GAAGAAAGGG TCTTATAACT TGAATAAATT

    TTCACATGGA GCAGAGAAAG CGCACCTCTG CGTTGGCGGC AATGTTAATT TGAGACCTAT ATAAATTGGA GCTTTCGTCA CCTTTTTTTG GCTTGTTCTG
201 AAGTGTACCT CGTCTCTTTC GCGGTGAGAC GCAACCGCCG TTACAATTAA ACTCTGCATA TATTTAACCT CGAAAGCAGT GGAAAAAAAC CGAACAAGAC

    TTGTCGGGTT CCTAATGTTA GTTTTATCCT TGTTTCATTC CCTTTTTTTT CCAGTGAAAA AGAAGTAACA ATG
301 AACAGCCCAA GAATTACAAT CAAAATAGGA ACTAAATAAG ACAAAGTAAG GGAAAAAAAA GGTCACTTTT TCTTCATTGT TAC
```

[a] The start of translation of the *INO1* gene occurs at the ATG at position 381. Transcription starts at position 375 marked by arrow (94). The positions of the 9-bp and 11-bp repeats are underlined by arrows showing their direction. The TATA box is located at position 257. The truncated *INO1* promotor, which exhibits constitutive expression and is insensitive to the presence of inositol, was constructed by deleting all sequences 5' to position 255 (marked by arrow).

DME or choline but not by ethanolamine or MME. In the case of each class of mutants (*cho1*, *cho2*, and *opi3*), the Opi⁻ phenotype is alleviated only by those soluble precursors that enter the pathway beyond the metabolic lesion permitting synthesis of PC (or PDME). Thus, ethanolamine suffices in the case of *cho1* mutants, since it enters the pathway downstream from the lesion in PS biosynthesis. In the case of the *cho2* mutants, the metabolic block occurs in the first phospholipid methylation, PE → PMME, and only those precursors entering beyond this point (i.e. MME, DME, or choline) restore regulation of I-1-P synthase (79). In the case of the *cho2* mutants, it has been demonstrated that the altered regulation of I-1-P synthase is due to altered expression of the *INO1* transcript (80). In *opi3* mutants, regulation of the I-1-P synthase subunit and *INO1* transcript abundance in response to inositol is restored only by the addition of DME or choline. Ethanolamine and MME do not restore the regulation (81). This result is particularly important in determining the nature of the metabolic signal responsible for triggering transcriptional regulation of phospholipid synthesis, since both of the final phospholipid methylations (i.e. PMME → PDME → PC) are blocked in *opi3* cells. In *opi3* cells exposed to DME, PDME is synthesized and PC is synthesized only when choline is present (Table 6). Thus, it can be concluded that regulation of the *INO1* gene in response to inositol occurs only when PDME or PC is being synthesized. Furthermore, it is clear that regulation in response to inositol occurs when one of these lipids is synthesized via either one of two routes: via the CDP-choline pathway as in *opi3* mutants, or via the methylation pathway as in *cho1* mutants exposed to ethanolamine or wild-type cells receiving no exogenous supplement.

How might the cell sense the presence of inositol when PC or PDME synthesis is ongoing? Kelley et al (36) have shown that inositol is a noncompetitive inhibitor of PS synthase. The addition of inositol, therefore, should lead to a decrease in substrate availability in the reaction sequence PS → PE → → → PC by drawing CDP-DG precursor away from PS biosynthesis and into PI biosynthesis. Presumably, this change in the pattern of phospholipid synthesis can only be detected by cells in which PC biosynthesis is ongoing. However, in *cho1*, *cho2*, and *opi3* cells, the reaction sequence PS → PE → → → PC is irreversibly blocked. The addition of soluble precursors beyond the block does not restore the continuity of the reaction sequence. Thus, yeast cells must have some mechanism for directly detecting the synthesis of end-product lipids and must be able to adjust the regulation of the intermediate steps accordingly. At present, the precise nature of the metabolic signal remains obscure, but the availability of mutants blocked in various parts of the CDP-choline pathway together with mutants in the methylation pathway should permit a systematic search for the ultimate metabolic signal.

CONCLUSIONS AND SUMMARY

Enormous progress has been made in the last several years in the analysis of phospholipid biosynthesis in yeast. This progress has included the purification to homogeneity and characterization of a number of enzymes. Furthermore, the structural genes encoding many of these enzymes have been identified, isolated, and sequenced, and work is in progress on many additional genes and proteins. The well-studied metabolic lesions in the large collection of mutants defective in phospholipid biosynthesis in yeast provide an unparalleled opportunity to dissect a complex metabolic signaling pathway. In addition, these mutants provide the basis for manipulation of phospholipid composition in vivo, which will permit an assessment of the cellular functions of the phospholipids, an assessment heretofore inaccessible in a eukaryotic organism.

ACKNOWLEDGMENTS

This work was supported by Public Health Service grants GM-28140 and GM-35655 to G.M.C., and GM-19629 to S.A.H. from the National Institutes of Health. This work was also supported by the Charles and Johanna Busch Memorial Fund and New Jersey State funds to G.M.C.

We are particularly grateful to J. Lopes and P. Chorgo for permitting us to quote their data prior to its publication. We thank P. McGraw, E. Summers, M. Kelley, A. Bailis & M. Johnson, and R. Bell & R. Hjelmstad for permitting us to quote data in press.

Literature Cited

1. Rothstein, R. J. 1983. *Methods Enzymol.* 101C:202–11
2. Hill, J. E., Chung, C., McGraw, P., Summers, E., Henry, S. A. 1989. In *Proceedings of the Harden Conference on Fungal Walls and Membranes,* ed. A. Trinci. In press
3. Henry, S. A. 1982. In *The Molecular Biology of the Yeast Saccharomyces: Metabolism and Gene Expression,* ed. J. N. Strathern, E. W. Jones, J. R. Broach, pp. 101–58. Cold Spring Harbor, NY: Cold Spring Harbor Lab.
4. Rattray, J. B., Schibeci, A., Kidby, D. K. 1975. *Bacteriol. Rev.* 39:197–231
5. Bae-Lee, M., Carman, G. M. 1984. *J. Biol. Chem.* 259:10857–62
6. Esko, J. D., Raetz, C. R. H. 1983. *The Enzymes* 16:207–53
7. Moore, T. S. 1982. *Annu. Rev. Plant Physiol.* 33:235–59
8. Lester, R. L., Steiner, M. R. 1968. *J. Biol. Chem.* 243:4889–93
9. Steiner, M. R., Lester, R. L. 1969. *Biochemistry* 9:63–69
10. Steiner, M. R., Lester, R. L. 1972. *Biochim. Biophys. Acta* 260:222–43
11. Waechter, C. J., Lester, R. L. 1973. *Arch. Biochem. Biophys.* 158:401–10
12. Atkinson, K., Fogel, S., Henry, S. A. 1980. *J. Biol. Chem.* 255:6653–61
13. Atkinson, K. D., Jensen, B., Kolat, A. I., Storm, E. M., Henry, S. A., Fogel, S. 1980. *J. Bacteriol.* 141:558–64
14. Nikawa, J., Yamashita, S. 1981. *Biochim. Biophys. Acta* 665:420–26
15. Kovac, L., Gbelska, I., Poliachova, V., Subik, J., Kovacova, V. 1980. *Eur. J. Biochem.* 111:491–501
16. Kennedy, E. P., Weiss, S. B. 1956. *J. Biol. Chem.* 222:193–214
17. Berridge, M. J. 1987. *Biochim. Biophys. Acta* 907:33–45
18. Kuchler, K., Daum, G., Paltauf, F. 1986. *J. Bacteriol.* 165:901–11
19. Hosaka, K., Yamashita, S. 1984. *Biochim. Biophys. Acta* 796:110–17
20. Morlock, K. R., Lin, Y.-P., Carman, G. M. 1988. *J. Bacteriol.* 170:3561–66

21. Donahue, T. F., Henry, S. A. 1981. *J. Biol. Chem.* 256:7077–85
22. Carter, J. R., Kennedy, E. P. 1966. *J. Lipid Res.* 7:678–83
23. Kelley, M. J., Carman, G. M. 1987. *J. Biol. Chem.* 262:14563–70
24. Hirabayashi, T., Larson, T. J., Dowhan, W. 1976. *Biochemistry* 15:5205–11
25. Larson, T. J., Hirabayashi, T., Dowhan, W. 1976. *Biochemistry* 15:974–79
26. Fischl, A. S., Carman, G. M. 1983. *J. Bacteriol.* 154:304–11
27. Belendiuk, G., Mangnall, D., Tung, B., Westley, J., Getz, G. S. 1978. *J. Biol. Chem.* 253:4555–65
28. Kepner, G. R., Macey, R. I. 1968. *Biochim. Biophys. Acta* 163:188–203
29. Kempner, E. S., Schlegel, W. 1979. *Anal. Biochem.* 92:2–10
30. Kanfer, J. N., Kennedy, E. P. 1964. *J. Biol. Chem.* 239:1720–26
31. Letts, V. A., Henry, S. A. 1985. *J. Bacteriol.* 163:560–67
32. Carman, G. M., Matas, J. 1981. *Can. J. Microbiol.* 27:1140–49
33. Poole, M. A., Fischl, A. S., Carman, G. M. 1985. *J. Bacteriol.* 161:772–74
34. Letts, V. A., Klig, L. S., Bae-Lee, M., Carman, G. M., Henry, S. A. 1983. *Proc. Natl. Acad. Sci. USA* 80:7279–83
35. Kiyono, K., Miura, K., Kushima, Y., Hikiji, T., Fukushima, M., et al. 1987. *J. Biochem.* 102:1089–100
36. Kelley, M. J., Bailis, A. M., Henry, S. A., Carman, G. M. 1988. *J. Biol. Chem.* 263:18078–85
37. Raetz, C. R. H., Carman, G. M., Dowhan, W., Jiang, R.-T., Waszkuc, W., et al. 1987. *Biochemistry* 26:4022–27
38. Paulus, H., Kennedy, E. P. 1960. *J. Biol. Chem.* 235:1303–11
39. Becker, G. W., Lester, R. L. 1977. *J. Biol. Chem.* 252:8684–91
40. Hanson, B. A., Lester, R. L. 1980. *J. Bacteriol.* 142:79–89
41. Henry, S. A., Atkinson, K. D., Kolat, A. I., Culbertson, M. R. 1977. *J. Bacteriol.* 130:472–84
42. Nikawa, J., Kodaki, T., Yamashita, S. 1987. *J. Biol. Chem.* 262:4876–81
43. Fischl, A. S. 1986. PhD thesis. *Reconstitution of purified phosphatidylinositol synthase from Saccharomyces cerevisiae into phospholipid vesicles.* Rutgers Univ. New Brunswick, NJ
44. Hromy, J. M., Carman, G. M. 1986. *J. Biol. Chem.* 261:15572–76
45. Colodzin, M., Kennedy, E. P. 1965. *J. Biol. Chem.* 240:3771–80
46. Belunis, C. J., Bae-Lee, M., Kelley, M.

J., Carman, G. M. 1988. *J. Biol. Chem.* 263:18897–903
47. Culbertson, M. R., Henry, S. A. 1975. *Genetics* 80:23–40
48. Homann, M. J., Henry, S. A., Carman, G. M. 1985. *J. Bacteriol.* 163:1265–66
49. Klig, L. S., Homann, M. J., Carman, G. M., Henry, S. A. 1985. *J. Bacteriol.* 162:1135–41
50. Bailis, A. M., Poole, M. A., Carman, G. M., Henry, S. A. 1987. *Mol. Cell. Biol.* 7:167–76
51. Carson, M. A., Atkinson, K. D., Waechter, C. J. 1982. *J. Biol. Chem.* 257:8115–21
52. Carson, M. A., Emala, M., Hogsten, P., Waechter, C. J. 1984. *J. Biol. Chem.* 259:6267–73
53. Poole, M. A., Homann, M. J., Bae-Lee, M., Carman, G. M. 1986. *J. Bacteriol.* 168:668–72
54. Yamashita, S., Oshima, A. 1980. *Eur. J. Biochem.* 104:611–16
55. Yamashita, S., Oshima, A., Nikawa, J., Hosaka, K. 1982. *Eur. J. Biochem.* 128:589–95
56. Waechter, C. J., Lester, R. L. 1971. *J. Bacteriol.* 105:837–43
57. Holland, K. M., Homann, M. J., Belunis, C. J., Carman, G. M. 1988. *J. Bacteriol.* 170:828–33
58. Homann, M. J., Poole, M. A., Gaynor, P. M., Ho, C.-T., Carman, G. M. 1987. *J. Bacteriol.* 169:533–39
59. Fischl, A. S., Homann, M. J., Poole, M. A., Carman, G. M. 1986. *J. Biol. Chem.* 261:3178–83
60. Kinney, A. J., Carman, G. M. 1988. *Proc. Natl. Acad. Sci. USA* 85:7962–66
61. Homann, M. J., Bailis, A. M., Henry, S. A., Carman, G. M. 1987. *J. Bacteriol.* 169:3276–80
62. Culbertson, M. R., Donahue, T. F., Henry, S. A. 1976. *J. Bacteriol.* 126:243–50
63. Greenberg, M. L., Hubbell, S., Lam, C. 1988. *Mol. Cell. Biol.* 8:4773–79
64. Henry, S. A., Klig, L. S., Loewy, B. S. 1984. *Annu. Rev. Genet.* 18:207–31
65. Greenberg, M. L., Reiner, B., Henry, S. A. 1982. *Genetics* 100:19–33
66. Greenberg, M. L., Goldwasser, P., Henry, S. A. 1982. *Mol. Gen. Genet.* 186:157–63
67. Loewy, B. S., Henry, S. A. 1984. *Mol. Cell. Biol.* 4:2479–85
68. Hosaka, K., Yamashita, S. 1981. *Eur. J. Biochem.* 116:1–6
69. Homann, M. J., 1987. *Regulation of CDP-diacylglycerol synthesis and utilization in Saccharomyces cerevisiae.* PhD thesis. Rutgers Univ. New Brunswick, NJ

70. Greenberg, M. L., Klig, L. S., Letts, V. A., Loewy, B. S., Henry, S. A. 1983. *J. Bacteriol.* 153:791–99
71. Lindegren, C., Shult, E., Hwang, Y. L. 1962. *Nature* 194:260–63
72. Kohlwein, S. D., Kuchler, K., Sperka-Gottlieb, C., Henry, S. A., Paltauf, F. 1988. *J. Bacteriol.* 170:3778–81
73. Johnson, M. J., Henry, S. A. 1988. *J. Biol. Chem.* 264:1274–83
74. Majumder, A. L., Duttagupta, S., Goldwasser, P., Donahue, T. F., Henry, S. A. 1981. *Mol. Gen. Genet.* 184:347–54
75. Nikawa, J., Yamashita, S. 1982. *Eur. J. Biochem.* 125:445–51
76. Nikawa, J., Yamashita, S. 1984. *Eur. J. Biochem.* 143:251–56
77. Klig, L. S., Homann, M. J., Kohlwein, S., Kelley, M. J., Henry, S. A., Carman, G. M. 1988. *J. Bacteriol.* 170:1878–86
78. Kodaki, T., Yamashita, S. 1987. *J. Biol. Chem.* 262:15428–35
79. Summers, E. F., Letts, V. A., McGraw, P., Henry, S. A. 1988. *Genetics* 120:909–22
80. Hirsch, J. P., Henry, S. A. 1986. *Mol. Cell. Biol.* 6:3320–28
81. McGraw, P., Henry, S. A. 1989. *Genetics.* In press
82. Hjelmstad, R. H., Bell, R. M. 1987. *J. Biol. Chem.* 262:3909–17
83. Hosaka, K., Yamashita, S. 1987. *Eur. J. Biochem.* 162:7–13
84. Nikawa, J., Yonemura, K., Yamashita, S. 1983. *Eur. J. Biochem.* 131:223–29
85. Tsukagoshi, Y., Nikawa, J., Yamashita, S. 1988. *Eur. J. Biochem.* 169:477–86
86. Hjelmstad, R. H., Bell, R. M. 1988. *J. Biol. Chem.* 263:19748–57
87. Klig, L. S., Hoshizaki, D. K., Henry, S. A. 1988. *Curr. Genet.* 13:7–14
88. Henry, S. A., Hoshizaki, D., Bailis, A., Homann, M. J., Carman, G. M. 1986. In *Enzymes in Lipid Metabolism II,* Proc. CNRS-INSERM Int. Symp. and NATO Workshop, ed. L. Freysz, H. Dreyfus, R. Massarelli, S. Gatt, pp. 623–32. New York: Plenum
89. Loewy, B. S., Hirsch, J., Johnson, M., Henry, S. A. 1986. *Proc. UCLA Symp. Yeast Cell Biol.* 33:551–65
90. Atkinson, K. D. 1984. *Genetics* 108:533–43
91. Atkinson, K. D. 1988. *Plant Membrane Structure Assembly and Function,* ed. J. Harwood, T. Walton, pp. 65–72. London: Biochem. Soc.
92. Nikawa, J., Tsukagoshi, Y., Kodaki, T., Yamashita, S. 1987. *Eur. J. Biochem.* 167:7–12
93. Guarente, L. 1988. *Cell* 52:303–5
94. Hirsch, J. 1987. Cis- *and* trans-*acting regulation of the* INO1 *gene of* Saccharomyces cerevisiae. PhD thesis. Albert Einstein Coll. Med.

Annu. Rev. Biochem. 1989. 58:671–717

ANIMAL VIRUS DNA REPLICATION[1]

Mark D. Challberg

Laboratory of Viral Diseases, National Institute of Allergy and Infectious Diseases, National Institutes of Health, Bethesda, Maryland 20892

Thomas J. Kelly

Department of Molecular Biology and Genetics, Johns Hopkins University School of Medicine, Baltimore, Maryland 21205

CONTENTS

PERSPECTIVES AND SUMMARY

Much of the impetus for studying the replication of animal virus genomes comes from a desire to understand the events that occur during the replication of eukaryotic chromosomes. Viruses offer many advantages for the study of

[1]The US government has the right to retain a nonexclusive, royalty-free license in and to any copyright covering this paper.

671

eukaryotic DNA replication. Viral genomes are relatively simple and can be readily manipulated by modern genetic methods. In addition, the replication of some viral genomes has proven amenable to analysis in cell-free systems. These facts significantly enhance the ability to analyze replication mechanisms at the molecular level. There are a number of potentially useful viral systems, and this review focuses on four of the best characterized: (a) adenovirus, (b) SV40, (c) herpes simplex virus, and (d) bovine papillomavirus. Each system has certain unique virtues that can be exploited to gain insight into different aspects of the replication process.

Adenovirus DNA replication occurs by a process that is significantly less complex than chromosomal DNA replication. Replication initiates by a novel protein priming mechanism, and all daughter strands are elongated by a continuous mode of synthesis such as occurs at the leading strand of a chromosomal replication fork. The biochemical dissection of a soluble in vitro system capable of faithfully replicating adenovirus DNA has led to the identification of most of the proteins involved. Adenovirus DNA replication requires the participation both of virus-encoded replication proteins and host-cell–encoded transcription factors. Since a linkage between replication and transcription has now been observed in other systems, it is likely that further analysis of the adenovirus system will provide insights that are of general importance.

The SV40 genome represents a more complete model system for studying cellular DNA replication. SV40 encodes only a single replication protein (T antigen) and relies predominantly on the host-cell replication machinery. In vivo studies have established that many of the details of SV40 DNA replication are closely similar to those of cellular DNA replication. Replication initiates at a fixed site on the viral genome and proceeds bidirectionally with continuous growth of leading strands and discontinuous growth of lagging strands. As in the case of adenovirus, an efficient cell-free replication system has been developed for SV40, and dissection of this system has identified several cellular replication proteins. A partial understanding of the mechanisms by which these proteins act is beginning to emerge, and it seems certain that this will be an area of continued rapid progress.

In contrast to SV40, herpes simplex virus (HSV) encodes many, if not all, of the proteins that are involved in the replication of its genome. Thus, HSV DNA replication has been studied by using a combination of genetics and biochemistry. The complete set of viral genes necessary for DNA synthesis has recently been identified, and the products of many of these genes have been purified and partially characterized. Several of these purified proteins have functions expected of replication proteins, including a DNA polymerase, a helicase, a primase, a single-stranded DNA binding protein, and an origin recognition protein. It seems likely that the availability of these purified proteins will soon lead to the development of an in vitro system capable of specifically replicating HSV DNA. Genetic and biochemical dissection of such a system should provide important new insights into the molecular mechanisms of eukaryotic DNA replication.

Adenovirus, SV40, and HSV are all examples of viruses that normally multiply by productive cytocidal infection. In all of these cases, viral DNA replication begins soon after infection and continues at a high rate until the death of the host cell. In contrast, bovine papillomavirus (BPV) represents an example of a virus that is capable of multiplying as a stable extrachromosomal element. In this case, viral DNA replication is controlled so that the number of viral genomes doubles only once per cell cycle, and under normal circumstances the host is not killed. As in the case of SV40, the BPV genome is relatively small and encodes only a small number of proteins involved in DNA replication; viral DNA synthesis depends heavily on host-cell replication proteins. The biochemical analysis of BPV DNA replication is in its infancy, but genetic analyses have provided evidence for a negative control system that apparently ensures that each viral genome is replicated once and only once during each cell cycle. Bovine papillomavirus therefore represents an excellent model for illuminating mechanisms involved in regulating DNA replication.

ADENOVIRUS DNA REPLICATION

Adenovirus DNA replication is better understood than the replication of the other animal virus genomes because it was the first to be established in vitro (1). Although many of the molecular details remain to be worked out, the basic features of the adenovirus DNA replication pathway are now reasonably clear, and most of the proteins involved in the replication process have been identified (2–5). On the basis of the information gathered to date, it is evident that the adenoviruses have evolved some very interesting and novel solutions to the replication problem, including the use of a "protein priming" mechanism for initiation and the diversion of cellular transcription factors to the viral replication machinery. It seems certain that further biochemical analysis of this system will illuminate and extend our understanding of DNA replication in the context of the animal cell.

The Adenovirus Chromosome

The genomes of the human adenoviruses are double-stranded linear DNA molecules containing approximately 35,000 base pairs. The 5' terminus of each strand of the viral genome is covalently attached to a virus-encoded protein (TP) with a molecular weight of about 55,000 (6, 7). In addition, the nucleotide sequences at the extreme ends of the genome are identical (8, 9).

Both of these structural features play important roles in the initiation of viral DNA replication.

Numerous in vivo studies have established that adenovirus DNA replication takes place in two stages (2–5). In the first stage, DNA synthesis is initiated at either terminus of the duplex viral genome by the protein priming mechanism (see below). The initiation process results in the establishment of a replication fork that moves from one end of the genome to the other. At each replication fork only one of the two parental DNA strands serves as a template for DNA synthesis. Thus, the products of the first stage of replication are a daughter duplex and a displaced single strand. In the second stage of DNA replication, the strand complementary to the displaced single strand is synthesized. It seems likely that the initial step in this process is the circularization of the single-stranded template by annealing of its self-complementary termini. The resulting duplex "panhandle" has the same structure as the terminus of the duplex adenovirus genome and is presumably recognized by the same initiation machinery that operates in the first stage of replication. Following a second initiation event, complementary strand synthesis proceeds from one end of the template to the other, generating a second daughter duplex. In both stages of adenovirus DNA replication there is only one priming event per nascent daughter strand, so all viral strands are synthesized in a continuous fashion from their 5' termini to their 3' termini.

The adenoviruses encode three proteins that play central roles in viral DNA replication: the terminal protein precursor (pTP), the adenovirus DNA polymerase (Ad pol), and the single-stranded DNA binding protein (DBP) (10–12). Together these three proteins consume approximately 25% of the coding capacity of the viral genome. The mRNAs for all three proteins are products of the same viral transcription unit and are produced by differential splicing of a common precursor (13). This genetic organization provides a simple mechanism for the coordinate regulation of the levels of replication proteins during infection. In addition to the virus-encoded proteins, adenovirus DNA replication requires the participation of several cellular proteins. Those identified to date include two cellular transcription factors (NF-I/CTF and NF-III/OTF-1) and a cellular topoisomerase activity (14–20). The biochemical roles of these viral and cellular replication proteins are discussed below.

Initiation of Adenovirus DNA Replication

THE PROTEIN PRIMING MODEL   Initiation of adenovirus DNA replication occurs by a novel mechanism in which the first nucleotide in the new DNA chain becomes covalently linked to a virus-encoded protein, the terminal protein precursor. This mechanism is unique among the DNA viruses of mammals, but a similar mechanism operates during the replication of other chromosomes, for example bacteriophage Φ29 (21). The protein priming model was first proposed following the discovery that the 5' ends of adenovirus DNA strands are covalently linked to the 55-kd terminal protein (22). Direct biochemical support for the model was obtained by analysis of the initiation reaction in the adenovirus cell-free replication system. Initial studies demonstrated that the adenovirus terminal protein is synthesized in the form of a larger 80-kd precursor (pTP), which is active in initiation (10, 13, 23, 24). The pTP is processed by proteolysis to the mature 55-kd form during packaging of the viral genome into virions (23). A series of isotope transfer experiments provided evidence that the critical first step in the replication reaction is the formation of an ester bond between the β-OH of a serine residue in the pTP and the α-phosphoryl group of dCMP, the first residue in the new DNA chain (10). The nascent strand then grows by extension from the 3' hydroxyl of the covalently bound dCMP residue (22). The subsequent development of a direct assay for the formation of a covalent complex between dCMP and the pTP (pTP-dCMP) made it possible to purify the pTP in functional form and to define the requirements for the initiation reaction (25–28). Work in a number of laboratories has shown that initiation is dependent upon the presence of specific nucleotide sequence elements at the termini of the viral genome and requires the participation of several viral and cellular proteins.

THE ADENOVIRUS ORIGIN OF DNA REPLICATION   The natural template for adenovirus DNA replication in vivo or in vitro is the viral chromosome with the covalently attached terminal protein (TP or pTP). However, as first demonstrated by Tamanoi & Stillman (28), plasmids containing the cloned adenovirus terminal sequence will support initiation of DNA replication in vitro, provided that the plasmid is cleaved with a restriction enzyme in such a way that the adenovirus terminus is located near the end of the resulting linear DNA molecule. This observation provided definitive evidence that specific nucleotide sequence elements in the viral genome are recognized by the initiation machinery. The efficiency of initiation with plasmid templates is considerably lower than that observed with adenovirus chromosomes isolated from purified virions. Moreover, recent studies have revealed that the protein and cofactor requirements for initiation are somewhat different for the two templates (29–32). However, analysis of plasmids with deletion and/or base substitution mutations has been very useful for defining the nucleotide sequence requirements for initiation.

Most in vitro studies of the nucleotide sequence requirements for adenovirus DNA replication have been conducted with serotypes 2 or 5. Analysis of a large number of deletion and base substitution mutations has revealed that the adenovirus origin of DNA replication is complex, containing at least three functionally distinct domains (17, 33–42). Domain A consists of the first 18 base pairs of the viral genome and represents the minimal origin of replication. The presence of domain A is absolutely required for initiation of adenovirus DNA replication, but templates containing only domain A initiate DNA synthesis at a very low efficiency. All adenovirus serotypes that have been examined share a common 10-base-pair sequence, ATAATATACC, within this region of the viral genome (34, 43, 44). It has been suggested that this conserved motif is important for the binding of the virus-encoded initiation proteins to the origin (32, 45). Domains B and C, while not absolutely required for initiation of adenovirus DNA replication, contribute significantly to the efficiency of the initiation reaction. Domain B consists of the segment between nucleotides 19 and 39 (36, 37, 40, 42). As in the case of domain A, there is considerable sequence conservation in this region among the various adenovirus serotypes. Many (but not all) adenovirus genomes contain a version of the consensus sequence TGG(A/C)NNNNNGCCAA. As described below, this motif is recognized by a cellular DNA-binding protein, nuclear factor I(CTF)(14–16). The presence of domain B increases the efficiency of initiation of adenovirus 5 DNA replication at least 10-fold. Domain C of the adenovirus origin includes nucleotides 40 to 51 and contributes an additional factor of three to the efficiency of initiation of viral DNA replication in vitro (17, 42). The consensus sequence, AT(G/T)N(A/T)AAT, has been identified in this region. A second cellular DNA-binding protein, nuclear factor III (ORP-C, OTF-1), recognizes this sequence (17–19). The spacing between the minimal origin and domain B appears to be critical for origin function (41, 42). The insertion or deletion of only a few base pairs between the two segments dramatically reduces the efficiency of initiation. This observation suggests that the initiation reaction may require relatively short-range interactions between the protein factors that bind to the various domains of the origin.

Analysis of the replication of deletion mutants in vivo is largely consistent with the general picture of the sequence organization of the adenovirus genome derived from the in vitro studies (46–51). Both the conserved sequence element in domain A and the nuclear factor I binding site in domain B have been shown to be essential for adenovirus 2 (or 5) DNA replication in cultured cells. The stimulatory effect of domain C has not yet been observed in vivo. In contrast to most adenovirus serotypes, the replication origin of adenovirus type 4 lacks a recognition site for nuclear factor I. It has been demonstrated that Ad4 DNA replication, both in vivo and in vitro, requires only the terminal 18 base pairs of the genome, which are identical to the minimal origin of Ad2 or Ad 5 (49, 52).

CELLULAR ORIGIN-BINDING PROTEINS   Nuclear factor I (NF-I) was originally identified as a cellular factor that stimulated formation of pTP-dCMP complexes by partially purified viral replication proteins (14). The stimulatory activity was subsequently purified to near homogeneity by recognition site affinity chromatography and shown to consist of a family of polypeptides with molecular weights between 52,000 and 66,000 (15). Recently, three human cDNA clones of NF-I have been isolated and characterized (53). The clones contain blocks of identical sequence interspersed with blocks of different sequence, suggesting that the corresponding mRNAs are generated by differential splicing. Analysis of the open reading frames in the clones suggests that each mRNA encodes a distinct protein. Thus, differential splicing may account, at least in part, for the multiplicity of NF-I polypeptides that have been observed. Although it has been demonstrated that the protein products of all three NF-I cDNA clones are active in stimulating adenovirus DNA replication in vitro, it is not yet clear whether all of the NF-I polypeptides are functionally equivalent (53).

The interaction between NF-I and its recognition sequence has been studied using chemical probes and in vitro mutagenesis (13, 39, 54). Taken together, these studies suggest that the optimal recognition site consists of the symmetrical sequence $TTGGCN_5GCCAA$. The principal contacts between protein and DNA are in the major groove, and nearly all of the contacts are accessible from one side of the helix (54). Given the symmetry of the protein-DNA contacts, it seems likely that NF-I binds as a dimer.

Purified nuclear factor I stimulates initiation of adenovirus DNA replication in vitro at least 30-fold (15, 42). The binding of NF-I to its recognition sequence is essential for the stimulatory effect, since base substitution mutations in the viral origin that abolish binding greatly reduce the efficiency of initiation (18, 36, 37, 39, 40, 42). The precise role of NF-I in the initiation reaction is not yet clear (see below).

Nuclear factor III (ORP-C) was identified as a stimulatory factor that recognized the sequence element TATGATAAT within domain C of the adenovirus 2 origin of replication (17, 18). The factor has been purified to homogeneity by recognition site affinity chromatography and shown to consist of a 92-kd polypeptide (55). Experiments with various chemical probes indicate that the protein makes both major and minor groove contacts with the DNA and that the contacts are not confined to one side of the helix (56). The binding site for NF-III is very close to that for NF-I; both proteins contact the same A/T base pair at position 39 in the adenovirus origin of replication (56). Despite their proximity, there is no evidence for cooperativity in binding (18, 56). Both NF-I and NF-III are required for optimal levels of DNA replication in vitro (17, 18, 55, 56), but the requirement for NF-III in vivo has not yet been demonstrated.

Interestingly, both nuclear factor I and nuclear factor III also appear to function as cellular transcription factors. A number of viral and cellular promoters contain functionally significant sequence elements that are related to the NF-I recognition sequence (57–67). A protein factor (CCAAT transcription factor or CTF) that recognizes the GCCAAT motif present in several such promoters has been purified to homogeneity and shown to be capable of stimulating transcription of the human α globin gene in vitro and in vivo (68). Purified CTF was found to consist of a series of polypeptides with molecular weights similar to those previously described for NF-I (15). Detailed comparison of the physical and biochemical properties of CTF with those of NF-I demonstrated that the two groups of proteins are indeed identical (16).

The recognition site for nuclear factor III is similar to the octamer sequence that has been implicated in the transcriptional regulation of several genes, including the histone H2b, immunoglobulin, and small nuclear RNA (U1 and U2) genes (17, 18). Binding studies have demonstrated that NF-III binds to the promoter/enhancer regions of several such genes (69). A 92-kd protein factor (octamer transcription factor or OTF-1) that recognizes the octamer sequence in the histone H2b promoter and markedly stimulates H2b transcription in vitro has been purified from HeLa cells (70). The purified OTF-1 protein is physically and biologically indistinguishable from NF-III (19). The implications of the finding that cellular sequence-specific DNA-binding proteins can participate in both transcription and DNA replication are not yet clear. On the one hand, adenovirus may have simply subverted cellular transcriptional factors for its own purposes. On the other hand, there is abundant circumstantial evidence for a fundamental relationship between transcription and replication in eukaryotic cells (5, 71–73). Indeed, several examples of transcriptional signals that significantly affect the efficiency of DNA replication are documented in other sections of this review. Further study of the roles of NF-I and NF-III in adenovirus DNA replication may lead to better understanding of the mechanistic role of transcriptional factors in DNA replication.

REQUIREMENTS FOR THE INITIATION REACTION   Initiation of adenovirus DNA replication is assayed in vitro by measuring the formation of a covalent complex between dCMP and the 80-kd pTP (25–28). The initiation reaction is absolutely dependent upon the presence of a DNA template. The most efficient template is the adenovirus chromosome containing the covalently attached 55-kd terminal protein, although other DNA molecules, such as linear plasmids or single-stranded DNA molecules, will support pTP-dCMP complex formation to some extent (see below). With adenovirus chromosomes as template, optimal initiation requires a minimum of four proteins: two cellular proteins, NF-I and NF-III (14, 15, 17, 18), and two virus-encoded proteins, pTP and the 140-kd adenovirus DNA polymerase (11, 74–76). The viral proteins copurify through several chromatographic steps, and their sedimentation behavior suggests that they exist in a 1:1 complex (11, 74, 75). The complex can be separated into the individual polypeptides by glycerol gradient sedimentation in the presence of urea (74, 75, 77). This has made it possible to demonstrate the absolute requirement for both the pTP and the DNA polymerase in the initiation reaction (74, 75, 77). Initiation also requires ATP, which appears to be serving an effector function, since nonhydrolyzable analogues also stimulate initiation, and no ATP hydrolysis has been detected (32). It has been reported that a third virus-encoded protein, the 72-kd single-stranded DNA-binding protein (DBP) (78), stimulates initiation several-fold, but the protein is clearly not an essential participant in the reaction (18, 32).

As described above, duplex templates that lack the covalently attached terminal protein (e.g. linear plasmids containing the adenovirus origin of DNA replication at one terminus) are capable of supporting the initiation reaction, albeit at lower efficiency than adenovirus chromosomes. With such templates, an additional protein, factor pL, is required for efficient formation of pTP-dCMP complexes (29, 30). Factor pL has been purified to near homogeneity and shown to be a 44-kd polypeptide with $5' \rightarrow 3'$ exonuclease activity (30). The pL exonuclease appears to activate adenovirus templates that lack the terminal protein by degrading the $5'$ end of the DNA strand that is normally displaced during adenovirus DNA replication (30, 31). This creates a short single-stranded region at the $3'$ end of the DNA strand that normally serves as the template for adenovirus DNA synthesis. Similar partially single-stranded templates, constructed using synthetic oligonucleotides, support the initiation reaction in the absence of factor pL (31). Thus, the presence of a single-stranded region at the $3'$ end of the template strand appears to allow the system to bypass the requirement for the 55-kd terminal protein on the input DNA. One possible interpretation of this result is that the 55-kd terminal protein attached to adenovirus chromosomes plays some role in opening the duplex during the early stages of the initiation process (see below). Except for the requirement for factor pL, the protein requirements for initiation on partially single-stranded templates are the same as those for adenovirus chromosomes (32). However, ATP is no longer required (32).

Single-stranded DNA molecules will also support formation of pTP-dCMP complexes in vitro (28, 29, 33, 35, 79). The sequence requirements for initiation with single-stranded templates appear to be somewhat less stringent than with adenovirus chromosomes or plasmid DNAs (35). A large number of single-stranded DNA molecules, including those lacking the specific adenovirus origin sequences, have been observed to support pTP-dCMP complex formation with varying degrees of efficiency. However, it has been reported that an oligonucleotide containing the template strand of the adenovirus origin is 5–20 times as active in initiation as other single-stranded DNA molecules

(32, 80). With single-stranded DNA templates, initiation is dependent on the virus-encoded replication proteins, but is not dependent on NF-I or NF-III (32). Both ATP and the adenovirus DBP are inhibitory.

MECHANISM OF INITIATION   The precise order of events during initiation of adenovirus DNA replication is not yet clear. The finding that single-stranded or partially single-stranded templates can support pTP-dCMP complex formation suggests that initiation is a two-step process (Figure 1). In the first step, the terminal region of the viral genome is unwound, exposing a short single-stranded region. In the second step, a dCMP residue is covalently linked to the pTP. Elucidation of the protein requirements for localized unwinding at the adenovirus origin must await the development of a direct unwinding assay. However, the available data suggest that the 55-kd terminal protein attached to the template DNA may play a role in unwinding, since the requirement for the terminal protein is obviated by presence of a single-stranded region at the end of the template. It is also possible that NF-I, NF-III, pTP, or the adenovirus polymerase participate in strand opening within the origin. The binding of the pTP and adenovirus DNA polymerase to the template strand of the origin presumably takes place following (or perhaps during) the unwinding reaction. There is some evidence that the complex of pTP and Ad polymerase may interact with sequence elements in the origin, but these interactions must be of relatively low specificity. It is possible that one role of NF-I and NF-III is to facilitate the binding or positioning of the pTP and adenovirus DNA polymerase on the DNA template. The final step in initiation, the formation of the covalent pTP-dCMP complex, probably occurs once the pTP-Ad pol complex is correctly positioned on the exposed template strand. Although both the pTP and the adenovirus DNA polymerase are required for the initiation reaction, an important unresolved question is whether the adenovirus DNA polymerase catalyzes the transfer of dCMP to the pTP or whether this is an autocatalytic process.

Elongation of Nascent DNA Chains

The synthesis of full-length adenovirus DNA strands in vitro requires the pTP, the Ad DNA polymerase, the Ad DBP, and nuclear factors I–III. At present there is good evidence that three of these proteins (Ad DNA polymerase, Ad DBP, and nuclear factor II) are directly involved in chain elongation (20, 81, 82). Although there is no direct evidence for the involvement of the other adenovirus replication proteins in elongation, this possibility cannot be completely ruled out. The adenovirus DNA polymerase is a 140-kd protein with physical and biochemical properties distinct from the other known eukaryotic DNA polymerases (11, 74, 81, 82). The enzyme is capable of utilizing a variety of deoxyribonucleotide homopolymer template-primers,

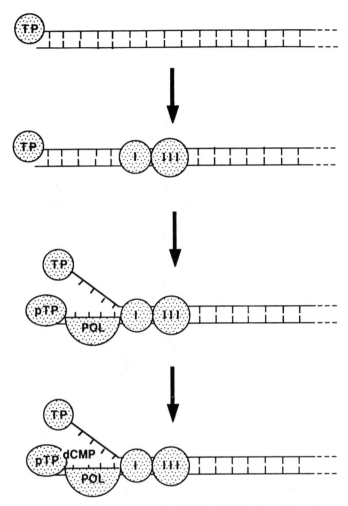

Figure 1 Diagrammatic representation of the initiation of adenovirus DNA replication. See text for details. Abbreviations used: I, nuclear factor I; III, nuclear factor III; TP, terminal protein; pTP, preterminal protein; POL, adenovirus DNA polymerase.

but is relatively inactive with RNA primers (82). Polymerase activity is inhibited by dideoxynucleotides and is resistant to aphidicolin. The purified polymerase contains an intrinsic 3'→5' exonuclease activity that is specific for single-stranded DNA and probably serves a proofreading function during polymerization (82). The Ad DBP is a 59-kd phosphoprotein that migrates in SDS-polyacrylamide gels with an apparent molecular weight of 72,000 (12, 83). The DBP binds tightly and cooperatively to single-stranded DNA in sequence-independent fashion (12, 83–87). At saturation approximately one molecule of DBP is bound for every seven nucleotides (85). With poly(dT) as template and oligo(dA) as primer the DBP stimulates DNA synthesis by the Ad DNA polymerase as much as 100-fold (82). The stimulatory effect is quite specific, since the *Escherichia coli* SSB has no effect on Ad DNA polymerase activity and the Ad DBP does not stimulate the activity of other DNA polymerases such as HeLa DNA polymerase α. Based upon these and other results (81, 82), it seems likely that there is a highly specific interaction between the DBP and the Ad DNA polymerase that increases the efficiency of polymerization; however, a stable complex of the two proteins has not yet been detected. In the presence of the DBP, the Ad polymerase is a highly processive enzyme, capable of synthesizing DNA chains at least 30,000 nucleotides in length from a single primer terminus (82). Moreover, under these conditions the polymerase appears to be able to translocate through long stretches of duplex DNA (81). Thus, it is possible that fork movement during adenovirus DNA replication does not require a separate helicase activity. Rather, unwinding of the parental strands may be mediated solely by the Ad DNA polymerase and DBP, and the energy required for unwinding may be provided by the hydrolysis of deoxynucleoside triphosphates. This possibility is consistent with the observation that little, if any, ATP hydrolysis occurs during adenovirus DNA replication in vitro (81).

In the presence of the pTP, the adenovirus DNA polymerase, the DBP and NF-I nascent adenovirus DNA chains are elongated to only about 25% of full length (20). Synthesis of complete adenovirus DNA strands requires an additional cellular protein, nuclear factor II (20). Nuclear factor II from HeLa cells has a native molecular weight of approximately 30,000 and copurifies with a DNA topoisomerase activity. Human or calf thymus topoisomerase I (but not *E. coli* topoisomerase I) will substitute for nuclear factor II in the adenovirus DNA replication reaction. The precise function of nuclear factor II in adenovirus DNA chain elongation is not yet clear. Since the protein has no significant effect on the synthesis of nascent strands up to 9 kb in length, it is presumably required to overcome the inhibitory effects of some DNA structure that appears only after extensive DNA synthesis.

SV40 DNA REPLICATION

SV40 has proven to be an excellent model system for studying the mechanisms of cellular DNA replication (88–91). The viral genome consists of a circular duplex DNA molecule of about 5000 base pairs and contains one origin of DNA replication. SV40 DNA replication takes place in the nucleus of the host cell where the viral genome is complexed with histones to form a nucleoprotein structure (minichromosome) indistinguishable from cellular chromatin. Since SV40 encodes only a single replication protein (T antigen), the virus makes extensive use of the cellular replication machinery. As a result there are many similarities between viral and cellular DNA replication. In both cases initiation of DNA synthesis results in the establishment of two replication forks that move in opposite directions. At each fork one of the two nascent strands (the leading strand) grows continuously, while the other strand (the lagging strand) grows discontinuously by joining together small (ca. 200 bp) segments of DNA that are independently initiated with RNA primers. Completion of replication occurs when two oppositely moving forks meet. In linear cellular chromosomes the two merging forks originate from adjacent origins, while in circular SV40 chromosomes they have a single origin.

Much has been learned about SV40 DNA replication from in vivo studies (see reviews in Refs. 88–91). However, the recent development of an efficient cell-free replication system has greatly accelerated progress in understanding the molecular mechanisms involved (92). An important dividend of the dissection of the cell-free system has been the identification and functional characterization of components of the cellular replication apparatus. Thus, this review focuses on in vitro studies.

Initiation of SV40 DNA Replication

THE ORIGIN OF DNA REPLICATION   The SV40 origin of replication is a 64-base-pair segment of the viral genome that contains all of the nucleotide sequence elements that are required for initiation of viral DNA replication in vitro and in vivo (93–106). Careful genetic analysis of base substitution mutations has revealed that the origin is complex, consisting of at least three functionally distinct sequence domains (104–106). At the center of the origin are four copies of a pentomeric sequence motif (GAGGC) organized as an inverted repeat. This sequence element is recognized by the viral initiation protein, T antigen (101, 106–120). On one side of the T antigen–binding site is a 17-base-pair segment containing A/T base pairs (105). It is suspected that this is the initial site of strand opening during initiation of SV40 DNA replication. On the other side of the T antigen–binding site is a 15-base-pair imperfect palindrome of unknown function. All three sequence domains of the origin are required for SV40 DNA replication, and there is some evidence that the spacing between them is critical for origin function (104).

Although the 65-base-pair core origin region is sufficient to support the initiation of SV40 DNA replication, sequences outside of the core can significantly influence the efficiency of initiation. A second T antigen–binding site located adjacent to the core origin increases replication efficiency several-fold both in vivo and in vitro (97, 101–103, 121, 122). Of even greater importance are elements previously associated with the activation of SV40 transcription, such as the SV40 enhancers or the binding sites for the transcriptional factor Sp-1. The presence of either of these sequence elements adjacent to the core origin increases the efficiency of DNA replication in vivo at least 10-fold (99–103, 121, 123). For maximal stimulation of SV40 DNA replication, the transcriptional elements must be relatively close to the core origin. Insertion of 180 base pairs between the core origin and the Sp-1 binding sites completely abolishes the effect (121, 124). As documented elsewhere in this review, the activation of DNA replication by enhancers and other transcriptional elements is not limited to SV40 and, in fact, appears to be a quite general feature of the replication of eukaryotic viruses. In the case of SV40, there is some evidence that the binding of transcriptional activator proteins affects replication indirectly by perturbing the local distribution of nucleosomes, so that the DNA in the adjacent core region is relatively nucleosome-free. This presumably facilitates the interactions of the core origin with T antigen and other initiation proteins. Direct analysis of SV40 chromatin isolated from infected cells has revealed that the core origin region is less likely to be packaged into a nucleosome than other regions of the viral genome (124–129). In addition, studies of viral mutants have demonstrated that the genetic determinants of the nucleosome exclusion effect reside in the SV40 enhancer elements and the Sp-1-binding sites (99, 130, 131). Thus, although other models have not been ruled out, it seems likely that the stimulatory effect of the transcriptional elements is due, at least in part, to effects on chromatin structure.

SV40 T ANTIGEN   The SV40 T antigen is a virus-encoded phosphoprotein with a polypeptide molecular weight of 82,000 (88–91, 132, 133). The protein plays the central role in initiation of viral DNA replication. Binding studies suggest that a T antigen molecule binds to each of the four pentamer repeats in the origin to form an organized nucleoprotein structure that is competent for initiation (134, 135). The precise number of T antigen monomers involved in formation of the T antigen/origin complex is not known. At physiologic temperatures complex formation is greatly facilitated by ATP (136, 137). Binding of T antigen appears to cause significant changes in the local DNA structure of the origin. Chemical probes have revealed destabilization of the helix in the region of the imperfect 15-base-pair palindrome and a structural deformation of unknown character in the AT-rich region (138).

Recently, considerable progress has been made in understanding the role of T antigen in the initiation of SV40 DNA replication. This is in large measure due to the discovery that T antigen has an intrinsic helicase activity (139). Once it is bound to the origin, T antigen is capable of entering the duplex and catalyzing the ATP-dependent unwinding of the two DNA strands (140–142). Unwinding appears to be a critical step that establishes the replication forks and generates the substrate that is required for the priming and elongation of nascent strands. The helicase activity of T antigen may also be involved in the elongation process itself (see below).

The unwinding reaction requires the presence of the AT-rich segment of the origin as well as the T antigen recognition site (143). Analysis of base substitution mutations in the AT-rich segment indicates that the precise nucleotide sequence of the segment, not just its base composition, is an important determinant of the efficiency of unwinding (105, 120, 143). One possible explanation for this observation is that T antigen makes sequence-specific contacts with the AT-rich domain during the course of the unwinding reaction. A second possibility is suggested by the finding that the AT domain induces a significant bend in the helix (105). Several single-base substitution mutations within the domain that reduce the efficiency of unwinding and replication also change the degree of net bending (105, 143). Thus, the conformation of the DNA in this domain may be critically important. For example, bending could contribute to the destabilization of the duplex in the AT domain or could facilitate the interaction of the domain with T antigen during entry or unwinding.

CELLULAR PROTEINS   In addition to specific nucleotide sequence elements, the T antigen–mediated unwinding reaction requires accessory proteins contributed by the host cell. For example, a single-stranded DNA-binding protein is required to prevent reassociation of the single strands exposed during unwinding (140–142). It seems likely that this function is normally fulfilled by a recently identified cellular protein designated replication protein A (RP-A or RF-A) (144, 145). RP-A has been purified to homogeneity, and the purified protein is absolutely required for SV40 DNA replication in the reconstituted cell-free system (144, 145). The protein consists of three tightly associated subunits of 70 kd, 32 kd, and 14 kd. The largest subunit binds specifically to single-stranded DNA (146). Heterologous single-stranded DNA binding proteins, such as E. coli SSB, will substitute for RP-A in the unwinding reaction (140, 141); however, E. coli SSB cannot replace RP-A in the complete DNA replication reaction, indicating that RP-A must play other roles in DNA replication (141). A single-stranded DNA-binding activity associated with polypeptides of 72 kd and 76 kd has also been identified in protein isolated from HeLa cells (147). This activity stimulates SV40 DNA replication in vitro and may be related to RP-A. Although T antigen and RP-A appear to be the only proteins that are absolutely required for origin-dependent unwinding, recent evidence indicates that the efficiency of unwinding is increased significantly by a second cellular protein, designated RP-C (146). Although it seems likely that RP-C interacts with T antigen during the early stages of the reaction, its precise function is not yet clear.

Elongation of Nascent DNA Chains

Fractionation of the cell-free SV40 system has yielded considerable information on the proteins involved in the elongation of nascent SV40 DNA chains.

DNA POLYMERASE α Mammalian cells contain four distinguishable DNA polymerase activities designated α, β, γ, and δ (148, 149). Of these, DNA polymerase α and probably DNA polymerase δ are required for SV40 DNA replication (92, 146, 150–156). DNA polymerase α has long been considered to be the major replicative polymerase in animal cells and has been purified from a variety of sources by several laboratories (148, 150, 157). There is general agreement that the enzyme is composed of four distinct subunits (148, 157). The largest subunit (180 kd) contains the polymerase active site. The large subunit of the *Drosophila* DNA polymerase α has also been shown to harbor a cryptic $3' \rightarrow 5'$ exonuclease that serves a proofreading function during polymerization (158). The exonuclease activity is normally masked by a second subunit of the polymerase with a molecular weight of 70,000. The exonuclease activity has not yet been detected in enzymes from species other than *Drosophila,* but its presence seems likely given the apparent structural conservation of DNA polymerase α during evolution. The proofreading exonuclease activity is probably an important general feature of DNA polymerase α that contributes significantly to the fidelity of DNA replication. The smallest subunits of DNA polymerase α (50–60 kd) constitute a primase enzyme capable of synthesizing short RNA transcripts that can serve as primers for subsequent DNA chain elongation by the catalytic subunit (148, 157, 159, 160). The mammalian DNA polymerase α is not a highly processive enzyme, as less than 100 nucleotides are polymerized per binding event under the usual assay conditions (161–164).

There is evidence that DNA polymerase α can form a specific complex with the SV40 T antigen. Thus, certain monoclonal antibodies against T antigen (or DNA polymerase α) will coprecipitate the two proteins from cell extracts (165, 166). The specificity of this interaction may play some role in determining the host specificity of SV40. For example, it has been observed that preparations of human DNA polymerase α, but not murine DNA polymerase α, can activate T antigen–dependent DNA replication in extracts of murine cells that are normally defective in SV40 DNA replication (151). In both lytically infected and transformed cells, T antigen is also associated with a cellular protein of unknown function referred to as p53 (167–169). Recent experiments suggest that murine p53 competes with DNA polymerase α for binding to T antigen (166). The significance of this observation is not yet clear.

DNA POLYMERASE δ AND PCNA   DNA polymerase δ was originally distinguished from DNA polymerase α because it contains a readily detectable 3'→5' exonuclease activity (170). The two polymerases also differ from one another in their template preferences, chromatographic properties, and sensitivities to various inhibitors (149, 170–174). Monoclonal antibodies against DNA polymerase α do not inhibit the activity of DNA polymerase δ, and DNA polymerase δ appears to lack an intrinsic primase activity. Thus, it appears likely that DNA polymerases α and δ are distinct molecular entities. However, since the structure of DNA polymerase δ has not been characterized in detail, it remains possible that the two polymerases share similar or identical subunits.

A 37-kd protein that dramatically increases the activity of DNA polymerase δ with template/primers containing long single-stranded regions was recently purified to homogeneity (175). Interestingly, the 37-kd protein proved to be identical to a previously described polypeptide found specifically in proliferating cells and referred to as proliferating cell nuclear antigen (PCNA) or cyclin (153, 154, 176). In the presence of PCNA, DNA polymerase δ is a highly processive enzyme capable of catalyzing the polymerization of at least 1000 nucleotides per binding event (164, 175). PCNA has no effect on the activity or processivity of DNA polymerase α (164).

It has been demonstrated by direct reconstitution that PCNA is required for efficient SV40 DNA replication in the cell-free system and is probably involved in DNA chain elongation (146, 152–154). In the absence of PCNA, initiation of DNA synthesis at the origin occurs, but only short nascent strands, containing a maximum of a few hundred nucleotides, are synthesized. The requirement for PCNA suggests the possibility that DNA polymerase δ may be involved in SV40 DNA replication, a view that is also supported by inhibitor studies. More direct evidence on this point has come from recent experiments suggesting that both DNA polymerase α and DNA polymerase δ are required to reconstitute efficient DNA replication in the SV40 cell-free system (146).

THE TWO POLYMERASE MODEL   If both DNA polymerase α and DNA polymerase δ are involved in DNA replication, it would seem likely that they fulfill different functions. One can envision several possible ways that two polymerases could divide the labor of replication, but one particularly interesting possibility is that DNA polymerase δ serves as the leading strand polymerase and DNA polymerase α serves as the lagging strand polymerase (164). This model is consistent with the known biochemical properties of the two enzymes. The leading strand polymerase would be expected to be highly processive and would derive little benefit from an associated primase activity.

The lagging strand polymerase, on the other hand, would require only moderate processivity, but would benefit enormously from a tightly associated primase activity. The model is also supported by recent studies indicating that PCNA, the accessory factor for DNA polymerase δ, is required for leading strand synthesis in the reconstituted SV40 DNA replication system (177).

TOPOISOMERASES   The roles of DNA topoisomerases in SV40 DNA replication have been investigated by depleting the cell-free system of topoisomerase activities with specific antibodies and then reconstituting with the purified enzymes (178). These experiments have demonstrated that DNA topoisomerase activity is absolutely required for SV40 DNA replication and have established two distinct roles for DNA topoisomerases in the replication process. One role is to act as a swivel to relieve superhelical tension that would otherwise hinder the unwinding of the parental strands as the replication forks advance. Either of the two known mammalian DNA topoisomerases, topoisomerase I or topoisomerase II, can provide this function. The second role is to mediate the separation of the newly synthesized daughter duplexes at the completion of DNA replication. In general, the links between the parental strands are not completely removed prior to termination of SV40 DNA synthesis, so the immediate products of replication consist of two circular DNA molecules that are multiply intertwined (179). The final act of replication is the segregation of these intertwined daughter molecules into two separate unlinked molecules. This reaction is catalyzed by topoisomerase II. It is of interest to note that genetic experiments in yeast indicate that the topoisomerases probably play these same two roles during the replication and segregation of cellular chromosomes (180–182).

HELICASE   Unwinding of the parental DNA ahead of the advancing replication fork presumably requires the action of a helicase. The identity of the helicase activity required for fork movement during SV40 DNA replication has not been established with certainty. One reasonable possibility is that T antigen fulfills this function in addition to its role in unwinding the origin region prior to initiation of DNA synthesis (183). In the presence of a single-stranded DNA-binding protein, the intrinsic helicase activity of T antigen is capable of unwinding long segments of duplex DNA in a highly processive manner (142, 184, 185). Moreover, it has been reported that monoclonal antibodies that inhibit the helicase activity of T antigen also inhibit ongoing chain elongation in a subcellular SV40 DNA replication system (183). Studies with model substrates indicate that the T antigen helicase translocates in the 3' to 5' direction on single DNA strands (184, 185). Thus, if T antigen is the helicase responsible for fork involvement, it would be expected to move along the leading strand. In contrast, all of the prokaryotic helicases that are known to be directly involved in DNA replication move along the lagging strand (185).

It remains possible that a cellular DNA helicase, rather than T antigen, is responsible for unwinding the parental strands at SV40 replication forks. While helicase activities have been identified in extracts of eukaryotic cells, it is not yet clear whether any of these activities are involved in SV40 (or cellular) DNA replication (186–188). It has also been reported that DNA polymerase δ is capable of some degree of strand displacement during DNA synthesis, suggesting the possibility that a separate helicase activity may not be required (163). Alternatively, the helicase activity required for chromosomal replication may reside in an as-yet-unidentified cellular protein.

HERPES SIMPLEX VIRUS DNA REPLICATION

The herpesviruses are DNA-containing viruses infecting a variety of animal species. Of the more than 80 different herpesviruses that have been isolated, the most extensively characterized are the human viruses herpes simplex virus type 1 (HSV-1) and the closely related virus, HSV-2 (collectively, HSV). HSV replicates lytically in epithelial cells in infected individuals and in a wide variety of cultured mammalian cells. The molecular biology of HSV has been studied extensively and a great deal is known about the structure of the genome, and the arrangement and regulated expression of viral genes (reviewed in Refs. 189–191). HSV DNA is a linear double-stranded molecule about 153 kb in size. The genome consists of two components, L and S, each of which is flanked by inverted repeat sequences (see Figure 2). Although the inverted repeats flanking the L and S components are different, they share a small terminal sequence (termed the "a sequence") so that HSV DNA is terminally redundant. During the course of viral replication, the two components invert relative to each other, so that purified viral DNA consists of an equimolar population of four isomers that differ in the relative orientations of L and S. It is clear that isomerization is not an obligatory feature of HSV DNA replication. Mutant viruses containing a deletion of sequences at the L/S junction are frozen in a single isomeric arrangement but nevertheless are viable and replicate DNA at normal levels (192). The HSV genome has now been completely sequenced (193). Analysis of the sequence suggests that HSV encodes a minimum of 72 proteins. Splicing of HSV genes is very uncommon, a fact that has greatly simplified the genetic analysis of this virus (189–191, 193).

Like all of the herpesviruses, HSV has the capacity to remain latent in an infected host. In the case of HSV, the cells that harbor the latent viral genome are neurons in sensory ganglia (reviewed in Refs. 194, 195). It is not yet known whether HSV, like some of the other herpesviruses that are latent in

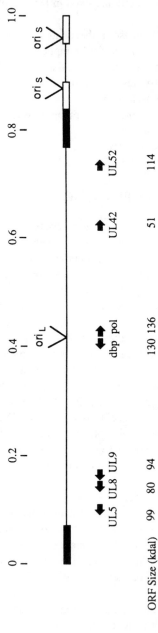

ORF Size (kdal)

Figure 2 Organization of the herpes simplex virus genome. Unique regions of the genome are indicated by thin lines. Filled boxes and open boxes indicate the inverted repeats flanking the long unique region and short unique region, respectively. Arrows indicate open reading frames (ORFs) encoding proteins involved in DNA replication.

Figure 3 Sequence of the herpes simplex virus origins of DNA replication. The nucleotides in ori$_L$ that differ from those in ori$_S$ are underlined.

dividing cells, has a regulated mode of DNA replication in the latent phase of infection; since neurons do not divide, however, there is no necessity for such a system. In this review we focus on the only documented mode of HSV DNA synthesis: the DNA synthesis that occurs during lytic growth of the virus.

In Vivo Studies

Studies of the overall mechanism of the HSV DNA replication in vivo have been hampered by the large size of the viral genome, but several important facts have emerged from analyses of the properties of replication intermediates and from electron microscopic examination of viral DNA isolated from infected cells (196–198). Replication intermediates have two characteristic properties. First, DNA pulse-labeled in vivo with 3H-thymidine sediments more rapidly than unit length viral DNA (196). Second, pulse-labeled DNA is "endless"; i.e. the molecular termini of mature viral DNA are fused together (197, 198). On the basis of these observations, it has been proposed that parental linear viral DNA is circularized shortly after entry into the host cell, and that replication takes place predominantly by a rolling circle mechanism, generating linear concatamers of tandemly repeated viral genomes (197). Although this remains an attractive hypothesis, it is largely untested.

Circularization of the incoming viral genome prior to the onset of DNA synthesis is certainly the simplest explanation for the complete lack of genomic termini in replicating DNA. In fact, unit-length circular genomes have been observed by electron microscopic analysis of viral DNA in infected cells (196). A comparison of the structure of the termini of mature DNA and L/S junction sequences suggests that circularization takes place by the direct ligation of the ends of linear viral DNA (199).

The mechanism(s) responsible for the very rapid sedimentation properties of replicating viral DNA is less clear. Certainly rolling circle–type replication would account for such intermediates. Other means of generating fast-sedimenting DNA, however, are plausible. For example, it is possible that replication takes place by a "Cairns" type mechanism and that segregation of daughter molecules is slow (198). Alternatively, daughter molecules may be rapidly joined together by homologous recombination. There is some indirect evidence that is consistent with the latter possibility. First, the structure of replicating viral DNA may not be simple. Electron microscopic studies have suggested that much of the viral DNA in infected cells is composed of complex networks (196). Second, several studies have shown that HSV DNA in infected cells undergoes high levels of homologous recombination (200–202). The rate of recombination has been estimated at about 1–3% per kb per infectious cycle (200). This high rate of recombination is closely linked to DNA synthesis (203). By analogy with bacteriophage T4, which produces replication intermediates with very similar properties (204, 205), it is possible that the networks are formed by a combination of DNA synthesis and recombination and that resolution of recombinational intermediates is a slow step. Clearly, this overall scheme alone could account for the sedimentation properties of pulse-labeled DNA, although recombination and rolling circle replication are, of course, not mutually exclusive. Recently developed techniques such as pulse-field gel elecrotrophoresis may provide a means to reevaluate some of these questions.

Replication Origins

The existence of *cis*-acting replication origins was first inferred from the structure of defective genomes (206–212), which arise during serial passage of HSV at high multiplicities of infection. Individual isolates of such defective genomes have several common features. They all consist of many tandemly repeated identical copies of small segments of viral DNA; the sequences present in a single repeat unit are not necessarily colinear with a single segment of viral DNA. Two classes of defectives were recognized (206–214). Class I defectives contain sequences from the short inverted repeat segment of the viral genome including the "a" sequence. Class II defectives contain sequences from near the middle of the long unique region, as well as the "a" sequence. It was proposed that the sequences present in defective genomes contain two *cis*-acting signals: an origin of replication, many copies of which would account for the selective advantage of defectives; and the "a" sequence, required for the formation of genomic termini from "endless" intermediates and for packaging of the DNA into infectious virions (212, 215, 216, 218, 219). This general picture of defectives has now been verified in several laboratories. Both the origin sequences and the cleavage/packing site have been cloned from the wild-type genome and analyzed in some detail. The analyses of origin sequences will be discussed below; since the cleavage/packaging signal is not required for DNA replication per se, it will not be discussed further. The interested reader is referred to several recent publications (221–223).

As implied from the existence of two classes of defective genomes, HSV contains two distinct origin sequences (212). As mentioned, both of the sequences have been cloned from the wild-type viral genome (211–219, 225–227). Plasmids containing either of these origin sequences are amplified when introduced into HSV-infected cells by transfection, and this transient plasmid amplification assay has formed the basis for most of the functional analyses of origin sequences. The origin present in class I–defective genomes has been designated ori_S (216, 218, 219). Since it is located within the inverted repeat sequence of the short component of viral DNA, there are two copies of ori_S in the HSV genome. The origin contained in class II–defective genomes, designated ori_L, is located in the middle of the long unique component of viral DNA (209, 214). The sequences of ori$_S$ and ori$_L$ are closely related (224, 227; see Figure 3). Both contain a rather extensive inverted repeat sequence, the central 18 base pairs of which are exclusively AT base pairs. The inverted repeat of ori$_L$ is considerably longer than that of ori$_S$; for this reason, plasmids containing ori$_L$, but not ori$_S$, are highly unstable in *E. coli*. Therefore, most of the functional analyses of HSV origin sequences have been done with ori$_S$. As described below, the minimum sequences required for the function of ori$_S$ correspond well to the region of highest similarity with ori$_L$ (228). Thus it is likely that ori$_S$ and ori$_L$ are functionally equivalent, although direct evidence for this supposition is currently lacking. The functional significance of three separate origins of replication in the HSV genome is not clear. Mutant viruses lacking ori$_L$ or one copy of ori$_S$ have been isolated and have no obvious defect in growth (229, 230). Attempts to construct a mutant virus lacking both copies of ori$_S$ have been unsuccessful to date; this seems to imply that virus replication requires at least one copy of ori$_S$ or, alternatively, at least two origin sequences, whether they be ori$_L$ or ori$_S$. Additional genetic experiments will be required to answer these questions.

Several laboratories have carried out deletion analyses of plasmids containing ori$_S$, with somewhat conflicting results (218–220, 228, 231). In the most extensive recent study (228), the left-hand boundary of the minimal or core ori$_S$ sequence was shown to lie between nucleotides 5 and 11 in the sequence shown in Figure 3, about 20 nucleotides to the left of the left arm of the palindrome. The right-hand boundary was located between nucleotides 74 and 77, corresponding almost precisely to the outer border of the right arm of the central palindrome. As detailed below, the two arms of the palindrome contain binding sites for a viral-encoded protein required for DNA synthesis, UL9. Deletion of the central AT-rich component of the palindrome completely eliminated DNA replication, as did replacement of the central AT region with an equal number of GC base pairs. Insertion of AT base pairs at the center of symmetry had an oscillating effect on function, with a periodicity of approximately 10 base pairs. Taken together with studies on the interaction between the origin and the HSV origin binding protein (see below), these results suggest that that ori$_S$ is composed of at least four distinct domains: the two arms of the palindrome, which serve as binding sites for UL9; the central AT-rich region, which serves as a spacer between the protein-binding sites and, by analogy with other origin sequences, potentially as a site of protein-induced unwinding; and the 20 or so base pairs to the left of the palindrome, to which no precise function has as yet been ascribed. As noted earlier, ori$_L$ is very similar to ori$_S$ throughout this core region (see Figure 3). Moreover, because of the greater degree of symmetry in ori$_L$, the left-most domain of ori$_S$ is represented twice in ori$_L$.

While there is general agreement among several laboratories that the left arm of the palindrome and the AT-rich region at the center of the palindrome are essential components of ori_S, there are conflicting results on the requirement for the right arm of the palindrome. In contrast to the results of Lockshon & Galloway (228), Deb & Doelberg (231) have reported that a plasmid completely lacking the right arm replicates equally as well as a control plasmid containing the intact origin. Studies on the origin of another alpha herpesvirus, varicella zoster virus (VZV), provide some additional support for this view (232). HSV-infected cells support the replication of plasmids containing the VZV origin of replication at a level about 10% that of plasmids containing ori_S. The VZV origin contains sequences homologous to the left arm of the HSV palindrome and a somewhat expanded AT-rich region to the right, but no sequences corresponding to the right arm of the ori_S palindrome. Clearly, more work needs to be done to resolve the discrepancies in the identification of the minimal HSV origin sequence.

The minimal HSV origin sequence, whether it includes the right arm of the origin palindrome or not, does not contain any obvious transcriptional regulatory sequences. Both ori_S and ori_L are located between divergently transcribed genes (233–237), but there is no evidence that either the promoters for these genes or any upstream regulatory elements have any effect on origin function. There is evidence that sequences flanking the core origin sequence do have a modest stimulatory effect on the extent of DNA replication in the plasmid amplification assay (218, 219, 228), but whether or not these stimulatory sequences are related in any way to transcriptional enhancer sequences, as is the case for other origins, remains to be determined. There is a transcript in HSV-infected cells that extends through ori_S (238). Neither the 5' nor 3' end of this transcript is located within sequences that have an effect on origin function. It has been suggested that this transcript plays a role in regulating DNA replication, but no evidence supports this speculation as yet.

Clearly, many questions remain concerning the HSV origins. By analogy with other better characterized systems, it seems likely that DNA replication begins at one or more of the origin sequences. There is no information as yet, however, on the molecular events by which HSV DNA replication is initiated. It seems likely that further insight into this process will await the development of a soluble, origin-dependent in vitro system.

Genetics of Viral DNA Replication

A large number of conditionally lethal mutants of HSV have been isolated and characterized over the past decade, and many of these mutants have defects in DNA synthesis (reviewed in Ref. 239). Some mutations, such as those occurring within the gene encoding an immediate early transcriptional regulatory protein (240, 241), affect viral DNA synthesis indirectly, but many mutants apparently have defects directly affecting DNA replication. There are seven complementation groups of such mutants now known (241–254), and for reasons that will be discussed below, it is unlikely that any more will be found.

The complete set of viral genes that are required for DNA replication were identified by means of a transient complementation assay in which cloned segments of HSV DNA were tested for the ability to support the replication of a cotransfected plasmid containing ori_S or ori_L (255). Seven genes were found to be both necessary and sufficient for origin-dependent DNA synthesis (256; see Figure 2). Detailed mapping of available ts mutants with clear DNA^- phenotypes has shown that all such mutants have defects in one of the seven genes identified using the transient assay system (241–249, 252, 254). In addition, viruses containing targeted null mutations (insertions or deletions) in five of these seven genes have now been isolated (251–253; P. A. Schaffer, personal communication; D. M. Knipe, personal communication). These viruses, which can be propagated in complementing cell lines expressing the wild-type gene, all fail to synthesize viral DNA in noncomplementing host cells. Thus, there is complete correspondence between the genes required for DNA synthesis in the transient system and the genes identified by mutational analysis to be directly involved in viral DNA replication.

The functions of the products of most of these genes are either unknown or only recently identified (see below). Thus, the use of virus mutants to analyze the properties of replication proteins has been limited. The exceptions are the genes encoding a DNA polymerase and a single-stranded DNA-binding protein, which have been known for some time and have been the subject of genetic analysis in several labs. Since, as mentioned above, directed mutagenesis of essential HSV genes is now possible, a great deal more information of this sort should be forthcoming in the near future.

Several temperature-sensitive mutants with amino-acid substitutions in the HSV DNA polymerase have been characterized (241–245). All of these mutants fail to synthesize viral DNA at the restrictive temperature. Since the DNA polymerase is the target of several antiviral drugs (deoxynucleotide and pyrophosphate analogues), it has also been possible to isolate virus mutants in which the encoded polymerase has altered sensitivities to these drugs (258–265). All such mutations that have been mapped carefully occur in the carboxy-terminal half of the DNA polymerase (from amino acid number 397 to 961). It has been proposed that these amino acids contain the nucleotide-binding domain of the DNA polymerase (266, 267, 267a). In this context it is also of interest that several regions of homology among the amino acid sequences of many DNA polymerases, both eukaryotic and prokaryotic, have been noted (266–268). Most of these regions of homology occur within the sequence of the HSV polymerase in the region proposed as the nucleotide-binding domain. Several polymerase mutants selected for resistance to the inhibitor phosphonoacetic acid were shown to also have an anti-mutator phenotype (269). The biochemical properties of the mutant polymerases that account for this phenotype have not yet been determined.

The other HSV DNA replication gene that has been extensively analyzed is that encoding a single-stranded DNA-binding protein (infected-cell polypeptide 8, or ICP8). Temperature-sensitive mutants with amino acid substitutions in the ICP8 coding sequence fail to synthesize viral DNA at the restrictive temperature (246–248). Some mutant ICP8 proteins are defective in nuclear localization, while others have been shown to have a defect in DNA binding (270–273). Fine mapping of mutants that encode ICP8 proteins that do localize to the nucleus has revealed that at least some of the residues between 346 and 450 are necessary for DNA binding (273). Several ts mutants with defects in ICP8 have also been shown to display altered sensitivities (at the permissive temperature) to drugs that inhibit the viral DNA polymerase (274). Finally, some ICP8 ts mutants overexpress certain late genes at the nonpermissive temperature (275, 276). Thus it is possible that ICP8 plays a role in the regulation of viral gene expression in addition to its role in viral DNA replication. It is not yet known whether such a regulatory function is related to the ability of ICP8 to bind to single- and double-standed nucleic acids, or whether it represents a completely unrelated activity.

In ending this section on the genetics of viral DNA replication, it is important to note that although HSV contains only seven genes that are required for viral DNA replication in cultured cells, the virus clearly encodes several other proteins that are also likely to be involved in DNA metabolism. These include a thymidine kinase (277), a ribonucleotide reductase (278, 279), a dUTPase (280), a uracil-N-glycosidase (281), and a nuclease (282, 283). Genetic studies have shown that while mutations in these genes have only minor or no effects on viral DNA synthesis in infected cells in culture, they may cause profound defects in the ability of the virus to replicate following experimental infection of animals (284–286; S. K. Weller, personal communication). It is likely, therefore, that rapidly dividing cultured cells provide many functions that are lacking in the cells encountered by the virus during a natural infection. Whether host cells provide any functions necessary for DNA replication that are not encoded by the virus remains to be determined.

Analysis of the Sequences of the HSV-1 Replication Proteins

The complete sequences of two other human herpesviruses in addition to HSV have been determined: VZV, like HSV an alphaherpesvirus (287); and EBV, a member of the more distantly related herpesvirus subgroup, the gammaherpesviruses (288). These viruses have biological properties that differ markedly from HSV, but the available evidence suggests that DNA replication during the lytic growth of these and all herpesviruses occurs by fundamentally the same mechanism. Thus it is of some interest to search the genomes of the less-well characterized herpesviruses for homologues to the HSV genes that are essential for DNA synthesis. A clear homologue of each of the seven HSV genes is present in the genome of VZV (257). Counterparts of four HSV genes (pol, ICP8, UL5, and UL52) with clear sequence similarity are present in the genome of the more widely diverged virus, EBV (257). Counterparts of the three remaining HSV genes can be proposed on the basis of genomic location alone, but for two of these, UL8 and UL9, they are markedly different in size. The functional significance of these similarities (or lack thereof) is not yet known. The lack of a clear counterpart in EBV of the HSV gene UL9, which encodes an origin-specific DNA-binding protein, is consistent with the fact that the lytic origin of EBV is not discernibly similar to the HSV origins (288a). It seems likely that as more information becomes available concerning the biochemical functions of the HSV replication proteins that these sequence comparisons will provide some useful insights into structure-function relationships. Conversely, knowledge of the function of the HSV genes should prove extremely useful in the analysis of the replication of the genomes of these other viruses.

The predicted sequences of the HSV DNA replication proteins have been examined for the presence of several consensus elements (257). Three of the proteins were found to contain a motif associated with nucleotide binding sites (289): the DNA polymerase, UL5, and UL9. This motif was also conserved in the VZV homologues of these genes and in the EBV homologues of the DNA polymerase and UL5. More recently, UL5 (and the VZV and EBV homologues of UL5) has been shown to contain several additional regions of similarity with a large group of proteins that all have helicase function (289a, 289b, 289c). It seems likely, therefore, that UL5 is a helicase.

Biochemical Analysis of Replication Proteins

The complete characterization of HSV DNA replication will ultimately depend on the development of a soluble, origin-dependent in vitro system. No such system has yet been described, but it has been possible to carry out some biochemical analyses of selected replication proteins. By analogy with other, better characterized systems, it is possible to predict certain biochemical activities that might be involved in HSV DNA replication. Extracts of infected cells can thus be analyzed for the presence of virus-induced proteins with these activities, and if found, the proteins can be purified and further characterized. In addition, since the complete set of viral genes required for DNA synthesis is now known and the products of these genes identified using specific antisera (290), it is possible to purify these proteins without any prior knowledge or assumptions concerning their functions. The purified proteins can then be assayed for pertinent biochemical activities.

DNA POLYMERASE   Extracts of HSV-infected cells contain a novel DNA polymerase activity (283, 291, 292). This activity is readily distinguished from the host DNA polymerases on the basis of its sensitivity to various inhibitors and by the fact that it is stimulated, rather than inhibited, by moderate concentrations of salt (292–295). As mentioned earlier, analysis of ts and drug-resistant mutants has clearly shown this enzyme to be virus-encoded. The HSV DNA polymerase has been extensively purified in several laboratories. As is the case with many other DNA polymerases, it has an intrinsic 3'-5' exonuclease activity that probably serves a proofreading function to increase the fidelity of DNA synthesis (294–296). The most highly purified preparations of the enzyme consist predominantly of a monomer of a single polypeptide chain of ~140 kd in size (294, 295), in good agreement with the size of the product of the polymerase gene predicted from DNA sequence analysis (266, 297). An unusual property of the HSV polymerase is that it is highly processive in the absence of accessory factors (295).

It is clear that the 140-kd polypeptide product of the polymerase gene itself has catalytic activity on simple primer templates such as activated DNA. As mentioned above, the most highly purified preparations of the polymerase do not contain any other polypeptide in stoichiometric amounts (but see below). In addition, catalytically active polymerase has been expressed by in vitro transcription/translation (295a) in yeast (295b) and in insect cells using a recombinant baculovirus (D. Coen, A. Marcey, P. Olivo, personal communication). It is not yet clear, however, whether there are other forms of the polymerase that contain additional accessory factors that increase the efficiency of the polymerase or modify its activity in some other way. It has been reported that the 140-kd polypeptide of HSV-2 co-purified with a 55-kd polypeptide (298). This 55-kd polypeptide has since been shown to be the product of the *UL42* gene, one of the HSV genes that is essential for DNA synthesis (see below); the effect of the UL42 protein on the activity of the polymerase, however, has not yet been determined.

SINGLE-STRANDED DNA-BINDING PROTEIN   ICP8 was recognized several years ago as an abundant HSV-induced protein of about 130 kd present in infected cells but not in virions (299, 300). ICP8 was shown to bind tightly to single-stranded DNA cellulose columns (301–305). Since, as indicated earlier, genetic evidence indicates that this protein is required for viral DNA synthesis, it seems reasonable to assume that the function of ICP8 is analogous to that of the gene 32 protein of bacteriophage T4 and the SSB protein of *E. coli:* namely, to bind to the single-stranded DNA formed at a replication fork by the unwinding of the parental duplex DNA, and to facilitate the use of these strands as templates for DNA polymerase. ICP8 in fact has many of the properties that are characteristic of this class of replication proteins: it binds more tightly to single-stranded DNA than to double-stranded DNA (301–306); binding to single-stranded DNA is cooperative and is independent of sequence (305). It has been difficult, however, to obtain direct biochemical evidence that single-stranded DNA bound to ICP8 is a better template for the HSV DNA polymerase than naked DNA. It has been reported that purified ICP8 has a small (no greater than twofold) stimulatory effect on the activity of purified HSV DNA polymerase using activated DNA as template (307). Surprisingly, however, ICP8 was shown to inhibit the activity of the polymerase on a long single-stranded DNA template (306). Obviously, a great deal more work needs to be done concerning the function of ICP8 and its interaction with other replication proteins.

The isolation several years ago of a monoclonal antibody against ICP8 (308) stimulated a detailed analysis of its intracellular localization (272, 309, 310). As expected from its essential role in DNA replication, ICP8 is localized predominantly in the nucleus (309–311). Under conditions in which viral DNA synthesis is inhibited, ICP8 is located in small discrete foci that contain newly replicated host-cell DNA (272, 272a). During DNA synthesis, ICP8 was found to move from these small foci to larger, more globular areas that are the sites for viral DNA synthesis (272). It has been suggested that one function of ICP8 may be to organize the structure and composition of these different compartments (272a). Whether all the viral replication proteins are found at these sites remains to be determined.

ORIGIN-BINDING PROTEIN   A protein that binds specifically to the HSV origins of DNA replication has been identified in extracts of HSV-infected cells (312). This protein has now been purified to near homogeneity by site affinity chromatography (313). It consists of a single major polypeptide of about 83 kd. The origin-binding protein has recently been shown to be the product of the HSV gene UL9, one of the seven genes that are essential for DNA replication. The UL9 protein was expressed in insect cells using a baculovirus expression system, and the recombinant UL9 protein was shown to interact with ori_S in a manner indistinguishable from that observed with the origin-binding protein purified from HSV-infected cells (314). Although the size of the UL9 protein predicted from DNA sequence analysis is 94 kd (257), the observed size of the protein expressed both by the recombinant baculovirus and in HSV-infected cells is 83 kd (313, 314). Sedimentation analysis of the purified recombinant protein suggests that it is a dimer in solution (M. D. Challberg, unpublished).

The purified UL9 protein binds to ori_S at two nearly identical sites, located on each arm of a palindrome (313, 314). Filter binding studies with synthetic double-stranded oligonucleotides corresponding to the two sites have shown that the intrinsic affinity for the site on the left arm is about 10 times greater than for the site on the right arm (313). Methylation interference experiments and a comparison of different binding sites suggest that the recognition sequence for UL9 is contained within the eight-base-pair sequence GTTCGCAC (315; M. D. Challberg, unpublished). It has been proposed (315) that the recognition sequence is GT (T/G)CG, which is contained twice within the eight-base-pair recognition domain as inverted repeats that share a two-base overlap. There is as yet no direct evidence in support of this proposal. It is consistent with the fact that purified UL9 does form a stable dimer, and with the fact that UL9 has a lower affinity for the binding site on the right arm of the ori_S palindrome, in which one of the inverted pentamers differs from the canonical sequence.

Although the intrinsic affinity of the binding sites for UL9 on the two arms of the ori_S palindrome differ, DNase I footprint analysis of the interaction of purified UL9 with the complete ori_S sequence suggests that the affinity of UL9 for the two sites is equal (313, 314). Moreover, deletion of the UL9-binding site on the right arm of the palindrome reduces the affinity of UL9 for the binding site on the left arm (M. D. Challberg, unpublished). These results suggest that there is a cooperative interaction between the protein bound at the two sites. It is not yet known whether this cooperativity has any functional significance, although the reported effect of insertions into the AT-rich region between the two UL9-binding sites is intriguing. As mentioned earlier, a series of mutant origins (228) have been constructed in which n copies of the AT dinucleotide were inserted into the center of the AT-rich region of ori_S. As n increases from 0 to 8, replication first sharply decreases to a minimum at $n=3$, then rises to a maximum at $n=5$ or 6, then decreases again. It is possible that this oscillation in activity reflects a requirement for the UL9 protein bound on each arm of the palindrome to be located on the same side of the DNA helix. This arrangement may be necessary to accommodate critical UL9-UL9 interactions. However, in view of the report that the right arm of the ori_S palindrome has no effect at all on origin function (231; see above), this question bears further investigation.

There is now good evidence to indicate that the carboxy-terminal portion of the UL9 protein contains the DNA-binding domain. The carboxy-terminal 37 kd of UL9 has been expressed in *E. coli* as a fusion protein. This truncated UL9 protein still binds specifically to DNA fragments containing ori_S (N. Stow, personal communication). There is also preliminary evidence that a truncated form of UL9 (of unknown size) containing the carboxy terminus and retaining ori_S-binding activity can be isolated from HSV-infected cells (N. Stow, personal communication; A. Koff, P. Tegtmeyer, personal communication). It is not yet known whether this truncated form has any biological function.

The role of UL9 binding in HSV DNA replication is not known. By analogy with other prokaryotic and eukaryotic replication origin recognition proteins, the binding of UL9 to ori_S and ori_L may initiate the assembly of a multiprotein replication complex. Alternatively, or additionally, UL9 may be involved in unwinding the two parental strands at the origin as a prelude to the initiation of daughter strand synthesis. This latter possibility is strengthened by the observation that the predicted amino acid sequence of UL9 contains a consensus ATP-binding sequence similar to that found in the SV40 T-antigen and the *E. coli* dna A protein (M. D. Challberg, unpublished). There is no convincing evidence as yet, however, that the UL9 protein binds or hydrolyzes ATP, or is capable of unwinding DNA at the HSV origins. Since the protein can now be produced in large quantities, the answer to some of these questions should be forthcoming.

HELICASE/PRIMASE   Infection of cells with HSV induces novel helicase and primase activities (316, 317). Recently, these two activities have been purified to homogeneity (J. Crute, I. R. Lehman, personal communication); both helicase and primase activities are components of a three-subunit enzyme composed of the products of the UL5, UL8, and UL52 genes. The helicase can utilize either ATP or GTP as a cofactor for unwinding. The activity of the helicase on model substrates suggests that it moves in the 5' to 3' direction on the strand to which it is bound. Thus, this enzyme may prime lagging strand synthesis as it unwinds DNA at a replication fork. The activities of the component polypeptides of the complex have not yet been determined, although as mentioned above, UL5 contains several sequence motifs that are shared by helicases.

DOUBLE-STRANDED DNA-BINDING PROTEIN   As mentioned above, purified preparations of the HSV-2 DNA polymerase were reported to contain two major polypeptides: the 140-kd product of the polymerase gene, and a 55-kd protein (294). It has since been demonstrated that this protein is the product of the UL42 gene (318–320), which genetic experiments have shown to be required for viral DNA synthesis (254, 256, 257). The UL42 protein has now been purified from HSV-1-infected cells in several laboratories, but its function remains obscure (320; C. Wu, M. D. Challberg, unpublished). As indicated earlier, this protein is not required for the catalytic activity of the HSV DNA polymerase. Immunoaffinity purification of the UL42 protein has provided evidence that there is an interaction between UL42 and the polymerase (320). The functional significance of this interaction remains to be determined. The UL42 protein binds strongly in a sequence-independent fashion to double-stranded DNA (320; C. Wu, M. D. Challberg, unpublished). The nature of this binding and its effects on DNA structure have not been detailed. It is not known whether DNA binding is an essential component of the function of UL42 in DNA synthesis.

BOVINE PAPILLOMAVIRUS DNA REPLICATION

The papillomaviruses are a group of small DNA-containing viruses that are associated with epithelial tumors in a variety of animal species (reviewed in 321–324). Although most of the cells in such tumors contain many copies of the viral genome in a latent state, the tumor cells contain no infectious virus; productive infection by these viruses appears to be confined to terminally differentiated keratinocytes. It has not yet been possible to propagate any of the papillomaviruses in a cell culture system. A subset of the papillomaviruses, however, of which bovine papillomavirus type 1(BPV-1) is the most extensively studied, also induce fibropapillomas in their natural hosts, and this subset readily transforms established rodent fibroblast cell lines in culture (325, 326). As in the case of naturally occurring tumors, transformed rodent cells contain the viral genetic information in a latent state. Only a subset of viral genes are expressed in transformed cells (327–329), and no infectious virus is produced. Most of the information on the molecular biology of the papillomaviruses has come from a study of this model in vitro cell culture system.

The BPV-1 genome is a covalently closed, circular, double-stranded DNA molecule of 7945 base pairs (330, 331). All of the open reading frames (ORFs) of at least 400 base pairs are located on one strand, and all of the mRNA species detectable both in transformed cells and in productively infected bovine fibropapillomas are homologous to that same strand (327–329). A 5.4-kb subgenomic fragment of BPV-1 DNA is sufficient for transformation of rodent cells in vitro (332). Within the transforming fragment there are eight open reading frames (E1 through E8) contained within a 4.5-kb segment (330, 331). On the 5' side of these ORFs there is an approximately 1-kb segment that contains no large open reading frames. This segment (the long control region, or LCR; also referred to as the upstream regulatory region) appears to contain a number of cis-acting regulatory elements, including several transcriptional promoters (328, 333), a transcriptional enhancer that is activated by one of the E2 gene products (334, 335), and the origin of DNA replication (333, 336, 337; see below).

Although the arrangement of ORFs deduced from DNA sequence analysis has provided a useful starting point for the analysis of viral gene products, there is clear genetic and biochemical evidence to suggest that the number and structure of viral proteins cannot be predicted solely from the DNA sequence of the genome. Analysis of viral mRNAs by electron microscopy and cDNA cloning has revealed a number of mRNAs formed by complex splicing patterns (328, 329). Thus, some viral proteins may correspond to a single ORF, others to only a portion of a single ORF, and yet others to combinations of ORFs. It has been possible to predict the primary sequence of some viral proteins from sequence analysis of cloned cDNAs. In addition, segments of several ORFs have been expressed in bacteria and the resulting proteins used to produce specific antisera for the direct analysis of viral proteins in transformed cells (338–340). Several viral proteins have now been identified, but much remains to be learned.

In cells transformed by BPV or cloned BPV DNA, the viral DNA is maintained extrachromosomally as a stable multicopy plasmid in the cell nucleus (341). As detailed below, the available evidence suggests that only a single viral gene product is directly involved in viral DNA synthesis; it seems likely, therefore, that viral DNA replication is carried out largely by host cell proteins. Moreover, the copy number of BPV plasmids appears to be tightly controlled, and there is evidence that each BPV genome replicates once and only once per cell cycle (342). Thus, BPV appears to represent a useful model for analyzing the mechanisms involved in regulating DNA replication in higher eukaryotes.

BPV DNA Replication In Vivo

The site at which DNA replication initiates has been determined from studies of the structure of replicative intermediates (336). Covalently closed viral DNA was isolated from hamster cells transformed by wild-type BPV and analyzed by electron microscopy. Circular molecules with two forks and no free ends were observed. By measuring molecules cleaved with various single-cut restriction enzymes, the position of the replication eye in such intermediates was localized to map position 6940 ± 5%. Thus, replication of latent BPV genomes initiates within the LCR and proceeds by way of Cairns-type intermediates. Since predominantly early intermediates were analyzed in this study, it was not possible to determine whether fork movement takes place in one or both directions from the origin following initiation.

Once the BPV genome is established as a plasmid in a transformed cell line, its copy number is maintained at a constant level. Hence, there must be some mechanism that ensures that there is an exact, or nearly exact, doubling of BPV DNA during each cell cycle. There are several different ways in which a constant average copy number could be maintained. One possibility is that viral DNA replication could be limited by the availability of some required factor. Viral genomes would replicate at random until this factor was exhausted, and then replication would cease until the factor was replenished during the next cell cycle. A prediction of this model is that some DNA molecules would replicate more than once during a cell cycle, and a similar fraction would not replicate at all. An alternative model is that each viral genome replicates once and only once per cell cycle. A requirement of such a model is that some mechanism exists that distinguishes between those viral genomes that have undergone a round of DNA replication and genomes that have yet to be replicated. The mode of replication of BPV DNA has been analyzed by means of density labeling experiments (342). Mouse cells transformed by BPV and containing an average of ~150 viral genomes per cell were labeled with bromodeoxyuridine during exponential growth. At intervals, plasmid DNA was isolated and analyzed by density gradient centrifugation. The results of this experiment clearly showed that during one cell cycle, nearly all of the viral DNA shifted to the hybrid density expected of molecules that had undergone a single round of semiconservative DNA synthesis; essentially no viral DNA with the density expected for molecules that had undergone more than one round of replication was observed until the cells were labeled for a period longer than a single cell division cycle. The simplest interpretation of this experiment is that the overwhelming majority of viral DNA molecules replicate once and only once per cycle. Very different results, however, were obtained in a variation of this experiment in which the density-labeled cells that had completed S-phase were selected by mitotic shake-off (343). In this experiment, up to 20% of the viral DNA labeled during a single S-phase banded at the density indicative of molecules that had undergone multiple rounds of DNA replication. The reason for the discrepancy between these results is not clear. It is possible that certain cell culture conditions may promote the transient appearance of cells in which viral DNA replication becomes unregulated. A small percentage of such cells could potentially result in the appearance of a relatively high proportion of viral DNA molecules that had undergone multiple rounds of DNA synthesis. Whatever the explanation for the latter results, it seems clear that under at least some conditions, BPV DNA replicates once and only once per cell cycle. Thus, in this respect, BPV DNA replication appears similar to chromosomal DNA replication.

Genetic Analysis of DNA Replication

CIS-ACTING ELEMENTS   Two distinct cis-acting sequences in the BPV genome have been found to allow autonomous replication of plasmid DNAs in the presence of BPV gene products. These elements were first identified by cloning fragments of BPV into a vector expressing the gene encoding neomycin resistance (333). The cloned DNAs were then introduced into mouse cells transformed by BPV, neomycin-resistant colonies were selected, and the physical state of the marker gene was analyzed. In most cases, the plasmid DNA was found to be integrated into the chromosomal DNA. Two small segments of BPV DNA (plasmid maintenance sequences; PMS) gave rise to neomycin-resistant colonies in which the marker gene was maintained as an extrachromosomal nuclear plasmid. When these same plasmids were introduced into untransformed mouse cells (not containing any additional BPV sequences), they failed to replicate extrachromosomally; hence, autonomous replication also requires trans-acting BPV gene products (see below).

One of the plasmid maintenance sequences, PMS-1, was mapped to a

521-bp segment (nucleotides 6945–7476) within the LCR. The other, PMS-2, was mapped to a 140-bp region (nucleotides 1515–1655) within the E1 ORF. The function of these sequences is not known with certainty. PMS-1 is located very near to the position at which the origin of replication of BPV was mapped (336). It seems likely, therefore, that PMS-1 is the site at which DNA replication is initiated. The sequences of PMS-1 and PMS-2 contain a region of extensive homology (333). Therefore, it seems reasonable to assume that PMS-2 may also function as an origin of replication. In the electron microscopic analysis of replication intermediates described earlier, no molecules with replication eyes centered on PMS-2 were observed. It is possible that initiation of replication at PMS-2 is much less efficient and, in the presence of PMS-1, does not occur at a measurable frequency. An analysis of replicative intermediates of plasmids that contain PMS-2 but lack PMS-1 has not been reported.

There is evidence to suggest that the plasmid maintenance sequences may serve other functions in addition to a role in the initiation of replication. Plasmids containing either PMS are stably maintained in cells at a constant copy number in the absence of selection. This observation implies that the PMS elements may contribute to the controlled partitioning of plasmids during cell division (333). In addition, the PMS may function as part of a system that suppresses integration of plasmid DNA. As mentioned, when recombinant plasmids containing PMS-1 or PMS-2 linked to a selectable marker are introduced into BPV-transformed cells, the marker gene is invariably found as an extrachromosomal plasmid. When the same recombinants are used to transfect untransformed cells, the marker gene is always integrated. Moreover, if a plasmid containing PMS-1 or PMS-2 is introduced into transformed cells together with an unlinked selectable marker gene, the marker gene is integrated, while the plasmid containing the PMS is again found as an extrachromosomal plasmid (333).

The sequences necessary for the function of PMS-1 have been analyzed in detail by looking at the effect of various mutations in PMS-1 on the ability of plasmids to replicate transiently following transfection into BPV-transformed cells (347). This transient replication system very likely corresponds to the amplification of viral DNA that must occur during the establishment of a stable final copy number of ~200 from a single infecting genome. Two distinct domains of PMS-1 have been identified by this means. Domain 2 (nucleotides 7116–7224) completely overlaps the region of homology with PMS-2. Domain 1 (nucleotides 6707–6848) appears to be a transcriptional enhancer: the function of domain 1 is independent of its orientation and exact distance away from domain 2, and can be replaced by known enhancer elements of other viruses. Whether the function of domain 1 in DNA replication relates in any way to transcriptional enhancer function is not known. In this context, however, it is also of interest that domain 2 also contains transcriptional regulato.y signals. An mRNA start site (called P1) has been located within domain 2 at nucleotide 7186 (344). Deletion analysis has revealed that the controlling elements required for transcription initiation at this site are located downstream of the site of initiation of transcription, also within the boundaries of domain 2. Deletion of 23 base pairs between nucleotides 7187 and 7234 was shown to abolish both PMS-1-dependent DNA replication and transcription from P1. On the basis of these results it has been suggested that a cellular transcription factor(s) required for transcription from P1 may also play a key role in the initiation of DNA replication. According to this model, the function of the domain 1 enhancer element is to potentiate the binding of the appropriate transcription/replication factors to domain 2. It is not known whether an enhancer element is also required for stable plasmid replication. It is also unclear whether PMS-2 will support transient replication, and if so, whether such replication also depends on the presence of an enhancer element. There is no known transcriptional start site in the vicinity of PMS-2. Therefore, it seems unlikely that there is an obligatory requirement for transcription per se in PMS function, although there is no evidence to rule out the possibility that transcription from PMS-2 takes place on plasmids that lack PMS-1. An evaluation of the role that transcription factors and/or transcription play in the various aspects of BPV DNA replication will almost certainly depend on the development and biochemical analysis of a PMS-dependent in vitro replication system.

TRANS-ACTING VIRAL GENE PRODUCTS   The BPV genes involved in DNA replication have also been investigated using both transient assays and assays involving the establishment of stable extrachromosomal plasmids. Deletion mutants lacking ORFs E2, E3, E4, and E5 were shown to be capable of autonomous replication (333). Mutations in other ORFs have now defined three different complementation groups of mutants that are defective in autonomous replication, and two additional groups of mutants in which the control of plasmid copy number is altered (337, 345–347).

Mutants with lesions in the 3' portion of the E1 ORF transform cells with the same efficiency as wild-type DNA, but the mutant DNA is invariably integrated into the chromosomal DNA of the transformed cell rather than replicating as an autonomous plasmid (345). These mutants (called rep⁻ or R⁻) also fail to replicate transiently following transfection, and cells transformed by R⁻ mutants do not support the transient replication of plasmids containing PMS elements. Autonomous replication by R⁻ mutants can be complemented in both transient assays and stable transformation assays by mutants in other complementation groups. Thus, the 3' portion of the E1 ORF encodes a protein that may be directly involved in BPV DNA replication. The product(s) of the R gene has not yet been identified.

Mutants with lesions in the 5' portion of the E1 ORF define a second complementation group (referred to as modulator⁻ or M⁻) that are defective in autonomous replication (348, 349). These mutants transform cells with very much lower efficiency than wild-type DNA. As in the case of R⁻ mutants, the mutant DNA is always integrated into the chromosomal DNA of transformed cells. Unlike R⁻ mutants, however, these mutants replicate as well as wild-type BPV DNA in transient assays. M⁻ mutants therefore define a function that is required for plasmid maintainance but not DNA synthesis per se. There is additional genetic evidence (see below) that the M function is a repressor that is required to prevent "runaway" replication of the BPV replicon. The available evidence strongly suggests that even though the M function and the R function are both encoded in the same open reading frame (E1), these two functions are the products of two distinct genes (348). First, the two groups of mutants complement each other fully. When R⁻ mutants are cotransfected with M⁻ mutants, both mutant genomes replicate in transient assays, and stably transformed lines derived from such transfections contain both genomes as extrachromosomal plasmids. Second, frame shift mutations in the M gene, at the 5' end of the E1 ORF, do not affect R function. Finally, the product of the M gene has been identified in BPV-transformed cells using antisera prepared against the amino-terminal portion of the E1 ORF expressed in *E. coli* (340); the apparent size of the M protein on SDS gels is 23 kd, significantly smaller than the predicted size of the complete E1 ORF (68 kd). Spliced mRNAs that could account for the synthesis of an amino-terminal truncation of the E1 ORF have been identified.

Recently, a third complementation group of mutants with properties intermediate between those of R⁻ and M⁻ mutants has been identified (M. Lusky, personal communication). These mutants have lesions in the E8 ORF, which is completely embedded within the E1 ORF in a different translational reading frame. Missense mutations in the E8 ORF that leave the E1 ORF unaltered have been constructed by in vitro mutagenesis. These mutants transform cells at the same efficiency as wild-type DNA. As in the case of both R⁻ and M⁻ mutants, the mutant DNA is not maintained as an extra-chromosomal plasmid. These mutants therefore define another viral product that is essential for plasmid maintainance. E8 mutants are able to replicate DNA in transient assays, although at a reduced rate compared to wild-type or M⁻ mutants. E8 mutants complement R⁻ mutants in transient assays and both R⁻ and M⁻ mutants in assays for stable autonomous replication. The protein responsible for the function lacking in these mutants has not yet been identified.

Mutations in the E6 and E7 ORFs do not affect the ability of BPV to replicate as an autonomous plasmid, but can result in a nearly 100-fold reduction in plasmid copy number, from about 200 to 1–5 genomes per cell (345). Analysis of cDNA clones has revealed two classes of mRNAs that contain sequences from this region. One class contains the E6 ORF intact, and the other class contains a splice that joins a portion of the E6 ORF to a portion of the E7 ORF to generate a putative E6/7 fusion protein. Mutants with lesions that specifically affect either the E6 product or the E6/7 product can complement each other and so define two distinct genes; however, both groups that display this altered copy number phenotype are collectively referred to as cop⁻ mutants (345, 347, 348). When cells are transfected with cop⁻ mutants, the mutant genomes initially replicate to high copy number, but with continued passage of the transformed cells the copy number gradually declines to 1–5 copies per cell (347). This copy number is then maintained stably for many generations. The mechanism by which the E6 and E6/7 genes exert an influence on copy number is not known. Analysis of many individual subclones derived from a single transformed cell has shown that this low average copy number of cop⁻ mutants is not due to a gross defect in segregation. It has been reported that the E6 and/or the E6/7 genes influence the activity of the enhancer element in PMS-1 (L. Turk, quoted in Ref. 347). Since the promoter for the M and R genes lies within PMS-1, it is possible that the cop genes affect the level of expression of the other genes that are involved in autonomous DNA replication. Alternatively, the level of expression from the promoter within PMS-1 may directly affect the initiation of DNA replication.

Complementation tests with E6 and E6/7 mutants have shed some light on the functions of some of these genes. When E6 or E6/7 mutants are cotransfected with wild-type BPV DNA or with mutants in other complementation groups, both DNAs replicate transiently and become established as high-copy-number plasmids (345, 347, 348). Hence, the E6 and E6/7 gene products function in *trans*. On the other hand, if cells are first transformed with a cop⁻ mutant and then supertransfected with wild-type DNA, the wild-type DNA is not amplified transiently and becomes established at low copy number (349). This result suggests that wild-type BPV encodes a function that represses plasmid amplification, and the cop gene products play some role in controlling the level of this repressor once stable plasmid copy number has been established. The available evidence suggests that the M gene encodes this repressor. When an M⁻ mutant is used to supertransfect cells carrying a low-copy-number mutant, the incoming M-DNA is transiently amplified but the resident low-copy mutant genome is not (347). The simplest explanation for this result is that the product of the M gene represses plasmid amplification, but in the steady state the protein is sequestered in some fashion so that it cannot readily work in *trans* on newly introduced DNA. In this context, it is also of interest to note that when cotransfected into cells along with a selectable marker, M⁻ mutants decrease the efficiency of transformation nearly 100-fold. This low transformation efficiency may well be due to a lethal effect of unconstrained replication of the M⁻ replicon, which lacks a critical negative element of the copy number regulatory system.

It is clear that BPV encodes a regulatory system that acts to repress amplification of viral DNA when plasmid copy number reaches a certain critical level (100–200 copies per cell). There is also evidence that the BPV repression mechanism can function with a heterologous replication origin (350–352). A hybrid replicon was constructed containing the 5.4-kb BPV-transforming fragment and the SV40 origin of replication. This chimeric molecule was introduced into cells expressing the SV40 T-antigen. Transient amplification of this hybrid plasmid was suppressed relative to a control plasmid containing the SV40 origin but lacking the BPV sequences, even when both plasmids were introduced simultaneously into the same cells. In addition, it was possible to establish stable cell lines in which the hybrid plasmid was maintained at a constant copy number. Density transfer experiments showed that as in the case of latent BPV genomes, once the steady state copy number of the hybrid plasmid was reached, every plasmid replicated once and only once per cell division cycle. Two BPV elements were required in *cis* for the suppressive effect on replication. One element coincided with PMS-1 and the other was near PMS-2. At least one BPV gene product is required in *trans*. Deletions removing the 5' portion of the E1 gene abolished the suppressive effect on replication, and supressed replication could be restored by cotransfection with wild-type BPV DNA. It therefore seems likely that the M gene is part of a BPV regulatory system that can repress DNA replication from a heterologous origin when it is linked to specific BPV *cis*-acting sequences. It will be of considerable interest to elucidate the mechanisms by which this regulatory system interacts with the cellular replication machinery.

ACKNOWLEDGMENTS

We thank our colleagues for many stimulating discussions. Work in TK's laboratory was supported by grants from the National Institutes of Health.

Literature Cited

1. Challberg, M. D., Kelly, T. J. 1979. *Proc. Natl. Acad. Sci. USA* 76:655–59
2. Stillman, B. W. 1983. *Cell* 35:7–9
3. Friefeld, B. R., Lichy, J. H., Field, J., Gronostajski, R. M., Guggenheimer, R. A., et al. 1984. *Curr. Top. Microbiol. Immunol.* 110:221–55
4. Kelly, T. J. 1984. In *The Adenoviruses*, ed. H. S. Ginsberg, pp. 271–308. New York: Plenum
5. Kelly, T. J., Wold, M. S., Li, J. 1988. *Adv. Virus Res.* 34:1–42
6. Robinson, A. J., Younghusband, H. B., Bellett, A. J. D. 1973. *J. Virol.* 56:54–69
7. Robinson, A. J., Bellett, A. J. D. 1974. *Cold Spring Harbor Symp. Quant. Biol.* 39:523–31
8. Wolfson, J., Dressler, D. 1972. *Proc. Natl. Acad. Sci. USA* 69:3054–957
9. Garon, C. F., Berry, K. N., Rose, J. A. 1972. *Proc. Natl. Acad. Sci. USA* 69:2391–95
10. Challberg, M. D., Desiderio, S. V.,

Kelly, T. J. 1980. *Proc. Natl. Acad. Sci. USA* 77:5105–9

11. Enomoto, T., Lichy, J. H., Ikeda, J.-E., Huwitz, J. 1981. *Proc. Natl. Acad. Sci. USA* 78:6779–83

12. van der Vliet, P. C., Levine, A. J. 1973. *Nature New Biol.* 246:170–74

13. Stillman, B. W., Lewis, J. B., Chow, L. T., Mathews, M. B., Smart, J. E. 1981. *Cell* 23:497–508

14. Nagata, K., Guggenheimer, R. A., Hurwitz, J. 1983. *Proc. Natl. Acad. Sci. USA* 80:6177–81

15. Rosenfeld, P. J., Kelly, T. J. 1986. *J. Biol. Chem.* 261:1398–408

16. Jones, K. A., Kadonaga, J. T., Rosenfeld, P. J., Kelly, T. J., Tjian, R. 1987. *Cell* 48:79–89

17. Pruijn, G. J. M., van Driel, W., van der Vliet, P. C. 1986. *Nature* 322:656–59

18. Rosenfeld, P. J., O'Neill, E. A., Wides, R. J., Kelly, T. J. 1987. *Mol. Cell. Biol.* 8:875–86

19. O'Neill, E. A., Fletcher, C., Burrow, C. R., Heintz, N., Roeder, R. G., Kelly, T. J. 1988. *Science* 241:1210–13

20. Nagata, K., Guggenheimer, R. A., Hurwitz, J. 1983. *Proc. Natl. Acad. Sci. USA* 80:4266–70

21. Salas, M. 1988. *Curr. Top. Microbiol. Immunol.* 136:71–88

22. Rekosh, D. M. K., Russell, W. C., Bellett, A. J. D., Robinson, A. J. 1977. *Cell* 11:283–95

23. Challberg, M. D., Kelly, T. J. 1981. *J. Virol.* 38:272–77

24. Smart, J. E., Stillman, B. W. 1982. *J. Biol. Chem.* 257:13499–13506

25. Lichy, J. H., Horwitz, M. S., Hurwitz, J. 1981. *Proc. Natl. Acad. Sci. USA* 78:2678–82

26. Pincus, S., Robertson, W., Rekosh, D. M. K. 1981. *Nucleic Acids Res.* 9:4919–35

27. Challberg, M. D., Ostrove, J., Kelly, T. J. 1982. *J. Virol.* 41:265–70

28. Tamanoi, F., Stillman, B. W. 1982. *Proc. Natl. Acad. Sci. USA* 79:2221–25

29. Guggenheimer, R. A., Nagata, K., Lindenbaum, J., Hurwitz, J. 1984. *J. Biol. Chem.* 259:7807–14

30. Guggenheimer, R. A., Nagata, K., Kenny, M., Hurwitz, J. 1984. *J. Biol. Chem.* 259:7815–25

31. Kenny, M. K., Balogh, L. A., Hurwitz, J. 1988. *J. Biol. Chem.* 263:9801–8

32. Kenny, M. K., Hurwitz, J. 1988. *J. Biol. Chem.* 263:9809–17

33. Tamanoi, F., Stillman, B. W. 1983. *Proc. Natl. Acad. Sci. USA* 80:6446–50

34. van Bergen, B. G. M., van der Ley, P. A., van Priel, W., van Mansfield, A. O.

M., van der Vliet, P. C. 1983. *Nucleic Acids Res.* 11:1975–89

35. Challberg, M. D., Rawlins, D. R. 1984. *Proc. Natl. Acad. Sci. USA* 81:100–4

36. Rawlins, D. R., Rosenfeld, P. J., Wides, R. J., Challberg, M. D., Kelly, T. J. 1984. *Cell* 37:309–18

37. Guggenheimer, R. A., Stillman, B. W., Nagata, K., Tamanoi, F., Hurwitz, J. 1984. *Proc. Natl. Acad. Sci. USA* 81:3069–73

38. Lally, C., Dorper, T., Groger, W., Antoine, G., Winnacker, E.-L. 1984. *EMBO J.* 3:333–37

39. Leegwater, P. A., van Driel, W., van der Vliet, P. C. 1985. *EMBO J.* 4:1515–21

40. de Vries, E., van Driel, W., Tromp, M., van Boom, J., van der Vliet, P. C. 1985. *Nucleic Acids Res.* 13:4935–52

41. Adhya, S., Shneidman, P. S., Hurwitz, J. 1986. *J. Biol. Chem.* 261:3339–46

42. Wides, R. J., Challberg, M. D., Rawlins, D. R., Kelly, T. J. 1987. *Mol. Cell. Biol.* 7:864–75

43. Tolun, A., Alestrom, P., Pettersson, U. 1979. *Cell* 17:705–13

44. Stillman, B. W., Topp, W. C., Engler, J. A. 1982. *J. Virol.* 44:530–37

45. Rijnders, A. W. M., van Bergen, B. G. M., van der Vliet, P. C., Sussenbach, J. S. 1983. *Nucleic Acids Res.* 24:8777–89

46. Stow, N. D. 1982. *Nucleic Acids Res.* 10:5105–19

47. Hay, R. T., Stow, N. D., McDougall, I. M. 1984. *J. Mol. Biol.* 175:493–510

48. Hay, R. T. 1985. *EMBO J.* 4:421–26

49. Hay, R. T. 1985. *J. Mol. Biol.* 186:128–36

50. Wang, K., Pearson, G. D. 1985. *Nucleic Acids Res.* 13:5173–87

51. Bernstein, J., Porter, J. M., Challberg, M. D. 1986. *Mol. Cell. Biol.* 6:2115–24

52. Harris, M. P. G., Hay, R. T. 1988. *J. Mol. Biol.* 201:57–67

53. Santoro, C., Mermod, N., Andrews, P. C., Tjian, R. 1988. *Nature* 334:218–242

54. de Vries, E., van Driel, W., van den Heuvel, S. J. L., van der Vliet, P. C. 1987. *EMBO J.* 6:161–68

55. O'Neill, E. A., Kelly, T. J. 1988. *J. Biol. Chem.* 263:931–37

56. van der Vliet, P. C., Claessens, J., de Vries, E., Leegwater, P. A. J., Pruijn, G. J. M., et al. 1988. In *Cancer Cells 6*, ed. T. J. Kelly, B. W. Stillman. New York: Cold Spring Harbor Lab.

57. Benoist, C., O'Hare, K., Breathnach, R., Chambon, P. 1980. *Nucleic Acids Res.* 8:127–42

58. McKnight, S. L., Kingsbury, R. C.,

Spence, A., Smith, M. 1984. *Cell* 37: 253–62
59. Siebenlist, U., Henninghausen, L., Battey, J., Leder, P. 1984. *Cell* 37:381–91
60. Henninghausen, L., Siebenlist, U., Danner, D., Leder, P., Rawlins, D., et al. 1985. *Nature* 314:289–92
61. Nowock, J., Borgmeyer, U., Puschel, A. W., Rupp, R. A., Sippel, A. E. 1985. *Nucleic Acids Res.* 13:2045–61
62. Efstratiadis, A., Posakony, J. W., Maniatis, T., Lawn, R. M., O'Connell, C., et al. 1980. *Cell* 21:653–68
63. Bienz, M., Pelham, H. R. B. 1986. *Cell* 45:753–60
64. Graves, B. J., Johnson, P. F., McKnight, S. L. 1986. *Cell* 44:565–72
65. Theisen, M., Srief, A., Sippel, A. E. 1986. *EMBO J.* 5:719–24
66. Cereghini, S., Raymondjean, M., Carranca, A. G., Herbomel, P., Yaniv, M. 1987. *Cell* 50:627–38
67. Morgan, W. D., Williams, G. T., Morimoto, R. I., Greene, J., Kingston, R. E., Tjian, R. 1987. *Mol. Cell. Biol.* 7:1129–38
68. Jones, K. A., Yamamoto, K. R., Tjian, R. 1985. *Cell* 42:559–72
69. Pruijn, G. J. M., van Driel, W., van Miltenburg, R. T., van der Vliet, P. C. 1987. *EMBO J.* 6:3771–78
70. Fletcher, C., Heintz, N. H., Roeder, R. G. 1987. *Cell* 51:773–84
71. DePamphilis, M. L. 1988. *Cell* 52(5): 635–38
72. Goldman, M. A., Holmquist, G. P., Gray, M. C., Caston, L. A., Nag, A. 1984. *Science* 224:686–92
73. Enver, T., Brewer, A. C., Patient, R. K. 1988. *Mol. Cell. Biol.* 8:1301–9
74. Lichy, J. H., Field, J., Horwitz, M. S., Hurwitz, J. 1982. *Proc. Natl. Acad. Sci. USA* 79:5225–29
75. Stillman, B. W., Tamanoi, F., Mathews, M. B. 1982. *Cell* 31:613–23
76. Ostrove, J. M., Rosenfeld, P., Williams, J., Kelly, T. J. 1983. *Proc. Natl. Acad. Sci. USA* 80:935–39
77. Friefeld, B. R., Lichy, J. H., Hurwitz, J., Horwitz, M. S. 1983. *Proc. Natl. Acad. Sci. USA* 80:1589–93
78. Nagata, K., Guggenheimer, R. A., Enomoto, T., Lichy, J. H., Hurwitz, J. 1982. *Proc. Natl. Acad. Sci. USA* 79:6438–42
79. Ikeda, J.-E., Enomoto, T., Hurwitz, J. 1982. *Proc. Natl. Acad. Sci. USA* 79:2442–46
80. Stillman, B. W., Tamanoi, F. 1983. *Cold Spring Harbor Symp. Quant. Biol.* 47:741–50
81. Lindenbaum, J. O., Field, J., Hurwitz, J. 1986. *J. Biol. Chem.* 261:10218–227

82. Field, J., Gronostajski, R. M., Hurwitz, J. 1984. *J. Biol. Chem.* 259:9487–95
83. Levine, A. J., van der Vliet, P. C., Rosenwirth, B., Rabek, J., Frenkel, G., Ensinger, M. 1975. *Cold Spring Harbor Symp. Quant. Biol.* 39:559–66
84. Schechter, N. M., Davies, W., Anderson, C. W. 1980. *Biochemistry* 19: 2802–10
85. van der Vliet, P. C., Keegstra, W., Jansz, H. S. 1978. *Eur. J. Biochem.* 86:389–98
86. Fowlkes, D. M., Lord, S. T., Linne, T., Pettersson, U., Philipson, L. 1979. *J. Mol. Biol.* 132:163–80
87. Nass, K., Frenkel, G. D. 1980. *J. Virol.* 35:314–19
88. DePamphilis, M. L., Bradley, M. K. 1986. In *The Papovaviridae*, ed. N. Salzman, pp. 99–150. New York: Plenum
89. Kelly, T. J., Wold, M. S., Li, J. 1988. *Adv. Virus Res.* 34:1–42
90. DePamphilis, M. L., Wasserman, P. M. 1982. In *Organization and Replication of Viral DNA*, ed. A. S. Kaplan, pp. 37–77. Boca Raton, Fla: CRC
91. Kelly, T. J. 1988. *J. Biol. Chem.* In press
92. Li, J., Kelly, T. J. 1984. *Proc. Natl. Acad. Sci. USA* 81:6973–77
93. Subramanian, K. N., Shenk, T. 1978. *Nucleic Acids Res.* 5:3635–42
94. Gutai, M. W., Nathans, D. 1978. *J. Mol. Biol.* 126:259–88
95. DiMaio, D., Nathans, D. 1980. *J. Mol. Biol.* 140:129–42
96. DiMaio, D., Nathans, D. 1982. *J. Mol. Biol.* 156:531–48
97. Myers, R. M., Tjian, R. 1980. *Proc. Natl. Acad. Sci. USA* 77:6491–95
98. Learned, R. M., Myers, R. M., Tjian, R. 1981. *ICN-UCLA Symp. Mol. Cell. Biol.* 21:555–66
99. Fromm, M., Berg, P. 1983. *Mol. Cell. Biol.* 3:991–99
100. Bergsma, D. J., Olive, D. M., Hartzell, S. W., Subramanian, K. N. 1982. *Proc. Natl. Acad. Sci. USA* 79:381–85
101. Jones, K. A., Tjian, R. 1984. *Cell* 36:155–62
102. Stillman, B. W., Gerard, R. D., Guggenheimer, R. A., Gluzman, Y. 1985. *EMBO J.* 4:2933–39
103. Li, J., Peden, K. W. C., Dixon, R. A. F., Kelly, T. J. 1986. *Mol. Cell. Biol.* 6:1117–28
104. Deb, S., DeLucia, A. L., Baur, C.-P., Koff, A., Tegtmeyer, P. 1986. *Mol. Cell. Biol.* 6:1663–70
105. Deb, S., DeLucia, A. L., Koff, A., Tsui, S., Tegtmeyer, P. 1986. *Mol. Cell. Biol.* 6:4578–84

106. Deb, S., Tsui, S., Koff, A., DeLucia, A. L., Parsons, R., Tegtmeyer, P. 1987. *J. Virol.* 61:2143–49
107. Tjian, R. 1978. *Cell* 13:165–79
108. Tjian, R. 1978. *Cold Spring Harbor Symp. Quant. Biol.* 43:655–62
109. Shalloway, D., Kleinberger, T., Livingston, D. M. 1980. *Cell* 20:411–22
110. Tegtmeyer, P., Anderson, B., Shaw, S. B., Wilson, V. G. 1981. *J. Virol.* 115:75–87
111. Tegtmeyer, P., Lewton, B. A., DeLucia, A. L., Wilson, V. G., Ryder, K. 1983. *J. Virol.* 46:151–64
112. Tenen, D. G., Taylor, T. S., Haines, L. L., Bradley, M. K., Martin, R. G., Livingston, D. M. 1983. *J. Mol. Biol.* 168:791–808
113. DeLucia, A. L., Lewton, B. A., Tjian, R., Tegtmeyer, P. 1983. *J. Virol.* 46:143–50
114. Prives, C., Covey, L., Scheller, A., Gluzman, Y. 1983. *Mol. Cell. Biol.* 3:1958–66
115. Lewton, B. A., DeLucia, A. L., Tegtmeyer, P. 1984. *J. Virol.* 49:9–13
116. Wright, P. J., DeLucia, A. L., Tegtmeyer, P. 1984. *Mol. Cell. Biol.* 4:2631–38
117. Ryder, K., DeLucia, A. L., Tegtmeyer, P. 1983. *Virology* 129:239–45
118. Ryder, K., Vakalopoulou, E., Mertz, R., Mastrangelo, I., Hough, P., et al. 1985. *Cell* 42:539–48
119. Gluzman, Y., Frisque, R. J., Sambrook, J. 1980. *Cold Spring Harbor Symp. Quant. Biol.* 44:293–99
120. Wasylyk, B., Wasylyk, C., Mattes, H., Wintzerith, M., Chambon, P. 1983. *EMBO J.* 2:1605–11
121. Lee-Chen, G.-J., Woodworth-Gutai, M. 1986. *Mol. Cell. Biol.* 6:3086–93
122. DeLucia, A. L., Deb, S., Partin, K., Tegtmeyer, P. 1986. *J. Virol.* 57:138–44
123. Hertz, G. Z., Mertz, J. E. 1986. *Mol. Cell. Biol.* 6:3513–20
124. Innis, J., Scott, W. A. 1984. *Mol. Cell. Biol.* 4:1499–507
125. Scott, W. A., Wigmore, D. J. 1978. *Cell* 15:1511–18
126. Varshavsky, A. J., Sundin, O. H., Bohm, M. J. 1978. *Nucleic Acids Res.* 5:3469–77
127. Waldeck, W., Rosl, F., Zentgraf, H. 1984. *EMBO J.* 3:2173–78
128. Cremisi, C., Pignatti, P. F., Croissant, O., Yaniv, M. 1976. *J. Virol.* 17:204–11
129. Saragosti, S., Moyne, G., Yaniv, M. 1980. *Cell* 20:65–73
130. Gerard, R. D., Montelone, B. A., Walter, C. F., Innis, J. W., Scott, W. A. 1985. *Mol. Cell. Biol.* 5:52–58
131. Jongstra, J., Reudelhuber, T. L., Oudet, P., Benoist, C., Chae, C.-B., et al. 1984. *Nature* 307:708–14
132. Tegtmeyer, P. 1980. In *Molecular Biology of Tumor Viruses,* ed. J. Tooze, Chapter 5. New York: Cold Spring Harbor Lab.
133. Rigby, P. W. J., Lane, D. P. 1983. *Adv. Viral Oncol.* 3:31–57
134. Mastrangelo, I. A., Hough, P. V. C., Wilson, V. G., Wall, J. S., Hainfeld, J. F., Tegtmeyer, P. 1985. *Proc. Natl. Acad. Sci. USA* 82:3626–30
135. Dean, F. B., Dodson, M., Echols, H., Hurwitz, J. 1987. *Proc. Natl. Acad. Sci. USA* 84:8981–85
136. Deb, S. P., Tegtmeyer, P. 1987. *J. Virol.* 61:3649–54
137. Borowiec, J. A., Hurwitz, J. 1988. *Proc. Natl. Acad. Sci. USA* 85:64–68
138. Borowiec, J. A., Hurwitz, J. 1988. *EMBO J.* 7:3149–58
139. Stahl, H., Droge, P., Knippers, R. 1986. *EMBO J.* 5:1939–44
140. Dean, F. B., Bullock, P., Murakami, Y., Wobbe, C. R., Weissbach, L., Hurwitz, J. 1987. *Proc. Natl. Acad. Sci. USA* 84:16–20
141. Wold, M. S., Li, J. J., Kelly, T. J. 1987. *Proc. Natl. Acad. Sci. USA* 84:3643–47
142. Dodson, M., Dean, F. B., Bullock, P., Echols, H., Hurwitz, J. 1987. *Science* 238:964–67
143. Dean, F. B., Borowiec, J. A., Ishimi, Y., Deb, S., Tegtmeyer, P., Hurwitz, J. 1987. *Proc. Natl. Acad. Sci. USA* 84:8267–71
144. Wold, M. S., Kelly, T. J. 1988. *Proc. Natl. Acad. Sci. USA* 85:2523–27
145. Fairman, M. P., Stillman, B. 1988. *EMBO J.* 7:1211–18
146. Wold, M. S., Weinberg, D. H., Virshup, D. M., Li, J. J., Kelly, T. J. 1988. *J. Biol. Chem.* In press
147. Wobbe, C. R., Weissbach, L., Borowiec, J. A., Dean, F. B., Murakami, Y., et al. 1987. *Proc. Natl. Acad. Sci. USA* 84:1834–38
148. Fry, M., Loeb, L. A. 1986. In *Animal Cell DNA Polymerases,* pp. 13–60. Boca Raton, Fla:CRC
149. So, A. G., Downey, K. M. 1988. *Biochemistry* 27:4591–95
150. Huberman, J. A. 1987. *Cell* 48:7–10
151. Murakami, Y., Wobbe, C. R., Weissbach, L., Dean, F. B., Hurwitz, J. 1986. *Proc. Natl. Acad. Sci. USA* 83:2868–73
152. Wold, M. S., Li, J. J., Weinberg, D.

H., Virshup, D. M., Sherley, J. L., et al. 1988. See Ref. 56, pp. 133–41
153. Prelich, G., Tan, C.-K., Kostura, M., Mathews, M. B., So, A. G., et al. 1987. *Nature* 326:517–19
154. Prelich, G., Kostura, M., Marshak, D. R., Mathews, M. B., Stillman, B. 1987. *Nature* 326:471–75
155. Dresler, S. L., Frattini, M. G. 1986. *Nucleic Acids Res.* 14:7093–100
156. Decker, R. S., Yamaguchi, M., Rossenti, R., Bradley, M. K., DePamphilis, M. L. 1987. *J. Biol. Chem.* 262:10863–72
157. Campbell, J. L. 1986. *Annu. Rev. Biochem.* 55:733–71
158. Cotterill, S. M., Reyland, M. E., Loeb, L. A., Lehman, I. R. 1987. *Proc. Natl. Acad. Sci. USA* 84:5635–39
159. Tseng, B. Y., Ahlem, C. N. 1982. *J. Biol. Chem.* 257:7280–90
160. Kaguni, L. S., Rossignol, J.-M., Conaway, R. C., Banks, G. R., Lehman, I. R. 1983. *J. Biol. Chem.* 258:9037–39
161. Hohn, K. T., Grosse, F. 1987. *Biochemistry* 26:2870–80
162. Villani, G., Fay, P. J., Bambara, R. A., Lehman, I. R. 1981. *J. Biol. Chem.* 256:8202–7
163. Hockensmith, J. W., Bambara, R. A. 1981. *Biochemistry* 20:227–32
164. Downey, K. M., Tan, C.-K., Andrews, D. M., Li, X., So, A. G. 1988. See Ref. 56, pp. 403–10
165. Smale, S. T., Tjian, R. 1986. *Mol. Cell. Biol.* 6:4077–79
166. Gannon, J. V., Lane, D. P. 1987. *Nature* 329:456–58
167. Lane, D. P., Crawford, L. V. 1979. *Nature* 278:261–64
168. Linzer, D. I. H., Levine, A. J. 1979. *Cell* 17:43–54
169. Harlow, E., Pim, D. C., Crawford, L. V. 1981. *J. Virol.* 37:564–73
170. Byrnes, J. J., Downey, K. M., Black, V. L., So, A. G. 1976. *Biochemistry* 15:2817–23
171. Lee, M. Y. W. T., Tan, C.-K., So, A. G., Downey, K. M. 1980. *Biochemistry* 19:2096–107
172. Lee, M. Y. W. T., Tan, C.-K., Downey, K. M., So, A. G. 1984. *Biochemistry* 23:1906–13
173. Byrnes, J. J. 1985. *Biochem. Biophys. Res. Commun.* 132:628–34
174. Crute, J. J., Wahl, A. F., Bambara, R. A. 1986. *Biochemistry* 25:26–36
175. Tan, C.-K., Castillo, C., So, A. G., Downey, K. M. 1986. *J. Biol. Chem.* 261:12310–16
176. Bravo, R., Frank, R., Blundell, P. A.,

MacDonald-Bravo, H. 1987. *Nature* 326:515–17
177. Prelich, G., Stillman, B. 1988. *Cell* 53:117–26
178. Yang, L., Wold, M. S., Li, J. J., Kelly, T. J., Liu, L. F. 1987. *Proc. Natl. Acad. Sci. USA* 84:950–54
179. Sundin, O., Varshavsky, A. 1980. *Cell* 21:103–14
180. DiNardo, S., Voelkel, K., Sternglanz, R. 1984. *Proc. Natl. Acad. Sci. USA* 81:2616–20
181. Brill, S. J., DiNardo, S., Voelkel-Meiman, K., Sternglanz, R. 1987. *Nature* 326:414–16
182. Uemura, T., Ohkura, H., Adachi, Y., Morino, K., Shiozaki, K., Yanagida, M. 1987. *Cell* 50:917–25
183. Wiekowski, M., Droge, P., Stahl, H. 1987. *J. Virol.* 61:411–18
184. Wiekowski, M., Schwartz, M. W., Stahl, H. 1988. *J. Biol. Chem.* 263:436–42
185. Goetz, G. S., Dean, F. B., Hurwitz, J., Matson, S. W. 1988. *J. Biol. Chem.* 263:383–92
186. LeBowitz, J. H., McMacken, R. 1986. *J. Biol. Chem.* 261:4738–48
187. Seki, M., Enomoto, T., Hanoka, F., Yamada, M. 1987. *Biochemistry* 26:2924–28
188. Hubscher, U., Stalder, H. P. 1985. *Nucleic Acids Res.* 13:5471–83
189. Spear, P. G., Roizman, B. 1980. See Ref. 132, Part 2, pp. 615–745
190. Wagner, E. K. 1985. In *The Herpesviruses*, ed. B. Roizman, 3:104. New York: Plenum
191. Roizman, B., Batterson, W. 1985. In *Virology*, ed. B. Fields, D. M. Knipe, R. M. Chanock, J. L. Melnick, B. Roizman, R. E. Shope, pp. 497–526. New York: Raven
192. Jenkins, F., Roizman, B. 1986. *J. Virol.* 59:494–99
193. McGeoch, D. J., Dalrymple, M. A., Davison, A. J., Dolan, A., Frame, M. C., et al. 1988. *J. Gen. Virol.* 69:1531–74
194. Klein, R. J. 1982. *Arch. Virol.* 72:143–68
195. Wildy, P., Field, H. J., Nash, A. A. 1982. In *Virus Persistence,* ed. B. W. J. Nahy, A. C. Minson, G. K. Darby, pp. 133–55. Cambridge, England: Cambridge Univ. Press
196. Jacob, R. J., Roizman, B. 1977. *J. Virol.* 23:394–411
197. Jacob, R. J., Morse, L. S., Roizman, B. 1979. *J. Virol.* 29:448–57
198. Jongeneel, C. V., Bachenheimer, S. L. 1981. *J. Virol.* 39:656–60

199. Mocarski, E. S., Roizman, B. 1982. *Cell* 31:89–97
200. Smiley, J. R., Wagner, M. J., Summers, W. P., Summers, W. C. 1980. *Virology* 102:83–93
201. Schaffer, P. A., Tevethia, M. J., Benyesh-Melnick, M. 1974. *Virology* 58:219–28
202. Honess, R. W., Buchan, A., Halliburton, I. W., Watson, D. H. 1980. *J. Virol.* 34:716–42
203. Weber, P. C., Challberg, M. D., Nelson, N. J., Levine, M., Glorioso, J. C. 1988. *Cell* 54:369–81
204. Huberman, J. A. 1968. *Cold Spring Harbor Symp. Quant. Biol.* 33:509–24
205. Mosig, G. 1983. In *Bacteriophage T4*, ed. C. Mathews, E. Kutter, G. Mosig, P. Berget, pp. 71–81. Washington, DC: Am. Soc. Microbiol.
206. Frenkel, N., Jacob, R. J., Honess, R. W., Hayward, G. S., Locker, H., Roizman, B. 1975. *J. Virol.* 16:153–67
207. Frenkel, N., Locker, H., Batterson, W., Hayward, G. S., Roizman, B. 1976. *J. Virol.* 20:527–31
208. Schroder, C. H., Stegmann, B., Lauppe, H. F., Kaerner, H. C. 1975. *Intervirology* 46:270–84
209. Frenkel, N., Locker, H., Vlazny, D. A. 1980. *Ann. NY Acad. Sci.* 354:347–70
210. Kaerner, H. C., Ott-Hartmann, A., Schatten, A., Schroder, C. H., Gray, C. P. 1981. *J. Virol.* 39:75–81
211. Denniston, K. J., Madden, N. M. J., Enquist, L. W., Vande Woude, G. 1981. *Gene* 15:365–78
212. Vlazny, D. A., Frenkel, N. 1981. *Proc. Natl. Acad. Sci. USA* 78:742–46
213. Vlazny, D. A., Kwong, A., Frenkel, N. 1982. *Proc. Natl. Acad. Sci. USA* 79:1423–27
214. Locker, H., Frenkel, N., Halliburton, I. 1982. *J. Virol.* 43:574–93
215. Spaete, R. R., Frenkel, N. 1982. *Cell* 30:295–304
216. Mocarski, E. S., Roizman, B. 1982. *Proc. Natl. Acad. Sci. USA* 79:5626–30
217. Barnett, J. W., Eppstein, D. A., Chan, H. W. 1983. *J. Virol.* 48:384–95
218. Stow, N. D. 1982. *EMBO J.* 1:863–67
219. Stow, N. D., McMonagle, E. C. 1983. *Virology* 130:427–38
220. Stow, N. D., McMonagle, E. C., Davison, A. J. 1983. *Nucleic Acids Res.* 11:8205–20
221. Deiss, L. P., Frenkel, N. 1986. *J. Virol.* 57:933–41
222. Varmuza, S. L., Smiley, J. R. 1985. *Cell* 41:793–802
223. Deiss, L. P., Chou, J., Frenkel, N. 1986. *J. Virol.* 59:605–18

224. Murchie, M. J., McGeoch, D. J. 1982. *J. Gen. Virol.* 62:1–15
225. Weller, S. K., Spadaro, A., Schaffer, J. E., Murray, A. W., Maxam, A. M., Schaffer, P. A. 1985. *Mol. Cell. Biol.* 5:930–42
226. Lockshon, D., Galloway, D. A. 1986. *J. Virol.* 58:513–21
227. Knopf, C. W., Spies, B., Kaerner, H. C. 1986. *Nucleic Acids Res.* 14:8655–67
228. Lockshon, D., Galloway, D. 1988. *Mol. Cell. Biol.* 8:4018–27
229. Polvino-Bodnar, M., Orberg, P. D., Schaffer, P. A. 1987. *J. Virol.* 61:3528–35
230. Longnecker, R., Roizman, B. 1986. *J. Virol.* 58:583–91
231. Deb, S., Doelberg, M. 1988. *J. Virol.* 62:2516–19
232. Stow, N. D., Davison, A. J. 1986. *J. Gen. Virol.* 67:1613–23
233. Cordingley, M. G., Campbell, M. E., Preston, C. M. 1983. *Nucleic Acids Res.* 11:2347–65
234. Mackem, S., Roizman, B. 1982. *Proc. Natl. Acad. Sci. USA* 79:4917–21
235. Mackem, S., Roizman, B. 1982. *J. Virol.* 44:939–49
236. Yager, D., Coen, D. M. 1988. *J. Virol.* 62:2007–15
237. Su, L., Knipe, D. M. 1987. *J. Virol.* 61:615–20
238. Hubenthal-Voss, J., Starr, L., Roizman, B. 1987. *J. Virol.* 61:3349–55
239. Schaffer, P. A., Preston, V. G., Wagner, E. K., Devi-Rao, G. B. 1987. *Genet. Maps Cold Spring Harbor Lab.* 4:93–98
240. Preston, C. M. 1979. *J. Virol.* 29:275–84
241. Dixon, R. A. F., Schaffer, P. A. 1980. *J. Virol.* 36:189–302
242. Purifoy, D. J. M., Lewis, R. B., Powell, K. L. 1977. *Nature* 269:621–23
243. Purifoy, D. J. M., Powell, K. L. 1981. *J. Gen. Virol.* 54:219–22
244. Chartrand, P., Crumpacker, C. S., Schaffer, P. A., Wilkie, N. M. 1980. *Virology* 103:311–26
245. Coen, D. M., Aschman, D. P., Gelep, P. T., Retondo, M. J., Weller, S. K., Schaffer, P. A. 1984. *J. Virol.* 49:236–47
246. Conley, A. J., Knipe, D. M., Jones, P. C., Roizman, B. 1981. *J. Virol.* 37:191–206
247. Weller, S. K., Lee, K. J., Sabourin, D. J., Schaffer, P. A. 1983. *J. Virol.* 45:354–66
248. Littler, E., Purifoy, D., Minson, A., Powell, K. L. 1983. *J. Gen. Virol.* 64:983–95

249. Weller, S. K., Carmichael, E. P., Aschman, D. P., Goldstein, D. J., Schaffer, P. A. 1987. *Virology* 161:198–210
250. Carmichael, E. P., Kosovsky, M. J., Weller, S. K. 1988. *J. Virol.* 62:91–99
251. Goldstein, D. J., Weller, S. K. 1988. *J. Virol.* 62:2970–77
252. Zhu, L., Weller, S. K. 1988. *Virology* 166:366–78
253. Carmichael, E. P., Weller, S. K. 1988. *J. Virol.* 62:2970–77
254. Marchetti, M. E., Smith, C. E., Schaffer, P. A. 1988. *J. Virol.* 62:715–21
255. Challberg, M. D. 1986. *Proc. Natl. Acad. Sci. USA* 83:9094–98
256. Wu, C. A., Nelson, N. J., McGeoch, D. J., Challberg, M. D. 1988. *J. Virol.* 62:435–43
257. McGeoch, D. J., Dalrymple, M. A., Dolan, A., McNab, D., Perry, L. J., et al. 1988. *J. Virol.* 62:444–53
258. Hay, J., Subak-Sharpe, J. H. 1976. *J. Gen. Virol.* 31:145–48
259. Honess, R. W., Watson, H. H. 1977. *J. Virol.* 21:584–600
260. Schaffer, P., Carter, A. V. C., Timbury, M. C. 1978. *J. Virol.* 27:490–504
261. Coen, D. M., Furman, P. A., Gelp, P. T., Schaffer, P. A. 1982. *J. Virol.* 41:909–18
262. Crumpacker, C. S., Chartrand, P., Subak-Sharpe, J. H., Wilkie, N. M. 1980. *Virology* 105:171–84
263. Furman, P. A., Coen, D. M., St. Clair, H. H., Schaffer, P. A. 1981. *J. Virol.* 40:936–41
264. Jofre, J. T., Schaffer, P. A., Parris, D. S. 1977. *J. Virol.* 23:833–36
265. Knipe, D. M., Ruyechan, W. T., Roizman, B. 1979. *J. Virol.* 29:698–704
266. Gibbs, J. S., Chiou, H. C., Hall, J. D., Mount, D., Retondo, M. J., et al. 1985. *Proc. Natl. Acad. Sci. USA* 82:7969–73
267. Larder, B. A., Kemp, S. D., Darby, G. 1987. *EMBO J.* 6:169–75
267a. Gibbs, J. S., Chiou, H. C., Bastow, K. F., Cheng, Y-C., Coen, D. M. 1988. *Proc. Natl. Acad. Sci. USA* 85:6672–76
268. Wong, S. W., Wahl, A. F., Yuan, P.-M., Arai, N., Pearson, B. E., et al. 1988. *EMBO J.* 7:37–47
269. Hall, J. D., Coen, D. M., Fisher, B. L., Weisslitz, M., Randall, S., et al. 1984. *Virology* 132:26–37
270. Lee, C. K., Knipe, D. M. 1983. *J. Virol.* 46:909–19
271. Leinbach, S. S., Casto, J. F., Pickett, T. K. 1984. *Virology* 137:287–96
272. Quinlan, M. P., Chen, L. B., Knipe, D. M. 1984. *Cell* 36:857–63
272a. de Bruyn Kops, A., Knipe, D. M. 1988. *Cell* 55:857–68

273. Gao, M., Bouchey, J., Curtin, K., Knipe, D. M. 1988. *Virology* 163:319–29
274. Chiou, H. C., Weller, S. K., Coen, D. M. 1985. *Virology* 145:213–26
275. Godowski, P. J., Knipe, D. M. 1983. *J. Virol.* 47:478–86
276. Godowski, P. J., Knipe, D. M. 1985. *J. Virol.* 55:357–65
277. Kit, S., Dubbs, D. 1963. *Biochem. Biophys. Res. Commun.* 11:55–66
278. Coen, G. H. 1972. *J. Virol.* 9:408–18
279. Wohlrab, F., Francke, B. 1980. *Proc. Natl. Acad. Sci. USA* 77:1872–76
280. Fisher, F. B., Preston, V. G. 1986. *Virology* 148:190–97
281. Caradonna, S., Worrad, D., Lirette, R. 1987. *J. Virol.* 61:3040–47
282. Preston, C. M., Cordingley, M. G. 1982. *J. Virol.* 43:386–94
283. Hay, J., Moss, H., Halliburton, I. W. 1971. *Biochem. J.* 124:64–76
284. Field, H. J., Wildy, P. 1978. *J. Hyg.* 81:267–77
285. Cameron, J. M., McDougall, I., Marsden, H. S., Preston, V. G., Ryan, D. M., Subak-Sharpe, J. H. 1988. *J. Gen. Virol.* 69:2607–12
286. Goldstein, D. J., Weller, S. K. 1988. *J. Virol.* 62:196–205
287. Davison, A. J., Scott, J. E. 1986. *J. Gen. Virol.* 67:1759–816
288. Baer, R., Bankier, A. T., Biggin, M. D., Deininger, P. L., Farrell, P. J., et al. 1984. *Nature* 310:207–11
288a. Hammerschmidt, W., Sugden, B. 1988. *Cell* 545:427–35
289. Walker, J. E., Saraste, M., Runswick, M., Gay, J. 1982. *EMBO J.* 1:945–51
289a. Gorbalenya, A. E., Koonin, E. V., Donchenko, A. P., Blinov, V. M. 1988. *Nature* 333:22
289b. Hodgman, T. C. 1988. *Nature* 333:22–23
289c. Lane, D. 1988. *Nature* 334:478
290. Olivo, P. D., Nelson, N. J., Challberg, M. D. 1989. *J. Virol.* 63:196–204
291. Keir, H. M., Hay, J., Morrison, J., Subak-Sharpe, J. H. 1966. *Nature* 210:369–71
292. Weissbach, A., Hong, S. L., Aucker, J., Muller, R. 1973. *J. Biol. Chem.* 248:6270–77
293. Ostrander, M., Cheng, Y.-C. 1980. *Biochem. Biophys. Acta* 609:232–45
294. Powell, K. L., Purifoy, D. J. M. 1977. *J. Virol.* 24:618–26
295. O'Donnell, M. E., Elias, P., Lehman, I. R. 1987. *J. Biol. Chem.* 262:4252–59
295a. Dorsky, D. I., Crumpacker, C. S. 1988. *J. Virol.* 62:3224–32

295b. Haffey, M. L., Stevens, J. T., Terry, B. J., Dorsdy, D. I., Crumpacker, C. S., et al. 1988. *J. Virol.* 62:4493–98

296. Knopf, K.-W. 1979. *Eur. J. Biochem.* 98:231–44

297. Quinn, J. P., McGeoch, D. J. 1985. *Nucleic Acids Res.* 13:8143–63

298. Vaughan, P. J., Banks, L. M., Purifoy, D. J. M., Powell, K. L. 1984. *J. Gen. Virol.* 65:2033–41

299. Honess, R. W., Roizman, B. 1973. *J. Virol.* 12:1346–65

300. Powell, K. L., Courtney, R. J. 1975. *Virology* 66:217–28

301. Bayliss, G. J., Marsden, H. S., Hay, J. 1975. *Virology* 68:124–34

302. Purifoy, D. J. M., Powell, K. L. 1976. *J. Virol.* 19:717–31

303. Powell, K. L., Littler, E., Purifoy, D. J. M. 1981. *J. Virol.* 39:894–902

304. Knipe, D. M., Quinlan, M. P., Spang, A. E. 1982. *J. Virol.* 44:736–41

305. Ruyechan, W. T. 1983. *J. Virol.* 46: 661–66

306. O'Donnell, M. E., Elias, P., Funnell, B., Lehman, I. R. 1987. *J. Biol. Chem.* 262:4260–66

307. Ruyechan, W. T., Weir, A. C. 1984. *J. Virol.* 52:727–33

308. Showalter, S. D., Zweig, M., Hampar, B. 1981. *Infect. Immun.* 34:684–92

309. Knipe, D. M., Spang, A. E. 1982. *J. Virol.* 43:314–24

310. Quinlan, M. P., Knipe, D. M. 1983. *Mol. Cell. Biol.* 3:315–24

311. Fenwick, M. L., Walker, M. J., Petkevich, J. M. 1978. *J. Gen. Virol.* 39:519–29

312. Elias, P., O'Donnell, M. E., Mocarski, E. S., Lehman, I. R. 1986. *Proc. Natl. Acad. Sci. USA* 83:6322–26

313. Elias, P., Lehman, I. R. 1988. *Proc. Natl. Acad. Sci. USA* 85:2959–63

314. Olivo, P. D., Nelson, N. J., Challberg, M. D. 1988. *Proc. Natl. Acad. Sci. USA* 85:5414–18

315. Koff, A., Tegtmeyer, P. 1988. *J. Virol.* 62:4096–103

316. Crute, J. J., Mocarski, E. S., Lehman, I. R. 1988. *Nucleic Acids Res.* 16:6585–96

317. Holmes, A. M., Wietstock, S. M., Ruyechan, W. T. 1988. *J. Virol.* 62: 1038–45

318. Marsden, H. S., Campbell, M. E. M., Haarr, L., Frame, M. C., Parris, D. S., et al. 1987. *J. Virol.* 61:2428–37

319. Parris, D. S., Cross, A., Haarr, L., Orr, A., Frame, M. C., et al. 1988. *J. Virol.* 62:818–25

320. Gallo, M. L., Jackwood, D. H., Murphy, M., Marsden, H. S., Parris, D. S. 1988. *J. Virol.* 62:2874–83

321. zur Hausen, H. 1981. In *DNA Tumor Viruses: Molecular Biology of Tumor Viruses,* ed. J. Tooze, pp. 371–82. Cold Spring Harbor, NY: Cold Spring Harbor Lab.

322. Lancaster, W. D., Olson, C. 1982. *Microbiol. Rev.* 46:191–207

323. Howley, P. M. 1982. *Arch. Pathol. Lab. Med.* 106:429–32

324. Shah, K. 1985. In *Virology,* ed. B. Fields, pp. 371–91. New York: Raven

325. Dvoretsky, I., Shober, R., Chattopadhayay, S. K., Lowy, D. R. 1980. *Virology* 103:369–75

326. Geraldes, A. 1970. *Nature* 226:81–82

327. Heilman, C. A., Engel, L., Lowy, D. R., Howley, P. M. 1982. *Virology* 119:22–34

328. Yang, Y.-C., Okayama, H., Howley, P. M. 1985. *Proc. Natl. Acad. Sci. USA* 82:1030–34

329. Stenlund, A., Zabielski, J., Ahola, H., Moreno-Lopez, J., Pettersson, U. 1985. *J. Mol. Biol.* 182:541–54

330. Chen, Y. E., Howley, P. M., Levinson, A. D., Seeburg, P. H. 1982. *Nature* 299:529–34

331. Danos, O., Engel, L. W., Chen, E. Y., Yaniv, M., Howley, P. M. 1983. *J. Virol.* 46:557–66

332. Lowy, D. R., Dvoretzky, I., Shober, R., Law, M.-F., Engel, L., Howley, P. M. 1980. *Nature* 287:72–74

333. Lusky, M., Botchan, M. R. 1984. *Cell* 36:391–401

334. Spalholz, B. A., Yang, Y.-C., Howley, P. M. 1985. *Cell* 42:183–91

335. Spalholz, B. A., Lambert, P. R., Yee, C. L., Howley, P. M. 1987. *J. Virol.* 61:2128–37

336. Waldeck, W., Rosl, F., Zentgraf, H. 1984. *EMBO J.* 32:2173–78

337. Lusky, M., Botchan, M. R. 1986. *Proc. Natl. Acad. Sci. USA* 83:3609–13

338. Androphy, E. J., Schiller, J. T., Lowy, D. R. 1985. *Science* 230:442–45

339. Schlegel, R., Wade-Glass, M., Rabson, M. S., Yang, Y.-C. 1986. *Science* 233:464–67

340. Thorner, L., Bucay, M., Choe, J., Botchan, M. 1988. *J. Virol.* 62:2474–82

341. Law, M. F., Lowy, D. R., Dvoretzky, I., Howley, P. M. 1981. *Proc. Natl. Acad. Sci. USA* 78:2727–31

342. Botchan, M. R., Berg, L. J., Reynolds, J., Lusky, M. 1986. In *Papillomaviruses,* ed. D. Evered, S. Clark, pp. 53–68. New York: Wiley

343. Gilbert, D. M., Cohen, S. N. 1987. *Cell* 50:59–68

344. Stenlund, A., Bream, G. L., Botchan, M. R. 1987. *Science* 236:1666–71
345. Lusky, M., Botchan, M. R. 1985. *J. Virol.* 53:955–65
346. Sarver, N., Rabson, M. S., Yang, Y.-C., Byrne, J. C., Howley, P. M. 1984. *J. Virol.* 53:377–88
347. Berg, L., Lusky, M., Stenlund, A., Botchan, M. R. 1986. *Cell* 46:753–62
348. Lusky, M., Botchan, M. R. 1986. *J. Virol.* 60:729–42
349. Berg, L. J., Singh, K., Botchan, M. R. 1986. *Mol. Cell. Biol.* 6:859–69
350. Roberts, J. M., Weintraub, H. 1986. *Cell* 46:741–52
351. Roberts, J. M., Weintraub, H. 1988. In *Eukaryotic DNA Replication,* ed. T. Kelly, B. Stillman, pp. 191–99. Cold Spring Harbor, NY: Cold Spring Harbor Lab.
352. Roberts, J. M., Weintraub, H. 1988. *Cell* 52:397–404

Annu. Rev. Biochem. 1989. 58:719–42

MOLECULAR BASIS OF FERTILIZATION

David L. Garbers

Howard Hughes Medical Institute, Departments of Pharmacology and Molecular Physiology and Biophysics, Vanderbilt University Medical Center, Nashville, Tennessee 37232-0295

CONTENTS

PERSPECTIVES AND SUMMARY

The molecular events that act in coordination to dictate a successful fertilization occur at many different levels, including animal behavior itself. Here, we will concentrate on the molecular basis for the specificity of interactions

719

0066-4154/89/0701-0719$02.00

between the gametes and their associated structures, the specificity that in large measure prohibits fertilization across the species.

A model, in simplest of forms, that summarizes known interactions between the gametes is given in Figure 1. Here, specific activation of sperm motility, attraction of spermatozoa to the egg, adhesion of sperm cells to the egg, induction of an acrosome reaction, membrane fusion between the gametes, and subsequent egg activation are identified. Evidence exists to support these specific interactions in a wide variety of animals, and examples will be discussed in detail later. As demonstrated in the model of Figure 1, eggs are generally enveloped in acellular matrices and/or adherent cells that the spermatozoon must first encounter. In much of the literature, the sites of specific interaction between the spermatozoon and the egg and/or the cellular and acellular components of the egg are not clear for a given species. Therefore, when we speak of molecules associated with the egg, the possible emanance of such molecules from cellular or acellular structures surrounding the egg or from the egg, itself, as well as the deposition of active molecules to the egg-associated structures by cells of the female reproductive tract are all included.

It is important to realize that fertilization may occur in the absence of many of the specific events described in Figure 1. However, fertilization rates certainly would be expected to be lower under natural conditions, especially

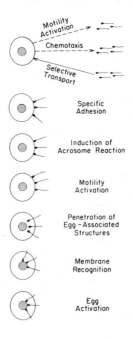

Figure 1   Sites of potential interactions between spermatozoa and the egg (including acellular and cellular matrices that surround the egg).

when stress is involved, and should range between zero and the normal value. Current evidence suggests specificity at the species level for each of the interactions depicted, although specificity is often not absolute but relative in terms of the concentration of effector molecule required to elicit a response. However, in some cases specificity is absolute and even the highest concentrations of effector molecule will not interact with a receptor molecule of another species. Examples of relative and absolute specificity can be observed with small peptides found in the egg-conditioned media of various animals that interact with cell surface receptors of spermatozoa. In closely related species the specificity may be based on relative potencies, but in more distantly related species, no cross-reaction is visible even at the highest peptide concentrations.

The biochemical and biological responses of sperm cells to a specific effector molecule are generally similar or identical. Therefore, it is reasonable to assume that the changed behavior of the spermatozoon is important and exerts pressure for the coevolution of effector molecules and receptor binding sites. The potential coevolution of effector and receptor molecules involved with fertilization raises the possibility that mutations at the germ cell level could, in some cases, mediate speciation itself. For this to occur, a mutation in an effector molecule (egg) and in a receptor molecule (spermatozoa) would occur such that fertilization between the mutant germ cells and spermatozoa or eggs of the general population would be markedly reduced or zero, but fertilization rates would be high between the two mutants. The offspring of this mating, then, could conceivably also not effectively fertilize except with each other, thereby resulting in the same phenomenon as geographic isolation.

Based on research on bacterial chemoreceptors (1–3) and other receptor families (4–7), a working hypothesis would be that cell surface receptors of spermatozoa remain highly conserved in the cytoplasmic domains but show considerable variation in the receptor-binding domains. Such models could provide reasonably simple explanations for what now appears to be considerable diversity in the structures of effector molecules but retention of the basic biological responses demonstrated in Figure 1 across the species.

Pertinent reviews on the molecular basis of fertilization include Shapiro (8), Trimmer & Vacquier (9), Vacquier (10), Wassarman (11, 12), and Garbers (13).

SELECTIVE TRANSPORT OF SPERMATOZOA

Ejaculated spermatozoa must often be first transported to the site of fertilization within the female. Selective transport within the female does not necessarily involve specific molecules associated with the egg, but may involve molecules released in follicular fluid at the time of ovulation, as well as local effector molecules released by the uterus or oviduct. In various mammals, it is well documented that spermatozoa may reside in restricted regions of the reproductive tract prior to final transport to the site of fertilization (14–18), and that sperm motility may be reduced within these regions (18). Several reports suggest that the transport of spermatozoa to the site of fertilization occurs coincident with the time of ovulation, and although the mechanism of transport activation is not fully understood, the stimulation of oviductal motility or ciliary motility by factors within the follicular fluid, and/or the release of sperm chemoattractants or sperm motility stimulants coincident with the release of the egg, are likely explanations. The potential for species-specific regulation of transport has apparently not yet been studied, although it represents one potential mechanism capable of preventing or diminishing cross-species fertilization.

ACTIVATION OF MOTILITY

Motility pattern changes and/or increased sperm velocity have been reported in both invertebrates and vertebrates in response to egg-associated molecules (19, 20). Often, however, it has not been clear whether or not the positive effects are due to a specific molecule involved with signaling or to a nonspecific effector (e.g. a metabolic substrate). Recently, a substance present in seminal plasma of pigs that stimulates adenylate cyclase activity was purified and identified as bicarbonate ion (21, 22); it also stimulated sperm motility. The authors have suggested that the bicarbonate effect on adenylate cyclase is specific for the enzyme found in spermatozoa. The stimulating effects of bicarbonate ion would normally be attributed to its role as a general metabolic substrate or possibly to its ability to elevate intracellular pH. However, in this case at least, a molecule that causes general effects in many different cells also appears to have specific effects in the spermatozoon; one therefore must be cautious with interpretations of the mechanisms of action of molecules that might be considered nonspecific. *Limulus polyphemus* spermatozoa represent an example of cells that are ejaculated and remain in an immotile state until egg-conditioned media are added to the cells (23, 24). Preliminary reports have suggested that the active molecule in the egg media is a peptide, since protease treatment of a partially purified principal destroys the activity (24). In *Orthopyxis*, however, spermatozoa appear to become immotile within the acellular matrix (jelly coat) of the egg until "active" molecules are released coincident with formation of the second polar body; subsequently, motility resumes and fertilization occurs (25). An example of motility enhancement, especially under conditions of slightly acidic extracellular pH, is represented by sea urchin spermatozoa, where egg-conditioned media can increase sperm respiration rates and motility (26, 27).

Identification of Active Molecules

Small peptides have been isolated from eggs that can stimulate sperm metabolism and motility under appropriate conditions in sea urchins (27–32). The two peptides studied in greatest detail are speract (Gly-Phe-Asp-Leu-Asn-Gly-Gly-Gly-Val-Gly), isolated from *Strongylocentrotus purpuratus* or *Hemicentrotus pulcherrimus,* and resact (Cys-Val-Thr-Gly-Ala-Pro-Gly-Cys-Val-Gly-Gly-Gly-Arg-LeuNH₂), obtained from *Arbacia punctulata* egg-conditioned media. Resact and speract do not cross-react detectably with spermatozoa of the species containing the opposite peptide even when very high concentrations are used. The specificity appears to be dictated by the primary structure of the carboxyl tail of the peptide. Considerable substitution is possible in the NH_3-terminal portion of speract with retention of biological activity, but deletion of the CO_2-terminal Gly or Val-Gly results in large or total losses of respiration-stimulating activity (33). Most chemically synthesized analogues have had an equal or decreased potency relative to speract, but one analogue (Gly-Phe-Asp-Leu-Ser-Gly-Gly-Gly-Val-Pro) appears to be 500 times more potent (34).

Nomura (35) also has studied the structure-activity relationships of Gly-Phe-Asp-Leu-Ser-Gly-Gly-Gly-Val-Gly, an egg peptide from *A. crassispina.* Elimination of the carboxyl-terminal Gly reduced activity to 1/3000 of the native peptide, whereas deletion of the amino-terminal Gly reduced activity by only 1/10. Tyr could replace Phe without loss of activity. The replacement of Phe by Leu, norleucine, Ala, Gly, or Pro, in contrast, considerably reduced activity. Valine also appeared important for activity, and it subsequently was suggested that hydrophobic residues at positions 2, 4, and 9 were important for optimal activity.

For the most part, structure-activity studies have concentrated on the relationship of structure to respiration stimulation; for a number of speract analogues, respiration-stimulating activity has coincided with cyclic nucleotide-elevating activity (30). Shimomura & Garbers (36) prepared various analogues of resact, however, and found that relative potencies varied dependent on the physiological parameter measured. Modification of the CO_2-terminal leucine-NH_2 of resact did not alter biological activity, but substitution of the two cystenyl residues by Ser or Tyr or methylation of the cystenyl residues resulted in divergent relative potencies dependent on whether respiration rates or cyclic nucleotide concentrations were measured.

The structures of various other sperm-stimulating peptides are now known (37–39); some of these resemble the structure of speract or resact, and others do not (Table 1).

Suzuki et al (40) have attempted to study the taxonomical distribution of the peptides in the sea urchin by testing the crude egg-conditioned media of a particular species on the spermatozoa of various other species. Their results suggest that egg-conditioned media of species within the same order will

Table 1 Structure of some of the peptides that are known to stimulate sperm metabolism and motility found in the egg-conditioned media of various sea urchins

Peptide name	Peptide structure
Speract	Gly-Phe-Asp-Leu-Asn-Gly-Gly-Gly-Val-Gly
	Gly-Phe-Asp-Leu-Ser-Gly-Gly-Gly-Val-Gly
	Gly-Phe-Asp-Leu-Thr-Gly-Gly-Gly-Val-Gly
	Gly-Phe-Asp-Leu-Thr-Gly-Gly-Gly-Val-Gln
	Gly-Phe-Ala-Leu-Gly-Gly-Gly-Gly-Val-Gly
	Gly-Phe-Ser-Leu-Asn-Gly-Gly-Gly-Val-Ser
	Ser-Phe-Ala-Leu-Gly-Gly-Gly-Gly-Val-Gly
	Gly-Phe-Ser-Leu-Ser-Gly-Ser-Gly-Val-Asp
	Gly-Phe-Ser-Leu-Ser-Gly-Ser-Gly-Val-Gly
	Asp-Ser-Asp-Ser-Ala-Gln-Asn-Leu-Ile-Gly
	Asp-Ser-Asp-Ser-Ala-His-Leu-Ile-Gly
Resact	Cys-Val-Thr-Gly-Ala-Pro-Gly-Cys-Val-Gly-Gly-Gly-Arg-LeuNH$_2$

stimulate the spermatozoa across species lines; this does not imply that the structures of the peptides are the same, but that the peptide structures are similar enough to cross-react with the same receptors. *L. pictus* and *S. purpuratus* fall within the same order, for example, and they both contain speract or speract-like molecules, but the structures of the peptides are different (37).

Identification of Egg Peptide Genes

One of the interesting observations made by Suzuki and coworkers (29, 32, 38, 39) was that various peptides with similar structures could be isolated from the egg-conditioned media of the same species. Since conditioned media were isolated from the eggs of many individuals, it was not known whether the different peptide structures reflected constituents from a single sea urchin, or whether individual sea urchins possessed one peptide and the variability in structures reflected individual variation in the population. C. S. Ramarao, D. J. Burks, and D. L. Garbers (personal communication) have now isolated cDNA clones from an ovarian cDNA library that encodes for speract and speractlike structures. Two different clones with insert sizes of 1.2 and 2.3 kb have been isolated; the open reading frame of the longer insert is shown in Figure 2. Eight speract or speractlike potential decapeptides are encoded for in one region, each separated by a single lysine residue. Four of the peptides are speract, and four contain structures that possess a Gly-Gly-Gly-Val-Gly carboxyl-terminus, suggestive that they therefore have motility-stimulating properties (see above). Another decapeptide, Gly-Thr-Met-Pro-Thr-Gly-Ala-Gly-Val-Asp, within the same region of the mRNA as well as a decapeptide

DEDUCED AMINO ACID SEQUENCE OF THE SPERACT PRECURSOR

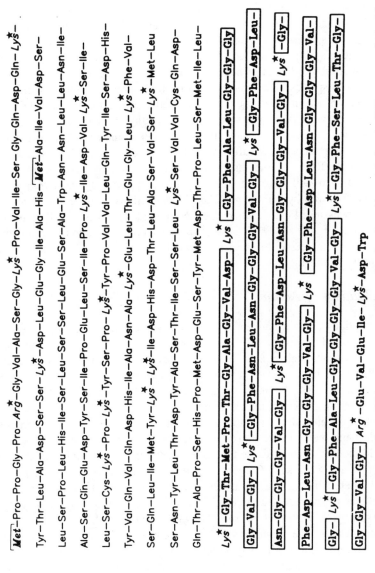

Figure 2 The deduced amino acid sequence of the speract precursor. The predicted amino acids are based on the DNA sequence obtained from an isolated 2.3 kb cDNA insert from a sea urchin ovary cDNA library. Two possible initiation sites (methionine) are marked with an overline. The lysine residues are starred and the nine potential decapeptides that are continuous with each other are boxed. One other potential decapeptide exists.

upstream (Ile-Asp-His-Asp-Thr-Leu-Ala-Ser-Val-Ser) also separated by single lysine residues, has yet to be tested for biological activity. The other isolated clone contained two additional speractlike structures: Ser-Phe-Asp-Leu-Asn-Gly-Gly-Gly-Val-Gly and Gly-Phe-Ser-Leu-Thr-Gly-Gly-Gly-Val-Gly. Hybridization of the cDNA clone with mRNA from ovaries revealed two major RNA species of 1.2 and 2.3 kb.

The gene(s) for speract are now being isolated, which will allow the determination of whether or not there are conserved regions of the gene that exist across species. The structure of the RNA, however, raises interesting questions. Since multiple speract or speractlike structures occur within the same RNA, what is the pressure to retain fidelity in multiple copies of the same biologically active peptide? Is a single speract molecule encoded within one mRNA molecule not sufficient? There are other potential peptides encoded by the RNA. Do they have a function? The processing of the predicted precursor also will be of interest. It can be speculated that a trypsinlike enzyme followed by carboxypeptidase B-like activity would result in appropriate processing. Finally, where are these peptides synthesized? Preliminary data of D. J. Burks and D. L. Garbers (personal communication) suggests that a precursor molecule for resact is found within the egg, suggesting its synthesis within the egg.

Mechanism of Action

The peptides cause a number of early biochemical responses that resemble those reported in somatic cells that respond to growth factors, hormones, and other environmental signals. Transient elevations of cyclic AMP, cyclic GMP, and of intracellular Ca^{2+} have been reported (20, 41, 42). A proton efflux occurs and the intracellular pH increases (41, 43). Although the temporal arrangement of the biochemical events has yet to be established, the elevation of intracellular pH appears to be the primary cause of the increased motility and metabolism. Hansbrough & Garbers (41) initially demonstrated that Monensin A could cause the same effects on sperm motility as the egg peptides, and Repaske & Garbers (43) later showed that weak acids and weak bases altered motility in a manner consistent with the hypothesis that intracellular pH represented the primary regulator of motility. The addition of 8-bromo-cyclic GMP but not of cyclic GMP, cyclic AMP, or various other nucleotides also can stimulate sea urchin sperm motility (44); the stimulation by 8-bromo-cyclic GMP may represent a property of this nucleotide not related to its structural similarity to cyclic GMP, since Shimomura & Garbers (36) demonstrated that elevations of sperm respiration rates could be observed in the absence of detectable changes in intracellular cyclic GMP.

CHEMOATTRACTION

Identification of Active Molecules

Sperm chemotaxis has been reported in various phyla and has been demonstrated in many different species (45) across both the animal and plant kingdom. The first and still only identified substance from animal eggs established as a chemoattractant is the peptide resact described in the previous section. The peptide was isolated based on its ability to stimulate sperm respiration rates and was subsequently shown to be a potent attractant (46). Miller (45) summarizes the chemical characteristics of attractants, a number of which strongly resemble peptides, from various animals. Further research on a starfish attractant has suggested it to be a peptide although its primary structure is not known (47). Resact has been tested on the spermatozoa of S. purpuratus and L. pictus as well as starfish spermatozoa and it fails to attract these spermatozoa; there is also no detectable specific binding of radiolabeled resact to spermatozoa from these animals. Consistent with these observations, Miller (45) has compiled extensive data demonstrating that species specificity of sperm attraction is a common phenomenon. The experiments to provide such information have involved the use of crude or partially purified extracts from eggs tested on the spermatozoa of the same or of different species. Although the structures of the active molecules remain to be determined, the experimental design clearly established species specificity to the attractant response.

Reports also exist that spermatozoa are attracted to peptides of the f-Met-Leu-Phe family (48, 49). In one case this has been reported for bovine and in the other case for human spermatozoa. Since prior to these reports chemotaxis had not been successfully demonstrated in mammalian spermatozoa, the results were at first exciting and suggested that a molecule related to this peptide family might be secreted by mammalian eggs. However, Miller (50) subsequently demonstrated that the work of Igbal et al (49) contained methodological errors, and the report of Gnessi et al (48) has yet to be confirmed. Following the apparent positive chemoattractant effects with human sperm cells, Gnessi et al (51) reported the specific binding of ^{35}S-fMet-Leu-Phe to human spermatozoa. Approximately 60,000 receptors/cell were estimated. Attempts to measure specific binding of ^{35}S-fMet-Leu-Phe to bovine or human spermatozoa in at least one other laboratory, however, have not been successful (S. E. Domino, D. L. Garbers, personal communication). One major concern with the previous studies is the possible contamination of spermatozoan preparations by neutrophils or neutrophil membranes. It now seems prudent to return to egg extracts, egg-conditioned media, follicular fluid, etc, to determine whether or not sperm chemotaxis in the mammal occurs. Studies are potentially more complicated in the mammal because of the process of capacitation and because of the normal storage of sperm cells in certain regions of the female reproductive tract prior to transport to the site of fertilization; it remains possible that chemoattractant responses of mammalian spermatozoa will not be observed unless the cell proceeds through these normal physiological events.

Mechanism of Action

The mechanism by which gradients of resact are detected by the sperm cell remains to be determined. However, it is known that Ca^{2+} in the extracellular medium is required in order for cells to respond to the peptide gradient (46). Since Ca^{2+} is not required for the binding of the peptides or for the stimulation of respiration rates, it would appear that Ca^{2+} movement from extracellular to intracellular compartments represents an important signal for sperm cell orientation. It has been demonstrated, in fact, that intracellular Ca^{2+} concentrations increase in a transient manner in response to speract or resact and that such elevations require the presence of extracellular Ca^{2+} (42). A theory that Ca^{2+} concentrations may lead to changes in the asymmetry of flagellar bending in response to resact has been proposed (52), and could explain the mechanism by which the cell moves in a peptide gradient. The signal for Ca^{2+} mobilization remains to be clarified, although cyclic AMP and/or cyclic GMP are potential candidates. In both cases there is prior evidence in other cells that these nucleotides can regulate ion channels (53–58).

Research of Lee & Garbers (59) demonstrated that the membrane potential of spermatozoa could be regulated by speract or resact, and Lee (60) later suggested involvement of a GTP-binding regulatory protein in a proposed speract receptor/K^+ channel coupled interaction. The model of Lee (60) is attractive, since guanine nucleotide regulatory protein(s) are known to regulate K^+ channels in other cells (61, 62). This model would suggest the regulation of the Na^+/H^+ antiporter by a primary effect on K^+ channels, since Lee (63) has previously demonstrated the existence of membrane potential–dependent Na^+/H^+ exchange.

RECEPTORS FOR EGG PEPTIDES

Identification of Receptor Molecules

Although an ^{125}I-labeled Bolton-Hunter adduct of speract was used in early studies of the speract receptor (33), the lack of a free amino group negated potential cross-linking studies to identify the receptor. Therefore, an analogue (Gly-Gly-Gly-Gly-Tyr-Asp-Leu-Asn-Gly-Gly-Gly-Val-Gly) was synthesized that retained respiration-stimulating activity equivalent to speract (64). The analogue also competed with ^{125}I-labeled Bolton-Hunter speract for receptor binding with equivalent potency to speract. The GGG[Y^2]speract was radiolabeled with ^{125}I and subsequently cross-linked to the apparent receptor with dissuccinimidyl suberate (64). The apparent receptor was identified as a glycoprotein with an estimated molecular weight of 77,000 (sodium dodecyl sulfate gels, reducing conditions). The specificity of the association of the peptide with the protein was determined by competition experiments with various speract analogues and by the failure of the radiolabeled peptide to covalently bind to spermatozoa from species that did not cross-react with speract. In the cross-linking experiments, high concentrations of GGG[Y^2]speract could not be used, and therefore a low-affinity receptor may not have been detected. In addition, there is the possible existence of receptor molecules that are not capable of being covalently coupled with GGG[Y^2]speract because of a lack of the necessary functional group at the binding site. The most definitive experiments that concern the determination of whether or not the cross-linked protein is the receptor are probably the isolation of the cDNA clone for the protein followed by the expression of the protein in cultured cells. L. J. Dangott et al (65) have managed to purify the cross-linked protein, to obtain amino acid sequence from the protein, and to isolate cDNA clones from testis cDNA libraries for the protein. A cDNA clone containing a 2.5 kb insert was isolated that hybridized to mRNA isolated from testis of the same size. An open reading frame containing 531 amino acids was contained within the insert. The predicted protein contains an amino-terminal domain relatively rich in cysteine residues and four repeat sequences of approximately 110 amino acids; the putative receptor resembles, then, the low-density lipoprotein and other receptors in this respect (66–68). In addition, proteins of the cellular adhesion supergene family contain similar repeat sequences (69).

The same approaches described above were used to identify the apparent resact receptor (70). An analogue of resact (Gly-Gly-Gly-Tyr-Gly-Cys-Val-Thr-Gly-Ala-Pro-Gly-Cys-Val-Gly-Gly-Gly-Arg-LeuNH$_2$) was chemically synthesized and shown to possess the same respiration-stimulating and receptor-binding activity as resact. The radioiodinated analogue was then used in cross-linking experiments, but instead of a major radiolabeled protein at a molecular weight of 77,000, the major radioactive band was present at an apparent molecular weight of 160,000. Speract failed to compete with the ^{125}I-labeled analogue in cross-linking experiments, whereas nonradioactive resact effectively competed. The radioactive, cross-linked protein was subsequently identified as the enzyme guanylate cyclase (70). As with the apparent resact receptor, amino acid sequences from this protein were obtained and a cDNA clone was subsequently isolated (71). The predicted amino acid sequence of guanylate cyclase showed an open reading frame of

986 amino acids with a 21-amino-acid signal peptide and a single transmembrane domain separating the protein into 478 extracellular and 459 intracellular amino acids. The enzyme was shown to possess significant identity in a carboxyl domain to the catalytic domains of all protein kinases, therefore making it a member of the protein kinase family.

After the isolation of the cDNA clone from *A. punctulata* spermatozoa, a cDNA clone from a *S. purpuratus* cDNA library was isolated and sequenced (72). A comparison of the predicted amino acid sequences between the two sea urchin clones is shown in Figure 3. High conservation of amino acids is observed, particularly in the cytoplasmic domain that is homologous to protein kinases. Greater variability in the amino terminal domain is apparent, and in the carboxyl-tail of the protein, the *S. purpuratus* sequence completely deviates from that of *A. punctulata*.

The finding of two apparent egg peptide receptor molecules (for speract and resact), whose structures showed no high degree of similarity, was unexpected. Since the biochemical responses of the sperm cells to the various peptides appear to be similar or identical, the simplest receptor model would be like that described for the bacterial chemoreceptors (1–3) where the amino-terminal domains are highly variable and the carboxyl domains remain highly conserved. The conservation of the cytoplasmic domains of the peptide receptors but variation of the extracellular, binding domain could easily account for the diversity of peptide specificity but retention of the same biological responses. A possible explanation of the above receptor data would be that the 77,000-dalton protein and guanylate cyclase are actually in close apposition in the membrane and possibly even subunits of the functional receptor. Since cross-linking studies relied on disuccinimidyl suberate as the coupling reagent, it remains possible that either protein could have been cross-linked dependent on the proximity of the amino-terminus of the peptide and a functional amino group in one of the two proteins; the specificity expected for the receptor would still be observed. Unpublished data of L. J. Dangott suggest that a speract receptor–like protein exists in *A. punctulata;* this evidence is based on the isolation of positive-hybridizing cDNA clones from a testis cDNA library. Since it is known that speract does not bind to *A. punctulata* spermatozoa, the homologous protein must be modified in the ligand-binding domain if it, in fact, represents the receptor. With cDNA clones available for both apparent receptors, it will now be possible to determine whether or not both molecules are, in fact, cell surface receptors whether they form subunits of a receptor, or whether one of the proteins is merely in close apposition to the other.

Comparison with Somatic Cell Receptors

At about the same time studies demonstrated the cross-linking of resact to guanylate cyclase of spermatozoa (70), reports appeared suggesting that atrial

```
SPGC  MAHARHLFLFMVAFTITMV-IARLDFNPTIINEDRGRTKIHVGLLAEWTTADGDQGTLGFPALGALPLAISLANQDSNILNGFDVQFEWVDTHCDINIGWH (100)
APGC          TTL LV VM   RS T HY  V L    PLIMTS N NS  -    SA F   QY NMD HYIN                          L  A (100)

SPGC  AVSDWWKRGFVGVIGPGCCTYEGRLASALNIPMIDYVCDENPVSDKSIYPTFLRTIPPSIQVVEAIILTLQRYELDQVSVVENITKYRNIFNTMKDKFD (201)
APGC       I L      F A          N EF                               DS L DM DWN T        V        EQ (201)

SPGC  ER-DYEILHEEYYAGFDPWDYEMDDFFSEIIQRTKETTRIYVFFGDASDLRQFAMTALDEGILDSGDYVILGAVVDLEVRDSQDYHSLDYILDTSEYLNQI (301)
APGC   M EEW      PDAAE TD K      SG           IS V N AV I I                                    ETEAD E (302)

SPGC  NPDYARLFKNREYTRSDNDRALEALKSVIIVTGAPVLKTRNWDRFSTFVIDNALDAPFNGELELRAEIDFASVYMFDATMQLLEALDRTHAAGGDIYDGEE (402)
APGC  QA EQM L T DE   M L E    RSQA HIY AI      E T KTD M I        TE K ALQ M               SQ (403)

SPGC  VVSTLLNSTYRSKTDTFYQFDENGDGVKPYVLLHLIPIPKGDGGATKDSLGMYPIGTFNR-ENGQWGFEEALDEDANVLKPVWHNRDEPPLDMPPCGFHGE (502)
APGC   NF TS  AKAD        S R V        MPPG P   VASHS ----NK PD-N        D        V (499)

SPGC  LCTNWALYLGASIPTFLIIFGGLIGEFIYRKRAYEAALDSLVWKVDWSEVQTKATDTNSQGFSMKNMVMSAISVISNAEKQQIFATIGTYRGTVCALHAVH (603)
APGC   G    TL A I      GL YY       K RESE       S L                                    I I (600)

SPGC  KNHIDLTRAVRTELKIMRDMRHDNICFFIGACIDRPHISILMHYCAKGSLQDILENDDIKLDSMFLSSLIADLVKGIVYLHSSEIKSHGHLKSSNCVVDNR (704)
APGC            L              C            M A                L (701)

SPGC  WVLQITDYGLNEFKKGQKQDVDLGDHAKLARQLWTSPEHLRQEGSMPTAGSPQGDIYSFAILLTELYSRQEPFHENEMDLADIIGRVKSGEVPPYRPILNA (805)
APGC   HR  E E K A   EGK HPG T K      S M         DLE A      SK           V (802)

SPGC  VNAAAPDCVLSAIRACWPEDPADRPNIMAVRTMLAFPLQKGLKPNILDNMIAIMERYTNNLEELVDERTQELQKEKTKTEQLLHRMLPPSIASQLIKGIAVL (906)
APGC   E T V  ME IE                                   A                                              S (903)

SPGC  PETFEMVSIFFSDIVGFTALSAASTPIQVVNLLNDLYTLFDAIIISNYDVYKVETIGDAYMLVSGLPLRNGDRHAGQIASTAHHLLESVKGFIVPHKPEVFL (1007)
APGC   D     LIHFSL SCRLFCSSQVLPLLVPWLHSLLTLPLHLPL WMNPLIS FAQPSWSAL SHSCSALHSS* (986)
```

Figure 3 Comparison of the deduced amino acid sequence of the putative guanylate cyclase from A. punctulata (APGC) or S. purpuratus (SPGC). The A. puntulata cDNA clone was isolated as described in Ref. 71, and this clone was then used as a probe to isolate the S. purpuratus cDNA clone. The predicted amino acid sequences of S. purpuratus are shown and those of A. punctulata are given only where different.

natriuretic peptides also could bind to guanylate cyclase (73–76). In both cases the peptides were shown to activate guanylate cyclase activity of membrane preparations (77–80) or of detergent-solubilized enzyme preparations (81, 82). These results not only suggested the conservation of a cell surface receptor with enzymatic activity in germ cells and somatic cells, but the conservation of this signaling system from echinoderms through mammals.

The deduced amino acid sequences of mammalian membrane forms of guanylate cyclase have now been obtained (83, 84), and a low-molecular-weight atrial natriuretic peptide (ANP) receptor that does not possess guanylate cyclase activity also has been cloned (85). The sea urchin sperm guanylate cyclases show some identity to the low-molecular-weight ANP receptor in the presumed peptide-binding domain, whereas the mammalian membrane forms are approximately 33% identical with this receptor in the presumed extracellular domain. It is now known that guanylate cyclase binds ANP and therefore serves as a cell surface receptor (83, 84). Of considerable interest has been the comparison of the membrane forms of guanylate cyclase to one of the subunits of the soluble form of guanylate cyclase from bovine lung. One of the subunits of the bovine lung soluble form of guanylate cyclase has been cloned by Koesling et al (86). As shown in Figure 4, the clone from *S. purpuratus* shows a high degree of identity to the bovine lung soluble form in the carboxyl region. The enzyme of *S. purpuratus*, in fact, resembles the soluble subunit within this region more so than the membrane form from *A. punctulata* (71). Since *S. purpuratus* is believed to have evolved much later than *A. punctulata*, the similarity in the carboxyl region of guanylate cyclase

```
S GC  379   T L R A L E D E K K T D T L L Y S V L P P S V A N E
P GC  872   R T Q E L Q K E K T K T E Q L L H R M L P P S I A S Q

S GC  406   L R H K R P V P A K R Y D N V T I L F S G I V G F N A
P GC  899   L I K G I A V L P E T F E M V S I F F S D I V G F T A

S GC  433   F C S K H A S G E G A M K I V N L L N D L Y T R F D T
P GC  926   L - - - - S A A S T P I Q V V N L L N D L Y T L F D A

S GC  460   L T D S R K N P F V Y K V E T V G D K Y M T V S G L P
P GC  949   I I - - - S N Y D V Y K V E T I G D A Y M L V S G L P

S GC  486   - E P C I H H A R S I C H L A L D M M E - I A G - Q V
P GC  973   L R N G D R H A G Q I A S T A H H L L E S V K G F I V

S GC  511   Q V D G E S - V Q I T I G I H T G E V V T G V I G Q R
P GC 1000   P H K P E V F L K L R I G I H S G S C V A G V V G L T

S GC  537   M P R Y C L F G N T V N L T S R T E T T G E K G K I N
P GC 1027   M P R Y C L F G D T V N T A S R M E S N G L A L R I H
```

Figure 4 Comparison of a region of the *S. purpuratus* (PGC) guanylate cyclase amino acid sequence with that of the soluble form of guanylate cyclase from bovine lung (SGC) (86). Identical residues are boxed.

from *S. purpuratus* and bovine lung may indicate the conservation of these residues, in general, in species that arose after *S. purpuratus*. This is supported by the observation that the mammalian membrane form of guanylate cyclase also displays a high degree of amino acid identity to the soluble enzyme subunit within this same region (83, 84).

In addition, the membrane forms of guanylate cyclase contain a region homologous to the protein kinases (71, 83, 84) and therefore they resemble, in part, the cell surface receptors of the protein tyrosine kinase family (87). They also resemble this family by virtue of possessing one predicted transmembrane domain. A schematic model of the guanylate cyclase family, then, is shown in Figure 5. Based on the sequence variability of the enzyme between sea urchin species, it also can be predicted that the guanylate cyclase family will include other members. Along these lines, it should be recalled that guanylate cyclase produces cyclic AMP at a low rate relative to cyclic GMP; in some cases, however, rates of cyclic AMP formation have approached 10% of the rates of cyclic GMP formation (88). Therefore, although there are no striking similarities between the bacterial or yeast adenylate cyclases and the sperm guanylate cyclase (71, 72), with sequence data on additional adenylate and guanylate cyclases, and subsequent closer examination, restricted conserved regions may become apparent. Some data, albeit slight, in fact, has been reported to suggest regions of identity between the yeast adenylate cyclase and the mammalian guanylate cyclase (83).

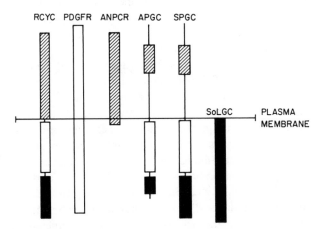

Figure 5   Members of the guanylate cyclase family. Presented are the deduced proteins for the rat brain membrane form of guanylate cyclase (83) (RCYC), PDGF receptor (PDGFR) (136), low-molecular-weight ANP receptor (ANPCR) (85), *A. punctulata* guanylate cyclase (APGC) (71), *S. purpuratus* guanylate cyclase (SPGC) (72), and the bovine soluble form of guanylate cyclase (SoLGC) (86). Regions of significant identity between the membrane forms of guanylate cyclase and the other proteins are boxed and shaded the same.

Regulation of Receptor Activity

Guanylate cyclase exists as a phosphoprotein in spermatozoa (89, 90), and when membranes are prepared such that the enzyme remains in a phosphorylated state, the egg peptides markedly activate the enzyme (77). The activation is transient, however, and coincident with a decrease in enzyme activity a loss of phosphate occurs (77). In the intact cell, elevations of cyclic GMP caused by egg peptide addition also are transient, and as with membranes a rapid dephosphorylation of guanylate cyclase coincides with subsequent decreases in cellular cyclic GMP (31, 89, 90). In the absence of peptide, sperm guanylate cyclase has been found to contain up to 17 mol phosphate/mol enzyme, all on serine residues (91, 92); after addition of resact only 1–3 mol phosphate/mol enzyme remain. Therefore, the egg peptide receptor:guanylate cyclase coupled reaction appears to be desensitized by dephosphorylation as opposed to desensitization by phosphorylation as has been reported in many other receptor models (93, 94).

GAMETE ADHESION

Identification of Active Molecules

In the sea urchin the spermatozoon first encounters an acellular structure called the jelly coat; after penetration of this material, the sperm cell interacts with the vitelline layer prior to its fusion with the plasma membrane of the egg. For the gametes to adhere to each other upon collision is a distinct advantage and, in fact, sperm cells adhere to the jelly coat of eggs. One of the mediators of this adhesion appears to be a fucose sulfate polymer (FSG, for fucose sulfate glycoconjugate; 10) that also induces an acrosome reaction (95–97). FSG is a highly sulfated polymer and when added to suspensions of either live or dead sea urchin spermatozoa causes marked agglutination. It seems likely that the large FSG molecule contains multiple recognition sites and therefore is able to bind to multiple spermatozoa resulting in agglutination. The highly charged nature of FSG, however, leads to relatively low specificity in terms of many proteins and cells adhering to the jelly coat.

Species-specific binding that demonstrates high fidelity is obtained at the level of the vitelline envelope; the structure of one of the interactive molecules is now known and is named bindin (98); it is a major protein of the acrosomal process and has been cloned and the amino acid sequence deduced (99). Bindin is present at an intracellular site and therefore mediates species-specific binding following the acrosome reaction. Its structure would be expected to vary between the species, and some preliminary data in support of this suggestion have been presented (99). Bindin is synthesized as a 51,000-dalton precursor molecule that is subsequently processed to yield a 24,000-dalton mature protein in the sea urchin, S. purpuratus (99). The synthesis of the protein appears specific to the testis. As pointed out by Gao et al (99), the specificity of the interaction of bindin with its receptor appears to represent one of the mechanisms that prevents gene flow between the species, since in at least *S. purpuratus* and *Strongylocentrotus franciscanus* it is known that viable hybrids between the species can be formed. However, the bindin of *S. purpuratus* fails to interact with the receptor of *S. franciscanus* eggs, demonstrating a barrier at the level of fertilization. The structure of the bindin receptor remains to be determined, although it has been partially characterized (100, 101).

In the mouse, the protein present in the zona pellucida responsible for adhesion has been isolated and its mRNA has been cloned (11, 12, 102). The protein, named ZP-3, is a glycoprotein; its O-linked oligosaccharide side chains appear required for adhesion, but another apparent property of ZP-3, the induction of an acrosome reaction, is lost upon removal of the protein component (11, 12). The precursor of ZP-3 has a molecular weight of 46,307 with a 22-amino-acid signal peptide that would result in a secreted protein of molecular weight equal to 43,943 (102). Although the structures of ZP-3-like molecules in other species are not yet known, the cDNA probe from mouse hybridizes with the RNA obtained from ovaries of various other mammalian species suggestive that considerable identity exists despite the relative species-specificity of adhesion. The cDNA probe from mouse does not cross-hybridize, however, with genomic DNA from various nonmammalian species *(Xenopus laevis, S. purpuratus, Drosophila melanogaster,* and Rainbow trout), thereby demonstrating, at least under the conditions of stringency used, that ZP-3 represents a mammalian-specific protein product (102). Clearly, the comparative structures of ZP-3 will be of great importance in the targeting of the active sites of adhesion as well as those sites involved in the induction of an acrosome reaction.

The sperm receptor for ZP-3 has not been unequivocally identified, although recent reports suggest galactosyltransferase as the receptor molecule (11, 12, 103). It has been suggested that another zona pellucida protein, ZP-2, mediates secondary adhesion of the gametes (104, 105). This protein, which would bind to a sperm receptor after completion of the acrosome reaction, would be analogous to the bindin/egg receptor mechanism reported in the sea urchin. The structure of ZP-2 has not yet been reported, but data demonstrating that proteinase inhibitors prevent this secondary adhesion suggest that a trypsinlike enzyme (possibly acrosin) is either directly or indirectly involved in the adhesion process (105–107).

INDUCTION OF THE ACROSOME REACTION

Identification of Active Molecules

In those animals whose spermatozoa possess an acrosome, the induction of an acrosome reaction is thought to be essential for fertilization. Although there were initial suggestions that this reaction did not normally proceed in response to extracellular signals, there is now compelling reason to suggest that specific molecules associated with the egg normally signal the spermatozoon when the time is appropriate for the reaction. There are also reports of follicular fluid and various constituents of the female reproductive tract or of molecules obtained from cumulus oophorous cells causing the induction of an acrosome reaction (108–112). In the sea urchin, starfish, and mouse considerable information about these signaling molecules is now known, although the sperm receptors remain to be identified.

In the sea urchin, a fucose-sulfate–rich molecule (FSG; 10) has been isolated that induces the acrosome reaction in a relatively species-specific fashion (95), and it has been suggested that the species specificity is due to the structure of the sugar-sulfate polysaccharides (96). SeGall & Lennarz (95) suggested in early studies that the removal of protein did not destroy the ability of the remaining sugar-sulfate to induce an acrosome reaction. However, it must be cautioned that in the face of no concentration-response data, it remains possible that sufficient remaining molecules with protein still attached existed that these molecules actually caused the acrosome reaction. In fact, Garbers et al (113) demonstrated that they could not remove all detectable protein from a preparation by NaOH or pronase treatment; the final preparation of enriched fucose-sulfate (an event that coincides with Ca^{2+} uptake and the acrosome reaction) still elevated cyclic AMP concentrations, but the relative potency decreased appreciably.

ZP-3 from the mouse, a zona pellucida protein discussed above with respect to its role in adhesion of the gametes, also has a proposed role in the induction of the acrosome reaction (11, 12). Unlike the situation with adhesion, however, the protein component of the molecule appears required for induction of the acrosome reaction (11, 12). Now that cDNA clones are available, it will be possible to define structure/activity relationships of the molecule. The apparent dual function of ZP-3 has been assigned under the assumption that the purified material is actually chemically pure; therefore, there still exists the possibility that a potent, small polypeptide contaminated the preparations and accounted for the acrosome reaction–inducing properties of ZP-3; this, however, seems unlikely since the same polypeptide would be expected to associate with other zona pellucida glycoproteins. Under the assumption that ZP-3 has a dual function then, it can be postulated that at least two receptors exist on the spermatozoon: one that recognizes the carbohydrate structure and mediates adhesion and another that recognizes a protein component of ZP-3 and upon binding transduces an intracellular signal(s) that culminates in the acrosome reaction.

On the surface there are similarities between ZP-3 and the FSG described above for the sea urchin. Both molecules induce the acrosome reaction and both appear to do so by primary effects on Ca^{2+} uptake. ZP-3 mediates cell adherence and FSG appears to possess the same property based on its ability to cause sperm agglutination when mixed with either live or dead cells and on the observed adhesion of sperm cells to the jelly coat of sea urchin eggs. Both FSG and ZP-3 appear to mediate their effects on the intact cell prior to an acrosome reaction, and both exhibit relative species specificity with respect to their ability to induce an acrosome reaction.

In the starfish the acrosome reaction apparently occurs in response to two components, a sulfated glycoprotein, and steroidal saponins (114, 115). It has been shown that a molecule, most likely an oligopeptide, also can participate in the acrosome reaction (116). The participation of steroidal saponins in the acrosome reaction of sea urchins or other species has not been reported, however, Yamaguchi et al (117) have suggested that speract can potentiate the acrosome reaction apparently induced by FSG. The mechanism of the potentiation is not yet understood but not surprising, since many of the biochemical changes caused by FSG (elevated intracellular pH, increased concentrations of cyclic AMP, mobilization of Ca^{2+}) are also induced by the egg peptides.

Mechanism of Action

The signal/transduction pathways of mammalian and nonmammalian spermatozoa may be similar if not identical. No study has yet dissociated a Ca^{2+} influx across the plasma membrane from the acrosome reaction, and presumably a primary event after receptor occupation, therefore, is Ca^{2+} mobilization. It remains to be determined by what mechanism the Ca^{2+} mobilization is activated, although it remains possible that the sperm receptor, itself, is an ion channel. The identification of the apparent receptor followed by the cloning of the mRNA for the receptor is now of paramount importance; little but continued speculation will likely occur prior to this accomplishment.

Nevertheless, presented here is a brief summary of current views on the molecular basis of the acrosome reaction. Gonzalez-Martinez & Darszon (118) have reported that a rapid hyperpolarization, possibly due to K^+ efflux, occurs transiently in response to FSG. The hyperpolarization is followed by a depolarization due principally to Ca^{2+} uptake (119, 120). It has been suggested that possibly a voltage-dependent Na^+/H^+ exchanger is regulated by FSG due to its effects on membrane potential (118); an increase in intracellular pH is also a primary response to FSG. Yanagimachi (121) has suggested that Ca^{2+} influx represents the primary signal for induction of an acrosome reaction. He has then postulated that the Na^+/K^+ ATPase would be inhibited by the elevated Ca^{2+} concentrations followed by elevated intracellular Na^+ concentrations that he suggests would then result in an efflux of H^+ via the Na^+/H^+ antiporter. However, the Na^+/H^+ antiporter in this scheme would transport Na^+ out and H^+ in and hence lower intracellular pH.

In the sea urchin cyclic AMP concentrations are markedly elevated in response to FSG, but FSG does not activate adenylate cyclase in broken cell preparations in the presence or absence of GTP or GTP analogues (19). Since the elevations of cyclic AMP can be reproduced by agents that cause Ca^{2+} entry into spermatozoa and since D-600 or verapamil can block cyclic AMP elevations in response to FSG, it appears that cyclic AMP is regulated by Ca^{2+}. However, in the sea urchin, direct regulation of sperm adenylate cyclase by Ca^{2+} has not been demonstrated. This is not the case for the spermatozoa of various other species where Ca^{2+}-regulated adenylate cyclases have been reported (122–124). Recently, Noland et al (125) have reported ZP-3–induced elevations of cyclic AMP in mouse spermatozoa, and these effects also appear to depend on the presence of extracellular Ca^{2+}.

It seems likely, based on these reports, that protein phosphorylation represents a component of the acrosome reaction. However, the role of this covalent modification in the reaction remains to be determined. Since added cyclic AMP or cyclic AMP analogues will not induce an acrosome reaction but may accelerate the time required for an acrosome reaction, protein phosphorylation is not sufficient, by itself, for this morphological event to occur.

With the realization that inositol polyphosphates could regulate Ca^{2+} concentrations of many cells, it became of interest whether or not one of these metabolites could mediate the FSG or ZP-3 effects on Ca^{2+} mobilization. Domino & Garbers (126) reported that inositol 1,4,5,-trisphosphate (IP_3) was elevated in sea urchin spermatozoa in response to FSG. However, the elevation of IP_3, like the elevation of cyclic AMP, appeared to require extracellular Ca^{2+}, and the mobilization of Ca^{2+} by other agents also could elevate IP_3 concentrations. Therefore, IP_3, IP_4, or other inositol polyphosphates probably do not mediate the Ca^{2+} movement across the plasma membrane. Since then, Domino et al (127) have demonstrated marked increases in phosphatidic acid after FSG addition to spermatozoa. The phosphatidic acid was subsequently shown to arise by virtue of an activation of phospholipase D, but the role, if any, of phosphatidic acid in the acrosome reaction remains to be determined.

Recent data of Endo et al (128) suggest that guanine nucleotide regulatory proteins are involved in the signal/transduction pathway in mouse spermatozoa. The sperm cells from both vertebrates and invertebrates contain G_o or G_i-like proteins (129, 130), and Van Dop et al (131) have recently isolated and sequenced a cDNA clone for one of the sperm guanine nucleotide regulatory proteins. The identification of these proteins in spermatozoa has relied on the use of toxins and their ability to ADP-ribosylate proteins of this family; however, the possible existence of guanine nucleotide regulatory proteins in spermatozoa that are insensitive to ADP-ribosylation remains to be investigated (132). In the mouse, pertussis toxin can block the ZP-3–induced acrosome reaction, and prior incubation of mouse spermatozoa with GTP analogues can block the pertussis toxin inhibition (128). These observations are enticing, since they would place the ZP-3 receptor into the guanine nucleotide binding protein-coupled family of receptors, a diverse family that is beginning to be understood at the molecular level.

As one compares the biochemical responses of spermatozoa to FSG, ZP-3, or the small egg peptides, many similarities are obvious. In the case of FSG and the egg peptides, a K^+ efflux appears to occur early that results in hyperpolarization; the change in membrane potential may in turn regulate a Na^+/H^+ antiporter that results in alkalinization of the cell interior. Ca^{2+} is mobilized in response to FSG, ZP-3, or the egg peptides, but the molecular mechanisms involved require further investigation. It is likely that the regulatory mechanisms for Ca^{2+} mobilization are different for ZP-3 and FSG compared to the egg peptides, or that receptors for the ligands exist in specific regions (e.g. peptide receptors in flagellum, FSG, and ZP-3 receptors in sperm head). The egg peptides do not induce an acrosome reaction, although they may potentiate the reaction caused by FSG (117). Similarly, FSG does not stimulate sperm respiration rates; after the acrosome reaction, in fact, respiration rates tend to decrease (133). FSG, ZP-3, and the egg peptides cause elevations of cyclic AMP; however, only the egg peptides significantly elevate cyclic GMP concentrations. It is likely, then, that cyclic GMP plays a key role in the mechanism of action of the egg peptides.

MEMBRANE RECOGNITION AND EGG ACTIVATION

Once gamete membranes come into contact, the probability of cross-species fertilization increases dramatically. The denuded hamster egg, for example, can be penetrated by the sperm cells of many, but not all, mammalian species (11, 12, 121). Therefore, recognition signals at this level are apparently highly conserved across the species.

Gould and colleagues (134, 135) have provided intriguing data on the activation of *Urechis* eggs by acrosomal extracts of spermatozoa. The extracts reproduce the effects of intact spermatozoa in terms of the egg activation. The ability to activate eggs with such extracts should allow the purification and structural determination of both the effector and receptor molecules, and the determination of whether or not significant conservation exists across the species by the use of recombinant DNA technology.

Literature Cited

1. Boyd, A., Kendall, K., Simon, M. I. 1983. *Nature* 301:623–26
2. Krikos, A., Conley, M. P., Boyd, A., Berg, H. C., Simon, M. I. 1985. *Proc. Natl. Acad. Sci. USA* 82:1326–30
3. Manoil, C., Beckwith, J. 1986. *Science* 233:1403–8
4. Kubo, T., Fukuda, K., Mikami, A., Takahashi, H., Mishina, M., et al. 1986. *Nature* 323:411–16
5. Yarden, Y., Rodriguez, H., Wong, S. K.-F., Brandt, D. R., May, D. C., et al. 1986. *Proc. Natl. Acad. Sci. USA* 83:6795–99
6. Dixon, R. A. F., Kobilka, B. K., Strader, D. J., Benovic, J. L., Dohlman, H. G., et al. 1986. *Nature* 321:75–79
7. Nathans, J., Hogness, D. S. 1984. *Proc. Natl. Acad. Sci. USA* 81:4851–55
8. Shapiro, B. M., Schackmann, R. W., Gabel, C. A. 1981. *Annu. Rev. Biochem.* 50:815–43
9. Trimmer, J. S., Vacquier, V. D. 1986. *Annu. Rev. Cell. Biol.* 2:1–26
10. Vacquier, V. D. 1986. *Trends Biochem. Sci.* 11:77–81
11. Wassarman, P. M. 1987. *Annu. Rev. Cell Biol.* 3:109–42
12. Wassarman, P. M. 1987. *Science* 235:553–60
13. Garbers, D. L. 1988. *ISI Atlas Sci.* 1:120–26
14. Hunter, R. H. F., Leglise, P. C. 1971. *J. Reprod. Fertil.* 24:233–45
15. Flechon, J. E., Hunter, R. H. F. 1981. *Tissue & Cell* 1:127–39
16. Hunter, R. H. F., Wilmut, I. 1984. *Reprod. Nutr. Dev.* 24:597–608
17. Smith, T. T., Koyanagi, F., Yanagimachi, R. 1987. *Biol. Reprod.* 37:225–34
18. Overstreet, J. W., Cooper, G. W. 1975. *Nature* 258:718–19
19. Garbers, D. L., Kopf, G. S. 1980. *Adv. Cyclic Nucleotide Res.* 13:251–306
20. Hansbrough, J. R., Garbers, D. L. 1981. *Adv. Enzyme Regul.* 19:351–76
21. Okamura, N., Sugita, Y. 1983. *J. Biol. Chem.* 258:13056–62
22. Okamura, N., Tajima, Y., Soegima, A., Masuda, H., Sugita, Y. 1985. *J. Biol. Chem.* 260:9699–705
23. Clapper, D. L., Brown, G. G. 1980. *Dev. Biol.* 76:341–49
24. Clapper, D. L., Brown, G. G. 1980. *Dev. Biol.* 76:350–57
25. Miller, R. L. 1978. *J. Exp. Zool.* 205:385–402
26. Ohtake, H. 1976. *J. Exp. Zool.* 198:303–12
27. Ohtake, H. 1976. *J. Exp. Zool.* 198:313–22
28. Hansbrough, J. R., Garbers, D. L. 1981. *J. Biol. Chem.* 256:1447–52
29. Suzuki, N., Nomura, K., Ohtake, H., Isaka, S. 1981. *Biochem. Biophys. Res. Commun.* 99:1238–44
30. Garbers, D. L., Watkins, H. D., Hansbrough, J. R., Smith, A., Misono, K. S. 1982. *J. Biol. Chem.* 257:2734–37
31. Suzuki, N., Shimomura, H., Radany, E. W., Ramarao, C. S., Ward, G. E., et al. 1984. *J. Biol. Chem.* 257:14874–79
32. Nomura, K., Suzuki, N., Ohtake, H., Isaka, S. 1983. *Biochem. Biophys. Res. Commun.* 117:147–53
33. Smith, A., Garbers, D. L. 1983. In *Biochemistry of Metabolic Processes*, ed. D. L. F. Lennon, F. W. Stratman, R. N. Zahlten, pp. 15–28. New York: Elsevier
34. Nomura, K., Isaka, S. 1985. *Biochem. Biophys. Res. Commun.* 126:974–82
35. Nomura, K. 1986. *Dev. Growth Differ.* 28:88
36. Shimomura, H., Garbers, D. L. 1986. *Biochemistry* 25:3405–10
37. Shimomura, H., Suzuki, N., Garbers, D. L. 1986. *Peptides* 7:491–95
38. Suzuki, N., Kajiura, H., Nomura, K., Garbers, D. L., Yoshino, K., et al. 1988. *Comp. Biochem. Physiol. B* 89:687–93
39. Suzuki, N., Kurita, M., Yoshino, K., Kajiura, H., Nomura, K., Yamaguchi, M. 1987. *Zool. Sci.* 4:649–56
40. Suzuki, N., Hoshi, M., Nomura, K., Isaka, S. 1982. *Comp. Biochem. Physiol. A* 72:489–95
41. Hansbrough, J. R., Garbers, D. L. 1981. *J. Biol. Chem.* 256:2235–41
42. Schackmann, R. W., Chock, P. B. 1986. *J. Biol. Chem.* 261:8719–28
43. Repaske, D. R., Garbers, D. L. 1983. *J. Biol. Chem.* 258:6025–29
44. Hansbrough, J. R., Kopf, G. S., Garbers, D. L. 1980. *Biochem. Biophys. Acta* 630:82–91
45. Miller, R. L. 1985. *Biol. Fertil.* 2:275–337
46. Ward, G. E., Brokaw, C. J., Garbers, D. L., Vacquier, V. D. 1985. *J. Cell Biol.* 101:2324–29
47. Tezon, J., Miller, R. H., Bardin, C. W. 1986. *Proc. Natl. Acad. Sci. USA* 83:3589–93
48. Gnessi, L., Ruff, M. R., Fraioli, F., Pert, C. B. 1985. *Exp. Cell Res.* 161:219–30
49. Igbal, M., Shivaji, S., Vijayasarathy,

S., Balaram, P. 1980. *Biochem. Biophys. Res. Commun.* 96:235–42
50. Miller, R. L. 1982. *Gamete Res.* 5:395–401
51. Gnessi, L., Fabbri, A., Silvestroni, L., Moretti, C., Fraioli, F., et al. 1986. *J. Clin. Endocrinol. Metab.* 63:841–46
52. Brokaw, C. J. 1987. *J. Cell. Biochem.* 35:175–84
53. Tonosaki, K., Funakoshi, M. 1988. *Nature* 331:354–56
54. Karpen, J. W., Zimmerman, A. L., Stryer, L., Baylor, D. A. 1988. *Proc. Natl. Acad. Sci. USA* 85:1287–91
55. Hanke, W., Cook, N. J., Kaupp, U. B. 1988. *Proc. Natl. Acad. Sci. USA* 85:94–98
56. Kolesnikov, S. S., Zhainazarov, A. B., Fesenko, E. E. 1988. *Biofizika* 33:101–8
57. Johnson, E. C., Robinson, P. R., Lisman, J. E. 1986. *Nature* 324:468–70
58. Paupardin-Tritsch, D., Hammond, C., Gerschenfeld, H. M., Nairn, A. C., Greengard, P. 1986. *Nature* 323:812–14
59. Lee, H. C., Garbers, D. L. 1986. *J. Biol. Chem.* 261:16026–32
60. Lee, H. C. 1988. *Dev. Biol.* 126:91–97
61. Codina, J., Yatani, A., Granet, D., Brown, A. M., Birnbaumer, L. 1987. *Science* 236:442–45
62. Neer, E. J., Clapham, D. E. 1988. *Nature* 333:129–34
63. Lee, H. C. 1985. *J. Biol. Chem.* 260:10794–99
64. Dangott, L. J., Garbers, D. L. 1984. *J. Biol. Chem.* 259:13712–16
65. Dangott, L. J., Jordan, J. E., Bellet, R. A., Garbers, D. L. 1989. *Proc. Natl. Acad. Sci. USA.* In press
66. Sudhof, T. C., Goldstein, J. L., Brown, M. S., Russell, D. W. 1985. *Science* 228:815–22
67. Sudhof, T. C., Russell, D. W., Goldstein, J. L., Brown, M. S., Sanchez-Pescandor, R., Bell, G. I. 1985. *Science* 228:893–95
68. Doolittle, R. F. 1985. *Trends Biochem. Sci.* 10:233–37
69. Montell, D. J., Goodman, C. S. 1988. *Cell* 53:463–73
70. Shimomura, H., Dangott, L. J., Garbers, D. L. 1986. *J. Biol. Chem.* 261:15778–82
71. Singh, S., Lowe, D. G., Thorpe, D. S., Rodriquez, H., Kuang, W. J., et al. 1988. *Nature* 334:708–12
72. Thorpe, D. S., Garbers, D. L. 1989. *J. Biol. Chem.* In press
73. Kuno, T., Andresen, J. W., Kamisaki, Y., Waldman, S. A., Chang, L. Y., et al. 1986. *J. Biol. Chem.* 261:5817–23
74. Paul, A. K., Marala, R. B., Jaiswal, R.

K., Sharma, R. K. 1987. *Science* 235:1224–26
75. Takayanagi, R., Inagami, T., Snajdar, R. M., Imada, T., Tamura, M., Misono, K. S. 1987. *J. Biol. Chem.* 262:12104–13
76. Takayanagi, R., Snajdar, R. M., Imada, T., Tamura, M., Pandey, K. N., et al. 1987. *Biochem. Biophys. Res. Commun.* 144:244–50
77. Bentley, J. K., Tubb, D. J., Garbers, D. L. 1986. *J. Biol. Chem.* 261:14859–62
78. Bentley, J. K., Shimomura, H., Garbers, D. L. 1986. *Cell* 45:281–88
79. Waldman, S. A., Rapoport, R. M., Murad, F. 1984. *J. Biol. Chem.* 259:14332–34
80. Leitman, D. C., Andresen, J. W., Catalano, R. M., Waldman, S. A., Tuan, J. J., Murad, F. 1988. *J. Biol. Chem.* 263:3720–28
81. Bentley, J. K., Garbers, D. L. 1988. *Biol. Reprod.* 39:639–47
82. Trembley, J., Gerzer, R., Pang, S. C., Cantin, M., Genest, J., Hamet, P. 1986. *FEBS Lett.* 194:210–14
83. Chinkers, M. C., Garbers, D. L., Chang, M. S., Lowe, D. G., Chin, H., et al. 1989. *Nature* 338:78–83
84. Lowe, D. G., Chang, M. S., Hellmiss, R., Singh, S., Chen, E., et al. 1989. *EMBO J.* In press
85. Fuller, F., Porter, J. G., Arfsten, A., Miller, J., Schilling, J. W., et al. 1988. *J. Biol. Chem.* 263:9395–401
86. Koesling, D., Herz, J., Gausepohl, H., Niroomand, F., Hinsch, K-D., et al. 1988. *FEBS Lett.* 239:29–34
87. Hanks, S. K., Quinn, A. M., Hunter, T. 1988. *Science* 241:42–53
88. Mittal, C. K., Murad, F. 1977. *J. Biol. Chem.* 252:3136–40
89. Ward, G. E., Vacquier, V. D. 1983. *Proc. Natl. Acad. Sci. USA* 80:5578–82
90. Ward, G. E., Garbers, D. L., Vacquier, V. D. 1985. *Science* 227:768–70
91. Vacquier, V. D., Moy, G. W. 1986. *Biochem. Biophys. Res. Commun.* 137:1148–52
92. Ramarao, C. S., Garbers, D. L. 1988. *J. Biol. Chem.* 263:1524–29
93. Strasser, R. H., Benovic, J. L., Caron, M. G., Lefkowitz, R. J. 1986. *Proc. Natl. Acad. Sci. USA* 83:6362–66
94. Cochet, C., Gill, G. N., Meisenhelder, J., Cooper, J. A., Hunter, T. 1984. *J. Biol. Chem.* 259:2553–58
95. SeGall, G. K., Lennarz, W. J. 1979. *Dev. Biol.* 71:33–48
96. SeGall, G. K., Lennarz, W. J. 1981. *Dev. Biol.* 86:87–93
97. Kopf, G. S., Garbers, D. L. 1980. *Biol. Reprod.* 22:1118–26

98. Vacquier, V. D., Moy, G. W. 1977. *Proc. Natl. Acad. Sci. USA* 74:2456–60
99. Gao, B., Klein, L. E., Britten, R. J., Davidson, E. H. 1986. *Proc. Natl. Acad. Sci. USA* 83:8634–38
100. Rossignol, D. P., Roschelle, A. J., Lennarz, W. J. 1981. *J. Supramol. Struct. Cell. Biochem.* 15:347–58
101. Ruiz-Bravo, N., Lennarz, W. J. 1986. *Dev. Biol.* 118:202–8
102. Rinquette, M. J., Chamberlin, M. E., Baur, A. W., Sobieski, D. A., Dean, J. 1988. *Dev. Biol.* 127:287–95
103. Lopez, L. C., Bayna, E. M., Litoff, D., Shaper, N. L., Shaper, J. H., Shur, B. D. 1985. *J. Cell Biol.* 101:1501–10
104. Bleil, J. D., Wassarman, P. M. 1986. *J. Cell Biol.* 102:1363–71
105. Bleil, J. D., Greve, J. M., Wassarman, P. M. 1988. *Dev. Biol.* 128:376–85
106. Saling, P. M. 1981. *Proc. Natl. Acad. Sci. USA* 78:6231–35
107. Benau, D. A., Storey, B. T. 1987. *Biol. Reprod.* 36:282–92
108. Lenz, R. W., Ax, R. L., Grimek, J. H., First, N. L. 1982. *Biochem. Biophys. Res. Commun.* 106:1092–98
109. Ball, G. D., Bellin, M. E., Ax, R. L., First, N. L. 1982. *Mol. Cell. Endocrinol.* 28:113–22
110. Parrish, J. J., Susko-Parrish, J. L., First, N. L. 1985. *Biol. Reprod.* 32:211
111. Meizel, S., Turner, K. O. 1986. *J. Exp. Zool.* 237:137–39
112. Siiteri, J. E., Dandekar, P., Meizel, S. 1988. *J. Exp. Zool.* 246:71–80
113. Garbers, D. L., Kopf, G. S., Tubb, D. J., Olson, G. 1983. *Biol. Reprod.* 29:1211–20
114. Ikadai, H., Hoshi, M. 1981. *Dev. Growth Differ.* 23:73–80
115. Ikadai, H., Hoshi, M. 1981. *Dev. Growth Differ.* 23:81–88
116. Matsui, T., Nishiyama, I., Hino, A., Hoshi, M. 1986. *Dev. Growth Differ.* 28:349–57
117. Yamaguchi, M., Niwa, T., Kurita, M., Suzuki, N. 1988. *Dev. Growth Differ.* 30:159–67
118. Gonzalez-Martinez, M., Darszon, A. 1987. *FEBS Lett.* 218:247–50
119. Schackmann, R. W., Christen, R., Shapiro, B. M. 1981. *Proc. Natl. Acad. Sci. USA* 78:6066–70
120. Schackmann, R. W., Christen, R., Shapiro, B. M. 1984. *J. Biol. Chem.* 259:13914–22
121. Yanagimachi, R. 1988. In *Physiology of Reproduction*, ed. E. Knobil, J. Neill, et al, pp. 135–85. New York: Raven
122. Kopf, G. S., Vacquier, V. D. 1984. *J. Biol. Chem.* 259:7590–96
123. Hyne, R. V., Garbers, D. L. 1979. *Biol. Reprod.* 21:1135–42
124. Gross, M. K., Toscano, D. G., Toscano, W. A., Jr. 1987. *J. Biol. Chem.* 262:8672–76
125. Noland, T. D., Garbers, D. L., Kopf, G. S. 1988. *Biol. Reprod.* 38:94
126. Domino, S. E., Garbers, D. L. 1988. *J. Biol. Chem.* 263:690–95
127. Domino, S. E., Bocckino, S. B., Garbers, D. L. 1989. *J. Biol. Chem.* In press
128. Endo, Y., Lee, M. A., Kopf, G. S. 1987. *Dev. Biol.* 19:210–16
129. Kopf, G. S., Woolkalis, M. J., Gerton, G. L. 1986. *J. Biol. Chem.* 261:7327–31
130. Bentley, J. K., Garbers, D. L., Domino, S. E., Noland, T. D., Van Dop, C. 1986. *Biochem. Biophys. Res. Commun.* 138:728–34
131. Van Dop, C., Stone, K., Apone, L. M. 1988. *Fed. Proc.* 2:A1686
132. Fong, H. K., Yoshimoto, K. K., Eversole-Circe, P., Simon, M. I. 1988. *Proc. Natl. Acad. Sci. USA* 85:3066–70
133. Kinsey, W. H., SeGall, G. K., Lennarz, W. J. 1979. *Dev. Biol.* 71:49–59
134. Gould, M., Stephano, J. L., Holland, L. Z. 1986. *Dev. Biol.* 117:306
135. Gould, M., Stephano, J. L. 1987. *Science* 235:1654–56
136. Yarden, Y., Excobedo, J. A., Kuang, W. J., Yang-Feng, T. L., Daniel, T. O., et al. 1986. *Nature* 323:226–32

Annu. Rev. Biochem. 1989. 58:743–64

GLUTATHIONE S-TRANSFERASES: GENE STRUCTURE, REGULATION, AND BIOLOGICAL FUNCTION

Cecil B. Pickett

Merck Frosst Centre for Therapeutic Research, P.O. Box 1005, Pointe-Claire—Dorval, Québec, Canada H9R 4P8

Anthony Y. H. Lu

Animal and Exploratory Drug Metabolism, Merck Sharp & Dohme Research Laboratories, P. O. Box 2000, Rahway, New Jersey 07065

CONTENTS

0066-4154/89/0701-0743$02.00

PERSPECTIVES AND SUMMARY

The glutathione S-transferases (GSTs) are a family of proteins that conjugate glutathione on the sulfur atom of cysteine to various electrophiles (1–5). In addition, GST binds with high affinity a variety of hydrophobic compounds such as heme, bilirubin, polycyclic aromatic hydrocarbons, and dexamethasone (6–9). Although GSTs have now been purified from a number of species (*Escherichia coli* to humans), those found in rat liver are the most extensively characterized (8, 10–15). GSTs are homodimers or heterodimers comprising at least seven subunits (see Table 1). Since specific GST subunits are induced by various xenobiotics (e.g. phenobarbitol, 3-methylcholanthrene, *trans*-stilbene oxide) and are expressed in a tissue-specific manner, the GST gene family is a useful model system to study induction and tissue-specific regulation of gene expression.

In the past few years, the application of molecular biological techniques to elucidate GST gene structure has begun to reveal the immense complexity of this multigene family. In this review, we focus most of our attention on structure, expression, and regulation of genes encoding the cytosolic GSTs. In addition, the catalytic mechanism and the role of GST in the activation and detoxication of xenobiotics are discussed. Since many of these studies use the rat as an animal model, the discussion focuses on this animal. For a detailed discussion on the purification and enzymology of various purified GSTs, the reader is referred to other recent reviews or book chapters (2–5).

QUANTITATION OF GST SUBUNITS

Although the occurrence of GST isozymes in various mammalian tissues is well-established, quantitation of each subunit in tissue extracts, prepared from

Table 1 Nomenclature for the cytosolic rat glutathione S-transferases

Subunit[a]	Nomenclature of Jakoby et al (19)	Nomenclature of Mannervik et al (11)	Other nomenclature (2, 6, 15, 17)
Y_aY_a	1-1	L_2	ligandin
Y_aY_c	1-2	BL	B
Y_cY_c	2-2	B_2	AA
$Y_{b1}Y_{b1}$	3-3	A_2	A
$Y_{b1}Y_{b2}$	3-4	AC	C
$Y_{b2}Y_{b2}$	4-4	C_2	D
Y_nY_n	6-6	—	—
Y_pY_p	7-7	—	P
Y_kY_k	8-8	—	K

[a] Nomenclature of Bass et al (10), Beale et al (16), Kitahara et al (17), and Hayes & Mantle (18).

untreated or induced animals, has not been achieved. Earlier attempts to quantitate tissue GST isozymes involved either immunochemical analysis (20), or the separation of GST isozymes followed by enzymatic assays (21). However, since most of the GST antibodies (either polyclonal or monoclonal) available are not monospecific, antibody-based assays generally quantitate a group of structurally related subunits. Likewise, because of the overlapping substrate specificities and the lack of complete separation and recovery of GST isozymes by conventional separation methods, catalytic-based assays also tend to quantitate more than one isozyme. Recently, a simple method has been described for the separation and quantitation of GST subunits (22). This method involves the purification of GST isozymes by affinity chromatography and the separation of subunits by reverse-phase HPLC. Subunits are quantitated from the area under the HPLC peaks and the known extinction coefficients of each subunit at 214 nm. Since the recovery of GST activities from the affinity column are high and the recovery of each subunit from the HPLC has been established, quantitation of each subunit in a given tissue can be easily achieved. For example, it is estimated that the cytosolic fraction from 80 mg rat liver contains 196, 63, 51, and 88 μg of subunits Y_a, Y_c, Y_{b1}, and Y_{b2} respectively. Although a few subunits are not completely separated, this method should be able to quantitate many of the major subunits in tissues and to examine the induction of specific subunits in tissues following drug treatment.

ELECTROPHILIC SUBSTRATES OF BIOLOGICAL IMPORTANCE

GST isozymes catalyze the nucleophilic addition of the thiol of GSH to electrophilic acceptors including aryl and alkyl halides, olefins, organic peroxides, quinones, and sulfate esters (1–5). New substrates are continuously being reported in the literature. Among the more interesting substrates are those toxic products generated from tissue damage. For example, lipid peroxidation of biological membranes yields reactive alkenes, epoxides, hydroperoxides, and aldehydes. Cholesterol α-oxide (23), arachidonic acid hydroperoxide, and linoleate hydroperoxide (24, 25) are substrates of GST. In addition, 4-hydroxyalkenals, the highly cytotoxic and mutagenic products of membrane lipid peroxidation (26, 27), are substrates of a variety of GST isozymes, especially excellent substrates for Y_kY_k (27).

Oxygen-centered free radicals can peroxidize not only membrane lipids but also DNA (28–30). For example, 5-hydroperoxymethyluracil is the thymine hydroperoxide generated by subjecting DNA to γ-radiation in aqueous solutions (31). Tan et al (25) have shown that GST isozymes Y_cY_c, Y_nY_n, and Y_pY_p have good activity toward 5-hydroperoxymethyluracil. Thus, GST may play an important role in protecting tissues from oxidative damage and oxidative stress.

CATALYTIC MECHANISM

The steady-state kinetic mechanisms of several GSTs have been extensively studied, and several mechanisms, including random, ping-pong, and sequential, have been proposed (2–4, 32). Interpretation of product inhibition patterns is complicated by the nonhyperbolic substrate saturation curve, nonlinear reaction rate, time-dependent conformational transitions of the enzyme, and the possible presence of other structurally related isozymes in a given purified enzyme preparation used in the study. However, more recent studies by Jakobson et al (33, 34) and Chen et al (35), using glutathione analogues and other analyses, suggest either a random sequential mechanism or an ordered sequential mechanism for isozymes $Y_{b1}Y_{b1}$ and $Y_{b2}Y_{b2}$.

Further evidence supporting a sequential mechanism is obtained from studying the stereochemical aspects of a variety of reactions catalyzed by the GST. In these studies, chiral compounds of known absolute configuration are used as substrates and the absolute configuration of the products is then analyzed. Mangold & Abdel-Monem (36) found that isozyme $Y_{b1}Y_{b1}$ catalyzes the reaction between glutathione and phenethyl halides with inversion of configuration at the benzylic carbon atom. Subsequent study by Ridgewell & Abdel-Monem (37) indicates that the conjugation reaction of (R,S)-2-iodooctane or (R,S)-2-bromooctane and glutathione, catalyzed by Y_aY_c, also proceeds with inversion of configuration at the chiral carbon of the substrate. These findings indicate a single displacement mechanism, in agreement with a ternary-complex sequential mechanism. In contrast, a ping-pong mechanism would be expected to operate with a double-displacement and retention of the stereochemical configuration.

GST-catalyzed addition of glutathione to arene oxides shows varying degrees of stereoselectivity (38, 39). Isozyme $Y_{b2}Y_{b2}$ is highly stereospecific and displays >99% attack at the oxirane carbon of R absolute configuration of phenanthrene 9,10-oxide, pyrene 4,5-oxide, benz[a]anthracene 5,6-oxide, and benzo[a]pyrene 4,5-oxide. Isozyme $Y_{b1}Y_{b1}$ exhibits a low degree of stereoselectivity toward arene oxides. All these data support a single-displacement sequential mechanism.

Although the amino acid sequence of many of the isozymes has been deduced from the nucleotide sequence of cDNAs (5), very little is known about the active site of the enzymes. GST shows very high specificity for the thiol substrate, indicating a strict structural requirement for the binding of glutathione to the active site. The substrate binding site is hydrophobic in nature and may be asymmetrically disposed with respect to the catalytic site

(32). Recently, Awasthi et al (40) reported that a histidine residue is essential for the catalytic activity of GST ψ from human liver based on protein modification studies. Whether or not the histidine is present at the active site remains to be determined. If this is indeed the case, then the imidazole of the histidine residue would facilitate the ionization of glutathione to the thiolate ion, the nucleophilic species of glutathione, and thus enhance the rate of the conjugation reaction.

The lack of posttranslational modification and the cytosolic localization of the GSTs have made it possible to express the enzymes in *E. coli* (41, 42). Site-directed mutagenesis and X-ray crystallographic analysis of these expressed proteins should be invaluable in defining the role of specific amino acids or domains in substrate binding, ligand binding, and in catalysis.

ROLE OF GST IN METABOLIC ACTIVATION AND DETOXICATION OF XENOBIOTICS

Fate of GSH Conjugates

For many toxic xenobiotics, including known carcinogens, glutathione S-conjugate formation represents a detoxication pathway (1–3). As shown in Figure 1, the glutathione conjugate is converted to the corresponding cysteine conjugate following sequential removal of glutamate and glycine. Cysteine conjugate is either metabolized to a mercapturate by acetylation or cleaved to a mercaptan by β-lyase (C-S lyase). In addition to the mercapturic acid pathway, methylation of the thiol to form the methylthio-containing metabolite and the glucuronydation of the mercaptan to form the thioglucuronide represent important metabolic steps for the biotransformation of the cysteine conjugate. Thus, excretion of mercapturate and the CH_3S-, CH_3SO-, and CH_3SO_2-containing metabolites are indicative of the in vivo formation of glutathione S-conjugates of xenobiotics.

Cysteine conjugate β-lyase is present in the liver and kidney, in gut microflora, and in some microorganisms (43, 44). It catalyzes the elimination of leaving groups from the β-carbon of cysteine conjugates to form pyruvate, ammonia, and thiols. The microfloral β-lyase appears to have a broader substrate specificity than the tissue β-lyase. Based on a number of observations, Bakke & Gustafsson (44) have suggested that the microfloral β-lyase–mediated reaction is quantitatively more important than the tissue enzyme in the in vivo metabolism of cysteine conjugates of xenobiotics.

Mechanism of Metabolic Activation

Although it is generally believed that glutathione and cysteine conjugates of many toxic xenobiotics are nontoxic and are eliminated from the body, either

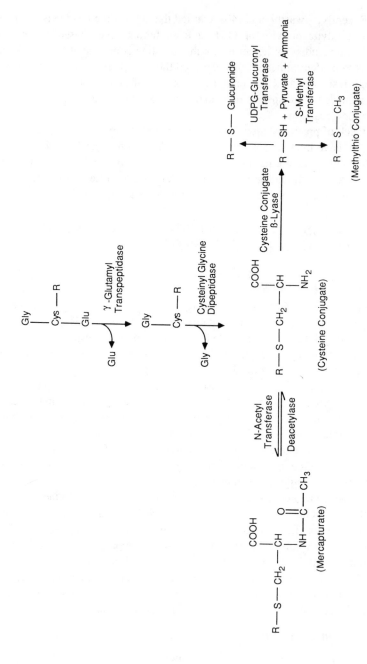

Figure 1 Metabolism of glutathione conjugates.

as conjugates or mercapturate or methylthio-containing metabolites (1–5), recent studies (43–46) have shown that some glutathione and cysteine conjugates are of toxicological concern. Some of these compounds, mostly conjugates of halogenated hydrocarbon, are listed in Table 2. The toxicity of these conjugates is associated with the formation of unstable thiols, which may be converted to alkylating agents or to stable but toxic metabolites, or both.

Two different mechanisms have been described to explain the toxicity of glutathione and cysteine conjugates. In the first mechanism, the conjugates act as direct alkylating agents and their toxic effects are dependent on the sulfur half-mustard moiety of the molecule (43–46). For example, GST catalyzes the conjugation between 1,2-dibromoethane and glutathione (Figure 2) to form a sulfur half-mustard. Displacement of the bromide ion by sulfur yields an electrophilic episulfonium ion that can react with nucleophiles to produce toxicity (51, 68, 69). The major DNA adduct formed from 1,2-dibromoethane has been identified as S-[2-(N^7-guanyl)ethyl]-glutathione (70). Covalent binding of the glutathione conjugate to DNA may explain the mutagenic and carcinogenic action of 1,2-dibromoethane. S-(2-chloroethyl)-DL-cysteine, the cysteine conjugate of 1,2-dichloroethane, is also a direct-acting hepatotoxic, nephrotoxic, and mutagenic agent. This cysteine con-

Table 2 Glutathione or cysteine S-conjugates of chemicals that are shown to be toxic

Conjugates of either glutathione or cysteine	Toxicity	Reference
1,1,2,3,4,4-Hexachloro-1,3-butadiene	nephrotoxic, mutagenic, carcinogenic	47–49
1,1,2,2-Tetrafluoroethylene	nephrotoxic	49
1,1,2,2,3,3-Hexafluoropropene	nephrotoxic	49
1,1,2-Trichloroethylene	nephrotoxic, mutagenic	49
1,2-Dibromoethane	mutagenic, carcinogenic	50–54
1,2-Dichloroethane	hepatotoxic, nephrotoxic, mutagenic	55–57
1,2-Dichloroethene	nephrotoxic, mutagenic, cytotoxic	58–61
2-Chloro-1,1,2-trifluoroethene	nephrotoxic, cytotoxic	58, 62
2-Bromohydroquinone	nephrotoxic	63
Allyl isothiocyanate	cytotoxic	64
Benzyl isothiocyanate	cytotoxic	64
Benzyl 1,2,3,4,4-pentachlorobuta-1,3-dienyl sulfide	cytotoxic	65
Benzyl 2-Chloro-1,1,2-trifluoroethyl sulfide	cytotoxic	65
3-Hydroxyamino-1-methyl-5H-pyrido [4,3-b]indole	mutagenic	66
1,1-Dichloro-2,2-Difluoroethylene	nephrotoxic, hepatotoxic	67

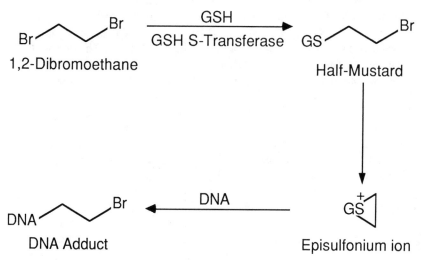

Figure 2 Proposed mechanism for the metabolic activation of 1,2-dibromoethane.

jugate can form a reactive episulfonium ion to alkylate macromolecules causing toxicity (71, 72), but it is not a substrate for the β-lyase (73).

The second mechanism by which glutathione conjugate exerts its toxicity involves the conversion of glutathione conjugate to its corresponding cysteine conjugate, followed by β-elimination of an electrophilic sulfur-containing fragment from the cysteine conjugate (43–46). This β-lyase-mediated reaction can lead to the formation of a reactive species, resulting in toxicity. For example, studies by Dekant et al (74) have shown that S-(2-chloro-1,1,2-trifluoroethyl)-L-cysteine is activated by the kidney β-lyase to form the unstable thiol 2-chloro-1,1,2-trifluoroethanethiol, which loses hydrogen fluoride to generate the acylating agent chlorofluorothionacetyl fluoride. Subsequent hydrolysis of the thionoacyl fluoride gives rise to the stable chlorofluoroacetic acid and inorganic fluoride. It is believed that the acylating agent and chlorofluoroacetic acid are responsible for the cytotoxicity of this cysteine conjugate.

The mechanism of cytotoxicity of several glutathione and cysteine conjugates has been studied in isolated kidney cells (75–77). These studies indicate that mitochondria may be the primary target of nephrotoxic cysteine conjugates. For example, S-(1,2-dichlorovinyl)-L-cysteine causes significant effects on kidney mitochondrial structural integrity, mitochondrial energy metabolism, and mitochondrial function, including the inhibition of respiration. Stevens et al (58) have shown that cysteine and glutathione conjugates of 1,2-vinyldichloride,1,2,3,4,4-pentachloro-1,3-butadiene and 2-chloro-1,1,2-trifluoroethane are toxic to LLC-PK1, a cell line derived from pig kidney.

Evidence has been obtained that S-(1,2-dichlorovinyl)-L-cysteine is metabolized to a reactive species that covalently binds to cellular macromolecules and that the binding is proportional to cellular toxicity. The use of α-methyl and homocysteine analogues, inhibitors, and activators of the renal β-lyase also supports the hypothesis that activation by the β-lyase is essential for the expression of toxicity of these cysteine conjugates (43, 75–77).

GST $Y_a Y_c$ GENE FAMILY

GST $Y_a Y_c$ cDNA Clones

Perhaps the first indication of the complexity of the GST multigene family became apparent shortly after the construction and characterization of cDNA clones complementary to the mRNAs encoding the Y_a and Y_c subunits (78–85). DNA sequence analysis of Y_a and Y_c clones revealed that the corresponding mRNAs are 75% identical in the protein-coding region; however, both the 5' and 3' untranslated regions of the two mRNAs are very divergent.

It also became clear from the DNA sequence analysis of two full-length Y_a cDNA clones (79, 80), pGTB38 and pGTR261, that these two clones differed by 15 nucleotides, which translated into eight amino acid differences. Furthermore, the 3' untranslated regions of the two clones were also divergent, indicating the two mRNAs are most likely encoded by distinct genes. A third Y_a cDNA (86), pGTB45, has also been isolated and is more similar to pGTR261 than pGTB38. However, pGTB45 contains a type-2 Alu repetitive element in the 3' untranslated region. The type-2 repetitive element contains two overlapping polyadenylation signals downstream from the polyadenylation signal found in pGTR261. The functional significance of the type-2 Alu repetitive element in the 3' untranslated region of a Y_a mRNA is unknown, but may play a role in the stability of this mRNA species.

The existence of two distinct classes of Y_a cDNA clones is not due to DNA sequencing mistakes or cloning artifacts. Numerous laboratories have constructed cDNA clones corresponding to one or the other clone of Y_a classes (78–83).

The amino acid sequences of the Y_a and Y_c subunits have an overall identity of 68%. The Y_a subunit comprises 222 amino acids with a molecular weight of 25,547; whereas the Y_c subunit comprises 221 amino acids with a molecular weight of 25,322.

Tu & Quian (87), Rhoads et al (88), and Board & Webb (89) have reported the nucleotide sequence of human GST cDNA clones. The open reading frame of these clones is 666 nucleotides long, encoding a polypeptide comprising 222 amino acids. The nucleotide sequences of the human clones are approximately 80% identical to the rat liver Y_a and Y_c clones. The predicted amino acid sequence of the human GST subunits is 75% identical to the rat sequence. Tu & Quian (87) have used computer analysis of the coding regions of a number of GST subunits, and found a small amino acid region that is conserved between the different families of subunits. This particular amino acid sequence corresponds to amino acids 70–95 of the rat Y_a subunit and is encoded by exon 4 of the Y_a structural gene. Exon 4 of the Y_a structural gene encodes an amino acid domain that is the most highly conserved between the Y_a and Y_c subunits (90). Therefore, it is possible that exon 4 encodes a domain of the GSTs, which is common between the various subunits. Common to all GSTs is the glutathione-binding domain.

Y_a Structural Gene

Early studies of southern blot analysis of rat genomic DNA using 5' and 3' regions of a Y_a clone, pGTB38, or a Y_c cDNA clone, pGTB42, indicated the presence of at least five Y_a genes and two Y_c genes in the rat genome (86). The presence of multiple genes in this family was confirmed by the isolation of four unique genomic fragments from a rat genomic library (86). One of the genomic fragments, λ GTB45-15, was characterized in detail (90) and was demonstrated to be a structural gene encoding a Y_a subunit. The Y_a gene comprises seven exons separated by six introns and is approximately 10 kb in length. Exon 1 is 43 bp in length and encodes the 5' untranslated region of the mRNA. The sizes of the remaining exons and all the introns are presented in Table 3. Interestingly, the amino acid sequence encoded by exon 3 represents the most divergent amino acids in the Y_a and Y_c polypeptides. Despite an overall identity of 66% between the Y_a and Y_c subunits, there is only a 36% sequence identity between the amino acids encoded by exon 3. Both exon 2 and, as mentioned earlier, exon 4 encode amino acid domains that are highly conserved between the Y_a and Y_c subunits. The amino acid sequences of the Y_a subunit encoded by exons 2 and 4 are 86% and 91% identical, respectively, to the corresponding amino acids in the Y_c subunits. Thus, it would appear that the genes encoding the Y_a and Y_c subunits comprise highly conserved as well as highly divergent exons, which may impart both similar and unique functional properties to the two subunits. The divergent exons may encode amino acids responsible for substrate specificity (i.e. Δ^5 androstene 3,17 dione for the Y_a subunit and cumene hydroperoxide for the Y_c subunit), whereas the exons encoding amino acids that are highly conserved between the two subunits may form the glutathione-binding domain.

Unfortunately, the precise number of genes encoding Y_a and Y_c subunits is unknown. In fact, no laboratory has characterized a Y_c subunit gene. As mentioned earlier, it is apparent from blotting data that more than one Y_a or Y_c gene exists; however, it is unknown how many genes are expressed in liver or peripheral tissues. Furthermore, it is unclear if unique Y_a or Y_c genes are

Table 3 Rat liver glutathione S-transferase Y_a exon and intron sizes

Exon	Size determined by DNA sequence analysis (base pairs)	Intron	Size determined by restriction endonuclease mapping (base pairs)
1	43	1	2350
2	109	2	3500
3	52	3	650
4	133	4	2100
5	142	5	1500
6	132	6	800
7	234	—	—

expressed in different tissues. It is clear from the above discussion that greater attention should be devoted to the isolation and characterization of additional GST genes in this gene family.

GST Y_b GENE FAMILY

GST Y_b cDNA Clones

At least three classes of Y_b cDNA clones have been isolated and characterized to date (91–95). These cDNA clones Y_{b1}, Y_{b2}, and Y_{b3} are more related, at the nucleotide and amino acid level, to each other than are the Y_a and Y_c cDNAs and subunits. In the protein-coding region, the Y_b clones are approximately 80% identical in nucleotide and amino acid sequence; whereas the 5' and 3' untranslated regions are very divergent.

Interestingly, the Y_b cDNAs and subunits share little sequence homology with other GST subunits. This has been somewhat surprising since all GST subunits have overlapping substrate specificity and all bind glutathione.

GST Y_b Structural Genes

Southern blots of rat genomic DNA with a Y_b probe also indicated the presence of multiple Y_b genes (93). To date, three Y_b genes have been characterized (96, 97; M. Morton, C. B. Pickett, unpublished results). These genes encode the Y_{b1}, Y_{b2}, and Y_{b4} subunits. Although Y_{b1} and Y_{b2} cDNA clones and the corresponding subunits have been characterized, there is no evidence that the Y_{b4} gene is expressed in the rat.

Similarly, a Y_{b3} cDNA clone has been utilized to demonstrate expression of the Y_{b3} mRNA primarily in the brain (95). However, no laboratory has isolated the Y_{b3} subunit gene. Sequence analysis demonstrates that the Y_b genes comprise eight exons separated by seven introns and span approximately 5 kb (Table 4; Figure 3). Interestingly, the nucleotide sequences of introns 3, 4, and 6 are also highly conserved between the Y_{b1}, Y_{b2}, and Y_{b4} genes

Table 4 Exon and intron sizes of rat liver glutathione S-transferase Y_{b1}, and Y_{b2} subunit genes[a]

Exon	Y_{b1} base pairs	Y_{b2}	Intron	Y_{b1} base pairs	Y_{b2}
1	73	69	1	295	268
2	76	76	2	349	318
3	65	65	3	247	285
4	82	82	4	90	97
5	101	101	5	~1000	864
6	96	96	6	82	81
7	111	111	7	~2000	1784
8	458	436			

[a] Introns 5 and 7 from the Y_{b1} gene are estimated based on restriction mapping, whereas the remaining exon and intron sizes of the Y_{b1} and Y_{b2} genes are based on nucleotide sequence analysis.

The data for the Y_{b1} gene is from M. Morton, and C. B. Pickett, unpublished. The data for the Y_{b2} gene is from Ref. 97.

(97). These latter data have led to the hypothesis that gene conversion may have played a role in the evolution of the Y_b genes (97).

GST Y_p GENE FAMILY

GST Y_p cDNA Clones

Suguoka et al (98) and Pemble et al (99) have isolated cDNA clones specific to the rat Y_p subunit. The Y_p cDNA is approximately 750 nucleotides in length with an open reading frame of 630 nucleotides encoding 210 amino acids. A comparison of the Y_p subunit with the Y_a and Y_c subunits revealed a modest amino acid sequence homology of 32%. More recently, Kano et al (100) reported the isolation and characterization of the human Y_p cDNA clone.

GST Y_p Structural Gene

Okuda et al (101) have reported the isolation and characterization of the rat Y_p structural gene. The rat gene consists of seven exons, separated by six introns, and spans approximately 3 kb (Figure 3). In addition to the Y_p gene, Okuda et al (101) reported the presence of several processed-type pseudogenes, which most likely originated by reverse transcription followed by insertion at specific sites.

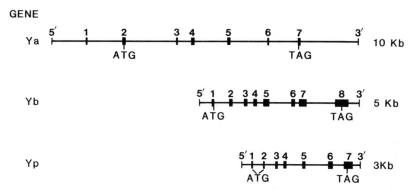

Figure 3   Exon-intron structures of rat glutathione S-transferase genes.

Microsomal Glutathione S-Transferase Gene(s)

DeJong et al (102) have recently isolated cDNA clones to rat and human liver microsomal GST. In both the rat and human sequences, there is a 154-amino-acid reading frame, encoding proteins with molecular weights of 17,430 and 17,450, respectively. The rat and human microsomal GSTs are 95% identical in their amino acid sequences, and the mRNAs share a 77% nucleotide sequence similarity in the protein-coding region. The protein sequence deduced from the rat cDNA clone is in good agreement with the sequence of the protein determined by conventional protein sequencing techniques (103). Southern blots with a microsomal GST cDNA suggest the presence of a single microsomal GST gene in the rat genome, which is approximately 12 kb long (102).

REGULATION OF GST GENE EXPRESSION

By Xenobiotics

When rats are administered various xenobiotics (e.g. 3-methylcholanthrene, phenobarbitol, and *trans*-stilbene oxide), some GST subunits are induced in the liver. The increase in enzyme content and/or activity is paralleled by an elevation in the translational activity of the Y_a mRNA (104, 105). Using in vitro translation and immunoprecipitation with specific antiserum, an increase in the translational activity of the Y_a mRNA could be detected as early as 4 h after a single administration of phenobarbitol, reaching maximal induction (5- to 7-fold) between 16 and 24 h (106). A blot analysis using GST cDNA clones confirmed that the increase in translational activity of the Y_a mRNA was due to an accumulation in the steady-state level of Y_a mRNA (80). The Y_c mRNA is only marginally affected by phenobarbitol or 3-methylcholanthrene. The use of specific regions of the Y_{b1} and Y_{b2} clones also demonstrated that the Y_{b1} and Y_{b2} mRNAs are elevated by phenobarbitol and 3-methylcholanthrene (80, 92). These observations led to studies that demonstrated that member(s) of the GST Y_a and Y_b gene families were transcriptionally activated in the rat (107).

The mechanisms that are responsible for transcriptional activation of the GSTs are currently being addressed. Telakowski-Hopkins et al (108) have constructed chimeric genes using various lengths of the 5' flanking region of a rat liver Y_a structural gene. The constructs contained 700 bp, 1600 bp, and 4000 bp of the 5' flanking region of the transferase promoter ligated to chloroamphenicol acetyl transferase (CAT). All of these constructs gave detectable CAT activity when transfected into human, rat, or mouse hepatoma cells. Interestingly, 10-fold differences in CAT activity was noted between the shortest construct pGTB.7 CAT, and a longer construct pGTB1.6 CAT in rat and human hepatoma cells. Transfection of the longest construct, pGTB4.0 CAT, gave CAT activity similar to pGTB1.6 CAT. When cells containing pGTB1.6 CAT were treated with the planar aromatic compound, β-napthaflavone, CAT activity was elevated 3–7-fold. The CAT activity directed by the pGTB.7 CAT construct was not affected by β-napthaflavone. These data suggest that a GST Y_a subunit structural gene contains *cis*-acting regulatory elements required for inducible expression by planar aromatic compounds such as 3-methylcholanthrene or β-napthaflavone, and the DNA sequence conferring the inducibility is located $\sim$ 650 to 1550 nucleotides upstream from transcription initiation. It was also concluded in the above studies that a second *cis*-acting regulatory element was present between $\sim$ 650 and 1550 nucleotides, which is required for maximal basal level expression.

The above studies also addressed the question of whether the Ah or dioxin receptor is involved in the induction process. In the cytochrome P-450 system, the gene encoding cytochrome P_1-450, as well as the gene encoding rat cytochrome P-450$_c$, contains a core promoter, TCDD dioxin-responsive elements, and a negative regulatory element (109–114). Presumably, the dioxin-responsive elements in the 5' flanking region of the cytochrome P_1-450 or P-450$_c$ gene is recognized by the Ah receptor–ligand complex, and this interaction results in the transcriptional activation of the cytochrome P_1-450 or P-450$_c$ gene. By deletion analysis of the 5' flanking region, the existence of these *cis*-acting DNA sequences in the 5' flanking region of the P_1-450 and P-450$_c$ genes have been confirmed. In addition, binding studies using gel shift and DNase protection assays support the hypothesis that specific *trans*-acting proteins bind to these DNA sequences (115, 116).

Using variant mouse hepatoma cells, defective in the number or translocation of the Ah receptor into the nucleus, Telakowski-Hopkins et al (108) demonstrated that functional receptors are required in order to get an induction in CAT activity after transient transfection in mouse cell lines. However, these data do not prove that the receptor directly interacts with specific sequences in the 5' flanking region of the GST Y_a gene.

Deletion analysis and protein-binding assays should provide definitive data on the exact role of the Ah receptor in regulating GST Y_a subunit gene expression. A working model describing the regulation of Y_a gene expression is presented in Figure 4.

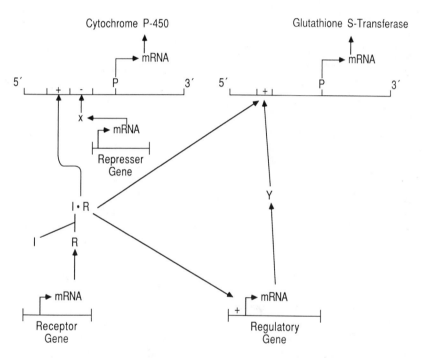

Figure 4   A model for the regulation of the glutathione S-transferase gene expression. A current model governing P_1-450 gene regulation by polycyclic aromatic hydrocarbons is also presented as a reference. In this model, the Ah or dioxin receptor (R) binds polycyclic aromatic hydrocarbons (I), forming a receptor ligand complex that is translocated to the nucleus and interacts with positive regulatory elements (polycyclic aromatic hydrocarbon-responsive elements), leading to the transcriptional activation of the P_1-450 gene and possibly the glutathione S-transferase Y_a gene. Alternatively, the receptor ligand complex (I.R.) could potentially interact with a regulatory gene, leading to the transcriptional activation of the regulatory gene, producing a *trans*-acting protein, which regulates expression of the glutathione S-transferase Y_a gene. To date, we have not been able to distinguish between these two possibilities. Finally, in the P-450 system, a repressor protein (X) is thought to regulate negatively the constitutive expression of the P_1-450 genes in the absence of ligand.

During Chemical Carcinogenesis

Solt & Farber (117) have developed a protocol to induce the formation of hepatocyte nodules in the liver of rats. A few of these nodules are persistent in liver and develop into hepatocellular carcinomas. The Solt-Farber model has been used to study chemically induced neoplastic transformation. The nodules have elevated levels of various phase II drug metabolizing enzymes (e.g. GSTs, NAD(P)H: quinone reductase, and epoxide hydrolase), however, cytochrome P-450 levels are depressed (118–125). The elevated levels of the phase II enzymes are consistent with the observation that persistent hepatocyte nodules are more resistant to the cytotoxic effect of carcinogens as compared to normal liver.

Using specific cDNA clones to the GSTs, it was found that the mRNAs specific for the Y_a and Y_b subunits were elevated in nodular tissue as compared to liver tissue surrounding the nodules (126). Similarly, it has been found that the mRNA specific for quinone reductase is markedly elevated in nodule tissue (127). Studies were also conducted to determine if the level of steady state Y_a and Y_b mRNAs could be regulated by 3-methylcholanthrene in rats that had undergone nodule induction. Surprisingly, it was found that only the Y_b mRNAs were inducible by 3-methylcholanthrene (126). These data indicate that the regulation of the Y_a mRNA in persistent hepatocyte nodules is aberrant; however, the precise regulatory defect is unknown.

Suguoka et al (98) have demonstrated that the GST Y_p mRNA is substantially elevated in acetylaminofluorene-induced hepatocellular carcinoma, and Satoh et al (14) have demonstrated that the GST-P subunit is elevated markedly in preneoplastic nodules. Pemble et al (99) have also demonstrated an elevation in Y_p mRNA in N,N-dimethyl-4-aminoazobenzene-induced rat hepatomas.

Overexpression of GST in Drug-Resistant Cells

Overexpression of GST has been reported in tumor cells resistant to anticancer drugs. Cowan and coworkers (128) have recently selected an adriamycin-resistant human breast cancer cell line that has a marked increase (45-fold) in GST activity. More than 90% of the increased activity is associated with an anionic species not found in the wild-type cells. This new anionic isozyme in adriamycin-resistant cells is immunologically related to the GST Y_p subunit (129). Overexpression of the Y_p subunit has also been observed in rat persistent liver nodules (130), which are resistant to various hepatotoxins, and in drug-resistant ovarian adenocarcinoma cells (131). Batist et al (129) have shown by RNA blot analysis that a marked elevation in Y_p mRNA occurred in DOX^R cells compared to the wild type.

The GST subunit compositions in nitrogen mustard-resistant and -sensitive Walker 256 rat mammary carcinoma cell lines have been examined by Buller et al (132). Overexpression of a Y_c class GST is observed in the resistant cells. The resistant cells show a 15-fold increase in resistance to the cytotoxic effects of chlorambucil, but ethacrynic acid and piriprost, substrates of GST, potentiate the cytotoxic activity of chlorambucil (133). In a Chinese hamster ovary cell line resistant to nitrogen mustard, the elevated GST activity is associated with the reduced levels of mechlorethamine-induced DNA cross-linking (134). These results suggest that overexpression of GST may contribute to the nitrogen mustard–resistant phenotype.

In another study, Lewis et al (135) studied two ovarian adenocarcinoma cell lines derived from a patient before and after the onset of drug resistance to cis-platinum, chlorambucil, and 5-fluorouracil. In drug-resistant cells, the GST activity is 3-fold higher, demonstrating that tumor cells obtained from a patient before and after the onset of drug resistance have very different GST activity.

As a way to explore the role of specific GST isozymes in the development of resistance to drugs, Manoharan et al (136) have ligated the coding region of the GST Y_a cDNA clone, pGTB38, to the SV40 early promoter of the Herpes simplex thymidine kinase gene. One construct containing the SV40 promoter produced significant levels of Y_a mRNA and GST activity in COS cells. Manoharan et al (136) treated a mixed population of COS cells containing Y_a subunit overexpressing cells with benzo(a)pyrene($\pm$)anti-diol epoxide, which is a cytotoxic agent and substrate for the Y_a homodimer. These investigators found a 20- to 30-fold enrichment in clones of overexpressing cells in the cells surviving treatment with the cytotoxic agent. These data provide direct evidence that overexpression of specific GSTs in mammalian cells can lead to significant biological resistance to known alkylating agents.

Although all these studies suggest a possible role for GST in drug resistance in various cell lines, they have not established that elevated GST activity in resistant cells correlates with higher rates of conjugation of the drug to glutathione. It is also important to establish that the drug that causes the resistance of the cells is a substrate of the overexpressed GST isozyme. Until these data become available, the biological significance of GST overexpression in drug resistance is unclear. Research is needed on expressing specific GST subunits in appropriate mammalian cells to examine their role in drug resistance.

Tissue-Specific Expression

The expression of GSTs has been reported to be tissue specific (8, 137–141). Expression of the Y_a and Y_c subunits does not occur in heart and testis, whereas the Y_b subunits are not expressed in kidneys (13). Similarly, the Y_p subunit is not expressed in normal rat liver. RNA blot analyses with specific cDNA clones have indicated that tissue-specific regulation of GST is pretranslational (140). Abramovitz & Listowsky (95) have isolated and characterized a novel Y_b cDNA clone, Y_{b3}, from a rat brain λgt 11 expression library. Using the 3' untranslated region of the Y_{b3} clone as a specific probe, these investigators found that the Y_{b3} mRNA is brain specific. These latter data provided the first direct proof for the existence of a third related subunit in the Y_b gene family and further confirmed the tissue-specific expression of specific glutathione S-transferases genes.

Although studies such as the one discussed above have suggested tissue-specific expression of GST genes, there have been no studies to date examining whether specific *cis*-acting regulatory elements in the 5' flanking region of GST genes are important for tissue-specific expression. A detailed analysis of both *cis*- and *trans*-acting elements that regulate tissue-specific GST gene expression is an important research area worthy of additional effort.

FUTURE DIRECTIONS

Although the above discussion focused primarily on GSTs from the rat, it is likely that other mammalian species will have a similar complexity of subunits and genes. The molecular mechanisms underlying induction of GSTs by planar aromatic compounds are just now beginning to receive attention. It should be possible, in the near future, to define *trans*- and *cis*-acting factors that regulate the expression of these genes. Of particular interest will be whether the Ah receptor, directly or indirectly, regulates expression of GST genes.

Although phenobarbitol transcriptionally activates GST genes, there appear to be no cell lines that are responsive to this xenobiotic. Consequently, experiments to define regulatory elements on GST genes that are required for transcriptional activation by phenobarbitol have not been done. The use of primary hepatocytes may be useful in examining phenobarbitol induction.

The role specific amino acid residues or domains play in substrate binding and ligand binding (e.g. heme, bilirubin, and catalytic activity) can now be approached using site-directed mutagenesis. The lack of posttranslational modification of GSTs and their cytosolic localization have made them ideal candidates for high-level expression in *E. coli*. Crystallographic analysis of GST subunits is already under way (142). These advances in the GST field should rapidly define residues required for catalysis.

One of the major problems in current chemotherapy is the establishment of cells that become resistant to the chemotherapeutic agent. Many investigators have found elevations in GST activity or protein level, but the physiological importance of this elevation remains obscure. Consequently, the identification and characterization of specific GSTs involved in the metabolism of these agents is critical. The molecular mechanisms underlying GST induction during chemotherapy will also be an important future area of research. The research directions described above should provide an exciting challenge for many investigators in the years ahead.

ACKNOWLEDGMENT

We thank Ms. Freda Dias for her assistance in the preparation of this review.

Literature Cited

1. Chasseaud, L. F. 1979. *Adv. Cancer Res.* 29:175–274
2. Jakoby, W. B., Habig, W. H. 1980. In *Enzymatic Basis of Detoxication,* ed. W. B. Jakoby, pp. 63–94. New York: Academic
3. Mannervik, B. 1985. *Adv. Enzymol.* 57: 357–417
4. Boyer, T. D., Kenney, W. C. 1985. In *Biochemical Pharmacology and Toxicology,* ed. D. Zakim, D. A. Vessey, 1:297–363. New York: Wiley
5. Mannervik, B., Danielson, U. H. 1988. *CRC Crit. Rev. Biochem.* 23:281–334
6. Litwack, G., Ketterer, B., Arias, I. M. 1971. *Nature* 234:466–67
7. Arias, I. M., Fleischner, G., Kirsch, R., Mishkin, S., Gatmaitan, Z. 1976. In *Glutathione: Metabolism and Function,* ed. I. M. Arias, W. B., Jakoby, pp. 175–88. New York: Raven
8. Bhargava, M. M., Ohmi, N., Listowsky, I., Arias, I. M. 1980. *J. Biol. Chem.* 255:724–27
9. Homma, H., Listowsky, I. 1985. *Proc. Natl. Acad. Sci. USA* 82:7165–69
10. Bass, N. M., Kirsch, R. E., Tuff, S. A., Marks, I., Saunders, S. J. 1977. *Biochem. Biophys. Acta* 492:163–75
11. Mannervik, B., Jensson, H. 1982. *J. Biol. Chem.* 257:9909–12
12. Reddy, C. C., Li, N.-Q., Tu, C.-P. D. 1984. *Biochem. Biophys. Res. Commun.* 121:1014–20
13. Hayes, J. D. 1984. *Biochem. J.* 224: 839–52
14. Satoh, K., Kitahara, A., Soma, Y., Inaba, Y., Hatayama, I., Sato, K. 1985. *Proc. Natl. Acad. Sci. USA* 82:3964–68
15. Hayes, J. D. 1986. *Biochem. J.* 233: 789–98
16. Beale, D., Meyer, D. J., Taylor, J. B., Ketterer, B. 1983. *Eur. J. Biochem.* 137:125–29
17. Kitahara, A., Satoh, K., Nishimura, K., Ishikawa, T., Ruike, K., et al. 1984. *Cancer Res.* 44:2698–703
18. Hayes, J. D., Mantle, T. J. 1986. *Biochem. J.* 237:731–40
19. Jakoby, W. B., Ketterer, B., Mannervik, B. 1984. *Biochem. Pharmacol.* 33:2539–40
20. Bass, N. M., Kirsch, R. E., Tuff, S. A., Saunders, S. J. 1977. *Biochim. Biophys. Acta* 494:131–43
21. Ketterer, B., Meyer, D. J., Coles, B., Taylor, J. B., Pemble, S. 1985. In *Microsomes and Drug Oxidations,* ed. A. R. Boobis, J. Caldwell, F. De Matteis, C. R. Elcombe, 6:167–77. London: Taylor & Francis
22. Farrants, A. K. O., Meyer, D. J., Coles, B., Southan, C., Aitken, A., et al. 1987. *Biochem. J.* 245:423–28
23. Meyer, D. J., Ketterer, B. 1982. *FEBS Lett.* 150:499–502
24. Danielson, U. H., Esterbauer, H., Mannervik, B. 1987. *Biochem. J.* 247:707–13
25. Tan, K. H., Meyer, D. J., Coles, B., Ketterer, B. 1986. *FEBS Lett.* 207:231–33
26. Esterbauer, H., Cheeseman, K. H., Dianzani, M. U., Poli, G., Slater, T. F. 1982. *Biochem. J.* 208:129–40
27. Esterbauer, H. 1982. In *Free Radicals, Lipid Peroxidation and Cancer,* ed. D. C. H. McBrien, T. F. Slater, pp. 101–28. New York: Academic
28. Totter, J. B. 1980. *Proc. Natl. Acad. Sci. USA* 77:1763–67
29. Slater, T. F. 1984. *Biochem. J.* 222:1–15
30. Emerit, I., Cerutti, P. A. 1982. *Proc. Natl. Acad. Sci. USA* 79:7509–13
31. Hahn, B. S., Wang, S. Y. 1973. *Biochem. Biophys. Res. Commun.* 54: 1224–31
32. Armstrong, R. N. 1987. *CRC Crit. Rev. Biochem.* 22:39–88
33. Jakobson, I., Warholm, M., Mannervik, B. 1979. *Biochem. J.* 177:861–68
34. Jakobson, I., Warholm, M., Mannervik, B. 1979. *J. Biol. Chem.* 254:7085–89
35. Chen, W.-J., Boehlert, C. C., Rider, K., Armstrong, R. N. 1985. *Biochem. Biophys. Res. Commun.* 128:233–40

36. Mangold, J. B., Abdel-Monem, M. M. 1983. *J. Med. Chem.* 26:66–71
37. Ridgewell, R. E., Abdel-Monem, M. M. 1987. *Drug Metab. Dispos.* 15:82–90
38. Cobb, D., Boehlert, C., Lewis, D., Armstrong, R. N. 1983. *Biochemistry* 22:805–12
39. Boehlert, C. C., Armstrong, R. N. 1984. *Biochem. Biophys. Res. Commun.* 121:980–86
40. Awasthi, Y. C., Bhatnagar, A., Singh, S. V. 1987. *Biochem. Biophys. Res. Commun.* 143:965–70
41. Board, P. G., Pierce, K. 1987. *Biochem. J.* 248:937–41
42. Wang, R. W., Pickett, C. B., Lu, A. Y. H. 1989. *Arch. Biochem. Biophys.* 269:536–43
43. Anders, M. W., Lash, L., Dekant, W., Elfarra, A. A., Dohn, D. R. 1988. *CRC Crit. Rev. Toxicol.* 18:311–41
44. Bakke, J., Gustafsson, J.-A. 1984. *Trends Pharmacol. Sci.* 5:517–21
45. Elfarra, A. A., Anders, M. W. 1984. *Biochem. Pharmacol.* 33:3729–32
46. Lash, L. H., Anders, M. W. 1988. *Rev. Biochem. Toxicol.* 9:29–67
47. Nash, J. A., King, L. J., Lock, E. A., Green, T. 1984. *Toxicol. Appl. Pharmacol.* 73:124–37
48. Jaffe, D. R., Hassall, C. D., Brendel, K., Gandolfi, A. J. 1983. *J. Toxicol. Environ. Health* 11:857–67
49. Green, T., Odum, J. 1985. *Chem. Biol. Interact.* 54:15–31
50. Van Bladeren, P. J., Breimer, D. D., Rotteveel-Smijs, G. M. T., de Jong, R. A. W., Buijs, W., et al. 1980. *Biochem. Pharmacol.* 29:2975–82
51. Rannug, U., Sundvall, A., Ramel, C. 1978. *Chem. Biol. Interact.* 20:1–16
52. Rannug, U. 1980. *Mutat. Res.* 76:269–95
53. Wong, L. C. K., Winston, J. M., Hong, C. B., Plotnick, H. 1982. *Toxicol. Appl. Pharmacol.* 63:155–65
54. Olson, W. A., Habermann, R. T., Weisburger, E. K., Ward, J. M., Weisburger, J. H. 1973. *J. Natl. Cancer Inst.* 51:1993–95
55. Webb, W. W., Elfarra, A. A., Webster, K. D., Thom, R. E., Anders, M. W. 1987. *Biochemistry* 26:3017–23
56. Elfarra, A. A., Baggs, R. B., Anders, M. W. 1985. *J. Pharmacol. Exp. Ther.* 233:512–16
57. Van Bladeren, P. J., van der Gen, A., Breimer, D. D., Mohn, G. R. 1979. *Biochem. Pharmacol.* 28:2521–24
58. Stevens, J., Hayden, P., Taylor, G. 1986. *J. Biol. Chem.* 261:3325–32
59. Elfarra, A. A., Jakobson, I., Anders,
60. Silber, P. M., Gandolfi, A. J., Brendel, K. 1986. *Drug Chem. Toxicol.* 9:285–303
61. Dekant, W., Vamvakas, S., Berthold, K., Schmidt, S., Wild, D., Henschler, D. 1986. *Chem. Biol. Interact.* 60:31–45
62. Dohn, D. R., Leininger, J. R., Lash, L. H., Quebbemann, A. J., Anders, M. W. 1985. *J. Pharmacol. Exp. Ther.* 235:851–57
63. Monks, T. J., Lau, S. S., Highet, R. J., Gillette, J. R. 1985. *Drug. Metab. Dispos.* 13:553–59
64. Bruggeman, I. M., Temmink, J. H. M., van Bladeren, P. J. 1986. *Toxicol. Appl. Pharmacol.* 83:349–59
65. Veltman, J. C., Dekant, W., Guengerich, F. P., Anders, M. W. 1988. *Chem. Res. Toxicol.* 1:35–40
66. Saito, K., Yamazoe, Y., Kamataki, T., Kato, R. 1983. *Carcinogenesis* 4:1551–57
67. Commandeur, J. N. M., Oostendorp, R. A. J., Schoofs, P. R., Xu, B., Vermeulen, N. P. E. 1987. *Biochem. Pharmacol.* 36:4229–37
68. Hill, D. L., Shih, T. W., Johnston, T. P., Struck, R. F. 1978. *Cancer Res.* 38:2438–42
69. Ozawa, N., Guengerich, F. P. 1983. *Proc. Natl. Acad. Sci. USA* 80:5266–70
70. Koga, N., Inskeep, P. B., Harris, T. M., Guengerich, F. P. 1986. *Biochemistry* 25:2192–98
71. Van Bladeren, P. J., Breimer, D. D., Rotteveel-Smijs, G. M. T., de Knijff, P., Mohn, G. R., et al. 1981. *Carcinogenesis* 2:499–505
72. Webb, W. W., Elfarra, A. A., Webster, K. D., Thom, R. E., Anders, M. W. 1987. *Biochemistry* 26:3017–23
73. Stevens, J., Jakoby, W. B. 1983. *Mol. Pharmacol.* 23:761–65
74. Dekant, W., Lash, L. H., Anders, M. W. 1987. *Proc. Natl. Acad. Sci. USA* 84:7443–47
75. Lash, L. H., Anders, M. W. 1986. *J. Biol. Chem.* 261:13076–81
76. Lash, L. H., Elfarra, A. A., Anders, M. W. 1986. *Arch. Biochem. Biophys.* 251:432–39
77. Lash, L. H., Anders, M. W. 1987. *Mol. Pharmacol.* 32:549–56
78. Taylor, J. B., Craig, R. K., Beale, D., Ketterer, B. 1984. *Biochem. J.* 219:223–31
79. Lai, H.-C. J., Li, N.-Q., Weiss, M. J., Reddy, C. C., Tu, C.-P. D. 1984. *J. Biol. Chem.* 259:5536–42
80. Pickett, C. B., Telakowski-Hopkins, C.

A., Ding, G. J.-F., Argenbright, L., Lu, A. Y. H. 1984. *J. Biol. Chem.* 259: 5182–88

81. Daniel, V., Sarid, S., Bar-Nun, S., Litwack, G. 1983. *Arch. Biochem. Biophys.* 227:266–71

82. Kalinyak, J. E., Taylor, J. M. 1982. *J. Biol. Chem.* 257:523–30

83. Tu, C.-P. D., Weiss, M. J., Karakawa, W. W., Reddy, C. C. 1982. *Nucleic Acids Res.* 10:5407–19

84. Tu, C.-P. D., Lai, H.-C. J., Li, N.-Q., Weiss, M. J., Reddy, C. C. 1984. *J. Biol. Chem.* 259:9434–39

85. Telakowski-Hopkins, C. A., Rodkey, J. A., Bennett, C. D., Lu, A. Y. H., Pickett, C. B. 1985. *J. Biol. Chem.* 260: 5820–25

86. Rothkopf, G. S., Telakowski-Hopkins, C. A., Stotish, R. L., Pickett, C. B. 1986. *Biochemistry* 25:993–1002

87. Tu, C.-P. D., Quian, B. 1986. *Biochem. Biophys. Res. Commun.* 141:229–37

88. Rhoads, D. M., Zarlengo, R. P., Tu, C.-P. D. 1987. *Biochem. Biophys. Res. Commun.* 145:474–81

89. Board, P. G., Webb, G. C. 1987. *Proc. Natl. Acad. Sci. USA* 84:2377–81

90. Telakowski-Hopkins, C. A., Rothkopf, G. S., Pickett, C. B. 1986. *Proc. Natl. Acad. Sci. USA* 83:9393–97

91. Ding, G. J.-F., Lu, A. Y. H., Pickett, C. B. 1985. *J. Biol. Chem.* 260:13268–71

92. Ding, G. J.-F., Ding, V. D.-H., Rodkey, J. A., Bennett, C. D., Lu, A. Y. H., Pickett, C. B. 1986. *J. Biol. Chem.* 261:7952–57

93. Lai, H. C.-J., Grove, G., Tu, C.-P. D. 1986. *Nucleic Acids Res.* 14:6101–14

94. Lai, H. C.-J., Tu, C.-P. D. 1986. *J. Biol. Chem.* 261:13793–99

95. Abramovitz, M., Listowsky, I. 1987. *J. Biol. Chem.* 262:7770–73

96. Tu, C.-P. D., Lai, H.-C. J., Reddy, C. C. 1987. In *Glutathione S-Transferases and Carcinogenesis*, ed. T. J. Mantle, C. B. Pickett, J. D. Hayes, pp. 87–110. London: Taylor & Francis

97. Lai, H. C. J., Qian, B., Grove, G., Tu, C. P. D. 1988. *J. Biol. Chem.* 263: 11389–95

98. Suguoka, Y., Kano, T., Okuda, A., Sakai, M., Kitagawa, T., Muramatsu, M. 1985. *Nucleic Acids Res.* 13:609–57

99. Pemble, S. E., Taylor, J. B., Ketterer, B. 1986. *Biochem. J.* 240:885–89

100. Kano, T., Sakai, M., Muramatsu, M. 1987. *Cancer Res.* 47:5626–30

101. Okuda, A., Sakai, M., Muramatsu, M. 1987. *J. Biol. Chem.* 262:3858–63

102. DeJong, J. L., Morgenstern, R., Jörn-vall, H., DePierre, J. W., Tu, C.-P. D. 1988. *J. Biol. Chem.* 263:8430–36

103. Morgenstern, R., DePierre, J. W., Jörnvall, H. 1985. *J. Biol. Chem.* 260: 13976–83

104. Pickett, C. B., Wells, W., Lu, A. Y. H., Hales, B. F. 1981. *Biochem. Biophys. Res. Commun.* 99:1002–10

105. Pickett, C. B., Telakowski-Hopkins, C. A., Donohue, A. M., Lu, A. Y. H., Hales, B. F. 1982. *Biochem. Biophys. Res. Commun.* 104:611–19

106. Pickett, C. B., Donohue, A. M., Lu, A. Y. H., Hales, B. F. 1982. *Arch. Biochem. Biophys.* 215:539–43

107. Ding, V. D.-H., Pickett, C. B. 1985. *Arch. Biochem. Biophys.* 240:553–59

108. Telakowski-Hopkins, C. A., King, R. G., Pickett, C. B. 1988. *Proc. Natl. Acad. Sci. USA* 85:1000–4

109. Jones, P. B. C., Galeazzi, D. R., Fisher, J. M., Whitlock, J. P. Jr. 1985. *Science* 227:1499–1502

110. Jones, P. B. C., Durrin, L. K., Galeazzi, D. R., Whitlock, J. P. Jr. 1986. *Proc. Natl. Acad. Sci. USA* 83:2802–6

111. Jones, P. B. C., Durrin, L. K., Fisher, J. M., Whitlock, J. P. Jr. 1986. *J. Biol. Chem.* 261:6647–50

112. Fujisawa-Sehara, A., Sogawa, K., Nishi, C., Fujii-Kuriyama, Y. 1986. *Nucleic Acids Res.* 14:1465–77

113. Gonzalez, F. J., Nebert, D. W. 1985. *Nucleic Acids Res.* 13:7269–88

114. Sogawa, K., Fujisawa-Sehara, A., Yamane, M., Fujii-Kuriyama, Y. 1986. *Proc. Natl. Acad. Sci. USA* 83:8044–48

115. Fujisawa-Sehara, A., Yamane, M., Fujii-Kuriyama, Y. 1988. *Proc. Natl. Acad. Sci. USA* 85:5859–63

116. Durrin, L. K., Whitlock, J. P. Jr. 1987. *Mol. Cell. Biol.* 7:3008–11

117. Solt, D. B., Farber, E. 1976. *Nature* 263:701–3

118. Farber, E. 1984. *Cancer Res.* 44:5463–74

119. Aström, A., DePierre, J. W., Eriksson, L. C. 1983. *Carcinogenesis* 4:577–81

120. Cameron, R., Sweeney, G. D., Jones, K., Lee, G., Farber, E. 1976. *Cancer Res.* 36:3888–93

121. Kitahara, A., Satoh, K., Nishimura, K., Ishikawa, T., Ruike, K., et al. 1984. *Cancer Res.* 44:2698–703

122. Kitahara, A., Satoh, K., Sato, K. 1983. *Biochem. Biophys. Res. Commun.* 112: 20–28

123. Levin, W., Lu, A. Y. H., Thomas, P. E., Ryan, D., Kizer, D. E., Griffin, M. J. 1978. *Proc. Natl. Acad. Sci. USA* 75:3240–43

124. Bock, K. W., Lilienblum, W., Pfeil, H., Eriksson, L. C. 1982. *Cancer Res.* 42:3747–52
125. Schor, N. A., Ogawa, K., Lee, G., Farber, E. 1978. *Cancer Lett.* 5:167–71
126. Pickett, C. B., Williams, J. B., Lu, A. Y. H., Cameron, R. G. 1984. *Proc. Natl. Acad. Sci. USA* 81:5091–95
127. Williams, J. B., Lu, A. Y. H., Cameron, R. G., Pickett, C. B. 1986. *J. Biol. Chem.* 261:5524–28
128. Cowan, K. H., Batist, G., Tulpule, A., Sinha, B. K., Myers, C. E. 1986. *Proc. Natl. Acad. Sci. USA* 83:9328–32
129. Batist, G., Tulpule, A., Sinha, B. K., Katki, A. G., Myers, C. E., Cowan, K. H. 1986. *J. Biol. Chem.* 261:15544–49
130. Rushmore, T. H., Sharma, R. N. S., Roomi, M. W., Harris, L., Satoh, K., et al. 1987. *Biochem. Biophys. Res. Commun.* 143:98–103
131. Wolf, C. R., Lewis, A. D., Carmichael, J., Ansell, J., Adams, D. J., et al. 1987. See Ref. 96, pp. 199–212
132. Buller, A. L., Clapper, M. L., Tew, K. D. 1981. *Mol. Pharmacol.* 31:575–78
133. Tew, K. D., Bomber, A. M., Hoffman, S. J. 1988. *Cancer Res.* 48:3622–25
134. Robson, C. N., Lewis, A. D., Wolf, C. R., Hayes, J. D., Hall, A., et al. 1987. *Cancer Res.* 47:6022–27
135. Lewis, A. D., Hayes, J. D., Wolf, C. R. 1988. *Carcinogenesis* 9:1283–87
136. Manoharan, T. H., Puchalski, R. B., Burgess, J. A., Pickett, C. B., Fahl, W. E. 1987. *J. Biol. Chem.* 262:3739–45
137. Scully, N. C., Mantle, T. J. 1981. *Biochem. J.* 193:367–70
138. Guthenberg, C., Mannervik, B. 1979. *Biochem. Biophys. Res. Commun.* 86:1304–10
139. Tu, C.-P. D., Weiss, M. J., Li, N.-Q., Reddy, C. C. 1983. *J. Biol. Chem.* 258:4659–62
140. Li, N.-Q., Reddanna, P., Thyagaraju, K., Reddy, C. C., Tu, C.-P. D. 1986. *J. Biol. Chem.* 261:7596–99
141. Hayes, J. D., Mantle, T. J. 1986. *Biochem. J.* 233:779–88
142. Sesay, M. A., Ammon, H. L., Armstrong, R. N. 1987. *J. Mol. Biol.* 197:377–78

Annu. Rev. Biochem. 1989. 58:765–98

MUTATIONAL EFFECTS ON PROTEIN STABILITY

Tom Alber

Department of Biochemistry, University of Utah, School of Medicine, Salt Lake City, Utah 84132

Department of Chemistry, University of Utah, Salt Lake City, Utah 84103

CONTENTS

PERSPECTIVES AND SUMMARY

The study of protein stability is changing dramatically. In the last decade, improvements in genetic and biophysical methods have facilitated measurements of the stabilizing contributions of specific interactions in proteins. Careful comparisons of closely related variants have been used to ask if traditional analogies to the properties of simple chemical model systems are applicable to the stability of proteins. Genetic identifications of critical stabilizing amino acids in several proteins have provided another important

765

0066-4154/89/0701-0765$02.00

advance. Despite the difficulty of partitioning mutational effects between the folded and unfolded states, several general conclusions are emerging:

1. The role of each amino acid depends on its structural context. Sensitivity to severe destabilizing substitutions is correlated with features of the folded state, implying that interactions in this state are often dominant. With the exception of charged residues, most amino acids that make critical interactions are rigid or buried in the folded structure.

2. Many different types of interactions—including disulfide bonds, hydrophobic forces, hydrogen bonds, electrostatic interactions, and dispersion forces—make quantitatively comparable contributions to stability.

3. Specific interactions of each type make a wide range of stabilizing contributions. The observed range of contributions is not adequately described by the behavior of the simple chemical model systems traditionally used to evaluate the strengths of noncovalent interactions. Model systems generally do not account for the unique environments of each residue in the folded and unfolded states or for the entropy changes associated with forming specific interactions.

4. Many amino acid substitutions do not have large effects on stability. Proteins tolerate substitutions because (*a*) some substitutions preserve critical interactions, (*b*) some interactions apparently do not make large contributions to stability, and (*c*) protein structures adjust to compensate for changes in sequence. The impact of an amino acid substitution is a combination of its intrinsic effects on the folded and unfolded states and the relative abilities of the two states to relax in response to the change. Relaxations minimize destabilizing effects.

In parallel with these experimental developments, a conceptual framework that accounts for the disparate contributions of specific interactions has been proposed by Creighton (1–5). Creighton's approach is based on the idea of "effective concentration," the ratio of the intramolecular and intermolecular association constants for two groups. Stabilizing substituents have higher effective concentrations (their interactions are more favored) in the folded state than in the unfolded state. Amino acids and water molecules that are solvated better in the unfolded state are destabilizing, even if they participate in observable interactions in the X-ray crystal structure of the folded protein.

The concept of effective concentration, since it includes all the determinants of association constants, does not necessarily provide a mechanistic understanding of stabilizing contributions. Unlike earlier approaches based on scalar contributions derived from studies of model systems, however, the idea of effective concentration rationalizes the wide range of contributions made by each type of interaction. The cooperativity of folding is also an integral part of the approach. Effective concentrations of specific groups in proteins have been measured experimentally in both the folded and unfolded states.

Several testable predictions follow from the theory, and one of these has recently been lent experimental support: many groups that are relatively rigid in the folded structure make important contributions to stability (1, 6).

This review summarizes recent studies that use protein variants to better understand the structural basis of protein thermodynamic stability. Because of the rapidly growing literature in this area, I have found it impossible to be comprehensive. A number of other reviews have been published recently (7–12). Several important topics are not treated here, including recent progress on the mechanisms of protein folding, the causes of irreversible protein inactivation, and the sources of ligand-binding energy. Discussions of the methods of genetics, thermodynamics, kinetics, X-ray crystallography, and calculations of free energy are also excluded. A theme of this review is the comparison of studies of variant proteins with traditional studies of model systems. The emphasis is on the search for structural patterns associated with stability that may provide rigorous criteria for assessing the plausibility of predicted structures and stringent tests for emerging theories of stabilization.

THE PROBLEM

The thermodynamic stability of proteins is modest and depends on environmental conditions in a complex way. The central problem is to quantitatively account for the small differences in free energy between the relatively small ensemble of folded conformations and the immense ensemble of rapidly interconverting unfolded alternatives. Qualitatively, the conformational entropy and hydration of the unfolded state are thought to be balanced by specific stabilizing interactions in the folded molecule. The physical forces that underlie this balance—such as the hydrophobic effect, van der Waals forces, electrostatic interactions, hydrogen bonds, covalent crosslinks, etc— were largely identified in the 1950s (13). A current challenge is to determine the contributions of individual amino acids to the stability of a specific protein as a function of environmental conditions. The structural basis of cooperativity and the mechanisms by which interactions are integrated are also not well understood.

The complexities of the folded state, the unfolded state, and the transition between them present major barriers to solving the stability problem. By now the structural complexity of folded proteins is generally appreciated. Discovery of structural patterns has led to proposals for simplifying architectural rules (14, 15) and to the identification of potentially stabilizing features. These include the tight packing of the protein interior (16), the α-helix dipole (17–19), helix caps (20, 21), and weakly polar interactions between aromatic groups (22, 23). The folded structure undergoes large and small motions that provide wells of conformational and vibrational entropy (24). Alternative folded conformations have been detected in a number of proteins (25–29). Interactions of the polypeptide chain with solvent and counterions are fluid and hard to model rigorously. Because of the modest resolution of most protein X-ray crystal structures, an extended atom representation that does not include hydrogens is commonly used. Yet incorrect placement of even a single hydrogen atom can thwart the computational analysis of stability (30, 31).

Since each amino acid influences the free energy of both the folded and unfolded states, insight into denatured proteins is crucial for understanding protein stability. The complexities of the folded state are only compounded when considering its alter ego. Major issues include the number of accessible conformations, the amount and nature of preferred structure, and the relationship between structure and free energy.

On unfolding, the polypeptide chain becomes less compact, more highly solvated, and much more flexible (32). Unfolded chains produced by heat and by guanidinium hydrochloride (GuHCl) are very nearly random coils, as judged by intrinsic viscosity measurements made at identical temperatures (25). Whatever residual structures may exist in unfolded states, comparisons of measurements made at the same temperature and pH show that changes in heat capacity, enthalpy, and free energy due to denaturation by heat and by GuHCl are very similar (25). Differences in local order do not result in changes in thermodynamic state.

The unfolded polypeptide, though, is not an ideal mathematically random chain. Excluded volume effects alone reduce the estimated number of allowed backbone conformations of a 100-residue chain to about 10^{16} compared to an estimated 10^{60} in a "random coil" (8, 33). There is also considerable experimental evidence for some local order in proteins unfolded by different denaturants (34–37). Using fluorescence energy transfer, Haas et al recently obseved a nonrandom distribution of conformations in reduced bovine pancreatic trypsin inhibitor denatured by GuHCl (38, 39). Distinguishable unfolded forms of staphylococcal nuclease were detected by NMR methods (29). Kinetic studies of folding have long provided evidence for slowly interconverting conformational distributions in the unfolded state that differ in isomerization of prolines (40, 41).

Preferred conformations in isolated short peptides in water have been detected under conditions that favor structure formation (42–44). Such local interactions may influence the unfolded state (36, 37). For groups distant in the amino acid sequence, effective concentrations of 10^{-4} to 10^{-2} M have been measured (45, 46). For adjacent residues, effective concentrations may exceed several molar (16). As a result, dilute aqueous solutions of model compounds may not adequately represent the unfolded state (25).

Structures in the unfolded state are likely to be very transient and consequently highly localized. Bond rotations (47), hydrogen exchange (48, 49), and helix propagation (50) are rapid. NMR spectra of denatured proteins generally show that side chains are found in similar average environments (48, 51). The simple kinetics of folding (excluding the effects of proline isomerization) imply that unfolded conformations are in rapid equilibrium (52). The immense conformational variability and flexibility of the unfolded state presents severe problems for establishing rigorous structural models whose interactions can be evaluated analytically or experimentally.

The thermodynamics of the transition between the folded and unfolded states are also proving to be complex. Thermodynamic studies provide the data that must be explained by structural theories of protein stability. This topic has been developed in a number of excellent papers (11, 25, 53–60) and will only be summarized here.

Protein stability depends on environmental conditions such as temperature, pressure, pH, ionic strength, and the concentration of specific ligands, stabilizers, and denaturants. Even under the most favorable conditions, the folded state is only stabilized by 5–15 kcal/mole. This narrow range of stabilization free energy is independent of molecular weight (25). Denaturation is highly cooperative. For single-domain proteins, intermediates are rarely detected at equilibrium. This property often allows the simplifying assumption that the polypeptide chain populates only two states, folded and unfolded.

A central feature of the energetics of protein denaturation is that changes in enthalpy and entropy are strongly dependent on temperature. In particular:

$$\Delta H_{\mathrm{T}} = \Delta H_0 + \int_{T_0}^{T} \Delta Cp \, dT \qquad\qquad 1.$$

$$\Delta S_{\mathrm{T}} = \Delta S_0 + \int_{T_0}^{T} \frac{\Delta Cp}{T} \, dT \qquad\qquad 2.$$

where ΔH_{T} and ΔS_{T} are the changes in enthalpy and entropy at temperature T, ΔH_0 and ΔS_0 are the changes in enthalpy and entropy at a reference temperature T_0, and ΔCp is the difference in heat capacity between the folded and unfolded states at constant pressure (55). When ΔCp is assumed to be constant with temperature, these expressions can be simplified to the more familiar:

$$\Delta H_{\mathrm{T}} = \Delta H_0 + \Delta Cp \, (T - T_0) \qquad\qquad 3.$$

$$\Delta S_{\mathrm{T}} = \Delta S_0 + \Delta Cp \, \ln \, (T/T_0) \qquad\qquad 4.$$

Privalov & Gill (55) recently found that ΔCp approaches zero at about 143°C but is nearly constant over the temperature range of most studies of protein denaturation (0–80°C). As a result, the common assumption of constant ΔCp introduces only small quantitative errors compared to the more rigorous treatment (55).

At normally accessible temperatures, ΔCp for unfolding is large and positive ($\sim$1–2 kcal/mol^{-1}K^{-1}), and ΔCp per g of protein is proportional to the number of nonpolar contacts per g in the folded state (25). This behavior has been taken as a confirmation of the central role of the hydrophobic effect in protein stabilization. The large value of ΔCp means that ΔH_T and ΔS_T are steep functions of temperature. A temperature change of 1°C, for example, causes changes in ΔH_T and in $T\Delta S_T$ of approximately 1–2 kcal/mole. ΔH_T and ΔS_T are zero near room temperature, and they have large compensating values at temperatures where proteins denature. The mechanisms of this compensation are not well understood (61, 62). In addition, the effects of changes in hydrophobic contacts on ΔH, ΔS, and ΔCp have not been systematically investigated.

The relationship between free energy of stabilization and temperature, $\Delta G_T = \Delta H_T - T\Delta S_T$, defines the protein stability curve (25, 54, 58). Measured protein stability curves can be used to estimate the change in ΔG associated with a change in the denaturation temperature, T_m, induced by mutations. The free energy of stabilization reaches a maximum near room temperature where $\Delta S_T = 0$, and the equilibrium constant for folding ([F]/[U]) reaches a maximum at a slightly lower temperature where $\Delta H_T = 0$. Denaturation can be induced by heat and, surprisingly, by cold as well (56). At high temperature, unfolding results in a large increase in entropy, presumably due to the added flexibility of the polypeptide chain, and a compensating increase in enthalpy, attributed to changes in interactions in the protein and solvent. In the cold, the system actually loses entropy (becomes more ordered) and releases heat on unfolding.

THERMODYNAMICS OF AMINO ACID SUBSTITUTIONS

Major barriers to understanding how the amino acid sequence gives rise to this behavior have been described above. They include the complexities of folded and unfolded structures and the critical role of the solvent in determining protein conformation. A fundamental practical problem is that thermodynamic measurements only provide information about the sum of all interactions in the system. The measurements do not directly identify features of a complicated structure that determine its stability.

Comparisons of closely related proteins offer the possibility of estimating the contributions of specific interactions to stability. Modern methods of directed mutagenesis have significantly increased the power of this strategy. Despite technical advances, at least three problems complicate the genetic approach. (*a*) Even chemically simple alterations and localized structural shifts can simultaneously change many different interactions in the folded protein (31, 63–69). As a result, changes in stability often cannot be ascribed to single interactions. This emphasizes the importance of detailed structural studies for understanding the properties of mutant proteins. (*b*) Amino acid substitutions can alter the distribution of unfolded conformations (29, 70, 71). (*c*) Thermodynamic measurements only evaluate the differences between the folded and unfolded states, not the absolute effect of the mutation on either state. This represents a fundamental limitation of the genetic approach, because the magnitudes of the energetic effects that must be accounted for in each state are unknown.

This problem is illustrated by the following thermodynamic cycle:

$$
\begin{array}{ccc}
U & \overset{K_1}{\rightleftharpoons} & F \\
K_3 \updownarrow & & \updownarrow K_4 \\
U_M & \underset{K_2}{\rightleftharpoons} & F_M
\end{array}
$$

where K_1 and K_2 are the equilibrium constants for folding a wild-type (U,F) and a mutant protein (U$_m$,F$_m$), respectively. K_1 and K_2 cannot be related directly because the two proteins are different compounds. K_3 and K_4 are the hypothetical equilibrium constants for making the amino acid substitution reversibly in the unfolded and folded states. Though free energy perturbation methods are being used to try to calculate K_3 and K_4, no general method of measuring the free energy cost of a substitution in either state is available. As a result, it is impossible to tell whether a change in stability due to a mutation is caused by effects on the folded state, on the unfolded state, or on a combination of the two.

Four approaches—measurement of disulfide stability, a kinetic test, an empirical test, and a thermodynamic test—have been put forward to overcome this problem. While none of these methods is completely satisfactory, the first three approaches suggest that many amino acid substitutions that cause large changes in stability have the largest relative effects on the folded state. This result is encouraging for attempts to design new proteins based on modeling reasonable folded conformations. Nonetheless, partitioning the free energy of mutation remains a primary limitation to interpretation of genetic experiments in structural terms. The four approaches are summarized here before going on to discuss the insights gained from studying protein variants.

For the specific case of the stability of disulfide bonds, the effects on the unfolded and folded states can be determined relative to an added disulfide exchange reagent (1, 3–5). The thiol reagent may serve as a reference to relate thermodynamic cycles for disulfide formation in two proteins that differ by a single amino acid (72).

$$RSSR + U\overset{SH}{\underset{SH}{\diagup}} \quad \overset{K_1}{\rightleftharpoons} \quad F\overset{SH}{\underset{SH}{\diagup}} + RSSR$$

$$K_3 \Big\Updownarrow \qquad\qquad \Big\Updownarrow K_4$$

$$2\ RSH + U\overset{S}{\underset{S}{\diagup|}} \quad \overset{K_2}{\rightleftharpoons} \quad F\overset{S}{\underset{S}{\diagup|}} + 2\ RSH$$

The equilibrium constants K_1 and K_2 describe the stability of the reduced and oxidized proteins, and K_3 and K_4 reflect the stability of the disulfide bond in the unfolded and folded states. Linkage between these equilibria requires that $K_1K_4 = K_2K_3$.

Measured effective concentrations for disulfide bonds in unfolded proteins range from 10^{-3} to 10^{-2} M, and the effective concentrations of the three disulfides in the folded conformation of bovine pancreatic trypsin inhibitor (BPTI) range between 10^2 M and 5×10^5 M (1).

Several amino acid substitutions in BPTI cause progressively larger changes in disulfide stability as folding proceeds (72a). Only two- to three-fold differences in disulfide stability were observed in the unfolded polypeptides, but mutations decreased the stability of disulfides in the folded state by factors of up to 10^4.

A second approach to assigning the effects of amino acid substitutions relies on the analysis of kinetic data. Using concepts from transition state theory, C. R. Matthews and coworkers (73) proposed that mutations that selectively alter the rate of folding shift the free energy of the unfolded state, mutations that alter only the rate of unfolding shift the free energy of the folded state, and mutations that alter both the rates of unfolding and folding shift the free energy of the transition state. This formalism describes the simplest interpretation of changes in the kinetics of reversible denaturation, but other effects are not ruled out. Mutations that selectively alter folding rates, for example, could reasonably affect the energies of both the folded state and the rate-limiting transition state. Nonetheless, the parsimonious interpretation of kinetic data has led to testable structural models for the folding process (73–76).

This kinetic test was applied to the denaturation of 28 variants of phage T4 lysozyme, and all 25 destabilizing substitutions examined speed the rate of unfolding (J. Klemm, T. Alber, D. P. Goldenberg, unpublished). The temperature-sensitive mutations tested destabilized the protein with respect to both the unfolded state and the transition state for denaturation. None of the substitutions exclusively stabilize the unfolded state. The increase in unfolding rate was qualitatively correlated with the reduction in thermodynamic stability. These results support the idea that destabilizing substitutions often disrupt important interactions in the folded state, but the possibility of changes in the unfolded state is not ruled out.

An empirical test for analyzing the effects of mutations has been used by several laboratories. This relies on measuring the stabilities of a series of mutant proteins, each containing a different amino acid at a given position. Correlations between stability and structure identify critical interactions (7, 69). A correspondence between transfer free energy and protein stability, for example, was used to support the importance of the hydrophobic effect at specific sites in the α-subunit of trp synthase (77), kanamycin nucleotidyl transferase (78), and phage T4 lysozyme (79). The importance of a particular hydrogen bond in the folded state of phage T4 lysozyme was inferred from the presence of this interaction in the X-ray crystal structures of the most stable of 14 variants with a different amino acid at position 157 (69). The overall correlation between the severity of destabilizing substitutions and the crystallographically determined mobility and solvent accessibility of the altered site has also emphasized the importance of mutational effects on the folded state (6).

A thermodynamic test for assigning selective effects of an amino acid substitution to the unfolded state has been suggested by Shortle & Meeker (70). They proposed that a change in the average structure of the unfolded state can be inferred from a change in the dependence of the logarithm of the equilibrium constant for folding on denaturant concentration. This idea is based on an interpretation of the thermodynamics of solvent denaturation in terms of the amount of exposed surface area available for binding in the unfolded state. This physical model, however, does not reflect all the interactions of denaturants with proteins and water (57, 80). As a result, the thermodynamic test does not rule out alternative interpretations. An increase in the slope of ln K versus denaturant concentration could reflect increased association of a component of the solvent with the native state. For example, specific binding of denaturants to several proteins has been observed crystallographically (81–83). A decrease in slope could reflect a higher population of intermediates in the transition region (70, 71).

Shortle and coworkers have described stabilizing and destabilizing substitutions in staphylococcal nuclease that change the dependence of the equilibrium constant on the denaturant concentration (70, 71). These data were interpreted to indicate the possibility of large effects on the structure and free energy of the unfolded state. Staph nuclease could present special complications, however, because multiple folded and unfolded states are populated (29). Mutationally induced shifts in the relative conformational distributions could increase or decrease the slope of the denaturation curve. Evidence has also been presented for equilibrium intermediates in the folding of some nuclease variants (70). These difficulties notwithstanding, the analysis of Shortle and coworkers has emphasized the critical importance of understanding more about changes in the unfolded state due to mutations. There is little information, for example, on the extent to which the energy of the unfolded state is influenced by changes in amino acid composition and shifts in the distribution of polypeptide conformations.

Despite the difficulty of partitioning the effects of mutations on the folded and unfolded states, the genetic approach to the protein stability problem is proving to be a rich source of new information. Recent progress in this area is highlighted in the following section.

GENETIC STUDIES OF PROTEIN STABILITY

Classical genetic studies of protein function have suggested that proteins are very tolerant of amino acid substitutions. In the *Escherichia coli lac* repressor, for example, a wild-type phenotype was produced by 58% of 323 substitutions generated by suppression of nonsense mutations at 90 sites (84). Repressor activity was drastically reduced by 15% of the substitutions, and only 11% of the sequence changes resulted in a temperature-sensitive (ts) phenotype. Analysis of almost 2000 missense mutants of the *lac* repressor showed that ts and inactivating lesions are clustered (85, 86). The protein apparently adapts to changes in large regions of the sequence.

Analyses of protein families emphasize similar conclusions. The globins, for example, contain few absolutely conserved residues (87, 88). In several families, the most variable residues are on the protein surface (88, 89).

Recent studies of mutant proteins have revealed three reasons for this pattern of variation. (*a*) Some amino acid substitutions preserve critical interactions. (*b*) Only a fraction of the residues in a protein make large contributions to stability (6, 90, 91). This limits the size of the target for destabilizing mutations. (*c*) Structural adjustments can mitigate the intrinsic effects of amino acid substitutions, even at critical sites (69, 92).

Randomly induced mutations that alter protein stability provide a relatively unbiased experimental identification of amino acids that make essential contributions. A large collection of mutant hemoglobins has been obtained by random screening of the human population (93). Collections of mutations have also been obtained by screening for temperature sensitivity of phage T4 lysozyme (6, 94) and for reduced activity of staphylococcal nuclease (94), cytochrome c (96), and the cI and cro repressors of phage λ (90, 91). Several methods for isolating thermostable variants have also been reported. These include screening for resistance to heat inactivation (97, 98), isolating pseudorevertants of destabilizing lesions (95, 99, 100), and selecting for gene function at elevated temperatures in a thermophilic bacterium (101–103). A new conditional phenotype, hypersensitivity to D_2O, promises to provide information about hydrogen bonds and about the roles of water in protein stabilization (104).

Saturation mutagenesis of a residue or a structural region has recently been used to probe the stabilizing contributions of specific amino acids (105, 106). After randomizing the target codon(s), functional variants are selected and sequenced. This approach has been used to study the dimer interface in λ repressor (106), the role of a loop in staph nuclease (R. Fox, personal communication), and the variability of residues in the arc and mnt proteins of phage P22 (R. T. Sauer, personal communication). In general, the number of substitutions allowed at each position and the most hydrophilic residue allowed at each position are often proportional to the solvent accessibility of the wide-type side chain in the folded structure (106).

Although the characterizations of stability differ from system to system, a number of general conclusions have been drawn. The most obvious finding is that amino acid substitutions that alter stability are chemically varied. Changes in charge, size, polarity, hydrophobicity, or hydrogen bonding capacity are found. This indicates that many different types of noncovalent interactions—including hydrogen bonds, van der Waals contacts, hydrophobic contacts, and ionic interactions—can make quantitatively comparable contributions to stability. No single type of interaction dominates; all the forces proposed to play a role can be important at specific sites.

It is also apparent that the effect of a substitution depends on the nature of the change and on its structural context. Analysis of the sites of temperature-sensitive mutations in λ repressor (90), λ cro (91), and phage T4 lysozyme (6) in terms of the X-ray crystal structures of the wild-type proteins showed that critical amino acids are generally relatively rigid (6) and inaccessible to solvent (6, 90, 91) in the folded protein. Substitutions at mobile and exposed sites usually have little effect on stability. The mutations in the cro protein highlight several interesting exceptions to this pattern. Destabilizing mutations can alter residues with relatively high mobility if the sites are charged or are undergoing segmental, rather than local, motion. Substitutions that would be expected to cause propagating shifts in the folded structure, such as insertion of prolines at inappropriate locations or replacement of conformationally special glycines, can also be destabilizing.

These studies suggest that only some of the interactions inferred from the X-ray crystal structure of a protein make large contributions to stability. Critical amino acids are generally rigid and buried in the folded protein. This pattern provides simple rules for using crystallographic data to predict the effects of amino acid substitutions on protein stability (6) and accounts for much of the silent variation observed in protein families and in genetic studies of protein function.

The correlations between low mobility, low solvent accessibility, and high sensitivity to destabilizing lesions also bear on current theories of protein stability. Side chains with a large amount of buried nonpolar surface area are sensitive to destabilizing substitutions, emphasizing the importance of van der Waals contacts and the hydrophobic effect. Even small buried residues, though, form part of the target for ts mutations. This is hard to rationalize on the basis of a constant linear relationship between surface area and hydrophobic stabilization, because small side chains have limited surface area. Instead, it supports the view that groups that become surrounded by protein atoms during folding can form specific stabilizing interactions. Uncharged substituents that are equally exposed to solvent in the folded and unfolded states may contribute less to stability, because their average environments in the two states are similar.

On the other hand, sites that are more exposed to solvent in the folded protein than in the average unfolded conformation may make substantial contributions to stability. Polar groups at maximally exposed sites would favor the folded structure; nonpolar side chains may stabilize the unfolded state. Consequently, the few amino acids that are more exposed in the folded structure than in an extended model may prove to be sensitive to both stabilizing and destabilizing substitutions.

The correlation between rigidity and sensitivity to destabilizing substitutions was predicted by Creighton's hypothesis that the relative effective concentrations of groups in the folded and unfolded states determines their contribution to stability (1). According to this view, rigid amino acids can make the largest contributions, because they can be constrained to interact productively. Their entropy is reduced by the folded structure in toto, so a greater fraction of the enthalpy of interaction is expressed in the free energy of stabilization. More flexible substituents may interact productively for a smaller fraction of time and lose more entropy in forming each stable contact.

A final intriguing conclusion drawn from the structural patterns of destabilizing substitutions is that severe mutations that primarily affect the free energy of the denatured state may be rare. In principle, mutations that affect the unfolded state could occur in any part of the sequence. Their positions should not correlate with features of the folded structure. Since a tight correlation is observed between the positions of severe destabilizing mutations and properties of the folded structure, large changes in the thermodynamic stability may generally involve significant effects on the folded protein. The rigid and buried residues in the folded state may comprise the part of the system that is least able to structurally relax to compensate for amino acid substitutions. The conformational diversity of the unfolded state may allow it to largely compensate for changes in sequence. As a result, the free energy of the unfolded state may be less sensitive to amino acid substitutions.

This argument notwithstanding, the possible magnitude of the effects of substitutions on the free energy of the unfolded state remains controversial. Shortle and coworkers (70, 71) have suggested that changes in interactions in the unfolded state can fully account for even large changes in stability. The kinetic formalism adopted by C. R. Matthews and coworkers (73, 75) also suggests that some mutations may differentially alter the free energy of unfolded forms.

COOPERATIVITY

A surprising finding from studies of thermal denaturation is that multiple amino acid substitutions generally have additive effects on stability (58, 95, 107; L. McIntosh, W. Becktel, W. Baase, D. Muchmore, and F. W. Dahlquist, in preparation). Nonadditive effects can occur if one of the substitutions alters ΔCp (108). Solvent denaturation studies carried out near room temperature also generally reflect the additivity of the effects of substitutions (70), although nonadditive effects have also been reported (109). The additivity of changes in free energy of stabilization is unexpected because of the cooperativity of protein folding. The theoretical treatment of Creighton discussed above (1, 2) also implies that substitutions could synergistically alter the contributions of neighboring interactions and lead to nonadditive effects even in the absence of structural shifts. The apparent additivity of amino acid substitutions implies that cooperative interactions are highly localized. This supports the simplifying assumption that protein stability may be understood as a sum of local interactions. Practically speaking, this means that multiple substitutions might be engineered to set the stability of a protein at any desired value. On the other hand, changes in the magnitudes of stabilizing contributions due to changes in cooperativity will be hard to measure using genetic approaches because substitutions usually cause structural shifts.

STRUCTURAL STUDIES OF PROTEIN VARIANTS

The growing number of structural studies of protein variants using X-ray crystallography and NMR emphasize that the native structure (and the unfolded state as well) can adjust to accommodate changes in sequence. The effects of a given substitution depend on its intrinsic impact on the free energies of the folded and unfolded states and on the relative abilities of the two states to structurally compensate for the new amino acid (92). Flexibility provides a mechanism of compensation. If proteins were completely rigid, changes in sequence would have larger effects on thermodynamic stability. Hypersensitivity to mutation would reduce the genetic variability important for evolution.

A range of conformational adjustments allow proteins to tolerate substitutions. Replacements of surface residues often result in small localized shifts. Positional changes of less than 1 Å in nearby side-chain atoms and 0.5 Å in main-chain atoms are common. Changes in the distribution of bound solvent and counterions have also been observed (31, 66, 69, 79, 110, 111). Even surface substitutions, however, can cause propagating structural adjustments. NMR data show that replacement of Pro 117 in staphylococcal nuclease changes the equilibrium between similar alternative conformations in both the folded and unfolded states (29). Positional adjustments propagate extensively through the folded conformation. In phage T4 lysozyme, replacement of Gly 156 by Asp (112) or of Pro 86 by any of seven amino acids (92) cause crystallographically observable shifts up to 20 Å apart on the protein surface. Surface substitutions generally do not cause structural changes to propagate into the protein interior.

Substitutions of interior residues sometimes cause only localized adjustments (64), but propagating shifts are more common. In phage T4 lysozyme, replacement of Ala 98 by Val in an interhelical contact causes structural changes in a 20 Å × 20 Å × 10 Å slice through the protein (T. Alber, B. W. Matthews, unpublished). The high-resolution X-ray crystal structure of the mutant protein shows extensive subtle repacking of the enzyme's hydrophobic core. This change is associated with a reduction in stability of ~4 kcal/mole at pH 2.0 and 42°C, but the activity of the protein is preserved. Replacements of Ala 146 by Thr, Val, Cys, Gly, and Ile also cause shifts to propagate over 10 Å into the interior, where a cavity in the wild-type phage lysozyme becomes filled (64; K. Wilson, D. Maki, T. Alber, D. Tronrud, M. Karpusas, J. Mendel-Hartvig, L. McIntosh, B. W. Matthews, unpublished results). These substitutions reduce stability even though many of the mutant proteins bury additional hydrophobic surface area in the folded state. Proteins with interior substitutions have proven to be the most difficult to crystallize and the most likely to produce new crystal forms (10). This suggests that structural changes can propagate to surface regions that make intermolecular crystal contacts.

Several other generalizations are emerging from structural studies of protein variants. (a) Positional shifts are channeled in specific directions and tend to be damped with distance from the site of the substitution. (b) Mutationally induced cavities are often filled by shifts in the protein or the solvent. (c) Even chemically and structurally simple substitutions can simultaneously alter several different kinds of interactions. As a result, careful comparisons of multiple substitutions may be required to evaluate the contributions of specific noncovalent bonds. (*d*) New side chains generally adopt commonly observed rotamer conformations (30, 68, 69, 92), indicating an absence of torsional strain. Side chains in rare unstrained rotamer conformations can also be stabilizing (69). (*e*) Examples of dramatic changes in solvent binding on the surface of mutant proteins have been reported (66, 69, 111). Binding of specific water molecules can be stabilizing. (*f*) Unexpected features are often found, and proteins can respond differently to seemingly conservative replacements. For example, functionally important differences in the position and mobility of leucine and isoleucine side chains at specific sites in hemoglobin and phage lysozyme have been reported (67, 69). At position 157 in phage lysozyme, Glu, Asp, and Asn are found in different average conformations, and Arg forms a new ion pair with a neighboring residue. These findings emphasize the importance of structural studies for understanding the properties of mutant proteins. Predicting structural shifts in protein variants is not yet a simple proposition (113–115).

FORCES THAT STABILIZE PROTEINS

As noted at the outset, it has been traditional to estimate the magnitudes of the interactions that stabilize proteins by considering simpler model systems. A theme of recent studies has been to use directed mutagenesis and chemical modification to determine if data from model systems quantitatively describe the contributions of specific interactions. While such studies are in their infancy, early indications are that proteins harbor complexities that are not accounted for in traditional model systems. Data on several classes of stabilizing interactions are summarized in this section.

Effects on Conformational Entropy

Covalent crosslinks are thought to stabilize proteins by reducing the conformational entropy of the unfolded chain. Statistical treatments have suggested that the destabilization of the unfolded state depends on the length of the loop formed by a single crosslink. Relating the observed effects of crosslinks to predictions based on calculations of loop entropy in the unfolded state is not straightforward. Most statistical analyses assume a value $(10–10^3$ M) for the effective concentration of the crosslink in the folded state and also assume that no entropy is lost on forming the crosslink in the folded state (5). Recent studies of proteins containing natural or engineered crosslinks indicate that effects on the folded state can also be crucial.

Chemical methods were used to join Gly 35 and Trp 108 in hen lysozyme (116), Lys 7 and Lys 41 in ribonuclease A (117), and the N- and C-termini of BPTI (72). Crystallographic and NMR studies indicated that these crosslinks do not cause large changes in the folded conformations of ribonuclease (118) and BPTI (119). The crosslinks in hen lysozyme and ribonuclease increased the melting temperature by 29°C and 25°C, respectively, at pH 2.0. Analysis of thermodynamic data suggested that the Lys 7-Lys 41 crosslink stabilizes ribonuclease by ~5 kcal/mole at 53°C. This effect is almost entirely entropic, consistent with a destabilization of the unfolded state (117). Kinetic data showing that the crosslink specifically reduces the rate of unfolding (120) led Matthews & Hurle, however, to suggest that the largest differential effect may be on the folded state (75).

Studies of crosslinks in BPTI emphasize the importance of effects on the folded state. A peptide bond joining the natural chain termini provided as much stabilization as the ion pair between the termini in the uncrosslinked protein (5, 72). This suggested that the crosslink destabilized both the folded and unfolded states equally. Strain and loss of entropy in the folded conformation were proposed as potential destabilizing effects.

Creighton has analyzed the stabilities of the three natural disulfides in BPTI relative to an added disulfide exchange reagent (1). The disulfides have similar stabilities in the unfolded state, with effective concentrations of about 10^{-2} M. In the folded state, however, the measured effective concentrations ranged from 2.3×10^2 M to 4.6×10^5 M. The effective concentration of each disulfide bond in the folded state correlated qualitatively with the change in the melting temperature of the protein caused by reduction of the disulfide.

Overall, these data suggest that the stabilizing contribution of a disulfide bond (or other crosslink) is determined by the size of the loop formed in the chain and by the compatibility of the crosslink with the folded structure. For a given loop size, the most effective crosslink may be formed between groups that are rigidly held in an optimum orientation by the folded structure (1, 5). Crosslinks that disrupt interactions or reduce the entropy of the folded state by reducing its flexibility may be proportionately less stabilizing.

Attempts to stabilize proteins with engineered disulfide bonds have produced mixed results. Sites for engineered disulfides have generally been selected to satisfy conformational restrictions observed in naturally occurring proteins (14). In the absence of shifts in the folded structure, very few pairs of α-carbons in any given protein can be joined by a disulfide bond with standard geometry (121). Single added disulfides in subtilisin BPN' and dihydrofolate reductase (DHFR) produce complex effects. Observation of an increase in stability depends on solution conditions (122, 123) or on the method of extrapolation to zero denaturant concentration (68). Crystallographic studies show that the engineered S-S bridge between residues 39 and 85 in DHFR reduces van der Waals contacts with nearby residues. The mobility of residues 80 to 90 also increases, perhaps reflecting destabilization of the native state (68).

An intermolecular disulfide bond joining residue 88 in adjacent subunits of λ repressor increases the T_m of the N-terminal domain by approximately 10°C (124). Formation of an engineered disulfide between residues 3 and 97 of phage T4 lysozyme increases T_m by 7°C at pH 2 (108, 125). The free energy of stabilization increases about 1.8 kcal/mole, which is less than expected from loop formation. In most cases reported to date, introduction of the cys residue(s) required to form a disulfide was destabilizing (68, 108). This reduces the net stabilization afforded by the S-S bridge.

Residues that do not introduce crosslinks may also influence the conformational entropy of the unfolded state. Simple estimates of the degrees of freedom of the polypeptide backbone suggest that removal of glycine (126) or introduction of proline (127) could destabilize the unfolded state by 0.8 and 1.4 kcal/mole, respectively, at 65°C. Substitutions that introduce β-branched or bulkier side chains may also restrict main-chain rotations in the unfolded state (126). Increases in stability of < 1 kcal/mole are caused by Gly to X and X to Pro substitutions that are compatible with the folded structures of phage lysozyme (127), phage λ repressor (100, 128), and the neutral protease of *Bacillus stearothermophilus* (129). In λ repressor and neutral protease, increases in stability were attributed to enhanced "helical propensity" at helical sites. Entropic effects, however, may contribute to the observed tendency of glycine to destabilize model helixes. In addition to possible effects on the unfolded state, these substitutions could contribute to stability by decreasing the flexibility of neighboring groups in the folded structure (6, 69).

The Hydrophobic Effect

One of the most important contributions to protein stability is made by the hydrophobic effect. Current estimates of amino acid hydrophobicity are based on the measured free energies of transferring side chains from water to organic solvents (130–134), to SDS micelles (135), and to the vapor phase (136). On the basis of these data, scalar multipliers of 22–25 cal per Å² have been proposed to estimate the free energy of stabilization from buried hydrophobic surface area (16). More sophisticated approaches based on summing the contributions of individual atoms have also been developed (133, 137). The free energy of transferring side chains from water to ethanol is correlated with their average extent of burial in known protein structures, supporting the applicability of the solvent transfer model to the problem of protein stability (138–140).

Baldwin (141) has analyzed the temperature dependence of the transfer of six liquid hydrocarbons to water and used the results to interpret thermodynamic data on the unfolding of hen lysozyme. For the unfolding of hen lysozyme, the temperature dependence of ΔH_T and ΔS_T could be largely accounted for by the hydrophobic effect. The fraction of ΔH_T and ΔS_T due to other contributions was independent of temperature. The temperature-independent part of ΔH_T was large and favored folding, presumably due to noncovalent interactions in the folded state. The temperature-independent part of ΔS_T was large and favored unfolding, presumably due to conformational entropy. From 10 to 100°C, the free energy of stabilization due to the hydrophobic effect increases with temperature, and it reaches a maximum at ~112°C if ΔCp is constant. This analysis suggested that cold denaturation is caused by the weakening of the hydrophobic effect. Unfolding occurs at high temperatures, because the destabilizing contribution of the conformational entropy of the unfolded state ($T\Delta S_{conf}$) increases more rapidly with temperature than the stabilizing contribution of the hydrophobic effect. Baldwin noted that this analysis was limited by assumptions made to determine ΔCp and by the possibility that the hydrocarbon transfer experiments may not be directly applicable to protein folding.

Privalov & Gill have undertaken a similar analysis also based on calorimetric studies of protein denaturation and on the temperature dependence of the solubility of nonpolar substances in water (55). The temperature dependence of ΔCp was described and was shown to produce quantitative but not qualitative differences in the thermodynamics of denaturation. The physical interactions proposed to account for denaturation, however, were different from Baldwin's model. Privalov & Gill suggested that, contrary to popular opinion, water solvation of hydrophobic groups is favorable at all accessible temperatures and becomes increasingly favorable as the temperature decreases. At sufficiently low temperatures, interactions between water and nonpolar side chains become stronger than van der Waals interactions within the protein, leading to cold denaturation. This model emphasizes interactions between water and protein atoms rather than interactions between water molecules.

Hydrophobicity is generally estimated using solvent transfer models, but the applicability of transfer experiments to protein stability has been questioned on theoretical and experimental grounds. (a) Measured hydrophobicities depend on the nature of the nonaqueous phase (134, 136), and organic solvents, detergent micelles, and gases are poor models of the protein interior. These models differ dramatically from proteins in packing density (16, 142), polarity (143), absolute size (144), water content (134), and in the energy of cavity formation. (b) The loss of entropy associated with fixing the position of a side chain during folding is likely to be significantly larger than the entropy change of transferring a side chain from water to another bulk phase (145). (c) Cooperativity may play a less significant role in solvent transfer (2, 145). (d) The transfer data may not account for the effects of pressure on protein stability (136–148). These criticisms suggest that the relationship between hydrocarbon transfer free energy in model systems and hydrophobic stabilization of proteins is undetermined. Because of differences in the energy of cavity formation (144), hydrophobicity may even vary from site to site and from protein to protein.

This problem has begun to be addressed in recent studies of multiple amino acid substitutions in the α-subunit of trp synthase (77, 149), kanamycin nucleotidyl transferase (78), and phage T4 lysozyme (79). Yutani et al (77) constructed 20 variants of the α-subunit of trp synthase, each with a different amino acid at position 49. The kinetics and thermodynamics of unfolding by guanidine hydrochloride were studied. With the exception of the bulky aromatic residues Trp, Tyr, Phe, and His and the negatively charged residues Asp and Glu, the stability of the protein was proportional to the free energy of transfer of side chains from water to ethanol. This result supports the qualitative validity of the transfer model.

Surprisingly, the constant of proportionality between protein stabilization and transfer free energy depended on the concentration of denaturant and on the pH. Based on extrapolations to zero denaturant concentration, the protein was apparently more hydrophobic than the ethanol model by factors of 2 at pH 5.5, 3.7 at pH 7, and 1.3 at pH 9. Yutani and coworkers (77) proposed that this variation may be due to differences in electrostatic interactions arising from differences in the local dielectric constant. The difficulty of extrapolating to zero denaturant concentration and the different degrees of extrapolation at each pH could also influence the apparent hydrophobicities.

For eight replacements of Asp 80 in kanamycin nucleotidyl transferase, Matsumura and coworkers (78) discovered a qualitatively similar relationship between transfer free energy and resistance to irreversible inactivation at 58°C. Although the structure of KNTase is not known, the observed correlation was taken as evidence that residue 80 is at least partially buried and that the solvent transfer model qualitatively applies to the hydrophobic stabilization of proteins.

A direct proportionality between protein stabilization and transfer free energy was observed for 11 substitutions of Ile 3 in phage T4 lysozyme (79). The constant of proportionality was independent of pH as expected and varied between 0.65 and 0.94, depending on the organic phase used in the transfer measurements. Side chains no larger than the wild-type Ile provided approximately 20 cal of stabilization for each $\mathring{A}^2$ buried in the folded structure. The large side chains of Phe, Tyr, and Trp could not be accommodated in the protein interior and, especially in the case of Trp, caused a large reduction in stability.

Additional studies of this nature will be necessary to identify the structural features that determine how hydrophobicity and protein stabilization are correlated. Considerably more effort will be required to understand how hydrophobic interactions contribute to ΔCp.

Hydrogen Bonds

One of the major contributions to the largely temperature-independent part of the enthalpy of stabilization is thought to come from hydrogen bonds. Hydrogen bonding groups are part of the target for destabilizing amino acid substitutions (6, 90, 91, 93). A survey of highly refined protein structures indicates that hydrogen bonding geometry is quite variable, and unpaired H-bond donors and acceptors are rare (150).

Model compounds that form detectable intermolecular hydrogen bonds in water have been used to estimate the contribution of hydrogen bonding to protein stability. Because of competition with water, model compounds generally associate weakly in aqueous solution (151–156). Different model systems have provided different estimates of the enthalpy of hydrogen bond formation. Studies of the dimerization of urea and δ-valerolactam in water led Schellman (151, 152) and Susi and coworkers (154) to conclude that the heat of formation of hydrogen bonds may be −1.5 kcal/mole or more. Klotz & Franzen (153), using aqueous N-methylacetamide as a model for the peptide bond, estimated that ΔH for hydrogen bond formation is close to zero.

The latter result has supported the popular view that, as long as equal numbers of hydrogen bonds are formed in the folded and unfolded states, hydrogen bonds do not contribute to the difference between the free energies of the states. Hydrogen bonds can help establish the specificity of folding, since unpaired donors or acceptors sequestered from solvent destabilize the folded conformation by reducing the relative number of hydrogen bonds. A central feature of this analysis is that hydrogen bonds in the folded and unfolded states are considered to be energetically equivalent. The concentration of donors and acceptors in water, ~110 M, is assumed to be the effective concentration of all hydrogen-bonding groups in both the folded and unfolded states.

Several authors have pointed out, however, that differences in the average geometry, the entropy of formation, and the average number of interacting partners can lead to significant energetic differences between hydrogen bonds (and other noncovalent interactions) in the folded and unfolded states (2, 3, 5, 155, 157). Groups with effective concentrations higher than 110 M in the folded state will be stabilizing; those with lower effective concentrations will be destabilizing. Obviously, if hydrogen bonds in a folded protein make a wide range of contributions to stability, data obtained from model systems cannot provide reliable estimates of the contributions of specific interactions.

Studies of protein-ligand complexes and of nucleic acids suggest that hydrogen bonds between uncharged donors and acceptors can contribute 0.5 to 1.8 kcal/mole to the observed binding energy (158–160). Hydrogen bonds between charged groups can contribute up to 6 kcal/mold. Unpaired donors or acceptors in protein-ligand complexes are destabilizing, reflecting a reduction in the relative number of hydrogen bonds in the bound state.

Evidence for intramolecular hydrogen bonds in the unfolded state has been obtained in NMR studies of short peptides. Based on the pH dependence of amide proton chemical shifts, Bundi & Wüthrich (43) concluded that ionized carboxylates in short peptides can hydrogen bond to nearby amides. Dyson and coworkers (44) found that a nonapeptide fragment of influenza virus hemagluttinin can form a hydrogen bonded β-turn in water. Surprisingly, this sequence is part of a strand of β-sheet in the folded protein, suggesting that the conformational preference of the fragment must be overcome during folding.

Evidence for a wide range of stabilizing contributions has been found by adding or removing specific hydrogen-bonding groups. In phage T4 lysozyme, for example, tertiary hydrogen bonds at position 157 are critical for stability, but a hydrogen bond involving residue 86 makes no contribution (31, 69, 92). The substitution of Thr 157 by Ile reduces the stability of phage lysozyme by about 3 kcal/mole at 42°C and pH 2.0 (31, 161). Twelve additional substitutions were engineered at this position to systematically evaluate the contributions of each atom in the Thr and Ile side chains (69). Crystallographic and thermodynamic studies showed that the five most stable lysozymes contained a hydrogen bond to the buried main-chain amide of Asp 159. This correlation between structure and stability suggested that the hydrogen bond in the folded state dominates the effects of all 13 amino acid substitutions on the free energies of both the folded and unfolded states.

A striking finding from this study is that bound water can stabilize proteins. A water molecule is localized by three hydrogen bonds in the crevice formed by Gly 157, and this mutant protein is almost as stable as the Ser 157 variant. Though the hydrogen bonds to residue 157 made the largest contribution to stability, loss of the Thr γ-methyl group (Ser 157) reduced stability by ~0.7 kcal/mole. The methyl group could stabilize the protein directly, through van der Waals and hydrophobic contacts, or indirectly, by increasing the probability of forming the hydrogen bonds to the Thr 157 hydroxyl group (69). Compensating structural changes made it impossible to determine the stabilizing contributions of specific hydrogen bonds.

In contrast, additional hydrogen bonds formed at position 86 in phage lysozyme are not associated with added stability (92). Of 10 substitutions for Pro 86, Ser and Cys form a new crystallographically inferred hydrogen bond with the side chain of Gln 122. This interaction, however, does not stabilize the protein; all 10 variants have essentially the same thermodynamic stability. Residues at position 86 have higher-than-average mobility and surface accessibility. As a result, the intramolecular hydrogen bond may be equivalent to the local interactions with water. The high mobility of the side chains at this site may reflect a significant loss of entropy in the formation of the intramolecular interaction.

Further studies will undoubtedly clarify the structural features that determine the contributions of specific hydrogen bonds to protein stability. Obvious parameters to consider include surface accessibility, mobility, average geometry, and number of partners in the folded state. In addition, the effects of competing structures in the unfolded state have not been quantified. Such information will help distinguish hydrogen bonds that are essential for protein stability from those that are neutral or destabilizing.

Dispersion Forces

The importance of dispersion or van der Waals forces for protein stability hinges on differences in packing in the folded and unfolded states. Unfolding usually involves very small changes in the volume of the system, so the average interatomic distance remains constant. This has been interpreted to imply that van der Waals forces do not change appreciably between the two states. According to this view, cavities in proteins are destabilizing because they reduce the relative contact surface of the folded state. Steric overlap is avoided in both states. This implies that van der Waals forces can contribute to the specificity of folding but not to the excess free energy of stabilization.

The overall volume of the system, however, is not an adequate measure of the sum of the interactions. During folding, water is transferred to the relatively open bulk phase, and the atoms that form the protein interior become as densely packed as molecules in crystals of small organic molecules (16, 142, 143). This changes the neighbors of solvent and protein atoms as well as the distribution of interatomic distances. Both of these effects can alter van der Waals interactions. Privalov & Gill have proposed that these changes can account for much of the temperature dependence of protein stability (55). A likely source of an increase in van der Waals forces is the interior of the folded protein, since this is the most densely packed phase of the system. Bello (162, 163) has argued from the similarity in densities that the average van der Waals contribution in the folded state can be estimated from the heat of fusion of small organic solids (~30 cal/g). This value, however, may not be appropriate because it does not take into account the changes in packing in the solvent, the greater flexibility of proteins, or the irregularities of packing of individual side chains with usually unlike neighbors (148, 164).

Analysis of the packing of proteins provides circumstantial evidence for the importance of dispersion forces (16). Contact surfaces of secondary structures are generally nonpolar (143), and cavities that are large enough are filled by water molecules or nonpolar solutes. Short nonbonded contacts are generally absent.

Amino acid substitutions that alter interior packing can reduce protein stability, but quantitative correlations have not been established. A reduction in tertiary contact surface is associated with the lack of stabilization of an engineered disulfide bond in DHFR (68). In phage T4 lysozyme, van der Waals and hydrophobic contacts of the γ-methyl group of Thr 157 may contribute up to 0.7 kcal/mole to stability at pH 2 and 42°C (69). Several destabilizing substitutions that alter interior packing of phage lysozyme have been analyzed crystallographically. The replacement of Ala 98 by Val in a helix-helix contact causes extensive subtle rearrangements in packing and destabilizes the protein by almost 4 kcal/mole at pH 2 and 42°C (T. Alber, S. Cook, B. W. Matthews, unpublished results). At residue 146, Gly, Val, Thr, Ile, and Phe are equally poor substitutes for the wild-type alanine. Stability is not correlated with side-chain volume at this largely buried site. A cavity in the interior of the protein is eliminated by the substitution of Ala 146 by Thr, and the same cavity is expanded by the substitution of Met 102 by Thr (64). Both amino acid substitutions destabilize the protein. The energetic costs of substitutions that decrease the volume of interior amino acids have also been explored using the enzyme barnase (165). Replacement of Ile 96 by Val and Ala decreased the extrapolated stability of the protein by 1.1 and 4.0 kcal/mole, respectively. In the absence of X-ray structural data, however, it is impossible to tell if these reductions are associated with the formation of interior cavities or are complicated by structural relaxations that fill cavities in the mutant proteins.

An algorithm for identifying combinations of amino acids that could pack similarly has been developed by Ponder & Richards (30). This list of sequences defines the "tertiary template" for a given region of main-chain conformation. The approach is based on the requirement that side chains fill space without steric overlap. Possible side-chain conformations are taken from a rotamer library derived from 19 highly refined protein crystal structures. Surprisingly, the conformations of 17 of the 20 amino acids can be represented by only 67 rotamers. The dihedral angles in the rotamer library cluster around minima of torsional energy, indicating that side chains generally adopt unstrained conformations (166).

Functional variants that have suffered larger changes in side chain volume than are normally allowed in the definition of a tertiary template have been described for several proteins. Nonetheless, determining the stabilization afforded by different members of a tertiary template provides a systematic approach to quantifying the effect of packing on stability. Modern genetic methods also allow proteins with all possible combinations of amino acids in a given tertiary cluster to be compared.

Electrostatic Interactions

The associations of acetate ions with guanidium or n-butylammonium cations in water have been used to model the contributions of ion pairs to protein stability (167, 168). Interactions between these ions are weak, fueling early skepticism that electrostatic interactions could be stabilizing. Like other simple model systems, however, these salt solutions do not account for unique changes in the solvation of specific groups during folding or for the differences in the entropy of formation of specific interactions. Folding alters the effective dielectric constant around each ion and creates a boundary between the high dielectric solvent and the lower dielectric of the protein interior. These effects can increase the interactions of ions and dipoles.

Ionizable groups are not distributed randomly over protein surfaces, reflecting structural and functional roles. Including the peptide dipoles, charges are, on average, surrounded by charges of the opposite sign (175-177). Examples of highly asymmetric charge distributions are found, and the resulting macrodipoles often point in the right direction to facilitate the binding of charged ligands (178). On average, only a third of the charged residues in proteins are involved in ion pairs, and 76% of these are between residues in different elements of secondary structure (177). Surprisingly, 17% of the ion pairs are buried, and these generally play clearly identifiable functional roles. Overall, however, ion pairs are poorly conserved in protein families, suggesting that at least some of them are not critical for folding and stability. Most unexpected was the finding that 20% of the buried charged groups do not form ion pairs and are solvated instead by hydrogen bonds (179). This highlights the importance of the self energy of ionic groups (180–182).

Data on protein variants suggests that ionic interactions can be important for folding stability. Even though charged residues are often mobile and exposed to solvent, they are sensitive to destabilizing mutations (90, 92, 169). Additional ion pairs distinguish several thermophilic proteins from their mesophilic counterparts (169, 170), and ion binding can be stabilizing (110). In the α-subunit of trp synthase, replacement of Arg 211 by Glu decreases the rate of folding by a factor of $\sim$70 (74; C. R. Matthews, unpublished results). Charged groups can also have dramatic effects on the stability of isolated helixes in solution (171–174).

The stabilizing contributions of several specific ion pairs have been measured. Analysis of the pH dependence of the activity of α- and δ- chymotrypsins suggested that the stabilization energy of the buried ion pair between Asp 194 and the α-amino group of Ile 16 is 2.9 kcal/mole (183). A solvent-exposed salt bridge between the termini of BPTI stabilizes the protein by $\sim$1 kcal/mole in 6 M guanidine hydrochloride at pH 6.0 and 75°C (184). In phage T4 lysozyme, a new ion pair formed when Thr 157 is replaced by Arg contributes about 1 kcal/mole to the stability of the mutant protein at pH 6.0 and 65°C (69; T. Alber, S. Cook, and B. W. Matthews, unpublished results). Based on studies of protein-ligand complexes, Fersht and coworkers have concluded that hydrogen-bonded ion pairs can contribute 3 to 6 kcal/mole to ligand-binding energies (159, 160).

The stabilizing contributions of charged groups that do not form ion pairs are also dependent on position. Increases in the positive charge density of BPTI and phage T4 lysozyme (pI's > 10) at solvent-exposed positions decrease stability by 0.25–1 kcal/mole (91, 185, 186). In the α subunit of trp synthase, a more neutral protein, replacement of Gly 211 by either Glu or Arg increases stability (74, 187). Isolated charges that cannot be well solvated in the folded protein are apparently very destabilizing (77, 188).

Ionic groups at helix termini also influence stability. The stability of ribonuclease S increases with increasing negative charge on the N-terminus of the S-peptide helix (173). Replacement of Arg 96 by His at the C-terminus of a helix in phage lysozyme reduces the free energy of stabilization by ~3.5 kcal/mole at low ionic strength (189). The guanidium group of Arg 96 forms hydrogen bonds to two carbonyl groups at the helix terminus, and the histidine side chain, though positively charged at low pH, is not long enough to make these interactions in the folded structure (63).

Electrostatic interactions can have very long-range effects in the folded state. Shifts of up to one unit in the pKa of His 64 in *Bacillus amyloliquefaciens* subtilisin were induced by substitutions of six charged residues 12–20 Å from the imidazole nitrogen (191–193). In phage T4 lysozyme, several replacements that increase positive charge over 20 Å from the catalytic Glu 11 carboxyl group caused a marked decrease in enzyme activity at neutral pH (92, 186). Such long-range interactions may account for the finding that mobile and solvent-exposed charged residues can make crucial contributions to protein stability.

Interactions of ionizable groups vary with pH, and the stability of a protein depends on the number of protons bound or released upon denaturation (25, 32, 58, 194). Surprisingly, many proteins are maximally stable near pH 6, regardless of their isoelectric points (58). This may reflect the ionization of histidines in the unfolded state. Stability varies smoothly with pH, suggesting the presence of compensating interactions in the constellation of ionizable groups. Consistent with this behavior, a calculation of the contribution of electrostatic energy to the stability of sperm whale myoglobin showed that stabilizing and destabilizing interactions were redistributed over the charge pairs as the pH was varied (195).

The strength of ionic interactions can be affected by changes in local effective dielectric constants. As a result, amino acid substitutions that alter neutral residues may alter electrostatic contributions to stability (77, 149, 196). The local dielectric within a protein can be quite high because of distribution of polar groups (197). Local differences in polarity have been proposed (182, 198) to contribute to the surprising 2.7 kcal/mole difference in the stability of an Asp-Arg ion pair compared to the inverted Arg-Asp pair in aspartate amino transferase (199). Published data, however, do not rule out the possibility that the reduction in binding is due to structural changes in the free enzyme.

Nonpolar amino acids may also participate in "weakly polar interactions" (22, 23). For example, aromatic groups in proteins tend to pack edge-to-face. This arrangement may be favored by short-range dispersion forces and by a longer-range electrostatic attraction between the electron-poor aromatic hydrogens and the electron-rich π orbitals (22, 23).

Theoretical approaches to calculating electrostatic energies are under rapid development (reviewed by 182, 200, 201). Whatever the computational method used, it is clear that experimental studies of mutant proteins can provide accurate benchmarks of electrostatic contributions to stability. Changes in stability due to shifts of pH and ionic strength as well as titrations of folded and unfolded proteins can be used to specifically evaluate electrostatic effects. As a result, ionic interactions may soon become the best understood contribution to protein stability.

Helix Stabilization

The helix-coil transition has long served as a model for protein folding, and considerable effort has been invested in understanding the balance of forces that promote helix formation. One aim of this approach is to understand the stabilizing contribution of protein helixes in terms of their intrinsic stability and their affinity for the remainder of the protein. The validity of this concept was directly demonstrated by Mitchinson & Baldwin (173). They found that addition of the ribonuclease S-protein to S-peptides of increasing intrinsic helical stability leads to concomitant stabilization of the resulting ribonuclease S variants.

The relative abilities of each of the 20 common amino acids to stabilize model α-helixes in water has been investigated by Scheraga and coworkers (202, 203). Equilibrium data on helix stability were analyzed using the theory of Zimm & Bragg (204) to derive nucleation constants, σ, and propagation constants, s, for each residue as a function of temperature. Helix formation in this system is highly cooperative. Values of σ range from 10^{-5} to 10^{-2}. The amino acids show small differences in helical propensity, with s values ranging from 0.59 to 1.35 at 20°C (203). Helix formation is favored when $s > 1$; helix termination is favored when $s < 1$. The values of σ and s indicate that isolated short helixes are likely to be unstable because both nucleation and termination are unfavorable. The relative helical propensities obtained from the analysis of random copolymers generally parallel the relative frequencies of occurrence of each amino acid in protein helixes (202, 205).

While the thermodynamic and statistical preferences for helix formation are weak, these measures of helix stability correlate qualitatively with the effects of amino acid substitutions on the thermodynamic stability of several proteins including staph nuclease, λ repressor, and T4 lysozyme (95, 100, 113, 128). This is consistent with the idea that the host-guest data can be used to describe both the stability of individual protein helixes and the contributions of helixes to the overall stability of a protein.

This conclusion has been challenged by investigations of the effects of amino acid substitutions on isolated short helixes (172, 206) and in a helix in phage T4 lysozyme (92). These studies indicate that the effects of changes in σ and s can be small or can be easily dominated by sequence-specific interactions. The decrease in helical propensity often associated with de-stabilizing substitutions may be a fortuitous consequence of the relatively high helical propensities of residues in actual protein helixes.

To examine the relationship between protein stabilization and helical propensity, 10 substitutions were made for Pro 86 in a helix in phage T4 lysozyme (93). The X-ray crystal structures of seven of the mutant lysozymes were determined, and in each case, the helix was extended by two residues. The amino acids at position 86 spanned a wide range of helical propensity. Surprisingly, the substitutions had little effect on protein stability. The strongest helix former studied, Ala, was no more stabilizing than the strongest helix breaker, Gly. This insensitivity of protein stability to the residue at position 86 is not simply explained by thermodynamic studies of model helixes.

Also contrary to expectations from the host-guest data, several short peptides have been found to form isolated helixes in water. The peptides include block copolymers of Ala_{20} and Glu_{20} (172), a designed 17-residue sequence with three Glu-Lys pairs (175) and variants of the S-peptide (residues 1–20) and C-peptide (residues 1–13) of ribonuclease (42, 207).

Helix formation in S-peptide and C-peptide analogues is promoted by low temperature and moderately acidic pH. The bell-shaped dependence of helix stability on pH suggested a crucial role for charged residues (207). Variants of C-peptide lacking Glu 2 and His 12 formed less helix under optimal conditions, and helix stability depended on the ionization of the remaining residue (172, 208). On this basis, the "charged group effect" was assigned to electrostatic interactions of Glu 2 and His 12 with the macrodipole resulting from the partial charges on the peptide groups at the helix termini. This explanation for the charged group effect is consistent with the finding that isolated helixes are stabilized by increasing N-terminal negative charge and destabilized by increasing N-terminal positive charge (171, 173, 174). Other residues in the S-peptide, including Phe 8 and Arg 10, may also be important for helix stability (208, 209).

While the importance of electrostatic interactions involving the partial charges on polarized helical peptide groups seems well established, the use of the dipole model is controversial. The macrodipole model was developed to estimate the effects of the parallel alignment of individual peptide dipoles (17, 18) and the polarization of peptide groups (19, 210) in a helix. Calculations assuming a low dielectric constant indicate that the effective charges at the helix termini may be as high as ± 0.5 electron (19, 210). Sheridan and coworkers (211) compared the calculated potentials of an α-helix modeled as a macrodipole and as an array of discrete charges. As expected, the models were essentially identical at distances that are large compared to the helix length. Long-range interactions of the helix dipole (such as those between secondary structural elements), however, may be very weak when the ends of the helixes are in a high dielectric medium (212). The dipole model broke down near the helix termini (211), resulting in large differences in the calculated potentials where binding sites are commonly located. In addition, the dipolar potential is cylindrically symmetric, while the calculated potential of the "all atom" model reflects the asymmetrical positions of the amide and carbonyl groups in the helix.

Helix termination in short peptides is also not expected on the basis of host-guest data on model helixes (202). Nonetheless, the helix in the isolated S-peptide stops near Met 13, the last helical residue in this region of ribonuclease S (206). Helix termination in the S-peptide does not require tertiary interactions. These studies indicate that specific interactions of the side chains can be as important for helix formation as the intrinsic helical propensities of the individual residues (172, 206, 208, 213).

Sequence-specific structural features that may stabilize helixes have recently been identified by analyzing detailed statistical preferences (91) and hydrogen-bonding patterns (20) near helix termini. Most helixes contain hydrogen bonding side chains that interact with the terminal main-chain amide and carbonyl groups. In addition, the frequency of Pro adjacent to the N-terminal boundary residue (N-cap) is almost three times higher than average, and Gly is the C-terminal boundary residue (C-cap) in fully one third of the helixes examined.

These patterns have been used to engineer stabilizing amino acid substitutions in phage T4 lysozyme. These include introduction of Asp near two helix termini [$\Delta\Delta G = 0.8$–1.0 kcal/mole (111)] and replacement of Ala with Pro at an N-cap $+1$ position ($\Delta\Delta G = 0.5$ kcal/mole (127)). Crystallographic studies of the Asp-containing variants suggest that the stabilization is due to electrostatic interactions rather than specific hydrogen bonding at the helix termini (111).

CONCLUDING REMARKS

Chemical model systems traditionally used to estimate the contributions of noncovalent interactions have not reliably predicted the effects of amino acid substitutions on protein stability. Model systems, while qualitatively applicable, do not account for the high local concentrations of substituents in proteins, for differences in the environments of interacting groups, or for differences in the entropies of formation of specific interactions. Analysis of mutational effects is also complicated by structural relaxations that tend to minimize destabilizing perturbations.

Nonetheless, a combination of genetic, thermodynamic, and structural data can be used to estimate the contributions of specific interactions to protein stability. This sets the stage for analysis of a large number of interactions that will define the range of stabilizing effects and identify structural patterns that characterize stabilizing groups. Genetic experiments have already suggested that critical amino acids are often buried or rigid in the folded structure.

The structural and thermodynamic effects of mutations will also improve tests for the plausibility of predicted protein structures. Even an empirical understanding of stability can be useful for rejecting predicted structures that do not contain sufficient stabilizing interactions. Residue hydrophobicity, for example, has provided a good criterion for identifying transmembrane helixes (132), and atom hydrophobicities have been used to distinguish misfolded from correctly folded structures (137). Data from mutational analyses may prove useful for parameterizing energy functions (115, 214), surface matching algorithms (215, 216), and packing algorithms (30) that are increasingly being used to rank predicted structures. The correlation between low mobility and sensitivity to destabilizing substitutions suggests a link between correct structural elements and molecular dynamics.

Physical studies of protein variants also provide benchmarks for mechanistic theories of stability. With technology currently outpacing theory, ideas about stability are increasingly being tested by engineering thermostable protein variants and by de novo protein design (217, 218). Coupling physical and genetic methods, these developments have brought studies of the structural basis of protein stability into an exciting period of rapid change.

ACKNOWLEDGMENTS

I am deeply grateful to Brian Matthews for my experiences in his laboratory working on the structural basis of the stability of phage T4 lysozyme. I would like to thank W. A. Baase, W. J. Becktel, J. Bell, B.-L. Chen, S. Cook, F. W. Dahlquist, D. Goldenberg, T. Gray, E. Haas, P. Kim, S. Marqusee, B. W. Matthews, C. R. Matthews, L. McIntosh, H. Nicholson, J. Nye, A. Pakula, G. A. Petsko, R. Sauer, C. Schellman, J. Schellman, G. Streisinger, D.-P. Sun, D. Tronrud, L. H. Weaver, K. Wilson, and J. A. Wozniak for many wonderful collaborations, discussions of the problem of protein stability, and communications of unpublished results. Joyce Eshleman prepared this manuscript. Thanks also to the American Cancer Society (MV-382) and the Pew Memorial Trust for financial support.

Literature Cited

1. Creighton, T. E. 1983. *Biopolymers* 22:49–58
2. Creighton, T. E. 1983. *Proteins.* New York: Freeman
3. Creighton, T. 1985. *J. Phys. Chem.* 89:2452–59
4. Creighton, T. E., Goldenberg, D. P. 1984. *J. Mol. Biol.* 179:497–526
5. Goldenberg, D. P. 1985. *J. Cell. Biochem.* 29:321–35
6. Alber, T., Sun, D.-P., Nye, J. A., Muchmore, D. C., Matthews, B. W. 1987. *Biochemistry* 26:3754–58
7. Ackers, G. K., Smith, F. R. 1985. *Annu. Rev. Biochem.* 54:597–629
8. Dill, K. A. 1987. *Protein Engineering,* ed. D. L. Oxender, C. F. Fox, pp. 187–92. New York: Liss
9. Baldwin, R. L., Eisenberg, D. 1987. See Ref. 8, pp. 127–48
10. Matthews, B. W. 1987. *Biochemistry* 26:6885–88
11. Schellman, J. A. 1987. *Annu. Rev. Biophys. Biophys. Chem.* 16:115–37
12. Goldenberg, D. P. 1988. *Annu. Rev. Biophys. Biophys. Chem.* 17:481–507
13. Kauzmann, W. 1959. *Adv. Protein Chem.* 14:1–63
14. Richardson, J. S. 1981. *Adv. Protein Chem.* 34:167–339
15. Chothia, C. 1984. *Annu. Rev. Biochem.* 53:537–72
16. Richards, F. M. 1977. *Annu. Rev. Biophys. Bioeng.* 6:151–76
17. Arridge, G. C., Cannon, C. G. 1963. *Proc. R. Soc. London Ser. A* 278:91–109
18. Brant, D. A., Flory, P. J. 1965. *J. Am. Chem. Soc.* 87:663–64
19. Hol, W. G. J., van Duijnen, P. T., Berendsen, H. J. C. 1978. *Nature* 273:443–46
20. Presta, L. G., Rose, G. D. 1988. *Science* 240:1632–41
21. Richardson, J. S., Richardson, D. C. 1988. *Science* 240:1648–52
22. Burley, S. K., Petsko, G. A. 1985. *Science* 229:23–28
23. Burley, S. K., Petsko, G. A. 1988. *Adv. Protein Chem.* 39:125–89
24. Sturtevant, J. M. 1977. *Proc. Natl. Acad. Sci. USA* 74:2236–40
25. Privalov, P. L. 1979. *Adv. Protein Chem.* 33:167–241
26. Dlott, D. D., Frauenfelder, H., Langer, P., Roder, H., DiIorio, E. E. 1983. *Proc. Natl. Acad. Sci. USA* 80:6239–43
27. Svensson, L. A., Sjölin, L., Gilliland, G. L., Finzel, B. C., Wlodawer, A. 1986. *Proteins* 1:370–75
28. Smith, J. L., Hendrickson, W. A., Honzatko, R. B., Sheriff, S. 1986. *Biochemistry* 25:5018–27
29. Evans, P. A., Dobson, C. M., Kautz, R. A., Hatfull, G., Fox, R. O. 1987. *Nature* 329:266–68
30. Ponder, J. W., Richards, F. M. 1987. *J. Mol. Biol.* 193:775–91
31. Grütter, M. G., Gray, T. M., Weaver, L. H., Alber, T., Wilson, K., et al. 1987. *J. Mol. Biol.* 197:315–29
32. Tanford, C. 1968. *Adv. Protein Chem.* 23:121–282
33. Dill, K. A. 1985. *Biochemistry* 24:1501–9
34. Aune, K. C., Salahuddin, A., Zarlengo, M. H., Tanford, C. 1967. *J. Biol. Chem.* 242:4486–89
35. Matthews, C. R., Westmoreland, D. G. 1975. *Biochemistry* 14:4532–38
36. Labhart, A. M. 1982. *J. Mol. Biol.* 157:331–55
37. Bierzynski, A., Baldwin, R. L. 1982. *J. Mol. Biol.* 162:173–86
38. Amir, D., Haas, E. 1987. *Biochemistry* 26:2162–75
39. Haas, E., Amir, D. 1987. *J. Cell. Biochem. Suppl.* 11C:214

40. Kim, P. S., Baldwin, R. L. 1982. *Annu. Rev. Biochem.* 51:459–89
41. Hurle, M. R., Matthews, C. R. 1987. *Biochim. Biophys. Acta* 913:179–84
42. Brown, J. E., Klee, W. A. 1971. *Biochemistry* 10:470–76
43. Bundi, A., Wüthrich, K. 1979. *Biopolymers* 18:299–311
44. Dyson, H. J., Cross, K. J., Houghten, R. A., Wilson, I. A., Wright, P. E., et al. 1985. *Nature* 318:480–83
45. Mutter, M. 1977. *J. Am. Chem. Soc.* 99:8307–14
46. Illuminati, G., Mandolini, L. 1981. *Acc. Chem. Res.* 14:95–102
47. Glushko, V., Lawson, P. J., Gurd, F. R. N. 1972. *J. Biol. Chem.* 247:3176–85
48. Roder, H., Wagner, G., Wüthrich, K. 1985. *Biochemistry* 24:7407–11
49. Loftus, D., Gbenle, G. O., Kim, P. S., Baldwin, R. L. 1986. *Biochemistry* 25:1428–36
50. Barksdale, A. D., Stuehr, J. E. 1971. *J. Am. Chem. Soc.* 94:3334–38
51. States, D. J., Creighton, T. E., Dobson, C. M., Karplus, M. 1987. *J. Mol. Biol.* 195:731–39
52. Creighton, T. E. 1988. *Proc. Natl. Acad. Sci. USA* 85:5082–86
53. Pace, C. N. 1975. *CRC Crit. Rev. Biochem.* 3:1–43
54. Schellman, J. A., Lindorfer, M., Hawkes, R., Grütter, M. 1981. *Biopolymers* 20:1989–99
55. Privalov, P. L., Gill, S. J. 1988. *Adv. Protein Chem.* 39:191–234
56. Privalov, P. L., Griko, Y. V., Venyaminov, S. Y., Kutyshenko, V. P. 1986. *J. Mol. Biol.* 190:487–98
57. Schellman, J. A. 1987. *Biopolymers* 26:549–59
58. Becktel, W. J., Schellman, J. A. 1987. *Biopolymers* 26:1859–77
59. Chen, B.-L., Schellman, J. A. 1989. *Biochemistry.* In press
60. Chen, B.-L., Baase, W. A., Schellman, J. A. 1989. *Biochemistry.* In press
61. Go, N. 1975. *Int. J. Pept. Protein Res.* 7:313–23
62. Ueda, Y., Go, N. 1976. *Int. J. Pept. Protein Res.* 8:551–58
63. Grütter, M. G., Hawkes, R. B., Matthews, B. W. 1979. *Nature* 277:667–68
64. Grütter, M. G., Weaver, L. H., Gray, T. M., Matthews, B. W. 1983. *Bacteriophage T4*, ed. C. K. Matthews, E. M. Kutter, G. Mosig, P. M. Berget, pp. 356–60. Washington, DC: Am. Soc. Microbiol.
65. Luisi, B. F., Nagai, K. 1986. *Nature* 320:555–56
66. Howell, E. E., Villafranca, J. E., Warren, M. S., Oatley, S. J., Kraut, J. 1986. *Science* 231:1123–28
67. Nagai, K., Luisi, B., Shih, D., Miyazaki, G., Imai, K., et al. 1987. *Nature* 329:858–60
68. Villafranca, J. E., Howell, E. E., Oatley, S. J., Xuong, N.-H., Kraut, J. 1987. *Biochemistry* 26:2182–89
69. Alber, T., Sun, D.-P., Wilson, K., Wozniak, J. A., Cook, S. P., et al. 1987. *Nature* 330:41–46
70. Shortle, D., Meeker, A. K. 1986. *Proteins* 1:81–89
71. Shortle, D., Meeker, A. K., Freire, E. 1988. *Biochemistry* 27:4761–68
72. Goldenberg, D. P., Creighton, T. E. 1984. *J. Mol. Biol.* 179:527–45
72a. Goldenberg, D. P., Frieden, R. W., Haack, J. A., Morrison, T. B. 1989. *Nature* 338:127–32
73. Beasty, A. M., Hurle, M. R., Manz, J. T., Stackhouse, T., Onuffer, J. J., et al. 1986. *Biochemistry* 25:2965–74
74. Beasty, A. M., Hurle, M., Manz, J. T., Stackhouse, T., Matthews, C. R. 1987. See Ref. 8, pp. 91–102
75. Matthews, C. R., Hurle, M. R. 1987. *Bioessays* 6:254–57
76. Perry, K. M., Onuffer, J. J., Touchette, N. A., Herndon, C. S., Gittelman, M. S., et al. 1987. *Biochemistry* 26:2674–82
77. Yutani, K., Ogasahara, K., Tsujita, T., Sugino, Y. 1987. *Proc. Natl. Acad. Sci. USA* 84:4441–44
78. Matsumura, M., Yahanda, S., Aiba, S. 1988. *Eur. J. Biochem.* 171:715–20
79. Matsumura, M., Becktel, W. J., Matthews, B. W. 1988. *Nature* 334:406–10
80. Schellman, J. A. 1978. *Biopolymers* 17:1305–22
81. Snape, K. W., Tjian, R., Blake, C. C. F., Koshland, D. E. 1974. *Nature* 250:295–98
82. Yonath, A., Podjarny, A., Honig, B., Sielecki, A., Traub, W. 1977. *Biochemistry* 16:1418–24
83. Hibbard, L. S., Tulinsky, A. 1978. *Biochemistry* 17:5460–68
84. Miller, J. H., Coulondre, C., Hofer, M., Schmeissner, U., Sommer, H., et al. 1979. *J. Mol. Biol.* 131:191–222
85. Miller, J. H., Schmeissner, V. 1979. *J. Mol. Biol.* 131:223–48
86. Miller, J. H. 1984. *J. Mol. Biol.* 180:205–12
87. Perutz, M. F., Kendrew, J. C., Watson, H. C. 1965. *J. Mol. Biol.* 13:669–78
88. Perutz, M. F., Lehmann, H. 1968. *Nature* 219:902–9

89. Go, M., Miyazawa, S. 1980. *Int. J. Pept. Protein Res.* 15:211–24
90. Hecht, M. H., Nelson, H. C. M., Sauer, R. T. 1983. *Proc. Natl. Acad. Sci. USA* 80:2676–80
91. Pakula, A. A., Young, V. B., Sauer, R. T. 1986. *Proc. Natl. Acad. Sci. USA* 83:8829–33
92. Alber, T., Bell, J. A., Sun, D.-P., Nicholson, H., Wozniak, J. A., et al. 1988. *Science* 239:631–35
93. Fermi, G., Perutz, M. F. 1981. *Haemoglobin and Myoglobin.* Oxford: Clarendon
94. Streisinger, G., Mukai, F., Dreyer, W. J., Miller, B., Horiuchi, S. 1961. *Cold Spring Harbor Symp. Quant. Biol.* 26:25–30
95. Shortle, D., Lin, B. 1985. *Genetics* 110:539–55
96. Hampsey, D. M., Das, G., Sherman, F. 1986. *J. Biol. Chem.* 261:3259–71
97. Alber, T., Wozniak, J. A. 1985. *Proc. Natl. Acad. Sci. USA* 82:747–50
98. Bryan, P. N., Rollence, M. L., Pantoliano, M. W., Wood, J., Finzel, B. C., et al. 1986. *Proteins* 1:326–34
99. Hecht, M. H., Sauer, R. T. 1985. *J. Mol. Biol.* 186:53–63
100. Hecht, M. H., Sturtevant, J. M., Sauer, R. T. 1984. *Proc. Natl. Acad. Sci. USA* 81:5685–89
101. Matsumura, M., Aiba, S. 1985. *J. Biol. Chem.* 260:15298–303
102. Hendrix, J. D., Welker, N. E. 1985. *J. Bacteriol.* 162:682–92
103. Liao, H., McKenzie, T., Hageman, R. 1986. *Proc. Natl. Acad. Sci. USA* 83:576–80
104. Bartel, B., Varshavsky, A. 1988. *Cell* 52:935–41
105. Schultz, S. C., Richards, J. H. 1986. *Proc. Natl. Acad. Sci. USA* 83:1588–92
106. Reidhaar-Olson, J. F., Sauer, R. T. 1988. *Science* 241:53–57
107. Matsumura, M., Yasumura, S., Aiba, S. 1986. *Nature* 323:356–58
108. Wetzel, R., Perry, L. J., Baase, W. A., Becktel, W. J. 1988. *Proc. Natl. Acad. Sci. USA.* 85:401–5
109. Hurle, M. R., Tweedy, N. B., Matthews, C. R. 1986. *Biochemistry* 25:6356–60
110. Pace, C. N., Grimsley, G. R. 1988. *Biochemistry* 27:3242–46
111. Nicholson, H., Becktel, W. J., Matthews, B. W. 1988. *Nature* 336:651–56
112. Gray, T. M., Matthews, B. W. 1987. *J. Biol. Chem.* 262:16858–64
113. Alber, T., Grütter, M. G., Gray, T. M., Wozniak, J. A., Weaver, L. H., et al. 1986. *UCLA Symp. Mol. Cell. Biol.* 39:307–18
114. Snow, M. E., Amzel, L. M. 1986. *Proteins* 1:267–79
115. Moult, J., James, M. N. G. 1986. *Proteins* 1:146–63
116. Johnson, R. E., Adams, P., Rupley, J. A. 1979. *Biochemistry* 17:1479–84
117. Lin, S. H., Konishi, Y., Denton, M. E., Scheraga, H. A. 1984. *Biochemistry* 23:5504–12
118. Weber, P. C., Sheriff, S., Ohlendorf, D. H., Finzel, B. C., Salemme, F. R. 1985. *Proc. Natl. Acad. Sci. USA* 82:8473–77
119. Chazin, W. J., Goldenberg, D. P., Creighton, T. E., Wüthrich, K. 1985. *Eur. J. Biochem.* 152:429–37
120. Lin, S. H., Konishi, Y., Nall, B. T., Scheraga, H. A. 1985. *Biochemistry* 24:2680–86
121. Pabo, C. O., Suchanek, E. G. 1986. *Biochemistry* 25:5987–91
122. Wells, J. A., Powers, D. B. 1986. *J. Biol. Chem.* 261:6564–70
123. Pantoliano, M. W., Ladner, R. C., Bryan, P. N., Rollence, M. L., Wood, J. F., et al. 1987. *Biochemistry* 26:2077–82
124. Sauer, R. T., Hehir, K., Stearman, R. S., Weiss, M. A., Jeitler-Nilsson, A., et al. 1986. *Biochemistry* 25:5992–98
125. Perry, L. J., Wetzel, R. 1984. *Science* 226:555–57
126. Nemethy, G., Leach, S. J., Scheraga, H. A. 1966. *J. Phys. Chem.* 70:998–1004
127. Matthews, B. W., Nicholson, H., Becktel, W. J. 1987. *Proc. Natl. Acad. Sci. USA* 84:6663–67
128. Hecht, M. H., Hehir, K. M., Nelson, H. C. M., Sturtevant, J. M., Sauer, R. T. 1985. *J. Cell. Biochem.* 29:217–24
129. Imanaka, T., Shibazaki, M., Takagi, M. 1986. *Nature* 324:695–97
130. Tanford, C. 1962. *J. Am. Chem. Soc.* 84:4240–47
131. Nozaki, Y., Tanford, C. 1971. *J. Biol. Chem.* 246:2211–17
132. Kyte, J., Doolittle, R. F. 1982. *J. Mol. Biol.* 157:105–32
133. Abraham, D. J., Leo, A. J. 1987. *Proteins* 2:130–52
134. Radzicka, A., Wolfenden, R. 1988. *Biochemistry* 27:1664–70
135. Wishnia, A. 1963. *J. Phys. Chem.* 67:2079–82
136. Wolfenden, R., Andersson, L., Cullis, P. M., Southgate, C. C. B. 1981. *Biochemistry* 20:849–55
137. Eisenberg, D., McLachlan, A. D. 1986. *Nature* 319:199–203
138. Rose, G. D., Geselowitz, A. R., Lesser, G. J., Lee, R. H., Zehfus, M. H. 1985. *Science* 229:834–38

139. Miller, S., Lesk, A. M., Janin, J., Chothia, C. 1987. *Nature* 328:834–36
140. Lawrence, C., Auger, I., Mannella, C. 1987. *Proteins* 2:153–67
141. Baldwin, R. L. 1986. *Proc. Natl. Acad. Sci. USA* 83:8069–72
142. Klapper, M. H. 1971. *Biochim. Biophys. Acta* 229:557–66
143. Chothia, C. 1976. *J. Mol. Biol.* 105:1–14
144. Lee, B. 1985. *Biopolymers* 24:813–23
145. Tanford, C. 1980. *The Hydrophobic Effect: Formation of Micelles and Biological Membranes.* New York: Wiley
146. Brandts, J. F., Oliveira, R. J., Westort, C. 1970. *Biochemistry* 9:1038–48
147. Zipp, A., Kauzmann, W. 1973. *Biochemistry* 12:4217–28
148. Hvidt, A. 1975. *J. Theor. Biol.* 50:245–52
149. Yutani, K., Ogasahara, K., Aoki, K., Kakuno, T., Sugino, Y. 1984. *J. Biol. Chem.* 259:14076–81
150. Baker, E. N., Hubbard, R. E. 1984. *Prog. Biophys. Mol. Biol.* 44:97–179
151. Schellman, J. A. 1955. *C.R. Trav. Lab. Carlsberg Ser. Chim.* 29:223–29
152. Schellman, J. A. 1955. *C.R. Trav. Lab. Carlsberg Ser. Chim.* 29:230–59
153. Klotz, I. M., Franzen, J. S. 1962. *J. Am. Chem. Soc.* 84:3461–66
154. Susi, H., Timasheff, S. N., Ard, J. S. 1964. *J. Biol. Chem.* 239:3051–54
155. Stahl, N., Jencks, W. P. 1986. *J. Am. Chem. Soc.* 108:4196–205
156. Rebek, J. Jr. 1987. *Science* 235:1478–84
157. Page, M. I., Jencks, W. P. 1971. *Proc. Natl. Acad. Sci. USA* 68:1678–83
158. Page, M. I. 1984. *The Chemistry of Enzyme Action,* ed. M. I. Page, pp. 1–54. Amsterdam: Elsevier Scientific
159. Fersht, A. R., Shi, J.-P., Knill-Jones, J., Lowe, D. M., Wilkinson, A. J., et al. 1985. *Nature* 314:235–38
160. Fersht, A. R. 1987. *Trends Biochem. Sci.* 12:301–4
161. Hawkes, R., Grütter, M. G., Schellman, J. 1984. *J. Mol. Biol.* 175:195–212
162. Bello, J. 1977. *J. Theor. Biol.* 68:139–42
163. Bello, J. 1978. *Int. J. Pept. Protein Res.* 12:38–41
164. Narayana, S. V. L., Argos, P. 1984. *Int. J. Pept. Protein Res.* 24:25–39
165. Kellis, J. T., Nyberg, K., Sali, D., Fersht, A. R. 1988. *Nature* 333:784–87
166. Janin, J., Wodak, S., Levitt, M., Maigret, B. 1978. *J. Mol. Biol.* 125:357–86
167. Tanford, C. 1954. *J. Am. Chem. Soc.* 76:945–46
168. Springs, B., Haake, P. 1977. *Bioorg. Chem.* 6:181–90
169. Perutz, M. F. 1978. *Science* 201:1187–91
170. Perutz, M. F., Raidt, H. 1975. *Nature* 255:256–59
171. Ihara, S., Ooi, T., Takahashi, S. 1982. *Biopolymers* 21:131–45
172. Shoemaker, K. R., Kim, P. S., Brems, D. N., Marqusee, S., York, E. J., et al. 1985. *Proc. Natl. Acad. Sci. USA* 82:2349–53
173. Mitchinson, C., Baldwin, R. L. 1986. *Proteins* 1:23–33
174. Marqusee, S., Baldwin, R. L. 1988. *Proc. Natl. Acad. Sci. USA* 84:8898–902
175. Wada, A., Nakamura, H. 1981. *Nature* 293:757–58
176. Thornton, J. M. 1982. *Nature* 295:13–14
177. Barlow, D. J., Thornton, J. M. 1983. *J. Mol. Biol.* 168:867–85
178. Barlow, D. J., Thornton, J. M. 1986. *Biopolymers* 25:1717–33
179. Rashin, A. A., Honig, B. 1984. *J. Mol. Biol.* 173:515–21
180. Russell, S. T., Warshel, A. 1985. *J. Mol. Biol.* 185:389–404
181. Gilson, M. K., Rashin, A., Fine, R., Honig, B. 1985. *J. Mol. Biol.* 183:503–16
182. Warshel, A. 1987. *Nature* 330:15–16
183. Fersht, A. R. 1971. *Cold Spring Harbor Symp. Quant. Biol.* 36:71–73
184. Brown, L. R., DeMarco, A., Richarz, R., Wagner, G., Wüthrich, K. 1978. *Eur. J. Biochem.* 88:87–95
185. Wagner, G., Kalb, A. J., Wüthrich, K. 1979. *Eur. J. Biochem.* 95:249–53
186. Grütter, M. G., Matthews, B. W. 1982. *J. Mol. Biol.* 154:525–35
187. Matthews, C. R., Crisanti, M. M., Gepner, G. L., Velicelebi, G., Sturtevant, J. M. 1980. *Biochemistry* 19:1290–93
188. Ahern, T. J., Casal, J. I., Petsko, G. A., Klibanov, A. M. 1987. *Proc. Natl. Acad. Sci. USA* 84:675–79
189. Sturtevant, J. M. 1987. *Annu. Rev. Phys. Chem.* 38:463–88
190. Deleted in proof
191. Thomas, P. G., Russell, A. J., Fersht, A. R. 1985. *Nature* 318:375–76
192. Sternberg, M. J. E., Hayes, F. R. F., Russell, A. J., Thomas, P. G., Fersht, A. R. 1987. *Nature* 330:86–88
193. Gilson, M. K., Honig, B. H. 1987. *Nature* 330:84–86
194. Tanford, C. 1970. *Adv. Protein Chem.* 24:1–95
195. Garcia-Moreno, B., Chen, L. X., March, K. L., Gurd, R. S., Gurd, F. R. N. 1985. *J. Biol. Chem.* 260:14070–82

196. Flanagan, M. A., Garcia-Moreno, B., Friend, S. H., Feldmann, R. J., Scouloudi, H., et al. 1983. *Biochemistry* 22:6027–37
197. Macgregor, R. B., Weber, G. 1986. *Nature* 319:70–73
198. Hwang, J.-K., Warshel, A. 1988. *Nature* 334:270–72
199. Cronin, C. N., Malcolm, B. A., Kirsh, J. F. 1987. *J. Am. Chem. Soc.* 109:2222–23
200. Warshel, A., Russell, S. T. 1984. *Q. Rev. Biophys.* 17:282–422
201. Matthew, J. B. 1985. *Annu. Rev. Biophys. Biophys. Chem.* 14:387–417
202. Scheraga, H. A. 1978. *Pure Appl. Chem.* 50:315–24
203. Sueki, M., Lee, S., Powers, S. P., Denton, J. B., Konishi, Y., et al. 1984. *Macromolecules* 17:148–55
204. Zimm, B. H., Bragg, J. K. 1959. *J. Chem. Phys.* 31:526–35
205. Lewis, P. N., Go, N., Go, M., Kotelchuck, D., Scheraga, H. A. 1970. *Proc. Natl. Acad. Sci. USA* 65:810–15
206. Kim, P. S., Baldwin, R. L. 1984. *Nature* 307:329–34
207. Bierzynski, A., Kim, P. S., Baldwin, R. L. 1982. *Proc. Natl. Acad. Sci. USA* 79:2470–74
208. Shoemaker, K. R., Kim, P. S., York, E. J., Stewart, J. M., Baldwin, R. L. 1987. *Nature* 326:563–67
209. Rico, M., Santoro, J., Bermejo, F. J., Herranz, J., Nieto, J. L., et al. 1986. *Biopolymers* 25:1031–53
210. Wada, A. 1976. *Adv. Biophys.* 9:1–63
211. Sheridan, R. P., Levy, R. M., Salemme, F. R. 1982. *Proc. Natl. Acad. Sci. USA* 79:4545–49
212. Rogers, N. K., Sternberg, M. J. E. 1984. *J. Mol. Biol.* 174:527–42
213. Scheraga, H. A. 1985. *Proc. Natl. Acad. Sci. USA* 82:5585–87
214. Novotny, J., Bruccoleri, R., Karplus, M. 1984. *J. Mol. Biol.* 177:787–818
215. Connolly, M. L. 1986. *Biopolymers* 25:1229–47
216. Connolly, M. L. 1986. *Int. J. Pept. Protein Res.* 28:360–63
217. Regan, L., Degrado, W. F. 1988. *Science* 241:976–78
218. Richardson, J. S., Richardson, D. C. 1987. See Ref. 8, pp. 149–63

Annu. Rev. Biochem. 1989. 58:799–839

EUKARYOTIC TRANSCRIPTIONAL REGULATORY PROTEINS

Peter F. Johnson[1] and Steven L. McKnight

Howard Hughes Research Laboratories, Carnegie Institution of Washington, Department of Embryology, 115 West University Parkway, Baltimore, Maryland 21210

CONTENTS

[1]Present address: Molecular Mechanisms of Carcinogenesis Laboratory, Basic Research Program, NCI-Frederick Cancer Research Facility, P.O. Box B, Frederick, Maryland 21701

0066-4154/89/0701-0799$02.00

PERSPECTIVES

The past five years have offered a virtual explosion of information in the field of eukaryotic gene regulation. The natural anticipation that a considerable degree of regulation would be imparted at the level of mRNA synthesis (transcription) has been satisfied. When the polypeptide product of a gene is absent from a eukaryotic cell, its absence is typically due to the fact that its encoding gene is transcriptionally inert. There are important and highly interesting exceptions to this general rule, such as the regulated splicing of transcripts synthesized from the various fruit fly genes that control sexual phenotype (1–2b). However, many lines of evidence identify transcription as the most common and immediate focal point of genetic regulation in eukaryotic cells.

As in prokaryotic organisms, transcriptional control in eukaryotes results from an interplay between regulatory DNA sequences and site-specific DNA-binding proteins. Transcription of eukaryotic genes is influenced by various regulatory elements, termed promoters, enhancers, and silencers (see 3–5 for reviews). These elements, in turn, are composed of discrete DNA sequence motifs, which constitute binding sites for sequence-specific DNA-binding proteins (reviewed in 6, 7). A major effort is now under way to identify sequence-specific DNA-binding proteins, to match them to their cognate sites within or around eukaryotic genes, and to elucidate how the binding of such proteins results in increased or decreased transcription of the associated gene. This review focuses on recent information regarding the nature of eukaryotic transcriptional regulatory proteins, primarily ones that bind to DNA in a sequence-specific manner. Albeit an exciting time, the past several years have also generated a considerable amount of information that has yet to fall into place. We have chosen to concentrate our discussion on a few examples that illustrate particular concepts. Most of these are cases where considerable biochemical information has been obtained, although many of these systems owe their origins to classical genetic studies.

The plethora of information concerning gene regulatory proteins can be largely attributed to several methodological advances that have occurred in the past decade. For example, cell-free extracts have been developed that recapitulate accurate transcription from RNA polymerase II promoters using cloned DNA templates (8–11). Also, detection of rare DNA-binding activities in crude cell extracts has become possible due to sensitive DNA protection (12) and gel retardation (13) assays. Finally, purification of these proteins has been expedited by the refinement of sequence-specific DNA affinity chromatography techniques (14–16).

It would be impossible to cover all of the developments in the diverse field of transcriptional regulators. In particular, we have circumvented two impor-

tant issues that have recently been reviewed elsewhere: how regulatory proteins communicate information to the transcriptional apparatus via activator
domains (17), and the biochemical events involving RNA polymerase II and
ancillary factors that occur during transcriptional initiation (18). We have
instead chosen to begin this review by outlining the basic structural motifs that
endow proteins the capacity to bind DNA in a sequence-specific manner, then
focus on results and observations that were not necessarily anticipated from
precedent studies on bacterial gene regulation. We hope to provide a
framework for the consideration of topics that qualify as looming enigmas,
and present speculations on the importance of protein:protein interactions.
Lastly, we address the problem of how transcriptional regulatory proteins are
themselves controlled.

DNA-BINDING MOTIFS

Proteins recognize DNA in much the same way as they recognize other
proteins. That is, they first form a tertiary shape or contour that is compatible
with the surface with which they must interact. They are then availed the
opportunity to come into sufficiently close contact, over a sufficiently broad
surface, to establish numerous atomic interactions. As in the interaction
between two proteins, the atomic contacts between protein and DNA are
varied, including hydrogen bonding, ionic interactions, and hydrophobic
interactions. As such, the problem of interaction specificity between protein
and DNA has profited most substantially from X-ray crystallographic studies.
Since this subject will not be reviewed in detail herein, we refer readers to the
excellent review of Pabo & Sauer (19) as well as to Ptashne's concise primer
on prokaryotic gene regulation (20).

Although any comprehensive understanding of how proteins recognize
DNA will require structural studies, the problem is substantially simplified by
the emerging indication that many different proteins recognize DNA by virtue
of common structural motifs. We start this review with descriptions of three
motifs that are utilized by eukaryotic regulatory proteins to bind specific DNA
sequences (Figure 1).

Helix-Turn-Helix

The structures of three sequence-specific DNA-binding proteins were solved
over a relatively short span of time in the early 1980s. Each was a gene
regulatory protein functional in bacterial cells; two were small DNA-binding
proteins encoded by bacteriophage lambda (the CRO and CI proteins), and the
other was the catabolite activator protein (CAP) of *Escherichia coli*. All three
proteins were known to bind DNA as dimers, and their binding sites on DNA
were known to be dyad symmetric. Upon resolution of their respective

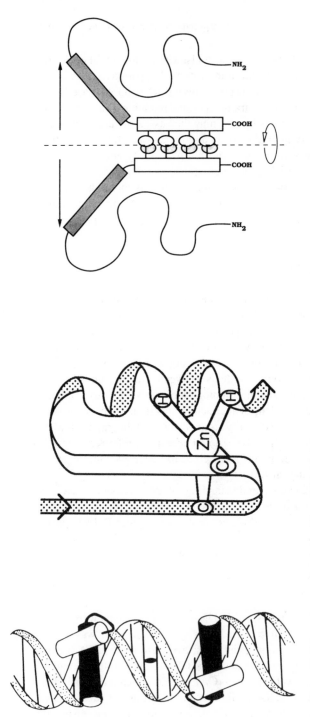

Figure 1 Schematic representations of three DNA-binding motifs.

Left panel shows a helix-turn-helix protein bound to a dyad-symmetric binding site on DNA. The recognition helix (helix 3) is represented as a dark barrel situated in the major groove of each half of the dyad-symmetric binding site. Helix 2 is situated above helix 3 in a position that helps lock the recognition helix into place. Figure kindly provided by Mark Ptashne.

Middle panel shows a zinc finger protein coordinated with a zinc ion through paired cysteine and histidine residues. Amino-to-carboxyl dipole is indicated by arrows. Descending and ascending polypeptide segments containing the two cysteine residues are predicted to exist as paired, antiparallel beta sheets. Descending polypeptide segment containing the two histidine residues is predicted to exist in an α helical conformation. Figure kindly provided by Jeremy Berg.

Right panel shows two polypeptide chains of a leucine zipper protein dimerized via hydrophobic interactions between two α helixes. Spherical projections at the dimerization interface represent interlocking leucine side chains. Stippled rectangles represent highly basic regions believed to establish direct contact with DNA. Dashed line represents axis of rotational symmetry. Opposing arrows at top of diagram represent dyad half-sites of recognition site on DNA. Figure kindly provided by Bill Landschulz.

three-dimensional structures (21–23), it became apparent that they shared a distinctive succession of two alpha helices separated by a relatively sharp beta turn (24). This structure is termed the helix-turn-helix motif.

In each case, the more carboxyl terminal helix from one subunit of the dimer, now termed helix 3, was observed to be juxtaposed in a unique manner relative to the analogous helix of the other subunit. This similarity of helix positioning rested on several criteria. First, the distances between the helices were approximately equivalent in the three proteins, and were matched to the distance separating successive major grooves on a single face of B form DNA. Second, the helices were aligned in an antiparallel conformation such that their orientations (amino-to-carboxyl dipoles) were matched relative to the DNA sequences on each half of their dyad-symmetric recognition sites. Third, in each case, helix 3 was locked into place via hydrophobic interactions with helix 2, which sat above helix 3 relative to its putative interface with DNA. According to this modeling, helix 3 was christened the recognition helix, owing to the expectation that it would contribute the atomic contacts with DNA that were most important for sequence-specific binding.

The results of protein engineering experiments, which were used to excise and exchange helix 3 between different repressor proteins (25, 26), coupled with extensive genetic information on the interaction between CI, CRO, and their three binding sites on the rightward operator of bacteriophage lambda (27, 28), are entirely consistent with the anticipated docking of helix 3 within the major groove of DNA. Moreover, recent X-ray diffraction studies on crystallized complexes between helix-turn-helix proteins and their cognate sites on DNA have begun to provide structural confirmation of this model (29, 29a, 29b, 29c).

The constancy of the helix-turn-helix structural motif in bacterial gene regulatory proteins is correlated with an underlying constancy of primary amino acid sequence (19). Although less striking than the uniformity of structure, the amino acid sequences of each protein exhibit distinctive regularity at certain positions. The fourth residue of helix 2 almost always contains a hydrophobic side chain, and the fifth residue is marked by either of the comparatively small side chains, glycine or alanine. The beta turn that occurs between the two helices begins most frequently with glycine, and is invariably followed by a residue bearing a hydrophobic side chain. Finally, residues four and seven of helix 3 are almost always occupied by hydrophobic amino acids.

These six positions offered a diagnostic constellation of reference points that, in turn, allowed tentative identification of the helix-turn-helix motif in several eukaryotic gene regulatory proteins (30–33). The amino acid sequences of two products of the mating type locus of *Saccharomyces cerevisiae*, MATa1 and MATα2, showed the distribution of hydrophobic residues at three of the four appropriate positions in helixes 2 and 3, and at the second position of the putative β turn. An exception to the glycine normally found at the first position of the β turn was observed in MATα2; however, its replacement (serine) was also noted in the sequence that specifies the helix-turn-helix motif of Tn3 resolvase (19). Both MATa1 and MATα2 were observed to contain a tryptophan at the seventh position of helix 3. Although many of the bacterial proteins contain classically defined hydrophobic residues at this position (leucine, isoleucine, valine, or alanine), at least three are known to contain tryptophan at position seven. The prediction from these amino acid sequence similarities, that the yeast mating type products would be capable of sequence-specific recognition of DNA, has been substantiated by the biochemical experiments of Johnson & Herskowitz (34). The product of the MATα2 locus was synthesized in bacterial cells and shown to be capable of sequence-specific recognition of a 32-base-pair operator site located upstream of one of the yeast genes known to be regulated by mating type.

Laughon & Scott (31), and Shepherd et al (33) noted that the sequence similarities between yeast and bacterial helix-turn-helix proteins could be further extended to the products of the *Drosophila* homeotic genes, *antennapedia, fushi tarazu,* and *ultrabithorax.* Earlier nucleic acid hybridization experiments had predicted an extensive degree of structural similarity between homeotic genes. DNA sequence analysis further defined these similarities which, when theoretically translated, typically exceeded 90% amino acid identity over a 60-amino-acid segment (termed the "homeobox"). A pattern of hydrophobic amino acids consistent with the helix-turn-helix motif occurs within a 30-amino-acid segment disposed toward the carboxyl half of the canonical homeobox sequence.

The realization that homeobox sequences share similarity with bacterial helix-turn-helix proteins led to the obvious prediction that these proteins would administer their regulatory function by sequence-specific interaction with DNA. Although the task of linking specific homeotic gene products with their appropriate sites on DNA has been problematic, it is clear that homeobox-containing proteins localize in the nuclear compartment (for example, 35–37), are capable of avid interaction with specific DNA sequences (38–41), and depend on the homeodomain for sequence-specific recognition of DNA (41).

The extreme degree of sequence relatedness between homeobox elements has facilitated direct cloning of frog (42), mouse (43), and human (44) genes that may also encode DNA-binding proteins. Conceptual translation of the DNA sequences of vertebrate and invertebrate homeobox-containing genes has revealed striking levels of amino acid sequence similarity. For example, two different human genes display amino acid sequences that share 90% and

91% identity with that of the antennapedia homeobox of *Drosophila melanogaster* (44). Moreover, different genes of a single species can share near identity within their homeobox regions. The observations of extreme sequence relatedness are inherently interesting and exciting. By allowing researchers to retrieve cloned copies of what will almost certainly turn out to be important regulatory genes, the extreme sequence relatedness between homeoboxes has facilitated an entirely new focus of attack on vertebrate development.

The high level of sequence similarity between different homeoboxes predicts that the target DNA sequences of many homeodomain proteins will be similar or identical. Recent in vitro footprint experiments using bacterially expressed *Drosophila* proteins (engrailed, fushi tarazu, ultrabithorax, even-skipped, zerknullt, and paired) support this prediction (39–41). These observations pose an obvious dilemma: how do proteins with similar DNA-binding specificities execute different regulatory roles? We return to this issue subsequently.

Zinc Finger

Discovery of the zinc finger motif came as a consequence of studies on the expression of *Xenopus laevis* 5S ribosomal RNA genes. Transcription of 5S genes is regulated by an internal promoter (45, 46). The internal promoter binds a protein termed transcription factor IIIA (TFIIIA; 47, 48). TFIIIA, in collaboration with at least two other proteins (TFIIIB and TFIIIC), assembles into a stable complex on the 5S gene, which is then the target for RNA polymerase III transcription (49, 50).

Critical insight into the nature of the DNA interaction interface of TFIIIA resulted from the discovery of a repetitive motif within its amino acid sequence (51, 52). A sequential and ordered occurrence of cysteine and histidine residues was observed to be repeated nine times in the primary amino acid sequence of TFIIIA ($Cys-N_2-Cys-N_{12}-His-N_3-His$). These observations, coupled with earlier experiments that had shown TFIIIA to be complexed with zinc (53), prompted Klug and colleagues to postulate that each repeat sequesters a single zinc ion via tetrahedral coordination with the spatially conserved cysteine and histidine residues. It was further hypothesized that the 12–14 amino acids intervening between the cysteines and histidines would loop out in a manner facilitating direct interaction with DNA, forming a distinct DNA interaction module termed the zinc finger (54, 55). Although X-ray crystallographic data have not yet been gathered to provide a rigorous test of the zinc finger model, evidence from a number of alternative avenues of investigation is consistent with its general tenets. For example, the tetrahedral coordination complex between zinc and the finger repeats of TFIIIA has been documented by X-ray adsorption spectroscopy

(56), and sequence-specific interaction of TFIIIA with its binding site on 5S DNA has been shown to be zinc dependent (57). Finally, mutagenized derivatives of the TFIIIA protein sequentially lose sequence-specific affinity for the internal promoter on 5S DNA in a manner that correlates with the sequential loss of zinc finger repeats (58).

The spatial distribution of cysteine and histidine residues within the TFIIIA repeats was sufficiently distinctive to allow rapid identification of other zinc finger proteins. Gene regulatory proteins from yeast (59, 60), fruit flies (61–64), and humans (65, 66) exhibit cysteine and histidine residues organized in constellations highly similar to that seen in TFIIIA (cys$_2$his$_2$). Moreover, the constancy of the zinc finger amino acid motif has allowed generation of synthetic oligonucleotides which, in turn, facilitated the isolation of genes that encode new zinc finger proteins of unknown function (67, 68).

An amino acid sequence motif highly related to, yet distinct from, the cys$_2$his$_2$ arrangement of TFIIIA-like zinc fingers has been observed in a wide variety of DNA-binding proteins, ranging from mammalian hormone receptors (69–75), to yeast gene regulatory proteins (76–80). This related motif is typified by clusters of 4 to 6 cysteine residues, with pairs of cysteines often separated by 2 or 4 amino acids. As in the TFIIIA paradigm, these cys$_x$ motifs occur in 25–30-residue clusters that tend to be repeated. Mutations that either delete the cys$_x$ cluster from a DNA-binding protein, or create amino acid substitutions at cysteines, eliminate sequence-specific DNA-binding activity (81–85). Moreover, X-ray atomic adsorption spectroscopy has been used to demonstrate that the cys$_x$ domain of the glucocorticoid receptor binds zinc via a tetrahedral coordination with cysteine residues (86). Genetic and biochemical experiments have also demonstrated that GAL4, a yeast regulatory protein bearing a single cys$_x$ finger, requires zinc in order to bind to DNA (87).

In the absence of detailed X-ray diffraction studies, it has been impossible to determine how zinc fingers establish sequence-specific contact with DNA. The model offered initially by Klug and colleagues, that the 12–14 residues intervening between the pairs of cysteines and histidines would loop out to establish sequence-specific contacts with DNA, is likely to be correct in a certain sense. That is, the aspect of the motif concerned with zinc coordination probably serves as a scaffold that facilitates appropriate disposition of intervening amino acids for interaction with DNA. However, as has been pointed out by Berg (88), the structure adopted by these intervening amino acid sequences may not be a simple loop. It will be exciting to watch the progress of crystallographic studies on zinc finger proteins. The compactness of the motif perhaps portends a rather simple fit between protein and DNA, an anticipation that we hope will expedite the process of understanding principles of recognition between proteins and DNA.

Leucine Zipper

The existence of a third category of DNA-binding motifs was first hinted by a study (89) that noted primary amino acid sequence similarities between the products of several oncogenes (FOS, MYC, and JUN) and a yeast regulatory protein (GCN4) known to be capable of sequence-specific recognition of DNA. The structure underlying these sequence similarities began to emerge upon evaluation of another related protein (termed C/EBP). C/EBP, a heat-stable, rat liver nuclear protein, was first characterized according to its capacity to bind to regulatory DNA sequences associated with viral genes (90, 91). The gene encoding C/EBP was cloned and sequenced, leading to the discovery that the amino acid sequence of its DNA-binding domain shares distinctive similarity with the FOS, MYC, and JUN transforming proteins (92). The regions of amino acid sequence similarity between these proteins are relatively free of residues that are incompatible with α helical structure (89, 93, 94). On the basis of purported helix "permissivity," the amino acid sequence of the C/EBP DNA-binding domain was displayed on an idealized α helix (94). This exercise offered two unanticipated observations. First, it showed that a 35-amino-acid region covering roughly half of the C/EBP DNA-binding domain contained a strict heptad repeat of leucine residues. Second, it revealed an exceedingly high density of oppositely charged amino acids (acidics and basics) juxtaposed in a manner suitable for intrahelical ion pairing. This high frequency of ion pairing predicted an unusually stable helical domain. Both of these properties were common to FOS, MYC, JUN, and GCN4 (94).

The heptad array of leucines common to each of these regulatory proteins, reminiscent of the heptad repeat common to proteins that adopt a coiled coil quaternary structure (e.g. keratins, lamins, and the tail of myosin heavy chain), led to the hypothesis that polypeptides of this class might dimerize via the leucine repeat helix (94). Unlike standard coiled coil proteins, however, the putative dimerization interface of C/EBP and its relatives appeared to rely on leucine to the exclusion of other hydrophobic amino acids. An evaluation of seven proteins that exhibited this motif (FOS, three members of the MYC protein family, JUN, GCN4, and C/EBP), revealed only a single deviation from leucine. This deviation constituted a substitution by methionine at an internal position within the leucine repeat of human L-MYC protein. Aside from general hydrophobicity, the side chains of leucine and methionine are common in two ways; they are long (extending at least three carbon atoms from the alpha carbon), and they bear no methyl groups extending laterally from their β carbon (as do valine and isoleucine). These commonalities led to the hypothesis that the leucines extending from the helix of one polypeptide would interdigitate intimately with those of the analogous helix of a second polypeptide, forming an interlock termed the leucine zipper (94).

The leucine zipper model was largely speculative when first hypothesized. Aside from amino acid sequence arrangements, the idea was supported only by the experiments of Hope & Struhl (95), which had shown that GCN4 binds DNA as a dimer, and preliminary experiments which indicated that C/EBP also exists as a stable dimer (W. Landschulz, P. Sigler, S. McKnight, unpublished). However, recently completed site-directed mutagenesis experiments on C/EBP strongly support the zipper hypothesis. When leucines of the putative C/EBP zipper are changed to either isoleucine or valine, C/EBP fails to dimerize and loses its capacity for sequence-specific recognition of DNA (W. Landschulz, P. Johnson, S. McKnight, unpublished). Similarly conceived mutagenesis experiments on MYC (95a), JUN (D. Bohmann, R. Tjian, personal communication), and FOS (R. Gentz, F. Rauscher, C. Abate, T. Curran, personal cummunication; R. Turner, R. Tjian, personal communication; 95b, 95c) have led to the conclusion that the leucine repeat regions of each of these proteins represent oligomerization interfaces.

The leucine zipper is an integral element of the C/EBP DNA-binding domain. When the leucine repeat of C/EBP is deleted, or otherwise mutated in such a way as to prevent dimerization, the protein is incapable of specific interaction with DNA. However, the zipper is not the only aspect of C/EBP that is required for sequence-specific interaction with DNA. A 30-amino-acid segment located immediately N-terminal to the leucine repeat must also remain intact in order for the protein to bind DNA. These additional sequences are characterized by a high degree of basicity, and distinct pockets of sequence similarity occur between this basic region of C/EBP and an analogous region of FOS (94). We have hypothesized that dimerization, facilitated by the leucine zipper, leads to a unique positioning of basic regions from two subunits, and that this bilateral display is critical for sequence-specific recognition of DNA. The proposed dimeric structure of leucine zipper proteins predicts that their binding sites on DNA will exist as rotationally symmetric dyads. Observations favoring this prediction have already been made for the binding sites of both GCN4 (96) and C/EBP (90).

If all zipper proteins share the same leucine repeat, is it possible that they will cross-dimerize "willy nilly," creating all sorts of heterodimers? Alternatively, will other specificities encoded within any given zipper protein restrict interaction to homodimeric forms? Although this issue has yet to be evaluated in a rigorous manner, we predict that the most likely answer will lie somewhere between these two extremes. On the one hand, C/EBP chromatographs during purification from rat liver as a discrete species, with properties matching those displayed by the purified homodimer. Likewise, it will not cross-dimerize with MYC protein under conditions suitable for redistribution of C/EBP and MYC subunits (W. Landschulz, C. Dang, S. McKnight, unpublished). On the other hand, compelling evidence is emerging in support of cross-dimerization of FOS and JUN via their respective leucine zippers. As is discussed in more detail subsequently, purified FOS protein is apparently incapable of sequence-specific interaction with DNA. However, a number of different studies have recently shown that FOS can interact stably with DNA when complexed with JUN (97–100c). The domain of JUN required for interaction with FOS has been localized to the leucine repeat (D. Bohmann, R. Tjian, personal communication). Likewise, the domain of FOS required for interaction with JUN has been localized to its leucine repeat (R. Gentz, F. Rauscher, C. Abate, T. Curran, personal communication; R. Turner, R. Tjian, personal communication; 95b, 95c). These observations lend support to initial predictions that the FOS/JUN complex might reflect heterodimer formation via the leucine zipper (94).

We speculate that cross-dimerization will occur between leucine zipper proteins, but that it will be limited and highly specific. Other amino acid residues within the leucine repeat helixes will almost certainly tailor "helix matching" by virtue of the same sorts of atomic interactions that are employed to form the tertiary structures of proteins. The code of specificities that dictate zipper formation constitutes a ripe target for X-ray crystallographic studies. Structural studies will also resolve the disposition of the adjacent basic region, which we anticipate to be the component of leucine zipper proteins that will make intimate contact with DNA. The basic regions of FOS, JUN, GCN4, and C/EBP tend to be free of prolines and glycines, and contain high densities of ion pairs (P. Sigler, W. Landschulz, P. Johnson, S. McKnight, unpublished). Will the basic regions occur in an α helical conformation? If so, how are these helixes disposed relative to the aligned zipper helixes? The fact that we do not even know whether the pairs of zipper helixes are oriented in a parallel or antiparallel manner emphasizes the need for detailed structural studies.

If it is indeed the case that certain leucine zipper proteins can cross-dimerize, one might imagine that the heterodimeric form would display a DNA-binding specificity different from either of the parental homodimers. This prediction stems from the expectation that the basic region of a leucine zipper protein will harbor the amino acids that make intimate contact with DNA. Heterodimer formation, by juxtaposing two different basic regions, might form a unique binding specificity. Ptashne and colleagues have succeeded in creating just this sort of situation by engineering an artificial, heterodimeric repressor protein that binds selectively to a correspondingly mixed dyad operator site (101). If our interpretations regarding heterotypic leucine zipper dimers are correct, a substantially expanded repertoire of DNA-binding specificities could be generated from a limited number of genetically encoded proteins.

In the examples of leucine zipper proteins that have so far been discussed, evidence is accumulating that the integrity of the zipper helix is required for DNA binding. On the other hand, inspection of the amino acid sequences of other regulatory proteins has revealed the presence of potential leucine zipper helixes in domains that are not required for interaction with DNA. A proposed role for this latter category of leucine zipper (found in certain hormone receptors, the *Drosophila* zeste protein, and the lymphoid cell–specific octamer-binding protein) is discussed in a subsequent section. We believe that the leucine zipper is not restricted, as was originally anticipated, to defining a particular class of DNA-binding proteins. Rather, this structure may also have been adapted for protein-protein interactions that are not directly involved in DNA recognition.

A recent amino acid sequence comparison of several regulatory proteins (*achete scute* and *daughterless* gene products of *Drosophila melanogaster;* mammalian MYOD, cMYC, lMYC, and nMYC proteins; and two immunoglobulin enhancer binding proteins) led Baltimore and colleagues to propose the existence of a new DNA binding motif termed "helix-loop-helix" (C. Murre, P. McCaw, D. Baltimore, personal communication). This hypothetical motif consists of two amphipathic helixes separated by a loop region of variable length. Kouzarides & Ziff (95b) noticed that this same region also exhibits sequence similarity to the DNA-binding domain of bacteriophage λ cII protein. In each of the MYC proteins, the helix-loop-helix motif resides immediately N-terminal to the leucine zipper motif, in a position equivalent to that occupied by the basic regions of proteins such as C/EBP, FOS, JUN, and GCN4. By inference, this region of the MYC proteins is hypothesized to confer the ability to recognize specific DNA sequences. If so, the MYC DNA-binding domain may represent a unique (thus far) combination of structural motifs. This may portend a variety of DNA-binding proteins constructed from different combinations of DNA-contacting and oligomerization domains.

Unclassified Binding Motifs

Genes that encode DNA-binding proteins are being cloned and characterized at a rapid pace (see Table 1). Geneticists are identifying, mapping, and isolating regulatory genes, many of which are turning out to encode proteins that bind to specific DNA sequences. Likewise, biochemists are identifying and purifying scores of sequence-specific DNA-binding factors. Such efforts, in an increasing number of cases, are leading to the acquisition of cloned genes. Finally, methods that allow the cloning of cDNAs encoding DNA-binding proteins by direct screening of expression libraries with radiolabeled DNA ligands have been recently developed (102, 103).

Given the reasonably large repertoire of these protein sequences now at hand, it is of interest to consider what proportion of characterized eukaryotic

Table 1 A partial list of genes encoding transcriptional regulatory proteins

regulatory protein[a]	DNA binding class[b]	transcriptional activator/genetic regulator[c]	in vitro DNA binding[d]	references[e]
Yeast				
GAL4	ZF	+	+	76
GCN4	LZ	+	+	265
HAP1	ZF	+	+	266
HAP2	n.c.	+	+	107
HAP3	n.c.	+	+	108
MATα1	HTH	+	+	267
MATa2	HTH	+	+	267
MATa1	HTH	+	+	267
ADR1	ZF	+	−	59
SWI5	ZF	+	+	60
PPR1	ZF	+	−	77
ARGRII	ZF	+	−	78
RAP1	n.c.	−	+	268
PHO4	n.c.	+	−	269
HSTF/HSF	n.c.	+	+	104, 105
LAC9	ZF	+	−	269a
Neurospora				
cys-3	LZ	+	+	270
qa-1F	ZF	+	+	79
cpc-1	LZ?	+	−	271
Drosophila				
antennapedia	HTH	+	−	32, 272
ultrabithorax	HTH	+	+	32, 272
fushi tarazu	HTH	+	+	32, 272
engrailed	HTH	+	+	273
paired	HTH	+	+	274
cut	HTH	+	−	275
snail	ZF	+	−	64
kruppel	ZF	+	−	62
hunchback	ZF	+	−	63
serendipity	ZF	+	−	276
suppressor of hairy wing	ZF	+	+	277
zeste	n.c.	+	+	220, 219
bsg 25D	LZ	−	−	278
achaete scute (T4 and T5)	n.c.[h]	+	−	279
daughterless	n.c.[h]	+	−	279a
Vertebrates				
v-FOS	LZ	+	+	280
c-JUN	LZ	+	+	205
Sp1	ZF	+	+	66

Table 1   (*continued*)

regulatory protein[a]	DNA binding class[b]	transcriptional activator/genetic regulator[c]	in vitro DNA binding[d]	references[e]
CTF/NF-1	n.c.	+	+	106
OTF-1(OCT1)	HTH	+	+	178
OTF-2(OCT2)	HTH	+	+	177, 177a
H2TF-1/NF-κB-like protein	ZF	—	+	f
SRF	n.c.	—	+	280a
PRDI	ZF	—	+	g
PIT-1	HTH	—	+	280b
C/EBP	LZ	—	+	92
TDF	ZF	+(?)	—	65
GLI	ZF	+	—	281
Evi-1	ZF	+	—	282
IRF-1	n.c.	+	+	283
glucocorticoid receptor	ZF	+	+	70, 284
estrogen receptor	ZF	+	+	72, 73, 285
progesterone receptor	ZF	+	+	71, 286
thyroid hormone receptor (c-erbA)	ZF	+	+	75
FRA-1	LZ	+(?)	—	287
JUN-B	LZ	+(?)	—	203
ZIF/268	ZF	+(?)	—	288
c-MYC	LZ[h]	+	—	289
MyoD	n.c.[h]	+	—	290
AP-2	n.c.	+	+	291
κE2	n.c.[h]	—	+	i

[a] This list includes only genes that have been cloned without bias toward a particular class of DNA-binding domain, in order to make a more valid estimate of the frequency with which the various DNA-binding classes are employed. Thus, genes that have been cloned by DNA cross-hybridization to existing DNA-binding motifs (for example many homeotic genes and receptor genes) have been excluded.

[b] Abbreviations used: HTH, helix-turn-helix; ZF, zinc finger; LZ, leucine zipper; n.c., not classified, meaning that no recognizable motif has been discerned from the deduced amino acid sequence.

[c] A plus (+) sign signifies that a protein (or its encoding gene) has been shown to act as a transcriptional activator or has a genetic regulatory function. These criteria include evidence of transcriptional stimulatory activity in vitro or in vivo, or genetic evidence that the protein either controls transcription of known genes, regulates developmental process (e.g. homeotic, pair rule, and gap genes), or is involved in regulating cell growth (e.g. nuclear oncogenes).

[d] Proteins that have been shown to bind specific sequences either autonomously, or in conjunction with other proteins (e.g. c-FOS with c-JUN) are indicated (+). Note that c-MYC has only been observed to bind DNA nonspecifically.

[e] The citations refer only to publications in which the gene sequences were reported.

[f] A. Baldwin, H. Singh, P. Sharp, personal communication

[g] A. Keller, T. Maniatis, personal communication

[h] The achaete scute, daughterless, MYC, MyoD1, and κE2 proteins form an apparently related family (279, 279a; C. Murre, P. McCaw, D. Baltimore, personal communication). From the sequence similarities, D. Baltimore and colleagues have proposed a structure, termed "helix-loop-helix," that constitutes a putative DNA-binding and/or dimerization domain for this family.

[i] C. Murre, P. McCaw, D. Baltimore, personal communication.

regulatory proteins bind DNA by one of the three classes of structural motifs (helix-turn-helix, zinc finger, or leucine zipper). Since each of the three established categories of DNA-binding motifs displays a diagnostic constellation of amino acids, it is possible to estimate the proportion of eukaryotic transcriptional regulatory proteins that fall within each category, as well as the proportion that appear to bind DNA via yet-undetermined structural motifs. The results of such an analysis are presented in Table 1. In hopes of preserving the validity of this survey, we have purposely omitted consideration of regulatory genes that have been cloned by virtue of nucleotide sequence similarity to known regulatory genes. According to our tally, roughly ⅘ of the eukaryotic DNA-binding proteins that have been sequenced can be placed into one of the three defined categories. We anticipate that the proteins falling outside of the three established classes, which include such interesting examples as heat-shock transcription factor (104, 105), serum response factor (R. Treisman, personal communication), CTF/NF1 (106), and HAP2 and HAP3 (107, 108), will found new families of DNA-binding motifs. However, given the calculation that a substantial proportion of the known examples fit into one of three established categories, it is possible that the entire spectrum of primary DNA-binding specificities will be encompassed within a handful of structural motifs.

PROTEINS THAT BIND TO DEGENERATE SEQUENCES

Prototypical repressor and activator proteins of bacterial cells bind with high affinity and specificity to discrete, easily recognized sites on DNA. Dissociation constants of 10^{-11} molar or lower commonly characterize the avidity of interaction. In contrast, it has become increasingly clear that eukaryotic gene regulatory proteins are not as simple and efficient as their prokaryotic counterparts. In this and the following two sections, we discuss three unexpected properties that have emerged from the study of eukaryotic protein/DNA interactions. First, in several cases purified proteins have exhibited the perplexing capacity to bind, with equal avidity, to sites on DNA that share only minimal nucleotide sequence similarity. Second, a given cis-regulatory sequence can frequently be the target of more than one DNA-binding protein. Finally, in an increasing number of cases, eukaryotic DNA-binding activities are turning out to require the collaboration of two different polypeptide chains. Although presently enigmatic, we anticipate that these subtleties may constitute early glimpses of the highly interdependent nature of eukaryotic gene control.

HAP1

One of the first examples of DNA-binding promiscuity was uncovered by Guarente and colleagues (109, 110) in studies of HAP1, a yeast protein that plays a regulatory role in the expression of genes that encode components of the respiratory chain (111). The *hap1* gene product is required for the heme-dependent expression of the *cyc1* and *cyc7* genes via their respective upstream activation sites (UASs), indicating that HAP1 behaves as a positive transcriptional regulator at both of these sites (110, 112). HAP1 binds to a site within the UAS of the *cyc7* gene of yeast, as well as to a site within the UAS of the *cyc1* gene. Paradoxically, however, the *cyc1*-binding site bears no recognizable DNA sequence similarity to the *cyc7* sequence site. Moreover, a HAP1 mutant, HAP1–18 (111), fails to bind to the *cyc1* UAS yet retains unaltered affinity for its recognition site within the *cyc7* UAS (110). Despite obvious DNA sequence dissimilarity, HAP1 binds with equivalent affinity to the two sites, makes comparable major groove contacts, and is cross-competed by either site in binding reactions performed in vitro. Since HAP1 does not appear to be capable of interacting simultaneously with both sites, it appears to have evolved in such a way as to accommodate two different binding specificities within a single DNA-binding domain. The proposed overlap of DNA binding specificities within the same domain of HAP1 differs from the organization of dual DNA binding specificities within the integrase protein of bacteriophage lambda (110a). In the latter case, Landy and colleagues have conclusively demonstrated, by proteolytic separation, that two distinct DNA binding specificities are encoded by the 40-kd integrase polypeptide.

Glucocorticoid Receptor

A second regulatory protein that exhibits binding promiscuity is the glucocorticoid receptor protein. After exposure to its cognate ligand, the glucocorticoid receptor acquires the capacity to bind to selective sites located within or around hormone-responsive genes. Binding sites that mediate the positive regulatory effect of the glucocorticoid receptor have been termed glucocorticoid response elements, or GREs (113). By evaluating nine high-affinity sites associated with murine mammary tumor virus (MMTV), Payvar et al (114) noted the occurrence of the degenerate octanucleotide AGAa/tCAGa/t. Related consensus sequences have been proposed from studies of glucocorticoid receptor binding to cellular genes (115–117).

Glucocorticoid receptor also acts as a hormone-dependent negative regulator of transcription. Sequences located upstream from the bovine prolactin gene confer glucocorticoid-dependent repression to a linked reporter gene (118). Sakai et al (119) mapped seven glucocorticoid receptor–binding sites in the upstream region of the prolactin gene, and by systematic mutagenesis, showed that at least three of the receptor-binding sites played a role in negative regulation of prolactin gene expression. Glucocorticoid receptor-binding sites that confer negative regulation (termed nGREs) bear minimal resemblance to the octanucleotide consensus sequence of positive GREs

(119). However, the region of the glucocorticoid receptor protein necessary for interaction with nGREs, as determined by binding assays carried out in vitro, corresponds to that required for interaction with GREs (120). Sakai et al (119) have speculated that nGREs might alter the conformation of DNA-bound glucocorticoid receptor, causing it to act as a repressor of transcription. An alternative interpretation for the negative regulatory effects of the gluco-corticoid receptor has been proposed by Akerblom and colleagues (120a). In studies on the human glycoprotein hormone gene, the latter authors concluded that the negative regulatory effects of the glucocorticoid receptor could be attributed to its occlusion of the binding site for another transcriptional activator protein.

These studies indicate that the glucocorticoid receptor, like HAP1, does not interact with its target sites on DNA in a rigid, unidimensional manner. In contrast to HAP1, however, the binding of glucocorticoid receptor to two different kinds of sequences can in some way effect opposite regulatory consequences.

C/EBP and OBP100

Binding site promiscuity has also been noted for several recently character-ized enhancer binding proteins. One such example comes from C/EBP, the rat liver protein that led to definition of the leucine zipper motif (94). C/EBP was originally detected according to its capacity for sequence-specific interaction with the CCAAT pentanucleotides of the herpesvirus thymidine kinase pro-moter and the murine sarcoma virus long terminal repeat (90). Unknowingly, Johnson et al (91) purified this same protein by biochemical fractionation of an activity that bound specifically to the "enhancer core homology," a de-generate nonamer sequence TGTGG$^{AAA}_{TTT}$G common to many animal virus enhancers (121). Since the protein bound avidly to both CCAAT homologies and enhancers, it was eventually termed CCAAT/enhancer binding protein (C/EBP). Studies of the interaction of C/EBP with genes that are expressed primarily in liver have underscored its binding promiscuity (122). Such studies have shown that C/EBP binds avidly to the albumin, α1-antitrypsin, and transthyretin genes at sites that share only minimal DNA sequence similarity. Although the molecular basis of C/EBP binding promiscuity re-mains to be resolved, footprint assays carried out with proteolyzed fragments of the protein have shown that both CCAAT and enhancer core specificities colocalize to a 14-kd domain containing the leucine zipper (92).

Another enhancer binding protein that displays relaxed interaction specifi-city is the 100 kd octamer binding protein (OBP100) studied by Herr and colleagues (123). This protein, which is considered in the context of two other issues in subsequent sections, was designated initially according to its ability to interact with the octamer sequence ATGCAAAT. More recently, Baumruker and colleagues (124) examined binding of OBP100 to a series of both natural and synthetic octamer sequence variants within the SV40 enhancer and within the enhancers of herpesvirus immediate early (IE) genes. One of the two sites of OBP100 interaction with the SV40 enhancer matched the canonical octamer sequence at only five out of eight positions. It was shown that OBP100 interaction at that site was dependent upon the integrity of DNA sequences that flanked the degenerate octamer. When internal residues were converted to match the canonical octamer consensus, dependence on flanking residues was relieved. Consistent with earlier observations that employed crude HeLa nuclear extracts (125), OBP100 was also observed to be capable of sequence-specific interaction with TAATGARAT, the nonanucleotide *cis*-regulatory sequence common to herpesvirus IE genes. OBP100 binding occurs at TAATGARAT, despite the fact that the herpesvirus IE sequence only matches the octamer consensus at four of eight positions. Despite the incongruence of sequence between the canonical octamer and the herpesvirus IE element, the two sites could be related by a smooth progression of base changes found in a series of OBP100-binding sites (124).

The examples of degenerate sequence binding that we have presented are not exceptional. Homeodomain proteins (engrailed, fushi tarazu, ultrabithorax, even-skipped, paired, and zerknullt) are each capable of binding to sets of DNA sequences whose members diverge considerably from a consensus (39–41). Likewise Davidson et al (126) have characterized a HeLa cell DNA-binding protein (TEF-1) that forms complexes with two different *cis*-regulatory elements of the SV40 enhancer (GT-IIC: TGGAATGTG and Sph: AAGc/tATGCA), which are identical at only four out of nine positions. Although we do not understand the significance of the relaxed stringency exhibited by many eukaryotic DNA-binding proteins, it is important to point out that these binary assays (using purified proteins and isolated target DNA sequences) could lack components or conditions necessary for optimal specificity.

ONE *cis*-REGULATORY SEQUENCE BINDS MULTIPLE PROTEINS

Octamer-Binding Proteins

In addition to the unanticipated flexibility of binding specificity exhibited by certain regulatory proteins, recent examples of the converse—unique DNA sequence elements that are capable of binding multiple regulatory proteins—have also appeared. Perhaps the clearest case of the latter phenomenon has emerged from the study of immunoglobulin genes. Lymphoid cell specificity of immunoglobulin gene transcription is conferred by both enhancers (127–131) and promoters (132–136) associated with these genes. A conserved sequence, usually referred to as the octamer element (ATGCAAAT), was found within the immunoglobulin heavy chain (IgH) enhancer and promoter, and within the kappa light chain (IgK) promoter (132, 137, 138). Deletion mutants of immunoglobulin gene promoters have implicated the octamer element as an important determinant of lymphoid cell–specific transcription (132, 133, 135, 136, 139, 140). It came as a surprise, therefore, to discover that the octamer motif is also associated with genes whose expression is not restricted to lymphoid cells. Examples of these include histone H2B (141, 142), herpesvirus thymidine kinase (143), and U1 and U2 snRNAs (144–147). Moreover, as mentioned in the preceding section, octamer-related elements occur within the enhancers of SV40 and herpesvirus IE genes.

Evidence for the binding of proteins to the immunoglobulin gene octamer elements was first demonstrated by Singh et al (148). Electrophoretic retardation assays were used to identify an activity, present in both B lymphoma and HeLa cell nuclear extracts, which bound several different octamer elements. Staudt and colleagues (149) extended the analysis by defining two octamer-binding activities, NF-A1 and NF-A2. The latter activity is lymphoid cell specific, whereas the former was observed in nuclear extracts derived from a number of different cultured cell types. Numerous subsequent studies have supported the notion of multiple octamer-binding proteins (150–159).

The lymphoid octanucleotide activity has been termed NF-A2 (149), OTF2 (160), or oct-B2 (158), whereas the ubiquitous protein has been variously termed NF-A1 (149), OTF1 (161), OBP100 (123), NFIII (176), and oct-B3 (158). For the sake of simplicity, we will adopt the terminology conventions proposed by P. Sharp and W. Herr, in which the ubiquitous and lymphoid proteins are designated OCT1 and OCT2, respectively. We recognize, however, that all "ubiquitous" octanucleotide-binding factors are not necessarily specified by the same gene product. There may indeed exist multiple ubiquitous or even multiple lymphoid-specific octamer-binding proteins (see, for example, Ref. 158).

The polypeptides specifying both lymphoid (160) and ubiquitous (123, 161) octamer-binding proteins have been purified. The intrinsic DNA-binding properties of the lymphoid and ubiquitous octamer-binding proteins are indistinguishable. They contact the octamer ligand in the same way according to methylation interference assays (149) and respond similarly to the effects of base substitution mutations in the octamer DNA sequence (149). Each protein behaves as a positive activator of transcription in cell-free systems. Roeder and colleagues have shown that the ubiquitous octamer-binding protein is capable of in vitro activation of the histone H2B promoter (161), and that the lymphoid-specific protein is capable of activation of the immunoglobulin promoter (160). However, reciprocal transcription tests were not included in this series of experiments. LeBowitz et al (162) tested the relative abilities of purified OCT1 and OCT2 to activate the Ig heavy chain promoter in vitro. In these experiments, both proteins elevated transcription when used to fortify extracts that had been depleted of octamer-binding proteins, although OCT2 was slightly more active than OCT1 in this assay.

In light of the minimal differences exhibited by OCT1 and OCT2 in vitro, it becomes difficult to understand why Ig promoters are inactive in nonlymphoid cells that express appreciable levels of OCT1. Although the resolution of this enigma is not yet clear, two properties have recently been discovered that may serve to distinguish OCT1 from OCT2. The first comes from studies of the involvement of octamer elements in herpesvirus immediate early (IE) gene transcription. The TAATGARAT sequence is a *cis*-regulatory element common to herpesvirus IE gene enhancers, and is a target for *trans*-activation by a viral gene product variously termed VP16 or Vmw65 (163–170). VP16 action at TAATGARAT is not direct (171), but rather appears to be mediated by cellular proteins that bind to this element (125, 169, 172–175). Consistent with this notion, a ternary complex involving TAATGARAT, cellular factors, and VP16 can be detected by gel retardation assays (173–175). TAATGARAT sequences resemble the octamer consensus (125, 176), suggesting that the cellular factor(s) that bind to this motif might include an octamer-binding protein (124, 125). Gerster & Roeder (175) have demonstrated that purified OCT1 and OCT2 bind with equivalent affinities to the TAATGARAT sequence, but that only OCT1 is capable of participating in the higher-order complex with VP16. This observation may be relevant to the ability of OCT1 and OCT2 to discriminate between different octamer sequences associated with cellular genes, and raises the possibility that there may be cellular analogues to VP16.

The genes that encode lymphoid (177, 177a, W. Schaffner, personal communication) and ubiquitous (178) octamer-binding proteins have recently been cloned and sequenced. The conceptually translated amino acid sequences of both proteins exhibit distinct regions of sequence similarity (177, 178). The most substantial region of amino acid sequence similarity between the two proteins coincides with an area that bears similarity to the helix-turn-helix motif (177). These findings probably account for the indistinguishable binding specificities of the lymphoid and ubiquitous proteins. Given the expectation that the proteins will play different regulatory roles in vivo, it is of interest to examine regions of sequence divergence between OCT1 and OCT2. Clerc and colleagues (177) and Scheidereit et al (177a) have pointed out a perfect heptad repeat of leucines located roughly 45 amino acids toward the carboxyl terminus from the putative helix-turn-helix motif of OCT2. This leucine repeat does not occur in OCT1 (178), and may therefore represent a component of OCT2 that helps tailor its activation specificity.

CCAAT-Binding Proteins

A second example of overlapping recognition specificity has emerged from studies of proteins that bind to the CCAAT elements. The pentanucleotide sequence CCAAT is commonly found 50 to 100 base pairs upstream from the start site of eukaryotic genes that encode mRNA. A variety of studies have demonstrated the involvement of CCAAT elements in promoter function (reviewed in 6, 7). The first two CCAAT-binding activities to have been identified, purified, and characterized at the amino acid sequence level were CTF/NFI (106, 179, 180) and C/EBP (90–92). Although early evidence indicated CTF/NFI and C/EBP were capable of sequence-specific interaction with CCAAT homologies of the same gene (the herpesvirus thymidine kinase gene; 90, 179), they exhibited certain distinguishing properties. For example, C/EBP was heat stable and CTF/NFI was not. Likewise, a C-to-G transversion mutation at the first residue of the CCAAT pentanucleotide had opposing effects on the binding of the two proteins. This mutation caused C/EBP to bind more avidly than normal (90), yet impeded binding by CTF/NFI (K. Jones, R. Tjian, unpublished observations). Taking these different properties into account, it was not surprising to find that CTF/NFI and C/EBP are specified by different polypeptides which, in turn, are encoded by entirely different genes (92, 106). What comes as some surprise, however, is the observation that CTF/NFI and C/EBP utilize entirely different DNA-binding motifs. As outlined earlier, C/EBP binds DNA via its leucine zipper and associated basic region. The amino acid sequence of CTF/NFI bears no resemblance to that of C/EBP, nor does it exhibit similarity to the constellation of residues diagnostic of either the helix-turn-helix motif or the zinc finger motif.

One might have imagined, a priori, that all CCAAT elements are equivalent in the sense that they bind a single cellular factor. On the contrary, biochemical evidence supports the view that there are many different CCAAT-binding proteins. In addition to C/EBP and CTF, CCAAT-binding factors have been described in numerous other studies. They have been detected in extracts from murine erythroleukemia cells (181), sea urchin embryos, and testis (CBF; 182), B myeloma cells (NF-Y; 183), NIH 3T3 and rat liver cells (also called CBF; 184). Moreover, Chodosh et al (185) observed three physically separable CCAAT-binding activities from HeLa nuclei (CP1, CP2, and CTF/NFI), and Raymondjean et al (186) reported that rat liver nuclei contain C/EBP, CTF/NFI, and a third CCAAT factor (probably NF-Y). In addition to their DNA-binding properties, CTF/NFI (179) and mammalian CBF (187) were shown to stimulate transcription in reconstituted extracts.

Due to many experimental variables (species differences, tissue-specific variations, differences in DNA-binding assays, etc), it is difficult to evaluate the relationships amongst those CCAAT proteins for which a pure polypeptide or cloned gene has not yet been obtained. However, the following general statements can be offered with reasonable confidence. First, there is no "universal" CCAAT-binding protein. Each factor binds to a subset of CCAAT boxes, and these specificities form partially overlapping sets. Second, the total number of different proteins is a minimum of three (yet probably greater). Third, multiple CCAAT-binding proteins can exist within the same cell. Finally, as discussed in the next section, many of these activities require two heterologous protein components.

HETERODIMERIC DNA-BINDING ACTIVITIES

CCAAT-Binding Proteins

The enigmatic nature of DNA protein recognition at the CCAAT pentanucleotide has been compounded by the recent definition of a class of DNA-binding activities that require two different polypeptides. Hatamochi et al (184) have described a DNA-binding activity from rat liver that is capable of stable interaction with the CCAAT element of the $\alpha2(I)$ collagen promoter. This activity, termed CBF, could be separated into two components (CBF-A and CBF-B) (188). CBF-A behaves as an anion at neutral pH and is heat labile, whereas CBF-B is a heat stable, cationic protein. By recovering and renaturing protein bands from SDS gels, it was possible to identify CBF-A as a 39-kd polypeptide, and CBF-B as a 41-kd polypeptide. Whether the heat stable CBF-B is related or identical to C/EBP is unclear. Chodosh et al (185) have also defined a multicomponent CCAAT-binding activity. This latter activity, discovered in extracts prepared from HeLa cells, was divided by conventional chromatography into two fractions, termed CP1A and CP1B. Neither fraction was alone capable of sufficiently stable interaction to retard a CCAAT DNA probe in a gel retardation assay. However, when the two fractions were mixed, binding activity was reconstituted. A second multicomponent CCAAT-binding activity, termed CP2, was also described in the Chodosh et al study (185). The four components that define CP1 and CP2 (CP1A and B, CP2A and B) were chromatographically separable, and only the homologous pairs could be mixed to give functional binding activities. CP1 and CP2 were further distinguished according to the capacity to interact with different CCAAT elements.

HAP2/HAP3

A similar, two-component DNA-binding dependency has been discovered via genetic studies on catabolite derepression of *cyc1* gene expression in yeast. As discussed earlier, *cyc1* can be activated by HAP1, a DNA-binding protein that acts via the UAS1 element located upstream of the *cyc1* gene. Activation of the *cyc1* gene can also be accomplished via a second *cis*-regulatory element, termed UAS2. Guarente and colleagues have shown that the products of two genes, termed *hap2* and *hap3*, are required for UAS2 activation (112, 189). Yeast extracts contain a DNA-binding activity specific for UAS2 that is detectable by the gel retardation assay (190). Strains containing either a HAP2/β-galactosidase, or HAP3/β-galactosidase fusion gene give rise to complexes of reduced electrophoretic mobility (relative to the complex established from extracts derived from wild-type yeast strains), indicating that both gene products are present in the DNA:protein complex. Moreover, Hahn & Guarente (191) have recently shown that the HAP2 and HAP3 proteins are capable of forming a complex in the absence of DNA.

During the process of characterizing the activation of *cyc1* by HAP2 and HAP3, it was noticed that the binding site within UAS2 was related to the CCAAT homology. The native sequence contained the pentanucleotide sequence CCAAC. A transition mutation that changed the last C residue to T, creating a perfect CCAAT pentanucleotide, was observed to increase UAS2 activating potential (112). In a remarkable series of experiments, Chodosh et al (192) demonstrated that CP1A is the HeLa equivalent of HAP3, and that CP1B is the equivalent of HAP2. That is, CCAAT-binding activity could be reconstituted by the mixing of CP1A and HAP2; likewise, activity could be regenerated by mixing CP1B and HAP3. These fascinating observations indicate that two distinguishable specificities have been preserved during the evolutionary divergence between yeast and humans. First, the shapes of the two polypeptide complexes that generate binding specificity have remained sufficiently similar to recognize the same DNA sequence. Second, the protein surfaces that lead to multimerization have remained sufficiently similar to allow subunit interchange. The amino acid sequences of HAP2 and HAP3 are known (107, 108), and neither exhibits relatedness to any of the classified DNA-binding motifs.

Yeast Mating Type Regulatory Proteins

Evidence for binding specificities that consist of multiple polypeptides has emerged, in three forms, from the yeast regulatory circuit that controls mating type. MATα1 is known to be an activator of α-specific genes (193, 194). UAS elements preceding three α-specific genes contain a related sequence. Bender & Sprague (195) have shown that extracts from yeast cells expressing MATα1 contain a binding activity capable of recognizing this regulatory DNA sequence, but that bacterially produced MATα1 does not bind to the sequence. The resolution to this puzzle came from the discovery of an activity, common to all sexual forms of yeast, that binds stably to the α-specific UAS when mixed with MATα1. This complementing activity, termed PRTF, can bind on its own to a palindromic version (P(PAL)) of the aforementioned α recognition sequence. However, most P sites cannot bind PRTF stably, and are found adjacent to a second conserved sequence (termed Q). PQ sites are recognized by PRTF only when MATα1 is present, suggesting that a MATα1:PRTF complex is the functional activity capable of activating α-specific genes (Figure 2).

MATα2 is a helix-turn-helix protein that represses a-specific genes in haploid yeast cells. In an elegant series of experiments, Johnson & Herskowitz (34) demonstrated that MATα2 binds to specific sites (α2 operators) located upstream of the genes that it represses. In diploid yeast cells, MATα2 collaborates with MATa1, another helix-turn-helix protein, to coordinately repress haploid-specific gene expression (196). Goutte & Johnson (197) found that the sequence specificity of MATα2 is altered in the presence of MATa1. Without a MATa1 subunit, MATα2 binds to α2 operators, whereas in the presence of MATa1 it loses this activity and acquires specificity for sites associated with haploid-specific genes (α2/a1 operators). It is not known whether MATα2, MATa1, or both proteins recognize the α2/a1 operator sites. However, α2 and α2/a1 sites are similar, and MATα2 binds weakly to α2/a1 operator sites on its own. Since both types of operator sites consist of two symmetrical half sites, and since both regulatory proteins exhibit helix-turn-helix motifs, it has been speculated that the MATα2/MATa1 complex exists as a heterodimer wherein each subunit interacts with its cognate half of the dyad binding site (198).

The regulatory circuit of yeast mating type has offered yet a third example of factor collaboration. Keleher et al (199) have found that transcriptional repression by MATα2 requires cooperative interaction with a second polypeptide. This auxiliary component, which is probably identical to PRTF, binds to a site on α2 operators located between widely spaced MATα2 dyad half sites. Repression in vivo requires the integrity of both dyad half sites, as well as the PRTF-binding site. Sauer et al (200) have shown that an amino-terminal domain of MATα2, fully distinguishable from its helix-turn-helix motif, specifies its ability both to form dimers, and to interact with PRTF. The MATα2 dimer is remarkably flexible. It can occupy half sites separated by variable amounts of intervening DNA, and binding does not require adjacent half sites to occur on successive faces of DNA.

FOS/JUN

Both FOS and JUN are nuclear proteins encoded by proto-oncogenes. That these transforming proteins might play a role in transcriptional regulation became apparent as a consequence of two recent discoveries: first, JUN is related to a set of sequence-specific DNA-binding proteins collectively

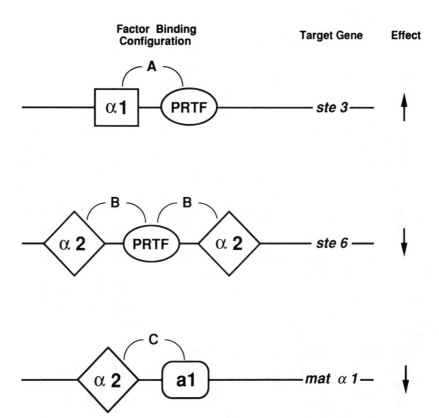

Figure 2 Schematic representation of protein:DNA interactions involved in the control of yeast mating type.

The consequences of transcription factor interaction with three yeast genes provide evidence of at least three protein:protein interaction interfaces. The *matα1* (α1) gene product binds in collaboration with PRTF to stimulate expression from the *ste3* gene, giving evidence of protein:protein interaction interface A. The *matα2* gene product (α2) binds in collaboration with PRTF to repress expression from the *ste6* gene, giving evidence of protein:protein interaction interface B. α2 also collaborates with the *mata1* gene product (a1) to repress expression from the *matα1* gene, giving evidence of protein:protein interaction interface C.

termed AP1; and second, JUN and FOS form a stable heteromeric complex in vivo. AP1 was originally defined as a HeLa transcription factor capable of sequence-specific interaction with the SV40 virus enhancer (201). R. Tjian and K. Struhl (personal communication) noticed similarity between the binding sites of AP1 and GCN4, a yeast protein known to play a pivotal role in general amino acid control. Point mutations within the DNA recognition site for GCN4 that inhibited binding also blocked binding of AP1 (201a). Thus, when Vogt and colleagues (89) reported amino acid sequence relatedness between the v-*jun* oncogene and GCN4, it was rapidly realized that AP1 might represent the product of the c-*jun* proto-oncogene (157, 202). The Recent work from several laboratories has indicated that AP1 activity in fact consists of several polypeptides, encoded by a family of highly related genes (99, 203), and that the JUN polypeptide is one of the members of this set. The synthesis or activity of AP1 proteins can be modulated by growth factors (203) and classically defined signal transducers (157, 204). As such, it is not surprising that an aberrant form of AP1 (v-JUN) leads to cellular transformation (205).

FOS is also a nuclear transforming protein (206). However, in its purified form, FOS has never been observed to make sequence-specific contact with DNA. Evidence that FOS might act as a direct regulator of gene expression, despite its inability to bind DNA on its own, came from three directions. First, it was shown to be capable of *trans*-activating a collagen gene promoter in transient transfection assays (207). Second, when linked to DNA by virtue of an artificially attached DNA-binding domain, it activated a *cis*-linked promoter (208). Third, it was found to be capable of forming a ternary complex with a promoter element of a gene expressed preferentially in adipocytes (97). A series of recent experiments have shown that FOS interaction sites on DNA are highly similar to AP1 recognition sites (97a, 98), and that the ternary complex observed in gel retardation assays involves either JUN or a highly related protein (99, 100). Evidence supporting the regulatory significance of these biochemically defined interactions has recently been presented by Chiu et al (209). In this study, a reporter gene containing an AP1 *cis*-regulatory element was transcribed only in cells that simultaneously express both the FOS and JUN proteins.

As outlined earlier, both FOS and AP1 are leucine zipper proteins (94), and the leucine repeat region of each protein is required to support complex formation (R. Gentz, F. Rauscher, C. Abate, T. Curran, personal communication; R. Turner, D. Bohmann, R. Tjian, personal communication; 95b, 95c). Landschulz et al (94) have hypothesized that dimerization of two polypeptides via the leucine zipper motif leads to a juxtapositioning of adjacent basic regions, and that it is these basic regions that provide sequence-specific contacts with DNA. Since the basic regions encoded in the FOS and JUN polypeptide sequences are different, one might predict that the binding properties of a FOS/JUN heterodimer would be distinguishable from a JUN/ JUN homodimer. Perhaps an indication of this comes from the observation that addition of FOS protein substantially increases the affinity of JUN for a variety of AP1 binding sites as measured by gel retardation assays (100a-c). Whether this improved binding to DNA is due to differences contributed by the FOS basic region, or to greater stability of the JUN/FOS heterodimer compared with the JUN/JUN homodimer, is currently unresolved.

THE JIGSAW PUZZLE HYPOTHESIS

We have pointed out several paradoxical properties of eukaryotic gene regulatory proteins. They frequently bind to their cognate recognition sites on DNA with substantially less avidity than bacterial repressors, and in numerous instances have been found to recognize binding sites that exhibit limited nucleotide sequence similarity. Moreover, it has become clear that polypeptides encoded by entirely different genes can, when assayed in vitro, bind with indistinguishable affinities to the same sites on DNA. How can regulatory proteins that exhibit these properties, such as *Drosophila* homeotic gene products, exert the distinguishable effects necessary to execute complex patterns of gene regulation? One view is that homeobox proteins compete for the same sites near target genes, in simple bimolecular reactions. Under this model, occupation of a particular *cis*-regulatory element is a dual function of the affinity of each protein for the *cis* element, compounded by the relative nuclear concentration of each homeodomain protein. An alternative explanation, which we tend to favor, posits that increased specificity or stability of binding can be achieved through protein:protein interactions. That is, a DNA-binding protein that exhibits promiscuous interaction specificity in vitro may acquire substantially improved specificity when complexed with one or more other proteins in vivo.

An ancillary factor that improves the binding specificity of a gene regulatory protein might operate in one of two ways. It could contribute, from its own polypeptide structure, additional contact surfaces with DNA. Alternatively, the role of an ancillary protein might be indirect. Its contacts with a second regulatory protein might prompt an allosteric change in the protein, causing it to bind certain DNA sequences more avidly. In the absence of extensive structural information, it may be difficult to distinguish between these two possibilities. For example, on its own, FOS protein exhibits little or no intrinsic affinity for specific DNA sequences. When mixed with JUN, FOS enters into a heterotypic protein:DNA complex that can be more stable than the complex formed by JUN alone. It is not known whether FOS donates DNA contact surfaces to the complex, as might be predicted by the leucine zipper model, or whether it enters into the complex solely by protein:protein contacts with JUN.

A particularly clear-cut example of the role of auxiliary factors has emerged from studies of TFIIIA and its formation of an active 5S gene transcription complex. On its own, TFIIIA binds to the 5S internal promoter with only moderate avidity (a K_d of approximately 10^{-9} M; 57). However, when abetted by an additional factor (TFIIIC), TFIIIA forms an exceedingly stable association with the 5S gene. Once formed, this complex is resistant to challenge by excess 5S DNA (46, 210, 211). Differences in stability of these transcription complexes appear to underlie the developmental switch during frog embryogenesis that turns off oocyte-type 5S genes and activates somatic-type 5S genes (reviewed by A. Wolffe, D. Brown; 212). The two types of 5S genes differ by only a few nucleotides in their promoter sequences and exhibit relatively minor differences in their affinity for purified TFIIIA. Thus, key regulatory features of 5S transcription have been revealed that would not have been predicted by the binding properties of TFIIIA alone.

Factors capable of improving the binding specificity of another protein can, in certain cases, exhibit an intrinsic DNA-binding activity of their own. The most refined examples of this category have emerged from studies of the regulatory circuit that controls sexual phenotype in yeast. Both MATα1 and MATα2 rely on PRTF in order to interact stably with certain regulatory sites on DNA (195, 199), and MATα2 somehow collaborates with MATa1 to repress haploid-specific genes in diploid yeast cells (197). All of these proteins (except MATa1) have been shown to exhibit intrinsic affinity for DNA, and it is only when their respective binding sites on DNA are appropriately juxtaposed that they can exert stabilizing influences on one another. In a conceptual sense, this phenomenon is no different from the cooperative association that stabilizes interaction between bacteriophage lambda repressor and its weaker binding sites on DNA (20). However, the novel contribution that has emerged from the yeast work is that cooperative interactions can be heterotypic. That is, PRTF can augment the interaction specificity of other proteins.

Cooperative interactions, at least in the case of lambda repressor, are thought to result from specific protein:protein contacts. Stabilization of repressor binding to adjacent operator sites is dependent on a carboxyl terminal domain of lambda repressor (20), which has recently been defined more precisely by the mapping of a single amino acid substitution mutation that blocks cooperativity (212a). If PRTF influence on MATα1 and MATα2 is comparable, in a mechanistic sense, to the cooperative behavior of lambda repressor, then one protein (PRTF) must have evolved the specificity to interact with at least two other proteins. These interactions may therefore begin to define a particular class of protein:protein interaction domains.

The story that is emerging from studies of mating type regulation in yeast may be compatible with interpretations generated from studies on viral enhancers. Enhancers consist of patchworks of cis-regulatory sequences, each of which is thought to specify the interaction site of a sequence-specific DNA-binding protein (reviewed in 5, 7). The combination of cis-regulatory sequences that generates a functional enhancer is almost certainly nonrandom. By evaluating the activating properties of synthetically constructed enhancers, as well as revertant forms of the SV40 enhancer generated by genetic selections initiated with a mutationally inactivated form, Herr and colleagues (213,

214) have shown that activity depends critically on the appropriate juxtaposi-
tion of *cis*-regulatory sequences. They propose that individual sequence
motifs can be either duplicated or combined in heterologous pairs to create an
active enhancer. Similar conclusions about enhancer architecture have been
reported by Chambon and colleagues (215). It is likely that the spatial
requirements necessary to assemble enhancers reflect cooperative interactions
between DNA-bound proteins.

Yamamoto (216) has outlined a simple, illuminating concept predicting
that the specificity of an enhancer will be generated through the varied
combinatorial arrangement of its *cis*-regulatory elements. The afore-
mentioned yeast and viral studies allow Yamamoto's combinatorial hypothesis to be refined. In order for a protein to participate in the formation of an
active enhancer complex, it must be suited for specific interactions not only
with a *cis*-regulatory DNA sequence within the enhancer, but also with
protein interfaces contributed by factors tethered at adjacent positions within
the enhancer. This concept can be compared to the properties of a jigsaw
puzzle. A regulatory protein that is to fit into an enhancer must be compatible
with multiple surfaces: these surfaces include both its cognate DNA sequence
motif and protein structural motifs presented by adjacently bound proteins.

A hypothetical example of the importance of protein:protein interaction in
establishing "jigsaw fit" comes from a comparison of OCT1 and OCT2, the
two octamer-binding proteins. Recall that the ubiquitous protein, OCT1, is
capable of forming a stable complex with the herpesvirus activator protein
VP16, but that the lymphoid cell–specific protein, OCT2, is not. By analogy
to the specificity of VP16 for OCT1, one might predict that OCT2 would also
be capable of some distinguishing protein:protein interaction. Although no
biochemical evidence has been generated to support this anticipation, recent
comparisons of the amino acid sequences of the two proteins (177, 178) have
revealed a potential leucine zipper structure in the sequence of OCT2 outside
the region thought to be necessary for DNA recognition. While OCT1 exhibits a helix-turn-helix motif that is exceedingly similar in sequence to that of
OCT2, it completely lacks this leucine repeat. It will be exciting to test
whether the leucine repeat segment of OCT2 plays any role in protein:protein
interactions that might qualify this factor for lymphocyte-specific gene expression.

Leucine repeats outside of the immediate DNA-binding domains of several
additional eukaryotic gene regulatory proteins have recently been observed.
The glucocorticoid, thyroid, and estrogen receptor proteins, which bind DNA
via cys_x zinc fingers (217), all contain two to four heptad repeats of leucines
(P. Godowski, K. Yamamoto, personal communication; Y. Yaoita, D.
Brown, personal communication; S. Tsai, B. O'Malley, personal communication). In each of these cases, the leucine repeats occur close to thei

respective carboxyl termini, and are not required for DNA binding as assayed in vitro. As in the case of OCT2, no role has been established for the leucine repeats associated with hormone receptors. Another intriguing correlation between leucine repeats and gene regulatory proteins has arisen in the case of the zeste gene product of *Drosophila*. The zeste protein binds DNA by virtue of a structural domain located close to its amino terminus (218). C. Dang (personal communication) pointed out to us that the carboxyl-terminal portion of zeste contains a helix-permissive region that encodes six heptad leucine repeats interrupted only once by histidine (LLHLLL; 219, 220). Since the zeste gene product is known to play a fundamental role in transvection (221), we offer the speculation that this carboxyl-terminal region might mediate the flow of regulatory information between synapsed chromatids.

Other types of multimerization interfaces are likely to be identified in the near future. For example, it should be possible to define the structural motif that allows PRTF to interact with both MATα1 and MATα2. Likewise, by combining gel retardation assays with in vitro mutagenesis, it should be possible to define the respective regions of VP16 and OCT1 that are required for their interaction. Finally, we make note of exciting recent discoveries showing that a related structural motif present in the transforming proteins of adenovirus and SV40 facilitates interaction with the retinoblastoma gene product (222, 223). Although this latter example does not necessarily relate directly to transcriptional regulation, it does reinforce our expectation that a limited number of multimerization interfaces may have evolved in the construction of genetic regulatory circuits.

LATENT ACTIVATORS

In the previous section we emphasized the involvement of multiple proteins in the generation of an active transcriptional regulatory complex. Although the interplay of multiple proteins is likely to be a common theme in eukaryotic gene regulation, the absence or inactivity of a single factor does not necessarily lead to dissolution of a transcription complex. That is, genes can be programmed to a state of "competence," awaiting only a single regulatory cue to activate transcription (e.g. 224). One straightforward explanation for the absence of the activating "cue" is the lack of an essential regulatory protein (due to the inactivity of its encoding gene, or to the regulated translation of its encoding mRNA). Instead of focusing on this category of control, which may be the mode of regulation used most often in developmental hierarchies of gene expression, we have chosen to devote this final section to "latent" activator proteins: proteins that already exist in a cell, yet are somehow masked in their activating function.

Cytogenetic studies first demonstrated that the gene expression program of a eukaryotic cell could be changed in the absence of new protein synthesis. This phenomenon, discovered in classical studies of puffing patterns in *Drosophila* salivary gland chromosomes (225–227), showed that transcriptional induction by both hormonal stimuli and heat shock could occur in the absence of protein synthesis. It is a tribute to the insight of these early studies that some of the most active and critical research remains focused on these same two systems.

Heat-Shock Transcription Factor

The new chromosome puffs that Ritossa (228) observed following heat treatment of *Drosophila* salivary chromosomes encode an evolutionarily conserved set of proteins (heat-shock proteins). The heat-shock genes are induced to be transcribed following exposure to increased temperature (reviewed in 229–231). The induction of heat-shock gene expression is mediated by a *cis*-regulatory sequence known as the heat-shock element, or HSE (232). The HSE has been found upstream of all heat-shock genes that have been studied, and represents the binding site for a transcription factor (HSTF) first defined in extracts of *Drosophila* cells (233, 234). HSTF proteins have also been isolated from yeast (235, 236) and HeLa cell cultures (237, 238).

The regulatory apparatus for the control of heat-shock gene expression has been highly conserved through evolution. HSEs associated with eukaryotic genes, regardless of origin, are nearly invariant. Moreover, HSTFs do not distinguish between HSEs from various species. However, the manner in which HSTF activity is regulated appears to differ between yeast and higher eukaryotic cells. In cultured *Drosophila* cells, it has been shown that the DNA-binding activity of HSTF is nearly undetectable when cells are grown at normal temperature. Upon heat shock, the affinity of HSTF for its binding site on DNA is rapidly unveiled, reaching a maximal level after five minutes. This induction occurs in the presence of cycloheximide, showing that new synthesis of HSTF does not account for the appearance of binding activity (239). Likewise, in HeLa cells, HSTF activity is only apparent after exposure of cells to elevated temperature (236, 238), and is independent of new protein synthesis.

In yeast, levels of HSE-binding activity are equivalent whether the cells are grown at 23°C or 39°C (236). This observation suggests that yeast HSTF may not be regulated at the level of DNA binding, but that its capacity to activate transcription is in some way regulated. Consistent with this interpretation is the observation that yeast HSTF adopts an altered electrophoretic mobility following heat treatment of cells, a transition that can be reversed by phosphatase treatment of the protein in vitro (105, 236). It thus appears that yeast HSTF can activate transcription only after heat shock–dependent phosphorylation. Sorger & Pelham (105) have hypothesized that the phosphorylation of yeast HSTF may create an acidic activation domain analogous to ones described recently on GAL4, GCN4, and VP16 (93, 169, 240). If HSTF is overexpressed in yeast, it boosts transcription of a heat-shock gene *(hsp70)* even at low temperature, suggesting that HSTF has a low basal activating capacity. This latter observation is consistent with the fact that the gene encoding yeast HSTF is essential, even at normal growth temperature (104, 105).

Hormone Receptors

Activation of gene expression in response to hormones, like the heat-shock response, can proceed in the absence of protein synthesis. Perhaps the clearest studies on hormone-mediated activation of gene expression have come from work on the glucocorticoid receptor protein. As outlined earlier, the glucocorticoid receptor binds to its cognate GRE via a cys_x zinc finger motif located in the central region of the protein. Studies making use of the cloned receptor gene have also localized the hormone-binding domain to the carboxyl-terminal half of the receptor (83, 84, 120, 241). Picard & Yamamoto (242) have identified an additional regulatory component within the glucocorticoid receptor. By fusing segments of the receptor to β-galactosidase, two nuclear localization signals were defined. One of these signals caused β-galactosidase to migrate to the nucleus constitutively. The other signal, which was mapped in close association with the hormone-binding domain of the protein, was observed to function as an active nuclear localization signal only in the presence of the cognate hormone. Since nuclear localization of the intact hormone receptor is also hormone dependent, it is possible that the regulated "unmasking" of a nuclear localization signal plays a fundamental role in hormone-mediated regulation of gene expression.

Properties relevant to hormone-dependent nuclear localization of glucocorticoid hormone receptor have also emerged from biochemical studies addressing its association with other cellular proteins. In the absence of hormone, the receptor is localized in the cytoplasm in a large multiprotein complex (243–246). One of the proteins of this complex is the 90-kd heat-shock protein (247, 248). The biological relevance of interaction between the glucocorticoid receptor and the 90-kd heat shock protein is supported by studies demonstrating that the association can be reversed by hormone treatment (249, 250). Furthermore, the receptor/hsp90 complex does not form on truncated versions of the receptor that lack the hormone-binding domain (251). "Transformation" of the latent glucocorticoid hormone receptor appears to be a surprisingly complex process, involving ligand binding, release from cytoplasmic proteins, unmasking of a nuclear localization domain, and direct association with target genes in the nucleus.

NF-κB

A second example of cytoplasmic sequestration of a latent gene activator protein has emerged from studies of NF-κB. Sen & Baltimore discovered NF-κB as a B-cell–specific DNA-binding protein that recognizes a *cis*-regulatory element within the enhancer of an immunoglobulin light chain gene (252). NF-κB binding to this enhancer is thought to be an important event in controlling the tissue specificity and developmental timing of immunoglobulin gene expression (253, 254).

Although its natural occurrence appears to be restricted to mature B lymphocytes, NF-κB activity can be induced in pre-B cells by treatment with bacterial lipopolysaccharide (LPS) or phorbol esters (255). The induction of NF-κB-binding activity does not require protein synthesis, suggesting that an inactive form of the protein exists prior to LPS stimulation. No NF-κB is detectable in extracts prepared from either cytoplasmic or nuclear fractions of pre-B cells (256). However, NF-κB activity could be generated from cytoplasmic extracts of pre-B cells if the extracts were cycled through a denaturation/renaturation regimen. Binding activity could not be unmasked by similar treatment of nuclear extracts. Moreover, phorbol ester treatment of cells prior to subcellular fractionation eliminated the latent activity from the cytoplasm, and led to the appearance of an "unmasked" form of NF-κB in the nuclear compartment. The simplest explanation for these results in that pre-B cells synthesize a form of NF-κB that is sequestered in the cytoplasm in an inactive state. It is possible that the active form is masked by an inhibitory molecule or protein that is released upon exposure to the appropriate intercellular signal and/or developmental cue.

GAL4/GAL80

Galactose metabolism in yeast is regulated by both positive and negative control (reviewed recently in 257). Positive control of many of the structural genes involved in galactose catabolism is effected by a common regulatory protein termed GAL4. As outlined in an earlier section, GAL4 binds DNA by virtue of a cys_x zinc finger located close to its amino terminus, and activates transcription by an entirely different domain located close to its carboxyl terminus (240, 258). The transcriptional activating domain of GAL4, like activating domains that have been localized in GCN4 (93) and VP16 (169), is characterized by a high degree of acidity. GAL4 is synthesized constitutively (259), and has been shown to interact with binding sites on target genes even when these genes are inactive (260).

Insight into the "masking" of GAL4 activating potential has come from studies of GAL80, the negative regulator of galactose metabolism. Genetic studies have shown that GAL80 repression acts through GAL4, and is selectively relieved by the natural inducer (galactose). Mutant forms of GAL4 that fail to be repressed by GAL80 have been identified (261, 262), and have been shown to be truncated at their carboxyl termini in a region close to the acidic activating domain (262a). Similarly impaired forms of GAL4 have been prepared by molecular genetic methods (240, 262a). Moreover, expression of high levels of the carboxyl terminus of GAL4 selectively blocks repression by GAL80 (240). These observations raise the possibility that GAL80 acts to repress GAL4 action by masking its acidic activating domain. If this is the case, galactose induction, an event known to occur with unusually rapid kinetics (263), might simply entail the dissociation of GAL80 from a GAL4:GAL80 complex already poised on target genes.

The use of latent activator proteins is advantageous in several ways. The presence of preexisting factors can obviously facilitate a rapid transcriptional response to an environmental stimulus. A prime example of this is the heat-shock response, where an expeditious reaction to elevated temperature may be crucial to the cell's survival. Dormant activator proteins may also be important in gene expression cascades that control cell differentiation. Sen & Baltimore (255) have pointed out that the use of postsynthetic control mechanisms to activate tissue-specific regulatory proteins (such as NF-κB) solves what would otherwise be a series of consecutive transcriptional regulation problems. If, for instance, control of NF-κB were at the level of mRNA synthesis, then another activator protein would be required, which itself would need to be regulated, and so on. The latent activator/inducer system overcomes this dilemma, reducing the problem to appropriate temporal and spatial control of the inducer.

RETROSPECTIVE

The field of eukaryotic gene regulation has come of age. Powerful technical approaches have begun to provide accurate information regarding DNA sequences and protein factors that mediate selective gene expression. In composing this review we have focused on experimental systems and observations that are breaking new ground. We realize that this course is accompanied by obvious risks of misinterpretation. We hope that the benefits of intellectual challenge outweigh the pitfalls of misconception, and trust that our composition provides adequate distinction between fact and fantasy. Finally, we predict that many of the observations on eukaryotic gene regulation that now appear to be so intriguing will fall into the more familiar context of bacterial paradigms once demystified by the acquisition of thorough, substantiated information. The overlapping binding specificities of CCAAT factors, for example, may simply reflect variations on a theme first discovered in comparisons between the binding specificities of the CI and CRO proteins of bacteriophage lambda. Regardless of the level of overlap between the mechanisms of gene regulation in prokaryotic and eukaryotic cells, it can safely be concluded that eukaryotic systems have begun to offer novel information. For example, long distance regulatory effects, first discovered from studies of viral enhancers, have now been observed in the control of bacterial gene expression. Perhaps Monod's adage might now be modified—studies of the elephant have begun to tell us something of the workings of a simple bacterial cell.

ACKNOWLEDGMENTS

We wish to thank the members of the McKnight laboratory (Kelly LaMarco, Bill Landschulz, Chuck Vinson, Alan Freidman, and Jon Shuman), our colleagues in the Department of Embryology of the Carnegie Institution of Washington (particularly Allan Spradling, Dennis McKearin, and Phil Beachy), and colleagues of other institutions (Mark Ptashne, Tom Curran, Harinder Singh, Keith Yamamoto, Robert Tjian, and David Baltimore) for their critical comments and assistance in researching this review. We are also indebted to the many scientists who provided discussion and information prior to publication, and to Jeremy Berg, Mark Ptashne, and Bill Landschulz for the provision of the figures. Finally, we thank Christine Pratt for heroic efforts in preparing the manuscript.

Literature Cited

1. Boggs, R. T., Gregor, P., Idriss, S., Belote, J. M., McKeown, M. 1987. *Cell* 50:739–47
2. Nagoshi, R. N., McKeown, M., Burtis, K. C., Belote, J. M., Baker, B. S. 1988. *Cell* 53:229–36
2a. Amrein, H., Gorman, M., Nöthiger, R. 1988. *Cell* 55:1025–35
2b. Bell, L. R., Maine, E. M., Schedl, P., Cline, T. W. 1988. *Cell* 55:1037–46
3. Khoury, G., Gruss, P. 1983. *Cell* 33:313–14
4. Maniatis, T., Goodbourn, S., Fischer, J. A. 1987. *Science* 236:1237–45
5. Muller, M. M., Gerster, T., Schaffner, W. 1988. *Eur. J. Biochem.* 176:485–95
6. McKnight, S., Tjian, R. 1986. *Cell* 46:795–805
7. Jones, N. C., Rigby, P. W. J., Ziff, E. B. 1988. *Genes Dev.* 2:267–81
8. Weil, P. A., Luse, D. S., Segall, J., Roeder, R. G. 1979. *Cell* 18:469–84
9. Manley, J. L., Fire, A., Cano, A., Sharp, P. A., Gefter, M. L. 1980. *Proc. Natl. Acad. Sci. USA* 77:5706–10
10. Dignam, J. D., Lebovitz, R. M., Roeder, R. G. 1983. *Nucleic Acids Res.* 11:1475–89

11. Gorski, K., Carneiro, M., Schibler, U. 1986. *Cell* 47:767–76
12. Galas, D. J., Schmitz, A. 1978. *Nucleic Acids Res.* 5:3157–70
13. Fried, M., Crothers, D. M. 1981. *Nucleic Acids Res.* 9:6505–25
14. Rosenfeld, P. J., Kelly, T. J. 1986. *J. Biol. Chem.* 281:1398–408
15. Kadonaga, J. T., Tjian, R. 1986. *Proc. Natl. Acad. Sci. USA* 83:5889–93
16. Franza, B. R. Jr., Josephs, S. F., Gilman, M. Z., Ryan, W., Clarkson, B. 1987. *Nature* 330:391–95
17. Ptashne, M. 1988. *Nature* 335:683–89
18. Mitchell, P. J., Tjian, R. 1989. *Science.* In press
19. Pabo, C. O., Sauer, R. T. 1984. *Annu. Rev. Biochem.* 53:293–321
20. Ptashne, M. 1986. *A Genetic Switch: Gene Control and Phage λ.* Cambridge, England: Blackwell Scientific and Cell Press. 128 pp.
21. Anderson, W. F., Ohlendorf, D. H., Takeda, Y., Matthews, B. W. 1981. *Nature* 290:754–58
22. McKay, D. B., Steitz, T. A. 1981. *Nature* 290:744–49

23. Pabo, C. O., Lewis, M. 1982. *Nature* 298:443–47
24. Steitz, T. A., Ohlendorf, D. H., McKay, D. B., Anderson, W. F., Matthews, B. W. 1982. *Proc. Natl. Acad. Sci. USA* 79:3097–100
25. Wharton, R. P., Brown, E. L., Ptashne, M. 1984. *Cell* 38:361–69
26. Wharton, R. P., Ptashne, M. 1985. *Nature* 316:601–5
27. Hochschild, A., Ptashne, M. 1986. *Cell* 44:925–33
28. Hochschild, A., Douhan, J., Ptashne, M. 1986. *Cell* 47:807–16
29. Anderson, J. E., Ptashne, M., Harrison, S. C. 1987. *Nature* 326:846–52
29a. Otwinowski, Z., Schevitz, R. W., Zhang, R.-G., Lawson, C. L., Joachimiak, A., et al. 1988. *Nature* 335:321–29
29b. Jordan, S. R., Pabo, C. O. 1988. *Science* 242:893–99
29c. Aggarwal, A. K., Rodgers, D. W., Drottar, M., Ptashne, M., Harrison, S. C. 1988. *Science* 242:899–907
30. Matthews, B. W., Ohlendorf, D. H., Anderson, W. F., Fisher, R. G., Takeda, Y. 1982. *Cold Spring Harbor Symp. Quant. Biol.* 67:427–33
31. Laughon, A., Scott, M. P. 1984. *Nature* 310:25–31
32. Scott, M. P., Weiner, A. J. 1984. *Proc. Natl. Acad. Sci. USA* 81:4115–19
33. Shepherd, J. C. W., McGinnis, W., Carrasco, A. E., De Robertis, E. M., Gehring, W. J. 1984. *Nature* 310:70–71
34. Johnson, A. D., Herskowitz, I. 1985. *Cell* 42:237–47
35. DiNardo, S., Kuner, J. M., Theis, J., O'Farrell, P. H. 1985. *Cell* 43:59–69
36. Carroll, S. B., Scott, M. P. 1985. *Cell* 43:47–57
37. Beachy, P. A., Helfand, S. L., Hogness, D. S. 1985. *Nature* 313:545–51
38. Desplan, C., Theis, J., O'Farrell, P. H. 1985. *Nature* 318:630–35
39. Hoey, T., Levine, M. 1988. *Nature* 332:858–61
40. Beachy, P. A., Krasnow, M. A., Gavis, E. R., Hogness, D. S. 1988. *Cell* 55:1069–81
41. Desplan, C., Theis, J., O'Farrell, P. H. 1988. *Cell* 54:1081–90
42. Carrasco, A. E., McGinnis, W., Gehring, W. J., De Robertis, E. M. 1984. *Cell* 37:409–14
43. McGinnis, W., Garber, R. L., Wirz, J., Kuroiwa, A., Gehring, W. J. 1984. *Cell* 37:403–8
44. Levine, M., Rubin, G. M., Tjian, R. 1984. *Cell* 38:667–73
45. Sakonju, S., Bogenhagen, D. F., Brown, D. D. 1980. *Cell* 19:13–25
46. Bogenhagen, D. F., Sakonju, S., Brown, D. D. 1980. *Cell* 19:27–35
47. Engelke, D. R., Ng, S.-Y., Shastry, B. S., Roeder, R. G. 1980. *Cell* 19:717–28
48. Sakonju, S., Brown, D. D. 1981. *Cell* 23:665–69
49. Segall, J., Matsui, T., Roeder, R. G. 1980. *J. Biol. Chem.* 255:11986–91
50. Shastry, B. S., Ng, S.-Y., Roeder, R. G. 1982. *J. Biol. Chem.* 257:12979–86
51. Miller, J., McLachlan, A. D., Klug, A. 1985. *EMBO J.* 4:1609–14
52. Brown, R. S., Sander, D., Argos, S. 1985. *FEBS Lett.* 186:271–74
53. Hanas, J. S., Hazuda, D. J., Bogenhagen, D. F., Wu, F. H.-Y., Wu, C.-W. 1983. *J. Biol. Chem.* 258:14120–25
54. Fairall, L., Rhodes, D., Klug, A. 1986. *J. Mol. Biol.* 192:577–91
55. Rhodes, D., Klug, A. 1986. *Cell* 46:123–32
56. Diakun, G. P., Fairall, L., Klug, A. 1986. *Nature* 324:698–99
57. Hanas, J. S., Bogenhagen, D. F., Wu, C.-W. 1983. *Proc. Natl. Acad. Sci. USA* 80:2142–45
58. Vrana, K., Churchill, M. E. A., Tullius, T. D., Brown, D. D. 1988. *Mol. Cell. Biol.* 8:1684–96
59. Hartshorne, T. A., Blumberg, H., Young, E. T. 1986. *Nature* 320:283–87
60. Stillman, D. J., Bankier, A. T., Seddon, A., Goenhout, E. G., Nasmyth, K. A. 1988. *EMBO J.* 7:485–94
61. Schuh, R., Aicher, W., Gaul, U., Cote, S., Preiss, A., et al. 1986. *Cell* 47:1025–32
62. Rosenberg, U. B., Schroeder, C., Preiss, A., Kienlin, A., Cote, S., et al. 1986. *Nature* 319:336–39
63. Tautz, D., Lehmann, R., Schnurch, H., Schuh, R., Seifert, E., et al. 1987. *Nature* 327:383–89
64. Boulay, J. L., Dennefeld, C., Alberga, A. 1987. *Nature* 330:395–98
65. Page, D. C., Mosher, R., Simpson, E. M., Fisher, E. M. C., Mardon, G., et al. 1987. *Cell* 51:1091–104
66. Kadonaga, J. T., Carner, K. R., Masiarz, F. R., Tjian, R. 1987. *Cell* 51:1079–90
67. Chowdhury, K., Deutsch, U., Gruss, P. 1987. *Cell* 48:771–78
68. Altaba, A., Ruiz, I., Perry-O'Keefe, H., Melton, D. A. 1987. *EMBO J.* 6:3065–70
69. Weinberger, C., Hollenberg, S. M., Rosenfeld, M. G., Evans, R. M. 1985. *Nature* 318:670–72
70. Miesfeld, R., Rusconi, S., Godowski, P. J., Maler, B. A., Okret, S., et al. 1986. *Cell* 46:389–99
71. Jeltsch, J. M., Krozowski, Z., Quirin-

Stricker, C., Gronemeyer, H., Simpson, R. J., et al. 1986. *Proc. Natl. Acad. Sci. USA* 83:5424–28

72. Green, S., Walter, P., Kumar, V., Krust, A., Bornert, J.-M., et al. 1986. *Nature* 320:134–39

73. Greene, G. L., Gilna, P., Waterfield, M., Baker, A., Hort, Y., et al. 1986. *Science* 231:1150–54

74. Arriza, J. L., Weinberger, C., Cerelli, G., Glaser, T. M., Handelin, B. L., et al. 1987. *Science* 237:268–75

75. Weinberger, C., Thompson, C. C., Ong, E. S., Lebo, R., Gruol, D. J., et al. 1986. *Nature* 324:641–46

76. Laughon, A., Gesteland, R. F. 1984. *Mol. Cell. Biol.* 4:260–67

77. Kammerer, B., Guyonvarch, A., Hubert, J. C. 1984. *J. Mol. Biol.* 180:239–50

78. Messenguy, R., Dubois, E., Descamps, F. 1986. *Eur. J. Biochem.* 157:77–81

79. Baum, J. A., Geever, R., Giles, N. H. 1987. *Mol. Cell. Biol.* 7:1256–66

80. Wray, L. V. Jr., Witte, M. M., Dickson, R. C., Riley, M. I. 1987. *Mol. Cell. Biol.* 7:1111–21

81. Giguere, V., Hollenberg, S. M., Rosenfeld, M. G., Evans, R. M. 1986. *Cell* 46:645–52

82. Kumar, V., Green, S., Staub, A., Chambon, P. 1986. *EMBO J.* 5:2231–36

83. Godowski, P. J., Rusconi, S., Miesfeld, R., Yamamoto, K. R. 1987. *Nature* 325:365–68

84. Hollenberg, S. M., Giguere, V., Segui, P., Evans, R. M. 1987. *Cell* 49:36–49

85. Green, S., Kumar, V., Theulaz, I., Wahli, W., Chambon, P. 1988. *EMBO J.* 7:3037–44

86. Freedman, L. P., Luisi, B. F., Korszun, Z. R., Basavappa, R., Sigler, P. B., et al. 1988. *Nature* 334:543–46

87. Johnston, M. 1987. *Nature* 328:353–55

88. Berg, J. 1988. *Proc. Natl. Acad. Sci. USA* 85:99–102

89. Vogt, P. K., Box, T. J., Doolittle, R. F. 1987. *Proc. Natl. Acad. Sci. USA* 84:3316–19

90. Graves, B. J., Johnson, P. F., McKnight, S. L. 1986. *Cell* 44:565–76

91. Johnson, P. F., Landschulz, W. H., Graves, B. J., McKnight, S. L. 1987. *Genes Dev.* 1:133–46

92. Landschulz, W. H., Johnson, P. F., Adashi, E. Y., Graves, B. J., McKnight, S. L. 1988. *Genes Dev.* 2:786–800

93. Hope, I. A., Struhl, K. 1986. *Cell* 46:885–94

94. Landschulz, W. H., Johnson, P. F., McKnight, S. L. 1988. *Science* 240:1759–64

95. Hope, I. A., Struhl, K. 1987. *EMBO J.* 6:2781–84

95a. Dang, C. V., McGuire, M., Buckmire, M., Lee, W. 1989. *Nature* 337:664–66

95b. Kouzarides, T., Ziff, E. 1988. *Nature* 336:646–51

95c. Sassone-Corsi, P., Ransone, L. J., Lamph, W. W., Verma, I. M. 1988. *Nature* 336:692–95

96. Hill, D. E., Hope, I. A., Macke, J. P., Struhl, K. 1986. *Science* 234:451–57

97. Distel, R. J., Ro, H-S., Rosen, B. S., Groves, D. L., Spiegelman, B. M. 1987. *Cell* 49:835–44

97a. Franza, B. R., Rauscher, F. J. III, Josephs, S. F., Curran, T. 1988. *Science* 1150–53

98. Rauscher, F. J. III, Sambucetti, L. C., Curran, T., Distel, R. J., Spiegelman, B. M. 1988. *Cell* 52:471–80

99. Rauscher, F. J. III, Cohen, D. R., Curran, T., Bos, T. J., Vogt, P. K., et al. 1988. *Science* 240:1010–16

100. Sassone-Corsi, P., Lamph, W. W., Kamps, M., Verma, I. 1988. *Cell* 54:553–60

100a. Rauscher, F. J. III, Voulalas, P. J., Franza, B. R. Jr., Curran, T. 1988. *Genes Dev.* 2:1687–99

100b. Nakabeppu, Y., Ryder, K., Nathans, D. 1988. *Cell* 55:907–15

100c. Halazonetis, T. D., Georgopoulos, K., Greenberg, M. E., Leder, P. 1988. *Cell* 55:917–24

101. Hollis, M., Valenzuela, D., Pioli, D., Wharton, R., Ptashne, M. 1988. *Proc. Natl. Acad. Sci. USA* 85:5834–38

102. Singh, H., LeBowitz, J. H., Baldwin, A. S. Jr., Sharp, P. A. 1988. *Cell* 52:415–23

103. Vinson, C. R., LaMarco, K. L., Johnson, P. F., Landschulz, W. H., McKnight, S. L. 1988. *Genes Dev.* 2:801–6

104. Wiederrecht, G., Seto, D., Parker, C. S. 1988. *Cell* 54:841–53

105. Sorger, P. K., Pelham, H. R. B. 1988. *Cell* 54:855–64

106. Santoro, C., Mermod, N., Andrews, P. C., Tjian, R. 1988. *Nature* 334:218–24

107. Pinkham, J. L., Olesen, J. T., Guarente, L. 1987. *Mol. Cell. Biol.* 7:578–87

108. Hahn, S., Pinkham, J., Wei, R., Miller, R., Guarente, L. 1988. *Mol. Cell. Biol.* 8:655–63

109. Pfeifer, K., Arcangioli, B., Guarente, L. 1987. *Cell* 49:9–18

110. Pfeifer, K., Prezant, T., Guarente, L. 1987. *Cell* 19–27

110a. deVargas, L. M., Pargellis, C. A.,

Hasan, N. M., Bushman, E. W., Landy, A. 1988. *Cell* 54:923–29

111. Verdiere, J., Greusot, F., Guarente, L., Slonimski, P. 1986. *Curr. Genet.* 10:339–42
112. Guarente, L., Lalonde, B., Gifford, P., Alanis, E. 1984. *Cell* 36:503–11
113. Chandler, V. L., Maler, B. A., Yamamoto, K. R. 1983. *Cell* 33:489–99
114. Payvar, F., DeFranco, D., Firestone, G. L., Edgar, B., Wrange, O., et al. 1983. *Cell* 35:381–92
115. Karin, M., Haslinger, A., Holtgreve, A., Richards, R. I., Krauter, P., et al. 1984. *Nature* 308:513–19
116. Moore, D. D., Marks, A. R., Buckley, D. I., Kapler, G., Payvar, F., et al. 1985. *Proc. Natl. Acad. Sci. USA* 82:699–702
117. Renkawitz, R., Schutz, G., von der Ahe, D., Beato, M. 1984. *Cell* 37:503–10
118. Camper, S. A., Yao, Y. A. S., Rottman, F. M. 1985. *J. Biol. Chem.* 260:12246–51
119. Sakai, D. D., Helms, S., Carlstedt-Duke, J., Gustafsson, J.-A., Rottman, F. M., et al. 1988. *Genes Dev.* 2:1144–54
120. Meisfeld, R., Godowski, P. J., Maler, B. A., Yamamoto, K. R. 1987. *Science* 236:423–27
120a. Akerblom, I. E., Slater, E., Beato, M., Baxter, J. D., Mellon, P. L. 1988. *Science* 241:350–53
121. Weiher, H., Konig, M., Gruss, P. 1983. *Science* 219:626–31
122. Costa, R. H., Grayson, D. R., Xanthopoulos, K. G., Darnell, J. E. 1988. *Proc. Natl. Acad. Sci. USA* 85:3840–44
123. Sturm, R., Baumruker, T., Franza, B. R. Jr., Herr, W. 1987. *Genes Dev.* 1:1147–60
124. Baumruker, T., Sturm, R., Herr, W. 1988. *Genes Dev.* 2:1400–13
125. O'Hare, P., Goding, C. R. 1988. *Cell* 52:435–45
126. Davidson, I., Xiao, J. H., Rosales, R., Staub, A., Chambon, P. 1988. *Cell* 54:931–42
127. Banerji, J., Olson, L., Schaffner, W. 1983. *Cell* 33:729–40
128. Gillies, S. D., Morrison, S. L., Oi, V. T., Tonegawa, S. 1983. *Cell* 33:717–28
129. Neuberger, M. S. 1983. *EMBO J.* 2:1373–78
130. Picard, D., Schaffner, W. 1984. *Nature* 307:80–82
131. Queen, C., Stafford, J. 1984. *Mol. Cell. Biol.* 4:1042–49
132. Falkner, F. G., Zachau, H. G. 1984. *Nature* 310:71–74
133. Foster, J., Stafford, J., Queen, C. 1985. *Nature* 315:423–25
134. Grosschedl, R., Baltimore, D. 1985. *Cell* 41:885–97
135. Mason, J. O., Williams, G. T., Neuberger, M. S. 1985. *Cell* 41:479–87
136. Picard, D., Schaffner, W. 1985. *EMBO J.* 4:2831–38
137. Parslow, T. G., Blair, D. L., Murphy, W. J., Granner, D. K. 1984. *Proc. Natl. Acad. Sci. USA* 81:2650–54
138. Falkner, F. G., Mocikat, R., Zachau, H. G. 1986. *Nucleic Acids Res.* 14:8819–27
139. Bergman, Y., Rice, D., Grosschedl, R., Baltimore, D. 1984. *Proc. Natl. Acad. Sci. USA* 81:7041–45
140. Ballard, D. W., Bothwell, A. 1986. *Proc. Natl. Acad. Sci. USA* 83:9626–30
141. Harvey, R. P., Robins, A. J., Wells, J. R. E. 1982. *Nucleic Acids Res.* 10:7851–63
142. Perry, M., Thomsen, G. H., Roeder, R. 1985. *J. Mol. Biol.* 185:479–99
143. Parslow, T. G., Jones, S. D., Bond, B., Yamamoto, K. R. 1987. *Science* 235:1498–1501
144. Mattaj, I. W., Lienhard, S., Jiricny, J., De Robertis, E. M. 1985. *Nature* 316:163–67
145. Ciliberto, G., Buckland, R., Cortese, R., Philipson, L. 1985. *EMBO J.* 4:1537–43
146. Krol, A., Lund, E., Dahlberg, J. E. 1985. *EMBO J.* 4:1525–35
147. Mangin, M., Ares, M., Weiner, A. M. 1986. *EMBO J.* 5:987–96
148. Singh, H., Sen, R., Baltimore, D., Sharp, P. A. 1986. *Nature* 319:154–58
149. Staudt, L. M., Singh, H., Sen, R., Wirth, T., Sharp, P. A., et al. 1986. *Nature* 323:640–43
150. Augereau, P., Chambon, P. 1986. *EMBO J.* 5:1791–97
151. Davidson, I., Fromental, C., Augereau, P., Wildeman, A., Zenke, M., et al. 1986. *Nature* 323:544–48
152. Hromas, R., Van Ness, B. 1986. *Nucleic Acids Res.* 12:4837–48
153. Landolfini, N. F., Capra, J. D., Tucker, P. W. 1986. *Nature* 323:548–51
154. Mocikat, R., Falkner, F. G., Mertz, R., Zachau, H. G. 1986. *Nucleic Acids Res.* 14:8829–44
155. Schlokat, U., Bohmann, D., Scholer, H., Gruss, P. 1986. *EMBO J.* 5:3251–58
156. Sive, H. L., Roeder, R. G. 1986. *Proc. Natl. Acad. Sci. USA* 83:6382–86
157. Bohmann, D., Keller, W., Dale, T., Scholer, H. R., Tebb, G., et al. 1987. *Nature* 325:268–72
158. Rosales, R., Vigneron, M., Macchi,

M., Davidson, I., Xiao, J. H., et al. 1987. *EMBO J.* 6:3015–25

159. Gerster, T., Matthias, P., Thali, M., Jiricny, J., Schaffner, W. 1987. *EMBO J.* 6:1323–30

160. Scheidereit, C., Heguy, A., Roeder, R. G. 1987. *Cell* 51:783–93

161. Fletcher, C., Heintz, N., Roeder, R. G. 1987. *Cell* 51:773–81

162. LeBowitz, J. H., Kobayashi, T., Staudt, L., Baltimore, D., Sharp, P. A. 1988. *Genes Dev.* 2:1227–37

163. Mackem, S., Roizman, B. 1982. *Proc. Natl. Acad. Sci. USA* 79:4917–21

164. Mackem, S., Roizman, B. 1982. *J. Virol.* 44:939–49

165. Kristie, T. M., Roizman, B. 1984. *Proc. Natl. Acad. Sci. USA* 81:4065–69

166. Preston, C. M., Cordingley, M. G., Stow, N. D. 1984. *J. Virol.* 50:708–16

167. Gaffney, D. F., McLauchlan, J., Whitton, J. L., Clements, J. B. 1985. *Nucleic Acids Res.* 13:7874–62

168. Bzik, D. J., Preston, C. M. 1986. *Nucleic Acids Res.* 14:929–43

169. Triezenberg, S. J., LaMarco, K. L., McKnight, S. L. 1988. *Genes Dev.* 2:730–42

170. Campbell, M. E. M., Palfreyman, J. W., Preston, C. M. 1984. *J. Mol. Biol.* 180:1–19

171. Marsden, H. S., Campbell, M. E. M., Haarr, L., Frame, M. C., Parris, D. S., et al. 1987. *J. Virol.* 61:2428–37

172. Kristie, T. M., Roizman, B. 1987. *Proc. Natl. Acad. Sci. USA* 84:71–75

173. McKnight, J. L. C., Kristie, T. M., Roizman, B. 1987. *Proc. Natl. Acad. Sci. USA* 84:7061–65

174. Preston, C. M., Frame, M. C., Campbell, M. E. M. 1988. *Cell* 52:425–34

175. Gerster, T., Roeder, R. G. 1988. *Proc. Natl. Acad. Sci. USA* 85:6347–51

176. Pruijn, G. J. M., van Driel, W., van der Vliet, P. C. 1986. *Nature* 322:656–59

177. Clerc, R. G., Corcoran, L. M., LeBowitz, J. H., Baltimore, D., Sharp, P. A. 1988. *Genes Dev.* 2:1570–81

177a. Scheidereit, C., Cromlish, J. A., Gerster, T., Kawakami, K., Balmaceda, C.-G., et al. 1988. *Nature* 336:551–57

178. Sturm, R. A., Das, G., Herr, W. 1988. *Genes Dev.* 2:1582–99

179. Jones, K. A., Yamamoto, K. R., Tjian, R. 1985. *Cell* 42:559–72

180. Jones, K. A., Kadonaga, J. T., Rosenfeld, P. J., Kelly, T. J., Tjian, R. 1987. *Cell* 48:79–89

181. Cohen, R. B., Sheffery, M., Kim, C. G. 1986. *Mol. Cell. Biol.* 6:821–32

182. Barberis, A., Superti-Furga, G., Busslinger, M. 1987. *Cell* 50:347–59

183. Dorn, A., Bollekens, J., Staub, A., Benoist, C., Mathis, D. 1987. *Cell* 50:863–72

184. Hatamochi, A., Paterson, B., de Crombrugghe, B. 1986. *J. Biol. Chem.* 261:11310–14

185. Chodosh, L. A., Baldwin, A. S., Carthew, R. W., Sharp, P. A. 1988. *Cell* 53:11–24

186. Raymondjean, M., Cereghini, S., Yaniv, M. 1988. *Proc. Natl. Acad. Sci. USA* 85:757–61

187. Maity, S. N., Golumbek, P. T., Karsenty, G., de Crombrugghe, B. 1988. *Science* 241:582–85

188. Hatamochi, A., Golumbek, P. T., Van Schaftingen, E., de Crombrugghe, B. 1988. *J. Biol. Chem.* 263:5940–47

189. Pinkham, J. L., Guarente, L. 1985. *Mol. Cell. Biol.* 5:3410–16

190. Olesen, J., Hahn, S., Guarente, L. 1987. *Cell* 51:953–61

191. Hahn, S., Guarente, L. 1988. *Science* 240:317–21

192. Chodosh, L. A., Olesen, J., Hahn, S., Baldwin, A. S., Guarente, L., et al. 1988. *Cell* 53:25–35

193. Fields, S., Herskowitz, I. 1985. *Cell* 42:923–30

194. Jarvis, E., Hagen, D. C., Sprague, G. F. Jr. 1987. *Mol. Cell. Biol.* 8:309–20

195. Bender, A., Sprague, G. F. Jr. 1987. *Cell* 50:681–91

196. Strathern, J., Hicks, J., Herskowitz, I. 1981. *J. Mol. Biol.* 147:357–72

197. Goutte, C., Johnson, A. D. 1988. *Cell* 52:875–82

198. Miller, J. H., McKay, V. L., Nasmyth, K. A. 1985. *Nature* 314:598–603

199. Keleher, C. A., Goutte, C., Johnson, A. D. 1988. *Cell* 53:927–36

200. Sauer, R. T., Smith, D. L., Johnson, A. D. 1988. *Genes Dev.* 2:807–16

201. Lee, W., Mitchell, P., Tjian, R. 1987. *Cell* 49:741–52

201a. Bohmann, D., Bos, T. J., Admon, A., Nishimura, T., Vogt, P. K., Tjian, R. 1987. *Science* 238:1386–92

202. Angel, P., Allegretto, E. A., Okino, S., Hattori, K., Boyle, W. J. 1988. *Nature* 332:166–71

203. Ryder, K., Lau, L. F., Nathans, D. 1988. *Proc. Natl. Acad. Sci. USA* 85:1487–91

204. Angel, P., Imagawa, M., Chiu, R., Stein, B., Imbra, R. J., et al. 1987. *Cell* 49:729–39

205. Maki, Y., Bos, T. J., Davis, C., Starbuck, M., Vogt, P. K. 1987. *Proc. Natl. Acad. Sci. USA* 84:2848–52

206. Bishop, J. M. 1987. *Science* 235:305–11

207. Setoyama, C., Frunzio, R., Liau, G., Mudryj, M., de Crombrugghe, B. 1986. *Proc. Natl. Acad. Sci. USA* 83:3213–17
208. Lech, K., Anderson, K., Brent, R. 1988. *Cell* 52:179–84
209. Chiu, R., Boyle, W. J., Meek, J., Smeal, T., Hunter, T., et al. 1988. *Cell* 54:541–52
210. Lassar, A. B., Martin, P. L., Roeder, R. G. 1983. *Science* 222:740–48
211. Setzer, D. R., Brown, D. D. 1985. *J. Biol. Chem.* 260:2483–92
212. Wolffe, A. P., Brown, D. D. 1988. *Science* 241:1626–32
212a. Hochschild, A., Ptashne, M. 1988. *Nature* 336:353–57
213. Herr, W., Clarke, J. 1986. *Cell* 45:461–70
214. Ondek, B., Gloss, L., Herr, W. 1988. *Nature* 333:40–45
215. Fromental, C., Kanno, M., Nomiyama, H., Chambon, P. 1988. *Cell* 54:943–53
216. Yamamoto, K. R. 1985. *Annu. Rev. Genet.* 19:209–52
217. Evans, R. M., Hollenberg, S. M. 1988. *Cell* 52:1–3
218. Mansukhani, A., Crickmore, A., Sherwood, P. W., Goldberg, M. L. 1988. *Mol. Cell. Biol.* 8:615–23
219. Pirrotta, V., Manet, E., Hardon, E., Bickel, S. E., Banson, M. 1987. *EMBO J.* 6:791–99
220. Mansukhani, A., Gunaratne, P. H., Sherwood, P. W., Sneath, B. J., Goldberg, M. L. 1988. *Mol. Genet.* 211:121–28
221. Judd, B. H. 1988. *Cell* 53:841–43
222. Whyte, P., Buchkovich, K. J., Horowitz, J. M., Friend, S. H., Raybuck, M., et al. 1988. *Nature* 334:124–29
223. DeCaprio, J. A., Ludlow, J. W., Figge, J., Shew, J.-Y., Huang, C.-M., et al. 1988. *Cell* 54:275–83
224. Burch, J. B. E., Weintraub, H. 1983. *Cell* 33:65–76
225. Clever, U. 1964. *Science* 146:794–95
226. Ashburner, M. 1970. *Chromosoma* 31:356–76
227. Ashburner, M. 1974. *Dev. Biol.* 39:141–57
228. Ritossa, F. M. 1964. *Exp. Cell. Res.* 35:601–7
229. Craig, E. A. 1985. *Crit. Rev. Biochem.* 18:239–80
230. Lindquist, S. 1986. *Annu. Rev. Biochem.* 55:1151–91
231. Pelham, H. 1985. *Trends Genet.* 1:31–35
232. Pelham, H. R. B. 1982. *Cell* 30:517–28
233. Parker, C. S., Topol, J. 1984. *Cell* 37:273–83
234. Wu, C., Wilson, S., Walker, B., Dawid, I., Paisley, T., et al. 1987. *Science* 238:1247–53
235. Wiederrecht, G., Shuey, D. J., Kibbe, W. A., Parker, C. S. 1987. *Cell* 48:507–18
236. Sorger, P. K., Lewis, M. J., Pelham, H. R. B. 1987. *Nature* 329:81–84
237. Morgan, W. D., Williams, G. T., Morimoto, R. I., Greene, J., Kingston, R. E., et al. 1987. *Mol. Cell. Biol.* 7:1129–1138
238. Kingston, R. E., Schuetz, T. J., Larin, Z. 1987. *Mol. Cell. Biol.* 7:1530–34
239. Zimarino, V., Wu, C. 1987. *Nature* 327:727–30
240. Ma, J., Ptashne, M. 1987. *Cell* 48:847–53
241. Rusconi, S., Yamamoto, K. R. 1987. *EMBO J.* 6:1309–15
242. Picard, D., Yamamoto, K. R. 1987. *EMBO J.* 6:3333–40
243. Holbrook, N. J., Bodwell, J. E., Jeffries, M., Munck, A. 1983. *J. Biol. Chem.* 258:6477–85
244. Sherman, M. R., Moran, M. C., Tuazon, F. B., Steven, Y.-W. 1983. *J. Biol. Chem.* 258:10366–77
245. Vedeckis, W. V. 1983. *Biochemistry* 22:1983–89
246. Raaka, B. M., Samuels, H. H. 1983. *J. Biol. Chem.* 258:417–25
247. Catelli, M. G., Binart, N., Jung-Testas, I., Renoir, J. M., Baulieu, E. E., et al. 1985. *EMBO J.* 4:3131–35
248. Sanchez, E. R., Toft, D. O., Schlesinger, M. J., Pratt, W. B. 1985. *J. Biol. Chem.* 260:12398–401
249. Sanchez, E. R., Meshinchi, S., Tienrungroj, W., Schlesinger, M. J., Toft, D. O., et al. 1987. *J. Biol. Chem.* 262:6986–91
250. Howard, K. J., Distelhorst, C. W. 1988. *J. Biol. Chem.* 263:3474–81
251. Pratt, W. B., Jolly, D. J., Pratt, D. V., Hollenberg, S. M., Giguere, V., et al. 1988. *J. Biol. Chem.* 263:267–73
252. Sen, R., Baltimore, D. 1986. *Cell* 46:705–16
253. Lenardo, M., Pierce, J. W., Baltimore, D. 1987. *Science* 236:1573–77
254. Pierce, J. W., Lenardo, M., Baltimore, D. 1988. *Proc. Natl. Acad. Sci. USA* 85:1482–86
255. Sen, R., Baltimore, D. 1986. *Cell* 47:921–28
256. Baeuerle, P. A., Baltimore, D. 1988. *Cell* 53:211–17
257. Johnston, M. 1987. *Microbiol. Rev.* 51:458–76
258. Keegan, L., Gill, G., Ptashne, M. 1986. *Science* 231:699–704

259. Matsumoto, K., Toh-e, A., Oshima, Y. 1978. *J. Bacteriol.* 134:446–57
260. Lohr, D., Hopper, J. E. 1985. *Nucleic Acids Res.* 13:8409–23
261. Douglas, H. C., Hawthorne, D. C. 1966. *Genetics* 54:911–16
262. Matsumoto, K., Adashi, Y., Toh-e, A., Oshima, Y. 1980. *J. Bacteriol.* 141:508–27
262a. Johnston, S. A., Salmeron, J. M. Jr., Dincher, S. S. 1987. *Cell* 50:143–46
263. Torchia, T. E., Hopper, J. E. 1986. *Genetics* 113:229–43
264. Gill, G., Ptashne, M. 1988. *Nature* 334:721–24
265. Hinnebusch, A. G. 1984. *Proc. Natl. Acad. Sci. USA* 81:6442–46
266. Pfeifer, C., Kim, K.-S., Kogan, S., Guarente, L. 1989. *Cell* 56:291–301
267. Astell, C. R., Ahlstrom-Johasson, L., Smith, M., Tatchell, K., Nasmyth, K. A., et al. 1981. *Cell* 27:15–23
268. Shore, D., Nasmyth, K. 1987. *Cell* 51:721–32
269. Legrain, M., De Wilde, M., Hilger, F. 1986. *Nucleic Acids Res.* 14:3059–73
269a. Wray, L. V. Jr., Witte, M. M., Dickson, R. C., Riley, M. I. 1987. *Mol. Cell. Biol.* 7:1111–21
270. Fu, Y.-H., Paietta, J. V., Mannix, D. G., Marzluff, G. A. 1989. *Mol. Cell. Biol.* 9:1120–27
271. Paluh, J. L., Orbach, M. J., Legerton, T. L., Yanofsky, C. 1988. *Proc. Natl. Acad. Sci. USA* 85:3728–32
272. McGinnis, W., Garber, R. L., Wirz, J., Kuroiwa, A., Gehring, W. J. 1984. *Cell* 37:403–8
273. Poole, S. J., Kauvar, L. M., Drees, B., Kornberg, T. 1985. *Cell* 40:37–43
274. Frigerio, G., Burri, M., Bopp, D., Baumgartner, S., Noll, M. 1986. *Cell* 47:735–46
275. Blochlinger, K., Bodmer, R., Jack, J., Jan, L. Y., Jan, Y. N. 1988. *Nature* 333:629–35
276. Vincent, A., Colot, N. V., Rosbash, M. 1985. *J. Mol. Biol.* 186:146–66
277. Parkhurst, S. M., Harrison, D. A., Remington, M. P., Spana, C., Kelley, R. L., et al. 1988. *Genes Dev.* 2:1205–15
278. Boyer, P. D., Mahoney, P. A., Lenguel, J. A. 1987. *Nucleic Acids Res.* 15:2309–25
279. Villares, R., Cabrara, C. V. 1987. *Cell* 50:415–24
279a. Caudy, M., Vässin, H., Brand, M., Tuma, R., Jan, L. Y., Jan, Y. N. 1988. *Cell* 55:1061–67
280. van Beveren, C., van Straaten, F., Curran, T., Mueller, R., Verma, I. M. 1983. *Cell* 32:1241–55
280a. Norman, C., Runswick, M., Pollock, R., Treisman, R. 1988. *Cell* 55:989–1003
280b. Ingraham, H. A., Chen, R., Managalam, H. J., Elsholtz, H. P., Flynn, S. E., et al. 1988. *Cell* 55:519–29
281. Kinzler, K. W., Ruppert, J. M., Bigner, S. H., Vogelstein, G. 1988. *Nature* 332:371–74
282. Morishita, K., Parker, D. S., Mucenski, M. L., Jenkins, N. A., Copeland, N. G., et al. 1988. *Cell* 54:831–40
283. Miyamoto, M., Fujita, T., Kimura, Y., Maruyama, M., Harada, H., et al. 1988. *Cell* 54:901–13
284. Hollenberg, S. M., Weinberger, C., Ong, E. S., Cerelli, G., Oro, A., et al. 1985. *Nature* 318:635–41
285. Krust, A., Green, S., Argos, P., Kumar, V., Walter, P., et al. 1986. *EMBO J.* 5:891–91
286. Conneely, O. M., Sullivan, W. P., Toft, D. O., Birnbaumer, M., Cook, R. G., et al. 1986. *Science* 233:767–70
287. Cohen, D. R., Curran, T. 1988. *Mol. Cell. Biol.* 8:2063–69
288. Christy, B. A., Lau, L. F., Nathans, D. 1988. *Proc. Natl. Acad. Sci. USA* 85:7857–61
289. Stanton, L. W., Pahrlander, P. D., Tesser, P. M., Marcu, K. B. 1984. *Nature* 310:423–25
290. Davis, R. L., Weintraub, H., Lassar, A. B. 1987. *Cell* 51:987–1000
291. Williams, T., Admon, A., Lüscher, B., Tjian, R. 1988. *Genes Dev.* 2:1557–69

Annu. Rev. Biochem. 1989. 58:841–74

GLYCOSYLATION IN THE NUCLEUS AND CYTOPLASM

Gerald W. Hart[1], Robert S. Haltiwanger, Gordon D. Holt, and William G. Kelly

Department of Biological Chemistry, The Johns Hopkins University School of Medicine, Baltimore, Maryland 21205

CONTENTS

PERSPECTIVES AND SUMMARY

It is widely accepted that glycosyl moieties on glycoproteins are almost exclusively localized at a cell's surface or lumenal compartments. This view has arisen from extensive studies by many laboratories of the topography of asparagine-linked glycosylation in the rough endoplasmic reticulum and of many types of glycosylation reactions in the Golgi apparatus (1–3). Numerous

[1]To whom correspondence should be addressed.

0066-4154/89/0701-0841$02.00

investigators have elucidated the biosynthetic and secretory pathways for both membrane and secretory glycoproteins (4–6), and have localized glycosyl-transferases to specific lumenal regions of the Golgi, endoplasmic reticulum (2, 3, 7), or cell surface (7–9). None of the currently accepted models of glycoprotein biosynthesis or transport predicted the existence of glycoproteins in either the nucleoplasmic or cytoplasmic compartments of the cell (1–5, 10, 11).

In contrast to this prevailing dogma, well over a hundred papers have presented evidence for the existence of glycoproteins in the nucleus and cytosol (for reviews, see 12–15). These reports are based upon lectin-binding studies, compositional analyses of purified nuclei or cytoplasm, or metabolic radiolabeling with sugar precursors followed by subcellular fractionation of labeled components. While many of these studies are provocative, their findings and conclusions have been largely ignored by the mainstream of the biochemical community for several reasons: (a) The putative glycoconjugates identified in many of these studies have not been further characterized, either in terms of sugar structure or in terms of type of saccharide-protein linkages. (b) In studies of purified organelles, it is very difficult to rule out contamination from plasma membrane or lumenal contents of broken endoplasmic reticulum or Golgi, especially since the amounts of glycoconjugates seen in the nucleus or cytoplasm are typically quite low (16). (c) No adequate theories exist that can both account for established models of glycoprotein synthesis and also explain how glycoproteins might be transported to, or synthesized in the nuclear or cytoplasmic compartments. Overall, there is currently no compelling structural evidence for the presence of the best-characterized types of glycoproteins [i.e. those with asparagine-linked oligosaccharides (4, 17) or "mucin-type" saccharides (18–21)] in the nucleoplasmic or cytoplasmic compartments of cells.

Convincing evidence for the existence of cytoplasmic or nucleoplasmic glycoproteins has recently been provided from the discovery of five novel types of glycoconjugates: (a) Single N-acetylglucosamine residues are glycosidically attached to serine or threonine hydroxyls (O-GlcNAc) on proteins found in the cytoplasm or nucleus, including proteins at the faces of nuclear pores (22–29), on well-characterized cytoskeletal proteins (30), and transcription factors (31). (b) Unusual types of O-linked mannose-containing proteoglycans are localized almost exclusively to cytoplasm in adult rat brain (32–37). (c) Unique types of heparan sulfates are greatly enriched in the nuclei of rat liver cells (38, 39). (d) Glucosyl residues attached to the hydroxyl moieties of tyrosine are found on the cytoplasmic "primer" for glycogen synthesis (40–44). (e) O-linked mannosyl residues on certain cytoplasmic proteins are substrates for a novel cytoplasmic glucose phosphotransferase (45–48).

In this review, we first critically examine the available data indicating the presence of cytoplasmic and nuclear glycoproteins. We then discuss in more detail the existence, localization, and possible functions of the new types of glycosylation that appear to be enriched in these subcellular domains. Finally, we speculate as to the significance of these findings with respect to future investigations in biochemistry and cell biology.

BINDING STUDIES USING LECTINS

Lectins, which are specific carbohydrate-binding proteins, are valuable probes for the presence of saccharides on cell surfaces and on purified glycoproteins (49–52). There are several disadvantages inherent in the use of lectin probes. These include the difficulty of biochemical characterization and unequivocal identification of the binding site, and the lack of strict ligand specificity. Examples of the overlapping binding specificity of lectins include the binding of concanavalin A (Con A) to both high-mannose oligosaccharides and inositol-containing compounds (53), and the interaction of wheat germ agglutinin (WGA) with both sialic acids and N-acetylglucosamine (54). Table 1 lists lectins, their specificies, and studies that have used lectin probes to detect glycoconjugates in nuclear membranes, nuclear matrix, and chromatin, and in the cytoplasm of various kinds of cells. Although the data gathered employing these techniques is provocative, it is nevertheless preliminary in nature and deserves further study.

Lectin Binding Sites in the Nucleus

NUCLEAR MEMBRANES   Several studies have demonstrated lectin binding to nuclear membranes (Table 1). These studies generally involve binding of labeled lectins (fluorescent, radiolabeled, ferritin-conjugated, etc) to isolated nuclei in the presence or absence of competing sugars. Using such techniques, one group has found evidence of mannose-, galactose-, GalNAc-, GlcNAc-, and fucose-containing structures on the cytoplasmic face of isolated bovine nuclei (55). These results are supported by several other experiments, including staining of the cytoplasmic face of nuclei from calf thymocytes with Con A (56), and the labeling of rat nuclei with both Con A and WGA (57, 58). Studies on the binding of the *Ricinus communis* agglutinin (RCA) to ascites hepatoma nuclei and normal rat liver nuclei suggested that hepatoma nuclear membranes have tenfold more RCA-binding sites than those from normal rat liver cells (57). Examination of lectin staining to rat liver nuclei at the electron microscopic level by one group (58) revealed what may be a major problem with several of these studies: ferritin–Con A appeared to stain only damaged nuclei. In several of the studies mentioned above, the integrity of the isolated nuclear membranes was not addressed, allowing for the possibility that lectins

Table 1 Demonstration of lectin binding sites in the nucleus or cytoplasm

Binding site	Lectin	Competing sugar	References
Nuclear membranes	Con A	α-D-glucosides & α-D-mannosides	(55[a], 57[a], 73) (56[a], 58–61, 75)
Nuclear membranes	WGA	D-GlcNAc & sialic acids	(55[a], 58[a], 60, 129[a]) (25[a], 62[a], 141[a], 28[a])
Nuclear membranes	RCA	D-galactose	(55[a], 57[a])
Chromatin & nuclear matrix	Con A	α-D-glucosides & α-D-mannosides	(68, 69, 72, 74, 92, 93) (63, 65, 70, 71, 75, 78)
Chromatin & nuclear matrix	WGA	D-GlcNAc & sialic acids	(28, 65, 69, 72, 75, 78, 92)
Chromatin & nuclear matrix	LCA	α-D-mannosides	(64, 67)
Chromatin & nuclear matrix	UEA-I	α-L-fucose	(68, 70, 72)
Chromatin & nuclear matrix	RCA	D-galactose	(72, 78)
Chromatin & nuclear matrix	RCA-II	D-GalNAc, D-galactose	(74)
Chromatin & nuclear matrix	APA	α-L-fucose	(92)
Chromatin & nuclear matrix	TTA	D-GlcNAc	(72)
Chromatin & nuclear matrix	SBA	D-GalNAc	(72)
Chromatin & nuclear matrix	PHA	oligosaccharides	(73)
Cytoplasmic sites	Con A	α-D-glucosides & α-D-mannosides	(80, 84–86)
Cytoplasmic sites	WGA	D-GlcNAc & sialic acids	(82, 84)
Cytoplasmic sites	RCA	D-Galactose	(83, 84)

[a] Binding observed on cytoplasmic and/or nucleoplasmic faces of nuclear membranes.

were interacting with glycoproteins normally present on the lumenal side of these membranes (4). In some cases, the purity of the isolated nuclei was not documented, and the possibility of contamination with membranes containing high concentrations of glycoproteins, such as the plasma membrane, was not addressed. The fact that ferritin–Con A stains the lumenal contents, but does not stain the cytoplasmic/nucleoplasmic faces of isolated nuclear envelopes from amphibian oocytes (59), supports these criticisms.

More recent work examining lectin binding sites at the electron microscopic level supports the traditional localization of glycosylation. Colloidal gold and ferritin conjugates of WGA and Con A were used to compare the distribution of lectin binding sites on the fracture faces of plasma membranes and intracellular membranes from rat exocrine and endocrine pancreatic cells (60). Unlike the external surface of the plasma membrane, which labeled with both lectins, the endoplasmic reticulum and nuclear envelopes labeled at their lumenal surfaces with Con A, but did not label with WGA (60), as expected according to conventional models of glycoprotein biosynthesis and processing. Lectin peroxidase-colloidal gold techniques applied to thin sections of rat liver detected Con A binding sites in plasma membranes, nuclear envelopes, rough endoplasmic reticulum, and smooth endoplasmic reticulum (61). However, gold particles were seen not only in cisternal spaces, as before, but also on membrane-bound ribosomes.

Although all of these lectin staining studies are intriguing, conclusions about the presence of glycoproteins in nuclear membrane are difficult to draw because of the contradictory results obtained by different laboratories. Definitive structural evidence of the carbohydrate moieties suggested by these studies is also lacking. In contrast, recent studies at both the light and electron microscopic levels have convincingly shown the presence of WGA-binding sites at the nucleoplasmic and cytoplasmic faces of nuclear pores (25, 62). As discussed below, other structural studies have identified these WGA-binding sites as O-GlcNAc moieties.

CHROMATIN AND NUCLEAR MATRIX   Specific lectin-binding sites have been demonstrated in isolated chromatin and nuclear matrix preparations by many different laboratories (Table 1). As with isolated nuclei, a major problem in many of these studies is the lack of chromatin purity, a problem compounded by the fact that purified chromatin is usually defined only operationally without the use of marker enzymes or proteins. Nonetheless, a number of intriguing studies have suggested the presence of glycosyl moieties in chromatin and nuclear matrix.

Scatchard analyses of the binding of Con A to purified rat liver chromatin indicates one type of binding site ($K_d = 3 \times 10^{-7}$) with one site per 1400 base pairs of DNA (63). Electroblots of nonhistone chromosomal fractions probed with ^{125}I-Con A detected three major proteins of 135 kd, 125 kd, and 69 kd. ^{131}I-LCA (lentil lectin, *Lens culinaris*), binding to sea urchin chromatin, isolated by detergent extraction of nuclei and sedimentation through sucrose to minimize membrane contamination, demonstrated saturable specific binding with a single type of binding site (64).

In studies of fluorescently labeled WGA or Con A binding to mitotic chromosomes or interphase nuclear substructures of cultured human fibroblasts, it was found that WGA brightly stained both mitotic and interphase chromatin structures, whereas Con A binding was not detectable (65). WGA staining of chromosomes produced a banding pattern similar to that of quinacrine staining. The interiors of nuclear "scaffolds" prepared by DNase treatment of HeLa cell chromosomes also stained intensely with WGA (65). The authors argued against contamination by nuclear membranes, which are absent during mitosis. Quantitative studies of ^{131}I-lentil lectin binding to sea urchin nuclei and nuclear matrix, prepared by the method of Berezney & Coffey (66), indicated that the majority of the nuclear-associated lectin binding sites were in the matrix residue (67). Histones isolated from the macronucleus of the protozoan, *Tetrahymena thermophila,* appear to contain fucose- and mannose-residues, based upon their specific binding to a fucose-specific lectin from *Ulex europeus* (UEA-I) and to Con A, respectively, as well as their metabolic incorporation of ^{3}H-fucose (68). H2A appears to be the most heavily fucosylated histone in these cells. The structure or linkage of the saccharides on these glycosylated histones has not been further elucidated. Fluorescein- and ferritin-WGA and –Con A were used to probe tissue sections of ovarian follicles of the lizard, *Lacerta vivipara,* in which the nuclei of the five major cell types have very distinctive morphologies (69). Fluorescent WGA intensely stained the nucleolus, whereas it only diffusely stained the nucleoplasm and the cytoplasm, with even weaker staining over the nuclear envelope. Fluorescent Con A yielded similar results, but stained much less intensely. At the electron microscopic level, ferritin-WGA binding sites appeared enriched over granular regions of the nucleoli, and binding was evident in the nucleoplasm, nuclear pores, and cytoplasm (69). In other studies using freeze-fracture electron microscopy and lectin probes conjugated to colloidal gold or horseradish peroxidase, glycoproteins were found in cross-fractured nuclei of duodenal columnar and exocrine pancreatic cells (70). Con A bound preferentially to euchromatin, and UEA-I was almost exclusively localized to cross-fractured zones of exposed euchromatin. Binding was abolished by trypsin and by the appropriate competitive sugar. The authors suggested that mannose- and fucose-containing glycoproteins appear to be enriched at sites of active chromatin (70).

Additional studies have shown that transcriptionally active regions of chromatin may contain glycosylated proteins. Fluorescein–Con A binding to polytene chromosomes from *Chironomus thummi* appeared to bind selectively to transcriptionally active regions termed "puffs" (71). The extent of specific Con A binding appeared proportional to the size of the puff. One study investigated the binding of several fluorescently derivatized lectins to isolated chromatin of sea urchin embryos at different developmental stages and divided these lectins into three distinct classes: (*a*) those whose binding increased at gastrulation (Con A and RCA-120); (*b*) one whose binding decreased at gastrulation, but increased afterward (TTA, *Tachypleus tridentatus*, Japanese horseshoe crab); and (*c*) those whose binding was low throughout development (WGA, SBA, UEA-I) (72). The authors suggested that this modification of chromatin constituents may be involved in the switching from maternal information to embryonic encoded RNA that occurs at gastrulation (72). Phytohemagglutinin (PHA) treatment of isolated lymphocyte nuclei increased their DNA-directed RNA synthesis twofold, and concomitantly increased actinomycin D binding, suggesting a loosening of chromatin structure (73). Direct binding of PHA to chromatin was not demonstrated. Partially purified human neuroblastoma DNA polymerase was shown to be inhibited by toxic lectins, Con A, RCA-II, and PHA-P (74). The significance of these inhibitions remains unclear, but these authors suggested that they may relate to the toxic properties of the lectins.

Since there appears to be a correlation between nuclear glycosylation and transcriptional activity, it is possible that nuclear glycoproteins might be glycosylated abnormally in neoplastically transformed cells. Lectin blotting of electrophoretically separated proteins from normal rat liver and several tumor cell types detected nuclear-associated glycoproteins (75). Most of these putative glycoproteins were removed from nuclei washed with nonionic detergent. However, a major WGA-binding protein of 62 kd and a major Con A–binding protein of 200 kd remained. These two glycoproteins likely represent subsequently identified nuclear pore glycoproteins (26, 76). Compositional analyses suggested that specific glycosylation of chromatin was associated with experimentally induced hepatocarcinogenesis (77). Several lectins were used to detect glycosylated proteins associated with residual fractions (insoluble in 4% SDS) or nuclear matrix prepared from normal rat liver, azo dye–induced rat hepatoma cells, and Walker 256 carcinosarcoma cells (78). Two lectin-binding proteins (95 kd and 55 kd) were present only in the tumor cells. A 62 kd WGA-binding protein was identified in normal rat liver and in the induced tumor cells, and a 213 kd Con A– and RCA-binding protein was identified in the Walker cell residual fraction. Based upon its lectin-binding properties, a 182 kd protein in the nuclear residual fraction appeared to be differentially glycosylated in the two tumors. These studies suggested that analogous to cell surface glycosylation (79), the glycosylation of intranuclear proteins might also correlate with the neoplastic phenotype.

During recent studies of nuclear pore glycoproteins, WGA binding sites on the inside of the nuclei were also detected (28). In support of these studies showing WGA binding to transcriptionally active chromatin, several well-studied transcription factors for RNA-polymerase II have been shown to bind WGA and to contain multiple O-GlcNAc residues (31).

Lectin Binding Sites in the Cytoplasm

Several papers have reported the existence of lectin binding sites in the cytoplasmic compartment (Table 1). Early work with lectins indicated that purified mitochondria were strongly agglutinated by WGA and RCA, but only weakly by Con A, and suggested that the mitochondrial-associated glycoproteins were cytosolic contaminants (55), although contamination by plasma membrane components was not excluded. [Acetyl-^{3}H]Con A was used to study lectin binding to purified rat liver mitochondria (80). Based upon detergent permeabilization studies, using adenylate kinase to measure latency, it was estimated that about 40% of the Con A binding sites were present on the outer cytoplasmic surface of the mitochondria. While the mitochondrial preparations used appeared to be fairly pure, a minor contamination by lysosomes could not be eliminated. Similar studies using ^{125}I-Con A binding to isolated chromaffin granules from bovine adrenal glands demonstrated binding to isolated granule membranes but not to intact granules, suggesting that the lectin binding sites were not exposed at the cytoplasmic face (81). In contrast, agglutination of "intact" chromaffin granules by WGA was taken as evidence that they contain complex carbohydrates at their cytoplasmic faces (82). In support of this contention, 50% of the granule-associated sialic acids and only 2% of the catecholamines were released upon treatment of intact granules with sialidase. It is interesting to note, however, that WGA-induced granule agglutination was not prevented by prior sialidase treatment (82). Similarly, ferritin-RCA was shown to bind to the cytoplasmic faces of isolated azureophil and specific granules of rabbit polymorphonuclear leukocytes (83), suggesting that the granule glycoconjugates face the cytoplasm. Recently, a 47 kd protein that binds to WGA and DBA (*Dolichos biflorus*; GalNAc-specific) has also been localized to the cytoplasm of *Leishmania braziliensis* by using fluorescently derivatized lectins (84).

Several studies have detected lectin binding sites on ribosomes (61, 85, 86). Ribosomes prepared from chicken liver or rabbit reticulocytes bound Con A in a molar ratio of 1:1 (85). The Con A appeared to bind to the large subunit in a saturable fashion ($K_d = 5 \times 10^{-7}$ at 0°C). The ribosome preparations used were judged to be free (<5%) of membrane contamination using GTP hydrolysis and alkaline phosphodiesterase assays. In these same studies, *Escherichia coli* ribosomes bound much less Con A. Metabolic radiolabeling studies, using ^{35}S-methionine and affinity chromatography, identified a single protein of 37 kd (~1.5% of the ribosomal proteins) that appeared to account for the binding of Con A to ribosomes (85). Unfortunately, subsequent studies have not characterized this putative ribosomal glycoprotein. In a similar study, free and membrane-bound ribosomes from wheat germ were agglutinated by Con A, even after extensive washing with detergents (86). Four putative glycoproteins (14 to 24 kd) were identified in both the free and membrane-bound ribosomes by periodic acid-Schiff staining of SDS-polyacrylamide gels. Three putative glycoproteins ranging from 15 kd to 25 kd were either underglycosylated or absent in free ribosomes, but detectable in membrane-bound ribosomes. It is important to note that contamination by membrane proteins was not carefully controlled in these studies. In a survey using lectin-gold complexes for electron microsopic localization of glycoconjugates, colloidal gold–Con A was seen to bind to ribosomes (61). If indeed certain ribosomal proteins are glycosylated, it would be of great interest to determine the structures of their glycan moieties as well as the role of their saccharides in either ribosomal structure or function. Perhaps the so-called "nonspecific" side effects on protein synthesis by the specific glycosylation inhibitor tunicamycin (87) are not so nonspecific after all. Only rigorous biochemical and functional characterization of these putative ribosomal glycoproteins will clarify these issues.

BIOCHEMICAL STUDIES

A few studies have used compositional analyses or metabolic radiolabeling with sugar precursors, in conjunction with subcellular fractionation, to argue for the existence of nuclear or cytoplasmic glycoconjugates. The majority of papers that contain the most convincing data describe novel types of glycoconjugates that appear to be selectively enriched in the nucleoplasmic or cytoplasmic compartments, thus defeating the arguments that these intracellular glycans might have arisen from contamination by other organelles.

General Compositional Analyses and Metabolic Labeling Studies

Carbohydrate compositional analyses on isolated rat liver nuclei indicated that glucosamine (GlcN) and mannose (Man) were the major components, with more than 60% of both sugars localized to the nuclear membranes (88). In nuclear membrane–derived glycopeptides, the GlcN:Man ratio was 1:3, whereas the nucleoplasmic sugars consisted almost entirely of GlcN. The kinetics of metabolic incorporation of tritiated glucosamine into nuclear components were rapid, leading these authors to suggest that the nuclear glycoproteins must be synthesized at or near the nucleus (88). Compositional analyses of chromatin from rat Walker Carcinoma 256 indicated that the

nonhistone and histone chromosomal proteins were 32% and <1% carbohydrate by weight, respectively (89). It was suggested that the high carbohydrate composition of the nonhistone fraction might have its origin in the
nuclear membrane where chromatin fibers converge and attach. Alternatively,
it was proposed that the carbohydrate might be acquired by and necessary for
some components of the nonhistone proteins during their transport from the
nucleus to the cytoplasm (89). Later studies of high salt–extracted chromatin
from the Walker 256 carcinosarcoma and the Novikoff hepatoma identified a
26 kd galactosamine-rich (15 moles/mole) glycoprotein that is absent in
chromatin from normal or regenerating rat liver (90). Another group argued
for the presence of cytosolic glycoproteins based upon compositional analyses
of protein-bound sugar residues in rat liver microsomal subfractions (91).
These authors reported that about 50% of the microsomal mannose and
galactose could be released from "intact" vesicles with trypsin (91). Unfortunately, these results are difficult to interpret because the integrity of the
microsomes was not well documented.

In a series of well-controlled fractionation studies, highly purified rat liver
nuclei and nuclear membranes were prepared and assayed for phospholipids,
glycolipids, cytochrome oxidase, protein-bound carbohydrate, Con A receptors, and histones (16). In these studies, nuclear membranes did not appear to
contain significant amounts of cardiolipin, gangliosides, cytochrome oxidase,
histones, or other basic proteins, but had neutral sugar compositions similar to
rough endoplasmic reticulum membranes. These workers found no indication
of carbohydrate in the nonmembraneous nuclear fraction, except for what
they termed as "unaccounted for" amounts of glucosamine and glucose (16).
PAS-staining of SDS-gels of 0.6 M NaCl extract of purified chromatin
identified six major putative glycoproteins (30 kd, 54 kd, 75 kd, 104 kd, and
127 kd) in Novikoff hepatoma ascites cells, whereas normal liver revealed
only four major bands (16 kd, 30 kd, 54 kd, and 75 kd) (92). In these same
studies, lectin rocket "affinoelectrophoresis" was used to identify chromatin-
associated glycoproteins; however, the method used to prepare chromatin did
not adequately eliminate nuclear membrane contamination (92). Metabolic
radiolabeling with [³H]glucosamine identified a major 130 kd glycoprotein in
mononucleosomes generated by digesting Ehrlich ascites nuclei with micrococcal nuclease (93). In these studies, [³H]fucose was used as a marker for
membrane contamination. The 130 kd putative glycoprotein was not extract-
able by Triton X-100 and, unlike the fucose-marker, remained associated with
the nucleosomes after sedimentation through sucrose (93). Polyvalent antisera
and monoclonal anibodies have identified a group of chromosomal protein
antigens that appear during azo dye–induced hepatocarcinogenesis (77).
These antigens are enriched in chromatin fractions that are insoluble in 6 M
guanidine-HCl:2% SDS. Compositional analysis suggests that these proteins

are glycosylated (1.4 and 2.0 moles of GalN and GlcN, per mole of protein, respectively).

In a recent paper, several lysosomotropic agents were claimed to induce the translocation of [2-^{3}H]mannose-labeled nonhistone glycoproteins into the nucleus (94). The authors suggested that these nuclear nonhistone glycoproteins contain N-linked oligosaccharides and reach the cytoplasmic compartment by vesicular traffic. While several radiolabeled glycoproteins did appear to become nuclear associated in these studies, contamination from other subcellular compartments or the identities, size, or number of such proteins were not well documented.

High Mobility Group Glycoproteins

High mobility group (HMG) proteins are fairly abundant (1/30th of histones) chromosomal proteins classified according to their relative electrophoretic mobilities (95). HMG proteins appear to be preferentially associated with regions of transcriptionally active chromatin. Although they have highly conserved amino acid sequences, they undergo a wide variety of posttranslational modifications, such as methylation, acetylation, phosphorylation, poly(ADP)ribosylation, as well as glycosylation (95). It is speculated that posttranslational modifications of HMG proteins in select regions of the chromosome might regulate gene activity or facilitate nuclear functions.

Purified HMG14 and HMG17 from mouse Friend erythroleukemic cells and calf thymus cells were found by direct composition analysis to contain N-acetylglucosamine, mannose, galactose, glucose, and fucose (96). They also bind a fucose-specific lectin from *Ulex europeus* (UEA-I). Furthermore, these HMG proteins can be metabolically labeled with sugar precursors, and the labeled saccharide moieties are insensitive to cleavage by dilute alkali, suggesting an N-linkage to protein. Surprisingly, removal of saccharides by mixed glycosidases abrogates the bindingof these HMG proteins to nuclear matrix (97). These workers suggested that the saccharides on HMGs might play an important role in the overall architecture of active domains of chromatin associated with the nuclear matrix. Unfortunately, virtually nothing is known about the structures or biosynthesis of the interesting saccharide moieties on HMG proteins.

Proteoglycans and Glycosaminoglycans

Over the years, there have been many reports suggesting that small amounts of proteoglycans and/or glycosaminoglycans are present in the nucleoplasmic or cytoplasmic compartments of cells (for reviews, see 13–15). These claims have met with almost universal skepticism, since the polysaccharide components of proteoglycans, the glycosaminoglycans, are clearly added in the lumenal compartments of the Golgi and endoplasmic reticulum (98). In subcellular fractionation experiments, the small amounts of these highly charged molecules typically observed in nuclei and cytoplasm have made it almost impossible to eliminate the possibility that they arose from contamination by cell-surface or lumenal components. Histochemical approaches have been complicated by the presence of large amounts of highly negative nucleic acids. Enzymatic confirmation of histochemical staining is weakened by the lack of product characterization and the uncertainty with respect to the relative purity or specificities of the enzymes employed. Nevertheless, there are several well-documented cases in which these obstacles have been overcome to demonstrate the occurrence of glycosaminoglycans and proteoglycans in these subcellular compartments.

NUCLEAR GLYCOSAMINOGLYCANS   As early as 1964, small amounts of hexosamines were found in purified rabbit liver nuclei, leading researchers to suggest that nuclei either contain connective tissue polysaccharides or are contaminated by them upon isolation (99). Kinoshita and colleagues have published a series of reports over the last 15 years that support a role for nuclear sulfated glycosaminoglycans or proteoglycans in sea urchin morphogenesis (100–105). An initial study had demonstrated that sulfate ion in sea water was essential in the gastrulation stage of sea urchin development (100). Incorporation of $^{35}SO_4$ into the embryo peaks at the gastrulation stage of development, and is largely into heparan sulfate-like molecules. Heparin was found to stimulate RNA synthesis in pregastrula sea urchin nuclei, but not in postgastrula nuclei, suggesting that nuclear-associated heparan sulfates might be involved in the switch from maternal gene-directed development to that directed by embryonic genes (100). Exogenously added chondroitin sulfate or hyaluronic acid had no effects in this system. Another study isolated chromatin nonhistone proteins from developing sea urchins at different stages and suggested the presence of hyaluronic acid, but no developmental changes were observed (106). In pulse-chase studies of a heparan sulfate proteoglycan in developing sea urchins, it was concluded that the proteoglycans were first present in a "cytoplasmic compartment" and were subsequently transferred to the nucleus (101). The "cytoplasmic compartment" was defined by its lack of sedimentation during high-speed centrifugation. This criteria, however, would not rule out a lumenal origin for the proteoglycans. In order to further support their hypothesis that nuclear heparan sulfates play a role in the dramatic increase in chromatin template activity that occurred at gastrulation, Kinoshita et al examined the effects of exogeneous heparin and sea urchin proteoglycans on chromatin thermal denaturation (102). Changes in the thermal denaturation of chromatin during sea urchin development were found to correlate with the onset of proteoglycan synthesis. In addition, heparin and purified proteoglycans from sea urchin nuclei were both found to "loosen" chromatin structure (102). Inhibitors of proteoglycan synthesis, such as β-xylosides and sodium selenate, block sea urchin development at the blastula stage when proteoglycan synthesis was reduced by 50% (103). Remarkably, this developmental arrest by β-xylosides was released by the exogenous addition of isolated proteoglycans from postgastrula but not pregastrula embryos (104). At higher concentrations, microinjection or direct addition of proteoglycans isolated from embryos, blocked development specifically at the same stage from which the molecules were isolated (104). In later studies with glycerol-permeabilized sea urchin cells, proteoglycans purified from more advanced stages of development accumulated rapidly in the nucleus, shifting the hyperchromicity of the chromatin to a lower temperature (105). While these studies are clearly out of the mainstream of current thinking, it is well established that glycosaminoglycans and proteoglycans can dramatically influence transcription in vitro (66). A weakness of these studies is in not providing unequivocal evidence for the presence of these molecules in the nucleus. Since these protein-polysaccharides are among the most negatively charged molecules in nature, it is difficult to establish that they are not simply binding to basic proteins in the nucleus, such as histones, when the embryos are disrupted.

In addition to the preceding functional studies, numerous groups have used metabolic radiolabeling together with subcellular fractionation to examine the distribution of these molecules in the cell. One such study isolated nuclei from B16 melanoma cells radiolabeled with $^{35}SO_4$ and [^3H]glucosamine, and found they contained a mixture of complex glycoconjugates, including proteoglycans (107). While glycosaminoglycans appeared to be minor components, compared to the glucosamine-containing glycoproteins, a family of high-molecular-weight chondroitin sulfates with different degrees of sulfation, and a minor amount of heparan sulfates were identified. Nuclei freed of outer membrane by washing in 2% Triton X-100 still retained 70–80% of the nuclear-associated $^{35}SO_4$-macromolecules. The nuclei were judged free of contamination by phase contrast and electron microscopy, as well as by marker enzymes (108). A brief, but provocative, report used similar techniques to argue for the presence of both glycoproteins and glycosaminoglycans in HeLa cell nuclei (14). Chromatin isolated from nuclei stripped of their outer membranes by detergent contained many high-molecular-weight glycoconjugates labeled with either ^3H-fucose or ^3H-glucosamine. Mixing experiments, in which radiolabeled cell-surface trypsinates were combined with unlabeled nuclear preparations prior to chromatin isolation, were used to argue against contamination by plasma membrane components. Chromatin purified through sucrose still retained >90% of the sugar-labeled macromolecules. Sensitivity to glycosidases identified the presence of chrondroitin sulfates in these chromatin fractions (14).

Studies on the composition, concentration, and sulfate radiolabeling of glycosaminoglycans and glycoproteins were performed on purified, Triton-extracted rat brain nuclei (32). Glycosaminoglycan concentration was estimated to be 0.142 μmole of hexosamine/100 mg of protein, with chondroitin 4-sulfates, chondroitin 6-sulfates, hyaluronate, and heparan sulfates accounting for 57%, 7%, 29%, and 7% of the total glycosaminoglycans, respectively. Premixing experiments suggested that <5% of the total glycosaminoglycans could be accounted for by nonspecific adsorption during nuclear isolation. As much as 24% of the total $^{35}SO_4$ incorporated into the nucleus was found in sulfated glycoproteins that were resistant to treatments that degrade glycosaminoglycans. Metabolic labeling studies indicated that the rat brain nuclear chondroitin 6-sulfates turned over almost five times faster than their 4-sulfated counterparts (32). A similar study of glycosaminoglycans from highly purified rat liver nuclei suggested that they were present at levels of 0.2–0.3 μg of hexuronic acid/mg of DNA, with the major component identified as hyaluronate, and minor components as chondroitin sulfates (109). A later paper suggested that these anionic polysaccharides increased in a fashion parallel to DNA synthesis in the nuclei of liver cells after a partial hepatectomy (110). Most of these glycosaminoglycans appeared to be in the chromatin fraction. Arguments against the putative nuclear glycosaminoglycans arising from other subcellular contamination, and supporting their intrinsic localization in the nucleus, are as follows: (*a*) Nuclear preparations were judged to be practically free from subcellular organelles and plasma membranes by phase-contrast and electron microscopy. (*b*) Repeated washing of nuclei with 2.2 M sucrose, or pretreatment of intact nuclei with hyaluronidase, did not change the quantitation. (*c*) Plasma membrane marker enzymes, such as 5'-nucleotidase, were present at <0.06% of their specific activity in plasma membranes. (*d*) Most of the nuclear-associated glycosaminoglycans are hyaluronic acids and chondroitin sulfates, whereas the majority at the plasma membrane are heparan sulfates. (*e*) Mixing unlabeled liver or crude nuclear fractions with [³H]glucosamine-labeled postnuclear supernatants did not result in the appearance of radiolabeled macromolecules in subsequently purified nuclei (109). While all of these arguments have limitations, together they provide a relatively strong case for the intrinsic localization of glycosaminoglycans in the nucleus.

Histochemical methods were used to demonstrate the presence of acidic glycosaminoglycans in the cell nuclei of human iris cells and other tissues (111). Staining was shown to be sensitive to hyaluronidases and chondroitinases, and simultaneous staining with alcian blue and Feulgen stain were used to distinguish glycosaminoglycans from nucleic acids. These nuclei were concluded to have three types of glycosaminoglycans, a nonsulfated glycosaminoglycan, chondroitin sulfates, and a testicular hyaluronidase-resistant species. Histochemical and metabolic radiolabeling studies suggested the existence of polysaccharides associated with mammalian chromosomes, and suggested that they changed with the cell-cyle (112). However, the specificity of the histochemical probes was not well documented, especially considering the large amounts of RNA and DNA present relative to the putative polysaccharides. Contamination by membranes and extracellular matrix was not ruled out in most metabolic labeling studies. High-resolution autoradiography of $^{35}SO_4$-labeled skin fibroblasts from patients with Hurler's or Sanfilippo's syndromes was used to identify nuclear localized glycosaminoglycans (113). Sixty-one percent of the nuclei had silver grains, mostly localized in chromatin-rich peripheral zones, and biochemical analyses showed that nearly all of the cell-associated radioactivity was in glycosaminoglycans (113). Computer-assisted statistical analyses argued that the β-particles seen in the autoradiographs originated in the nucleus; however, the numbers of grains in each sample were quite small, making the data unconvincing.

The majority of the previously mentioned studies have remained controversial, largely because of questions about the specificities of probes used, or because of the difficulties associated with proving the glycosaminoglycans did not arise from contamination by other cellular fractions. Recently, unequivocal evidence for the presence of a structurally unique form of heparan sulfate in the nuclei of rat hepatocytes was obtained (38). Nuclear heparan sulfate accounts for ~6% of the total heparan sulfates in these cells, and exists mostly in the form of a free polysaccharide. More importantly, the nuclear-specific heparan sulfate has a unique structure not seen in other cellular compartments or in the extracellular environment. Nuclear heparan sulfates contain large amounts of an unusual sulfated glucuronosyl residue, whereas in non-nuclear heparan sulfates the uronic acid residues are generally not sulfated until they are converted into iduronic acid (38). Subsequent studies have used pulse-chase metabolic radiolabeling to study the biosynthesis and uptake of these unusual nuclear heparan sulfates (39). Heparan sulfates appear in the nucleus as early as 2 h after synthesis, and turn over with a half-life of ~8 h. Chloroquine did not affect the nuclear uptake, suggesting that lysosomes are not involved in the processing of heparan sulfate proteoglycans to the forms found in the nucleus, but the drug did increase the half-life of the polysaccharides to more than 20 h. When added to hepatocytes at 37°C, 10% of the ^{35}S-heparan sulfate proteoglycan accumulated in the nucleus, where it appears as free polysaccharide chains with substantially increased levels of sulfated glucuronosyl moieties (39). The uptake into the nucleus still occurs at 16°C, a temperature that stops lysosomal transport, and ammonium chloride does not block the appearance of the free heparan sulfate chains. These data suggest that heparan sulfate proteoglycans, which are synthesized by the conventional endoplasmic reticulum and Golgi pathways, are specifically endocytosed via a nonlysosomal pathway, and a small amount is degraded to release free heparan sulfate side-chains, which are transported to the nucleus. It is not yet clear where the sulfation of the glucuronosyl residues occurs, however. The discovery of a nuclear-specific structure circumvents many of the arguments against earlier claims that these types of glycoconjugates are present in the nucleoplasmic compartment.

CYTOPLASMIC PROTEOGLYCANS   In contrast to the nuclear studies, few studies suggest the existence of cytoplasmic proteoglycans or glycosaminoglycans. Most of these have come from groups working on the nervous system (33, 35–37, 114–118). Previous studies had demonstrated that most of the chondroitin sulfates in brain exist on soluble proteoglycans (119–121). The fact that the proteoglycans were soluble but not released during cell isolation, or by treatment of intact cells with trypsin, suggested that they might be cytoplasmic rather than a component of the extracellular matrix. The distribution of glycosaminoglycans was studied in the cytoplasmic and particulate fractions of neurons isolated from rat cerebrum, using the release of lactate dehydrogenase as a marker for cell integrity (35). Hypotonic lysis of neurons released 20% of protein, 90% of lactate dehydrogenase, 82% of chondroitin sulfates, 55% of heparan sulfates, and 25% of hyaluronate and glycoproteins. These data suggest that in adult brain, chondroitin sulfate proteoglycans are mostly cytoplasmic constituents rather than being primarily components of the extracellular matrix. Surprisingly, subsequent analyses of these cytoplasmic proteoglycans indicated that they contain more than half of their carbohydrate-protein linkages as novel O-linked mannose oligosaccharides, including both neutral and sialylated species (114, 117, 118). The cytoplasmic localization in neurons and astrocytes was confirmed by using peroxidase-coupled monoclonal antibodies to the chondroitin sulfate proteoglycans together with immuno-electron microscopy (33). In contrast, similar studies in developing rat brain indicated that nearly all of the chondroitin sulfates in 7 d postnatal brain are extracellular (34). At 10 to 14 days of development, there is a decrease in extracellular staining, and by day 21 of postnatal development, distinct cytoplasmic staining is evident in both neurons and astrocytes. By day 33, the localization is the same as that for adult brain. Biochemical analyses suggested that the extracellular proteoglycans in day 7 neonates are similar to the intracellular proteoglycans seen in adult brains (34). Histochemical studies, together with the use of glycosaminoglycan-specific glycosidases, have also suggested the presence of cytoplasmic glycosaminoglycans in human brain cells (116).

Using a biotinylated hyaluronate-binding region of chondroitin sulfate proteoglycan from Swarm rat chondrosarcoma, together with avidin-peroxidase and light or electron microscopy, hyaluronic acids were localized in developing rat cerebellum (37). Similar to chondroitin sulfate pro-teoglycans, hyaluronates were found to be predominantly extracellular through the third postnatal week. Cytoplasmic hyaluronate appeared about day 20 of postnatal development. While it seems likely that these cytoplasmic proteoglycans and glycosaminoglycans are synthesized and secreted by well-characterized pathways (98) and subsequently endocytosed, no data is yet available. Alternatively, it is possible that they are synthesized in the cyto-plasm by brain cell–specific pathways that are yet to be described. Further-more, the functions of these cytoplasmically localized glycoconjugates in the brain are not at all clear.

Glycogenin

Classical work by the Coris on glycogen biosynthesis suggested that a "prim-er" must be required and would likely be a small oligosaccharide (reviewed in 122). Recent work has identified a 37–42 kd protein, called glycogenin, which appears to be the long sought-after primer for glycogen elongation. Glycogenin has glucosyl residues covalently attached at tyrosine residues and appears to be closely associated with glycogen synthase (40, 41, 43). It therefore appears that glycogen is in fact a type of cytoplasmic proteoglycan (40). Subsequent studies identified the 38 kd subunit of rabbit muscle gly-cogen synthase as glycogenin, which exists as a tight complex in a 1 : 1 molar ratio with the 86 kd catalytic subunit (44). Glycogen synthase is able to glycosylate the purified glycogenin molecule, which appears to lose its affinity for the complex when the glycogen polymer reaches a certain length. Recently, an antibody to glycogenin from rabbit skeletal muscle was used to show that the glycogen primer is present in heart, liver, and in high-speed supernatants containing glycogen synthase "acceptor protein" (42). While we tend not to consider glycogen as a proteoglycan or to categorize it with the more complex glycoconjugates, it is at present the only clear-cut example of cytoplasmic glycosylation of a polypeptide that has a known function.

Viruses

Many viruses have coat proteins that are heavily glycosylated. It is clear that for the most part, these viruses utilize the host endoplasmic reticulum or Golgi glycosylation pathways. However, certain viral antigens localized to nuclear membranes, nucleoplasm, or cytoplasm also appear to be glycosylated (123–125). Simian virus 40 (SV-40) large T antigen appears to contain an unusual form of glycosylation (124). T antigen is metabolically labeled by galactose and glucosamine, but not by fucose or mannose. The incorporated sugar is resistant to endoglycosidase H and tunicamycin, suggesting that its saccha-rides might be O-linked to protein. T antigen is found in the nucleus, cytoplasm, and on the cell surface. Subsequent studies have shown that T antigen specifically binds soybean agglutinin (a galactose-specific lectin), and cell fractionation studies have shown that [³H]glucosamine-labeled T antigen is preferentially associated with the nucleus (125). In contrast, the majority of the T antigen labeled with ³⁵S-methionine is found in the cytoplasm. While the T antigen glycosylation sites have been localized to the same peptide as the T antigen phosphorylation sites (amino acids 653 to 691), nothing is known with respect to the linkage or types of saccharide structures present on the molecule.

O-Linked N-Acetylglucosamine-Bearing Glycoproteins

During investigations in which bovine milk galactosyltransferase was used to probe the surface saccharide topography of living murine lymphocyte sub-populations, a novel O-linked N-acetylglucosamine saccharide-protein linkage (O-GlcNAc) was discovered (126). The linkages to protein and saccharide structure were confirmed by kinetic analyses of alkali-induced β-elimination reactions, resistance of the linkage to peptide: N-glycosidase A, sensitivity of galactosyltransferase labeled products to β-galactosidase, and chromatographic analyses of β-elimination products. O-GlcNAc appeared to occur as clustered monosaccharide residues on several different integral membrane proteins. Unexpectedly, latency experiments indicated that the vast majority of these monosaccharides were attached to intracellular proteins. B lymphocytes were found to contain more than 150 million O-GlcNAc moieties per cell. Metabolic radiolabeling indicated that O-GlcNAc accounted for from 5 to 12% of the total macromolecular [³H]glucosamine incorporation.

Subsequently, an abstract (127) reported the presence of single GlcNAc residues on proteins in rat liver nuclei. Other studies using highly purified, detergent solubilized subcellular fractions from rat liver indicated that O-GlcNAc-bearing proteins were particularly enriched in nuclear envelope, nucleus, and in cytosol (22). Rough endoplasmic reticulum and Golgi also were found to contain substantial amounts of these glycoproteins, whereas mitochondria were totally lacking in O-GlcNAc-bearing proteins. Several integral membrane proteins of the smooth and rough endoplasmic reticulum of rat liver have also been demonstrated to contain terminal GlcNAc moieties at their cytoplasmic faces (128). Linkage analysis of one major 55 kd protein of smooth endoplasmic reticulum demonstrated the presence of cytoplasmic O-GlcNAc. The smooth endoplasmic reticulum appears to contain larger amounts of these saccharides on a different set of proteins than the rough endoplasmic reticulum (128). Resistance to peptide: N-glycosidase F digestion and chromatographic analyses of β-elimination products indicated that nearly all of the protein-bound terminal GlcNAc moieties within the cytoplasmic and nuclear compartments are in the form of O-GlcNAc (22).

Even though serum contains a myriad of N-linked and "mucin-type" O-linked glycoproteins, it was found to completely lack O-GlcNAc-bearing proteins. Soluble cytosolic fractions from perfused livers contained nearly all of their terminal GlcNAc residues as O-GlcNAc (22).

NUCLEAR PORE GLYCOPROTEINS   A monoclonal antibody was shown to bind to a 62 kd (p62) nuclear pore protein by immunofluorescence and immuno-ferritin electron microscopy (26). Pulse-chase studies suggested that this p62 is first synthesized in the cytoplasm as a 61 kd precursor. Based upon earlier descriptions of wheat germ agglutinin–binding proteins of the same apparent size (129), and on the pronounced enrichment of O-GlcNAc in the nuclear envelope (22), Davis & Blobel suggested that maturation of p62 might involve glycosylation by O-GlcNAc (26). This suggestion was supported by the fact that p62 bound WGA, but not Con A on nitrocellulose blots. Further analyses indicated that the p62 pore protein has 10–12 sites of O-GlcNAc (mostly at serine and to a much lesser extent threonine residues) (23). Concomitantly, several different monoclonal antibodies that recognize different subsets of the proteins associated with the nuclear pore complex were isolated (24). These monoclonal antibodies identified a group of 8–10 major polypeptides associated with the nuclear pore complex. While these proteins were tightly associated with the nuclear pores, only one appeared to be an integral membrane protein. The antibodies stained the nucleus in a characteristic punctate fashion by immunofluorescence and stained the cytoplasmic and nucleoplasmic faces of the pore complexes by immuno-gold electron microscopy (24). Also, the antibody staining demonstrated that the pore polypeptides were dispersed throughout the cytoplasm at mitosis, when the nuclear envelope breaks down. It was demonstrated that all of these major nuclear pore proteins are multiply glycosylated by O-GlcNAc residues, and that these pore glycoproteins account for the majority of the O-GlcNAc in the nuclear envelope (23). Surprisingly, masking of O-GlcNAc residues by galactosylation or removing O-GlcNAc by hexosaminidase abrogated the binding of all of the monoclonal antibodies to varying extents, indicating that the saccharide moieties are major epitopes recognized by all of these "pore-specific" antibodies. These experiments suggested that clustered O-GlcNAc moieties are highly immunogenic. Together with the afore-mentioned biochemical studies, these data strongly support the cytoplasmic and nucleoplasmic orientation of O-GlcNAc moieties.

Recently, these studies were extended to further confirm the presence of O-GlcNAc on nuclear pore glycoproteins (27). Using WGA-blotting and galactosyltransferase as a probe, it was demonstrated that the p62–p64 increased to an apparent molecular mass of 65-66 kd after galactosylation. Schindler et al suggested that O-GlcNAc might protect the proteins from proteolysis or serve as a targeting signal for nuclear proteins. The cytoplasmic orientation of O-GlcNAc was also supported by a series of studies using mannose-6-phosphatase (a known lumenal enzyme) as a marker for latency (25). In intact rat liver nuclei, O-GlcNAc residues were found to be fully accessible to external galactosylation even when greater than 70% of mannose-6-phosphatase was latent. Detergent permeabilization of the nuclei did not expose more O-GlcNAc. Further, it was demonstrated that rhodamine-WGA intensely stained sealed nuclei and that WGA-ferritin selectively bound to the cytoplasmic and nucleoplasmic faces of the nuclear pore complexes. The apparently high immunogenicity of O-GlcNAc was made more evident, since a series of independently isolated monoclonal antibodies to nuclear pores also appeared to recognize epitopes dependent upon O-GlcNAc (29). Pulse-chase radiolabeling of Buffalo rat liver cells and immunoprecipitation with a monoclonal antibody specific for peptide epitopes on at least three nuclear pore proteins, including p62, suggested that most of the O-GlcNAc residues are added to the p62 protein within five minutes of its synthesis while it is still in the cytoplasm (28). Using an antibody that recognizes O-GlcNAc-dependent epitopes, immunofluorescence and subcellular fractionation suggested that some of these glycoproteins are present within the nucleus (28). The available evidence suggests that O-GlcNAc is added in the cytoplasm most likely to proteins synthesized on free polysomes. Very recently, a specific assay for O-GlcNAc:protein glycosyltransferase has been developed (130). Surprisingly, the enzyme seems to be membrane associated and remains in the particulate fractions after a high-speed centrifugation of homogenates. Latency studies have directly demonstrated that the enzyme's active site is cytoplasmic. The sites of glycosylation by O-GlcNAc have been identified on three proteins: p62 nuclear pore protein of rat liver, a 65 kd polypeptide of human erthrocytes, and Band 4.1 of human erthrocytes (131). The glycosylated hydroxyamino acid is adjacent to an acidic amino acid and within two residues of a proline on each of the glycopeptides investigated. The sequence data around the glycosylation site of p62 was recently used to obtain a cDNA clone encoding this nuclear pore glycoprotein (132).

O-LINKED N-ACETYLGLUCOSAMINE AND NUCLEAR TRANSPORT   Nuclear structure and transport processes have recently been reviewed elsewhere (133, 134). Early studies demonstrated that antisera to a 67 kd nuclear envelope protein of clam oocytes *(Spisula solidissima)* and the lectin WGA specifically blocked the ATP-dependent RNA transport from isolated nuclei (129). A heterologous in vitro nuclear transport systems, composed of *Xenopus* oocyte extract, purified rat liver nuclei, and fluorescein-derivatized nucleoplasmin, was used to demonstrate WGA's inhibition of nucleoplasmin uptake into isolated nuclei (62). Nucleoplasmin is a well-characterized *Xenopus* nuclear protein that is often used in nuclear transport studies (135). Fluorescent-

WGA stained the nuclei in a punctate fashion, and multiple molecules of WGA-ferritin were shown to bind to the cytoplasmic and nucleoplasmic faces of the pore complex (62). Lectin blotting showed that ^{125}I-WGA predominantly bound to a 63–65 kd protein. Among 10 lectins tested, only WGA and Con A stained the nuclear periphery, with Con A staining only "damaged" nuclei. Earlier studies had found that WGA did not block the influx of 64 kd dextrans, suggesting that WGA's effect might not be due to simple blockage of the nuclear pores (136). Subsequently, microinjection of WGA into *Xenopus* oocytes was used to demonstrate that WGA blocks the nuclear uptake of labeled nuclear proteins (137). Similarly, WGA blocks the transport of fluorescein-nucleoplasmin into the nucleus when they are coinjected into the cytoplasm of rat hepatoma cells (137). The uptake of all karyophilic proteins appeared to be similarly affected by the lectin. Dose response measurements suggested that WGA slowed the rate of nuclear uptake but not its final extent.

An unusual D-mannose-specific lectin from the coral, *Gerardia savaglia*, blocked the ability of rat liver nuclear envelope to translocate mRNA in vitro (138). On the other hand, Con A, which supposedly has a similar specificity, had no effect. The coral lectin bound to nuclear pore proteins in a manner similar to WGA, reacting with six major proteins in rat liver nuclear envelope (190 kd, 110 kd, 78 kd, 62 kd, 56 kd, and 52 kd). In these studies, Con A primarily bound to endoplasmic reticulum and Golgi, whereas the coral lectin primarily bound to nuclear structures. Before the relationship of these data to the rest of the field can be evaluated, the specificities of the *Gerardia* lectin(s) need to be better defined.

Microinjection of non-nuclear proteins that had been covalently conjugated to the SV-40 T-antigen nuclear localization sequence into cultured human cells together with a variety of lectins demonstrated that WGA specifically blocked nuclear uptake of these hybrid, karyophilic proteins (139). The WGA effect persisted for at least one hour after injection and could be prevented by coinjection with 0.3 M GlcNAc. Simultaneously, WGA had no effect on the passive diffusion of fluorescein-dextrans ($M_r = 17,900$), thus suggesting the lectin is affecting a step involved in active transport. To further demonstrate the involvement of GlcNAc residues, the GlcNAc-specific lectin DSA *(Datura stamonium)* was also shown to block nuclear transport in this system (139). WGA also had no effect on RNA synthesis, as determined by measuring ^{3}H-uridine incorporation 12 hours after injection of the lectin. Very recently, workers in two laboratories have independently shown that nuclear translocation involves two separable steps: (*a*) binding that does not require energy, and (*b*) translocation that requires ATP hydrolysis (140, 141). Surprisingly, WGA appeared to specifically block the translocation step of nuclear transport, but did not affect the binding of the proteins to the pore complex (141).

OTHER O-LINKED N-ACETYLGLUCOSAMINE–CONTAINING PROTEINS   Not only are O-GlcNAc-bearing proteins widely distributed intracellulary, but they also appear to be present throughout eukaryotic phylogeny. To date, they have been detected in viruses (142), yeast (G. D. Holt, R. S. Haltiwanger, G. W. Hart, unpublished), trypanosomes (W. G. Kelly, G. W. Hart, unpublished), shistosomes (143), rodents (22, 23, 25–27, 126), frogs (G. D. Holt, G. W. Hart, unpublished), fruit flies (31, 144), as well as in humans (30, 31). O-GlcNAc was not detected in *E. coli,* suggesting that it may not exist in prokaryotes (G. D. Holt, R. S. Haltiwanger, G. W. Hart, unpublished). The human parasitic blood fluke, *Schistosoma mansoni,* has two major forms of simple O-linked sugar chains: a minor type linked to protein by O-GalNAc, and the major type containing O-GlcNAc residues (143). O-GlcNAc, which appears to be highly clustered on the polypeptide chains, accounts for more than 10% of the total radioactive glucosamine incorporated into macromolecules by this parasite. The vast majority of all terminal N-acetylglucosamine residues occur cytoplasmically in human erthrocytes in the form of protein-bound O-GlcNAc (30). The majority of the erythrocyte O-GlcNAc occurs on a single cytosolic, unidentified polypeptide of 65 kd. However, several other proteins are also modified by O-GlcNAc addition, including the well characterized cytoskeletal protein Band 4.1 (30), which is important in anchoring the cytoskeleton to the plasma membrane (145). Other major erythrocyte proteins, such as hemoglobin, actin, and spectrin, are not glycosylated.

Similar analyses of isolated human cytomegalovirus have demonstrated O-GlcNAc modification of the 149 kd virion basic phosphoprotein (142). Tryptic mapping suggested only one site of glycosylation. Interestingly, the basic phosphoprotein of the virus is localized in the tegument between the capsid and the viral envelope, an unprecedented localization for a glycosylated viral protein. Basic phosphoproteins, which account for up to 20% of the total viral proteins, contain O-GlcNAc in all four human cytomegalovirus strains. However, cytomegalovirus strain Colburn and herpes simplex virus-1 did not appear to contain O-GlcNAc moieties on their basic phosphoproteins, which is surprising since they are produced in the same cell and are thought to mature through similar pathways (142).

Recently, bovine milk galactosyltransferase and the lectin WGA were used to demonstrate that several transcription factors for RNA polymerase II are multiply glycosylated by O-GlcNAc (31). Sp1, a human protein that activates the transcription of viral and cellular genes containing a GC-rich decanucleotide transcriptional control sequence (the GC-Box; 146), appears to contain at least nine sites of O-GlcNAc addition per protomer. Preincubation of Sp1 with WGA resulted in a three- to four-fold inhibition of transcriptional stimulation. The lectin-induced inactivation of Sp1 was prevented by concomitant exposure of the factor to 0.3 M GlcNAc. In contrast to human Sp1, the form synthesized in *E. coli* was not affected by WGA. DNase I protection assays demonstrated that WGA had no substantial effect on Sp1 binding to DNA. Of eight RNA polymerase II transcription factors studied (five from HeLa cells and three from *Drosophila)*, all appeared to be modified by O-GlcNAc. Unlike Sp1, most appeared to have only three to five glycosylation sites. However, factor AP-2 (147) seemed to contain only 0.1 GlcNAc per mole (31). In contrast, RNA polymerase I transcription factor UBF1, RNA polymerase III transcription factor TFIIIA (148), and SV-40 large T antigen did not contain detectable levels of O-GlcNAc accessible to galactosylation (31). One potentially important finding is that only a subset of the family of proteins constituting a particular RNA polymerase II transcription factor appear to contain O-GlcNAc. For example, the proteins labeled by galactosylation of AP-1 family appeared to be the c-Jun protein, the c-Fos protein, and a "Fos-related antigen" protein (31). While the functional importance of the glycosylation of these transcription factors remains to be elucidated, the saccharide moieties are already proving valuable in their purification.

NUCLEAR AND CYTOPLASMIC SACCHARIDE-BINDING PROTEINS

If intracellular glycosyl moieties function as recognition signals or in the assembly of multiprotein complexes, there must be specific proteins inside the cell that bind to and/or modify these carbohydrate structures. There are now several studies providing clear evidence for the existence of nucleoplasmic and cytoplasmic endogenous lectins, glycosidases, or glycosyltransferases.

Endogenous Lectins

Several laboratories have identified endogenous carbohydrate-binding proteins in the nucleoplasmic or cytoplasmic compartments of cells (149–161). Preliminary studies of sea urchin nonhistone chromosomal proteins demonstrated their ability to agglutinate erythrocytes in a manner that was inhibited by high concentrations of free D-galactose (149). A comprehensive study of lectins isolated from calf tissues and chicken hearts by asialo-fetuin-Sepharose affinity chromatography compared the biochemical characteristics, binding properties, antigenic cross-reactivities, and cellular localization of a 12 kd calf and a 13 kd chicken galactoside-binding lectin (150). These lectins all have acidic isoelectric points, and require thiol groups and divalent cations for binding. Immunofluorescence studies on several cultured cell lines indicated that both the bovine and chicken lectins have primarily intracellular cytoplasmic localization. Similar lactose-binding lectins have been characterized in chicken muscle, where they occur as dimers with a subunit molecular mass of 15 kd (151). In developing muscle and brain, immunofluorescence studies detected a small amount of this lectin on the cell surface, but most appeared to be intracellular. Using rabbit antisera against bovine heart and human skeletal muscle β-galactoside lectins, indirect immunofluorescence and immunoperoxidase staining of cryostat sections not only detected extracellular lectins, but also demonstrated staining in sections of nuclei (152). Fluorescein-neoglycoproteins (galactose-BSA-FITC) also specifically stained the nucleus. This staining was inhibited by galactose-BSA but not by mannose-BSA (152). Monoclonal antibody to the commonly occurring 13 kd mammalian β-galactoside lectin was used to demonstrate the presence of nuclear and cytoplasmic proteins (130 kd, 80 kd, 65 kd, and 13 kd) that are antigenically related (156). Interestingly, the levels of these lectins and their patterns of expression change with the transformation state of the cell. The authors suggested that these proteins might be a family of lectin-related proteins that are involved in the growth regulatory systems of transformed and activated cells (156).

Asialo-fetuin-Sepharose affinity chromatography of detergent extracts from murine fibroblasts identified three β-galactoside-specific lectins (CBP35, CBP16, CBP13.5)(162, 163) with properties very similar to those described from several other sources, including calf and chicken, as discussed above (150). Anti-CBP35 antibodies were used to screen for similar proteins in various cultured cells and in murine tissues and organs. Cross-reactive proteins of the same size were not detected in adult mouse liver, but were identified in mouse, human, and chicken fibroblasts, and also in a murine macrophage cell line. Rabbit antibodies to CBP35 detected a small amount of this lectin on the surfaces of murine fibroblasts by immunofluorescent staining, by agglutination of live cells, and by isolation of ^{125}I-lectin after cell surface iodination (158). However, staining of fixed and permeabilized cells showed substantial quantities in the nucleus and cytoplasm. Immunoblotting of subcellular fractions demonstrated the presence of the lectin in nuclear pellet, soluble fraction, and plasma membrane fraction of the postnuclear supernatant (158). Moutsatsos et al have also shown that the levels of CBP35 (high affinity for Galβ1-4GlcNAc) are elevated in proliferating cells, where it is predominantly localized to the nucleus (160). Nuclei are stained in a punctate fashion, indicating clustering of the lectin. In quiescent cells, nuclear staining was greatly diminished with an overall decrease in fluorescent intensity. Stimulation of serum-starved cells causes the fluorescent anti-CBP35 staining of the nucleus to return to high levels. Very recently, CBP35 was shown by several lines of evidence to be a component of heterogeneous nuclear ribonucleoprotein complexes (161). In these studies, N-ϵ-aminocaproyl-D-galactosamine-Sepharose affinity columns were used to isolate a complex containing RNA, as well as a set of polypeptides matching the core particle peptides of hnRNP, one of which was CBP35. In contrast, CBP35 isolated from the cytoplasm by the same method did not contain RNA or the hnRNP core polypeptides (161). The complete cDNA sequence of cloned CBP35 suggested an amino acid sequence of two major domains: an amino terminal portion homologous to hnRNP proteins and a carboxyl terminal portion homologous to β-galactoside-binding lectins isolated from a variety of tissues (164).

Skin of *Xenopus laevis* also contains a soluble β-galactoside-specific lectin of 16 kd similar to that found in many other sources (159). This skin lectin is concentrated in the cytoplasm of granular and mucous gland cells. Upon injection of epinephrine, there is a massive secretion of the cytoplasmic lectin from the granular gland cells (165). This phenomenon is striking because it provides a dramatic example of secretion of a cytoplasmic protein without involving secretory vesicles. Secretion appears to involve the rupture of plasma membranes at the base of the glandular ducts, induced by contraction of surrounding muco-epithelial cells, which squeeze out the cytoplasm. The cell membranes subsequently re-seal (159). Although representative of a special case, similar more subtle mechanisms might operate in other cells.

Fluorescein-labeled neoglycoproteins were used to detect lectins in isolated nuclei from baby hamster kidney (BHK) or murine L1210 leukemia cells (155). Fluorescence was seen throughout the nucleus, but the brightest staining occurred over the nucleoli. Of nine neoglycoproteins tested, α-rhamnose-, α-glucose, β-GlcNAc-, and α-Man-6-P-BSA strongly labeled nuclei, while reacting little with intact cells. Nuclei bound from 300 to 800 fg of neoglycoprotein per nucleus. Staining at lower intensity was also evident in the cytoplasm. Nuclei appeared to contain a large amount of Man-6-P-binding protein. The entire nucleus stained brightly with GlcNAc-BSA-FITC, again with strongest staining over the nucleus. Seve et al suggested that these nuclear lectins might function in transport, modulation of enzymes within the nucleus, or in maintaining chromatin architecture (155). Similar studies detected specific mannose-binding sites in the nucleus and cytoplasm of lizard *(Lacerta vivapara)* granulosa cells using mannose-ferritin conjugates (154). The strongest labeling was observed over nucleoli, chromatin, and the external leaflet of the nuclear envelope. These studies suggested that mannose-binding lectins were associated with ribosomal precursors and with cytoplasmic ribosomes (154).

Quantitative flow microfluorometry of fluorescein-BSA neoglycoproteins was used to demonstrate the presence of endogenous lectins in isolated baby hamster kidney cell nuclei (157). Similar labeling intensities were obtained using two different methods of nuclear isolation, suggesting that the lectins were not derived from cytoplasmic or membrane-derived contaminants. Most lectins appeared to be associated with nucleoli and nucleoplasmic ribonucleoprotein elements. Interestingly, nuclei from growing cells bound much greater amounts of neoglycoprotein than nuclei from contact-inhibited cells. Permeabilized nuclei from exponentially growing cells bound more than tenfold more fucose-BSA, GlcNAc-BSA, and lactose-BSA $(20+/-5/\text{mole})$ than did nuclei from contact-inhibited cells. In these studies, mannose-BSA and BSA alone did not bind significantly to nuclei (157).

Glycosidases

While several laboratories have investigated "cytosolic" glycosidases, in many cases it is not clear whether the enzymes are actually localized in the cytosol, are situated in easily broken vesicles, or are present from serum contamination of homogenized tissues. Operationally, the term "cytosolic" has often been applied to mean an enzyme with a relatively high pH optimum that is soluble in the supernatant of a high-speed sedimentation of cell homogenates. Early studies described both soluble and lysosomal sialidases from rat liver with pH optima of 5.8 and 4.4, respectively. These two enzymes had different cation requirements, as well as different substrate specificities (166). The "cytosolic" enzyme remained soluble when isolated from nonperfused livers homogenized in isotonic potassium chloride. In a separate study, "cytosolic" or "neutral" α-mannosidase has been purified 12,000-fold from rat liver (167). This enzyme has a pH optimum of 5.5 to 5.9, is PAS-positive (suggesting that it is glycosylated), and it is in the soluble fraction of nonperfused rat livers homogenized in low-salt/Tris buffers. Shoup & Touster suggested that this enzyme might be involved in the catabolism of cytoplasmically localized glycoproteins. Later studies suggested that this soluble α-mannosidase is immunologically related to the endoplasmic reticulum α-mannosidase, and that they are two forms of the same enzyme (168). Bischoff & Kornfeld suggested that the soluble mannosidase is derived from the endoplasmic reticulum form by limited proteolysis during tissue preparation. However, inhibition of the soluble mannosidase by high concentrations of swainsonine was reported to cause accumulation of oligosaccharides in the cytosolic fractions obtained from a mild homogenization procedure. These data gave rise to the suggestion that the "cytosolic" mannosidase is important in glycoprotein catabolism (169). Clearly, better in situ localization data are required to determine if the enzyme or oligosaccharides are actually cytosolic in a living cell. Similar studies have described a "cytosolic" sialidase from rat liver (170). This sialidase has a pH optimum of 6.5, and is thought to be associated with the cytoplasmic faces of microsomal membranes. Again, contamination by lumenal enzymes, serum, or secreted enzymes needs to be ruled out, and careful in situ localization studies are required to determine if this enzyme is indeed cytosolic in vivo. Until fairly strong evidence is obtained, these enzymes should probably be referred to as "soluble" rather than as cytosolic.

Glycosyltransferases

As mentioned earlier, it is clear that most glycosyltransferases are localized lumenally (1, 2) or on cell surfaces (7, 8). Clearly, if cytoplasmic glycoconjugates exist, then either certain transport pathways between lumenal/extracellular compartments and the cytoplasm must be present, or alternatively, some glycosyltransferases must be cytoplasmically localized. A few reports have suggested the presence of glycosyltransferases within nucleoplasmic or cytoplasmic compartments (48, 171–178). Five glycosyltransferases were identified in purified rat liver nuclei using endogenous glycoconjugates or glycosidase-treated fetuin as substrates (mannosyl- galactosyl-, N-acetylglucosaminyl-, N-acetylgalactosaminyl-, and sialyltransferases) (171). In an attempt to control for outside contamination, only perfused livers were used and the nuclear purity was checked by electron microscopy and several enzyme markers. Slight contamination of the nuclei by plasma membranes and endoplasmic reticulum was indicated. Other studies have demonstrated a N-acetylglucosaminyl transferase that uses endogenous substrates in purified nuclei of the cellular slime mold, *Dictyostelium* (172). To argue against contamination, nuclei were isolated without detergents, purity was checked by electron microscopy, and marker enzymes were followed throughout isolation. Using radiolabeled UDP-GlcNAc with isolated nuclei and autoradiography, heavy labeling was seen in the nuclei. Products were susceptible to both mild base and mild acid hydrolysis, but labeled oligosaccharides were not structurally characterized (172). Recent preliminary studies have incubated UDP-[^{14}C]GlcNAc with purified rat hepatocyte nuclei and demonstrated the incorporation of GlcNAc into several endogenous nuclear proteins of low molecular weight (178). Incorporation into these endogenous substrates was only inhibited 30% by tunicamycin, and the suggestion was made that some of the GlcNAc appeared to be O-linked. Unfortunately, the endogenously labeled saccharides were not characterized.

Other studies have described a β1-3GalNAc-O-protein galactosyltransferase in human erythrocytes, which appears to have a cytosolic orientation (174). In the absence of detergents, no enzyme activity was detectable in sealed ghosts, but detergent permeabilization of the cells exposed the enzyme. The galactosyltransferase activity was also resistant to proteases in sealed ghosts but sensitive in leaky ghosts or inside-out vesicles. Based upon classical arguments that have been used to describe the orientation of several erythrocyte membrane proteins (179), this enzyme appears to be cytosolically localized in erythrocytes (174). Surprisingly, recent studies using cDNA probes to identify mRNA encoding murine β1-4 galactosyltransferase found two forms of mRNA that differ in length by about 200 base pairs (177). The long form of the enzyme contains an additional N-terminal 13 amino acids, which appears to be a cleavable signal sequence. The more abundant short mRNA does not contain any such sequence. Based upon the known orientation of many proteins, which strongly correlates with the types of signal sequences they contain (180), Shaper et al speculate that the short abundant form encodes an enzyme with its active site in the lumen of the Golgi, while the longer less abundant mRNA may encode an enzyme with its active site in the cytosol (177). If the larger galactosyltransferase is indeed cytosolic, it may play a role in the elongation of cytoplasmically localized saccharides, such as O-GlcNAc-bearing proteins. However, currently there are no data available suggesting the elongation of this form of glycosylation.

Mannose was shown to be incorporated into nonhistone proteins of monkey liver upon incubation of purified nuclei with GDP-[^{14}C]mannose (175). The major mannosylated protein appeared to have a molecular mass of 13 kd by SDS-PAGE. The purity of the nuclei and the type of oligosaccharide structure containing the mannose were not further characterized. Another study suggested the existence of a retinylphosphate-independent mannosyltransferase in rat liver nuclear membranes (176), but no information on the topology of the enzyme is available. A recent study has demonstrated the presence of a mannose-specific α-glucose phosphotransferase in the cytoplasm of rat liver (48). This novel enzyme appears to attach α-glucose-1-phosphate residues to O-linked mannosyl residues. The major substrate in liver is a 62-kd protein. The transferase appears to be cytoplasmically oriented but membrane associated by the following criteria: (*a*) Endogenous labeling with a UDP-^{35}S-phosphorothioate nonhydrolyzable substrate showed that 85% of the product was in the high-speed supernatant, while in the same preparations 94% of the CMP-NeuAc incorporated remained microsomal. (*b*) Optimal activity of the glucose phosphotransferase did not require detergent permeabilization of sealed vesicles as did the latent activities of mannose-6-phosphatase or galactosyltransferase. (*c*) The transferase is also susceptible to proteases in sealed vesicles. Given the striking similarities to the well-studied lysosomal mannose-6-phosphate targeting system (181), this cytoplasmic phosphotransferase may play an important role in cytoplasmic targeting of proteins.

FUTURE DIRECTIONS

It is clear from available data that several types of glycoproteins are present in the cytoplasm and/or nucleoplasm. Little is known about how these cytosolic glycoproteins are synthesized and localized. They could be synthesized in the lumenal spaces of the endoplasmic reticulum and the Golgi apparatus, much like conventional asparagine-linked and "mucin-type" O-linked glycoproteins. If this is the case, which it appears to be for nuclear heparan sulfates (39), then novel mechanisms for the transport of macromolecules from either the lumenal or extracellular environments to the cytosol must exist. The fact that no definitive evidence exists for the presence of classical asparagine-linked and "mucin-type" O-linked oligosaccharide structures in the nucleoplasm or cytoplasm implies that alternative biosynthetic pathways are likely. Cytosolic glycoproteins, for instance, could be synthesized on free ribosomes and glycosylated in the cytoplasm or nucleoplasm. As discussed above, the demonstration of several glycosyltransferase activities in these subcellular compartments indicates that this latter biosynthetic pathway is probable for at least some of these glycoproteins. The number and different types of cytoplasmic or nucleoplasmic glycosyltransferases present is a major question requiring further study. Perhaps these enzymes are the result of differential splicing of mRNA from genes that predominantly encode well-studied lumenal glycosyltransferase, as might be the case for murine galactosyltransferase (177).

As with most well-studied lumenal and cell-surface forms of glycosylation, the actual role of the glycosyl moieties on cytoplasmic and nuclear glycoproteins has yet to be determined. One possible role for these saccharides may be that they are signals involved in targeting proteins to specific intracellular compartments, much like phosphorylated mannosyl residues on N-linked oligosaccharides target proteins to lysosomes (181). Since many nuclear proteins contain O-GlcNAc, this saccharide could play a similar role in nuclear targeting (27). The sugar could also play a more active role in functional aspects of proteins, as suggested by lectin inhibition of nuclear pore transport (62) and initial studies of glycosylated transcription factors (31). In addition, O-GlcNAc might be involved in the formation or stabilization of multiprotein or multienzyme complexes (23, 30). Perhaps O-GlcNAc blocks phosphorylation sites and is regulated by a specific *N*-acetylglucosaminidase (30)? Also, why do O-GlcNAc residues appear to occur in organized clusters on proteins, and why are these moieties so immunogenic? We know nothing about the relationship between intracellular lectins and cytosolic or nucleoplasmic glycoconjugates. For example, why are there apparently so many nuclear β-galactoside-binding proteins, when prevailing evidence suggests that of the more common sugar linkages found in the nucleus, β-galactosides have rarely been seen? What functions do proteoglycans in the cytoplasm of adult nerve cells carry out (37)? Why does the cell synthesize CMP-NeuAc within the nucleus when it is predominantly utilized in the lumen of the Golgi (182–184)?

Now that certain types of glycoconjugates have been established to exist in the nucleoplasmic and/or cytoplasmic spaces, the door has been unlocked for an open-minded and critical evaluation of the distribution and possible functions of glycoproteins within the cell. What will be required are careful studies at the structural level that characterize both the linkage and structures of these types of glycoconjugates. Not until nuclear or cytoplasmic glycoproteins are structurally well characterized can we begin to address their modes of biosynthesis or transport, and functions in a meaningful way. It is likely that we will continue to discover unique saccharide structures and glycosyltransferases in these compartments that may play critical roles in many important intracellular processes.

ACKNOWLEDGMENTS

We thank Drs. Sidney S. Whiteheart and Antonino Passaniti for critically reading the manuscript, and Dr. Richard Marchase for providing a manuscript prior to publication. The authors' research is supported by NIH grants HD 13563 and CA 42486, and was performed during the tenure of an Established Investigatorship of the American Heart Association (to G. W. H.). R.S.H. is a Postdoctoral Fellow of the Arthritis Foundation and W.G.K. is a Predoctoral Fellow of the March of Dimes.

Literature Cited

1. Lennarz, W. J. 1987. *Biochemistry* 26:7205–10
2. Hirschberg, C. B., Snider, M. D. 1987. *Annu. Rev. Biochem.* 56:63–87
3. Roth, J. 1987. *Biochim. Biophys. Acta* 906:405–36
4. Kornfeld, R., Kornfeld, S. 1985. *Annu. Rev. Biochem.* 54:631–64
5. Hanover, J. A., Lennarz, W. J. 1981. *Arch. Biochem. Biophys.* 211:1–19
6. Beyer, T. A., Sadler, J. E., Rearick, J. I., Paulson, J. C., Hill, R. L. 1981. *Adv. Enzymol.* 52:23–175
7. Shur, B. D., Roth, S. 1975. *Biochim. Biophys. Acta* 415:473–512
8. Pierce, M., Turley, E. A., Roth, S. 1980. *Int. Rev. Cytol.* 65:1–47
9. Lopez, L. C., Bayna, E. M., Litoff, D., Shaper, N. L., Shaper, J. H., Shur, B. D. 1985. *J. Cell Biol.* 101:1501–10
10. Garoff, H. 1985. *Annu. Rev. Cell Biol.* 1:403–45
11. Farquhar, M. G. 1985. *Annu. Rev. Cell Biol.* 1:447–88
12. Hart, G. W., Holt, G. D., Haltiwanger, R. S. 1988. *Trends Biochem. Sci.* 13:380–84
13. Stoddart, R. W. 1979. *Biol. Rev.* 54:199–235
14. Stein, G. S., Roberts, R. M., Davis, J. L., Head, W. J., Stein, J. L., et al. 1975. *Nature* 258:639–41
15. Stein, G. S., Roberts, R. M., Stein, J. L., Davis, J. L. 1981. In *The Cell Nucleus,* ed. H. Busch, 9:341–56. New York: Academic
16. Franke, W. W., Keenan, T. W., Stadler, J., Genz, R., Jarasch, E-D., Kartenbeck, J. 1976. *Cytobiologie* 13:28–56
17. Lennarz, W. J., ed. 1980. *The Biochemistry of Glycoproteins and Proteoglycans.* New York: Plenum
18. Pigman, W. 1977. In *The Glycoconjugates,* ed. M. I. Horowitz, W. Pigman, Vol. I, pp. 137–52. Orlando, Fla: Academic
19. Horowitz, M. I. 1977. See Ref. 18, pp. 15–34
20. Holden, K. G., Griggs, L. J. 1977. See Ref. 18, pp. 215–37
21. Doehr, S. A. 1977. See Ref. 18, pp. 239–57
22. Holt, G. D., Hart, G. W. 1986. *J. Biol. Chem.* 261:8049–57
23. Holt, G. D., Snow, C. M., Senior, A., Haltiwanger, R. S., Gerace, L., Hart, G. W. 1987. *J. Cell Biol.* 104:1157–64

24. Snow, C. M., Senior, A., Gerace, L. 1987. *J. Cell Biol.* 104:1143–56
25. Hanover, J. A., Cohen, C. K., Willingham, M. C., Park, M. K. 1987. *J. Biol. Chem.* 262:9887–94
26. Davis, L. I., Blobel, G. 1986. *Cell* 45:699–709
27. Schindler, M., Hogan, M., Miller, R., DeGaetano, D. 1987. *J. Biol. Chem.* 262:1254–60
28. Davis, L. I., Blobel, G. 1987. *Proc. Natl. Acad. Sci. USA* 84:7552–56
29. Park, M. K., D'Onofrio, M., Willingham, M. C., Hanover, J. A. 1987. *Proc. Natl. Acad. Sci. USA* 84:6462–66
30. Holt, G. D., Haltiwanger, R. S., Torres, C. R., Hart, G. W. 1987. *J. Biol. Chem.* 262:14847–50
31. Jackson, S. P., Tjian, R. 1988. *Cell* 55:125–33
32. Margolis, R. K., Crockett, C. P., Kiang, W. L., Margolis, R. U. 1976. *Biochim. Biophys. Acta* 451:465–69
33. Aquino, D. A., Margolis, R. U., Margolis, R. K. 1984. *J. Cell Biol.* 99:1117–29
34. Aquino, D. A., Margolis, R. U., Margolis, R. K. 1984. *J. Cell Biol.* 99:1130–39
35. Margolis, R. K., Thomas, M. D., Crockett, C. P., Margolis, R. U. 1979. *Proc. Natl. Acad. Sci. USA* 76:1711–15
36. Margolis, R. U., Aquino, D. A., Klinger, M. M., Ripellino, J. A., Margolis, R. K. 1986. *Ann. NY Acad. Sci.* 481:46–54
37. Ripellino, J. A., Bailo, M., Margolis, R. U., Margolis, R. K. 1988. *J. Cell Biol.* 106:845–55
38. Fedarko, N. S., Conrad, H. E. 1986. *J. Cell Biol.* 102:587–99
39. Ishihara, M., Fedarko, N. S., Conrad, H. E. 1986. *J. Biol. Chem.* 261:13575–80
40. Rodriguez, I., Whelan, W. J. 1985. *Biochem. Biophys. Res. Commun.* 132:829–36
41. Blumenfeld, M. L., Krisman, C. R. 1986. *Eur. J. Biochem.* 156:163–69
42. Rodriguez, I., Fleisler, S. 1988. *Arch Biochem. Biophys.* 260:628–37
43. Aon, M. A., Curtino, J. A. 1985. *Biochem. J.* 229:269–72
44. Pitcher, J., Smythe, C., Campbell, D. G., Cohen, P. 1987. *Eur. J. Biochem.* 169:497–502
45. Marchase, R. B., Saunders, A. M., Rivera, A. A., Cook, J. M. 1987. *Biochim. Biophys. Acta* 916:157–62
46. Hiller, A. M., Koro, L. A., Marchase, R. B. 1987. *J. Biol. Chem.* 262:4377–81
47. Koro, L. A., Marchase, R. B. 1982. *Cell* 31:739–48
48. Srisomsap, C., Richardson, K. L., Jay, J. C., Marchase, R. B. 1988. *J. Biol. Chem.* 263:17792–97
49. Goldstein, I. J., Hayes, C. E. 1978. *Adv. Carbohydr. Chem. Biochem.* 35:127–340
50. Osawa, T., Tsuji, T. 1987. *Annu. Rev. Biochem.* 56:21–42
51. Sharon, N. 1983. *Adv. Immunol.* 34:213–98
52. Lis, H., Sharon, N. 1986. *Annu. Rev. Biochem.* 55:35–67
53. Wassef, N.M., Richardson, E.C., Alving, C. R. 1985. *Biochem. Biophys. Res. Commun.* 130:76–83
54. Monsigny, M., Roche, A. C., Sene, C., Maget-Dana, R., Delmotte, F. 1980. *Eur. J. Biochem.* 104:147–53
55. Nicolson, G., Lacorbiere, M., Delmonte, P. 1972. *Exp. Cell Res.* 71:468–73
56. Monneron, A., Segretain, D. 1974. *FEBS Lett.* 42:209–13
57. Kaneko, I., Sato, H., Ukita, T. 1972. *Biochem. Biophys. Res. Commun.* 48:1504–10
58. Virtanen, I., Wartiovaara, J. 1976. *J. Cell Sci.* 22:335–44
59. Feldherr, C. M., Richmond, P. A., Noonan, K. D. 1977. *Exp. Cell Res.* 107:439–44
60. Pinto da Silva, P., Torrisi, M. R., Kachar, B. 1981. *J.Cell Biol.* 91:361–72
61. Roth, J. 1983. *J. Histochem. Cytochem.* 31:987–99
62. Finlay, D. R., Newmeyer, D. D., Price, T. M., Forbes, D. J. 1987. *J. Cell Biol.* 104:189–200
63. Rizzo, W. B., Bustin, M. 1977. *J. Biol. Chem.* 252:7062–67
64. Sevaljevic, L., Petrovic, S. L., Petrovic, M. 1979. *Experientia* 35:193–94
65. Hozier, J., Furcht, L. T. 1980. *Cell Biol. Int. Rep.* 4:1091–99
66. Berezney, R., Coffey, D. S. 1977. *J. Cell Biol.* 73:616–37
67. Sevaljevic, L., Poznanovic, G., Petrovic, M., Krtolica, K. 1981. *Biochem. Int.* 2:77–84
68. Levy-Wilson, B. 1983. *Biochemistry* 22:484–89
69. Seve, A. P., Hubert, J., Bouvier, D., Masson, C., Geraud, G., Bouteille, M. 1984. *J. Submicrosc. Cytol.* 16:631–41
70. Kan, F., Pinto da Silva, P. 1986. *J. Cell Biol.* 102:576–86
71. Kurth, P. D., Bustin, M., Moudrianakis, E. N. 1979. *Nature* 279:448–50
72. Kinoshita, S., Yoshii, K., Tonegawa, Y. 1988. *Exp. Cell Res.* 175:148–57

73. Rubin, A. D., Davis, S., Schultz, E. 1972. *Biochem. Biophys. Res. Commun.* 46:2067–74
74. Bhattacharya, P., Simet, I., Basu, S. 1979. *Proc. Natl. Acad. Sci. USA* 76:2218–21
75. Glass, W. F., Briggs, R. C., Hnilica, L. S. 1981. *Anal. Biochem.* 115:219–24
76. Gerace, L., Ottaviano, Y., Kondor-Koch, C. 1982. *J. Cell Biol.* 95:826–37
77. Schmidt, W. N., McKusick, K. B., Schmidt, C. A., Hoffman, L. H., Hnilica, L. S. 1984. *Cancer Res.* 44:5291–304
78. Burrus, G. R., Schmidt, W. N., Briggs, J. A., Hnilica, L. S., Briggs, R. C. 1988. *Cancer Res.* 48:551–55
79. Dennis, J. W., Laferte, S. 1987. *Cancer Metastasis Rev.* 5:185–204
80. Glew, R. H., Kayman, S. C., Kuhlenschmidt, M. S. 1973. *J. Biol. Chem.* 248:3137–45
81. Eagles, P. A., Johnson, L. N., Van Horn, C. 1975. *J. Cell Sci.* 19:33–54
82. Meyer, D. I., Burger, M. M. 1976. *Biochim. Biophys. Acta* 443:428–36
83. Feigenson, M. E., Schnebli, H. P., Baggiolini, M. 1975. *J. Cell Biol.* 66:183–88
84. Nagakura, K., Tachibana, H., Kaneda, Y., Nakae, T. 1986. *Exp. Parasitol.* 61:335–42
85. Howard, G. A., Schnebli, H. P. 1977. *Proc. Natl. Acad. Sci. USA* 74:818–21
86. Yoshida, K. 1978. *J. Biochem.* 83:1609–14
87. Elbein, A. D. 1987. *Annu. Rev. Biochem.* 56:497–534
88. Kawasaki, T., Yamashina, I. 1972. *J. Biochem.* 72:1517–25
89. Tuan, D., Smith, S., Folkman, J., Merler, E. 1973. *Biochemistry* 12:3159–65
90. Yeoman, L. C., Jordan, J. J., Busch, R. K., Taylor, C. W., Savage, H. E., Busch, H. 1976. *Proc. Natl. Acad. Sci. USA* 73:3258–62
91. Bergman, A., Dallner, G. 1976. *Biochim. Biophys. Acta* 433:496–508
92. Goldberg, A. H., Yeoman, L. C., Busch, H. 1978. *Cancer Res.* 38:1052–56
93. Miki, B. L., Gurd, J. W., Brown, I. R. 1980. *Can. J. Biochem.* 58:1261–69
94. Polet, H., Molnar, J. 1988. *J. Cell Physiol.* 135:47–54
95. Einck, L., Bustin, M. 1985. *Exp. Cell Res.* 156:295–310
96. Reeves, R., Chang, D., Chung, S. C. 1981. *Proc. Natl. Acad. Sci. USA* 78:6704–8
97. Reeves, R., Chang, D. 1983. *J. Biol. Chem.* 258:679–87
98. Roden, L. 1980. See Ref. 17, pp. 241–371
99. Yamashina, I., Izumi, K., Naka, H. 1964. *Biochem. J.* 55:652–58
100. Kinoshita, S. 1971. *Exp. Cell Res.* 64:403–11
101. Kinoshita, S. 1974. *Exp. Cell Res.* 85:31–40
102. Kinoshita, S. 1976. *Exp. Cell Res.* 102:153–61
103. Kinoshita, S., Saiga, H. 1979. *Exp. Cell. Res.* 123:229–36
104. Kinoshita, S., Yoshii, K. 1979. *Exp. Cell Res.* 124:361–69
105. Kinoshita, S., Yoshii, Y. 1983. *Experientia* 39:189–90
106. Ljiljana, S., Koviljka, K. 1973. *J. Biochem.* 4:345–48
107. Bhavanandan, V. P., Davidson, E. A. 1975. *Proc. Natl. Acad. Sci. USA* 72:2032–36
108. Bhavanandan, V. P. 1979. *Glycoconjugates Res.* 1:499–502
109. Furukawa, K., Terayama, H. 1977. *Biochim. Biophys. Acta* 499:278–89
110. Furukawa, K., Terayama, H. 1979. *Biochim. Biophys. Acta* 585:575–88
111. Sames, K. 1979. *Acta Anat. (Basel)* 103:74–82
112. Ohnishi, T., Yamamoto, K., Terayama, H. 1973. *Histochemie* 35:1–10
113. Fromme, H. G., Buddecke, E., von Figura, K., Kresse, H. 1976. *Exp. Cell Res.* 102:445–49
114. Finne, J., Krusius, T., Margolis, R. K., Margolis, R. U. 1979. *J. Biol. Chem.* 254:10295–300
115. Vannucchi, S., Fibbi, G., Cappelletti, R., Del Rosso, M., Chiarugi, V. 1982. *J. Cell Physiol.* 111:149–54
116. Alvarado, M. V., Castejon, H. V. 1984. *J. Neurosci. Res.* 11:13–26
117. Krusius, T., Reinhold, V. N., Margolis, R. K., Margolis, R. U. 1987. *Biochem. J.* 245:229–34
118. Krusius, T., Finne, J., Margolis, R. K., Margolis, R. U. 1986. *J. Biol. Chem.* 261:8237–42
119. Margolis, R. U., Margolis, R. K., Atherton, D. M. 1972. *J. Neurochem.* 19:2317–24
120. Margolis, R. K., Margolis, R. U., Preti, C., Lai, D. 1975. *Biochemistry* 14:4797–804
121. Kiang, W. L., Crockett, C. P., Margolis, R. K., Margolis, R. U. 1978. *Biochemistry* 17:3841–48
122. Whelan, W. J. 1976. *Trends Biochem. Sci.* 1:13–15
123. Lyles, D. S., McConnell, K. A. 1981. *J. Virol.* 39:263–72
124. Jarvis, D. L., Butel, J. S. 1985. *Virology* 141:173–89

125. Schmitt, M. K., Mann, K. 1987. *Virology* 156:268–81
126. Torres, C. R., Hart, G. W. 1984. *J. Biol. Chem.* 259:3308–17
127. Schindler, M., Hogan, M. 1984. *J. Cell Biol.* 99:99a (Abstr.)
128. Abeijon, C., Hirschberg, C. B. 1988. *Proc. Natl. Acad. Sci. USA* 85:1010–14
129. Baglia, F., Maul, G. G. 1983. *Proc. Natl. Acad. Sci. USA* 80:2285–89
130. Hart, G. W., Haltiwanger, R. S., Holt, G. D., Kelly, W. G. 1988. *J. Cell Biol.* 107(6, pt. 3):11a
131. Hart, G. W., Haltiwanger, R. S., Holt, G. D., Kelly, W. G. 1989. *Ciba Found. Symp.* 145: In press
132. D'Onofrio, M., Starr, C. M., Park, M. K., Holt, G. D., Haltiwanger, R. S., et al. 1988. *Proc. Natl. Acad. Sci. USA* 85:9595–99
133. Dingwall, C., Laskey, R. A. 1986. *Annu. Rev. Cell Biol.* 2:367–90
134. Newport, J. W., Forbes, D. J. 1987. *Annu. Rev. Biochem.* 56:535–65
135. Dingwall, C. 1985. *Trends Biochem. Sci.* 10:64–66
136. Jiang, L. W., Schindler, M. 1986. *J. Cell Biol.* 102:853–58
137. Dabauvalle, M. C., Schulz, B., Scheer, U., Peters, R. 1988. *Exp. Cell Res.* 174:291–96
138. Kljajic, Z., Schroder, H. C., Rottmann, M., Cuperlovic, M., Movsesian, M., et al. 1987. *Eur. J. Biochem.* 169:97–104
139. Yoneda, Y., Imamoto-Sonobe, N., Yamaizumi, M., Uchida, T. 1987. *Exp. Cell Res.* 173:586–95
140. Richardson, W. D., Mills, A. D., Dilworth, S. M., Laskey, R. A., Dingwall, C. 1988. *Cell* 52:655–64
141. Newmeyer, D. D., Forbes, D. J. 1988. *Cell* 52:641–53
142. Benko, D. M., Haltiwanger, R. S., Hart, G. W., Gibson, W. 1988. *Proc. Natl. Acad. Sci. USA* 85:2573–77
143. Nyame, K., Cummings, R. D., Damian, R. T. 1987. *J. Biol. Chem.* 262:7990–95
144. Kelly, W. G., Hart, G. W. 1989. *Cell* 57. In press
145. Bennett, V. 1985. *Annu. Rev. Biochem.* 54:273–304
146. Briggs, M. R., Kadonaga, J. T., Bell, S. P., Tjian, R. 1986. *Science* 234:47–52
147. Mitchell, P. J., Wang, C., Tjian, R. 1987. *Cell* 50:847–61
148. Engelke, D. R., Ng, S. Y., Shastry, B. S., Roeder, R. G. 1980. *Cell* 19:717–28
149. Sevaljevic, L., Konstantinovic, M., Tomovic, M., Pavlovic, Z. 1977. *Mol. Biol. Rep.* 3:265–67
150. Briles, B., Gregory, W., Fletcher, P.,

151. Kornfeld, S. 1979. *J. Cell Biol.* 81:528–37
151. Beyer, E. C., Barondes, S. H. 1980. *J. Supramol. Struct.* 13:219–27
152. Childs, R. A., Feizi, T. 1980. *Cell Biol. Int. Rep.* 4:775–81
153. Raz, A., Meromsky, L., Carmi, P., Karakash, R., Lotan, D., Lotan, R. 1984. *EMBO J.* 3:2979–83
154. Hubert, J., Seve, A. P., Bouvier, D., Masson, C., Bouteille, M., Monsigny, M. 1985. *Biol. Cell* 55:15–20
155. Seve, A. P., Hubert, J., Bouvier, D., Bouteille, M., Maintier, C., Monsigny, M. 1985. *Exp. Cell Res.* 157:533–38
156. Carding, S. R., Thorpe, S. J., Thorpe, R., Feizi, T. 1985. *Biochem. Biophys. Res. Commun.* 127:680–86
157. Seve, A. P., Hubert, J., Bouvier, D., Bourgeois, C., Midoux, P., et al. 1986. *Proc. Natl. Acad. Sci. USA* 83:5997–6001
158. Moutsatsos, I. K., Davis, J. M., Wang, J. L. 1986. *J. Cell Biol.* 102:477–83
159. Bols, N. C., Roberson, M. M., Haywood-Reid, P. L., Cerra, R. F., Barondes, S. H. 1986. *J. Cell Biol.* 102:492–99
160. Moutsatsos, I. K., Wade, M., Schindler, M., Wang, J. L. 1987. *Proc. Natl. Acad. Sci. USA* 84:6452–56
161. Laing, J. G., Wang, J. L. 1988. *Biochemistry* 27:5329–34
162. Roff, C. F., Rosevear, P. R., Wang, J. L., Barker, R. 1983. *Biochem. J.* 211:625–29
163. Roff, C. F., Wang, J. L. 1983. *J. Biol. Chem.* 258:10657–63
164. Jia, S., Wang, J. L. 1988. *J. Biol. Chem.* 263:6009–11
165. Cerra, R. F., Haywood-Reid, P. L., Barondes, S. H. 1984. *J. Cell Biol.* 98:1580–89
166. Tulsiani, D. R., Carubelli, R. 1970. *J. Biol. Chem.* 245:1821–27
167. Shoup, V. A., Touster, O. 1976. *J. Biol. Chem.* 251:3845–52
168. Bischoff, J., Kornfeld, R. 1986. *J. Biol. Chem.* 261:4758–65
169. Tulsiani, D. R., Touster, O. 1987. *J. Biol. Chem.* 262:6506–14
170. Miyagi, T., Tsuiki, S. 1985. *J. Biol. Chem.* 260:6710–16
171. Richard, M., Martin, A., Louisot, P. 1975. *Biochem. Biophys. Res. Commun.* 64:109–14
172. Rogge, H., Neises, M., Passow, H., Grunz, H., Risse, H. J. 1975. In *New Approaches to the Evaluation of Abnormal Embryonic Development,* ed. D. Neubet, H. J. Mecker, pp. 772–91. Stuttgart: Thieme
173. Berthillier, G., Benedetto, J. P., Got, R.

1980. *Biochim. Biophys. Acta* 603:245–54

174. Hesford, F. J., Berger, E. G. 1981. *Biochim. Biophys. Acta* 649:709–16
175. Berthillier, G., Got, R. 1982. *Mol. Cell Biochem.* 44:39–43
176. Fayet, Y., Galland, S., Degiuli, A., Got, R., Frot-Coutaz, J. 1988. *Biochem. Int.* 16:429–38
177. Shaper, N. L., Hollis, G. F., Douglas, J. G., Kirsch, I. R., Shaper, J. H. 1988. *J. Biol. Chem.* 263:10420–28
178. Galland, S., Degiuli, A., Frot-Coutaz, J., Got, R. 1988. *Biochem. Int.* 17:59–67
179. Steck, T. L. 1974. *J. Cell Biol.* 62:1–19
180. Geisow, M. J. 1986. *Bioessays* 4:149–51
181. Kornfeld, S. 1987. *FASEB J.* 1:462–68
182. Kean, E. L. 1970. *J. Biol. Chem.* 245:2301–8
183. Gielen, W., Schaper, R., Pink, H. 1971. *Z. Physiol. Chem.* 352:1291–96
184. van den Eijnden, D. H. 1973. *J. Neurochem.* 21:949–58

Annu. Rev. Biochem. 1989. 58:875-911

THE MULTI-SUBUNIT
INTERLEUKIN-2 RECEPTOR[1,2]

Thomas A. Waldmann

Metabolism Branch, National Center Institute, National Institutes of Health, Bethesda, Maryland 20892

CONTENTS

[1]The US government has the right to retain a nonexclusive, royalty-free license in and to any copyright covering this paper.

[2]Abbreviations used in this paper: ADF, ATL-derived factor; AIDS, acquired immune deficiency syndrome; ATL, adult T-cell leukemia; CAT, chlorophenicol acetyl transferase; HIV, human immunodeficiency virus; HTLV-I, human T-cell lymphotrophic virus I; IL, interleukin; LAK, lymphokine-activated killer; LGL, large granular lymphocytes; LTR; long terminal repeat; NK, natural killer; PHA, phytohemagglutinin; PMA, phorbol myristate acetate; SCID, severe combined immunodeficiency disease; SDS-PAGE, sodium dodecyl sulfate-polyacrylamide gel electrophoresis.

PERSPECTIVES AND SUMMARY

The human body defends itself against foreign invaders such as bacteria and viruses by a defense system that involves antibodies and thymus-derived lymphocytes (T cells). T cells mediate important regulatory functions such as help or suppression, as well as effector functions such as the cytotoxic destruction of antigen-bearing cells and the production of lymphokines. The success of these responses requires that T cells change from a resting to an activated state. The sequence of events involved in the activation of T cells begins with the interaction of appropriately processed and presented antigen with antigen receptors on the surface of the T cells. This binding then triggers the T cell to enter the activated state. Following antigen activation, T cells synthesize and secrete interleukin-2 (IL-2), a 15.5-kd glycoprotein. To exert its biological effect, IL-2 must interact with specific high-affinity receptors. Resting cells do not express high-affinity IL-2 receptors, but receptors are rapidly expressed on T cells after interaction with antigen or mitogen. IL-2 binding to the high-affinity form of its receptor then induces T-cell proliferation and, ultimately, the generation of specific regulatory and effector cells. Thus, both the growth factor and its receptor are absent in resting T cells, but following antigen activation, the genes for both proteins become expressed. Failure of this amplification process results in failure of the immune response. Thus, both the production of IL-2 and the display of IL-2 receptors are pivotal events in the full expression of the immune response.

High-affinity (K_d of ~ 10 pM) and low-affinity (K_d of ~ 10 nM) forms of the human IL-2 receptor have been identified. The study of the human IL-2 receptor was markedly facilitated by the identification of a monoclonal anti-receptor peptide antibody, anti-Tac. Utilizing this antibody, this IL-2 receptor peptide has been biochemically characterized as a 55-kd transmembrane glycoprotein. cDNAs encoding the Tac peptide have been isolated and the primary sequence of 251 amino acids determined. The 55-kd Tac peptide has been shown to participate in the formation of both affinity classes of IL-2 receptors. Until recently, the structural basis for these two classes of receptors remained unclear. Recently, using cross-linking methodology, two IL-2 binding peptides have been identified: the 55-kd peptide reactive with the anti-Tac monoclonal antibody, and a novel 75-kd non-Tac IL-2-binding peptide. We have proposed a multisubunit model for the high-affinity IL-2 receptor in which independently existing p55 (Tac) or p75 peptides would create low- or intermediate-affinity receptors, whereas high-affinity receptors would be created when both receptors are expressed and noncovalently associated in a receptor complex. There is evidence for a yet more complex structure that involves a 90–100-kd peptide in addition to the p55 and p75 peptides in the multisubunit high-affinity IL-2 receptor. The p55 Tac and p75 proteins have been shown to interact with different regions of the IL-2 molecule, with amino acids 11–20 of IL-2 required for binding to the p75 and amino acids 33–56 required for binding to the p55 protein. Kinetic binding studies with radiolabeled IL-2 have provided a perspective on how the p55 and p75 IL-2-binding proteins cooperate to form the high-affinity receptor. Each chain reacts with IL-2 with distinct kinetic and equilibrium binding constants. The on and off rate for IL-2 binding to the p55 protein is very rapid (i.e. 5–10 sec), while the on and off rates for IL-2 binding to the p75 protein is markedly slower (>20 min). The kinetic data obtained when high-affinity receptors are analyzed show that the association rate of this receptor is dependent upon the fast-reacting p55 Tac chain, whereas the dissociation rate is derived from the slow-releasing p75 chain. Since the affinity of binding at equilibrium is determined by the ratio of the dissociation and the association rate constants, this kinetic cooperation between the low- and intermediate-affinity ligand binding sites results in a receptor with a very high affinity for IL-2.

In contrast to the lack of p55 peptide expression in normal resting mononuclear cells, this receptor peptide is expressed by a proportion of the activated cells in certain forms of lymphoid neoplasia and select autoimmune diseases, and in individuals rejecting allografts. A proportion of the abnormal cells in these diseases express the Tac antigen on their surface. Furthermore, the serum concentration of a soluble released form of the Tac peptide is elevated. Specifically, human T-cell lymphotrophic virus type I (HTLV-I)-associated adult T-cell leukemia cells constitutively express very large numbers of IL-2 receptors. Recent studies have linked this disordered expression of IL-2 receptors in this retrovirus-induced T-cell leukemia with the action of a *trans*-activator gene product (*tax*-1) encoded by this virus. It seems possible that the Tax protein plays an important role in HTLV-I-induced malignancy by both augmenting viral transcription and by deregulating the expression of the cellular genes encoding IL-2 and the 55-kd IL-2 receptor peptide, which control T-cell proliferation.

To exploit the fact that IL-2 receptors are present on abnormally activated T cells but not on normal resting cells, clinical trials have been initiated involving patients with neoplastic or autoimmune disorders as well as those receiving organ allografts. These patients are being treated with unmodified anti-Tac, with anti-Tac conjugated to toxins, with toxin-IL-2 conjugates, with isotopic chelates of anti-Tac, and with recombinant "humanized" antibodies to the IL-2 receptor. Thus, the development of monoclonal antibodies and toxin-lymphokine conjugates directed toward the IL-2 receptors expressed on leukemia and lymphoma cells, autoreactive T cells of certain patients with autoimmune disorders, and host T cells responding to foreign histocompatibility antigens of organ allografts is leading to the development of rational, novel therapeutic approaches for these clinical conditions.

INTRODUCTION—OVERVIEW OF INTERLEUKIN-2 AND THE INTERLEUKIN-2 RECEPTOR

The vertebrate immune system has a virtually unlimited capacity to recognize and distinguish different molecular configurations and thus to bind to millions of potential antigens. Two broad lineages of cells that react specifically with antigens, B cells and T cells, are involved in this immune response. B cells are precursors of the antibody-secreting cells of the humoral immune system. T cells consist of an array of different subtypes of cells, including those that serve critical regulatory and effector functions. The receptor molecules that are involved in the specific immune response have a heterodimeric structure for both B and T lymphocytes. Immunoglobulins, with their light and heavy chain polypeptides, serve as the cell membrane antigen receptors for B cells. The T-cell antigen receptors are heterodimeric structures that are composed of α and β or, alternatively, γ and δ chains (1–7). The interaction of appropriately presented antigen with the heterodimeric receptor initiates the immune response. This interaction is not, however, sufficient to signal the initiation of effector functions, but provides the signal required for the induction of lymphokines and their receptors. These lymphokines play a major role in intercellular communication. They are involved in proliferative clonal expansion and in differentiation to specialized functions.

The failure of T cells to function normally may lead to neoplastic, autoimmune, or immunodeficiency diseases such as AIDS. Successful T-cell immune responses require that T cells change from a resting to an activated state, and this activation requires two sets of signals from cell surface receptors to the nucleus. The first signal is initiated when appropriately processed and presented foreign antigen interacts with the 90-kd polymorphic heterodimeric T-cell surface receptor for the specific antigen. After the antigen presented in the context of products of the major histocompatibility locus and interleukin (IL)-1 or IL-6 interacts with the antigen receptor, T cells synthesize a series of lymphokines, including IL-2 (8–13). IL-2 is a 15.5-kd glycoprotein encoded by a gene localized to human chromosome 4 (14). To exert its biological effect, IL-2 must interact with specific high-affinity membrane receptors. Resting cells do not express high-affinity IL-2 receptors, but receptors are rapidly expressed on T cells after interaction with antigen or mitogen. Thus, the interaction of antigen with its receptor on resting T cells induces the expression of both the lymphokine IL-2 and its ligand-binding receptor. IL-2 binding to the high-affinity form of its receptor then provides the signals that lead to T-cell proliferation and ultimately to the generation of specific regulatory and effector cells. Both IL-2 synthesis and IL-2 receptor expression are transient events, and the decline in expression of these proteins plays an important role in the normal termination of the T-cell immune response. Thus, although the interaction of appropriately presented antigen with its specific receptors confers specificity for an immune response, the interaction of IL-2 with high-affinity IL-2 receptors determines its magnitude and duration.

Utilizing purified, biosynthetically labeled IL-2, Robb and coworkers (11) demonstrated specific, saturable high-affinity binding sites on IL-2-dependent T-cell lines, as well as on mitogen- and alloantigen-activated T cells. Further progress in the analysis of the structure, function, and expression of the human IL-2 receptor was facilitated by our production of an IgG2a mouse monoclonal antibody (termed anti-Tac) that is now known to recognize the human IL-2 receptor (15–20). The data that support the view that anti-Tac recognizes a human IL-2 receptor peptide include the following observations: anti-Tac blocks IL-2-dependent T-cell proliferation; anti-Tac blocks the binding of radiolabeled IL-2 to activated T cells; and IL-2 at high concentration blocks the binding of radiolabeled anti-Tac to activated cells (17–20). The IL-2 receptor peptide identified by anti-Tac is composed of a 33-kd (251-amino-acid) peptide precursor that is posttranslationally glycosylated, sulfated, and phosphorylated in its mature 55-kd form (17, 18, 21–26).

Initial receptor-binding studies with radiolabeled IL-2 and anti-Tac suggested that activated normal T cells and leukemic T-cell populations express 5–20-fold more binding sites for anti-Tac than for IL-2; that is, while only 4000 high-affinity (10^{-11} M) IL-2 receptors were detected on PHA-activated lymphocytes in binding assays performed with radiolabeled IL-2, 30,000–60,000 sites were demonstrable with anti-Tac (27, 28). However, Robb and coworkers (29) repeated these studies utilizing greater quantities of IL-2 than had been used previously and showed that, in terms of affinity, there are two classes of IL-2 receptors. In general, most cell populations display 5–20% of their IL-2 receptors with an apparent affinity of 10^{-11} M, whereas the remaining receptors bind IL-2 with an affinity three logs lower (10^{-8} M).

The Tac peptide was shown to participate in both the high- and low-affinity forms of the IL-2 receptor. Isolation of cDNAs encoding the Tac peptide did not provide an explanation for the great difference in affinity between high- and low-affinity receptors. In studies of this issue, Tac-peptide cDNA was shown to reconstitute high-affinity receptors only when transfected into lymphoid cells but not when transfected into nonlymphoid cells (30–32). Furthermore, there was a conversion of low-affinity receptors to the high-affinity form following the fusion of membranes from human T cells with cell membranes from L cells transfected with cDNA encoding the p55 murine counterpart of the Tac peptide (33). These two observations supported the view that a cofactor or "converter" protein from T cells somehow imparted high affinity to the low-affinity binding protein (30–34). However, no one considered it likely that a separate IL-2-binding peptide might account for the observations.

Certain additional observations led us to experiments that helped resolve this issue. It had been shown that certain Tac-nonexpressing cells, including large granular lymphocytes (LGL), which are the precursors of natural killer (NK) and lymphokine-activated killer (LAK) cells, could make nonproliferative responses to IL-2 (35, 36). Furthermore, we identified a cell line, MLA-144, that did not express the Tac peptide yet manifested IL-2 binding sites. Using cross-linking methodology with radiolabeled IL-2, we demonstrated the existence of a novel IL-2-binding peptide (p75) with a M_r of 68,000–76,000 on the MLA-144 line that did not express the Tac antigen but manifested intermediate-affinity IL-2 binding sites (37–40) (Figure 1). Sharon and coworkers (41, 42) provided parallel evidence for this novel peptide. All cell lines expressing low- or intermediate-affinity IL-2 receptors express the p55 Tac or p75 peptides alone, whereas cell lines bearing high-affinity receptors express both the p55 Tac and p75 peptides. Therefore, we proposed a multichain model for the high-affinity IL-2 receptor in which an independently existing p55 or p75 peptide would create a low- or intermediate-affinity receptor, respectively, whereas high-affinity receptors would be created when both peptides are expressed and associated noncovalently in a receptor complex (39, 41). The p55 and p75 differ in the site of the IL-2 molecule that they bind, the kinetics of their association and dissociation with IL-2, and in their potential role in signal transduction (11, 12, 43–45).

A number of malignant and autoimmune diseases are associated with disorders of IL-2 receptor expression. In contrast to normal resting cells, the Tac peptide of the IL-2 receptor is expressed by a proportion of the abnormal cells in certain forms of lymphoid neoplasia—especially those associated with the retrovirus HTLV-I—in select autoimmune diseases, and in individuals rejecting allografts (10, 46–50). The observation that T cells in patients with these disorders express IL-2 receptors identified by the anti-Tac monoclonal antibody, whereas normal resting cells and their precursors do not, provided the scientific basis for therapeutic trials using agents to eliminate IL-2 receptor-expressing cells. In these studies, monoclonal antibodies and toxin-lymphokine conjugates directed toward the IL-2 receptor on abnormal Tac-expressing cells are being used for the treatment of patients with leukemia or autoimmune disorders, as well as those receiving organ allografts (4, 40, 51).

CHEMICAL CHARACTERIZATION OF THE MULTIPLE INTERLEUKIN-2 RECEPTOR SUBUNITS

Biochemical Properties of the p55 Tac Peptide of the Human Interleukin-2 Receptor

The IL-2 binding peptide identified by the anti-Tac monoclonal antibody was characterized as a 55-kd glycoprotein. Leonard and coworkers (17, 18) defined the posttranslational processing of this 55-kd glycoprotein by using a combination of pulse-chase and tunicamycin experiments. On the basis of these studies, it was shown that the IL-2 receptor was composed of a 33-kd peptide precursor after cleavage of the hydrophobic leader sequence. This precursor was cotranslationally N-glycosylated to 35- and 37-kd forms. One hour after the addition of unlabeled amino acids, the 55-kd mature form of the receptor was observed, suggesting that O-linked carbohydrate was added to the IL-2 receptor. Furthermore, the IL-2 receptor was shown to be sulfated and phosphorylated on a serine residue (21–23).

The Interleukin-2 Receptor (p55 Peptide) Gene

The Tac peptide was purified utilizing an anti-Tac affinity column, and its NH_2-terminal amino acid sequence was determined. Appropriate oligonucleotide probes based on this sequence were used to isolate full-length cDNAs encoding this protein. The deduced amino acid sequence of this peptide indicates that it is composed of 251 amino acids, as well as a 21-amino-acid signal peptide (24–26). The 219 NH_2-terminal amino acids make up an extracellular domain. This domain contains two potential N-linked glycosylation sites (Asn-X-Ser/Thr) and multiple possible O-linked carbohydrate sites. Multiple cysteine residues have been identified that participate in the formation of intramolecular disulfide bonds required for the IL-2 binding site. A second domain contains a single hydrophobic region near the COOH-terminal of the protein of 19 amino acids, which is presumably a membrane-spanning region. The third and final domain is a very short (13-amino-acid) cytoplasmic domain that contains several positively charged amino acids, presumably involved in anchoring of the protein. Potential phosphate acceptor sites (serine and threonine, but not tyrosine) are present within this intracytoplasmic domain. However, the cytoplasmic domain of the IL-2 receptor peptide identified by anti-Tac appears to be too small for enzymatic function. Thus, this receptor differs from many other known growth factor receptors that have large intracytoplasmic domains with tyrosine kinase activity. This finding raised the issue of how the Tac peptide could effectively transduce the intracellular signals required for T-cell proliferation and differentiation.

   A single gene encodes the Tac peptide of the IL-2 receptor (52). This gene consists of eight exons and seven introns spanning a minimum distance of 25

882   WALDMANN

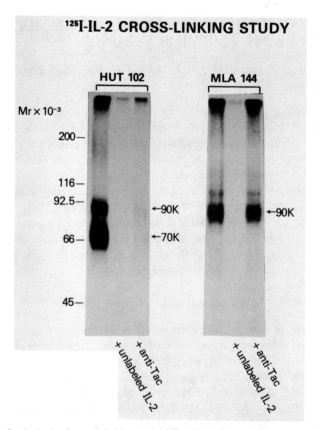

Figure 1 On the basis of a cross-linking study, MLA-144 cells were shown to express a novel non-Tac IL-2-binding peptide (p70–75), whereas HUT 102 cells express both the p55 and p75 peptides. Cross-linking studies using ^{125}I-labeled IL-2 and disuccinimidyl suberate (DSS) were performed to define the size of the surface molecule on MLA-144 cells responsible for IL-2 binding. With MLA-144, a labeled band with a M_r 90,000 was observed. This labeled band was abolished by addition of unlabeled IL-2, whereas the addition of anti-Tac did not affect the labeling. Thus, the IL-2-binding peptide on MLA-144 is a non-Tac peptide with a calculated M_r of 75,000 that is cross-linked with the M_r 15,000 IL-2 yield the M_r 90,000 band observed. The labeling of two bands was observed with HUT 102; a more heavily labeled lower band migrating at M_r 70,000 representing the Tac peptide (M_r 55,000) cross-linked with ^{125}I-IL-2 and the upper band (M_r 90,000) representing the p75 peptide cross-linked with ^{125}I-IL-2. The addition of anti-Tac antibody abolished not only the lower band but also the upper band, suggesting that the p70–75 peptide is closely associated with the p55 peptide on HUT 102 cells.

kb. In situ hybridization studies localized the receptor gene to chromosome 10p band 14 → 15. The first exon contains the 5' untranslated region and signal peptide. Exon 2 begins with the first amino acid of the mature protein. Limited homology with the Ba fragment of complement factor B was noted. Exons 2 and 4 (required for the functional molecule) contain sequences related to each other and to the duplicated mouse exons 2 and 4 (53). Furthermore, a similar pattern of repeating units, each approximately 61 amino acids in length, has been identified in β2 glycoprotein I and a series of proteins of the complement cascade, including C4b binding proteins, the C3b receptor (CR1), decay accelerating factor (DAF), complement factors B and C2, and the murine complement protein H (54). The fourth exon of this IL-2 receptor gene is involved in some cases in a form of posttranscriptional splicing, resulting in the joining of message from exon 3 to that of exon 5 (24, 52). This aberrantly sliced message does not yield a cell surface protein capable of binding IL-2 or anti-Tac. It has been suggested that sequences within exon 2 and 4 are involved in the IL-2 binding site. The sixth exon contains multiple potential sites for O-linked glycosylation. The 19-residue hydrophobic transmembrane domain is encoded predominantly by the seventh exon, with a minor contribution from the sixth exon. The eighth exon encodes part of the 13-amino-acid region that makes up the intracytoplasmic domain, as well as the 3' untranslated region of the IL-2 receptor. This 3' untranslated region can differ markedly in length, yielding mRNAs of approximately 1500 and 3500 bases due to the use of two or more polyadenylation signals (24, 52). The 5' end of the IL-2 receptor mRNAs has also been found to vary, with receptor gene transcription initiated at two principal sites in activated T lymphocytes (24).

Additional Interleukin-2 Receptor Peptides (p75 and p95) Participate in the High-Affinity Receptor

Several observations challenged the view that the p55 Tac peptide was the only molecule involved in IL-2 binding and triggering an immunological response. The unresolved questions concerned: (*a*) the structural explanation for the great difference in affinity ($>10^3$ in kd) between high- and low-affinity receptors (29); (*b*) how, in light of the short cytoplasmic tail of 13 amino acids of the Tac peptide, the receptor signals are transduced to the nucleus (24–26); (*c*) why the transfection of nonlymphoid mouse cells (fibroblasts and L cells) with cDNAs encoding human p55 generates only low-affinity receptors, whereas similar transfectants of mouse T-cell lines display both high- and low-affinity receptors (30–32); (*d*) how IL-2 can up-regulate the expression of Tac mRNA in multiple cell types, including some cells that did not initially express the Tac peptide (55–58; see below); and (*e*) how certain Tac-negative cells (e.g. LGL that are precursors of the NK and LAK cells) make nonproliferative responses to IL-2 (35, 36). These questions led us to consider the possibility that the high-affinity IL-2 receptor was not a single peptide but rather a receptor complex that included the Tac peptide as well as novel non-Tac peptides.

In studies initially presented at the sixth International Immunology Congress in July 1986, we utilized radiolabeled IL-2 in cross-linking to define the size of the IL-2 receptor peptides on various cell lines (37–39). In these studies, ^{125}I-labeled IL-2 was cross-linked using the bifunctional agent disuccinimidyl suberate to activated T cells and HTLV-I-infected leukemic T cells that express both high- and low-affinity IL-2 receptors. The labeling of two bands was observed in these cases. A heavily labeled lower band migrated at 69,000–72,000; the calculated size of this IL-2 binding peptide after subtracting the M_r of IL-2 (15,500) agreed with the reported molecular weight of the Tac peptide. There was an additional upper band that was frequently a doublet with a M_r of 84,000–91,000. Both excess unlabeled IL-2 and anti-Tac abolished the labeling of both bands.

On the basis of these initial studies, it was difficult to distinguish between two alternatives. Theoretically, two IL-2-binding sites could have been present per Tac peptide that bound to two homologous regions of the peptide encoded by the second and fourth exons of the Tac gene. Alternatively, the band at M_r 90,000 could have reflected a novel non-Tac IL-2-binding protein. We resolved this issue by performing radiolabeled IL-2 cross-linking studies on MLA-144, a T-cell line that expresses approximately 6000 IL-2 receptors of low to intermediate affinity yet neither displays the Tac peptide nor expresses Tac mRNA. In these studies, we identified a 75-kd IL-2-binding protein on this cell line (37–40). The binding of IL-2 to this peptide was blocked by excess unlabeled IL-2 but not by the anti-Tac antibody, confirming the presence of a novel 70–75-kd IL-2-binding protein (p75). When a variety of T-cell lines were examined for IL-2 binding and were subjected to IL-2 cross-linking studies, we demonstrated that there was a correlation between the affinity of IL-2 binding and the IL-2-binding peptides expressed (40). In these studies, cell lines bearing either the p55 Tac or the p75 peptide alone manifested low- or intermediate-affinity IL-2 binding, whereas cell lines bearing both peptides manifested both high- and low-affinity receptors. In light of these observations, we proposed a multichain model for the high-affinity IL-2 receptor in which an independently existing p55 or p75 peptide would create low- or intermediate-affinity receptors, whereas high-affinity receptors would be created when both receptors were expressed and noncovalently associated in a receptor complex (38–40). We have shown that fusion of cell membranes from a low-affinity IL-2 receptor cell line bearing the Tac peptide alone (MT1) with membranes from a line with intermediate-affinity receptors bearing the p75 peptide alone (MLA-144) generated hybrid membranes bearing high-affinity receptors (39). These findings support the proposed multichain model for the high-affinity receptor.

In independent studies, Sharon and coworkers (41, 42) proposed a similar model. This group, using similar cross-linking studies with radiolabeled IL-2 on activated and HTLV-I-infected T-cell lines, identified bands at 68–72 kd and 85–92 kd. This latter band was not present when IL-2 was cross-linked to cells transfected with the Tac peptide gene. In subsequent studies, these authors have partially purified the p70–75 peptide. Partial proteolysis with V8 protease followed by analysis of the fragments by SDS-PAGE yielded different peptide maps with the p55 and p70–75 complexes. These studies, demonstrating a novel IL-2-binding peptide p70–75 and its association with the p55 Tac peptide in the high-affinity receptors, have been confirmed and extended in subsequent studies by others (59–64).

Dukovich and coworkers (60) identified the p70–75 IL-2-binding peptide in the absence of the Tac peptide on the surface of a B-cell line, SKW6.4B, that secretes immunoglobulin in response to IL-2. In addition, Robb (61) and Teshigawara (59) and their coworkers demonstrated that the subclones of the human leukemic NK line YT, which do not express the Tac antigen yet express intermediate-affinity IL-2 receptors, express the p70–75 IL-2-binding polypeptide. Furthermore, these groups showed that YT cells induced to display high-affinity receptors by various treatments, including the addition of forskolin, concomitantly express both p55 and p70–75. Furthermore, Dukovich and coworkers (60) showed that the introduction of cDNA encoding p55 into MLA-144 cells by protoplast fusion led to the generation of high-affinity IL-2 receptors (K_d of 35–70 pm) in a proportion of these cells. These reconstituted high-affinity receptors included the Tac antigen, since the addition of anti-Tac blocked the high-affinity but not the intermediate-affinity component of IL-2 binding.

Kinetic binding studies with IL-2 have provided an interesting perspective on how the two separate IL-2-binding proteins cooperate to form the high-affinity receptor. Each chain reacts with IL-2 very differently, with distinct kinetic and equilibrium binding constants. The on and off rate for IL-2 binding to the Tac protein is very rapid (5–10 sec), while the on and off rate for IL-2 binding to the p70–75 protein is markedly slower (20–50 min) (62–64). The kinetic binding data obtained when high-affinity receptors are analyzed show that the association rate of this receptor is dependent upon the fast-reacting p55 Tac chain, whereas the dissociation rate is derived from the slow-reacting p75 chain. Since the affinity of binding at equilibrium is determined by the ratio of the dissociation constant and the association rate constant, this kinetic cooperation between the low- and intermediate-affinity ligand binding sites results in a receptor with a very high affinity for IL-2 (62–64).

There is evidence for a more complex subunit structure that involves other peptides in addition to p55 and p75 (65–68). Two monoclonal antibodies, OKT27 and OKT27b, were produced that react with distinct epitopes of a 95-kd peptide expressed on activated T cells. The OKT27b antibody inconsistently coprecipitated the 55-kd Tac peptide as well as the 95-kd peptide, indicating some association between the two (65). In collaboration with Szöllösi, we investigated this relationship in intact cells by labeling each peptide with an appropriate fluorescent monoclonal antibody and then measuring the average distance between them with a flow cytometric energy transfer technique (65). This technique permits the determination of intermolecular distances at 2–10 nm on a cell-by-cell basis. The technique was used to show proximity of the p95 (T27) and Tac peptides, as well as proximity of the HLA class I antigen. Appropriate specificity controls showed that the p95 and Tac associations were not fortuitous. These results indicated that the p95 and Tac peptides are associated on the surface of at least some activated T cells. Furthermore, fluorescence photobleaching recovery measurements performed in conjunction with Edidin indicated that the Tac and p95 peptides are not only in proximity on HUT-102 lymphocyte membranes but also interact physically in situ (66).

In independent chemical cross-linking studies with radiolabeled IL-2, Herrmann & Diamanstein (67) and Saragovi & Malek (68) presented evidence for a 90–100-kd IL-2-binding peptide in mice associated with the 55- and 70–75-kd chains of the high-affinity form of the IL-2 receptor. Herrmann & Diamanstein (67), using IL-2 in high-affinity conditions, demonstrated bands corresponding to those associated with the 55- and 70–75-kd peptides of humans and, in addition, 90-kd and 100-kd peptides associated with the 15-kd IL-2 molecule on mouse T-cell blasts or the CTLL-16 cell line. When the radiolabeled cross-linked molecules were incubated with an antibody to the p55 component coupled to protein A Sepharose, all of the 55-kd peptides were removed, whereas the materials corresponding to peptides of 70–75 and 90–100 kd were not completely immunoprecipitated. The authors proposed that a third IL-2-binding chain is associated with the p55 and p70–75 chains in the high-affinity IL-2 receptor. In parallel studies, Saragovi & Malek (68) cross-linked ^{125}I-labeled IL-2 to the mouse EL-4 line that had been transfected with 55-kd IL-2 receptor cDNA. Furthermore, these authors showed that the 90- and 115-kd bands contained IL-2 cross-linked to peptides that were not reactive with an anti-p55 IL-2 receptor antibody. Thus, these authors also raised the possibility that the 115-kd band demonstrable by cross-linking the 15-kd radiolabeled IL-2 to the high-affinity receptor represents a new receptor subunit of approximately 100 kd. Taken together, these studies suggest that three IL-2-binding chains (p55, p70–75, and p90–100) are associated in a multisubunit high-affinity IL-2 receptor. In addition, cross-linked components of 22, 35, 40, 135, and 180 kd have been observed on occasion in association with the 55- and 75-kd components.

The regions of IL-2 involved in binding to these receptor peptides are being defined (43–45, 69) (Figure 2). The three-dimensional structure of the 133-amino-acid lymphokine IL-2 has been determined to 3.0 Å resolution (70).

IL-2 has been shown to have a novel α-helical structure with no segments of β structure. One portion of the molecule forms a structural scaffold that underlies the receptor-binding facets of the molecule. Using monoclonal antibodies directed toward defined regions of IL-2, deletional analysis, and site-specific mutagenesis of IL-2 in neutralization and binding assays has aided in the analysis of the sites of human IL-2 involved in receptor binding. The p55 Tac and p70–75 proteins have been shown to interact with different regions of the IL-2 molecule (Figure 2). Kuo & Robb (43), using antibodies for specific regions of the IL-2 molecule, have demonstrated that residues 8–27 and

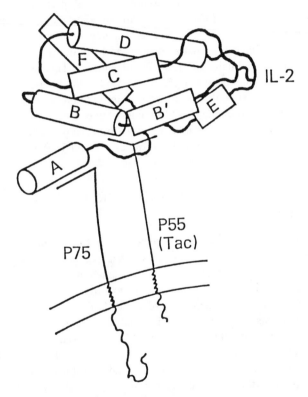

Figure 2   Schematic drawing of IL-2; helixes are represented as cylinders and are lettered sequentially from the NH$_2$-terminus. Helix A (amino acids 11–19) and the neighboring amino acid 20 are required for effective binding to the p75 protein. The second helix, B and B', involving residues 33–56, is most likely required for binding to the p55 Tac peptide. An additional α helix (E amino acids 107–113) is also positioned on the binding scaffold plane and could theoretically bind an additional IL-2-binding peptide. However, it must be emphasized that no evidence for this exists and that no extensive studies of this region are available. [Reproduced with modifications from B. J. Brandhuber, T. Boone, W. C. Kenney, and D. B. McKay (70).]

33–54 are involved in high-affinity receptor interactions. Furthermore, they have suggested that amino acids 8–25 interact with p70–75 receptor peptide but not with the Tac protein.

Ju (44) and Collins (45) and their coworkers have used deletional analysis and site-directed mutagenesis to define the regions of the IL-2 molecule involved in receptor binding. A short helical segment (helix A, amino acids 11–19) is required for biological activity, and its normal conformation is required for binding to the p75 IL-2-binding peptide. Collins and coworkers (45) have shown that the neighboring amino acid 20 (Asp) is required for effective binding to the p70–75 protein. Specific mutations at this site exhibit a very low level of bioactivity yet retain normal overall protein conformation as assessed by circular dichroism and are still bound by the p55 peptide. The second helix on the structural scaffold is an extended loop involving residues 33–56 that form an interrupted helix "bent" near the middle by Pro^{45}. The two segments of the second helix are designated B and B'. This segment is most likely required for binding to the p55 Tac peptide. An additional α helix E (amino acids 107–113) is also positioned on the binding scaffold and could theoretically bind an additional IL-2-binding peptide. However, no extensive studies of this region are available. Finally, the carboxy terminal residues 131–133 and two of the three cysteine residues (58 and 105) are required for full biological activity and binding (44).

REGULATION OF INTERLEUKIN-2 RECEPTOR EXPRESSION

The expression of high-affinity IL-2 receptors on resting T cells is initiated by the interaction of antigen with antigen-specific receptors. This interaction involves the presentation of appropriately processed antigen by macrophages in the context of products of the major histocompatibility locus. Antigen binding in the presence of macrophage-derived IL-1 or IL-6 then induces the coordinate synthesis and secretion of IL-2 and the expression of high-affinity membrane receptors for this lymphokine (28, 71). Activation through the T-cell receptor complex is associated with an increase in intracellular calcium, activation of the phosphoinositol cycle, and the translocation of protein kinase C from the cytoplasm to the membrane (72, 73). IL-2 receptors may be alternatively induced by the action of mitogens, by antibodies that interact with either the antigen-specific heterodimer or the T3 elements (CD3) of the complex T-cell antigen receptor, or by alternative pathways through the action of other mitogenic antibodies such as CD2 (anti-T11) or CD28 (9.3p44) (74, 75). IL-2 receptors may also be induced without interaction with T-cell antigen receptors by the use of the calcium ionophore A23187, which increases intracellular calcium, or by phorbol myristate acetate (PMA), which acts on protein kinase C (28).

Using radiolabeled anti-Tac to evaluate cell surface expression of the 55-kd IL-2 receptor peptide, little or no specific binding is demonstrable with freshly isolated human lymphocytes. However, after activation with PHA, IL-2 receptors are demonstrable within 4 to 8 hr and reach a peak of approximately 30,000 to 60,000 sites/cell 48 to 96 hr after activation (12, 28). Following this period, there is a progressive decline in the number of receptors so that, after 10 to 21 days in culture, the number declines by 80–90% from that observed during peak expression. The decline in receptor number parallels a decline in mRNA transcription for the Tac peptide. Tac mRNA is demonstrable at 4 hr after activation with PHA, peaks at 8–12 hr, and then declines (12, 28, 76). The proliferative rate of cells follows and parallels the rise and fall in IL-2 receptor expression. The decline in IL-2 receptor expression in T cells maintained in culture may be responsible for the crisis period occurring after 3 to 6 weeks in culture, at which time T cells no longer proliferate in response to IL-2. Furthermore, these data suggest that IL-2-dependent T-cell proliferative immune responses may be regulated by both the amount of IL-2 secreted and the magnitude of IL-2 receptor expression. These findings also suggest that, in vivo, the magnitude and duration of IL-2 receptor expression may well define the magnitude of T-cell clonal expansion and the resultant immune responses.

A series of signals can reactivate IL-2 receptor gene transcription and cell proliferation on PHA-activated lymphoblasts that had lost most of their receptors during culture. The expression of the Tac receptor can be reactivated by the addition of the initial stimulus, mitogenic lectins, or antigens. Furthermore, Depper and coworkers (28) demonstrated that the addition of phorbol diesters, phospholipase C, or the diasolglycerol congeners that activate protein kinase C but not IL-1 also resulted in augmented Tac expression. Farrar & Ruscetti (73) demonstrated that agents that induced Tac expression stimulated transient redistribution and activation of protein kinase C. However, Mills (77) and Valge (78) and their coworkers demonstrated that T-cell clones lacking protein kinase C can still proliferate in response to IL-2.

A series of lymphokines has been shown to play a role in the regulation of Tac antigen expression. For example, IL-1 has been shown to be required for the expression of receptors for IL-2 in a cloned helper T-cell line (79). Furthermore, IL-1 augments the expression of the Tac antigen on human NK cells and the NK-like cell line YT. Wakasugi and coworkers (80) have shown that an IL-1-like factor derived from an Epstein-Barr virus-transformed B-cell line, 3B6, induces Tac expression on YT cells. Resting T cells treated with either IL-4 or PMA alone show little, if any, up-regulation of the expression of the Tac receptor; however, treatment of cells with IL-4 plus PMA resulted in up-regulation of this receptor by the majority of cells (81, 82). In addition, IL-4 plus PMA pretreatment of resting cells prepared those cells to respond to either IL-4 or IL-2 alone (81, 82). IL-5, a T-cell-derived differentiation factor, induces IL-2 receptor expression on peanut agglutinin-binding thymocytes and on activated B cells (84). The induction of T-cell responsiveness to IL-2, T-cell proliferation, and Tac expression stimulated by the addition of an anti-T cell (anti-CD28) monoclonal antibody required monocytes or monocyte culture supernatants. The active compound responsible for the helper signal in the monocyte culture supernatant has been identified to be IL-6, not IL-1 (84).

HTLV-I-associated adult T-cell leukemia (ATL) cells constitutively express the Tac peptide of the IL-2 receptor system. Yodoi and coworkers (85, 86) have identified an ATL-derived factor (ADF) that induces the expression of the high-affinity IL-2 receptor on a variety of cells, including HTLV-I-positive T cells, myeloid leukemia cells, and YT cells. The ADF protein has been purified and the partial NH_2-terminal amino acid sequence determined (86). The 20-amino-acid sequence had no homology to IL-1, IL-2, or other cytokines derived, with the exception of the IL-1-like factor from the Epstein-Barr virus-transformed B-cell line 3B6 (85, 86).

Several observations have indicated that IL-2 plays a major role in the regulation of its own receptors. We demonstrated that the addition of IL-2 to a cloned normal IL-2 receptor-expressing B-cell line led to the induction of mRNA and the expression of receptors identified by anti-Tac (51, 57, 58). In parallel studies, Reem & Yeh (56), using dexamethasone to inhibit IL-2 production during lectin activation of human lymphocytes, observed that IL-2 receptor expression was augmented by the addition of IL-2. Welte et al (87), using anti-T3 antibodies, also demonstrated that IL-2 augmented IL-2 receptor expression. In subsequent studies, the addition of IL-2 was shown to augment the expression of the 55-kd IL-2 receptor peptide when cultured with thymocytes, cloned T-cell lines, freshly obtained peripheral blood lymphocytes, and LGL that are the precursors of activated NK and lymphokine-activated killer effector cells. The in vivo administration of purified human IL-2 to patients with cancer led to the development of circulating Tac-positive T cells and monocytes (88). Furthermore, it led to a dramatic increase in the serum levels of the soluble Tac receptor (88). Smith & Cantrell (71, 89) added purified IL-2 to activated T cells brought to rest by IL-2 deprivation and observed an 8–10-fold increase in the expression of the Tac peptide but a simultaneous 20–30% decrease in detectable high-affinity IL-2-binding sites.

The discovery of the 75-kd IL-2-binding peptide (p75) has helped in our understanding of these latter observations as well as the observations that IL-2 can induce the Tac peptide in cells that initially do not express this IL-2-binding peptide. We have demonstrated that the addition of IL-2 to p75-expressing Tac-negative LGL leukemic cells augmented transcription of the Tac gene and induced the expression of the Tac peptide (90). This up-regulation of Tac gene expression was not inhibited by the addition of anti-Tac. The results strongly suggest that the p75 peptide is responsible for IL-2-stimulated induction of the transcription of the Tac gene and the expression of the Tac peptide. Thus, the interaction of IL-2 with the p75 peptide may play an important role in the function of LGL by leading to the induction of the Tac peptide, thus yielding a cell that expresses both peptides required for the high-affinity IL-2 receptor.

The slight reduction in the number of high-affinity receptors observed by Smith & Cantrell (71, 89) following the addition of IL-2 to activated T cells that initially display both receptors can be understood when the signals required for receptor-mediated endocytosis are considered. The interaction of IL-2 with the p75 but not the p55 Tac IL-2 receptor peptide conveys the signal for rapid receptor-mediated endocytosis of radiolabeled IL-2. Thus, the addition of IL-2 to a cell expressing the p75 peptide leads, on the one hand, to increased expression of the Tac peptide by augmenting transcription and translation for this chain, while simultaneously leading to a reduction in p75 chain expression by IL-2-stimulated internalization. When the number of high-affinity receptors and the capacity for T-cell proliferation and function are considered, different patterns are observed when IL-2 is added to activated T cells that already express both receptor peptides than is observed when one adds IL-2 to resting T cells and LGL that initially express the p75 peptide alone. In the former case, the addition of IL-2 leads to a net reduction in high-affinity IL-2 receptor expression and thus a movement toward the resting state. In the latter case with cells initially expressing p75 alone, the addition of IL-2, by inducing the expression of Tac peptide, leads to conversion of cells bearing only the intermediate-affinity receptor to ones expressing the high-affinity IL-2 receptor, which are then capable of responding to the low levels of IL-2 present in biological fluids.

Much less is known about the factors regulating the expression of the 75-kd IL-2-binding peptide, since antibodies to this peptide are just being generated and the gene encoding this peptide has just been cloned. Quantitation of the number of p75 receptor peptides expressed has depended on radiolabeled IL-2 binding studies in the presence and absence of anti-Tac. As noted above, the addition of IL-2 has differential effects on the expression of p55 Tac and p75 IL-2 receptor peptides. Furthermore, the addition of PHA has a more dramatic effect on the expression of Tac than p75 receptors. Resting T cells have been reported to manifest approximately 600 p75 receptor sites per cell and virtually no Tac sites; however, following PHA activation, there is a dramatic increase in the number of Tac sites to 30,000 to 60,000, exceeding the 1500–2000 high-affinity IL-2-binding sites bearing p75 peptides by a factor of 20 (91). Furthermore, forskolin, a potent inducer of the Tac molecule on YT cells, failed to enhance the expression of p75 molecules on such cells, supporting the view that the regulation mechanism of p75 is different from that of Tac (91). The ADF factor, which induces Tac peptide expression, also did not significantly augment p75 expression (59). Finally, the transfection of a fragment of the HTLV-I gene containing a *trans*-acting, transcriptional activator gene into YT cells led to an increased level of the expression of the Tac peptide but did not affect p75 expression (86). This may explain the predominant expression of low-affinity receptors in adult T-cell leukemia.

Several studies have been directed toward defining the *cis*-acting sequences and the *trans*-acting factors that bind to them that regulate the expression of the Tac receptor gene (92–97). The elements associated with the IL-2 receptor promoter have been related to those in the long terminal repeat (LTR) of type 1 human immunodeficiency virus (HIV-1), since both of these regulatory elements are induced by comparable signals, including PMA, PHA, and the *tax* gene product of HTLV-I. Plasmids containing the upstream flanking sequences of IL-2 receptor linked to the chlorophenicol acetyl transferase (CAT) reporter gene introduced into T lymphocytes, are induced by PMA, anti-T3 monoclonal antibodies, and the *trans*-activator protein of the HTLV-I retrovirus (Tax). Cross et al (92) and Lowenthal et al (93) have demonstrated *cis*-acting regulatory elements required for the mitogenic and PMA induction of Jurkat T cells between nucleotides −327 and −179. In Jurkat cells, sequences between −317 and −271, as well as those 3' of −271 were required. In contrast, sequences 3' of base −271 were sufficient for promoter induction in YT1 cells stimulated with PMA, IL-1, or forskolin, or in Jurkat cells expressing the Tax protein of HTLV-I. In the studies, Cross (92) showed, with transiently transfected HTLV-I-transformed T cells, that although deletions with end points of −1242, −499, and −481 possessed promoter activity, a 10–15-fold increase was seen in levels of CAT conversion as 5' sequences were progressively removed up to −317. Similarly, in the studies of Lowenthal et al (93) using YT1 cells stimulated with PMA, IL-1, or forskolin, −317 pTac-CAT plasmids were 3–4-fold more active than −471 pTac-CAT plasmids. These studies, as well as those of Suzuki et al (94), suggest that a region with negative regulatory activity for the IL-2 receptor-55-kd promoter may be present 5' of 317 in the promoter fragment.

Leung & Nabel (95), as well as Böhnlein et al (96), have identified a sequence upstream (−267 to −256) of the IL-2 receptor-55-kd gene, which is the same at 9 of 11 base pairs as the binding site of the transcriptional factor called NF-κB. A similar site has been identified in the 5' region of the IL-2 gene as well. This site, an enhancer region for the IL-2 receptor, was initially identified within the enhancer sequence of the immunoglobulin κ light chain gene. Furthermore, 9 of 11 and 8 of 11 base pairs were shared with the downstream and upstream elements, respectively, of the HIV-1-directed repeats. The active form of the NF-κB protein is not expressed in resting T cells but is induced upon T-cell activation (95). NF-κB has been shown to bind to the HIV-1 enhancer (97). In addition, this protein and hiven86a, a nuclear protein whose relationship to NF-κB remains uncertain, have been shown to be capable of specifically associating with both the HIV enhancer and the 55-kd IL-2 receptor promoter. It has also been shown that the cellular and viral binding sites for this factor confer mitogen inducibility on the promoters. These findings suggest that the HIV provirus is able to utilize a nuclear protein involved in T-cell activation and the regulation of IL-2 receptor gene expression to permit activation of the retroviral LTR, augment viral gene transcription, and ultimately to cause high-level viral replication and cytopathic effects within the infected T cells. As will be discussed more extensively below, the retrovirus HTLV-I produces a *tax* gene product that probably induces an NF-κB-like effect that underlies the constitutive IL-2 receptor expression in HTLV-I-associated ATL. Thus, it has been suggested than an NF-κB-like factor is a target of HTLV-I infection activated by the *tax* product (98). In the case of T cells infected with HIV, NF-κB is a cellular activator of HIV gene expression. In uninfected T cells, it may normally regulate the expression of the Tac IL-2 receptor peptide.

HIV-1 and the promoter region encoding the IL-2 receptor share negative as well as positive regulatory *trans*-acting factors. Patarca et al (99) have purified and molecularly cloned the *rpt*-1 (for regulatory protein, T-lymphocyte 1) gene and have shown that this gene selectively expressed by resting but not activated CD4 inducer T cells, encodes an intracellular protein (Rpt-1, M_r 41,000). This protein down-regulates gene expression directed by the promoter region of the gene encoding the Tac peptide and that directed by the LTR of HIV-1. These studies suggest that Rpt-1 levels may be inversely correlated with activation of CD4 T cells and HIV replication and that the expression of this protein may be related to the relatively long latency period of HIV in resting T cells.

DISTRIBUTION OF INTERLEUKIN-2 RECEPTORS

The distribution of the 55-kd Tac peptide of the IL-2 receptor has been defined using monoclonal antibodies (CD25), including the anti-Tac mono-clonal antibody. Antibodies to the p55 peptide of other species have been prepared (99–102). Since antibodies to the p75 peptide are just being pro-duced, the distribution of this peptide has been defined using radiolabeled IL-2 in cross-linking studies, as well as radiolabeled or biotinylated IL-2 in binding studies. In these latter studies, antibodies blocking the binding site of the p55 protein can be added to prevent IL-2 from binding to this receptor peptide. Most resting T cells, B cells, or monocytes in the circulation do not display the p55 peptide of the IL-2 receptor. Specifically, less than 5% of freshly isolated, unstimulated human peripheral blood T cells react with anti-p55 monoclonal antibodies (15). A proportion of immature thymocytes display this peptide (103). The majority of T cells, however, can be induced to express IL-2 receptors by interaction with lectins, by antigen or alloantigen stimulation, or by the addition of monoclonal antibodies to the T-cell antigen receptor complex, appropriate pairs of CD2 antibodies, or CD28 antibodies (15, 74, 84).

The p55 peptide has also been demonstrated on activated B cells. Korsmeyer and coworkers (104) showed that hairy cell leukemia cells of the B-cell series expressed the peptide identified by the anti-Tac monoclonal antibody. Subsequently, the p55 peptide has been shown to be expressed on many B-cell lines from patients with Burkitt's lymphoma and on B-cell lines containing the HTLV-I genome (55, 57). Normal peripheral blood B cells activated by *Staphylococcus aureus* Cowan I organisms, pokeweed mitogen, phorbol myristate acetate, or anti-μ immunoglobulins can be induced to express the p55 antigen (57, 105–107). In addition, cloned Epstein-Barr virus–transformed B-cell lines derived from p55-expressing, activated normal B cells continue to express this antigen in long-term culture (57). Such p55-expressing cells manifested both high- and low-affinity IL-2 receptors at a ratio comparable to that observed with IL-2-dependent T-cell lines and activated T lymphocytes. The size of the IL-2 receptors on activated normal B cells was comparable (53–57 kd) to that of receptors on PHA-stimulated T lymphoblasts. Furthermore, p55-expressing B cells transcribe 1500- and 3500-base mRNAs for the p55 IL-2 receptor peptide (57). Thus, certain malignant as well as activated normal B cells display the p55 peptide and manifest high-affinity receptors for IL-2.

IL-2 receptors identified with the anti-Tac monoclonal antibody have been detected on activated cells of the monocyte/macrophage series, including the Reed-Sternberg cells in Hodgkin's disease, Kupffer's cells of the liver, cultured lung macrophages, Langerhans' cells of the skin, and normal human peripheral blood monocytes stimulated with lipopolysaccharide or interferon-γ (108, 109). IL-2 receptors have also been identified on cells other than those of the lymphocyte/monocyte series. Granulocytic cells from the promyelocyte stage onward lacked p55 IL-2 receptor expression before and after in vitro stimulation (110–113). In contrast, a proportion of the blast cells from approximately half of the cases with acute myelogenous leukemia studied expressed this IL-2 receptor peptide. Furthermore, p55 expression could be induced on such cells with various agents, including TPA, γ-interferon, and colony-stimulating factor—suggesting a potential role for IL-2 in early granulopoiesis.

In addition to the membrane-associated form of 55-kd IL-2 receptor, a soluble form of this receptor is released by activated normal peripheral blood mononuclear cells and by HTLV-I-infected leukemic cell lines. Rubin, Nelson, and coworkers (114) showed that the soluble form of the receptor is smaller (45 kd) than its cellular counterpart (55 kd). Like its cellular counterpart, the soluble receptor is capable of binding IL-2 with an affinity of 10^{-8} M. Using an ELISA with two monoclonal antibodies that recognize distinct epitopes on the human IL-2 p55 receptor, they showed that normal individuals have measurable amounts of this peptide in their plasma and that certain lymphoreticular malignancies, autoimmune disorders, and allograft rejections are associated with elevated plasma levels of this receptor (114). The release of soluble IL-2 receptors appears to be a consequence of activation of various cell types, especially T cells, that may play a role in the regulation of the immune response. The analysis of plasma levels of IL-2 receptors appears to provide a very valuable noninvasive method for the analysis of both normal and disease-associated lymphocyte activation in vivo.

The p75 IL-2-binding peptide is expressed along with the 55-kd Tac peptide on activated T cells, B cells, and monocytes. Specifically, T cells stimulated by PHA or phorbol myristate acetate and those infected with HTLV-I express the p75 peptide. The number of such IL-2-binding peptides on PHA-stimulated T cells, as estimated by Skatchard analysis of radiolabeled IL-2 binding in the presence of anti-Tac, is approximately 2000 molecules per cell, a value approximately one tenth that of the Tac peptide (91). Resting B cells stimulated with *S. aureus* Cowan-I and monocytes stimulated by lipopolysaccharide or a combination of lipopolysaccharide and γ-interferon also express the p75 peptide. In addition, the p75 peptide is expressed on certain circulating cells and cell lines that do not express the p55 antigen. It is expressed alone on the B-cell line SKW6.4B, which does not express the Tac peptide yet responds to high concentrations of IL-2 with increased IgM synthesis. It is also expressed on resting T cells of especially the CD8$^+$ subpopulations. These cells express approximately 600–700 p75 molecules per cell, values at least 10-fold greater than the expression of the p55 peptides on these cells (115).

The demonstration of p75 on resting T cells may provide an explanation for the ability of these cells, which are initially p55 nonexpressing, to exhibit proliferative responses to high concentrations of IL-2. Tac-nonexpressing LGL can be stimulated by IL-2 to enhance NK and LAK activity. Using cross-linking methodology with radiolabeled IL-2, we demonstrated that clonal populations of LGL from patients with CD2$^+$/CD3$^-$/CD4$^-$/CD8$^-$ leukemic LGL as well as from patients with Tγ lymphoproliferative disease with the CD2$^+$/CD3$^+$/CD8$^+$ phenotype expressed the p75 IL-2-binding peptide but did not express the Tac peptide (90). Furthermore, in studies performed in our laboratory (90) and those by Siegal and coworkers (116), normal LGL were also shown to express the p75 but not the p55 IL-2-binding peptide.

LYMPHOCYTE FUNCTIONS THAT ARE REGULATED BY THE INTERACTION OF INTERLEUKIN-2 WITH ITS RECEPTOR

Extensive studies have attempted to define the contributions to lymphocyte functions made by the p55 and p75 peptides when expressed alone and when associated in the high-affinity IL-2 receptor complex. Certain of these studies have been directed toward defining the contributions of the two peptides to the receptor-mediated internalization of the ligand IL-2. Weissman, Fujii, and their coworkers (117, 118) demonstrated that IL-2 binding to the high-affinity receptor results in rapid receptor-mediated endocytosis of the ligand-receptor complex, while ligand binding to the low-affinity p55 site does not lead to internalization of the complex. Using cell lines that express either the p55 or the p75 proteins, IL-2 internalization was shown to occur when the ligand is bound to the isolated p75 peptide but not when it is bound to isolated p55 (119). The kinetics of IL-2 internalization mediated by p75 alone were nearly identical to those of the high-affinity heterodimer (half-life of 10–15 min), and each type of receptor targeted the bound IL-2 for intracellular degradation in lysosomes. Although the p75 plays the dominant role in the rapid internalization of IL-2, evidence exists that at least some internalization can follow the interaction of IL-2 with the p55 peptide. Lorberboum-Galski and coworkers (120) studied the internalization of IL-2 via the individual p55 and p75 subunits of the IL-2 receptor using IL-2-PE40, a chimeric protein composed of human IL-2 genetically fused to the NH_2-terminus of a modified form of *Pseudomonas* exotoxin. Internalization was assessed by measuring inhibition of protein synthesis caused by the toxin moiety of IL-2-PE40. This internalization was studied on several mouse and human cell lines expressing either the p55, the p75, or both IL-2 receptor subunits. IL-2 internalization was mediated by either the p55 receptor subunit or by the p75 subunit but was much more efficient when high-affinity receptors composed of both subunits were present.

One of the major questions that is not yet fully answered is how the growth signal is transduced following interaction of IL-2 with the IL-2 receptor peptides. The p55 IL-2-binding molecule is not sufficient to transduce the IL-2 signal. This 55-kd molecule has a 13-amino-acid intracytoplasmic domain with no homology to tyrosine kinases. In addition, either deletions or mutations of the transmembrane or intracellular portions of this IL-2-binding molecule, including replacement of this total region with the transmembrane portion of another receptor molecule, do not inhibit triggering of cell proliferation. Furthermore, cell lines expressing the Tac peptide alone or non-lymphoid cells transfected with the Tac gene are not induced to meaningful proliferation by IL-2. It therefore seems unlikely that the p55 peptide on its own is sufficient for signal transduction. In contrast, it appears that the p75 peptide without the p55 peptide is sufficient for signal transduction when large concentrations of IL-2 are used. As noted above, LGL express the p75, but not the p55, IL-2-binding peptide (90). These p55-nonexpressing LGL made proliferative responses to IL-2 but required a much higher concentration than that required for the proliferation of normal PHA-stimulated T lymphoblasts that express high-affinity receptors (90). Furthermore, up-regulation of p55 mRNA and peptide expression by IL-2 has been demonstrated for a number of cell types that express p75 alone, including LGL (90). Neither the IL-2-induced activation of LGL nor the up-regulation of p55 expression in such cells was inhibited by the addition of anti-Tac. These results strongly suggested that the p75 peptide is responsible for IL-2-induced activation of LGL and that this peptide alone can mediate an IL-2 signal. Siegal and coworkers (116) also concluded that the interaction of IL-2 with the p75 peptide on LGL mediates the initial phase of induction of LAK, NK, and proliferative activities. Subsequently, p55 was induced and functional high-affinity receptors consequently expressed. Others (121) have confirmed these observations concerning LGL and have extended them to other cells, including thymocytes and acute lymphoblastic leukemia cells that express the p75 peptide alone. The ability of the p75 peptide to transduce signals was examined further in the gibbon ape cell line MLA-144 that expresses this peptide alone and that proliferates in an autocrine fashion in response to IL-2 that it produces itself. The addition by Smith (122) of glucocorticoids to this cell inhibited IL-2 production. Addition of exogenous IL-2 to glucocorticoid-treated MLA-144 cells supported a proliferative response. Furthermore, although an array of antibodies to the p55 peptide have been produced, none have been shown to have agonist activity. In contrast, Inamoto and coworkers (123) have produced a series of monoclonal antibodies recognizing a p75 peptide on LGL that was mitogenic for peripheral blood lymphocytes. Furthermore, the antibody rapidly induced the Tac antigen on YT cells that initially expressed the p75 peptide alone. One of the earliest events following the interaction of IL-2 with its receptor on the surface of cells is an increase in intracellular pH due to activation of the Na^+/H^+ antiport. Mills & May (124) have demonstrated that the interaction of IL-2 with the p75 IL-2-binding molecule is sufficient to activate the Na^+/H^+ antiport, whereas IL-2 does not yield a similar response in cells expressing only the 55-kd IL-2-binding molecule. Furthermore, Ishii and coworkers (125) demonstrated that IL-2 rapidly induced phosphorylation of a 67-kd and four 63-kd cellular proteins in various T cells. Using various cell lines, including the MLA-144 cell line that expresses only the p75 IL-2-binding peptides, these authors have shown that the protein phosphorylation is mediated by the interaction of IL-2 with the p75 IL-2 receptor peptide. Finally, the interaction of IL-2 with the p75 IL-2 receptor peptide is sufficient to induce the expression of c-*myc* and c-*myb* (115). Taken together, it is clear from these studies that the p55 IL-2-binding peptide is not required for the transduction of the signals initiated by high concentrations of IL-2. As a more complete biochemical and molecular description of the p75 protein becomes available, several of the critical questions concerning the mechanism of signal transduction may be answered. It should be emphasized that other peptides, especially the p95 and p35 peptides, that have been shown to be associated with the p55 and p75 peptides of the IL-2 receptor, may play a role in the transduction of the IL-2 signal.

Although the Tac peptide is not required for a cellular response to high concentrations of IL-2, it does play a critical role in the high-affinity receptor. In contrast to the constitutive nature of p75 expression on certain lymphoid cells, the *tac* gene is highly regulated. As noted above, IL-2 binds to and dissociates from the p55 peptide very rapidly, whereas it binds to and dissociates from the p75 chains relatively slowly. The association rate of the heterodimer is contributed by the fast-reacting p55 chain. Furthermore, the affinity of the p55, p75 heterodimer is 100-fold greater ($K_d = 10^{-11}$ M) than that of the p75 peptide alone ($K_d = 10^{-9}$ M). This increase in affinity is required for a cellular response to the exceedingly low levels of IL-2 present in vivo. The importance of IL-2 interaction with the Tac peptide is underscored by the effects on lymphocyte functions observed following addition of the anti-Tac monoclonal antibody. This antibody disrupts the interaction of IL-2 with the p55 peptide and thus disrupts IL-2 interaction with the high-affinity p55/p75 receptor. The addition of anti-Tac to cultures of human peripheral blood mononuclear cells inhibited a variety of immune reactions. Anti-Tac profoundly inhibited the proliferation of T lymphocytes stimulated by soluble antigens and by cell surface antigens (e.g. autologous and allogeneic mixed lymphocyte reactions) (16, 126). Upon activation, human T cells acquire other surface structures that, in large measure, are growth factor receptors that are not easily detectable during their resting stage (127–129). The addition of anti-Tac during initiation of cultures of T cells stimulated by mitogens, antigens, or the T3 antibody inhibited the expression of the late-appearing activation proteins examined, the insulin and transferrin receptors, and the Ia proteins (128, 129). Anti-Tac was also shown to inhibit a series of T-cell functions, including the generation of both cytotoxic and suppressor T lymphocytes in allogeneic cultures but did not inhibit their action once generated (16, 126, 130). Furthermore, anti-Tac inhibited B-lymphocyte immunoglobulin production activated by polyclonal activators (57, 126, 131). In functional studies, peripheral blood B cells from normal individuals activated with *S. aureus* Cowan I organisms could be induced to proliferate and synthesize immunoglobulin molecules by the addition of recombinant IL-2 (131). These responses were abolished when anti-Tac was added to the cultures. As noted above, in contrast to the action on T cells and B cells, anti-Tac did not inhibit the IL-2-induced activation of LGL into effective NK and LAK cells (90). Thus, the p75 peptide alone contributes to the initial triggering of LGL, whereas both the p55 and p75 peptides participate in the formation of the high-affinity receptor complex on activated T and B cells that is required for functional responses to the low concentrations of IL-2 and the normal regulation of humoral and cellular responses in vivo.

DISORDERS OF INTERLEUKIN-2 RECEPTOR EXPRESSION IN IMMUNODEFICIENCY, MALIGNANT, AND AUTOIMMUNE DISEASES

Immunodeficiency States Associated with Deficient Interleukin-2 Receptor Expression

Diminished production of IL-2, or the response to this lymphokine, has been described in association with a number of primary and acquired immunodeficiency diseases, as well as in drug-induced immunosuppression (132–140). Diminished IL-2 production and IL-2 receptor Tac expression following mitogen stimulation of lymphocytes in vitro has been observed in patients with gliomas and in individuals with lepromatous leprosy (133, 134). In addition, Trypanosoma cruzi inhibited IL-2 receptor expression and suppressed T-cell proliferation (135). Furthermore, the activated lymphocytes from patients with systemic lupus erythematosus expressed normal proportions of Tac-positive cells but manifested impaired expression of the high-affinity IL-2 receptor and reduced responsiveness to exogenous IL-2 (136). Although the production of IL-2 and its receptor may be diminished in some patients with primary and secondary immunodeficiency disorders, the T cells from most patients can be induced to express IL-2 disorders. However, in rare cases, IL-2 receptor expression cannot be induced in the circulating lymphocytes of patients with severe combined immunodeficiency disease (SCID). Doi and coworkers (137) showed that a patient with SCID and his father lacked a proliferative response to PHA. These individuals were defective in IL-2 production and IL-2 receptor expression (less than 1% of normal), whereas the production of other lymphokines (including B-cell differentiation factor and interferon-γ) were not impaired significantly. There were no abnormalities demonstrable by Southern blot analysis using DNA probes of IL-2 and IL-2 receptor, whereas transcription of DNA coding for the IL-2 receptor was lacking in the patient. The authors propose that the inherent defects of IL-2 production and IL-2 receptor expression in this family do not reside in the DNA coding for IL-2 and its receptor but are due to a transcriptional deficiency.

In certain cases, drugs that act to induce immunosuppression do so by inhibiting IL-2 production. For example, glucocorticoids such as dexamethasone prevent IL-2 production but do not prevent IL-2 receptor expression (56, 138). Similarly, it has been shown that cyclosporin A does not prevent IL-2 receptor expression but inhibits IL-2 production by inhibiting IL-2 gene expression at the level of mRNA transcription (140).

Increased Interleukin-2 Receptor Expression in Association with Lymphoid Neoplasia, Autoimmune Diseases, and Allograft Rejection

In contrast to the lack of Tac peptide expression in normal resting mononuclear cells, this receptor peptide is expressed by a proportion of the activated cells in certain forms of lymphoid neoplasia, select autoimmune diseases, and in individuals rejecting allografts (46–50, 141–149). A proportion of the abnormal cells in these diseases express the Tac antigen on their surface. Furthermore, the serum concentration of the soluble form of the Tac peptide is elevated. In terms of the neoplasias, certain T-cell, B-cell, monocytic, and even granulocytic leukemias express the Tac antigen (108–113, 141–149). Specifically, virtually all of the abnormal cells of patients with HTLV-I ATL express the Tac antigen (49, 145). Similarly, a proportion of patients with cutaneous T-cell lymphomas, including the Sézary syndrome and mycosis fungoides, express the Tac peptide (49, 142–144). Furthermore, the malignant B cells of virtually all patients with hairy cell leukemia and a proportion of patients with large and mixed cell diffuse lymphomas are Tac positive (104). The Tac antigen is also expressed on the Reed-Sternberg cells of patients with Hodgkin's disease and on the malignant cells of patients with true histiocytic lymphoma (138, 139, 141). Finally, a proportion of the leukemic cells of patients with chronic and acute myelogenous leukemia express the Tac antigen (110–113). In addition to these Tac-expressing leukemias and lymphomas, there are certain leukemias (e.g. acute lymphoblastic leukemia and LGL leukemia) that do not express the Tac peptide but express the p75 peptide of the IL-2 receptor (90).

Autoimmune diseases may also be associated with disorders of Tac-antigen expression. A proportion of the mononuclear cells in the involved tissues express the Tac antigen, and the serum concentration of the soluble form of the Tac peptide is elevated. Such evidence for T-cell activation and disorders of Tac-antigen expression are present in patients with rheumatoid arthritis, systemic lupus erythematosus, pulmonary sarcoidosis, subsets of patients with aplastic anemia, and individuals with HTLV-I-associated tropical spastic paraparesis (46, 50, 146–149). Disorders of IL-2-receptor expression have also been demonstrated in animal models of these diseases, including adjuvant arthritis of rats, the rodent models of type I diabetes (NOD mouse and

BB rat), experimental allergic encephalomyelitis (EAE) of mice, and in rodent models of systemic lupus erythematosus (47, 49). Modest elevations of the soluble IL-2 receptor have been reported in patients with AIDS. The Tac peptide is also expressed by the activated lymphocytes in recipients of renal, hepatic, and cardiac allografts that are reacting to the foreign histocompatibility antigens expressed on the donor organs (151, 152).

Disorders of Interleukin-2 Receptor Expression in Adult T-Cell Leukemia

A distinct form of mature T-cell leukemia was defined by Takasuki and coworkers (153) and termed adult T-cell leukemia (ATL). ATL is a malignant proliferation of mature T cells that have a tendency to infiltrate the skin. Cases of ATL are associated with hypercalcemia and an immunodeficiency state that usually has a very aggressive course. The ATL cases are clustered within families and geographically, occurring where HTLV-I is endemic in the southwest of Japan, the Caribbean basin, and sub-Saharan Africa. HTLV-I has been shown to be a primary etiologic agent in ATL (154). The leukemic cells are usually of the CD4-positive phenotype yet functionally do not manifest T helper activity but function as suppressor-effector cells and inhibited the immunoglobulin synthesis on cocultured pokeweed mitogen-stimulated mononuclear target populations (49). The leukemic cells we and others have examined from patients with HTLV-I-associated ATL expressed high- and low-affinity IL-2 receptors, including the Tac peptide (49, 145). The expression of the p55 IL-2 receptor peptide on ATL cells differs from that of normal T cells (40, 145). First, unlike normal T cells, ATL cells do not require prior activation to express IL-2 receptors. Furthermore, using a ^{3}H-labeled anti-Tac receptor assay, HTLV-I-infected leukemic T-cell lines characteristically expressed 5–10-fold more receptors per cell (270,000–1,000,000) than did maximally PHA-stimulated T lymphoblasts (30,000–60,000). In addition, whereas normal human T lymphocytes maintained in culture with IL-2 demonstrate a rapid decline in receptor number, adult ATL lines do not show a similar decline (145). Tsudo and coworkers (155) have reported abnormal regulation of the Tac antigen on ATL cells. Furthermore, the IL-2 receptor on the ATL line MT-1 is spontaneously phosphorylated, whereas the IL-2 receptor on PHA-stimulated normal cells is IL-2 dependently phosphorylated. Although in vitro HTLV-I-immortalized lines from patients with ATL express the Tac peptide in all cases, they release detectable IL-2 and transcribe mRNA for this lymphokine in only a minority of cases (156). In those cases where the malignant cells do not produce IL-2, the IL-2/IL-2 receptor-mediated autocrine model is not possible. However, there is another subset of patients in which the available evidence supports an IL-2-mediated autocrine model for the proliferation of ATL cells (157).

Furthermore, IL-2-mediated autocrine or paracrine cell proliferation may be involved in the early phases of leukemogenesis. Even in later phases, it is conceivable that the constant presence of high numbers of IL-2 receptors on ATL cells may play a role in the pathogenesis of the uncontrolled growth of these malignant T cells.

As noted above, T-cell leukemias caused by HTLV-I, as well as all T-cell and B-cell lines infected with HTLV-I, universally express large numbers of IL-2 Tac receptor peptides (46). An analysis of this virus and its protein products suggests a potential mechanism for this association between HTLV-I and constitutive IL-2 receptor expression. The complete sequence of HTLV-I has been determined by Seiki and colleagues (158). In addition to the presence of typical LTRs, *gag, pol,* and *env* genes, retroviral gene sequences common to other groups of retroviruses, HTLV-I and -II were shown to contain an additional genomic region between *env* and the 3' LTR referred to as pX that encodes at least three peptides of 21, 27, and 40–42 kd. Sodroski and colleagues (159) demonstrated that one of these, a 42-kd protein they termed the Tat (transactivator of transcription) protein (now termed Tax), is essential for viral replication. The mRNA for this protein is produced by a double splicing event. The Tax protein requires the presence of three 21-bp enhancer-like repeats within the LTR of HTLV-I to stimulate transcription (160, 161).

In studies involving the transfection of cDNA encoding the *tax* product of human HTLV-I into Jurkat T cells, it has been shown that the Tax protein plays a central role in increasing the transcription of host genes governing human IL-2 and IL-2 p55 receptor peptide expression (94, 98, 162–165). In these transient expression studies, the Tax protein stimulated an increase in IL-2 receptor promoter activity. In contrast, Tax alone had little effect on IL-2 promoter activity in Jurkat T cells but markedly synergized with other mitogenic stimuli (PHA or phorbol myristate acetate), which alone were ineffective. A direct interaction of the Tax protein with specific DNA sequences seems unlikely, since the promoters of IL-2 and IL-2 receptor genes do not share strong sequence homologies with the 21-bp enhancer-like regions of the LTR of HTLV-I that have been shown to be involved in Tax-induced *trans*-activation. Ruben and coworkers (98) demonstrated that the 12-base sequence motif in the −255 to −267 5' region to the IL-2 receptor *tac* gene are required for activation by the *tax* gene. This is the site of action of the nuclear factor NFκB. The authors suggest that the activation of the IL-2 receptor *tac* gene expression by HTLV-I Tax protein occurs through an interaction with, or activation of, a host transcription factor with properties similar, if not identical, to those of NFκB. Fujii and coworkers (165b) reported that the protooncogene *fos* is transactivated by *tax* in a variety of cell types. They have identified at least two regions in the *cis*-acting elements that mediate transactivation of the c-fos promoter by *tax*, including one that is similar to the 21-base-pair element in HTLV-I required for transactivation by *tax*. Another element is a consensus sequence C-C-(A+T)rich-G-G that is a protein binding site common to transcription regulatory regions of *fos* and the IL-2 receptor genes. These workers suggest the scenario that *tax* activates c-fos which activates NFκB or AP-1 to regulate transcription of IL-2 and the IL-2 receptor. As noted above, an uncontrolled autocrine T-cell growth model has been proposed for the early events of HTLV-I-induced T-cell transformation. Thus, it seems possible that the Tax protein plays an important role in HTLV-I-induced malignancy by deregulating the expression of the cellular genes encoding IL-2 and the IL-2 receptor, which are involved in the normal control of T-cell proliferation.

THE INTERLEUKIN-2 RECEPTOR AS A TARGET FOR THERAPY IN PATIENTS WITH TAC-EXPRESSING LEUKEMIA, PATIENTS WITH AUTOIMMUNE DISORDERS, AND INDIVIDUALS RECEIVING ORGAN ALLOGRAFTS

The observation that T cells in patients with certain lymphoid malignancies, select autoimmune diseases, and individuals rejecting allografts express IL-2 receptors identified by the anti-Tac monoclonal antibody, whereas normal resting cells and their precursors do not, provides the scientific basis for therapeutic trials using agents to eliminate the IL-2 receptor-expressing cells (10, 40, 47, 48, 51, 166). Such agents could theoretically eliminate Tac-expressing leukemic cells or activated T cells involved in other disease states while retaining the Tac-negative mature normal T cells and their precursors that express the full repertoire of antigen receptors for T-cell immune responses. The agents that we have used include (*a*) unmodified anti-Tac monoclonal; (*b*) toxin conjugates of anti-Tac (e.g. A chain of ricin toxin, *Pseudomonas* exotoxin [PE] and truncated PE [PE40]); (*c*) IL-2 truncated toxin fusion proteins (e.g. IL-2 PE40); (*d*) α- and β-emitting isotopic (e.g. ^{212}Bi and ^{90}Y) chelates of anti-Tac; and (*e*) hybrid "humanized" anti-Tac with mouse light and heavy chain variable or hypervariable regions joined to the human constant κ light chain and IgG$_1$ or IgG$_3$ heavy chain regions.

We have initiated a clinical trial to evaluate the efficacy of intravenously administered anti-Tac monoclonal antibody in the treatment of patients with ATL (51, 166). None of the nine patients treated suffered any untoward reactions, and only one, a patient with anti-Tac-induced clinical remission, produced antibodies to the mouse immunoglobulin or to the idiotype of the anti-Tac monoclonal. Three of the patients had a temporary mixed, partial, or complete remission following anti-Tac therapy (51).

These therapeutic studies have been extended by examining the efficacy of toxins coupled to anti-Tac to selectively inhibit protein synthesis and viability of Tac-positive ATL lines (167, 168). The addition of anti-Tac antibody coupled to PE inhibited protein synthesis by Tac-expressing HUT-102-B2 cells, but not that by the acute T-cell leukemia line MOLT 4, which does not express the Tac antigen (168). However, the initial PE-anti-Tac conjugates were hepatotoxic when administered to patients with ATL. Functional analysis of deletion mutants of the PE structural gene has shown that domain I of the PE molecule is responsible for ubiquitous cell recognition, whereas the remainder of the molecule is actually responsible for cell death (169). A PE molecule from which domain I has been deleted (PE40) was conjugated to anti-Tac. These anti-Tac-PE40 conjugates inhibited protein synthesis of Tac-expressing lines but not that of Tac-nonexpressing lines. Furthermore, when administered to mice, these conjugates were approximately 75-fold less hepatotoxic than conjugates with unmodified PE.

IL-2-PE40, a chimeric protein composed of human IL-2 genetically fused to the NH_2-terminal of PE40, was constructed to provide an alternative (lymphokine-mediated) method of delivering PE40 to the surface of IL-2 receptor-positive cells. The addition of this recombinant IL-2-PE40 led to the inhibition of protein synthesis when added to human cell lines expressing either the p55, p75, or both IL-2 receptor subunits, but did not inhibit the protein synthesis or curtail the viability of cell lines that did not express any IL-2 receptor peptide (120).

The action of toxin conjugates of monoclonal antibodies depends on their ability to be internalized by the cell and released into the cytoplasm. Anti-Tac bound to IL-2 receptors on leukemic cells is internalized slowly. Furthermore, the toxin conjugates do not pass easily from the endosome to the cytosol, as required for their action on elongation factor 2. To circumvent these limitations, alternative cytotoxic agents were developed that could be conjugated to anti-Tac and were effective when bound to the surface of Tac-expressing cells. In one case, we showed that bismuth-212 (^{212}Bi), an α-emitting radionuclide conjugated to anti-Tac by use of a bifunctional chelate, was well suited for this role (170). Activity levels of 0.5 μCi targeted by ^{212}Bi-labeled anti-Tac eliminated greater than 98% of the proliferative capacity of the HUT-102 cells, with only a modest effect on IL-2 receptor-negative lines. This specific cytotoxicity was blocked by excess unlabeled anti-Tac but not by IgG. In parallel studies, the β-emitting yttrium-90 (^{90}Y) was bound to anti-Tac using chelates that did not permit elution of the radiolabeled yttrium from the monoclonal antibody (171). As noted below, Rhesus monkeys receiving a xenograft of a cynomolgus heart showed a marked prolongation of xenograft survival following the administration of ^{90}Y-labeled anti-Tac. Thus, ^{212}Bi-labeled anti-Tac and ^{90}Y-labeled anti-Tac are potentially effective and specific immunocytotoxic agents for the elimination of Tac-expressing cells.

In addition to their use in the therapy of patients with ATL, these IL-2 receptor antibodies are being evaluated for their ability to eliminate IL-2 receptor-expressing T cells in other clinical states, including certain autoimmune disorders. In these disorders, the lymphocytes infiltrating the affected organs express the Tac antigen, and the soluble form of the IL-2 receptor in the serum is elevated. Murine anti-IL-2 receptor antibodies were shown to suppress murine diabetic insulitis, lupus nephritis, experimental allergic encephalomyelitis, and adjuvant arthritis (47, 48). Weckerle & Diamanstein (172) demonstrated that the administration of ART-18, an antibody to the 55-kd IL-2 receptor peptide, protects Lewis rats from lethal experimental autoimmune encephalomyelitis induced by the injection of myelin basic protein-specific permanent T-cell lines. Furthermore, a similar treatment with this antibody protected rats from polyarthritis caused by the passive transfer of lymphocytes from mycobacterium adjuvant-sensitized rats. In contrast to the inhibition by anti-IL-2 receptor monoclonal antibody therapy of the development of autoimmune disease caused by the passive transfer of antigen-specific T cells, this anti-IL-2 receptor monoclonal antibody approach failed to effect actively induced diseases. In parallel studies, Kelley and coworkers (173) used an anti-IL-2 receptor (M7/20) monoclonal antibody that recognizes the 55-kd subunit of the murine IL-2 receptor in two models of autoimmunity, the nonobese diabetic mouse and the (NZB × NZW)F$_1$ hybrid with lupus. Treatment with the anti-IL-2 receptor monoclonal antibody reduced the autoimmune insulitis characteristic of the nonobese diabetic (NOD) mouse and protected (NZB × NZW)F$_1$ hybrid mice from the development of renal injury in this model of systemic lupus erythematosus. With these encouraging results in animal models, IL-2 receptor-directed therapy is being initiated in patients with autoimmune disorders.

Monoclonal antibodies that recognize the IL-2 receptor have been used to inhibit the graft-versus-host reaction and organ allograft rejection. Antibodies to the p55 IL-2 receptor (Tac) were shown to inhibit the proliferation of T cells to foreign histocompatibility antigens expressed on the donor organ and to prevent the generation of cytotoxic T cells in allogeneic cell cocultures (48, 174). Volk and coworkers (174) demonstrated that the acute graft-versus-host reaction across a strong MHC barrier in mice can be suppressed by AMT-13, a monoclonal antibody directed against the 55-kd IL-2 receptor on activated mouse lymphocytes. Furthermore, in studies by Kirkman and coworkers (175), the survival of cardiac allografts was prolonged in some cases to indefinite survival in rodent recipients treated with an anti-IL-2 receptor monoclonal antibody. In parallel studies, the administration of anti-Tac for the initial 10 days after transplantation prolonged the survival of renal allografts in cynomolgus monkeys (176). Furthermore, Bacha and coworkers (177) have achieved prolongation of allograft survival and suppression of delayed type hypersensitivity with genetically engineered diphtheria toxin linked to the NH_2-terminus of human IL-2. However, unmodified anti-Tac did not lead to a prolongation of graft survival in heterotopic cardiac xenografts in which Rhesus monkeys received cardiac xenografts from cynomolgus donors. In contrast, animals receiving ^{90}Y-labeled anti-Tac showed a prolongation of graft survival from a value in untreated animals of 6–8 days to a mean graft survival of 40 days in animals receiving a dose of radioactivity that had acceptable toxicity (171). In light of these encouraging results, human recipients of cadaver donor renal allografts are receiving different anti-IL-2 receptor monoclonal antibodies as adjunctive immunotherapy (47, 178, 179). Antibody treatment has been well tolerated, and 50 of 53 recipients treated retain a functioning allograft. Thus, the development of monoclonal antibodies and toxin-lymphokine conjugates, directed toward the IL-2 receptors expressed on leukemia and lymphoma cells, on autoreactive T cells of certain patients with autoimmune disorders, and on host T cells responding to foreign histocompatibility antigens of organ allografts, is leading to the development of rational, novel therapeutic approaches for these clinical conditions.

CONCLUSIONS

Antigen- or mitogen-stimulated activation of resting T cells induces the synthesis of IL-2, as well as the expression of high-affinity cell surface receptors for this lymphokine. Failure of the production of either IL-2 or its receptor results in a failure of the T-cell immune response. There are two forms of cellular receptors for IL-2, one with a very high affinity and the other with a lower affinity. At least two IL-2-binding peptides, a 55-kd peptide reactive with the anti-Tac monoclonal antibody and a 70–75-kd novel non-Tac peptide, have been identified. We propose a multisubunit model for the high-affinity IL-2 receptor in which both the p55 Tac and p75 peptides are associated in a receptor complex. The two chains form a hybrid receptor complex that exploits the rapid association rate contributed by the p55 chain and the slow dissociation rate provided by the p75 chain. Recently, evidence has been developed for a more complex subunit structure that involves a 95-kd peptide (p95) associated with the p55 and p75 peptides in a multisubunit high-affinity IL-2 receptor complex. Investigative efforts in the future will certainly focus on the molecular cloning of the genes encoding the p75 and p95 peptides. Such studies might help to elucidate the mechanism of growth factor signal transduction through the membrane and the biochemical pathways activated by IL-2 that ultimately lead to DNA replication, cell proliferation, and differentiation. Furthermore, IL-2 receptor-directed therapy will be expanded to take advantage of the observation that T cells in patients with certain lymphoid malignancies or select autoimmune diseases, and individuals rejecting allografts, express IL-2 receptors identified by the anti-

Tac monoclonal antibody, whereas normal resting cells and their precursors do not. Thus, the development of monoclonal antibodies and toxin-lymphokine conjugates directed toward IL-2 receptors is leading to the development of rational, novel therapeutic approaches for these benign and malignant clinical conditions.

Literature Cited

1. Hedrick, S. M., Neilsen, E. A., Kavaler, J., Cohen, D. I., Davis, M. M. 1984. *Nature* 308:153–58
2. Yanagi, Y., Yoshikai, Y., Leggett, K., Clark, S. P., Aleksander, I., et al. 1984. *Nature* 308:145–49
3. Saito, H., Kranz, D. M., Takagaki, Y., Hayday, A. C., Eisen, H. N., et al. 1984. *Nature* 312:36–40
4. Murre, C., Waldmann, R. A., Morton, C. C., Waldmann, T. A., Bongiovanni, K. F., et al. 1985. *Nature* 316:549–52
5. Sim, G. K., Yagüe, J., Nelson, J., Marrack, P., Palmer, E., et al. 1984. *Nature* 312:771–75
6. Chien, Y. H., Iwashima, M., Kaplan, K. B., Elliott, J. F., Davis, M. M. 1987. *Nature* 327:677–82
7. Waldmann, T.A. 1987. *Adv. Immunol.* 40:247–321
8. Morgan, D. A., Ruscetti, F. W., Gallo, R. C. 1976. *Science* 193:1007–8
9. Smith, K. A. 1980. *Immunol. Rev.* 51:337–57
10. Waldmann, T. A. 1986. *Science* 232:727–32
11. Robb, R. J., Munck, A., Smith, K. A. 1981. *J. Exp. Med.* 154:1455–74
12. Smith, K. A. 1988. *Science* 240:1169–76
13. Greene, W. C., Leonard, W. J. 1986. *Annu. Rev. Immunol.* 4:69–95
14. Taniguchi, T., Matsui, H., Fujita, T., Takaoka, C., Kashima, N., et al. 1983. *Nature* 302:305–10
15. Uchiyama, T., Broder, S., Waldmann, T. A. 1981. *J. Immunol.* 126:1393–97
16. Uchiyama, T., Nelson, D. L., Fleischer, T. A., Waldmann, T. A. 1981. *J. Immunol.* 126:1398–403
17. Leonard, W. J., Depper, J. M., Uchiyama, T., Smith, K. A., Waldmann, T. A., et al. 1982. *Nature* 300:267–69
18. Leonard, W. J., Depper, J. M., Robb, R. J., Waldmann, T. A., Greene, W. C. 1983. *Proc. Natl. Acad. Sci. USA* 80:6957–61
19. Miyawaki, T., Yachie, A., Uwadana, N., Ohzeki, S., Nagaoki, T., et al. 1982. *J. Immunol.* 129:2474–78
20. Robb, R. J., Greene, W. C. 1983. *J. Exp. Med.* 158:1332–37
21. Leonard, W. J., Depper, J. M., Waldmann, T. A., Greene, W. C. 1984. *Receptors and Recognition,* ed. M. Greaves, 17:45–66. London: Chapman & Hall
22. Shackelford, D. A., Trowbridge, I. S. 1984. *J. Biol. Chem.* 259:11706–12
23. Gaulton, G. N., Eardley, D. D. 1986. *J. Immunol.* 136:2470–77
24. Leonard, W. J., Depper, J. M., Crabtree, G. R., Rudikoff, S., Pumphrey, J., et al. 1984. *Nature* 311:626–31
25. Cosman, D., Cerretti, D. P., Larsen, A., Park, L., March, C., et al. 1984. *Nature* 312:768–71
26. Nikaido, T., Shimizu, A., Ishida, N., Sabe, H., Teshigawara, K., et al. 1984. *Nature* 311:631–35
27. Depper, J. M., Leonard, W. J., Drogula, C., Krönke, M. J., Waldmann, T. A., et al. 1985. *J. Cell Biochem.* 27:267–76
28. Depper, J. M., Leonard, W. J., Krönke, M., Noguchi, P. D., Cunningham, R. E., et al. 1984. *J. Immunol.* 133:3054–61
29. Robb, R. J., Greene, W. C., Rusk, C. M. 1984. *J. Exp. Med.* 160:1126–46
30. Hatakeyama, M., Minamoto, S., Uchiyama, T., Hardy, R. R., Yamada, G., et al. 1985. *Nature* 318:467–70
31. Sabe, H., Kondo, S., Shimizu, A., Tagaya, Y., Yodoi, J., et al. 1984. *Mol. Biol. Med.* 2:379–96
32. Greene, W. C., Robb, R. J., Svetlik, P. B., Rusk, C. M., Depper, J. M., et al. 1985. *J. Exp. Med.* 162:363–68
33. Robb, R. J. 1986. *Proc. Natl. Acad. Sci. USA* 83:3992–96
34. Kondo, S., Shimizu, A., Saito, Y., Kinoshita, M., Honjo, T. 1986. *Proc. Natl. Acad. Sci. USA* 83:9026–29
35. Ortaldo, J. R., Mason, A. T., Gerard, J. P., Henderson, L. E., Farrar, W., et al. 1984. *J. Immunol.* 133:779–83
36. Grimm, E. A., Rosenberg, S. A. 1984. *Lymphokine,* pp. 279–311. New York: Academic
37. Waldmann, T. A., Kozak, R. W., Tsu-

do, M., Oh-ishi, T., Bongiovanni, K.
F., Goldman, C. K. 1986. *Progress in
Immunology VI,* ed. B. Cinidar, R. G.
Miller, pp. 553–62. Orlando: Academic
38. Tsudo, M., Kozak, R. W., Goldman, C.
K., Waldmann, T. A. 1986. *Proc. Natl.
Acad. Sci. USA* 83:9694–98
39. Tsudo, M., Kozak, R. W., Goldman, C.
K., Waldmann, T. A. 1987. *Proc. Natl.
Acad. Sci. USA* 84:4215–18
40. Waldmann, T. A. 1988. *Harvey Lect.
Ser.* 82:1–17
41. Sharon, M., Klausner, R. D., Cullen,
B. R., Chizzonite, R., Leonard, W. J.
1986. *Science* 234:859–63
42. Sharon, M., Siegel, J. P., Tosato, G.,
Yodoi, J., Gerrard, T. L., et al. 1988. *J.
Exp. Med.* 167:1265–70
43. Kuo, L., Robb, R. J. 1986. *J. Immunol.*
137:1538–43
44. Ju, G., Collins, L., Kaffka, K. L.,
Tsien, W.-H., Chizzonite, R., et al.
1987. *J. Biol. Chem.* 262:5723–31
45. Collins, L., Tsien, W.-H., Seals, C.,
Hakimi, J., Weber, D., et al. 1988.
Proc. Natl. Acad. Sci. USA 85:7709
46. Nelson, D. L., Rubin, L. A., Kurman,
C. C., Fritz, M. E., Boutin, B. 1986. *J.
Clin. Immunol.* 6:114–20
47. Williams, J. M., Kelley, V. E., Kirkman, R. L., Tilney, N. L., Shapiro, M.
E., et al. 1988. *Immunol. Invest.*
16:687–723
48. Diamantstein, T., Osawa, H. 1986. *Immunol. Rev.* 92:5–27
49. Waldmann, T. A., Greene, W. C.,
Sarin, P. S., Saxinger, C., Blayney, D.
W., et al. 1984. *J. Clin. Invest.*
73:1711–18
50. Waldmann, T. A. 1988. *J. Autoimmun.*
In press
51. Waldmann, T. A., Goldman, C. K.,
Bongiovanni, K. F., Sharrow, S. O.,
Davey, M. P., et al. 1988. *Blood*
72:1805–16
52. Leonard, W. J., Depper, J. M., Krönke,
M., Peffer, N. J., Svetlik, P. B., et al.
1985. *Science* 230:633–39
53. Miller, J., Malek, T. R., Leonard, W.
J., Greene, W. C., Shevach, E. M., et
al. 1985. *J. Immunol.* 134:4212–17
54. Kristensen, T., Tack, B. F. 1986. *Proc.
Natl. Acad. Sci. USA* 83:3963–67
55. Waldmann, T. A., Goldman, C. K.,
Leonard, W. J., Depper, J. M., Robb,
R. J., et al. 1984. *Curr. Top. Microbiol.
Immunol.* 133:96–101
56. Reem, G., Yeh, N.-H. 1984. *Science*
255:429–30
57. Waldmann, T. A., Goldman, C. K.,
Robb, R. J., Depper, J. M., Leonard,
W. J., et al. 1984. *J. Exp. Med.*
160:1450–66
58. Depper, J. M., Leonard, W. J., Drogula, C., Krönke, M., Waldmann, T. A.,
et al. 1985. *Proc. Natl. Acad. Sci. USA*
82:4230–34
59. Teshigawara, K., Wang, H.-M., Kato,
K., Smith, K. A. 1987. *J. Exp. Med.*
165:223–38
60. Dukovich, M., Wano, Y., Thuy, L. T.
B., Katz, P., Cullens, B. R., et al.
1987. *Nature* 327:518–21
61. Robb, R. J., Ruck, C. M., Yodoi, J.,
Greene, W. C. 1987. *Proc. Natl. Acad.
Sci. USA* 84:2002–11
62. Lowenthal, J. W., Greene, W. C. 1987.
J. Exp. Med. 166:1156–61
63. Wang, H.-M., Smith, K. A. 1988. *J.
Exp. Med.* 166:1055–69
64. Smith, K. A. 1988. *Adv. Immunol.*
42:165–97
65. Szöllösi, J., Damjanovich, S., Goldman, C. K., Fulwyler, M., Aszalos, A.
A., et al. 1987. *Proc. Natl. Acad. Sci.
USA* 84:7246–51
66. Edidin, M., Aszalos, A., Damjanovich,
S., Waldmann, T. A. 1988. *J. Immunol.*
141:1206–10
67. Herrmann, F., Diamantstein, T. 1987.
Immunobiology 175:145–58
68. Saragovi, H., Malek, T. R. 1987. *J.
Immunol.* 139:1918–26
69. Liang, S.-M., Thatcher, D. R., Liang,
C.-M., Allet, B. 1986. *J. Biol. Chem.*
261:334–37
70. Brandhuber, B. J., Boone, T., Kenney,
W. C., McKay, D. B. 1987. *Science*
238:1707–9
71. Cantrell, D. A., Smith, K. A. 1984.
Science 224:1312–16
72. Weiss, A., Imboden, J., Shoback, D.,
Stobo, J. 1984. *Proc. Natl. Acad. Sci.
USA* 81:4169–73
73. Farrar, W. L., Ruscetti, F. W. 1986. *J.
Immunol.* 136:1266–73
74. Meuer, S. C., Hogdon, J. C., Hussey,
R. E., Protentis, J. P., Schlossman, S.
F., et al. 1983. *J. Exp. Med.* 158:988–
93
75. Hara, T., Fu, S. M., Hansen, J. A.
1985. *J. Exp. Med.* 161:1513–24
76. Krönke, M., Leonard, W. J., Depper, J.
M., Greene, W. C. 1985. *Science*
228:1215–17
77. Mills, G. B., Girard, P., Grinstein, S.,
Gelfand, E. W. 1988. *Cell* 55:91–
100
78. Valge, V. E., Wong, J. G. P., Datlof,
B. M., Sinskey, A. J., Rao, A. 1988.
Cell 55:101–12
79. Kaye, J., Gillis, S., Mizel, S. B., Shevach, E. M., Malek, T. R., et al. 1984.
J. Immunol. 133:1339–45
80. Wakasugi, N., Rimsky, L., Mahe, Y.,
Kamel, A. M., Fradelize, D., et al.

1987. *Proc. Natl. Acad. Sci. USA* 84: 804–8

81. Brown, M., Hu-Li, J., Paul, W. E. 1988. *J. Immunol.* 141:504–11
82. Hu-Li, J., Shevach, E. M., Mizuguchi, J., Ohara, J., Mosmann, T., et al. 1987. *J. Exp. Med.* 165:157–72
83. Takatsu, K., Kikuchi, Y., Takahashi, T., Honjo, T., Matsumoto, M., et al. 1987. *Proc. Natl. Acad. Sci. USA* 84:4234–38
84. Baroja, M. I., Couppens, J. L., Van Danna, J., Billiau, A. 1988. *J. Immunol.* 141:1502–7
85. Okada, M., Maeda, M., Tagaya, Y., Taniguchi, Y., Teshigawara, K., et al. 1985. *J. Immunol.* 135:3995–4003
86. Yodoi, J., Tagaya, Y., Shendo, T., Sugie, K., Nakamura, Y. 1988. *Gann Monogr. Cancer Res.* 34:177–89
87. Welte, K., Andreeff, M., Platzer, E., Holloway, K., Rubin, B. Y., et al. 1984. *J. Exp. Med.* 160:1390–403
88. Lotze, M. T., Custer, M. C., Sharrow, S. O., Rubin, L. A., Nelson, D. L., et al. 1987. *Cancer Res.* 47:2188–95
89. Smith, K. A., Cantrell, D. A. 1985. *Proc. Natl. Acad. Sci. USA* 82:864–68
90. Tsudo, M., Goldman, C. K., Bongiovanni, K. F., Chan, W. C., Winton, E. F., et al. 1987. *Proc. Natl. Acad. Sci. USA* 84:5394–98
91. Thuy, L. T. B., Dukovich, M., Peffer, N. J., Fauci, A. S., Kehrl, J. H., et al. 1987. *J. Immunol.* 139:1550–55
92. Cross, S. L., Feinberg, M. B., Wolf, J. B., Holbrook, N. J., Wong-Staal, F., et al. 1987. *Cell* 49:47–56
93. Lowenthal, J. W., Böhnlein, E., Ballard, D. W., Greene, W. C. 1988. *Proc. Natl. Acad. Sci. USA* 85:4468–72
94. Suzuki, N., Matsunami, N., Kanamori, H., Ishida, N., Shimizu, A., et al. 1987. *J. Biol. Chem.* 262:5079–86
95. Leung, K., Nabel, G. J. 1988. *Nature* 333:776–78
96. Böhnlein, E., Lowenthal, J. W., Siekevitz, M., Ballard, D. W., Franza, B. R., et al. 1988. *Cell* 53:827–36
97. Nabel, G., Baltimore, D. 1987. *Nature* 326:711–13
98. Ruben, S., Poteat, H., Tan, T.-H., Kawakami, K., Roeder, R., et al. 1988. *Science* 241:89–92
99. Patarca, R., Schwartz, J., Singh, R. P., Kong, Q.-T., Murphy, E., et al. 1988. *Proc. Natl. Acad. Sci. USA* 85:2733–37
100. Malek, T. R., Robb, R. J., Shevach, E. M. 1983. *Proc. Natl. Acad. Sci. USA* 80:5694–98
101. Osawa, H., Diamantstein, T. 1983. *J. Immunol.* 130:51–55
102. Gaulton, G. N., Bangs, J., Maddock,

S., Springer, T., Eardley, D. D., et al. 1985. *Clin. Immunol. Immunopathol.* 36:18–29
103. Raulet, D. H. 1985. *Nature* 314:101–3
104. Korsmeyer, S. J., Greene, W. C., Cossman, J., Hsu, S. M., Jensen, J. P., et al. 1983. *Proc. Natl. Acad. Sci. USA* 80:4522–26
105. Maraguchi, A., Kehrl, J. H., Longo, D. L., Volkman, D. L., Smith, K. A., et al. 1985. *J. Exp. Med.* 161:181–97
106. Tsudo, M., Uchiyama, T., Uchino, H. 1984. *J. Exp. Med.* 160:612–17
107. Jung, L. K. I., Hara, T., Fu, S. M. 1984. *J. Exp. Med.* 160:1597–602
108. Herrmann, F., Cannistra, S. A., Levine, H., Griffin, J. D. 1985. *J. Exp. Med.* 162:1111–16
109. Holter, W., Goldman, C. K., Casabo, L., Nelson, D. L., Greene, W. C., et al. 1987. *J. Immunol.* 138:2917–22
110. Armitage, R. J., Lai, A. P., Roberts, P. J., Cawley, J. C. 1986. *Br. J. Haematol.* 64:799–807
111. Visani, G., Delwel, R., Touw, I., Bot, F., Lowenberg, B. 1987. *Blood* 69:1182–87
112. Yamamoto, S., Hattori, T., Matsuoka, M., Ishii, T., Asou, N., et al. 1986. *Blood* 67:1714–20
113. Colamonici, O. R., Rosolen, A., Cole, D., Kirsch, I., Felix, C., et al. 1988. *J. Immunol.* 141:1202–5
114. Rubin, L. A., Kurman, C. C., Biddison, W. E., Goldman, N. D., Nelson, D. L. 1985. *Hybridoma* 4:91–102
115. Greene, W. C. 1987. *Clin. Res.* 35:439–50
116. Siegel, J. P., Sharon, M., Smith, P. L., Leonard, W. J. 1987. *Science* 238:75–78
117. Weissman, A. M., Harford, J. B., Svetlik, P. B., Leonard, W. L., Depper, J. M., et al. 1986. *Proc. Natl. Acad. Sci. USA* 83:1463–66
118. Fujii, M., Sugamura, K., Sano, K., Nakai, M., Sugita, K., et al. 1986. *J. Exp. Med.* 163:550–62
119. Robb, R. J., Greene, W. C. 1987. *J. Exp. Med.* 165:1201–6
120. Lorberboum-Galski, H., Kozak, R., Waldmann, T., Bailon, P., FitzGerald, D., et al. 1988. *J. Biol. Chem.* 263:18650–56
121. Colamonici, O. R., Quinones, R., Rosolen, A., Trepel, J. B., Sausville, E., et al. 1988. *Blood* 71:825–28
122. Smith, K. A. 1982. *Immunobiology* 161:157–73
123. Inamoto, T., Nakamura, Y., Sugie, K., Masutani, H., Shindo, T., et al. 1989. *Proc. Natl. Acad. Sci. USA.* In press

124. Mills, G. B., May, C. 1987. *J. Immunol.* 139:4083–87
125. Ishii, T., Takeshita, T., Numata, N., Sugamura, K. 1988. *J. Immunol.* 141: 174–79
126. Depper, J. M., Leonard, W. J., Waldmann, T. A., Greene, W. C. 1983. *J. Immunol.* 131:690–96
127. Cotner, T., Williams, J. M., Christenson, L., Shapiro, H. M., Strom, T. B., et al. 1983. *J. Exp. Med.* 157:461–72
128. Tsudo, M. T., Uchiyama, K., Takatsuki, K., Uchino, H., Yodoi, J. 1982. *J. Immunol.* 129:592–95
129. Neckers, L. M., Cossman, J. 1983. *Proc. Natl. Acad. Sci. USA* 80:3494–98
130. Oh-ishi, T., Goldman, C. K., Misiti, J., Waldmann, T. A. 1988. *Proc. Natl. Acad. Sci. USA* 85:6478–82
131. Mingari, M. C., Gerosa, F., Carra, G., Accolla, R. S., Moretta, A., et al. 1984. *Nature* 312:641–43
132. Welte, K., Mertelsmann, R. 1985. *Cancer Invest.* 3:35–49
133. Elliott, L., Brooks, W., Roszman, T. 1987. *J. Natl. Cancer Inst.* 78:919–22
134. Mohagheghpour, N., Gelber, R. H., Larrick, J. W., Sasaki, D. T., Brennan, P. J., et al. 1985. *J. Immunol.* 135:1443–49
135. Beltz, L. A., Sztein, M. B., Kierszenbaum, F. 1988. *J. Immunol.* 141:289–94
136. Ishida, H., Kumagai, S., Umehara, H., Sano, H., Tagaya, Y., et al. 1987. *J. Immunol.* 139:1070–74
137. Doi, S., Saiki, O., Tamaka, T., Hakawa, K., Igarashi, T., et al. 1988. *Clin. Immunol. Immunopathol.* 46:24–36
138. Larsson, E. L. 1980. *J. Immunol.* 124: 2828–33
139. Hess, A. D., Tutschka, P. J., Santos, G. W. 1982. *J. Immunol.* 128:355–59
140. Krönke, M., Leonard, W. J., Depper, J. M., Arya, S. K., Wong-Staal, F., et al. 1984. *Proc. Natl. Acad. Sci. USA* 81:5214–18
141. Lawrence, E. C., Brousseau, K. P., Berger, M. B., Kurman, C. C., Marcon, L., et al. 1988. *Annu. Rev. Respir. Dis.* 137:759–64
142. Ralfkiaer, E., Wantzin, G. L., Stein, H., Thomsen, K., Mason, D. Y. 1987. *J. Am. Acad. Dermatol.* 15:628–37
143. Sheibani, K., Winberg, C. D., Van de Velde, S., Blayney, D. W., Rappaport, H. 1987. *Am. J. Pathol.* 127:27–37
144. Schwarting, R., Gerdes, J., Stein, H. 1985. *J. Clin. Pathol.* 38:1196–97
145. Uchiyama, T., Hori, T., Tsudo, M., Wano, Y., Umadome, H., et al. 1985. *J. Clin. Invest.* 76:446–53
146. Keystone, E. L., Snow, K. M., Bom-

bardier, C., Chang, C.-H., Nelson, D. L., et al. 1988. *Arthritis Rheum.* 31: 844–49
147. Marcon, L., Greenberg, S. J., Nelson, D. L., Jacobson, S., McFarlin, D. E., et al. 1988. *Clin. Res.* 34:443A
148. Zoumbos, N. C., Gascon, P., Djeu, J. Y., Trost, S. R., Young, N. S. 1985. *New Engl. J. Med.* 312:257–65
149. Pizzolo, G. 1988. *Immunol. Clin.* 7:13–21
150. Kloster, B. E., John, P. A., Miller, L. E., Rubin, L. A., Nelson, D. L., et al. 1987. *Clin. Immunol. Immunopathol.* 45:4440–61
151. Colvin, R. B., Fuller, T. C., Mackeen, L., Kung, P. C., Ip, S. H., et al. 1987. *Clin. Immunol. Immunopathol.* 43: 1273–76
152. Lawrence, E. C., Brousseau, K. P., Kurman, C. C., Nelson, D. L., Young, J. B., et al. 1986. *Chest* 89(Suppl.): S526
153. Takatsuki, K., Uchiyama, T., Sagawa, K., Yodoi, J. 1977. *Topics in Hematology*, ed. S. Seno, F. Takau, S. Irino, pp. 73–77. Amsterdam: Excerpta Medica
154. Poiesz, B. J., Ruscetti, F. W., Gazdar, A. F., Bunn, P. A., Minna, J. D., et al. 1980. *Proc. Natl. Acad. Sci. USA* 77: 7415–19
155. Tsudo, M. T., Uchiyama, T., Uchino, H., Yodoi, J. 1983. *Blood* 61:1014–16
156. Arya, S. K., Wong-Staal, F., Gallo, R. C. 1984. *Science* 223:1086–87
157. Arima, N., Daitoku, Y., Ohgaki, S., Fukumori, J., Tanaka, H., et al. 1986. *Blood* 68:779–82
158. Seiki, M., Hattori, S., Hirayama, Y., Yoshida, M. 1983. *Proc. Natl. Acad. Sci. USA* 80:3618–22
159. Sodroski, J. G., Rosen, C. A., Haseltine, W. A. 1984. *Science* 225:381–85
160. Paskalis, H., Felber, B. K., Pavlakis, G. N. 1986. *Proc. Natl. Acad. Sci. USA* 83:6558–62
161. Rosen, C. A., Sodroski, J. G., Haseltine, W. A. 1985. *Proc. Natl. Acad. Sci. USA* 82:6502–6
162. Shimotohno, K., Miwa, M., Slamon, D. J., Chen, I. S. Y., Hoshino, H. O., et al. 1985. *Proc. Natl. Acad. Sci. USA* 82:302–6
163. Inoue, J., Seiki, M., Taniguchi, T., Tsuru, S., Yoshida, M. 1986. *EMBO J.* 5:2883–88
164. Maruyama, M., Shibuya, H., Harada, H., Hatakeyama, M., Seiki, M., et al. 1987. *Cell* 48:343–50
165. Siekevitz, M., Feinberg, M. B., Holbrook, N., Wong-Staal, F., Greene, W. C. 1987. *Proc. Natl. Acad. Sci. USA* 84:5389–93

165b. Fujii, M., Sassone-Corsi, P., Verma, I. M. 1988. *Proc. Natl. Acad. Sci. USA* 85:8526–30
166. Waldmann, T. A., Longo, D. L., Leonard, W. J., Depper, J. M., Thompson, C. B., et al. 1985. *Cancer Res.* 45:4559s–62s
167. Krönke, M., Depper, J. M., Leonard, W. J., Vitetta, E. S., Waldmann, T. A., et al. 1985. *Blood* 65:1416–21
168. FitzGerald, D., Waldmann, T. A., Willingham, M. C., Pastan, I. 1984. *J. Clin. Invest.* 74:966–71
169. Hwang, J., FitzGerald, D. T. J., Adya, S., Pastan, I. 1987. *Cell* 48:129–36
170. Kozak, R. W., Atcher, R. W., Gansow, O. A., Friedman, A. M., Hines, J. J., et al. 1986. *Proc. Natl. Acad. Sci. USA* 83:474–78
171. Cooper, M. M., Robbins, R. C., Gansow, O. A., Clark, R. E., Waldmann, T. A. 1988. *Surg. Forum* 39:353–55
172. Weckerle, H., Diamantstein, T. 1986. *Ann. NY Acad. Sci.* 475:401–3

173. Kelley, V. E., Gaulton, G. N., Hattori, M., Ikegami, H., Eisenbarth, G., et al. 1988. *J. Immunol.* 140:59–61
174. Volk, H. D., Brocke, S., Osawa, H., Diamantstein, T. 1986. *Clin. Exp. Immunol.* 66:126–31
175. Kirkman, R. L., Barrett, L. V., Gaulton, G. N., Kelley, V. E., Ythier, A., et al. 1985. *J. Exp. Med.* 162:358–62
176. Shapiro, M. E., Kirkman, R. L., Reed, M. H., Puskas, J. D., Mazoujian, G., et al. 1987. *Transplant. Proc.* 19:594–98
177. Bacha, P., Williams, D. P., Waters, C., Williams, J. M., Murphy, J. R., et al. 1988. *J. Exp. Med.* 167:612–22
178. Soulillou, J. P., Peyronnel, P., LeMauff, B., Hourmant, M., Olive, D., et al. 1987. *Lancet* 1:1339–42
179. Shapiro, M. E., Kirkman, R. L., Carpenter, C. B., Milford, E. L., Tilney, N. L., et al. 1988. *Transplantation.* In press

Annu. Rev. Biochem. 1989. 58:913–49

DYNAMIC, STRUCTURAL, AND REGULATORY ASPECTS OF λ SITE-SPECIFIC RECOMBINATION

Arthur Landy

Division of Biology and Medicine, Brown University, Providence, Rhode Island 02912

CONTENTS

913

0066-4154/89/0701-0913$02.00

PERSPECTIVES AND SUMMARY

On the 25th anniversary of Robin Holliday's model for single-strand exchange intermediates in homologous recombination (1), it is especially fitting to review the most recent developments in the site-specific recombination system of *Escherichia coli* bacteriophage λ. This article focuses on developments since the most recent reviews of λ recombination (2, 3), and has a different emphasis than a concurrent review (4) that explores the regulatory aspects of λ recombination in more detail. The present state of the field has its roots in the wealth of elegant λ genetics that rallied around Allan Campbell's model for viral integration (5). The purification of essential proteins by Howard Nash and his collaborators (6–8), and the definition of recombination target sites (9–11), allowed this system to be the first recombination pathway eligible for rigorous biochemical analysis. New results to be discussed in this review concern the dynamic, structural, and regulatory aspects of λ recombination. They are summarized below.

The dynamic aspects of λ recombination (such as mechanisms of strand exchange) prominently feature the biochemical analysis of Holliday junction recombination intermediates. Artificial DNA constructs designed to model putative Holliday-junction recombination intermediates are efficiently and specifically resolved by purified λ Integrase into normal recombinant products (12). Holliday intermediates can also be trapped as the major product in a λ recombination reaction using "suicide recombination substrates" (13). The pair of reciprocal strand exchanges that first form and then resolve the Holliday junctions proceed in a strictly prescribed order that is the same for integrative and excisive recombination (13, 14). This order is determined by DNA sequences and protein binding sites distant from the region of strand exchange (13, 14). Formation and resolution of the Holliday junction proceeds via a high-energy intermediate containing a covalent linkage between the DNA and Tyr-342 of Integrase protein (15). The requirement for a full 7 bp of DNA homology between recombining partners (16, 17) appears not to be for synapsis but rather for the purpose of branch migration between the staggered strand exchange sites within the Holliday intermediate (13, 18) (B. DeMassy, R. A. Weisberg, personal communication). The synapsis of *att* sites is most likely mediated by protein-protein and protein-DNA interactions, and in integrative recombination this may involve a protein-decorated supercoiled "donor" (*att*P) pairing with a protein-free "recipient" (*att*B) (19).

The structural aspects of λ recombination are becoming interesting as models for other complex protein–nucleic acid systems, such as those found in regulation of transcription and DNA replication. The *att* site DNAs function as part of higher-order protein-DNA complexes called "intasomes" (20, 21). These are formed by a set of cooperative and competitive protein-protein interactions involving four proteins and 15 binding sites. Formation of one of these intasomes (*att*P) requires supercoiled DNA (22). The intasome structure is shaped largely by three sequence-specific "accessory" proteins (IHF, Xis, and FIS) that induce extremely sharp bends in DNA (23). The Integrase protein has the potential to tether distant sequences and form DNA loops by virtue of two autonomous DNA-binding domains with different recognition specificities (24). Integrase executes recombination (7) by a mechanism of strand cleavage and ligation that is shared with more than 15 related recombinases (Int Family) (15, 25).

The extreme directionality and regulation of λ recombination are among its most striking features. Directionality is the consequence of two distinctly different pathways for integrative and excisive recombination. Each pathway uses a unique, but overlapping, set of the 15 protein binding sites that comprise *att* site DNAs (26, 27). Cooperative and competitive interactions involving all four recombination proteins determine the direction of recombination. They also bestow upon the reactions sensitivity to host and viral physiology and to environmental conditions. Regulation of recombination at the level of gene expression is augmented by a mechanism-based response in which host-encoded proteins are incorporated as integral elements of the recombination reaction. One of the host-encoded proteins (IHF) that is required for both integrative and excisive recombination has one binding site that must be occupied for integration and must be vacant for excision (28). The other host-encoded protein (FIS) dramatically stimulates excisive recombination when levels of the phage-encoded Xis protein are low (29). Intracellular levels of FIS, which is also an accessory protein for a different class of recombination reactions, drop by 70-fold as cells go from exponential to stationary phase (29).

In addition to the recent developments outlined above, reference will also be made to FLP and Cre, two well-characterized members of the Int family. Although they are similar to Int in the basic mechanisms of strand exchange, they are quite different in terms of structural complexity, directionality, and regulation.

BACKGROUND

Conservative site-specific recombination is distinguished from homologous recombination and transposition by a high degree of specificity for both partners; the strand exchange mechanism involves the cleavage and rejoining of specific DNA sequences in the absence of DNA synthesis. Two major families comprise this class of reactions. The Resolvase-Invertase family is distinguished by the use of a conserved serine to form covalent intermediates with DNA, a constraint to intramolecular reactions, and an orientation preference for recombination sites that can be either direct repeats (e.g. $\gamma\delta$ and Tn-3 Resolvase) or inverted repeats (e.g. the Hin, Gin, and Cin inversion systems).

The other family of reactions (Int Family) (15, 25) is distinguished by the use of a conserved tyrosine to form covalent intermediates with DNA, the ability to execute both inter and intra molecular reactions, and the capacity to recombine sites with either direct or inverted orientation. Members of the Int family can be further subdivided into the minimal systems (e.g. FLP of yeast 2μ circle and Cre of bacteriophage P1) and the complex systems, such as bacteriophage λ and related viruses.

The λ recombination system consists of four recombination sites (*att* sites) and four proteins that carry out the integrative and excisive recombination reactions as shown in Figure 1. The Integrase (Int) protein cuts and reseals DNA to carry out strand exchange. The other phage-encoded protein (Xis) is required only for excisive recombination. Of the two host-encoded proteins, IHF is required for both reactions, while FIS enhances only excisive recombination and only under special circumstances. Integrative recombination between specific sites on the viral (*att*P) and bacterial (*att*B) chromosomes generates prophage bounded by the recombinant products *att*L and *att*R. An excisive recombination between *att*L and *att*R regenerates *att*B and *att*P in the bacterial and excised viral chromosomes. Of the four *att* sites, *att*P (POP') is the most complex and *att*B (BOB') is the simplest (see Figures 1 and 2 for the coordinates and protein binding sites). The prophage sites *att*L (BOP') and *att*R (POB') are "hybrids" of *att*P and *att*B (see Figure 1).

The reader is referred to other closely related, or overlapping reviews on regulation of λ recombination (4), conservative site-specific recombination pathways (30, 31), Resolvases (32, 33), Invertases (34), enhancers (35), resolution of Holliday junctions (36, 37), FLP (38), topology (39–41), and to many related chapters in the books *Transposition* (42), *Genetic Recombination* (43), and *Mobile DNA* (44).

PROTEINS AND THEIR GENES

Int

Int is a basic protein of 40,330 M_r (45, 46) that has type I topoisomerase activity (6, 7, 15, 47, 48) and binds specifically to two different families of DNA sequences (49, 50). During recombination, it carries out the cutting and resealing of *att* site DNA via a covalent Int-DNA intermediate in the absense of any high-energy cofactors (8, 51, 52). Int can also transfer covalently bound DNA to the 5' OH of another DNA molecule in a sequence-independent reaction (S. E. Nunes-Düby, A. Landy, unpublished results). Int is encoded in the λ genome adjacent to the phage *att* site, reflecting the

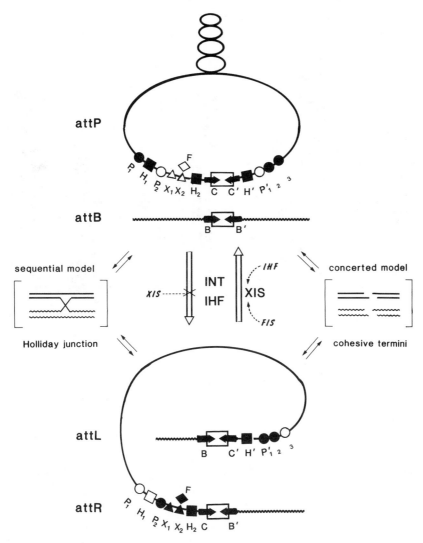

Figure 1 Integrative and excisive recombination pathways. The protein binding sites for arm-type Int (○), core-type Int (➤), IHF (☐), Xis (△), and FIS (◇) are indicated by filled symbols when that site is occupied by its cognate protein to make a competent recombination partner for integrative (⇓) or excisive (⇑) recombination. Proteins required for each reaction (Int, IHF, and Xis) are in bold, proteins that inhibit (Xis and IHF) or enhance (FIS) the indicated reactions are in italics. Of the two models for strand exchange ([ ]), the sequential model (*left*) is correct.

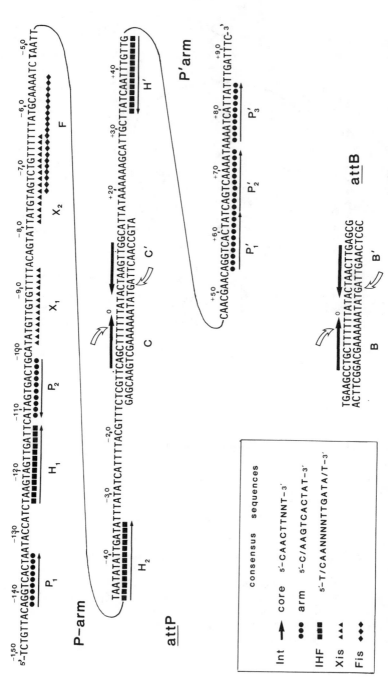

Figure 2 DNA sequence, coordinates, and protein binding sites for *att*P and *att*B. Recognition sequences for each protein are denoted with the symbols indicated lower left. Relative orientation of the binding sites is shown (——). Curved arrows (⤸) indicate the staggered sites of strand exchange and the boundaries of the 7 bp overlap regions.

common motif of recombinase genes that are very close to their sites of action. The specific arrangement of recombination genes often reflects their respective strategies for regulation of gene expression, which, in the case of λ *int,* involves retroregulation (53–57), its own cII-dependent promoter (58), and transcription from the distant P_L promoter [reviewed in (4)]. In phages P22 and P2, the *int* genes are oriented as in λ, but in ø80 and P4 they have the opposite orientation (59–61). The P2 *int* gene, which was once hypothesized to be split by the *att* site (62), has recently been shown to terminate its coding region just before *att* (61).

Thus far only one classical *int* mutant (resulting in a Glu to Lys change at position 174) (63) has been well characterized biochemically. The mutation was isolated in two independent experiments: (*a*) by selection for λ bacteriophage that could undergo site-specific recombination in an *E. coli* host mutant for IHF (called *int*-h) (64); (*b*) by selection for an Int partially independent of Xis in excisive recombination (called *xin*) (C. Gritzmacher, L. W. Enquist, R. A. Weisberg, unpublished data). In the absence of IHF, where Int^+ activity is depressed approximately 500-fold, Int-h activity is only depressed approximately 10-fold (52). The fact that this reduced, but significant, Int-h activity is identical for supercoiled and relaxed substrates indicates that the normal supercoiling requirement for *att*P is determined largely by IHF. The two most plausible explanations for this phenotype are that Int-h/ Xin may have a higher affinity than Int^+ for core-type binding sites; or that it works differently on the synaptic intermediate, either to stabilize it or accelerate its conversion to recombinant product (65). A number of other *int* mutants that are interesting because they are recombination proficient (66) or dominant (63) have been studied but not thoroughly characterized.

A comparative study of the primary structures of proteins from seven different bacteriophages pointed to an "Integrase family" of recombination proteins that now includes more than 15 members (15, 25). Within this family of proteins there is one region of approximately 40 amino acids near the carboxy-terminus where at 25 positions more than 50% of the residues belong to the same amino acid exchange group (67). Particularly striking are three perfectly conserved residues within this region: His-308, Arg-311, and Tyr-342 (using the numbering of λ Int).

The biochemical significance of this region was determined using a family of suicide recombination substrates designed to accumulate transient recombination intermediates (13) (see below, SYNAPSIS AND STRAND EXCHANGE, and Figure 3). It was shown that Tyr-342 of Int forms a covalent bond with DNA at the sites of strand exchange. A mutant Int in which Tyr-342 is changed to phenylalanine is devoid of both topoisomerase and recombinase activity but still binds to both classes of Int DNA-binding sites with an affinity comparable to wild-type Int (15). The applicability of these

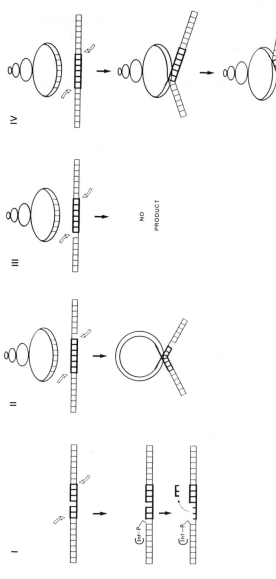

Figure 3 Suicide and canonical recombination substrates, and their respective Int-promoted products. I: medial nicked substrate when incubated with Int yields covalent Int-DNA product. II: bottom strand exchange site nick, after reaction with supercoiled *att*P and exchange of the top strands, yields a Holliday junction intermediate in the form of an α-structure with a nick. III: top strand exchange site nick yields no product when reacted with supercoiled *att*P. IV: in recombination between canonical *att* sites a left side (*top*) strand exchange leads to a Holliday junction; homology in the overlap region is necessary for the second strand exchange and would allow branch migration of the crossed strands; recombination is completed by a right side (*bottom*) strand exchange.

results to the Int family as a whole is supported by studies on FLP and Cre (68–70). In contrast, the Resolvase-Invertase family of recombinases forms a covalent linkage with DNA via a Ser(p) (71) located very close to the amino-terminus (35, 71–73) (see also BACKGROUND).

The recombinases within the Int family are found in a rather wide range of biological functions, such as phage integration and excision, fimbrial phase variation, and plasmid maintenance in *E. coli* and yeast. Some members rely on accessory proteins while others are self-sufficient. Some members, like Cre and FLP, are recombinases with a single DNA recognition site, while others, like λ Int, have two different DNA-binding domains. This intrafamily heterogeneity is reflected by the fact that only a single highly conserved 40-amino-acid region is common to all members of the Int family. It has been proposed that the unifying feature of the Int family is the protein domain for the DNA cutting-ligating activity; the identification of Tyr-342 as the active nucleophile in Int lends strong support to this view (15).

The fact that Int recognizes two classes of DNA-binding sites (49, 50) (see below and Figure 2), along with the biochemical differences in binding to each class (51, 74), led to the proposal for a new kind of DNA-binding protein (50). The existence of two functionally distinct DNA recognition domains in the Integrase protein was demonstrated by the fact that core-type and arm-type DNA sequences do not compete with each other for binding in nuclease protection experiments (24). These results rule out the proposal (75) that all of the Int binding sites are recognized by Int on the basis of similar structural DNA features. They also indicate that an Int monomer is capable of binding to both classes of sites simultaneously. The "bivalency" of Int provides a simple structural explanation for the ability of some *int* mutants to negatively complement (63, 76–79).

The identity and structural autonomy of the two DNA-binding domains was established by proteolytic cleavage of Int and footprinting analysis of the resulting two major peptides (24). A chymotryptic cleavage between Leu-64 and Thr-65 generates a 32-kd carboxy-terminal fragment that binds core-type sites and a 7-kd amino-terminal fragment that binds arm-type sites. The amino-terminal fragment that binds exclusively to arm-type sites is not required for catalytic function, while the carboxy-terminal peptide retains topoisomerase function and resolves synthetic *att* site Holliday junctions (24). This is consistent with earlier observations that Int can efficiently resolve Holliday structures, even in the absence of arm-type binding sites (12).

Xis

Xis is a small (M_r of 8630), basic, phage-encoded protein that is required for excision but not integration (45, 46, 80). Xis has no significant homologies with other proteins, including analogous excisionases from other lambdoid phage, and nothing is known about its structure. While Xis is relatively thermostable in vitro, it is rapidly degraded intracellularly by an unknown protease, so that 80% of the activity decays within five minutes (80–92). Int, on the other hand, is quite stable intracellularly.

In the presence of Int and IHF, Xis promotes efficient recombination between attL and attR. With greater than 100 mM salt, Xis is required for this reaction, while at low salt it is dispensable, although still stimulatory (83). Several lines of evidence indicate that the low-salt Xis-independent reaction proceeds by a different pathway than the Xis-stimulated reaction (84). A similar difference in pathways has been proposed for integrative recombination at high and low salt with supercoiled versus relaxed attP (85).

Xis has been found to confer enhanced thermostability on Int protein in vitro (80), but there is no evidence to indicate whether this reflects an interaction relevant to recombination. No catalytic activity has been associated with Xis, and its function is exerted by sequence-specific cooperative binding to two adjacent sites in the P arm (26, 86). As discussed below, binding at these sites introduces a very sharp bend in the DNA and is also associated with cooperative interactions with DNA-bound Int and FIS (23, 29, 87, 88).

IHF

Integration Host Factor (IHF), as the name implies, was discovered as a cellular function essential for λ site-specific recombination, both integration and excision (89–92). IHF is a heterodimer composed of two subunits, with predicted molecular weights of 11,200 and 10,580 for the α and β subunits respectively (8, 93–95). Each of the subunits is very similar in sequence to the type II DNA-binding proteins, a family that includes the major histone-like proteins of E. coli (HU) and other bacteria [reviewed in (96)]. The only known (primary) function of IHF is its ability to bind (97–99) and to bend (23, 100–103) DNA at specific sites.

It may be valid to infer some basic features of IHF structure from a comparison with the HU protein from Bacillus stearothermophilus (HBs), whose crystal structure has been solved to 3 Å resolution (104). Both of the IHF subunits show significant homology to the subunit of the HBs homodimer (94). The HBs structure has two intertwined subunits that form a hydrophobic wedge-shape body and two long flexible arms extending from the wide portion of the wedge. It is proposed that the two flexible arms form a right-handed helix that could bind within the major or minor groove of double-stranded B DNA.

Craig & Nash (97) have proposed that IHF binds in the minor groove of DNA based on the patterns of protection against modification by dimethylsulfate (DMS). H. A. Nash (personal communication) has found additional support for this proposal in the patterns of protection against cleavage of the sugar-phosphate backbone by hydroxyl radicals. In these experiments, which were done on the three IHF sites of *att* DNA, as well as in nuclease protection experiments on other IHF sites (98, 99), IHF is found to protect three to four helical turns of DNA. Despite this relatively large protected region, preliminary stoichiometric data indicate that only a single IHF dimer binds per site (H. A. Nash, personal communication). It is also interesting to note that while IHF has high sequence specificity, the closely related HU and HBs proteins have little or no sequence specificity.

The genes for IHF have been mapped (89, 90), cloned (93–95, 105), and overexpressed (106). The α-encoding gene, *him*A, maps to minute 37, and the β-encoding gene, which is referred to as both *hip* and *him*D, maps to minute 20 on the *E. coli* chromosome. *Hip/him*D is part of a gene cluster with at least two promoters (E. L. Flamm, R. A. Weisberg, personal communication). The *him*A gene is part of a more complex cluster whose transcription involves at least seven promoters and three terminators, with the regulation being influenced by growth rate and the SOS response. For a recent review of the regulation see (4).

Some evidence suggests that levels of IHF may vary as a function of cellular physiology, but this point is not clear. A large increase in IHF levels is detected in stationary-phase cells, but this depends on the conditions used to extract the protein (27, 28). Attempts to resolve this question by in vivo chemical modification have not been definitive because all three of the IHF sites in *att*P that were monitored were fully occupied under the growth conditions tested (29).

The lambdoid phages ø80, P22, and λ all have IHF-binding sites within their very different *att* site DNA sequences (98). ø80, like λ, is unable to lysogenize *him*A or *hip/him*D mutants (107), while P22 has not been tested. Since the number of IHF-binding sites is different in each of the phage *att* sites, it will be interesting to compare the three recombination reactions at a mechanistic level. (Cre and FLP are two members of the Int family that do not require IHF.) Transposons are another class of recombination elements where a role for IHF has been demonstrated or implicated (99, 101, 108, 109). In some recombination systems an IHF requirement or HU requirement can be replaced with low efficiency by the other protein (108, 110). The roles of IHF extend well beyond recombination and are discussed thoroughly in a recent review (111).

FIS

A second host-encoded protein (in addition to IHF) involved in the λ site-specific recombination reaction was first discovered by testing *E. coli* extracts for proteins that would bind to *att* site DNA in a gel mobility shift assay. Such an activity was found and subsequently shown to be FIS (29), a protein that had been identified (112–114), and purified (115, 116), on the basis of its ability to stimulate three recombination inversion reactions from the Resolvase-Invertase family. In the λ system, FIS stimulates in vitro excision approximately 20-fold when Xis levels are limiting, however it cannot replace Xis (29). FIS has no effect when Xis is at saturating levels and it does not influence integrative recombination. The in vivo relevance of FIS has been established by demonstrating occupancy of its binding site in the P arm (29), as well as dramatic differences in excision efficiency for FIS$^+$ and FIS$^-$ cells (R. Johnson, personal communication). When bound to its single binding site in the P arm, FIS induces a DNA bend of approximately 90° (23).

FIS, like IHF, is a small, basic, heat-stable DNA-binding protein (115, 116). Unlike IHF, which is a heterodimer, FIS is a homodimer in solution. The protomer, based on its gene sequence, has 98 amino acids and a calculated pI of 9.5 (117, 118). There is no sequence homology between FIS and the type II DNA-binding proteins such as HU or IHF, despite their gross similarity in physical properties and somewhat analogous roles as bending and accessory proteins in several different recombination pathways. FIS has no cysteines or histidines so that a "zinc-finger" mode of DNA recognition is unlikely (117). However, within the carboxy-terminal portion, at amino acids 74–93, there is a sequence with a high probability of adopting the helix-turn-helix DNA binding motif (117, 118) that has been characterized for repressors and other proteins (119).

The Hin system of *Salmonella typhimurium,* the Gin system of bacteriophage Mu, and the Cin system of bacteriophage P1 constitute a closely related family of FIS-dependent site-specific recombination systems. These systems specifically invert a segment of DNA and thereby mediate expression of two alternate forms of a flagellar antigen or phage tail fiber protein (34, 35, 120). Their requirement for the bacterial protein FIS is mediated by a binding site on the DNA that has been called a "recombinational enhancer." The function of recombinational enhancers, which contain two or three FIS binding sites, is relatively independent of position or orientation relative to the sites of recombination (112–114, 118, 121, 122).

The *fis* gene has been mapped to approximately 72.5 min on the *E. coli* chromosome, between *fab*E and *aro*E (117, 118). The fact that the initiating methionine is preceded by a poor Shine-Delgarno translation initiation site, coupled with the high number of rare codons (11%), may contribute to the low abundance of FIS in *E. coli*. A null mutation in *fis* leads to a 10^3–10^4-fold reduction of Hin and Gin inversion in vivo and a complete loss of detectable FIS activity when extracts are assayed for ability to stimulate inversion in vitro (117, 118). These null mutants also exhibit a 10^3–10^4-fold reduction in λ excisive recombination efficiency, suggesting that under these experimental conditions, intracellular levels of Xis are limiting (R. Johnson, personal communication). Strains carrying these same null mutations show no reduction in viability, even when this mutation is combined with a mutation in *hip/himD* (encoding one of the subunits of IHF). The inability to find major effects of *fis* mutations on cell physiology, or on the expression of other cellular genes, may reflect the short time these mutants have been available for study.

One reason for predicting some interesting effects of FIS on cell physiology, and a possible role in global aspects of regulation, is that its intracellular concentration drops more than 70-fold as *E. coli* goes from exponential growth to stationary phase (29). It is not known what controls the degradation of FIS or whether gene expression is also involved.

ATT SITE STRUCTURE AND PROTEIN BINDING SITES

The most prominent feature of the four *att* site DNA sequences is a 15 bp "core sequence" they have in common. Individual sequence elements of functional significance lie within, outside, and across the boundaries of this common core, which itself has no functional significance. Controlled resection experiments established that *att*P extends from -150 to $+85$ from the center of the core, while *att*B extends from approximately -15 to $+15$ (see Figure 2) (10, 11, 123). The prophage sites generated by integrative recombination, *att*L and *att*R, are fully competent for excisive recombination, but they contain more than the minimal sequences necessary for excision (27) (see below, DIRECTIONALITY AND REGULATION). It is interesting to note that the *att*B sites of six different systems are located within the 3'-terminal regions of various tRNA genes (60, 98) (W. Reiter, P. Palm, S. Yeats, personal communication). While this association with tRNA genes does not hold for the majority of *att*B sites, it is significant and may offer some useful insights about evolution and/or mechanisms.

The *att*P sequence is 73% A+T, with one 48 bp region being 90% A+T (9, 10). As might be expected from this base composition, there are several runs of up to six consecutive adenines, a sequence feature known to generate intrinsic curvature within the DNA (124–126). The center of *att*P curvature has been mapped to -45 to -50 in the P arm (127). This same region is also hypersensitive to modification by bromoacetaldehyde, but only if the DNA is supercoiled (128). While the intrinsic curvature in this region of the P arm may enhance the efficiency of recombination, it is not absolutely required (127). Its function may be to position the *att* site to an external, easily accessible loop in plectonemically supercoiled DNA (129, 130).

In vitro integrative recombination reactions traced the observed requirement for negative supercoiling to the *att*P partner (131). Below 50 mM NaCl the reaction does not require supercoiled *att*P (132); however, several lines of evidence suggest that this low-salt reaction with linear *att*P proceeds by a different recombination pathway (85). In contrast to integration, excision does not require supercoiling (133), although supercoiling of either *att*L or *att*R stimulates the reaction at limiting protein concentrations. Excision is also less sensitive than integration to salt concentration (132).

Overlap Region

The 7 bp between the staggered cuts responsible for strand exchange is referred to as the "overlap region" (3), since in recombinant molecules this region receives one DNA strand from each partner. The *att* site mutants of Shulman & Gottesman (134) that segregate to both recombinant products have been located within the overlap region by DNA sequencing (135). These mutants greatly impair recombination efficiency by deleting one bp from the overlap region. Insertions of one bp are not as detrimental (136), but a systematic test of spacing has not been carried out. Because this region receives one DNA strand from each partner during recombination, one would predict that it must have the same sequence in both partners. Indeed, many sequence changes can be accommodated within the overlap region as long as the same changes are introduced into both partners (17, 81). The role of DNA:DNA homology will be discussed below (SYNAPSIS AND STRAND EXCHANGE).

An overlap region, defined both by the position of staggered cuts and by the requirement for DNA:DNA homology, is also found in the *lox* sites (Cre system) and FRT sites (FLP system), 6 bp in the former (137–141) and 8 bp in the latter (142–145). In both of these systems some specific sequences within the overlap region have been observed to reduce recombination despite the preservation of homology between the partners. The weaker effect of specific overlap region sequences in λ *att* sites may be due to the dominating effects of the proteins bound to the P and P' arms. Similarly, in the absence of the asymmetry conferred by the P and P' arms of λ, the orientation of *lox* and FRT recombination partners relative to one another is determined wholly by the overlap region sequence (139, 143, 145, 146). The Resolvase-Invertase systems also make use of staggered cuts during strand exchange, but the size of the overlap region is only 2 bp (71, 73, 112, 147, 148).

Int Binding Sites

There are seven Int-binding sites in *att*P, two of which define the outermost limits of *att*P (10, 49). Sequence inspection suggested that these sites could be grouped into two distinct families of sequences: "core-type" (also called "junction-type") sites that are adjacent to the points of strand exchange and "arm-type" sites that are distal to the region of strand exchange. This distinction was confirmed, and reliable consensus sequences were obtained for each family, by analyzing a number of fortuitous Int binding sites in non-*att* DNA (49, 50). As described above, each family is recognized by one of two autonomous DNA-binding domains of Int (24).

CORE-TYPE SITES   The core-type sites (C and C' in *att*P and B and B' in *att*B) are found as inverted repeats at the two ends of the overlap region and include the phosphodiester bonds that are cut by Int during strand exchange. The two base pairs flanking this phosphodiester bond are not specified in the consensus recognition sequence of seven bp (50). This lack of specificity may be a consequence of protein structural features in the catalytic region of Int. Int bound at the core-type sites protects DNA against attack by dimethyl sulfate in both the major and minor grooves along one face of the DNA helix; the inverted repeats face each other across the central major groove. The opposite face of the DNA helix appears to be devoid of strong protein contacts (50). The appropriate configuration of core-type sites is the only requirement for the resolution of synthetic Holliday junctions by Int, or its carboxy-terminal domain (12) (C. Pargellis, B. Franz, and A. Landy, unpublished results).

Although affinity constants for Int binding have not been measured, they are considerably lower for the core-type sites than for the arm-type sites, as estimated by their respective behavior in nuclease protection (19, 49) and gel mobility shift (23) experiments. The relative order of binding affinity for core-type sites is C' > C = B > B' (50). However, the differential in affinities does not appear to be essential, since different sites can be substituted for one another with little or no effect on recombination (13, 26). The differences in affinities of the individual sites may be partially masked by the significant amount of cooperativity between Int proteins bound at two core-type sites (50).

The amount of Int-induced DNA bending at the COC' core-type sites has been estimated to be approximately 17° (23). However, this value must be viewed cautiously because of the relatively low affinity of Int for these sites (50). Additionally, measurements on the isolated core-type sites may not be relevant to the functional recombinogenic complexes (see below, COOPERA-TIVE INTERACTIONS).

Two core-type sites flanking the overlap region comprise *att*B. This configuration is very similar to the minimal recombination sites of both partners in the Cre and FLP systems. The simplicity of *att*B explains the existence of secondary *att* sites in the *E. coli* chromosome. These sites are utilized for λ integration (at greatly reduced efficiencies) when *att*B is deleted (149–151). Sequence comparisons of the secondary *att* sites indicates that they are degenerate facsimiles of the simple *att*B site. The efficiency of a particular secondary *att* site is presumably determined by the extent to which it emulates the canonical spacing and sequence of the overlap region and core-type Int binding sites (3, 50). Similarly, fortuitous facsimiles of *att*B, or core-type sites, could comprise the targets for Int topoisomerase activity on supercoiled DNA not containing an *att* site.

ARM-TYPE SITES   The initial characterization of a distinct class of essential Int-binding sites well removed from the region of strand exchange was both intriguing and challenging in terms of recombinogenic structures and mechanisms (49, 74). Evidence for how they function in site-specific recombination is beginning to emerge. The five arm-type Int-binding sites have a consensus recognition sequence of seven contiguous base pairs (49). The approximate binding affinities for individual sites indicates a hierarchy of P'1 = P1 > P2 (49, 87, 88). P'2 and P'3 have not been examined in isolation and all three of the P' sites are subject to cooperative binding effects amongst themselves (L. Moitoso de Vargas, A. Landy, unpublished results). In addition to these cooperative effects, sequences outside of the binding site can also influence the binding affinities. [However, the low affinity of Int for P2, which has the best match to the consensus sequence, is not due to context effects (87).]

Classical genetic techniques have produced only one protein-binding mutant in the *att* sites. This mutation, *hen,* is in the P'3 site and changes a completely conserved C to T (152). It is especially interesting because it abolishes integrative recombination without affecting excisive recombination (152). To examine the roles of other arm-type sites, the single base transition of the *hen* mutation has been introduced into each of the five arm-type sites (153). Multi-site changes have also been introduced into a subset of the arm-type sites (27, 88). From these studies and resection experiments (10, 11), it is clear that the P1 and P'2, and P'3 sites are required for integrative recombination, while the P2 site is not. The apparent dispensability of P'1 for integrative recombination is subject to the same reservation that applies to all point mutants without a phenotype in highly cooperative systems.

In excisive recombination, none of the point mutations has as large an effect as seen with integrative recombination (153). The largest effect is observed with the P'1 mutation, suggesting that this site is required for excision, with some of the defect being compensated by cooperative interactions with other sites. P1 is clearly not required for excision, because when it is deleted the *att*R works as well or better than wild-type *att*R (27). While P2 is also dispensable under some circumstances, it does appear to be important for excision. In the absence of the P2 site, more Xis is required to obtain the same level of recombination as with a wild-type *att*R (27, 87). Additionally, the reaction between an *att*R-*hen* P2 and an *att*L-*hen* P'2 or *att*L-*hen* P'3 is reduced relative to a wild-type *att*L (153). Both of these observations are reflective of cooperative interactions to be discussed below. See Figure 1 for a summary of the site requirements for integrative and excisive recombination.

Xis Sites

Xis protein protects a single region of approximately 40 bp in the P arm (86). Gel mobility shift assays have shown that highly cooperative binding between at least two Xis molecules takes place (26). Mutagenesis of the binding region showed that each half could bind Xis independently, although with lower affinity than the canonical site (26). Protection patterns against DMS modification also provide evidence for two binding sites, denoted X1 and X2, and further suggest that they are arranged as a direct repeat (86). Determining a consensus recognition sequence will require the analysis of fortuitous Xis recognition sites and/or a mutational study of the canonical sites.

The Xis mutations isolated thus far include a single base deletion between the two sites and a series of resected *att* sites in which varying extents of the P arm have been replaced with heterologous DNA. In the first case, excisive recombination is abolished. In the second, it was shown that both X1 and X2 must be occupied; defects in X1 are only overcome when sufficient Xis is present to bind to the mutated as well as the normal site (27).

Xis binding to X1 or X2 bends the DNA approximately 45° and 95°, respectively. When X1 and X2 are both occupied, as is normally the case, the bending angle is greater than 140° (23). This very sharp bend is undoubtedly a key feature in the role of Xis during recombination.

IHF Sites

There are three IHF sites in *att*P: H1 and H2 in the P arm and H' in the P' arm (97). IHF binding to multiple sites on the same DNA fragment is not cooperative (28, 88, 97, 101, 154). Derivation and refinement of a consensus recognition sequence has been aided by the analysis of IHF binding sites from a variety of phage, plasmid, and bacterial sources (97–100, 154–156). IHF protection patterns against modification by dimethyl sulfate further delimit those positions involved in binding (97, 100) (see Figure 2).

As discussed above, and similar to Xis, the only known function of IHF is its ability to bend DNA (23, 100–103). The extent of IHF-induced bending is estimated to be greater than 140° at each of the three *att*P binding sites (23). The strong IHF-induced bend is consistent with the observations and analyses of DNA bending induced by CAP protein (157). The large effects (greater than 100-fold) on IHF binding affinity by sequence changes outside the consensus region (101) (L. Moitoso de Vargas, A. Landy, unpublished results) could be the result of changes in the bendability, and/or intrinsic curvature, of the DNA. Similarly, variation in the amount of DNA protected by IHF at each of several different binding sites (97, 98) might reflect differences in the length of DNA involved in making the curve around IHF protein. The H2 and H' IHF binding sites, as well as some non-*att* IHF sites (100), occur in regions where the DNA has a substantial intrinsic curvature in the absence of bound protein (23, 127), e.g. as determined by gel mobility shift permutation analyses (125). While intrinsic curvature may have the potential to favor IHF binding if it is in phase with the IHF-induced bend (157), it is not required. A number of IHF sites, such as H1, do not show any evidence of intrinsic curvature (23, 102).

Despite the gross similarity of H1, H2, and H' in the extent of IHF induced bending, the three sites are not functionally equivalent in recombination. Whereas H1 is required for integrative but not excisive recombination, H2 and H' are required for both reactions (26–28, 154). Additionally, it appears that the IHF requirement at H2 and H', but not at H1, can be partially substituted by another protein, possibly HU (154). Progress toward answering some of the questions about IHF should come from the recent IHF-DNA co-crystals made with synthetic oligomer (K. Appelt, personal communication).

FIS Sites

FIS binds to a single site, called F, that overlaps the X2 site in the P arm; the region protected against nuclease digestion is slightly larger than X2 (29). In contrast, the recombinational enhancers of the Hin and Gin family consist of two independent FIS binding sites separated by approximately 48 bp (113, 121, 122). The FIS site in λ *att* also seems to differ in sequence from the enhancer FIS sites (122). This lack of similarity suggests that the FIS recognition determinants are not fully understood and may involve specific DNA conformations not readily apparent in a simple sequence alignment.

Even though FIS can substitute for Xis binding at X2, the two proteins do not recognize the same features in the DNA: their respective DMS-modification protection patterns and the boundaries of the regions protected against nuclease digestion are different, and FIS does not bind to the X1 site (29, 86). In order to separate the effects of FIS and Xis, the X1-X2 site was replaced with an X1-X1 site in the P arm, thereby eliminating FIS binding while retaining the Xis binding (29, 127). This FIS$^-$ mutant made it possible to show that, in contrast to its stimulation of excision, FIS has no effect on integrative recombination.

The effect of adding FIS to a nuclease protection or gel mobility shift assay is equivalent in many respects to simply adding more Xis. If Xis is limiting, the addition of FIS leads to occupancy of the X1 and F sites with the same cooperativity seen for the X1 and X2 sites when Xis is increased in the absence of FIS. FIS binding to F is also similar to Xis binding to X2 in the extent of induced DNA bending, approximately 90° in both cases (23). Furthermore, the increase in the total apparent bend to more than 140° is similar for X1-F occupancy and for X1-X2 occupancy. Finally, as discussed below, all of the cooperative interactions in which Xis participates with Int and IHF occur equally well when FIS is bound in combination with Xis.

COOPERATIVE INTERACTIONS AND HIGHER-ORDER STRUCTURES

The first evidence for the wrapping of DNA in a putative higher-order structure was the appearance of a 10-base-pair repeat in the nuclease protection patterns of Int-DNA complexes (10, 49, 158). Such patterns of enhanced cutting suggest that one face of the helix is oriented toward the solvent, while the other face is less accessible to nuclease because it lies along a protein surface or forms the inside of a loop (159–162). Electron microscopy graphically revealed large protein complexes at the phage *att* site (20, 163) that suggested the involvement of approximately 230 bp of *att*P DNA and 4–8 Int monomers (20). The term *intasome* was coined to describe these complexes that were considered to contain only Int and DNA (20). It is now used as a generic term for protein–*att* site DNA complexes that also contain the other proteins involved in recombinogenic structures. While electron microscopy was not successful in revealing the importance of IHF in complex formation, it did show that Xis protein favored the formation of complexes involving *att*R and *att*P, and that many combinations of complexes could pair with one another (20, 21).

The first evidence that the functional recombinogenic complex involves DNA wrapped in a nucleosome-like structure came from the topological studies of Pollock & Nash (132). Intramolecular recombination between two *att* sites (either *att*P × *att*B or *att*L × *att*R) oriented in opposite directions on a circular supercoiled molecule inverts one segment of the circle with respect to the other and produces knotted products (164, 165). For integrative recombination, under conditions that minimized the knotting due to random interwrapping of the superhelical DNA, approximately one half of the recombinant products are simple circles and the rest are knotted. For excisive recombination, all of the recombinants are simple circles. The excess knotting of the integrative recombination products is postulated to reflect the wrapping of DNA in the *att*P complex (132). Subsequent topological analyses of knotted and catenated recombinant products have confirmed this result and indicate that the *att*P "intasome" is wrapped with a negative sign, which in the nucleosome cores of chromatin produces a left-handed solenoid (130, 166–168).

Several lines of evidence suggest that the role of the supercoiling requirement for integrative recombination is to facilitate formation of the *att*P intasome. Since only *att*P requires supercoiling, it is not likely that any step subsequent to synapsis, such as strand exchange, is responsible for the requirement (169). The singularity of *att*P is evidenced by the fact that supercoiling is not required for excisive recombination, although it stimulates the reaction (170). Any degree of DNA wrapping in the *att*L or *att*R intasomes should also have a negative sign, since negative supercoiling of either one enhances the efficiency of recombination (133) (W. Bushman, J. Thompson, A. Landy, unpublished results).

In a more direct biochemical analysis, it was shown that binding of Int to P1, and to a lesser extent to P'3, is more resistant to challenge by excess salmon sperm DNA when *att*P is negatively supercoiled (22). This effect requires the presence of IHF, and is seen only on *att*P (i.e. not when the arm binding sites are separated as in *att*L and *att*R). The good correlation between the sign and degree of supercoiling needed to promote these effects, and that needed to promote recombination, supports the notion that the primary role of supercoiling is to favor formation of the intasome (22).

The importance of the spatial orientation of the P1 and H1 sites within the intasome is seen when the helical phase of these sites is altered with respect to the remainder of *att*P by making insertions or deletions in a nonessential intervening region (85). Under normal reaction conditions (where supercoiling is required), substrates with changes of integral multiples of the DNA helical repeat recombine, while substrates with nonintegral changes do not. With nonsupercoiled (linear) *att*P, in low salt conditions, both the helical phasing and the P1 and H1 sites are much less important (although the overall efficiency is lower than with supercoiled *att*P) (85).

The specific interactions responsible for the intasome structures have been deduced by correlations of recombination with multiprotein nuclease protection assays, competition binding studies, site-specific mutagenesis of protein-binding sites, and results from suicide recombination substrates. These interactions are summarized below and illustrated schematically in Figure 4.

The pairwise cooperative interactions between Int molecules bound at the core-type sites was demonstrated by the fact that Int binds much better to core-type sites positioned as inverted repeats 7 bp apart (i.e. the canonical configuration) than to isolated core-type sites (50, 135). Similar cooperativity is also inferred from studies with artificial Holliday junctions, but this configuration includes the potential interactions among Int protomers bound at

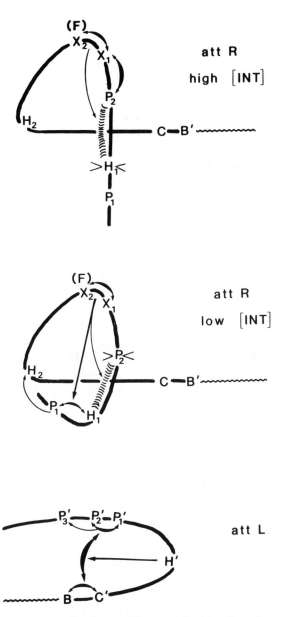

Figure 4   Schematic representation of some of the cooperative interactions observed with linear recombinogenic complexes. In some cases, cooperative interactions (⌒) and competitive interactions ( ))))₩₩₩₩₩) are dependent upon binding of another protein (⟶). The outcome of a competition at high or low Int is indicated by a vacated (✕) protein binding site. For more information about the individual protein binding sites see Figures 1 and 2.

each of the four core-type sites (B. Franz, A. Landy, unpublished results). Cooperative interactions among Int molecules bound at the core-type sites have been difficult to study because Int has a low affinity for these sites.

The most thoroughly characterized of the pairwise cooperative interactions involves Xis binding at the adjacent X1 and X2 sites (26, 86). This rather strong cooperativity is of special interest for two reasons. First, it involves a head-to-tail arrangement of sites, which is not so common for cooperatively binding proteins. Second, the Xis-Xis cooperativity can be substituted by Xis-FIS cooperativity. Keeping in mind that Xis and FIS do not have similar amino acid sequences, and that they bind to different DNA sequences, it is all the more surprising that for every interaction tested, the X1-F pair works as well as the X1-X2 pair (29). In the following discussion, all Xis interactions refer to Xis bound simultaneously at X1 and X2, and are probably equally applicable to simultaneous binding of Xis at X1 and FIS at F.

Xis bound at X1-X2 (or X1-F binding) cooperatively assists Int binding at P2. In nuclease protection experiments there is a 16-fold reduction in the amount of Int required to fill P2 if X1-X2 is also filled (26). This cooperativity is also observed in recombinations involving attR with a defective P2 or X1 site (27, 87, 88).

In addition to the cooperative interactions within the intasome, there are also several competitive interactions. These, however, are not simple pairwise competitions, since any single protein can fill all of its binding sites, and any pairwise combination of proteins can fill all of their respective binding sites (26, 29, 49, 86, 88, 97). Int binding at P2 and IHF binding at the neighboring H1 site are perfectly compatible in the absence of Xis. However, when X1-X2(F) is filled, then binding at H1 is competitive with binding at P2 (29, 88). Clearly, this competition is not due to simple occlusion of a binding site by neighboring proteins, given the pairwise compatibility of protein binding. Rather, the favored interpretation is a competition for adjoining space within a higher-order structure.

In order to better understand the dynamics of the H1-P2 competition, it is necessary to also refer to a complex cooperative interaction involving the P1 site (88). In linear molecules, Int binding to P1 in attR is strongly enhanced by the simultaneous occupancy of H1, H2, and X1-X2(F). However, in attP this cooperativity is not observed (the P' arm is involved in a different set of cooperative interactions that includes the core-type Int binding sites) (L. Moitoso de Vargas, A. Landy, unpublished). Thus, the binding affinity of Int for P1 in a higher-order complex is determined by interactions occurring on DNA at least 150 bp distant (see Figure 4). Richet et al (22) have found IHF-dependent cooperative interactions involving the P1 site in attP that require supercoiling. Thus, two major features that distinguish integrative and excisive recombination (the supercoiling requirement for attP and the Xis requirement for *att*R) are also needed for the long-range interactions involving the P1 site.

The above discussion indicates that each partner in the H1 vs P2 competition is involved in a separate set of cooperative interactions. Therefore, the outcome of the competition is determined by the configuration of *att* site arms, i.e. *att*R vs *att*P, by the relative concentrations of each protein, and by the topological state of the DNA. As will be discussed (DIRECTIONALITY AND REGULATION), the cooperative transition in binding of proteins from the P1, H1, X1/X2(F), H2 set to the P2, X1/X2(F), H2 set (see Figure 5) occurs over the same range of protein concentrations that is required to overcome IHF inhibition of excision and to favor efficient excisive recombination (28, 88).

In contrast to the complexity of the P arm, the relative simplicity of the P' arm highlights one of the primary interactions responsible for intasome structure. In *att*L, IHF bound at H' enhances binding of Int to core-type sites as long as the P' arm-type sites are present. This same cooperative interaction is reflected in an (H'-P')-enhanced nicking of *att*L suicide recombination substrates. In *att*L, both the IHF-enhanced Int binding at core-type sites and recombination proficiency depend upon the correct helical phasing of the P' arm-type sites with respect to the core region (L. Moitoso de Vargas, S. H. Kim, A. Landy, unpublished). These cooperative interactions lead to the architectural element shown in Figure 6 and discussed below.

The number and complexity of the intasome-forming interactions stands in sharp contrast to the compact "minimized" reactions of some other Int family members such as Cre and FLP. Since these are highly efficient site-specific recombinases, it is appealing to ascribe much of the additional complexity of the λ system to the requirements for directionality and regulation, as discussed below.

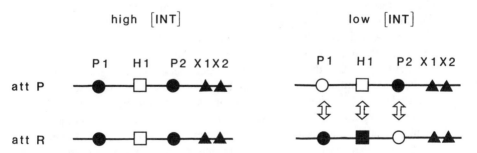

Figure 5 Coordinate occupancy of protein binding sites in the P arm of *att*R and *att*P as a result of cooperative and long-range interactions. Filled symbols indicate occupancy of the site by its cognate protein (P1 and P2 by Int, H1 by IHF, X1X2 by Xis). Double-headed arrows highlight the differences between *att*P and *att*R.

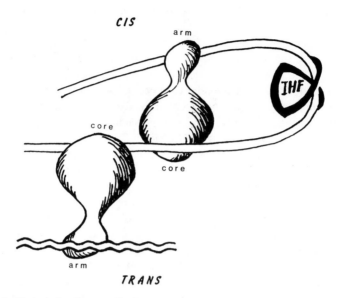

Figure 6 Basic design element of an intasome. IHF induces a sharp (≥140°) bend in DNA to bring the distant arm-type and core-type sites close enough to be bound by an Int protomer (*cis* interaction). The bivalent Int can also be involved in *(trans)* interactions between the arm-type and core-type sites on different partners. Xis and FIS are postulated to play a similar role to IHF.

SYNAPSIS AND STRAND EXCHANGE

Covalent Intermediates

The sites of strand exchange, i.e. the locus of cutting and resealing individual DNA strands, was narrowed down to two adjacent phosphodiester bonds, in an elegant isotope transfer experiment by M. Mizuuchi and K. Mizuuchi (171). Resolution of the remaining ambiguity came from sequence comparisons of many secondary *att* sites (3, 50) and from the demonstration by Craig & Nash (51) that Int makes a covalent link with the 3' phosphate at the site(s) of strand exchange (reviewed in Ref. 3).

In order to carry the analysis of transient recombination intermediates further, two families of suicide recombination substrates proved useful (see Figure 3) (13, 15). These *att* sites, which contain a strategically placed nick within the overlap region, initiate recombination normally and become deviant only after Int has acted on them. They are unable to complete or reverse the reaction. The particular intermediate trapped by such substrates depends upon where the nick is placed within the overlap region. One family of suicide substrates contains a medial nick in the overlap region (centered between the two sites of strand exchange) and leads to the formation of covalent Int-DNA complexes in high yield. This work showed that Tyr-342 of Int forms a covalent link with DNA as an intermediate during strand cleavage and religation (15). The other family of suicide substrates is discussed below.

The Order of Strand Exchange

The existence of covalent Int-DNA intermediates does not distinguish between the two possible mechanisms of strand exchange: a concerted mechanism, in which both DNA strands in each *att* site are cleaved prior to strand exchange; and a sequential mechanism, in which one DNA strand in each *att* site is cleaved and exchanged followed by a second cycle of cleavage and exchange (see Figure 1).

From genetic crosses, it was found that Int-promoted recombination in vivo can generate the progeny predicted by a sequential strand exchange mechanism (172, 173). However, the low yield of these progeny was cited as evidence against their being on the primary recombination pathway (128). In vitro data that were interpreted to favor a concerted mechanism included the observation of low levels of double strand breakage by Int protein (128, 174). Support for a sequential strand exchange mechanism came from the specific resolution of synthetic *att* site Holliday junctions by purified Int protein to yield completed recombination products (12). The high efficiency and specificity of the resolution reaction was persuasive but not compelling.

Two different experimental devices finally led to the isolation of the elusive single-strand exchange recombination intermediates. One of these is the second family of suicide recombination substrates discussed above. In this family of suicide substrates, the preexisting *att* site nick is located at either the left (top strand) or right (bottom strand) Int cleavage site of the *att*B partner (see Figure 3). When the nick is in the bottom strand, single-strand exchange intermediates are produced under normal reaction conditions with almost the same efficiency as recombinant products from normal substrates. However, when the nick is in the top strand, no intermediates or recombinant products are formed (13). Thus, recombination proceeds via a sequential mechanism of strand exchange and the order is fixed such that the top strands must exchange first. Blocking the top strand exchange with a nick prevents the reaction from initiating. Using suicide recombination substrates for an *att*L × *att*R reaction, it was shown that excisive recombination has the same prescribed order of top strand exchange followed by bottom strand exchange (13, 14).

Similar conclusions have been drawn for integrative recombination (excision was not tested) from experiments in which substitution by phosphorothioate nucleotides into *att* site DNA was used to block Int cleavage (14, 65). Substitution in the top strand blocked strand exchange, while substitution in the bottom strand generated Holliday intermediates. In these experiments, however, very low levels of the recombination intermediate were seen, even when using the Int variant known as Int-h (see above) to increase their yield. This difference between the analogue substitution and suicide recombination substrates is informative, especially when coupled with the low levels of Holliday structure obtained in normal reactions, or those blocked by a heterology in the overlap region (see below) (13, 14, 18). These differences have led to the suggestion that the first strand exchange during integrative recombination is reversible. Thus, in order to obtain high yields of recombination intermediate, it is not sufficient to block forward progression of the reaction, but it is also necessary to block its reversal (13). Understanding which of several possible mechanisms is responsible for the ability of the suicide substrate to trap Holliday junction intermediates is likely to provide useful insights into the normal reaction. Thus far, results obtained with a variety of different suicide recombination substrates indicate that Int protein, when provided with an appropriate 5' OH acceptor, will efficiently execute a single DNA strand transfer reaction (13) (S. E. Nunes-Düby, A. Landy, unpublished results).

The strict order of strand exchanges does not arise from asymmetry in the core region, but rather from the asymmetric arrangement of proteins bound to the P and P' arm sequences: (a) the core regions in the suicide recombination substrates were constructed to be virtually symmetric (13); (b) the same order of strand exchanges is obtained with att sites having different core region sequences (13, 14); (c) inversion of the overlap (13) or the full core region (14) does not change the order of strand exchange relative to the P and P' arms.

Holliday recombination intermediates have also been detected in the Cre and FLP systems (141, 145, 175). Thus, it is likely that this mechanism of sequential strand exchanges is common to members of the Int family.

DNA Homology

Although the specific sequence of the overlap region is not critical (within a wide latitude) for the order of strand exchange, or for the overall efficiency of recombination, it is required that the overlap region have the same sequence in both recombination partners (16, 17, 136). Two models have been proposed for the role of DNA:DNA homology in the overlap region. In the first, homology is required for synapsis of the recombining partners, and more specifically for the formation of a four-stranded DNA structure (7, 176, 177). From this view of recombination, considerable attention has been focused on whether the interstrand interactions are likely to be of the type proposed by McGavin (178) or by Wilson (179), and the likely direction in which strands rotate around the four-sided block during recombination (166, 177). In the second model, which is strongly favored by recent data, DNA:DNA homology is not required for synapsis but rather for some later step, such as branch migration of a Holliday junction (16). Experiments with *att* sites that are heteroduplex within the overlap region argue against mechanisms involving a homology-dependent annealing of cohesive ends (180), however they do not distinguish between these two models.

A distinction between the two models is afforded by the experimental devices described above for obtaining single-strand exchange intermediates containing Holliday junctions. Whereas the formation of single-strand exchange intermediates is completely blocked by a nonhomology on the left side of the overlap region, it is not affected by a nonhomology on the right (13, 18). When the center of the overlap region is tested, single-strand exchange intermediate formation is blocked by a nonhomology at position 0 but not at position +1 (18). These results rule out that class of models, including that of four-stranded DNA, in which homology over all (or most) of the overlap region is required for synapsis. However, they do not exclude a requirement for a local homology of several base pairs in the immediate vicinity of a strand exchange site.

Whether or not strand exchange depends on some very short local homologies, a proposed requirement for a homology-dependent branch migration step is attractive (16). Artificial *att* site Holliday junctions have been made with sequence nonhomologies in the overlap region designed to restrict branch migration. For some of these constructions, resolution is strongly biased in favor of parental-type products, in contrast to wild-type Holliday structures that resolve to parental and recombinant-type products without pronounced bias. These results suggest that heterology prevents the Holliday junction from migrating from one Int cleavage site to the other and that resolution is executed by Int if the junction is at, or near, the Int cleavage site (B. DeMassy, L. Dorgai, R. A. Weisberg, personal communication). However, some data from these experiments suggest that Int may not be able to cleave when the junction is constrained very close to the cleavage site. The DNA branch migration model also gains further credibility from its ability to explain the results of recombination in which one partner is heteroduplex in the overlap region (17, 180).

Synapsis

Additional evidence that DNA homology is not required for synapsis comes from studies on both integrative and excisive recombination. Heteroduplex *att*B was incubated with a supercoiled homoduplex *att*P under recombination conditions; although the two *att* sites cannot recombine, because of nonhomologies between their overlap regions, double-strand cleavage of *att*B is promoted by the *att*P intasome. Thus, capture of *att*B by the *att*P intasome does not depend upon DNA homology between the sites (19). In addition, chemical footprinting assays establish that under recombination conditions, *att*B cannot stably bind Int in competition with other DNAs. Taken together, these experiments suggest that *att*B may obtain its Int by binding to a preformed *att*P intasome in a synapsis that is governed primarily by protein-protein and protein-DNA interactions and is independent of DNA homology (19). Thus the asymmetry in structural complexity is indeed reflected in functional asymmetry. For the analysis of excisive recombination, half-*att* sites were constructed by cutting at a restriction site within the overlap region. These half-*att* sites are proficient for synapsis and strand transfer to an intact *att* site partner despite the lack of homology between their overlap regions (S. E. Nunes-Düby, A. Landy, unpublished results).

A particularly challenging aspect of synapsis is how two recombining *att* sites initially come together and are productively oriented with respect to one another. The fact that recombination can occur intermolecularly, intramolecularly, and between directly repeated or inverted sites suggests that *att* sites come together by random collision rather than by some form of tracking along the DNA. This general view is substantiated by topological analyses of the knots and catenanes generated by recombination: the topological complexity of recombinant products increases with both the density of supercoils and the distance between the *att* sites (130, 132, 164, 166, 181).

In contrast to the Int family, the Resolvase-Invertase family has more stringent orientation requirements and yields topologically invariant recombination products (33, 34, 39, 40). These features are consistent with a class of models in which the two recombination sites are interwrapped in a very specific manner, a requirement that is equivalent to a "topological filter" (182–185). From the topological experiments, it is clear that λ att sites do not have such an interwrapping requirement for productive synapsis. [Although an interesting exception may be the Xis-independent attL × attR recombination in low salt (84).] The usefulness of topological analyses of recombinant products lies mainly in its ability to set limits on acceptable models (40, 130, 166) when the number of biochemical unknowns is small. However, any acceptable mechanism for synapsis and strand exchange must be validated by its ability to explain the topological consequences of recombination.

Several features of the λ system are quite similar to those found in FLP and Cre, while others appear to be influenced by the unique arms of λ *att* sites. Both FLP (68, 186–188) and Cre (189–191) share with λ all those properties that suggest a random collision-type of synapsis. Cre, however, appears to exhibit an additional constraint in synapsis, suggesting some interwrapping of *lox* sites (189, 191, 192). However, it is unlikely that this aspect of synapsis reflects a fundamental difference in mechanism, since relatively small mutational changes in either Cre or *lox* can make the Cre system topologically similar to FLP and Int (192).

DIRECTIONALITY AND REGULATION

One of the unique and intriguing aspects of the site-specific recombination of λ (and related phages) is its extreme directionality. Although the pathway is commonly diagrammed as a reversible integration/excision reaction, and the normal products of one reaction are the normal substrates of the other, such notation belies the fact that there are in fact two distinct pathways. Integrative recombination normally depends upon supercoiling for assembly of the attP intasome (22, 169), while the attL and attR intasomes do not have this requirement (although they are stimulated by it) (133). Integrative recombination requires higher levels of Int than excisive recombination, both in vivo (193) and in vitro (when attR is supercoiled) (27, S. E. Nunes-Düby, A. Landy, unpublished results). Xis protein is required for excisive recombination under physiological conditions, while it is inhibitory for integrative recombination (80, 83, 194). The H1 site must be occupied by IHF for integrative recombination and it must be vacant for excision (27, 28, 88, 154). Of a total constellation of 13 protein binding sites, P1, H1, and P'3 are uniquely required for integrative recombination, and P2, X1, X2, F, and P'1 are uniquely required for, or involved in, excisive recombination (27–29, 87, 88, 153). Indeed, deletion of the P1 and H1 sites, which are required for attP function, makes an attR that can be more efficient than wild type (27). It is clear that the λ att sites have evolved to efficiently execute one reaction or the other in a mutually exclusive manner.

The regulation of phage integration and prophage excision consists of a sophisticated network of pathways involving multiple phage and host genes. A detailed description of how these pathways direct recombination, and some models for their regulation, have recently been presented (4). Several features of these models are summarized below.

In contrast to the analysis of gene expression, regulation at the level of the recombination reaction itself has been difficult to study in vivo. The in vitro data, and resulting models pointing to mechanism-based regulation, have to be tested by in vivo experiments specifically designed to detect such effects. For example, the large effect of FIS on in vivo excision, discussed above, was not detected until specifically looked for. The other mechanism-based regulation by a host-encoded protein involves IHF and the H1 binding site. The effect of IHF inhibition of excision is to greatly sensitize the reaction of Int concentration via the cooperative and competitive interactions described above (28, 88). Under one set of in vitro conditions, IHF inhibition at H1 sharpens the Int concentration dependence of excision by more than 10-fold (28).

The two host-encoded proteins involved in λ site-specific recombination both exert regulatory effects on excision; FIS stimulates and IHF inhibits. It is significant that both regulatory responses are overridden by high levels of phage-encoded proteins: high Xis levels abolish FIS stimulation and high Int levels overcome IHF inhibition. These mechanisms endow the prophage excision reaction with a responsiveness to host physiology, viral physiology, and a range of environmental conditions (4, 28, 29). When there is a strong inducing stimulus, such as heavy UV irradiation, death of the host cell is imminent and it is advantageous for the prophage to excise and initiate a lytic cycle without delay or modulation. These are the conditions when levels of Int and Xis are expected to be high. When the inducing stimulus is weak and cell death is not certain, it is advantageous for viral excision and lytic development to be influenced by other factors, such as the likelihood of completing a lytic cycle. These are the conditions when Int and Xis levels may be low. Additionally, in the unfavorable conditions of stationary phase, FIS levels are low and excision is more difficult (requires higher levels of Xis).

The ability of FIS to stimulate excision at low levels of Xis may also be relevant in the phenomenon of spontaneous phage production, which occurs at a frequency of one in 10^3 to 10^5 cell generations. Some "spontaneous" phage production probably results from DNA damage and a strong induction in a small fraction of cells. Additionally, some cells may produce phage as a consequence of random fluctuations in repressor concentration, leading to a weak induction. It has been suggested (4) that the possible role of FIS in assisting this low-frequency pathway is not unlike the FIS stimulation of recombination of other low-frequency recombinations (112, 113, 115, 116). Both the *hin* inversion system of *Salmonella* flagellar antigens (195) and the *gin* inversion system of bacteriophage Mu tail fibers (196) also involve fluctuations in the synthesis of a poorly expressed recombinase. The stimulation of all three low-frequency reactions by FIS suggests that they will occur preferentially in exponentially growing cells.

COMMENTS AND SPECULATIONS

Cooperative Systems

In retrospect we can see why it was so difficult to isolate *att* site defective mutants (despite the existence of many potential targets for disruption). In highly cooperative complexes like intasomes, it is likely that small changes in specific protein-binding sites will be masked by the remaining functional sites. Indeed, even with the ability to make site-directed mutations, the phenotypes can be misleading. A two-base change in the consensus recognition sequence, which reduced IHF binding to the H1 site by 500-fold, yielded little change in phenotype. Only after the introduction of five base changes could it be determined that the H1 site is required for integrative recombination (28). Thus, in highly cooperative systems, considerable caution must be exercised in interpreting null phenotypes. Another example of confusing or misleading phenotypes was the observation that, under certain recombination conditions, deletion of the P2 Int binding site produced much less of a defect than a multisite mutant that destroyed its function. In fact, the interplay of competitive and cooperative binding in the P arm explains this surprising result, and is substantiated by biochemical analysis of the two mutants (87). A final example of considerations relevant to complex cooperative systems concerns the conditions under which apparently independent binding sites are actually used. The most prominent instances of this are the P arm binding site interactions (88), and the recent finding that some (or all) of the core-type sites do not obtain Int from solution (19), but through delivery by arm-type sites (L. Moitoso de Vargas, S. H. Kim, A. Landy, unpublished results).

We have also learned that optimization of an in vitro system can result in overlooking some of the regulatory and biochemical complexity that is important under nonoptimal conditions. Indeed, nonoptimal conditions are prevalent in nature. The effect of FIS protein could not have been detected in an in vitro system optimized for Xis concentration.

Dynamic Aspects

In considering how two *att* sites might synapse, it is useful to summarize the features of Int binding in the region of strand exchange. Both of the phosphodiester bonds that are cut during exchange of each strand, and all of the contacts for Int binding, are on the same face of the DNA helix, straddling the central major groove of the overlap region (50). The recent demonstration that synapsis does not require DNA homology across the entire 7 bp overlap region (13, 18, 19) rules out models of four-stranded DNA and eliminates the motivation for proposing extensive DNA-DNA interactions in the early steps of synapsis. This makes even more appealing the view that the Int proteins lie on the inside faces of a synaptic complex where they can interact with one another. One particularly attractive configuration is that of two helixes lying across one another (as opposed to a side-by-side alignment). This would serve to maximize the interaction between all of the bound Int proteins, as suggested by some preliminary data on the resolution of synthetic Holliday junctions and the properties of certain suicide recombination substrates (B. Franz, S. H. Kim, S. E. Nunes-Düby, A. Landy, unpublished results). It would also minimize the repulsive forces between the two DNA helixes. This view of the synaptic complex, with an "interior" cluster of Int molecules in the region of strand exchange, is similar to that suggested for γδ Resolvase on the basis of recent mutant and crystallographic data. These indicate a tetrameric complex of Resolvase monomers with the active-site serines on the outer surface (T. Steitz, N. Grindley, personal communication).

Structural Aspects

The sharp ($\cong$ 140°) bends induced at the loci of binding by IHF, Xis, and FIS, along with the bivalent DNA-binding capacity of Int, are probably the major determinants of intasome structure. Intrinsic DNA curvature, as a consequence of A tracts (124–126), is also likely to contribute to the overall structures. Within the intasome, these elements are likely to work in conjunction to form DNA loops that are tethered by an Int, binding simultaneously to arm-type and core-type sites. The delivery of Int to the low-affinity core-type sites, by the higher-affinity arm-type sites, would be facilitated by the protein-induced DNA bending. Binding at core-type sites could also be assisted through secondary (protein-protein) interactions. For example, one Int bound to both arm- and core-type sites could bring another Int to the core region, either on the same *att* site or on the synapsed partner. This view is also consistent with the cooperativity of Int binding in the core regions of all four *att* sites (50) (B. Franz, S. H. Kim, S. Nunes-Düby, L. Moitoso de Vargas, A. Landy, unpublished results).

This model of promoting binding to the core-type sites by the formation of DNA loops that deliver Int bound at arm-type sites predicts specific pairwise interactions between each core-type site and an arm-type site, or between two core-type sites. These interactions could involve sites within the same partner (*cis* interactions), and/or sites on different partners (*trans* interactions) (Figure 6). Similarly, the cooperative interactions between proteins bound at core-type sites could be *cis* and/or *trans*. Identifying these pairwise interactions is one of the challenges of understanding the intasome structures. A related challenge is characterization of the interactions between DNA and the site-specific DNA bending proteins IHF, Xis, FIS.

Regulatory Aspects

Bacteriophage λ has long served as a divining rod for uncovering host functions. Yet another example may be the host-encoded IHF and FIS proteins. It is tempting to speculate that the mechanism-based regulation of recombination incorporates IHF and FIS as integral elements of the reaction, because these two proteins reflect global signals of *E. coli* physiology. This suggestion is appealing because of the striking variation in FIS levels as a function of growth phase (29) and the wide range of reactions in which IHF is involved (111).

The role of FIS in stimulating excision has led to speculation about a new phage regulatory pathway that could maintain derepression even under weak inducing conditions (4, 29). After excision, new genes from the *b* region are expressed by N-dependent transcription through *att*P. One of these (*ben*) encodes a double-stranded DNA endonuclease (197, 198) that has been purified and found to cleave only supercoiled DNA (199). This endonuclease, or one of the others encoded in the *b* region (22, 200, 201), could play a role in prolonging (inducing) an SOS response by cleaving the bacterial genome and generating new sites that activate RecA protein. The activated RecA would promote continuous cleavage of λ, thus preventing repression of the newly excised phage.

The model proposes a function for the enigmatic *b* region (202) and, more importantly, it suggests how FIS could enhance excision, even under conditions that might otherwise lead to repression of the excised phage. According to this model, spontaneous phage production does not reflect the lower limits of prophage repression, but rather is a programmed strategy in maintaining phage populations. In the absence of a pathway devoted to maintaining derepression (under weak inducing conditions), the dramatic effects of FIS on prophage excision would be primarily relevant to normal induction conditions.

CLOSING THE CIRCLE

Analysis of the λ recombination system has benefitted greatly from Robin Holliday's provocative insight about possible intermediates in the exchange (or rearrangement) of genetic information. In return, studies of the λ pathway have contributed significantly to understanding more about the formation, properties, and resolution of these genetic intermediates. The Holliday structures were conceived in the context of homologous, not site-specific, recombination—but such extrapolations are commonplace. One example is the possible relevance of intasome structures to the complex organizations involved in DNA replication, transcription, and other recombination systems. Understanding how the accessory proteins IHF, Xis, and FIS bend and shape DNA and how the bivalent Int protein tethers two distant DNA sites is likely to be of interest well beyond the field of site-specific recombination. Another example is the possible window into global regulatory devices that is afforded by the discovery of IHF and FIS proteins and their role in mechanism-based regulation of recombination.

The circle is always closing but it is never closed.

ACKNOWLEDGMENTS

I thank the many colleagues who provided information and manuscripts prior to publication. I am greatly indebted to the members of our research group, past and present, for their ideas, criticism and enthusiasm, especially to John Thompson, Simone Nunes-Düby, and Lina Moitoso de Vargas, who made significant contributions to this review. I also thank Joan Boyles for her dedicated and good-natured assistance in preparing the manuscript.

Literature Cited

1. Holliday, R. 1964. *Genet. Res.* 5:282–304
2. Nash, H. A. 1981. *Annu. Rev. Genet.* 15:143–67
3. Weisberg, R. A., Landy, A. 1983. In *Lambda II*, ed. F. W. Stahl, J. Roberts, R. A. Weisberg, pp. 211–50. New York: Cold Spring Harbor Lab.
4. Thompson, J. F., Landy, A. 1989. See Ref. 44, pp. 1–22
5. Campbell, A. M. 1962. *Adv. Genet.* 11:101–45
6. Kikuchi, Y., Nash, H. A. 1978. *J. Biol. Chem.* 253:7149–57
7. Kikuchi, Y., Nash, H. A. 1979. *Proc. Natl. Acad. Sci. USA* 76:3760–64
8. Nash, H. A., Robertson, C. A. 1981. *J. Biol. Chem.* 256:9246–53
9. Landy, A., Ross, W. 1977. *Science* 197:1147–60
10. Hsu, P.-L., Ross, W., Landy, A. 1980. *Nature* 285:85–91
11. Mizuuchi, M., Mizuuchi, K. 1980. *Proc. Natl. Acad. Sci. USA* 77:3220–24
12. Hsu, P.-L., Landy, A. 1984. *Nature* 311:721–26
13. Nunes-Düby, S. E., Matsumoto, L., Landy, A. 1987. *Cell* 50:779–88
14. Kitts, P. A., Nash, H. A. 1988. *J. Mol. Biol.* 204:95–108
15. Pargellis, C. A., Nunes-Düby, S. E., Moitoso de Vargas, L., Landy, A. 1988. *J. Biol. Chem.* 263:7678–85
16. Weisberg, R. A., Enquist, L. W., Foeller, C., Landy, A. 1983. *J. Mol. Biol.* 170:319–42
17. Bauer, C. E., Gardner, J. F., Gumport, R. I. 1985. *J. Mol. Biol.* 181:187–97
18. Kitts, P. A., Nash, H. A. 1987. *Nature* 329:346–48
19. Richet, E., Abcarian, P., Nash, H. A. 1988. *Cell* 52:9–17
20. Better, M., Lu, C., Williams, R. C., Echols, H. 1982. *Proc. Natl. Acad. Sci. USA* 79:5837–41
21. Better, M., Wickner, S., Auerbach, J., Echols, H. 1983. *Cell* 32:161–68
22. Richet, E., Abcarian, P., Nash, H. A. 1986. *Cell* 46:1011–21
23. Thompson, J. F., Landy, A. 1988. *Nucleic Acids Res.* 16:9687–705
24. Moitoso de Vargas, L., Pargellis, C. A., Hasan, N. M., Bushman, E. W., Landy, A. 1988. *Cell* 54:923–29
25. Argos, W., Landy, A., Abremski, K., Egan, J. B., Haggård-Ljungquist, E., et al. 1986. *EMBO J.* 5:433–40
26. Bushman, W., Yin, S., Thio, L. L., Landy, A. 1984. *Cell* 39:699–706
27. Bushman, W., Thompson, J. F., Vargas, L., Landy, A. 1985. *Science* 230:906–11
28. Thompson, J. F., Waechter-Brulla, D., Gumport, R. I., Gardner, J. F., Moitoso de Vargas, L., Landy, A. 1986. *J. Bacteriol.* 168:1343–51
29. Thompson, J. F., Moitoso de Vargas, L., Koch, C., Kahmann, R., Landy, A. 1987. *Cell* 50:901–8
30. Sadowski, P. 1986. *J. Bacteriol.* 165:341–47
31. Craig, N. L. 1988. *Annu. Rev. Genet.* 22:77–106
32. Hatfull, G. F., Salvo, J. J., Falvey, E. E., Rimphanitchayakit, V., Grindley, N. D. F. 1988. See Ref. 42, pp. 149–81
33. Hatfull, G. F., Grindley, N. D. F. 1988. See Ref. 43, pp. 357–96
34. Glasgow, A. C., Hughes, K. T., Simon, M. I. 1989. See. Ref. 44, pp. 637–59
35. Johnson, R. C., Simon, M. I. 1987. *Trends Genet.* 3:262–67
36. West, S. C. 1989. In *Nucleic Acids and Molecular Biology,* Vol. 3, ed. F. Eckstein, D. M. J. Lilley. Berlin: Springer Verlag. In press
37. West, S. C. 1988. *Trends Genet.* 4:8–13
38. Cox, M. M. 1989. See Ref. 44, pp. 661–70
39. Gellert, M., Nash, H. 1987. *Nature* 325:401–4
40. Benjamin, H. W., Cozzarelli, N. R. 1985. In *Proc. Robert A. Welch Found. Conf. Chem. Res. XXIX. Genetic Chemistry and the Molecular Basis of Heredity,* pp. 107–26. Houston: Robert A. Welch Found.
41. Wasserman, S. A., Cozzarelli, N. R. 1986. *Science* 232:951–60
42. Kingsman, A. J., Kingsman, S. M., Chater, K. F., eds. 1988. *Transposition.* Cambridge, England: Cambridge Univ. Press
43. Kucherlapati, R., Smith, G. R., eds. 1988. *Genetic Recombination.* Washington, DC: Am. Soc. Microbiol.
44. Berg, D. E., Howe, M. M., eds. 1989. *Mobile DNA.* Washington, DC: Am. Soc. Microbiol.
45. Hoess, R. H., Foeller, C., Bidwell, K., Landy, A. 1980. *Proc. Natl. Acad. Sci. USA* 77:2482–86
46. Davies, R. W. 1980. *Nucleic Acids Res.* 8:1765–82
47. Kotewicz, M., Grzesiuk, E., Courschesne, W., Fischer, R., Echols, H. 1980. *J. Biol. Chem.* 255:2433–39
48. Nash, H. A. 1981. *Enzymes* 14:471–80
49. Ross, W., Landy, A. 1982. *Proc. Natl. Acad. Sci. USA* 79:7724–28
50. Ross, W., Landy, A. 1983. *Cell* 33:261–72
51. Craig, N. L., Nash, H. A. 1983. *Cell* 35:795–803

52. Lange-Gustafson, B. J., Nash, H. A. 1984. *J. Biol. Chem.* 259:12724–32
53. Belfort, M. 1980. *Gene* 11:149–55
54. Schindler, D., Echols, H. 1981. *Proc. Natl. Acad. Sci. USA* 78:4475–79
55. Guarneros, G., Montanez, C., Hernandez, T., Court, D. 1982. *Proc. Natl. Acad. Sci. USA* 79:238–42
56. Schmeissner, U., McKenney, K., Rosenberg, M., Court, D. 1984. *J. Mol. Biol.* 176:39–53
57. Montanez, C., Bueno, J., Schmeissner, U., Court, D. L., Guarneros, G. 1986. *J. Mol. Biol.* 191:29–37
58. Shimada, K., Campbell, A. 1974. *Proc. Natl. Acad. Sci. USA* 71:237–41
59. Leong, J., Nunes-Düby, S., Oser, A. B., Lesser, C., Youderian, P., et al. 1986. *J. Mol. Biol.* 189:603–16
60. Pierson, L. S. III, Kahn, M. L. 1987. *J. Mol. Biol.* 196:487–96
61. Yu, A., Bertani, L. E., Haggård-Ljungquist, E. 1989. *Gene.* In press
62. Ljungquist, E., Bertani, L. E. 1983. *Mol. Gen. Genet.* 192:87–94
63. Bear, S. E., Clemens, J. B., Enquist, L. W., Zagursky, R. J. 1987. *J. Bacteriol.* 169:5880–83
64. Miller, H. I., Mozola, M. A., Friedman, D. I. 1980. *Cell* 20:721–29
65. Kitts, P. A., Nash, H. A. 1988. *Nucleic Acids Res.* 16:6839–56
66. Enquist, L. W., Weisberg, R. A. 1984. *Mol. Gen. Genet.* 195:62–69
67. Dayhoff, M. O., Schwartz, R. M., Orcott, B. L. 1978. In *Atlas of Protein Sequence and Structure,* ed. M. O. Dayhoff, pp. 345–52. Washington, DC: Natl. Biomed. Res. Found.
68. Gronostajski, R. M., Sadowski, P. D. 1985. *Mol. Cell. Biol.* 5:3274–79
69. Prasad, P. V., Young, L.-J., Jayaram, M. 1987. *Proc. Natl. Acad. Sci. USA* 84:2189–93
70. Wierzbicki, A., Kendall, M., Abremski, K., Hoess, R. 1987. *J. Mol. Biol.* 195:785–94
71. Reed, R. R., Moser, C. D. 1984. *Cold Spring Harbor Symp. Quant. Biol.* 49:245–49
72. Hatfull, G. F., Grindley, N. D. F. 1986. *Proc. Natl. Acad. Sci. USA* 83:5429–33
73. Klippel, A., Mertens, G., Patschinsky, T., Kahmann, R. 1988. *EMBO J.* 7:1229–37
74. Ross, W., Landy, A., Kikuchi, Y., Nash, H. 1979. *Cell* 18:297–307
75. Nussinov, R., Weisberg, R. A. 1986. *J. Biomol. Struct. Dyn.* 3:1133–44
76. Enquist, L. W., Weisberg, R. A. 1977. *J. Mol. Biol.* 111:97–120
77. Gingery, R., Echols, H. 1967. *Proc. Natl. Acad. Sci. USA* 58:1507–14
78. Gottesman, M. E., Yarmolinsky, M.

79. Zissler, J. 1967. *Virology* 31:189
80. Abremski, K., Gottesman, S. 1982. *J. Biol. Chem.* 257:9658–62
81. Weisberg, R. A., Gottesman, M. E. 1971. In *Bacteriophage Lambda,* ed. A. D. Hershey, pp. 489–500. New York: Cold Spring Harbor Lab. Press
82. Gottesman, S., Gottesman, M., Shaw, J. E., Pearson, M. L. 1981. *Cell* 24:225–33
83. Abremski, K., Gottesman, S. 1981. *J. Mol. Biol.* 153:67–78
84. Craig, N. L., Nash, H. A. 1983. In *Mechanism of DNA Replication and Recombination,* ed. N. R. Cozzarelli, pp. 617–36. New York: Liss
85. Thompson, J. F., Snyder, U. K., Landy, A. 1988. *Proc. Natl. Acad. Sci. USA* 85:6323–27
86. Yin, S., Bushman, W., Landy, A. 1985. *Proc. Natl. Acad. Sci. USA* 82:1040–44
87. Thompson, J. F., Moitoso de Vargas, L., Nunes-Düby, S. E., Pargellis, C., Skinner, S. E., Landy, A. 1987. In *Mechanisms of DNA Replication and Recombination,* ed. T. Kelly, R. McMacken, pp. 735–44. New York: Liss
88. Thompson, J. F., Moitoso de Vargas, L., Skinner, S. E., Landy, A. 1987. *J. Mol. Biol.* 195:481–93
89. Miller, H. I., Friedman, D. I. 1980. *Cell* 20:711–19
90. Kikuchi, A., Flamm, E., Weisberg, R. A. 1985. *J. Mol. Biol.* 183:129–40
91. Miller, H. I., Kikuchi, A., Nash, H. A., Weisberg, R. A., Friedman, D. I. 1979. *Cold Spring Harbor Symp. Quant. Biol.* 43:1121–26
92. Miller, H. I., Nash, H. A. 1981. *Nature* 290:523–26
93. Miller, H. I. 1984. *Cold Spring Harbor Symp. Quant. Biol.* 49:691–98
94. Flamm, E. L., Weisberg, R. A. 1985. *J. Mol. Biol.* 183:117–28
95. Mechulam, Y., Fayat, G., Blanquet, S. 1985. *J. Bacteriol.* 163:787–91
96. Drlica, K., Rouviere-Yaniv, J. 1987. *Microbiol. Rev.* 51:301–19
97. Craig, N. L., Nash, H. A. 1984. *Cell* 39:707–16
98. Leong, J. M., Nunes-Düby, S., Lesser, C. F., Youderian, P., Susskind, M. M., Landy, A. 1985. *J. Biol. Chem.* 260:4468–77
99. Gamas, P., Chandler, M. G., Prentki, P., Galas, D. J. 1987. *J. Mol. Biol.* 195:261–72
100. Stenzel, T. T., Patel, P., Bastia, D. 1987. *Cell* 49:709–17
101. Prentki, P., Chandler, M., Galas, D. J. 1987. *EMBO J.* 6:2479–87
1968. *J. Mol. Biol.* 31:487–505

102. Robertson, C. A., Nash, H. A. 1988., *J. Biol. Chem.* 263:3554–57
103. Kosturko, L. D., Daub, E., Murialdo, H. 1989. *Nucleic Acids Res.* 17:17–34
104. Tanaka, I., Appelt, K., Dijk, J., White, S. W., Wilson, K. S. 1984. *Nature* 310:376–81
105. Friedrich, M. J., DeVeaux, L. C., Kadner, R. J. 1986. *J. Bacteriol.* 167:928–34
106. Nash, H. A., Robertson, C. A., Flamm, E., Weisberg, R. A., Miller, H. I. 1987. *J. Bacteriol.* 169:4124–27
107. Miller, H. I., Friedman, D. I. 1977. In *DNA Insertion Elements,* ed. A. I. Bukhari, J. A. Shapiro, S. L. Adhya, pp. 349–56. New York: Cold Spring Harbor Lab.
108. Morisato, D., Kleckner, N. 1987. *Cell* 51:101–11
109. Way, J. C., Kleckner, N. 1986. *Proc. Natl. Acad. Sci. USA* 81:3452–56
110. Surette, M. G., Chaconas, G. 1989. *J. Biol. Chem.* 264:3028–34
111. Friedman, D. I. 1988. *Cell* 55:545–54
112. Johnson, R. C., Simon, M. I. 1985. *Cell* 41:781–91
113. Kahmann, R., Rudt, F., Koch, C., Mertens, G. 1985. *Cell* 41:771–80
114. Huber, H. E., Iida, S., Arber, W., Bickle, T. A. 1985. *Proc. Natl. Acad. Sci. USA* 82:3776–80
115. Johnson, R. C., Bruist, M. F., Simon, M. I. 1986. *Cell* 46:531–39
116. Koch, C., Kahmann, R. 1986. *J. Biol. Chem.* 261:15673–78
117. Johnson, R. C., Ball, C. A., Pfeffer, D., Simon, M. I. 1988. *Proc. Natl. Acad. Sci. USA* 85:3484–88
118. Koch, C., Vandekerckhove, J., Kahmann, R. 1988. *Proc. Natl. Acad. Sci. USA* 85:4237–41
119. Pabo, C., Sauer, R. 1984. *Annu. Rev. Biochem.* 53:293–321
120. Plasterk, R. H. A., van de Putte, P. 1984. *Biochim. Biophys. Acta* 782:111–19
121. Johnson, R. C., Glasgow, A. C., Simon, M. I. 1987. *Nature* 329:462–65
122. Bruist, M. F., Glasgow, A. C., Johnson, R. C., Simon, M. I. 1987. *Genes Dev.* 1:762–72
123. Mizuuchi, M., Mizuuchi, K. 1985. *Nucleic Acids Res.* 13:1193–208
124. Marini, J. C., Levene, S. D., Crothers, D. M., Englund, P. T. 1982. *Proc. Natl. Acad. Sci. USA* 79:7664–68
125. Wu, H.-M., Crothers, D. M. 1984. *Nature* 308:509–13
126. Diekmann, S., Wang, J. C. 1985. *J. Mol. Biol.* 186:1–11
127. Thompson, J. F., Mark, H. F., Franz, B., Landy, A. 1988. In *DNA Bending and Curvature,* ed. W. K. Olson, M. H. Sarma, R. H. Sarma, M. Sundaralingam, pp. 119–28. Guilderland, NY: Adenine
128. Kitts, P., Richet, E., Nash, H. A. 1984. *Cold Spring Harbor Symp. Quant. Biol.* 49:735–44
129. Laundon, C. H., Griffith, J. D. 1988. *Cell* 52:545–49
130. Spengler, S. J., Stasiak, A., Cozzarelli, N. R. 1985. *Cell* 42:325–34
131. Mizuuchi, K., Gellert, M., Nash, H. 1978. *J. Mol. Biol.* 121:375–92
132. Pollock, T. J., Nash, H. A. 1983. *J. Mol. Biol.* 170:1–18
133. Abremski, K., Gottesman, S. 1979. *J. Mol. Biol.* 131:637–49
134. Shulman, M., Gottesman, M. 1973. *J. Mol. Biol.* 81:461–82
135. Ross, W., Landy, A. 1982. *J. Mol. Biol.* 156:505–29
136. DeMassy, B., Studier, F. W., Dorgai, L., Appelbaum, E., Weisberg, R. A. 1984. *Cold Spring Harbor Symp. Quant. Biol.* 49:715–26
137. Hoess, R. H., Abremski, K. 1985. *J. Mol. Biol.* 181:351–62
138. Hoess, R., Abremski, K., Sternberg, N. 1984. *Cold Spring Harbor Symp. Quant. Biol.* 49:761–68
139. Hoess, R. H., Wierzbicki, A., Abremski, K. 1986. *Nucleic Acids Res.* 14:2287–300
140. Abremski, K., Wierzbicki, A., Frommer, B., Hoess, R. H. 1986. *J. Biol. Chem.* 261:391–96
141. Hoess, R., Wierzbicki, A., Abremski, K. 1987. *Proc. Natl. Acad. Sci. USA* 84:6840–44
142. Andrews, B. J., Beatty, L. G., Sadowski, P. D. 1987. *J. Mol. Biol.* 193:345–58
143. Senecoff, J. F., Cox, M. M. 1986. *J. Biol. Chem.* 261:7380–86
144. Umlauf, S. W., Cox, M. M. 1988. *EMBO J.* 7:1845–52
145. Meyer-Leon, L., Huang, L.-C., Umlauf, S. W., Cox, M. M., Inman, R. B. 1988. *Mol. Cell. Biol.* 8:3784–96
146. Andrews, B. J., McLeod, M., Broach, J. R., Sadowski, P. D. 1986. *Mol. Cell. Biol.* 6:2482–89
147. Iida, S., Hiestand-Nauer, R. 1987. *Mol. Gen. Genet.* 208:464–68
148. Reed, R. R., Grindley, N. D. F. 1981. *Cell* 25:721–28
149. Shimada, K., Weisberg, R. A., Gottesman, M. E. 1972. *J. Mol. Biol.* 63:483–503
150. Shimada, K., Weisberg, R. A., Gottesman, M. E. 1972. *J. Mol. Biol.* 80:297–314
151. Shimada, K., Weisberg, R. A., Gottesman, M. E. 1975. *J. Mol. Biol.* 93:415–29

152. Winoto, A., Chung, S., Abraham, J., Echols, H. 1986. *J. Mol. Biol.* 192:677–80
153. Bauer, C. E., Hesse, S. D., Gumport, R. I., Gardner, J. F. 1986. *J. Mol. Biol.* 192:513–27
154. Gardner, J. F., Nash, H. A. 1986. *J. Mol. Biol.* 191:181–89
155. Krause, H. M., Higgins, N. P. 1986. *J. Biol. Chem.* 261:3744–52
156. Filutowicz, M., Appelt, K. 1988. *Nucleic Acids Res.* 16:3829–32
157. Gartenberg, M. R., Crothers, D. M. 1988. *Nature* 333:824–29
158. Davies, R. W., Schreier, P. H., Kotewicz, M. L., Echols, H. 1979. *Nucleic Acids Res.* 7:2255–73
159. Rhodes, D., Klug, A. 1980. *Nature* 286:573–78
160. Klug, A., Lutter, L. C. 1981. *Nucleic Acids Res.* 9:4267–83
161. Liu, L. F., Wang, J. 1978. *Cell* 15:979–84
162. Drew, H. R., Travers, A. A. 1985. *J. Mol. Biol.* 186:773–90
163. Hamilton, D., Yuan, R., Kikuchi, Y. 1981. *J. Mol. Biol.* 152:163–69
164. Mizuuchi, K., Gellert, M., Weisberg, R. A., Nash, H. A. 1980. *J. Mol. Biol.* 141:485–95
165. Mizuuchi, K., Fisher, L. M., O'Dea, M. H., Gellert, M. 1980. *Proc. Natl. Acad. Sci. USA* 72:1847–51
166. Griffith, J. D., Nash, H. A. 1985. *Proc. Natl. Acad. Sci. USA* 82:3124–28
167. Germond, J. E., Hirt, B., Oudet, P., Gross-Bellard, M., Chambon, P. 1975. *Proc. Natl. Acad. Sci. USA* 72:1843–47
168. Richmond, T. J., Finch, J. T., Klug, A. 1983. *Cold Spring Harbor Symp. Quant. Biol.* 47:493–501
169. Mizuuchi, K., Mizuuchi, M. 1979. *Cold Spring Harbor Symp. Quant. Biol.* 43:1111–14
170. Pollock, T. J., Abremski, K. 1979. *J. Mol. Biol.* 131:651–54
171. Mizuuchi, K., Weisberg, R., Enquist, L., Mizuuchi, M., Buraczynska, M., et al. 1981. *Cold Spring Harbor Symp. Quant. Biol.* 45:429–37
172. Echols, H., Green, L. 1979. *Genetics* 93:297–307
173. Enquist, L. W., Nash, H., Weisberg, R. A. 1979. *Proc. Natl. Acad. Sci. USA* 76:1363–67
174. Bauer, C. E., Hesse, S. D., Gardner, J. F., Gumport, R. I. 1984. *Cold Spring Harbor Symp. Quant. Biol.* 49:699–705
175. Jayaram, M., Crain, K. L., Parsons, R. L., Harshey, R. M. 1988. *Proc. Natl. Acad. Sci. USA* 85:7902–6
176. Nash, H. A., Mizuuchi, K., Enquist, L. W., Weisberg, R. A. 1981. *Cold Spring Harbor Symp. Quant. Biol.* 45:417–27
177. Nash, H. A., Pollock, T. J. 1983. *J. Mol. Biol.* 170:19–38
178. McGavin, S. 1971. *J. Mol. Biol.* 55:293–98
179. Wilson, J. H. 1979. *Proc. Natl. Acad. Sci. USA* 76:3641–45
180. Nash, H. A., Bauer, C. E., Gardner, J. F. 1987. *Proc. Natl. Acad. Sci. USA* 84:4049–53
181. Bliska, J. B., Cozzarelli, N. R. 1987. *J. Mol. Biol.* 197:205–18
182. Boocock, M. R., Brown, J. L., Sherratt, D. J. 1985. *Biochem. Soc. Trans.* 14:214–16
183. Craigie, R., Mizuuchi, K. 1986. *Cell* 45:795–800
184. Wasserman, S. A., Cozzarelli, N. R. 1985. *Proc. Natl. Acad. Sci. USA* 82:1079–83
185. Wasserman, S. A., Dungan, J. M., Cozzarelli, N. R. 1985. *Science* 229:171–74
186. Meyer-Leon, L., Gates, C. A., Attwood, J. M., Wood, E. A., Cox, M. M. 1987. *Nucleic Acids Res.* 15:6469–88
187. Vetter, D., Andrews, B. J., Roberts-Beatty, L., Sadowski, P. D. 1983. *Proc. Natl. Acad. Sci. USA* 80:7284–88
188. Beatty, L. G., Babineau-Clary, P., Hogrefe, C., Sadowski, P. D. 1986. *J. Mol. Biol.* 188:529–44
189. Abremski, K., Hoess, R., Sternberg, N. 1983. *Cell* 32:1301–11
190. Abremski, K., Frommer, B., Hoess, R. H. 1986. *J. Mol. Biol.* 192:17–26
191. Abremski, K., Hoess, R. 1985. *J. Mol. Biol.* 184:211–20
192. Abremski, K., Frommer, B., Wierzbicki, A., Hoess, R. H. 1988. *J. Mol. Biol.* 201:1–8
193. Enquist, L. W., Kikuchi, A., Weisberg, R. A. 1979. *Cold Spring Harbor Symp. Quant. Biol.* 43:1115–20
194. Nash, H. A. 1975. *Proc. Natl. Acad. Sci. USA* 72:1072–76
195. van de Putte, P., Cramer, S., Giphart-Gassler, M. 1980. *Nature* 286:218–22
196. Zieg, J., Hillman, M., Simon, M. 1978. *Cell* 15:237–44
197. Benchimol, S., Lucko, H., Becker, A. 1982. *J. Biol. Chem.* 257:5201–10
198. Sumner-Smith, M., Benchimol, S., Murialdo, H., Becker, A. 1982. *J. Mol. Biol.* 160:1–22
199. Benchimol, S., Lucko, H., Becker, A. 1982. *J. Biol. Chem.* 257:5211–19
200. Becker, A. 1970. *Biochem. Biophys. Res. Commun.* 41:63–70
201. Chowdbury, M. R., Dunbar, S., Becker, A. 1972. *Virology* 49:314–18
202. Kellenberger, G., Zichichi, M. L., Weigle, J. 1961. *J. Mol. Biol.* 3:399–408

Annu. Rev. Biochem. 1989. 58:951–98

CRYSTAL STRUCTURES OF THE HELIX-LOOP-HELIX CALCIUM-BINDING PROTEINS[1]

Natalie C. J. Strynadka and Michael N. G. James

Medical Research Council of Canada Group in Protein Structure and Function, Department of Biochemistry, University of Alberta, Edmonton, Alberta, Canada T6G 2H7

CONTENTS

PERSPECTIVES AND SUMMARY

Calcium ions have a number of diverse functions in biological systems, from biomineralization in bones, teeth, and shells, to a complex role as an intracellular messenger (1). The calcium-binding proteins that have been sub-

---

[1] Abbreviations used: TnC, troponin C; CaM, calmodulin; Parv, parvalbumin; ICaBP, intestinal calcium-binding protein; MLCK, myosin light chain kinase; HLH, helix-loop-helix; EF-hand, a convenient mnemonic for the helix-loop-helix motif; rms, root mean square; NMR, nuclear magnetic resonance.

0066-4154/89/0701-0951$02.00

jected to high-resolution crystal structure analyses fall into two general categories (Table 1). One group includes many extracellular enzymes and proteins that have enhanced thermal stability or resistance to proteolytic degradation as a result of binding Ca^{2+} ions. For some of the enzymes, Ca^{2+} may play an additional role in catalysis. The other group comprises a family of intracellular proteins that reversibly bind Ca^{2+} ions and thereby modulate the action of other proteins or enzymes. This second group is distinguished from the first in that its members have a common Ca^{2+}-binding motif consisting of two helices that flank a "loop" of 12 contiguous residues from which the oxygen ligands for the calcium ion are derived. The structure of a single HLH motif has been likened to an index finger (E-helix), a curled second finger (the loop), and a thumb (F-helix) of a right hand, so that the term "EF-hand" has been widely applied to describe such a Ca^{2+}-binding site (29). A second distinguishing characteristic of the HLH family is the fact that the Ca^{2+}-binding motifs occur in intimately linked pairs (Table 1). With the exception of thermolysin, all other proteins in Table 1 contain only one calcium-binding site that has no associated helices. Thermolysin has a double site (two Ca^{2+} ions separated by 3.8 Å) in which one Ca^{2+} ion is six coordinate and the other seven coordinate (13, 14).

There have been a number of reviews concerning various structural aspects of calcium-binding to small molecules and to proteins (44–47). Subsequently, additional high-resolution crystal structure analyses or refinements of earlier structures have been completed. In writing this review, we have been extremely fortunate to have had access to the refined atomic coordinates of bovine brain CaM (38), turkey skeletal muscle TnC (41), chicken skeletal muscle TnC (43), carp Parv (31), and bovine ICaBP (34). This has allowed us to make direct structural comparisons among the members of this family.

There are several new deductions and generalizations regarding Ca^{2+}-binding sites that can be made from an analysis of the data base of new structures (Table 1). The majority of Ca^{2+}-binding sites contribute seven oxygen ligands to the metal ion. There are only three instances in which the Ca^{2+} coordination number is six. For those with seven ligands, the oxygen atoms are located approximately at the seven vertices of a pentagonal bipyramid ~2.4 Å from the central Ca^{2+} ion. Not all of the seven ligands are from the protein; many of the binding sites have one or more water molecules in the coordination sphere of the Ca^{2+} ion. The HLH family has a very characteristic and consistent deviation of one of the five ligands from the pentagonal plane. This deviation may be important in providing for both Ca^{2+} and Mg^{2+} binding.

The calcium-binding sites of the HLH proteins involve a segment of polypeptide chain with 12 contiguous residues, whereas the sites in the other proteins are discontinuous with the Ca^{2+} ligands coming from residues on

Table 1 Calcium-binding proteins crystal structures

Protein	Resolution (Å)	No. Ca^{2+} sites	Ca^{2+}-binding characteristics			Role of Ca^{2+} ion	References
			Coordinating peptide[a]	Coordination number	$K_d^{Ca^{2+}}$ (M)		
bovine β-trypsin	1.9	1	C	6	4×10^{-4}	stabilizing	2–4
Streptomyces griseus trypsin	1.7	1	D	7	—	stabilizing	5–7
subtilisin Novo	2.1	1	D	7	—	stabilizing	8
subtilisin Carlsberg	1.2, 1.8	1	D	7	—	stabilizing	9–12
thermolysin	1.6	4	D	6, 7	2×10^{-5}, 1×10^{-6}	stabilizing	13–16
phospholipase A$_2$	1.7	1	D	7	2.5×10^{-4}	stabilizing (catalytic)	17, 18
staphylococcal nuclease	1.5	1	D	6	$\sim10^{-3}$	(catalytic)	19, 20
concanavalin A	1.75	1	D	7	3×10^{-4}	cell binding	21–23
α-lactalbumin	1.7	1	C	7	10^{-6}–10^{-7}	stabilizing cofactor	24–27
D-galactose binding protein	1.9	1	D	7	—	—	28
carp Parv	1.6	2	C	7	$\sim10^{-9}$	intracellular calcium buffer	29–31
pike Parv	1.9	2	C	7	10^{-8}	intracellular calcium buffer	32
bovine ICaBP	2.3	2	C	7	10^{-8}	calcium transport	33–35
bovine CaM	2.2	4	C	7	10^{-5}, 10^{-6}	intracellular enzyme regulation	36–38
avian TnC	2.0	4	C	7	10^{-5}, 10^{-7}	muscle contraction and regulation	39–43

[a] C denotes a continuous polypeptide that contains the Ca^{2+} ligands.
D indicates that the Ca^{2+} ligands come from several separated segments of polypeptide.

widely separated segments of polypeptide chain (Table 1). There are two exceptions in this last group; bovine trypsin (2) and α-lactalbumin (24) have continuous calcium-binding segments of 11 and 10 residues, repectively.

Each of the 12 residues of the loop in an HLH motif plays an important role in defining the structure of the calcium-binding site. Invariably, five of them are involved directly in providing oxygen ligands to the Ca^{2+} ion. The other residues provide hydrogen bonding via main-chain NH groups to stabilize the geometry of the loop required for Ca^{2+} binding.

The majority of the helical segments in the HLH proteins display normal α-helical geometry with the expected n to $n + 4$ hydrogen bonding pattern. However, there are several helices (one in each protein) that are considerably bent or distorted, allowing more favorable hydrophobic interactions. Helix crossing angles have often been used as indicators of the overall molecular conformations (34, 36, 39, 44). However, detailed comparisons of the several structures show that these parameters are not particularly informative indicators of conformational differences. The interhelical packing between HLH structural units is strongly conserved. Comparison of the Ca^{2+}-free N-terminal domain of TnC with its C-terminal domain indicates that conformational changes that occur on Ca^{2+} binding probably involve coupled movements of a pair of helices (48). Such movements on Ca^{2+} binding would expose an extensive hydrophobic patch on the molecular surface. A variety of biophysical studies on TnC and CaM support this model for the Ca^{2+}-mediated conformational change.

FUNCTIONAL OVERVIEW

The helix-loop-helix Ca^{2+}-binding proteins are a family of highly homologous, intracellular proteins whose activities are regulated by the Ca^{2+}-binding event. Binding Ca^{2+} to these proteins induces in them a conformational change that is subsequently transmitted to their respective target molecules. For this reason they are often termed Ca^{2+}-modulated. The two most well-characterized examples of the family in terms of structure and function are TnC and CaM.

The cellular functions of two other members of the HLH family are less clear. Although the strong primary and tertiary structural homology of Parv and ICaBP warrant their inclusion in the TnC/CaM family, the term Ca^{2+}-modulated cannot be applied to them. There are as yet no firm biochemical data that would indicate whether Parv and ICaBP directly modulate the activity of any specific secondary target molecules. Most evidence thus far suggests a more passive role for Parv and ICaBP, in that binding of calcium to these molecules aids in the regulation of calcium concentrations in the cell.

Because protein structure and function are such an inseparable marriage, one cannot fully discuss one aspect without referring to the other. For this reason, a brief summary of the current views of the functionality of these four HLH proteins is presented.

Troponin C

TnC is a key player in the Ca^{2+}-mediated regulation of muscle contraction. The cross-bridge model proposes that the S1 heads of myosin in the thick filaments cyclically attach and detach from the actin-containing thin filaments (49). With myosin bound to actin, the complex forms a Mg^{2+}-activated ATPase, actomyosin. As ATP is hydrolyzed, the two filaments slide past one another and the contractile force is generated. Regulation of the contraction/relaxation cycle in skeletal muscle is Ca^{2+}-mediated at the level of the thin filament through the protein complex of troponin and tropomyosin (50–52). TnC is the acidic (pI 4.25) 18,000-dalton Ca^{2+}-binding component of troponin, a tripartite complex consisting of TnC, TnI, and TnT (53, 54). Troponin binds at regular intervals along the actin polymer of the thin filament. TnT is primarily responsible for binding the complex to the long coiled-coil molecule of tropomyosin. TnI inhibits the Mg^{2+}-activated ATPase of actomyosin (53). TnC binds to specific regions of TnI, and the strength of this interaction is increased in the presence of Ca^{2+} (55–60). Several studies suggest that TnI binding occurs on hydrophobic surfaces of TnC that become accessible when Ca^{2+} binds (61–64). Whatever the case, Ca^{2+} binding to TnC induces a conformational change that directly affects its interaction with TnI (65, 66). This results in a change in the disposition of the troponin/tropomyosin complex relative to actin such that either a steric hindrance to the approach of myosin heads to actin is removed (49, 67, 68) or the release of inorganic phosphate (Pi) from the actin/myosin head/ADP:Pi complex, normally blocked in the relaxed state, is allowed (69).

Skeletal TnC contains two high-affinity Ca^{2+}-binding sites ($K_d \sim 10^{-7}$ M). These sites also bind Mg^{2+} competitively ($K_d \sim 10^{-3}$ M) and are therefore called the Ca^{2+}-Mg^{2+} sites (54, 63, 70). Skeletal TnC also contains two sites of lower Ca^{2+} affinity ($K_d \sim 10^{-5}$ M) that are essentially specific for Ca^{2+} at physiological Mg^{2+} concentrations (71).

Calmodulin

CaM is an intracellular calcium receptor found in all eukaryotic cells from yeast to higher mammals (72). It is a small, acidic protein of molecular weight 16,700 whose cellular function encompasses the Ca^{2+}-mediated activation of a number of different intracellular enzymes (73) including phosphodiesterase (74, 75), myosin light chain kinase (MLCK) (76, 77), calcineurin (78), erythrocyte Ca^{2+}-ATPase (79, 80), brain adenylate cyclase (81, 82), phosphorylase kinase (83), and nicotinamide dinucleotide kinase (84). The Ca^{2+}-

saturated molecule is the active form of CaM. As in the case of TnC, conformational changes produced by Ca^{2+} binding are thought to result in exposure of hydrophobic surfaces with which target enzymes or inhibitory drugs interact (85–95).

CaM has four binding sites with association constants for Ca^{2+} falling within one order of magnitude of one another. It is generally accepted that the four Ca^{2+}-specific sites of CaM can be divided into the lower-affinity sites I and II with $K_d \sim 10^{-5}$ M and sites III and IV with slightly higher affinity ($K_d \sim 10^{-6}$ M) (89, 96–100). Biochemical characterization of CaM is extensive, and has been discussed recently elsewhere (85, 101, 102).

Parvalbumin

Parvalbumins are a large group of Ca^{2+}-binding proteins that have been isolated mainly from fast twitch muscles of fishes, amphibians, and mammals (103). Members of the family exhibit a variety of molecular weights and isoelectric points (pI). Carp Parv is small (11,000 daltons) and acidic (pI $\sim$ 4.25). Although their function is not defined, several kinetic and equilibrium Ca^{2+}/Mg^{2+}-binding studies suggest that they may be involved in the relaxation event following muscle contraction (104–107). It has been proposed that Parv is usually in the Mg^{2+} bound form in muscle. At physiological levels of Mg^{2+} (1 mM) and K^+ (80 mM), and at levels of Ca^{2+} corresponding to those of resting muscle ($\sim 10^{-8}$ M), Parv binds two Mg^{2+} ions and no Ca^{2+}. When Ca^{2+} is released into the cell following a nerve impulse, the Ca^{2+} binds first to TnC and CaM, presumably because of the very slow off rate of Mg^{2+} from Parv. Upon muscle relaxation, Parv could take up the Ca^{2+} released by TnC and CaM, as Mg^{2+} is released. Therefore it could act as a calcium buffer, by quickly reducing the calcium concentration so contraction is not reinitiated. That parvalbumins are found in greatest quantities in fast twitch muscles may support this conclusion (108). It should be noted that parvalbumins are essentially skeletal muscle proteins; they are usually not found in cardiac or smooth muscle and so are not indispensable components of the contractile mechanism.

Parvalbumins bind Ca^{2+} very tightly with impressive K_ds of up to 10^{-9} M (106, 108–110). Mg^{2+}, Na^+, and K^+ ions compete with Ca^{2+} for the metal ion binding sites of Parv only at high concentrations (109, 111). Under normal physiological conditions, Parv is not found in the apo metal-free state because of the exceptionally high affinity for this ion.

Intestinal Calcium-Binding Protein

The vitamin D–dependent ICaBP is also known as calbindin 9K. It is a soluble protein located primarily in the cytoplasm of the absorptive cells of mammalian intestinal tissue (112). As with Parv, the function of ICaBP is not certain. Various studies support its role as a Ca^{2+} buffer for vitamin D–stimulated Ca^{2+} absorption (113, 114); others suggest it may be an aqueous intracellular Ca^{2+} transporter (115). Preliminary biophysical studies indicate that ICaBP does not undergo a large conformational change upon Ca^{2+} binding (116, 117).

ICaBP binds two Ca^{2+} ions with moderate affinity ($K_d \sim 10^{-6}$ M to 10^{-8} M) (118–121). It does not appear to bind Mg^{2+} at physiological concentrations.

STRUCTURAL OVERVIEW

Primary Structures

Table 2 presents the aligned amino acid sequences of chicken skeletal TnC, bovine brain CaM, carp Parv, and bovine vitamin D–dependent ICaBP. By far the greatest sequence identity occurs between CaM and TnC (51% identity). Parv and ICaBP show considerably less homology with each other and with other members of the family.

In Table 2 the helices of each HLH unit are labeled alphabetically and the loops are denoted with Roman numerals. The nomenclature of the Ca^{2+}-binding loop derives from the original octahedral description of the Ca^{2+}-coordinating ligand geometry (29). In that description, six residues in positions 1(X), 3(Y), 5(Z), 7(−Y), 9(−X), and 12(−Z) provided oxygen ligands to the Ca^{2+} ion. In all known HLH units, the residue at −Z is a glutamate; it contributes both of its side-chain oxygen atoms to the metal ion coordination. This bidentate coordination at −Z, along with the oxygen ligands at Y, Z, and −Y, defines an approximately planar pentagonal arrangement. The invariant aspartate at X contributes one oxygen to the pyramidal apex; the opposite apex (−X) is occupied generally by a water molecule (39). Only in loop I of Parv is there a residue in position 9 that has a side chain sufficiently long to coordinate directly to the Ca^{2+}. Therefore, since there are seven oxygen ligands, the true Ca^{2+}-coordination is pentagonal bipyramidal, not octahedral as previously described. In spite of this we will use the established octahedral nomenclature given in Table 2. Also shown in Table 2 is the three-residue overlap between the Ca^{2+}-binding loops (positions 10, 11, and 12) and the following helices.

All four proteins have N-terminal extensions of variable lengths that precede the first HLH folding unit (TnC, 15 residues; CaM, 5 residues; Parv, 39 residues; and ICaBP, 2 residues). The linking peptides between HLH domains in TnC and CaM are 5 amino acids long. Those of Parv and ICaBP are 8 and 10 residues long, respectively.

The HLH proteins are characterized by relatively high percentages of acidic residues (TnC, 29%; CaM, 25%; Parv, 18%; and ICaBP, 23%). There are

Table 2 Primary sequences[b]

```
                      N
         1       5        10        15
TnC   Ac A S M T D Q Q A E A R A F L S
CaM      - - - - - - - - - - A D Q L T
                             1       5
```

```
              Helix                 Loop              Helix                Linker
                A                     I                  B
                                X   Y   Z  -Y  -X    -Z
         16    20        25      30        35       40        45        50
                                 *   *   *   *            *
TnC    E E M I A E F K A A F D M F D A D G G G D I S T K E L G T V M R M L G Q N P T
CaM    E E Q I A E F K E A F S L F D K D G D G T I T T K E L G T V M R S L G Q N P T
       6      10        15        20        25        30        35        40
```

```
                 C                    II                 D
         55        60        65        70        75        80        85        90
                                       *   *   *   *            *
TnC      K E E L D A I I E E V D E D G S G T I D F E E F L V M M V R Q M K E D A K G K S
CaM      E A E L Q D M I N E V D A D G N G T I D F P E F L T M M A R K M K D T D - - - S
         45        50        55        60        65        70        75        80
```

```
                 E                   III                 F
         95        100       105       110       115       120       125       130
                                       *   *   *   *            *
TnC      E E E L A N C F R I F D K N A D G F I D I E E L G E I L R A T G E H V T
CaM      E E E I R E A F R V F D K D G N G Y I S A A E L R H V M T N L G E K L T
         82      85        90        95        100       105       110       | 117
                                                                              Me3
```

```
                 G                    IV                 H
         131       135       140       145       150       155       160
                                       *   *   *   *            *
TnC      E E D I E D L M K D S D K N N D G R I D F D E F L K M M E G V Q
CaM      D E E V D E M I R E A D I D G D G Q V N Y E E F V Q M M T A K -
         118 120     125       130       135       140       145
```

```
                        A                                     B
         1       5        10        15        20        25        30        35
Parv  Ac A F A G V L N D A D I A A A L E A C K A A D S F D H K A F F A K V G L T S K S
```

```
                 C                    I                  D
         40        45        50        55        60        65        70        75
                                       *   *   *   *            *
Parv     A D D V K K A F A I I D Q D K S G F I E E D E L K L F L Q N F K A D A R A L T
```

```
                 E                    II                 F
         79 80      85        90        95        100       105
                                       *   *   *   *
Parv     D G E T K T F L K A G D S D G D G K I G V D E F T A L V K A
```

```
                 A                    I                  B
         1       5        10        15        20        25        30        35        40        45
                                     *A   *     *P   *                  *
ICaBP    K S P E E L K G I F E K Y A K E G D N Q L S K E E L K L L L Q T E F P S L L K G P S T
```

```
                 C                    II                 D
         50        55        60        65        70        75
                                       *   *   *   *            *
ICaBP    L D E L F E E L D K N G D G E V S F E E F Q V L V K K I S Q
```

---

[a] Helix A extends from Asp8 to Ala17; helix B from His26 to Val33 (31).

[b] The residues that form the calcium coordination sphere, labeled X, Y, Z, -Y, -X, -Z, are denoted by the astertisk above those residues in each loop.

The amino acid sequences are given in the one-letter code. The sources of the sequences are: chicken skeletal TnC (122, 123); bovine brain CaM (124, 125); carp Parv (126); bovine ICaBP (127).

also a number of conserved hydrophobic amino acids, many of which are aromatic phenylalanine residues. None of the four proteins has a tryptophan; chicken TnC has no tyrosine; bovine brain CaM has no cysteine; carp Parv has no methionine, tyrosine, or proline; bovine ICaBP has no methionine, arginine, cysteine, or histidine.

The conserved sequences of the Ca^{2+}-binding loops and their flanking helices have allowed for the derivation of a consensus sequence (34, 64). Successful prediction of HLH binding sites in newly determined protein sequences has been possible with these sequences (128–130).

The chicken sequence of TnC originally determined by Wilkinson (122) has been completed by DNA sequencing methods (123). DNA sequencing methods have corrected the original report that in CaM residue 129 was an asparagine (124). It has subsequently been confirmed as aspartic acid (125, 131). The crystal structure of CaM that has been reported (38) still retains the originally assigned asparagine at position 129.

Tertiary Structures

CRYSTALLOGRAPHIC ANALYSES   In this section we review briefly the reliability of the five crystal structure determinations. Table 3 presents a summary of the crystallographic data. All five structures have been refined to final R factors of 0.155 to 0.187. These values indicate well refined crystal structures at medium resolution.

Both avian skeletal TnC structures were refined at 2.0 Å resolution and are comparable determinations (41, 43). Analysis of the coordinates by a least-squares minimization procedure reveals that the two molecules are very similar. For the 628 common main-chain atoms, the root mean square (rms) difference is 0.29 Å; for the 1210 common atoms of whole TnC, the rms difference is 0.66 Å. The larger number is primarily a reflection of the greater inaccuracy in coordinates for the more mobile side chains, like glutamates and lysines. Other than these residues, there are only two major conformational differences in the side-chain positions between turkey and chicken TnC at Phe-105 and Ile115. Overall, the differences in these two completely independent determinations of similar molecules are minimal and compare favorably to results on dual determinations of other protein structures [e.g. bovine trypsin (135) and bovine chymotrypsin (136–138)]. Four residues were not seen in the final electron density map of turkey TnC, the first two, the last one, and the side chain of Glu67.

The CaM structure has been refined at 2.2 Å resolution (38). The positions of five of its residues were not defined also due, most likely, to disorder. The refinement of the Ca^{2+} and Cd^{2+} complexes of carp Parv have been done at the highest resolution, 1.6 Å (31). Since these two structures were

Table 3 Summary of crystallographic data

	Turkey sTnC	Chicken sTnC	CaM	Parv	ICaBP
Crystallization conditions					
	40% $(NH_4)_2SO_4$	43% $(NH_4)_2SO_4$	45–60% MPD	70–85% $(NH_4)_2SO_4$	70–80% $(NH_4)_2SO_4$
	100 mM NaOAc 2.5–5% PEG200 10 mM $CaCl_2$	50 mM NaOAc 3% PEG200 5 mM $MgCl_2$	50 mM cacodylate 4 mM $CaCl_2$	2 mM phosphate	20 mM Tris
	pH 4.9–5.0	pH 5.1	pH 5.2–5.8	pH 6.8	pH 8.9
	Ref. 132	Ref. 43	Ref. 133	Refs. 29, 30	Ref. 33
Crystallographic refinement information					
Final R[a]	Ref. 41	Ref. 43	Ref. 38	Ref. 31	Ref. 34
	.155	.172	.175	.187	.178
Resolution (Å)	10.0–2.0	8.0–2.0	5.0–2.2	10.0–1.6	9.0–2.3
Cutoff	$I \geq 2\sigma(I)$	$I \geq 2\sigma(I)$	$I \geq 2.5\sigma(I)$	$I \geq 3\sigma(I)$	not given
No. reflections	8054	8100	6685	10879	2350
No. protein atoms	1259	1248	1121	808	600
Ca^{2+} ions	2	2	4	2	2
No. solvent atoms	158	68	69	75	37 ($1SO_4 =$)
Residues not seen	1, 2, 162, 67	1, 2	1–4, 148	—	
Deviations from ideal geometry[b]					
Bond distances	.019 (.017)	.04 (.03)	.016 (.020)	.020 (.02)	.016 (.02)
Angle distances	.045 (.028)	.07 (.05)	.034 (.030)	.037 (.04)	.040 (.03)
Planar 1–4 distances	.049 (.034)	.11 (.10)	.044 (.040)	.060 (.05)	.061 (.06)
Transpeptide ω, (°) (deviations from 180°)	2.5 (3.5)	4.9 (5.0)	2.3 (5.0)	3.9 (3.0)	2.1 (3.0)
Estimated coord. error (Å)	~.30	~.17	~.15	~.15	~.25

[a] The agreement factor R is defined as $\Sigma||F_o|-|F_c||/\Sigma|F_o|$, where $|F_o|$ and $|F_c|$ are the observed and calculated structure factor amplitudes, respectively.
[b] The values given correspond to the root mean square deviations from ideal stereochemistry. The numbers in parentheses are the final input parameters to the refinement program that determine the relative weights of the corresponding restraints (134).

isomorphous and have essentially identical tertiary structures, we have included only the results for the Ca^{2+} Parv in the present discussion. The previously reported structures of Parv were of lower resolution and not completely refined (29, 30). Finally, ICaBP has been refined at 2.1 Å (34).

Solvent molecules are an integral component of protein structure. Analysis of the results from high-resolution (better than 2.0 Å) protein structures indicates that the ratio of the number of ordered solvent sites to the number of residues in the protein is roughly 1:1. This would indicate (see Table 3) that for chicken TnC, CaM, Parv, and ICaBP, the description of ordered solvent may still be incomplete, thus precluding detailed comparisons. Nonetheless, it is likely that all the strongly bound solvent molecules have been selected, and among them there are many conserved solvent sites common to the five protein structures.

Table 3 also gives the rms deviations of each of the model structures from ideal stereochemistry. The size of these numbers, in conjunction with the amount of data included in the refinement and the value of the agreement factor R, give a general indication of how complete the structure refinements have been. Chicken skeletal muscle TnC (43) has rms deviations that are approximately twice those of the other four structures. Other parameters are comparable among all of them.

Estimates of the error in the refined atomic coordinates range from 0.15 Å to 0.30 Å. The lower estimates for CaM and chicken TnC are overly optimistic. The low value of the coordinate errors (0.15 Å) for the Parv structure reflects the much higher resolution of that study. For turkey TnC, the more conservative estimate for the coordinate accuracy was determined from a σ_A plot (139).

OVERALL ARCHITECTURE   Analysis of the crystallographic structures of TnC, CaM, Parv, and ICaBP reveals a conserved structural arrangement of each Ca^{2+}-binding domain. It consists of a pair of HLH structural motifs joined by linker peptides of 5 to 10 residues in length. Figure 1 shows a diagrammatic representation of one domain (the HLH units are generically denoted as helix A–loop I–helix B and helix C–loop II–helix D). The helices are approximately 10 to 12 residues long and they flank a 12-residue loop. The HLH structures are related to each other by an approximate intradomain 2-fold rotation axis centrally located between the two loops (Figure 1). Single HLH units that bind calcium have not been observed thus far. All Ca^{2+}-binding domains analyzed to date involve a pair of HLH motifs, even though in some cases one of the sites may have lost its Ca^{2+}-binding ability [cardiac TnC (140); crayfish TnC(141)].

In all determined structures, the two HLH motifs are intimately associated.

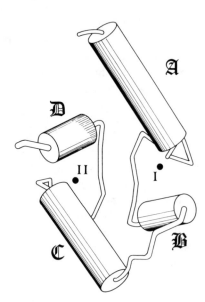

Figure 1 A diagrammatic representation of a typical HLH Ca^{2+}-binding domain. There are two HLH structural units (helix A–loop I–helix B and helix C–loop II–helix D) that are related by a pseudo-two-fold rotation axis between the loops. Ca^{2+} ions are represented by filled circles (●).

The inter-motif interactions occur via the loops and the helices. The two loops of adjacent HLH motifs interact via two antiparallel β-sheet hydrogen bonds. This is the only β-sheet secondary structure in the molecules. The nature and arrangement of the helices also play a critical role in the interaction of HLH motifs. All four helices of the domain are amphipathic. They pack with their hydrophobic faces inward, away from solvent, forming a central core that consists of a number of intra- and inter-helical hydrophobic interactions. On their solvent-exposed faces, charged residues form favorable electrostatic interactions by binding metal ions, bulk solvent, or more rarely, by forming intramolecular ion pairs. The Ca^{2+}-binding domain can be thought of as cup-shaped (Figure 1); the inside of the cup is lined with relatively exposed hydrophobic residues, the bottom of the cup contains the Ca^{2+}-binding loops, and the outside and rim of the cup is comprised of charged, hydrophilic residues. The similarity in tertiary structure of the Ca^{2+}-filled domains can be appreciated by comparing them pairwise using a least-squares procedure (Table 4). The comparison excluded the ICaBP structure because it has major conformational differences with the other domains. The smallest rms difference is between the C-terminal domain of TnC and C-terminal domain of CaM. On the other hand, the largest rms difference is between TnC-C and CaM-N. While the Ca^{2+}-binding domain is conserved in both sequence and architecture, significant structural differences among the four molecules arise mainly from modifications to the linker regions that bind adjacent HLH motifs together, and from structural additions to the N-termini of the Ca^{2+}-binding domains.

Table 4 Ca^{2+}-binding domain comparisons[a]

	TnC-C	CaM-N	CaM-C	Parv
TnC-C	—	0.951[b]	0.817	0.827
CaM-N	197	—	0.751	0.865
CaM-C	238	238	—	0.855
Parv	228	203	223	—

[a] The Ca^{2+}-binding domains are defined in Table 2. The comparisons of the corresponding pairs of HLH units were carried out using a least-squares fitting computer program originally written by W. Bennett. Only main-chain atoms (N, C^{α}, C, O) were included in the calculations; the atoms of the linking peptides were omitted. For TnC and CaM, the N- and C-terminal domains are denoted by an N or C following the name.

[b] The numbers in the upper triangle of this matrix are the rms deviations (in Å units) for each pairwise comparison. Those in the lower triangle are the number of pairs of atoms used in the least-squares superposition. Atom pairs with a final deviation larger than 1.9 Å were considered nonequivalent and omitted from the corresponding comparison.

Troponin C   TnC is a 70 Å long dumbbell-shaped molecule. The globular N- and C-terminal domains are connected by a 31-residue α helix (Figure 2a). The domains have mean radii of approximately 17 Å; their centers are separated by 44 Å (41). Each domain of TnC consists of a pair of HLH motifs. The pair in the N-terminal domain are helix A–loop I–helix B and helix C–loop II–helix D. The N-terminal domain has an additional helical structure, the N-helix, that is unique to the TnCs. Helix E–loop III–helix F and Helix G–loop IV–helix H form the EF-hand pair in the C-terminal domain. The central connecting peptide between the two domains is formed by the continuous helical sweep from the D-helix, through the D-E helical linker region, to the E-Helix (Figure 2a and Table 2).

In both X-ray crystallographic structures of chicken and turkey TnC discussed here (41, 43), the C-terminal high-affinity sites (III and IV) are occupied by metal ions, whereas the two Ca^{2+}-specific low-affinity sites (I and II) in the N-terminal domain have no bound Ca^{2+} ions. This lack of Ca^{2+}-binding is attributed to the low pH at which the crystals were obtained (Table 3). Comparison of the structures of the pair of HLH motifs in the N- and C-terminal domains indicates marked tertiary structural difference between the Ca^{2+}-free and Ca^{2+}-bound forms.

Calmodulin   From the very high sequence identity and similar biochemical properties of TnC and CaM, one would expect extensive structural similarities between the two molecules (142). Comparison of Figures 2a and 2b shows that they do exhibit similar portraits. There are three major structural differences between TnC and CaM. Both the N- and C-terminal domains of

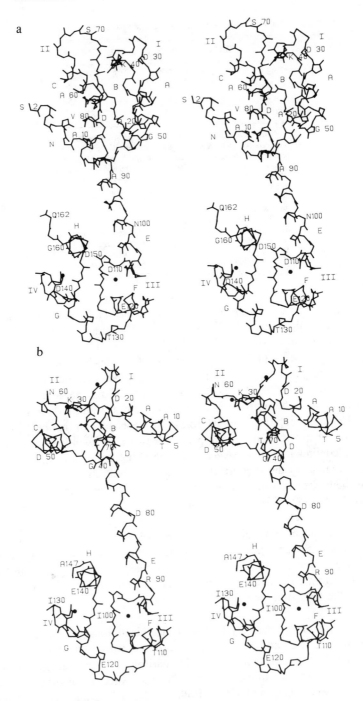

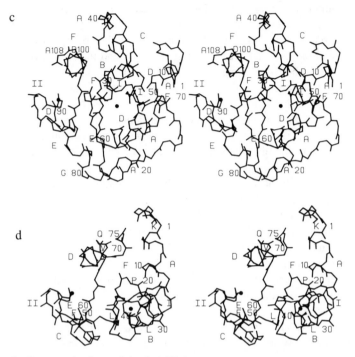

Figure 2 Stereoscopic views of the four HLH proteins discussed in this review. Only the polypeptide chain atoms, N, C$^\alpha$, C, O, and the Ca^{2+} ions of each protein are represented. The Ca^{2+}-binding domains of CaM (C-terminal domain), Parv, and ICaBP were structurally aligned to the C-terminal Ca^{2+}-binding domain of TnC and all four displayed from the same vantage point. (*a*) turkey skeletal muscle TnC. (*b*) bovine brain CaM. (*c*) carp Parv. (*d*) bovine ICaBP.

CaM are in the Ca^{2+}-bound form; the N-terminal domain of TnC is metal-free. The rms difference for the main-chain atom comparison of the N- and C-terminal domains of CaM (0.751 Å, Table 4) indicates the very similar conformations. The N-terminal helical arm of TnC has no counterpart in CaM. There is a three-residue deletion in the central D/E helical linker of CaM (see also Table 2). As a result of the deletion, CaM is approximately 5 Å shorter than TnC, with a length of 65 Å (38) and the N- and C-terminal domains are oriented differently. In Figures 2*a* and 2*b* the C-terminal domains of TnC and CaM are displayed from the same vantage point. It can be seen that relative to the fixed C-termini, the orientation of the N-terminal domain of TnC differs by ∼ 60° from that of CaM.

The different lengths and resulting orientation of domains in TnC and CaM may play a role in differentiating their function. A recent study has shown that a three-residue insertion into the central helix of CaM causes a marked decrease in the activation of some of the molecule's target enzymes (143).

The dumbbell shapes of TnC and CaM were totally unexpected (144). The refined atomic coordinates from the crystal structure studies show interatom pair distribution functions [P(r)] that are bimodal. The first peak corresponds to the mean radius of the Ca^{2+}-binding domains and the second peak, to the mean distance between domains (41, 145, 146). In addition, several small-angle X-ray scattering studies (SAXS) have evaluated the shapes of these molecules in solution (145–148). These studies showed P(r) distributions that are similar to that of the X-ray crystal structures, but with the minimum between the peaks not so pronounced. The maximum lengths of both CaM and TnC determined in solution agree well with the lengths from the single crystal X-ray studies. The results are thus consistent with the presence of a population of molecules in solution that have the extended dumbbell shape. In order to explain the larger number of intermediate-length pair distances (between 25 and 40 Å) on the solution P(r) function, two alternatives have been put forward. One of them proposes a population of molecules in solution with a bent central helix (145). A bend of ~ 65° produces good agreement with the values of the P(r) function in solution. The alternative explanation involves bound water molecules between the domains. This hydration layer would tend to raise the minimum in the interatomic distance plot (at ~ 30 Å) thereby accounting for the shape of the experimental curves (146). Most likely both explanations are valid. No clear-cut experiments that would differentiate between these alternatives have yet been done.

Parvalbumin   Unlike the dumbbell-shaped TnC and CaM, Parv is a globular molecule (Figure 2c) that can be described as a prolate ellipsoid with dimensions of $36 \times 30 \times 30$ Å (30). Parv contains six helices A through F (Table 2). Helices C and D flank loop I; E and F flank loop II. Collectively, they constitute the familiar cup-shaped Ca^{2+}-binding domain. In addition, Parv has a 39-residue N-terminal extension comprised of two helices, A and B, that flank an 8-residue loop. It has been proposed that this region is a defunct Ca^{2+}-binding site that has arisen by gene triplication (149). Unlike TnC and CaM, whose N- and C-terminal Ca^{2+}-binding domains are linked by a long central helix, the single N-terminal HLH motif of Parv folds over, packing its hydrophobic face into the hydrophobic cup formed by helices C, D, E, and F of the Ca^{2+}-binding domain. As a result, Parv has none of the exposed hydrophobic surfaces that are thought to be the sites of target molecule binding in TnC and CaM. Thus, Parv depicts a globular shape, with a buried central core of hydrophobic residues and an outer shell of hydrophilic ones.

Intestinal calcium-binding protein   ICaBP had 75 amino acid residues and is the smallest of the known HLH proteins. It consists of only four helices (A,

B, C, and D) and two loops (I and II) that make up the two HLH motifs of the Ca^{2+}-binding domain (Table 2 and Figure 2d). ICaBP is also an approximate prolate ellipsoid with dimensions of 30 Å long and 25 Å in diameter (34). The most distinguishing features of ICaBP are the altered Ca^{2+}-binding loop I [a result of a two-residue insertion (Table 2)], and the 10-residue linking peptide that joins the two HLH structural units between helices B and C. This extended peptide linker has hydrophobic residues that contribute their side chains to the hydrophobic interior of the cup formed by the inner surfaces of helixes A to D. ICaBP, like Parv, has no significant hydrophobic patches on the surface that might serve as interaction sites for target molecules.

THE CA^{2+} COORDINATION   The coordination of Ca^{2+} ions in the calcium-binding proteins that have been studied crystallographically has often been referred to as octahedral, thereby implying six ligands. The more highly refined crystal structures now show that there are seven oxygen atoms at an average distance of 2.4 Å from the Ca^{2+} ion. Five of them (in an approximate pentagonal arrangement) lie close to a common plane that includes the central Ca^{2+} ion. The vector joining the other two oxygen atoms passes near to the Ca^{2+} ion and is approximately perpendicular to this pentagonal plane (Figure 3). This defines the coordination geometry as pentagonal bipyramidal. Figure 4 shows that this is the case for all of the HLH proteins. In fact, almost all of the presently determined and highly refined crystal structures of proteins that bind Ca^{2+} ions do so in this seven-coordinate fashion (Table 1). The three exceptions seem to be bovine trypsin (2), in which the Ca^{2+} is six coordinate, thermolysin, which has a six-coordinate and a seven-coordinate double site (14), and staphylococcal nuclease, details of which have not been fully published (20, 47).

Many Ca^{2+}-binding motifs have been observed in proteins (Table 1). Some have all seven ligands originating from atoms in the protein (Parv, subtilisin). Others, including most members of the EF-hand family, have one water molecule in their ligand sphere; some have two water ligands (e.g. SGT), and others have three water ligands (e.g. site 3 of thermolysin). One extraordinary calcium-binding geometry is that from rhizopus chinensis aspartic proteinase (150); it has a single protein ligand (a main-chain carbonyl oxygen atom) and six water molecules that complete the seven-coordinate pentagonal bipyramidal geometry.

Figure 3a shows the idealized pentagonal bipyramid with its seven vertices at a distance of 2.40 Å from the central Ca^{2+} ion. Superimposed upon that are the oxygen ligands of loop III from the refined CaM structure (38). The equatorial ligands are Asp95 $O^{\delta2}$, Asn97 $O^{\delta1}$, Tyr99 O, and the two carboxyl group oxygen atoms of Glu 104. The bidentate coordination of this side chain forces the oxygen atoms at positions Y and −Y toward the oxygen atom of the

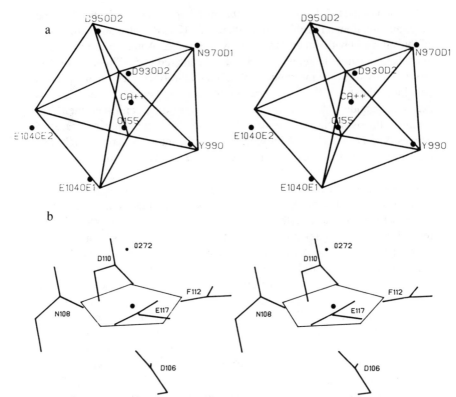

Figure 3 (*a*) A stereoscopic representation of the oxygen atom ligands of Ca^{2+}-binding site III in CaM superimposed onto an ideal pentagonal bipyramid by a least-squares procedure. (*b*) The residues of the same Ca^{2+}-binding site showing the deviation of Glu104 $O^{\epsilon 1}$ from coplanarity with the pentagonal equatorial plane. The other four oxygen ligands and the Ca^{2+} ion lie in the plane of the superposed pentagon.

ligand at Z. We have done a least-squares fitting of the seven oxygen ligands of each of the HLH calcium-binding loops to the idealized pentagonal bipyramid. The rms deviations from the ideal for the 10 refined loops range from 0.36 Å to 0.50 Å

The HLH Ca^{2+}-binding proteins depart from strict coplanarity of the Ca^{2+} ion and the five oxygen ligands that determine the equatorial pentagonal plane (Figure 3*b*). The deviation is consistently exhibited by all of the metal-binding sites in these proteins. The glutamate in position 12 (–Z) has one carboxylate oxygen that is in the plane, but the carboxylate group has been rotated so that the other oxygen atom is out of the plane by ~1.2 Å. Thus, the pentagonal arrangement of oxygen ligands has an envelope pucker that is reminiscent of one of the commonly observed conformations of a ribose ring (151).

Figure 3b also indicates that the carboxylate or amide planes of the ligands in positions X, Y, and Z are not directed toward the central Ca^{2+} ion. Rather, the Ca^{2+} is displaced by ~ 1.0 to 1.5 Å out of each of these planes (Figures 4b–f). This is not the case for most of the Ca^{2+}-binding proteins outside the HLH family (Table 1). For these, the metal ion lies close to the carboxylate or amide planes.

It is tempting to speculate that the coordination geometry of the HLH proteins is designed to accommodate, separately, both Ca^{2+} and Mg^{2+} ions. The average ligand distance for Mg^{2+} is shorter, ~ 2.1 Å (152) and its coordination number is predominantly six. One of the main determinants of the Ca^{2+} seven-fold coordination is the bidentate nature of the glutamate carboxylate and the conformation of this side chain. In order to adapt this site so that it would bind Mg^{2+}, subtle changes in the torsional angles χ_2 and χ_3 would rotate the carboxylate so that its plane would be approximately perpendicular to the equatorial plane. In this configuration only one of the oxygen atoms of the carboxylate would still be on the equatorial plane and coordinating to the metal ion, allowing the other five coordinating ligands to cluster more closely around a smaller central Mg^{2+} ion with the appropriate octahedral coordination. This could be achieved through rotations of $\sim 30°$ about χ_1 in the residues at positions X, Y, and Z, in such a way that the carboxylate or amide planes of these residues would now be directed toward the Mg^{2+} ion. This subtlety may be required for those sites that adjust to both Mg^{2+} and Ca^{2+} ions.

The seven-coordinate geometry seen for the Ca^{2+} ions holds equally well for the Cd^{2+} ions in the refined Cd parvalbumin (31). The average $Cd^{2+}\cdots O$ distance in that structure was also 2.40 Å. The seven coordinating oxygen atoms that have been confirmed in the Cd-Parv refinement agree well with the [113]Cd NMR work (153), a study that indicated that the coordination number of the metal had to be greater than six for both metal-binding sites of parvalbumin.

Comparison of loop conformations   In order to assess the similarity of conformation of the loops in these several proteins (Figure 4), we have superimposed the 48 main-chain atoms of each loop (4 × 12 residues) with one another and in pairs (96 atoms in each pair) via a least-squares procedure. The results of the comparisons of the pairs of loops are summarized in Table 5. For the C-terminal domain of TnC for both domains of CaM and for Parv, the rms deviations are approximately 0.50 Å. The pair of loops in ICaBP have a slightly different conformation. This is due to the altered binding site I. For this protein only those residues of loop I deemed to be topologically similar to the residues of the common loop were used in the comparisons (Gln22 to Glu27).

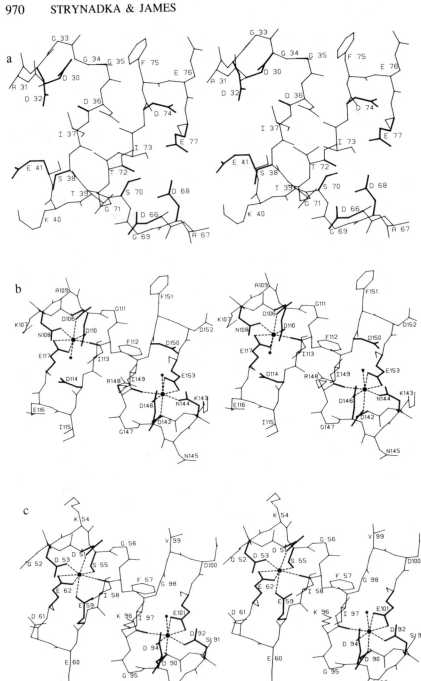

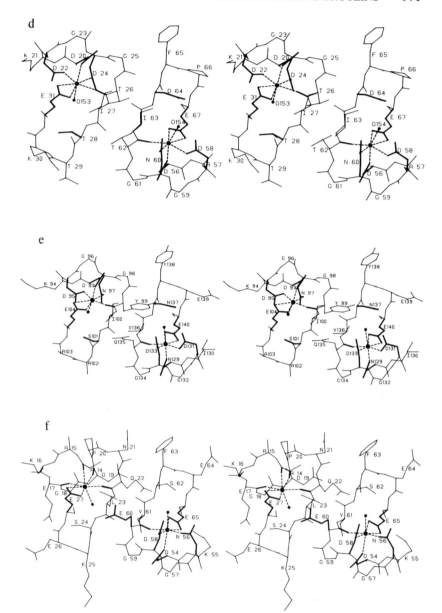

Figure 4   Stereoscopic views of the pairs of 12-residue loops in each of the Ca²⁺-binding loops of the four proteins. The Ca²⁺ ions are connected to their coordinating ligands by dashed lines. Residues in positions 1, 3, 5, 7 (C = O), 9, and 12 are represented by thick lines. (*a*) The Ca²⁺-free loops I and II of TnC. (*b*) Loops III and IV of TnC. (*c*) Loops I and II of Parv. (*d*) Loops I and II of CaM. (*e*) Loops III and IV of CaM. (*f*) Loops I and II of ICaBP. There are two insertions in the first part of loop I in ICaBP that contribute to its altered conformation.

The comparisons of the individual loops among themselves showed two important features. Firstly, the rms deviations for the 36 comparisons ranged from 0.28 Å (TnC IV vs Parv II) to 0.68 Å (CaM III vs CaM IV). These individual values are similar in magnitude to the values that result when the loops were compared as pairs (Table 5). Therefore, in the presence of Ca^{2+}, the two β strands within the loops interact with one another in a common fashion (twist angle is $\sim -34°$). Secondly, comparisons of loops I with other loops I or III and loops II with other loops II or IV show significantly smaller rms deviations than the values resulting from comparisons of loop I with loops II or IV and loops III with loops II or IV. Thus, loop I is structurally more similar to loop III and loop II is more similar to loop IV. Since all of the loop regions have highly homologous sequences (Table 2), the reason for the structural differentiation between I, III and II, IV must arise from the nature of the flanking helices. Indeed, the greatest sequence homology is between helices flanking loops I and III and those flanking loops II and IV. This EF-hand homology was first noted by Weeds & McLachlan (154). In particular, the conserved pair of aromatic residues on helices A and E, preceding loops I and III, is not present on the helices C and G, that precede loops II and IV. Similarly, the aromatic residues on helices D and H that follow loops II and IV are not on helices B and F.

The standard Ca^{2+}-binding loop   Perhaps the most convenient way to discuss the conformation of a generic Ca^{2+}-binding loop is to proceed from the N-terminus to the C-terminus and to discuss each residue in turn regarding its contribution to the metal coordination and to the loop conformation. The values of ϕ, ψ, and χ_n that are given for each position are the averages for the 11 loops that have Ca^{2+} bound in the five structures we are discussing. The spread of values of conformational angles is small for most residues of the loops. The variant loop I of ICaBP and the Ca^{2+}-free loops of TnC will be discussed separately.

Table 5   Superpositions of pairs of loops

Protein	TnC-N	CaM-N	CaM-C	Parv	ICaBP[a]
TnC-C	3.068[b]	0.512	0.491	0.428	0.765
CaM-N		—	0.510	0.495	0.878
CaM-C			—	0.594	0.994
Parv				—	0.864

[a] For loop I of ICABP the least-squares superposition used only residues Gln22 to Glu27 plus all the residues in loop II.

[b] The numbers listed are the rms deviations in Ångstrom units. They result from the least-squares fitting of the 96 main-chain atoms (N, C^α, C, O) of the loop regions of pairs of EF-hands (see footnote to Table 2).

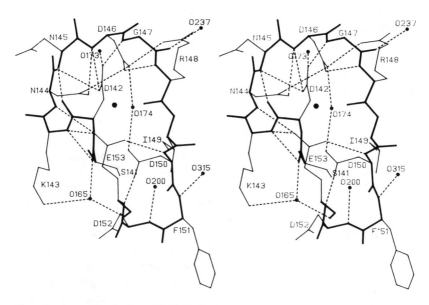

Figure 5   A stereoscopic view of Ca²⁺-binding loop IV in turkey skeletal TnC. The backbone is represented by a thick line; the side chains by a thin line. The hydrogen-bonding pattern shown (dashed lines) is preserved amongst the standard Ca²⁺-binding loops of the HLH family. Conserved water molecules are depicted as small filled circles, the Ca²⁺ ion by a larger filled circle. The β-sheet hydrogen-bonding interactions of Ile149 are not shown.

1.   Position 1 ($<\phi,\psi> = -75°, 84°; < \chi_1, \chi_2 > = -176°, 21°$).

The first residue of the loop is invariant; it is an aspartate. The incoming helix is quite regular up to this residue, but it finishes here (ψ is $+84°$). The ϕ, ψ angles are characteristic of those of the middle residue of a γ turn (155) in which there is a main-chain hydrogen bond between the first and third residues (156). This hydrogen bond, from the NH of the residue in position 2, to the CO of the last residue in the helix (position -1) is present in all loops. The turn does not reverse the chain direction, but is the secondary structural feature that terminates the incoming helix and initiates the loop.

The carboxylate group of the side chain plays a prominent role in Ca²⁺ binding (position X in Table 2) and in providing a focus around which the first six residues of the loop fold. One of the oxygen atoms of the carboxylate is a direct Ca²⁺ ion ligand (Figure 4), and it is one of the pyramidal apices of the pentagonal bipyramid. The other oxygen atom is the recipient of a strong hydrogen bond from the main-chain NH of the conserved glycine in position 6

(Figure 5). It also forms a hydrogen bond to a conserved water. Additional main-chain NH groups at positions 4 and 5 form hydrogen bonds to the coordinating oxygen atom. In light of the extensive hydrogen bonding to both oxygen atoms of this aspartate side chain, it is clear that even the conservative change to an asparagine would not be tolerated. It would result in considerable reorganization of the loop with concomitant loss of Ca^{2+} binding affinity.

2. Position 2 ($<\phi,\psi> = -65°, -37°$).

The nature of the residue in this position is not highly conserved. However, in approximately 50% of the loops, it is a basic residue, a lysine or less frequently an arginine (Table 2). The side chain of lysine 143 on loop IV of TnC extends into the solvent (Figure 5) and is directed toward the C-terminus of the incoming helix so that the positive charge on this basic residue has a strong stabilizing effect upon the helix dipole. It is curious that in loop II of the TnC molecules from different species this residue is a conserved glutamate, a negatively charged side chain that should have a pronounced destabilizing effect on the C helix. In fact, in turkey TnC there is no electron density for the side chain of Glu67, indicating that this residue is disordered and highly mobile. In the C-terminal domain of TnC (the high-affinity Ca^{2+}-binding sites) both loops have a basic residue in position 2. In addition to the γ-turn hydrogen bonding, the NH of this residue forms a bifurcated H-bond with the carboxylate oxygen of the glutamate in position 12 (Figure 5).

3. Position 3 ($<\phi,\psi> = -98°, 7°; < \chi_1, \chi_2 > = 67°, 11°$).

The residue in position 3 (Y) is most commonly an aspartate or an asparagine. It forms a direct coordination to the Ca^{2+} ion and contributes to the pentagonal plane. The main-chain nitrogen atom forms a relatively long hydrogen bond to the carboxylate of the invariant glutamate at position 12 (Figure 5). In those cases where the residue at position 3 is an asparagine, the side-chain amide can form a hydrogen bond to either the water molecule in the $-X$ position or to the side chain of the coordinating residue in position 5. This latter interaction between Asn144 $N^{\delta 2}$ and Asp146 $O^{\delta 2}$ of loop IV of turkey TnC can be seen in Figure 5.

4. Position 4 ($<\phi,\psi> = 58°, 36°$).

The most common residue in this position is a glycine. This is commensurate with the ϕ, ψ angles. Other residues that are found in position 4 include asparagine, alanine, and lysine. A bifurcated hydrogen bond from the main-chain NH is made to the carbonyl oxygen atom of the aspartate in position 1 (a Type I 3_{10} turn) and to the Ca^{2+}-coordinating oxygen atom of the same aspartate. Both distances are long, $\sim$ 3.3 Å on average for the several loops analyzed. In loop two of CaM the distances are especially long, but the directions are still such that a favorable electrostatic interaction could result.

5.   Position 5 ($<\phi,\psi>$ = $-90°$, $-3°$, $< \chi_1, \chi_2 >$ = $65°$, $10°$).

Position 5 is also a Ca^{2+}-binding ligand (Z in Table 2). The common residue is an aspartate or asparagine, but serine and glycine have also been observed. Serine at this position is particularly interesting. In the parvalbumins almost all Ca^{2+}-binding loops I have a serine at this position. Not only does the O^γ atom form a coordinating ligand to the Ca^{2+} ion, but also it forms a strong hydrogen bond to the carboxylate group of the glutamate in position 9 (distance $O\gamma\cdots O^{\epsilon 2}$ 2.8 Å, Figure 4c). A glutamate is the only amino acid with a sufficiently long side chain to coordinate directly to the Ca^{2+} from position 9 (157). For those Ca^{2+} ions that have a water molecule in the $-X$ position, the side chain of the residue in position 5 is favorably disposed to accept (aspartate) or donate (asparagine) a hydrogen bond with the water. The main-chain NH of residue 5 forms, on average, a bifurcated hydrogen bond with the oxygen atoms that coordinate to the Ca^{2+} ion of residues in positions 1 and 3 (Figure 5). The interaction with the side chain of position 3 is an n to $n + 2$ type (156). For those loops in which adjacent positions 1 and 3 both contain a charged aspartate, this favorable electrostatic interaction is highly stabilizing.

6.   Position 6 ($<\phi,\psi>$ = $90°$, $2°$).

This position is occupied by an invariant glycine. The main-chain NH forms the important hydrogen bond to the noncoordinating oxygen atom of the carboxylate of aspartate in position 1 (Figure 5). This is a relatively strong hydrogen bond with an average N to O distance of 2.7 Å in the 11 loops. The main-chain conformation for position 6 is a common conformation for a glycine residue. It assists the chain to make a 90° turn in direction so that the remaining Ca^{2+} ligands are in coordinating positions. The carbonyl-oxygen atom of glycine 6 forms a hydrogen bond with a conserved water molecule in 7 of the 11 loops. This water, in those structures for which it was selected as a well-ordered solvent, forms a hydrogen-bonded bridge to the adjacent loop of the pair at the main-chain NH of the residue in position 10, the first residue of the exiting helix. It is not present in the Ca^{2+}-free loops of the avian TnC molecules.

7.   Position 7 ($<\phi,\psi> = -126°$, 157°).

From the ϕ,ψ values it is apparent that the residue in this position initiates the small β strand that extends for three residues. Of all positions in the calcium-binding loop, this is the most variable in terms of the nature of the amino acid. The side chains extend into the solvent. However, there does seem to be a fairly common electrostatic interaction between aromatic groups (phenylalanine or tyrosine) in the first loop of a high-affinity pair and a basic residue (arginine or lysine) in the second loop. This interaction is typified by loops III and IV of TnC and loops I and II of Parv (Figure 4b,c). Loops III and IV of CaM have a tyrosine and glutamine respectively that are similarly disposed (Figure 4e).

Position 7 also provides its main-chain carbonyl oxygen atom to the coordination sphere of the Ca^{2+} ion at $-Y$ (see Table 2 and Figure 4). The β conformation allows both of the NH and the CO groups of this residue to point in toward the center of the loop. The NH forms an n to $n-2$ hydrogen bond with the Ca^{2+}-coordinating oxygen atom in the side chain of the residue in position 5. This hydrogen-bonding interaction, along with the others mentioned above, serves to stabilize partially the close proximity of negatively charged oxygen atoms in the coordination sphere of the calcium.

8.   Position 8 ($<\phi,\psi> = -108°$, 120°).

Position 8 is the central residue of the short β strand. In all of the loops this residue has a hydrophobic character with a preponderance for isoleucine. The side chains of the residues in position 8 penetrate the protein core and seem to form a central point around which the loops can adjust their conformations (158, 159). The main-chain NH and CO groups of the residue in position 8 face away from the central cavity of the loop toward the neighboring loop. Antiparallel β-sheet type interactions are formed with the main-chain amide and carbonyl group of the adjacent strand also at position 8. The pseudo-two-fold axis relating the two loops in the pair of EF-hands passes approximately between these two residues at position 8. The cooperativity in Ca^{2+} binding (101, 160, 161) may involve this region of the loop and likely includes these residues at position 8.

9.   Position 9 ($<\phi,\psi> = -94°$, 171°; $<\chi_1,\chi_2> = 68°$, 16°).

It appears that the residue in position 9 has several functions relative to the stability of the loop in a Ca^{2+}-binding conformation. It was originally proposed as one of the coordinating ligands to Ca^{2+} (29) and indeed in the parvalbumin loop I, the glutamate at this position ($-X$) does perform such a role (Figure 4c). On the other hand, in the majority of cases when the side chain at position 9 is a Ser, Thr, Asp, or Asn, its side chain is too short to coordinate directly to the Ca^{2+} and the ligand at the $-X$ position is always a water molecule (157). This is even true in the case of the less common variant loop I of ICaBP (Figure 4f). The side chain of the residue in position 9 does not always form a hydrogen bond to the coordinating water molecule. However, it does seem to be the residue that initiates the exiting helix. This initiation (or stabilization) involves the side chain of the residue in position 9 forming a hydrogen bond to the NH of the invariant glutamate in position 12 [an n to n + 3 interaction detailed by Baker & Hubbard (156)]. Several of the loops also have n to n + 2 hydrogen bonds from the position 9 side chain to the NH of the residue at position 11, but the most common interaction is the n to n + 3 (Figure 4 and 5). The majority of helices in the CaM family of proteins are initiated by similar side-chain (Ser, Thr, Asp, or Asn) to main-chain NH hydrogen-bonding intractions (see Table 2). Parvalbumin is a notable exception with Glu59 and Gly98 in the initiating positions.

Another very important interaction is provided by the main chain of position 9. The NH forms a hydrogen bond to the carboxylate group of the invariant glutamate (Figure 5). Thus, in the Ca^{2+}-bound form of these loops, the two invariant residues, aspartate at position 1 and glutamate at position 12, are the recipients of many stabilizing, favorable electrostatic interactions. The carbonyl oxygen of the residue in position 9 is involved with a normal 1-5 α-helix type hydrogen bond in the exiting helix.

10.   Position 10 ($<\phi,\psi> = -60°, -41°$).

The residue in this position is the first one of the exiting helix with helical ϕ,ψ values and normal α-helical 1–5 hydrogen bonding for its carbonyl oxygen atom. The amino nitrogen of this residue is often hydrogen bonded to a conserved water molecule (Figure 5). There is a preponderance of aromatic residues at position 10 in loop II or loop IV of a pair of EF-hands (Table 2, Figure 4). In these cases, the aromatic side chain forms part of the aromatic cluster that involves the two aromatic residues at the C-terminus of the incoming helix (at -1 and -4) of loops I and III and the two aromatics of the exiting helix of loops II and IV [position 10 of the loops and the residue following the invariant glutamate (Table 2)].

11.   Position 11 ($<\phi,\psi> = -63°, -44°$).

The residue in position 11 is most commonly a negatively charged residue, an aspartate or glutamate, although in some loops it is a basic residue, lysine in loop I of TnC and CaM. The negatively charged carboxylate groups are electrostatically stabilizing for the amino-terminus of the exiting helix. Often the NH of this residue forms a hydrogen bond to a water molecule (Figure 5).

12. Position 12 ($<\phi,\psi> = -66°, -41°$; $<\chi_1,\chi_2,\chi_3> = -72°$, 166°, -25°).

The last position in the Ca^{2+}-binding loop is also the third position of the exiting helix. Position 12 is the invariant glutamate that has both oxygen atoms of its carboxylate group coordinating to the Ca^{2+} ion in a bidentate manner (Figure 3b and Figures 4a–f). As described above, this residue could be the key residue for adapting the metal-binding loop toward the binding of either Mg^{2+} or Ca^{2+} ions. The long-range hydrogen bonds to the carboxylate oxygen atoms from main-chain NH groups of the residues in positions 2, 3, and 9 have already been described.

Until recently, it was thought that the Ca^{2+}-binding loop conformation described above only occurred in the HLH family. However, the crystal structure of the galactose-binding protein revealed a very similar tertiary structure for the Ca^{2+}-binding loop (28) in spite of the different secondary structures of the flanking segments of polypeptide chain. The loop consists of nine residues (Asp134 to Gln142), each alternate residue contributing a ligand to the Ca^{2+} ion. The final two ligands come from both oxygen atoms of the carboxylate of a neighboring glutamate (Glu205). The seven oxygen ligands are arranged in an approximate pentagonal bipyramidal fashion analogously to the one found in EF-hand loops. A least-squares overlap of the 44 common main-chain atoms in the galactose-binding protein with the main-chain atoms of the EF site of parvalbumin yielded an rms deviation of 0.60 Å, indicating the remarkable structural equivalence (28). Hydrogen-bonding interactions within the loop are similar to those shown in Figure 5 for loop IV of TnC. The similarity extends even to the stabilizing main-chain hydrogen bonds from Asn136NH and Gln142NH to both oxygens of the Glu205 carboxylate group. Although not explicitly stated in the paper, it appears that the isoleucine at position 8 in the galactose-binding protein loop forms β-type hydrogen-bonding interactions with an adjacent strand. However, this β strand is parallel and not antiparallel as in the pair of loops in the HLH motif proteins. The branched aliphatic side chain contributes to the hydrophobic interior of the domain, as in the other proteins.

N-terminal loops of TnC   The N-terminal loops of TnC are metal-free in both the turkey and chicken crystal structures (41, 43). As a result, they have quite a different conformation to the loops with Ca^{2+} bound (Figure 4a) (157). The rms difference between the N- and C-terminal loops of TnC (96 main-chain atoms in the comparison) is 3.07 Å (Table 5). On the other hand, if one compares individual loops (i.e. I with III and IV and II with III and IV), the rms deviations (48 atoms compared) are ~1.9 Å (43, 157). Furthermore, if one limits the comparisons to the first six residues of the loops or the last six residues of the loops, then rms deviations of ~0.6 Å result. This suggests that the Ca^{2+} ligands in each half of the loop are close to their optimal position for Ca^{2+} binding and only need to come together for it to take place (40, 157, 162). However, it is not a simple hinge motion; rather, cumulative small differences in ϕ and ψ over several residues in the central portion of the loop bring the two parts of the loop together to form the completed site. These Ca^{2+}-dependent differences in conformation of the loop would have attendant changes in the interhelical angles of the whole domain when Ca^{2+} ions bind.

In addition to the two antiparallel β-sheet hydrogen-bonding interactions at position 8 of the Ca^{2+}-filled loops, the Ca^{2+}-free loops I and II have the third hydrogen bond. This bond, from the NH of the residue at position 10 direct to the CO of the invariant glycine in the adjacent strand, extends the β-sheet. In the Ca^{2+}-filled conformation, the equivalent residues at positions 10 and 6 are also hydrogen bonded but through a conserved water molecule (Figure 5).

The variant loop of ICaBP   ICaBP has the common HLH motif for binding site II, but the loop of binding site I departs from the normal structure (33, 34). This structural difference has also been inferred from solution NMR studies (163). In spite of the difference, this loop also has seven ligands that coordinate to the Ca^{2+} ion. Table 2 shows that there is no sequence homology with the other EF-hand structures for the first part of loop I. The Ca^{2+}-binding ligands at X, Y, and Z do not come from side-chain oxygen atoms as in the common EF-hands; the ligands are carbonyl oxygen atoms of the main chain at Ala14, Glu17, and Asp19. Therefore the negative charges arising from aspartate residues in the coordination sphere of Ca^{2+} are replaced by peptide dipoles with the negative pole directed toward the Ca^{2+} ion. There are two insertions in the first part of the loop such that the right-angled bend in the loop normally provided by the invariant glycine at position six is now at the eighth amino acid of the loop (Asn21). In addition, the conformational angles (ϕ, ψ) of this asparagine are not the same as those of the invariant glycine.

There are stabilizing hydrogen bonds for the first part of the loop (Figure 4f). The equivalent residue to the invariant glutamate (Glu27) is the recipient of a favorable electrostatic interaction from the main-chain NH of Glu17 (position Y) and a hydrogen bond from the NH of Ser24, the equivalent to position 9 in the commonly observed loop. The side chain of Asp19, which does not provide the ligand, is involved with an n to n+2 hydrogen bond to the main-chain NH of Asn21 and in a hydrogen bond to the side chain of

Gln22 (equivalent to position 7). There is a long favorable interaction from the NH of Gln22 to the CO of Asp19 in the Z position; this has an analogous interaction in the previously described structures.

The last six residues of loop I in ICaBP (Gln22 to Glu27) are structurally similar to those equivalent residues in the standard loops. The $-X$ position of the Ca^{2+}-coordination sphere is also a water molecule. The only differences are the side-chain interactions of the residues in position 7 of loop I and loop II. Each one (Gln22 and Glu60) reaches across to the adjacent loop (Figure 4*f*) and forms a hydrogen bond to the water molecule coordinating the Ca^{2+} ion ($-X$). In spite of the very different tertiary structure that loop I has from the standard loop, the pK_d for Ca^{2+} is very similar to that of loop II and to other EF-hand loops, ~6.6 (163, 164).

HELICES

Helix regularity   Figures 2*a–d* clearly show that the predominant secondary structure in the Ca^{2+}-modulated proteins is α- or 3_{10}-helix. Analysis of the main-chain dihedral angles and hydrogen-bonding patterns indicates the following percentages of residues in helical conformations: TnC, 68%; CaM, 66%; Parv, 56%; and ICaBP, 59%. For the most part the helices are amphipathic and display the regular n to $n + 4$ hydrogen-bonding pattern typical of α-helices. However, there are notable exceptions in various helices of all four proteins. The central helix of CaM is irregular and curved. The B helix of TnC, the D helix of parvalbumin, and the D helix of ICaBP are markedly kinked. The latter three forego their α-helical conformations in specific regions in order to maximize hydrophobic interactions with adjacent helices.

Figure 6(*a*) shows a comparison of the conformations of the central parts of the two helices that connect the N- and C-terminal domains of TnC and CaM. These long helices are comprised of the relatively hydrophobic D helix, the highly charged D/E linker, and the amphipathic E helix; in total, continuous helical segments of 31 and 28 residues, respectively. In TnC, the entire length of the helix is relatively straight [a slight bend with a radius of curvature of 137 Å (41)]. Average values of ϕ and ψ are $-60°$ and $-45°$ and the hydrogen bonding is the normal n to $n + 4$ of an α helix. The only aberration in α-helical parameters occurs at the central Lys91-Gly92-Lys93 triplet where ϕ, ψ values are $-47°$ and $-56°$ (Figure 6*a*). Also in this region, Ser94 O^γ provides an additional hydrogen bond to the main-chain carbonyl oxygen of Lys91.

The central helix of CaM is much less regular in its conformational angles (38). There are two places in which the helical parameters are disrupted, at Thr70 and at Thr79. The environment of the second threonine is visible in Figure 6*a*. For both of them the n to $n + 4$ hydrogen bond is long (3.53 Å) and the ϕ, ψ values are atypical of α-helical parameters, Thr70 ($-74°$, $-34°$), Thr79 ($-100°$, $-12°$). In the region from Lys77 to Glu84, there are several bifurcated hydrogen bonds. These residues are therefore in a mixed conformation between 4_{13} and 3_{10} helices. There are also some unambiguous 3_{10} turns with n to $n + 3$ hydrogen bonding. As a result, the carbonyl oxygen atoms in this region are splayed out toward solvent (Asp78 O, Thr79 O, Asp80 O). The side-chain O^γ atom of Ser81 forms a hydrogen bond to Lys77 O; the main chain has conformational angles ϕ, ψ of $-93°$, $-63°$ and its carbonyl oxygen, Ser81 O, accepts two hydrogen bonds, one from Ile85 N (2.99 Å) and the other from Glu84 N (2.82 Å) (see Figure 6a).

The residues of the central long helix in both CaM and TnC exhibit some of the highest temperature factors in these molecules (38, 41, 43). These crystallographic studies suggest conformational flexibility in the central helix. Neither in TnC nor in CaM does this segment of the helix have close intermolecular packing contacts in the crystals that might warrant its distortions.

Some biochemical and biophysical studies have also suggested flexibility in the central helices of TnC and CaM (145, 146, 165, 166). Circumstantial support for flexibility of the central helix comes from the site of proteolytic cleavage in both CaM and TnC. The major sites of hydrolysis by bovine trypsin in these two molecules are at Lys77 and Lys87, respectively (167, 168). The conformational specificity of trypsin for substrates would require that the central helices of TnC and CaM would partly unfold in order for them to fit into its active site. In addition, one study has shown that intramolecular crosslinking of the N- and C-terminal domains of CaM could be achieved by replacing Gln3 and Thr146 with cysteine residues (169). The authors suggest that the major bend in the central helix required for the crosslinking could occur by adjusting the backbone ϕ, ψ angles of Thr79. Furthermore, they have used the concept of helix bending to formulate a model for the interaction of CaM with its target enzymes (170). However, the issue of central helix bending in the modulation of TnC and CaM activity is still very controversial (1, 51). Two recent site-specific mutagenesis studies of the central helices on TnC and CaM have provided interpretations that are contrary to helix bending in the function of these proteins. Replacement of Thr79 in CaM and Gly92 in TnC by proline residues did not significantly alter the activities of these proteins toward their target molecules (123, 143).

The B helix of TnC has a pronounced bend (Figure 6b) that is accommodated by the insertion of two solvent molecules into the helix path (43). Helix B is the exiting helix from loop I in TnC. Normally the exiting helices are initiated by the side chain of the residue in position 9 of the loop (see above). In the B helix however, Ser38 O^γ forms an n to $n + 2$ hydrogen bond with Lys40 N rather than with the NH of Glu41, as it would when initiating a helix.

In addition, the NH of Glu41 forms an n to $n + 3$ hydrogen bond to the carbonyl oxygen of Ser38, in a type III β-turn. This is followed by a type I β-turn in which the NH of Leu42 forms an n to $n + 3$ hydrogen bond to the CO of Thr39. It is at this point that the bend occurs. Solvents O181 and O183 assume the structural positions of the carbonyl oxygen atoms of Thr39 and Lys40 in an undistorted helix. These solvent atoms are involved in extensive hydrogen-bonding interactions (Figure 6b) that presumably stabilize the irregular helix. From Gly43 on, the remaining residues of helix B adopt a regular α-helical structure. A major consequence of the bend in helix B is to augment the hydrophobic interactions made by Leu42, Val45, and Met46 with the hydrophobic surfaces of helices A and D. Only in the Ca^{2+}-free form are these three helices in sufficiently close proximity to form the hydrophobic interactions. Therefore the lack of Ca^{2+} binding in loop I may be directly related to this helix distortion.

Helix D in Parv is decidedly kinked with a bend angle of 124° (Figure 6c). As for the B helix of TnC, an ordered water molecule is found in a compensating position where distortion of the helix precludes normal main-chain hydrogen bonds. Glu59, because of its involvement with Ca^{2+} coordination in loop

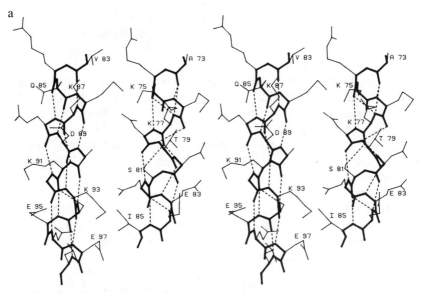

Figure 6   (*a*) A stereoview of the two central helices in TnC (*left*) and CaM (*right*). In this and in parts (*b*) and (*c*), the main chain is shown in thick lines, the side chains in thin lines, and the hydrogen bonds are dashed. (*b*) A stereoview of the kink in helix B of TnC. Two water molecules (O181 and O183) are inserted into the helix at the point where it bends. (*c*) A view of helix D in parvalbumin. The hydrogen bonding in this helix is predominantly that of a 3_{10} helix. Water O113 forms hydrogen bonds to backbone atoms of the helix.

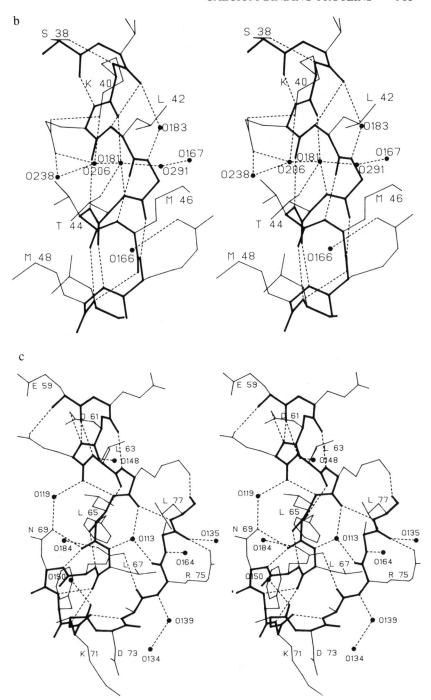

I, does not initiate the helix. From Glu60 to Lys64, the main-chain conformation is essentially α-helical. There is a type III β-turn formed by residues Glu61 to Leu65. This is followed by a type I β-turn with the NH of Phe66 forming a hydrogen bond to the CO of Leu63. The distortion in the path of the polypeptide chain occurs at Leu65 (ϕ, ψ; $-97°$, $13°$). The water molecule (O118) bridges the main-chain NH of Leu67 and the carbonyl oxygen atom of Lys64 that would normally be directly hydrogen-bonded in 3_{10} helical conformation (Figure 6c). The remaining residues of the D-helix adopt a 3_{10} helical conformation. Water O118 also forms a hydrogen bond with the main-chain carbonyl oxygen atom of Arg75.

Enhanced hydrophobic interactions are facilitated by the bending of helix D. Phe66, Leu67, and Phe70 form a hydrophobic array with which the four phenylalanine residues on the A/B domain interact.

Helix D of ICaBP begins as a regular α-helix that is initiated normally by the hydrogen bond from O^γ of Ser62 to the NH of Glu65 (34). At residue Phe66, the helix bends to an angle of 150°, adopts a 3_{10} conformation for one turn, makes a further bend to 130°, and finishes with the final six residues in an α-helical conformation. The bends allow helix D to pack tightly against the rest of the molecule and close in over the hydrophobic surface. As in the other cases, the favorable hydrophobic contacts formed presumably stabilize the distorted helix.

The above description of these several distorted helices suggests that α-helices are highly adaptable to specific environments. Large distortions in which the helix bends by up to 60° (from 180°) can be achieved by altering the main-chain conformational angles of the involved residues by only a few degrees, with the concomitant change of the hydrogen bonding interactions from a 4_{13} to a 3_{10} pattern. Solvent molecules such as water are ideally suited to compensate for the loss of main-chain hydrogen-bonding that result from such rearrangements.

Helix packing   The four helices of the Ca^{2+}-binding domains pack against one another in a highly conserved fashion (Table 6, Figure 1). Adjacent helices within a single HLH motif, i.e. E/F or G/H, are approximately perpendicular (range, 84° to 109°). They contact each other only at the ends, where they are joined to the Ca^{2+} binding loop (Figures 1, 2a–d). Adjacent helices from two separate HLH motifs (F/G and E/H) pack together with an average crossing angle of 118° (range, 110° to 124°). The main interhelix axis contact distance for these helical pairs is 9.9 Å (Table 6). The overall outcome of the above packing scheme for the Ca^{2+}-bound domains is that there is a much larger, more favorable interaction area between helices from neighboring HLH motifs (E/H and F/G) than for those within a single HLH motif.

The large variation in the interhelix axis crossing angles between E/F or

Table 6 Interhelix angles and contact distances

Helices	TnC Angle (°)	TnC Dist. (Å)[b]	CaM Angle (°)	CaM Dist. (Å)[b]	Parv[a] Angle (°)	Parv[a] Dist. (Å)[b]	ICaBP[a] Angle (°)	ICaBP[a] Dist. (Å)
N/D	61	8.3	—	—				
A/B	134	10.0	96	*				
B/C	122	9.7	112	9.4				
C/D	146	11.8	84	*				
A/D	115	11.2	116	10.4				
B/D	53	11.3	—	—				
E/F	109	*	105	*	97	*	129	8.4
F/G	124	10.2	111	9.3	121	9.1	116	10.4
G/H	108	*	91	*	106	*	119	8.7
E/H	123	10.8	120	10.4	110	9.5	123	10.9

[a] For Parv and ICaBP, the interhelix angle and distance data have been aligned with the values for the C-terminal domains of TnC and CaM only for convenience of presentation. The standardized nomenclature for these helices is for Parv CDEF and for ICaBP ABCD.

[b] An asterisk indicates that the interhelix distances have been omitted because the helix pairs concerned make contact only at the points where they enter or exit the Ca^{2+}-binding loop. Therefore such distances are not representative of the overall interhelical interactions and would be misleading (see Figure 1).

G/H pairs suggests major conformational differences in the Ca^{2+}-binding domains. However, the data in Table 4 show that this is not the case and the Ca^{2+}-bound domains are very similar. The exception is ICaBP (34), which has much larger interhelical angles within HLH motifs (129° and 119°). These angles are a result of several key hydrophobic contacts between helices B, C, and their linking peptide with helix D that curves in toward them (Figure 2d). ICaBP is the only Ca^{2+}-bound domain in which all four helices interact with one another.

The interhelical angles between helices on adjacent HLH motifs (A/D and B/C) of the Ca^{2+}-free N-terminal domain of TnC maintain the average packing angle of 118° seen in the Ca^{2+}-bound domains (48, 162) (Table 6). On the other hand, the crossing angles between helices within a single HLH motif are much larger than those of the Ca^{2+}-filled domains.

Interhelical interactions are much more favorable in the Ca^{2+}-free form. Adjacent helices within an HLH motif become nearly antiparallel instead of roughly perpendicular. This increases energetically favorable interactions in several ways. The contact area amongst all four helices in the N-terminal domain increases relative to that of the corresponding Ca^{2+}-bound domain. The helix dipoles are more antiparallel and therefore provide favorable electrostatic interactions along their length. In the Ca^{2+}-filled form, adjacent helices cross only at their ends; therefore, helix dipole interactions are limited. Furthermore, interhelix hydrophobic interactions are maximized and there are no exposed hydrophobic surfaces as in the C-terminal domain. Thus, in the absence of the favorable forces associated with Ca^{2+} binding, the helices seem to adopt a lower energy conformation with increased stabilizing helix-helix interactions (171, 172).

HYDROPHOBIC INTERACTIONS   All four proteins have a hydrophobic core located at the interface of the two HLH motifs.[2] The two conserved hydrophobic residues in position 8 of the adjacent Ca^{2+}-binding loops form the nucleus of this core (Table 2, Figures 7a–f). These two residues not only interact with each other to stabilize the loops, but also form several key contacts with conserved hydrophobic residues in each of the four helices.

The interface between helices E and H consists of a conserved set of hydrophobic interactions that involve adjacent phenylalanine residues (Figures 7b–f). The aromatic side chains of these residues (Phe102 and Phe154) pack against one another in the same manner in all four proteins. In both domains of CaM and TnC and in ICaBP, a third aromatic group (Phe151) stacks against the first two. The characteristic aromatic ring stacking, edge to face and roughly perpendicular, is observed in the cluster. In Parv, the analogous position is occupied by Val99, which also interacts with the adjacent phenylalanine side chains (Figure 7d).

Whereas the E/H interface is predominantly aromatic, the interface between helices F and G is aliphatic in nature (Figures 7b–f). The consistently larger interhelical distances between helices E and H (Table 6) may be a reflection of this phenomenon. Unlike the extensive and highly conserved packing between the E and H helices, the hydrophobic interactions between the F and G helices are more variable and often involve the methylene carbon atoms of charged residues. The conserved hydrophobic interactions between the HLH motifs extending from the E/H helices, past the loop and to the F/G helices, form the structural interior of all four molecules regardless of variation in primary sequence, interhelical angles, or distortions in secondary structure. Some of these contacts had been suggested earlier based on solution studies using NMR methods (174).

In addition to the hydrophobic core, the C-terminal domain of TnC and both domains of CaM have large expanses of hydrophobic surface exposed to solvent (38, 48, 162). Some of the residues in these regions have been implicated in the binding of target molecules (38, 61, 92, 94, 142, 175–180). The homologous residues of the Ca^{2+}-free N-terminal domain of TnC, of Parv, and of ICaBP are also hydrophobic but not accessible to solvent.

---

[2]In the following discussion, we refer to the four helices of the Ca^{2+}-binding domains in the generic sense as shown in Figure 7a (E, F, G, H). For convenience, the nomenclature given in Figure 7a is for the C-terminal domain of TnC; the equivalent positions for the other domains can be gleaned from Table 2.

Figures 7b and 7c display the hydrophobic surfaces of the C-terminal domains of TnC and CaM. The sequence (Table 2) and structural homologies of these surfaces are evident. There are many methionine residues in this region, four in each of the domains of CaM and two in the C-terminal domain of TnC. NMR studies have shown that the chemical shifts of methionine and phenylalanine residues are especially sensitive to complexation of CaM with the CaM-binding domain of MLCK (177) and with trifluoperazine, a potent CaM antagonist (175). Furthermore, chemical modification of several methionine residues in the N-terminal domain of CaM (e.g. Met71, Met72, and Met76) inhibited its activation of cyclic nucleotide phosphodiesterase (178, 146), adenylate cyclase (179), erythrocyte Ca^{2+}-ATPase (176), as well as inhibiting the binding of trifluoperazine (180). This latter study also showed that nitration of Tyr99 and Tyr138 (Figure 4e), which are on the surface of the molecule totally opposite to the hydrophobic patch (38, 142), has no effect on the ability of CaM to activate phosphodiesterase.

Conserved hydrophobic interactions involving the surface residues of TnC and CaM are shown in Figure 7a. A residue involved in one of these contacts, Phe105 of turkey TnC (Figure 7b), exhibits a major difference in conformation from the homologous residues in CaM (Figure 7c) and chicken TnC. In turkey TnC, this residue has a high energy conformation with $\chi_2 \sim 5°$. On the other, the homologous phenylalanine residues in chicken TnC and CaM adopt the minimum energy conformation with $\chi_2 \sim 90°$. It appears that chicken and turkey TnC have sampled conformational space in a different manner; neighboring residues Cys101 and Ile121 also adopt alternative conformations. In turkey TnC the mean B-factors for Phe105 are ~ 10 Å^2 (41); in chicken, the equivalent atoms have B-factors of 40 Å^2 (43).

In the Ca^{2+}-free domain of TnC, homologous hydrophobic residues are present (Figure 7f, Table 2). The N-terminal helix (N) is stabilized by the presence of several additional hydrophobic residues on helices A (Met18, Ile19) and D (Val80 and Val83). There are significant differences in the spatial distributions of some homologous hydrophobic residues due to the changed disposition of the helices in the N-terminal domain (cf. Figures 7b and 7f). As a result, this domain exhibits many more interhelical hydrophobic interactions and no large exposed hydrophobic surfaces.

Like the N-terminal domain of TnC, Parv does not have a large exposed hydrophobic surface. In the case of Parv, this is not the result of changed helix disposition, but is due to the folding of the additional 39 residues of the A/B domain to fill the hydrophobic cup formed by helices C, D, E, and F of the Ca^{2+} binding domain (Figure 2c, Figure 7d). The hydrophobic residues of helices A and B are accommodated by residues lining the hydrophobic cup unique to the parvalbumins (Table 2). Phe66, 70, and 85, and Leu77 interact with Phe24, 29, and 30 in the classical herringbone fashion. Perhaps the

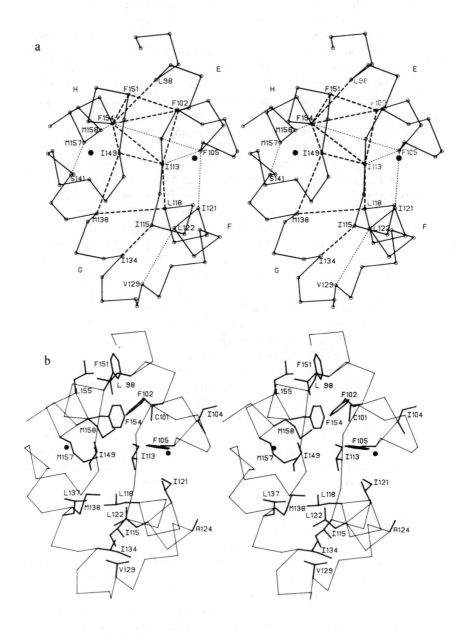

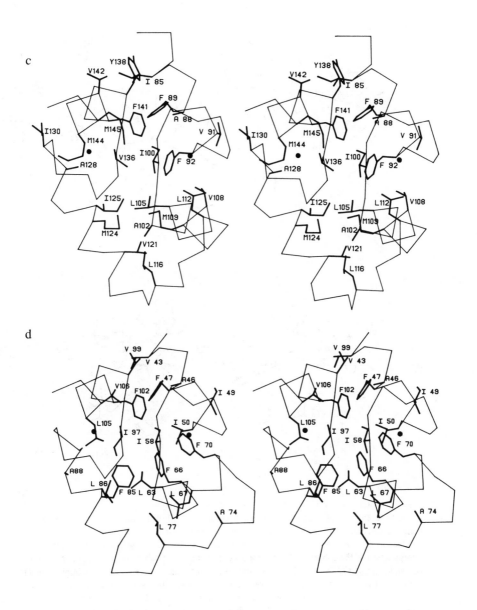

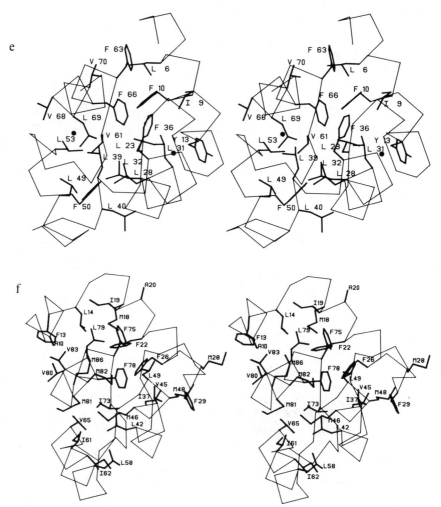

Figure 7 Stereoscopic views of the disposition of hydrophobic residues in the HLH proteins. (*a*) A representation of hydrophobic contacts common to all of the Ca^{2+}-binding domains. For convenience, the C^{α}-backbone of the C-terminal domain of TnC is displayed. The analogous residues in the Ca^{2+}-binding domains of the other proteins can be gleaned from Table 2. Our interpretation of the conserved hydrophobic contacts differs in substance from that of Sekharudu & Sundaralingam (173). Thick dashed lines between C^{α}-atoms represent hydrophobic contacts between HLH motifs that form the core of all four proteins; thin dotted lines represent common contacts among "surface" hydrophobic residues. In this view and part (*b*), Val161 and Gln162 have been omitted. In parts (*b*) to (*f*), in addition to the C^{α}-backbone, the complete side chains for the hydrophobic residues are indicated by thick lines. (*b*) C-terminal domain of TnC. (*c*) C-terminal domain of CaM. (*d*) Parv [the N-terminal 41 amino acids that fold over the hydrophobic cup have been omitted for clarity (c.f. Figure 2*c*)]. (*e*) ICaBP. (*f*) N-terminal domain of TnC (the first five residues of the N-helix are omitted).

presence of the N-terminal 39 amino acids in Parv aids in stabilizing the Ca^{2+}-bound conformation of the domain by burying the hydrophobic surface.

Like the N-terminal domain of TnC, the disposition of helices in ICaBP buries the potential hydrophobic surface (34). The distortion of the D-helix along with the hydrophobic residues on the extended peptide linker (Phe36, Leu39, and Leu40) contribute to the stabilization of the more "closed" conformation (Table 6, Figure 7e).

ELECTROSTATIC INTERACTIONS   The charged residues in the four HLH Ca^{2+}-binding proteins form electrostatic interactions with the metal ions, the bulk solvent or, less frequently, with one another. Negatively charged carboxylate groups provide many of the oxygen ligands for the Ca^{2+} ions. In addition, there are two other regions in which they are clustered together. Both TnC and CaM exhibit these clusters on the N-terminal portions of helices A, C, E, and G (Table 2). These negatively charged patches surround the hydrophobic surfaces discussed in the previous section (Figures 7a–f). The conserved nature of these charged regions in TnC and CaM suggest that they may play an important role, along with the hydrophobic surfaces, in interacting with target molecules. Indeed, the proposed TnC-binding site on TnI (51, 60), the CaM-binding site on MLCK (181), the CaM-binding site on phosphofructokinase (182), and several of the antagonists of TnC and CaM (183, 184) exhibit clusters of positive charge complementary to the negatively charged clusters on TnC and CaM. Several computer studies that have modeled the interactions of target molecules and drugs to CaM have also implicated the negatively charged regions (142, 185–187). More direct evidence of the importance of the charge cluster on helix E in CaM comes from site-directed mutagenesis in which the three glutamates were changed to lysine residues (188). The resulting mutant CaM was unable to activate the myosin light chain kinase effectively.

Positively charged amino acids do not form specific clusters as do the negatively charged groups. The most conserved lysine is in position 2 of the loops (see below). In each of the domains of CaM there is a positively charged NH_3^+-group from Lys75 and from Lys148 that extends over the hydrophobic surface. Differential labeling of these lysine residues was important in pinpointing the sites of interactions of the target enzymes with Ca^{2+}-calmodulin (189–193).

The contribution of electrostatic interactions among charged side chains to protein structure and function is difficult to evaluate. A complete description of the solvation state, the presence of metal ions, the net charge of the group, and the mobilities are needed in conjunction with the results of the crystal structure determinations in order to define what constitutes a significant charge-charge interaction. Some of these problems have been addressed in other systems (194, 195). Until similar calculations will have been done with the HLH family of Ca^{2+}-binding proteins, a meaningful description of the ion pairs is of necessity incomplete.

Examination of the structurally aligned sequences in Table 2 shows many conserved sites for the charged residues. This implies that there would also be a number of conserved ion pair interactions among them. In general, though, this is not the case. For example, the ion pair interaction between Arg84 and Glu64 is present in both avian TnC molecules. In CaM, the side chains of the homologous residues are 10 Å apart, and in ICaBP, Lys72 and Glu52 interact through a bridging water molecule. Parv has neither of these residues.

The sequences also show the presence of oppositely charged amino acids two or three residues apart in some of the helices. It has been suggested that these salt bridges provide stabilization of the long central helices of TnC and CaM (196). Only two of the seven originally predicted are confirmed as direct hydrogen-bonded ion pairs in the refined TnC structures (41, 43). The others, although too far apart to be considered direct salt bridges, could still provide favorable electrostatic interactions and enhance the level of ordered solvent in this helical region (197).

CONFORMATIONAL CHANGE OF TNC   There is ample evidence that TnC undergoes a conformational change when it binds Ca^{2+} ions (50–52). A model for this conformational change has been proposed (48). It was based upon the extensive structural similarity of the Ca^{2+}-filled domains (Table 4) and the high degree of sequence homology between these domains and the Ca^{2+}-free N-terminal domain of TnC (Table 2). In that proposal, the loops and helices of the N-terminal domain would adopt the disposition of those of the C-terminal domain upon Ca^{2+}-binding (48, 162).

Figures 8 and 9 show the essence of this proposal. The Ca^{2+}-free conformation, represented by the N-terminal domain of the TnC crystal structures, has a closed conformation, in which the hydrophobic residues of the surface are inaccessible to solvent (Figure 8a). Upon Ca^{2+}-binding, the N-terminal domain would adopt a more open conformation (Figure 8b). The transition requires a reorientation of the helices within a single HLH domain (A/B, C/D). Residues that comprise the linker peptide between helix B and helix C move by up to 14 Å (48, 162). Adjacent helical pairs A/D and B/C retain their relative orientations and interhelix crossing angles (Table 6). This is a reflection of the more extensive hydrophobic interactions that bind these pairs together. The proposal also specifies that helix N does not change its orientation relative to helix D. Such a model for the conformational change involves only a rearrangement of existing secondary structural units.

The model also proposes that Ca^{2+}-binding to the N-terminal domain would not have much effect on the conformation of the C-terminal domain.

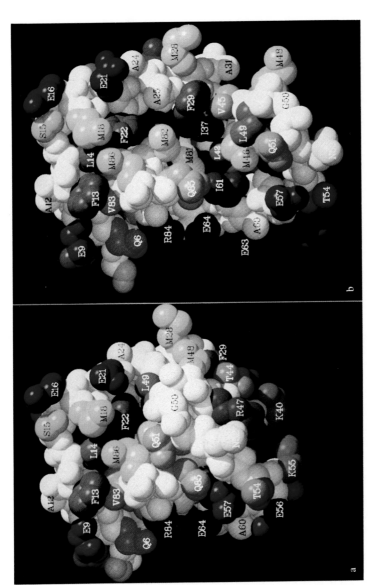

Figure 9 Surface representation of the two conformational extremes of the N-terminal domain of TnC. Atoms are shown with their appropriate van der Waals radii. These views correspond to the two views in Figure 8. This computer graphics rendering was produced with software (RASTER3D) developed by David Bacon. The color coding for the side chains is: shades of green, aliphatic residues; brown, aromatic residues; blue, basic amino acids; red, acidic; oranges and pinks, hydrophilic; yellow, methionine. (*a*) Ca^{2+}-free conformation (39,41). (*b*) Proposed Ca^{2+}-bound conformation (48).

a   b

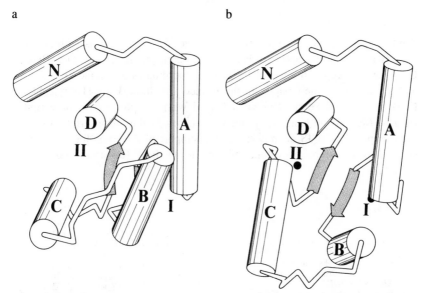

Figure 8   Diagrammatic representation of the proposed Ca^{2+}-induced conformational change in the N-terminal domain of TnC (48). In this model helixes N, A, and D retain their relative dispositions. Helixes B and C and the linker peptide move by up to 14 Å when Ca^{2+} binds. The relative dispositions of helixes B and C also remain constant. (*a*) Ca^{2+}-free conformation of the N-terminal domain of TnC. (*b*) Proposed Ca^{2+}-bound conformation of this domain.

This is consistent with the results from numerous circular dichroism and ${}^1H$ NMR studies done on whole TnC and on each of the isolated N- and C-terminal domains (50, 158, 198). Additionally, the similarity of the overall dimensions of the proposed 4-Ca^{2+}-model with the 2-Ca^{2+} crystal structures is strongly supported by the recent solution X-ray scattering data (146).

The opening of the N-terminal domain upon Ca^{2+} binding exposes a prominent hydrophobic surface analogous to those of the TnC C-terminal domain and both domains of CaM (Figures 9 and 7*b* and *c*, respectively). Exposing these hydrophobic surfaces is attended by large changes in the environments of Phe22, Phe29, Val45, Leu49, and Glu57 (Figures 9*a*,*b*). These movements are supported by an earlier NMR study on Ca^{2+} titration of the N-terminal sites of TnC that showed large changes in the chemical shifts of valine, leucine, and two or more phenylalanine and glutamate residues (63). Figure 9 also shows that the side chains of Met82 and Glu85 on helix D, and of Val45, Met46, Met48, and Leu49 on helix B have dramatically increased solvent accessibilities (48, 162). These residues, as well as Met18, Phe29, and Met86, constitute the exposed hydrophobic surface that, along with the negatively charged clusters of Glu16, Glu21, Glu56, Glu57, and Asp59 are implicated in Ca^{2+}-mediated target molecule binding.

ACKNOWLEDGMENTS

We thank Drs. Swain, Amma, and Kretsinger for supplying us with atomic coordinates of the refined carp Parv prior to their public release; Drs. Babu, Cook, and Bugg for the use of the refined coordinates of bovine brain CaM and a preprint of their paper on the refined structure. Tony Hawrylechko, Mark Israel, and Perry D'obrenan helped with the preparation of various figures. Mae Wylie handled the many revisions of this manuscript with her usual expertise. NCJS acknowledges the support of AHFMR for a studentship. This research was supported by the Medical Research Council of Canada.

Literature Cited

1. Forsén, S. 1989. In *Inorganic Biochemistry*, ed. I. Bertini. Mill Valley, Calif: Univ. Sci. Books. In press
2. Bode, W., Schwager, P. 1975. *J. Mol. Biol.* 98:693–717
3. Bier, M., Nord, F. F. 1951. *Arch. Biochem. Biophys.* 33:320–32
4. Cliffe, S. G. R., Grant, D. A. W. 1981. *Biochem. J.* 193:655–58
5. Read, R. J., James, M. N. G. 1988. *J. Mol. Biol.* 200:523–51
6. Russin, D. J., Floyd, B. F., Toomey, T. P., Brady, A. H., Awad, W. M. Jr. 1974. *J. Biol. Chem.* 249:6144–48
7. Olafson, R. W., Smillie, L. B. 1975. *Biochemistry* 14:1161–67
8. McPhalen, C. A., Svendsen, I., Jonassen, I., James, M. N. G. 1985. *Proc. Natl. Acad. Sci. USA* 82:7242–46
9. McPhalen, C. A., Schnebli, H. P., James, M. N. G. 1985. *FEBS Lett.* 188:55–58
10. McPhalen, C. A., James, M. N. G. 1988. *Biochemistry* 27:6582–98
11. Bode, W., Papamokos, E., Musil, D. 1987. *Eur. J. Biochem.* 166:673–92
12. Matsubara, H., Hagihara, B., Nakai, M., Komaki, T., Yonetani, T., Okunuki, K. 1958. *J. Biochem.* 45:251–55
13. Matthews, B. W., Weaver, L. H., Kester, W. R. 1974. *J. Biol. Chem.* 249:8030–44
14. Holmes, M. A., Matthews, B. W. 1982. *J. Mol. Biol.* 160:623–39
15. Feder, J., Garrett, L. R., Wildi, B. S. 1971. *Biochemistry* 10:4552–56
16. Voordouw, G., Roche, R. S. 1974. *Biochemistry* 13:5017–22
17. Dijkstra, B. W., Kalk, K. H., Hol, W. G. J., Drenth, J. 1981. *J. Mol. Biol.* 147:97–123
18. De Haas, G. H., Bonsen, P. P. M.,
Pieterson, W. A., van Deenen, L. L. M. 1971. *Biochim. Biophys. Acta* 239:252–66
19. Arnone, A., Bier, C. J., Cotton, F. A., Day, V. W., Hazen, E. E. Jr., et al. 1971. *J. Biol. Chem.* 246:2302–16
20. Cotton, F. A., Hazen, E. E. Jr., Legg, M. J. 1979. *Proc. Natl. Acad. Sci. USA* 76:2551–55
21. Hardman, K. D., Ainsworth, C. F. 1972. *Biochemistry* 11:4910–19
22. Becker, J. W., Reeke, G. N. Jr., Wang, J. L., Cunningham, B. A., Edelman, G. M. 1975. *J. Biol. Chem.* 250:1513–24
23. Kalb, A. J., Levitzki, A. 1968. *Biochem. J.* 109:669–72
24. Stuart, D. I., Acharya, K. R., Walker, N. P. C., Smith, S. G., Lewis, M., Phillips, D. C. 1986. *Nature* 324:84–87
25. Mitani, M., Harushima, Y., Kuwajima, K., Ikeguchi, M., Sugai, S. 1986. *J. Biol. Chem.* 261:8824–29
26. Hiraoka, Y., Sugai, S. 1984. *Int. J. Peptide Protein Res.* 23:535–42
27. Musci, G., Berliner, L. J. 1985. *Biochemistry* 24:6945–48
28. Vyas, N. K., Vyas, M. N., Quiocho, F. A. 1987. *Nature* 327:635–38
29. Kretsinger, R. H., Nockolds, C. E. 1973. *J. Biol. Chem.* 248:3313–26
30. Moews, P. C., Kretsinger, R. H. 1975. *J. Mol. Biol.* 91:201–28
31. Swain, A. 1988. *Restrained least-squares refinement of native (Ca) and Cd-substituted carp parvalbumin using X-ray data to 1.6 Å resolution.* PhD thesis. Univ. S. Carolina
32. Declercq, J.-P., Tinant, B., Parello, J., Etienne, G., Huber, R. 1988. *J. Mol. Biol.* 202:349–53
33. Szebenyi, D. M. E., Obendorf, S. K., Moffat, K. 1981. *Nature* 294:327–32

34. Szebenyi, D. M. E., Moffat, K. 1986. *J. Biol. Chem.* 261:8761–77
35. Szebenyi, D. M. E., Moffat, K. 1987. *Methods Enzymol.* 139:585–610
36. Babu, Y. S., Sack, J. S., Greenhough, T. J., Bugg, C. E., Means, A. R., Cook, W. J. 1985. *Nature* 315:37–40
37. Babu, Y. S., Bugg, C. E., Cook, W. J. 1987. *Methods Enzymol.* 139:632–42
38. Babu, Y. S., Bugg, C. E., Cook, W. J. 1988. *J. Mol. Biol.* 203:191–204
39. Herzberg, O., James, M. N. G. 1985. *Nature* 313:653–59
40. Herzberg, O., Moult, J., James, M. N. G. 1987. *Methods Enzymol.* 139:610–32
41. Herzberg, O., James, M. N. G. 1988. *J. Mol. Biol.* 203:761–79
42. Sundaralingam, M., Bergstrom, R., Strasburg, G., Rao, S. T., Roychowdhury, P., et al. 1985. *Science* 227:945–48
43. Satyshur, K. A., Rao, S. T., Pyzalska, D., Drendel, W., Greaser, M., Sundaralingam, M. 1988. *J. Biol. Chem.* 263:1628–47
44. Kretsinger, R. H. 1980. *CRC Crit. Rev. Biochem.* 8:119–74
45. Einspahr, H., Bugg, C. E. 1980. *Acta Crystallogr., Sect.* B36:264–71
46. Einspahr, H., Bugg, C. E. 1981. *Acta Crystallogr., Sect.* B37:1044–52
47. Einspahr, H., Bugg, C. 1984. In *Metal Ions Biological Systems*, ed. H. Sigel, 17:51–97. New York/Basel: Dekker
48. Herzberg, O., Moult, J., James, M. N. G. 1986. *J. Biol. Chem.* 261:2638–44
49. Huxley, H. E. 1973. *Cold Spring Harbor Symp. Quant. Biol.* 37:361–76
50. Leavis, P. C., Gergely, J. 1984. *CRC Crit. Rev. Biochem.* 16:235–305
51. Zot, A. S., Potter, J. D. 1987. *Annu. Rev. Biophys. Biophys. Chem.* 16:535–59
52. Kay, C. M., McCubbin, W. D., Sykes, B. D. 1987. *Biopolymers* 26:S123–S144
53. Greaser, M. L., Gergely, J. 1973. *J. Biol. Chem.* 248:2125–33
54. Potter, J. D., Gergely, J. 1975. 250:4628–33
55. Grabarek, Z., Drabikowski, W. 1981. *J. Biol. Chem.* 256:13121–27
56. Leavis, P. C., Rosenfeld, S. S., Gergely, J. 1978. *J. Biol. Chem.* 253:5452–59
57. Talbot, J. A., Hodges, R. S. 1981. *J. Biol. Chem.* 256:2798–802
58. Syska, H., Wilkinson, J. M., Grand, R. J. A., Perry, S. V. 1976. *Biochem. J.* 153:375–87
59. Weeks, R. A., Perry, S. V. 1978. *Biochem. J.* 173:449–57
60. van Eyk, J. E., Hodges, R. S. 1988. *J. Biol. Chem.* 263:1726–32
61. Drabikowski, W., Dalgarno, D. C.,

Levine, B. A., Gergely, J., Grabarek, Z., Leavis, P. C. 1985. *Eur. J. Biochem.* 151:17–28
62. Cachia, P. J., Gariépy, J., Hodges, R. S. 1985. In *Calmodulin Antagonists and Cellular Physiology*, ed. H. Hidaka, D. J. Hartshorne, Chapter 5, pp. 63–88. New York: Academic
63. Levine, B. A., Coffman, D. M. D., Thornton, J. M. 1977. *J. Mol. Biol.* 115:743–60
64. Gariépy, J., Hodges, R. S. 1983. *FEBS Lett.* 160:1–6
65. Ingraham, R. H., Swenson, C. A. 1984. *J. Biol. Chem.* 59:9544–48
66. Wang, C.-K., Cheung, H. C. 1985. *Biophys. J.* 48:727–39
67. Haselgrove, J. C. 1973. *Cold Spring Harbor Symp. Quant. Biol.* 37:341–52
68. Wakabayashi, T., Huxley, H. E., Amos, L. A., Klug, A. 1975. *J. Mol. Biol.* 93:477–97
69. Chalovich, J. M., Chock, P. B., Eisenberg, E. 1981. *J. Biol. Chem.* 256:575–78
70. Potter, J. D., Seidel, J. C., Leavis, P., Lehrer, S. S., Gergely, J. 1976. *J. Biol. Chem.* 251:7551–56
71. Ellis, P. D., Strang, P., Potter, J. D. 1984. *J. Biol. Chem.* 259:10348–56
72. Means, A. R., Tash, J. S., Chafouleas, J. G. 1982. *Physiol. Rev.* 62:1–39
73. Klee, C. B., Crouch, T. H., Richman, P. G. 1980. *Annu. Rev. Biochem.* 49:489–515
74. Lin, Y. M., Liu, Y. R., Cheung, W. Y. 1974. *J. Biol. Chem.* 249:4943–54
75. Teo, T. S., Wang, J. H. 1973. *J. Biol. Chem.* 248:5950–55
76. Hathaway, D. R., Adelstein, R. S. 1979. *Proc. Natl. Acad. Sci. USA* 76:1653–57
77. Dabrowska, R., Sherry, J. M. F., Aromatorio, D. K., Hartshorne, D. J. 1978. *Biochemistry* 17:253–58
78. Stewart, A. A., Ingebritsen, T. S., Manalan, A., Klee, C. B., Cohen, P. 1982. *FEBS Lett.* 137:80–84
79. Lynch, T. J., Cheung, W. Y. 1979. *Arch. Biochem. Biophys.* 194:165–70
80. Hanahan, D. J., Taverna, R. D., Flynn, D. D., Ekholm, J. E. 1978. *Biochem. Biophys. Res. Commun.* 84:1009–15
81. Brostrom, M. A., Brostrom, C. O., Wolff, D. J. 1978. *Arch. Biochem. Biophys.* 191:341–50
82. Westcott, K. R., La Porte, D. C., Storm, D. R. 1979. *Proc. Natl. Acad. Sci. USA* 76:204–8
83. Cohen, P., Burchell, A., Foulkes, J. G., Cohen, P. T. W., Vanaman, T. C., Nairn, A. C. 1978. *FEBS Lett.* 92:287–93

84. Anderson, J. M., Cormier, M. J. 1978. *Biochem. Biophys. Res. Commun.* 84: 595–602
85. Klee, C. B., Vanaman, T. C. 1982. *Adv. Protein Chem.* 35:213–321
86. La Porte, D. C., Wierman, B. M., Storm, D. R. 1980. *Biochemistry* 19: 3814–19
87. Tanaka, T., Hidaka, H. 1980. *J. Biol. Chem.* 255:11078–80
88. Tanaka, T., Hidaka, H. 1981. *Biochem. Biophys. Res. Commun.* 101:447–53
89. Klevit, R. E., Dalgarno, D. C., Levine, B. A., Williams, R. J. P. 1984. *Eur. J. Biochem.* 139:109–14
90. Krebs, J., Buerkler, J., Guerini, D., Brunner, J., Carafoli, E. 1984. *Biochemistry* 23:400–3
91. Follenius, A., Gerard, D. 1984. *Biochem. Biophys. Res. Commun.* 119: 1154–60
92. Dalgarno, D. C., Klevit, R. E., Levine, B. A., Scott, G. M. M., Williams, R. J. P., et al. 1984. *Biochim. Biophys. Acta* 791:164–72
93. Burger, D., Cox, J. A., Comte, M., Stein, E. A. 1984. *Biochemistry* 23: 1966–71
94. Jarrett, H. W. 1984. *J. Biol. Chem.* 259:10136–44
95. Zimmer, M., Hofmann, F. 1987. *Eur. J. Biochem.* 164:411–20
96. Seamon, K. B. 1980. *Biochemistry* 19: 207–15
97. Crouch, T. H., Klee, C. B. 1980. *Biochemistry* 19:3692–98
98. Wolff, D. J., Poirier, P. G., Brostrom, C. O., Brostrom, M. A. 1977. *J. Biol. Chem.* 252:4108–17
99. Martin, S. R., Andersson Teleman, A., Bayley, P. M., Drakenberg, T., Forsén, S. 1985. *Eur. J. Biochem.* 151:543–50
100. Teleman, A., Drakenberg, T., Forsén, S. 1986. *Biochim. Biophys. Acta* 873: 204–13
101. Forsén, S., Vogel, H. J., Drakenberg, T. 1986. In *Calcium and Cell Function,* ed. W. Y. Cheung, 6:113–57. New York: Academic
102. Means, A. R. 1988. *Recent Prog. Horm. Res.* 44:223–86
103. Baron, G., Demaille, J., Dutruge, E. 1975. *FEBS Lett.* 56:156–60
104. Blum, H. E., Lehky, P., Kohler, L., Stein, E. A., Fischer, E. H. 1977. *J. Biol. Chem.* 252:2834–38
105. Pechère, J.-F., Derancourt, J., Haiech, J. 1977. *FEBS Lett.* 75:111–14
106. Haiech, J., Derancourt, J., Pechère, J.-F., Demaille, J. G. 1979. *Biochemistry* 18:2752–58
107. Gillis, J. M., Thomason, P., Lefevre,

I., Kretsinger, R. H. 1982. *J. Muscle Res. Cell. Motil.* 3:377–98
108. Wnuk, W., Cox, J. A., Stein, E. A. 1982. In *Calcium and Cell Function, ed.* W. Y. Cheung, 2:243–78. New York: Academic
109. Pechère, J. F. 1977. In *Calcium Binding Proteins and Calcium Function, ed.* R. H. Wasserman, R. A. Carradino, E. Carafoli, R. H. Kretsinger, D. H. Maclennan, F. L. Siegal, pp. 212–21. New York: North Holland
110. Permyakov, E. A., Yarmolenko, V. V., Emelyanenko, V. I., Burstein, E. A., Closset, J., Gerday, C. 1980. *Eur. J. Biochem.* 109:307–15
111. Grandjean, J., Laszlo, P., Gerday, C. 1977. *FEBS Lett.* 81:376–80
112. Taylor, A. N. 1983. In *Calcium Binding Proteins,* ed. B. de Bernard, G. L. Sottocasa, G. Sandri, E. Carafoli, A. W. Taylor, pp. 207–13. Amsterdam: Elsevier Science
113. Wasserman, R. H., Fullmer, C. S. 1982. See Ref. 108, pp. 175–216
114. Wasserman, R. H., Shimura, F., Meyer, S. A., Fullmer, C. S. 1983. See Ref. 112, pp. 183–205
115. Levine, B. A., Williams, R. J. P. 1982. See Ref. 108, pp. 1–38
116. Dalgarno, D. C., Levine, B. A., Williams, R. J. P., Fullmer, C. S., Wasserman, R. H. 1983. *Eur. J. Biochem.* 137:523–29
117. Chiba, K., Mohri, T. 1987. *Biochemistry* 26:711–15
118. Shelling, J. G., Sykes, B. D., O'Neil, J. D. J., Hofmann, T. 1983. *Biochemistry* 22:2649–54
119. O'Neil, J. D. J., Dorrington, K. J., Hofmann, T. 1984. *Can. J. Biochem. Cell Biol.* 62:434–42
120. Bryant, D. T. W., Andrews, P. 1984. *Biochem. J.* 219:287–92
121. Shelling, J. G., Sykes, B. D. 1985. *J. Biol. Chem.* 260:8342–47
122. Wilkinson, J. M. 1976. *FEBS Lett.* 70:254–56
123. Reinach, F. C., Karlsson, R. 1988. *J. Biol. Chem.* 263:2371–76
124. Watterson, D. M., Sharief, F., Vanaman, T. C. 1980. *J. Biol. Chem.* 255: 962–75
125. Watterson, D., Burgess, W. H., Lukas, T., Iverson, D., Marshak, D. R., et al. 1984. *Adv. Cyclic Nucleotide Protein Phosphorylation Res.* 16:205–26
126. Coffee, C. J., Bradshaw, R. A. 1973. *J. Biol. Chem.* 248:3305–12
127. Fullmer, C. S., Wasserman, R. H. 1981. *J. Biol. Chem.* 256:5669–74
128. Tufty, R. M., Kretsinger, R. H. 1975. *Science* 187:167–69

129. Kobayashi, T., Takagi, T., Konishi, K., Ohnishi, K., Watanabe, Y. 1988. *Eur. J. Biochem.* 174:579–84
130. Hardin, S. H., Keast, M. J., Hardin, P. E., Klein, W. H. 1987. *Biochemistry* 26:3518–23
131. Putkey, J. A., Ts'ui, K. F., Tanaka, T., Lagacé, L., Stein, J. P., et al. 1983. *J. Biol. Chem.* 258:11864–70
132. Herzberg, O., Hayakawa, K., James, M. N. G. 1984. *J. Mol. Biol.* 172:345–46
133. Cook, W. J., Sack, J. S. 1983. *Methods Enzymol.* 102:143–47
134. Hendrickson, W. A., Konnert, J. H. 1980. In *Computing in Crystallography*, ed. R. Diamond, S. Ramaseshan, K. Venkatesan, pp. 13.01–13.23. Bangalore, India: Indian Acad. Sci., Int. Union Crystallography
135. Chambers, J. L., Stroud, R. M. 1979. *Acta Crystallogr., Sect.* B35:1861–74
136. Blevins, R. A., Tulinsky, A. 1985. *J. Biol. Chem.* 260:4264–75
137. Tsukada, H., Blow, D. M. 1985. *J. Mol. Biol.* 184:703–11
138. Fujinaga, M., Sielecki, A. R., Read, R. J., Ardelt, W., Laskowski, M. Jr., James, M. N. G. 1987. *J. Mol. Biol.* 195:397–418
139. Read, R. J. 1986. *Acta Crystallogr., Sect.* A42:140–49
140. van Eerd, J.-P., Takahashi, K. 1976. *Biochemistry* 15:1171–80
141. Wnuk, W., Schoechlin, M., Stein, E. A. 1984. *J. Biol. Chem.* 259:9017–23
142. Strynadka, N. C. J., James, M. N. G. 1988. *Proteins: Struct. Funct. Genet.* 3:1–17
143. Putkey, J. A., Ono, T., VanBerkum, M. F. A., Means, A. R. 1988. *J. Biol. Chem.* 263:11242–49
144. Schutt, C. 1985. *Nature* 315:15
145. Heidorn, D. B., Trewhella, J. 1988. *Biochemistry* 27:909–15
146. Hubbard, S. R., Hodgson, K. O., Doniach, S. 1988. *J. Biol. Chem.* 263:4151–58
147. Seaton, B. A., Head, J. F., Engelman, D. M., Richards, F. M. 1985. *Biochemistry* 24:6740–43
148. Fujisawa, T., Ueki, T., Inoko, Y. 1987. *J. Appl. Crystallogr.* 20:349–55
149. Kretsinger, R. H. 1972. *Nature* 240:85–88
150. Suguna, K., Bott, R. R., Padlan, E. A., Subramanian, E., Sheriff, S., et al. 1987. *J. Mol. Biol.* 196:877–900
151. Saenger, W. 1984. In *Principles of Nucleic Acid Structure*, ed. C. R. Cantor, pp. 55–69. New York: Springer-Verlag

152. Shannon, R. D., Prewitt, C. T. 1988. *Acta Crystallogr., Sect.* B25:925–46
153. Rodisiler, P. F., Amma, E. L. 1982. *J. Chem. Soc. Chem. Commun.*, pp. 182–84
154. Weeds, A. G., McLachlan, A. D. 1974. *Nature* 252:646–49
155. Nemethy, G., Printz, M. P. 1972. *Macromolecules* 5:755–58
156. Baker, E. N., Hubbard, R. E. 1984. *Prog. Biophys. Mol. Biol.* 44:97–179
157. Herzberg, O., James, M. N. G. 1985. *Biochemistry* 24:5298–302
158. Dalgarno, D. C., Klevit, R. E., Levine, B. A., Williams, R. J. P., Dobrowolski, Z., Drabikowski, W. 1984. *Eur. J. Biochem.* 138:281–89
159. Ikura, M., Minowa, O., Hikichi, K. 1985. *Biochemistry* 24:4264–69
160. Teleman, O., Drakenberg, T., Forsén, S., Thulin, E. 1983. *Eur. J. Biochem.* 134:453–57
161. Linse, S., Brodin, P., Drakenberg, T., Thulin, E., Sellers, P., et al. 1987. *Biochemistry* 26:6723–35
162. Herzberg, O., Moult, J., James, M. N. G. 1986. In *Calcium and the Cell, Ciba Found. Symp. 122*, ed. D. Evered, J. Whelan, pp. 120–44. Chichester, UK: Wiley
163. Vogel, H. J., Drakenberg, T., Forsén, S., O'Neil, J. D. J., Hofmann, T. 1985. *Biochemistry* 24:3870–76
164. Fullmer, C. S., Wasserman, R. H. 1980. In *Calcium Binding Proteins: Structure and Function*, ed. F. L. Siegel, E. Carafoli, R. H. Kretsinger, D. H. MacLennan, R. H. Wasserman, pp. 363–70. New York: Elsevier/North-Holland
165. Wang, C.-L. A., Gergely, J. 1986. *Eur. J. Biochem.* 154:225–28
166. Grabarek, Z., Leavis, P. C., Gergely, J. 1986. *J. Biol. Chem.* 261:608–13
167. Newton, D. L., Oldewurtel, M. D., Krinks, M. H., Shiloach, J., Klee, C. B. 1984. *J. Biol. Chem.* 259:4419–26
168. Drabikowski, W., Grabarek, Z., Barylko, B. 1977. *Biochim. Biophys. Acta* 490:216–24
169. Persechini, A., Kretsinger, R. H. 1988. *J. Biol. Chem.* 263:12175–78
170. Persechini, A., Kretsinger, R. H. 1989. *J. Cardiovasc. Pharmacol.* In press
171. Weber, P. C., Salemme, F. R. 1980. *Nature* 287:82–85
172. Sheridan, R. P., Levy, R. M., Salemme, F. R. 1982. *Proc. Natl. Acad. Sci. USA* 79:4545–49
173. Sekharudu, Y. C., Sundaralingam, M. 1988. *Protein Eng.* 2:139–46
174. Aulabaugh, A., Niemczura, W. P.,

Blundell, T. L., Gibbons, W. A. 1984. *Eur. J. Biochem.* 143:409–18

175. Krebs, J., Carafoli, E. 1982. *Eur. J. Biochem.* 124:619–27

176. Guerini, D., Krebs, J., Carafoli, E. 1987. *Eur. J. Biochem.* 170:35–42

177. Klevit, R. E., Blumenthal, D. K., Wemmer, D. E., Krebs, E. G. 1985. *Biochemistry* 24:8152–57

178. Walsh, M., Stevens, F. C. 1978. *Biochemistry* 17:3924–30

179. Thiry, P., Vandermeers, A., Vandermeers-Piret, M.-C., Rathe, J., Christophe, J. 1980. *Eur. J. Biochem.* 103:409–14

180. Tanaka, T., Ohmura, T., Hidaka, H. 1983. *Pharmacology* 26:249–57

181. Blumenthal, D. K., Takio, K., Edelman, A. M., Charbonneau, H., Titani, K., et al. 1985. *Proc. Natl. Acad. Sci. USA* 82:3187–91

182. Buschmeier, B., Meyer, H. E., Mayr, G. W. 1987. *J. Biol. Chem.* 262:9454–62

183. Prozialeck, W. C., Weiss, B. 1982. *J. Pharmacol. Exp. Ther.* 222:509–14

184. Erickson-Viitanen, S., O'Neil, K. T., DeGrado, W. F. 1987. In *Protein Engineering*, ed. D. L. Oxender, C. F. Fox, pp. 201–11. New York: Liss

185. Gariépy, J., Hodges, R. S. 1983. *Biochemistry* 22:1586–94

186. O'Neil, K. T., DeGrado, W. F. 1985. *Proc. Natl. Acad. Sci. USA* 82:4954–58

187. Gresh, N. 1987. *Mol. Pharmacol.* 31:617–22

188. Craig, T. A., Watterson, D. M., Prendergast, F. G., Haiech, J., Roberts, D. M. 1987. *J. Biol. Chem.* 262:3278–84

189. Jackson, A. E., Carraway, K. L. III, Puett, D., Brew, K. 1986. *J. Biol. Chem.* 221:12226–32

190. Manalan, A. S., Klee, C. B. 1987. *Biochemistry* 26:1382–90

191. Winkler, M. A., Fried, V. A., Merat, D. L., Cheung, W. Y. 1987. *J. Biol. Chem.* 262;15466–71

192. Faust, F. M., Slisz, M., Jarrett, H. W. 1987. *J. Biol. Chem.* 262:1938–41

193. Mann, D. M., Vanaman, T. C. 1988. *J. Biol. Chem.* 263:11284–90

194. Warshel, A. 1978. *J. Phys. Chem.* 83:1640–52

195. Gilson, M. K., Honig, B. 1988. *Proteins: Struct. Funct. Genet.* 4:7–18

196. Sundaralingam, M., Drendel, W., Greaser, M. 1985. *Proc. Natl. Acad. Sci. USA* 82:7944–47

197. Sundaralingam, M., Sekharudu, Y. C., Yathindra, N., Ravichandran, V. 1987. *Proteins: Struct. Funct. Genet.* 2:64–71

198. Drakenberg, T., Forsén, S., Thulin, E., Vogel, H. J. 1987. *J. Biol. Chem.* 262:672–78

Annu. Rev. Biochem. 1989. 58:999–1027
Copyright © 1989 by Annual Reviews Inc. All rights reserved

TOPOGRAPHY OF MEMBRANE PROTEINS

Michael L. Jennings

Department of Physiology and Biophysics, University of Texas Medical Branch, Galveston, Texas 77550

CONTENTS

PERSPECTIVES AND SUMMARY

Nearly two decades ago Bretscher (1) presented experimental evidence that a single polypeptide chain can span the lipid bilayer of a biological membrane. In the intervening years, considerable effort has been devoted to determining

999

0066-4154/89/0701-0999$02.00

the topographical arrangement in the lipid bilayer of integral membrane polypeptides. The classical approaches to the determination of membrane protein topography are vectorial labeling and in situ proteolysis; these methods are still in use and continue to provide valuable information. More recently there have been advances in the use of immunological methods for the localization of exposed portions of membrane proteins. Another important recent development is a method for studying membrane protein topography by constructing fusion proteins with marker enzymes that are active only on one side of the membrane. The increase in the availability of complete amino acid sequences for membrane proteins has led to computational methods for predicting topography from sequence. Although these methods are becoming increasingly sophisticated, there are several examples in which predictions disagree with experimental data. The purpose of this review is to summarize the methods, both computational and experimental, that are currently being used to study membrane protein topography.

In the context of the larger problem of protein structure determination, it should be emphasized that the topographical information reviewed here is essentially two-dimensional and can never provide a high-resolution, three-dimension image of a membrane protein. Even if all the membrane-spanning segments and hydrophilic loops were identified, and if the boundaries of these regions relative to the lipid bilayer were known, the structure of the protein would still be seriously underdetermined. Given the limitations of the con-clusions, and the difficulty of the work itself, why not just forego the two-dimensional picture and devote more effort to getting high-resolution crystallographic information? The answer is that for many membrane proteins it will be years before high-resolution crystal structures are determined, and the relatively crude two-dimensional models are the best current hope for providing concrete, testable hypotheses about structure-function relations. In addition, the topographical maps place major constraints on the possible folding patterns of a given protein, and those constraints should be of value in interpreting intermediate-resolution diffraction data and structural data obtained from other physical techniques.

SCOPE AND LIMITATIONS

Many hundred recent papers have been devoted to the subject of membrane protein structure and function. In a review of this kind it is necessary to limit the subject matter, and there are several topics that will not be covered in depth.

The sequences of dozens of membrane proteins have been determined from the use of recombinant DNA technology (e.g. 2–10). The availability of complete sequence data represents a major advance in the study of membrane proteins, because conventional sequence determination had proven to be difficult for integral membrane proteins. Knowledge of the sequence provides important information about the possible arrangements of a protein in the membrane, and several computational methods have been described for interpreting sequence data in terms of transmembrane topography (11–17). These methods will be reviewed but not discussed in detail.

This article will be confined mainly to the relatively small number of membrane proteins for which a large body of biochemical information is available (e.g. the nicotinic acetylcholine receptor, rhodopsins, Na,K-ATPase, Ca-ATPase, *Escherichia coli* β-galactoside transporter, red cell band 3, glucose transporter, sucrase-isomaltase, glycophorin, ADP-ATP exchanger). For these systems no attempt has been made to present current models of each structure. The emphasis is on approaches to the study of topography, and specific systems will be discussed as examples of how various approaches have been used.

A subject that is very closely related to the arrangement of membrane proteins is the insertion of the proteins during (or in some cases after) synthesis. In fact, several excellent discussions of membrane protein topography have been written from the standpoint of explaining how a particular topography came about (18–23). A consideration of membrane protein insertion is far beyond the scope of this article and is mentioned only very briefly. Another limitation of this article is that it considers only membrane proteins that are inserted into a membrane in the cell in which the protein was synthesized. Accordingly, the topographies of self-inserting proteins such as diphtheria toxin, mellitin, colicins, or components of the complement system will not be discussed. There is also no mention of fatty acylation or phosphoinositide anchoring of membrane proteins.

Yet another major subject that has been omitted is the quaternary structure of membrane proteins. Subunit associations are undoubtedly important in determining the structure and function of membrane proteins, and there is a large literature on the subject. There are many examples of associations among integral subunits of membrane proteins that take place by lateral arrangements in the plane of the membrane. This article includes only very cursory mention of the lateral associations between subunits; rather, the emphasis is on the arrangement of an individual polypeptide chain relative to the lipid bilayer.

A central reason for studying membrane protein topography is to provide insight into the function of the protein. The transmembrane segments of proteins are very important in such functions as active transport, coupled antiport and cotransport, transmembrane signaling, and ion conduction. Any consideration of structure-function relations in specific proteins in this article would necessarily be so superficial as to be worthless. Therefore, an effort has been made to confine discussion to topography alone, with minimal reference to the mechanisms of action of particular proteins.

ESTABLISHED CLASSES OF MEMBRANE PROTEIN TOPOGRAPHY

Topology and Topography

The words *topography* and *topology* are both commonly used to refer to the arrangement of membrane polypeptides relative to the two sides of a bilayer. There is little reason to make a major distinction between them, but *topology* refers mainly to the number and orientation of the membrane-spanning segments, and *topography* has a somewhat broader meaning. A protein's topography is defined by not only the number of membrane crossings but also the shape of the protein surfaces. Much of this review concerns methods for defining topology rather than actually specifying the three-dimensional topography of protein surfaces. However, to avoid switching back and forth between similar terms, *topography* will be used throughout.

The definition of an integral membrane has not changed significantly since 1972 (24): an integral membrane protein requires disruption of the lipid bilayer in order to be released from the membrane. A commonly used criterion is that an integral protein or protein fragment is not released from the membrane by exposure to very alkaline pH (e.g. 25–27), indicating that it interacts strongly and hydrophobically with the lipid bilayer. The goal of study of the topography of a membrane protein is to determine all the portions of the protein that are buried in the bilayer and to determine the "sidedness" of the sequences that are exposed to the aqueous media.

Monotopic and Bitopic Proteins

Blobel (18) classified integral membrane proteins on the basis of the number of times the polypeptide spans the membrane. Monotopic proteins are hydrophobically associated with the membrane but do not pass all the way across the bilayer. Bitopic proteins cross the membrane exactly once, and polytopic proteins cross the membrane more than once.

Monotopic proteins are rare. Cytochrome b5 assumes a monotopic configuration, with a hairpin loop that does not pass all the way through the bilayer (28–30), although some evidence suggests that the polypeptide may actually cross the membrane (31) when the protein is in the "tight-binding" form. It should be recognized that a hairpin loop with a beta turn in the middle of the bilayer is energetically unfavorable because of backbone polar groups that are not hydrogen bonded. Indeed, in a survey of soluble proteins of known structure, Rose et al (32) found that interior turns are always associated with one or more water molecules in a hydrogen-bonded complex. It is unknown whether water is associated with the mid-bilayer loop of cytochrome b5.

Bitopic membrane proteins are quite common and include spike proteins of enveloped viruses (33, 34), glycophorin (35), HL-A antigens of human cells (36), and intestinal sucrase-isomaltase (37), all of which have most of their mass extracellular. Other bitopic proteins have major cytoplasmic domains, e.g. the liver asialoglycoprotein receptor (38). The epidermal growth factor receptor, the insulin receptor beta subunit, and related receptors and oncogene products (39, 40) are also very likely to be bitopic proteins, with major domains on both sides of the membrane. Some bitopic proteins are oriented with the N-terminus facing the cytoplasm and others have the opposite orientation.

Polytopic Proteins

The number of possible classes of polytopic membrane proteins is quite large. For the purpose of illustrating the variety of known or suspected topographies of membrane proteins, six categories of polytopic proteins are defined (Figure 1). The definitions are based on the number of membrane crossings and the dispositions of the two termini. Figure 1 shows that at least one example of each of these classes of polytopic proteins exists or is believed to exist. Given that essentially all imaginable topographies of membrane proteins have been found in nature, there are few a priori constraints on the possible arrangements of a protein in a bilayer. Accordingly, the task of defining the number and positions of the membrane-spanning segments is formidable.

INFERENCES FROM SEQUENCE

Membrane-Spanning Helices

It is well known that the hydrophobic interior of a lipid bilayer is an energetically very unfavorable environment for peptide bonds, unless the polar groups are hydrogen bonded (e.g. 12). For this reason, and because of the tendency of hydrophobic amino acid side chains to partition into lipid, a membrane-spanning alpha helix is an energetically favorable structure for a sequence of hydrophobic amino acid residues. An antiparallel pair of helices should also be stabilized by the opposing electric dipoles associated with the backbones (56). The thickness of the hydrocarbon portion of membranes varies somewhat among cells and organisms, but it is commonly accepted that an alpha helix of about 20 residues is sufficiently long to span a lipid bilayer if the helix is oriented perpendicular to the bilayer plane.

Given the theoretical tendency for a hydrophobic sequence to form alpha helix in a lipid bilayer, and the experimental fact that many integral membrane proteins are high in alpha helix content (57–61), it is not surprising that

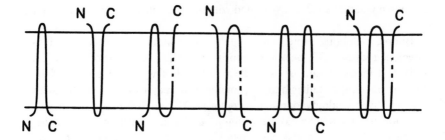

CYTOPLASM

Figure 1 Diagrammatic representation of possible transmembrane arrangements of polytopic membrane proteins. The following categories are defined: two crossings; odd number (>1) of crossings; even number (>2) of crossings. For each of these groups, there are two possible arrangements of the termini. Each group is represented by at least one known or strongly suspected example (see Table 1).

major efforts have been made toward identifying amino acid sequences that are likely to exist as membrane-spanning alpha helices. In 1982, Kyte & Doolittle (11) published a method for representing the average hydrophobicity of each sequence of N residues in a larger sequence. The hydrophobicity, or "hydropathy" is plotted as a function of distance along the polypeptide chain, and sequences of high hydropathy are more likely to exist as membrane-spanning helix. As the authors stated, "although the method is not unqiue and embodies principles that have long been appreciated, its simplicity and its graphic nature make it a very useful tool for the evaluation of protein structures." Indeed, it has become very common for publications of membrane protein sequence from cDNA sequence to include a Kyte-Doolittle hydropathy plot.

Engelman et al (12) have recently reviewed methods, including that of Kyte & Doolittle, for identifying membrane-spanning helices from protein sequence data. The Goldman-Engelman-Steitz (GES) method for calculating hydropathy is significantly different from that of Kyte & Doolittle; the former assigns more polarity to glutamate, aspartate, lysine, and arginine, and it also includes a contribution from the entropy of immobilization. For some proteins, both methods give the same result, in that the same segments would be identified as transmembrane helices. However, when polar or charged groups are present in an otherwise nonpolar segment, different methods can give somewhat different profiles.

It has been noted previously (see 11, 12) but is worth repeating that hydropathy profiles can be misleading. For example, one of the most hydrophobic segments known is a stretch of 19 residues in dogfish lactate dehydrogenase, a water-soluble protein. Similarly, trypsinogen has a hydrophibic stretch of residues that could be mistaken for a transmembrane

Table 1 Examples of the different classes of polytopic membrane proteins that are represented pictorially in Figure 1

Number of crossings	Sidedness of N-terminus	Examples (ref)
Two	Cytoplasmic	Bacterial aspartate receptor (41)
Two	Extracytoplasmic	Signal peptidase (42, 43)
Three or more (odd)	Cytoplasmic	L, M subunits of photosynthetic reaction center (44, 45)
Three or more (odd)	Extracytoplasmic	Rhodopsins (46–49)
Four or more (even)	Cytoplasmic	Glucose transporter (10)
		Band 3 (50, 51)
		β-galactoside transporter (52, 53)
Four or more (even)	Extracytoplasmic	Delta subunit of AChR?? (see 54, 55)

segment (12). Klein et al (13) have developed a method based on discriminant analysis for the detection of membrane-spanning polypeptides and for distinguishing a hydrophobic segment of a soluble protein from a membrane-spanning helix. The method successfully distinguishes between integral and soluble proteins for nearly all the proteins in a large (NBRF) data base; however, several lactate dehyrogenases would be classified as an integral by this method, and two integral proteins (out of 66) would be classified as peripheral. The method successfully predicts the positions of the boundaries of the seven membrane-spanning helices of bacteriorhodopsin.

Amphipathic Helices

Computational approaches to identifying transmembrane helices can usually detect the single membrane-spanning helix of bitopic proteins. The problem is more difficult for polytopic proteins, many of which function as transport pathways for hydrophilic solutes and therefore are not likely to have an entirely hydrophobic interior. Such proteins might be expected to have amphipathic helices with one polar face and one apolar face; the polar face would interact with the polar solute and the apolar face would be in contact with the bilayer. Argos et al (14) devised a method for detecting amphipathic helices based on recognition of periodicities in hydrophobicity. This approach has been criticized on the grounds that a random sequence of polar and nonpolar amino acids has a significant probability of having most of the polar residues on one side of a helix that is 20 residues long (15).

An improved method for detecting amphipathic helices is the hydrophobic moment plot of Eisenberg et al (16, 17). The hydrophobic moment is calculated from the hydrophobicity and orientation of successive side chains arranged in an alpha helix. Sequences in which hydrophobic and hydrophilic residues tend to orient in opposite directions have a high hydrophilic moment. The sequences most strikingly indicative of amphipathic helices are in surface-seeking proteins such as mellitin and delta-hemolysin (16). In permanently anchored membrane proteins with multiple crossings there are certain segments that, if helical, would have one face that is more polar than the other. However, it remains to be seen whether the sequences identified from sequence as potential amphipathic helices actually assume such structures in the proteins themselves. For example, the sequences of the subunits of the nicotinic acetylcholine receptor include some apparently amphipathic helices, but direct evidence indicates that these segments may not be embedded in the bilayer (see below).

How polar can a transmembrane alpha helix be? Not enough is known about membrane protein structure to answer this question. Transmembrane helix B of the L subunit of the bacterial photosynthetic reaction center has five charged residues concentrated in half the length of the helix (44). The proposed helix 4 in rhodopsin (62) has a rather low hydropathy score because of the presence of charged residues. In the absence of other information, neither of these helices would be easily predicted from sequence.

Because of the known presence of charged residues in transmembrane helices, it is worthwhile to ask how these charges can be stabilized. Honig & Hubbell (63) have pointed out that the removal of an ion pair from water to a region of low dielectric constant is not as energetically unfavorable as had been previously believed; they predict that ion pairs may be common in membrane proteins. Such pairs could be between adjacent helices or between two residues (with two or three intervening uncharged residues) in the same helix. Ion pairing of course would be expected to have an important role in the function of many transport proteins.

In discussing the placement of charged residues in a transmembrane protein, it would be very useful to know how much, if any, water is associated with the transmembrane portion of the protein. A recent study of the red cell glucose transporter by Alvarez et al (64) has shown that more than 80% of the protein backbone is accessible to deuterium exchange, indicating considerably more solvent water contact than had been expected from the sequence, which is rather hydrophobic. Other proteins, such as the calcium ATPase and rhodopsin, do not exchange backbone hydrogen as readily, indicating less contact with water (65, 66). It is of considerable interest to continue to determine solvent water accessibility for various membrane proteins, especially those suspected of forming hydrophilic channels.

Hydrophilic Connector Loops

The identification of the sidedness of hydrophilic loops between transmembrane helices places major restriction on the possible arrangements of the polypeptide, and a major goal of much chemical labeling, proteolysis, and immunological work is to establish the sidedness of such loops. The study of membrane protein synthesis and insertion has led to some insights regarding the characteristics of intracellular and extracellular sequences that flank transmembrane helices. Von Heijne (21, 21a) has compiled a large amount of sequence data on both bitopic and polytopic transmembrane proteins with the goal of identifying common features of either extracellular or intracellular sequences that flank transmembrane helices. There are theoretical arguments, at least for prokaryotes, that hydrophobic membrane-spanning segments of polypeptide should be flanked by positively charged residues on the cytoplasmic side and negatively charged residues on the extracytoplasmic side (19–23; 67). In keeping with this idea, in the data base examined by von Heijne (21, 21a), short extracytoplasmic hydrophilic loops of polytopic proteins contain only about 25% as many positively charged side chains as do hydrophilic cytoplasmic loops of similar length. A cluster of positive charges adjacent to a hydrophobic stretch is therefore indication that the positive charges are cytoplasmic.

The "positive inside" correlation of von Heijne is not claimed to be conclusively diagnostic, and there are already examples of short extracellular segments that have no negatively charged residues and two positively charged residues. The sequence for the mouse erythroid anion exchanger between residues 571 and 583 is believed to be a connector loop between two transmembrane helices. The loop contains two lysines and no aspartates or glutamates (6). The corresponding portion of human band 3 is undoubtedly extracellular; it contains sites for extracellular proteolysis of intact cells, as established in at least four laboratories (68–71). Therefore, the "positive in" rule definitely does not apply to this hydrophilic loop of band 3. As knowledge of sequence and topography of polytopic membrane proteins increases, the number of positively charged extracellular loops will probably increase, especially for eukaryotic proteins that are inserted into a membrane (the rough endoplasmic reticulum) that has very little transmembrane electrical potential difference. For prokaryotes, the "positive in" correlation may hold up better because of the negative (inside) potential difference that would tend to keep positive charges cytoplasmic. For example, the L and M subunits of the photosynthetic reaction center of purple bacteria are arranged according to the "positive in" rule (44).

Nonhelical Structures

The emphasis of this and most other discussions of integral membrane protein topography has been on alpha helices. This emphasis is a consequence of the known high helix content of many membrane proteins (57–61) and of the existence of many known sequences that are suggestive of transmembrane alpha helix. It is probable that beta turns are common elements of hydrophilic loops between transmembrane helices. In addition to beta turns, however, there is evidence that beta strands exist in several integral membrane proteins (64, 72–77). The best examples are the porin proteins of the outer membrane gram-negative bacteria (72, 73). Recent applications of circular dichroism and Fourier transfer infrared spectroscopy (74) have provided evidence for substantial amounts of beta strand in the Ca-ATPase of sarcoplasmic reticulum (75), the red cell glucose transporter (64), myelin proteolipid (76), and even bacteriorhodopsin (75, 77). The existence of beta strands in bacteriorhodopsin, however, is controversial (see 78).

The evidence for beta strands in the membrane-bound portions of some membrane proteins makes it necessary to consider the possibility of nonhelical membrane-spanning sequences. Indeed, some models of well-studied membrane proteins include such segments. For example, the adenine nucleotide translocator of the mitochondrial inner membrane is believed to have a beta strand that spans the membrane (79, 80). One recent model of the alpha subunit of the AChR includes a nonhelical membrane-crossing sequence (81). The thermodynamic arguments against placing a beta strand in a hydrophobic environment are of course still valid (12). Beta sheets, unless closed into a cylindrical barrel, have hydrogen bonding requirements for the backbones at each edge of the sheet. Despite this theoretical argument in favor of alpha helix, membrane-associated beta strands may be common (but not universal) elements of integral membrane proteins.

Inferences from Intron-Exon Boundaries

The sequence of genomic DNA is known for several eukaryotic membrane proteins, including hamster 3-hydroxy-3-methylglutaryl coenzyme A reductase (82), bovine rhodopsin (62), mouse anion transporter (83), and *Drosophila* sodium channel (84). In these genes the intron-exon boundaries are mainly in the sequences coding for hydrophilic connector loops rather than in the middle of transmembrane helices. One apparent exception is in rhodopsin, where the first intron is at a position that is in the interior of the third helix (62). In the case of HMG CoA reductase, each membrane-spanning segment is coded by its own exon (79), but in bovine rhodopsin there are seven transmembrane helices coded by four exons (59). For the anion transporter, the topography is not known precisely, but there is a reasonable correlation between exons and membrane-spanning segments (83). In the gene for the *Drosophila* sodium channel (84), no introns are in three highly conserved membrane-spanning sequences, but there are introns in two other sequences that are believed to be membrane-spanning. In summary, it is too early to tell how well intron-exon boundaries correlate with membrane protein topography, and it will be worthwhile to continue such comparisons as the data base for genomic sequence increases.

Negative Inference

Throughout the above discussion, caution has been advised against over-interpretation of computations of secondary structure and topography from sequence alone. The same cautions have been expressed by the authors of the methods cited (11–17). Most of the caveats, however, concern *positive* inferences, e.g. that a particular sequence is a membrane-spanning helix. Although they are not as often discussed, *negative* inferences from sequence, e.g. that there is no membrane-spanning segment in a particular sequence, can be important. For example, a large body of careful work by Semenza and coworkers (85–87) had led to the conclusion that the intestinal sucrase-isomaltase complex was anchored to the extracellular side of the brush border membrane by a hydrophobic hairpin loop that crossed the membrane twice. However, the cDNA sequence reveals that there is one hydrophobic segment near the N-terminus; this sequence is of the appropriate length to span the membrane once, but the flanking sequences are much too hydrophilic to be suggestive of a membrane-spanning helix (37). The knowledge of the sequence led to further experiments that showed that the previous topographical studies of the protein had been misleading, in part because of leakiness of the membrane preparation (see 37). Thus, the sequence provides powerful evidence for the absence of a second transmembrane helix. This kind of negative inference is most convincing for proteins that do not have a large number of membrane-spanning segments.

EXPERIMENTAL APPROACHES

Vectorial Labeling

The concept of "vectorial" labeling is well established and is in principle quite simple. Vectorial labeling requires a sealed membrane preparation (either inside-out or right-side-out) and also requires a labeling agent that only has access to one side of the membrane. Labeling agents can be covalently reacting radioactive, fluorescent, or spin-labeled small molecules, or they can be antibodies or proteolytic enzymes. Before considering any of these methods in detail, it is worthwhile to emphasize the importance of the membrane preparation itself.

THE MEMBRANE PREPARATION   Many differnt kinds of membrane preparations have been used in studies of protein topography, including intact cells, right-side-out plasma membrane vesicles (87–91), inside-out vesicles (92–99), whole organelles (100, 101), membranes sealed around latex (102) or silica particles (103), and lipid bilayer membranes containing reconstituted protein (e.g. 28, 29, 55, 98). The simplest kind of vectorial labeling experiment is to use a membrane-impermeant reagent to label sites that are outside the permeability barrier of a sealed membrane preparation of known sidedness. Successful application of vectorial labeling depends critically on the integrity of the membrane preparation. Numerous criteria for sealing have been used, including flotation on density gradients; electron microscopic observation of the exclusion of a marker; latency of intravesicular marker enzyme; accessibility of antibody to intravesicular marker; direct measurements of permeability; direction of active ion transport; toxin binding; and extent of labeling of known *trans* proteins, e.g. cytoplasmic proteins of intact cells or proteins sealed inside vesicles.

Unfortunately, it is sometimes very difficult to be certain that the preparation is really impermeable to the labeling reagent, and direct measurements of permeation are often not performed because of limited intravesicular volume or because external binding can mimic permeation. The issue of permeability of the reagent is nonetheless very important, and it may not be sufficient to have a preparation that is mainly resealed, with a small fraction of leaky membranes. For example, if a particular site on the *trans* side of the membrane is much more reactive than any site on the *cis* side, the *trans* site could be the main labeled site even if 90% of the membranes are completely impermeable to the reagent.

The working assumption in nearly all vectorial labeling experiments is that a charged reagent cannot cross the permeability barrier. However, there are exceptions. For example, isethionyl acetimidate was shown by Whitely & Berg (104) not to penetrate the red cell membrane on the time scale of the labeling, but the reagent has recently been shown to permeate rapidly the membrane of the glycosome of *Trypanosoma brucii* (105). The glycosome membrane very likely has unusual permeability properties, but the example illustrates the importance of estimating the permeability of the labeling reagent in the membrane system under study.

SURFACE LABELING   In discussing methods for labeling exposed groups on one side or the other of a membrane protein, the term *surface* refers to all sites that are accessible from the aqueous medium on a given side of the membrane. By this definition, the "surface" includes sites that are not literally in the plane of the outer surface of the lipid bilayer but rather may lie in a hydrophilic cleft that is some distance from the actual membrane surface. For example, the stilbenedisulfonate site (106) of the red cell anion exchange protein is believed to be in a hydrophilic pocket that is part way through the membrane but still outside the main permeability barrier (107). Amino acid residues at this site are defined as extracellular because they can be reached by extracellular reagents that cannot penetrate the membrane. There may be instances in which a reagent can move slowly through a particular transmembrane protein, and there could be multiple barriers to the permeation of the reagent. In these cases the notion of an "extracellular" or "cytoplasmic" site becomes somewhat unclear, because there may be sites that are rapidly accessible from neither compartment. As labeling methods become more sophisticated, it may be possible to label sites that are beyond an initial permeation barrier but not all the way across the membrane. For the present purposes we define "surface" labeling as including all sites with which an impermeant reagent has rapid diffusional access, even if the site lies in a hydrophilic cleft.

LABELING STOICHIOMETRY   Bretscher (108) argued that labeling of membranes should be with agents of high sensitivity, so that labeling is easily detectable at very low stoichiometries. The reasoning was that low levels of incorporation of label are less likely to cause perturbation of the structure. This argument has merit, but it should also be recognized that labeling at very low stoichiometries may exaggerate the effect of a highly reactive group on the *trans* side of the membrane in a leaky subset of the membranes. Suppose again that 10% of the membranes are leaky, and in the leaky membranes the labeling reagent has access to a group that reacts 20 times as rapidly as does any group on the outer surface of the sealed membranes. If the labeling is performed under conditions in which the total incorporation is much less than one mole/mole, then the reactive *trans* group will be labeled more than the less reactive *cis* group. On the other hand, if the labeling is at a higher stoichiometry, the *trans* group cannot be labeled to higher than 10% stoichiometry, because only 10% of the membranes are unsealed. This example is hypothetical but is intended to illustrate the potential importance of stoichiometry in the interpretation of labeling experiments.

LABELING THE *TRANS* SIDE OF A MEMBRANE   It would of course be desirable to have, for a given protein, a choice of sealed membranes of either orientation, as is possible with certain preparations (92–99). Even when an inside-out preparation is unavailable, it is possible to draw inferences from labeling experiments in which intact membranes are compared with membranes that have been rendered leaky. The following methods can be used to disrupt the permeability barrier: freeze-thaw cycles, sonication, alkaline pH, saponin, or detergents. A typical approach is to compare the labeling of intact membranes with that of disrupted membranes. Sites that can be labeled only in disrupted membranes are inferred to reside on the *trans* side of the membrane. The interpretation of the experiment is correct as long as the disruption does not expose sites that were not originally exposed to water on either side of the membrane. It is very difficult to evaluate this assumption. The safest ways to try to gain access to the inside of a sealed vesicle are mechanical disruptions such as freeze-thaw cycles and sonication. There is no proof that mechanical disruption exposes no new sites to water, but it is reasonable to expect that, after the sonication or freeze-thaw cycle, most hydrophobic surfaces that were exposed (e.g. in a torn bilayer) would reanneal.

It is harder to say whether saponin permeabilization exposes new sites to water. Kyte et al (109) have argued that saponin does not expose sites in the Na,K-ATPase that had not been previously exposed, because saponin, unlike detergents, leaves an intact bilayer, and saponin does not inhibit the enzymatic activity of the Na,K-ATPase. These are reasonable arguments, but it is too early to say whether saponin is a "safe" permeabilizing agent for all plasma membranes.

LOCALIZATION OF LABELED SITES IN THE SEQUENCE   Once a given site has been identified as being labeled from one side or the other of a membrane, or the interior of the bilayer in the case of hydrophobic probes (see below), the labeled site must be localized in the sequence in order for the labeling to provide topographical information. Prior knowledge of the sequence of the protein makes an enormous difference in this regard. If the sequence is not known, large fragments must be prepared and aligned in order to make any inference about where the labeled site is in the sequence. With knowledge of the sequence, relatively small peptides can be prepared by digesting the isolated protein in the presence of detergents. The small labeled peptide is often sufficiently hydrophilic to be isolated by liquid chromatography using the standard methods that have been developed for peptides from soluble proteins. Edman degradation and knowledge of the sequence then makes it possible to position the labeled peptide (and often the labeled amino acid itself) in the sequence.

Although the localization of labeled sites is simple in principle, the work itself is difficult because of the difficulty of generating and separating the labeled peptides. Membrane proteins sometimes aggregate in detergent solution, thereby limiting accessibility of cleavage sites. In denaturing detergent such as SDS, the protein may not be aggregated, but the effectiveness of the proteolytic enzyme is compromised. There is not yet a standard formula for the generation of small peptides from a solubilized, purified membrane protein. Some recent examples of the successful preparation and sequence analysis of peptides from solubilized membrane proteins are in Refs. 110–115.

SOME USEFUL HYDROPHILIC LABELING AGENTS   Numerous agents are known to react covalently with membranes under conditions that are sufficiently mild that membrane integrity is preserved. No attempt is made here to assemble a complete list of reagents. However, there are some methodologies that are of sufficiently broad applicability that they deserve mention, even though many of the methods are not new. The red blood cell, because of its simplicity, has been a proving ground for new chemical labeling methods; most methods were in fact first demonstrated in studies with red cells. In extrapolating the behavior of a given reagent with red cells to other cells, it is worth remembering that the red cell membrane is less permeable to cations and has a much higher rate of monovalent anion exchange than most membranes.

Radioiodination is one of the most common methods for labeling the *cis* side of a sealed membrane preparation. The original lactoperoxidase methods developed by Phillips & Morrison (116) and Hubbard & Cohn (117) are still used. Although lactoperoxidase itself is certainly membrane-impermeant, it can catalyze the formation of I_2, which can penetrate membranes rapidly and itself act as an iodinating agent; for a good discussion of precautions to be used in surface iodinations with lactoperoxidase, see Morrison (118). The nonenzymatic method of Markwell & Fox (119) provides vectorial labeling under some conditions, with higher specific activity than the lactoperoxidase method. An alternative to iodinating tyrosines is the Bolton-Hunter (120) reagent, which is itself iodinated and reacts covalently with amino groups. The reagent itself is uncharged and penetrates membranes, but Thompson et al (121) recently showed that a sulfonated analogue of the Bolton-Hunter reagent can be used as a membrane-impermeant iodinating reagent. This reagent should be of considerable use for labeling external amino groups to high specific activity.

Pyridoxal phosphate reacts with amino groups in proteins to form a Schiff's base adduct that can be reduced and labeled with $[^3H]BH_4$. PLP cannot cross most membranes because of its negative charge. It is, however, a very slowly permeating substrate for the anion exchanger (122). Because of the mild conditions under which PLP reacts, and the ease and economy of using $[^3H]BH_4$ as the reductant, PLP has been successfully applied to topographical studies of a variety of membrane proteins, including the Na,K-ATPase (109), the ATP-ADP exchanger of the mitochondrial inner membrane (80, 100), the red cell band 3 protein (110), and the photosynthetic membranes of *Rhodobacter sphaeroides* (123).

An alternative to comparing intact membranes with permeabilized membranes is to compare the labeling by two reagents that differ in charge such that one can cross the membrane and the other cannot. Whitely & Berg (104) compared the labeling of red cell membranes with ethylacetimidate and isethionylacetimidate. The former is uncharged and can cross the membrane readily; the latter is charged and impermeant under the labeling conditions. A similar approach was used by Abbott & Schachter (124), who introduced glutathionemaleimide as a membrane-impermeant sulfhydryl reagent (as compared with *N*-ethylmaleimide, which is membrane-permeant). In this kind of experiment it is valid to conclude that sites labeled with the impermeant reagent are outside the permeability barrier, but it is not safe to conclude that all sites labeled only by the permeant reagent are on the *trans* side of the membrane. The permeant reagent, by its nature, may have access to sites that are not in contact with the water on either side of the membrane.

A useful class of membrane-impermeant active ester crosslinking reagents has been developed by Staros (125, 126). These agents, which are derivatives of N-hydroxy sulfosuccinimide, are impermeant or slowly permeant by virtue of the sulfonic acid moiety; they react with amino groups under very mild temperatures and pH. In addition to the crosslinking reagents, monofunctional active esters (127) can be readily prepared by the method described by Staros; the single negative charge on the monofunctional active esters slows the permeation of these active esters, but some of the monofunctional agents penetrate the red cell membrane as substrates for the anion exchanger.

The modification of carboxyl groups with the membrane-impermeant car-bodiimides has been used in recent studies of the red cell band 3 protein (128, 129); impermeant carbodiimides, in with radioactive nuceophiles, may in principle be used as a vectorial label. It is also possible to use a carbodiimide to activate radioactive citrate, which then reacts covalently with exofacial amino groups (130). Another novel use of carbodiimides as a vectorial label is to use a membrane-impermeant nucleophile (taurine) as a secondary radioactive label (29).

A negatively charged reagent, Woodward's reagent K, with borohydride as a secondary label, was recently used to convert extracellular carboxyl groups of the red cell membrane to labeled alcohols under very mild conditions (131). The lack of labeling of inward-facing proteins indicates that the labeling is extracellular; moreover, direct uptake measurements of the in-activated reagent indicates negligible permeation (132). The Woodward's reagent K labeling may be an example of the vectorial labeling of a carboxyl group in band 3 that is outside the permeability barrier in one conformation of the protein, but inside the barrier in another (see 132).

Most of the above labeling reagents exhibit considerable group selectivity. For some applications it is desirable to label as many surface-exposed residues as possible, irrespective of side chain. For this purpose the most useful agents are charged, membrane-impermeant photolabels such as NAP-taurine (133, 134), diazotized sulfanilic acid (89), and 3-azido-2,7-naphthalene disulfonate (135). These agents, like the hydrophobic photolabels (see below), have some group selectivity but nonetheless can label a variety of residues.

A new impermeant labeling agent was recently developed by Hundle & Richards (136), who synthesized the guanidinating reagent 2-S-thiuroniu-methanesulfonate and showed that it can be used as a vectorial label for membrane proteins in chromatophores.

There are examples of vectorial labeling even by reagents (formaldehyde/borohydride and phenyglyoxal) that can cross membranes rapidly (137, 138). The labeling is confined to the extracellular side of the membrane by keeping the intracellular pH much lower than the extracellular pH. Both examples of this approach have used the red cell, which has a huge intracellular hydrogen ion buffer power, but the method could in principle be applied to other cells or membrane vesicles enclosing a high concentration of buffer.

NATURAL MARKERS   No summary of hydrophilic labels would be complete without mention of two "natural markers" that provide information about the sidedness of a site in a membrane protein. The first is glycosylation, which is universally on the extracytoplasmic side of the membrane. (Nonenzymatic glucosylation of lysine residues is an exception.) The second is phosphorylation, which is almost certain to be cytoplasmic under physiological conditions. It should be emphasized that the presence of a consensus sequence for glycosylation does not imply that the site is glycosylated; consensus sequences for glycosylation that are almost certainly cytoplasmic are known (6).

Hydrophobic Labels

If all the exposed amino acid residues on both sides of a membrane protein could be identified by surface labeling with hydrophilic reagents, then all the membrane-spanning sequences would be specified. A complementary approach is to try to identify the membrane-spanning segments directly using hydrophobic labels. Most such labels are derivatives of the photoactivated nitrenes or carbenes that were developed about 10 years ago in several laboratories (e.g. 139–143), although new photoactivated hydrophobic labels continue to be produced (144–146). A special class of hydrophobic photolabels are derivatives of phospholipids, with a photoreactive probe on either a fatty acid or on the head group (142–144). The lipid derivatives can either be used in reconstituted preparations or introduced into existing membranes with a phospholipid exchange protein (see 143).

CHEMICAL SELECTIVITY   Both nitrenes and carbenes can react with a variety of amino acid side chains, but both have some group selectivity, e.g. for tryptophan in gramicidin (143). Cysteine residues are preferentially labeled by nitrene-generating arylazides (147, 148). The carbene-generating reagent TID [3-(Trifluoromethyl)-3(-m-iodophenyl)diazirine] labels several residues in the F_1F_0 ATPase of mitochondria (149), but the labeling of cysteine and methionine is more extensive than that of other residues. There is no recovery of labeled aspartate and glutamate in this system, probably because the adduct is an ester that is hydrolyzed during deformylation of the protein. Carbenes definitely do react with carboxylates; the major residue labeled in human glycophorin A is glutamate (150). Carbenes can add to methylene groups of aliphatic hydrocarbons but apparently do not react with the methyl side chain of alanine (149). In summary, nitrenes and especially carbenes have a broad specificity, but even carbenes do not react equally well with all side chains.

NONLIPID PHOTOLABELS   Ideally, hydrophobic photolabels are confined entirely to the lipid bilayer, and they label only the portions of a membrane protein that are embedded in the bilayer. The most common indication that the label does not contact water is the lack of effect of the water-soluble scavenger glutathione (140). In some instances glutathione does affect labeling, indicating that the photoactivated label has access at least to the interface between lipid and water. For example, in a thorough study of the labeling of ovine opsin by an aryl aide probe, Davison & Findlay (147) showed that a major labeled residue was Cys-316, which is also labeled by hydrophilic reagents. The combination of hydrophobic and hydrophilic labeling reagents may be of general use in identifying interfacial residues.

The very hydrophobic carbene-generating reagent TID appears to be less likely to label interfacial residues, and it was originally believed that this agent labels only integral membrane proteins (141). However, it has become clear in the past few years that labeling with TID or adamantine diazirine does not imply that the labeled amino acid resides in a bilayer. Soluble proteins such as calmodulin (151) and blood coagulation factor Va (152), as well as the self-inserting protein diphtheria toxin (153), are labeled by TID in the absence of membranes. The labeling is presumably at hydrophobic sites in these proteins, but the presence of a lipid bilayer is not required. Thus, TID and adamantane diazirine cannot distinguish a membrane-spanning segment from a hydrophobic sequence in a globular part of a membrane protein.

Although hydrophobic probes do label soluble proteins, they may still be of considerable value in detailed studies of well-characterized membrane proteins. Recent work by Brunner et al (154) suggests that TID can be used to identify the amino acid residues facing the lipid in a transmembrane protein that has multiple crossings. Bacteriorhodopsin was labeled with [^{125}I] TID and the radioactivity localized in the sequence by Edman degradation of fragments. One CNBr fragment was analyzed in detail, and the labeling pattern for this fragment showed a periodicity of 3–4 residues, as expected if the residues on one face of the helix were preferentially labeled. Although there were some confusing aspects of the labeling results (seemingly similar labeled peptides eluted in distinct pools), the work indicates that TID may used to identify lipid-facing sides of membrane-spanning helices.

PHOSPHOLIPID PROBES   The original reason for attaching a photoactivatable probe to the acyl chain of phospholipids was not only to identify the parts of the protein that are in contact with lipid but also to sample various depths in the bilayer by positioning the activatable groups at various distances from the head group. Unfortunately, the phospholipid probes do not appear to be able to distinguish the depth of a labeled residue, at least not at high resolution. For example, the same glutamate residue of glycophorin A was labeled by probes with five-carbon and ten-carbon spacers separating the aryl diazirine from the fatty acid ester carbonyl (150). Thus, because of the flexibility of the fatty acid, the long-chain probe can label residues that are embedded in the bilayer but are nearer the interface and not in the center of the bilayer.

ISOLATION OF HYDROPHOBIC PEPTIDES   The use of hydrophobic labeling reagents necessitates the isolation of hydrophobic peptides. It is well known that hydrophobic peptides are difficult to purify, and there have been no real breakthroughs in this area. Despite some examples of hydrophobic peptide purification by reversed phase liquid chromatography (e.g. 155), it remains true that high-resolution reversed phase separations of such peptides, especially larger peptides and fragments, are difficult. A useful method for separating medium-to-large fragments is gel filtration in either formic acid–ethanol (156, 157), chloroform-methanol (158), or 80% formic acid (159). Another common method is gel filtration in the presence of sodium dodecyl sulfate. These methods, although they do not have the high resolution of reversed phase HPLC, are sometimes necessary for the isolation of medium-to-large hydrophobic peptides.

In Situ Proteolysis

ENDOPEPTIDASES   Proteolytic enzymes have been used extensively to identify sites that are exposed at the surfaces of membrane proteins. As with chemical labeling methods, the impermeability of the membrane preparation is crucial to the interpretation of proteolysis studies. The most common use of in situ proteolysis is to generate and isolate relatively large hydrophobic fragments (160–168). End group determination (and in recent years prior knowledge of the sequence) makes it possible to localize the cleavage site in the primary structure. Proteolytic enzymes are usually used on the outer surface of sealed membranes, or for bilateral cleavage of an unsealed preparation. There are examples, however, of sealing proteolytic enzymes inside red blood cell ghosts to establish intracellular cleavage sites (168, 169).

Many different enzymes have been used for in situ proteolysis studies, often at concentrations that are far higher than are necessary to produce cleavage of soluble proteins. The high concentrations are necessary for high yields, presumably because of hindered access to potentially susceptible sites. The use of in situ proteolysis for generating relatively large, hydrophobic fragments has not changed dramatically in the past decade. It is a still a valuable approach, especially if an added tool, such as an antibody (e.g. 168), is available to facilitate isolation of fragments of interest.

An alternative approach is to recover the soluble peptides that are released from the membrane by proteolysis. These peptides are then sequenced, and prior knowledge of the protein sequence specifies their location. Again, the release of these peptides from a given side of the membrane unambiguously identifies the sidedness of that portion of the sequence. This approach has recently been applied to the Na,K-ATPase (170) and the red cell glucose transporter (171). A specific protease such as trypsin is preferable for this kind of study because a less specific enzyme would degrade the released peptides. A definite advantage of this approach is that the small peptides are generally very well behaved and can be isolated easily by reversed phase HPLC.

POSSIBLE DESTABILIZATION OF TRANSMEMBRANE HELIXES   An assumption inherent in the proteolysis approach, as in the chemical labeling approach, is that the treatment does not cause gross changes in the structure of the protein. For one-sided treatment of membranes with even very aggressive enzymes, there is little evidence that major changes are induced in the topography of the protein. However, bilateral proteolysis can potentially cause major changes. Dumont et al (172) recently digested unsealed membrane fragments from *Halobacterium halobium* with proteinase K and found that, under vigorous digestion conditions, even the membrane-spanning segments of bacteriorhodopsin were degraded. The explanation for this somewhat surprising finding is that, once the protein is digested down to its membrane-spanning segments, the removal of one or two more residues destabilizes the peptide and causes it to oscillate out of the plane of the membrane. This leads to further digestion and further destabilization until there is not enough protein left to partition into the membrane.

The importance of this finding is that it shows that very vigorous bilateral proteolysis can cause loss of membrane-spanning helices and therefore does not provide reliable information on the number and identities of the helices. This destabilization phenomenon may explain why Ramjeesingh et al (173) concluded that there are only five membrane-associated transmembrane fragments of red cell band 3 following vigorous pepsin digestion, even though there actually appear to be at least eight membrane crossings in this protein (168).

It remains to be seen whether the proteolytic destabilization of transmembrane helices is a general phenomenon. There are examples of proteins in which transmembrane secondary structure persists even after vigorous digestion. The purified and reconstituted glucose transporter from red cells retained much of its original secondary structure even after digestion with papain for three days (171). The cautions indicated by the work of Dumont et al (172), however, may apply to many membrane proteins, and vigorous bilateral proteolysis cannot be assumed to preserve transmembrane helices.

EXOPEPTIDASES   A special case of in situ proteolysis is the use of carboxypeptidases or, less commonly, aminopeptidases, to try to determine the sidedness of the C-terminus of a membrane protein. A good early example is bacteriorhodopsin, which is susceptible to carboxypeptidase digestion in membrane sheets or inside-out vesicles, but not in right-side-out vesicles (165). Carboxypeptidase has also been used to localize the C-terminus of halorhodopsin (46). The localization of the C-termini of other polytopic membrane proteins has proven to be more difficult. Even in such a well-characterized protein as red cell band 3, the cytoplasmic location of the C-terminus was only recently established (50, 51). In the delta subunit of the nicotinic acetylcholine receptor and the alpha subunit of the Na,K-ATPase, the location of the C-terminus is a matter of controversy (54, 55, 170). In the red cell glucose transporter, carboxypeptidase Y treatment of resealed ghosts led to the conclusion that the C-terminus is extracellular (164), which disagrees with recent data using C-terminal-specific antibodies (174); it is possible that the presence of some unsealed ghosts accounted for the carboxypeptidase Y effect (164).

In summary, carboxypeptidases have been used to establish the sidedness of C-termini of proteins, but the C-terminus is not accessible in all proteins even when the enzyme is on the correct side of the membrane (50). It is important in this kind of experiment to obtain a good yield of cleaved protein to rule out the possibility of an artifact resulting from a small portion of leaky membranes.

Immunological Methods

Immunological tools are being used increasingly in the study of membrane protein topography. In order to be useful for such purposes, the antibody should be directed against a specific portion of the primary structure. To obtain such antibodies, one approach is to prepare a panel of monoclonal antibodies, many of which have continuous epitopes. Once the epitopes are mapped in the sequence (a nontrivial accomplishment), the antibodies can be used in topographical work, e.g. by determining which side of a sealed membrane preparation binds each antibody. The work of Lindstrom et al (81, 175–177) on the nicotinic acetylcholine receptor and the work of Ovchinnikov et al (178) on bacteriorhodopsin exemplify this approach.

A method that has become more common in the past few years is to make synthetic peptides from portions of the protein sequence that are considered likely to be exposed on one or the other surface of the membrane. The peptides are coupled to albumin or keyhole limpet hemocyanin by any of several methods, and antisera are raised. The antisera are then used as specific markers for the chosen sequence. The use of synthetic peptides has the attraction that the study can be designed with only the knowledge of the sequence, without the necessity of isolating the protein or its fragments. The use of synthetic peptides as immunogens has been discussed in several reviews (179–181), and strategies for the choice of peptide and the technology of synthesizing and coupling the peptides will not be considered here. Instead, some recent examples of the use of antibodies in the study of membrane protein topography are briefly summarized.

Ball & Loftice (182) raised antibodies against a series of synthetic peptides

(11–15 residues each) from the alpha subunit of the Na,K-ATPase. Five of six peptides were immunogenic when used without coupling to albumin or KLH; however, none of the antibodies reacted with native Na,K-ATPase. When peptides coupled to KLH were used as immunogens, four of seven antisera recognized the native holoenzyme, thus establishing several segments of sequence that are exposed. The antibody against N-terminal peptide reacted with holoenzyme, but that against the C-terminal peptide did not, suggesting that the C-terminus is sequestered either because of its interaction with the bilayer or with the β subunit.

Seckler et al (183) raised antisera against seven synthetic hydrophilic peptides (8–13 residues) from the sequence of the *E. coli* lactose transporter. The C-terminal peptide reacted well with the native protein. Earlier studies had used antibodies against this peptide to show that the C-terminus is cytoplasmic. Of the other six antisera, only one reacted with the intact protein. This antibody was used in combination with proteolysis of right-side-out or inside-out vesicles to show that residues 125–135 include a cytoplasmic site.

Antibody against C-terminal peptide has been used to localize the C-terminus of the large subunit of the *Electrophorus electricus* sodium channel (184). Antibody was found to bind only to saponin-permeabilized right-side-out vesicles, and immunoelectronmicroscopy showed that most of the membrane-associated colloidal gold particles were on the cytoplasmic side of postsynaptic membranes. As mentioned above, antibodies against C-terminal peptides have also been used to establish that the C-terminus of the glucose transporter (174) and band 3 (51) are cytoplasmic.

Partly because of interest in autoimmune myasthenia gravis, several laboratories have performed topographical and functional studies with monoclonal and polyclonal antibodies against the nicotinic acetylcholine receptor (81, 175–177, 185–190). These data have led to conflicting conclusions, and no attempt is made here to resolve all the disagreements. It is notable, however, that two different kinds of antibody studies indicate that a sequence previously thought to be an amphipathic transmembrane helix is not actually buried in the membrane (186, 190).

Molecular Genetic Approaches

In 1986 Manoil & Beckwith (191) published a new method for assessing membrane protein topography that is based on the sidedness of fusion proteins formed from portions of the membrane protein of interest attached to marker enzyme alkaline phosphatase, which is active only when exported to the periplasm (192). The fusion protein is constructed with the membrane polypeptide upstream from alkaline phosphatase, and the eventual location of the phosphatase will reflect the sidedness of the C-terminus of the membrane polypeptide. For example, if the C-terminus of the membrane protein is periplasmic, then the phosphatase will also be periplasmic and therefore active. The method was initially tested with a protein of relatively simple topography: the Tsr protein of E. coli, which acts as a chemoreceptor. Like the aspartate chemoreceptor (41), the Tsr protein crosses the inner membrane twice, and both termini are cytoplasmic. In keeping with this topography, in-frame fusions of alkaline phosphatase to presumed periplasmic sites resulted in high phosphatase activity, and fusions at cyoplasmic sites resulted in low activity (191).

Boyd, Manoil, & Beckwith (193) extended this approach to the E. coli maltose transport protein, which has more than two membrane-spanning segments. Nine fusions at sites predicted from hydropathy plots to be periplasmic had very high alkaline phosphatase activity. Six of these fusions were in a major periplasmic domain, but three were at sites that are believed to be in relatively short connector loops between transmembrane helices. This indicates that each "in to out" membrane-spanning segment can support the export of alkaline phosphatase. The fusions at sites believed to be intracellular resulted in low alkaline phosphatase activity, with three interesting exceptions. Fusions immediately following an "out to in" transmembrane segment had fairly high activity, but fusions at the C-terminal end of a cytoplasmic loop had much lower activity than at the N-terminal end of same loop, implying that the cytoplasmic loops are important contributors to intracellular anchoring.

The same laboratory (194) developed a complementary method that employs fusions with β-galactosidase, which is active only when it is cytoplasmic. Accordingly, fusions to periplasmic C-termini of a membrane polypeptide give low β-galactosidase activity. Membrane fractionation experiments also allowed soluble fusion proteins to be distinguished from those that are membrane bound. Thus, a membrane-bound, active fusion protein has an even number of membrane-spanning segments. The results obtained for the maltose transporter were consistent with those derived from the phosphatase fusions (193).

In the short time since the method has been developed, it has been applied successfully in other laboratories. Chun & Parkinson (195) used the fusion technique to examine the topography of a membrane protein (MotB) involved in bacterial motility. The sequence of this protein had suggested a model in which there are two transmembrane segments with most of the mass cytoplasmic. However, the fusion proteins with alkaline phosphatase give high activity even for fusions near the C-terminus of the MotB protein. This indicates that most of the protein is periplasmic, with a single membrane-spanning segment. The single-spanning topography was confirmed by proteolysis of spheroplasts.

Akihama & Ito (196) used the alkaline phosphatase fusion method in combination with in situ proteolysis to study the topography of the SecY protein, which is involved in protein export in *E. coli*. The proteolysis studies were made possible by the existence of overproducing strains. The results indicate that there are 10 membrane-spanning segments, with both termini cytoplasmic.

In addition to phosphatase and β-galactosidase, it has been shown recently that fusion proteins with β-lactamase can be used for topographical studies (197). If the lactamase is fused with a C-terminus that is extracellular, the cells exhibit the ampicillin resistance conferred by the periplasmic lactamase. The method was applied to a protein, EnvZ, involved in osmoregulation.

CONCLUSIONS

In almost every membrane protein that has been studied in detail by more than one method, conflicting information has emerged regarding its topography. The reasons for these disagreements are not clear in many cases, and an obvious conclusion to be drawn from the disagreements is that it is very difficult to establish the topography even of relatively simple proteins such as cytochrome b5 (28–31) and the β subunit of the Na,K-ATPase (26, 198, 199). For more complex proteins such as the nicotinic acetylcholine receptor, it is perhaps less surprising that different techniques can lead to different conclusions. However, it is also remarkable that two sets of carefully obtained and convincing data can lead to opposite conclusions (see 54, 55, 81). Perhaps one hope of resolving conflicting conclusions is to use different methods (e.g. immunological, vectorial reduction) on exactly the same membrane preparation.

No one approach to the study of topography is inherently more reliable than the others. The labeling and proteolysis methods reviewed here give reliable results as long as sufficient attention is paid to three important parameters: the homogeneity of the membrane preparation, the yield of labeled or cleaved product, and the permeability of the labeling agent. A small fraction of leaky or wrong-sided vesicles can have a large effect on labeling experiments if insufficient attention is paid to labeling stoichiometry. Similarly, proteolysis can be misleading if the yield of cleaved product is low and if the preparation does not have uniform sidedness. Methods that rely on "permeabilizing" membranes should be scrutinized carefully, because it is so difficult to determine whether previously buried sites become exposed.

Immunological methods for the study of topography will continue to evolve. While it is true that many antibodies against synthetic peptides do not turn out to be useful in topographic studies, it is also true that such antibodies are a major way of applying sequence information to the study of the protein itself. In applying immunological methods it should be recognized that antibody specificity is not absolute; an instructive example is mAb 10 of Lindstrom and coworkers (see 81), which affects receptor channel function from the outside but also binds to a sequence that is believed to be cytoplasmic. It is important, then, to be aware of the possibility of alternate binding sites, especially when high concentrations of antibody are used.

The fusion protein methods of Beckwith and coworkers (191, 193, 194) will undoubtedly continue to provide topographical information about prokaryotic membrane proteins. The methods should be especially useful for proteins that are not available in abundance. It is also probable that the method can be applied to eukaryotic membrane proteins.

Literature Cited

1. Bretscher, M. S. 1971. *Nature New Biol.* 231:229–32
2. Noda, M., Takahashi, H., Tanabe, T., Toyasato, M., Furutani, Y., et al. 1982. *Nature* 299:793–97
3. Ballivet, M., Patrick, J., Lee, J., Heinemann, S. 1982. *Proc. Natl. Acad. Sci. USA* 79:4466–70
4. Schull, G. E., Schwartz, A., Lingrell, J. B. 1985. *Nature* 316:691–95
5. Schull, G. E., Lane, L. K., Lingrell, J. B. 1986. *Nature* 321:429–31
6. Kopito, R. R., Lodish, H. F. 1985. *Nature* 316:234–38
7. MacLennan, D. H., Brandl, C. J., Korczak, B., Green, N. M. 1985. *Nature* 316:696–700
8. Noda, M., Ikeda, T., Kayano, T., Suzuki, H., Takeshima, H., et al. 1986. *Nature* 320:188–92
9. Tanabe, T., Takeshima, H., Mikami, A., Flockerzi, V., Takahashi, H., et al. 1987. *Nature* 328:313–18
10. Mueckler, M., Caruso, C., Baldwin, S. A., Panico, M., Blench, I., et al. 1985. *Science* 229:941–45
11. Kyte, J., Doolittle, R. F. 1982. *J. Mol. Biol.* 157:105–32
12. Engelman, D. M., Steitz, T. A., Goldman, A. 1986. *Annu. Rev. Biophys. Biophys. Chem.* 15:321–53
13. Klein, P., Kanehisa, M., DeLisi, C. 1985. *Biochim. Biophys. Acta* 815:468–76
14. Argos, P., Rao, J. K. M., Hargrave, P. A. 1982. *Eur. J. Biochem.* 128:565–75
15. Flinta, C., von Heijne, G., Johansson, J. 1983. *J. Mol. Biol.* 168:193–96
16. Eisenberg, D., Schwarz, E., Komaromy, M., Wall, R. 1984. *J. Mol. Biol.* 179:125–42
17. Eisenberg, D. 1984. *Annu. Rev. Biochem.* 53:595–623
18. Blobel, G. 1980. *Proc. Natl. Acad. Sci. USA* 77:1496–500
19. Wickner, W. T., Lodish, H. F. 1986. *Science* 230:400–6
20. Zimmerman, R., Watts, C., Wickner, W. 1982. *J. Biol. Chem.* 257:6529–36
21. von Heijne, G. 1986. *EMBO J.* 5:3021–27
21a. von Heijne, G. 1988. *Biochim. Biophys. Acta* 947:307–33
22. Neville, D. M. Jr., Hudson, T. H. 1986. *Annu. Rev. Biochem.* 55:195–224
23. Engelman, D. M., Steitz, T. A. 1981. *Cell* 23:411–22
24. Singer, S. J., Nicholson, G. 1972. *Science* 175:720–31
25. Steck, T. L., Yu, J. 1973. *J. Supramol. Struct.* 1:220–32
26. Farley, R. A., Miller, R. P., Kudrow, A. 1986. *Biochim. Biophys. Acta* 873: 136–42
27. Audigier, Y., Friedlander, M., Blobel, G. 1987. *Proc. Natl. Acad. Sci. USA* 84:5783–87
28. Dailey, H. A., Strittmatter, P. 1981. *J. Biol. Chem.* 256:3951–55
29. Arinc, E., Rzepecki, L. M., Strittmatter, P. 1987. *J. Biol. Chem.* 262:15563–67
30. Takagaki, Y., Radhakrishnan, R., Gupta, C. M., Khorana, H. G. 1983. *J. Biol. Chem.* 258:9128–35
31. Takagaki, Y., Radhakrishnan, R., Wirtz, K. W. A., Khorana, H G. 1983. *J. Biol. Chem.* 258:9136–42
32. Rose, G. D., Young, W. B., Gierasch, L. M. 1983. *Nature* 304:654–57
33. Wiley, D. C., Skehel, J. J. 1987. *Annu. Rev. Biochem.* 56:365–94
34. Rose, J. K., Welch, W. J., Sefton, B. M., Esch, F. S., Ling, N. C. 1980. *Proc. Natl. Acad. Sci. USA* 77:3884–88
35. Marchesi, V. T., Furthmayr, H., Tomi-

ta, M. 1976. *Annu. Rev. Biochem.* 45:667–98

36. Lopez de Castro, J. A., Barbosa, J. A., Krangel, M. S., Biro, P. A., Strominger, J. L. 1985. *Immunol. Rev.* 85:149–68

37. Hunziker, W., Spiess, M., Semenza, G., Lodish, H. F. 1986. *Cell* 46:227–34

38. Chiacchia, K. B., Drickamer, K. 1984. *J. Biol. Chem.* 259:15440–46

39. Ullrich, A., Gray, A., Tam, A. W., Yang-Feng, T., Tsubokawa, M., et al. 1986. *EMBO J.* 5:2503–12

40. Yarden, Y., Ullrich, A. 1988. *Annu. Rev. Biochem.* 57:443–78

41. Russo, A. F., Koshland, D. E. Jr. 1983. *Science* 220:1016–20

42. Moore, K. E., Miura, S. 1987. *J. Biol. Chem.* 262:8806–13

43. Dalbey, R. E., Kuhn, A., Wickner, W. 1987. *J. Biol. Chem.* 262:13241–45

44. Michel, H., Weyer, K. A., Gruenberg, H., Dunger, I., Oesterhelt, D., Lottspeich, F. 1986. *EMBO J.* 5:1149–58

45. Deisenhofer, J., Epp, O., Miki, K., Huber, R., Michel, H. 1985. *Nature* 318:618–24

46. Schobert, B., Lanyi, J. K., Oesterhelt, D. 1988. *EMBO J.* 7:905–11

47. Henderson, R., Unwin, P. N. T. 1975. *Nature* 257:28–32

48. Stoeckenius, W., Lozier, R. H., Bogomolni, R. A. 1979. *Biochim. Biophys. Acta* 505:215–78

49. Hargrave, P. A., McDowell, J. H., Curtis, D. R., Wang, J. K., Juszczak, K. E., et al. 1983. *Biophys. Struct. Mech.* 9:235–44

50. Lieberman, D. M., Nattriss, M., Reithmeier, R. A. F. 1987. *Biochim. Biophys. Acta* 903:37–47

51. Lieberman, D. M., Reithmeier, R. A. F. 1988. *J. Biol. Chem.* 263:10022–28

52. Kaback, H. R. 1986. *Annu. Rev. Biophys. Biophys. Chem.* 15:279–319

53. Wright, J. K., Seckler, R., Overath, P. 1986. *Annu. Rev. Biochem.* 55:225–48

54. McCrea, P. D., Engelman, D. M., Popot, J.-L. 1988. *Trends Biochem. Sci.* 13:289–90

55. McCrea, P. D., Popot, J.-L., Engelman, D. M. 1987. *EMBO J.* 6:3619–26

56. Sheridan, R. P., Levy, R. M., Salemme, F. R. 1982. *Proc. Natl. Acad. Sci. USA* 79:4545–49

57. Mao, D., Wallace, B. A. 1984. *Biochemistry* 23:2667–73

58. Foster, D. L., Boublik, M., Kaback, H. R. 1983. *J. Biol. Chem.* 258:31–34

59. Vogel, H., Wright, J. K., Jähnig, F. 1985. *EMBO J.* 4:3625–31

60. Oikawa, K., Lieberman, D. M., Reith-

meier, R. A. F. 1985. *Biochemistry* 24:2843–48

61. Chin, J. J., Jung, E. K. Y., Jung, C. Y. 1986. *J. Biol. Chem.* 7101–4

62. Nathans, J., Hogness, D. S. 1983. *Cell* 34:807–14

63. Honig, B., Hubbell, W. L. 1984. *Proc. Natl. Acad. Sci. USA* 81:5412–16

64. Alvarez, J., Lee, D. C., Baldwin, S. A., Chapman, D. 1987. *J. Biol. Chem.* 262:3502–9

65. Osborne, H. B., Nabedryk-Viala, E. 1977. *FEBS Lett.* 84:217–20

66. Downer, N. W., Bruchman, T. J., Hazzard, J. H. 1986. *J. Biol. Chem.* 261:3640–47

67. Weinstein, J. N., Blumenthal, R., van Renswoude, J., Kempf, C., Klausner, R. D. 1982. *J. Membr. Biol.* 66:203–12

68. Drickamer, L. K. 1976. *J. Biol. Chem.* 251:5115–23

69. Steck, T. L., Koziarz, J. J., Singh, M. K., Reddy, R., Köhler, H. 1978. *Biochemistry* 17:1216–22

70. Jenkins, R. E., Tanner, M. J. A. 1977. *Biochem. J.* 161:139–47

71. Jennings, M. L., Adams, M. F. 1981. *Biochemistry* 20:7118–23

72. Rosenbusch, J. P. 1974. *J. Biol. Chem.* 249:8019–29

73. Kleffel, B., Garavito, R. M., Baumeister, W., Rosenbusch, J. P. 1985. *EMBO J.* 4:1589–92

74. Surewicz, W. K., Mantsch, H. H. 1988. *Biochim. Biophys. Acta* 952:115–30

75. Lee, D. C., Hayward, J. A., Restall, C. J., Chapman, D. 1985. *Biochemistry* 24:4364–73

76. Surewicz, W. K., Moscarello, M. A., Mantsch, H. H. 1987. *J. Biol. Chem.* 262:8598–602

77. Jap, B. K., Maestre, M. F., Hayward, S. B., Glaeser, R. M. 1983. *Biophys. J.* 43:81–89

78. Wallace, B. A., Teeters, C. L. 1987. *Biochemistry* 26:65–70

79. Aquila, H., Link, T. A., Klingenberg, M. 1987. *FEBS Lett.* 212:1–9

80. Bogner, W., Aquila, H., Klingenberg, M. 1986. *Eur. J. Biochem.* 161:611–20

81. Ratnam, M., Nguyen, D. L., Rivier, J., Sargent, P. B., Lindstrom, J. 1986. *Biochemistry* 25:2633–43

82. Liscum, L., Finer-Moore, J., Stroud, R. M., Luskey, K. L., Brown, M. S., Goldstein, J. L. 1985. *J. Biol. Chem.* 260:522–30

83. Kopito, R. R., Andersson, M., Lodish, H. F. 1987. *J. Biol. Chem.* 262:8035–40

84. Salkoff, L., Butler, A., Wei, A., Scavarda, N., Giffen, K., et al. 1987. *Science* 237:744–49

85. Frank, G., Brunner, J., Hauser, H., Wacker, H., Semenza, G., Zuber, H. 1978. *FEBS Lett.* 96:183–88
86. Sjöström, H., Norén, O., Christiansen, L., Wacker, H., Spiess, M., et al. 1982. *FEBS Lett.* 148:321–25
87. Bürgi, R., Brunner, J., Semenza, G. 1983. *J. Biol. Chem.* 258:15114–19
88. Forbush, B. III. 1982. *J. Biol. Chem.* 257:12678–84
89. O'Connell, M. A. 1982. *Biochemistry* 21:5984–91
90. St. John, P. A., Froehner, S. C., Goodenough, D. A., Cohen, J. B. 1982. *J. Cell. Biol.* 92:333–42
91. Resch, K., Schneider, S., Szamel, M. 1981. *Anal. Biochem.* 117:282–92
92. Steck, T. L. 1974. In *Methods in Membrane Biology,* Vol. 2, ed. E. D. Korn, pp. 245–81. New York: Plenum
93. Kondo, T., Dale, G. L., Beutler, E. 1980. *Biochim. Biophys. Acta* 602:127–30
94. Walsh, F. S., Crumpton, M. J. 1977. *Nature* 269:307–11
95. Bürgi, R., Suter, F., Zuber, H. 1987. *Biochim. Biophys. Acta* 890:346–51
96. Akerlund, H.-E., Andersson, B. 1983. *Biochim. Biophys. Acta* 725:34–40
97. Velema, J., van Amsterdam, F. T. M., Zaagsma, J. 1987. *Int. J. Biochem.* 19:467–70
98. Seckler, R., Wright, J. K. 1984. *Eur. J. Biochem.* 142:269–79
99. Grouzis, J.-P., Gibrat, R., Rigaud, J., Grignon, C. 1987. *Biochim. Biophys. Acta* 903:449–64
100. Bogner, W., Aquila, H., Klingenberg, M. 1982. *FEBS Lett.* 146:259–61
101. Abbs, M. T., Phillips, J. H. 1980. *Biochim. Biophys. Acta* 595:200–21
102. Hunt, R. C., Brown, J. C. 1975. *J. Mol. Biol.* 97:413–22
103. Shiozawa, J. A., Jelenska, M. M., Jacobson, B. S. 1987. *Biochemistry* 26:4884–92
104. Whitely, N. M., Berg, H. C. 1974. *J. Mol. Biol.* 87:541–61
105. Patthey, J.-P., Deshusses, J. 1987. *FEBS Lett.* 210:137–41
106. Cabantchik, Z. I., Rothstein, A. 1974. *J. Membr. Biol.* 15:207–26
107. Rao, A., Martin, P., Reithmeier, R. A. F., Cantley, L. C. 1979. *Biochemistry* 18:4505–16
108. Bretscher, M. S. 1971. *J. Mol. Biol.* 58:775–81
109. Kyte, J., Xu, K., Bayer, R. 1987. *Biochemistry* 26:8350–60
110. Kawano, Y., Okubo, K., Tokunaga, F., Miyata, T., Iwanaga, S., Hamasaki, N. 1988. *J. Biol. Chem.* 263:8232–38
111. Farley, R. A., Tran, C. M., Carilli, C.

T., Hawke, D., Shively, J. E. 1984. *J. Biol. Chem.* 259:9532–35
112. Oberthür, W., Muhn, P., Baumann, H., Lottspeich, F., Wittmann-Liebold, B., Hucho, F. 1986. *EMBO J.* 5:1815–19
113. Giraudat, J., Dennis, M., Heidmann, T., Haumont, P.-Y., Lederer, F., Changeux, J.-P. 1987. *Biochemistry* 26:2410–18
114. Aquila, H., Misra, D., Eulitz, M., Klingenberg, M. 1982. *Hoppe-Seyler's Z. Phys. Chem.* 363:345–49
115. Filoteo, A. G., Gorski, J. P., Penniston, J. T. 1987. *J. Biol. Chem.* 262:6526–30
116. Phillips, D. R., Morrison, M. 1971. *Biochemistry* 10:1766–71
117. Hubbard, A. L., Cohn, Z. A. 1972. *J. Cell Biol.* 55:390–405
118. Morrison, M. 1980. *Methods Enzymol.* 70:214–20
119. Markwell, M. A. K., Fox, C. F. 1978. *Biochemistry* 17:4807–17
120. Bolton, A. E., Hunter, W. M. 1973. *Biochem. J.* 133:529–39
121. Thompson, J. A., Lau, A. L., Cunningham, D. D. 1987. *Biochemistry* 26:743–50
122. Cabantchik, Z. I., Balshin, M., Breuer, W., Rothstein, A. 1975. *J. Biol. Chem.* 250:5130–36
123. Francis, G. A., Richards, W. R. 1980. *Biochemistry* 19:5104–11
124. Abbott, R. E., Schachter, D. 1976. *J. Biol. Chem.* 251:7176–83
125. Staros, J. V. 1982. *Biochemistry* 21:3950–55
126. Anjaneyulu, P. S. R., Staros, J. V. 1987. *Int. J. Peptide Protein Res.* 30:117–24
127. Donovan, J. A., Jennings, M. L. 1986. *Biochemistry* 25:1538–44
128. Bjerrum, P. J. 1983. In *Structure and Function of Membrane Proteins,* ed. E. Quagliariello, F. Palmieri, pp. 107–15. Amsterdam: Elsevier
129. Craik, J. D., Reithmeier, R. A. F. 1985. *J. Biol. Chem.* 260:2404–8
130. Werner, P. K., Reithmeier, R. A. F. 1988. *Biochim. Biophys. Acta* 942:19–32
131. Jennings, M. L., Anderson, M. P. 1987. *J. Biol. Chem.* 262:1691–97
132. Jennings, M. L., Al-Rhaiyel, S. 1988. *J. Gen. Physiol.* 92:161–78
133. Staros, J. V., Richards, F. M. 1974. *Biochemistry* 13:2720–26
134. Staros, J. V., Richards, F. M., Haley, B. E. 1975. *J. Biol. Chem.* 250:8174–78
135. Moreland, R. B., Docktor, M. E. 1980. *Anal. Biochem.* 103:26–32
136. Hundle, B. S., Richards, W. R. 1987. *Biochemistry* 26:4505–11

137. Jennings, M. L., Nicknish, J. S. 1984. *Biochemistry* 23:6432–36
138. Bjerrum, P. J., Wieth, J. O., Borders, C. L. Jr. 1983. *J. Gen. Physiol.* 81:453–84
139. Bayley, H., Knowles, J. R. 1978. *Biochemistry* 17:2414–19
140. Bayley, H., Knowles, J. R. 1978. *Biochemistry* 17:2420–23
141. Brunner, J., Semenza, G. 1981. *Biochemistry* 20:7174–82
142. Gupta, C. M., Radhakrishnan, R., Khorana, H. G. 1977. *Proc. Natl. Acad. Sci. USA* 74:4315–19
143. Brunner, J., Richards, F. M. 1980. *J. Biol. Chem.* 255:3319–29
144. Montecucco, C., Schiavo, G. 1986. *Biochem. J.* 237:309–12
145. Pradhan, D., Lala, A. K. 1987. *J. Biol. Chem.* 262:8242–51
146. Middlemas, D. S., Raftery, M. A. 1987. *Biochemistry* 26:1219–23
147. Davison, M. D., Findlay, J. B. C. 1986. *Biochem. J.* 234:413–20
148. Giraudat, J., Montecucco, C., Bisson, R., Changeux, J.-P. 1985. *Biochemistry* 24:3121–27
149. Hoppe, J., Brunner, J., Jørgensen, B. 1984. *Biochemistry* 23:5610–16
150. Ross, A. H., Radhakrishnan, R., Robson, R. J., Khorana, H. G. 1982. *J. Biol. Chem.* 257:4152–61
151. Krebs, J., Buerkler, J., Guerini, D., Brunner, J., Carafoli, E. 1984. *Biochemistry* 23:400–3
152. Krieg, U. C., Isaacs, B. S., Yemul, S. S., Esmon, C. T., Bayley, H., Johnson, A. E. 1987. *Biochemistry* 26:103–9
153. Dumont, M. E., Richards, F. M. 1988. *J. Biol. Chem.* 263:2087–97
154. Brunner, J., Franzusoff, A. J., Lüscher, B., Zugliani, C., Semenza, G. 1985. *Biochemistry* 24:5422–30
155. Nicholas, R. A. 1984. *Biochemistry* 23:888–98
156. Bayley, H., Huang, K.-S., Radhakrishnan, R., Ross, A. H., Takagaki, Y., Khorana, H. G. 1981. *Proc. Natl. Acad. Sci. USA* 78:2225–29
157. Mawby, W. J., Findlay, J. B. C. 1982. *Biochem. J.* 205:465–75
158. Fillingame, R. H. 1976. *J. Biol. Chem.* 251:6630–37
159. Brock, C. J., Tanner, M. J. A., Kempf, C. 1983. *Biochem. J.* 213:577–86
160. Steck, T. L., Ramos, B., Strapazon, E. 1976. *Biochemistry* 15:1154–61
161. Castro, J., Farley, R. A. 1979. *J. Biol. Chem.* 254:2221–28
162. Chin, G., Forgac, M. 1983. *Biochemistry* 22:3405–10
162a. Jennings, M. L., Adams-Lackey, M.,

Denney, G. H. 1984. *J. Biol. Chem.* 259:4652–60
163. Jenkins, R. E., Tanner, M. J. A. 1977. *Biochem. J.* 161:131–38
164. Shanahan, M. F., D'Artel-Ellis, J. 1984. *J. Biol. Chem.* 259:13878–84
165. Gerber, G. E., Gray, C. P., Wildenauer, D., Khorana, H. G. 1977. *Proc. Natl. Acad. Sci. USA* 74:5426–30
166. Deziel, M. R., Rothstein, A. 1984. *Biochim. Biophys. Acta* 776:10–20
167. Wennogle, L. P., Changeux, J. P. 1980. *Eur. J. Biochem.* 106:381–93
168. Jennings, M. L., Anderson, M. P., Monaghan, R. 1986. *J. Biol. Chem.* 261:9002–10
169. Lepke, S., Passow, H. 1976. *Biochim. Biophys. Acta* 455:353–70
170. Ovchinnikov, Y. A., Arzamazova, N. M., Arystarkhova, E. A., Gevondyan, N. M., Aldanova, N. A., Modyanov, N. N. 1987. *FEBS Lett.* 217:269–74
171. Cairns, M. T., Alvarez, J., Panico, M., Gibbs, A. F., Morris, H. R., et al. 1987. *Biochim. Biophys. Acta* 905:295–310
172. Dumont, M. E., Trewhella, J., Engelman, D. M., Richards, F. M. 1985. *J. Membr. Biol.* 88:233–47
173. Ramjeesingh, M., Gaarn, A., Rothstein, A. 1984. *Biochim. Biophys. Acta* 769:381–89
174. Andersson, L., Lundahl, P. 1988. *J. Biol. Chem.* 263:11414–20
175. Conti-Tronconi, B., Tzartos, S., Lindstrom, J. 1981. *Biochemistry* 20:2181–91
176. Gullick, W. J., Lindstrom, J. M. 1983. *Biochemistry* 22:3312–20
177. Ratnam, M., Sargent, P. B., Sarin, V., Fox, J. L., Nguyen, D. L., et al. 1986. *Biochemistry* 25:2621–32
178. Ovchinnikov, Y. A., Abdulaev, N. G., Vasilov, R. G., Vturina, I. Y., Kuryatov, A. B., Kiselev, A. V. 1985. *FEBS Lett.* 179:343–50
179. Berzofsky, J. A. 1985. *Science* 229:932–40
180. Sutcliffe, J. G., Shinnick, T. M., Green, N., Lerner, R. A. 1983. *Science* 219:660–66
181. Van Regenmortel, M. H. V. 1987. *Trends Biochem. Sci.* 12:237–40
182. Ball, W. J. Jr., Loftice, C. D. 1987. *Biochim. Biophys. Acta* 916:100–11
183. Seckler, R., Möröy, T., Wright, J. K., Overath, P. 1986. *Biochemistry* 25:2403–9
184. Gordon, R. D., Fieles, W. E., Schotland, D. L., Hogue-Angeletti, R., Barchi, R. L. 1987. *Proc. Natl. Acad. Sci. USA* 84:308–12

185. Strader, C. B. D., Revel, J.-P., Raftery, M. A. 1979. *J. Cell Biol.* 83:499–510
186. Roth, B., Schwendimann, B., Hughes, G. J., Tzartos, S. J., Barkas, T. 1987. *FEBS Lett.* 221:172–78
187. Criado, M., Sarin, V., Fox, J. L., Lindstrom, J. 1986. *Biochemistry* 25:2839–46
188. Atassi, M. Z., Mulac-Jericevic, B., Yokoi, T., Manshouri, T. 1987. *Fed. Proc.* 46:2538–47
189. Lennon, V. A., McCormick, D. J., Lambert, E. H., Griesmann, G. E., Atassi, M. Z. 1985. *Proc. Natl. Acad. Sci. USA* 82:8805–9
190. Dwyer, B. P. 1988. *Biochemistry* 27: 5586–92
191. Manoil, C., Beckwith, J. 1986. *Science* 233:1403–8
192. Hoffman, C. S., Wright, A. 1985. *Proc. Natl. Acad. Sci. USA* 82:5107–11
193. Boyd, D., Manoil, C., Beckwith, J. 1987. *Proc. Natl. Acad. Sci. USA* 84:8525–29
194. Froshauer, S., Green, G. N., Boyd, D., McGovern, K., Beckwith, J. 1988. *J. Mol. Biol.* 200:501–11
195. Chun, S. Y., Parkinson, J. S. 1988. *Science* 239:276–78
196. Akihama, Y., Ito, K. 1987. *EMBO J.* 6:3465–70
197. Forst, S., Comeau, D., Norioka, S., Inouye, M. 1987. *J. Biol. Chem.* 262:16433–38
198. Chin, G. J. 1985. *Biochemistry* 24: 5943–47
199. Zibirre, R., Hippler-Feldtmann, G., Kühne, J., Poronnik, P., Warnecke, G., Koch, G. 1987. *J. Biol. Chem.* 262:4349–54

Annu. Rev. Biochem. 1989. 58:1029–49

tRNA IDENTITY

Jennifer Normanly[1] and John Abelson

Department of Biology, California Institute of Technology, Pasadena, California 91125

CONTENTS

BACKGROUND

In the translation of genetic information from nucleic acid to protein, transfer RNA plays a crucial intermediate role. Each codon is read by a unique tRNA, which has been aminoacylated with the appropriate amino acid. Because of degeneracy in the genetic code there are, in general, more than one tRNA for each amino acid (1). Although there are some exceptions, there is a single aminoacyl tRNA synthetase (AAS) for each amino acid. An AAS must correctly recognize and activate an amino acid and then join it to each of the cognate set of tRNAs. By virtue of the fact that they must function interchangeably on the ribosome during protein synthesis, all tRNAs have similar primary (2), secondary (3), and tertiary (4, 5) structures (Figure 1). Within these constraints each cognate set of tRNAs contains distinct elements

---

[1]Present address: The Whitehead Institute, Cambridge, Massachusetts 02142

0066-4154/89/0701-1029$02.00

that mediate correct recognition by the AAS. These elements comprise the *identity* of a tRNA. Identity elements can be both positive and negative. Positive elements are those features of the tRNA that the cognate AAS recognizes directly, and negative elements are those features that block the recognition by other AASs.

Although tRNAs comprise a homogeneous set of molecules, AASs are remarkably diverse. The quaternary structures α, α_2, α_4, and $\alpha_2\beta_2$ have all been observed (6, 7). AAS subunit sizes range from 300 to 900 amino acid residues. The details of the interaction between these diverse enzymes and their tRNA substrates will not be known until high-resolution X-ray structures of tRNA-AAS complexes have been solved. Fortunately, there is the prospect that several such complexes will be solved in the near future (8, 9). Already, we have a remarkably detailed description of the amino acid and ATP binding sites for the *Bacillus stearothermophilus* tyrosine AAS (10) and the *Escherichia coli* methionine AAS (11). It is not, however, our purpose to review the large body of information on the structure and properties of the AAS, or on the indirect efforts that have been employed to map their interactions with tRNA. These have been well reviewed before in these volumes (6, 7). Rather, we wish to describe the results of fresh and energetic approaches to the problem of tRNA identity, which have appeared in the last few years.

PAST AND PRESENT APPROACHES TO DEFINING tRNA IDENTITY

We describe two new approaches to the determination of tRNA identity. Both have their roots in previous work but are now more powerful because of advances in gene synthesis technology. In the first, attempts are made to alter with the fewest changes, the identity of a tRNA while retaining complete biological function. Those changes responsible for the redirection of AAS recognition constitute identity elements. We term this type of experiment an identity swap. In the second, tRNA variants are generated by altering the sequence of a tRNA, and their properties as a substrate for the AAS are determined in vitro. This approach has been greatly facilitated by the ability to synthesize tRNA in vitro with bacteriophage T7 RNA polymerase.

The identity swap has its origins in the study of amber suppressors. An amber suppressor is a mutant tRNA that can recognize and suppress the chain-terminating effects of a UAG codon, by virtue of a change in the tRNA's anticodon (12, 13). Suppressors have greatly facilitated the application of genetics to the study of tRNA, and have been utilized extensively in bacteria (14), T4 bacteriophage (15, 16), and yeast (17). Such work has led to a formal proof of the secondary structure of tRNA (14) and to information about the mechanism of tRNA processing (18–20). Suppressors have also

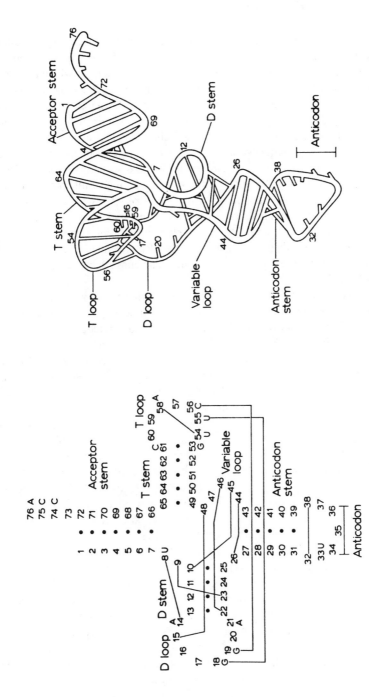

Figure 1 Secondary and tertiary structure of tRNA. In the cloverleaf structure (*left*) invariant nucleotides that are involved in stabilizing tertiary interactions are indicated; the lines indicate tertiary interactions between bases. Dots indicate base pairing. (*Right*) The L-model for tRNA tertiary structure (4, 5).

been used to address the tRNA identity problem. Attempts were made to alter the specificity of the *E. coli* tRNA^{Tyr} suppressor, su$_3^+$, by selecting mutants of this tRNA that could suppress amber mutations at positions in which tyrosine was not an acceptable amino acid. This approach was taken independently by J. D. Smith & S. Brenner in Cambridge and by H. Ozeki & Y. Shimura in Kyoto nearly 20 years ago (21, 22). Both groups obtained the same results. The su$_3^+$ mutants that were isolated inserted glutamine instead of tyrosine into amber codons (23). Sequence analysis revealed changes in the fourth base from the 3' end, A$_{72}$ → G (22), and in the first and second base pairs of the acceptor stem 1-72 and 2-71 (21). This was an exciting result, but it did not prove to be universally applicable to the study of tRNA identity. Despite many attempts, su$_3^+$ mutants that inserted an amino acid other than glutamine or tyrosine were never found (23). It is likely that such transformations would have required more than one change in the tRNA sequence and therefore could not be obtained by genetic selection. This approach was abandoned for 15 years.

We have returned to this approach armed with an automated DNA synthesizer, which allows the rapid synthesis of altered suppressors de novo, as in the pioneering work of Khorana (24). The synthetic genes are assembled from oligonucleotides and inserted in cassette form into a plasmid that supplies a promoter, either the inducible *lac* promoter (25) or the constitutive *lpp* promoter (26, 27), and a transcription terminator. Usually a screen is employed, in which the suppression of an amber mutation in a gene requires the insertion of a specific amino acid to allow growth. The amino acid specificity of the suppressor is determined by sequencing the protein product of a suppressed amber mutant gene. For this purpose the *E. coli* dihydrofolate reductase (DHFR) gene with an amber mutation at codon 10 is used. DHFR is easily purified by affinity chromatography on a methotrexate resin. X-ray crystallographic data (28, 29) revealed that residue 10 of DHFR lies on the surface of the protein, well removed from the active site, and it appears that any amino acid is acceptable at this position. If more than one amino acid is inserted by the suppressor, they can be readily detected by this method. Protein sequence data acquired in this manner reveal the outcome of the competition among all 20 AASs in the cell for the suppressor. The drawbacks to this method are twofold. First, the anticodon of the tRNA must be CUA to allow recognition of amber codons; so the contribution of the anticodon to tRNA identity cannot be assessed. As described below, the anticodon is a strong identity element for some tRNAs. Second, the effectiveness of the suppressor tRNA as a substrate for the AAS can only be measured by the efficiency of suppression. Multiple factors, such as the ability of the tRNA to be transcribed, processed, modified, and accepted by the ribosome, can affect the observed suppression efficiency, and in this method cannot be distinguished from the efficiency of aminoacylation.

In the second method developed by Uhlenbeck (30), tRNA genes and their variants are synthesized and placed under the control of the bacteriophage T7 promoter so that the first nucleotide of the in vitro transcript is the mature 5' end of the tRNA. Alternatively, synthetic tRNAs with 5' precursor sequences can be processed correctly by incubation with the RNA component of RNAse P (31; L. Schulman, unpublished). A Bst N1 restriction endonuclease recognition site (5'CC·AGG) is encoded at the 3' end of the gene. Run-off transcription of Bst N1–cleaved DNA generates the correct CCA terminus. Prior to the availability of this methodology, the study of tRNA was hindered in that most variants were not processed or were unstable in vivo. This technique not only overcomes such obstacles, but also allows the easy purification of the transcript. Reactions can be scaled up, and milligrams of some transcripts have been produced for NMR studies (K. Hall et al, unpublished). These synthetic tRNAs do not contain the normal base modifications found in native tRNA (and are therefore excellent substrates for the study of any modifying enzyme). They are nonetheless, good substrates for the AAS. Typically, the K_m values for the substrates are one- to four-fold higher than for fully modified tRNA. V_{max} values for the unmodified substrates are at most, two-fold lower, resulting in values of the specificity constant (k_{cat}/K_m) that are 5–17-fold lower. In contrast, the k_{cat}/K_m for a typical misacylation is orders of magnitude lower (6). These results indicate that the modified nucleotides are not likely to play a large role in recognition by the AAS, as their absence does not significantly diminish the efficiency of aminoacylation.

In order to determine that the synthetic tRNAs are properly folded, Sampson & Uhlenbeck (32) have exploited the phenomenon of lead-mediated cleavage of tRNA. This reaction is mediated by a tightly bound Pb^{2+} ion whose position is determined by the tertiary interactions between the D-loop and T-loop (33, 34). In native yeast tRNA^{Phe}, cleavage takes place at residue 17. Unmodified tRNA^{Phe} is cleaved at the same position, with similar kinetics, indicating that the synthetic molecule has folded properly. The in vitro approach is attractive in that one can quantitate the effects of individual nucleotide changes on the recognition of the tRNA by cognate and noncognate AAS. The method does not allow assessment of the effect of competition between one AAS and the other 19 AASs. In the end, it will be desirable to apply both the in vivo and in vitro approaches to determine the identity elements of any tRNA.

THE ROLE OF THE ANTICODON IN tRNA IDENTITY

Since the anticodon is directly correlated with the identity of any tRNA species, it is the most logical identity element. Indeed, one of the earliest recognition hypotheses states that the "coding triplet (anticodon) plays an important, perhaps decisive role in the interaction with the synthetase" (35). Early evidence to support this hypothesis came from a combination of chemical modification and oligonucleotide interference experiments as well as from analysis of mutant tRNAs, and suggested that for a number of tRNAs, the anticodon can be implicated in recognition. Among them are *E. coli* tRNA^{Gly}, tRNA^{Lys}, tRNA^{Arg}, tRNA^{Glu}, tRNA$^{Met(f)}$, tRNA^{Trp}, yeast tRNA^{Gly}, tRNA^{Val}, and beef pancreatic tRNA^{Trp} (reviewed in 36).

The opposing notion, that the anticodon could not be a recognition element, was derived from the observation that the anticodon could be altered to create amber suppressors with no effect upon amino acid specificity in at least four tRNAs. Additionally, translation of the codons (six each) for serine, leucine, and arginine requires a set of tRNAs with different nucleotides in two out of three positions for leucine and arginine, and in all three for serine. The idea that the anticodon is not an identity element seems to have held sway, in some quarters at least, because a few textbooks still state this unequivocally. Nonetheless, it is now clear that the anticodon is a recognition element for tRNA identity in perhaps a majority of tRNAs.

The strongest evidence in support of this view comes from L. Schulman's work on *E. coli* tRNA^{Met}. Experiments conducted over a period of 15 years have demonstrated conclusively that the CAU anticodon of *E. coli* tRNA^{Met} is an essential recognition element for the methionine AAS. Chemical modification experiments have shown that the anticodon nucleotides C_{34} and A_{35} cannot be modified without loss of methionine acceptor activity in vitro (37). The anticodon of tRNA^{Met} was altered by the judicial insertion of oligoribonucleotides into the tRNA, which had been cleaved specifically by RNAse. These experiments revealed that substitution of C_{34} with a U, A, or G lead to a decrease in the aminoacylation rate by four to five orders of magnitude (38). Altering the other two positions of the anticodon (A_{35} and U_{36}) affects AAS recognition as well, decreasing the rate of aminoacylation one to four orders of magnitude to that of normal (39). Substitutions of any position in the anticodon with nucleotides containing functional groups in common with those of the wild-type nucleotides were tolerated more than if nucleotides, lacking similar functional groups, were substituted. The conclusion was that the synthetase must be interacting with specific functional groups in the anticodon (40). In addition, chemical crosslinking experiments demonstrated that C_{34} of the anticodon is uniquely crosslinked to Lys 465 in the methionine AAS (41).

More recently, Schulman has conducted identity swap experiments in the in vitro system (42). In these experiments genes for both tRNA^{Trp} and tRNA^{Val} of *E. coli* were synthesized with the methionine (CAU) anticodon and transcribed in vitro, and the kinetics of acylation by the methionine AAS and valine AAS were determined for the synthetic tRNAs (Table 1). tRNA^{Val}

(CAU) was aminoacylated by the methionine AAS with a V_{max}/K_m similar to that of native tRNA^{Met}. tRNA^{Trp} (CAU) was also acylated by the methione AAS, but the V_{max}/K_m was down 10-fold. Conversely, tRNA^{Met} with the valine anticodon (UAC) was acylated by the valine AAS with a V_{max}/K_m value 1/10 of that for normal tRNA^{Val}. In comparison, the value for acylation of tRNA^{Met} by the valine AAS was four orders of magnitude lower than that for tRNA^{Val}. These results clearly show that the anticodon is a major, if not the sole, recognition element for tRNA^{Met} and is important for tRNA^{Val} identity as well.

Theories about the role of the anticodon in AAS recognition have generally been influenced by the existence and the properties of amber suppressors. On the one hand, the demonstration that suppressors for tRNA^{Tyr}, tRNA^{Ser}, tRNA^{Gln}, and tRNA^{Leu}, were correctly charged, indicated that not all AASs recognize the anticodon nucleotides of their cognate tRNAs. A more complete picture of the effects of the amber anticodon on AAS recognition has been obtained from our efforts in collaboration with J. Miller and coworkers to construct a complete bank of 20 amber suppressors (27, 43; J. Normanly, et al, unpublished). These suppressors have been generated by synthesizing the genes of isoacceptors from the 20 species of tRNA in *E. coli,* each containing a CUA anticodon. As in the identity swap experiments described above, the tRNA genes are inserted in cassette form into a plasmid containing the constitutive *lpp* promoter and a terminator. The amino acid specificity of each suppressor was determined as described above by sequencing the suppressed product of the amber mutation at position 10 in DHFR. Tables 2 and 3 give a

Table 1   Kinetic parameters for aminoacylation of tRNAs with *E. coli* methionine and valine AAS

tRNA	Anticodon	V/K_m	Relative V/K_m
Methionine AAS			
Met (f)[a]	CAU	1.3	7×10^5
Met (m)[a]	CAU	3.7	2×10^6
Met (m)	CAU	1.7	9×10^5
Val	CAU	1.3	7×10^5
Val	UAC	2×10^{-6}	1.0
Trp	CAU	0.1	5×10^4
Trp	CCA	2×10^{-6}	1.0
Valine AAS			
Val[a]	UAC	4.0	5×10^5
Val	UAC	1.5	2×10^5
Met (m)	UAC	0.16	2×10^4
Met (m)	CAU	8×10^{-6}	1.0

[a] Native tRNA
Data from (42).

Table 2   Amber suppressor tRNA genes in *E. coli*[a]

Derived by genetic means	Derived by synthetic means	
Ser (su$_1^+$)[b]	Ala2	Ile1
Gln (su$_2^+$)[b]	Arg	Ile2
Tyr (su$_3^+$)[b]	AspM	Leu5[i]
Leu (su$_6^+$)[c]	Asn[g]	Lys
Trp (Su$_7^+$)[d]	Cys[h]	Met (m)
GlyU[e]	GlyU	Phe[h]
GlyT[f]	GlyT	Pro
	GluA	Thr2
	HisA	Val

[a] The specificity of these tRNAs is not indicated. From (J. Normanly, et al, unpublished) unless otherwise indicated.
[b] (13)
[c] (70)
[d] (45)
[e] (71)
[f] (72, 73)
[g] Not a functional suppressor, specificity unknown.
[h] (27)
[i] (25)

summary progress report of this project. Table 2 lists the species of tRNA that exist as amber alleles, derived either genetically or by synthetic means. Table 3 classifies the suppressors by their amino acid specificity. Seven of the synthetic suppressors insert the correct amino acid. Together with the four original suppressors, 11 of 20 tRNAs in *E. coli* can tolerate the amber anticodon without misacylating. In some cases suppressor tRNAs are aminoacylated by the glutamine AAS. The synthetically derived tRNA$_{CUA}^{GluA}$ inserts 20% glutamine. From previous work (44), we know that tRNA$_{CUA}^{Met(f)}$ is charged by the glutamine AAS in vitro, and the su$_7^+$ allele of tRNA^{Trp} inserts a mixture of glutamine and tryptophan in vivo (45). The amber allele of one isoacceptor of tRNA^{Gly}, tRNA$_{CUA}^{GlyU}$, was correctly charged in vivo, while tRNA$_{CUA}^{GlyT}$ was aminoacylated by the glutamine AAS (46; J. Normanly, et al, unpublished). We were surprised to find that six of the suppressors generated in this work, tRNA$_{CUA}^{Arg}$, tRNA$_{CUA}^{AspM}$, tRNA$_{CUA}^{Ile2}$, tRNA$_{CUA}^{Met(m)}$, tRNA$_{CUA}^{Thr2}$, and tRNA$_{CUA}^{Val}$, were aminoacylated by the lysine AAS. There are also examples of suppressors that are aminoacylated by both the glutamine and lysine AASs. Such is the case for tRNA$_{CUA}^{Ile1}$, as well as for some mutant suppressor tRNAs (46–48).

Apparently, only the glutamine AAS and the lysine AAS misacylate amber suppressors. Each tends to misacylate a separate set. Since the CUA anti-

Table 3 tRNAs classified by the effect of an altered anticodon upon AAS recognition

Class I: No affect upon charging	Class II: Glutamine AAS mischarging	Class III: Lysine AAS mischarging
Ala2	Ile1	Ile2
GlyU	GlyT	Arg
Cys	Met (f)[a]	Met (m)
Phe	GluA	AspM
ProH	Trp (su$_7^+$)	Thr2
HisA		Val
Lys		
Ser (su$_1^+$)		
Gln (su$_2^+$)		
Tyr (su$_3^+$)		
Leu (su$_6^+$; also synthetic Leu5)		

[a] Demonstrated in vitro only (44)

codon is recognized by both the glutamine AAS and the lysine AAS, there must be negative elements that prevent aminoacylation by these enzymes. Some suppressors have both elements, some only one, and others have neither. It is interesting to note that there are two lysine AAS genes in *E. coli*, one of which is preferentially expressed during heat shock (49). Preliminary experiments designed to detect which of the enzymes is responsible for the observed misacylation indicated that both are.

A SINGLE BASE PAIR IS A MAJOR IDENTITY DETERMINANT FOR tRNA^{Ala}

Hou & Schimmel set out to isolate mutants of a tRNA^{Ala} amber suppressor that could not be recognized by the alanine AAS using the in vivo approach (50). An impressive collection of synthetic mutants was generated with multiple changes throughout the tRNA. Only those mutants having in common a change in the G-U pair at position 3-70 failed to insert alanine. This finding is consistent with two previous observations. First, earlier work by Murgola had indicated that a G-U pair at this position in a tRNA^{Lys} missense suppressor caused the insertion of either alanine or glycine (51). Second, the G-U pair at this position is conserved in virtually all of the tRNA^{Ala} that have been sequenced. To test the significance of the G$_3$-U$_{70}$ pair, Hou & Schimmel performed two identity swap experiments. tRNA^{Cys} and tRNA^{Phe} amber suppressors were synthesized, each containing the G-U pair at position 3-70. The altered cysteine suppressor inserted entirely alanine (Table 4). The

Table 4 Conversion to tRNA^{Ala} identity

Starting tRNA identity (E. coli)	Nucleotide changes	tRNA identity (% amino acid inserted)
Cys	$(C_3-G_{70})\rightarrow(G-U)$	Ala, 100[a]
Gly	$C_{70}\rightarrow U$	Gln, 95; Gly, 5[b]
Phe	$(C_3-G_{70})\rightarrow (G-U)$	Ala, 63; Phe, 37[a]
		Phe, 76; Ala, 24[b]
Phe	$(C_3-G_{70})\rightarrow(G-U)$; $U_{16}\rightarrow C$; $C_{17}\rightarrow U$; $U_{20}\rightarrow G$; $U_{60}\rightarrow C$; $U_{51}\rightarrow C$	Ala, 94; Lys, 6[b]
Lys	$C_{70}\rightarrow U$	Ala, 95[c]

[a] (50)
[b] (46)
[c] (48)

altered phenylalanine suppressor inserted predominantly alanine (63%) and some phenylalanine (37%). This result is certainly one of the most dramatic identity shift experiments to date in that it implicated a single G-U pair as the major identity element for the alanine AAS.

The ability of the G_3-U_{70} pair to direct alanine AAS recognition has been tested in other tRNAs with mixed results. Superimposing the G-U pair at position 3-70 of a lysine suppressor results in complete conversion to alanine identity (48). However, the same alteration to a glycine suppressor tRNA (GlyT) results in the insertion of predominantly glutamine, a small amount of glycine, and no detectable levels of alanine (46). It should be noted that the glycine amber suppressor with the normal G-C base pair at position 3-70 inserts predominantly glutamine (46; J. Normanly, et al, unpublished). In the case of tRNA^{GlyT}, simply superimposing the G-U pair at position 3-70 is not sufficient to counteract the identity determinants for the glutamine AAS. The glycine-to-alanine identity swap experiment might be more successful if the amber allele of another isoacceptor of tRNA^{Gly} that is not misacylated, were used instead.

McClain & Foss approached the problem of tRNA^{Ala} from a different angle (46). McClain has developed an algorithm (52, 53) that compares the sequence of a tRNA with its isoacceptors as well as isoacceptors of all 19 other tRNAs. This algorithm selects those positions in a tRNA that are correlated with its isoacceptors and anticorrelated with the other tRNA sequences. Subjecting tRNA^{Ala} to this type of analysis resulted in the prediction that four nucleotides were significant to tRNA^{Ala} in addition to the anticodon. These were the G_3-U_{70} pair, as well as G_{20} and C_{60} (Figure 2). The phenylalanine suppressor was altered to include these nucleotides, as well as changed at position 51 in order to increase transcriptional efficiency (54). Changes at positions 16 and 17 that were designed to remove phenylalanine identity were also made. In the DHFR assay, the resulting suppressor inserted 96% alanine and 4% lysine, indicating that the phenylalanine identity determinants had been removed.

McClain & Foss examined the contribution of C_{60} to alanine identity, and found that altering this putative identity element in $tRNA^{Ala}$ did not abolish alanine acceptance, nor did it give rise to aminoacylation by the alanine AAS when superimposed upon $tRNA^{Phe}$. In a similar fashion, the contribution of G_{20} to alanine identity was determined to be minimal. Alanine acceptance was not abolished when G_{20} was altered in $tRNA^{Ala}$, nor did G_{20} transplanted into $tRNA^{Phe}$ result in alanine insertion. It would appear that these elements of alanine identity predicted by the algorithm do not have a strong effect upon alanine AAS recognition.

$tRNA^{Phe}$ containing the G_3-U_{70} base pair alone, inserted 24% alanine and 76% phenylalanine. This result was the inverse of that obtained by Hou & Schimmel (63% alanine and 37% phenylalanine). The reason for this discrepancy is not clear, since both groups used the same expression system and the same specificity assay; namely, suppression of an amber mutation in DHFR. In any case, the G_3-U_{70} pair is clearly the dominant determinant of alanine identity, and is the strongest contributor to alanine identity predicted by McClain's algorithm.

The G-U pair as an identity element raises an interesting question. Does the alanine AAS recognize the functional groups of these bases via interactions in the deep groove of the A-Helix, or is it the perturbations to the helix stability and structure imposed by the G-U "wobble" pair that are recognized? To answer this question, McClain has now constructed alanine suppressors that contain each of 16 possible base combinations at position 3-70 (48). All the

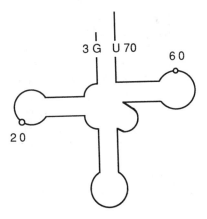

Figure 2   Cloverleaf secondary structure of $tRNA^{Ala}$. The G-U base pair at position 3-70, along with bases 20 and 60 (indicated by open circles), were predicted to be important alanine identity elements by statistical analysis (46).

variants had substantially lowered suppression efficiencies, and some of the base combinations yielded inactive suppressors. The four Watson-Crick base pairs at this position yielded suppressors that inserted lysine and glutamine. tRNA^{Ala} containing the mispairs G-A, C-A, and U-U at position 3-70 all inserted 90% alanine and, although less efficient, were similar to G-U at this position. Other variants, containing A-C, C-C, C-U, or U-G at position 3-70, inserted ~80% alanine, along with lysine and glutamine. The mispairs, G-G, A-A, and A-G, led to inactive suppressors, and the variant with a U-C mispair inserted predominantly glutamine. These results suggest that the alanine AAS is recognizing a structural perturbation imparted by the G-U pair and not specific bases. However, not all mispairs are acceptable; so the perturbation that is recognized must be sequence dependent. (It should be noted that NMR and X-ray crystallographic data reveal that a G-U base pair does not have a dramatic effect upon the overall structure of the acceptor stem helix.) Translocation of the G-U pair to position 4-69 led to partial alanine insertion, but only if the 3-70 pair was A-U and not G-C. G-U at position 2-71 had no activity. More questions are raised by these observations than are answered. The results emphasize the fact that there is insufficient data on the sequence dependence of RNA structure. Among the structural variations that one can imagine is a bending of the helix or even a separation of the strands.

THE tRNA$^{Leu \rightarrow Ser}$ IDENTITY SWAP

We have used the suppressor approach to define the elements of tRNA^{Ser} identity (25). Serine was chosen for several reasons. There are six serine codons, AGU/C and UCN (where N denotes any nucleotide) recognized by at least four different tRNAs. One of the original amber suppressors (su$_1^+$) inserted serine; so in all, the serine AAS can recognize tRNAs with at least five anticodons. It is unlikely that there are any serine recognition elements in the anticodon, although this assumption has not yet been tested in vitro. We set out to change the identity of a tRNA^{Leu} amber suppressor to serine. An examination of the six *E. coli* tRNA^{Ser} sequences revealed 14 nucleotides conserved in all tRNA^{Ser} sequences but not found in the tRNA$_5^{Leu}$ starting sequence. Additionally, the base pair at position 3-70, although not conserved (A-U or U-A), was found by McClain's statistical analysis to be correlated with serine and anticorrelated with other sequences (52, 53). This base pair was included as a potential element of serine AAS recognition (Figure 3a). The conserved nucleotides in the acceptor stem and in the D-stem-and-loop were included in the original trial construction. In addition, the size of the D-loop is highly conserved in the six tRNA^{Ser} sequences. Changes were made in that region so as to conform to the serine D-loop size. The first trial construction contained 12 changes from the starting tRNA$_5^{Leu}$ amber suppres-

a

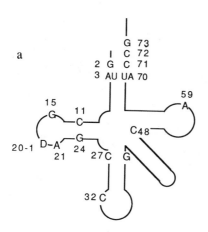

b

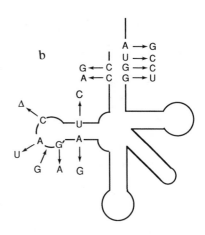

c

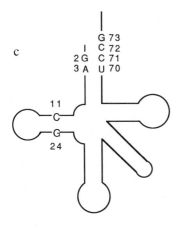

Figure 3 (*a*) Composite *E. coli* tRNA^{Ser}. Bases highly conserved in serine tRNAs and not found in tRNA$^{Leu}_{5CUA}$ are indicated. Adapted from (25). (*b*) Mutations introduced into tRNA$^{Leu}_{5CUA}$ to effect the conversion from leucine to serine identity result in the mutant tRNA$^{Leu \rightarrow Ser}$. Adapted from (25). (*c*) The minimum number of nucleotides required to effect the leucine-to-serine conversion (J. Normanly, J. Abelson, unpublished).

sor (Figure 3*b*). The resultant suppressor, tRNA$^{Leu \rightarrow Ser}$, was able to suppress an amber mutation in the β-lactamase gene, which requires the insertion of serine to produce an active enzyme. However, the suppression efficiency of tRNA$^{Leu \rightarrow Ser}$, 1%, was quite low compared to the 60–100% efficiency of the starting tRNA^{Leu} suppressor. Using the DHFR assay, the amino acid specificity of the tRNA$^{Leu \rightarrow Ser}$ suppressor was determined to be serine.

Following our initial successful attempt at the identity swap, we then set out to determine the minimum number of changes required to effect the transformation. This has now been accomplished and the results are shown in Figure 3*c*. The minimum number of base changes required is eight: base pairs 2-71, 3-70, nucleotides 72 and 73 of the acceptor stem, and the 11-24 base pair in the D-stem. The resulting suppressor containing these minimal changes is approximately 40% efficient. The efficiency was regained when the unnecessary changes in the D-loop were discarded. Interestingly, we find that only an A-U is acceptable at position 3-70 even though some serine tRNAs contain U-A at this position. We speculate that the A-U is acting as a deterrent to the leucine AAS. We are now using this information to conduct further identity swaps. Perhaps other serine identity elements exist that were already present in tRNA$_5^{Leu}$. By this iterative procedure we hope to arrive at a clear understanding of tRNA^{Ser} identity. The *E. coli* serine AAS has been purified and crystallized by R. Leberman et al (55). The ultimate understanding of tRNA^{Ser} recognition will come when the manner in which these identity elements are recognized by the enzyme has been defined.

YEAST tRNA^{Phe} RECOGNITION

As in the case of *E. coli* tRNA^{Met}, the anticodon has been implicated in the recognition of yeast tRNA^{Phe} by the yeast phenylalanine AAS. Early work by Zachau, in which the anticodon of yeast tRNA^{Phe} was removed, resulted in a tRNA that was aminoacylated by the phenylalanine AAS at a level 20% that of wild-type tRNA^{Phe} (55a). Using recombinant RNA technology, Uhlenbeck and coworkers have demonstrated that substitution of any of the yeast tRNA^{Phe} anticodon nucleotides diminishes aminoacylation (k_{cat}/K_m was decreased 3–10-fold) (56). In addition, transplanting the phenylalanine anticodon onto tRNA^{Tyr} leads to aminoacylation of the hybrid molecule by the phenylalanine AAS (57). As the aminoacylation of the hybrid tRNA was fairly weak, they speculated that there must be phenylalanine recognition elements outside of the anticodon as well.

Using the in vitro system, transcription of synthetic tRNA genes with T7 RNA polymerase, a large number of variants have been constructed and their properties as substrates have been tested (30, 58). This is by far the largest and most comprehensive set of tRNA variants ever constructed. At this point

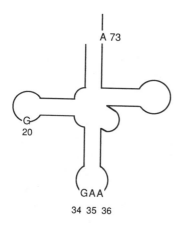

34 35 36

Figure 4   Nucleotides postulated to be important for yeast tRNA^{Phe} identity (58).

nearly all of the nonuniversal bases in tRNA^{Phe} have been changed and the variants have been tested as substrates. Experiments in which the nucleotides involved in tertiary interactions were altered revealed that changes in these bases can be tolerated as long as the overall tertiary interactions are not disrupted. For example, the G_{19}-C_{56} tertiary Watson-Crick base pair can be altered to C_{19}-G_{56} without affecting recognition by the phenylalanine AAS. Either of the single mutations, $G_{19} \rightarrow C$ or $C_{56} \rightarrow G$, leads to a fivefold drop in k_{cat}/K_m, indicating that the tertiary structure of the tRNA must be maintained but that neither G_{19} nor C_{56} is recognized directly. When G_1-C_{72} is altered to either G-U or A-U, the k_{cat}/K_m drops four- and seven-fold, respectively (58a). It is not clear yet whether G_1-C_{72} is important as a recognition element for the phenylalanine AAS, or as a structural feature. The results of this study led to the proposal that five nucleotides, all in single strand regions, are recognition elements for the phenylalanine AAS. They are the three anticodon bases, G_{34}, A_{35}, and A_{36}, G_{20} in the D-loop, and A_{73}, the fourth base from the 3' end (Figure 4). Several tests of the model were undertaken. First, the *E. coli* tRNA^{Phe} (Table 5), a rather poor substrate for the yeast phenylalanine AAS, contained all of the putative identity elements save the one at position 20. Changing U_{20} to G brought k_{cat}/K_m to within a factor of two for that of yeast tRNA^{Phe}. Identity swaps based on the conserved bases were also carried out[2], changing yeast tRNA^{Met}, tRNA^{Arg}, and tRNA^{Tyr} to good substrates for the phenylalanine AAS. It appears that each of the five bases contributes independently to the specificity. Thus, a tRNA containing four of the five recognition elements has a k_{cat}/K_m one order of magnitude lower than wild type, and a substrate with only three recognition elements has

[2]Changes were incorporated elsewhere as well, either to facilitate the run-off transcription reaction, or to introduce tRNA^{Phe} structural elements where they differed from the starting tRNA.

a k_{cat}/K_m that is two orders of magnitude lower. Some years ago Dudock and collaborators reported the mischarging of a number of tRNAs by the phenylalanine AAS (59). With knowledge of the recognition elements and thus relative contributions to specificity, Sampson & Uhlenbeck were able to reconcile that data nicely. Of the yeast tRNAs that have been sequenced, none has more than three of the recognition elements so that the in vivo specificity can be explained if one also assumes a contribution to overall specificity due to competition with the cognate AAS. The five recognition elements are located far apart from one another in the tRNA tertiary structure so the phenylalanine AAS must contact the entire surface of the tRNA molecule.

McClain & Foss have used the suppressor approach to investigate E. coli tRNA^{Phe} identity. From their results it appears that U_{20} and A_{73} are implicated in tRNA^{Phe} identity. In addition, small effects are observed in changing the anticodon stem base pairs 27-43, and 28-42 (47).

tRNA^{Gln} IDENTITY

It is pertinent to discuss what is known about the recognition of E. coli tRNA^{Gln}. In a short time we will know more about the interaction of this tRNA with its AAS than for any other system. At this point (October 1988) Steitz & Söll have generated tRNA^{Gln}-AAS co-crystals that refract to high resolution. Data has been collected and a map is being constructed. Where do we expect the contacts will be made?

First, as we have already mentioned, it is likely that the glutamine AAS recognizes the anticodon. The su$_7^+$ allele of E. coli tRNA^{Trp}, which inserts

Table 5 Aminoacylation kinetics of wild-type and mutant tRNA^{Phe} transcripts

tRNA	Nucleotide alterations	Relative k_{cat}/K_m
yeast Phe	wild-type	1.0
E. coli Phe	wild-type	.04
E. coli Phe	$U_{20} \rightarrow G$	.52
yeast Met→Phe	$A_{20} \rightarrow G$; $C_{34} \rightarrow G$; $U_{36} \rightarrow A$; $(G_{49}-C_{65}) \rightarrow (C-G)$; $A_{59} \rightarrow U$;	.68
yeast Arg→Phe	$C_{20} \rightarrow G$; $U_{34} \rightarrow G$; $C_{35} \rightarrow A$; $U_{36} \rightarrow A$; $C_{59} \rightarrow U$; $C_{73} \rightarrow A$; $U \rightarrow_{17}$[a]	.64
yeast Tyr→Phe	$(U_{20}, U_{20-1}, U_{20-2}) \rightarrow G$; $A_{13} \rightarrow C$; $A_{22} \rightarrow G$; $(C_{12}-G_{23}) \rightarrow (U-A)$; $(C_1-G_{72}) \rightarrow (G-C)$; $U_{35} \rightarrow A$; $A_{46} \rightarrow G$; $(U_2-A_{71}) \rightarrow (C-G)$; $(C_3-G_{70}) \rightarrow (G-C)$	1.5

[a] Represents an insertion
Data from (58).

both glutamine and tryptophan in vivo, has a single change in the anticodon from CCA to CUA, implicating the U in the middle position (45). In vitro, the su_7^+ tRNA is a better substrate for the glutamine AAS by about six orders of magnitude (60). Changing the anticodon of *E. coli* tRNA$^{Met(f)}$ from CAU to CUA leads to aminoacylation by the glutamine AAS in vitro, at levels similar to that seen for su_7^+ tRNA (44). The su_3^+ allele of *E. coli* tRNA^{Tyr} is correctly charged in vivo, but elevated levels of the glutamine AAS can lead to mischarging in vitro (61).

We also anticipate that the glutamine AAS will interact with the 3' end of the tRNA. Mutants of su_3^+ tRNA^{Tyr} with alterations at the 3' end result in recognition by the glutamine AAS. One of these mutants, $A_{73} \rightarrow G$, is particularly effective (21–23). tRNA^{Gln} has a G at position 73. Rogers & Söll have also described mutational changes of the tRNA^{Ser} in the acceptor base pairs 1-72 and 2-73 that convert this tRNA to glutamine specificity in the DHFR assay (62). Note that these are also positions recognized by the serine AAS, so competition between the two enzymes was decreased by making these changes.

THE VARIABLE POCKET: A POTENTIAL SITE IN tRNA FOR AAS RECOGNITION

In discussing the structure of yeast tRNA^{Phe}, Klug pointed out that the cluster of nucleotides formed by the interaction of the D-loop and T-loop forms a patch that arches out from the surface of the molecule (63). The nucleotides involved in this patch are 16, 17, 20, 59, and 60 (Figure 1). The patch is variable in that these nucleotides are not conserved among different tRNAs, and there can be insertions or deletions in the D-loop that could potentially change the configuration of this pocket. Klug explicitly suggested that this variable pocket may "form part of a recognition system for different tRNAs, perhaps for sorting into classes for synthetase discrimination."

We have already seen that G_{20} in yeast tRNA^{Phe} is a recognition element for the phenylalanine AAS (58). In an analysis of the six *E. coli* tRNA^{Arg} sequences, McClain observed a strong correlation of A_{20} with arginine identity. Consequently, a phenylalanine-to-arginine identity swap was attempted (64). The A_{20} was inserted into the phenylalanine D-loop, causing U_{20} to become $U_{20.1}$. This suppressor was not active. However, with the variable pocket model in mind, residues 16, 17, and 59 were altered in an attempt to create the variable pocket characteristic of tRNA$_{ACG-1}^{Arg}$. This suppressor inserts 75% arginine, 5% lysine, and 6% tyrosine. It is a significantly better arginine suppressor than one constructed simply by changing the anticodon of tRNA$_{ACG-1}^{Arg}$ to CUA, which inserts 37% Arg and 55% lysine. Insertion of A_{20} and A_{59} alone had virtually the same effect. Taken together with the tRNA^{Phe}

results, this finding certainly highlights residue 20 as an important synthetase recognition element. Given the exposed configuration of the variable pocket, other synthetases may well use different residues in this group as recognition elements.

tRNA SPECIFICITY DEFINED BY NUCLEOTIDE MODIFICATION

In general, tRNA modifications do not play a large role in tRNA identity. That this is so is most clearly demonstrated by the near normal substrate activity of unmodified transcripts. Indeed, even a tDNA^{Phe} can be specifically aminoacylated (65). There is, however, a very interesting case in which a tRNA modification plays a crucial role not only in AAS recognition, but in codon recognition as well. *E. coli* tRNA$_2^{Ile}$ recognizes the codon AUA. It was therefore a surprise when it was discovered that the anticodon in the DNA sequence of the tRNA$_2^{Ile}$ gene is CAT (66), the anticodon for methionine. The same sequence was found in the bacteriophage T4 tRNA^{Ile} gene (67, 68), and in a spinach chloroplast tRNA^{Ile} gene (69). Recall that C_{34} in tRNA^{Met} is a crucial recognition element. Thus, we have the dilemma that this tRNA^{Ile} should not only miscode, but it should also mischarge. The dilemma was resolved by the discovery that the C in tRNA$_2^{Ile}$ is modified to lysidine (Figure 5). This alteration affects recognition by the synthetase; the unmodified tRNA is recognized by the methionine AAS and the modified tRNA by the isoleucine AAS (66). Lysidine also changes the coding capacity of the tRNA so that it recognizes AUA and not AUG. Why is this bizarre route taken, rather than simply making that anticodon UAU? Muramatsu et al suggest that the evolution of the lysidine mechanism may have allowed a change in the genetic code converting AUA from methionine to isoleucine specificity. Once the mechanism was fixed, it was retained, except in some mitochondria where AUA has reverted to methionine specificity. Muramatsu et al suggest that this was due to the loss of the lysidine biosynthetic mechanism (66).

CONCLUDING REMARKS

Much progress has been made recently with the application of two powerful methods to the study of tRNA identity. The in vitro approach assesses the contribution of specific nucleotides to tRNA identity by examining the aminoacylation kinetics of specifically altered tRNAs, providing a very clear picture of recognition elements for a given tRNA. Adding to this approach, determination of the in vivo specificity of altered tRNAs provides clues into the physiological relevance of specific nucleotides, i. e. what effect they have upon the competition between all 20 AASs. The experiments reviewed here

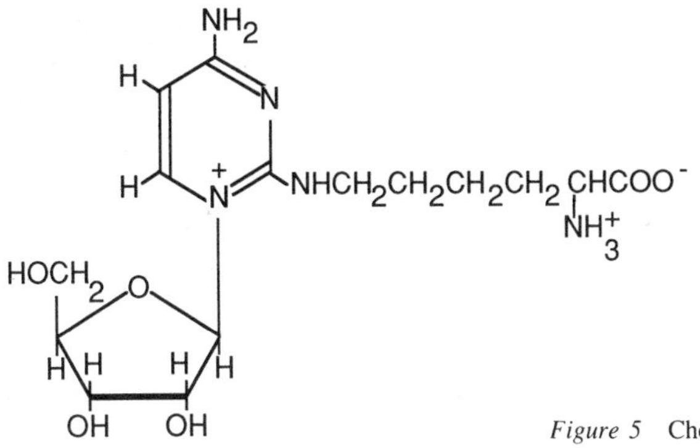

Figure 5   Chemical structure of lysidine.

have revealed the following: (*a*) tRNA identity is defined by a relatively small number of elements, and (*b*) the anticodon appears to be an important identity element for a majority of tRNAs.

Identity elements may vary in the extent of their evolutionary conservation. For example, the G_3-U_{70} base pair so important for alanine identity is present in virtually every tRNA^{Ala} sequenced to date, over a wide phylogenetic spectrum. The same appears to hold true for the A_{20} found to be involved in arginine identity. The determinants of serine tRNA identity, however, are conserved only in bacterial genes.

There are likely to be classes of tRNAs in regard to AAS recognition. On the simplest level, tRNAs can be classified according to the effect that an alteration in the anticodon has upon AAS recognition. Cognate recognition is either (*a*) normal, or (*b*) diminished/abolished. Within these subsets further divisions can be made. For example, both *E. coli* tRNA$^{Met(m)}$ and yeast tRNA^{Phe} fall into the latter class. Yet for tRNA$^{Met(m)}$ the anticodon appears to be the sole identity element, while for yeast tRNA^{Phe} two other nucleotides are important for complete phenylalanine acceptance. Similarly, *E. coli* tRNA^{Ser} and tRNA^{Ala} fall into class one, i.e. the anticodon does not appear to play a role in cognate AAS recognition. While eight identity elements have been identified for tRNA^{Ser}, a single base pair is primarily responsible for tRNA^{Ala} identity.

The phenomena of lysine and glutamine mischarging remain a mystery. The composition of the anticodon must in large part direct aminoacylation by these two AASs (e.g. a U at position 35 is a recurring theme in glutamine aminoacylation), but, as they do not misacylate universally, there must be identity determinants or deterrents outside of the anticodon that dictate which tRNAs can be readily misacylated. Both of these AASs may have rather loose specificity, and, subsequently, behave as "default" AASs for tRNAs that are not well charged by their cognate AAS. Sorting out identity determinants from deterrents will no doubt prove difficult; however, a systematic application of both the in vitro and in vivo approaches will go a long way toward solving the problem.

Recent progress in the crystallography of tRNA-AAS complexes (8, 9) promises to give insight into the interaction between the two molecules, and will undoubtedly suggest new pathways to take in the study of the recognition problem.

ACKNOWLEDGMENTS

We wish to thank Jeff Sampson, Olke Uhlenbeck, LaDonne Schulman, and Bill McClain for sharing their results prior to publication, and to David Horowitz, Olke Uhlenbeck, and Jeff Sampson for critical reading of the manuscript. The authors were supported by grants from NIH (GM32637) and ONR (N00014-86-K-0755).

Literature Cited

1. Crick, F. H. C. 1966. *J. Mol. Biol.* 19:548–55
2. Sprinzl, M., Hartmann, T., Meissner, F., Moll, J., Vorderwülbecke, T. 1987. *Nucleic Acids Res.* 15:r53–r188
3. Holley, R. W., Apgar, J., Everett, G. A., Madison, J. T., Marquisee, M., et al. 1965. *Science* 147:1462–65
4. Kim, S. H., Sussman, J. L., Suddath, F. L., Quigley, G. J., McPherson, A., et al. 1974. *Proc. Natl. Acad. Sci. USA* 71:4970–74
5. Klug, A., Robertus, J. D., Ladner, J. E., Brown, R. S., Finch, J. T. 1974. *Proc. Natl. Acad. Sci. USA* 71:3711–15
6. Schimmel, P. R., Söll, D. 1979. *Annu. Rev. Biochem.* 48:601–48
7. Schimmel, P. R. 1987. *Annu. Rev. Biochem.* 56:125–58
8. Podjarny, A., Rees, B., Thierry, J. C., Cavarelli, J., Jesior, J. C., et al. 1987. *J. Biomol. Struct. Dyn.* 5:187–98
9. Perona, J. J., Swanson, R., Steitz, T. A., Söll, D. 1988. *J. Mol. Biol.* In press
10. Fersht, A. R., Shi, J. P., Knill-Jones, J., Lowe, D. M., Wilkinson, A. J., et al. 1985. *Nature* 314:235–38
11. Zelwer, C., Risler, J. L., Brunie, S. 1982. *J. Mol. Biol.* 155:63–81
12. Goodman, H. M., Abelson, J., Landy, A., Brenner, S., Smith, J. D. 1968. *Nature* 217:1019–24
13. Steege, D. A., Söll, D. 1979. In *Biological Regulation and Development,* ed. R. F. Goldberger, I:433–85. New York: Plenum
14. Smith, J. D. 1979. In *Nonsense Mutations and tRNA Suppressors,* ed. J. E. Celis, J. D. Smith, pp. 109–25. London: Academic. 349 pp.
15. Wilson, J. H., Abelson, J. N. 1972. *J. Mol. Biol.* 69:57–73
16. McClain, W. H., Wilson, J. H., Seidman, J. G. 1988. *J. Mol. Biol.* 203:549–53
17. Kurjan, J., Hall, B. D., Gillam, S., Smith, M. 1980. *Cell.* 20:701–9
18. Altman, S. 1979. See Ref. 14, pp. 173–89
19. Willis, I., Hottinger, H., Pearson, D., Chisholm, V., Leupold, U., et al. 1984. *EMBO J.* 3:1573–80
20. Nishikura, K., Kurjan, J., Hall, B. D., DeRobertis, E. M. 1982. *EMBO J.* 1:263–68
21. Hooper, M. L., Russell, R. L., Smith, J. D. 1972. *FEBS Lett.* 22:149–56
22. Shimura, Y., Aono, H., Ozeki, H., Sarabhai, A., Lamfrom, H., Abelson, J. 1972. *FEBS Lett.* 22:144–48
23. Ghysen, A., Celis, J. E. 1974. *J. Mol. Biol.* 83:333–51
24. Khorana, H. G. 1979. *Science* 203:614–25
25. Normanly, J., Ogden, R. C., Horvath, S. J., Abelson, J. 1986. *Nature* 321:213–19
26. Masson, J.-M., Miller, J. H. 1987. *Gene* 47:179–83
27. Normanly, J., Masson, J.-M., Kleina, L. G., Abelson, J., Miller, J. H. 1986. *Proc. Natl. Acad. Sci. USA* 83:6548–52
28. Bolin, J. T., Filman, D. J., Matthews, D. A., Hamlin, R. C., Kraut, J. 1982. *J. Biol. Chem.* 257:13650–62
29. Filman, D. J., Bolin, J. T., Matthews,

D. A., Kraut, J. 1982. *J. Biol. Chem.* 257:13663–72
30. Sampson, J. R., Uhlenbeck, O. C. 1988. *Proc. Natl. Acad. Sci. USA* 85: 1033–37
31. Samuelsson, T., Boren, T., Johansen, T.-I., Lustig, F. 1988. *J. Biol. Chem.* 263:13692-99
32. Sampson, J. R., Sullivan, F. X., Behlen, L. S., DiRenzo, A. B., Uhlenbeck, O. C. 1987. *Cold Spring Harbor Symp. Quant. Biol.* 52:267–79
33. Brown, R. S., Dewan, J. C., Klug, A. 1985. *Biochemistry* 24:4785–801
34. Dirheimer, G., Ebel, J.-P., Bonnet, J., Gangloff, J., Keith, G., et al. 1972. *Biochimie* 54:127–44
35. Engelhardt, W. A., Kisselev, L. L. 1966. In *Current Aspects of Biochemical Energetics*, ed. N. O. Kaplan, E. O. Kennedy, pp. 213–15. New York: Academic
36. Kisselev, L. L. 1985. In *Progress in Nucleic Acids Research and Molecular Biology*, ed. W. E. Cohn, K. Moldave, 32:237–66. Fla: Academic
37. Schulman, L. H. 1979. In *Transfer RNA: Structure, Properties and Recognition*, ed. P. R. Schimmel, D. Söll, J. N. Abelson, pp. 311–24. New York: Cold Spring Harbor Lab. Press
38. Schulman, L. H., Pelka, H., Susani, M. 1983. *Nucleic Acids Res.* 11:1439–55
39. Schulman, L. H., Pelka, H. 1983. *Proc. Natl. Acad. Sci. USA* 80:6755–59
40. Schulman, L. H., Pelka, H. 1984. *Fed. Proc.* 43:2977–80
41. Valenzuela, D., Leon, O., Schulman, L. H. 1984. *Biochem. Biophys. Res. Commun.* 119:677–84
42. Schulman, L. H., Pelka, H. 1988. *Science* 242:765–68
43. Miller, J. H., Normanly, J., Masson, J.-M., Kleina, L. G., Abelson, J. 1987. In *Integration and Control of Metabolic Processes: Pure and Applied Aspects*, pp. 571–81. England: Cambridge Univ. Press
44. Schulman, L. H., Pelka, H. 1985. *Biochemistry* 24:7309–14
45. Celis, J. E., Coulondre, C., Miller, J. H. 1976. *J. Mol. Biol.* 104:729–34
46. McClain, W. H., Foss, K. 1988. *Science* 240:793–96
47. McClain, W. H., Foss, K. 1988. *J. Mol. Biol.* 202:697–709
48. McClain, W. H., Foss, K., Chen, Y.-M., Schneider, J. 1988. *Science* 242: 1681–84
49. Hirshfield, I. N., Bloch, P. L., Van Bogelen, R. A., Neidhardt, F. C. 1981. *J. Bacteriol.* 146:345–51
50. Hou, Y.-M., Schimmel, P. R. 1988. *Nature* 333:140–45

51. Prather, N. E., Murgola, E. J., Mims, B. H. 1984. *J. Mol. Biol.* 172:177–84
52. Atilgan, T., Nicholas, H. B., McClain, W. H. 1986. *Nucleic Acids Res.* 14:375–80
53. McClain, W. H., Nicholas, H. B. 1987. *J. Mol. Biol.* 194:635–42
54. McClain, W. H., Guerrier-Takada, C., Altman, S. 1987. *Science* 238:527–30
55. Leberman, R., Berthet-Colominas, C., Cusack, S., Härtlein, M. 1987. *J. Mol. Biol.* 193:423–25
55a. Thiebe, R., Harbers, K., Zachau, H. G. 1972. *Eur. J. Biochem.* 26:144–52
56. Bruce, A. G., Uhlenbeck, O. C. 1982. *Biochemistry* 21:3921–27
57. Bare, L., Uhlenbeck, O. C. 1985. *Biochemistry* 24:2354–60
58. Sampson, J. R., DiRenzo, A., Behlen, L., Uhlenbeck, O. C. 1989. *Science.* 243:1363–66
58a. Uhlenbeck, O. C., Wu, H.-N., Sampson, J. R. 1987. In *Molecular Biology of RNA, New Perspectives*, ed. M. Inouye, B. S. Dudock, pp. 285–94. Academic
59. Roe, B., Sirover, M., Dudock, B. 1973. *Biochemistry* 12:4146–54
60. Yarus, M., Knowlton, R., Soll, L. 1977. In *Nucleic Acid–Protein Recognition*, pp. 391–408. New York: Academic
61. Hoben, P., Uemura, H., Yamao, F., Cheung, A., Swanson, R., et al. 1984. *Fed Proc.* 43:2972–76
62. Rogers, M. J., Söll, D. 1988. *Proc. Natl. Acad. Sci. USA* 85:6627–31
63. Ladner, J. E., Jack, A., Robertus, J. D., Brown, R. S., Rhodes, D., et al. 1975. *Proc. Natl. Acad. Sci. USA* 72:4414–18
64. McClain, W. H., Foss, K. 1988. *Science* 241:1804–7
65. Khan, A. S., Roe, B. 1988. *Science* 241:74–79
66. Muramatsu, T., Nishikawa, K., Nemoto, F., Kuchino, Y., Nishimura, S., et al. 1988. *Nature* 336:179–81
67. Fukada, K., Abelson, J. 1980. *J. Mol. Biol.* 139:377–91
68. Mazzara, G. P., Plunkett, G. III, McClain, W. H. 1981. *Proc. Natl. Acad. Sci. USA* 78:889–92
69. Kashdan, M. A., Dudock, B. S. 1982. *J. Biol. Chem.* 257:11191–94
70. Thorbjarnardottir, S., Dingermann, T., Rafnar, T., Andresson, O. S., Söll, D., et al. 1985. *J. Bacteriol.* 161:219–22
71. Murgola, E. J. 1985. *Annu. Rev. Genet.* 19:57–80
72. Murgola, E. J., Hijazi, K. A. 1983. *Mol. Gen. Genet.* 191:132–37
73. Murgola, E. J., Prather, N. E., Pagel, F. T., Mims, B. H., Hijazi, K. A. 1984. *Mol. Gen. Genet.* 193:76–81

Annu. Rev. Biochem. 1989. 58:1051–77

MOLECULAR MECHANISMS OF TRANSCRIPTIONAL REGULATION IN YEAST

Kevin Struhl

Department of Biological Chemistry, Harvard Medical School, Boston, Massachusetts 02115

CONTENTS

PERSPECTIVES AND SUMMARY

Regulation of gene expression is a fundamental aspect of many biological phenomena such as cell growth, morphology, the development of multi-cellular organisms, the response to environmental conditions, and disease. The processes of decoding genes and synthesizing appropriate amounts of gene products are complex, and regulation can occur at one or more of the

1051

various steps along the pathway. Nevertheless, a major point of gene control occurs at the first step, namely the initiation of messenger RNA synthesis.

Over the past several years, many observations have increasingly pointed to common molecular mechanisms of transcriptional regulation in eukaryotic organisms ranging from humans to yeasts (for previous reviews, see 1–4). The subunit structure and catalytic properties of RNA polymerase II as well as the posttranscriptional modifications of the primary transcripts are conserved throughout the eukaryotic kingdom. A sequence resembling TATAAA is found in most eukaryotic promoters at a position near the mRNA initiation site. Genetic analyses of yeast and higher eukaryotic promoters have revealed enhancer or UAS elements that activate transcription when located at long distances from the initiation site. The largest subunit of RNA polymerase II contains an unusual repeating heptapeptide tail at the C-terminus and is otherwise highly conserved (5, 6). More recently, it has been shown that yeast and higher eukaryotic cells contain structurally similar and functionally analogous transcription factors that recognize essentially identical sequences (7–14). Moreover, yeast proteins can activate transcription in a wide variety of eukaryotic organisms (15–18), and vertebrate proteins can stimulate transcription in yeast cells (19–22). Such functional interchangeability makes it possible to study molecular mechanisms in vivo and in vitro by using mixtures of yeast and mammalian components.

Much of our knowledge of the molecular mechanisms involved in transcriptional regulation has come from studies of the baker's yeast *Saccharomyces cerevisiae*. The relative simplicity and rapid growth rate of the organism, the availability of powerful genetic selections and screens to identify important genes and gene products and to obtain mutant strains, the ease of cloning essentially any gene, and the ability to alter the genome at will by performing exact replacements of normal chromosomal sequences with mutated derivatives constructed in vitro have been crucial for the relatively advanced state of knowledge. At present, many yeast promoters have been analyzed in detail, and a number of specific transcription factors have been identified and characterized for their DNA-binding and transcriptional activation functions. In some cases, the regulatory circuits for altering the level of transcription under appropriate conditions have been outlined.

YEAST PROMOTER ELEMENTS

The yeast genome specifies approximately 5000 protein-coding genes that are densely clustered on 16 linear chromosomes. The average yeast gene is transcribed about 5–10 times during each cell cycle, which results in a steady-state level of 1–2 mRNA molecules/cell. Although many genes are transcribed constitutively at the average level, RNA levels of different genes vary over 2–3 orders of magnitude. Moreover, some genes are transcribed at variable rates depending on the physiological conditions, and groups of genes are regulated coordinately. Despite the different transcriptional regulatory properties, essentially all yeast promoters contain three basic kinds of DNA sequence elements (Figure 1). The properties of these upstream (UAS), TATA, and initiation (I) elements have been reviewed previously (1–3) and are briefly summarized below.

Upstream elements are short DNA sequences, typically 10–30 bp in length, that are located relatively far from the mRNA start site and are required for transcription. Depending on the gene, upstream elements can be located anywhere from 100 to 1500 bp upstream of the initiation site. For genes whose transcription rates vary according to the physiological conditions, the upstream element is usually the major determinant of the particular regulatory properties of a given promoter. Genes subject to a common control mechanism contain upstream elements that are similar in DNA sequence, whereas noncoordinately regulated genes contain upstream elements with different DNA sequences. Upstream elements are analogous to mammalian enhancer sequences in that they function in both orientations and at long and variable distances with respect to other promoter elements and the mRNA initiation site. In contrast to enhancers, UAS elements do not activate transcription when located downstream of the mRNA initiation site (24, 25).

In addition to upstream elements, TATA elements are necessary for transcriptional initiation of most yeast genes; deletion of these elements greatly reduces the mRNA level (26–29). TATA elements, which have been historically defined by their similarity to the sequence TATAAA, are typically located anywhere between 40 and 120 bp upstream of the mRNA initiation

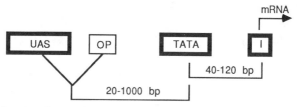

Figure 1 *cis*-acting elements for a simple yeast promoter.
    The initiator element (I) is important for determining where transcription begins (arrow). The TATA element, located 40 to 120 base pairs upstream from the initiation site, is required for transcription. The upstream element (UAS), which is located at variable and sometimes very long distances from the other elements, is important for transcription and usually determines the particular regulatory properties of the promoter. Some promoters contain operator elements (OP), which mediate negative control of transcription; operators are generally located at variable distances upstream of TATA elements, but their position with respect to UAS elements depends on the promoter. See text for details.

sites. This variability in location differs from the situation in most other eukaryotes, in which TATA sequences are almost always located 25–30 bp away from the initiation site. The presence of the conserved TATAAA sequence in many different kinds of promoters has led to the view that TATA elements have a general role in the transcription process much like RNA polymerase. However, more recent observations have argued against this view and instead have suggested the existence of functionally distinct classes of TATA elements.

The initiator element, located near the actual mRNA start site, is the primary determinant of where transcription begins. In yeast, accurate initiation is still observed when the distance to the TATA or upstream element is varied or when transcription depends on "foreign" promoter elements located at different positions from the natural elements (29–32). In contrast, selection of higher eukaryotic mRNA start sites is determined not by specific DNA sequences but rather by distance (25–30 bp) from the TATA element. The spacing between yeast initiation sites and TATA elements is much more flexible and somewhat larger, although there are limits to the distance over which a TATA element can act (roughly 40–120 bp). However, the nonstringent spacing requirements account for why some yeast genes have many more initiation sites than TATA elements. The DNA sequence requirements for yeast initiator elements are poorly understood, although it is clear that many different sequences can carry out the function. Unlike upstream or TATA elements, the initiator element is relatively unimportant for determining the rate of transcriptional initiation.

Some yeast promoters contain a fourth class of element, the operator, that represses the level of transcriptional initiation (reviewed in 33). The properties of yeast operators are similar to those of upstream elements in that they are short sequences that function bidirectionally and at variable distances upstream of TATA elements. Like upstream elements, similar operator sequences in different promoters provide the basic mechanism for coordinate regulation of transcription (34, 35). Operators can repress transcription when located upstream of upstream elements, but repression is generally much more efficient when the operator lies between the upstream and TATA element (28, 34, 36). One major exception to this rule is the mating-type silencer that efficiently represses transcription when located at distances as far as 2 kb upstream or downstream from the mRNA initiation sites (37–39).

SPECIFIC DNA-BINDING PROTEINS

It is almost an axiom of molecular biology that the function of promoter DNA sequences is to act as targets for the specific DNA-binding proteins that actually activate or repress transcription. As might be expected from the wide variety of upstream promoter elements, yeast cells contain many specific transcription factors that recognize different DNA sequences. Unlike higher eukaryotic factors that were identified by direct DNA-binding assays, yeast regulatory proteins were first identified by mutations that alter the transcription of a specific gene or set of genes. Using various experimental procedures, it has been shown in a number of cases that these mutations define genes encoding specific DNA-binding proteins. The genetic approach is particularly valuable because the mutant strains provide direct information about the function(s) of the DNA-binding protein in vivo and facilitate the cloning of the genes.

The availability of the cloned genes has allowed for a detailed analysis of the structures and functions of these yeast DNA-binding proteins. For several cases, analyses of deletion proteins indicate that small autonomous domains containing less than 100 amino acid residues are sufficient for specific DNA-binding activity. Although detailed structural information is not yet available, it is clear from the primary sequences and biochemical properties of the DNA-binding domains that different structural motifs are employed for DNA sequence recognition. These motifs include the helix-turn-helix of bacterial repressors and activators (40), the presumptive zinc finger (41), the putative leucine zipper (42), and possibly others that have yet to be described. Studies of yeast DNA-binding proteins are most advanced for the GAL4, GCN4, MATα2, and HAP1 proteins.

GAL4   The GAL4 protein (881 amino acids) is required for transcription of the *GAL* (galactose metabolizing) genes that occurs in galactose medium. GAL4 binds to four sites within the upstream regulatory region required for bidirectional activation of the *GAL1* and *GAL10* genes as well as to sites within several other promoters (43–45). Sequence comparison of the binding sites indicates that a 17-bp sequence of imperfect dyad symmetry is important for recognition by GAL4. In vivo footprinting experiments indicate that in galactose medium the GAL4-binding sites in the genome are occupied most if not all of the time (44, 46), and in certain cases it appears that GAL4 can bind cooperatively to adjacent sites (47). Interestingly, GAL4 is bound to its genomic sites even when *GAL* transcription is not induced (glycerol medium), although it is not bound in glucose medium when *GAL* transcription is catabolite repressed (44, 46).

The GAL4 DNA-binding domain is localized to the N-terminal 73 amino acids (48), and it contains two pairs of cysteine residue sequences that resemble coordination sites for zinc ions (41, 50). An important role for zinc is inferred from gal4 mutants that are functional only in the presence of high concentrations of zinc ions (50, 51). In addition, mutations of the cysteine residues abolish DNA-binding activity (51–53).

The yeast ARGR2 and PPR1 activator proteins contain amino acid sequences that resemble the region of GAL4 that contains the pairs of cysteine residues even though these proteins recognize very different DNA sequences (54, 55). In contrast, LAC9, a protein from the related yeast *Kluveromyces lactis* that can functionally substitute for GAL4 in *S. cerevisiae* cells, has an additional sequence in its DNA-binding domain that is very similar to that found in GAL4. Thus, it is likely that this additional sequence is involved in recognition of the DNA target, whereas the region that includes the cysteine pairs is more involved in maintaining overall structure and perhaps nonspecific interactions to DNA. Consistent with this hypothesis, a synthetic peptide containing the zinc finger of ADR1 can form a discrete structure in the presence of zinc that binds DNA nonspecifically, but it fails to recognize its normal target sequence (56).

GCN4   GCN4 protein (281 amino acids) binds specifically to the promoter regions of many amino acid biosynthetic genes and induces their transcription in response to amino acid starvation (57, 58). Saturation point mutagenesis of the *HIS3* regulatory site indicates that GCN4 recognizes a 9-bp region with optimal binding to the dyad-symmetric sequence ATGA(C/G)TCAT (59). The optimal binding site binds to GCN4 with higher affinity than the native *HIS3* site, and it induces transcription to higher levels, thus suggesting that GCN4 protein levels are limiting in vivo. The optimal GCN4 binding site also represents the consensus of presumptive recognition sequences from 15 genes subject to coordinate induction by GCN4 (59). Interestingly, none of these naturally occurring sites are identical to the consensus sequence but instead have 1–2 deviations. Thus, it appears that yeast cells maintain a fine balance between the level of GCN4 protein and the affinities of the many binding sites in the genome such that transcription of the coregulated genes can be induced even though GCN4 does not fully occupy the sites.

The dyad-symmetric nature of the optimal GCN4 binding site strongly suggests that GCN4 binds as a dimer to adjacent half sites. GCN4 does indeed bind as a dimer because synthetic mixtures of wild-type and deleted GCN4 proteins yield functional heterodimers at the expected frequency (60). The GCN4 recognition site is unusual because it is very short and because mutation of the central C:G base pair, even to the symmetric G:C counterpart, significantly reduces binding to DNA. This suggests that the central base pair is part of the half site recognized by a GCN4 monomer and consequently that the half sites overlap (60). Moreover, it follows that the optimal binding site must contain two nonequivalent half sites, ATGAC and ATGAG. Interestingly, GCN4 binds well to a 10-bp site containing adjacent ATGAC half sites, whereas it fails to bind to a 10-bp site containing adjacent ATGAG half sites (J. W. Sellers, and K. Struhl, unpublished). Thus, GCN4 binds to overlapping half sites whose optimal sequence is ATGAC, and the protein is surprisingly flexible in that it can bind DNA even when a base pair is inserted at the middle of the recognition sequence.

Deletion analysis of GCN4 reveals that the 60 C-terminal amino acids are fully sufficient for specific DNA-binding and for dimerization (60, 61). Strikingly, the GCN4 DNA-binding domain shows 45% amino acid identity to the C-terminal region of the jun oncoprotein that causes fibrosarcomas in chickens (62). Moreover, the GCN4 DNA-binding domain can be functionally replaced by the homologous jun region to generate a protein that possesses DNA-binding properties indistinguishable from those of GCN4 on a series of target sites (7). The fact that GCN4 and jun recognize identical DNA sequences suggests that the amino acid residues involved in direct contacts to DNA are located within the 30-residue stretch that is most highly conserved.

Recently, it has been proposed that GCN4 belongs to a class of proteins that utilizes a new motif for DNA binding, the leucine zipper (42). These proteins, which include the jun, fos, and myc oncoproteins and the C/EBP enhancer–binding protein, all contain 4–5 leucine residues that could be viewed as being periodically repeated every two turns of an α helix. The hypothetical model is that the leucine residues are important for interdigitating two α helixes, one from each monomer unit, that provide the structural basis for the dimer formation. However, a variety of mutations of the conserved leucines have little if any effect on DNA binding (J. Sellers, K. Struhl, unpublished).

HAP1   The HAP1 protein binds to the upstream regulatory elements of two cytochrome c genes, *CYC1* and *CYC7,* and induces their transcription (63, 64). Specific DNA-binding in vitro and transcriptional induction in vivo are stimulated by heme. Although both binding sites compete for HAP1 binding, activation through the *CYC1* site is about 10-fold more efficient. A derivative of the protein, HAP1-18, appears to have an altered specificity because it fails to bind the *CYC1* site while retaining the ability to bind the *CYC7* site. Strains containing HAP1-18 show increased levels of *CYC7* expression and decreased levels of *CYC1* expression.

Surprisingly, the *CYC1* and *CYC7* binding sites have no obvious sequence similarity, thus leading to the suggestion that HAP1 recognizes two qualitatively different DNA sequences, possibly by utilizing a single DNA-binding domain (64). However, in accord with proposed sequence relationships between these two sites (65), a particular base-pair substitution of the *CYC7* site generates a new site with *CYC1*-like properties; high expression in the presence of HAP1 and low expression dependent on HAP1-18 (66). The possibility that the *CYC1* and *CYC7* sites might be related, though divergent, forms of the same sequence has not been eliminated.

MATα2   The MATα2 protein (210 amino acids) regulates yeast cell type by binding specifically to operator sequences of several a-specific genes and repressing their transcription (34). The α2 operators are unusual in that they are very large (roughly 30 bp), having highly conserved sequences at both ends with an approximate twofold symmetry (CATGTAA) and lacking sequence similarities in the middle. In accord with this unusual arrangement, α2 protects the two end portions from DNase I, dimethylsulfate, and hydroxyl radical, while leaving the middle accessible to attack by either of these reagents (67). A 7-bp substitution in the center of the operator has little effect on repressor binding. These observations suggest that α2 binds as a dimer to half sites whose centers are two and one-half helical turns apart and does not contact the middle of the operator. Surprisingly, α2 also binds with similar affinity to an operator derivative with the central 13 bp deleted such that the half sites are immediately adjacent. Thus, α2 is remarkably flexible in that it can bind to differently spaced operator half sites even when located on the same or opposite side of the DNA helix (67).

Proteolytic fragmentation of α2 indicates that the protein is composed of two independently folding structural domains (67). The 79-residue C-terminal fragment interacts specifically with operator DNA although with a reduced affinity that is characteristic of half-site binding. The 102-residue N-terminal segment does not bind DNA itself, but instead is necessary for the dimerization that permits the protein to bind simultaneously to both half sites at high affinity. The purified α2 dimers are very unusual in that the monomer subunits are covalently linked through the disulfide bonds between cysteine residues in the N-terminal region. Although α2 dimers might form in vivo by such a mechanism, it now appears more likely that the cysteine residues serve as zinc-binding sites that hold the monomer subunits together (67).

Interestingly, the DNA-binding domain of α2, a protein that specifies cell type in yeast, shows strong primary sequence similarities to the homeobox motif of *Drosophila* proteins that selects the choice of developmental pathway (68). Amino acid substitutions within the α2 homeo-like region abolish function, presumably by virtue of a loss of DNA-binding activity (69). These proteins also show some sequence similarity to the helix-turn-helix motif of bacterial repressors and activators (70), suggesting that this structure may be involved in specific DNA-binding by α2.

TATA-BINDING PROTEIN   Unlike the above activator and repressor proteins that affect relatively few genes of related function, a protein binding to TATA elements should be important for the transcription of many genes. As the genetic approach is restricted to transcription factors that are nonessential for growth and can be mutated to produce simple phenotypes, it is not useful for identifying a TATA-binding protein. Initial evidence for a TATA-binding activity was obtained by photofootprinting in vivo, which revealed physical changes in the *GAL1,10* TATA elements only under conditions when the genes were transcriptionally active (71).

More recently, such a protein has been identified and purified by its ability to substitute for the mammalian TATA-binding transcription factor (TFIID) in a reconstituted in vitro transcription system (13, 14). The yeast factor binds to two of the three genetically identified TATA elements in the *CYC1* promoter, is surprisingly small (25 kd by sedimentation and gel filtration), and appears to be a basic protein (13). Interestingly, transcription stimulated in vitro by the yeast TATA-binding protein is initiated 30 bp downstream from a *CYC1* or a mammalian TATA element, a distance typical of mammalian promoters (13). This observation suggests that the difference between yeast and mammalian promoters regarding the spacing between the TATA element and mRNA initiation site is due not to the TATA-binding proteins, but rather to differences in the basic transcription machinery that interacts with the TATA factor.

TRANSCRIPTIONAL ACTIVATION

The binding of activator proteins to upstream promoter elements is necessary but not sufficient for stimulating transcription in vivo. GAL4 and GCN4 deletion proteins containing only the intact DNA-binding domain do not activate transcription, but instead can actually repress transcription when their recognition sites are located between a heterologous upstream element and TATA sequence (48, 61). Conversely, regions of GAL4 or GCN4 lacking their own DNA-binding domains can activate transcription when fused to a heterologous DNA-binding domain such as that of the *Escherichia coli* LexA repressor (61, 72). These LexA hybrid proteins activate a promoter having a LexA operator as the upstream element by binding via the LexA repressor-operator interaction and stimulate transcription utilizing the GAL4 and GCN4 activation function. Thus, DNA binding and transcriptional activation are distinct functions that are located in separate parts of the protein.

Deletion analyses have localized the GCN4 and GAL4 transcriptional activation functions to short regions that contain a relatively high proportion of acidic residues (61, 73). In both cases, surprisingly large portions of the protein-coding sequence can be removed without significantly affecting the transcriptional activation function. The GCN4 transcriptional activation function is localized to a short acidic region in the center of the protein, whereas GAL4 contains two separate activation regions. In the case of GCN4, derivatives retaining as few as 35–40 amino acids from the acidic region are sufficient for wild-type levels of transcriptional activation when fused directly to the DNA-binding domain (74). The distance and orientation of the GCN4 and GAL4 activation regions with respect to their DNA-binding domains is functionally unimportant. This indicates that the activation region encodes an autonomous function and that there is no requirement for a spacer between the activation region and DNA-binding domain.

Several lines of evidence indicate that yeast transcriptional activation functions are defined by short acidic regions with little sequence homology. First, different portions of the GCN4 acidic region are equally capable of activating transcription even though their primary sequences are dissimilar (61, 74). Second, the GCN4 and GAL4 transcriptional activation regions are all acidic but have no other noticeable sequence similarities (61, 73, 74). In this regard, other yeast activator proteins also have acidic regions that may be required for transcriptional stimulation in vivo. Third, acidic character is the common feature of functional transcriptional activation regions selected from short *E. coli* DNA segments (75). Fourth, single amino acid changes in a GAL4 derivative that increase or decrease the level of activation are almost always associated with an increased or decreased negative charge, respectively (76). These results strongly suggest that transcriptional activation regions are not defined by a specific primary sequence but rather a more general property such as net negative charge.

Despite the clear importance of acidic character, it appears that functional transcriptional activation regions must have additional structural features. An acidic 15-residue peptide whose sequence could form two turns of an α helix confers some transcriptional activity when fused to the GAL4 DNA-binding domain (77). However, a related peptide with the exact length and composition but a different sequence fails to stimulate transcription. Progressive fine-scale deletions of the GCN4 transcriptional activation region yields a series of small, stepwise reductions in activity (74). GCN4 activity appears directly related to the size of the transcription activation region remaining, whereas there is no such precise relationship of transcriptional activity to the number of acidic residues. These experiments strongly suggest that transcriptional activation regions do not have a defined tertiary structure such as found in active sites or domains in a protein, a view that is confirmed by proteolysis experiments involving GCN4 (74).

Nevertheless, the strong correlation between the length of the GCN4 activation region and level of transcriptional activity is strongly suggestive of a repeating structure consisting of units that act additively. Independent evidence for some kind of structure comes from an unusual chymotrypsin digestion pattern observed for GCN4 and deleted derivatives that depends on the presence of a functional transcriptional activation region (74). A clue to the structure of the GCN4 activation region is that the boundaries defining the stepwise levels of activation may occur every seven amino acid residues, a repeat unit consistent with two turns of an α helix (74). The GCN4 activation

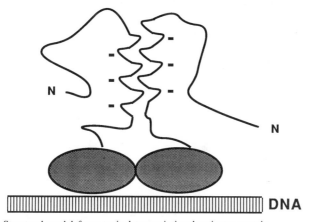

Figure 2 Structural model for a typical transcriptional activator protein.
The regions of an activator protein include a dimeric DNA-binding domain (shaded ovals) interacting with a dyad symmetric sequence (arrows), a transcriptional activation region that is hypothesized to be a dimer of interacting α helices (wavy lines) with acidic residues exposed ($-$), and a nonessential region shown as unstructured. See text.

region could form three amphipathic α helices with acidic and hydrophobic residues tending to be clustered along separate surfaces. Similarly, the sequence of the synthetic peptide described above that functions as an activation region is consistent with forming two turns of an amphipathic helix (77). However, as the major GAL4 and other activation regions are unlikely to form amphipathic helices, a simple relationship between this structure and function appears unlikely.

A structural model has been proposed in which the activation region is a dimer of intertwined α helices, one helix from each monomer (74) (Figure 2). The formation/stability of this structure should be facilitated by the stability of dimeric DNA-binding domains. Moreover, as the LexA domain binds very poorly to its operator because of weak dimerization (78), it is likely that transcriptional stimulation through the LexA domain also requires that the activation region facilitate dimerization (7, 61, 75). By this model, amphipathic helices could form functional activation regions as they would easily permit a structure involving interacting hydrophobic residues that are protected from solvent and exposed acidic residues.

The short acidic regions that are sufficient for activation are likely to be surfaces that are used for interactions with other proteins of the transcription machinery. It seems extremely unlikely that these short acidic regions of limited homology could encode catalytic activities such as topoisomerases, nucleases, methylases, etc that might be involved in transcription. Acidic regions of DNA-binding proteins are likely to be general requirements for transcriptional activation in all eukaryotic organisms. GAL4 activates transcription from appropriate target promoters in mammalian cells, and in flies and plants; the acidic activation region is required in all cases (15–18). Conversely, transcriptional activation in yeast cells by the jun oncoprotein and the glucocorticoid receptor requires acidic sequences in addition to the DNA-binding domains (20, 22). The implication of these results is that the acidic activation regions contact some part of the transcription machinery that is conserved functionally throughout the eukaryotic kingdom. The obvious candidates for such an interaction are TATA-binding proteins, RNA polymerase II, or histones.

REGULATION

In all forms of transcriptional regulation, there must be at least two distinct physiological conditions that are defined operationally by different RNA levels of a particular gene or set of genes. Furthermore, there must be a mechanism by which one of these physiological states can be converted to another state by some molecular signal. Such a signal can be caused either by a change in the external environment (the presence or the absence of a particular compound), or by an internal change governed by a particular developmental program. Although transcriptional control is executed by DNA-binding proteins, there must be additional regulatory molecules that govern when these regulatory proteins execute their roles in transcription. Thus, the DNA-binding proteins interact not only with specific nucleotide sequences, but also (directly or indirectly) with signal molecules that distinguish between the two physiological states.

Yeast cells use a variety of molecular mechanisms to regulate the activity of specific transcription factors. In one mechanism, the activity of a DNA-binding protein is altered by the binding of a small molecule. For example, HAP1 protein requires heme for efficient DNA binding in vitro and transcriptional activation in vivo (63). In another mechanism, the activity of a DNA-binding protein is affected by the binding of another protein. The GAL80 protein inhibits transcriptional activation by interacting directly with GAL4 such that it masks the primary acidic transcriptional activation region; it does not affect the ability of GAL4 to bind DNA (79–81). The GAL4-GAL80 interaction is eliminated when cells are grown in galactose medium, presumably by an interaction between GAL80 and galactose (or related metabolite). A third mechanism involves alteration by covalent modification of the DNA-binding protein. The heat-shock transcription factor is present under all growth conditions, but during heat shock it appears to be phosphorylated (82, 83). As the target genes are expressed only during heat-shock conditions, presumably the phosphorylated form represents the active transcription factor.

By analogy with the acidic transcriptional activation regions of GAL4 and GCN4, phosphorylation of the heat-shock factor might be required for activation simply by increasing the negative charge of the protein. Another potential example of phosphorylation involves the SNF1 protein kinase that is required for transcriptional activation by GAL4 and ADR2 (84); however it is not clear if the effects of SNF1 are direct.

In some cases, yeast cells regulate transcription by controlling the amount of a specific DNA-binding protein rather than by modulating the activity of the protein. GCN4 activates transcription only during amino acid starvation because it is not synthesized under normal growth conditions. During starvation, GCN4 protein levels increase 30–50-fold, while GCN4 mRNA levels remain relatively unchanged (85, 86). This translational regulation involves four small open reading frames present in the extremely large 5' untranslated leader of GCN4 mRNA (87, 88). As expected, the open reading frames prevent translation in normal circumstances, whereas it appears that the basic rules of translational initiation are circumvented under starvation. As GCN4 effectively regulates protein synthesis by controlling the amount of amino acid precursors, the unusual translational control mechanism is sensible because the effector protein is sensitive to the process it controls.

A different method for regulating transcription by controlling the amount of a DNA-binding protein is exemplified by the proteins that determine yeast mating type. Unlike GCN4, whose level changes in response to environmental conditions, the MATα2, MATα1, and MATa1 proteins regulate transcription of their target genes by being present only in the relevant cell types (reviewed in 89). The segregation of transcriptional regulatory proteins into different cell types probably represents the major mechanism for developmental regulation of genes in multicellular organisms.

MORE COMPLEX PHENOMENA

In previous sections, I discussed the basic components required for transcriptional regulation. For the simplest promoters, an upstream activator protein binds to its target site and stimulates transcription through the acidic activation region in a process that also requires the binding of a common TATA factor to its target site. Other promoters may contain multiple upstream, TATA, initiator, or operator elements that allow for distinct regulatory properties. Although some of these may be viewed as a set of independently acting simple promoters, it is becoming increasingly clear that there are additional complexities that represent new principles.

Multiple Proteins Recognizing the Same Target Sequence

The existence of a protein distinct from GCN4 but recognizing similar DNA sequences was initially inferred from the observation that the GCN4 binding site in the *HIS3* promoter activates low levels of transcription in the absence of GCN4 protein (90). Such a protein, yAP-1, was identified (10, 11) by virtue of its ability to bind a site recognized by AP-1, the mammalian transcription factor when DNA-binding domain is structurally and functionally related to GCN4 (8, 9). yAP-1 and GCN4 show about 35% sequence identity in their DNA-binding domains and they bind very similar sequences, although with an altered specificity (91). Genetic experiments indicate that GCN4 and yAP-1 activate transcription from promoters whose upstream elements contain related sequences. However, their in vivo roles must differ because yAP-1 cannot compensate for the defects caused by *gcn4* mutations. The role of yAP-1 is unclear because mutant strains lacking the protein have nearly normal growth properties (91). Perhaps yAP-1 is important for low-level, constitutive expression of some of the amino acid biosynthetic genes that are induced by GCN4 under starvation conditions.

Several lines of evidence suggest that there are additional yeast proteins with DNA-binding properties similar to those of GCN4 and yAP-1. First, proteins interacting with the AP-1 binding site are present in extracts lacking both GCN4 and yAP-1 (91). Second, a yeast protein(s) has been isolated by virtue of its ability to bind a sequence recognized by the mammalian ATF/CREB factor (92). This sequence is similar to the functional 10-bp GCN4-binding site that contains an additional G:C base pair in the middle of the site. Third, this 10-bp GCN4-binding site represses *HIS3* transcription in the absence of GCN4; in the presence of GCN4, *HIS3* transcription is partially induced probably due to competition between GCN4 and the putative repressor protein for the binding site (J. Sellers, K. Struhl, unpublished).

Another example involves an upstream element in the *CYC1* promoter that interacts both with HAP1, which induces transcription in response to heme, and with RC2, a distinct protein found in crude extracts (63, 93). The protein-DNA contacts for these two interactions are indistinguishable, suggesting that binding by these proteins is mutually exclusive (63). The role of RC2 in vivo is unclear.

The existence of multiple proteins that recognize related sequences increases the sensitivity and flexibility for coordinately and independently regulating subclasses of genes. Given the increase in complexity, it seems likely such families of proteins will be important for regulatory processes of basic importance for the organism. For example, the bacteriophage λ repressor and cro proteins recognize similar but not identical sequences that control the developmental decision between lysis and lysogeny (94). In the case of the GCN4-related proteins, one might imagine that it is crucial for yeast cells to precisely regulate amino acid biosynthesis due to its crucial role in protein synthesis and hence cell growth.

Multiple Proteins Necessary for a Single DNA-Binding Event

It is generally assumed that a given DNA-binding protein is fully capable of interacting with its target site. However, there are instances in which two proteins together can bind DNA whereas neither protein can bind alone. Both the HAP2 and HAP3 proteins are required for binding to an upstream element in the *CYC1* promoter and are present in the final protein-DNA complex (95). The HAP2-HAP3 complex forms in the absence of DNA and can be purified to near homogeneity through a number of chromatographic steps (96). The HAP2-HAP3 complex has a remarkable functional relationship to the mammalian CP1 binding factor that itself is a complex between the CP-1A and CP-1B proteins (97). Both complexes recognize the same DNA sequences, and functional yeast-mammalian hybrid complexes can be created; HAP2 is functionally equivalent to CP-1B and HAP3 is equivalent to CP-1A (12). It is unknown whether binding by the heteromeric complex involves specific DNA contacts being mediated through HAP2, HAP3, or both together.

Two other examples can be found among the proteins that regulate mating type. In one case, regulatory sites in three α-specific genes are bound by a combination of the MATα1 transcriptional activator and a second protein PTRF; neither protein binds alone (98). PRTF itself appears to bind some promoter elements whose consensus sequence is dyad symmetric. The α-specific genes contain mutated and functionally defective versions of the PTRF element as well as an additional element termed Q (99). It has been suggested that α1 and PTRF interact respectively with the Q and defective PTRF elements in the α-specific genes, with both proteins being necessary for high-affinity binding (98, 99).

The second example involves the MATα2 and MAT**a**1 proteins that are both required to bind promoter sites in diploid-specific genes (100). The target sequence for this combination of proteins is large and includes a sequence resembling an α2-binding site (35, 101). However, the **a**1-α2 combination does not bind to a simple α2 operator sequence, thus suggesting that **a**1 protein alters the binding specificity of α2 repressor (100). Different portions of α2 are required for the different binding activities because removal of the 62 N-terminal residues abolishes **a**1-α2 binding, but does not effect simple α2 binding (100, 102). It is likely the N-terminal region of α2 is not required for specific contacts to DNA, but rather for protein-protein contacts to **a**1. Although many amino acid substitutions in the α2 DNA-binding domain abolish both binding functions, some mutations specifically affect α2 activities, whereas others specifically affect **a**1-α2 activities (69, 103). It has been suggested that **a**1-α2 might be a heterodimer that recognizes a different sequence from the presumptive homodimeric α2 repressor (100).

The principle that the combined action of two proteins can be necessary for

DNA-binding increases the precision of transcriptional regulation. Such combinatorial regulation makes it possible to influence gene expression only when two specific physiological conditions occur; e.g. regulation of diploid-specific genes only in **a**/α cells or in complex cell-type regulation in higher organisms. Combinatorial DNA-binding might also increase the flexibility of regulation by having a given protein associate with a variety of different proteins to yield heteromeric species with distinct sequence recognition properties.

Multiple Proteins Necessary for Activation or Repression

In the above examples where two proteins are necessary to bind a target sequence, it follows that both proteins are necessary for the transcriptional induction or repression that is mediated through the site. However, in the situations described below, it appears that two proteins are necessary for activation or repression even though each protein can bind independently and simultaneously to the target. As discussed for combinatorial DNA-binding, synergy between activators or repressors increases both the precision and flexibility of transcriptional regulation. In addition, synergy is economical because it minimizes the number of distinct transcription factors that are necessary to achieve the wide variety of regulatory responses.

Transcriptional activation mediated by the UAS1 element in the *CYC1* promoter occurs in the presence of heme and requires the HAP1 activator protein. However, more detailed analysis of UAS1 indicates that it is actually composed of two parts, only one of which is bound by HAP1 (104). Thus, HAP1 is necessary but not sufficient for activation of the *CYC1* promoter. The RAF1 protein identified in crude extracts that binds to the other part of UAS1 may also be required for this activation (63).

Binding of the MATα2 protein to operators of **a**-specific genes is necessary but not sufficient for transcriptional repression (105). As mentioned earlier, α2 binds to the ends of the operator, leaving the middle free. Although the middle of the operator is not important for α2 binding, it is crucial for repression in vivo. The GRM protein, which is present in all cell types, binds to the center of the operator even at the same time as when α2 is bound at the ends. Moreover, GRM and α2 bind cooperatively to the operator, and the N-terminal domain of α2 is necessary for cooperative binding in vitro (105) and repression in vivo (102). These observations have suggested that simultaneous, and possibly cooperative, binding of GRM and α2 is required for repression.

Another example of synergy is represented by the mating-type silencer that efficiently represses transcription when located at distances as far as 2 kb upstream or downstream from the mRNA initiation sites (37–39). Three distinct elements, two of which interact with known DNA-binding proteins, are involved in silencer function. Surprisingly, no individual element is either necessary or sufficient for silencer function, but transcriptional repression occurs with any two of the elements (106). Presumably, any two of the proteins that bind these elements can act together to repress transcription, whereas no individual protein is capable.

There are three basic models to explain synergy. First, one protein that carries out the activation or repression might bind efficiently to DNA only in the presence of the other protein. Such cooperative binding would imply a very specific interaction between the two proteins, and hence seems unlikely to be a general explanation. Second, the proteins may form a relatively nonspecific interaction once both are bound to DNA. By the model invoking dimeric acidic activation regions, one might imagine that acidic regions of two distinct proteins might associate to yield a heterodimer whose structure results in a higher level of function. Third, the two proteins might not interact directly, but may both contact a common target. In this view, the target would require a threshold of contacts, which neither protein alone would be able to achieve.

Activation and Repression by the Same Protein

Several instances have been described that are consistent with the idea that a single protein can serve either as a transcriptional activator or repressor depending on the promoter. Based on the consensus sequences for binding and the properties of *mcm1* mutant strains, it has been suggested that GRM, the protein that binds cooperatively with the $\alpha2$ repressor to a-specific promoters (105), might be identical (or related) to PTRF, the protein that binds in combination with the $\alpha1$ activator to α-specific promoters (98). If true, a single non-cell-type-specific protein could act as a co-activator or a co-repressor depending on the cell-type-specific protein with which it acts in combination. In addition, sequences that bind these proteins appear to act on their own as weak upstream promoter elements in a-cells (107), suggesting that PTRF/GRM can also act as an independent activator.

Proteins initially identified by their ability to bind to elements in the mating type silencer many also represent other examples of the same phenomenon (108, 109). Two of the silencer elements act as transcriptional activating elements when fused upstream of TATA elements (106, 109). The RAP1 protein, also known as GRF, binds to the upstream promoter elements of some ribosomal protein genes and to the telomers of chromosomes in addition to binding a silencer element (108, 109). In the appropriate context, the ribosomal promoter element functions as a silencer element (108). It will be interesting to determine if distinct regions of RAP1 are involved in transcriptional activation or repression.

Functional Distinctions Between TATA Elements

It is commonly assumed that TATA sequences are general promoter elements that are recognized by a common transcription factor that is part of the basic transcriptional machinery. However, the *HIS3* promoter region contains two distinct classes of TATA elements, constitutive (T_C) and regulatory (T_R), that are defined by their interactions with upstream promoter elements, selectivity of initiation sites, and chromatin structure (110). Transcription dependent on T_C is initiated equally from two sites, $+1$ and $+12$, whereas transcription dependent on T_R initiates preferentially from the $+12$ site; this selectivity is determined primarily by the distance between T_R and the initiation sites. Transcriptional activation by GCN4 and GAL4 occurs only in combination with the T_R element, not T_C. As assayed by its ability to activate transcription in combination with GAL4, the sequence TATAAA in the *HIS3* promoter is fully sufficient for T_R function. Saturation mutagenesis of this sequence revealed that 17 out of the 18 possible single mutations abolish T_R function (TATATA being the only exception) (111).

The high sequence specificity of the *his3* T_R element provides strong genetic evidence for a T_R-binding protein. It is very likely that this T_R protein is the yeast TATA-binding protein identified by in vitro transcription using reconstituted mammalian factors (13, 14). However, the T_C element necessary for constitutive *HIS3* expression does not have a sequence that fits the T_R rules. Thus, it is almost certain that the T_R and T_C elements interact with different proteins, an explanation that easily accounts for the function distinctions between T_R and T_C elements with regard to their interactions with upstream activator proteins (110, 111). This idea also accounts for why overproduction of GAL4 "squelches" *HIS3* transcription from the $+12$ but not the $+1$ initiation site; presumably GAL4 is titrating out the T_R protein but not the T_C protein (112).

By analogy with bacterial σ factors that interact with the core RNA polymerase to generate distinct holoenzymes that recognize different promoter sequences, yeast cells may contain multiple proteins that carry out a related "downstream element function" but have different specificities for DNA-binding. Two additional lines of evidence support this view. First, two TATA point mutations have the novel property of activating transcription in combination with GCN4 but not with GAL4 (113). The simplest interpretation of these results is that there are two "TATA-binding" proteins; T_R, which recognizes TATAAA and interacts functionally with both GCN4 and GAL4, and a distinct protein that recognizes a related sequence but interacts only with GCN4. Second, the T_R element was replaced by random-sequence oligonucleotides, and functional promoter elements were selected by virtue of their ability to activate transcription in combination with GAL4. In addition to the expected T_R-like sequences, other functional elements having no sequence resemblance to T_R were obtained (V. Singer, K. Struhl, unpublished).

Poly(dA-dT) Sequences and the Effect of Chromatin

Many yeast promoters contain poly(dA-dT) homopolymer sequences greater than 10 bp in length located upstream of the mRNA initiation site. For several constitutively expressed genes these poly(dA-dT) sequences act as the upstream promoter elements required for transcription (114). In addition, adr2 promoter mutations that lengthen the natural poly(dA-dT) tract cause high constitutive levels of transcription (115). Interestingly, the DED1 poly(dA-dT) element stimulates transcription by bacteriophage T7 RNA polymerase in yeast cells by a factor similar to its ability to stimulate the natural DED1 promoter (116). These similar enhancement effects on two very different transcription machineries suggests that this poly(dA-dT) element mediates its effects through the chromatin template. As poly(dA-dT) sequences inhibit nucleosome formation in vitro (117, 118), it has been proposed that this particular class of upstream elements might act by excluding nucleosomes, not by binding specific proteins (114).

Consistent with this idea, several observations indicate that transcription from some promoters can be stimulated by direct alterations of chromatin structure. Inhibition of nucleosome formation by preventing the synthesis of histone H4 leads to an increase of PHO5 transcription (119). Altering the balance of histones by gene dosage can influence transcription from certain promoters (120). Small N-terminal deletions of histone H4 do not affect cell growth, but they increase transcription from the silent mating type genes, presumably by interfering with silencer function (121). These experiments provide direct evidence that normal chromatin structure represents a transcriptionally repressed state.

Such chromatin structural models have been weakened recently by the identification of proteins binding to yeast poly(dA-dT) elements. The DED1 poly(dA-dT) element stimulates transcription in vitro under conditions where nucleosome formation does not occur (122). Moreover, this effect is blocked by an excess of competing oligonucleotide, suggesting that transcriptional stimulation depends on a protein(s) binding to the element. In a separate line of experiments, a protein that specifically recognizes homopolymer (dA-dT) sequences at least 9 bp in length has been purified, and the gene encoding this protein has been cloned (E. Winter, A. Varshavsky, unpublished). Surprisingly, this poly(dA-dT)–binding protein appears to act as a transcriptional repressor because a mutant strain lacking the protein shows increased levels of transcription from an artificial promoter containing a poly(dA-dT) upstream element (E. Winter, A. Varshavsky, unpublished). However, the existence of specific binding proteins does not exclude the possibility that poly(dA-dT) sequences might also influence transcription by virtue of their unusual DNA structures.

Upstream Activator Proteins Function When Bound at the TATA Position

It is generally assumed that upstream activator proteins such as GAL4 and GCN4 stimulate transcription only when bound upstream of a TATA element. However, GCN4 can efficiently activate transcription from a *GAL-HIS3* promoter in which the conventional TATA element is replaced by a GCN4-binding site (123). In other words, GCN4 can activate transcription in the absence of a TATA element when bound close to the mRNA initiation site. Transcription occurs from wild-type initiation sites and requires both the GCN4 DNA-binding domain and the acidic activation region, but it is not affected by changing the spacing between the GCN4-binding site and the mRNA start sites. GCN4 is not sufficient for this TATA-independent activation; a sequence in the GAL region distinct from the GAL4 binding sites is also required. Thus, GCN4 functions both upstream of a TATA element and in place of a TATA element, suggesting that there might not be an intrinsic difference between an upstream and TATA activator protein (123).

The idea that the downstream element function can be carried out by multiple proteins, including conventional upstream activator proteins, can easily account for transcriptional activity from promoters lacking the conserved TATAAA sequence. For example, the *TRP3* promoter lacks the conserved TATAAA sequence and instead contains a GCN4-binding site 28 bp upstream of the mRNA start site that presumably mediates induction in response to amino acid starvation. It is important to note that even when GCN4 replaces the TATA function, at least two distinct promoter elements are necessary for transcription. However, not all combinations of upstream and downstream elements result in functional promoters.

MOLECULAR MECHANISM OF ACTIVATION AND REPRESSION

Although RNA polymerase II can catalyze RNA synthesis on a variety of artificial templates in vitro, it does not bind to specific DNA sequences and does not initiate transcription, even randomly, on normal double-stranded templates. A view of the transcriptional initiation process is that RNA polymerase II initiates mRNA synthesis at discrete sites upon recognizing a transcription complex composed of upstream activator proteins, TATA factors, and the DNA (Figure 3). The complex is formed/stabilized by specific interactions between the proteins and cognate DNA sequences, and by pro-

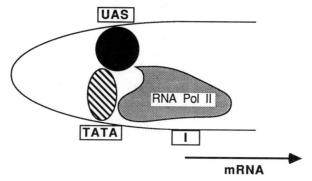

Figure 3 Molecular model for transcriptional activation.

An upstream activator protein (black circle), TATA-binding protein (striped oval), and RNA polymerase II are shown as interacting with their target promoter elements (boxes) and with each other. The protein-protein interactions and the RNA polymerase II association with the initiator element are hypothetical. The DNA (line) is illustrated as bending to allow for the protein-protein interactions. See text.

tein-protein interactions between the various components. In this sense, the DNA serves as a specific scaffold for the assembly of an active transcription complex. The crucial mechanistic questions are which proteins are in direct contact, and how is RNA polymerase II activated.

Assuming that the acidic transcriptional activation regions are surfaces that contact other proteins, several observations suggest that the TATA-binding protein is a likely target. From the functional distinctions between the *HIS3* transcription T_R and T_C elements (110), it has been suggested that the GCN4 and GAL4 activation regions stimulate transcription by associating with a protein that binds T_R, whereas they are unable to interact with the protein that recognizes T_C (1, 113). The proposed interaction between GAL4 and the T_R proteins also explains why overproduction of GAL4 selectively inhibits *HIS3* transcription that depends on the T_R element (112). Finally, the interaction of the mammalian TATA factor and its target DNA site appears to be altered by GAL4 derivatives that contain a functional activation region (124).

The proposed interaction between upstream activators and TATA factors does not explain how RNA polymerase II is activated for transcription. The fact that transcription can be initiated accurately, though inefficiently, in vitro from promoters containing only a "typical" TATA element has led to the belief that the major TATA factor is part of the basic transcription machinery and hence associated with RNA polymerase II. In principle, this presumptive interaction could either increase the binding of the TATA factor to its target site, or allosterically affect the TATA protein such that it would be able to promote transcription more efficiently.

An important issue is whether upstream activator proteins directly contact

RNA polymerase II or whether they affect transcription indirectly through the TATA factor. The fact that LexA hybrid proteins activate transcription when bound upstream of a TATA element argues against a strict requirement for a direct contact to RNA polymerase II. However, GCN4 can activate transcription when its binding site replaces the TATA element (123), and affinity chromatography indicates that GCN4 interacts directly with RNA polymerase II in vitro (125). Surprisingly, the GCN4 DNA-binding domain is necessary and sufficient for this interaction in vitro; the acidic activation region does not appear to be involved. Thus, GCN4 might facilitate the formation of an active initiation complex by utilizing different regions of the protein for contacting RNA polymerase II and a TATA factor (Figure 3). However, the significance of the GCN4-RNA polymerase II interaction for transcriptional activation in vivo remains to be determined.

By any model, RNA polymerase II must be the ultimate target for the activities of TATA and upstream activator proteins. In this regard, the largest subunit of RNA polymerase II from yeast to human contains a conserved seven-amino-acid sequence that is repeated many times at the C-terminus (126, 127). This "tail" is required for transcription in vitro (128) and in vivo (129, 130), although it is not important for basic RNA polymerizing activity. Although the yeast tail contains 26 repeats, derivatives containing only 13 repeats have minimal effects on cell growth, and those containing 10–12 repeats are viable at normal (but not high or low) temperatures (129, 130). Interestingly, the yeast tail can be functionally replaced by the analogous tail from hamster, but not by the more divergent tail from flies (130). It has been proposed that the tail might be an interaction site for TATA and or upstream activator proteins. The idea is attractive because of the repeating nature of both the tail and the acidic activation region (74), and because it provides an explanation for the synergism between upstream activator proteins that is often observed.

A basic property of upstream activator proteins is their ability to function when bound at long and variable distances from TATA elements and mRNA initiation sites. By analogy with well-documented examples of prokaryotic regulatory proteins, it is now believed that action at a distance is explained by looping out of the intervening DNA such that relevant proteins can directly interact (reviewed in 131). However, unlike the prokaryotic examples, which involve highly specific contacts between identical protein molecules, looping involved in eukaryotic transcriptional activation would presumably involve relatively nonspecific contacts between different proteins. Looping does not easily explain how activators can function at relatively short and variable distances because of the high energy involved in generating short loops. In this case, such flexibility might reflect variable conformations of the large nonessential parts of the activator protein such that its critical acidic surface can interact with the other protein (1).

Given the above ideas about activation, repressor proteins could inhibit transcription by a variety of molecular mechanisms. The simplest model is steric hindrance, in which the binding of a repressor protein to its operator prevents binding of an upstream activator or TATA protein to its cognate promoter element. Steric hindrance might account for regulation of the *BAR1* gene (107) and be involved in the examples where multiple proteins bind very relaxed sequences. However, a bacterial repressor protein (132) or derivatives of yeast activator proteins lacking the acidic activation region (48, 61) can inhibit transcription when bound at various positions between an upstream and TATA element. In these cases, it is likely that the repressors interfere with the putative interaction between the transcriptional activator proteins rather than preventing DNA binding. Such interference might simply reflect steric constraints between the relevant proteins, or an increased difficulty of DNA looping due to the presence of a bound protein within the loop.

Neither of these steric hindrance models can explain how some repressors can function when bound upstream of promoter elements. By analogy with transcriptional activators, it seems that repressor proteins that function at a distance must require a transcriptional repression function that is distinct from DNA binding. Consistent with this analogy, N-terminal deletions of $\alpha2$ behave as negative control mutants in that they do not repress transcription even though they can bind DNA (102). Perhaps, the putative negative regulatory regions of repressors interact with the same target as that recognized by the acidic activation regions, thus forming a distinct and stable complex that prevents the formation of an active initiation complex. The idea that activators and repressors may interact with a common target also suggests a mechanism for how a single protein could act as a positive or negative regulatory protein depending on the promoter context.

EVOLUTION

It is now clear that the mechanism of transcription is remarkably conserved throughout the eukaryotic kingdom. Yeast upstream activator proteins function in a variety of eukaryotic organisms, vertebrate transcription factors function in yeast cells, and the yeast and mammalian TATA factors are functionally interchangeable. Such functional conservation undoubtedly indicates that the basic mechanism of transcriptional initiation has existed since the first eukaryotic organisms.

Surprisingly, yeast and mammalian cells have structurally related transcription factors that recognize essentially identical sequences even though the homologues perform different functions in their respective organisms. For example, GCN4 and HAP2,3 activate the amino acid biosynthetic and oxygen-regulated genes, respectively, in yeast, whereas their evolutionary counterparts, jun/AP-1 and CP-1, activate a variety of genes whose functions appear unrelated. Assuming that the original eukaryotic organisms contained multiple genes that utilized these conserved binding sites, it would be difficult to alter the sequence recognition properties of the regulatory protein without affecting the transcription of many genes. Similar arguments have been used to explain why the genetic code is essentially universal and why eukaryotic TATA elements and prokaryotic -10 sequences are extremely similar even though transcriptional and translational mechanism are quite different (133).

Despite all the similarities, there are a few significant differences between yeast and mammalian transcriptional mechanisms. First, unlike mammalian activators, yeast proteins cannot activate transcription when bound downstream of the mRNA initiation site. However, this property does not reflect an intrinsic difference between activator proteins because yeast activators can function at a downstream position in mammalian cells, and mammalian factors cannot function downstream in yeast cells. Second, the distance between TATA element and mRNA initiation sites is rather large and variable in yeast promoters when compared to the short fixed distance in mammalian promoters. Again this does not reflect basic differences between the yeast and mammalian TATA factors; in mammalian in vitro transcription systems, the yeast TATA factor stimulates initiation at the distance expected for a mammalian promoter. In both these cases, it is clear that there is some difference in the basic transcription machineries of yeast and higher eukaryotes such that they respond differently to upstream activator proteins and TATA factors. It is tempting to speculate that this difference involves the initiator element. Perhaps, a factor(s) involved in correct initiation in yeast also blocks the ability of upstream activator proteins to function in a downstream position. In any event, it is clear that transcriptional regulatory mechanisms in yeast and higher eukaryotes, though not identical, are extremely similar, and that continuing analysis of yeast genes will continue to uncover new and general principles that are relevant to all eukaryotic organisms.

ACKNOWLEDGMENTS

I thank colleagues in the yeast gene regulation field for many valuable discussions over the years. This work was supported by grant GM 30186 from the National Institutes of Health.

Literature Cited

1. Struhl, K. 1987. *Cell* 49:295–97
2. Guarente, L. 1987. *Annu. Rev. Genet.* 21:425–52
3. Guarente, L. 1988. *Cell* 52:303–5
4. McKnight, S. L., Tjian, R. 1986. *Cell* 46:795–805
5. Allison, L. A., Moyle, M., Shales, M., Ingles, C. J. 1985. *Cell* 42:599–610
6. Corden, J. L., Cadena, D. L., Ahearn, J. M. Jr., Dahmus, M. E. 1985. *Proc. Natl. Acad. Sci. USA* 82:7934–38
7. Struhl, K. 1987. *Cell* 50:841–46
8. Bohmann, D., Bos, T. J., Admon, A., Nishimura, T., Vogt, P. K., Tjian, R. 1987. *Science* 238:1386–92
9. Angel, P., Allegretto, E. A., Okino, S. T., Hattori, K., Boyle, W. J., et al. 1988. *Nature* 332:166–71
10. Harshman, K. D., Moye-Rowley, W. S., Parker, C. S. 1988. *Cell* 53:321–30
11. Jones, R. H., Moreno, S., Nurse, P., Jones, N. C. 1988. *Cell* 53:659–67
12. Chodosh, L. A., Olesen, J., Hahn, S., Baldwin, A. S., Guarente, L., Sharp, P. A. 1988. *Cell* 53:25–35
13. Buratowski, S., Hahn, S., Sharp, P. A., Guarente, L. 1988. *Nature* 334:37–42
14. Cavallini, B., Huet, J., Plassat, J.-L., Sentenac, A., Egly, J.-M., Chambon, P. 1988. *Nature* 334:77–80
15. Kakidani, H., Ptashne, M. 1988. *Cell* 52:161–67
16. Webster, N., Jin, J. R., Green, S., Hollis, M., Chambon, P. 1988. *Cell* 52:169–78
17. Fischer, J. A., Giniger, E., Maniatis, T., Ptashne, M. 1988. *Nature* 332:853–56
18. Ma, J., Przibilla, E., Hu, J., Bogorad, L., Ptashne, M. 1988. *Nature* 334:631–33
19. Lech, K., Anderson, K., Brent, R. 1988. *Cell* 52:179–84
20. Struhl, K. 1988. *Nature* 332:649–50
21. Metzger, D., White, J. H., Chambon, P. 1988. *Nature* 334:31–36
22. Schena, M., Yamamoto, K. R. 1988. *Science* 241:965–67
23. Struhl, K. 1983. *Nature* 305:391–96
24. Guarente, L., Hoar, E. 1984. *Proc. Natl. Acad. Sci. USA* 81:7860–64
25. Struhl, K. 1984. *Proc. Natl. Acad. Sci. USA* 81:7865–69
26. Struhl, K. 1982. *Proc. Natl. Acad. Sci. USA* 79:7385–89
27. Guarente, L., Mason, T. 1983. *Cell* 32:1279–86
28. Siliciano, P. G., Tatchell, K. 1984. *Cell* 37:969–78
29. Nagawa, F., Fink, G. R. 1985. *Proc. Natl. Acad. Sci. USA* 82:8557–61
30. Chen, W., Struhl, K. 1985. *EMBO J.* 4:3273–80
31. Hahn, S., Hoar, E. T., Guarente, L. 1985. *Proc. Natl. Acad. Sci. USA* 82:8562–66
32. McNeil, J. B., Smith, M. 1986. *J. Mol. Biol.* 187:363–78
33. Brent, R. 1985. *Cell* 42:3–4
34. Johnson, A. D., Herskowitz, I. 1985. *Cell* 42:237–47
35. Miller, A. M., MacKay, V. L., Nasmyth, K. A. 1985. *Nature* 314:598–603
36. Struhl, K. 1985. *Nature* 317:822–24
37. Abraham, J., Nasmyth, K. A., Strathern, J. N., Klar, A. J. S., Hicks, J. B. 1984. *J. Mol. Biol.* 176:307–31
38. Feldman, J. B., Hicks, J. B., Broach, J. R. 1984. *J. Mol. Biol.* 178:815–34
39. Brand, A. H., Breeden, L., Abraham, J., Sternglanz, R., Nasmyth, K. 1985. *Cell* 41:41–48
40. Pabo, C. O., Sauer, R. T. 1984. *Annu. Rev. Biochem.* 53:293–321
41. Miller, J., McLachlan, A. D., Klug, A. 1985. *EMBO J.* 4:1609–14
42. Landschulz, W. H., Johnson, P. F., McKnight, S. L. 1988. *Science* 240:1759–64
43. Bram, R., Kornberg, R. 1985. *Proc. Natl. Acad. Sci. USA* 82:43–47
44. Giniger, E., Varnum, S. M., Ptashne, M. 1985. *Cell* 40:767–74
45. Bram, R. J., Lue, N. F., Kornberg, R. D. 1986. *EMBO J.* 5:603–8
46. Selleck, S. B., Majors, J. 1987. *Mol. Cell. Biol.* 7:3260–67
47. Giniger, E., Ptashne, M. 1988. *Proc. Natl. Acad. Sci. USA* 85:382–86
48. Keegan, L., Gill, G., Ptashne, M. 1986. *Science* 231:699–704
49. Berg, J. 1986. *Science* 232:485–87
50. Johnston, M. 1987. *Nature* 328:353–55
51. Johnston, M., Dover, J. 1988. *Genetics* 120:63–74
52. Johnston, M., Dover, J. 1987. *Proc. Natl. Acad. Sci. USA* 84:2401–5
53. Witte, M. M., Dickson, R. C. 1988. *Mol. Cell. Biol.* 8:3726–33
54. Salmeron, J. M., Johnston, S. A. 1986. *Nucleic Acids Res.* 14:7767–81
55. Wray, L. V., Witte, M. M., Dickson, R. C., Riley, M. I. 1987. *Mol. Cell. Biol.* 7:1111–21
56. Parraga, G., Horvath, S. J., Eisen, A., Taylor, W. E., Hood, L., et al. 1988. *Science* 241:1489–92

57. Hope, I. A., Struhl, K. 1985. *Cell* 43:177–88
58. Arndt, K., Fink, G. 1986. *Proc. Natl. Acad. Sci. USA* 83:8516–20
59. Hill, D. E., Hope, I. A., Macke, J. P., Struhl, K. 1986. *Science* 234:451–57
60. Hope, I. A., Struhl, K. 1987. *EMBO J.* 6:2781–84
61. Hope, I. A., Struhl, K. 1986. *Cell* 46:885–94
62. Vogt, P. K., Bos, T. J., Doolittle, R. F. 1987. *Proc. Natl. Acad. Sci. USA* 84:3316–19
63. Pfeifer, K., Arcangioli, B., Guarente, L. 1987. *Cell* 49:9–18
64. Pfeifer, K., Prezant, T., Guarente, L. 1987. *Cell* 49:19–27
65. Zitomer, R. S., Sellers, J. W., McCarter, D. W., Hastings, G. A., Wick, P., Lowry, C. V. 1987. *Mol. Cell. Biol.* 7:2212–20
66. Cerdan, M. E., Zitomer, R. S. 1988. *Mol. Cell. Biol.* 8:2275–79
67. Sauer, R. T., Smith, D. L., Johnson, A. D. 1988. *Genes Dev.* 2:807–16
68. Shepherd, J. C. W., McGinnis, W., Carrasco, A. E., DeRobertis, E. M., Gehring, W. J. 1984. *Nature* 310:70–71
69. Porter, S. D., Smith, M. 1986. *Nature* 320:766–68
70. Laughon, A., Scott, M. P. 1984. *Nature* 310:25–30
71. Selleck, S. B., Majors, J. 1987. *Nature* 325:173–77
72. Brent, R., Ptashne, M. 1985. *Cell* 43:729–36
73. Ma, J., Ptashne, M. 1987. *Cell* 48:847–53
74. Hope, I. A., Mahadevan, S., Struhl, K. 1988. *Nature* 333:635–40
75. Ma, J., Ptashne, M. 1987. *Cell* 51:113–19
76. Gill, G., Ptashne, G. 1987. *Cell* 51:121–26
77. Giniger, E., Ptashne, M. 1987. *Nature* 330:670–72
78. Hurstel, S., Granger-Schnarr, M., Schnarr, M. 1988. *EMBO J.* 7:269–75
79. Lue, N. F., Chasman, D. I., Buchman, A. R., Kornberg, R. D. 1987. *Mol. Cell. Biol.* 7:3446–51
80. Ma, J., Ptashne, M. 1987. *Cell* 50:137–42
81. Johnston, S. A., Salmeron, J. M. Jr., Dincher, S. S. 1987. *Cell* 50:143–46
82. Sorger, P. K., Lewis, M. J., Pelham, H. R. B. 1987. *Nature* 329:81–84
83. Sorger, P. K., Pelham, H. R. B. 1988. *Cell* 54:855–64
84. Celenza, J. L., Carlson, M. 1986. *Science* 233:1175–80
85. Thireos, G., Penn, M. D., Greer, H.

1984. *Proc. Natl. Acad. Sci. USA* 81:5096–100
86. Hinnebusch, A. G. 1984. *Proc. Natl. Acad. Sci. USA* 81:6442–46
87. Mueller, P. P., Hinnebusch, A. G. 1986. *Cell* 42:201–7
88. Tzamarias, D., Alexandraki, D., Thireos, G. 1986. *Proc. Natl. Acad. Sci. USA* 83:4849–53
89. Nasmyth, K. A., Shore, D. 1987. *Science* 237:1162–70
90. Struhl, K., Hill, D. E. 1987. *Mol. Cell. Biol.* 7:104–10
91. Moye-Rowley, W. S., Harshman, K. D., Parker, C. S. 1989. *Genes Dev.* 3:283–92
92. Lin, Y.-S., Green, M. R. 1989. *Proc. Natl. Acad. Sci. USA.* In press
93. Arcangioli, B., Lescure, B. 1985. *EMBO J.* 4:2627–33
94. Johnson, A. D., Poteete, A. R., Lauer, G., Sauer, R. T., Ackers, G. K., Ptashne, M. 1981. *Nature* 294:217–23
95. Olesen, J., Hahn, S., Guarente, L. 1987. *Cell* 51:953–61
96. Hahn, S., Guarente, L. 1988. *Science* 240:317–21
97. Chodosh, L. A., Baldwin, A. S., Carthew, R. W., Sharp, P. A. 1988. *Cell* 53:11–24
98. Bender, A., Sprague, G. F. Jr. 1987. *Cell* 50:681–91
99. Jarvis, E. E., Hagen, D. C., Sprague, G. F. Jr. 1988. *Mol. Cell. Biol.* 8:309–20
100. Goutte, C., Johnson, A. D. 1988. *Cell* 52:875–82
101. Siliciano, P. G., Tatchell, K. 1986. *Proc. Natl. Acad. Sci. USA* 83:2330–34
102. Hall, M. N., Johnson, A. D. 1987. *Science* 237:1007–12
103. Strathern, J., Shafer, B., Hicks, J., McGill, C. 1988. *Genetics* 120:75–81
104. Lalonde, B., Arcangioli, B., Guarente, L. 1986. *Mol. Cell. Biol.* 6:4690–96
105. Keleher, C. A., Goutte, C., Johnson, A. D. 1988. *Cell* 53:927–36
106. Brand, A. H., Micklem, G., Nasmyth, K. 1987. *Cell* 51:709–19
107. Kronstad, J. W., Holly, J. A., MacKay, V. L. 1987. *Cell* 50:369–77
108. Shore, D., Nasmyth, K. 1987. *Cell* 51:721–32
109. Buchman, A. R., Kimmerly, W. J., Rine, J., Kornberg, R. D. 1988. *Mol. Cell. Biol.* 8:210–25
110. Struhl, K. 1986. *Mol. Cell. Biol.* 6:3847–53
111. Chen, W., Struhl, K. 1988. *Proc. Natl. Acad. Sci. USA* 85:2691–95
112. Gill, G., Ptashne, M. 1988. *Nature* 334:721–24

113. Struhl, K., Brandl, C. J., Chen, W., Harbury, P. A. B., Hope, I. A., Mahadevan, S. 1988. *Cold Spring Harbor Symp. Quant. Biol.* 53:701–9
114. Struhl, K. 1985. *Proc. Natl. Acad. Sci. USA* 82:8419–23
115. Russell, D. W., Smith, M., Cox, D., Williamson, V. M., Young, E. T. 1983. *Nature* 304:652–54
116. Chen, W., Tabor, S., Struhl, K. 1987. *Cell* 50:1047–55
117. Kunkel, G. R., Martinson, H. G. 1981. *Nucleic Acids Res.* 9:6869–88
118. Prunell, A. 1982. *EMBO J.* 1:173–79
119. Han, M., Kim, U-J., Kayne, P., Grunstein, M. 1988. *EMBO J.* 7:2221–28
120. Clark-Adams, C. D., Norris, D., Osley, M. A., Fassler, J., Winston, F. 1988. *Genes Dev.* 2:150–59
121. Han, M., Grunstein, M. 1988. *Cell* 55:1137–45
122. Lue, N. F., Buchman, A. R., Kornberg, R. D. 1989. *Proc. Natl. Acad. Sci. USA* 86:486–90
123. Chen, W., Struhl, K. 1989. *EMBO J.* 8:261–68
124. Horikoshi, M., Carey, M. F., Kakidani, H., Roeder, R. G. 1988. *Cell* 54:665–69
125. Brandl, C. J., Struhl, K. 1989. *Proc. Natl. Acad. Sci. USA.* In press
126. Allison, L. A., Moyle, M., Shales, M., Ingles, C. J. 1985. *Cell* 42:599–610
127. Corden, J. L., Cadena, D. L., Ahearn, J. M. Jr., Dahmus, M. E. 1985. *Proc. Natl. Acad. Sci. USA* 82:7934–38
128. Dahmus, M. E., Kedinger, C. 1983. *J. Biol. Chem.* 258:2303–7
129. Nonet, M., Sweetser, D., Young, R. A. 1987. *Cell* 50:909–15
130. Allison, L. A., Wong, J. K., Fitzpatrick, V. D., Moyle, M., Ingles, C. J. 1988. *Mol. Cell. Biol.* 8:321–29
131. Ptashne, M. 1986. *Nature* 322:697–701
132. Brent, R., Ptashne, M. 1984. *Nature* 312:612–15
133. Struhl, K. 1986. *J. Mol. Biol.* 191:221–29

AUTHOR INDEX

1079

SUBJECT INDEX

A

Acetylcholine receptor
  topography and, 116, 1008,
    1019-20
Acetyl-CoA carboxylase
  biotin and, 196-97, 202
  protein phosphatases and, 480
Aconitase
  purification of, 5
Acidic transcriptional regulators,
    1059-62
Acridines
  intercalative antitumor drugs
    and, 358-61
Acrosomes
  active molecules in, 735-36
  signal transduction and, 737-
    39
Actinomyces naeslundii
  glycolipid binding and, 337
Actinomyces pneumoniae
  glycolipid binding and, 316
Actinomyces viscosus
  glycolipid binding and, 337
Actinomycin D, 358
Adamantane diazirine, 1016
Adenosine deaminase
  multiple isotope effects and,
    397
Adenosylcobalamin
  ribonucleotide reductase and,
    264
  rearrangement reactions and,
    267-70
Adenovirus
  DNA replication and, 673-82
ADR1
  transcriptional regulation and,
    1056
ADR2
  transcriptional regulation and,
    1063
Adrenalin
  protein phosphatases and,
    477, 483, 486-89
β-Adrenergic agonists
  protein phosphatases and,
    477, 483-84
β-Adrenergic receptor
  reovirus infection and, 561
Adriamycin, 358
Adult T-cell leukemia
  interleukin-2 receptors and,
    901-3
Agglutinization
  fertilization and, 734
Ah receptor
  glutathione S-transferases and,
    756

AIDS
  interleukin-2 receptors and,
    878
Alfalfa mosaic virus, 540, 548
Alkaloids
  P-glycoproteins and, 138
Alzheimer's disease, 287-304
  amyloid deposits and, 288-
    97
  See also amyloid proteins
Amber mutations
  tRNA identity and, 1030-48
Amide hydrogens
  two-dimensional NMR and,
    226-30
Amine oxidases
  protein radicals and, 280-83
Amino acid radicals
  ribonucleotide reductase and,
    256-67
  tyrosyl, 259-63
Aminoacyl tRNA
  tRNA identity and, 1029-47
Aminoacyl tRNA synthase
  tRNA identity and, 1029-48
Amphipathic α-helices
  membrane proteins, 1005-6
  transcriptional regulation and,
    1060-62
AMP nucleosidase
  multiple isotope effects and,
    384
α-Amylase
  protein phosphatases and,
    458
α-Amylase inhibitor
  two-dimensional NMR and,
    244
Amyloid proteins, 287-304
  amyloid plaque cores and,
    293-95
  gene conversion and, 511
  molecular pathology of, 288
  See also amyloid A4 protein
Amyloid A4 protein
  cDNA precursors of, 297-300
  cerebrovascular amyloid and,
    295-97
  paired helical filaments and,
    291-93
  plaque cores and, 293-95
  precursor gene of, 301-2
  precursor proteins of, 300-1
Anemonia sulca
  antiviral protein BDS-1 and,
    244
Antibody diversity
  gene conversion and, 509-31
  avian immunoglobulin
    genes and, 512-15

germline corrections and,
    517-22
  mammalian im-
    munoglobulin genes
    and, 515-16
  minigene hypothesis and,
    516-17
  somatic conversions and,
    522-27
Anticodon
  tRNA identity and, 1029-48
Antioxidant defenses, 86-90
Antiviral compounds, 565-69
Antiviral protein BDS-1
  two-dimensional NMR and,
    244
AP1
  transcriptional regulation and,
    823-24, 1064, 1074
  glycosylation and, 863
Arachidonic acid
  PKC activation and, 37
  radical-mediated prostaglandin
    synthesis and, 274-77
Arbacia punctulata
  fertilization and, 723, 730-
    33
ARGR2
  transcriptional regulation and,
    1056
Arm-type sites
  λ site-specific recombination
    and, 928
Ascorbate
  dopamine β-hydroxylase and,
    417
  hydroperoxide metabolism
    and, 96-97
Ascorbic acid
  antioxidant defenses and, 88
Asparaginylglucose,
  isolation of, 180-81
Aspergillus
  protein phosphatases and,
    457
ATF/CREB factor
  transcriptional regulation and,
    1064
Athracyclines
  intercalative tumor drugs and,
    358-61
ATP-binding sequences
  P-glycoproteins and, 156-57
  PKC and, 35
ATP hydrolysis
  biotin-dependent carboxylation
    and, 205-6
  P-glycoproteins and, 138
ATP: phosphatidylinositol 4-
    phosphotransferase

1139

CUMULATIVE INDEXES

CONTRIBUTING AUTHORS, VOLUMES 54–58

Travers, A., 58:427–52
Tsuji, T., 56:21–42
Turk, J., 55:69–102
Tzagoloff, A. A., 55:249–86

U

Ueda, K., 54:73–100
Ullrich, A., 57:443–78

V

Varmus, H. E., 56:651–94
Vignais, P. V., 54:977–1014
von Figura, K., 55:167–94

W

Waldmann, T. A., 58:875–912
Walker, G. C., 54:425–57
Wang, J., 54:665–97
Wassarman, P. M., 57:415–42
Wehrenberg, W. B., 54:403–23
Wewer, U. M., 55:1037–58
Wickner, R. B., 55:373–96
Wieland, F., 58:173–94
Wikström, P. M., 56:263–88
Wiley, D. C., 56:365–94
Williams, A. F., 57:285–320
Williams, K. R., 55:103–36
Wimmer, E., 57:701–54
Wistow, G. J., 57:479–504

Wood, H. G., 54:1–41; 57:235–60
Wright, C. E., 55:427–54
Wright, J. K., 55:225–48
Wysocki, L. J., 58:509–31

Y

Yarden, Y., 57:443–78
Yeates, T. O., 58:607–33

Z

Zeytin, F., 54:403–23
Ziegler, D. M., 54:305–29
Zoon, K. C., 56:727–78

CHAPTER TITLES, VOLUMES 54–58

1159

1162   CHAPTER TITLES

Annual Reviews Inc.

A NONPROFIT SCIENTIFIC PUBLISHER

4139 El Camino Way
P.O. Box 10139
Palo Alto, CA 94303-0897 ● USA

Annual Reviews Inc. publications may be ordered directly from our office by mail, Telex, or use our Toll Free Telephone line (for orders paid by credit card or purchase order*, and customer service calls only); through booksellers and subscription agents, worldwide; and through participating professional societies. **Prices subject to change without notice.** ARI Federal I.D. #94-1156476

- **Individuals:** Prepayment required on new accounts by check or money order (in U.S. dollars, check drawn on U.S. bank) or charge to credit card—American Express, VISA, MasterCard.
- **Institutional buyers:** Please include purchase order number.
- **Students:** $10.00 discount from retail price, per volume. Prepayment required. Proof of student status must be provided (photocopy of student I.D. or signature of department secretary is acceptable). Students must send orders direct to Annual Reviews. Orders received through bookstores and institutions requesting student rates will be returned. You may order at the Student Rate for a maximum of 3 years.
- **Professional Society Members:** Members of professional societies that have a contractual arrangement with Annual Reviews may order books through their society at a reduced rate. Check with your society for information.
- **Toll Free Telephone orders:** Call 1-800-523-8635 (except from California) for orders paid by credit card or purchase order and customer service calls only. California customers and all other business calls use 415-493-4400 (not toll free). Hours: 8:00 AM to 4:00 PM, Monday-Friday, Pacific Time. ***Written confirmation** is required on purchase orders from universities before shipment.
- **Telex: 910-290-0275**

Regular orders: Please list the volumes you wish to order by volume number.
Standing orders: New volume in the series will be sent to you automatically each year upon publication. Cancellation may be made at any time. Please indicate volume number to begin standing order.
Prepublication orders: Volumes not yet published will be shipped in month and year indicated.
California orders: Add applicable sales tax.
Postage paid (4th class bookrate/surface mail) **by Annual Reviews Inc.** Airmail postage or UPS, extra.

ANNUAL REVIEWS SERIES		Prices Postpaid per volume USA & Canada/elsewhere	Regular Order Please send:	Standing Order Begin with:
			Vol. number	Vol. number
Annual Review of ANTHROPOLOGY				
Vols. 1-14	(1972-1985)	$27.00/$30.00		
Vols. 15-16	(1986-1987)	$31.00/$34.00		
Vol. 17	(1988)	$35.00/$39.00		
Vol. 18	(avail. Oct. 1989)	$35.00/$39.00	Vol(s). _____	Vol. _____
Annual Review of ASTRONOMY AND ASTROPHYSICS				
Vols. 1, 4-14, 16-20	(1963, 1966-1976, 1978-1982)	$27.00/$30.00		
Vols. 21-25	(1983-1987)	$44.00/$47.00		
Vol. 26	(1988)	$47.00/$51.00		
Vol. 27	(avail. Sept. 1989)	$47.00/$51.00	Vol(s). _____	Vol. _____
Annual Review of BIOCHEMISTRY				
Vols. 30-34, 36-54	(1961-1965, 1967-1985)	$29.00/$32.00		
Vols. 55-56	(1986-1987)	$33.00/$36.00		
Vol. 57	(1988)	$35.00/$39.00		
Vol. 58	(avail. July 1989)	$35.00/$39.00	Vol(s). _____	Vol. _____
Annual Review of BIOPHYSICS AND BIOPHYSICAL CHEMISTRY				
Vols. 1-11	(1972-1982)	$27.00/$30.00		
Vols. 12-16	(1983-1987)	$47.00/$50.00		
Vol. 17	(1988)	$49.00/$53.00		
Vol. 18	(avail. June 1989)	$49.00/$53.00	Vol(s). _____	Vol. _____
Annual Review of CELL BIOLOGY				
Vol. 1	(1985)	$27.00/$30.00		
Vols. 2-3	(1986-1987)	$31.00/$34.00		
Vol. 4	(1988)	$35.00/$39.00		
Vol. 5	(avail. Nov. 1989)	$35.00/$39.00	Vol(s). _____	Vol. _____

ANNUAL REVIEWS SERIES	Prices Postpaid per volume USA & Canada/elsewhere	Regular Order Please send:	Standing Order Begin with:
		Vol. number	Vol. number

Annual Review of **COMPUTER SCIENCE**

Vols. 1-2	(1986-1987)................$39.00/$42.00		
Vol. 3	(1988) ....$45.00/$49.00		
Vol. 4	(avail. Nov. 1989)...........$45.00/$49.00	Vol(s). _____	Vol. _____

Annual Review of **EARTH AND PLANETARY SCIENCES**

Vols. 1-10	(1973-1982)................$27.00/$30.00		
Vols. 11-15	(1983-1987)................$44.00/$47.00		
Vol. 16	(1988) ....$49.00/$53.00		
Vol. 17	(avail. May 1989)...........$49.00/$53.00	Vol(s). _____	Vol. _____

Annual Review of **ECOLOGY AND SYSTEMATICS**

Vols. 2-16	(1971-1985)................$27.00/$30.00		
Vols. 17-18	(1986-1987)................$31.00/$34.00		
Vol. 19	(1988) ....$34.00/$38.00		
Vol. 20	(avail. Nov. 1989)...........$34.00/$38.00	Vol(s). _____	Vol. _____

Annual Review of **ENERGY**

Vols. 1-7	(1976-1982)................$27.00/$30.00		
Vols. 8-12	(1983-1987)................$56.00/$59.00		
Vol. 13	(1988) ....$58.00/$62.00		
Vol. 14	(avail. Oct. 1989)...........$58.00/$62.00	Vol(s). _____	Vol. _____

Annual Review of **ENTOMOLOGY**

Vols. 10-16, 18	(1965-1971, 1973)		
20-30	(1975-1985)................$27.00/$30.00		
Vols. 31-32	(1986-1987)................$31.00/$34.00		
Vol. 33	(1988) ....$34.00/$38.00		
Vol. 34	(avail. Jan. 1989)...........$34.00/$38.00	Vol(s). _____	Vol. _____

Annual Review of **FLUID MECHANICS**

Vols. 1-4, 7-17	(1969-1972, 1975-1985).......$28.00/$31.00		
Vols. 18-19	(1986-1987)................$32.00/$35.00		
Vol. 20	(1988) ....$34.00/$38.00		
Vol. 21	(avail. Jan. 1989)...........$34.00/$38.00	Vol(s). _____	Vol. _____

Annual Review of **GENETICS**

Vols. 1-19	(1967-1985)................$27.00/$30.00		
Vols. 20-21	(1986-1987)................$31.00/$34.00		
Vol. 22	(1988) ....$34.00/$38.00		
Vol. 23	(avail. Dec. 1989)...........$34.00/$38.00	Vol(s). _____	Vol. _____

Annual Review of **IMMUNOLOGY**

Vols. 1-3	(1983-1985)................$27.00/$30.00		
Vols. 4-5	(1986-1987)................$31.00/$34.00		
Vol. 6	(1988) ....$34.00/$38.00		
Vol. 7	(avail. April 1989)...........$34.00/$38.00	Vol(s). _____	Vol. _____

Annual Review of **MATERIALS SCIENCE**

Vols. 1, 3-12	(1971, 1973-1982)...........$27.00/$30.00		
Vols. 13-17	(1983-1987)................$64.00/$67.00		
Vol. 18	(1988) ....$66.00/$70.00		
Vol. 19	(avail. Aug. 1989)...........$66.00/$70.00	Vol(s). _____	Vol. _____

Annual Review of **MEDICINE**

Vols. 9, 11-15	(1958, 1960-1964)		
17-36	(1966-1985)................$27.00/$30.00		
Vols. 37-38	(1986-1987)................$31.00/$34.00		
Vol. 39	(1988) ....$34.00/$38.00		
Vol. 40	(avail. April 1989) ....$34.00/$38.00	Vol(s). _____	Vol. _____